DESIGN OF
CONCRETE STRUCTURES

(In S.I. System of Units as per Code IS: 456-2000)

[S.I. UNITS]

3rd Revised and Enlarged Edition

DR. RAMCHANDRA

BE, M.E. (Hons.), Ph.D. (Roorkee at present I.I.T., Roorkee)
Professor (Formerly Head)
Department of Structural Engineering
M.B.M. Engineering College J.N.V. University,
JODHPUR (Rajasthan)

VIRENDRA GEHLOT

B.E. (Civil), ME (Civil Structures)
Consultant Engineer
JODHPUR (Rajasthan)
[E-mail: viren224@yahoo.co.in]

Published by:

Scientific Publishers (India)
5 A, New Pali Road, P.O. Box 91
Jodhpur 342 001 (India)

E-mail: info@scientificpub.com
Website: www.scientificpub.com

Branch Office
Scientific Publishers (India)
4806/24, Ansari Road, Daryaganj
New Delhi - 110 002 (India)

3rd Revised Reprint: 2013

ISBN: 978-81-7233-659-2

ISBN:9788172336592
9 788172 336592
Price: 590/-

Lasertype set: Rajesh Ojha
Printed in India

Dedicated to

OUR TEACHERS AND PARENTS

LIST OF THE PAPERS PUBLISHED BY

Prof. Dr. Ram Chandra and Virendra Gehlot***

I. PAPERS PUBLISHED IN INTERNATIONAL JOURNALS

*1. "Non-Linear Analysis of Steel Space Structures," Journals of American Society of Civil Engineers, Volume 116, No. 4, April, 1990, Paper no. 24525 (pp. 898-909) New York.

*2. "Elastic-Plastic Analysis of Steel Space Structures", Journal of American Society of Civil Engineers, Volume 116, No. 4, April, 1990 Paper no. 24528 (pp. 939-955) New York.

II. PAPERS PUBLISHED IN NATIONAL JOURNALS, CONFERENCES & SEMINARS

*3. "Non-Linear Elastic-Plastic Analysis of Skeletal Steel Plane Frames Hinged at supports", proceedings of the seminar on Modern trend in structural analysis and design, 25-26 February, 1984, Department of Civil Engineering, Banara Hindu University, Varanasi (U.P.)

*4. "Non-Linear Elastic Plastic Analysis of Skeletal Plane Steel Frame", International conference on computer applications in Civil Engineering, October 23-25, 1979, University of Roorkee, Roorkee.

*5. "Stub Column Tests on Indian Standard H-Beams", Journal of the Institution of Engineers (India), Volume 51, No. 11, C.I. 6, July, 1971.

*6. "Kani's Iteration Method I-Analysis of Continuous Beams" The Indian Engineer, Volume XIV, No. 6 June, 1970.

*7. "Kani's Iteration Method II Analysis of Frames", The Indian Engineer", Volume XIV, No. 8 August, 1970.

*8. Hinge Formation in Plane Frames Considering Finite Deformations All India Conference on "Tall Buildings" The Institution of Engineers (1) Roorkee Local Centre, March 1-3, 1993.

**9. "Design, Construction and Behaviour Aspects of Concrete and Steel Structures' Recent Trends in Civil Engineering National Seminar, Jodhpur, February 22-23, 2002, Department of Civil Engineering MBM Engineering College, Faculty of Engineering, JN Vyas University, Jodhpur pp 229.

**10. 'Corrosion of Reinforcing Steel Bars in Concrete' Futuristic Construction Materials and Technique (FCMT-2004, Feb. 21-22, 2005) Department of Structural Engineering Faculty of Engineering, JN Vyas University Jodhpur.

**11. 'Stresses in Reinforcing Steel Bars, HYSD of Grade Fe 415' Futuristic Construction Materials and Technique (FCMT 2004, Feb. 21-22, 2005) Department of Structural Engineering Faculty of Engineering, JN Vyas University, Jodhpur.

ACKNOWLEDGEMENT

The author highly acknowledges Bureaue of the Indian Standards for reproducing in this book some of the 'Tables and Clauses' from the Indian Standard Specifications.

It is desirable that for complete detail, reference be made to the latest versions of the Standards which are available from Indian Standards Institution, Manak Bhavan, 9, Bahadur Shah Zafar Marg, New Delhi-1, or from its branch offices at Bombay, Calcutta, Kanpur and Madras.

SYSTEM INTERNATIONAL d' UNITES (SI-System of Units)

In order to avoid the conversion of results obtained by engineers working with the foot-pound second system (gravitational) of units in terms of centimetre-gram second absolute system of units used by the scientists, a need of common system of units was realised. The General Conference on Weights and Measures held at Paris is 1960 finalised the System International *d'* Unites (SI). It is an absolute system of units. The mass is considered as fundamental unit and not the force. ISI has included a comment of transition in IS 3616-1966. 'Recommendation on the International System (SI) Units' that this system has begun to replace older systems of units in several branches of science and technology. The SI is a universal system of units and it has been adopted in France as a legal system and it is likely to become common in many countries. SI Units have the following six basic units.

Units of length (metre, m)

The length equal to 1,650,763.73 wave lengths, in vacuum, of the radiation corresponding to the transition between $2p^{19}$ and $5d^5$ levels of the krypton n atom of mass 86 is known as *one metre*.

Linear distances are expressed in metres and multiples or division of 10^3 of metres (i.e., one kilo-meter (km) = 10^3 m, one metre (m) = 1m, and one millimetre (mm) = 10^{-3} m.

Unit of mass (kilogram, kg)

The mass of planium-iridium cylinder deposited at the International Bureau of Weights and Measures and declared as the international prototype of the kilogram by the First General Conference of Weights and Measures is called as *one kilogram*.

Units of time (second, s)

1131, 566, 925, 974.7 of the length of the tropical year for 1900, the year commending at 1200 hours universal time on the first day of January, 1900 is termed as *one second*.

Units of electric current (Ampere, A)

The constant current which flows in two parallel straight conductors of infinite length of negligible circular cross-section and placed at a distance of one metre from each other in vacuum producing a force 2×10^{-7} Newtons per metre length between the conductors is defined as *one ampere.*

Unit of thermo-dynamic temperature (degree Kelvin, °K)

The degree interval of the thermo-dynamic scale on which the temperature of triple point of water is 273.16 degrees, is known as *one degree Kelvin.*

For temperatures, conventional degrees centigrade (°C) may be used, since, temperature changes are concerned rather than absolute temperatures.

Units of luminous intensity (candela, cd)

One sixtieth of luminous intensity normally emitted by one centimetre square of integral radiator (black body) at the temperature of solidification of platinum is called as *one candela.*

Following are the derived units (derived from the basic SI units and these are relevant for the study of design of steel structures).

Force	Newton (N)	$kg.m.sec^{-2}$
Work, energy	Joule (J)	$kg.m^2$, sec^{-2} = (N – m)
Power	Watt (W)	$kg.m^2$. sec^{-3} = (J . s^{-1})
Frequency	Hertz (Hz)	Cycles /sec.

Accerelation due to gravity used is 9.81 ms^{-2}.

The force is a derived quantity and physical law connecting the quantity of the fundamental quantities or previously obtained derived quantities is force = mass x acceleration. It is defined as that force which produces unit acceleration (i.e., 1 m per sec^2) in a unit mass of 1 kg. Its unit is Newton (N). Though, the Newton is a small unit, a still larger unit kN may be used. The intensity of force (viz., stress) due to 1 Newton over a unit area of one metre square is known as one *pascal*. It is denoted by symbol, Pa (1 Pa = 1 N/m^2 and 10^6. Pa = 1 N/mm^2, viz., 1 MPa = 1 N/mm^2).

The unit of force, (the Newton N), is the force required to develop unit acceleration (m s^{-2}) to unit mass (1 kg). In terms of Newtons, the common force units in the foot-pound-second system (with g = 9.81 m s^{-2}) are

1 lb-wt = 4.45 Newtons (N)

1 ton-wt = 9.96×10^{-3} Newtons (N)

1 ton-wt = 9.96 kN

The unit of force, the Newton (N) is used for the external loads and the internal forces, such as the shear force. Torque and bending moments are expressed as Newton-metres (N-m).

Another important unit is stress. In the foot-pound-second system the stresses are commonly expressed in lb-wt/in^2 and tons/in^2. In the SI system of units, these are taken as

$$1 \text{ lb. wt/in}^2 = 6.89 \text{ x } 103 \text{ N/mm}^2 \ 6.89 \text{ kN/m}^2$$

$$1 \text{ ton. wt/in}^2 = 15.42 \text{ x } 10\text{-}6 \text{ N/mm}^2 = 15.42 \text{ MN/m}^2$$

Yield stresses of the common metallic materials are in the range [200 MN/m^2 to 750 MN/m^2]. Young's modules for steel,

$$E_{STEEL} = 30 \times 10^6 \text{ lbs. wt/in}^2 = 207 \times 10^5 \text{ N/mm}^2 = 207 \text{ GN/m}^2$$

The SI units make the use of multiples and sub-multiples 1000 times or 1/1000 times the unit quantity and in powers of 10^3 *(kilo)* or 10^{-3} *(milli)* in respect of still larger and smaller quantities respectively. The lengths are measured usually in kilometre (1 km = 1000 m). metre and millimetre (1 mm = 10^{-3} m). The symbols of units are not to be suffixed with `s' for plural.

[(kilo k 10^{+3}), (mega M 10^{+6}), (giga G 10^{+9})].

SI system of units have many advantages. The units are very handy. The burden of non-decimal coefficients in foot-pound second system is avoided. It has relatively large main units in contrast to centimetre-gram-second system. At the same time, it is closely related to centimetre-gram-second system of units. In practice, it results in perfectly reasonable number when the value of g = 10 m/sec^2 is used instead of 9.806 m/sec^2.

References

1. Professor V.S. Mokashi, Visvesvaraya Regional College of Engineering, Nagpur in his paper titled as International System (SI) Units and their Application to Engineering, Published in Journal of Institution of Engineers, India, Vol. 19, March 1970 has highlighted the advantages and discussed SI units. A reference has been made to this paper.
2. John, Case and A.H. Chilver, `Strength of Materials and Structures' Second Edition (ELBS)/Edward Arnold, 1986, Printed of the Bath Press, Avon, Great Britain.

PREFACE (Thirteenth Edition)

This book `Design of Concrete Structures' in S.I. Units is based on working stress method as per code IS: 456-2000.

All the chapters of the book have been revised and re-arranged in eight parts (32 thirty two chapters) separate aspects of design of one structrual member have been described in different subsequent chapters.

In addition to above (i) the service life of concrete structures, (ii) Non-destructive tests/ Evaluation of strength (NDT/NDE) of materials and (iii) futuristic construction materials and Technique (FCMT) likely to be used for the concrete are new topics. Text for these topics (rarely, available in current books by other authors) have been first time given to familiarize the readers.

Authors are thankful to their colleagues and friends for their liking and sending useful suggestions. Authors express their thanks to the publishers of this book, M/s Scientific Publishers (India), Jodhpur for preparing make-up and printing this book in the latest technology and bring-out the same in a nice-getup.

Authors wish to express their thanks to Shri Rajesh Ojha for nice lasertype setting.

Suggestions from the students and affectionate teachers of various Institutions shall be highly appreciated.

12 February, 2010

Ramchandra
Virendra Gehlot

PREFACE (First Edition)

In this book, the author with his long teaching experience in the subject has made an attempt to present the subject mater of `Design of Concrete Structures' in a way which lays emphasis on the fundamentals, keeping in view, the difficulties experienced by the students. Every basic principle, method, equation or theory has been presented in a simplified manner. The text has been described in a sequence most naturally desired and appealed to the student. SI system of units has been used throughout the text. Indian Standard Specifications (specially, IS: 456-1978) have been used. The symbols and notations used in the text match with those of present code of practice. The book is intended for Degree, Diploma and AMIE students in various branches of engineering. The book deals with design of structural members.

The complete text in this book has been divided in **eight parts** [Part I: Introduction, Part II: Loads and Stresses, Part III: Design of Beams, Part IV: Design of Slabs, Part V: Design of Columns, Part VI: Design of Footings, Part VII: Design of Other Concrete Structures, and Part VIII: Design of Concrete Mix and Laboratory Tests] and **twenty-one** chapters (detailed as in contents).

A number of design problems has been solved to illustratre the theory and practice. The chapters have been so arranged that it facilitates self-understanding of the subject during study. In spite of careful scrutiny of the manuscript, it is possible that some typographical and computational errors are still left. The author shall be highly obliged to all those who will bring these errors in his notice.

The author is thankful to his colleagues for their suggestions. The author is also thankful to the publishers M/s Standard Book House for bringing the book in a nice get up and in a short time. Suggestions from the students and affectionate teachers of various Institutions shall be highly appreciated.

January 1, 1990
Jodhpur

RAM CHANDRA

CONTENTS

Part - I.
Concrete Structures, its Service Life, Concrete, its Constituents and Design Codes and Specification

1. Introduction ... 1-54

1.1 Introducton ... 1
1.2. Concrete ... 3
1.3. Cements ... 4
1.4. Cement Tests ... 6
1.5. Aggregates ... 9
1.6. Other Types of Aggregates ... 14
1.7. Fitness modulus of aggregates ... 15
1.8. Water ... 16
1.9. Admixtures ... 18
1.10. Reinforcement ... 21
1.11. Storage of Materials ... 24
1.12. Properties of Concrete ... 24
1.13. Requirements of Durability (IS: 456-2000) ... 28
1.14. Concrete Mix Proportions ... 28
1.15. Limits for Deleterious Mix Constituents ... 30
1.16. Concrete in Aggressive Soils and Water ... 31
1.17. Compaction, Finishing and Curing ... 31
1.18. Concrete in Sea-Water ... 32
1.19. Shape and Size of Component Members ... 32
1.20. Environmental exposure conditions ... 33
1.21. Nominal Concrete cover to reinforcement ... 35
1.22. Quality of Fine and Coarse Aggregates ... 36
1.23. Quality of Water ... 36
1.24. Permeability of Concrete ... 36
1.25. Curing of Concrete ... 37
1.26. Compaction of Concrete ... 37
1.27. Concrete in Foundations ... 38
1.28. Durability of Steel ... 38
1.29. Fire Resistance of Concrete Structures ... 42
1.30. Requirements and Aim of Design ... 43
1.31. Service Life of Concrete Structures ... 45
1.32. Design Codes and Specification. ... 47

Part - II.
Loads, Strength, Factors of Safety and Methods of Design

2. Loads (Actions), Strength of Materials, Factors of safety and Methods of Design ... 55-106

2.1. Introduction ... 55
2.2. Dead Lodes (Permanent Loads) (D.L.) ... 57
2.3. Live Loads (Variable Loads/Imposed (Live) Loads) (LL) ... 61
2.4. Wind Loads (WL) Transient Loads ... 68
2.5. Snow Load (Exceptional Loads) ... 71
2.6. Seismic Loads (Exceptional Loads) ... 71
2.7. Temperature effects ... 72
2.8. Shrinkage ... 72
2.9. Creep ... 76
2.10. Other Forces and Effects ... 81
2.11. Structural Safety and Factor of Safety (F.S.) . 83
2.12. Structural Failures ... 84
2.13. Characteristic Strength of Concrete ... 85
2.14. Factors influencing strength of Concrete ... 92
2.15. Stress-Strain Relationship for Concrete ... 95
2.16. Characteristic Strength of Steel Reinforcement ... 99
2.17. Stress-Strain Relationship for Steel ... 99
2.18. Methods of Design ... 102
2.19. Elastic (Working Stress) Method of Design ... 103
2.20. Ultimate Load Method (ULM). ... 104

Part - III.
Design of Beams

3. Bending (Flexural) Strength and Design of Singly Reinforced Rectangular Beams (SRRB-I) ... 107-142

3.1. Introduction ... 107

3.2. Behaviour of Reinforced Concrete Beam 108
3.3. Bending (Flexural) Strength of Singly Reinforced Beam 110
3.4. Balanced, under-reinforced and over reinforced Beam Sections 115
3.5. Permissible Stresses in Bending in Concrete and Steel 119
3.6. Design of Singly Reinforced Beam Sections ... 128
3.7. Bending of Singly Reinforced Beams of Varying Depth 138

4. Shear Strength of Beams and Design of Singly Reinforced Rectangular Beams (SRRB-II) 143-175

4.1. Introduction 143
4.2. Shear Stresses in Beams of Varying Depth 148
4.3. Diagonal Tension in Homogeneous Elastic Beams 150
4.4. Beams without Shear Reinforcement 152
4.5. Behaviour of Diagonally Cracked Beams 153
4.6. Shear Strength of Reinforced Concrete Beams 154
4.7. Reinforced Concrete Beams with Shear Reinforcement 156
4.8. Reinforced Concrete Beams with Vertical Stirrups 156
4.9. Reinforced Concrete Beams with Bent-up Bars 159
4.10. Reinforced Concrete Beams with inclined Bars (Stirrups) 160
4.11. Function of Shear Reinforcement 161
4.12. Critical Section for Shear Force 165
4.13. Permissible Shear Stresses in Concrete 167
4.14. Design of Shear Reinforcement. 168

5. Bond and Development Length in Singly Reinforced Rectangular Beams (SRRB-III) 176-189

5.1. Introduction 176
5.2. Bond Stresses 177
5.3. Development length of Reinforcement Bars in Tension 179
5.4. Anchorage of Plain Steel Bars in Tension 180
5.5. Curtailment of Tension Reinforcement in Beams 182
5.6. Conditions for Curtailment of Flexural Reinforcement in Tension Zone 184
5.7. Special Requirements near points of Zero Moment for Curtailment of Tension Reinforcement 185
5.8. Bar Splices. 187

6. Deflections and their limitations in Singly Reinforced Rectangular Beams (SRRB-IV) 190-192

6.1. Introduction 190
6.2. Slenderness Limits for Beams 192

7. Bending (Flexural) Strength and Design of Doubly Reinforced Rectangular Beams (DRRB-I) 193-214

7.1. Introduction 193
7.2. Location of Neutral Axis 195
7.3. Bending Strength of Doubly Reinforced Beam 196
7.4. Stresses in Concrete and Steel 197
7.5. Design of Doubly Reinforced Beam 204
7.6. Steel Beam Theory 208

8. Shear strength and Design of Doubly Reinforced Rectangular Beams (DRRB-II) 215-219

8.1. Introduction 215

9. Bond and Development length in Doubly Reinforced Rectangular Beams (DRRB-III) 220-223

9.1. Introduction 220
9.2. Development Length of Reinforcement Bars in Compression 220
9.3. Anchorage of Plain Bars in Compression 221
9.4. Lap Splices in Compression 224

10. Bending (Flexural) Strength and Design of Flanged (TEE) Beams (FB-I) 224-255

10.1. Introduction 224
10.2. Effective Width of Flagne 225
10.3. Locations of Neutral Axis 226
10.4. Lever Arm for T-Beam Section 228
10.5. Bending Strength of T-Beams 230
10.6. Stresses in Concrete and Steel 239

10.7. Effective Depth of T-Beam 243
10.8. Design of T-Beams, Problems 245

11. Shear Strength and Design of Flanged (TEE) Beams (FB-II) 256-259

11.1. Introduction 256

12. Torsional Strength and Design of Flanged (L) Beams (FB-III) 260-270

12.1. Introduction 260
12.2. Torsion in Plain Concrete Members 261
12.3. Torsion in Reinforced Concrete Members 264
12.4. Members without shear stirrups subjected to combined torsion and shear 269
12.5. Members with shear stirrups subjected to combined torsion and shear 269

13. Development length of Main Reinforcement Bars for Flanged (Tee) Beams (FB-IV) 271

13.1 Introduction 271

14. Bending (Flexural) Strength and Design of Continuous Beams (CB-I) 272-296

14.1. Introduction 272
14.2. Effective Span 273
14.3. Critical Section for Moment 274
14.4. Coefficients for moment for continuous beams 275
14.5. Concrete Building Frames 276
14.6. Arrangement of Live Load for Bending Moment 279
14.7. Compression Steel at Support 281
14.8. Tension Steel at Support 283
14.9. Tension Steel at Centre of Span 285
14.10. Haunches in Continuous Beams 285
14.11. Extension of Reinforcementy 286
14.12. Design of Continuous Beams 286
14.13. Bending of Bars 287

15. Shear Strength and Design of Continuous Beams (CB-II) 297-301

15.1. Introduction 297
15.2. Coefficients for Shear Forces for Continuous Beams 297
15.3. Arrangement of Live Load for Shear Force .. 298

Part IV - Design of Slabs

16. Design of Slabs spanning in one direction (SSOD) 302-332

16.1. Introduction 302
16.2. Design of Slabs spanning in one direction 304
16.3. Temperature and Shrinkage (Distribution) Reinforcement 308
164. Concentrated Loads over Slabs 310
16.5. Slabs Continuous over Walls 320
16.6. Shear in Slabs 331

17. Design of Slabs spanning in Two Directions (SSTD) 333-368

17.1. Introduction 333
17.2. Behaviour of Slabs Spanning in Two Directions 333
17.3. Grashoff-Rankine Method 334
17.4. Restrained Slabs 337
17.5. Marcus Method 339
17.6. Pigeaud's Method 344
17.7. Westergaard's Method 346
17.8. IS: 456-2000 Method 349
17.9. Tension Reinforcement for Two-way Supported Slabs 352
17.10. Torsion Reinforcement for Two-Way Edge Supported Slabs 354
17.11. Design of Slab Spanning in Two Directions 356

18. Design of Flat slabs 369-426

18.1. Introduction 369
18.2. Development of Flat Slabs 371
18.3. Elements of Flat Slabs 372
18.4. Statics of a Flat Slab 373
18.5. Behaviour of Flat Slabs in Bending 376
18.6. Direct Design Method 377
18.7. Effects of Pattern Loading (Direct Design Method) 383
18.8. Equivalent Frame Method 384
18.9. Relative Stiffness of Slabs 385
18.10. Relative Stiffness of Columns 390
18.11. Equivalent Column 394
18.12. Torsional Member 395

18.13. Loading patterns (Equivalent Frame Method) 398
18.14. Critical Sections for Negative Design Moments 400
18.15. Modification of Maximum Moment 400
18.16. Bending Moments in Panels with Marginal Beams or Walls 401
18.17. Transfer of Bending Moments to Columns 402
18.18. Moments in Columns 402
18.19. Slab Reinforcement 403
18.20. Behaviour of Flat Slab in Sher 407
18.21. Shear in Flat Slabs 408
18.22. Openings in Flat Slabs 426

19. Design of Circular slabs 427-452

19.1. Introduction 427
19.2. Circular Slab (Simply Supported at Edge) carrying uniformly distributed load 428
19.3. Circular Slab (Fixed at Edges) carrying uniformly distributed load 430
19.4. Circular slab (Simply Supported at Edge) carrying uniformly distributed total load W along circumference of a concentric circle 434
19.5. Circular Slab (Simply Supported at Edge) carrying uniformly distributed load inside a circle 435
19.6. Circular slab (Simply supported at Edge with a central hole) carrying total uniformly distributed load W along circumference of a concentric circle 436
19.7. Circular Slab (Simply Supported at Edge with Central Hole) Carrying uniformly distributed load 437
19.8. Circular slab carrying a concentrated load at centre 438

Part V -

Design of Columns

20. Design of Axially loaded columns and compression members 453-496

20.1. Introduction 453
20.2. Axially Loaded Columns 454
20.3. Effective Length of Compression Members 456
20.4. Equivalent (Transformed) Cross-sectional Area of Columns 461
20.5. Radius of Gyration of a Column Section 462
20.6. Slenderness ratio of Compression Members 463
20.7. Unsupported Length of Compression Member 465
20.8. Lateral Ties and Spirals 468
20.9. Minimum Eccentricity 469
20.10. Clear Cover to Reinforcement 470
20.11. Permissible Stresses in Concrete and Steel Bars 470
20.12. Strength of Axially Loaded Columns and Compression Members 471
20.13. Design of Axially Loaded reinforced concrete columns and compression members 481
20.14. Requirement of Longitudinal (Main) Reinforcement for Columns 483
20.15. Requirements of Transverse (Lateral) Reinforcement 485

21. Design of columns and compression members subjected to combined Axial loads and moments 497-556

21.1. Introduction 497
21.2. Eccentrically loaded columns and compression members 498
21.3. Resultant compressive and tensile stresses in columns 500
21.4. Uncracked Concrete Section of Columns 502
21.5. Cracked Concrete Section of Columns 504
21.6. Distribution of Stress over column section (Eccentricity of Load Less than 0.25 D) 511
21.7. Distribution of Stresses over Column section (Eccentricity of Load more than 1.5D) 513
21.8. Distribution of Stress over column section (Eccentricity of Load more than 0.25 D and less than 1.5 D) 516
21.9. Eccentrically loaded circular columns (Spirally reinforced) 529
21.10. Distribution of Stress over member section [Eccentricity of tensile load (Less than 0.25 D)] 542
21.11. Distribution of Stress over member section (Eccentricity of Tensile Load is more than 1.5 D) 544

21.12. Distribution of Stress over Member section (Eccentricity of Tensile Force more than 0.25 and Less than 1.5 D)....547

Part - VI.
Design of Footings and Foundations

22. Design of Independent Footing 557-639

22.1. Introduction....557
22.2. Types of independent footings....558
22.3. Bearing pressure distributional in soil....559
22.4. Bearing capacity of Soil....559
22.5. Depth of Foundation....562
22.6. Thickness at Edge of Footing....562
22.7. Moments and Bending moments in footings....563
22.8. Shear forces in Footings....567
22.9. Transfer of Load at base of Column (Bearing under columns)....569
22.10. Basic requirements for columns....571
22.11. Wal Footings....572
22.12. Square Footing....575
22.13. Rectangular Footing....577
22.14. Stepped Footings....580
22.15. Sloped Footings....582
22.16. Circular Footing....611
22.17. Footings on Rock....621
22.18. Balancing Footing Pressures....621
22.19. Eccentrically Loaded independent spread footings....622
22.20. Kern distance....625
22.21. Symmetrical Footings subjected to Axial Load and Moment....628
22.22. Unsymmetrical Footings subjected to Eccentric Load....629

23. Design of Combined footing 640-716

23.1. Introduction....640
23.2. Cantilever or Strap Footing....642
23.3. Continuous Strip Footings....644
23.4. Grid and Mat (RAFT) Foundations....647
23.5. Rectangular Footing Supporting Two Columns Load....648
23.6. Trapezoidal footing supporting two Columns Loads....651
23.7. Approximate method of design of footing supporting two columns....653
23.8. Footings on Piles....654
23.9. Number of Piles....654
23.10. Evaluation of Pile Reaction....655
23.11. Reinforced Concrete Piles....695
23.12. Load carrying capacity of Piles....696
23.13. Stresses in Precast Piles....700
23.14. Pile Cap....703
23.15. Analysis of Pile Caps....704
23.16. Design of Piles, Problems....707

Part - VII.
Design of Other Concrete Structures

24. Design of Stairs 717-747

24.1. Introduction....717
24.2. Types of Stairs....718
24.3. Effective Span of Stairs....720
24.4. Distribution of Loads on Stairs....723
24.5. Loads on Stairs....725
24.6. Structural System of Stair Slab....727

25. Design of Retaining walls 748-809

25.1. Introduction....748
25.2. Types of Retaining Walls....748
25.3. Active Earth Pressure....749
25.4. Passive Earth Pressure....752
25.5. Effect of Surcharge on Earth Pressure....753
25.6. Earth Pressures for Common Conditional of Backfill....754
25.7. Forces Acting on Retaining Wall....757
25.8. Strength of Retaing Wall....758
25.9. Stability of Retaing Wall....758
25.10. Behaviour of Cantilever retaining Wall....763
25.11. Preliminary Proportions of Cantilever Retaining Wall....765
25.12. Design of Cantilever Retaining Wall....769
25.13. Counterfort Retaining Walls....785
25.14. Behaviour of Counterfort Retaining Walls....786
25.15. Design of Counterfort Retaining Wall....787

26. Design of Domes 810-832

26.1. Introduction....810
26.2. Definitions of Terms Related with Domes....811

26.3. Various Loads Acting on Domes 812
26.4. Spherical Dome Carrying uniformly distributed load 813
26.5. Spherical Dome Carrying Concentrated Load at Crown 817
26.6. Design of Domes 820
26.7. Conical Dome 827

27. Design of Reinforced Brick Slabs and Hollow tiles 833-847

27.1. Introduction 833
27.2. Design of Reinforced Brick Slab 834
27.3. Hollow Clay Tiles 839
27.4. Design of Ribbed Slab 839

28. Design of Form-Works 848-866

28.1. Introduction 848
28.2. Requirements of Form-Work 849
28.3. Stripping Time 850
28.4. Economy of Form-Work 851
28.5. Materials for Form-Works 852
28.6. Loads on Form-Work 854
28.7. Form-Work for Beam and Slab 855
28.8. Form-Work for Columns 857

Part VIII.
Concrete Mix and Laboratory Tests

29. Design of Concrete Mixes 867-930

29.1. Introduction 867
29.2. Concrete Mix Proportioning 868
29.3. Nominal Mix Concrete 870
29.4. Common Terms 871
29.5. Statistical Quality Control of Concrete 872
29.6. Standard Deviation 872
29.7. Target Strength 874
29.8. Acceptance Criteria 875
29.9. Methods of Concrete Mix Design 877
29.10. Minimum Voids Method 878
29.11. Maximum Density Method 879
29.12. Fineness Modulus Method 881
29.13. Road Research Laboratory Method 887
29.14. American Concrete Institute Method 901
29.15. British Mix Design Method 909
29.16. USBR Mix Design Method 910
29.17. Indian Road Congress Mix Design Method 912
29.18. Indian Standard Mix Design Method 920

30. Laboratory Tests for cement and concrete 931-951

30.1. Introduction 931
30.2. Fineness of Cement 932
30.3. Soundness of Cement 937
30.4. Consistency of Standard Cement Paste 938
30.5. Initial and Final Setting Times of Cement 939
30.6. Compressive Strength of Cement 940
30.7. Tensile Strength of Cement 941
30.8. Slump Test 943
30.9. Compacting Factor Test 945
30.10. Compressive Strength of Concrete 946
30.11. Flexural Strength of Concrete 949

31. Non-Desctructive Tests/Evaluation (NDT/NDE) [Non-mechanial Test Methods] 952-987

31.1. Introduction 952
31.2. Non-Destructive Tests/Evaluation (NDT/NDV) [Non Mechanical Tests Methods] 953
31.3. Comprehensive (Broad) Classification of NDT Techniques 954
31.4. Historical Background (Methods to Estimate in-Place Strength) 956
31.5. Visual Inspection 957
31.6. Sonic Tests 958
31.7. Some terms (Electric-Grating and Electron diffractions) used in Ultra Sonic Method 959
31.8. Ultrasonic (Supersonic) Pulse Velocity (UPV) Test 961
31.9. Schmidt Rebound Hammer 965
31.10. Combined use of Rebound Hammer and Ultra-Sonic Pulse Velocity Method 967
31.11. Electrical and Magnetic Methods for Reinforcement 968
31.12. Half Cell Potential Method 971
31.13. Maturity Method (Temperature Time Factor) 973
31.14. Permeability Test for Concrete 974
31.15. Resonance Frequency Method 977
31.16. Pull-Out Test 978

31.17. Break-Off Test 980
31.18. Probe-Penetration Method 981
31.19. Nuclear (Radio active) Methods 982
31.20. Radiometric Methods 983
31.21. Back-Scatter Method 984
31.22. Radiographic Methods 984
31.23. Comments on Nuclear Methods 985
31.24. Comments for use and Applications of Nondestructive Tests/Evaluations. 985

32. Futuristic construction materials and techniques (FCMT) 988-1018

32.1. Introduction 988
32.2. Designed Requirements for Futuristic Construction Materials (FCMT) 990
32.3. Self-Compacting Concrete (SCC) 991
32.4. Properties of Self-Compacting Concrete 993
32.5. High Performance Concrete (HPC) 994
32.6. Concrete Nano-Technology 998
32.7. Polymer-Modified Cement, Mortar and Concrete 1000
32.8. Polymer Mortar (PM) and Polymer Concrete (PC) 1004
32.9. Polymer impregnated Mortar (PIM) and Polymer Impregnated Concrete (PIC) 1007
32.10. Polymer Composites in Civil Engineering Construction 1009
32.11. Fibre Reinforced Polymer (FRP) and Alternative Construction Material 1009
References 1013

Appendix - A. Keep Awareness about earthquakes 1018-1037

Appendix - B. Service Life of Concrete Structures 1038-1042

Index 1043-1054

1

PART I

CONCRETE STRUCTURES, ITS SERVICE LIFE CONCRETE, ITS CONSTITUENTS AND DESIGN CODES AND SPECIFICATIONS

INTRODUCTION

1.1 INTRODUCTION

Word 'Engineer' is originated from the Latin-word 'Ingener'. In our Indian culture, (i.e., Hindu Sanskrati), a person who creates artifacts (i.e., Brahmaji/ Vishwakarma/Vidhata). An engineer uses his power of imagination and capacity, constructs basically structures ingenu-ously and with simple and pure thoughts. An engineer understands others constraints and their capacity to spend. The construction of such structures which is possible and practi-cable for use, commonly it may be said, an engineer is one who designs and manufactures engines (a machine composed of simple parts). Long back, there used to be only two catego-ries of engineers (military engineer and civil engineer) who used to look after all expects. Now-a-days, there are different engineers for minute disciplines and their capacity and expertisation in their fields are immense.

An engineer understands the importance of finance, significance of project management and information technology. Engineer improves and updates his knowledge and quality of his design. It is essential to understand and keep collaboration with other disciplines. To-day's time is the time of inter-discipline, to work together. It is necessary to maintain comm-unication across professional limits. Engineer is to ensure that his design is proper and suitable for use, economical, durable and safe (both for users and neighbours). Engineer ensures that his construction is aethetically pleasing.

Civil engineering is the art and science of planning, design, construction, operation, maintenance and rehabilitation of structures. Term structure includes bridges, multi-storey buildings and all types of civil engineering structures (towers, shells, ...etc.) composed of any structural material. In common language it may be said, that engineering is an art of const-ructing and manufacturing structures. Structural engineering is a branch of civil engineering.

The **structural engineering** is that branch of engineering which deals with struct-ural analysis and structural design. The structural engineering covers a wide range. The structural engineering plays an important and necessary part in mechanical engineering, electrical engineering, civil engineering, naval engineering, aeronautical engineering and in all the specialized phases of engineering. The **structural analysis** deals with the development of suitable arrangement of structural elements for the structures to support the external loads or the various critical combinations of the loads which are likely to act on the structure. The analysis also deals with the determination of internal forces in the various members (viz., axial forces, bending moments and shear forces), state of stresses or critical combination of the stresses at the various points (which includes the nature, magnitude and direction of these stresses) and the external reactions due to the worst possible combination of the loads. The internal force per unit area of the original cross-section is called **stress.** The external reactions are transmitted to the foundation. The methods of structural analysis and the principles involved in them remain independent of the materials used for all types of structures, whether the structures are built of plastics, aluminium, timber, reinforced concrete or steel.

The analysis of stress, strain and the deflection characteristics of a structure is referred as **structural analysis.**

Structural design is an art and science combining the experienced engineers, feelings for the behaviours of a structure with a sound knowledge of principles of statics, dynamics, mechanics of materials and structural analysis to produce a safe economical structure which will serve its intended purpose. In the design, certain criteria must be established to evaluate whether or not an optimum has been obtained. The typical criteria may be (i) minimum cost, (ii) minimum weight, (iii) minimum construction time, (iv) minimum labour, (v) minimum cost of manufacture of owner's product and maximum efficiency of operation to owner. In most practical situations, the evaluation must be qualitative. Only simple objective criteria, such as weight or cost must be kept in mind.

The **structural design** deals with the selection of proper material, proper sizes, proportions and shape of each member and its connecting details. The selection is such that it is economical and safe. It satisfies all the stress requirements imposed by the most severe combination of the loads to which the structure is required to transmit or resist including its self- weight. The structural design further deals with the preparation of final layout of the structure. The finished design drawings with all pertinent or controlling dimensions for all the members, parts and connections are made. The finished design drawings are necessary for fabrication and construction.

The structural design is governed with the standard specifications. The hand-books are used as working tools in the structural design. When the reinforced cement concrete is used for the construction of the structures, the structural design is known as **design of reinforced concrete structures.** All the structural reinforced concrete components are designed as per IS: 456-2000, Code of Practice for Plain and Reinforced Concrete. The provisions of code are thoroughly followed, intelligently interpreted and carefully applied.

In reinforced concrete, reinforcing steel bars and concrete are used together with advantageous combination of their physical and mechanical properties. The steel and concrete deform together under the load due to the strong grip between them. The steel encased in dense concrete gets protection against corrosion and direct exposure to fire. The coefficients of linear thermal expansion and contraction of steel and concrete are practically equal. As such, the steel does not slip from the concrete and the initial stresses in two materials remain insignificant, even if, the ambient temperature varies with ±100°C. Reinforced concrete is extensively used in construction owing to its durability, weather and fire resistance, good compressive strength, high dynamic resistance, excellent formability and relative low cost of concrete and high tensile strength, much greater ductility and toughness of steel. Reinforced concrete is used in multi-storey buildings, industrial buildings, thermal power stations, nuclear power stations and reactors, hydroelectrical power plants, large span thin-walled shell structures, bunkers, tanks and smoke stacks. Reinforced concrete is also used for highway and railway bridges, tunnels, irrigation works, structures for underground works, etc.

1.2 CONCRETES

Concrete is a non-homogeneous manufactur state composed of graded, granular inert material which are held together by the action of cement and water.

A carefully proportioned mixture of cement (used as binder), fine aggregate (sand) and coarse aggregate (gravel) and water, which hardens to a stone-like mass, is called as *concrete.* It is a composite material [consists essentially of a binding medium (cement)] within which particles of aggregate are embedded. This hardened mass of concrete appears like stone as regards its weight, hardness and strength. A plastic mixture (wet concrete) is formed by mixing these four ingredients. The wet concrete is poured in forms (moulds) of the shape and dimensions of the desired structure or structural component and it is cured to hard mass. All the four ingredients have their separate functions. The coarse aggregates act as main filler. The fine aggregates fill in the voids in the coarse aggregates. The coarse aggregates and fine aggregates increase the volume of material. The cement and water act as binder. The chemical interaction of cement and water binds the aggregate particles into a solid mass. The wet concrete is made workable by using additional water over and above required for the chemical reaction. The wet concrete solidifies and attains useful strength slowly. The completion of chemical reaction of cement and water needs favourable temper-ature and occurs in three distinct stages of time. The freshly mixed concrete decreases in plasticity and develops marked resistance to flow. This first stage of setting, known as **initial set** takes *thirty minutes* to *sixty minutes* after the original mixing operation. After this, the concrete becomes relatively soft solid without surface hardness. This second stage of setting, called as **final set** takes *five hours* to *six hours* after the original mixing operations. There afterwards, the concrete continues to harden and gains strength progre-ssively and indefinitely. The major portion of its potential hardness and strength is gained rapidly practically within one month after mixing. Concrete behaves very well when subjected to compressive forces.

The concrete used to be designed since long in terms of volume of cement, fine aggregate and coarse aggregate used in their proportions (e.g., $1 : 2 : 4$, $1 : 1\frac{1}{2} : 3$, . . . etc.). Now proportions are specified by weight (the ingredients are in the same order : cement first, fine aggregate next and coarse aggregate last) and the total water to be used is also mentioned. By appropriate adjustment of the proportions of the ingredients, the concrete in a wide range of strength properties may be obtained. An even wider variety of strength properties may be obtained by using special cements (such as high early strength cement), special aggregates (such as lightweight or heavyweight aggregates), admixtures (such as plasticizers and air-entertaining agents) and special curing methods (such as steam- curing). The process of controlling the conditions of humidity and temperature in which the concrete mix is maintained from the time, it is placed in the forms (moulds) until it is fully hardened. The process known as **curing** influences the strength property of the concrete, greatly.

The high compressive strength of concrete makes the concrete suitable for columns and arches (members primarily subjected to compression). The small tensile strength of concrete as compared with its compressive strength prevents its economical use in structural members that are subjected to tension either entirely (e.g., the rods) or over part of their cross-section (e.g., beams or other flexual members).

Concrete ruptures suddenly when the tensile forces are applied. The steel bars are used to reinforce the concrete. The steel bars completely surrounded by the hardened concrete form internal part of the structural member. In such a manner that the steel bars and concrete act together in resisting forces. The concrete reinforced with the steel bars is known as **reinforced concrete**. The reinforcement (a word from the police and military and means to increase the existing strength).

The combination of steel and concrete permits the almost unlimited uses of reinforced concrete.

1.3 CEMENT

Cement is a cementitious material, which has the adhesive and cohesive properties necessary to bond inert aggregates into a solid mass of adequate strength and durability. Cement is the most important constituent of concrete. In earlier time, crushed material from natural rocks used to be burnt and ground and used as cement. Such cement is referred as natural cement Water is needed for the chemical process (**hydration**) in which the cement powder forms soft paste, which sets and hardens into one slid mass. As such the cements are referred as **hydraulic cements**. An appropriate type of cement is used for a particular requirement. The different types of cements are commonly made by adjustment in relative proportions of chemical compounds and fineness. The various types of cements may be classified in three categories (viz., portland cements, special cements and other cements). The brief description of different Portland cements (conforming to respective Indian Standards, referred by appropriate IS : numbers) used for reinforced concrete are as follows:

Ordinary portland cement (IS : 269)

Ordinary portland cement is the most commonly used cement. *Ordinary portland cement* is a finely powdered grayish material obtained by intimately mixing together calcareous material (such as lime stone or chalk) and argillaceous material and/or other silica, alumina, or iron oxide bearing materials (e.g., clay and shale), burning them at a clinkering temperature 1400°C -1450°C) and grinding the resulting clinker. The gypsum or air entertaining agent is only added after burning. Ordinary cement is the basic cement commercially manufactured in large quantities. *The cement made with ordinary portland cement shrinks while hardening and setting and the concrete has a tendency to develop cracks.* In case there is no exposure to sulphates in the soil or in underground water, the ordinary portland cement is suitable for plain and reinforced constructions. This cement resembles in its properties with stone quarried at Portland, United kingdom and as such, it is named as portland cement. This cement was discovered by brick-layer Joseph Aspin of Yorkshire.

Lime stone (which provides CaO), clays or sales (which furnishes silica, SiO_2) and alumina (Al_2O_3) are the three major and common raw materials used for the manufacture of this cement. Two parts of calcareous material (lime stone $CaCO_3$) and one part of argillaceous material (clays and alumina, viz., SiO_2 and Al_2O_3) are mixed, crushed and ground in ball mills. In addition to these, this mixture contains iron oxide, magnesia, sulphur trioxide and alkalies in small proportions. This finely ground powder is burnt in rotary kiln to white heat and fused to clinkers. The clinkers are cooled and ground to the required fineness. The gypsum or air entraining agent is the only material added after the burning. The percentages of main chemical constituents of portland cement are as under :

Chemical Constituents of Portland Cement

1	Lime	$Ca.O$	59-64
2	Silica	$Si.O_2$	19-24
3	Alumina	$Al_2.O_3$	3-6
4	Iron-oxide	$Fe_2.O_3$	10-4
5	Magnesia	$Mg.O$	0.5-4
6	Sulphur trioxide	SO_3	1.2-7.5
7	Alkalies (Soda and Potash)	$Na_2.O+K_2O$	0.5-1

During burning and fusion, the constituents of portland cement form the following mineral compounds due to chemical reaction. The reactions of these mineral compounds with water were explained by Professor P.H. Bogue and W. Lerch in their paper titled as "Hydration of Portland Cement Compounds" published in PCA Fellowship paper 27, August, 1934. Later on, these mineral compounds were also called as **Bogue compounds.**

Mineral Compounds of Portland Cement

S.No.	Mineral compounds	Conventional formula	Abbreviated form
1	Tri-calcium silicate	$3\ CaO.SiO_2$	C_3S
2	Di-calcium silicate	$2\ CaO.SiO_2$	C_2S
3	Tri-calcium aluminate	$3\ CaO.Al_2O_3$	C_3A
4	Tetra-calcium alumino-ferrite	$4\ CaO.Al_2O_3.Fe_2O_3$	C_4AF

When the cement is mixed with water, the *cement paste* is formed. During the process of **setting** and **hardening,** the cement paste gradually stiffens until it becomes a solid. In setting, the cement paste changes its plastic state to a stiff solid state and gains only small compressive strength. In *hardening*, the mass becomes harder and its compressive strength increases as hydration proceeds with time. The **initial setting** and the early hardening of concrete occur due to the hydration of tri-calcium aluminate, C_3A. The vigorous reaction and rapid setting effect of C_3A is checked by adding adequate amount of retardent (gypsum, $CaSO_4$). It seems irregular that the sulphates are used to control the setting. The sulphates are destructive to hardened concrete. C_3A provides early effect. Its maximum strengthening effect is within 24 hours. The strengthening of concrete due to C_3A continues upto 28 days only. C_3A combines with C_4AF actively with water and forms hydrated products. These products contribute moderate strength to concrete within the few hours or days of their existence.

C_3S and C_2S compounds together form *seventy to seventy- five percent* portion of the cement. C_3S hydrates rapidly and provides early strength as well as ultimate strength. The reaction of C_2S with water takes place slowly and provides the strength after *seven days*. The prolonged reaction of C_2S with water provide progressive increase in strength of portland cement concrete. The concrete resists the attacks of acids and alkalies and becomes more durable due to C_3S and C_2S compounds.

33-grade ordinary Portland Cement conforming to IS 269 is increasingly used.

The various other types of cements are

(i) Low heat portland cement (IS : 269),

(ii) Portland pozzolana cement (IS : 1489),

(a) (fly ash based) conforming to IS : 1489 (Part I)

(b) (calcined clay based) conforming to IS : 1489 (Part II)

(iii) Rapid hardening portland cement (IS : 8041),

(iv) High strength portland cement (IS : 8112),

43-grade ordinary Portland cement conforming to IS : 8112

(v) Portland slag cement (IS : 455),

(vi) Hydrophobic cement (IS : 8043),

(vii) Sulphate resisting portland cement,

(viii) Water-proof portland cement,

(ix) White and coloured portland cements (IS: 8042),

(x) High alumina cements (IS : 6452),

(xi) Super sulphated cement (IS : 6909),

(xii) Quick setting cement,

(xiii) Expansive cement,

(xiv) Low heat portland cement conforming to IS : 12600,

(xv) Sulphate resisting portland cement conforming to IS : 12330,

53 grade ordinary Portland cement conforming to IS : 12269.

These cements have been described in detail in author's book (Design of Concrete Structures, Vol. 1).

The type of cement used may be any of the mentioned above and the type selected should be appropriate for the intended use.

1.4 CEMENT TESTS

Indian Standard Institution has recommended following specifications and tests in respective IS : codes in order to maintain the quality of different varieties of cements. The various laboratory tests for cement namely, **fineness of cement** (Dry sieve analysis method, Blaine's air permeability method), **Soundness of cement, consistency of cement, initial and final setting times of cement, tensile strength of cement**, and **compressive strength of cement** have been described in detail in this book [Chapter 30]. The cement is tested and it is seen that the cement satisfies the following chemical and physical requirements.

1. Chemical requirements

The chemical analysis of cement to be used is carried out as per IS : 4032 -1968 (Methods of chemical analysis of hydraulic cement) and it is seen that the cement satisfies the following chemical requirements.

(i) The percentage loss of *ignition* (by mass) and percentage of *insoluble residues* do not exceed the values given in Table 1.1 (chemical requirements of cements).

(ii) The percentage of lime, PL = ($CaO - 0.7\ SO_3$) and percentage of silica, alumina and iron oxide, PSAF = ($2.8\ SiO_2 + 1.2\ Al_2O_3$) + $0.65\ Fe_2O_3$) are determined. It is seed that the ratio of PL and PSAF for ordinary portland cement and rapid hardening portland cement is not more than 1.02 and it is not less than 0.66.

(iii) The percentage of alumina, PA and percentage of iron oxide PF are found. It is checked that the ratio of PA and PF for ordinary portland cement and rapid hardening portland cement is not less than 0.66.

(iv) The percentage of *magnesia* oxide and *mananese* present in the cement (by mass) do not exceed the value given in Table 1.1.

(v) Total *sulphuric content* calculated as sulphuric anhydride, SO_3 shall not exceed percentages given in Table 1.1.

TABLE 1.1. CHEMICAL REQUIREMENTS OF CEMENTS

Chemical characteristics	Portland cements				Special cements				Other cements
	Ordinary	Pazzalana	Rapid hardening	Low heat	Slag	High strength	Hydrophobic	Sulpher sulphated	Masonary
1 Ignition loss	>5	5	>5	>5	>4	>5	>5	–	–
2 Insoluble residues	>2	*	>2	>2	>2.5	>2	>2	>4	–
3 Magnesia (percent by mass)	>6	6	>6	>6	>8	>6	>6	>10	–
4 Sulphur content (Sulphuric anhydride percent by mas)	2.75	2.75	2.75	2.75	3	–	–	6	–
5 Manganese oxide (Mn_2O_2)	–	–	–	–	2	–	–	–	–

*Percentage value = [x+ 2.0 (100 - x)/100] where x is the actual percentage of pozzalana.

The percentage of *lime* for low heat portland cement (after deducting the quantity necessary to combine with sulphuric anhydride, SO_3 present is found. It should not be more than

$$[2.4\ (SiO_2) + 1.2\ (Al_2O_3) + 0.65\ (Fe_2O_3)]$$

It should not be less than

$$[1.9\ (SiO_2) + 1.2\ (Al_2O_3) + 0.65\ (Fe_2O_3)]$$

It is to note that a symbol within a small pair of parenthesis to the percentage (by mass of total cement). The percentage contained in the insoluble residue is excluded.

2. Physical requirements

The cement is tested for the physical requirements which include **compressive strength, fineness, heat of hydration, setting time, soundness and tensile strength.**

(i) Compressive strength. At least twelve cubes (area of face 0.50 m^2) are prepared with one part of cement and three parts of Standard Leighton Buzzard Sand conforming to IS: 650-1966 by mass and water. The percentage of water mixed shall be [(Pa/4+3.0)] percent by mass, where Pa is the percentage of water needed for normal consistency. The crushing strengths of one set of three cubes (tested in accordance to IS : 4031-1968) are found after one day (24 hrs) from time of preparing cubes, and then for second set of cubes after three days (72 hrs), third set of cubes after *seven days* (168 hrs) and fourth set of cubes *twenty eight days* (672 hrs) are found. The average of three values of each set tested should not be less than that given in Table 1.2 (physical requirements of cements).

(ii) Fineness. The fineness of cement is found in terms of specific surface (i.e., the surface per unit mass (m^2/gm) of the cement. The rate of hydration and hydrolysis of cement depends upon its fineness. Finer cement acts more quickly with water and develops high early strength. The method of sieve and method of Blaine's permeability as described in IS : 4031-1968 are two methods to determine the fineness of cement. The specific area found for the cement should not be less than the values given in Table 1.2.

(iii) Heat of hydration. The heat of hydration for low heat portland cement and for super sulphate cement (calories per gm) found after 7 days and 28 days should not exceed the values given in Table 1.2.

TABLE 1.2. PHYSICAL REQUIREMENTS OF CEMENTS

Chemical characteristics	Portland cements				Special cements				Other cements
	Ordinary	Pazzalana	Rapid hardening	Low heat	Slag	High strength	Hydrophobic	Sulpher sulphated	Masonary
1 Compressive strength min (N/mm^2)									
(24 hrs ± 0.5 hrs)	–	–	<15	–	–	–	–	–	–
(72 hrs ± 1 hr)	<16	–	<27	<10	<16	<23	<21	<22	–
(168 hrs ± 2 hrs)	<22	<22	–	<16	<22	<33	<21	<22	<2.5
(672 hrs ± 4 hrs)	–	<31	–	<35	–	<43	<30	<30	<5.0
2 Fineness (m^2/gm)	<0.2250	<0.3000	<0.3250	<0.3200	<0.2250	<0.3500	<0.3500	<0.4000	–
3 Heat of hydration									
7 days (cal/gm)	–	–	–	–	–	–	–	–	–
28 days (cal/gm)	–	–	–	–	–	–	–	–	–

4 Setting time									
Initial (minutes)	<30	<30	<30	<30	<30	<30	<30	<30	<90
Final (minutes)	>600	>600	>600	>600	>600	>600	>600	>600	>1400
5 Soundness (mm)	>10	>10	>10	>10	>10	>10	>10	>5	>10
Tensile strength (N/mm²)									
6 24 hrs	<-	–	<2	–	–	–	–	–	–
72 hrs	<2	–	<3	–	–	–	–	–	–
168 hrs	<2.5	–	–	–	–	–	–	–	–

(iv) Setting time. The main aim of determining setting times is to find the distinction between normal setting and quick setting types of cements and also to know the deterioration of cement in storage. The initial and final setting times as found by Vickat apparatus and these values for a cement should not be less than that given for that type of cement in Table 1.2.

(v) Soundness. The main aim of finding the soundness of cement is to know the amount of overburnt uncombined lime present in the cement. Due to presence of unburnt uncombined lime and magnesia, the volumetric changes occur in the cement. After setting of cement, these ingredients continue to slack very slowly. The volumetri expansion is called unsoundness and distortion, disintegration and cracks in concrete may occur. It is found by Lechatelier's method in accordance with IS : 4031-1968 and its value should not be more than that in Table 1.2.

(vi) Tensile strength. Six standard briquettes are made of one part cement and three parts of Standard Leighton Buzzard Sad. The tensile strength of briquettes after *on day* (24 hrs), *three days* (72 hrs) and seven days (168 hrs) immersion respectively are found. The respective tensile strength found should not be less than the value give in Table 1.2.

(vii) Consistency of cement paste: The standard consistency of a cement paste is defined as that consistency which allows the vicat plunger top penetrate to points 5 mm to 7 mm from the bottom of the vicat method. The quantity of water needed to produce a cement paste of standard consistency vicat mould method is described in this book, Chapter 30.

Design of Concrete structures: The unit of consistency is percentage of water by mass of dry cement. It is denoted by P. The vicat apparants (as per IS: 5513-1976) is used. In which a plunger 50 long and 10 mm diameter used to penetrate the cement mortar.

The flexural and splitting tensile strength shall be obtained as described in IS : 5516 & IS: 5816, respectively. The tensile strength may be estimated from the compressive strength whenever, it is desired) by using the following formula flexural strength for = 0.7 $(f_{ck})^{1/2}$ N/mm².

1.5 AGGREGATES

Aggregate is inert material (chemically inactive) such as natural sand, slag, air cooled iron blast slag, pebbles, gravel, broken stone, broken bricks, industrial bye products etc., which is mixed with cement and water and concrete or mortar is prepared. The aggregate is added to cement paste to increase the bulk. *Seventy to seventy-five* percent of volume of the hardened mass of concrete consists of aggregates. The remaining *thirty to twenty-five* percent consists of hardened cement paste, uncombined water (i.e. water not involved in the hydration of the cement) and air voids. Aggregate must be clean, dense, hard, durable, structurally sound, capable of developing good and with cement, weather resistance and unaffected by water. Sand gravel and broken stone are natural available materials most commonly used as aggregates. Blast furnace slag (air-cooled and crushed) and broken bricks are the artificial and processed aggregates used in concrete. Natural aggregates are generally classified as coarse and fine aggregates. Natural materials having uniform shear strength in all directions are ideal for use as aggregates. Other natural materials having pronounced planes of weakness or cleavage, (e.g., slate, shale, and micaceous materials) are usually undesirable. The aggregates used for concrete should confer to IS: 383-1970 (specifications for coarse and fine aggregate from natural sources for concrete).

1. Coarse aggregate

Broken stones, pebbles and gravel retained on IS : 4.75 mm sieve is called as *coarse aggregate*. Coarse aggregate is prepared by crushing igneous stone (e.g., granite, gneiss), sedimentary stone (e.g., sand stones), and metamorphic stone (e.g., crystalline lime). Coarse aggregates prepared from very fine-grained granite stone is used for high strength concrete. Coarse aggregates obtained from crushing of hard sand stone are also suitable for use in concrete. Coarse aggregates made from the soft sand stone cracks when there is high degree of shrinking in concrete, and it should be used with care. Coarse aggregates (if prepared from close-grained crystalline lime stone) is also suitable for use in plain cement concrete. Coarse aggregate made from broken bricks is also used for plain cement concrete only. It makes the concrete weak in strength. In structural concrete, the reinforcement is likely to be rusted when the broken brick aggregates are used. Coarse aggregates are separated in two or three size groups for use in concrete for large and important constructions. Each size group of aggregate is separately weighed according to grade charts and then combined to result in densely packed concrete. For construction of less importance, aggregates brought from the pits or river beds (called as **all-in-aggregate**) are directly used.

2. Fine aggregate

Sand and any other aggregates which pass through IS : 4.75 mm sieve is called *fine aggregate*. Sand is brought from lakes, pits, rivers and sea-shores. Sand is thoroughly washed so that there remain no clay films over the surface of the aggregate. It prevents adhesion of cement to the aggregate. Clay and silt are fine materials which pass through 75-micron IS : sieve. Clay and silt materials (if present) will increase water content. Dust is also detrimental material to concrete. As such the fine aggregates should be free from dust, organic impurities, and soils. Chlorides (present in sea-shore sand) cause efflorescence and rusting and corrosion of reinforcement in structure concrete.

In order to prepare strong concrete, angular grained sand (having good interlocking property) is used. Round grained particles of sand lack in interlocking property.

In order to prepare the concrete to have good strength, weather resistance and durability, the aggregates are closely packed. For this reason the gradation of particle sizes of the aggregates is of considerable importance.

3. **All-in aggregate** material composed of a mixture of coarse and fine aggregate.

Grading of aggregates

Gradation of particle sizes of the aggregate vary with the nature and course of material and the requirements in this respect depend upon the type of structure. The particle sizes of fine and coarse aggregates are properly graded. Use of well graded aggregates reduces the air voids. The cement fills up the air voids of fine aggregates. The cement and sand both together fill up the air voids of the coarse aggregate.

The finishing, workability and uniformity of concrete are considerably affected by the grading of *fine aggregates*. In the grading of fine aggregates, most of the natural sands of this country are divided into *four zones* as given in IS : 383-1970. Following *square hole perforated plate type and fine mesh wire cloth type* IS : Sieve are used:

(i) Square hole perforated plate type IS : Sieves

80 mm, 63 m, 50 mm, 40 mm, 31.5 mm, 25 mm, 20 mm, 16 mm, 12.5 mm, 10 mm, 6.3 mm, 4.75 mm.

(ii) Fine mesh wire cloth type IS : Sieves

3.35 mm, 2.36 m, 1.18 mm, 600-micron, 300 micron, 150-micron, 75- micron.

Natural sands correspond to one or the other out of these four zones. Sands which may fall in grading zone-I, II or III and IV are used in concrete. Typical good sands fall in grade zone-II. However, finer or coarser sand may be used with suitable adjustments in the ratio of quantities of coarse to fine aggregate.

If a lean concrete mix is desired, the coarse sand falling in grading zone-I should be avoided. Very fine sands falling in grading zone-IV should not be used.

Single size *coarse aggregates* (ungraded) used in concrete may have nominal size of 63 mm, 40 mm, 20 mm, 16 mm, 12.5 mm and 10 mm. The grading limits for single size coarse aggregate are given in Table 1.3 (A) as per IS : 383-1970. Coarse aggregates of two or more sizes (Graded coarse aggregates), 40 mm, 20 mm, 16 mm and 12.5 mm mixed together give densely packed aggregates. The graded limits for graded aggregate are given in Table 1.3(B) as per IS:383-1970.

The grading limits for *fine aggregates* are given in Tale 1.4. Grading zone-I corresponds to coarse sands. Grading zone-II corresponds to typical good sands. Grading zone-III corresponds to little fine sands and grading zone-IV corresponds to very fine sands.

The grading limits for all-in-aggregates are given in Table 1.5.

Table 1.3(A). GRADING LIMITS
For Single-sized coarse aggregate of nominal size (IS: 383-1970)

IS: Sieve Designation	Percentage passing					
	63 mm	40 mm	20 mm	16 mm	12.5 mm	10 mm
80 mm	100	–	–	–	–	–
63 mm	85-100	100	–	–	–	–

40 mm	0-30	85-100	100	–	–	–
20 mm	0-5	0-20	85-100	100	–	–
16 mm	–	–	–	85-100	100	
12.5 mm	–	–	–	–	85-100	100
10 mm	0-5	0-5	0-20	0-30	0-45	85-100
4.75 mm	–	–	0-5	0-5	0-10	0-20
2.36 mm	–	–	–	–	–	0-5

Table 1.3(A). GRADING LIMITS
For graded coarse aggregate of nominal size (IS: 383-1970)

IS: Sieve Designation	Percentage passing			
	40 mm	20 mm	16 mm	12.5 mm
80 mm	100	–	–	–
63 mm	–	–	–	–
40 mm	95-100	100	–	–
20 mm	33-70	95-100	100	100
16 mm	–	–	90-100	–
12.5 mm	–	–	–	90-100
10 mm	10-35	25-55	30-70	40-85
4.75 mm	0-5	0-10	0-10	0-10
2.36 mm	–	–	–	–

Table 1.4. GRADING LIMITS
For fine aggregates (IS: 383-1970)

IS: Sieve Designation	Percentage passing			
	Grading zone-I	Grading zone-II	Grading zone - III	Grading Zone - IV
10 mm	100	100	100	100
4.7 mm	90-100	90-100	90-100	95-100
2.36 mm	60-95	75-100	85-100	95-100
1.18 mm	30-70	55-90	75-100	90-100
600 miccron	15-34	35-59	60-79	80-100
300 miccron	5-20	8-30	12-40	15-50
150 micron	0-10	0-10	0-10	0-15

Table 1.5. GRADING LIMITS
For all-in aggregates (IS: 383-1970)

IS: Sieve Designation	Percentage passing (Nominal size)	
	40 mm	20 mm
80 mm	100	–

40 mm	95-100	100
20 mm	45-75	95-100
4.75 mm	25-45	30-50
600 micron	8-30	10-35
150 micron	0-6	0-6

Table 1.6. LIMITS OF DELETERIOUS MATERIALS

Deleterious substances	Percentage limits by weight			
	Coarse aggregate		Fine aggregate	
	Crushed	Uncrushed	Crushed	Uncrushed
Coal, lignite	1.00	1.00	1.00	1.00
Clay lumps	1.00	1.00	1.00	1.00
Material passing 75 micron sieve	3.00	3.00	15.00	3.00
Soft fragments	–	3.00	–	–
Shale	–	–	–	1.00
Total limit	5.00	5.00	2.00	5.00

Impurities in aggregate

In order to have good strength, durability and weather resistance, the surface of aggregate is kept free from impurities such as iron pyrites, coal, mica, shale, loam, silt, clay, alkali, soft fragments, sea shell, organic matter and other and other deleterious (harmful) materials, which may weaken the bond with cement paste; and that no unfavourable reaction takes place between the aggregate and cement. Use of aggregate containing reactive silica (such as, chert and chalcedony) should be avoided. Weak aggregate (obtained from soft lime stone, soft sand stone or other porous stone) should not be used for concrete in sea water. Table 1.6 gives the limits of impurities as per IS : 383-1970.

Size of aggregate

For mass concrete (such as in dams) and plain concrete, the maximum size of coarse aggregate may be 200 mm and 63 mm respectively.

Plums above 160 mm and upto any reasonable size may be used in plain concrete work upto a maximum limit of 20 percent by volume of concrete. The plums should be distributed uniformly. The plums should be placed at depths 150 m or more from the surface.

For heavily reinforced concrete work, the nominal size of coarse aggregate should be as large as possible within the specified limits but in no case, it should be greater than *one-fourth* of the minimum thickness of the member, and it should be possible that the concrete can be placed without any difficulty so as to surround all the reinforcement thoroughly and fill the corners of forms (moulds) and honey comb pockets are avoided. The largest possible size and properly graded aggregates are used, which reduce the water demand. A lower maximum size of aggregate is adopted for high compressive strength.

The nominal maximum size of aggregate is usually limited to 5 mm less than the minimum clear cover to the reinforcement. The minimum clear cover specified in IS : 456-2000 for columns of dimensions more than 200 mm and for columns of minimum dimension 200 mm are 40 mm and 25 mm, respectively. As such, the nominal sizes of coarse aggregates adopted are (40-5 = 35 mm) and (25-5=20 mm), respectively. The minimum clear cover recommended in IS:456-1978 for beams is 25 mm. Hence, the nominal size of coarse aggregate used is (25-5 = 20 mm).

In the solid slabs, the minimum cover mentioned is 15 mm. Therefore, the nominal size of coarse aggregate used is (15-5 = 10 mm).

1.6. OTHER TYPES OF AGGREGATES

Some unusual and unprocessed aggregates such as blast furnance slag, pumice, calcined clay, diatomaceous silica, asbestos, sawdust, vegetable fibre such as seaweed and others are used for thermal insulating and acoustical concretes.

Slag and crushed overburnt brick or tile aggregates may be found suitable with regard to strength, durability and free from harmful materials and effects may be used for secondary members. Such aggregates should not contain more than *one percent* of sulphates and should not absorb more than *ten percent* of its own mass of water.

1. Light weight aggregates

The aggregates having a dry loose density of 11 kN/m³ or less. Expanded shales, clays, slates, slags, pelletized fly ash aggregates are light in weight because of porous and cellular structure of the individual aggregate particle. Such structure is obtained by gas or steam formation in processing the aggregates in rotary kilns at high temperatures (commonly in excess of 1000°C. Light weight aggregate is further distinguished in three classes namely *low density* (having a unit weight 8 kN/m³ and used for insulation) *moderate strength aggregates* (having unit weight 9-10 kN/m³ and chiefly used as fill) and *over light weight* (having unit weight 14.40 - 19.20 kN/m³).

2. Normal weight aggregates

Usual aggregates described in Art. 1.5 are referred as normal weight aggregates. IS:383-1970 classifies the normal aggregates on the basis of particle shape and surface textures of the aggregates as in Table 1.7 (A) and (B).

3. Heavy weight aggregates

For nuclear reactor and similar installations, heavy weight aggregate are essentially needed for shielding against gamma and X-ray radiation. For special purposes, (e.g., counterweights of lift bridges) the heavy weight aggregates are used. Heavy iron ores or barite (barium sulphate) rock crushed to suitable sizes are used as heavy weight aggregates. Steel in the form of scrap, punchings or shots as fines) are also used. Unit weight of heavy concretes with natural heavy weight aggregates is 32 kN/m³. In case, ores are used as fines only and steel for coarse aggregate, unit weight is 53 kN/m³.

Table 1.7 (A). CLASSIFICATION OF AGGREGATE
Based on particle shape (IS : 383-1970)

Classification	Description	Examples
Rounded	Fully waterworn, or completely shaped by attrition	River or sea shore gravels; desert, seashore and wind-blown sands.
Irregular or partially rounded	Naturally irregular or partly shaped by attrition, and having rounded edges.	Pit sand and gravels: land or dug flints: cuboids rock.
Angular	Having well defined edges.	Crushed rocks of all types: talus, scress.
Flaky	Usually angular whose thickness is small relative to width and length	Laminated rocks.

Table 1.7(B). CLASSIFICATION OF GGREGATE
Based on surface texture (IS : 383 - 1970)

Surface Texture	Examples
Glassy	Black flint
Smooth,	Chert, slate, marble, some rhyolite
Granular	Sand stone, colites
Crystaline	
Fine	Basalt, trachyte, keratophyte
Medium	Delerite, granophyte, granulite, microgranite, some lime stone, many dolomites
Coarse	Gabbro, gneiss, granite, granodiorite, syemite
Honey combed and porous	Scoriae, pumica, trass

1.7 FINENESS MODULUS OF AGGREGATES

Fineness modulus indicates the relative fineness of the aggregate.

Fineness modulus of an aggregate is an index number used to indicate the average size of particles in the aggregate. The fineness modulus is found as the sum of percentage of the masses of aggregate retained on a set of ten IS : sieves (80 mm, 40 mm, 20 mm, 10 mm, 4.75 mm, 2.36 mm, 1.18 mm, 600 micron and 150 micron) and divided by one hundred. It is commonly abbreviate as F.M. Its value is an index of the average surface area of the aggregate.

Aggregates of different particle-size distribution may have similar fineness moduli. For aggregates of a specific type, it is reasonably certain that the aggregates of similar fineness moduli will, (if used in similar quantities) produce concrete of similar plasticity. It is worthwhile to see the aggregate from a particular source may be seen readily-from the fineness modulus.

It may be desired to obtain a specific fineness modulus, p for an aggregate mix of fine and coarse aggregates. The fineness moduli of fine and coarse aggregates p_1 and p_2 are found by passing the aggregates separately through set of nine standard sieves. The quantity by mass, m of fine aggregate to be mixed with one part of coarse aggregate is found by

$$m = \left(\frac{p_2 - p}{p - p_1}\right) \quad (1.1)$$

1.8 WATER

Water is needed for chemical process (**hydration**) in which the cement powder forms a soft paste, which sets and hardens into one solid mass. In the chemical process involved in setting and hardening, heat (known as heat of hydration) as liberated. For complete hydration of a given quantity of cement, quantity of water equal to 25 percent of that of cement by mass is needed. In addition to this, water is also required for wetting the surface of aggregate and for distributing the cement paste evenly so that every particle of coarse and fine aggregates is coated and brought into intimate contact with each other. Water acts as lubricant. As water is added, the plasticity and fluidity of mix *(i.e., workability)* increases but the strength decreases because of the larger volume of voids created by free water. Water is also required later on for curing the concrete after it is casted in moulds. Thus. water has both chemical and physical functions. Water used for drinking (potable water) is generally considered satisfactory for mixing concrete.

Water used for mixing and curing shall be clean and free from such materials which are harmful to concrete or steel. Oils, acids, alkalies, salts, sugar organic and inorganic materials if present in quantities more than their permissible quantities given in Table 1.8 (as per IS:456-2000 are injurious for concrete or steel.

Marsh waters, mine and colliery waste waters and several industrial waste waters are not likely to meet the requirements of IS : 456, and therefore, should be used only after careful consideration. Mineral oil (not mixed with animal or vegetable oils) in concentrations more than 2 percent by mass of cement may decrease the concrete strength by more than 20 percent.

Acids and alkalies (if present) should not exceed in their concentrations as their maximum permissible values as per IS : 456.

(i) To neutralise 200 ml sample of water, using phenolphthalein as an indicator, it should not require more than 2 ml of 0.1 normal NaOH.

(ii) To neutralise 200 ml sample of water, using methyl orange as an indicator, it should not need more than 10 ml of 0.1 normal HCl.

Sodium and potassium carbonates and bicarbonates affect the setting time of cements. In case, these salts are present in excess, the strength of concrete reduces. *Calcium and magnesium carbonates* are generally insoluble in water and therefore, these salts do not remain present in significant extents. However, *their bicarbonates* are soluble in water. Contents of such salts should not exceed 400 ppm.

Sodium chlorides (if present) do not have any harmful effects and its contents should not exceed 20,000 ppm. *Magnesium chlorides* also do not have any harmful effects. Its contents even upto 40,000 ppm do not influence the strength of concrete. *Calcium chloride* upto 2 percent by mass of cement accelerates setting and hardening. Total quantity of chlorides as

Chloride-ions (as Cl and not as Cl_2) in the concrete at the time of placing should be limited to 0.15 percent. The chloride ions are responsible for corrosion of embedded metal in concrete.

Table 1.8. PERMISSIBLE LIMITS FOR SOLIDS
(If present in water)

Substances	Tested as per	Maximum permissible limits (mg per litre)
Organic	IS3025-part 18	200
Inorganic	IS3025-part 18	3000
Sulphates (SO_4)	IS3025-part 24	400
Chlorides (Cl)	IS3025-part 32	
(i) (for plain concrete)	for concrete not containing embeded steel and 500 mg/litre for reinforced concrete work 2000 mg/litre	2000
(ii) (for reinforced concrete)		1000
Suspended matter	IS3025-part 17	2000

Sodium sulphates (if present) do not have harmful effects and its contents should not exceed 10,000 ppm. *Magnesium sulphates* also do not have harmful effects. Its contents even upto 40,000 ppm do not influence the strength of concrete. Total amount of soluble sulphates (referred as SO_3 and not as SO_4 followed in conventions used in cement chemistry) in the concrete at the time of placing should be limited to 4 percent by mass of cement. The concentration of sulphates expressed as SO_3 in soil and in underground water should not exceed those given in IS : 456 (Appendix-A).

Sugar if present in water may not have any adverse effect on strength of concrete so long as its contents do not exceed 500 ppm. However, it is likely that the setting time is affected due to presence of sugar.

Organic material such as, algae if present in water may cause large reduction in the strength of concrete.

For plain cement concrete, the *sea water* containing upto 35000 ppm salts (including sodium chloride and other salts) may be used for mixing and curing under unavoidable circumstances. However, in reinforced concrete, sea water should not be used. Sea water contains harmful salts. Sea water may cause corrosion of the reinforcement. Concrete prepared by mixing sea water will show efflorescence.

In remote areas and in areas where available water from some sources is not used for domestic purposes. The suitability of such water remains doubtful. In such cases, the compressive strength and initial setting are found. Salts of zinc, copper and lead and in particular, sodium sulphide may cause a considerable reduction in strength. These substances are not listed in that of harmful materials. In case contents of these salts present are very little, then, such water my be used.

In order to determine the compressive strength and initial setting time sample of water is taken from the water proposed to be used for concrete, however, the seasonal variation is kept in view. This sample of water is not treated other than envisaged in the regular supply of water proposed for use in concrete. A clean container previously rinsed out with similar water is taken to store the sample of water.

At least three cubes of 150 mm size are prepared from the sample water. Average 28 days, compressive strength of these cubes is found. It should not be less than 90 percent of the average of strength of three similar cubes prepared with distilled water.

The initial setting time with sample water and appropriate cement shall not be less than 30 minutes. At the same time, it should not differ ± 30 minutes for same cement and distilled water.

The pH value of water shall generally be not less than six.

1.9 ADMIXTURES

Admixtures are the materials other than portland cement, aggegate or water. The seam added to concrete to modify the properties of concrete.

An admixture is an extra component added to the main components of concrete mix during mixing in order to create a special property or to neutralize one or more normal characteristic of the fresh or hardened concrete or to correct some deficiency of the mixtures to improve the workability, to accelerate or retard setting and hardening, to aid in curing, to improve durability, to add colour and to give other properties. Some admixtures have established beneficial effects. These admixtures are manufacture's products and have consistent behaviour. Some admixtures are industrial waste products and have little sale values and their performance as concrete ingredients is inconsistent. The claims for such admixtures should be examined carefully. The distinguished admixtures may be arranged in four main groups *(i) accelerating admixtures (ii) retarding admixtures (iii) water-reducing or plasticizing admixtures and (iv) air-entraining admixtures.*

1. Accelerating admixtures

In order to increase the early strength development, accelerating admixtures are added. These admixtures react chemically with one or more constituents of ordinary port-land cement. These admixtures facilitate the early removal of form work. These admixtures are also used to reduce period of curing and to accelerate the concerting in cold weather and repairs. Calcium chloride, fluosilicates and tri-ethanolmine are commonly used as accelerating admixtures for ordinary portland cement.

If calcium chloride is used during warm weather or in heated concrete, the setting is accelerated to large extent and impairs finishing. The heat of hydration also increases to a degree that causes cracking. It aggregates the dry shrinkage of hardened concrete and it increases expansion created due to alkali- aggregate reactivity.

2. Retarding admixtures

In order to slow down the rate of setting of ordinary portland cement, the retarding admixtures are mixed. These admixtures also react chemically with one or more constitue-nts of ordinary portland cement. Retarding admixtures are useful in hot water concreting, for avoiding cold joints in mass concrete works and for special treatment of concrete surfaces.

Starches and cellulose products, sugars and hydroxyl-carboxyl acids and their salts (e.g., sodium tartrate) are generally used as retarding admixtures. Except sugars and carbohydrates, many of these retarding admixtures reduce water requirements.

3. Water reducing or plasticing admixtures

In order to permit more workability for a given water-cement ratio or alternatively to retain the workability while reducing water content, plasticizing admixtures (simply called as *plasticizers*) are added. *Plasticizers* and so called *super plasticizers* are used increasingly, particularly for high strength concretes because these admixtures allow considerable water reduction while maintaining high slumps needed for proper placement and compaction of concrete. *Super plasticizers* are also suitable where an extremely high workability is desired (e.g., where, the concrete is to be pumped).

These admixtures act mechanically during the mechanical life of the concrete. They provide cohesion and plasticity to poorly workable mixtures and act in two ways. Least significantly, there are fillers of small voids. Most importantly, there are spacers. They mechanically separate particles and reduce internal friction. They provide space for more cement paste and more surface areas to be wet by cement paste. Because of their small size, they stiffen and add skeleal structure to the paste. As such, the bleeding of water and sedimentation of cement are reduced.

3 (i) Lignosulphanate salts are the basic ingredients in these water-reducing admixtures. These contain sugar due to which retardation may occur. Alternatively a refined sugar reduced product may be used which will avoid excessive retardation. *Polyhdroxy compounds* are also basic ingredients of these admixtures. These compounds are sensitive to over-dosing, in which case, excessive retardation may be caused.

3 (ii) Superplasticizers are based on *synthesized condensates* (e.g., melamine formaldehyde and napthalene sulphonate-formaldehyde). In order to be totally effective, it is ensured that super plasticizers give high fluidity to concrete. At the same, it is seen that these admixtures do not cause undesirable side effects such as bleeding and segregation.

4. Air-entraining admixtures

In order to improve the workability and durability, air- entraining admixtures are added to intentionally entrain a controlled quantity of air in the form of small (microscopic) bubbles without considerably altering the setting or hardening of the concrete. These dispersed minute bubbles are stabilized so that they cannot escape or collapse. These admixtures are chiefly resistant to freezing and thawing (i.e., resistance to frost damage) and reduce sagregation during placing. These admixtures decrease the density of concrete because of increased void ratio and thereby strength is also reduced; however this reduction can be partially offset by decreasing mixing water without loss of workability. At present air-entraining admixtures are most improvement and most widely used admixtures. Air entraining admixtures are chiefly used in pavements, but these are also used for structures, particularly for exposed components.

The quantity of air-entrained in concrete depends on the quantity of such admixtures and as well as on other factors also. Animal and vegetable oils and fats, natural wood resins and their sodium salts (e.g., vinsol resin) and alkali-salts of sulphated and sulphonated organic compounds are mainly used also air- entraining admixtures.

Water reducing admixtures known as surfactants are also used. Surfactants need not necessarily produce-stable form of bubbles.

Pozzolona is also used as admixture, where its physical property such as fineness is important as compared to its chemical reactivity. Fly ash (a finely divided material) is also used as mineral admixture. Fly ash may be added only when the concrete mixtures are deficient in *fines*. When, fly ash in used, the water demand is increased and strength of concrete is reduced.

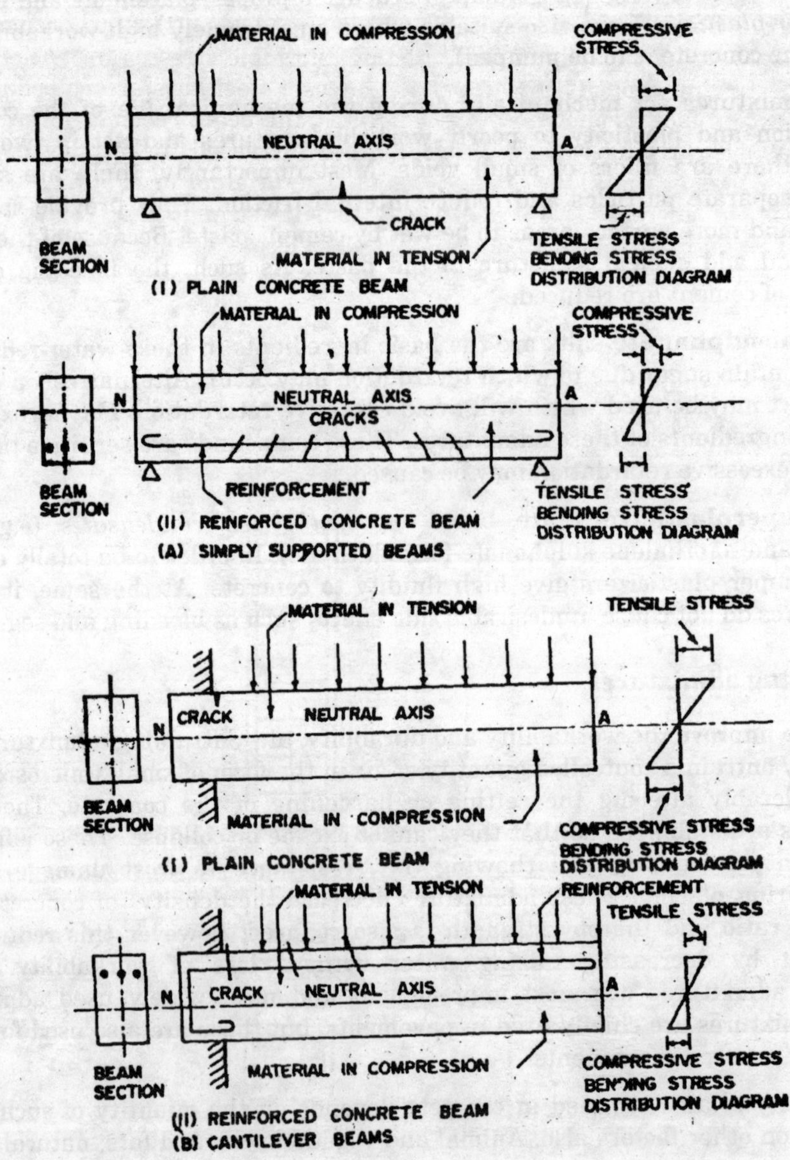

Fig. 1.1

Silica fume, rice-husk ash, metakaoline (having fireness between 700 to 900 m^2/kg), and ground granulated blast furnance slag are mineral admixtures which have been recommended in IS: 456-2000 (fourth edition) for use keeping in view the requirements mentioned.

1.10 REINFORCEMENT

Reinforcement is a term from military and police organisations. It means to increase the existing strength by additional force. Concrete is strong in compression and weak in tension.

A simply supported plain concrete beam subjected to loads acting downward carries bending compressive stresses in the material above neutral axis and bending tensile stresses in the materrial below neutral axis as shown in Fig. 1.1(A) (i). Since, concrete is weak in tension, its load-carring capacity is controlled by allowable stresses in tension and it is seen that the load-carrying capacity is very small. In case, steel bars are embedded in concrete close to the tension face as shown in Fig. 1.1(A) (ii), the beam becomes a reinforced concrete beam and its load carrying capacity is very much increased. It becomes as high as about *twenty-times* of that of the plain concrete beam. The original or existing strength of plain concrete beam is increased by additional tensile strength by using steel bars as reinforcement. Hence, in general, in reinforced concrete beams, the concrete resists the compressive force and longitudinal steel bars resist tensile force. Fig. 1.1 B (i) and (ii) show plain and reinforced cantiever beams.

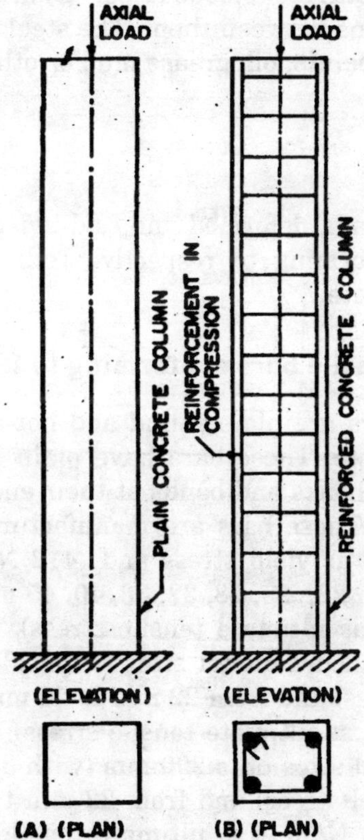

Fig. 1.2

In case, plain concrete columns are used, as shown in Fig. 1.2 (A) it is seen that their load carrying capability is also small, though the concrete carries compression only. In case the steel bars are embedded in the column of same size, of columns as shown in Fig. 1.2 (B), since the steel also has high compressive strength, the original (existing) load carrying capacity of plain cement concrete column is very much increased by providing additional strength by using steel bars. The columns become reinforced concrete columns. Thus, reinforcement is also used for resisting compressive forces. The cross-sectional dimensions of columns and compression members may be reduced, if desired.

Steel is high strength material as compared with concrete. The compressive yield strength of ordinary reinforcing steel bars is about 25 *times* the compressive strength of structural concrete and the tensile strength of steel bars is about 70 *times* the tensile strength of concrete. However, steel is a high cost material as compared with concrete. Two materials are best used in combination of their physical and mechanical properties.

General requirements

For most effective reinforcing action, there should be a sufficient strong bond between the hardened concrete and steel bars and two materials deform together and there is no relative movements between them. Concrete surrounding the reinforcing steel bars provides *excellent corrosion protection* and *fire resistance*. The *coefficients of linear thermal expansions* of steel (0.000012 per °C) and concrete (0.000011 per °C) are practically equal and there is no differential thermal deformations between them. The steel bars must be free from loose mill scales, loose rust, and coats of paints, oil, grease mud or other coatings which may destroy or reduce the bond.

Types of reinforcement

Following types of steel plain, deformed and cold-twisted bars; hard-drawn wire fabrics and rolled steel sections conforming to respective Indian standards codes are used for reinforcing the structural concrete :

(i) Mild steel and medium tensile bars conforming to IS : 432 (Part I) 1982

Mild steel (grade I) bars are plain round and hot-rolled. These are most commonly used for reinforcing the concrete. These bars have plain surface and have sufficient bond with concrete These bars need hooks and bends at their ends for anchorages. These bars are bent easily and are weldable. These bars are manufactured in 5, 6, 8, 10, 12, 16, 20 mm nominal size (with 255 N/mm^2 as yield stress and 412 N/mm^2 as ultimate tensile stress). These bars are also produced in 22, 25, 28, 32, 36, 40, 45 mm nominal size (with 236 N/mm^2 yield stress and 412 N/mm^2 as ultimate tensile stress). **Mild steel (grade II) bars** are manufactured in above mentioned nominal sizes (with 231 N/mm^2 as yield stress and 373 N/mm^2 as ultimate tensile stress) and from 22 mm to 50 mm nominal sizes (with 211 N/mm^2 as yield stress and 373 N/mm^2 as ultimate tensile stress). **Medium tensile steel bars** are manufactured in above nominal sizes upto 200 mm (with 353 N/mm^2 as yield stress and 538 N/mm^2 as ultimate tensile steel stress) and from 22 mm to 40 mm nominal sizes (with 348 N/mm as yield stress and 538 N/mm^2 as ultimate tensile stress) and 45 mm and 50 mm nominal sizes (with 323 N/mm^2 as yield stress and 538 N/mm^2 as ultimate tensile stress).

(ii) Hot rolled deformed bars IS : 1139-1966

A reinforcing bar with manufactured surface deformations which provides a locking anchorage with surrounding concrete.

These bars are also most commonly used for reinforcing the concrete. The yield stress for this quality of steel is 412 N/mm^2 and the ultimate tensile stress is 15 *percent* higher than the measured yield stress.

(iii) High strength deformed steel bars (hysd-steel bars) conforming to IS : 1786-1979

Cold-twisted deformed bars have deformations on their surface produced by cold-twisting a hot rolled bars and which have lugs, ribs or deformations on their surfaces.

Cold-twisted bars are usually mild steel bars. The yield point of this quality of steel is eliminated by cold working. Generally, twisting under controlled conditions results in a higher yield stress and consequently higher permissible (working) stress. These bars have lugs, ribs or deformations on the surface, so that the bond strength with concrete is very high (at least 40 percent more than that of plan bar of the same size). Because of high yiled stress (415 N/mm^2) and deformations this variety of steel is commonly referred as **high yield strength deformed bars (briefly referred as HYSD bars)**, Cold-twisted bars are manufactured in 6, 8, 10, 12, 16, 18, 20, 22, 25, 28, 32, 36, 40, 45 and 50 mm nominal sizes. Cold-twisted bars are also produced weith high yield stress of 500 N/mm^2. These two varieties of this steel are referred as Fe 415 and Fe 500, respectively. Fe 415 variety of this steel is very commonly used. Because of high bond with concrete, these bars do not need hooks and bends at their ends for anchorage. Cold twisted deformed bars are shown in Fig. 1.3.

The nominal size of the deformed bar is equivalent to the diameter or side of a plain bar having the same weight per metre run as that of the deformed bar.

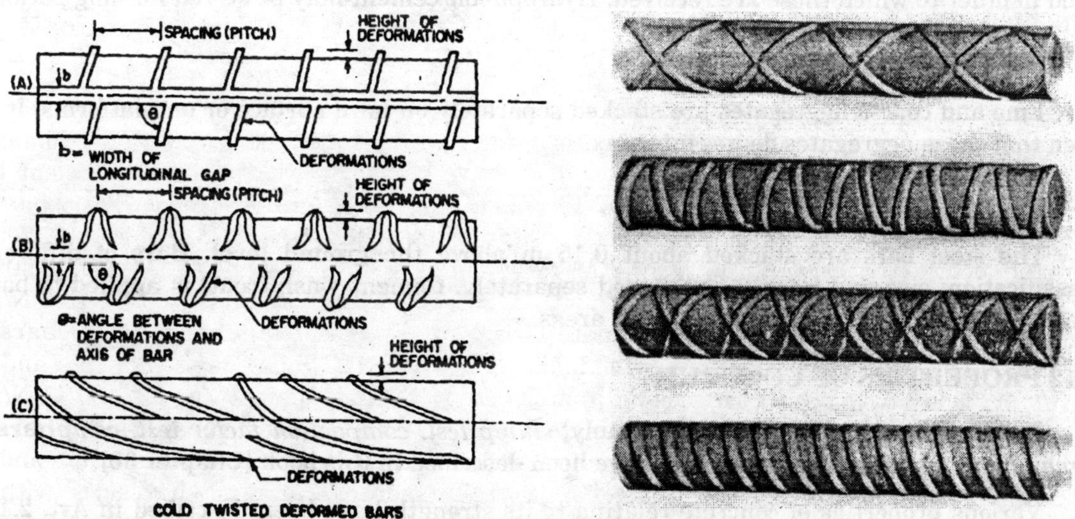

Fig. 1.3. Cold-twisted deformed bars.

(iv) Hard-drawn steel wire fabrics conforming to IS : 1566-1982

Hard-drawn steel wire fabrics are occasionally used for reinforcing floor slabs, flat and shell roof, etc. These are also used as secondary reinforcement in developing fire resistance and in some precast concrete like pipes. Square-mesh and oblong-mesh are also made by welding wire fabrics. Square-mesh fabrics have wires of the same size and spacing in both directions. Oblong-mesh fabrics have transverse wires that are smaller and more widely spaced than the main longitudinal wires.

(v) Rolled steel made from structural steel conforming to IS : 226-1975

Mild steel bars of grade-I with all nominal sizes mentioned above and grade-II with nominal sizes upto 20 mm have the chemical composition of steel conforming to IS : 226-1975.

The modulus of elasticity of steel shall be taken as 200 kN/mm^2 (i.e, 2 x 10^5 N/mm^2). The characteristic yield strength of different steel shall be assumed as the minimum yield stress/0.2 percent proof stress specified in the relevant codes.

1.11 STORAGE OF MATERIALS

Materials such as cement, fine aggregate, coarse aggregate, steel and other construc-tion materials are stacked and stored at site as per general guidance specified in IS : 4082-1977. Following instructions briefly described may be followed for storing the above mentioned materials :

(i) Cement

Cement bags are stored in dry and water proof sheds on platforms raised about 0.2 m above ground level and about 0.30 m clear-off the walls. These cement bags are taken and used in order to which these are received. Hydrophobic cement may be stored for long period.

(ii) Aggregate

Fine and coarse aggregates are stacked separately on hard surface or on platforms. It is seen that these aggregates do not intermix.

(iii) Reinforcing steel bars

The steel bars are stacked about 0.15 m above the ground level. Bars of different classification, size and lengths are stored separately. Cement wash (coat) is applied to bars stored for long period or stored in coastal areas.

1.12 PROPERTIES OF CONCRETE

The various tests for concrete, namely, *slump test, compaction factor test, compressive strength test and flexual strength test* have been described in this book [Chapter 30].

Various properties of concrete relating to its strength have been described in Art. 2.14. Other most important properties of concrete are *workability* for concrete in its fresh state) and *durability* (for concrete in its hardened state). These two properties of concrete are described here are as follows :

1. Workability

Workability of concrete is scientifically defined as that property of concrete which decides the quantum of useful internal work necessary to produce full compaction. IS : 6461 (Part VII) - 1973 defines workability. It is that property of freshly mixed concrete which determines the ease and homogenity with which it can be mixed, placed in the formwork, compacted and finished. A fresh concrete should satisfy number of requirements. A concrete mix should be stable (viz., it should not segregate during transportation and compaction). The tendency of bleeding (excess of water coming up and resulting in small pores in concrete) should be minimum. The concrete mix should be cohesive and mobile sufficient to be placed around the reinforcement. It should be possible to cast into the required shape. A concrete mix may be placed easily in large sections with minimum reinforcement. At the same time, it is difficult to place concrete mix in thin sections with heavy reinforcement. The proper and thorough compaction of mix concrete should be possible in placing the concrete and with the available facilities of compaction. Sometimes, it is difficult to compact by hand. In that case, the concrete mix may be compacted satisfactorily by mechanical vibrator. The workability of a concrete mix may be controlled by measurement of water.

The workability of a fresh concrete mix may be measured by different methods (viz., by carrying out slump test, compacting factor test, and Vee-Bee test).

(i) Slump test

A slump test is essentially a measure of **consistency** or the **wetness** of a concrete mix. A vessel having the shape of frutsum of a cone and open at the both ends is laced on a flat surface and filled with freshly mixed concrete in four layers of equal thickness. Each layer is compacted by 25 strokes with a standard tampering rod 16 mm in diameter and 600 mm in length. The vessel is carefully raised so that the concrete filling is not disturbed. Free concrete is allowed to subside or settle. A vertical settlement of concrete mix noted in mm is called as **slump.**

A concrete mix used for road work or mass concrete needs a slump (20-30) mm or (25-30) mm, respectively. Whereas for beam and slabs, and for columns and retaining wall, the values of slump adopted are (50-100) mm, and (75 to 150 mm) mm, respectively. For vibrated concrete a slump of (12 to 25) mm is adopted. A slump of (25 to 125) mm shows that the concrete mix is having medium to high workabilities. A slump of zero shows that the concrete mix is very stiff. It does not indicate any difference in concrete of different workabilities.

In slump test the apparatus and test procedure are simple. As such this test is widely used for concrete works.

(ii) Compaction factor test

Compaction factor test depends on the definition of workability expressed as internal work. The compability, (that is, the quantum of work necessary to compact the given mass of concrete) is an important aspect of workability. An indirect procedure is used in this test to measure the workability. The degree of compaction obtained is found for known quantum of work.

Two conical hoppers with quick release doors at the bottom of each hopper placed one above the other are used. The concrete mix is filled in the uppermost hopper and its door is opened and the concrete falls in the middle hopper. The door of middle hopper is also opened and the concrete is allowed to fall in a third cylindrical vessel. The concrete mass, m_1 in third vessel is measured. Later on, the concrete in third cylindrical vessel is filled in 50 mm layers and compacted each time and its mass, m_2 is again found. The ratio of (m_1/m_2) termed as **compaction factor** is determined.

The compaction factors of 0.9 and 0.8 show that the concrete mixes have medium and low workability. This method is more accurate and used in laboratory conditions. However, it is not suitable for the concrete mix having very low workability (i.e., compacton factor of 0.7 and below

(iii) Vee-bee test

The stiff concrete mixes having low or very low workability is tested by Vee-bee test. In this test, the advantage is, a similar treatment is given to the concrete mix, as it would receive in actual practice. In this test, an index (termed as Vee-bee time defined as the time taken for the concrete surface to uniformly adhere to the glass disc or rider of the apparatus) is found. It is observed by visual inspection.

A source of error is introduced in this test, due to difficulty of establishing the end point. If the time is very short (2 to 5) seconds, it becomes more difficult. This test is not suitable for concrete mixes of high workability having a value of slump as 75 mm or above.

TABLE 1.9

Conditions of placing	Degree of workability	Values of workability		
		Slump	Compaction factor	Vee-bee time
Concrete in shallow sections with vibration	Very low	–	0.75-0.80	20-10 sec
Concrete in lightly reinforced sections with vibrations	Low	–	0.80-0.85	10-5 sec
Concrete in lightly reinforced sections without vibration or heavily reinforced sections with vibration	Medium	5-75 mm for 20 min aggregate	0.85-0.92	5-2 sec
Concrete in heavily reinforced sections without vibration.	High	75-125 mm for 20 mm aggregate	Above 0.92	–

Note : For smaller aggregate the slump values will be lower.

The quantity of water, the material and mix proportions of aggregate and the environmental conditions are the various factors which influence the workability of a fresh concrete mix. The workability increases with increase in quantity of water. Coarse aggregate of large size gives higher workability as the total surface area and particles interference reduce. Use of fine sand with corresponding increase in specific surface area needs more water for the same workability.

For some conditions of placing of concrete, the concrete mix may have values of workability measured in accordance with IS : 1199-1959, in the range given Table 1.9.

2. Durability

A durable concrete is one that performs satisfactorily in the working environment during its anticipated exposure conditions during service. The materials and mix proportions specified and used should be such as to maintain its integrity and, if applicable, to protect embedded reinforcement from corrosion.

One of the main characteristics which influences the durability of concrete is its permeability to the ingress of water, oxygen, carbon dioxide, chloride, sulphate and other potential deleterious substances. Impermeability is governed by the constituently and workmanship used in making the concrete. With normal weight aggregates, a suitable low permeability is obtained by having an adequate cement content, sufficiently low free water/ cement ratio, by ensuring complete compaction of the concrete and by adequate curing.

Durability of concrete is defined and interpreted as its resistance to deteriorating influences which may be present inside the concrete itself by negligence or which are inherent in the environment in which the concrete is placed. Sometimes, the ingradients of concrete may have deleterious material and it is not known before hand. And sometimes, it may not be imagined that the concrete is likely to placed in harmful environment. Under such circumstances the problems use to arise for durability of concrete.

The factors which influence durability include (i) the environment (ii) the cover to the embedded steel, (iii) the type and quality of constituent materials, (iv) the cement content and water/cement ratio of the concrete, (v) workanship, to obtain full compaction and efficient curing and (vi) the shape and size of the concrete.

The degree of exposure anticipated for the concrete during its service life together with other relevant factors relatng to mix composition, workmanship, design and detailing should be considered. The concrete mix to provide adequate durability under these conditions should be so selected taking account of the accuracy of concern testing regimes for control and compliance as described in this code, IS: 456-2000.

External forces (weathering, attack by natural or industrial liquids and gases, bacterial growth, etc.) or internal forces (harmul alkali aggregate reactions, volume change due to non-compatibility of thermal and mechanical properties of aggregate and cement paste, presence of sulphates and chlorides in ingradients of concrete reduce the durability of concrete.

The possibility of corrosion of steel in reinforced concrete increases when the moisture or air enters the concrete. It leads to an increase in volume of steel. The concrete cover may crack and spill.

3. Thermal/Expansion

The coefficient of thermal expansion depends on nature of cement, the aggregate, the cement content, the relative humidity. and the size of sections. The value of coefficient of thermal expansion for concrete with different aggregates may be adopted as below.

Type of Aggregate	Coefficient of Thermal expansion for concrete/°C
Quartz	1.2 to 1.3 x 10^{-5}
Sandstone	0.9 to 1.2 x 10^{-5}
Granite	0.7 to 0.95 x 10^{-5}
Basalt	0.8 to 0.95 x 10^{-5}
Limestone	0.6 to 0.9 x 10^{-5}

1.13 REQUIREMENTS OF DURABILITY (IS : 456-2000)

Concrete may be made durable by satisfying its requirements specified in IS : 456-2000 for (i) shape and size of member (ii) control of environment exposure conditions (iii) concrete surfaces exposed to abrasive action and (iv) exposure to sulphate attack. Concrete may be made durable also by satisfying the requirement of concrete cover, concrete mix proportions, maximum cement content, mix constituents (i.e., chlorides, sulphates, alkali-aggregate reaction. Concrete may be made still more durable by adjusting the minimum cement contents, and maximum water-cement ratio. Concrete in aggressive soils and water, and concrete in sea-water need special attention as regards drainage, compaction, finishing and curing. By satisfying the limits and requirements specified in IS :456-2000 taken together, a dense, workable, placeable and durable concrete may be prepared. The concrete will also have as low a permeability as possible.

In IS: 456-2000 detailed guidance has been given concerning the factors affecting durability. These factors have been described in subsequent sections. Structure is checked for the durability as a part of design. In long term performance the durability is equally important as the strength of concrete structure. The durability of concrete and steel has to be checked seperately.

1.14 CONCRETE MIX PROPORTIONS

Free water-cement ratio is an important factor which control the durability of concrete. The value of water-cement ratio should always be the lowest. Appropriate values for minimum cement content and the maximum free water cement ratio are given in Table 1.10 (A) and 1.10(B) for different exposure conditions. The minimum cement content and the maximum water-cement ratio apply to 20 mm nominal maximum size aggregate.

TABLE 1.10 (A). MINIMUM CEMENT CONTENT, MAXIMUM WATER-CEMENT RATIO, AND MINIMUM GRADE OF CONCRETE FOR DIFFERENT EXPOSURES WITH NORMAL WEIGHT AGGREGATE OF 20 MM NOMINAL MAXIMUM SIZE (IS : 456-2000)

S.No.	Exposure	Plain concrete		
		Minimum cement content (kg/m^3)	Maximum free-water cement ratio	Minimum grade of concrete
1	Mild	220	0.60	–
2	Moderate	240	0.60	M15

3	Severe	250	0.50	M20
4	Very severe	260	0.45	M20
5	Extreme	280	0.40	M25

TABLE 1.10(B)

S.No.	Exposure	Reinforced concrete		
		Minimum cement content (kg/m³)	Maximum free water cement ratio	Minimum grade of concrete
1	Mild	300	0.55	M20
2	Moderate	300	0.50	M25
3	Severe	320	0.45	M30
4	Very severe	340	0.45	M35
5	Extreme	360	0.40	M40

It is to note that cement content prescribed as above is irrespective of the grades of cement. The additions such as fly- ash or ground granulated blast furnance slag may be taken into account in the concrete composition with respect to the cement content and water-cement ratio in case the suitability is established and as long as the maximum quantities are taken into account, the limit of pozzolona and slag should not be exceeded.

Further the minimum grade for plain concrete under mild exposure condition is not specified.

For sizes of aggregate other than 20 mm, the values given in Tables 1.10 (A) and 1.10 (B) are changed as given in Table 1.11.

Maximum cement content

It is recommended in IS : 456-2000 that the cement content not including fly ash and ground granulated blast furnace slag in excess of 450 kg/m³ should not be used unless special consideration has been given in design to the increased risk of cracking due to drying shrinkage in thin sections, or to early thermal cracking and to increased risk of damage due to alkali silica reactions.

TABLE 1.11. ADJUSTMENT TO MINIMUM CEMENT CONTENTS, (FOR AGGREGATE OTHER THAN 20 MM NOMINAL MAXIMUM SIZE) AS PER IS : 456-2000

	Nominal maximum aggregate size (mm)	Adjustments to minimum cement content in Tab.e 1.10 kg/m³
1	10	+40
2	20	0
3	40	– 30

1.15 LIMITS FOR DELETERIOUS MIX CONSTITUENTS

Careful selection of the mix and materials is necessary for concrete to be durable, so that the deleterious constituents do not exceed the limits. Chlorides, sulphates and alkalies (if present) increases risk of corrosion of reinforcement, cause expansion and disruption of concrete produce an expansive reactions, respectively. IS: 456-2000 specifies limits, which should not be exceed for the following chemicals.

1. Chlorides in concrete Whenever, chlorides are present in concrete, the risk of corrosion of reinforcement is increased. The higher chloride content, or if, subsequently exposed to warm moist conditions increases the risk of corrosion of reinforcement largely.

All constituents may contain chlorides and concrete may be contaminated by chlorides from the external environment. The chances of deterioration of concrete from harmful chemical salts may be minimized by keeping their levels low. Such harmful salts in concrete come from concrete materials (i.e., cement, aggregate, water and admixtures) as well as by diffusion from the environment. The quantities of chlorides should be limited. Total quantity of chlorides content (as chlorine) in the concrete at the time of placing should be as given in Table 1.12 as recommended in IS : 456-2000.

The total acid soluble chloride content should be found from the mix proportions and the measured chloride contents of the constituents. Whenever possible, the total chloride content of the concrete should be calculated.

2. Sulphates in concrete. Sulphates remain present in most cements and in some aggregates. Excessive quantities of water soluble sulphate from these or other mix constituents may cause expansion and disruption of concrete. This may be prevented. Total water-soluble sulphate content of the concrete mix expressed as SO_3 should not exceed 4 percent by mass of the cement in the mix. The sulphate content may be found as the total from the various constituents of the mix.

This 4 percent limit is not applicable to concrete made with super-sulphated cement complying with IS : 6909.

TABLE 1.12. LIMITS OF CHLORIDE CONTENTS OF CONCRETE (as per IS 456-2000)

S.No.	Types or use of concrete	Maximum total acid soluble chloride content expressed at kg/m³ of concrete
1	Concrete containing metal and steam cured at elevated temperature and pre-stressed concrete	0.4
2	Reinforced concrete or plain concrete containing embedded metal	0.6
3	Concrete not containing embedded or any material requiring protection from chloride	1.0

3. Alkali-aggregate reaction. Some aggregates may contain particular varieties of silica, which may be susceptible to attack by alkalies. (Na_2O and K_2O) originating from cement or other sources, producing an expansive reaction which may cause cracking and disruption of concrete. Damage to concrete from the reaction will normally only occur when all the following are present together (i) high moisture level within the concrete, (ii) a cement with high alkali content (iii) aggregate containing an alkali reactive constituent.

Where the service records of particular cement/aggregate combination are well established, and do not include any instances of cracking due to alkali-aggregate reaction, no further precautions should be necessary. When the materials are unfamiliar, precautions should be taken in one or more of the following forms.

(a) Non-reactive aggregate from the alternate sources should be used

(b) Low alkali ordinary portland cement having total alkali content not more than 0.6 percent (as Na_2O equivalent) should be used.

Further advantage may be obtained by use of fly ash (grade I) conforming to IS : 3812 or granulated blast fumance slag conforming to IS : 12089 as part replacement of ordinary portland cement (having total alkali content as Na_2O equivalent not more than 0.6 percent) provided fly ash content is at least 20 percent or slag is at least 50 percent.

(c) Measures to reduce the degree of saturation of the concrete during service such as use of impermeable membrance

(d) Limiting the cement content in the concrete mix and thereby limiting total alkali content in the concrete mix.

1.16 CONCRETE IN AGGRESSIVE SOILS AND WATER

In case, the concrete remains in contact with aggressive soils and water, these agents have progressive destruction action. The rate of deterioration decreases as the concrete is made stronger and more impermeable. This rate increases as the salt content of water incre-ases. Where the structures are only partially immersed or are in contact with aggressive soils or water is on one side only, the evaporation may cause serious concen-tration of salts with subsequent deterioration, even where the original salt content of the soil or water is not high. Following factor influences the durability of concrete as recomme-nded in IS : 456- 2000.

Drainage. The ground water may be lowered by drainage at sides where alkali concen-trations are high or may become very high. Then, the alkalies will not come in direct contact with the concrete.

Chemically resistant stone facing the concrete may be used for additional protection. Alternatively, a layer of plaster of paris covered with suitable fabric, (e.g. jute) may be applied thoroughly impregnated with bituminous material.

1.17 COMPACTION, FINISHING AND CURING

Adequate compaction without segregation should be ensured by providing suitable workability. This may be ensured by using appropriate placing and compacting equipment and and procedure. Full compaction is particularly important in the vicinity of construction and movement joints and of embedded water bars and reinforcement.

Good finishing practices are essential for durable concrete. Over working the surface and the addition of water/cement to aid finishing should be avoided. The resulting laitance will have impaired strength and durability and will be particularly vulnerable to freezing and thawing under wet conditions.

Adequate curing techniques may be used to reduce the permeability of the concrete. This enhances the durability of concrete by extending the hydration of the cement, particularly in its surface zone.

Consistency of the concrete Wetness of freshly mixed wet concrete. It is a property of freshly mixed net concrete virtue of which freshly mixed wet concrete must flow freely around the corners and between reinforcing steel bars. The standard method of determining the relative consistency of the concrete is to conduct slump test described in author's `Design of Concrete Structures'. It is also measured in terms of V.B. seconds a Vee-Bee consistometer is used.

1.18 CONCRETE IN SEA-WATER

In case of plain concrete, at least M 20 grade of concrete shall be used where it remains in sea-water or exposed directly along the sea-coast. In case of reinforced concrete, M 30 grade of concrete shall be provided. The use of slag or pozzolana cement is advantageous under such conditions.

Special attention shall be given to the design of the mix, to obtain the densest possible concrete. Slag, broken bricks, soft lime stone or other porous or weak aggregates shall not be used.

As far as possible, the preference shall be given to precast members unreinforced, well cured and hardened, without sharp corners and having trowel smooth finished surfaces free from crazing, cracks or other defects. Plaster should not be done.

No construction joints shall be allowed within 600 mm below low water-level or within 600 mm of the upper and lower planes of wave action. Where usually severe conditions or abrasion are anticipated such parts of the works shall be protected by bituminous or silico-fluride coatings or stone facing bedded with bitumen.

In reinforced concrete structures, care shall be taken to protect the reinforcement from exposure to saline atmosphere during storage, fabrication and use. It may be activated by heating the surface of reinforcement with cement wash or by suitable methods.

1.19 SHAPE AND SIZE OF COMPONENT MEMBERS

The shape and size of the component members and design details of exposed structures should be such as to promote good drainage of water. Standing pools and rundown of water should be avoided. The cracks should be minimized so that the water should not be collected and transmitted.

Adequate curing is essential to avoid the harmful effects of early loss of moisture.

The member profile (i.e., shape and size) and there intersections with other members shall be designed and detailed in a way to ensure easy flow of concrete and proper compaction during concrete.

Concrete is more vulnerable to deterioration due to chemical or climatic attack when it is thin in cross-sections, in sections under hydro-static pressure from one side only, in partially immersed sections and at corners and edges of elements

The life of the structure may be enhanced by providing extra cover to steel by chamfering the corners or by using surface coating which prevent or reduce the entrance of water, carbon dioxide or aggressive chemicals.

1.20 ENVIRONMENTAL EXPOSURE CONDITIONS

IS: 456-2000 recommends that the following factors (viz, exposure conditions, abrasive action of surfaces, freezing and thawing under wet conditions and exposure to sulphate attack) influence the durability of concrete.

1. Exposure conditions

The general environment to which the concrete will be exposed during its working life is classified into following five levels in IS : 456-2000

(i) **Mild environment**, In this category, the concrete surfaces are protected against water or aggressive conditions, except those located in coastal area.

(ii) **Moderate environment**. In this category the concrete surfaces are sheltered from severe rain or freezing whilst wet concrete exposed to condensation and rain, concrete continuously under water, concrete in contact or buried under non-aggressive soil/ground water, concrete surfaces sheltered from saturated salt air in coastal area are included in this category.

(iii) **Severe environment** : The concrete surfaces exposed to severe rain, alternate wetting and drying or occasional freezing whilst wet or sever condensation are considered as exposure to severe environment concrete completely immersed in sea-water and the concrete exposed to coastal environment are also considered as exposed to severe environment.

(iv) **Very severe environment** : The concrete surfaces exposed to sea-water spray corrosive fumes or severe freezing conditions whilst wet are included in this category. The concrete in contact with or buried under aggressive sub-soil/ground water are also considered as exposed to very severe exposure.

(v) **Extreme environment** : Surfaces of members in tidal zone, members in direct contact with liquid/solid aggressive chemicals are included in the category of extreme environment exposure.

2. Concrete surfaces exposed to abrasive action

Special literatures dealing with abrasive action of concrete surfaces may be referred to for durability requirements of concrete (e.g., in case of machinery and metal tyres).

3. Concrete surfaces exposed to freezing and thawing

Where freezing and thawing actions under wet conditions exist, enhanced durability may be obtained by the use of suitable air entraining admixtures. Where concrete lower than grade M 50 is under these conditions, the mean total air content by volume of the fresh concrete at the time of delivery into the construction should be as below :

Corresponding to nominal maximum sizes of aggregate 20 mm and 40 mm, the percentage of entrained air should be (5 ± 1) and (4 ± 1), respectively.

The air entrainment reduces the strength. Therefore, suitable adjustments may be made in the mixed design for achieving required strength.

4. Exposure to sulphate attack

As per IS : 456-2000, the recommendations for the type of cement, maximum free water/cement ratio and minimum cement content, which are required at different sulphate concentrations in near-neutral ground water having pH of 6 to 9 have been specified as given Tables 1.13 (A) and (B).

TABLE 1.13 (A). REQUIREMENTS FOR CONCRETE EXPOSED TO SULPHATE ATTACK (AS PER IS : 456-2000

S.N.	Class	Concentration of sulphates (expressed as SO_3)		
		(in soil)		(in gourd)
		Total SO_3	SO_3 in 2 : 1 water soil extract	g/litre
1	1	Traces (<<0.2	Less than 1.0	Less than 0.3
2	2	(0.2 to 0.5)	(1.0 to 1.9)	(0.3 to 1.2)
3	3	(0.5 to 1.0)	(1.9 to 3.1)	(1.2 to 2.5)
4	4	(1.0 to 2.0)	(3.1 to 5.0)	(2.5 to 5.0)
5	5	More than 2.0	More than 5.0	More than 5.0

TABLE 1.13(B). REQUIREMENTS FOR CONCRETE EXPOSED TO SULPHATE ATTACK

(AS PER IS : 456-2000) [Following abbreviations have been used in this table for different types of cements]

1. OPC : Ordinary portland cement, 2. PSC : Portland slag cement; 3. PPC : Portland pozzolana cement, 4. SSC : Super-sulphate cement; 5. SRPC : Sulphate resisting portland cement, 6. PC : Portland cement

S.No.	Class	Type of cement	Dense fully compacted concrete with 20 mm nominal max. size aggregate complying with IS:383	
			Minimum cement content (kg/m^3)	Maximum face water cement ratio
1	1	OPC PSC PPC	280	0.55
2a	2a	OPC PSC PPC	330	0.50
2b	2b	SSC SRPC	310	0.50
3a	3a	SSC SRPC	330	0.50

3b	3b	PPC PSC	350	0.45
4	4	SSC SRPC PC	370	0.45
5	5	SRPC SSC	400	0.40

1.21 NOMINAL CONCRETE COVER TO REINFORCEMENT

The protection of steel reinforcement is provided by adequate thickness of good quality concrete. It is essential to protect the steel bars against corrosion.

The nominal cover is defined as the design depth of concrete cover to all steel reinforcement bars, including links. It is the dimension used in design and indicated in the drawings. It shall not be less than the diameter of the bar.

For the requirement of durability, the minimum values for the nominal cover of normal-weight concrete which should be provided to all reinforcement, including links depending on the conditions of exposures described in Art. 1.20. Specific values have been given in IS:456-2000 as given in Table 1.14.

TABLE 1.14. NOMINAL COVER FOR DURABILITY REQUIREMENT (AS PER IS 456-2000)

S.N.	Environment exposure conditions	Nominal concrete cover not less than (in mm)	Grades of concrete
1	Mild	20	M20 and higher
2	Moderate	30	M25 and higher
3	Severe	45	M30 and higher
4	Very severe	50	M35 and higher
5	Extreme	75	M40

Notes

1. For mild exposure, for main reinforcement steel bars upto 12 mm diameter, the nominal cover may be reduced by 5 mm
2. Unless specified otherwise, actual concrete cover should not deviate from the required nominal cover by 10 mm.
3. For severe and very severe exposure conditions, 5 mm reduction may be made, where concrete grade is M 35 and above.

In order to meet specified period of **fire resistance**, the minimum values of nominal cover of normal weight aggregate concrete for all reinforcement steel bars including link in order to meet specified period for resistance has been recommended in IS : 456-2000. The values of nominal cover to meet specified period of fire resistance have been described in the respective chapters of beams, slabs and columns.

1.22 QUALITY OF FINE AND COARSE AGGREGATES

Some harmful chemicals such as chlorides and sulphate may come from concrete materials (i.e., cement, fine aggregate, coarse aggregate, water and admixtures). The chemicals deteriorate the durability of the concrete. The chances of coming of these chemicals may be minimized by careful selection of these concrete materials and level of these chemicals may be kept limited (i.e., below the maximum values specified in third edition of code IS : 456-1978.

Total quantity of chlorides as (Cl) and total quantity of soluble sulphates (as SO_3) in the concrete at the time of placing should be limited 0.15 percent by mass of cement and 4 percent by mass of cement, respectively. IS : 456-2000 recommends that total acid soluble chloride content should be less than 0.4 kg/m^3.

1.23 QUALITY OF WATER

Water used for mixing and curing shall be clean. Oils, acids, alkalies, salts, sugar, organic materials or other substances may be deleterious to concrete and steel. Water used must be free from such injurious substances.

Potable water is considered satisfactory for mixing concrete. IS : 456-2000 recommends the maximum permissible values as a guide.

(i) Phenolphthalene is used as an indicator. 100 ml sample of water should not require more than 5 ml of 0.02 normal NaOH to neutralize it.

(ii) In case, the mixed indicator is used, then, 100 ml sample of water should not require more than 25 ml of 0.02 normal $H_2 SO_4$.

(iii) The pH value of water shall not be less than 6.

(iv) The permissible limits for dissolved solids per litre of water as specified in IS : 456-2000 are as under.

Organic solids not more than 200 mg/litre, inorganic solids not more than 3000 mg/litre, sulphate as SO_3 not more than 400 mg/litre and chlorides not more than 2000 mg/litre. According to standards of WHO only 500 mg of dissolved solids are allowed.

1.24 PERMEABILITY OF CONCRETE

The permeability of concrete is one of the main characteristic which influences the durability of concrete. The permeability is entrance of water, oxygen, carbon dioxide, chloride, sulphate and other potential deleterious substances. Impermeability is controlled by the constituents and workmanship used in preparing the concrete. A suitable low permeability with normal-weight aggregate is obtained by having an adequate cement content sufficiently low free water/cement ratio, by ensuring complete compaction of the concrete and by adequate curing. The recommendations as regards minimum cement content (in kg/m^3) and maximum free water-cement ratio for the different grades of concrete have been specified in IS : 456-2000 and these have been listed in Table 1.10 (B). The maximum cement content should not exceed 450 kg/m^3 exclusive of fly-ash and ground granulated blast furnace slag.

1.25 CURING OF CONCRETE

Curing is the process of preventing the loss of moisture from the concrete whilst maintaining a satisfactory temperature regime. The prevention of moisture loss from the concrete is particularly important in case the water-cement ratio is low if the cement has a high rate of strength development, and if the concrete contains granulated blast furnace slag or pulverised fuel ash. The curing regime should also prevent thc development of high temperature gradients within the concrete. Proper hydration of cement in the freshly placed concrete by proper curing is done.

The rate of strength development at early ages of concrete made with supersulphated cement is significantly reduced at lower temperatures. Super-sulphated cement concrete is seriously affected by inadequate curing and the surface has to be kept moist for atleast seven days.

Following two types of curing have been described in IS : 456- 2000.

1. Moist curing : The exposed surfaces of concrete are kept continuously in a damp or wet condition by ponding (i.e., by adding) water or by covering with a layer of sacking, canvas, hessian (i.e, strong coarse cloth of mixed hemp and jute), or similar materials and kept constantly wet for at least seven days from the date of placing concrete in case of ordinary portland cement and at least ten days where mineral admixture or blended cements are used. The period of curing shall not be less than ten days for concrete exposed to dry and hot weather conditions. In the case of concrete where mineral admixtures or blended cements are used, it is recommended in IS : 456-2000 that above minimum periods may be extended to fourteen days.

2. Memberance curing

Approved curing compounds may be used in lieu of moisture curing with the permission of the engineer incharge. Such compounds shall be applied to all exposed surfaces of the concrete as soon as possible after the concrete has set. Impermeable memberances such as polythelene sheeting covering closely the concrete surface may also be used to provide effective barrier against evaporation.

For concrete containing portland pozzolana cement, portland slag cement or mineral admixtures, period of curing may be increased.

Alternate wetting and drying should not occur. Frequently the water should be poured over the concrete.

The curing methods are considered good, average and poor depending upon whether concrctc is protected from direct sun and relative humidity is maintained more than 80 percent, in between 80 and 50 percent and less than 50 percent respectively.

1.26 COMPACTION OF CONCRETE

Adequate compaction without segregation should be ensured by providing suitable workability and using appropriate placing and compacting equipment and procedure. Full compaction is particularly important in the vicinity of construction and movement joints. The proper compaction of concrete near the outer face (where there is reinforcement) is necessary as regards durability. The segregation and bleeding near these portions should be avoided.

Concrete should be thoroughly compacted and fully worked around the reinforcement and around the embedded fixtures and into the corners of the framework.

Mechanical vibrators should be used for thorough compaction of concrete. Over vibrations and under vibrations of concrete, both are harmful and should be avoided. The vibrations of very wet mixes should also be avoided.

Whenever, the vibration has to be done externally, the design of framework and the deposition of the vibrators should receive special consideration to get efficient compaction and to avoid surface blemishes.

1.27 CONCRETE IN FOUNDATIONS

The concrete in foundations remains below the ground and in contact with the aggressive soils. The sulphates and chlorides in the aggressive soil cause the deterioration of concrete. The rate of deterioration decreases as the concrete is made stronger and more impermeable where the structures are only partially immersed or are in contact with the aggressive soils or waters on one side only, evaporation may cause serious concentrations of salts with subsequent deterioration, even where the original salt content of the soil or water is not high.

The concrete in sea-water or exposed directly along the sea-coast shall be at least be of M20 grade in case of plain cement concrete and M30 in case of reinforced concrete. The use of slag cement or pozzolana cement is advantageous under such conditions Supersulphated cement gives an acceptables life provided that the concrete is dense and prepared with water-cement ratio 0.4 or less in mineral acids down to pH 3.5. Supersulphated cement gives better durability in comparison with the portland cement.

The design of mix to obtain possible dense concrete needs special attention. Slag, broken brick, soft lime stone or other porous or weak aggregate shall not be used. Precast members unreinforced, well cured and hardened, without sharp corners and having towel smooth finished surfaces from crazing, cracks or other defects, plastering should be avoided.

No construction joints should be allowed within 600 mm below low water level or within 600 mm of upper and lower planes of wave action. There are chlorides in sea-water which restricts the action of sulphates.

1.28 DURABILITY OF STEEL

In IS456-2000, the necessity of durability has been emphasized. The durability is the ability of the material to resist deterioration over long periods of time. A durable steel material for the reinforcing steel bars is one that performs satisfactorily the desired function in the working environment under the anticipated exposure condition during its service life, without deterioration of the cross-sectional area and loss of the strength due to corrosion.

For the durability of reinforced concrete, the durability of steel is also essential. The main factor which influences the durability of steel is corrosion. (i.e., rusting).

Requirements for durability

The design, fabrication and erection details of exposed structure should be such that good drainage of water is ensured. Standing pool of water, moisture accumulation and rundown of water for the extended duration are to be avoided.

All the exposed surfaces are easily accessible for inspection and maintenance. Such surfaces which are not easily accessible are completely sealed against-ingress of moisture.

The general environment, to which the reinforcing steel bars in concrete structure is exposed during its working life, is classified into following five categories.

Exposure conditions

1. Mild exposure: The reinforcing steel bars normally protected against exposure to weather or aggressive codnition as in interior of buildings, except when located in coasted areas.

2. Moderate exposure: The reinforcing steel bars exposed to (i) condensation and rain, (ii) continously under water, (iii) non-aggressive soil/ground water, (iv) sheltered from saturated salt air in coastal areas.

3. Severe exposure: The reinforcing steel bars exposed to (i) severe frequent rain, (ii) alternate wetting and drying, (iii) severe condensation (iv) completely immersed in sea-water, (v) exposed to saturated salt air in coastal area.

4. Very severe exposure: The reinforcing steel bars exposed to (i) sea water spray, (ii) corrosive fumes, (iii) aggressive sub-soil or ground water.

5. Extreme exposure: The reinforcing steel bars exposed to (i) tidal zones and splash zones in the sea (ii) aggressive liquid or solid chemicals.

Appropriate coatings may be used when surfaces of the reinforcing steel bars are exposed to concentration of sulphates (SO_3) in soil, ground water etc.

In case, the reinforcing steel bars are exposed to very high sulphate concentration, of more than 2 per cent in soil and 5 per cent in water, some form of living such as polythene, polychloroprene sheet or surface coating is used on asphalt, chlorinated rubber, epoxy or polymethane material should be used to completely avoid access of the solution to the reinforcing steel bars.

I. Corrosion of steel Corrosion of steel is an electro-chemical process. Anodic and cathodic zones are formed on the surface of steel due to differences in electro-chemical potential. In the presence of moisture (which acts like electrolyte) and oxygen, an electric current is produced and the anode is dissolved. The ferrous ions from the anodic areas are released and move towards the cathode and combine with water and oxygen and from hydroxyl ion which react with the ferrous ions. Areas of steel surface covered in dirt or rust scales acts like anode, whereas the areas readily exposed to atmospheric oxygen acts like cathode. As the corrosion reactions proceed, these areas on the surface of steel shift and change. This corrosion reaction produces hydrated ferrous oxide, commonly known rust (yellowish-brown loose powder). This hydrated ferrous oxide further gets oxidized into ferric oxide which is called as red rust. As rust builds up on portion of the body it becomes less anodic with respect to previously rusted area. The electric batteries are formed and reformed and corrode the entire surface. This rust becomes two and a half times in volume than its original volume as steel.

Both, moisture and oxygen are essential for corrosion to take place. The rate of corrosion depends on the impurities in the water. In relatively clean environments (such as rural areas) the corrosion rate remains low. In polluted atmospheres (such as bad industrial areas having smoke and soot), the rate of corrosion becomes high. The rate at which surface is lost may reach 0.075 mm/year, more in case particularly harmful agents such as sulphur dioxide are present. In the marine environment, the chloride ions remain present. These aggressive environment quicken the process.

Different types of corrosion. There are following different types of corrosion seen.

1. Pitting corrosion : Corrosion pits are formed in anodic areas of the steel surface in mild steel immersed in water or soil. Due to the presence of moisture specially aided by improper detailing or due to constant exposure to alternate wetting and drying, this type of corrosion occurs. It is easily handled by providing quick drainage in proper detailing and allowing free flow of air. The surface becomes dry.

2. Crevice corrosion : In this type of corrosion, long narrow openings (chinks) and fissures (thin splits) occur again due to improper detailing. The tops of splits become local anodes with respect to surrounding metal and corrosion takes place at this point. The oxygen content of water trapped in splits is less than that of water, which is exposed to air. The corrosion begins inside the split.

3. Bimetallic corrosion : When two different metals (for e.g. iron and aluminium) are in electrical contact put together in an electrolyte (normal water) an electrical current passes in between them and corrosion occurs.

The basen metal becomes anodic and the nobler metals becomes cathodic. The nobler metal tends to be protected at the expense of the base material.

Depending on the electric-potential, the metals may be arranged into a table, known as the *`galvanic series'*. Therefore, this bimetallic corrosion is also called **as galvanic corrosion**. The electrical potential difference shall be large in this case. The distance between two metals becomes large.

The metallic path between these two metals may be broken by insulating the dissimilar materials from each other and this bimetallic corrosion may be prevented. Otherwise the electro-type path may be broken by preventing the formation of a continuous bridge of electrolyte solution between two metals. Insulating bushes and washers may be fitted in former procedure while the metals may be painted in the alternative procedure.

4. Stress corrosion : Under the simultaneous influence of a static tensile stress and a specific corrosive atmosphere, at the stress concentration areas, the stress makes some spots in a metal more anodic compared with the remaining. The tip of a split (or crack) becomes anodic part and it corrodes and the split becomes more wider. It is not common in ferrous metals. Some stainless steels are susceptible to this type of corrosion.

5. Fretting corrosion : In case, two oxide coated films or rusted surfaces are rubbed together, the rust may be mechanically removed from projecting high spots (i.e., the contacting surfaces). These high spots become active anode and the rest of the surfaces act as cathode, and corrosions begin. In mechanical components, this type of corrosion occurs.

6. Bacterial corrosion : The pipelines, buried structures, and offshore structures remain in soils and water and bacterial corrosion occurs in them due to microbiological activity.

Above types of corrosions may be grouped in atmospheric corrosion, which occurs in normal environment. Others may be grouped as chlorides corrosions.

III. Deterioration of concrete and steel by corrosion : In concrete structures, rusted steel reinforcement bars swell. Due to this, the cracks develop in concrete. As rusting is progressive phenomenon, the width of cracks increases and pushes the concrete at the sides. The concrete cover breaks and falls down. The rate of corrosion accelerates by moisture. Along with the fall of concrete cover the cross-sectional areas of steel bars (as these are eaten-up) reduce and become unable to take up forces for which these are designed. With progress of time, the structures may fail due to progressive corrosion alone.

IV. Protection from corrosion: The steel reinforcement bars should be protected from moisture and oxygen to minimise the risk of corrosion. The reinforcement bars need proper protection. Most protective treatments used to protect the steels bars from corrosion are as follows :

The methods of corrosion protection are controlled by the actual conditions as per IS: 9077 and IS: 9172. The main corrosion protction methods include.

(i) Controlling the electro-potential

(ii) Inhibitors

(iii) Inorganic/metal coating or organic/painting system.

In the case of mild exposure, a coat of primer after removal of any loose will scale may be adequate. As the exposure condition becomes more severe elaborate surface preparations and coatings becomes necessary.

The reinforcing steel bars shall be provided with atleast coat of primer immediately after its surface preparation, such a sand blasting to remove all will scale and rust to expose the bars.

1. Paint coating - Various effective paints are available, which are used to provide paint coating on the steel bars. However, the paints are also permeable Commonly, a zinc, aluminium or lead based priming coat is applied first so as to provide a good base for the paints coat.

2. Metallic coating : Metallic coating including galvanizing and sheradizing (in both the process, zinc is used), is done for sheets. Electro-plating is mainly confined to small items like fasteners.

3. Corrosion resistant steel. Special corrosion resistant steel are used as protective treatment. Such steel rapidly forms its own protective layer of oxide film. The rate of corrosion reduces to negligible level after few years. In Britain, such steel is known as **weatherng steel.** Commercially, such steel is available as **Cor-Ten** It is originally developed by the United States Steel Corporation. Elements such as copper, nickle and chromium are added in this variety of steel. There is stainless steel, which does not corrode.

4. Coated reinforcement. In IS: 9077-1979, a special method of coating the individual reinforcement bars with a phosphate jelly to prevent corrosion has been described. This method has been developed at the Central Electro Chemical Research Institute (CECRI) at

Karaikudi, Tamilnadu. The individual bars are treated at site with derusting solution. The bars are washed with an alkaline cleaning powder. Then, the phosphate jelly is applied along with prohibitory solution, cement slurry, sealing solution and cement slurry again. Soon after coating, the bars are used. This process is tedious and time consuming.

Other methods of preventing the corrosion have been described alongwith mentioning different types of corrosion.

1.29 FIRE RESISTANCE OF CONCRETE STRUCTURES

Fire resistance is defined as ability of an element component or structure, to fulfil for a stated period of time, the required stability integrity, thermal insullation and/or other expected performance in a stand fire test.

Adequate fire resistance is necessary in buildings to allow the escape of the occupants and the reasonable safety for the fire fighters engaged in fire under control.

Fire resistance and protection are important aspects for the design in the event of fire inside the structures. An understanding of the behaviour of materials in elevated temperature is needed to ensure safety. Adequate fire resistance to structural elements is necessary to safe guard the stability of the structure Essentially, it is intended to ensure public safety than to safe guard the structure itself. The main criteria are to prevent premature collapse and the main sketelon of the building to stand up long enough so that to allow escape of the inside occupants and reasonable safety for fighters engaged in extinguishing the fire and bringing the fire under control and to limit the spread of fire, thus reducing the risk to surrounding property and their occupants. Smaller the area in which the fire may be contained, the smaller the escape and fire fighting times will need to be, although the different materials stored in the building will affect this time.

Fire resistance is obtained by ensuring that the cross-sections are of minimum thickness and that the cover reinforcement is sufficient.

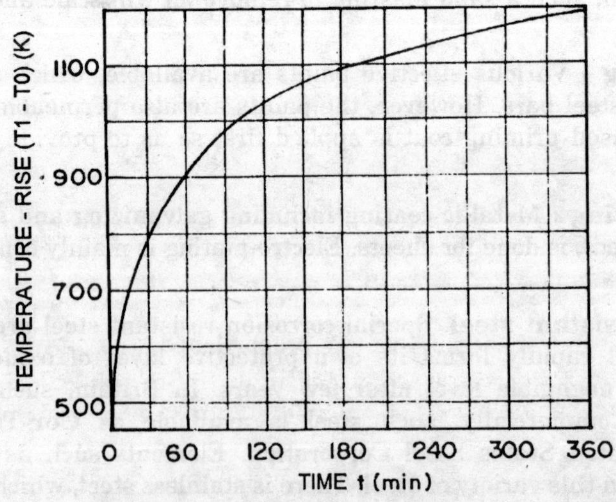

STANDARD CURVE (TIME v/s TEMPERATURE RISE.)

Fig. 1.4.

In Britain, the necessary requirements form part of the Building Regulations. The minimum periods for the inside occupants to escape range from 30 minutes for a small residential buildings to 4 hours for a large store. And it is specified in building regulations. The period for which fire resistant is needed is calculated by the volume of the enclosed section. For a larger volume a greater period of fire resistance is required than for a lesser volume. The fire resistance of a concrete structural element is expressed in terms of time period in hours in accordance with IS : 1641.

All materials lose strength with increase in temperature. Some idea of the of the temperatures likely to meet may be gained by reference to an internationally - accepted standard time-temperature curve shown in Fig. 1.4.

The fire resistance of a building is usually specified in hours (minutes). The fire-resistance grading period for a structural elements or system in minutes, whcih required to be attended in a standard fire test is referred as Fire Resistance Level.

A reinforced concrete structure or a structural element required to have fire resistance should be designed to possess an appropriate degree of resistance to flame penetration, heat transmission and failure. The fire resistance of reinforced concrete depends to a large extent on the control of temperature of the reinforcement and hence on the degree of insulation provided, which is obtained by concrete cover or by mixture of concrete cover, plaster and other finishes. The thickness of section also enters into the calculations. The fire resistance of concrete elements depends upon details of member size, cover to steel reinforcement and type of aggregate (normal weight or light weight). The general requirements for fire protection are given in IS: 1642. The beams should have minimum width, the floors should have minimum thickness and the columns should have minimum lateral dimensions.

The cover provided for the protection against corrosion and for the durability may not be sufficient for protection against fire. The nominal cover to meet specified period of fire resistance have been specified in IS : 456-2000 for beams, slabs, ribs and columns and these have been given in respective portions.

1.30 REQUIREMENTS AND AIM OF DESIGN

The structures and the structural members are designed to satisfy the three basic requirements (viz. *strength, stability and stiffness*).

1. Strength requirements

According to the *Strength requirements*, all the sections of the structures and the structural members should have its actual strength, which must be large enough to resist, with some margin to spare, the structural effects of all foreseen loads and forces that occur during the construction and use throughout the life of the structure, without failure or other distress. The structural members are so proportioned, (i.e., the dimensions of concrete sections and reinforcement are selected), so that, the strength of member is sufficient to resist forces resulting from even certain imaginary overload stages (significantly above loads expected actually to occur in service).

2. Stability requirements

According to the *stability requirements*, the structures and structural members are safe guarded against *overturning* and *sliding* (horizontal forward movement). The live loads having favourable effect are not considered for these purposes. The possibility of soil being permanently or temporarily removed is also kept into consideration.

2 (i) Overturning

The stability of a structure as a whole against overturning is verified so that *the restoring moment* is more than the sum of 1.2 times the maximum overturning due to characteristic dead load and 1.4 times the maximum overturning moment due to the characteristics imposed loads. In case, where the dead loads provide the restoring moments only 0.9 times the characteristic dead load are considered. The restoring movement due to imposed loads (live loads) are not considered. The anchorages or counter weights provided for overhanging members (during construction and service) should be such that static equilibrium remains even when the overturning moment is doubled.

2. (ii) Sliding

The factor of safety against sliding of structures should be more than 1.4 under the most adverse combination of the applied characteristic forces. In this case only 0.9 times the characteristic dead load should be considered. IS : 1904-1978 states that the factor of safety against sliding of structures which resist lateral forces (e.g., retaining walls) should be more than 1.5 when dead load, live load and earth pressures are considered together with wind or seismic forces. When the dead load, live load and earth pressures are only considered, the factor of safety should be more than 1.75

3. Stiffness requirements

In order to meet the *stiffness requirements* for the structure as a whole and for the structural members, the limitations for deflections are specified. The deflection of the structure or part thereof should not adversely affect the appearance or efficiency of the structure or finishes or partitions. **The relative stiffness** of structural members are found for the analysis of indeterminate structures. IS : 456-1978 permits three assumptions, for calculating the relative stiffness of members. The moment of inertia of a reinforced concrete section may be found on the basis of following :

1. The gross cross-section of a member neglecting the reinforcement may be used. It may be considered for convenience and but it is not reasonable.

2. The transformed or equivalent section (viz., cross-sectional area of concrete plus the area of steel reinforcement transformed by using modular ratio) may be used. It is most rational.

3. The cracked section (viz., area of concrete in compression plus the area of reinforcement transformed as equivalent area by using modular ratio) is used. It is also rational, when the section is a cracked section.

The aim of design is the achievement of an acceptable probability that structures being designed will perform satisfactorily, during their intended life, with an appropriate degree of safety, the structures should sustain all the loads and deformations of normal construction and use and have adequate durability and adequate resistance to the effects of misuse and fire.

1.31 SERVICE LIFE OF CONCRETE STRUCTURES

Service life of concrete structures is the time in service rendered by the structures *Functional service life* is the time in service until the structure becomes out of date due to change in functional requirements. *Economic service life* is the time in service until the replacement of the structure is economically more advantageous than keeping the structure in service *Technical service life* is that after which the performance of the structure may be considered to reduce with time.

Definition of functional service life and economic service life arise from the recognition that, in fact, structures commonly attain the end of their life (i.e., demolished and site is redeveloped) because the structures become out of date, then because of an inability to perform in the way envisaged at the design stage.

While, as a discussion point, the definition serves some purpose, functional service life may probably be considered to fall within the definition of economic service life. Indeed, except for certain special circumstances, the technical service life would also fall within this definition.

Mostly technical short-comings may be rectified by repair and whether repairs are done, or the structures replaced will depend upon the relative economics of the two options. In fact, the only circumstance that comes easily to mind where service life is not defined by economic conditions is if the structure collapses without warning. In all other circumstances where time exists to make a decision about the fate of a structure, a balance will strike between the costs of repair, renovation, modification, the value of the upgraded structure and costs of replacement. Value may not be exclusively monetary, as in the case of ancient monuments where the value is highly subjective quantity.

End of life is thus, practically, the point in time when it is judged to no longer be economically viable to keep the structure alive. This concept may be employed usefully to establish a required life.

In principle, it may, through studies of whole life costing. The basics of such an approach are given by White, having defined a suitable design life from economic consideration. It would then be necessary to ensure that the technical design life exceeded this figure. In practice, detail studies to define a design life would only rarely be appropriate and what would be provided are arbitrary figure such as the 120 years givem in BS : 5400 for the bridges.

Above has been concerned with the actual life of a strucutre or the life required by the client, inorder to design to achieve a given life, a different definition is required, based on the technical service life. Following may be a possible definition :

Design Service life is the anticipated time in service in the anticipated environment until a defined acceptable probability that the structure will have reached an unacceptable state is exceeded, for a defined level of maintenance.

Any definition must contain the following elements which are essential to the definition of a logical design approach. These are as under :

(i) The definition must make clear that a specific environment is assumed.

(ii) There must be a definite criterion of failure (attainment of unacceptable state)

(iii) The statement may express in specific terms (in probabilistic terms)

(iv) Mention must be made of an assumed level of maintenance.

End of life of the concrete structures

Practical problem which now arises before the idea of designing to a specified design life may become a reality is to clarify the end point of the technical design life. This has been stated to be the point when a defined unacceptable state is reached.

First idea that comes to mind is that this must relate to some specified reduction in safety; which will decide the reduced level of safety factor that may allowed.

Acceptable level of safety has been considered in relation to the appraisal of existing structures by the Institution of Structural Engineers. Their report allows reduced safety factors to be used in assessment situations, but the reductions in the partial factors are each justified on the grounds that more is known about the structure being assessed than was known at the design stage. Thus, the factor of safety on the dead loads may be reduced in case the member sizes are actually measured so that a more reliable estimate of the dead load may be found. Similarly, the factor on the concrete strength may be reduced when the actual strength of the concrete has been assessed from cores.

An essential feature of design is that the constructions will last without undue or unexpected deterioration for its projected useful life.

The question which must now be considered is whether it is useful to introduce a life time or design life explicitly into the design process. Looking at the whole of construction, it is clear that sometimes it is and sometimes it is not. Paint, for example, clearly has an expected life and maintenance schedules are organised to ensure repainting at appropriate intervals.

Structural elements may very occasionally, also be treated as replaceable and the maintenance and life time costing arranged to take accounts of this.

It has not been found useful to introduce design life into the design process.

Stonework or brickwork or mass concrete are considered to last effectively indefinitely if properly specified and constructed other materials, such as steel or soft wood are protected and, provided the protection is maintained, life is not threatend.

Reinforced or prestressed concrete seems unique for a long life (50-60 years) and the suggestion only relates to one specific cause of deterioration corrosion of reinforcement or prestressing tendons. Other forms of deterioration are treated by avoidance of the problem. For example, mix ingredients should be so selected so that deleterious alkali-silica reactions do not occur. A similar line is adopted for frost attack and sulphate attack. If these forms of deterioration do occur then it is reliability or system failure. Even for design against corrosion of prestressing tendons, the philosophy is beginning to develop that, where a sudden failure could result from corrosion, the tendons should be inspectable and replaceable, thus removing design life consideration from the design.

For reinforced concrete and prestressed concrete, it is foreseen that in the relatively near future, it will be possible to select appropriate covers and mix designs to give a chosen design life, however this may be defined. The discussion on reliability issues above suggests that the durability failures due to corrosion may generally be system failures and that careful selection of covers and mix designs may not, in fact, lead to any expectable changes in actual durability.

It may be more reasonable to approach the improvement of design for durability in the following two stages.

Stage 1 would address what is clearly the most serious problem. This is the reduction in the workmanship and design errors. This can only be achieved by improved education and defining of all involved in the construction process.

Once stage 1 is complete and the incidence of errors is greatly reduced, **Stage 2**, the introduction of design methods which relate directly to design life may be attempted. The introduction of **Stage 2** before successful completion of stage 1, will simply divert the attention from more important factors and will probably inhibit, rather than help, the achievement of acceptable standards of durability.

[Reference : Professor A.W. Beeby `Design for Durability' Report presented at BRE Seminar on Design for Durability, December, 1990]

Professor Beeby acknowledged that the content of this paper is largely drawn from a report produced by the British Cement Association for the Building Research Establi-shment.

Looking to the significance of the paper, this paper is given in an appendix of this book in the wider interest of the persons studying, working and involved in this studies. It will give an easy access. With courtesy and permission in anticipation from professor Beeby.

1.32 DESIGN CODES AND SPECIFICATIONS

Codes are set of technical specifications drafted to control the design and to provide guidance to learning students and newly passed engineering graduates. National building codes are formulated in different countries to lay down guidelines for the design and construction of structures. Codes are drafted by the export structural engineers, using the joint knowledge, gained over the years.

The construction of concrete structures involves four parties, (viz., the owners, the designers, engineers and architects, the fabricators and the constructors). The acceptable standard of quality of materials and workmanship in fabrication are defined in the specifications. The designers prepare three types of specifications, (viz., *project specific-ations, material standards and design codes*). The object of drafting specifications and preparing codes is to achieve economy in the use of materials and the cost of construction without sacrificing the safety. The *project specifications* describe complete information about the owner's and the engineer's requirements for the completed structures. The *material standards* describe the quality of materials and their economic use. The design codes incorporate the following :

1. General guidance as regards the various loads

2. Working (allowable) stresses

3. Methods of design and structural analysis and procedures

4. Generally accepted uniformity throughout the country.

For complex and time consuming formulae and provisions of designs, the tables are included in the code, which reduce the design calculations. Various professional associa-tions, institutes and government agencies prepare codes and specifications.

United States does not have an official or national code governing institute. The responsibility for preparing and maintaining design specifications rests with various professional groups, trade associations and technical institutes. The America Concrete Institute (ACI) has long been a leading institute and it has published the widely recognized `Building Code Requirements for Reinforced Concrete'. It serves as a guide in the design and construction of reinforced concrete buildings. The American Concrete Institute also publishes important journals and standards, as well as recommendations for the analysis and design of special structures such as tanks. Most of highway bridges in the United States are designed according to the requirements of AASHTO (American Association of State Highway and Transportation Officials) bridge specifications. The design of railway bridges is done according to specifications of AREA (American Railway Engineering Association) Manual of Railway Engineering.

ACI 319-1989 is a Building Code requirement for reinforced concrete.

The changes in the various respective codes shall be made as a regular fea- ture, which incorporated theoretical and laboratory resarches made in future time to time also.

In United Kingdom, the design of reinforced concrete is based largely on British Standards and BS codes of practice. The structural use of concrete (CP 110 : Parts 1, 2 and 3), the structural use of normal reinforced concrete in buildings (CP 114), the structural use of concrete for retaining aqueous liquids (BS 5337), steel, concrete and composite bridges (BS 5400), Part 2 : Specifications for loads, Part 4 : Design of concrete bridges, loading (CP 3: Chapter V)... etc. are British Standards and Codes in practice. In addition to these, National Building Regulations and the London Building (constructional) By- laws are also used.

Following codes are also followed in UK : These codes have been published by British standard Institution

CP114: 1957, 1965, 1972;
CP115; CP110

BS : 5328-1991 Part 1.	Guide to specifying concrete.
BS : 5328-1991 Part 2.	Methods of specifying concrete mixes.
BS : 5328-1990 Part 3.	Specification for the procedures to be used in producing and transporting concrete.
BS : 5950	The structural use of steel was in building (There are nine parts of this code)
BS : 5328-1990 Part 4.	Specification for the procedure to be used in sampling, testing and assessing compliance of concrete.
BS : 8110 - 1985 Part 1.	Structural use of concrete. Code of practice for the design and construction.
BS : 8110-1985 Part 2. EC	Structural use of concrete; Code of Practice for special

(Euro codes) circumstances Standard Association of Australia make the use of AS3600-1998, ``Concrete structures".

In our country Indian Standard Institute (Bureau of Indian Standards), Indian Road Congress (IRC) and Ministry of Railways and Railways Board have prepared and published various codes and specifications. IS : 456-1953 (Indian Standard Code of Practice for Plain and Reinforced Concrete for General Building Construction) was published in 1953. This codes was then revised in 1957 and referred as IS : 456-1957. It was again revised and issued in 1964 as IS : 456-1964, under modified title Code of Practice for Plain and Reinforced Concrete. The scope of use of this code was thus enlarged. This code mainly deals with the working stress method.

The codes serve at least follwoing four distinct functions.

1. Code ensure adequate structural safety, by specifying certain essential minimum requirements for the design.

2. Codes provide the designers relatively simple procedures.

3. Codes ensures a measure of consistence among different designers practising in the differents parts of the country.

4. Codes ensure some legal validity. The structural engineers (designers) from any liability in case of some structural failures which are caused by inadequate supervision and/or use of defective materials and improper construction.

Important benefits of standardization are improvements of the suitability of materials, products, processes and services for their intended purposes, prevention of barrier trade and facility of technological co-operation. Standardization is an activity of establishing with regard to actual or potential problems, provisions for common and repeated use, aimed at the achivement of the optimum degree of order in a given context. Standard cover almost all aspects of economic activity of a person-engineering industry, construction and many others. Correct perspective of aims of standardization is found by defining them at the following levels.

1. **International standardization:** Relevant bodies of all countries have their involvement at this level. International standards are formulated and adopted and made available for use.

2. **Regional standardization:** Relevant bodies of some countries (e.g. from geographical, political or economic area of the world) have their involvement. These are formulated and adopted made available for use.

3. **National standardization:** These are done by one country. These are formulated by national standard body and adopted and made available for use.

In addition to above, relevant bodies of different associations and companies may have their own involvement for formulation of standards.

These standards may differ in form and type depending on the particular aspect of a subject that may be covered. These aspects may include (i) a set of nomenclature or definition of terms, (ii) a specification for the quality, composition, or performance of a material, an instrument, a machine or a structure, (iii) a method of sampling or inspection to find conformity to a specified requirement of a large batch of material by inspection of a smaller sample, (iv) a method of test or analysis to evaluate specifiied characteristics of a material or a chemical. (v) a scheme of simplification or rationalization, (i.e., limitation of variety of sizes, shapes or grades) designed to meet most economically the needs. This also includes dimensional freezing of component design to ensure interchangeability as also method of grading and grade definitions for natural products, such as timber, minerals, etc.

In United states, most of the reinforced concrete buildings and related constructions are designed in accordance with the current ACI code. It has served as a model document for the many other countries. It is regarded as authorative statement of current good practice in the field of reinforced concrete. As a result it has been incorporated into the international building code.

International Building Code, International Code Council Falls Church, VA2000 is an example of a consensus code governing structural design and it has been adopted by local bodies. ACI 318-02, Building Code Requirement for structural concrete American Concrete Institute, Farmington Hills, MI, 2002 is an example of good concensus Commentary on Building Code Requirement for structural concrete, ACI 318R-0, Concrete Institute, Farmington Hills, MI, 2002 provides background material and rationale for the provisisions of code.

The various Indian Standards for aggregates and apparatus for testing aggregates, cements, cement sampling and cement tests, concrete, concrete design and construction and concrete testing and for reinforcement have been grouped and listed as below. The special publications for reinforced concrete are also listed.

AGGREGATES AND APPARATUS FOR TESTING AGGREGATES

1	9103-1979	Admixtures for concrete
2	383-1970	Aggregates (coarse and fine), from natural sources for concrete
3	5640-1970	Method of test for determining aggregates impact value of soft coarse aggregates
4	2430-1986	Methods for sampling of aggregates for concrete
5	2386 (Part 1) - 1963	Methods of test for aggregates for concrete (Particle size and shape)
6	2386 (Part 2) - 1963	Estimation of deleterious materials and organic impurites
7	2386 (Part 3) - 1963	Specific gravity, density, voids, absorption and buckling
8	2386 (Part 4) - 1963	Mechanical properties
9	2386 (Part 5) - 1963	Soundness
10	2386 (Part 6) - 1963	Measuring mortar making properties of fine aggregates
11	2386 (Part 7) - 1963	Alkalies aggregates reactivity
12	650-1966	Standard sand for testng of cement
13	9377-1979	Apparatus for aggregate impact value
14	9376-1979	Apparatus for measurng aggregate crushing value and ten percent fines value
15	10079-1982	Cylindrical metal measures for use in tests of aggregate and concrete
16	10070-1982	Machine for abrasion testing of coarse aggregate

17	10086-1982	Moulds for use in tests of cement and concrete
18	3812 Part III	Fly ash for use as fine aggregate for mortar and concrete
19	2430-1969	Methods of sampling of aggregates for concrete
20	6925-1973	Methods of test for determination of water soluble chlorides as concrete admixtures.

CEMENTS, CEMENT SAMPLING AND CEMENT TESTS

1	4845-1968	Definitions and terminology relating to hydraulic cement.
2	6452-1972	High alumna cement for structural use
3	8112-1976	High strength ordinary Portland cement
4	3535-1986	Hydraulic cements, methods of sampling
5	4031-1968	Hydraulic cement, physical tests for methods
6	8043-1978	Hydraulic Portland cement
7	4032-1985	Method of chemical analysis of hydraulc cement
8	8229-1986	Oil-well cement
9	269-1976	Ordinary and low heat Portland cement
10	1489-1976	Portland-pozzolana cement
11	8041-1978	Rapid hardening Portland cement
12	455-1976	Slag cement, Portland
13	650-1966	Standard sand for testng of cement
14	6909-1973	Super sulphated cement
15	8042-1978	White Portland cement
16	5514-1969	Apparatus used in Lechatelier's test
17	11262-1985	Calorimeter for determination of heat of hydration of hydraulc cement
18	5536-1969	Constants flow type air-permeability apparatus (Lea and Nurse Type)
19	11263-1985	Cylinder measurs for determination of air content of hydraulc cement mortar
20	10078-1982	Jolting apparatus used for testing of cement
21	10890-1984	Planetary mixer used in tests of cement and pozzalana
22	5516-1969	Variable flow type air-permeability apparatus (Blaine type)
23	10510-1983	Vee-bee consistometer
24	10080-1982	Vibrator machine
25	5513-1976	Vicate apparatus
26	7509-1974	Thermal insulating cements (Type 750)
27	7510-1974	Thermal insulating cements (Tyle 350)
28	2645-1975	Integral cement water proofing compounds
29	5724-1970	Methods of tests for thermal insulating cement
30	4031-1968	Method of physical test for Hydraulc cement

CONCRETE CONCRETE DESIGN & CONSTRUCTION AND CONCRETE TESTING

1	7861-1975	Extreme weather concreting, code of Practice
	(A) 7861 (Part 1) - 1975	Recommended practice for hot weather concreting

	(B) 7861 (Part 2)	Recommended practice for cold weather concreting
2	6461 (Part 1) - 1973	Glossary of terms relating to cement concrete
3	6241-1971	Methods of test for determination of stripping value of road aggregate
4	7325-1974	Apparatus for determining constituents or fresh concrete
5	9399-1979	Apparatus for flexural testing of concrete
6	9459-1980	Apparatus for use in measurement of length change of hardened cement paste, mortar and concrete
7	5515-1983	Compaction factor apparatus
8	7320-1964	Concrete slump test apparatus
9	456-2000	Plain and reinforced concrete code of practice
10	1343-1980	Prestressed concrete
11	10262-1982	Recommended guidelines for concrete mix design
12	9012-1978	Recommended practice for short concreting
13	9284-1979	Abraison resistance of concrete method test
14		Autoclaved cellular concrete products,, Methods of test
	6441 (Part 1) - 1972	Determination of unit weight or bulk density and moisture content
	6441 (Part 2) - 1972	Determination of drying shrinkage
	6441 (Part 4) - 1972	Corrosion protection of stress reinforcement in autoclave cellular concrete
	6441 (Part 5) - 1972	Determination of compressive strength
	6441 (Part 6) - 1973	Strength, deformation and cracking of flexural members subjected to bending - short duration loading test
	6441 (Part 7) - 1973	Strength deformation and cracking of flexural members subjected to bending sustained loading
	6441 (Part 8) - 1973	Load tests for flexural members in diagonal tension
	6441 (Part 9) - 1973	Jointing of autoclaved cellular concrete elements
15	8142-1976	Determining setting time of concrete by penetration resistance test
16	9013-1978	Making, curing and determining compressive strength of accelerated cured concrete test specimens, method
17	3085-1965	Permeability of cement mortar and concrete method of test
18	2514-1963	Concrete vibrating tables
19	7246-1974	Recommendations for use of table vibrators for consolidating concrete
20	5816-1970	Method of test for spliting tensile strength of concrete cylinders
21	2770 (part I) - 1967	Methods of testng bond in reinforced concrete
22	516-1959	Method of test for strength of concrete
23	4925-1968	Concrete mixing and Rotating Plant
24	1791-1968	Batch type concrete mixer
25	5891-1970	Hand Operated Concrete Mixer

26	5892-1970	Concrete transit mixers and Agitators
27	2505-1968	Concrete vibrators immersion type
28	2506-1964	Screed Board Concrete Vibrators
29	3558-1966	Code for Practice for use of immersion vibrations for consolidation concrete
30	5452-1969	Autoclaved cellular concrete board
31	5751-1970	Precast concrete copies blocks
36	2185-1967	Hollow Cement Concrete Blocks
37	6041-1971	Code of Practice for construction for autoclaved cellular concrete block masonry
38	6042-1969	Code of Practice for construction of light weight concrete block masonry
39	2572-1963	Code of Practice for construction of hollwo concrete block masonry
40	456-2000	Code of Practice Plain and Reinforced Concrete, Code of Practice (Fourth Revision)
41	3370 (Part I to VI)	Code of Practice for Concrete Structures for the storage of liquids
42	7861-1971	Code of Practice for extreme weather concrete
43	785-1964	Reinforced concrete poles for overhead power and telecommunication lines
44	2193-1962	Prestressed Concrete street lighting columns
45	1332-1959	Reinforced Concrete street lighting columns
46	2905-1966	Methods of test for poles for overheat power and telecommunication lines
47	1199-1959	Methods of Sampling and Analysis of Concrete
48	6003-1970	Intended wire for prestressed concrete
49	1343-1960	Code of Practice for prestressed Concrete
50.	10262-1982	Recommended guidelines for concrete mix-design

REINFORCEMENT

1	1786-1966	Cold twisted steel bars for concrete reinforcement
2	1139-1966	Hot rolled mild steel, medium tensile steel and high yield strength deformed bars for concrete reinforcement
3	432-1982 Part 1, Part 2, Part 1. Steel bar Part 2. Hand drawn steel wire.	Mild steel and medium tensile steel bars and hard drawn steel wire for concrete reinforcement
4	1566-1982	Fabric for concrete reinforcement Hard drawn steel wire
5	5525-1969	Reinforcement for detailing of reinforcement in reinforced concrete works
6	2502-1963	Code of practice for bending and fixing of bars for Concrete Reinforcement
7	6006-1983	Uncoated stress relieved strand for prestressed concrete.
8	6003-1983	Indented wire for prestressed concrete

9	1785-1983 Parts 1, 2	Plain and Hard drawn steel wire for prestressed concrete
10	6461 Part (3) - 1972	Glossary of terms relating to cement concrete Part 3.
11	1786-1985	High strength deformed steel bars and wires for Concrete Reinforcement
12	2090-1983	High tensile steel bars used in prestressed concrete
13	10790-1984 Part -I	Methods of sampling of steel prestressed and reinforced concrete
	10790-1984 Part- 1	Prestressing
	10790-1984 Part-2	Reinforcing

LOADS

	IS : 875-1987	Code of Practice for Design Loads (Other than Earthquake) for Buildings and structures
1	IS : 875 (Part 1) 1987	Dead Loads- Unit Weight of Building Materials and Stored Materials
2	IS : 875 (Part 2) 1987	Imposed Loads
3	IS : 875 (Part 3) 1987	Wind Loads
4	IS : 875 (Part 4) 1987	Snow Loads
5	IS : 875 (Part 5) 1987	Special Loads and Load Combinations

In addition to above codes, Bureau of Indian Standards have published following Special Publications (SP) which are useful as explanations for various provisions of codes and as an aid to design concerning to this subject.

1. SP : 22 (S & T) - 1982 Explanatory Handbook on Codes for Earthquake Engineering (IS:1893 and IS : 4326-1976)
2. SP : 23-1982 Handbook of Concrete Mixes
3. SP : 22-1983 Explanatory Handbook on Indian Standards for Plain and Reinforced Concrete, (IS : 456-1978).
4. SP : 25 (S & T) - 1984 Handbook on Causes and Prevention of cracks in Buildings
5. SP : 34 (S & T) - 1987 Handbook on Concrete Requirement and Detailing.

2

PART - II
LOADS, STRENGTH, FACTORS OF SAFETY AND METHODS OF DESIGN

LOADS (ACTIONS), STRENGTH OF MATERIALS, FACTORS OF SAFETY AND METHODS OF DESIGN

2.1. INTRODUCTION

The basic aim of a structural engineer is to design the structures and structural members which satisfy the *functional* and *structural aspects*. Both the aspects are inter-related. **The functional aspect** takes into consideration the purpose for which the structure is designed and built-up. This aspect includes the determination of amount and arrangement of space necessary for the structure. The structure is able to perform satisfactorily the service for which it is designed. The functional aspect includes the determination of location and arrangement of operating utilities, special equipment, machinery or other features, incident to the proper functioning of the structures The preliminary layout of the structure is made. The preliminary layout of structure is useful in determining the type of structure to be used, the general size, shape and arrangement of the various parts of the structure. The approximate dimensions of structures are established. These dimensions are used for the structural design.

In **the structural aspect**, it is ensured that the building or the structure is structurally safe, strong, durable and serviceable throughout its life time, aesthetically good in appearance and economical. The safety of structures means that the structures should not fail under the loads unless exceeded by a given margin technically termed as *overload* safety factors. The serviceability means that the structures serve throughout their service life in both appearance and comfort. That is, the cracking (if any) and the deflections due to the working loads should not be in excess of their values specified in the code. The design should give the saving of materials used (that will provide economy). Many structures are built to selter human beings and to protect them from the forces of nature. Occupants of the buildings or the structure should feel safe with regard to their lives and properties. The safety of many individuals at one time depends on the structural safety of the structures. The structures are designed between two limits (viz., the structural safety and economy).

A substatial economy in cost of a structure is obtained by appropriate choice of structural systems (for example, load bearing walls, frames, shear walls and their various combinations).

All the sections of the structures should have strength at least equal to the structural effects of design loads and forces that occur during the construction and use. All the loads should be taken into consideraton which a building or a structure would be required to bear. The term *load* includes both applied forces and imposed deformations, such as those caused by restraint of movements due to changes in temperature. The **nominal loads** are very similar to those which appear as **working loads**. Loads are externally aplied forces (actions). A structure may be required to withstand any load out of the various loads or the combination of the various loads. These loads include overhead and other cranes, equipments ... etc., in industrial buildings occupancy etc., in buildings; vehicles on bridges; water in hydraulic structures; loose materials for bins and bunkers and so on. For the purpose of determining the maximum stresses in any sructure or a structural member, the various loads which are likely to act on the structure are given below :

When the loads act on a structure, the stresses develop in the various members and the deformations occur due to the action of loads. A new term **ACTION** has been introduced by EURO-CODES for loads. Therefore, the loads are referred by this word `action'. The **actions** are the primary cause for developing the stresses and deformations.

1. Dead load
2. Live load (Imposed load)
3. Wind load
4. Snow load
5. Seismic load
6. Temperature effects
7. Shrinkage
8. Creep.

In addition to the above loads and effects, following forces and effects are also considered while designing the structures if these are liable to affect materially the safety and serviceability of the structures.

1. Foundation movements
2. Elastic axial shortening
3. Soil and fluid pressure
4. Vibrations
5. Fatigue
6. Impact (Dynamic effects)
7. Erection loads
8. Stress concentration effect due to point load and the like.

Similar to the new term **`action'** for the loads, other new term **`ACTION EFFECTS'** [or **LOAD EFFECTS'**] has also been introduced by the EURO-CODES for the inerval forces (e.g., axial forces, shear forces, bending moments and twisting moments) developed in the various members of a structure due to **action** loads.

The nominal loads are divided into three groups namely **permanent loads, transient loads** and **exceptional loads.**

1. Permanent loads

Permanent loads are defined as dead loads, superimposed dead loads, loads due to filling materials, different settlement and loads derived from the nature of the structural material. In concrete structures, it refers to shrinkage and creep.

2. Transient loads

All loads other than the permanent loads referred to above are transient loads. These loads include wind loads, temperature effects, exceptional loads, erection loads,... etc.

3. Exceptional loads

These loads include the loads due to otherwise unaccounted effects, such as earthquakes, stream flows, ice packs,...etc.

2.2 DEAD LOADS (PERMANENT LOADS) (D.L.)

The dead loads (viz., the self-weight of the structural members), superimposed dead loads and loads due to filling materials are referred as permanent dead loads.

Dead loads are loads which are constant in magnitude and fixed in position throughout the life time of the structure. Dead loads in a building comprises of the *self weight* of the structure and all other *super-imposed loads* (viz., all permanent constructions and installations including weight of walls, partitions, floors, and roofs). The self-weight of the structure being unknown in the beginning of the design, it is assumed by the designer in the light of experience and thumb rules and formula available. The super-imposed dead loads are determined for the various materials specified for the construction and unit-weight of materials. The unit weights of plain concrete made with sand and gravel or crushed natural stone aggregate may be taken as 24 kN/mm³, and that of the reinforced concrete as 25 kN/m³, respectively. On completion of design, total dead load is calculated and checked with the dead load assumed for design and the design is revised, if necessary. Table 2.1 gives the unit-weights of materials used in building construction. The dead loads are also called as permanent loads as regards to the duration of their action.

Dead loads act due to gravity. Therefore, these are also referred as gravity loads. These are loads which arise due to gravity.

TABLE 2.1. UNIT WEIGHT OF BUILDING MATERIALS, BUILDING PARTS AND COMPONENTS IS : 875 (PART I) - 1987 (Extracts from Tables I and II of IS : 1911-1967, schedule of unit-weight of materials)

No.	Materials	Nominal size or thickness	Weight/mass kN	kg	per
	Building materials				
1.	Asbestos natural raw	—	29.8	3040	m³
			5.98 to 8.85	600 to 900	m³
2.	Asbestos cement sheeting	6 mm	0.118 to	12	m²
	corrugated (146 mm)	6 mm	0.130	13.3	m²
	Semi-corrugated (240 mm)	6 mm	0.118 to	12	m²
		6 mm	0.127	13	m²
	Plain	5 mm	0.09	9.16	m²
3.	Bitumen	-	0.102	10.4	m³
4.	Boards (cork)				
	Compressed	10 mm	0.04	4	m²
	Ordinary	10 mm	0.02	2	m²
5.	Bricks		chips & broken bricks		

	Fine	—	14.2	1450	m^3
	Coarse	—	9.9	1010	m^3
6.	Brick dust (Surkhi)	—	9.9	1010	m^3
7	Bricks (Engineering)	—	21.2	2160	m^3
8	Cement				
	Ordinary and Aluminious	—	14.10	1440	m^3
	Rapid hardening	—	12.55	1280	m^3
9.	Plain cement concrete with sand and gravel or crushed natural stone	—	22 to 23.5	2240-2400	m^3
10.	Reinforced cement concrete sandstone aggregate with				
	1 percent steel	—	22.75-24.20	2310-2470	m^3
	2 percent steel	—	23.25-24.80	2370-2530	m^3
	5 percent steel	—	24.80-26.50	2530-2700	m^3
11.	Glass	—	24.384	2480	m^3
	Glass bricks	—	24.304	2480	m^3
	Window glass and		24.304	2480	m^3
	Looking glass		24.304	2480	m^3
	Sheet	2.0 mm	0.049	5.0	m^3
		2.5 mm	0.062	6.3	m^3
		3.0 mm	0.074	7.5	m^3
		4.0 mm	0.098	10.0	m^3
		5.0 mm	0.123	12.5	m^3
		5.5 mm	0.134	13.7	m^3
		6.5 mm	0.167	17.0	m^3
12	Gypsum				
	Mortar	—	11.75	1200	m^3
	Powder	—	13.89-17.25	1410-1760	m^3
13	Iron				
	Pig	—	70.60	7200	m^3
	Gray cast	—	68.95-69.90	7030-7130	m^3
	White cast	—	74.30-75.70	7530-7720	m^3
	Wrought	—	75.50	7700	m^3
14	Masic asphalt	10 mm	0.215	22	m^3
15	Metal sheeting Galvanised steel sheets (plain and corrugated)				

	Case 1	1.60 mm	0.131	13.31	m^2
		1.25 mm	0.104	10.50	m^2
		1.00 mm	0.084	8.60	m^2
		0.80 mm	0.069	7.03	m^2
		0.63 mm	0.056	5.70	m^2
	Case 2	1.60 mm	0.129	13.16	m^2
		1.25 mm	0.102	10.41	m^2
		1.00 mm	0.083	8.45	m^2
		0.80 mm	0.067	6.88	m^2
		0.63 mm	0.054	5.55	m^2
	Case 3	1.60 mm	0.128	13.01	m^2
		1.25 mm	0.101	10.26	m^2
		1.00 mm	0.081	8.30	m^2
		0.80 mm	0.066	6.73	m^2
		0.63 mm	0.053	5.40	m^2
	Case 4	1.60 mm	0.127	12.94	m^2
		1.25 mm	0.100	10.19	m^2
		1.00 mm	0.081	8.22	m^2
		0.80 mm	0.065	6.66	m^2
		0.63 mm	0.052	5.32	m^2
16.	Mortar				
	Cement	—	20.40	2080	m^3
	Lime	—	15.70-18.05	1600-1840	m^3
17	Plaster				
	Cement	—	20.40	2080	m^3
	Lime	—	17.25	1760	m^3
18	Soils and gravels				
	Alluvial ground (undisturbed)	—	15.60	1600	m^3
	Clay (dry and lump)	—	10.20	1040	m^3
	Dry (compact)	—	14.10	1440	m^3
	Damp (compact)	—	17.25	1760	m^3
	Clay (wet and compact)	—	20.40	2080	m^3
	Undisturbed	—	18.85	1920	m^3
	Undisturbed bed gravelly	—	20.40	2080	m^3
	Earth				
	Dry	—	13.85-18.05	1410-1840	m^3

	Moist	—	15.70-19.60	1600-2000	m^3
	Gravel				
	Loose	—	15.70	1600	m^3
	Rammed	—	18.85-21.20	1920-2160	m^3
19	Stone				
	Agate	—	25.50	2600	m^3
	Aggregate	—	15.70-18.85	1600-1920	m^3
	Basalt	—	27.95-29.05	2850-2960	m^3
	Granite	—	25.90-27.45	2640-2800	m^3
	Lime	—	23.55-25.90	2400-2640	m^3
	Marble	—	26.70	2720	m^3
	Sandstone	—	27.45	2800	m^3
20.	Tiles clay				
	Mangalore pattern	—			
21.	Timber				
	Hard wood (e.g. Babul)	—	6.27-9.41	640-960	m^3
	Light wood (e.g. Fir)	—	3.92-4.71	400-480	m^3
	Medium (e.g. Deodar)	—	4.71-6.7	480-640	m^3
	Building parts and Components				
22	Ceiling (on wire netting)	10 mm	0.265	27	m^3
23	Flooring				
	Asphalt	10 mm	0.22	22	m^2
	Terrazo paving	10 mm	0.24	24	m^2
24	Roofing				
	Allahabad tiles (including battens)				
	Single	—	0.83	85	m^2
	Double	—	1.67	170	m^2
	Country tiles (with battens)				
	Single	—	0.69	70	m^2
	Double	—	1.18	120	m^2
	Mangalore tiles with battens tiles		0.64	65	m^2
	with mortar		1.08	110	m^2
	Slates (on battens)	—	0.34-0.49	35-50	m^2
	Glazed with aluminium alloy bars for span upto 3m	6.4 mm	0.19	19.50	m^2
	Glazed with lead covered steel bars at 0.6 m centres	6.4 mm	0.25-0.28	26-29	m^2

25	Wall				
	Engg. bricks (in masonry)	—	23.55	2400	m^3
	Stone (masonry)				
	Granite ashlar	—	25.90	2640	m^3
	rubble	—	23.55	2400	m^3
	Lime ashler	—	25.70	2560	m^3
	Marble dressed	—	26.50	2700	m^3
	Sand stone	—	22.00	2240	m^3

2.3. LIVE LOADS (VARIABLE LOADS/IMPOSED (LIVE) LOADS) (LL)

A companying live imposed loads are the live (imposed) of lower magnitude which may act together with a leading imposed. Imposed load having a larger magnitude is known as leading imposed load.

Live loads are the loads which vary in magnitude and/or in positions. Live loads are also known as *imposed* or *transient loads*. Live loads include any external loads imposed upon the structure when it is serving its normal purpose. Live loads are assumed to be produced by the intended use of occupancy in buildings including distributed, concenrated, impact and vibration and snow loads. Live loads are expressed as *unformly distributed static loads (U.D.L.)*. Live loads include the weight of materials stored, furniture and movable equipments. For buildings in most cities, the loads imposed on floors, stairs and roofs are specified in codes. For design purpose, certain live loads on floors have been assumed by many countries and which are almost in agreement. Some efforts have been made at the international level to decide live loads on floors and these have been specified in the International Standards 2103 (imposed floor loads in residental and public building and 2633 (Determination of imposed floor loads in production buildings and warehouses). These codes have been published in the International Organisation of Standardization.

Code IS : 875 (Part 2) - 1987 defines the **principal occupancy** for which a building or part of a building is used or intended to be used. The buildings are classified according to occupancy as under as per IS : 875 (Part 2) - 1987. *An occupancy shall be deemed to include subsidiary occupancies which are contingent upon*.

1. Assembly buildings. The assembly buildings including any building or part of a building where groups of people gather together for amusement, recreation, social, religious, patriotic, civil, travel and similar purposes (e.g., theatres, motion picture houses, assembly halls, city halls, marriage halls, townhalls, auditoria, exhibition halls, museums, skating rinks, gymnasiums, restaurants, places of worship, dance halls, club rooms, passenger stations and terminals of air, surface and other public transportation services, recreation piers and stadia, etc.).

2. Business buildings. The business buildings include any building or part of a building, which is used to condut business (other than the covered in 6) for maintaining of accounts and records for similar purpose, offices, banks, professional establishments, court houses, and libraries shall be classified in this group so far as principal function of these is to deal with public business and the keeping of books and records.

The office buildings are primarily used for office purposes (e.g., purpose of administration, clerical work, handling money, telephone and telegraph operating and operating computers, calculating machines). The clerical work includes writing, book-keeping, sorting papers, typing, filing, duplicating, punching cards or tapes, drawing of matter for publication and the editorial preparation of matter for publication).

3. Educational buildings. The educational buildings include any building used for school, college or day-care purposes involving assembly for instructions, education or recreation and which is not covered in 1.

4. Industrial buildings. The industrial buildings include any building or a part of a building or structure in which products or materials of various kinds and properties are fabricated, assembled or processed like assembly plants, power plants, refineries, gas plants, mills, diaries, factories, workshops, etc.

5. Institutional buildings. Institutional buildings include any building or a part thereof, which is used for purpose, such as medical or other treatment in case of persons suffering from physical and mental illness, disease or infirmity; care of infants, convalescents of aged persons and for penal or correctional detention in which the liberty of the inmates is restricted. Institutional buildings ordinarily provide sleeping accommodation for the occupants. These include hospitals, sanitoria, custodial institutions or panel institutions (e.g., jails, prisons and reformations).

6. Mercantile buildings These buildings include building or a part of a building which is used as shops, stores, market for display and sale of merchandise either wholesale or retail. Office, storage and service facilities incidental to the sale of merchandise and located in the same building shall be included under this group.

7. Residential buildings The residential buildings include any building in which sleeping accommodation is provided for normal residential purposes with or without cooking, a dining or both facilities (except building covered in 5). These buildings include one or multi-family dwellings, apartment houses (flats), lodging or rooming houses, restaurants, hostels, dormitories ad residential hotels.

The *dwellings* include any building or part occupied by members of single/ multi- family units with independent cooling facilities. These also include apartment houses (flats).

8. Storage buildings The storage buildings inlcude any building or part of building used primarily for the storage or sheltering of goods, wares or merchandize, (e.g., warehouses, cold storages, freight depots, transity sheds, store houses, garages, hangers, truck terminals, grain elevators, barns and stables).

The **imposed loads** are the loads assumed to be produced by the intended use or occupancy of a building, including the weight of movable partitions, distributed loads, concentrated loads, loads due to impact and vibration, and dust load but excluding wind, seismic, snow and other loads due to temperature changes, creep, shrinkage, differential settlement, etc. The imposed loads are the largest loads those probably will be produced by the intended use or occupancy, but shall not be less than the equivalent minimum loads specified in Table 2.2 subjected to any reductions allowed by code IS : 875 (Part 2) - 1987.

The floors are investigated for both, the uniformly distributed loads (UDL) and the corresponding cocentrated loads specified in Table 2.2 and designed for the most adverse

effects but these shall not be considered to act simultaneously. The concentrated loads specified in Table 2.2 may be assumed to act over an area of 0.3m x 0.3 m. However, the concentrated loads may not be considered where the floors are capable of effective lateral distribution of this load.

Table 2.2. IMPOSED FLOOR LOADS FOR DIFFERENT OCCUPANCIES (As per IS : 875 (Part 2) - 1987

S.N.	Occupancy Classification	Uniformly Distributed Load (UDL) (kN/m^2)	Concentrated Load (kN)
(i)	RESIDENTIAL BUILDINGS		
	(a) Dwelling houses:		
	(1) All rooms and kitchen	2.0	1.8
	(2) Toilet and bath rooms	2.0	—
	(3) Corridors, passages, staircases including fire escapes and store rooms	3.0	4.5
	(4) Balconies	3.0	1.5 per metre run concentrated at the outer edge
	(b) Dwelling units planned and executed in accordance with IS: 8888-1989* only:		
	(1) Habitable rooms, kitchens, toilet and bathrooms	1.5	1.4
	(2) Corridors, passages and staircases including fire escapes	1.5	1.3
	(3) Balconies	3.0	1.5 per metre run concentrated at outer edge
	(c) Hotels, hostels, boarding houses, lodging houses, dormitories, residential clubs:		
	(1) Living rooms, bed rooms and domitories	2.0	1.8
	(2) Kitchens and laundries	3.0	4.5
	(3) Billiards room and public lounges	3.0	2.7
	(4) Store rooms	5.0	4.5
	(5) Dining rooms, cafeterias and restaurants	4.0	2.7
	(6) Office rooms	2.5	2.7
	(7) Rooms for indoor games	3.0	1.8
	(8) Bathrooms and toilets	2.0	—
	(9) Corridors, passages, staircases including fire escapes, lobbies - as per the floor serviced (excluding stores and the like) but not less than	3.0	4.5
	(10) Balconies	Same as rooms to which they give access but with a minimum of 4.0	1.5 per metre run concentrated at the outer edge

	(d) Boiler rooms and plant room- to be calculated but not less than	5.0	6.7
	(e) Garages		
	(1) Garage floors (including parking area and repair workshop) for passenger cars and vehicles not exceeding 25 kN gross weight, including access ways and ramps to be calculated but not less than	2.5	9.0
	(2) Garage floors for vehicles not exceeding 40 kN gross weight (including access ways and ramps) - to be calculated but not less than	5.0	9.0
(ii)	EDUCATIONAL BUILDINGS		
	(a) Class rooms and lecture rooms (not used for assembly purposes)	3.0	2.7
	(b) Dining rooms, cafterias and restaurants	3.0+	2.7
	(c) Offices, lounges and staff rooms	2.5	2.7
	(d) Domitories	2.0	2.7
	(e) Projection rooms	5.0	—
	(f) Kitchens	3.0	4.5
	(g) Toilets and bathrooms	2.0	—
	(h) Store rooms	5.0	4.5
	(i) Libraries and archieves:		
	(1) Stack rooms/stack area	6.0 kN/m^2 for a min. height of 2.2 m + 2.0 kN/m^2 per metre height beyond 2.2m	4.5
	(2) Reading rooms (without separate storage)	4.0	4.5
	(3) Reading rooms (with separate storage)	3.0	4.5
	(j) Boiler rooms and plant rooms - to be calculated but not less than	4.0	4.5
	(k) Corridors, passages, lobbies, staircases including fire escapes - as per the floor services (without accounting for storage and projection rooms) but not less than	4.0	4.5
	(l) Balconies	Same as rooms to which they give access but with a minimum of 4.0	1.5 per metre run concentrated at the outer edge
(iii)	INSTITUTIONAL BUILDINGS		
	(a) Bed rooms, wards, dressing rooms, dormitores and lounges	2.0	1.8
	(b) Kitches, laundries and laboratories	3.0	4.5

	(c) Dining rooms, cafeterias and restaurants	3.0+	2.7
	(d) Toilets and bathrooms	2.0	—
	(e) X-ray rooms, operating rooms, general sto- rage areas - to be calculated but not less than	3.0	4.5
	(f) Office rooms and OPD rooms	2.5	2.7
	(g) Corridors, passages , lobbies and staircases including fire escapes - as per the floor services but not less than	4.0	4.5
	(h) Boiler rooms and plant rooms - to be calculated but not less than	5.0	4.5
	(i) Balconies	Same as the rooms to which they give access but with a minimum of 4.0	1.5 per metre run concentrated at the outer edge
(iv)	ASSEMBLY BUILDINGS		
	(a) Assembly areas:		
	(1) with fixed seats+	4.0	—
	(2) without fixed seats	5.0	3.6
	(b) Restaurants (subject to assembly), museums and art galleries and gymnasia	4.0	4.5
	(c) Projection rooms	5.0	—
	(d) Stages	5.0	4.5
	(e) Office rooms, kitchens and laundries	3.0	4.5
	(f) Dressing rooms	2.0	1.8
	(g) Lounges and billiards rooms	2.0	2.7
	(h) Toilets and bathrooms	2.0	—
	(i) Corridors, passages, staircases including fire escapes	4.0	4.5
	(j) Balconies	Same as the rooms to which they give access but with a minimum of 4.0	1.5 per metre run concentrated at the outer edge
	(k) Boiler rooms and plant rooms including weight of machinery	7.5	4.5
	(l) Corridors passages subject to loads greater than from crowd such as wheeled vehicles, trolleys and the like. Corridors, staircases and passages in grandstands	5.0	4.5
(v)	BUSINESS AND OFFICE BUILDINGS (see also 3.1.2)		
	(a) Rooms for general use with separate storage	2.5	2.7

	(b) Rooms without separate storage	4.0	4.5
	(c) Banking halls	3.0	2.7
	(d) Business computing machine rooms (with fixed computers or similar equipment)	3.5	4.5
	(e) Records/files store rooms and storage space	5.0	4.5
	(f) Vaults and strong room - to be calculated but not less than	5.0	4.5
	(g) Cafeterias and dining rooms	3.0	2.7
	(h) Kitchens	3.0	2.7
	(i) Corridors, passages, lobbies and staircases including fire escapes - as per the floor serviced (excluding stores) but not less than	4.0	4.5
	(j) Bath rooms and toilet rooms	2.0	
	(k) Balconies	Same as the rooms to which they give access but with a minimum of 4.0	1.5 per metre run concentrated at the outer edge
	(l) Stationery stores	4.0 for each metre of storage height	9.0
	(m) Boiler rooms and plant rooms - to be calculated but not less than	5.0	6.7
	(n) Libraries	see Sl. No. (ii)	
(vi)	MERCANTILE BUILDINGS		
	(a) Retails shops	4.0	3.6
	(b) Wholesale shops - to be calculated but not less than	6.0	4.5
	(c) Office rooms	2.5	2.7
	(d) Dining rooms, restaurants and cafeterias	3.0	2.7
	(e) Toilets	2.0	—
	(f) Kitchens and laundries	3.0	4.5
	(g) Boiler rooms and plant rooms - to be calculated but not less than	5.0	6.7
	(h) Corridors, passages, staircases including fire escapes and lobbies	4.0	4.5
	(i) Corridors, passages, staircases subject to loads greater than from crowds, such as wheeled vehicles, trolleys and the like	5.0	4.5
	(j) Balconies	Same as the rooms to which they give access but with a minimum of 4.0	1.5 per metre run concentrated at the outer edge

(vii) INDUSTRIAL BUILDINGS		
(a) Work areas without machines/equipment	2.5	4.5
(b) Work areas with machinery/equipment		
1. Light duty (To be calculated but not %-2)	5.0	4.5
2. Medium duty less than)	7.0	4.5
3. Heavy duty	10.0	4.5
(c) Boiler rooms and plant rooms - to be calculated but not less than	5.0	6.7
(d) Cafeterias and dining rooms	3.0	2.7
(e) Corridors, passages and staircases including fire escapes	4.0	4.5
(f) Corridors, passages, staircases subject to machine loads, wheeled vehicles - to be calculated but not less than	5.0	4.5
(g) Kitchens	3.0	4.5
(h) Toilets and bathrooms	2.0	—
(viii) STORAGE BUILDINGS		
(a) Storage rooms (other than cold storage) warehouses - to be calculated based on the bulk density of materials stored but not less than	2.4 kN/m² per each metre of storage height with a minimum of 7.5 kN/m²	7.0
(b) Cold storage - to be calculated but not less than	5.0 kN/m² per each metre of storage height with a minimum of 15 kN/m²	9.0
(c) Corridors, passages and staircases including fire escapes - as per the floor serviced but not less than	4.0	4.5
(d) Corridors, passages subject to loads greater than from crowds, such as wheeled vehicles, trolleys and the like	5.0	4.5
(e) Boiler rooms and plant rooms	7.5	4.5

*Guide for requirements of low income housing.

Where unrestricted assembly of persons is anticipated, the value of UDL should be increased to 4.0 kN/m²

`With fixed seats' implies that the removal of the seating and the use of the space for other purposes is improbable. The maximum likely load in this case is, therefore, closely controlled.

The loading in industrial buildings (workshops and factories) varies considerably and so three loadings under the terms `light', `medium', and `heavy' are introduced in order to allow for more economical designs but the terms have no special meanings in themselves other than the imposed load for which the relevant floor is designed. It is, however, important particularly in the case of heavy weight loads, to assess the actual loads to ensure that they are not in excess of 10 kN/m², in case where they are in excess, the desing shall be based on the actual loadings.

For various mechanical handling equipment which are used to transport goods, as in warehouses, workshops, store rooms, etc, the actual load coming from the use of such equipment shall be ascertained and design should cater to such loads.

Note 1. Where no values are given (in this table) for concentrated load, it may be assumed that the tabulation distributed load is adequate for design purposes.

Note 2. The loads specified in above table are equivalent uniformly distributed loads on the plan area and provide for normal effect of impact and acceleration. These loads do not take into consideration special concentrated loads and other loads.

Note 3. Where the use of an area or floor is not provided in the above table, the imposed load due to the use and occupancy of such an area shall be determined from the analysis of the loads resulting from :

(i) weight of the probably assembly of persons : (ii) weight of the probably accumulation of equipment and furniture; (iii) weight of the probable storage materials; and (iv) impact factor, if any.

While designing columns, abutments, piers, walls, their supports and foundations, the live loads on floors are reduced as in Table 2.3 as per IS : 800-84.

Table 2.3. PERCENT REDUCTION OF TOTAL LIVE LOAD

Number of floors carried by member under consideration	Percent reduction in total live load on all floors above the member under consideration
1	0
2	10
3	20
4	30
5 or more	40

Reduction in live load shall not be made in the case of ware houses, garages and other buildings used for storage purposes and for factoris and workshops designed for 5 kN/m². However, above reductions are made for buildings such as factories and workshops designed for a live load more than 5 kN/m² provided that the loading assumed for any column, etc., is not less it would have been if the floors had been designed for 5 kN/m² with no reduction.

As per IS : 875-1984, where a single span of beam or grider supports not less than 50m² of floor at one general level, the live load may be reduced in the design of beam or grider by 5 percent, for each 50 m² supported floor, subjected to a maximum reduction of 25 percent. This reduction or that given in Table 2.3 above, whichever is greater, may be taken into account in the design of columns, supporting such a beam. These reductions are not applicable for the floors used for storage purposes, in the weight of any plant or machinery which is specifically allowed for.

2.4 WIND LOADS (W.L.) TRANSIENT LOADS

Wind loads are loads experienced by a member, or a structure due to wind pressure acting on a structure.

The wind loads are the transient loads. Relative to the surface of earth, the air remains in motion. The difference in solar and terrestrial radiations setting up irregularities in temperatures (which give rise to convection either upwards or downwards) is the primary cause of wind. Near the surface of earth, the vertical component of atmospheric motion is small as compared to its horizontal component. **This horizontal component of atmospheric motion is referred as wind.** The force due to wind is called as *wind load.* The wind load acting over a structure depends on the velocity of wind and the shape and size of the exposed structural member.

The velocity of wind depends on the districts in which the structure is erected, the height of the structure, and the shelter afforded by the buildings or hills in the neighbourhood. The wind velocities are either measured by anemometers or found from anemographs which are installed at meteorological observatories at heights varying from 10 to 30 m. During cyclonic storms, thunderstorms, dust storms or vigorous monsoons, the velocity of wind may be more than 80 kmph. In our country, the cyclonic storms rapidly weaken after crossing the coasts and move as depressions inland. Sometimes, the velocity of wind may even be more than 120 kmph. In general, it may not exceed 60 to 70 kmph.

Usually at any time, the wind used to be steady. However, the velocity of wind recorded at any locality may be extremely variable. Sometimes, the gusts of small duration (say, few seconds) cause locally increased in air pressure and momentary decreased in air pressure elsewhere and the walls, panels and the roof sheeting may be more seriously influenced. However, because of inertia of the building, there may not be appreciable increase in stress in the main structural component of the building.

The intensity of wind pressure is given by $p = k.v^2$

where, p is pressure due to wind in kN/m^2 and v is velocity of wind in kmph.

k = a coefficient (adopted as 0.00006)

The wind pressure is expressed in terms of basic wind pressure (defined as equivalent static pressure in the direction of flow of wind). Our country has been divided into a number of zones, and maps have been prepared for wind pressure including and excluding winds of short durations. The basic wind pressure acting (in kN/m^2) at any height over structure may be noted from these maps given in IS : 875-(PART 3) - 1987. Upto a height of 30 m, the wind pressure is considered to act uniformly. Above 30 m height, the wind pressure increases.

The wind speeds are assessed with the aid of anemometers or anemograph which are installed at meterological observatories at heights generally varying from 10 m to 30 m above ground. Very weak winds (i.e., winds at low soppeds in the range of 5 kmph to 10 kmph) are called breeze. The strong winds (i.e. winds at speed say 50 kmph) are generally associated with cyclonic storms. Very strong winds (i.e., winds say 80 kmph) are also associated with cyclonic storms, thunder-storms, dust storms or vigorous monsoon's. The cyclonic storms weaken rapidly after the coast and move to depressions in land. A severe storem may have wind of 60 kmph to 120 kmph. The wind of very high speeds and very short duration (hurricanes) are called **Kal Baisaki or Norwestors** occur fairly frequently during summer months over North East of India.

In addition to wind at any time, there are effects of **gusts** (sudden blast of wind, which may fast for few seconds) which should be considered. These gusts cause increase in air pressure but their effect on the stability of the building may not be so important. Often the gusts influence only part of the building and the increased local pressures may be more than balanced by a momentary reduction in pressure elsewhere short period gusts may cause any appreciable increase in stress in main components of buildings (because of inertia of the building) although the walls, root sheetings and individual cladding units (glass panels) and their supporting members such as purlins, sheeting rails and glazing bars may be more seriously affected. Gusts may also be extremely important for design of structures with slenderness ratios.

The modern practice is to use `peak gust loading' as the basis for design, instead of mean load average over one minute, which was until recently used. It gives more realistic assessment of wind peak gust is likely to be observed over an average time 3 seconds, 5 seconds or 15 seconds. In USA, the wind speed is expressed as average velocity. The average velocity of wind is the fastest for 1.609 km, which passes through the instrument in t seconds. That is

$$\text{Average velocity} = \left(\frac{1.609 \times 3600}{t}\right) = \left(\frac{5792.4}{t}\right) \text{kmph}$$

The effects of wind on the structures are found on two approaches.

1. Peak-wind approach (PWA) with gust related factors.

2. The mean-wind-approach (MWA) with gust related factors.

Peak-wind-appraoch (PWA) is recommended in the code of our country.

Our country, as per IS: 875-1987 (Part 3), at present, there is only one map and whole country is divided in siz zones (with basic wind speeds 33, 39, 44, 47, 50 and 55 metre per second which correspond to 120, 140, 160, 170, 180 and 200 kmph. the peak gust velocity averaged over short time interval of 3 seconds duration represents the basic maximum wind speed.

A **hurricane** is a classification for a particularly strong tropical cyclone. Tropical disturbances, tropical depressions, tropical typhoons and hurricanes are all specific names for tropical cyclones depending on strength and location. These are an important part of the atmospheric circulation system which moves heat from the equatorial region towards high latitudes. Structurally, it is alarge rotating systems of clouds, wind and thunder-storm activity. Its primary energy source is releaseof heat of condensation from water condensing at high altitudes. Tropical cyclones are classified as tropical depressions, tropical storms and a third group whose name depends on the region.

A tropical depression is an organised system of clouds and thurder-storms with a defined surface circulation and maximum sustained winds of less than 62 kmph. It does not typically have the spiral shape of more powerful storms. A tropical storm is a similar, but has sustained winds between 62-117 kmph. At these sppeds, the distinctive cyclonic shape starts to develop at hurricane intensity about 16 km to 80 km wide, in which the strongest thunder storms and winds circulate around the storm's centre. The circulation of clouds around a cyclone's centre imparts a distinct spiral shape to the system.

Bands or arms may extend over great distances as clouds are drawn toward the cyclone. The direction of the cyclonic circulation is counter-clockwise in the Northern hemisphere and clockwise in the Southern hemisphere.

2.5 SNOW LOAD (EXCEPTIONAL LOADS)

The snow loads are also referred as **exceptional loads**. IS : 875 - PART 4) - 1987 may be used to note values of snow loads.

The *snow load* depends upon latitude of the place and atmospheric humidity and accordingly snow fall at the place (if any). The snow load acts vertically and it is expressed in kN per square metre of plan area. Actual load due to snow depends upon the shape of roof and its capacity to retain the snow. When actual data for snow load is not available, snow load is assumed as 0.0025 kN/m² per mm depth of snow. It is likely that there may be total snow load or partial snow load. *One-half* of the roof may carry design snow load and the *remaining half* may carry *half* the design snow load. When the slope of roof is more than 50°, the snow load is not considered. It is likely that snow pockets may be formed & in that case, it is considered.

The roof is designed for actual snow load or the live loads, which ever is more severe.

2.6 SEISMIC LOADS (EXCEPTIONAL LOADS)

The inertia forces produced in a structure due to the movement and vibrations due to earthquake.

The seismic loads are also referred as **exceptional loads.**

It becoms essential to consider *seismic loads* in the design of structures, if the structure is situated in the seismic areas. The seismic areas are the regions which are geological young and unstable parts and which have experienced earthquakes in the past and are likely to experience earthquakes in furture. The Himalayan region, Indo Gangetic Plain, Western India, Cutch and Kathiawar are the places in our country which experience earthquaks frequently. Sometims, these earthquakes are violent also. Seismic loads are caused by the shocks due to earthquakes. The earthquakes range from small tremors to severe shocks. The earthquake shocks cause movement of ground, as a result of which the structure vibrates, which may be resolved into three perpendicular directions. One horizo-ntal direction dominates over other two directions. In some cases, structures are designed for horizontal seismic forces only and in some cases, horizontal and vertical both the components are considered. The seismic accelerations for the design may be arrived at from **seismic coefficient**, which is defined as the ratio of acceleration due to earthquakes and acceleration due to gravity. Our country has been divided into seven zones for determining seismic coefficients. The seismic coefficients have also been recommended for different types of soils for the guidance of designers. IS: 1893-2000 (Indian Standard Recommendations for Earthquake Resistance Design of Structures) may be referred for the actual design, for guidelines for earthquake resistant construction of buildings. IS: 4326-1976 may be also consulted.

For less important and relatively small structures, certain construction rules and precautions are observed, and the analysis for earthquake forces is not required. For buildings upto 40 m in height, the seismic coefficient method is used for estimating the earthquake forces. For building from 40 m upto 90 m height, it is advised to have model analysis. However, for earthquake zones I to III the seismic coefficient may also be used. For tall buildings of height more than 90 m, for zones I and II, the analysis may be carried by model analysis. For other zones, detailed dynamic analysis should be done, which one is based on expected ground motion.

2.7 TEMPERATURE EFFECTS

The temperature effects are considered as **transient loads**. IS : 875 (Part 5) - 1987 may be used to note the values.

Similar to other materials, concrete expands with increase in temeprature and contracts with decrease in temperature. The range of variation in temperature varies from localities to localities, season to season and day to day. The objectionable cracks may occur in concrete due to contraction in concrete combined with the effect of shrinkage. Occasi-onally, large and harmful stresses may develop due to deformations because of temperature changes. The coefficient of thermal expansion and contraction depends on type and quantity of cement, the type of aggregate, the relative humidity and the size of sections. For calculating stresses and deformations due to change in temperature, the coefficient of thermal expansion and contraction may be adopted satisfactorily as 0.0001) mm/mm per °C. As per IS : 456-1978, the values of coefficient of thermal expansion and contraction for concrete for quartz, sand stone and granite aggregate are 1.2 to 1.3 x 10^{-5}, 0.9 to 1.2 x 10^{-5} and 0.7 to 0.95 x 10^{-5}, respectively. And those for basalt and limestone aggregate are 0.8 to 0.95 x 10^{-5} and 0.6 to 10^{-5}, respectively.

2.8 SHRINKAGE

Total shrinkage of concrete depends upon the constituents of concrete, size of the member and the environmental conditions. For a given humidity and temperature, total shrinkage of concrete is most influenced by the total quantity of water present in the concrete at the time of mixing and to a lesser extent by the cement content.

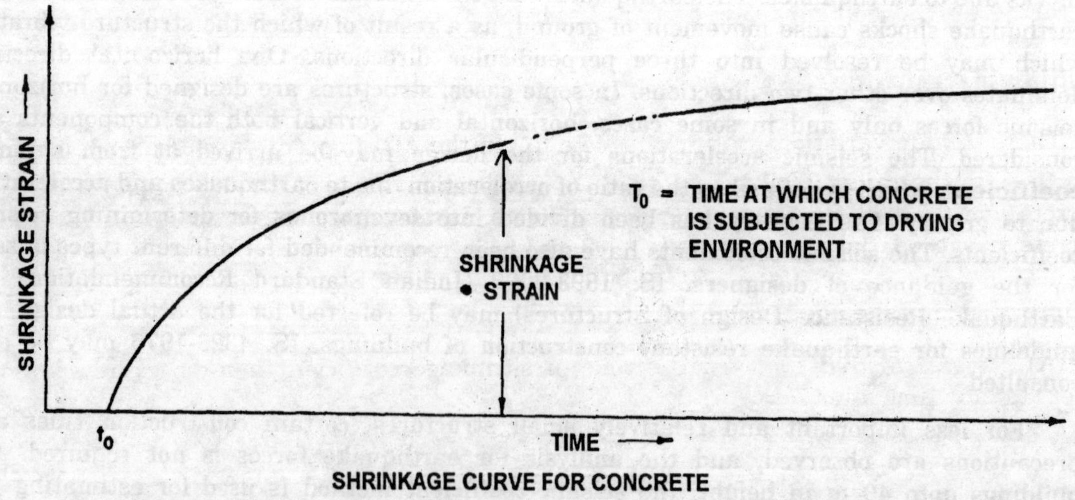

Fig. 2.1. Shrinkage curve for concrete

The shrinkage of concrete occurs due to loss of moisture by evaporation. The strains due to shrinkage are independent of the conditions of stress in concrete. In case, the concrete is restrained, the strains due to shrinkage may result in cracking of concrete. It will result in deflection of the structural members to increase with time.

Fig. 2.1 shows the increase in strains due to shrinkage with time. The shrinkage occurs at a decreasing rate with time. The total strains due to shrinkage vary largely, being generally in range 0.0002 to 0.0006 and sometimes as much as 0.0010.

In the absence of test data the approximate value of the total shrinkage strain for design may be taken as 0.0003.

To a large extent, the shrinkage is a reversible phenomenon. In case, the concrete is saturated with water after shrinkage has occurred, the concrete expands to almost its original volume. Alternating volume changes of concrete may occur due to alternating dry and wet conditions. The fluctuating deflections occur partly due to phenomenon of shrinkage. For example, concrete bridges exposed to seasonal changes each year.

As a rule, the concrete that exhibits a high creep also displays high shrinkage. The composition of concrete and the environment are the factors which influence the magnitude of the strains due to shrinkage. Empirial methods for estimation of strains due to shrinkage have been proposed in the ACI Committee 209 [`*Prediction of Creep, shrinkage and temperature in Concrete Structures*', `*Designing for the effects of creep, shrinkage and temperature in Concrete structures*' SP -27, American Concrete Institute, Detroit, 1971 (pp. 51-93)] and CEB-FIP [`*International Recommendations for Design and Construction of Concrete Structures*', Comite Europeen du Benton - Federation Internale de 19 Precontrainte, Paris 1970, (English translation available from Cement and Concrete Association, London, (pp. 88)]. ACE approach is described as below.

The unrestrained strains due to shrinkage at any time t is given by –

$$\varepsilon_{sh} = \left(\varepsilon_{shu} \cdot S_t \cdot S_h \cdot S_{th} \cdot S_s \cdot S_f \cdot S_e \cdot S_c\right) \tag{2.1}$$

where, these eight coefficients are described as below :

1. Ultimate strain due to shrinkage, ε_{shu}

There is wide variation in the value of ε_{shu}. In the ACI Committee 209 review, it was seen to be in the range 0.000415 to 0.000107, with mean values of 0.00080 for moist-cured concrete and 0.00073 for the steam-cured concrete. In absence of more exact data for the concrete to be used, the mean values may be used.

2. Time of shrinkage coefficient

(A) *For moist-cured concrete*, the time of shrinkage coefficient may be expressed as under : (at any time after 7 days)

$$S_t = \left(\frac{t}{(35+t)}\right) \tag{2.2}$$

where, t is the time in days from age of 7 days. The values of time shrinkage coefficients are as under for different times.

Time	Value of S_t
1 month	0.46
3 months	0.72
6 months	0.84
1 year	0.91
5 years	0.98

(B) ***For steam-cured concrete*****, time of shrinkage coefficient may be expressed as under : (at any time after 3 days)**

$$S_t = \left(\frac{t}{(55+t)}\right) \tag{2.3}$$

where, t is the time in days from age 1 to 3 days. The values of time shrinkage coefficients are as under for different times.

To account for shrinkage from greater ages than given above, the difference may be calculated using either of the two expressions given above for any period after that time. ***For example*****, the shrinkage for moist-cured concrete between say, 1 month and 1 year would be equal to the 7 days to 1-year shrinkage minus the 7 days to 1-month shrinkage. It is assumed in the foregoing procedure that the moist-cured concrete has been cured for 3 to 7 days. For the shrinkage of moist-cured concrete from 1 day, the shrinkage requires to be multiplied by 1.2; a linear interpolation between 1.2 at 1 day and 1.0 at 7 days may be used.**

3. Relative humidity coefficient, S_h

(A) For [40 < (Humidity = H) < 80 percent]

$$S_h = (1.4 - 0.01\,H) \tag{2.4}$$

(B) For [180 < (Humidity = H) < 100 percent]

$$S_h = (3.0 - 0.03\,H) \tag{2.5}$$

The values of coefficients, S_h for different percentage of relative humidity are as under:

Relative humidity, H percent	Values of S_h
40	1.00
60	0.80
80	0.60
100	0.00

4. Minimum thickness of member coefficient, S_{th}

For 152.4 mm thickness of the structural member, the value of S_{th} is 1.00 and for 228.6 mm, the value cf S_{th} is 0.84.

5. Slump of concrete coefficient, s_s

The values of coefficient, s_s for different values of slump are as below :

Stump	Value of S_h
50.80 mm	0.97
68.58 mm	1.00
76.20 mm	1.01
101.60 mm	1.05
127.00 mm	1.09

6. Fine coefficients, s_f

The values of coefficients, s_f for different values of fines by weight are as under:

Fines by weight (per cent)	Value of S_h
40	0.86
50	1.00
70	1.04

7. Air content coefficient, s_e

The values of coefficient, s_e are 0.98, 1.00 and 1.03 for 4 percent, 6 percent and 10 percent, respectively.

8. Cement content factor, s_c

The values of factor, s_c for different contents of cement are as follows :

Contents of cement (kg/m³)	Values of factor S_c
223.00	0.87
334.45	0.95
418.06	1.00
557.42	1.09

Total shrinkage of concrete depends upon the ingredients of concrete (cement, aggregates, water), size of the member and atmospheric conditions. For given atmospheric conditions, the total shrinkage is very much influenced by the total quantity of water present in the concrete at the time of mixing.

The quantity of water in any workable concrete mix used to be more than that required for hydration. The large quantity of water evaporates in time from the concrete surfaces exposed to air. The rate of drying and total drying of concrete depends on the ambient temperature and humidity. The concrete shrinks in volume as it dries The shrinkage in concrete is probably due to the capillary tension, (which develops in the remaining water in concrete). Shrinkage continues in concrete at reducing rate for several months. Shrinkage damages concrete in several respects. When the shrinkage is not properly controlled,

unsighty and harmful cracks develop in concrete (e.g., in slabs and beams). In indeterm-inate structures, large and harmful stresses may develop in concrete.

Unit water content in the fresh concrete is the main factor, which ascertains the extent of final shrinkage. In order to reduce the shrinkage, the content of water in the fresh concrete may be reduced. The content of water may be kept minimum compatible with the required workability. Further, the shrinkage may be controlled to the beneficial extent by prolonged and careful curing.

The shrinkage of concrete is influenced to a very small extent by the cement content.

The extent of shrinkage also depends upon water absorption property of the aggregate. The shrinkage in concrete having less water absorptive aggregate (e.g., granite and lime stone) is even half or still less than that in concrete having more water absorptive aggregate (e.g., some sandstones and slates). The extent of shrinkage is also more in some light weight aggregates because of porosity.

The approximate value of total shrinkage strain for design for ordinary concrete is taken 0.003 mm/mm.

2.9 CREEP

Professor Rusch, H (*'Researches towards a General Flexural Theory for Structural Concrete'* Journal ACI, Vol. 57, No. 1, July, 1960, (pp. 1-28)] conducted long term loading tests on unconfined concrete. It was observed that the sustained load compressive strength is approximately 80 percent of the short- term strength. The short-term strength is the strengths of an identically old and identically cast specimen, that is loaded to failure over a period of *ten minutes* when the specimen under sustained load has collapsed. In practice, the strength of concrete considered in the design of structure are usually based on the anticipated short-term strength at 28 days. The decrease in strength due to long term loading will be atleast partly offset by the property of concrete to attain a higher strength at greater ages. The strains due to creep due to long-term loading cause modification in the shape of the stress-strain curve. Professor Rusch obtained some curves as shown in Fig. 2.2 for various rates of loadig. These curves show with a reducig rate of strain, the value of maximum stress attained gradually reduces but the descending branch of curve falls less rapidly, the strain (at which the maximum stress is reached) is increased.

Fig. 2.2 shows that the stress-strain relation of concrete is function of time. A gradual increase in strain with time occurs in the concrete under stress because of creep deformations of the concrete. The total creep strain may be several *times* as large as the initial elastic strain. Generally, the creep has very small effect on the stregth of a structure, but it will result a redistribution of stress in reinforced concrete members at the service load deflections. Creep deformations are beneficial in some cases (*for example*, stress in concrete due to differential settlement of structures) are decreased due to creep. The creep in tension also delays shrinkage cracking in concrete. Fig. 2.3 shows the deformation due to creep under constant axial compressive stress.

Fig. 2.3 shows that the creep proceeds at a reducing rate with time. In case, the load is removed, the elastic strain is immediately recovered. However, this recovered elastic strain is less than the initial elastic strain because the elastic modulus increases with time. The elastic recovery is followed by the creep recovery, which is a small proportion of the total creep strain.

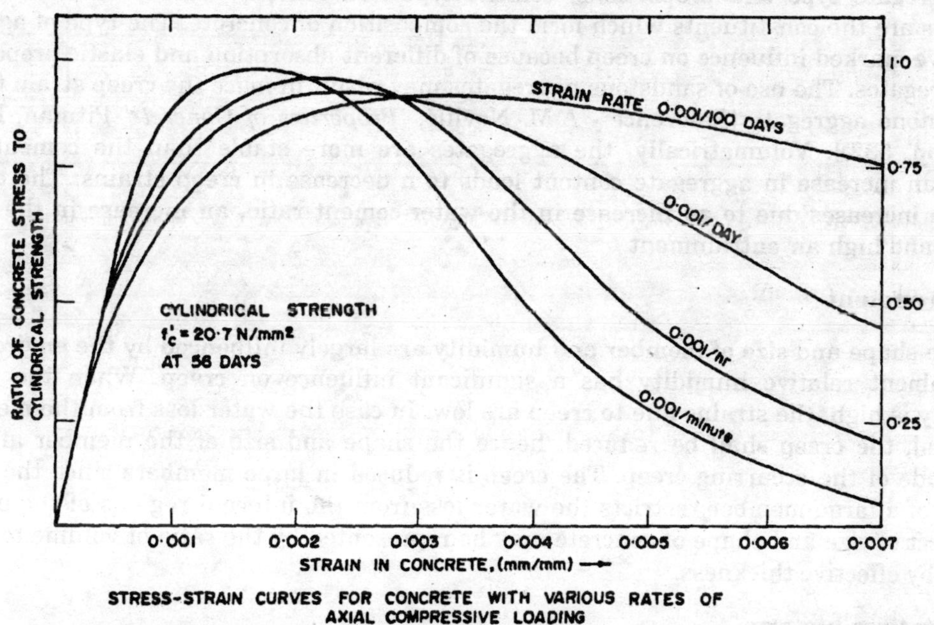

Fig. 2.2. Stress-strain curves for concrete with various rates of axial compressive loading.

The experiment tests show that the creep strain occurring over a given period is proportional to the applied stress, provided level of stress is not high. The composition of the concrete, the environment and the stress-time history are the factors which influence the magnitude of strain due to creep as below :

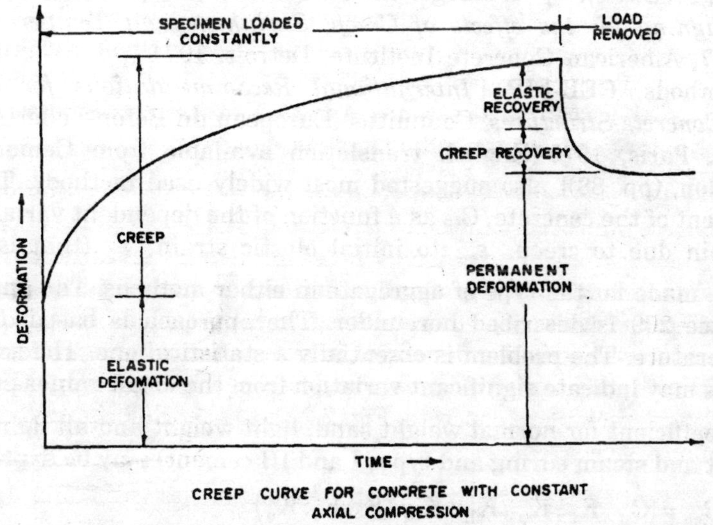

Fig. 2.3. Creep curve for concrete with constant axial compression.

1. Composition of concrete

Aggregate type and proportions, cement type and content, water- cement ratio and additives are the constituents which form the composition of concrete. The type of aggregate may have marked influence on creep because of different absorption and elastic properties of the aggregates. The use of sandstone aggregates may result in *twice* the creep strain than for a lime stone aggregate [Reference - A.M. Neville, `*Properties of Concrete*' Pitman, London, 1963, (pp. 532)]. Volumetrically, the aggregates are more stable than the cement paste. Hence, an increase in aggregate content leads to a decrease in creep strains. The creep in concrete increases due to an increase in the water-cement ratio, an increase in the cement content and high air entrainment.

2. Environment

The shape and size of member and humidity are largely influenced by the environment. The ambient relative humidity has a significant influence on creep. When the relative humidity is high, the strains due to creep are low. In case the water loss from the member is restricted, the creep shall be reduced, hence the shape and size of the member affect the magnitude of the occurring creep. The creep is reduced in large members since the outside regions of a large member restricts the water loss from the internal regions of the member. The effect of size and shape of concrete may be represented by the ratio of volume to surface area or by effective thickness.

3. Stress-time history

At the stage of first loading and time under loading are factors which describe the stress-time history. The high strains due to creep results due to loading at an early age and there is a marked decrease in strain due to creep in case there is increase in age at first loading. The strain due to creep increases with duration of loading.

The strains due to creep may be calculated from the several empirical methods. ACI Committee 209 [`*Prediction of Crinkage, Shrinkage and Temperature Effects of Concrete Structures*', `*Designing for the effects of Creep, Shrinkage and Temperature in Concrete Structures*', SP-27, American Concrete Institute, Detroit, 1971 (pp. 51-93)] suggested most widely used methods. CEB-FIP [`*International Recommendations for the Design and Construction of Concrete Structures*' Committee European du Beton-Federation Internale de la Precontrainte, Paris, 1970-(English translation available from Cement and Concrete Association London, (pp. 88)] also suggested most widely used methods. The methods give the creep coefficient of the concrete, C_{cp} as a function of the dependent variables, where C_{cp} is the ratio of strain due to creep, ε_{cp} to initial elastic strain, ε_i (that is $C_{cp} = (\varepsilon_{cp}/\varepsilon_i)$). Allowance is not made for the type of aggregate in either methods. The approach suggested in ACI Committee 209 is described hereunder. The approach is based on comprehensive review of the literature. The problem is essentially a statistical one. The scatter of test data and actual values may indicate significant variation from the mean values proposed.

The creep coefficient for normal weight sand, light weight and all light weight concrete (using both moist and steam curing and types I and III cement) may be expressed as under :

$$C_{cp} = \left(C_u \,.\, K_t \,.\, K_a \,.\, K_h \,.\, K_{th} \,.\, K_s .\, K_f .\, K_e\right) \tag{2.6}$$

where, these eight coefficients are as follows :

1. Ultimate creep coefficient, C_u

There is wide variation in the values of C_u. The values of C_u was seen to be in the range of 1.30 to 4.15 with an average value of 2.35. In absence of more exact data for the concrete, the average value of C_u may be used.

2. Time under load coefficient, K_i

It is expressed as follows : $K_1 = \left(\frac{t^{0.6}}{10 + t^{0.6}}\right)$ (2.7)

where, t is the time in days after application of load. For t = 1 month, 3 months, 6 months, 1 year and 5 years, the values of K_t are 0.44, 0.60, 0.69, 0.78 and 0.90, respectively).

3. Age when loaded coefficient, K_a

(A) For moist-cured concrete, $K_a = \left(1.25\, t_1^{-6.118}\right)$ (2.8)

[For 7, 10, 30 and 90 days, the values of K_a are 1.00, 0.95, 0.83 and 0.74, respectively].

(B) For steam-cured concrete, $K_a = \left(1.13\, t_1^{-0.095}\right)$ (2.9)

where, t_i is the age of concrete in days, when the load is first applied [For 1 to 3, 10, 30 and 90 days, the values of K_a are 1.00, 0.90, 0.82 and 0.74, respectively].

4. Relative humidity coefficient, K_h

It is expressed as below :

$K_h = (1.27 - 0.0067\, H)$ (for H more than 40 percent) (2.10)

(For H, less than or equal to 40, 60, 80 and 100 percent, the values of K_h are 1.00 0.87, 0.73 and 0.60, respectively).

5. Minimum thickness of member coefficient, K_{th}

For the thickness of member 152.4 mm or less, the value of K_{th} is unity. And for the thickness of member 305 mm, the value of K_{th} is 0.82.

6. Slump of concrete coefficient, K_s

The values of K_s for different values of slump of the concrete are as below :

Values of slump	Value of K_s
50 mm	0.95
8.58 mm	1.00
76.2 mm	1.02
101.6 mm	1.09
127 mm	1.16

7. Fines coefficient, K_f

The values of fines coefficients are as under :

Fines by weight percent	Value of K_f
30	0.95
50	1.00
70	1.05

8. Air content coefficient K_e

The values of air content coefficient are as under :

Air (percent)	Value of K_s
up to 6	1.00
7	1.09
8	1.17

For cement content between 278.7 kg/m^3 and 444.75 kg/m^3, the cement content may not be taken into consideration.

Rate and duration of loading have the pronounced effect on the behaviour of the concrete. Concrete creeps under the effect of long-time loading. **Creep** *is defined as continued deformation, over considerable length of time under additional elastic contraction exhibited by slow concrete, sustained (constant) stress or load.* The creep in concrete depends on applied stress, age of concrete at loading and duration of loading. The deformations due to creep in concrete is proportional to the magnitude of the applied stress. Low strength concretes show more creep as compared to high strength concretes. With passage of time, the deformations due to creep continue at reducing rate. After 2 to 5 years time, the deformations due to creep cease.

Characteristic values of creep, expressed in deformation per unit length, for 1:2:4 concrete loaded at 28 days with sustained stress of 4 N/mm^2 are 0.0003 and 0.0006 at one year.

Besides the magnitude of applied stress, the creep also depends upon average ambient relative humidity. As the part of reduction in volume under sustained load is caused due to evaporation of free pore water in the surrounding atmosphere. Type of cement and aggregate, age of concrete when first loaded and concrete strength are some important factors which influence the extent of creep.

The effect of creep of concrete is not often considered directly in reinforced concrete design. It is, however, taken into consideration when determining the deflections according to the rigorous method by modifying the elastic modulus of concrete.

Creep of concrete depends on the stress in the concrete, age at loading and duration of loading in addition to the constituents of concrete, size of member and environmental conditions. As long as the stress in concrete does not exceed one third of its characteristic compressive strength may be assumed to be proportional to the stress. The ultimate creep strain may be estimated from the following values of creep coefficients (i.e. ultimate creep strain/elastic strain at the age of loading); for long span structures, it is advisible to find actual creep strain likely to take place. The values of **creep coefficient** at the age at loading

7 days, 28 days and one year (as per IS : 456-2000) may be adopted as 2.2, 1.6 and 1.1, respectively. It is to note that the ultimate creep strain estimated as mentioned above does not include the elastic strain.

Shrinkage and creep do normally affect the ultimate strength of a reinforced concrete section. However, the shrinkage and creep affect the deflection of structures at working loads. For normal purposes, the ratios of span/depth recommended in IS : 456-1978 takes into consideration the shrinkage and creep in beams and slabs. Shrinkage and creep are not likely to affect the strength of columns designed by the ultimate strength methods. For tall frames (more than 30 storeys), the shrinkage and creep combined with axial shortening are likely to substantially change the forces in columns and beams.

The ratio of ultimate creep strain (excluding elastic strain) in the elastic strain at the age of loading is called as **creep coefficient**. As per IS : 456-1978, the values of creep coefficients at 7 days, 28 days and 1 years' age of loading are 2.2, 1.6 and 1.1, respectively.

2.10 OTHER FORCES AND EFFECTS

In addition to the loads described in Arts, 2.2 to 2.9, following forces and effects are also considered while designing the structural components and the structures in case these are likely to influence significantly the safety and the serviceability of the structure :

1. Foundation movement

Settlement, average tilt and angular distortions may occur due to *movement of foundation*. *Angular distortion* is angular change due to differential (unequal) settlement. Major cracks in buildings occur mainly due to differential settlement (viz., angular distortion). IS : 1904-1978 (code of practice for structural safety of buildings : shallow foundations) gives the allowable values of total settlement, differential settlement and angular distortion. The superstructures may accommodate themselves to the movements without harmful distortions in case these deformations are within the specified limits. In practice, the foundations are designed such that the angular distortion is limited instead of resisting the forces developed by the differential settlement.

2. Elastic axial shortening

In tall buildings carrying service loads, the columns and beams become curved due to presence of end moments in addition to the axial forces.

The difference between curved (original) length of the member and its chord length is known as *axial shortening*. The effect of elastic axial shortening is taken into account along with the effects of shrinkage and creep in tall buildings having 30 storeys or more height.

3. Soil and fluid pressure

The soil and fluid pressure exerted by the soil or water or both is taken into consideration on the basis of established theory in the design of structure or parts of the structures (e.g., basement floors and walls). Allowance is also made for the surcharge due to stationary or moving loads. The lateral soil pressure on slender structural members, such as pillars which rest in sloping soils is calculated by taking the width of member *twice* its actual width. In general, the relieving pressure of soil in front of the structural member con-cerned is not considered. IS : 875 (PART 5)-1987 may be used to note details for this effect.

4. Vibrations

In general, the concrete structures carry service loads. To comply with the limitations for the deflections, stiffeners are provided such that the further consideration of vibration is not necessary. However, in case, columns support crane girders, the effect of vibration is taken into account.

5. Fatigue

Fatigue is the damage caused by the repeated fluactions of stress, leading to progressive and consequent cracking of a structural element.

The phenomenon of decrease of resistance (strength) of material due to repeated applied stresses is known as *Fatigue*. Due to repeated stresses, the structural members are subjected to either **variation of stresses** or **reversals of stresses** (i.e., **fluctuation of stresses**), which may result in fatigue failure. The members may be subjected to low cycle fatigue or high cycle fatigue.

In low cycle fatigue, the number of reversal of stresses is not large. The failure mechanism is similar to those as under static loading. The strength and ductility of the material are not badly influenced.

In high cycle fatigue, (viz., proper fatigue), the fluctuation of stresses occurs in millions of cycles. Considerably small stresses (as compared to yield stress) may cause failure of the structure or a structural member. The degree of fatigue differs for direct compression, direct tension and bending. Fatigue due to tensile stress controls fatigue of concrete in bending. Proper fatigue may have its significant effects on the widths of cracks and on deflections, which may increase by *twenty five percent* compared to equivalent static load. IS : 875 (PART 5) - 1987 may be used to note the details.

The values of fatigue factors may be noted from IS:1893-1975 code of practice for the design of steel bridges.

6. Impact

Effect of sudden hit of machinery during its movement, precast structural members during their handling or **dynamic effect of moving loads is called as impact**. It is possible that lift trucks may strike the columns. Impact in such cases should be avoided. As a measure of protection, the rails may be provided to surround the columns during the phase of construction. Accidental impacts should also be avoided. Sometimes, the **impact factors** are used and the fraction of live loads in bridges are considered as **impact loads**. IS : 875 (PART 5) - 1987 may be used to note the details.

7. Erection loads

The action (loads and deformations) experienced by the structure during erection.

During construction and erection of a structure, construction materials are stored on the floors and equipments needed for erection are also brought and kept on the floor. The structure is to carry the weight of such items and loads caused due to the operation of such equipments. All such loads are called as **erection loads**. The stresses due to erection loads may be opposite in nature from those due to the structural members carry during actual working. The proper provisions (e.g., temporary bracings) shall be made to take care of all such stresses likely to develop during erection. IS : 875 (PART 5) - 1987 may be used to note the details.

8. Stress concentration

At the points of application of concentrated loads, abrupt changes in cross-section, re-entrant angles, opening in slabs, the stresses are far above their average value, and it is referred as *stress concentration*. In the design of concrete structures and the structural members, the concentration of stress is properly taken into consideration.

2.11 STRUCTURAL SAFETY AND FACTOR OF SAFETY (F.S.)

The factor of safety is the factor by which the yield stress of the material is divided to obtained the permissible stress in the material.

It is aimed to design structures which are structurally safe, strong, durable and serviceable throughout its life time.

The meaning of safety and to ensure safety are not so straight forward. It is difficult to define the concept of safety. For the purpose of structural design, one needs a precise, mathematically accepted definition. A balance in a rational and consistent manner is essential to be clear about the meaning of the safety. One should express and give the definition of safety as a quantity, in the absence of which one continues to balance a highly quantified concept (economy, measured in money) against a vague quantitative concept of safety. The main structural safety requirement is to ensure that there is only acceptably low risk of failure during its design life. Another important requirement is that the structure is sufficiently stiff. The excessive deflection and vibrations do not influence the in-service performance of the structure.

There are many uncertainties (e.g., uncertainty about the loading, uncertainty about the strengths of materials which will be used, uncertainty about the quality of construction, uncertainty about one's ability to predict the behaviour). Inspite of all these uncertainties, one has to design and construct safe and economical structures.

Until the beginning of the nineteenth century, the safety was assured by experience. A high value of factor of safety (around three) used to be arbitrarily adopted, which allows, a margin and considers all the uncertainties in the estimation of the loads and the strength of materials. As a result of large value of factor of safety, the allowable stresses calculated remained small and the size of members (components of the structures) adopted in design remained large.

There were no means for the measurement of the strength of materials used for the construction. There was no theoretical knowledge of the behaviour of the structures developed. The load carrying capacity of the structure could not be predicted. These conditions, continued until the industrial revolution.

The safety of structures depends upon many factors such as the type of building, fire rating of the fire proofing material, construction details, durability, probability of failure in the structural members and connections, inspection of quality control, and the amount of supervision. **Lower the probability of failure, higher is the safety.**

2.12 STRUCTURAL FAILURES

Structural failures may be caused by unsatisfactory material, fabrication or errection or faulty designs. Most common causes of structural failures are as under.

Structural failure does not always mean collapse. Commonly excessive distorsion of the structural frame prevents the structure from proper functioning and constitutes a failure as serious as collapse. Failure in the structure occurs when some of the principal members or connections fail by slip (shear), cleavage (tension), buckling, or crushing. Complete failure may occur under severe blast or impact.

1. Foundation movement

There are numerous failures due to movement or unequal settlement of foundation. The structural designers should be particularly careful to appraise the effect of settlement on the behaviour of the structures, especially indeterminate structures such as continuous structures.

2. Earthquake response

The structures which are not adequately designed for earthquake forces were severely damaged. Failures of walls, particularly windows, and local damage to main members or connections were commonly seen. The displacements may occur as a result of earthquake.

The extent of possible damage depends on the design as well as on the estimation of maximum intensity of probable earthquake during the life of the structure. These damages may be reduced or prevented by a rational analysis of dynamic response and by calculating properly the earthquake forces.

3. Dynamic response and dynamic instability

Dynamic response and dynamic instability are distinct phenomena which may occur under dynamic loading. When the periodic motions are imposed on the structures and the frequency of vibrations becomes approximately equal to the one of the natural modes of the structures, then, there occurs **resonance** and vibrations increase. When the frequency of wind gusts becomes equal to nearly the same as the natural frequency of the cables, then, the resonance occurs in electric power transmission lines.

The gradual or sudden build-up of catastrophic oscillatory motion in a structure subjected to wind is known as the phenomenon of **aerodynamic instability.** It occurs commonly in slender flexible structures exposed to wind.

Failure of the Tacoma Narrows suspension Bridge on November 7, 1940 was probably the greatest bridge failure in history.

4. Inadequate connections

Some of the forces acting on the connections are not properly considered. The connections may be designed for the axial loads, where as the loads are acting eccentrically. These connection fail due to inadequate connections.

5. Incorrect appraisal of buckling strength

The component members are parts of a structure as a whole. In such cases, it is very difficult to establish load eccentricity, the rigidity of the end supports, and the effectiveness of lateral bracing. The buckling of the elements and the members under various conditions of loading and support, and torsional buckling of some of the compression members are common causes of serious failures. The Quebec Bridge failed while it was under construction in 1907.

6. Lack of adequate bracing

There is need for adequate bracing against lateral sway. During even in construction and lifting, there should be temporary bracing in both the directions. Due to lack of bracing, serious accident may occur.

7. Overloading

In the 1930s a number of old bridges failed largely because the traffic became heavier and desenser than that assumed in the original design. Building converted from one occupancy to another without proper investigation of structural safety often are overloaded and as a result of which the failures occured.

8. Fatigue failures

In bridges, in addition to over loading, the number of loading cycles are increased resulting in fatigue failures.

In 1953, the Manhattan Bridge in New York failed due to fatigue. The failure occurred due to failure of lower chord member of the stiffening truss.

All the failures cannot be prevented and it is not humanly possible. However, major disasters may be reduced and prevented in future, by learning a lesson from each failure. The study of reasons of failures and their discussions in technical literature may be as useful to the profession as discussions of great achievements.

2.13 CHARACTERISTIC STRENGTH OF CONCRETE

The characteristic strength of concrete is defined as the specific strength of concrete below which not more than a small percentage (five percent) of the test results may be expected to fall.

A concrete mix is prepared by mixing cement, fine aggregate, coarse aggregate and water. When these ingredients of concrete are mixed in different proportions, the concrete mixes so prepared shall have different strengths. The strength of a concrete mix is specified as **characteristic strength of concrete** and it is denoted by f_{ck}. The characteristic strength of a concrete mix is defined as the compressive strength of 150 mm cube at 28 days expressed in N/mm^2 below which not more than 5 percent of the test results are expected to fall as shown in Fig. 2.4.

The different concrete mixes are graded and each mix is identified by its **grade designation**. In the designation of a concrete, mix, a letter M is used to denote the mix and a number is specified, which is numerically equal to characteristic strength at 28 days.

Different grades of concrete are grouped in three groups. M10, M15 and M20 grades of concrete are grouped for ordinary concrete. M25 and above upto M55 grades of concrete are grouped for standard concrete. M60 and above upto M80 grades of concrete are grouped for high strength concrete. The characteristic compressive strength, f_{ck} of these grades of concrete (M10 to M80) are given in Table 2.4 as per IS : 456 - 2000.

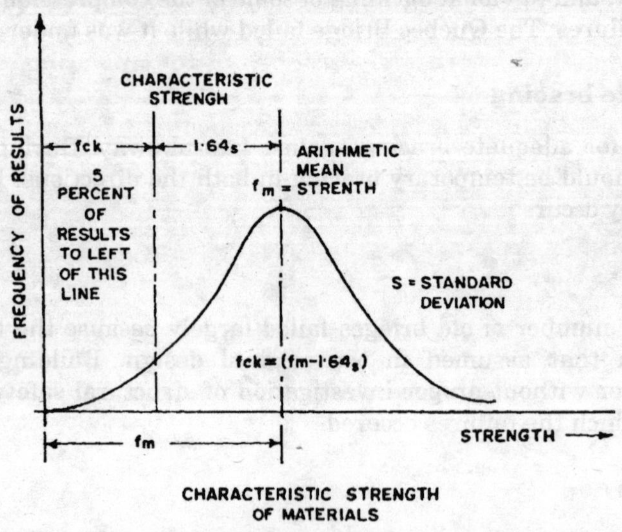

Fig. 2.4.

Characteristic strength= (Mean strength – 1.64) times standard deviation for strength).

where, S is the standard deviation and 1.64 is the value of the constant required to comply with 5 percent of test results below the characteristic strength.

The characteristic strength of a concrete mix may be obtained as under :

(Characteristic strength) = (Arithmetic mean strength) – (k x standard deviation)

$$f_{ck} = (T - K.S) \quad \text{(i)}$$

The factor K ensures that the probability of the actual values is to have smaller value than the characteristic value. As per code, IS 456-2000 recommends that the characteristic strength is the compressive strength below which not more than (five) percent of test results are expected to fall (i.e., K = 1.64). That is

$$f_{ck} = (\bar{f} - 1.64 \times S) \quad (2.13)$$

where, S = Standard deviation.

The strength of concrete and the imposed load that is likely to act over a specified area cannot be provided with certainty. Such variables are referred to as random variables in the theory of probability. The behaviour of random variables is described by its probability law which may be characterised in a number of ways. The most common procedure is through the probability distribution of random variables. The Norman or gaussian probability distribution shown in Fig. 2.4 is used in limit state of design. It is expressed by the following expressions

$$f(u) = \frac{1}{S.\sqrt{2\pi}} . e\left[-0.5\left(\frac{f-\bar{i}}{S}\right)^2\right] \quad (\sigma < x < \sigma) \qquad \text{(i)}$$

where, $\bar{x}$ is the arithmetic mean and S in the standard deviation.

Table 2.4. GRADES OF CONCRETE
(Characteristic compressive strength of 150 mm cube at 28 days, f_{ck})

Group	Grade Designation	f_{ck} (N/mm²)
Ordinary concrete	M10	10
	M15	15
	M20	20
Standard concrete	M25	25
	M30	30
	M35	35
	M40	40
	M50	50
	M55	55
High strength concrete	M60	60
	M65	65
	M70	70
	M75	75
	M80	80

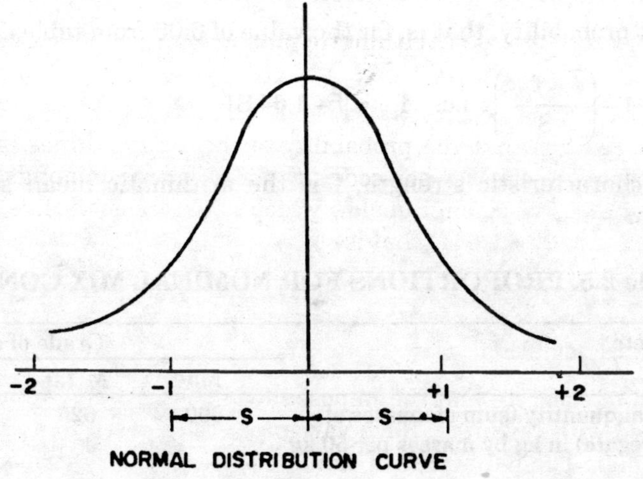

Fig. 2.5.

M20 grade of concrete is used for the beams (flexural members). This grade of concrete is also used for columns. M25 and M30 grades of concrete are used for water storage structures.

M35 and M40, M50 and M55 grades of concrete are used for *prestressed concrete* and standard concrete. M60 and above grades of concrete are used for high strength concrete. It is to note that in IS : 456-2000 (fourth revision) the minimum grade of concrete shall be not less than M20 in reinforced concrete work. It is to note than M5 and M 7.5 are used for lean concrete simple foundations for masonry walls and other simple or temporary concrete construction for masonry walls. It is not necessary to design M5 and M7.5 mixes.

It is preferred to design the proportions of all the ingredients for all the grades of concrete mix. However, **nominal mix concrete** (viz., proportions or quantities by mass of materials) as specified in IS : 456-2000 may be used for concrete of grades M5, M7.5, M10, M15 and M20 as in Table 2.5.

The normal distribution is listed in standard mathematical hand books in the form of standardized variable z, such that z = $(\bar{x} - x)/s$, that is

$$f(z) = \frac{1}{2\pi}\left(e^{-5\ z^2}\right) \qquad \text{(ii)}$$

The area under the curve may be found by numerically integrating the above expression within appropriate limits. From the standard tables for the curve, it may be seen that 95 per cent of the area lies between - 1.64 and +. ∞ This mean that there is only a 5 per cent probability that results will fall below $(\bar{f} - 1.64\ s)$.

The normal distribution curve is symmetrical about the vertical- axis passing through (z = 0) and its highest point is at $(0, 1/\sqrt{2\pi})$. This curve is concave downward between (z = ±1) and concave upward for values of Z outside this range. The total area under the curve and z-axis is unity.

For calculating f_{ck} it is known that $z = \left(\frac{f - f_{ck}}{S}\right)$ (iii)

For 5 per cent probability, that is, for the value of 0.05 from tables, z = 1.64.

$$1.64 = \left(\frac{\bar{i} - f_{ck}}{S}\right). \text{ i.e. } f_{ck} = \left[\bar{f} - 1.64\ S\right] \qquad \text{(iv)}$$

where, f_{ck} is the characteristic strength, f is the arithmatic mean strength and S is the standard deviation.

Table 2.5. PROPORTIONS FOR NOMINAL MIX CONCRETE

Ingredients of concrete	Grade of concrete				
	M5	M 7.5	M 10	M 15	M20
1. Dry aggregate total quantity (sum of masses of fine and coarse aggregate) in kg by masses per 50 kg of cement.	800	625	480	350	250
2. Proportions of fine and coarse aggregates by mass	Generally 1 : 2 but subject to an upper limit of $1:1\frac{1}{2}$ and lower limit of $1:2\frac{1}{2}$				

3. Water quantity per 50 kg of cement (litres)	60	45	34	32	30

The proportions of fine to coarse aggregates are adjusted from upper limit to lower limit progressively as the grading of fine aggregates becomes finer and the maximum size of coarse aggregate becomes larger. Graded coarse aggregates are used. For an average fine aggregate, the proportions are $1 : 1\frac{1}{2}$, $1 : 2$ and $1 : 2\frac{1}{2}$ for maximum size of aggregates 10 mm, 20 mm and 40 mm, respectively.

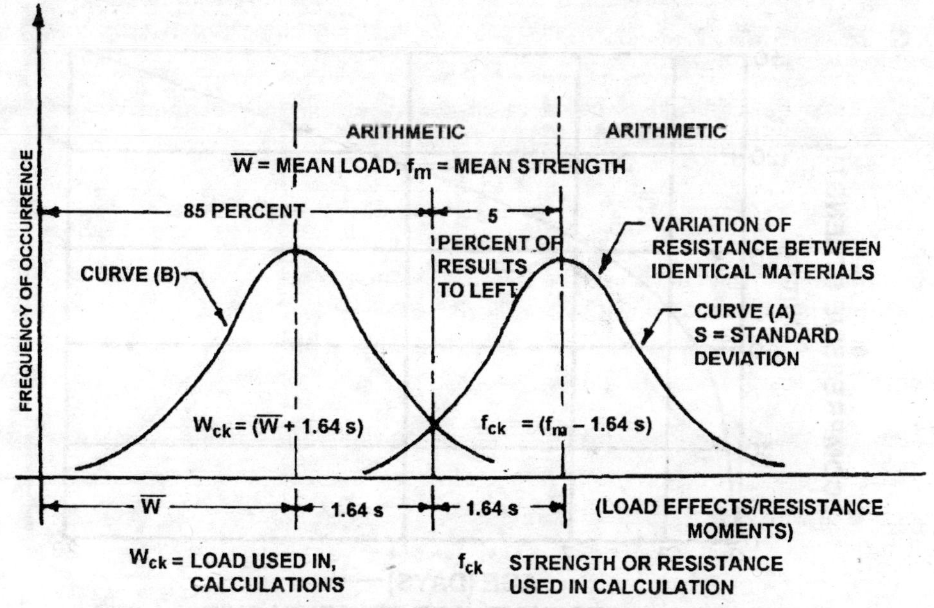

Fig. 2.6.

As an appropriate guidance, the nominal mix proportions by volume for M10, M 15 and M20 grades are $1 : 3 : 6$, $1 : 2 : 4$ and $1 : 1\frac{1}{2} : 3$, respectively and nominal maximum size of aggregate is 20 mm. The nominal mix concrete is used for works of minor nature in which the strength of concrete is not critical.

The strength of concrete and other properties may be adjusted to some extent during its manufacture. As such, the concrete is considered as an ideal material for foundation and other large or complicate parts of the structures. Those must be rigid and resistant to compressive forces.

The various properties of concrete related with its strength are as follows :

1. Increase in strength with age

The strength of a fresh concrete mix (after mixing appropriate quantity of water and beginning of hydration) increases with time in days. The strength of concrete at any age, may be expressed by an approximate formula

$$f_t = \left(\frac{t}{(a + b.t)}\right) f_{28} \qquad (2.14)$$

Where f_{28} is the strength of concrete at 28 days and t is the time in days, a and b are empirical constants and there values are approximately 4.7 and 0.833, respectively. The strength of concrete after 3 days, 7 days or 14 days or at any age in days may be found using above formula. The increase in strength for concrete mix prepared by using ordinary portland cement is shown in Fig. 2.7.

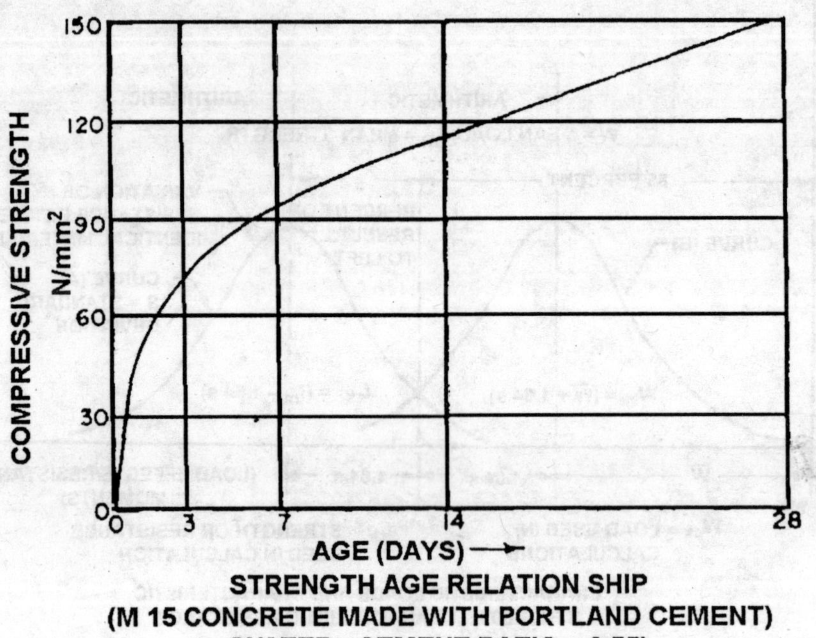

STRENGTH AGE RELATION SHIP
(M 15 CONCRETE MADE WITH PORTLAND CEMENT)
(WATER - CEMENT RATIO = 0.55)

Fig. 2.7.

The increase in strength in concrete mix prepared using different types of cements is shown in Fig. 2.8. The rate of gain of initial strength for different cement is different. However, the strength of concrete at later ages tend to become similar, except that high alumina cement.

(Reference: Orchard, D.F. concrete technology, Vol.1, Fourth edition, 1979 published by Applied Science Publishers Limited, London).

As in characteristic strength of concrete, a cube of 150 mm size is prepared and its compressive strength is found by crushing the cube after 28 days. Sometimes, when the test results for 28 days, strength of concrete mix is found to be lower than that specified, the concrete mix may be approved for use for construction (e.g., in foundation and lower columns in multistorey buildings) where the structure or structural member will not receive its full design load/stress within a period of 28 days after the casting of concrete. The concrete gains sufficient strength due to continuing of hydration of cement by the full time loads are applied.

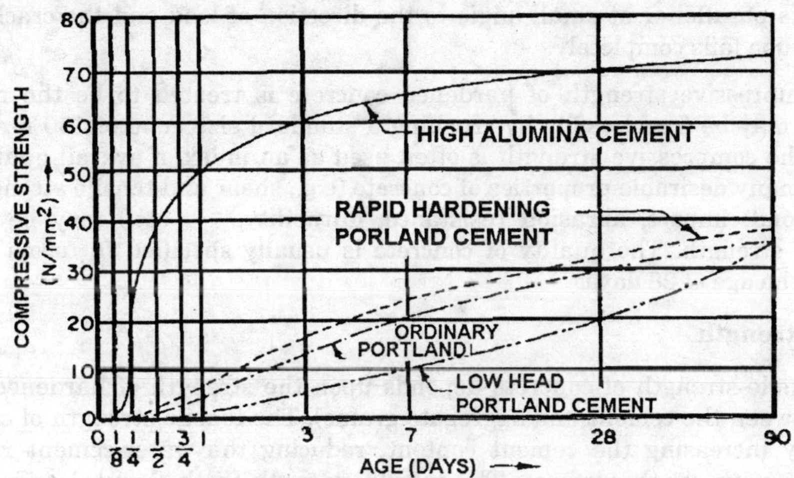

STRENGTH-AGE RELATIONSHIP FOR CONCRETE 1:2:4 BY WEIGHT MADE WITH DIFFERENT CONTENTS

Fig. 2.8.

There is normally a gain of strength beyond 28 days. The amount of increase depends upon the grade and type of cement, curing and environmental conditions.

In case, full design load/stress is expressed to be applied after 1 month or 3 months, the characteristic compressive strength of concrete mix (M15, M20. . .) may be increased by multiplying by age factor 1.0 or 1.10, respectively and in case, the load/stress is expected to be applied after 6 months or 12 months, then, this strength may be increased by multiplying by the age factors 1.15 or 1.20, respectively.

However, in case, high alumina cement is used for concrete mix, its strength is not increased for age or period of loading as such a mix gains its major strength in the beginning itself. In case the members are subjected to lower direct load during construction, such members are checked for stresses resulting from combination of direct load and bending during construction.

This increase in permissible stresses is also applicable to bending, shear and direct compression etc.

2. Compressive strength

Concrete is a non-uniform material. It is non-uniformly stressed when it is subjected to external loads. The stresses in a cube in compression test concentrate at the harder particles having a higher modulus of elasticity. The stresses also concentrate near pores and voids in a concrete cube. The stresses at one pore overlap with those at an adjacent hole. As a result of this, an axially loaded concrete cube carries compressive stresses parallel to the direction of load (longitudinal direction) tensile stresses in the lateral direction). The concrete cube in compression test fails due to lateral rupture of the concrete. As the load increases, the microscopic cracks develop throughout due to the bond failure, and these cracks merge into visible cracks parallel or at small angle to the direction of load and the cracks open up and finally the cube fails completely.

The compressive strength of hardened concrete is treated to be the most important property. It may be found easily by crushing a standard size cube of 150 mm or cylindrical specimen. The compressive strength is often used as an index of overall quality of concrete. In general, many desirable properties of concrete (e.g., shear and tensile strength, modulus of elasticity, bond, impact, abrasion resistance, durability,. . . etc.) may be related to the compressive strength. The quality of concrete is usually specified in terms of compressive strength at the age of 28 days.

3. Tensile strength

The tensile strength of concrete depends upon the strength of hardened cement (bond strength between the cement and aggregate grains). The tensile strength of concrete may be enhanced by increasing the cement content, reducing the water-cement ratio and using broken stone with rough surfaces. The tensile strength (both flexural and split tensile) is closely related to the compressive strength of concrete. However, there is no direct proportionality between them. As the compressive strength of concrete increases, tensile strength also increases but at a reducing rate.

4. The flexural tensile strength of concrete corresponds to modulus of rupture (the moment at first crack) which is needed for calculating the deflection of beams. The flexural tensile strength of concrete may be estimated from the compressive strength from the following formula as per 456-2000

$$f_{cr} = 0.7\left(f_{ck}\right)^{1/2} \tag{2.15}$$

where, f_{cr} is the flexural tensile strength of concrete and f_{ck} is the characteristic compressive strength of concrete.

5. The split tensile strength is found by splitting concrete cylinders by application of the compressive line load along two lines diametrically opposite to each other. The splitting tensile strength of concrete is considered while calculating the shear strength of beams with unreinforced webs, and in the design of cylindrical liquid containing structures. It is also taken into account in the design of slabs.

The tensile strength of concrete is more sensitive to insufficient curing than the compressive strength. It is possibly because of the serious effects of non-uniform shrinkage as in flexural test beams. The value of ratio of tensile strength and compressive strength of air cured concrete is low as compared to that of concrete cured in water. The grading of aggregate also affects the ratio of these two strengths.

2.14 FACTORS INFLUENCING STRENGTH OF CONCRETE

The compressive strength of concrete depends upon characteristics of cement, water-cement ratio, characteristic and proportions of aggregate, age, degree of compaction, efficiency of curing, temperature during curing period... etc. The effect of age and characteristics and contents of cement have been discussed in Art. 2.19. The water- cement ratio and aggregate cement ratio are most important factors which influence the compressive strength of concrete. The influence of each factor on the strength of concrete is as follows :

1. Water-cement ratio

The water cement ratio is the most important single factor involved in mixing concrete using average materials, the strength of concrete and all the desireable properties are directly related to the water-cement ratio.

Water is needed for the chemical processes (viz., **hydrolysis and hydration**) which occur simultaneously during the chemical reaction of cement and water. **Hydrolysis** is defined as the change of compound into others as a result of the chemical action of water. Whereas, **hydration** is the combination of material with water. When water is mixed with cement to form a soft paste, it gradually stiffens and it becomes one solid mass. This process is termed as **setting and hardening**. In **setting,** the cement stiffens and gains adequate rigidity to support an arbitrary specified pressure. In **hardening,** the cement continues to gain further strength for a long time. The water in the paste dissolves the material at the surfaces of the cement grains and forms a gel. This gel gradually increases in volume and stiffens. The process of *hydration* continues to proceed deep and deep into the cement grains, at decreasing speed, with continued stiffening and hardening.

As a result of research sponsored by Portland Cement Association and performed at Lewis Institute, in 1918. Professor Duff A. Abrahm reported that the compressive strength of concrete at a given age and under normal temperature depends primarily on *water- cement ratio.* **The water-cement** *ratio is defined as the weight of water to be used for mixing (correction for absorption of water by aggregate is made) to weight of cement for concrete mix.* The compressive strength of concrete is more when the water-cement ratio is small. Professor Abrahm enunciated this as **law of water cement ratio** and *with given concrete materials and conditions of test, the quantity of mixing water used determines the strength of concrete, so long as the mix is of workable plasticity.* This law is expressed by Eqs. 2.16 and 2.17.

$$CS = K_1/(K_2)^{w/c} \tag{2.16}$$

or

$$\log_e = (CS) = \log_e K_1 - (w/c)\log_e K_2 \tag{2.17}$$

where, CS is the compressive strength and (w/c) is the water- cement ratio of a fully compacted concrete mix by weight, and K_1 and K_2 are the empirical constants. In day to day practice, the constants K_1 and K_2 are not found. However, the relationships between the compressive strength and water-cement ratio are adopted. These relationships are assumed to be valid for a wide range of conditions. The values of constants K_1 and K_2 depend upon age of concrete, characteristics of cement and aggregate, climatic condition, . . .etc. Fig. 2.9 shows the relationship or compressive strength of concrete and water-cement ratio for ordinary portland cement and rapid hardening portland cement for fully compacted concrete and time of curing in days.

According to Professor H. Rusch (Researches towards a general flexural theory for structural concrete, Journal of ACI Vol. 32, No. 1, July 1960), the quantity of water needed for complete hydration of a given quantity of cement is equal to 25 per cent of that of cement by weight. In order to provide the mobility for the water in the cement paste during process of hydration, an additional 10 to 15 percent is added. Thus, the total minimum water-cement ratio becomes about 0.35 to 0.40 by weight. Any amount of water above 25 percent required for the chemical reaction produces pores in the cement paste.

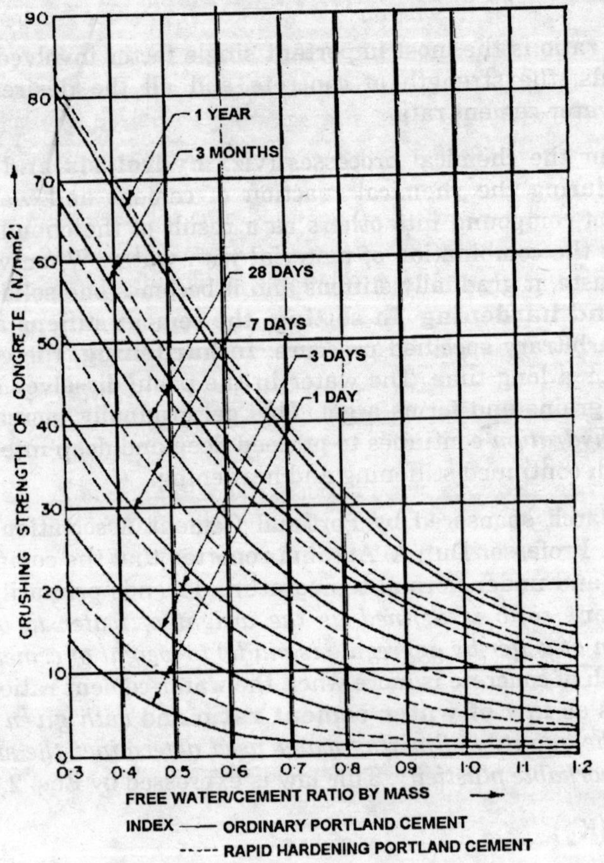

CRUSHING STRENGTH AND WATER CEMENT RATIO RELATIONSHIP (FOR FULLY COMPACTED CONCRETE)

Fig. 2.9

2. Aggregate-cement ratio

The physical properties of aggregates (e.g., strength, surface texture, particle shape, flakiness and elongation indices, maximum size, grading, deterious ingredients,. . . etc.) influence the compressive strength and workability of the concrete. Neville, A.M. (`Properties of concrete', 1972 Edition, published by Pitman) has written that as long as the workability of concrete is maintained at a satisfactory level and the water-cement ratio is kept constant, the compressive strength of concrete increases with increase in aggregate- cement ratio. It is not correct for high-strength concrete mixes of lower workability or in such situations where due to increase in aggregate-cement ratio, the workability is decreased to such an extent that the fresh concrete cannot be placed properly and compacted thoroughly.

Macintosh, J.D. Concrete mix design' 1966 second-edition, Cement and Concrete Association, London) has mentioned that there is small increase in comprehensive strength (in high-strength mixes of low workability) with a decrease in aggregate-cement ratio provided the water content in the mix is also proportionately decreased.

2.15 STRESS-STRAIN RELATIONSHIP FOR CONCRETE

The compressive stress-strain relationship for concrete is shown in Fig. 2.10. The rate of loading has pronounced effect on the behaviour of concrete. Professor H. Rusch has published stress-strain curves at various strain rates in his paper. `Research towards a general flexural theory for structural concrete', in Journal of ACI. Vol. 32, No. 1, July 1960 and it is seen that the curves are concentric. In case a specimen is loaded with slow application of loading (0.001 mm/mm per day), the internal disintegration of the material is less pronounced and that with relatively fast loading (0.001 mm/mm per minute), the internal disintegration of material is more pronounced. The maximum strength reached is somewhat smaller at slow rate of strain and somewhat more at fast rate of loading.

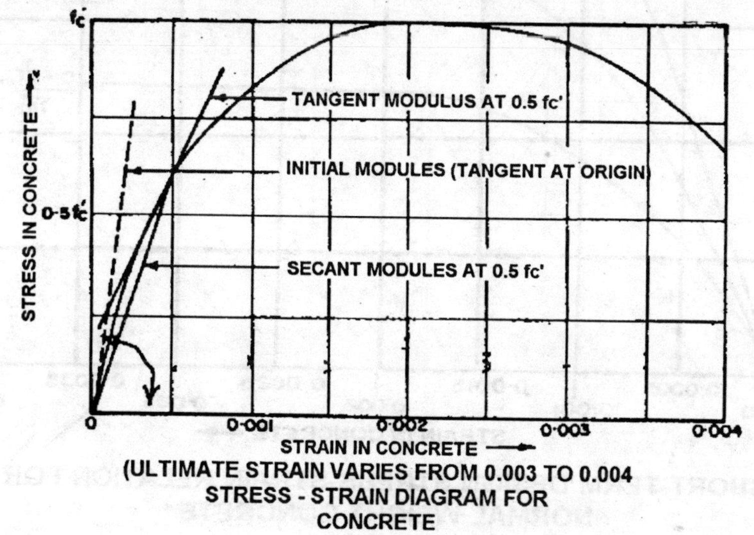

Fig. 2.10.

The shape of the stress-strain curve for the concrete is influenced by a number of factors (e.g., creep, type of aggregate, strength of concrete, age of concrete, curing condition, Fig. 2.10 (A) shows the effect of creep on the shape of the stress-strain curve for a particular concrete mix.

Initial modulus of elasticity (modulus of elasticity), tangent modulus and secant modulus of concrete may be noted from Fig. 2.10(B). These modulii of elasticity of concrete may be defined as follows :

1. Modulus of elasticity of concrete

The modulus of elasticity is primarily influenced by the elastic properties of the aggregate and to a lesser extent by the conditions of curing, age of the concrete, the mix proportions and type of cement. The modulus of elasticity is normally related to the compressive strength of concrete. It is to note that the instantaneous strain is 0.06 percent. The maximum strain is 0.30 percent. The maximum strain due to creep is 0.4 percent. Further it is to note the maximum stress that could be applied for an indefinite period without failure is 0.82 times the cylinder strength.

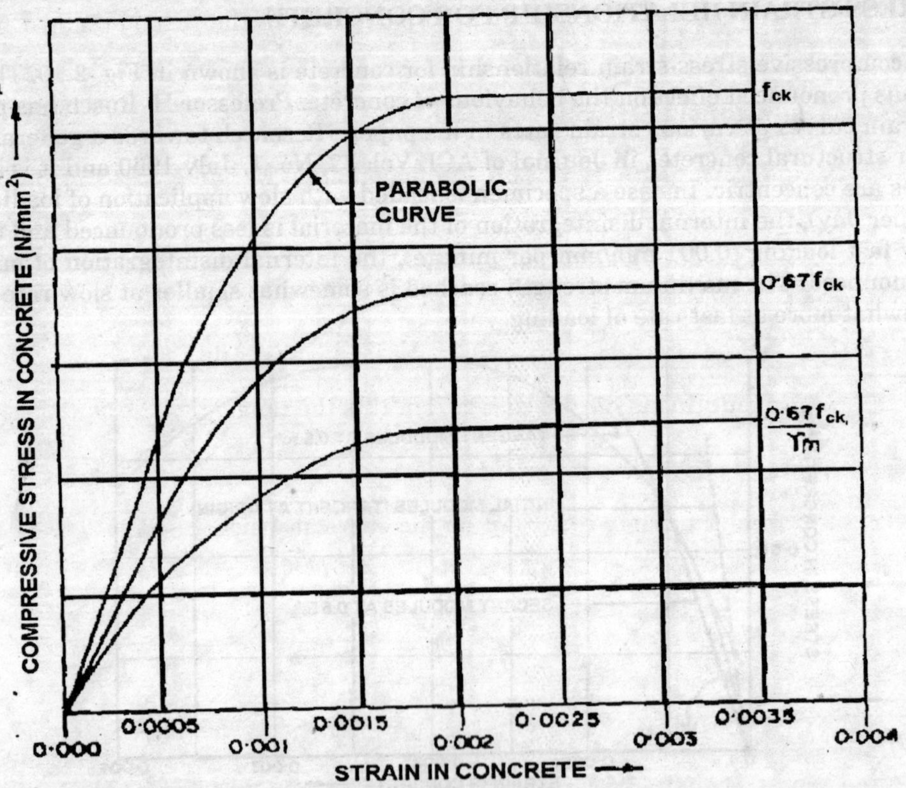

SHORT-TERM DESIGN STRESS-STRAIN RELATION FOR NORMAL WEIGHT CONCRETE

Fig. 2.11

The modulus of elasticity of concrete in compression, E_c is the ratio of unit stress expressed as N/mm^2 to the elastic deformation mm/mm of gauge length. It is the slope of the initial straight portion of the stress-strain curve. It is also termed as **initial modulus**. It is tangent at the origin. The modulus of elasticity varies with the strength. It is primarily influenced by the elastic properties of aggregate and to a lesser extent by the conditions of curing and the age of concrete, the mix proportions, type of cement, rate of loading and type and size of specimen. The concrete exhibits some permanent set even under small loads. In different codes, the various definitions are given for the modulus of elasticity. The modulus of elasticity for structural concrete is assumed as per IS : 456-2000 as follows :

$$E_c = 5000 \, . (f_{ck})^{1/2} \tag{2.17}$$

where, E_c is the **short term static modulus of elasticity** in N/mm² and f_{ck} is the characteristic cube strength of concrete in N/mm².

1. Elastic limit is the limit of stress beyond which the strain is not wholly recoverable. The stress-strain relationship beyond which no longer continues to be linear (i.e., it becomes curvilinear).

Actual measured value may differ by ± 20 percent from the value obtained from above expression.

2. Tangent modulus of elasticity of concrete

Tangent modulus of elasticity of concrete denoted by E'_c, describes the total strain in concrete (including creep) and it is a variable quantity. In Fig 2.4, *it is defined as the slope of the curve at any given point*

$$E'_c = \left(df_c/d\varepsilon_c\right) \tag{2.18}$$

where, df_c is small increase in stress for small increase in strain $d\varepsilon_c$.

Initial tangent modulus of elasticity, Eci. It is the tangent to the curve at its origin.

3. Secant modulus of elasticity of concrete

A line drawn from the origin to a point on the curve at which $f_c = 0.45\,f_c\,(0.5\,f_c)$ is called secant modulus of elasticity.

In some codes, in design calculations, an average quantity termed as *secant modulus of elasticity*, E_c is used. *It is defined as the slope of the chord drawn through a point on the stress-strain curve corresponding to a given stress.* It is also a variable quantity. It is smaller than the initial tangent modulus of elasticity.

When the concrete is compressed in one direction, it expands in the direction transverse to that of applied stress. The ratio of *transverse strain to longitudinal strain* is termed as **Poisson's ratio.** It depends some what on strength, composition and other factors. The value of Poisson's ratio for concrete for small stresses is about 0.15 to 0.20.

Idealised stress-strain curve for concrete

A stress-strain curve as shown in Fig. 2.11 has been accepted as design stress- strain curve for concrete in IS : 456-2000. The peak stress occurs at a strain of about 0.002. It is a parabolic-rectangular relationship for the concrete.

First portion of this curve from zero strain to 0.0020 strain, the curve is a second-degree parabola. The parabolic part of the curve may be represented by the expression

$$f_c = f_d\left[2\left(\frac{\varepsilon_c}{\varepsilon_d}\right) - \left(\frac{\varepsilon_c}{\varepsilon_d}\right)^2\right] \tag{2.19}$$

where $f_d = 0.446\,f_{ck}$ and $\varepsilon_d = 0.002$. Therefore

$$f_c = 0.446\,f_{ck}\left[2\left(\frac{\varepsilon_c}{0.002}\right) - \left(\frac{\varepsilon_c}{0.002}\right)^2\right]$$

$$f_c = 0.446\,f_{ck} \times 1000\ \left[\varepsilon_c - 250\,\varepsilon_c^2\right] \tag{2.20}$$

Later portion of this curve from 0.0020 strain to 0.0035 strain, is a straight line and the stress remains constant. The visible cracks in the concrete does not occur until the strain attains the value of 0.0035. The maximum compressive stress at collapse in a member subjected to short-term load is equal to the cylinder strength of concrete f_{cyl}. And

$$f_{cyl} = 0.8^* f_{cube} \text{ (Assumed)} \tag{i}$$

Considering the factors influencing the strength of concrete and mainly the effect of sustained load, the maximum compressive stress equal to 0.85 fcyl. Therefore,

$$0.85\, f_{cyl} = (0.85 \times 0.8\, f_{cube}) = 0.67\, f_{cube} = 0.67\, f_{ck}$$

The maximum stress on the character curve as shown in Fig. 2.11 has been recommended in the code IS : 456-2000.

$$(0.67\, f_{ck} = 0.67\, f_{cube}) \tag{iii}$$

For the purpose of design (in limit state method) the compressive strength of concrete in the structures is assumed to be 0.67 f_{ck}. For the material strength, a partial factor of safety, r_{ms} = 1.5 is applied in addition to this. The maximum compressive stress in concrete for design purpose becomes $(0.67\, f_{ck}/1.5) = 0.446\, f_{ck}$.

The short-term design stress-strain curve for concrete is shown in Fig. 2.11. By substituting the relevant value of γ_{ms} depending on the limit state being considered, the appropriate design stress-strain curve may be obtained. It is worthwhile to note the following:

1. Design strength is defined as the characteristic strength divided by γ_{rs} and yet the maximum stress value is given as $(0.67\, f_{cu}/\gamma_{ms})$. Since, the characteristic strength is derived from cube-tests. From these tests, it is well established that the maximum compressive stress at failure in a member of the same concrete as a cube, has a value in the region of $(0.8\, f_{cu})$. This is a peak value and as an additional safety factor against compressive failure, this value is reduced to $(0.67\, f_{cu})$ [which agrees with the design methods using ultimate load]. In case, one uses cylinders in finding the characteristic strength, the factor would be of the order 0.85, as the cylinder strength is near the actual behaviour and is approximately (0.8 x cube strength).

2. It is suggested in the code, that for the serviceability limit states, the short-term elastic modulus may be taken from the code (i.e., a linear stress-strain relationship is assumed with a specified value of Ec depending as f_{cu}.

The initial tangenet mudulus for serviceability limit state will seldom be used unless a rigorous is being carried out. Where the sustained loading is being considered however, allowance for shrinkage and creep be made. For the serviceability limit state poisson's ratio is adopted as 0.2.

For the reinforcement (steel bars) the short-term relationship is considered as bilinear (i.e., the value of E is same in compression and tension). The value of elastic modulus remains as 2×10^5 N/mm² (200 kN/mm²).

2.16 CHARACTERISTIC STRENGTH OF STEEL REINFORCEMENT

At present the concept of characteristic strength for steel reinforcement has not been included into any relevant Indian Standard.

As per IS : 456-2000, the characteristic strength of steel reinforcement shall be assumed as the minimum value of yield stress, below which not more than 5 percent of the test results may be expected to fall.

In otherwards, this strength is expected to be excoaded by 95 percent of the cases.

In case, any quality of steel does not show the distinct value or position of yield, IS : 456-2000 recommends that the characteristic strength of steel reinforcement shall be equal to 0.2 per cent proof stress (in the case of cold-worked steel bars) specified in the relevant Indian Standard specifications below which not more than 5 percent of the test materials may be expected to fall.

Proof stress is the stress which is just sufficient to produce under load, a non-proportioned elongation equal to 0.2 percent of the original gauge length (i.e., strain of 0.002 for most structural steel used in absence of well-defined yield points as in high strength steels).

2.17 STRESS-STRAIN RELATIONSHIP FOR STEEL

Yield stress, f_y (generally identical in tension and compression) and modulus of *elasticity,* E_s of steel are two main characteristics of reinforcing steel bars. The *yield stress,* f_y for reinforcing steel bars is quoted as the *characteristic strength of steel*. For all types of reinforcing steel bars, the value of modulus of elasticity, E_s is adopted as 2 x 10^5 N/mm^2. The shape of stress-strain curve and particulary of its initial portion has significant influence on the performance of reinforced concrete members. It is also defined as the characteristic stress of the material in tension.

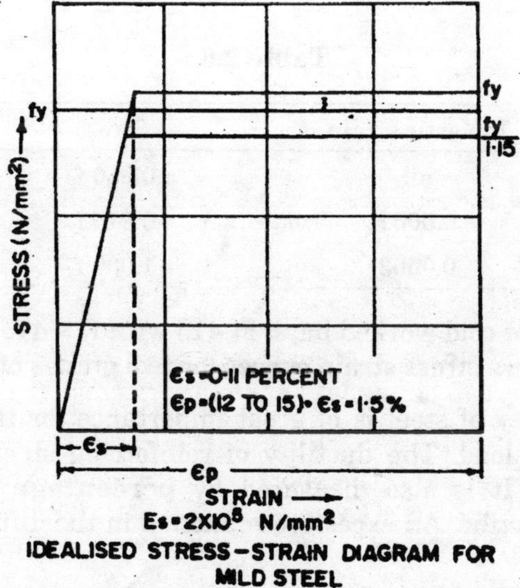

Fig. 2.12

For mild steel (hot-rolled) reinforcing bars, the stress is proportional to the strain upto yield stress and there after the strain increases at constant stress as shown in Fig. 2.12. Fig. 2.12 shows *an elastic portion* followed by *a* **yield plateau**. A yield plateau is a horizontal portion of curve where the strain continues to increase at constant stress. For further increase in strain, the stress begins to increase again, though at a slower rate. This phenomenon is known as **strain-hardening.** In idealised stress-strain curve, Fig. 2.12, the portion of curve beyond strain hardening is dropped from the complete shape of stress-strain curve. The portion of stress strain curve which shows upper yield stress is also dropped in order to simplify the calculation in the analysis.

The mild steel has a definite yield point. The partial factor of safety for strength of material is 1.15. Therefore, a stress of $(f_y/1.15) = 0.87\ f_y$ is used for the mild steel. The elastic strain in steel varies along a straight line. It is directly proportional upto stress 0.87 f_y. The elastic strain upto 0.87 f_y is 0.11 per cent (viz. 0.0011). The plastic strain commences at constant stress 0.87 f_y represented by a horizontal line. The plastic strain ε_p is about 12 to 15 times the elastic strain, ε_s. The plastic strain extends upto 1.5 per cent (viz., 0.015).

Heat treatment and hardening by drawing, increase the strength of hot-rolled reinforcing steel several times. Heat treatment includes hardening (heating to 800°C - 900°C temperature and rapid cooling) following by partial tempering (heating to 300°C -400°C temperature and slow cooling. In hardening by drawing, hot-rolled reinforcing steel is drawn until the stress exceeds the yield stress. Cold drawing or cold working changes the crystal lattice and the reinforcing steel hardens. The cold worked reinforcing steel bars have artificially raised yield stress (high yield strength).

Fig. 2.13 shows stress-strain curves for cold-worked steel as accepted by IS : 456-2000. The cold worked high strength steel bars do not have a definite yield point. The characteristic strength refers to 0.2 percent proof stress. The design yield stress (or 0.2 percent proof stress of steel), $f_{yd} = f_y/r_{ms}$ (with $r_{ms} = 1.15$, $f_{yd} = f_y/1.15 = 0.87\ f_y$). As per IS: 456-2000, therefore the stress-strain curve is defined as given below :

Table 2.6

Stress	Inelastic strain	Stress	Inelastic strain
0.80 f_y	nil	0.950 f_y	0.0007
0.85 f_y	0.0001	0.975 f_y	0.0010
0.90 f_y	0.0003	1.000 f_y	0.0020

There are two grades of cold worked bars, F_e 415 with f_y = 415 N/mm^2 and F_e 500 with f_y = 500 N/mm^2. Fig. 2.13 shows stress strain curves for two grades of steel.

The ductility property of steel is of great importance for the behaviour of reinforced concrete structures under load. The ductility of reinforcing steel is expressed in terms of **percentage elongation**. It is also measured by **percentage reduction of area**. The reinforcing steel is very ductilie. An excessive reduction in the difficulty may result in brittle fracture under load.

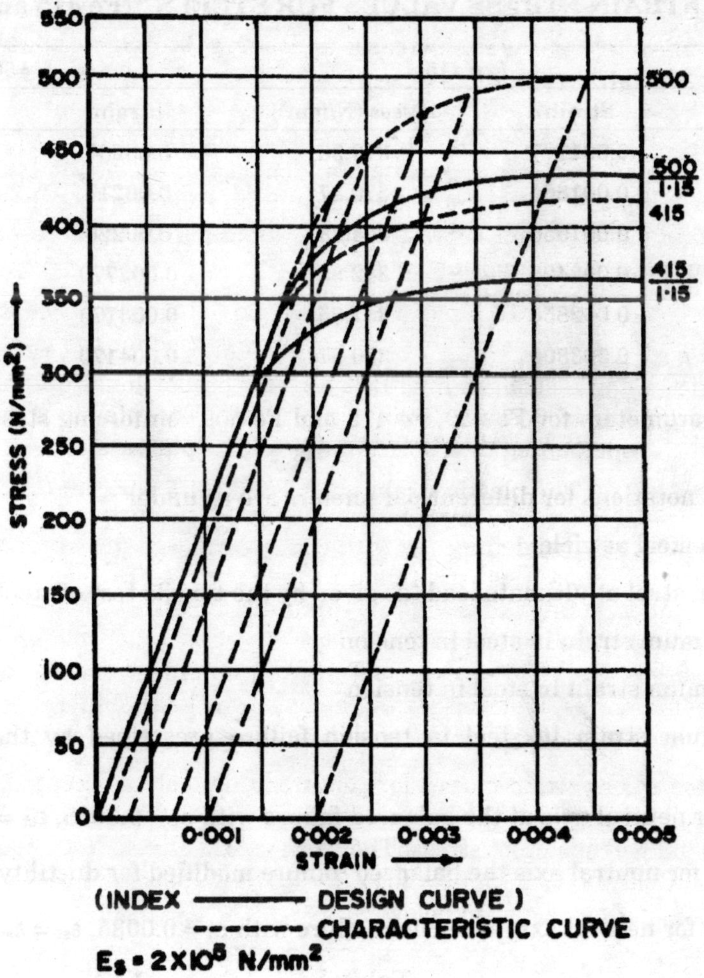

Fig. 2.13

It is the property of the materials indicating the extent to which it can deform beyond limit of yield deformation before the failure or fracture. The ratio of the ultimate to yield deformation is also termed as ductility.

In cold deformed steel, the stress remains proportional upto a stress 0.8 f_y. After this, the stress strain curve becomes non-linear and reaches the yield stress at a strain of $\left[\left(f_y/E_s\right)+0.002\right]$. After this, the strain increases further and the stress remains constant.

The partial safety factor for the material strength of steel is same as that of mild steel, (viz., 1.15 the design stress is equal to (f_y/1.15) = 0.87 f_y). In design stress-strain curve, the relationship remains straight line upto 0.8 x 0.87 f_y. Beyond this, the relationship is non-linear and the design stress reaches 0.87 f_y at strain [0.87 f_s/E_s) + 0.002].

The values of strains and stresses at different levels from strain-stress diagrams. Fig. 2.13 for steels Fe 415 and Fe 500 are given in Table 2.7.

Table 2.7. STRAIN-STRESS VALUES FOR STEELS (Fe 415 and Fe 500)

Stress levels $[f_{yd} = (f_y/1.15)]$	$f_y = 415$		$f_y = 500$	
	Strain	Stress (N/mm²)	Strain	Stress (N/mm²)
0.870 f_{yd}	0.001777	313.96	0.00207	378.26
0.885 f_{yd}	0.001866	319.37	0.00217	384.78
0.900 f_{yd}	0.001956	324.78	0.00226	391.30
0.950 f_{yd}	0.002420	342.83	0.002770	413.04
0.975 f_{yd}	0.002856	351.85	0.003775	423.91
1.000 f_{yd}	0.003800	360.87	0.004170	434.78

The design parameters for Fe 250, Fe 415 and Fe 500 reinforcing steel bars as given in Table 2.8.

The different notations for different parameters are as under :

ε_{sy} = strain in steel at yield

ε_{stu} = strain in steel at ultimate load (as given by the tensile test of steel sample)

$\varepsilon_{st,\,max}$ = maximum strain in steel in tension

$\varepsilon_{st,\,min}$ = minimum strain in steel in tension

ε_{std} = minimum strain in steel in tension failure prescribed by the code to ensure ductility

η_b = factor for neutral axis at the balanced failure with $\varepsilon_1 = 0.0035$, $\varepsilon_{st} = \varepsilon_{sy}$.

η_{max} = factor for neutral axis the balanced failure modified for ductility

η_{min} = factor for neutral axis at tension failure with $\varepsilon_1 = 0.0035$, $\varepsilon_{st} = \varepsilon_{st,\,max}$.

Table 2.9. DESIGN PARAMETERS FOR Fe 250, Fe 415 and Fe 500 STEEL ($E_s = 2 \times 10^5$ N/mm²)

Type of steel reinforcement	Parameters								
	f_y (N/mm²)	$\left(\dfrac{f_y}{1.15\,E_s}\right)$	ε_{sy}	ε_{std}	ε_{stu}	$\varepsilon_{st.mx}$	η_b	η_{max}	η_{min}
Fe250	250	0.0011	0.0011	0.0031	0.2000	0.0200	0.76	0.53	0.15
Fe415	415	0.0018	0.0038	0.0038	0.1450	0.0145	0.48	0.48	0.19
Fe500	500	0.0022	0.0042	0.0042	0.1200	0.0120	0.46	0.46	0.23

2.18. METHODS OF DESIGN

There are three basic methods for the design of reinforced concrete structures and structural members namely *working stress method, ultimate load method* and *limit state method*. All the three metods use certain common basic assumptions *for example* the distribution of strain across a section is considered as linear and the strength of concrete in tension is usually neglected (except in certain cases in the design of liquid containing structures). Other assumptions differ from method to method.

1. In 1900, the working stress method of Coignet and Tedsco became commonly accepted. In this method, *the straight line (elastic) theory* was the conventional method of design for the other materials. The straight-line distribution of stress led to mathematical simplication. The tests had also shown that the use of elastic theory with carefully adopted values for the allowable working stresses led to a structure displaying satisfactory behaviour at the service loads. This method has an adequate margin of safety against collapse. Thus, this theory has been the basis of design of reinforced concrete structures and the strcutural members for many years.

The method of analysis and design of reinforced concrete structures and the structural members should be such that, it satisfies functional, aesthetic and economic requirement in the most efficient manner. It should be capable of predicting the realistic behaviour of structures when subjected to various design loads and gives satisfactory design which remains safe and servicable throughout its life.

2. The ultimate strength theories, for example, Thullie' flexural theory of 1897 and the parabolic stress-distribution theory of Ritter in 1899 remained the basis for the several of the early studies of reinforced concrete members. Recently, there has been renewed interest in *ultimate strength theory* as a basis of design. The practical experience of more than half a century and laboratory tests have vastly enhanced the knowledge of the behaviour of the structural concrete. The deficiencies of the working stress method (elastic theory) became evident. It became increasingly apparent that a method of design should be based on the actual inelastic properties of the concrete and steel. Thus, the *ultimate strength method of design* became accepted as an alternative to working stress method in the buildng codes for reinforced concrete of the American Concrete Institute (ACI) in 1956 and of the United Kingdom in 1957.

3. Limit State Method: IS: 456-1978 (Third revision) introduced limit state method for design of plain and reiforced concrete.

These three methods for the design of reinforced concrete structures and members have been described in the subsequent articles.

2.19 ELASTIC (WORKING STRESS) METHOD OF DESIGN

The working stress method is deterministic in nature. In this method, the behaviour of the structure is assumed to act as linearly elastic body under the action of service loads.

The *elastic (working) stress method of design* of reinforced concrete member is also known as *elastic method of design or* **modular ratio method**. *The method of elastic design of a structure is defined as a method which limits the structural usefulness of the material of the structure upto a certain load at which the maximum stress in extreme fibre reaches the characteristic strength of material in bending*. This method is a conventional method of design and in use since early 1900's. The behaviour of structure is considered under working or service loads only. It is assumed that both concrete and steel behave perfectly elastically and the relationship between stress and strain is linear. It is also assumed that there is no slip between the concrete and the reinforcing steel bars (i.e., there exists a perfect grip or adhesion between concrete and steel). The strain in steel, ε_s and strain in concrete, ε_c at the level where the concrete completely surrounds the steel bars are equal. Therefore,

$$\varepsilon_s = \varepsilon_c \ (\sigma_s/E_s) = (\sigma_c/E_c)$$

$$\text{or} \quad \sigma_s = \sigma_c \ (E_s / E_c) = m \, . \, \sigma_c \qquad (2.)$$

where, ε_s and ε_c are the stresses in steel bar and concrete, respectively and E_s and E_c are moduli of elasticity of steel and concrete, respectively. The ratio of E_s and E_c is known as **modular ratio** and it is denoted by *m*.

The structures are analysed by the *linear elstic theory* and the members are proportioned such that the stresses in concrete and steel resulting from the normal service loads are within the specified limits (limits of *allowable* or *working or permissible stresses*). The allowable stresses are only some fractions of the failure stresses of the material. The allowable stresses are adopted as fixed proportions of the yield strength of the material; *for example*, for compression in bending, it is assumed 0.45 times the cylindrical strength of the concrete.

Concrete responds reasonably elastically upto compression stresses not exceedng about half its strength. Steel remains elastic practically upto the yield stress. The normal service loads are the maximum loads, which are likely to arise in normal use throughout the life-time of the structure. The factor of safety accounts for unpredictable over loads. The bending moment and other forces that act on statically indeterminate structures are calculated assuming linear-elastic behaviour.

The working stress method of design has many difficulties and deficiencies. This method is physically unreal and unnecessarily, it leads to conservative design. The maximum stress may occur at one critical section in the structure and the safety is being evaluated for a very local effect, where as the structure may take higher loads. The linear stress-strain relationship for reinforced concrete structure assumed is not correct. The degree of safety against collapse varies with the end conditions and type of loading. The elastic analysis gives moments and shears generally on the safe side, but their use involves at least a philosophical inconsistecy. Although the width of cracking is not usually calcu-lated, but it should be noted that the influence of the cracking of some of the members and the influence of joints where beam and column have a common section may modify the usual (based on members with a constant EI, often based on gross concrete section) calculated moments more than most designers realize. There is no simple way to account for different degree of various kinds of loads. Commonly the dead load (gravity load due to self-weigth of the structural elements and the permanently attachments) is known more correctly. But the live loads as prescribed in code may have known and variable distribution. The creep and shrinkage, which are time-dependent effects on a structure are not easily accountd for by the elastic calculations of the stresses. The stress in concrete is not proportional to strain upto its crushing strength The inherent safety provided remains unknown.

2.20 ULTIMATE LOAD METHOD (ULM)

There are two principal drawbacks in the working stress method, the distribution of stress over a section at failure differs markedly from that under service loads. And the sections having large cross-sectional area of compression steel are uneconomic as the complete section fails under the collapse or ultimate load before the full strength of its components is utilised. The *ultimate load method* came in practice in 1957 to overcome the short comings of working stress method.

In the ultimate load method of design, the reinforced concrete structures are designed to have the required ultimate load carrying capacity with appropriate load factors for the operating loads and the structural members are designed to develop the corresponding ultimate strength with appropriate load factors. In case of *statically determinate structures*, the ultimate load carrying capacity is controlled by the ultimate strength of the section subjected to the critical combination of loads and moments. However, in case of *statically indeterminate structures*, the load carrying capacity of the structure does not come to an end by failure of one section. The redistribution of moments takes place as a result of formation of plastic hinges. The methods of determining the collapse mechanism of an indeterminate structure for different combination of ultimate loads and for the given properties of the structure and its members are not covered by Code IS : 456-1964. In ultimate load method, the service loads are raised by multiplying by *load factors* to get the ultimate loads. The structures are analysed for the ultimate loads. The non-linear behaviour of the structure and non-linear stress-strain relationship are considered. *The method of ultimate design of a structure is defined as a method which limits the structural usefulness of the material of the structure upto ultimate load.* The ultimate load method of analysis is also called as **method of limit analysis**. It is a means of assessing the ultimate (collapse) load of a structure.

In ultimate strength design of reinforced concrete structures and the structural members, besides the common assumptions, it is further assumed that at ultimate strength, stresses and strains are not proportional and the distribution of compressive stress in a section subjected to bending is non-linear. The diagram of compressive stress distribution in concrete may be assumed as a rectangle, trapezoidal, parabola or any other shape which results in ultimate strength in reasonable agreement with test. The maximum fibre stress in concrete does not exceed 0.68 times ultimate cube stregth of concrete in compression. The members are proportioned such that they are capable of carrying without failure the critical load combinations. It is ensured that there is ample of safety against increased in super-imposed service loads. The strains under working loads should not be so large as to cause excessive cracking.

The reinforced concrete structural members are designed considering the inelastic strains to arrive the ultimate (maximum) strength (i.e., the concrete at maximum strength and usually the steel at yielding) when the ultimate load, equal to the sum of each service load multiplied by its respective load factor is applied to the structure. The bending moments and forces developed at the ultimate load are found assuming linear-elastic behaviour of the structure upto the ultimate load. Alteratively, the bending moments and forces developed are obtained by considering the redistribution of action that may occur because of the non-linear relationships that exist between the actions and deformations in the members at high loads.

Many ultimate load theories came in practice. Different idealised stress-strain curves have been assumed in different theories. Whitney's ultimate load theory became most popular due to simplification. The actual parabolic compressive stress distribution from neutral axis upto the extreme fibre in compression in bending was replaced by rectangular stress distribution diagram. The areas of two stress distribution diagrams were considered equal and the line of action of resultant compressive force through centre of gravity was kept unaltered.

The trend towards the ultimate load design continued for some of the reasons as next page:

At high loads, the reinforced concrete sections behave inelastically. The ultimate strength of the members cannot be reliably predicted by the elastic theory as inelastic strains are not considered. A more rational selection of the load factors may be made in the ultimate strength design. For exactly known values of the loads, a low factor may be used (e.g., in dead loads and for less certain loads (e.g., live loads), a higher load factor is used.

The stress-strain relationship for concrete is non-linear. It is also time dependent. The creep strains for concrete under constant sustained loads may be *many times* the initial elastic strain. The creep strains cause a substantial redistribution of stress in reinforced concrete. The reserve strength is utilised in the ultimate strength design. The high strength reinforcement is efficiently used in the ultimate strength. The ductility of the structure in the post-elastic range may be evaluated in the ultimate strength design.

The ultimate load method of design has serious limitations and drawbacks. This method cannot predict the response of the structure to service loads. The serviceability considerations are equally important as strength requirements. Excessive deflections and cracking are the common causes of damage in reinforced concrete structures and the structures may become unserviceable and unsafe. The variations which are likely to occur in the loads and in the strength of materials are not considered in this method.

REFERENCES

1. IS: 456-1978. Code of Practice for Plain and Reinforced Concrete (Third - revision) First reprinted in 1981.
2. IS: 456-2000. Code of Practice for Plain and Reinforced Concrete (Fourth revision).
3. CP110 : 1972. Elementary Strucural Design in Concrete, by A.W. Astill and L.H. Martin, Pubished by Arnold, 1975.
4. BS : 8110. Reinforced Concrete Design Simply Explained by A.H. Allen, Published by E & F.N. SPON London, New York, 1988.

3

PART III DESIGN OF BEAMS

BENDING (FLEXURAL) STRENGTH AND DESIGN OF SINGLY REINFORCED RECTANGULAR BEAMS (SRRB-I)

3.1 INTRODUCTION

A beam is defined as the structural member subjected to transverse loads. The plane of transverse loads is parallel to the plane of symmetry of the cross-section of the beam and it passes through the shear centre, so that the simple bending of beam occurs. The transverse loads produce bending moments and shear forces (called as the internal forces) in the beams at all the sections of the beams. The beams are called as *simply supported, overhanging, cantilever, fixed and continuous* depending upon nature of supports and end conditions. The cross- sections of most simple reinforced concrete beams are rectangular in shape.

Plain concrete beams are inefficient as flexural members. The tensile strength of such beams in bending is very small as compared to their compressive strength. As such, these beams fail on the tension side at small loads much earlier than the strength of concrete on compression side is fully utilised. Reinforced concrete beams are non-homogeneous. These beams are made of entirely two different materials (viz., cement concrete and steel). These beams differ considerably in their internal action from those made of homogeneous materials, (such as steel, timber or any other structural material). The fundamental principles involved in the analysis are however, essentially the same. In reinforced beams, the steel reinforcing bars are placed on the tension side as near the extreme tension fibre as it is necessary for fire and corrosion protection for the steel. The reinforced concrete beams in which steel reinforcement is placed on tension only, are known as *singly reinforced beam*. In singly reinforced beams, the tension developed due to the bending moment is mainly resisted by the steel reinforced and the compression is resisted by concrete alone. In simply supported beams as shown in Fig. 1.1(A) (ii), the steel bars are placed near bottom face (tension side) and in cantilever beams as shown in Fig. 1.1B (ii), the steel bars are placed near top face (tension side).

A joint action of concrete and steel in reinforced concrete design is assumed in case there is no relative slip between them. The steel reinforcing bars are perfectly gripped by the surrounding concrete. A perfect grip is termed as bond between the steel bars and concrete. This bond is achieved as the additive effects of adhesion between steel bars and concrete, resistance to sliding due to friction and further gripping of steel bars as a result of shrinkage of concrete.

3.2 BEHAVIOUR OF REINFORCED CONCRETE BEAM

A singly reinforced rectangular beam is shown in Fig. 3.1 (i) with dimensions of its cross-section. When the load is applied to a simply supported beam having such a cross-section and the load is gradually increased from zero to failure load. The beam passes through different and distinct stages and it shows three characteristic stages of stress-strain behaviour.

At small loads, the maximum tensile stress in bending in concrete remains smaller than the modulus of rupture. The complete concrete cross-section remains effective in resisting the bending of the beam. There is compressive stress in bending in concrete on one side (above) and tensile stress in bending in concrete on the other side (below) the neutral axis. The steel reinforcing bars deform equally as the surrounding concrete. There is also tensile stress in bending in steel bars. Upto this stage, the values of all the stresses in concrete are small. The distribution of strains and stresses across the depth of section is essentially same as in an elastic beams of homogeneous material as shown in Fig. 3.1 (ii) and (iii), respectively. The cracks do not develop along the length of the beam. *The stress in beam are elastic and the section of beam remains uncracked.*

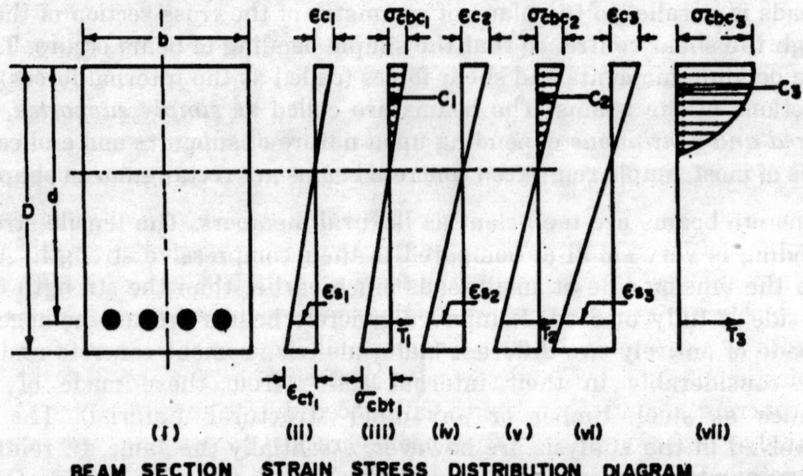

Fig. 3.1.

The strains in surrounding concrete, ε_c and in steel ε_s are equal. The stress in steel, σ_{st} is m (modular ratio) times the stress in surrounding concrete, σ_{ibc}. Therefore,

$$\sigma_{st} = (m \, . \, \sigma_{ibc}) \tag{3.1}$$

The complete cross-section of beam having concrete and steel may be replaced by an imaginary cross-section of beam called as **transformed section** or **equivalent section** which consists of concrete only as shown in Fig. 3.2 (A) and (B). The area of steel in tension, A_{st} is replaced by equivalent concrete area equal to ($m.A_{st}$) positioned at the level of steel. The analysis of beam upto this stage may be done as that of a homogeneous beam. The neutral axis of beam section may be located and moment of inertia and modulus of section of beam cross-sections may be calculated as usual to determine the stresses in extreme fibres.

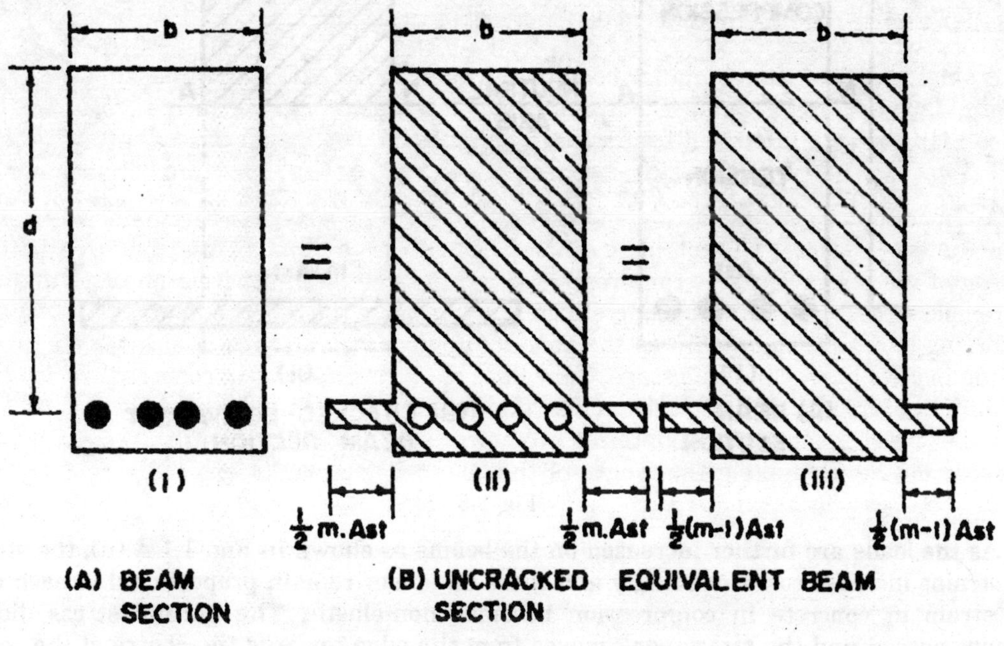

Fig. 3.2.

As the loads are increased gradually on the beam, Fig. 1.1 A (ii), the tensile strength of concrete is soon arrived. The tensile stress in bending in concrete exceeds the modulus of rupture at the bottom face. At this stage, the cracks develop at the bottom face. These cracks propagate rapidly upward to or near to the level of neutral plane. The neutral plane also shifts upward with progressing cracking. The distribution of these cracks is shown in Fig. 1.1 A (ii). The width of these cracks remain very small. These thin (hair line) cracks are not considered objectionable for fire and corrosion protection or for appearance. However, the behaviour of concrete beam is very much influenced due to presence of these cracks. The rectangular singly reinforced beam section beams cracked from the bottom face upto neutral axis. The concrete below neutral axis does not resist any tensile stress. The steel bars resist total tension. This situation commonly prevails in structures under service loads. The loads upto this stage may be referred as moderate loads and these loads correspond to service loads. The concrete and steel continuc to bchavc clastically, or very nearly so. The strains and stresses also continue to be closely proportional and their distributions are as shown in Fig. 3.1 (iv) and (v), respectively. For simplicity and with little (if any) error, it may be assumed that the cracks have travelled upto the neutral axis. The reinforcing steel bars retain its bond with the concrete.

The actual beam section of concrete and steel may be replaced by imaginary beam section consisting of concrete alone. The area of steel, A_{st} in tension may be replaced by its equivalent area $m.A_{st}$ of concrete. This imaginary section is called as **cracked equivalent section** as shown in Fig. 3.3. The method of analysis of such beams subjected to service loads shall be described in subsequent articles.

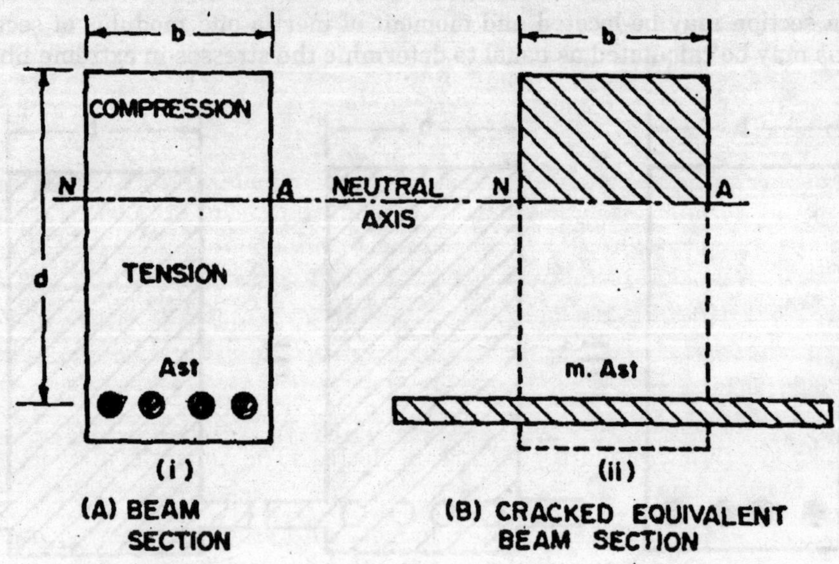

Fig. 3.3

As the loads are further increased on the beams as shown in Fig. 1.1 A (ii), the stresses and strains increase correspondingly and these no longer remain proportional to each other. The strain in concrete in compression becomes non-elastic. The normal stress diagram becomes curved and the stress peak moves from the edge towards the centre of the section. The distribution of strains and concrete stresses on the compression side close to the ultimate load are as shown in Fig. 3.1 (vi) and (vii), respectively. Finally, the load carrying capacity of the beam is reached and the failure of beam occurs in one of the following two ways :

When the cross-sectional area of steel used is relatively moderate, at some values of load, the stress in reinforcing steel bars reaches the physical or proof yield point and the reinforcement yields suddenly and stretches by a large length. The cracks on the tension face widen and become visible and propagate upwards. Simultaneously, the beam deflects significantly. At slightly larger load, the strains in remaining compression zone of concrete increase to such a degree that the crushing of concrete (the secondary compression failure) occurs. This type of failure of beam is called **yield failure**. The yield failure is gradual. It proceeds with visible widening and lengthening of cracks, and marked increase of deflection.

On the other hand, the cross-sectional area of steel bars may be large or the normal cross-section of steel of very high strength is used. The compressive strength of concrete exhausts earlier than the yielding of the steel bars. The crushing of concrete occurs and the beam fails. It has been seen that the rectangular beams fail in compression. The strains in concrete becomes about 0.0030 to 0.0040. The **crushing failure** of concrete occurs suddenly and without warming.

3.3 BENDING (FLEXURAL) STRENGTH OF SINGLY REINFORCED BEAM

In practice, the beams are analysed by the *linear elastic theory* (viz., by working stress method). In order to analyse the beams, following fundamental assumptions are necessary.

Fundamental assumptions

1. At any cross-section, **plane sections before loading** (bending) remain plane after loading (bending). In other words the strains vary lineary with the distance from the neutral axis.

It means that transverse sections (the sections perpendicular to the longitudinal axis of beam) are plane and normal to the longitudinal axis of beam before the loads are applied to the beam. These sections remain plane and normal to the longitudinal axis after loads are applied. The effect of shear is neglected. *The unit strains in a beam above and below the neutral axis are proportional to the distance from that axis.*

2. All the **tensile stresses are taken up by reinforcement** and the concrete does not resist any tensile stress.

It means that upto service (working) loads, the cross-section of beam acts as a cracked section. The portion of concrete in tension is neglected. The concrete in tension zone is cracked upto the neutral axis.

3. The **stress-strain relationships for concrete and steel under working** loads are linear. (that is, the stresses in steel and concrete are proportional to strains).

It means that, the behaviour of concrete and steel is assumed elastic upto service loads.

4. The value of modular ratio of concrete, m is found as $[280/(3.\sigma_{cbc})]$, where σ_{cbc} is the permissible compressive stress in bending in concrete in N/mm^2.

This expression for modular ratio, m partially takes into consideration long-term effects such as creep. It is different from **that calculated as (Es/Ec)**.

5. The modulus of **elasticity of concrete is constant**.

6. The adhesion (grip) of concrete to steel is perfect within the elastic limit of the steel.

7. There are no initial stresses in the steel.

The beam may be analysed by locating the neutral axis of the cross-section and by calculating the stresses and deflections (if desired). Consider a singly reinforced rectangular beam as shown in Fig. 3.4 with dimensions of its shape. Let b and d be the breadth (width) and effective depth of the section and A_{st} is the area of steel in tension. When the beam is simply supported at its both the ends, the portion of concrete above neutral axis is in compression and that below the neutral axis is in tension.

1. Depth of Neutral axis

The neutral axis is the axis which passes through the centroid of the beam section. At the neutral axis, the bending stress is zero. Let the distance of neutral axis from the extreme fibre of concrete in compression be n. It is also expressed conventionally as kd, as a fraction of the effective depth, d where, **k is the neutral axis depth factor**. The neutral axis is located by taking the moment of area of concrete in compression and the moment of equivalent area of steel in tension about the neutral axis. Therefore,

$$\left[(b\,.\,n\,).\frac{n}{2} - m\,A_{st}\,(d-n)\right] = 0 \qquad (3.2)$$

$$(b.n)\frac{n}{2} = (m.A_{st}).(d-n) \quad \text{(i)}$$

or $$\left[b.kd.\frac{kd}{2} - m\,A_{st}\,(d-kd)\right] = 0 \quad (3.3)$$

By solving this quadratic equation, the depth of neutral axis n or kd may be found.

It is to note that the rectangular section of the beam is symmetric about both the axes. Therefore, shear centre and centroid coincide.

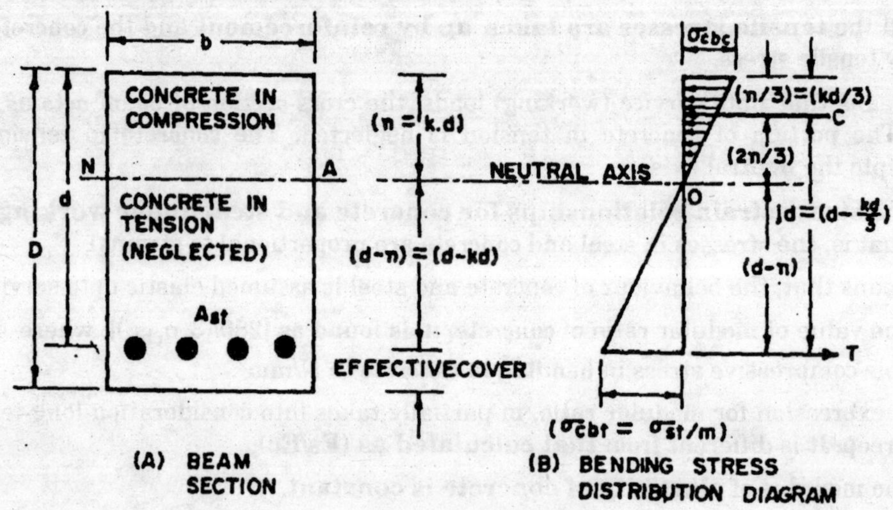

Fig. 3.4.

2. Factor k for depth of NA (Stresses in concrete and steel)

Let σ_{cbc} and σ_{st} be the compressive stress in bending in concrete at the extreme fibre and tensile stress in bending in steel, respectively. From Eq. 3.1, σ_{st} is equal to m. σ_{cbc}, as such, the stress in concrete surrounding the steel at the level of centre of steel is (σ_{st}/m). From the bending stress distribution diagram. Fig. 3.4(B), (triangles above and below the neutral axis are similar), therefore,

$$\left(\frac{\sigma_{cbc}}{kd}\right) = \left(\frac{(\sigma_{st}/m)}{(d-kd)}\right) \quad \text{(ii)}$$

or $$k = \left(\frac{m.\sigma_{cbc}}{(m\,\sigma_{cbc} + \sigma_{st})}\right) \quad (3.4)$$

where $$m = \left(\frac{280}{3\,\sigma_{cbc}}\right)$$

Note **: It takes in consideration long time effects. viz. shrinkage and creep in.**

As such, in case, the **actual stresses** σ_{cbc} and σ_{st} are known, even then the depth of neutral axis may be found from Eq. 3.4.

In case, the dimension of beam sections b and d, and the area of steel A_{st} are given, the actual stresses developed in the beam section for the moment M, at the section may be found as under. Total compressive force in concrete, C and total tensile force in steel T are as follows :

$$C = (\text{Average stress}).(\text{Area of concrete in compression})$$

$$C = \frac{1}{2}(\sigma_{cbc}.b.kd)$$

and $$T = (\text{Stress in steel}).(\text{Area of steel bars})$$

$$T = (\sigma_{st}.A_{st})$$

The beam is in equilibrium. Therefore, (C = T)

$$\left(\frac{1}{2}\sigma_{cbc}.b.kd\right) = (\sigma_{st}.A_{st})$$

$$A_{st} = \frac{1}{2}\left(\frac{\sigma_{cbc}}{\sigma_{st}}\right)(b.kd) \tag{3.7}$$

3. Percentage of steel

$$\text{Ratio } p = \left(\frac{A_{st}}{b.d}\right) = \left(\frac{1}{2}\frac{\sigma_{cac}}{\sigma_{st}}.k\right) \tag{3.8}$$

or $$p = \frac{1}{2}.\frac{\sigma_{cbc}}{\sigma_{st}}\left(\frac{m.\sigma_{cbc}}{m\,\sigma_{cbc} + \sigma_{st}}\right) \tag{3.9}$$

The ratio p is also expressed as the percentage of steel. Therefore

$$p = \frac{1}{2}\left(\frac{\sigma_{cbc}}{\sigma_{st}}.k\right) \times 100 \tag{3.10}$$

In case A_{st} is substituted as p.bd in Eq. (3.3) and Eq. 3.4 is solved for k, one obtains

$$k = [2mp + (mp)^2]^{1/2} - mp \tag{3.11}$$

4. Stresses

These two forces C and T are numerically equal, parallel and act in the opposite directions. These two forces form a couple and resist the external moment, M. By taking the moment of compressive force, C about the centre of steel bars

$$C.(\text{lever arm}) = M$$

or $$\left(\frac{1}{2}\sigma_{cbc}\right).(b.kd).jd = M$$

$$\sigma_{cbc} = \left(\frac{2.M}{k.j.bd^2}\right) \tag{3.12}$$

where, jd is the distance between two parallel forces C and T and it is called as lever arm, and *j* is termed as lever arm factor.

4(B). Stress in steel in tension in bending

By taking the moment of tensile force, T about the line of action of compressive force, C,

$$T. (\text{lever arm}) = M$$

or $$(\sigma_{st} . A_{st}). jd = M$$

$$\sigma_{st} = \left(\frac{M}{A_{st} . jd}\right) \tag{3.13}$$

From Fig. 3.4 (B) jd = (d – kd/3)

$$j = (1 - k/3) \tag{3.14}$$

For a given moment at a section, M, the actual stress, σ_{cbc} at the extreme fibre of concrete in compression, σ_{st} the actual stress in the steel bars may be found from Eqs. 3.12 and 3.13.

5. Bending (Flexural) strength of a beam section

In case, the area of steel, A_{st} in the singly reinforced beam is such that, the tensile stress in steel bars, σ_{st} becomes equal to the permissible stress in steel reinforcement, when the beam is loaded, the position of neutral axis of beam section shall be such that the value of σ_{cbc} shall be less than the permissible stress in compression in bending in concrete. The bending strength of beam is found from the steel in tension. Then

$$MR_1 = T. \text{ lever arm}$$

or $$MR_1 = (\sigma_{st} . A_{st} . jd) \tag{3.15}$$

where, MR_1 is the **moment of resistance** or the **bending strength** of beam sections and σ_{st} is permissible stress in steel in tension.

In case, the area of steel, A_{st} on the other hand is such that, the compressive stress in bending in concrete of the extreme fibre σ_{cbc} becomes equal to its respective permissible stress, when the beam is loaded, the position of neutral axis shall be such that the value of σ_{st} shall be less than its respective permissible stress, then

$$MR_2 = (C. \text{ lever arm})$$

or $$MR_2 = \left[\frac{1}{2}\sigma_{cbc} . (b . kd). jd\right] \tag{3.16}$$

where, MR_2 is the **moment of resistance or the bending strength** of beam section, σ_{cbc} is the permissible compressive stress in bending in concrete Eqs. 3.15 and 3.16 are the fundamental equations in the analysis and design of singly reinforced rectangular beams. Upto this stage, without knowing that the beam section is under-reinforced or over reinforced, the bending strength (viz., moment of resistance) of a beam section moment of resistance of a beam section shall be MR_1 and MR_2 whichever is minimum.

3.4 BALANCED, UNDER-REINFORCED AND OVER REINFORCED BEAM SECTIONS

For the given grades of concrete and steel, the position of neutral axis of reinforced concrete beams changes with the percentage of steel within the beam section. Depending upon the percentage of steel in a beam section, a beam section is called as *balanced, under reinforced or over reinforced* described as follows:

1. Balanced section. A reinforced concrete beam section is called as *balanced section,* in case the area of steel within the section is such that the **maximum compressive stress in bending in concrete,** σ_{cbc} at the extreme fibre on compression side and tensile stress in bending in steel, σ_{st} attain the values of their respective permissible stresses, simultaneously, when the beam is loaded. A balanced beam section is also called as *critical beam section* or an *ideally reinforced beam section.* For a balanced beam section,

$$\text{Ratio of}\left(\frac{\text{Actual } \sigma_{cbc}}{\text{Actual } \sigma_{st}}\right) = \text{Ratio of}\left(\frac{\text{Allowable } \sigma_{cbc}}{\text{Allowable } \sigma_{st}}\right)$$

For a balanced beam section, using M 15 grade of concrete and mild steel bars of diameter less than or equal to 20 mm, the values of depth factor for neutral axis k and ratio of steel, p may be found from Eqs. 3.4 and 3.8, respectively.

For M 15 grade of concrete and mild steel bars

$$\sigma_{cbc} = 5 \text{ N/mm}^2,\ \sigma_{st} = 140 \text{ N/mm}^2,\ m = \left(\frac{280}{3\times 5}\right) = 18.667$$

$$k_B = \left(\frac{m\,\sigma_{cbc}}{(m\,\sigma_{cbc} + \sigma_{st})}\right) = \left(\frac{18.667\times 5}{(18.667\times 5 + 140)}\right) = 0.4 \qquad \text{(i)A}$$

$$p_B = \frac{1}{2}\left(\frac{\sigma_{cbc}\cdot k}{\sigma_{st}}\right) = \left(\frac{5\times 0.4}{2\times 140}\right) = 0.7143 \text{ per cent} \qquad \text{(ii)A}$$

For M 20 grade of concrete and hysd steel bars,

$$\sigma_{cbc} = 7 \text{ N/mm}^2, \text{ and } \sigma_{st} = 230 \text{ N/mm}^2,\ m = \left(\frac{280}{3\times 7}\right) = 13.33$$

$$k_B = \left(\frac{13.33\times 7}{13.33\times 7 + 230}\right) = 0.2886$$

$$p_B = \frac{1}{2}\left(\frac{\sigma_{cbc}\,.\,k}{\sigma_{st}}\right) = \frac{1}{2}\times\left(\frac{7\times0.2886}{230}\right)\times100 = \mathbf{0.4392} \text{ per cent}$$

For the balanced section, depth factor for neutral axis, *k* is denoted by k_B or k_C lever arm factor *j* is denoted by j_B, neutral axis, *n* is devoted by n_B or n_C, and percentage of steel is denoted by p_B. The bending strength or moment of resistance of balanced beam section when worked out from Eqs. 3.15 and 3.16, MR_1 and MR_2 will be found to be equal

From Eq. 3.15, $$MR_1 = \left[\frac{1}{2}\sigma_{cbc}\,.(b\,.\,kd)\,.\,jd\right]$$

or $$MR_1 = \left(\frac{1}{2}\sigma_{cbc}\,.\,k\,.\,j\right).\,bd^2 = R.b.d^2 \qquad (3.17)$$

where, $$R = R_B = \left(\frac{1}{2}\sigma_{cbc}\,.\,k\,.\,j\right) \qquad (3.18)$$

R is called as **factor for moment of resistance**, and it is denoted by R_B for balanced section.

For Eq. 3.16 $$MR_2 = (\sigma_{st}\,A_{st}\,.\,jd) \text{ and } MR_1 = MR_2 \qquad (iii)$$

2. Under reinforced section. A reinforced concrete beam section is called as *under reinforced section*, when the percentage of steel in the beam section, p is less than the percentage of steel in the balanced beam section, p_B when the loads are applied to the beam and gradually increased, the tensile stress in bending in steel, σ_{st} attains the value equal to permissible stress in steel. At this stage of loads, the value of compressive stress in bending in concrete at the extreme fibre in compression, σ_{cbc1} remains less than its respective permissible stress. For the under reinforced beam section,

$$\text{Ratio}\left(\frac{\text{Actual } \sigma_{cbc}}{\text{Actual } \sigma_{st}}\right) < \text{Ratio}\left(\frac{\text{Allowable } \sigma_{cbc}}{\text{Allowable } \sigma_{st}}\right)$$

An under-reinforced beam section and its respective bending stress distribution diagram are shown in Fig. 3.5 (A) and (B), respectively. In case, the beam section would have been a balanced section, its neutral axis would pass through point *O* and σ_{st} and σ_{cbc} would be the values of permissible stresses for steel and concrete, respectively. The position of actual neutral axis of a beam section is located by equating the moment of equivalent areas about its neutral axis. From Fig. 3.5,

$$(b\,.\,n).\,\frac{n}{2} = [(m.\,A_{st})\,.\,(d-n)] \qquad (v)$$

Since, the area of steel, A_{st} is less than that for the balanced section, then, (d – n) is greater than (d – n_B) as the product of two values of left side has to be equal to that on right side. Therefore, the value of n, depth of actual neutral axis is less than n_B that for balanced section, and actual neutral axis shifts upward. The moment of resistance or bending strength of beam section is found from Eq. 3.16, (i.e., from tension side). As such

$$MR_2 = (\sigma_{st} . A_{st}) . jd$$

or $$MR_2 = \left(\frac{1}{2}\sigma_{cbc} . kj\right) bd^2 = \left(R.bd^2\right) \quad \text{(vi)}$$

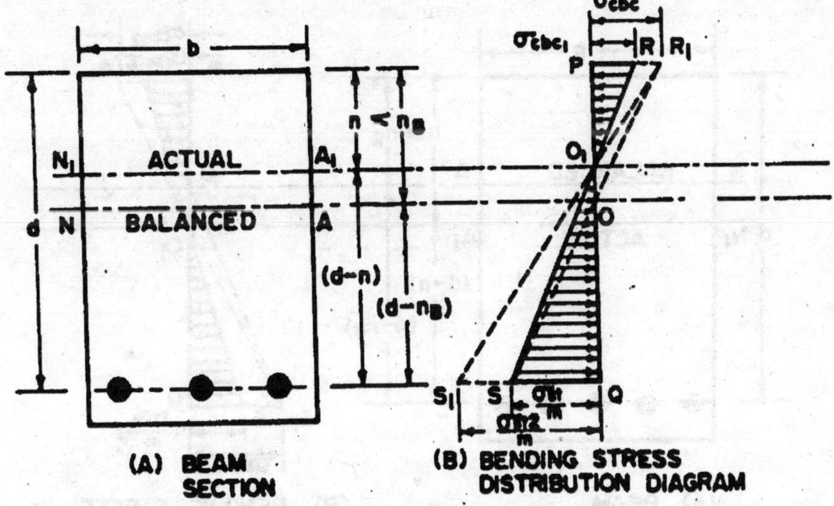

Fig. 3.5

In case, the applied loads on beam increases such that σ_{cbc} may increase upto σ_{cbc} then the actual value of tensile stress, σ_{st2}, in steel will exceed its permissible value σ_{st} and then the beam would be considered, as if, it had failed. However, the steel being ductile, the yielding of bars occurs and widening of cracks take place. The cracks become visible (at the bottom surface for a simply supported beam) and the measures may be taken to avoid total collapse. The beam gives warning and allows time to the occupants to escape. Therefore, it is desirable to design the beam as under reinforced beam.

3. Over reinforced section. A reinforced concrete beam section, is called as *over reinforced section*, when the percentage of the steel in the beam section, p is more than the percentage of steel in the balanced section p_B. When the loads are applied to the beam and gradually increased, the compressive stress in bending in concrete at the extreme fibre of concrete in compression, σ_{cbc} attains the value equal to the permissible stress in concrete. At this stage of loads, the value of tensile stress in steel, σ_{st} remains less than its respective permissible stress. For the over reinforced beam section,

$$\text{Ratio}\left(\frac{\text{Actual } \sigma_{cbc}}{\text{Actual } \sigma_{st}}\right) > \text{Ratio}\left(\frac{\text{Allowable } \sigma_{cbc}}{\text{Allowable } \sigma_{st}}\right)$$

An over reinforced beam section and its respective bending stress distribution diagram are shown in Fig. 3.6 (A) and (B), respectively. In case, the beam section would have been a balanced section, its neutral axis would pass through O and σ_{cbc} and σ_{st} would be the values of permissible stresses for concrete and steel, respectively. The position of actual neutral axis of a beam section is located by equating the moment of equivalent areas about its neutral axis. From Fig. 3.6.

$$(b.n).\frac{n}{2}=[(m\,A_{st})\,(d-n)] \qquad \text{(v)}$$

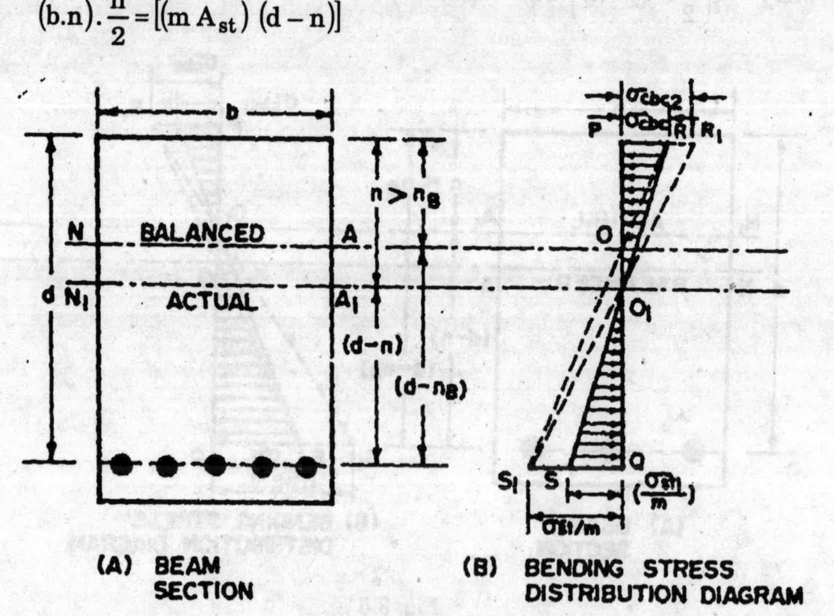

Fig. 3.6.

Since, the area of steel, A_{st} is more than that for the balanced section, then, (d – n) is less than (d – n_B), as the product of two values of left side has to be equal to that on right side. The value of n depth of actual neutral axis is more than n_B that for balanced section and actual neutral axis shifts downward. The moment of resistance or bending strength of beam section is obtained from Eq. 3.15 (i.e., from compression side). As such

$$MR_1=\frac{1}{2}\sigma_{cbc}\,.\,(b\,.\,kd)\,.\,jd$$

or
$$MR_1=\left(\frac{1}{2}\sigma_{cbc}\,.\,k\,.\,j\right).\,bd^2=\left(R\,.\,bd^2\right) \qquad \text{(vi)}$$

In case, the applied loads of the beam increases, such that σ_{st1} may increase upto σ_{st} then, the actual value of compressive stress in bending in concrete σ_{cbc2} will exceed its permissible value, σ_{cbc} and the beam would be considered, as if, it had failed. The failure occurs by crushing of concrete in the compression zone (at the top surface in simply supported beam, where, it is not visible also). The failure of over-reinforced beam is always brittle in nature and it occurs suddenly, and of an almost explosive nature. It occurs without warning. For this reason, it is not desirable to have an over-reinforced beam.

Many times, more number of beams are used. The actual reinforcement in a beam is kept more than the theoretically calculated reinforcement in order to simplify the details, the bars in different beams are made alike. Sometimes, more concrete is used than needed for strength alone in order to have the thickness or general dimensions desired. In general, excessive variety of sizes are avoided, which used to increase formwork.

It is worthwhile to note that a balanced design of a beam may not be an economical design. The economy of any construction depends upon relative costs of concrete and steel at a place.

3.5 PERMISSIBLE STRESSES IN BENDING IN CONCRETE AND STEEL

Permissible stresses in compression in bending for various grades of concrete are adopted from Table 3.1 as specified in IS : 456-1978.

TABLE 3.1. PERMISSIBLE STRESSES IN BENDING IN CONCRETE (σ_{cbc}, N/mm²)

Grading of concrete	Bending stress σ_{cbc} (N/mm²)	Grade of concrete	Bending stress σ_{cbc} (N/mm²)
M 15	5.0	M 30	10.0
M 20	7.0	M 35	11.5
M 25	8.5	M 40	13.0

Permissible stresses (σ_{st} or σ_{sc}) in steel reinforcement in tension for mild steel bars conforming to Grade I of IS: 432 (Part I)-1966 or deformed mild steel bars conforming to IS: 1139-1966 are as follows

(i) upto and including 20 mm diameter bars, $\sigma_{st} = 140$ N/mm²

(ii) over 20 mm diameter bars $\sigma_{st} = 130$ N/mm²

Permissible stresses (σ_{st} or σ_{sv}) in steel reinforcement in tension for medium tensile steel conforming to IS : 432 (Part I)-1966 or deformed medium tensile steel bars conforming to IS : 1139-1966 is adopted as half the guaranteed yield stress subject to a maximum of 190 N/mm².

Permissible stresses (σ_{st} or σ_{sv}) in steel reinforcement in tension for high yield strength deformed bars conforming to IS : 1139-1966 or IS : 1786-1979 (Grade F_e 4.15) is 230 N/mm².

For high yield strength deformed bars of Grade F_e 500, the permissible stress in flexural tension is 0.55 f_y.

Example 3.1. A singly reinforced rectangular beam is 400 mm wide. The effective depth of beam section is 560 mm and its effective cover is 40 mm. A steel reinforcement consisting of 4 mild steel 18 mm diameter has been used in the beam section. The grade of concrete is M 15. Locate the neutral axis of the beam section.

Solution : The grade of concrete is M 15. From IS : 456-1978, the permissible stress in compression in bending in concrete σ_{cbc} is 5 N/mm².

Step 1 : Modular ratio of concrete

$$m = \left(\frac{280}{3 . \sigma_{cbc}}\right) = \left(\frac{280}{3 \times 5}\right) = 18.667 \quad \text{(i)}$$

Step 2 : Area of steel reinforcement

$$A_{st} = \left(4 \times \frac{\pi}{4} \times 18^2\right) = 1017.87 \text{ mm}^2 \quad \text{(ii)}$$

Step 3 : Neutral Axis for beam

A rectangular beam section is shown in Fig. 3.4. In order to locate its neutral axis, equate moment of equivalent areas of concrete and steel about neutral axis. From Eq. 3.2

$$(b . n)\ \frac{n}{2} = [(m.\ A_{st})\ .\ (d - n)] \quad \text{(iii)}$$

$$\left[400 \times n \times \frac{n}{2} + (18.667 \times 1017.87)n\right] - (18.667 \times 1017.87) \times 560 = 0$$

$$n = 187.9937 \text{ mm } \textbf{(say 187.994 mm)} \quad \text{(iv)}$$

Alternatively. Ratio of area of steel in tension to *bd*

$$p = \left(\frac{A_{st}}{b . d}\right) = \left(\frac{1017.87}{400 \times 560}\right) = 4.544 \times 10^{-3}$$

From Eq. 3.11 $k = [2mp + (mp)]^{1/2} - mp$

$k = [2 \times 18.667 \times 4.544 \times 10^{-3} + (18.667 \times 4.544 \times 10^{-3})] - 18.667 \times 4.544 \times 10^{-3}$

Therefore, k = 0.3357 and, kd = (0.3357 x 560) = 187.99 mm.

Example 3.2. In Example 3.1, the bending moment at a transverse section of a beam is 70 kN-m. Determine the compressive stress in bending in concrete, σ_{cbc} at the extreme fibre and tensile stress in steel, σ_{st}.

Solution : In example 3.1, the depth of neutral axis n = kd is 187.994 mm.

Step 1 : Design constant

$$\therefore \quad k = \left(\frac{187.994}{560}\right) = 0.3357 \quad \text{and} \quad j = \left(1 - \frac{k}{3}\right) = 0.881 \quad \text{(i)}$$

Step 2 : Stresses in materials of beam

Moment of resistance from compression side

M = Compression force x lever arm

$$M = \left[\left(\frac{1}{2}\sigma_{cbc}\right)(b.kd)\ .\ jd\right]$$

Step 3 : Stress in concrete in compression, σ_{cbc}

$$\sigma_{cbc} = \left(\frac{2M}{k.j.bd^2}\right) = \left(\frac{2\times70\times10^6}{0.3357\times0.881\times400\times560^2}\right) = 3.7435\ \text{N/mm}^2$$

Step 4 : Stress in steel in tension

Moment of resistance from tension side

$$M = \text{Tensile force x lever arm} = (\sigma_{st}.A_{st}).jd$$

$$\sigma_{st} = \left(\frac{70\times10^6}{1017.87\times0.8881\times560}\right) = 138.278\ \text{N/mm}^2$$

Alternatively, from bending stress distribution diagram Fig. 3.4 (B)

$$\left(\frac{(\sigma_{st}/m)}{(d-n)}\right) = \left(\frac{\sigma_{cbc}}{n}\right)$$

$$\sigma_{st} = m\,\sigma_{cbc}\ .\ \frac{(d-n)}{n} = \left(\frac{18.667\times3.7435\times(560-187.994)}{187.994}\right)$$

$$\sigma_{st} = 138.279\ \text{N/mm}^2$$

Example 3.3. In Example 3.1, the compressive stress in bending in concrete, σ_{cbc} at the extreme fibre in compression is 3.7435 N/mm² and tensile stress in bending in steel is 138.279 N/mm². Locate the neutral axis of the beam section. Also determine the moment of resistance (flexural strength) of the beam section.

Solution :

Step 1 : Design constants

From Eq. 3.4, the factor for depth of neutral axis

$$k = \left(\frac{m\,\sigma_{cbc}}{(m\,\sigma_{cbc}+\sigma_{st})}\right) = \left(\frac{18.667\times3.7435}{18.667\times3.7435+138.279}\right) = 0.3357 \quad \text{(i)}$$

Step 2 : Depth of neutral axis

$$n = kd = (0.3357 \text{ x } 560) = 187.992 \text{ mm}$$

$$j = \left(1-\frac{k}{3}\right) = \left(1-\frac{0.3357}{3}\right) = 0.8881 \quad \text{(ii)}$$

Step 3 : Moment of resistance of beam

In case, it is not known that the beam is under-reinforced or over-reinforced, then, the moment of resistance MR_1 or MR_2 both are found. The flexural strength of beam shall be MR_1 or MR_2, whichever is less. From, tension side, from Eq. 3.15

$$MR_1 = (\text{T. lever arm}) = (\sigma_{st} \cdot A_{st}\, jd)$$

or $$MR_1 = \left(\frac{140 \times 1017.87 \times 0.8881 \times 560}{1000 \times 1000}\right) = 70.87 \text{ kN - m}$$

From, compression side, from Eq. 3.16

$$MR_2 = (\text{C. lever arm}) = \left(\frac{1}{2}\sigma_{cbc} \cdot (b \cdot kd)\, jd\right)$$

or $$MR_2 = \left(\frac{\frac{1}{2} \times 5 \times (400 \times 0.3357 \times 560) \times 0.8881 \times 560}{1000 \times 1000}\right) = 93.495 \text{ kN - m} = 93.495 \text{ kN}$$

It is to note that σ_{cbc} and σ_{st} are the permissible values. Flexural strength of beam section is 70.87 kN-m.

Example 3.4. A singly reinforced rectangular beam is 400 mm wide. The effective depth of beam is 560 mm. Determine the percentage of steel and cross-sectional area of reinforcement for the balanced beam section. The grade of concrete is M 15. The grade of steel is (i) mild steel, (ii) hysd-steel. Keep diameter of steel bars equal to or less than 20 mm.

Solution : For M 15 grade of concrete, permissible stress in compression in bending in concrete from IS : 456-1978 $\sigma_{cbc} = 5 \text{ N/mm}^2$

Step 1 : Design constants

$$\therefore\ m = \left(\frac{280}{3 \times \sigma_{cbc}}\right) = \left(\frac{280}{3 \times 5}\right) = 18.667 \qquad \text{(i)}$$

(i) **For mild steel,** diameter of steel bars less than or equal to 20 mm, from IS : 456-1978

$$\sigma_{st} = 140 \text{ N/mm}^2 \qquad \text{(ii)}$$

For balanced section, k = $k = k_B$, $n = n_B$

$$k_B = \left(\frac{m\ \sigma_{cbc}}{(m\ \sigma_{cbc} + \sigma_{st})}\right) = \left(\frac{18.667 \times 5}{(18.667 \times 5 + 140)}\right) = 0.4 \qquad \text{(iii)}$$

$$n_B = k_B \cdot d = (0.4 \times 560) = 224 \text{ mm.}$$

Step 2 : Area of steel reinforcement

Equate moment of equivalent areas of concrete and steel about neutral axis

$$b.n_B \;.\; \frac{n_g}{2} = m\,.\,A_{st}\,(d - n_B)$$

$$\left(\frac{400 \times 224 \times 224}{2}\right) = [18.667 \times A_{st} \times (560 - 224)]$$

A_{st} = 1599.97 mm²

Step 3 : Percentage of steel

$$p = p_B = \left(\frac{A_{st}}{bd}\right) \times 100 = \left(\frac{1599.97 \times 100}{400 \times 560}\right) = 0.7143 \textit{ percent}$$

(ii) **For hysd-steel,** from IS:456-1978, σ_{st} = 230 N/mm² (iv)

Step 1 (A) : Design constants

For balanced beam section,

$k = k_B.\ n = n_B$

$$k_B = \left(\frac{m\,\sigma_{cbc}}{(m\sigma_{cbc} + \sigma_{st})}\right) = \left(\frac{18.667 \times 5}{(18.667 \times 5 + 230)}\right) = 0.28866$$

$n_B = k_B.d = 0.28866 \times 560 = 161.65$ mm

Step 2 (A) : Area of steel reinforcement

Equate moment of equivalent areas of concrete and steel about neutral axis

$$\left(b\,.\,n_B\,.\frac{n_B}{2}\right) = m\,.\,A_{st}\,(d - n_B)$$

$$\left(\frac{400 \times 161.65 \times 165.65}{2}\right) = [18.667 \times A_{st}\ (560 - 161.65)]$$

A_{st} = 702.82 mm²

Step 3 (A) : Percentage of steel

$$p = \left(\frac{A_{st}}{b.d}\right) \times 100 = \left(\frac{702.82 \times 1000}{400 \times 560}\right) = 0.3137 \textit{ percent.}$$

Example 3.5. In example 3.4, determine the moment of resistance (flexural strength) of the beam section when (i) mild steel reinforcement is used, (ii) hysd steel bars are used.

Solution : The moment of resistance (flexural strength) of beam section is found as follows :

Step 1 : Design constants

(i) *When wild steel reinforcing bars less than or equal to 20 mm diameter are used.* From Example 3.4, design constants are as below:

$$\sigma_{cbc} = 5\ N/mm^2, \sigma_{st} = 140\ N/mm^2, k_B = 0.4$$

$$j_B = \left(1 - \frac{k_B}{3}\right) = \left(1 - \frac{0.40}{3}\right) = 0.8667$$

$$R_B = \left(\frac{1}{2}\sigma_{cbc} \cdot k_B \cdot j_B\right) = \left(\frac{1}{2} \times 5 \times 0.4 \times 0.8667\right) = 0.8667$$

Step 2 : Moment of resistance (flexural strength) of beam

From compressive side, From Eq. 3.16

$$MR_2 = C \cdot lever\ arm = \left(\frac{1}{2}\sigma_{cbc} \cdot k_B \cdot j_B\right) \cdot bd^2$$

or $$MR_2 = \left(R_B \cdot bd^2\right) = \left(\frac{0.8667 \times 400 \times 560^2}{1000 \times 1000}\right) = 108.719\ kN-m$$

Moment of resistance (flexural strength) from tension side, from Eq. 3.15

$$MR_1 = T.\ (lever\ arm) = (\sigma_{st}\ A_{st}) \cdot jd$$

or $$MR_1 = \left(\frac{140 \times 1599.97 \times 0.8667 \times 560}{1000 \times 1000}\right) = 108.717\ kN-m$$

It is to note that for the balanced beam section, the flexural strength of beam found from compression side and that obtained from the tension side shall be equal.

Step 1 (A) : Design constants

(ii) When hysd-steel bars are used, from Example 3.4, design constants are as below :

$$\sigma_{cbc} = 5\ N/mm^2, \sigma_{st} = 230\ N/mm^2, k_B = 0.28866$$

$$j_B = \left(1 - \frac{k_B}{3}\right) = \left(1 - \frac{0.28866}{3}\right) = 0.90378$$

$$R_B = \left(\frac{1}{2}\sigma_{cbc} \cdot k_B \cdot j_B\right) = \left(\frac{1}{2} \times 5 \times 0.28866 \times 0.90378\right) = 0.6522$$

Step 2 (A): Moment of resistance (flexural strength) of beam

From compressive side, from Eq. 3.16

$$MR_2 = (C\,.\,lever\ arm) = \left(\frac{1}{2}\sigma_{cbc}\,.\,k_B\,.\,j_B\right).bd^2$$

or $$MR_2 = (R_B\,.\,bd^2) = \left(\frac{0.6552 \times 400 \times 560^2}{1000 \times 1000}\right) = 81.812\ kN - m$$

Moment of resistance (flexural strength) from tension side, from Eq. 3.15

$$MR_1 = T.\,(lever\ arm) = (\sigma_{st}\,.\,A_{st}).\,j_d$$

or $$MR_1 = \left(\frac{230 \times 702.82 \times 0.90378 \times 560}{1000 \times 1000}\right) = 81.813\ kN - m$$

It is to note that the flexural strength for balanced beam section found compression side and that obtained from the tension side shall be equal.

Example 3.6. A singly reinforced rectangular beam is 400 mm wide. The effective depth of beam is 560 mm and its effective cover is 40 mm. The steel reinforcement in the beam section consists of 4 mild steel bars, 18 mm diameter. The grade of concrete is M 15. Determine whether the beam section is under- reinforced or over-reinforced. Also determine the flexural strength of the beam section.

Solution : For M 15 grade of concrete, σ_{cbc}, permissible compressive stress in bending in concrete from IS : 456-1978 is 5 N/mm². For mild steel bars less than or equal to 20 mm diameter, σ_{st} = 140 N/mm².

Step 1 : Design constants

Modular ratio of M 15 grade of concrete

$$m = \left(\frac{280}{3\,\sigma_{cbc}}\right) = \left(\frac{280}{3 \times 5}\right) = 18.667.$$

For balanced beam section, k= kB, n = nB,

$$k_B = \left(\frac{m\,\sigma_{cbc}}{(m\sigma_{cbc} + \sigma_{st})}\right) = \left(\frac{18.667 \times 5}{(18.667 \times 5 + 140)}\right) = 0.4 \qquad \text{(i)}$$

$$n_B = k_B\,.\,d = (0.40 \times 560) = 224\ mm \qquad \text{(ii)}$$

Step 2 : Area of steel

$$A_{st} = \left(4 \times \frac{\pi}{4} \times 18^2\right) = 1017.87\ mm^2$$

In order to locate the actual neutral axis, equate moment of equivalent areas concrete and steel about neutral axis. From Eq. 3.2

$$\left(b\,.\,n\,.\,\frac{n}{2}\right) = m\,A_{st}\,(d-n)$$

or $$\left(400 \times n \times \frac{n}{2}\right) = [18.667 \times 1017.87\,(560-n)]$$

$$n = 187.9937 \text{ mm},\ k = \left(\frac{187.9937}{560}\right) = 0.3357,\ j = \left(1 - \frac{k}{3}\right) = 0.8881$$

Since, n is less than n_B, the beam section is under-reinforced. Actual neutral axis is above the neutral axis for balanced section as shown in Fig. 3.5.

Step 3 : Moment of resistance of beam

For under-reinforced beam, Fig. 3.5, stress in steel attains its permissible stress value earlier than concrete. The flexural strength (moment of resistance) of beam section is governed by tension in steel. From Eq. 3.15

$$MR_1 = T.\,(lever\ arm) = (\sigma_{st}\,A_{st}).\,jd$$

or $$MR_1 = \left(\frac{140 \times 1017.87 \times 0.8881 \times 560}{1000 \times 1000}\right) = 70.87 \text{ kN-m}$$

Whether a beam section is under reinforced or over reinforced. It may also be confirmed by comparing the areas of steel (or percentage of steel) for balance section and actual beam section. For balanced section, area of steel and percentage area of steel, from Example 3.4 (i)

$$(A_{st})_{balanced} = 1599.97\ mm^2\ and\ p_B = 0.7143 \text{ percent}$$

For actual beam section A_{st} = 1017.87 mm²,

and $$p = \left(\frac{1017.87 \times 100}{400 \times 560}\right) = 0.4544\ per\ cent$$

p is less than p_B. Hence, the beam is under-reinforced.

Example 3.7. A singly reinforced rectangular beam is 400 mm wide. The effective depth of beam is 560 mm and its effective cover is mm. The steel reinforcement in the beam section consists of 4 hysd steel bar, 18 mm diameter. The grade of concrete is M 15. Determine whether the beam section is under reinforced or over-reinforced. Also determine the flexural strength of the beam section.

Solution :

For M 15 grade of concrete, permissible compressive stress in bending in concrete, σ_{cbc} from IS : 456-1978 is 5 N/mm². For hysd-steel, σ_{st} is 230 N/mm².

Step 1 : Design constants

Modular ratio of concrete

$$m = \left(\frac{280}{m \times \sigma_{cbc}}\right) = \left(\frac{280}{3 \times 5}\right) = 18.667 \qquad \text{(i)}$$

For balanced beam section, $k = k_B$, $n = n_B$

$$k_B = \left(\frac{m\ \sigma_{cbc}}{m\ \sigma_{cbc} + \sigma_{st}}\right) = \left(\frac{18.667 \times 5}{18.667 \times 5 + 230}\right) = 0.28866 \qquad \text{(ii)}$$

$$j_B = \left(1 - \frac{k_B}{3}\right) = \left(1 - \frac{0.28866}{3}\right) = 0.90378 \qquad \text{(iii)}$$

$$R_B = \left(\frac{1}{2}\sigma_{cbc}\ .\ j_B\ .\ k_B\right) = \left(\frac{1}{2} \times 5 \times 0.90378 \times 0.28866\right) = 0.6522$$

$n_B = k_B.d = (0.28866 \times 560) = 161.65$ mm

Step 2 : Area of hysd steel reinforcement in tension

$$A_{st} = \left(4 \times \frac{\pi}{4}\right) \times 18^2 = 1017.87\ mm^2$$

Step 3 : Depth of neutral axis

In order to locate the actual neutral axis, equate moment of equivalent areas of concrete and steel about neutral axis

$$\left(b\ .\ n\ .\ \frac{n}{2}\right) = m\ A_{st}\ (d - n)$$

$$\left(400 \times n \times \frac{n}{2}\right) = [18.667 \times 1017.87\ (560 - n)]$$

$n = 187.9937$ mm (iv)

Since, n is greater than n_B, the beam section is over- reinforced. Actual neutral axis is below the balanced neutral axis as shown in Fig. 3.6. For over-reinforced beam, Fig. 3.6, the compressive stress in bending in concrete attains its permissible value earlier than the steel.

Step 4 : Moment of resistance of beam

The flexural strength (moment of resistance of beam is controlled by compression in concrete). From Eq. 3.16

$$MR_2 = C\ .\ lever\ arm = \left(\frac{1}{2}\sigma_{cbc}\ .\ k_B\ .\ j_B\right).\ bd^2$$

or $$MR_2 = \left(R_B \,.\, bd^2\right) = \left(\frac{0.6552 \times 400 \times 560^2}{1000 \times 1000}\right) = 81.812\ kN - m$$

Step 5 : Area of steel reinforcement

Whether a beam section is under reinforced or over reinforced, it may also be seen by comparing the areas of steel (or percentage of steel) for balanced section and actual beam section. For balanced section, area of steel and percentage area of steel, from Example 3.4(ii)

$$\left(A_{st}\right)_{balanced} = 702.82\ mm^2\ and\ p_B = 0.3137\ percent$$

For actual beam section, A_{st} = 1017.87 mm²

and $$p = \left(\frac{1017.87}{400 \times 500}\right) \times 100 = 0.4544\ per\ cent$$

p is greater than p_B. Hence, the beam is over-reinforced.

3.6 DESIGN OF SINGLY REINFORCED BEAM SECTIONS

The design of singly reinforced beam deals with the selection of proper grade of concrete mix and proper quality of steel, proper shape, sizes and proportions of beams. M15 and M 20 grades of concrete mix are generally used for beams. Mild steel or hysd steel bars are used as reinforcement. After deciding the grade of concrete and type of steel reinforcement, the various design constants, m, k_B, j_B and R_B for the balanced beam section are calculated. The reinforced concrete beam sections are commonly made rectangular in shape, though other shapes are also used. For the design of beams, the dead and live loads to be carried by the beams and the clear span of the beams are usually known. Following are the usual steps in the design of singly reinforced beam sections.

Step 1. For determining the maximum bending moment and maximum shear force occurring at any sections, the effective span is needed.

Effective span

(i) For simply supported beams, the effective span of a beam that is not constructed integrally with its supports is adopted as below :

Effective span = (Clear span + Effective depth)

ES = (C . S + d) (3.19)

Effective span = (Centre to centre of supports) (3.20)

Before the design of beam is done, the size of beam is not known. In the beginning, the effective depth, *d* is assumed depending upon intensity of loads and length of spans. The effective depth of beam is assumed as one-tenth (1/10)th of span for beams to carry light loads and beams of moderate span and one-eighth (1/8)th of span for beams to carry heavy loads and beams of long span. In order to find the effective span from Eq. 3.20, the widths of supports at the two ends of the beam are assumed as 300 mm to 600 mm. The effective span for simply supported beam shall be minimum of the values calculated from Eqs. 3.19 and 3.20.

(ii) For cantilever beams, the effective span is taken as the length from free end of the cantilever to the face of support (viz., clear projection) plus half its effective depth. That is,

$$ES = \left(\textit{Clear projection} + \frac{1}{2} d \right) \tag{3.21}$$

Step 2. For calculating the maximum bending moment and shear force, it is also necessary to know the types and magnitudes of loads to be supported by the beams.

Loads. The dead loads of materials attached and supported by the beams are found by knowing the unit weight of materials used and their quantities. The self-weight of beam is calculated by estimating the size (viz., proportions of width and depth) of beam.

Proportions of beam. Reinforced concrete beams may be wide and shallow, or relatively narrow and deep. Consideration of maximum materials economy often leads to proportions with width of beam in the range from about half to two-third of the effective depth. The effective depth of beam, d is estimated as per the guide-lines described above along with effective span. The width of beam, b may be assumed as half of the effective depth in the beginning. However, constraints (if any) may dictate other choices. In building construction, it is common to use group of beams of equal widths, b and equal depths, d, which will also permit the use of a single flat bottom form, resulting in fact, economical construction and allowing levelled ceilings. The width and depth of reinforced concrete beams must be sufficient to be permissible stress in concrete in bending compression. After fixing the width of beam, b from the estimated depth, d, the self-weight of beam may be found. Exact depth necessary to resist the resulting maximum bending moment is found in the subsequent steps.

The live loads are calculated by noting the corresponding intensity of live loads per square metre from IS : 875-1984 for the type and use of floor (e.g., residential, class room, restaurant or workshop). Thus, the total loads to be carried by the beam is found.

Step 3. Required effective depth of beam is found after calculating the maximum bending moment, M resulting due to the loads and effective span of beam. *The beam section is designed as balanced section.* As such, the bending moment due to external loads, M is equated to moment of resistance of the beam section, calculated from compressive force in concrete. Therefore,

$$M = R \,.\, bd^2 \tag{i}$$

$$d = \left(\frac{M}{R\ b} \right)^{1/2} \tag{3.22}$$

where R is the factor of moment of resistance for the balanced beam section.

Step 4. In order to provide adequate protection to steel against fire and corrosion due to harmful elements present in the atmosphere, a certain thickness of concrete cover is provided outside of the outermost surface of the steel reinforcement bars. Thus, there remains concrete all round the steel bars.

Cover to reinforcement. The cover to reinforcement at each end along the length of the beam shall not be less than 25 mm, nor less than twice the diameter of such bar. The cover to reinforcement measured from outermost surface of main reinforcement (Also referred as clear cover) as shown in Fig. 3.7(A) upto surface of concrete in tension, shall be not less than 25 mm or less than the diameter of such bar. When the concrete surface remains exposed directly to the ground or weather, the harmful elements (e.g., some chemicals, acids, vapour, salts, sulphrous smoke, etc.) are likely to damage the concrete, the clear cover recommended above is further increased between 15 mm and 50 mm. The reinforced concrete beams may remain totally immersed in sea water. In such a case, the clear cover shall be 50 mm more than specified above. In case, the reinforced concrete beam remains periodically immersed in sea water, the clear cover shall be 50 mm more than the specified above. For concrete of grade M 25 and above, the additional thickness of cover specified above is made half. However, in all such cases, the concrete cover shall not exceed 75 mm.

The cover to reinforcement measured form centre of main reinforcement upto surface of concrete in tension as shown in Fig. 3.7(A) is called effective cover. The side cover to reinforcement measured from outermost face of reinforcement bar upto side face as shown in Fig. 3.7 (B) shall be not less than **25 mm nor less than the diameter of bar.**

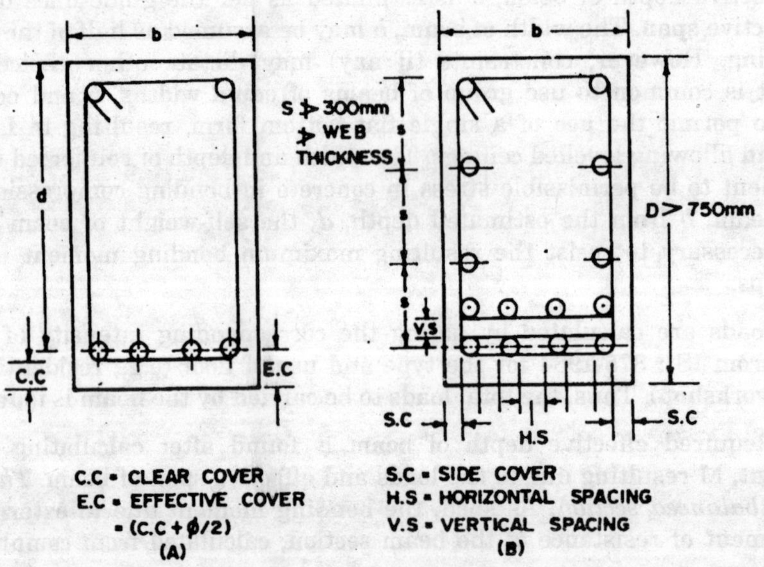

Fig. 3.7.

The overall depth of beam, D is found by adding effective cover (say, 40 mm + $\phi/2$ = 50 mm) to the effective depth, d calculated from Eq. 3.22 and then rounding upward to the nearest 5 mm, 10 mm or 20 mm. As a result, the actual effective depth, d is obtained by subtracting effective cover from overall depth of beam, D. As actual effective depth of beam becomes more than that found from Eq. 3.22, then design of beam becomes that for under-reinforced section for beam.

Step 5. Area of steel reinforcement in tension, A_{st} is calculated from Eq. 3.23 below :

$$A_{st} = \left(\frac{M}{\sigma_{st} \cdot j_d}\right) \qquad (3.23)$$

where σ_{st} is the permissible stress in steel in tension and *jd* is the lever arm and *d* is the actual effective depth of the beam section. Since, actual effective depth of beam has been adopted more than that calculated by Eq. 3.22, this area of steel, A_{st} found from Eq. 3.22 is definitely less than that if it would have been determined by using calculated effective depth. As such, the beam section being designed corresponding to design of under reinforced beam (which one is desirable). Area of steel provided is kept slightly more than that calculated from Eq. 3.23. 16 mm, 18 mm, 20 mm and 22 mm diameter bars are common used and arranged symmetrically about the vertical centre line of the beam cross-section. The bars of diameter more than 25 mm may also be used if necessary. The area of steel provided should be more than the minimum and less than the maximum limits specified in IS: 456-1978 (described below):

Following are the **requirements of main longitudinal (tensile) reinforcement.**

1. Minimum reinforcement in tension

The minimum reinforcement in tension shall not be less than given by Eq. 3.24 below:

$$\left(\frac{(A_{st})_{.min}}{b \, . \, d}\right) = \left(\frac{0.85}{f_y}\right) \qquad (3.24)$$

where, f_y is the characteristic (yield) strength of reinforcement in N/mm². For mild steel f_y is 250 N/mm² and for hysd steel Fe 415 and Fe 500, the values of f_y are 415 N/mm² and 500 N/mm², respectively.

For mild steel

$$(A_{st})_{.min} = \left(\frac{0.85}{250} \times bd\right) = (0.34 \text{ percent})$$

For hysd steel Fe 415

$$(A_{st})_{.min} = \left(\frac{0.85}{415} \times bd\right) = (0.2048 \text{ per cent})$$

For hysd steel Fe 500

$$(A_{st})_{min} = \left(\frac{0.85}{500} \times bd\right) = (0.17 \text{ per cent})$$

Sometimes for architectural or other purposes, the cross-sections of beam are kept larger than those necessary for the considerations of strength alone. The area of steel calculated for such beams for the requirement of strength will be very small. The strength of such beam by considering the beam section as cracked becomes even less than that for corresponding plain concrete section determined from its modulus of rupture. In such beams, the failure takes place suddenly. In order to avoid such eventuality, it is necessary to provide area of steel more than minimum.

2. Maximum reinforcement in tension. The maximum area of reinforcement in tension shall not exceed *0.04. bD* (viz. 4% of the cross-section, so that practical difficulties in placing and compacting of concrete may not arise. In the singly reinforced beams, it may not become necessary to use 4 percent of reinforcement.

3. Side face reinforcement. When the overall depth of beam becomes more than 750 mm, the side face reinforcement shall be provided along the two faces of the beam section as shown in Fig. 3.7(B) to take into consideration of crack width limitation and lateral buckling of the web in beam. The side face reinforcement shall not be less than 0.1% of the web area and shall be distributed equally in two faces at a spacing, S not to exceed 300 mm or web thickness whichever is less.

4. Spacing of reinforcement. In order to ensure proper placement of concrete around the main reinforcement bars and to achieve full surface contact between the bars and the concrete (*viz.*, to get maximum bond strength), it is necessary to maintain a certain minimum distance between adjacent bars. The air pocket likely to be left below the bars are to be avoided.

5. Minimum distance between individual bars. The clear horizontal distance between two adjacent parallel main reinforcement bars as shown in Fig. 3.7 (B) shall be not less than maximum of the following :

(i) The diameter of the bar (in case the diameters are equal)

(ii) The diameter of larger bar (in case the diameters are unequal), and

(iii) 5 mm more than the nominal maximum size of coarse aggregate.

The nominal maximum size of the aggregate is usually kept 5 mm less than the minimum cover to the reinforcement (viz., 25 mm minus 5 mm). As such aggregates having a nominal size of 20 mm are generally considered satisfactory.

Wherever possible, the horizontal distance between adjacent bars, Fig. 3.7(B) is provided more than the minimum specified above. In case needle vibrators are used for compaction of concrete, this horizontal distance between group of bars may be reduced to two-third the nominal maximum size of the coarse aggregate provided that the sufficient space is left between the groups of bars to enable the vibrator to be immersed.

Sometimes, the bars of main reinforcement are placed in two or more layers as shown in Fig. 3.7(B) when the bars cannot be accommodated in one layer. In such a case, the bars are arranged vertically in line. The minimum clear vertical distance between the bars as shown in Fig. 3.7(B) shall be not less than the maximum of the following:

(i) 15 mm

(ii) Two thirds the maximum size of aggregate

(iii) The maximum diameter of main bar.

6. Maximum distance between main bars in tension. When the beams are exposed to mild or moderate environment, the surface width of cracks should not, in general, exceed 0.3 mm, the maximum horizontal distance between adjacent bars for the exposed to normal internal or external conditions, shall not exceed beams that specified in IS : 456-1978 (described as under).

The horizontal distance between two adjacent parallel reinforcement bars, or group of bars, near the tension face of a beam shall not exceed 300 mm, 180 mm, or 150 mm for the characteristic strength of steel as 250 N/mm^2, 415 N/mm^2 or 500 N/mm^2 respectively, where the percentage of redistribution to or from section is zero.

In addition to the design of beams for bending strength and main reinforcement, the shear strength, transverse reinforcement and development length considerations (discussed later) are also made.

Example 3.8. The floor of a class room 5 m × 9 m consisting of 120 mm thick reinforced concrete slab is supported by singly reinforced rectangular beams. Design the beams. Adopt the permissible stresses as recommended in IS : 456-1978.

Design : The beams are placed over width of room at 3 m centre to centre distance as shown in Fig. 3.8. The clear span, CS of beam is 5 m. The thickness of supporting wall is assumed 600 mm. Overall depth of beam may be estimated as one-eighth to one-tenth of clear span, (viz., D = 620 mm). The effective depth, d may be taken as 580 mm. The beams are simply supported over the supports.

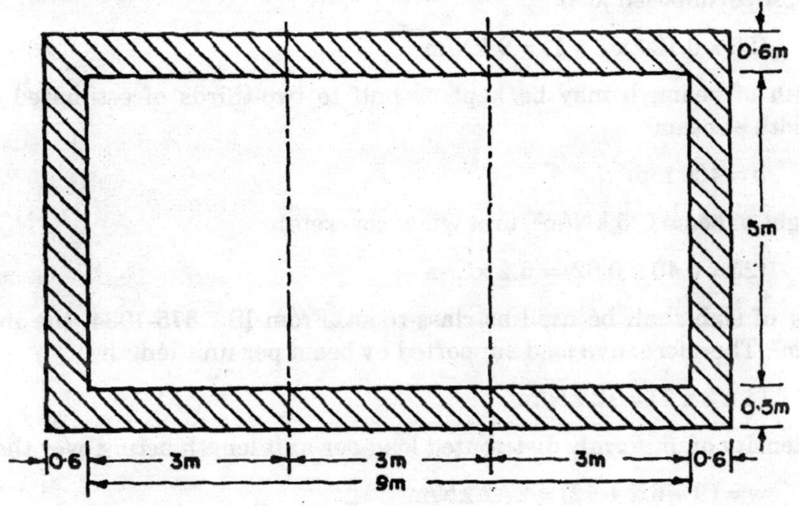

Fig. 3.8

Step 1 : Effective span of beams

The effective span of beams, ES shall be as follows :

$$ES = \text{centre to centre of supports} = (5 + 0.3 + 0.3) = 5.60 \text{ m} \quad \text{(i)}$$

$$ES = (CS + d) = (5 + 0.58) = 5.58 \text{ m} \quad \text{(ii)}$$

The effective span is whichever is less from the expression (i) and (ii). Hence, the effective span of beam, L is 5.58 m.

Step 2 : Design constants. M 15 grade of concrete mix and mild steel reinforcement bars of diameter less than or equal to 20 mm shall be used. From IS: 456-2000, the permissible stresses are σ_{cbc} = 5 N/mm^2 and σ_{st} = 140 N/mm^2. The modular ratio of concrete

$$m = \left(\frac{280}{3 \times \sigma_{cbc}}\right) = \left(\frac{280}{3 \times 5}\right) = 18.67 \quad \text{(iii)}$$

Various other design constants are as below :

$$k_B = \left(\frac{m\,\sigma_{cbc}}{m\,\sigma_{cdc} + \sigma_{st}}\right) = \left(\frac{18.67 \times 5}{18.67 \times 5 + 140}\right) = 0.4 \quad \text{(iv)}$$

$$j_B = (1 - k_B/3) = (1 - 0.4/3) = 0.867 \quad \text{(v)}$$

$$R_B = \left(\frac{1}{2}\sigma_{cbc} \cdot k_B \cdot j_B\right)$$

$$R_B = \left(\frac{1}{2} \times 5 \times 0.4 \times 0.867\right) = 0.867 \quad \text{(vi)}$$

Step 3 : Total load coming over each beam per unit length. From reinforced concrete slab, superimposed load

$$(25 \text{ x } 0.120 \text{ x } 3 \text{ x } 1) = 9 \text{ kN/m} \quad \text{(vii)}$$

The width of beam, b may be kept as half to two-thirds of estimated overall depth. Therefore, width of beam

$$b = 400 \text{ mm} \quad \text{(viii)}$$

Self-weight of beam (25 kN/m³, unit wt. of concrete)

$$(25 \text{ x } 0.40 \text{ x } 0.62) = 6.2 \text{ kN/m} \quad \text{(ix)}$$

The floor of slab shall be used as class-room. From IS : 875-1984, the intensity of live load is 4 kN/m². Therefore, live load supported by beam per unit length

$$(4 \text{ x } 3 \text{ x } 1) = 12 \text{ kN/m} \quad \text{(x)}$$

Total intensity of uniformly distributed load per unit length acting over the beam

$$w = (9 + 6.2 + 12) = 27.2 \text{ kN/m} \quad \text{(xi)}$$

Step 4 : Maximum bending moment. The beams act as simply supported at the ends. The maximum bending moment, M occurs at the centre of span

$$M = \left(\frac{w \cdot L^2}{8}\right) = \left(\frac{27.2 \times 5.58^2}{8}\right) = 105.864 \text{ kN - m}$$

Step 5 : Effective depth. The section of beam is designed as balanced section. The effective depth, *d* necessary for the beam

$$d = \left(\frac{M}{R_B\, b}\right)^{1/2} = \left(\frac{105.864 \times 10^6}{0.867 \times 400}\right)^{1/2} = 552.50 \text{ mm}$$

Overall depth,

D = (d + effective cover) = (552.50 + 50)

= 602.5 mm say 620 mm.

Actual effective depth

d = (620 – 50) = 570 mm

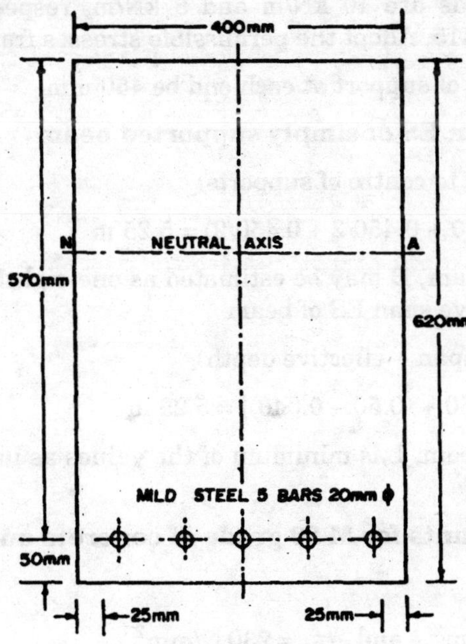

Fig. 3.9.

Step 6 : Necessary area of steel in tension

$$A_{st} = \left(\frac{M}{\sigma_{st} \times jd}\right) = \left(\frac{105.86 \times 10^6}{140 \times 0.867 \times 570}\right) = 1530.12 \text{ mm}^2$$

$(A_{st})_{min}$ necessary

$$(A_{st})_{min} = \left(\frac{0.85 \times 400 \times 570}{250}\right) = 775.2 \text{ mm}^2 < 1530.12 \text{ mm}^2$$

Provide 5 bars of 20 mm diameter. Actual area of steel provided

A_{st} = [5 x (π/4) x 20²] = 1570.8 mm²

Actual horizontal spacing between two adjacent main reinforcement bars

(400 – 5 x 20 – 2 x 25)/4 = 62.5 mm

It is more than the diameter of bar and (5 mm + nominal size of aggregate, 20 mm). Five bars of 20 mm diameter mild steel are provided as main tension reinforcement and bars are symmetrically placed about the central vertical line of the beam cross-section as shown in Fig. 3.9.

The design of beams for transverse shear reinforcement to resist shear force and for development length for steel bars has been done in Example 3.11. The details of complete design is shown in Fig. 3.35.

Example 3.9. Design a singly reinforced rectangular beam simply supported at its two ends. The clear span of beam is 4.80 m. The intensities of uniformly distributed super-imposed dead and live loads are 10 kN/m and 8 kN/m, respectively. Use M 20 grade of concrete and hysd steel Fe 415. Adopt the permissible stresses from IS : 456-2000.

Design : Let the width of support at each end be 450 mm.

Step 1 : Effective span ES of simply supported beam

ES = (centre to centre of supports)

or ES = (4.80 + 0.450/2 + 0.450/2) = 5.25 m (i)

The overall depth of beam, D may be estimated as one-eighth to one-tenth of clear span. ($\therefore$ D = 540 mm). The effective span ES of beam

ES = (clear span + effective depth)

or ES = [4.80 + (0.50 – 0.040)] = 5.29 m (ii)

The effective span of beam, L is minimum of the values as in the expressions (i) and (ii). (As such, L = 5.25 m).

Step 2 : Design constants for M 20 grade of concrete and hysd steel Fe 415

From IS : 456-1978

$$\sigma_{cbc} = 7\ \text{N/mm}^2, \text{ and } \sigma_{st} = 230\ \text{N/mm}^2$$

Modular ratio for the concrete of grade M 20

$$m = \left(\frac{280}{3 \times \sigma_{cbc}}\right) = \left(\frac{280}{3 \times 7}\right) = 13.33$$

Depth factor for the neutral axis

$$k_B = \left(\frac{m\,\sigma_{cbc}}{(m\,\sigma_{cbc} + \sigma_{st})}\right) = \left(\frac{13.33 \times 7}{13.33 \times 7 + 230}\right) = 0.2886$$

Lever arm factor for the beam

$$j_B = (1 - k_B/3) = (1 - 0.2886/3) = 0.9038$$

Factor for the moment of resistance

$$R_B = \left(\frac{1}{2} \times \sigma_{cbc} \times k_B \times j_B\right) = \left(\frac{1}{2} \times 7 \times 0.2886 \times 0.9038\right) = 0.9129$$

Step 3 : Total load coming over the beam

Superimposed load = 10 kN/m (iii)

The width of beam, b may be kept as half to two-thirds of the overall depth D

b = 360 mm

Self-weight of the beam

(25 x 0.36 x 0.540 x 1) = 4.86 kN/m **(iv)**

Live load = 8 kN/m **(v)**

Total intensity of uniformly distributed load acting over the beam

Total load, w = (10 + 4.86 + 8) = 22.86 kN/m **(vi)**

Step 4. Maximum bending moment, M for simply supported beam carrying uniformly distributed load maximum bending moment occurs at the centre.

$$M = \left(\frac{wL^2}{8}\right) = \left(\frac{22.86 \times 5.25 \times 5.25}{8}\right) = 78.76 \text{ kN-m} \quad \text{(vii)}$$

Step 5. Effective depth. The section of beam is designed as balanced section. The effective depth, d necessary for the beam

$$d = \left(\frac{M}{R_B \times b}\right)^{1/2} = \left(\frac{78.76 \times 10^6}{0.9129 \times 360}\right)^{1/2} = 489.54 \text{ mm} \quad \text{(viii)}$$

Overall depth,

D = (d + effective cover)

D = (489.54 + 50) = 539.54 mm, say D = 540 mm **(ix)**

Actual effective depth

d = (550 – 50) = 500 mm **(x)**

Step 6 : Necessary area of steel in tension

$$A_{st} = \left(\frac{M}{\sigma_{st} \cdot j_{Bd}}\right) = \left(\frac{78.76 \times 10^4}{230 \times 0.9038 \times 500}\right) = 757.76 \text{ mm}^2$$

$(A_{st})_{min}$ necessary

$$(A_{st})_{min} = \left(\frac{0.85 \times 369 \times 500}{415}\right) = 368.675 \text{ mm}^2$$

Necessary area of steel in tension calculated is more than minimum area of steel. Hence, it is satisfactory.

Provide 4 bars of 16 mm diameter. Actual area of steel provided.

$$A_{st} = \left(4 \times (\pi/4) \times 16^2\right) = 804.25 \text{ mm}^2$$

Actual horizontal spacing between adjacent main reinforcement bars

(360 – 4 x 16 – 2 x 25)/3 = 82 mm

it is more than the diameter of bar and (5 mm + nominal size of aggregate, 20 mm).

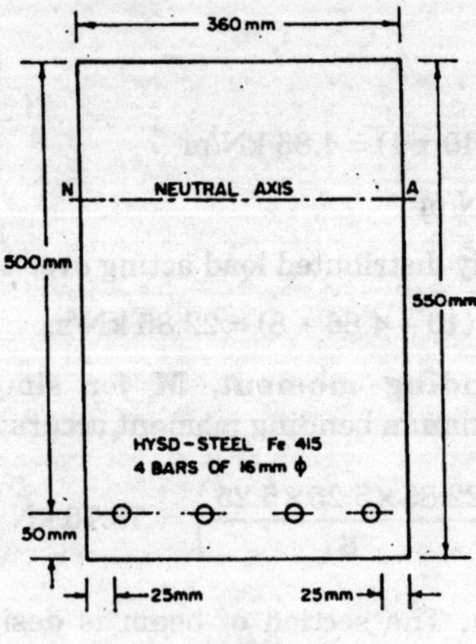

Fig. 3.10.

Four bars of 16 mm diameter hysd steel Fe 415 are provided as main tension reinforcement and bars are placed symmetrical about the central line of the beam cross-section as shown in Fig. 3.11.

The design of beam for transverse reinforcement to resist shear force and development length for steel bars has been done in Example 3.12. The details of complete design is shown in Fig. 3.11.

3.7 BENDING OF SINGLY REINFORCED BEAMS OF VARYING DEPTH

Many times, the hauches are provided with beams at the supports and such portions of beams have varying depth (viz either one or both the faces of beam remain inclined with the horizontal as shown in Fig. 3.11 (A), in which the faces of beams are not perpendicular to the section of bending moments). The retaining walls, footings and dams using reinforced cement concrete are the examples of such beams. These beams may be referred as wedge shaped beams. Professor William Cain suggested the following formulae in his studies related to earth pressure, retaining walls and bins.

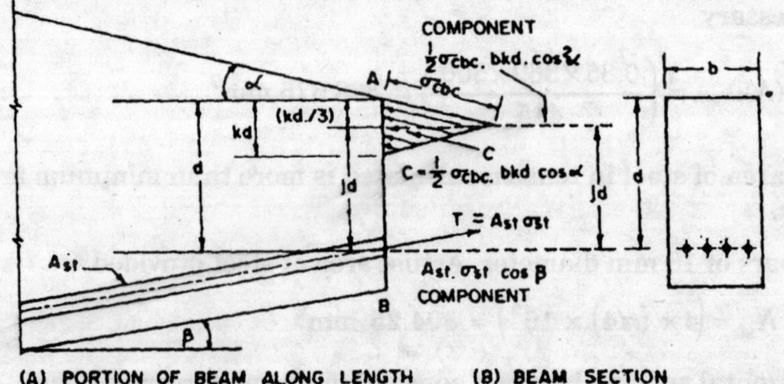

Fig. 3.11.

The stress developed due to bending of beams carrying loads shall not be at right angles to the section AB, Fig. 3.11 (A). In concrete, the stresses are parallel to the compression face while, in steel, the stresses act in the direction of reinforcement provided parallel to tension face. In bending stress distribution diagram, the depth of neutral axis measured perpendicular to compression face $0_1 D = (kd.\cos\alpha)$ upto centre of steel reinforcement (in the direction perpendicular to tension face from O_1) O_1E = (d – kd) . cos β. Various expressions for the following are as below : When the angles α and β exceed 10°, then, these expressions are used.

1. Factor, *k* for the depth of neutral axis

By equating the moment of area of concrete and equivalent area of steel about neutral axis

$$b.(kd\cos\alpha)\frac{(kd\ \cos\alpha)}{2} = m.\ A_{st}\ .\ (d-kd)\ \cos\beta \qquad (i)$$

$$k=\left(\frac{\cos\beta}{\cos^2\ \alpha}\right)\left[-mp+\left\{(2mp\cos^2\alpha/\cos\beta)+(mp)^2\right\}^{1/2}\right] \qquad (3.23A)$$

where p = (A_{st}/bd)

2. Lever arm factor, *j*

$$j = (1 - k/3) \qquad (3.24A)$$

3. Ratio of area of steel to beam section, *p*

Total compression in concrete, C acting at the vertical depth equal to kd/3 parallel to the compression face

$$C = \left(\frac{1}{2}\sigma_{cbc}\ .\ b\right).(kd\ \cos\ \alpha) \qquad (ii)$$

Horizontal component of the compressive force C

$$C_1 = (C.\cos\alpha) = \left(\frac{1}{2}\sigma_{cbc}\ b\right).\ (kd\cos^2\alpha) \qquad (iii)$$

Total tensile force in steel reinforce parallel to the tension face

$$T = (A_{st}\ .\ \sigma_{st}) \qquad (iv)$$

Horizontal component of the tensile force, T

$$T_1 = T.\cos\beta\ =\ (A_{st}\ .\ \sigma_{st})\ .\cos\beta \qquad (v)$$

Two horizontal components C_1 and T_1 are equal, since the beam section is in equilibrium. Therefore,

$$\left(\frac{1}{2}\sigma_{cbc} \cdot b \cdot kd \cos^2 \alpha\right) = \left(\sigma_{st} A_{st} \cos \beta_1\right)$$

or $$p = \left(\frac{1}{2}\frac{\sigma_{cbc}}{\sigma_{st}} \cdot \frac{k \cdot \cos^2 \alpha}{\cos \beta}\right) = \left(\frac{A_s}{b \cdot d}\right) \tag{3.25A}$$

4. Stress in reinforcement in tension

When the bending moment at the section of beam is M, then

$$\sigma_{st} = \left(\frac{M}{A_{st} \cdot jd \cdot \cos \beta}\right) \tag{3.26A}$$

5. Stress in concrete

$$\sigma_{cbc} = \left(\frac{2M}{b \cdot kd \cos \alpha \cdot jd \cos \alpha}\right) \tag{3.27A}$$

When the compression face of beam is inclined and the tension face is horizontal

When the tension face of beam is horizontal, then, angle B shall be zero and cos B shall be unity. Above expression Eq. 3.23 (A) to Eq. 3.27 (A) reduce as follows :

$$k = \left(\frac{1}{\cos^2 \alpha}\right)\left[-mp + \left\{\left(2mp \cos^2 \alpha\right) + (mp)^2\right\}^{1/2}\right] \tag{3.23B}$$

$$j = (1 - k/3) \tag{3.24B}$$

$$p = \left(\frac{1}{2}\frac{\sigma_{cbc}}{\sigma_{st}} \cdot k\right) \cdot \cos^2 \alpha \tag{3.25B}$$

$$\sigma_{st} = \left[M/\left(A_{st} \cdot j \cdot d\right)\right] \tag{3.26B}$$

$$\sigma_{cbc} = \left[2M/\left(b \cdot kd \cdot \cos \alpha \cdot jd \cos \alpha\right)\right] \tag{3.27B}$$

When the tension face of bema is inclined and the compression face is horizontal

When the compression face of beam is horizontal, angle a shall be zero and cos a shall be unity. Above expresses Eqs. 3.23(A) to 3.27(A) reduce as follows :

$$k = k = \cos \beta \left[-mp + \left\{\left(2mp / \cos^2 \alpha\right) + (mp)^{1/2}\right\}^{1/2}\right] \tag{3.23C}$$

$$j = (1 - k/3) \tag{3.24C}$$

$$p = \left[\frac{1}{2}\frac{\sigma_{cbc}}{\sigma_{st}} \cdot \frac{k}{\cos \beta}\right] \tag{3.25C}$$

$$\sigma_{st} = \left(\frac{M}{A_{st} \cdot jd}\right) \tag{3.26C}$$

$$\sigma_{cbc} = \left(\frac{2M}{b \cdot kd \cdot jd}\right) \tag{3.27C}$$

When the compression and tension (both) faces of beam are horizontal.

When both the faces of beam are horizontal, both the angles a and B shall be zero and the value of cos a shall be unity and that of cos B shall also be unity. In such a case, Eqs. 3.23(A) to 3.27(A) reduce as follows.

$$k = \left[2\,mp + (mp)^2\right]^{1/2} - mp \tag{3.23D}$$

$$j = (1 - k/3) \tag{3.24D}$$

$$p = \left(\frac{1}{2} \cdot \frac{\sigma_{cbc}}{\sigma_{st}} \cdot k\right) \tag{3.25D}$$

$$\sigma_{st} = \left[M/(A_{st} \cdot jd)\right] \tag{3.26D}$$

$$\sigma_{cbc} = \left[2M/(b \cdot kd \cdot jd)\right] \tag{3.27D}$$

Eq. 3.29 (D) to 3.33 (D) are same as Eqs. (3.11), (3.14), (3.8), (3.13) and (3.12) derived for theory of beams for usual singly reinforced beams.

PROBLEMS

3.1. A singly reinforced rectangular beams is 360 mm wide. The effective depth of beam section is 480 mm and its effective cover is 40 mm. A steel reinforcement consisting of 4 mild steel 16 mm diameter has been used in the beam section. The grade of concrete is M 15. Locate the neutral axis of the beam.

3.2. In Problem 3.1, the bending moment at a transverse section of a beam is 40 kN-m. Determine the compressive stress in bending in concrete, σ_{cbc} at the extreme fibre and tensile stress in steel, σ_{st}.

3.3. In Problem 3.1, determine the moment of resistance (flexural strength) of the beam section.

3.4. A singly reinforced rectangular beam is 360 mm wide. The effective depth of beam is 480 mm. Determine the percentage of steel and the cross-sectional area of reinforcement for the balanced beam section. The grade of concrete is M 20. The quality of steel is (i) mild steel, Fe 250 (ii) hysd-steel, Fe 415. Diameter of steel bars shall be kept less than 20 mm.

3.5. In Problem 3.4, determine the moment of resistance (flexural strength) of the beam section, when (i) the mild steel reinforcement is used (ii) the hysd-steel bars are used.

3.6. A singly reinforced rectangular beam is 360 mm wide. The effective depth of beam is 480 mm and its effective cover is 70 mm. The steel reinforcement in the beam section consists of 4 mild-steel bars, 16 mm diameter. The grade of concrete is M15. Determine whether the beam section is under reinforced or over reinforced. Also determined the flexural strength of the beam section.

3.7. A singly reinforced rectangular beam is 360 mm wide. The effective depth of beam is 480 mm and its effective cover is 40 mm. The steel reinforcement in the beam section consists of 4 hysd-steel bars, 16 mm diameter. The grade of concrete is M20. Determine whether the beam section is under reinforced or over reinforced. Also determine the flexural strength of the beam section.

4

SHEAR STRENGTH OF BEAMS AND DESIGN OF SINGLY REINFORCED RECTANGULAR BEAMS (SRRB-II)

4.1 INTRODUCTION

When the beams carry transverse loads, the bending of beams takes place. The bending of beams create a tendency in the particles to slide upon each other within the beam. This tendency of the particles to slide upon each other is called shear. The resistance tc. sliding offered by the material of beam is called shear strength. The algebraic sum of the external loads to the left or right (only on one side) of an imaginary transverse section of a beam called shear force, V also develops along with the bending M at the transverse section. The bending stresses develop at the section due to bending moment. Similarly, the shear stresses also develop at the section. The magnitude and distribution of the shear stresses at the various sections of the beam depends upon the external loads, the shape and the dimensions of beam. Due to vertical relative sliding on either side of a transverse section, the vertical shear stresses develop at the section. Due to horizontal sliding on either of a horizontal or longitudinal section of a beam, the horizontal shear stresses develop at any point in the transverse section are equal. The magnitude of shear stress at a height y from the neutral axis for a beam of homogenous material and its distribution across the cross-section of beam may be found as follows :

A small portion of a simply supported beam (say Fig. 1.1 A(i)) of homogenous material and of rectangular cross-section is shown in Fig. 4.1. AD and BC are two transverse section (sections perpendicular to length) δx distance apart measured parallel to the neutral surface GH. Let the width of beam section at any height y above GH be denoted by *z*. Let M and V, and (δM>) and V be the bending moments and shear forces at the sections AD and BC, respectively. The longitudinal or horizontal bending stress at the section AD at any height, y is σ_b = (M.y/I) where *I* is the moment of inertia of the beam cross-sections. Consider the equilibrium of portion ABEF between two sections. The longitudinal force at AE on any element of cross-section, of area zdy is

$$F_{AE} = (\sigma_b \,.\, z.dy) = \left(\frac{M}{I} y \,.\, z \,.\, dy\right) \qquad \text{(i)}$$

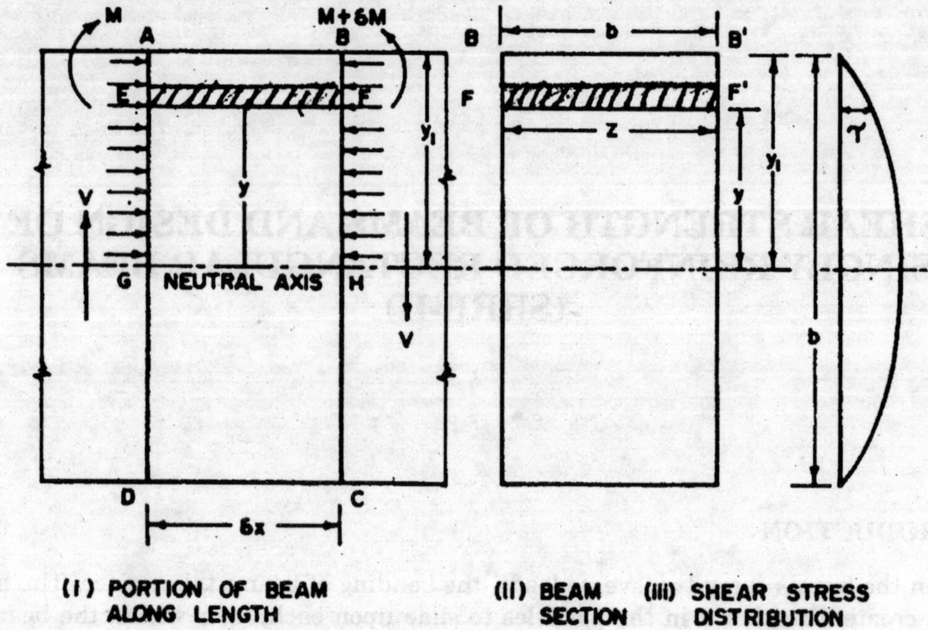

Fig. 4.1

The longitudinal force at BF on the same element at the same height y

$$F_{BF} = \left[\frac{(M + \delta M)}{I} y \,.\, z \,.\, dy\right] \tag{ii}$$

The difference between the two longitudinal forces

$$(F_{BF} - F_{AE}) = \left(\frac{\delta M}{I} \,.\, y \,.\, z \,.\, dy\right) \tag{iii}$$

The total difference between the two longitudinal forces

$$\Sigma\left(F_{BF} - F_{BA}\right) = \int_y^{y1} \frac{\delta M}{I} y.z.\ dy = \frac{\delta M}{I} \int_y^{y1} y \,.\, z \,.\, dy \tag{iv}$$

The portion of beam is in equilibrium. Hence, the total difference of longitudinal force on the portion ABFE is zero. It is balanced, by the horizontal shear force on the surface EF. Let τ represent the shear stress at a height y (neglecting any change in τ in length δx), the shear force

$$\tau \,.\, z \,.\, \delta x = \frac{\delta M}{I} \int_y^{y_1} y \,.\, z \,.\, dy \tag{v}$$

or $$\tau = \frac{\delta M}{\delta x} \cdot \frac{1}{I \,.\, z} \,.\, \int_y^{y_1} (y \,.\, z \,.\, dy) \tag{vi}$$

or $$\tau = \frac{V}{I.z}\int_{y}^{y_1}(y.z.dy) \qquad \left[\because \frac{\delta M}{\delta x} = V\right] \tag{4.1}$$

It may be noted that $\int_{y}^{y_1} y.z.dy$ is the moment of area BFF`B' about the neutral axis, (which is equal to product of area BFF`B = A and distance of its centroid from neutral axis, $\bar{y}$). Therefore,

$$\tau = \left(\frac{V}{I.z}A\,\bar{y}\right) = \left(\frac{V.Q}{I.z}\right) \tag{4.2}$$

where Q is the static moment of area above the surface considered about the neutral axis of the beam section. The stresses at the outer fibres are zero.

For the rectangular beam section, substituting the values of I and Z in Eq. 4.1, the shear stress, τ is found as under:

$$\tau = \frac{V}{\left(\frac{1}{12}bh^3.b\right)} \times \int_{y}^{h/2} b.y\,dy = \frac{6V}{b.h^3}\left[\frac{h^2}{4} - y^2\right] \tag{vii}$$

The various values of τ plotted as ordinates with height of beam, h as the base line, the distribution of shear stress τ is seen as parabolic, with y = 0.

$$\tau = \left(\frac{1.5\,V}{b.h}\right) \text{ (maximum at neutral axis)}$$

In case of reinforced concrete beam, the distribution of shear stress is different from that for a beam of homogenous material. In reinforced concrete beam, the beam section is considered as cracked section. The portion of concrete in tension is neglected. However, a perfect bond exists between concrete and steel in tension. Consider a small portion of a simply supported beam reinforced with steel bars in tension only and of rectangular section as shown in Fig. 4.2 AD and BC are two transverse section, δx distance apart. Due to the moments shown in Fig. 4.2, C and (C + δC) and T and (T + δT) are the compressive forces and tensile forces developed at sections AD and BC, respectively. Let *n* be the distance of top fibre in concrete in compression and (d-n) be the distance of centre of steel in tension from the neutral axis.

The distribution of shear stresses in concrete above neutral axis is similar to that in beam of homogenous material. The magnitude of shear stress may be written from 4.1 as below for the reinforced concrete beam section for the layer at a height *y* above the neutral axis upto the top fibre. Therefore,

$$\tau = \frac{V}{I.z}\int_{y}^{y_1}(y.z.dy) = \frac{V}{I.z}\int_{y}^{n}(y.z.dy) \tag{ix}$$

or $$\tau = \frac{V}{I.b}.b.\frac{1}{2}\left[n^2 - y^2\right] \tag{4.3}$$

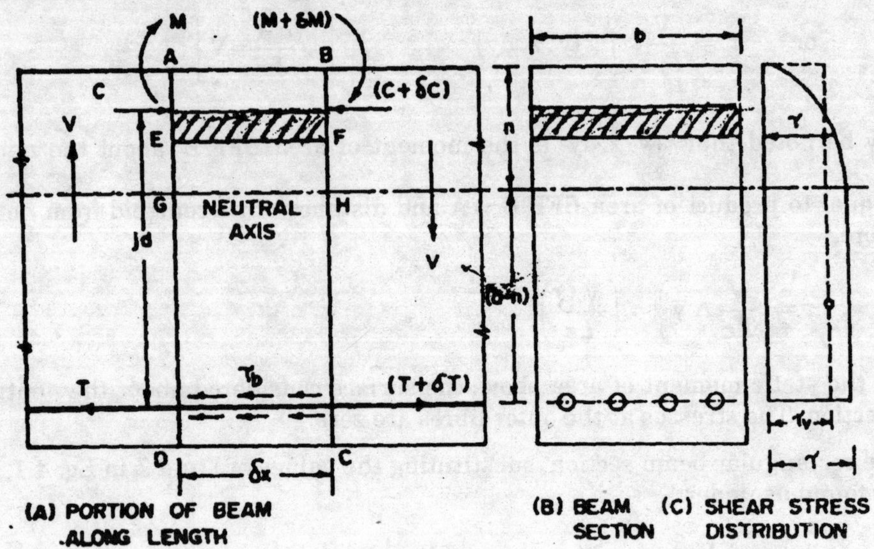

Fig. 4.2

From Fig. 4.3 it is seen that, the distribution of stress is parabolic from top of the beam upto neutral axis as shown in Fig. 4.2(C). The difference between longitudinal or horizontal forces below the neutral axis [T + δT – T = δT] is balanced by the shear force. Therefore,

$$(\tau \, . \, b \, . \, \delta x) = \delta T \qquad \text{(x)}$$

The portion of beam ABCD is in equilibrium and hence, the sum of moments is zero. Therefore, taking moment of forces about the line of action of compressive forces

$$(T + \delta T) \, . \, jd = T \, . \, jd + V \, . \, \delta x \qquad \text{(xi)}$$

$$\delta T = \left(\frac{V \, . \, \delta x}{j \, . \, d} \right) \qquad \text{(xii)}$$

From the expression, (x) and (xii),

$$\tau \, . \, b \, . \, \delta x = \left(V \, \frac{\delta x}{jd} \right)$$

or $$\tau = \left(\frac{V}{b \, . \, jd} \right) \qquad (4.4)$$

From Eq. (4.3), it is seen that the shear stress, τ below neutral axis remains constant and its distribution is rectangular as shown in Fig. 4.3(C).

Nominal shear stress

A concept of nominal shear stress has been newly introduced in IS : 456-2000. The nominal shear stress in beams of uniform depth is represented by τ_v and it is calculated as below :

$$\tau_v = \left(\frac{V}{b.d}\right) \tag{4.5}$$

The distribution of nominal shear stress, τ_v along the full effective depth, d is considered as uniform as shown by dotted line, Fig. 4.2(C).

The stress developed due to bending of beams carrying loads shall not be at right angles to the section AB, Fig. 4.3(A). In concrete, the stresses at parallel to the compression face while, in steel, the stresses act in the direction of reinforcement provided parallel to tension face. In bending stress distribution diagram, the depth of neutral axis measured perpendicular to compression face O_1D = (kd . cos α) and the distance measured upto centre of steel reinforcement (in the direction perpendicular to tension face from O_1) O_1E = (d – kd). cos B. Various expressions for the following are as below: When the angles a and B exceed 10°, then, these expressions are used.

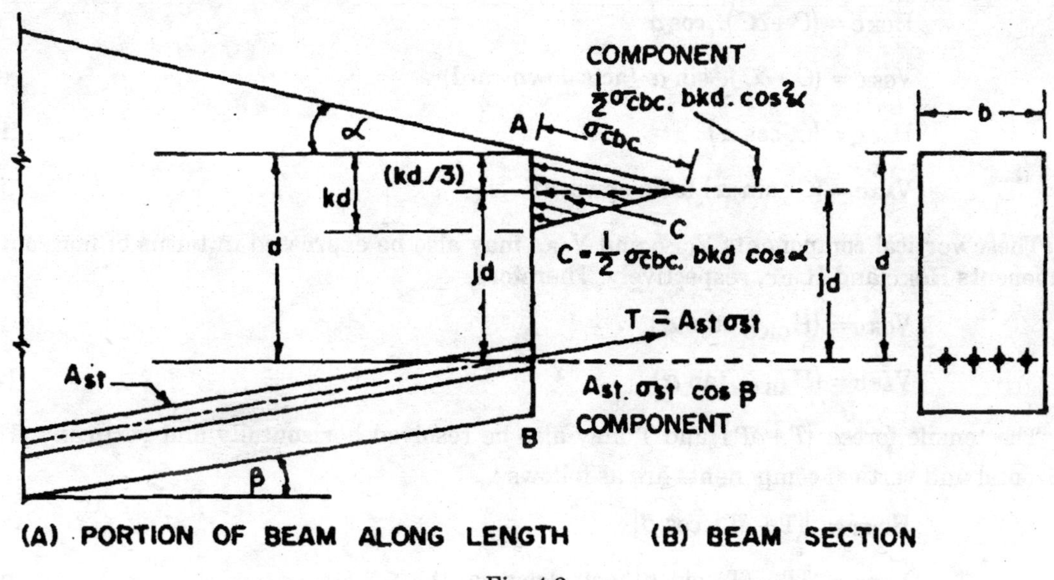

Fig. 4.3

1. Factor, k for the depth of neutral axis

By equating the moment of area of concrete and equivalent area of steel about neutral axis

$$b.\,(kd\cos\alpha)\,\frac{(kd\cos\alpha)}{2} = m.\,Ast.\,(d-kd)\,\cos\beta \tag{i}$$

$$k = \left(\frac{\cos\beta}{\cos^2\alpha}\right)\left[-mp + \left\{\left(2\,mp\cos^2\alpha/\cos\beta\right) + (mp)^2\right\}^{1/2}\right] \tag{4.6}$$

where, p = (A_{st})/bd)

2. Lever arm factor, *j*, $j = (1 - k/3)$ (4.7)

4.2 SHEAR STRESSES IN BEAMS OF VARYING DEPTH

In practice, the beam sections of uniform width and varying depth are commonly used. For example, in cantilever beams, the depth of beam section increases from its free end towards fixed support. The bending moment, M also increases numerically in the same direction as the effective depth, d increases. Consider a small portion of this wedged shape beam of length δx as shown in Fig. 4.3(A). Let d and d_1 be the effective depths of beam section at the faces AB and GH respectively. Let V be the shear force due to design loads at the face GK.

Let $(T+\delta T)$ and T be the tensile forces in steel reinforcement (provided parallel to the face inclined at an angle βwith the horizontal, and $(C+\delta C)$ and C be the compressive forces (parallel to the face inclined at an angle a with the horizontal) in concrete at sections GH and AB, respectively developed due to bending of beam. The compressive forces α and C may be resolved horizontally and vertically. Their horizontal and vertical components are as follows :

$$H_{GK.C} = (C+\delta C).\ \cos\alpha \qquad \text{(i)}$$

$$V_{GK.C} = (C+\delta C).\ \sin\alpha \text{ (acts downward)} \qquad \text{(ii)}$$

$$H_{AB.C} = (C.\cos\alpha) \qquad \text{(iii)}$$

$$V_{AB.C} = (C.\sin\alpha) \text{ (acts upward)} \qquad \text{(iv)}$$

These vertical components $V_{GK.C}$ and $V_{AB.C}$ may also be expressed in terms of horizontal components $H_{GK.C}$ and $H_{AB.C}$, respectively. Therefore,

$$V_{GK.C} = (H_{GK.C}.\tan\alpha) \qquad \text{(v)}$$

$$V_{AB.C} = (H_{AB.C}.\tan\alpha) \qquad \text{(vi)}$$

The tensile forces $(T+\delta T)$ and T may also be resolved horizontally and vertically. The horizontal and vertical components are as follows :

$$H_{GK.T} = [(T+\delta T).\cos\beta] \qquad \text{(vii)}$$

$$V_{GK.T} = [(T+\delta T).\sin\beta] \text{ (acts downward)} \qquad \text{(viii)}$$

$$H_{AB.T} = (T.\ \cos\beta) \qquad \text{(ix)}$$

$$V_{AB.T} = (T.\sin\beta) \text{ (acts upward)} \qquad \text{(x)}$$

These vertical components may also be expressed in terms of horizontal components. Therefore,

$$V_{GK.T} = (H_{GK.T}.\ \tan\beta) \qquad \text{(xi)}$$

$$V_{AB.T} = (H_{AB.T}\ \tan\beta) \qquad \text{(xii)}$$

Shear force at the face GK

Net shear force at the face GK shall be as below : ($\therefore H_{GK.C} = H_{GK.T}$)

$$[V - H_{GK.C} \tan\alpha - H_{GK.T} \tan\beta] = [V - H_{GK}.(\tan\alpha + \tan\beta)] \quad \text{(xiii)}$$

The horizontal component, H_{GK} may be approximately written as under :

$$H_{GK} = (M_{GK}/d_1) \quad \text{(xiv)}$$

Therefore, the net shear force may be written as

$$= [V - M_{GK} (\tan\alpha + \tan\beta)] \quad \text{(xv)}$$

The difference between the forces in steel reinforcement

$$[(T + \delta T) - T = \delta T] \quad \text{(xvi)}$$

is balanced by the shear force. Therefore, $\tau . b . \delta x \sec\beta = \delta T$

The portion of beam, Fig. 4.3(A) is in equilibrium and hence, the sum of moments is zero. Therefore, taking moment about a point on the line of action of compressive forces at the section GK

$$(T + \delta T) jd_1 = T . jd_1 + [V - M_{GK} (\tan\alpha + \tan\beta)/ d_1]\ \delta x . \sec\beta$$

$$\therefore \delta T = [V - M_{GK} (\tan\alpha + \tan\beta)/d_1]\ . \delta x \sec\beta/jd_1 \quad \text{(xvii)}$$

From the expression (xvi) and (xvii)

$$\tau . b . \delta x . \sec\beta = [V - M_{GK} . (\tan\alpha + \tan\beta)/d_1] \left(\frac{\delta x . \sec\beta}{jd_1}\right)$$

$$\tau = \left(\frac{V - M_{GK} . (\tan\beta)/d_1}{b . jd_1}\right) \quad (4.8)$$

When the compression face of beam is horizontal and the tension face only is inclined

When the compression face of beam is horizontal, the value of tan a shall be zero. Then Eq. 4.8 (A) reduces as below :

$$\tau = \left(\frac{V - M_{GK} . (\tan\beta)/d_1}{b . jd_1}\right) \quad (4.9)$$

When the tension face of beam is horizontal and the compression face is inclined

When the tension face of beam is horizontal, the value of tan B shall be zero. Then, Eq. 4.8 (A) reduced as below :

$$\tau = \left(\frac{V - M_{GK} (\tan\alpha)/d_1}{b\ .\ jd_1}\right) \quad (4.10)$$

For beams of varying depth, Eq. 4.5, 4.6, and 4.7 reduce as below for nominal shear stress

$$\tau_v = \left(\frac{V \mp M(\tan \alpha + \tan \beta)/d}{b \cdot d} \right) \tag{4.11}$$

$$\tau_v = \left(\frac{V \mp M(\tan \beta)/d}{b \cdot d} \right) \tag{4.12}$$

$$\tau_v = \left(\frac{V \mp M(\tan \alpha)/d}{b \cdot d} \right) \tag{4.13}$$

where M and d are the bending moment and the effective depth at the section.

The negative sign in Eq. 4.11 is used when the bending moment, M increases numerically in the same direction as the effective depth, *d* increases. And the positive sign when the moment decreases numerically in this direction.

4.3 DIAGONAL TENSION IN HOMOGENEOUS ELASTIC BEAMS

A beam of homogeneous material (say, steel or timber) simply supported at its both the ends and carrying transverse loads is shown in Fig. 4.4(A). Let V and M be the shear force and the bending moment at any section XX of the beam. The portion of beam above neutral axis carries compressive bending stresses in the horizontal direction (normal to section XX) and that below the neutral axis carries tensile bending stresses in the horizontal direction due to bending moment. The bending stresses

$$\sigma_b' = (M' \, y/I) \tag{i}$$

The section of beam also carries shear stresses

$$\tau = [V \cdot (A\bar{y})/(I \cdot b)] \tag{ii}$$

where $A\bar{y}$ is the statical moment of area A above/below a line (parallel to neutral axis) upto the face of the beam (upper or lower) and $\bar{y}$ is the distance of its centroid from neutral axis. I is the moment of inertia of the beam cross-section about neutral axis and b is the width of beam at the line being considered.

The shear stresses, τ_v as in expression (ii) act vertically (parallel to section XX) and these shear stresses always accompany **complementary shear** stresses, τ_v' which act horizontally and *b* are equal.

In case any elementary portion ABCD of beam shown on ex-aggregated scale in Fig. 4.4 (B) located anywhere (but neither at the neutral axis nor at the outer edges. This elementary portion ABCD carries six stresses as shown in Fig. 4.4 (B). These six stresses may be combined to principal stresses, (compressive and tensile) which act at right angles to each other. The principal stresses are given by

$$\sigma = \frac{\sigma_b}{2} \pm \left[\left(\frac{\sigma_b}{2} \right)^2 + (\tau_v)^2 \right]^{1/2} \tag{iii}$$

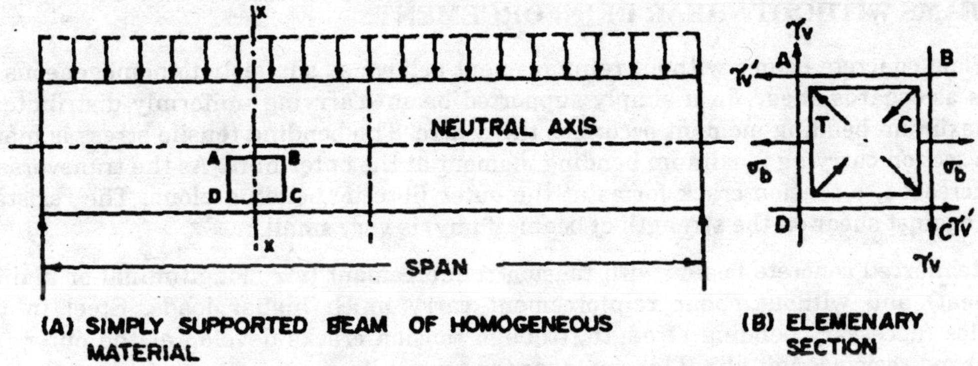

Fig. 4.4

The inclination of principal stresses a is given by

$$\tan 2\alpha = (2\tau_v/\sigma_b) \qquad \text{(iv)}$$

The magnitudes of bending stress, σ_b and the shear stress, τ_v change both along the length of beam and vertically with the distance from the neutral axis. Therefore, the magnitudes and inclination of principal stress, σ also change for the element depending upon its location.

At the neutral axis or at the support of beam it is seen that the bending stresses are zero. Therefore, from expression (iv)

$$\tan 2\alpha = \left(\frac{2\,\tau_v}{\text{(zero)}} = \infty\right), \qquad (\therefore\ \text{a} = 45°) \qquad \text{(v)}$$

The inclination of principal stresses, α is 45°. At support, the principal stress σ is tensile along the diagonal AC and it is compressive along the diagonal BD. The diagonal tensile stress is called as diagonal tension. Its magnitude is obtained from expression (iii), as below

$$|\sigma| = |\tau_v| \qquad \text{(v)}$$

At the outer faces or at the centre of beam, the shear stresses τ_v are zero. Therefore, from expression (iv)

$$\tan 2\alpha = \left(\frac{2.\text{(zero)}}{\sigma_b}\right) = \text{(zero)}, \qquad (\therefore\ \alpha = 0) \qquad \text{(vi)}$$

The inclination of principal stress, a is zero. The principal stress acts horizontally.

It is to note that the principal tensile stress obtained by combining the bending and shear stresses, is also referred as diagonal tension. Its inclination varies from zero at mid span to 45° at support. These principal tensile stresses exist at all the locations of a beam and it may damage or harm the integrity of beam, if it is not adequately provided for.

Since, the tensile strength of concrete is low, this principal tensile stress (known as diagonal tension also must be carefully considered in the reinforced concrete beam design).

4.4 BEAMS WITHOUT SHEAR REINFORCEMENT

Plain concrete beams without reinforcement behave as similarly to homogeneous elastic beams as regards shear. In a simply supported beam carrying uniformly distributed load, the maximum bending moment occurs at the centre. The bending tensile stress is maximum at the section carrying maximum bending moment at the outer fibre. As the transverse loads are increased, a tension crack forms at the outer fibre by bending alone. The resistance of beam against shear on the strength of beam (if any) is very small.

Reinforced concrete beams with tension reinforcement (viz., longitudinal or main reinforcement) and without shear reinforcement carry much higher loads. Steel in tension provides necessary bending strength, (though tension cracks develop at the outer fibre as above) and resists longitudinal tension near the tension face.

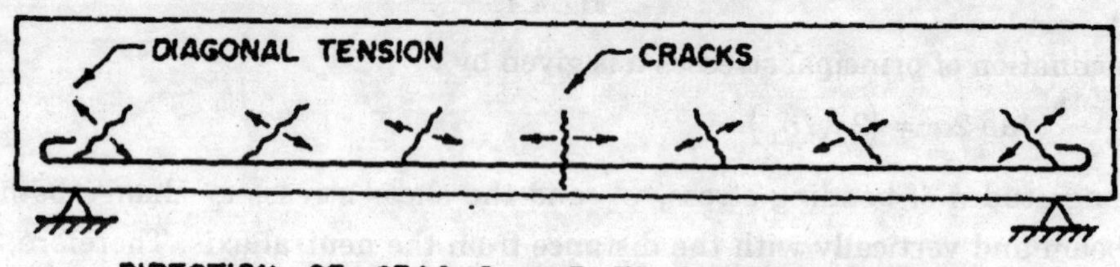

Fig. 4.5

Shear stresses develop in beam along with bending stresses. The shear stresses increase as the loads are increased. The shear stresses combined with bending stresses result in principal stresses. The principal tensile stresses exist in all the regions of beam. The longitudinal reinforcement does not resist the principal tensile stresses existing at other locations. The concrete is weak in tension. The principal tensile stresses increase as the loads are increased and cause cracks in the direction perpendicular to these stresses as shown in Fig. 4.5. These cracks are called as **diagonal cracks** in distinction to the **vertical flexural cracks. Vertical flexural cracks** occur in regions of large bending moments and where the shear force is very small or zero. These cracks are due to elongation of the bars and inability of concrete to stretch equally. These cracks are usually normal to the axis of beam or nearly so. Diagonal cracks occur in regions in which the shear forces are very high. After formation of a crack, the shear and diagonal tensile stresses cannot be transmitted across the opening by the concrete alone, even though the wind of crack may be only hair thin.

Three types of cracks are seen in simply supported beams. In beams of long span, the magnitudes of bending moments and thus the flexural stress remain very high in the middle portion of the span and flexural cracks occur in the direction normal to the tension face as shown in Fig. 4.6. The tension reinforcement yields and flexural cracks only open out (as the loads are increased) in the bottom fibres. In the beams of intermediate spans, vertical flexural cracks form near the section of maximum bending moment first. But in the regions where both the shear force and bending moment have comparatively large values, the diagonal tensile stress at the upper end of one or more of these cracks exceeds the tensile strength of the concrete, the cracks bend in diagonal direction and their lengths and widths increase. These cracks are called as flexural-shear cracks as shown in Fig. 4.6. In the beams of short spans, the flexural stresses are negligibly small and the shear stresses are very high near the supports. The diagonal tensile stresses are inclined 45° and are numerically equal to the shear stresses which are maximum at the neutral axis. The diagonal cracks (which are also called as web-shear cracks) form mostly at or near the neutral axis when the tensile stresses exceeds the tensile strength of concrete as shown in Fig. 4.6.

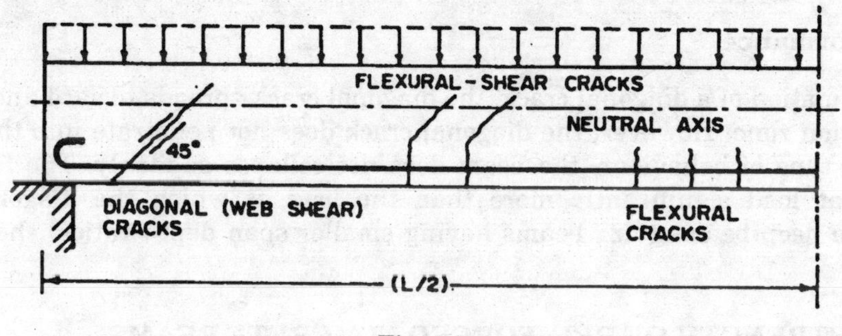

Fig. 4.6

In order to counteract the effect of diagonal tension, it is necessary to provide shear reinforcement in the reinforced concrete beams.

4.5 BEHAVIOUR OF DIAGONALLY CRACKED BEAMS

In reinforced concrete beams without shear reinforcement, the vertical flexural cracks are allowed to occur on the tension side. These cracks are not harmful to the strength of beam because, the beams have sufficient longitudinal reinforcement to resist the bending tensile stresses. However, the principal tensile stresses (diagonal tension) exists at all regions and these is no shear reinforcement. When the loads are increased on the beams, the diagonal tension (principal tensile stress) also increases and it exceeds the tensile strength of the concrete and the diagonal cracks are formed in the concrete in the direction perpendicular to the diagonal tension. These diagonal cracks govern the behaviour and strength of reinforced beams without shear reinforcement. Present knowledge is based on many tests conducted on such beams, which show that the performance of such beams is of following two types.

Type-1 performance

After formation of a diagonal crack, the diagonal crack spreads either immediately at the same loads or at increasing the load slightly. From the tension reinforcement, the diagonal crack transverses the beam to the compression face. The beam splits in two portions and the beam fails.

This process occurs suddenly without warning. It happens mainly in shallow beams (i.e., beams with span depth ratios of about 8 or more). The reinforced concrete beams without shear reinforcement become vulnerable to accidental large over loads and such beams are likely to have catastrophic failure without warning.

Even if the beams do not need shear reinforcement, it is good practice to provide a minimum shear reinforcement. The shear reinforcement counteracts the growth of diagonal cracks. The shear reinforcement increases the ductility of the beam. These shall be warning before actual failure.

In some slabs and footings, the actual shear stresses are very small as compared to their critical values and in such some situations, an unusual large factor of safety is used. For such structural members, it is not necessary to provide shear reinforcement and it is omitted.

Type-2 performance

After formation of a diagonal crack, the diagonal crack spreads toward and partially into the compression zone. However, the diagonal crack does not penetrate into the compression zone. In this type of behaviour, the beam does not collapse suddenly. The failure of beam takes place at load significantly more than the load at which the diagonal crack first develops. The deep beams (viz., beams having smaller span depth ratios) show this type of performance.

4.6 SHEAR STRENGTH OF REINFORCED CONCRETE BEAMS

When the beams are subjected to transverse loads, bending moment M and shear force V develop at every transverse section of a beam. When the loads are increased, such that the diagonal tension in beam exceeds the tensile strength of concrete, a diagonal tension is formed as shown in Fig. 4.7. The external shear force, V is resisted by internal shear strength of the beam, V_{int}. The internal shear strength V_{int} consists of shear strength of uncracked concrete, V_{cc} in compression zone, the vertical component of aggregate interlocking force, V_{yi} and the dowel action, V_{da} (the shear force) acting across the longitudinal steel bars. Therefore, the resistance of concrete beam with longitudinal reinforcement only.

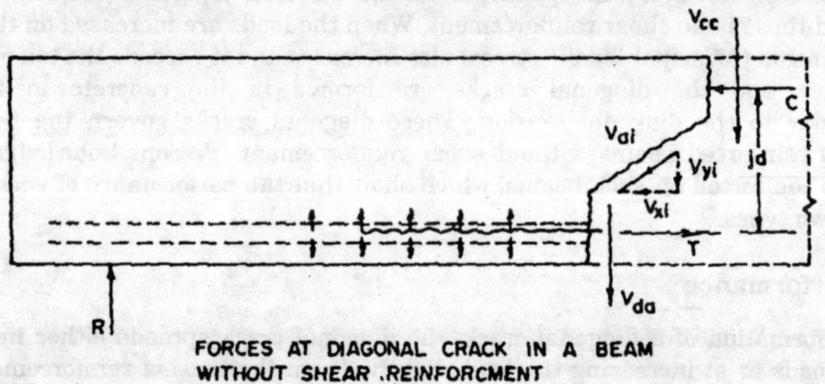

Fig. 4.7

$$V_{int} = (V_{cc} + V_{ai} + V_{da}) \quad \text{(i)}$$

The beam is in equilibrium, and therefore V_{int} is equal to V_{ext}. Therefore,

$$V_{cc} = (V_{ext} - V_{ai} - V_{da}) \quad \text{(ii)}$$

The shear strength of uncracked concrete, V_{cc} increases with the quality of concrete. The interlocking force, V_{ai} depends upon type of aggregate. For light weight aggregates, the interlocking force, V_{ai} shall be comparatively small. This interlocking force, V_{ai} is one-third and more of the total shear force. The dowel action force, V_{da} across the longitudinal reinforcement is usually very small. The longitudinal reinforcement is supported against vertical displacement mainly by the thin layer of concrete. The vertical tensile stresses develop in this concrete due to bearing pressure due to V_{da} as shown in Fig. 4.7. These vertical tensile stresses split the concrete along the longitudinal reinforcement and decrease the dowel action force V_{da} and widen the diagonal crack. In this process, the aggregate interlocking force, V_{ai} also decreases and vanishes completely and frequently leads to immediate failure due to rapid transfer of large shear force to concrete in compression.

The resistance of concrete against external shear is given by

$$V_c = (\tau_c \, . \, b \, . \, d) \tag{4.14}$$

where, τ_c = permissible shear stress in concrete in beam with longitudinal reinforcement and without shear reinforcement discussed in Art. 3.22.

Professor B.V. Rangan has specified a criterion in his paper titled as `Diagonal cracking strengths in shear of reinforced concrete beams' published in Civil Engineering Transactions, Institution of Engineers, Australia, Vol. C.E. 14, No. 1, 1972 that the diagonal cracking strength in shear is reached whenever the principal tensile stress at the neutral axis is reached whenever the principal tensile stress at the neutral axis of flexurally cracked beam attains the tensile strength of concrete. Professor B.V. Rangan has specified a semi-emperical formula τ_c based on this criterion as below

$$\tau_c = \left(\frac{0.85\,(0.8 f_{ck})^{1/2} \left[(1+5\beta)^{1/2} - 1 \right]}{6\beta} \right) \tag{4.15}$$

where, $\beta = 0.8\, f_{ck}/(6.89\, p_t)$, but not less than unity, and

$p_t = (100\, A_s/b.d)$

The cylinder strength has been converted to cube strength by using a factor 0.8 in the above formula. A reduction factor of 0.85 similar to the partial safety factor for the material is used.

The permissible shear stress in concrete, τ_c serves two purposes. Firstly it prescribes a limit for the shear carried by beams without shear reinforcements. Secondly, it gives a measure of shear carried by concrete in a member with shear reinforcement.

When the resistance of concrete, V_c in shear becomes insufficient, then, the shear reinforcement is provided to keep the two portions of splitted beam together. The shear reinforcement is also referred as web reinforcement or transverse reinforcement. Total external shear force, V_{ext} is resisted by the resistance of concrete to shear, V_c and the resistance of shear reinforcement to shear, V_s as shown in Fig. 4.8. Therefore

$$V_{ext} = [(V_{cc} + V_{yi} + V_{da}) + V_s] \tag{iii}$$

$$\text{or } V_{ext} = (V_c + V_s) \tag{iv}$$

$$\text{or } V_s = (V_{ext} - V_c) \tag{v}$$

$$\text{or } V_s = [V_{ext} - \tau_c \,).\, bd)] \tag{4.13}$$

where, V_s = Shear force carried by shear reinforcement.

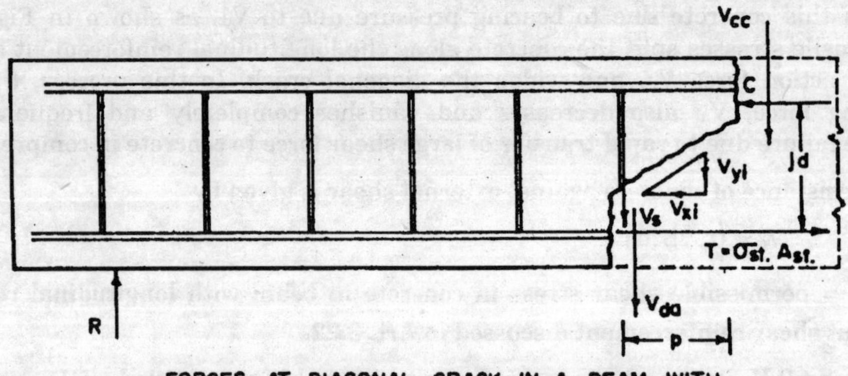

Fig. 4.8

Practically, the shear reinforcement remains free of stress prior to the development of diagonal cracks. The shear reinforcement increases the shear resistance of a reinforced concrete beam after the formation of diagonal cracks as follows :

A portion of shear force, viz., V_s is resisted by the vertical/inclined bars which cross a particular crack. The growth of diagonal cracks is checked by the presence of these bars. The possibility of penetration of diagonal crack into the compression zone is also reduced due to these bars. Two cracked faces of a diagonal crack is kept in close contact by the shear reinforcement. It increases the interlocking force, V_{ai}. The longitudinal reinforcement is tied into the main mass of the concrete by the shear reinforcement. The longitudinal reinforcement gets support in addition to that from concrete in cover. The possibility of development of a crack along the longitudinal reinforcement is reduced and the portion of shear force resisted by the dowel action of concrete, V_{da} increases.

4.7 REINFORCED CONCRETE BEAMS WITH SHEAR REINFORCEMENT

When the diagonal tensile stress in the surrounding of neutral axis of a beam attains the value of tensile strength of the concrete, a diagonal crack develops in the beam. In order to resist the diagonal tensile stress and to check the development of diagonal cracks, the shear reinforcement (also referred as web reinforcement or transverse reinforcement) is provided in the beam having longitudinal or main reinforcement. Following types of shear reinforcement are provided in the reinforced concrete beams.

Vertical stirrups placed at varying spacing along the length of the beam are very commonly used as shear reinforcement. Alternatively, the bent-up bars (a part of longitudinal tensile reinforcement where it is no more required to resist tension due to bending) are provided in the reinforced concrete beams. The bent-up bars along with vertical stirrups are also used as shear reinforcement. Formerly, inclined stirrups in the direction perpendicular to the possible diagonal cracks used to be provided.

4.8 REINFORCED CONCRETE BEAMS WITH VERTICAL STIRRUPS

Vertical stirrups (stirrups in the direction perpendicular to the longitudinal reinforcement are most frequently used as shear reinforcement in reinforced concrete beams.

Although, the vertical stirrups are not placed theoretically in the direction of diagonal tension, these stirrups resist the diagonal tensile stresses very effectively. These stirrups are secured against slipping in a horizontal direction. These are easily kept in position during construction. Stirrups are made of steel bars of comparatively small diameter, (e.g., 5 mm, 6 mm, 8 mm, 10 mm or 12 mm). Small sizes of bars are used for shear stirrups as these are usually strong enough when provided with closed spacing. The bars may be bent more easily and sharply to the shape as desired. These vertical stirrups are independent of the main reinforcement. The vertical stirrups are taken round the main reinforcement in tension and these bars extend from tension zone independent of the main reinforcement in tension and these bars extend from tension zone into the compression zone and are anchored there. Holding bars also termed as (anchored bars) which run in longitudinal direction in compression zone, are used to keep the stirrups in position. Fig. 4.9 shows one-legged, two-legged, four legged and multi-legged vertical stirrups used as shear reinforcement.

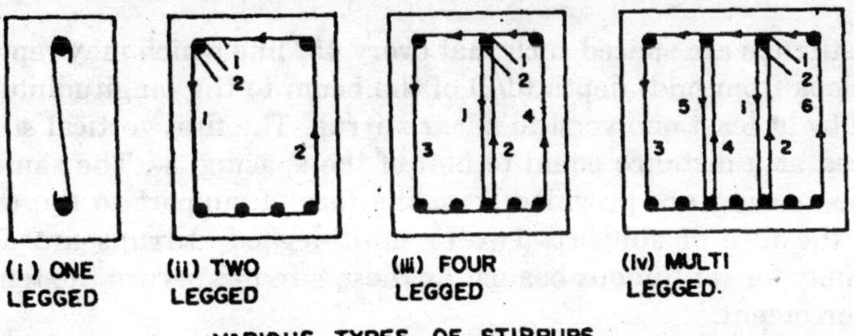

Fig. 4.9

The steel bars of vertical stirrups are bent round the holding bars of at least its own diameter through an angle of 135° and are continued beyond the end of curve for a length of at least six times the diameter of bar. The shear reinforcement is thus anchored with the holding bars in compression zone of the concrete.

One vertical stirrups traversing a diagonal crack exerts a force $A_v . \sigma_{sv}$ on the given portion of beam, where A_v is the cross-sectional area of the stirrup bar and σ_{sv} is the tensile stress in the stirrups, $(\sigma_{sv} = \sigma_{st})$. In case of two-legged stirrups, A_v is twice the cross-sectional area of one bar. In case V_s is the portion of the shear force resisted by the shear reinforcement (*viz.*, vertical stirrups) and *n* is the number of vertical stirrups traversing a diagonal crack, then

$$V_s = (n . A_{sv} . \sigma_{sv}) \quad \text{(i)}$$

The external shear force, V varies along the length of a beam. The spacing of vertical stirrups, s_v also varies along the length of beam. The spacing of vertical stirrups is found as follows :

Spacing of vertical stirrups

Fig. 4.8 shows the horizontal projection, p for a diagonal crack. The number of stirrups, n traversing the diagonal crack spaced, s_v distance depends on the horizontal projection p of the diagonal crack. The number of stirrups, n is given by the following expression,

$$n = (p/s_v) \quad \text{(ii)}$$

The diagonal crack used to develop at inclination somewhat flatter than 45. As such, the horizontal projection length, *p* may be assumed equal to the effective depth of the beam. Therefore,

$$n = (d/s_v) \quad \text{(iii)}$$

Hence, the shear resistance of the vertical stirrups may be found by substituting the value of n in the expression (i). Therefore,

$$V_s = (A_{sv} \cdot \sigma_{sv} \cdot d/s_v) \quad (4.16)$$

Then, the spacing of vertical stirrups is found as below :

$$s_v = (A_{sv} \cdot \sigma_{sv} \cdot d/V_s) \quad (4.17)$$

The vertical stirrups are spaced such that every 45° line which may represent a diagonal crack which develops from mid- depth (d/2) of the beam to the longitudinal tension reinforcement is crossed by at least one vertical shear stirrup. The first vertical stirrups at the end of a beam is placed at a distance equal to half of the spacing, s_v. The same size of vertical stirrups and same spacing are provided from the face of support to the critical section at distance, d from the face of support. Two or multi-legged stirrups are advantageous for erection and specially for continuous beams. As these stirrups secure the tendency as well as compressive reinforcement.

Following are the requirements of shear (transverse) reinforcement in beams

1. Maximum spacing of vertical stirrups

IS : 456-1978 recommends that the maximum spacing of vertical stirrups measured along the axis of the beam shall not exceed 0.75 times the effective depth, d. This spacing ensures that a potential shear crack is crossed by atleast one stirrup as mentioned above. In no case, this maximum spacing shall exceed 450 mm.

2. Minimum shear reinforcement

The minimum cross-sectional area of shear reinforcement, $(A_{sv})_{minimum}$ shall be provided even if, it is seen by the calculations that shear reinforcement is not needed

$$\left(\frac{(A_{sv})_{,min}}{b \cdot s_v}\right) \geq \left(\frac{0.4}{f_y}\right) \quad (4.18)$$

$$\text{or} \quad (A_{sv})_{min} \geq 0.4 \cdot \left(\frac{b \cdot s_v}{f_y}\right) \quad (4.19)$$

where f_y is the characteristic strength of yield stress of the steel used for stirrup reinforcement. The value of f_y for mild steel is 250 N/mm^2 and that for hysd steel of grade Fe 415 is 415 N/mm^2.

4.9 REINFORCED CONCRETE BEAMS WITH BENT-UP BARS

The cross-section of a reinforced concrete beam is designed to resist maximum bending moment. In a simply supported beam carrying uniformly distributed transverse loads, the maximum bending moment occurs at the centre. The area of steel bars (size and their number) in tension (*viz.,* longitudinal or main reinforcement) is also calculated to resist the maximum bending moment. The bending moment at the transverse section anywhere in between support and the centre of beam is smaller than its maximum value at the centre. All the steel bars at such transverse section are not needed. In case these steel bars are not cut-off and these excess longitudinal steel bars are bent-up (45° or more with respect to the longitudinal reinforcement) into the compression zone, as shown in Fig. 4.10, these bent-up bars when properly distributed and inclined serve as excellent shear or web reinforcement. These bent-up bars have the advantage resulting from closer coincidence with the direction of diagonal tensile stress. These bars also have the advantage of being rigidly connected with the main reinforcement. These bars are bent-up two at a time for symmetry. These bars are shifted from the bottom of the beam to the top, where these bars resist negative bending moment, if any.

There is also a disadvantage of bent-up bars. Their number in a reinforced concrete beam is small and the point/points of bending up is controlled by the requirement of bending moment and which one in turn do not coincide with the requirements of diagonal tension reinforcement.

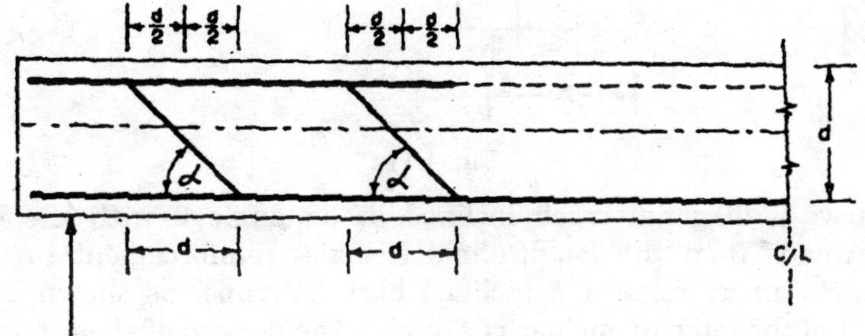

Fig. 4.10

The bent-up bars are made fully effective as diagonal tension reinforcement, by keeping the spacing of these bars measured along the axis of beam not larger than the effective depth of beam. Where only one bar is bent up or the spacing is large, the region of effectiveness of the bent-up bar as diagonal tension reinforcement extends, on both the sides, from the point of intersection of the bars with the neutral axis for a distance along the beam equal to one-half of the depth of bcam.

The bent up bar resist the shear force, V_s equal to the vertical component of force in these bars. Therefore,

$$V_s = (\sigma_{sv} \cdot A_{sv} \cdot \sin\alpha) \qquad (4.20)$$

where, A_{sv} = total cross-sectional area of bent up bars at the cross-section

σ_{sv} = permissible tensile stress in shear reinforcement

α = angle between bent up bars and axis of member not less than 45°

4.10 REINFORCED CONCRETE BEAMS WITH INCLINED BARS (STIRRUPS)

Bars or stirrups inclined at an angle a (not less than 45°) with respect to the longitudinal reinforcement placed in the direction of diagonal tensile stress is the most appropriate shear reinforcement to resist these type of stresses. The inclination of diagonal tension changes in different parts of the beam. The inclination of diagonal tension in case of simply supported beam carrying uniformly distributed transverse loads is maximum near the support and it flattens out towards the centre of beam. In order to follow the direction of diagonal tensile stress, the inclination of inclined stirrups would have to be changed throughout the beam and it is impracticable. The inclined bars (stirrups) have advantage of its superiority which results from a closer coincidence between the direction of the stress and the direction of the bar. In case, the inclined stirrups are used, these are placed inclined 45° or more with the longitudinal reinforcement as shown in Fig. 4.11. For practical reasons, (viz., it is difficult to attach these bars with the main reinforcement and to keep them in place), the inclined bars are rarely used as shear reinforcement.

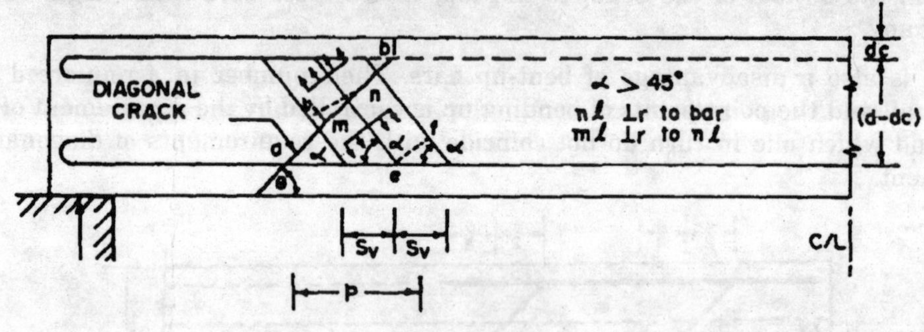

Fig. 4.11

Fig. 4.11 shows a diagonal crack inclined at an angle θ with the horizontal. This diagonal crack extends from the longitudinal (tensile) reinforcement to the compression surface. It intersects an average of n inclined bars (stirrups) as shown in Fig. 4.11. The vertical component of the force in one bar is Asv. σ_{sv}. The portion of shear force resisted by all the bars which cross the crack is given by the following expression :

$$V_s = \left(n \,.\, A_{sv} \,.\, \sigma_{sv} \,.\, \sin\alpha\right) \quad \text{(i)}$$

The number of inclined bars, *n* spaced at a distance, *sv* apart, horizontally crossing the diagonal crack may be found as under :

Let the horizontal projection of the crack ab be p and its inclined length *ab* be *i*,

$$i = \left(p/\cos\theta\right) \quad \text{(ii)}$$

The distance between two adjacent bars measured parallel to the direction of crack, a is equal to *mn*, the distance between two middle points *m* and *n*. In order to find the length *mn*, draw *nl* perpendicular to the middle bar and *ml* perpendicular to the *nl*.

$$a = mn = [nl.\sec(\text{angle } mnl)] = nl.\sec[90 - (\alpha - \theta)] = nl.\operatorname{cosec}(\alpha + \theta) \quad \text{(iii)}$$

The perpendicular distance *nl* is equal to *ef* perpendicular distance between two adjacent bars

$$nl = ef = s_v \,.\, \sin \alpha \qquad \text{(iv)}$$

Therefore, $a = mn = s_v \,.\, \sin\alpha \,.\, \text{cosec}\,(\alpha + \theta)$ (v)

or $a = s_v / [\sin\theta\,(\cot\theta + \cot\alpha)]$ (vi)

The number of bars crossing the crack

$$n = \left(\frac{i}{a}\right) = \left(\frac{p}{a \,.\, \cos\theta}\right)$$

$$n = [p \sin\theta\,(\cot\theta + \cot\alpha)/\cos\theta \,.\, s_v] = \frac{p}{s_v}(1 + \tan\theta \,.\, \cot\alpha) \qquad \text{(vii)}$$

Therefore, the portion of shear force resisted by the inclined bars (stirrups) from the expressions (i) and (vii)

$$V_s = \frac{p}{s_v}(1 + \tan\theta \,.\, \cot\alpha)\, A_{sv} \,.\, \sigma_{sv} \,.\, \sin\alpha$$

or $$V_s = A_v \,.\, \sigma_{sv} \,.\, \frac{p}{s_v}(\sin\alpha + \cos\alpha \,.\, \tan\beta) \qquad \text{(viii)}$$

The inclination of diagonal crack, θ, which varies somewhat depending on various influences, is commonly assumed to be 45. The expression (viii) reduces as follows :

$$V_s = \left(\frac{A_{sv} \,.\, \sigma_{sv} \,.\, p}{s_{sv}}\right)(\sin\alpha + \cos\alpha) \qquad \text{(ix)}$$

The horizontal projected distance, *p* of the diagonal crack may be assumed as the effective depth of beam, *d* as in case of the vertical stirrups. Therefore,

$$V_s = \left(\frac{A_{sv} \,.\, \sigma_{sv} \,.\, d}{s_{sv}}\right)(\sin\alpha + \cos\alpha) \qquad (4.21)$$

or $$s_v = \left(\frac{A_{sv} \,.\, \sigma_{sv}\, d}{V_s}\right)(\sin\alpha + \cos\alpha) \qquad (4.22)$$

It is to note that Eqs. 4.16 and 4.17 developed for the vertical stirrups, are only special case of Eqs. 4.21 and 4.22, for a = 90°. α 4.21 and 4.22 are more general expressions.

4.11 FUNCTION OF SHEAR REINFORCEMENT

The function of shear reinforcement in a reinforced concrete beam may also be explained by analogy with simply steel trusses. The expressions as in Eqs. 4.16, 4.17, 4.20, 4.21, 4.22 are based in truss analogies IS: 456-2000 also recommends Eqs. 4.16, 4.20, 4.21 to determine the portion of shear force resisted by vertical stirrups, bent up bars and inclined stirrups, respectively, placed as shear reinforcement in reinforced concrete beams.

Truss analogy - 1

A simple steel truss is shown in Fig. 4.12(A). The top and bottom chords of this steel truss are in compression and tension, respectively. The web members of this truss are diagonal members, which are alternately in compression and tension.

In a reinforced concrete beam with shear reinforcement (inclined stirrups) as shown in Fig. 4.12(B), the concrete above neutral axis resists the compression and the longitudinal reinforcement at the bottom resists the tension. The inclined stirrups used as shear reinforcement resist the diagonal tensile forces and the concrete resists diagonal compressive forces. Thus, it is seen that there is truss analogy between the simply steel truss shown in Fig. 4.12(A) and reinforced concrete beam with shear reinforced concrete beam with shear reinforcement shown in Fig. 4.12(B). The crushing of concrete in web is avoided be limiting the maximum shear stress $\tau_{c.max}$, which depends upon the grade of concrete used.

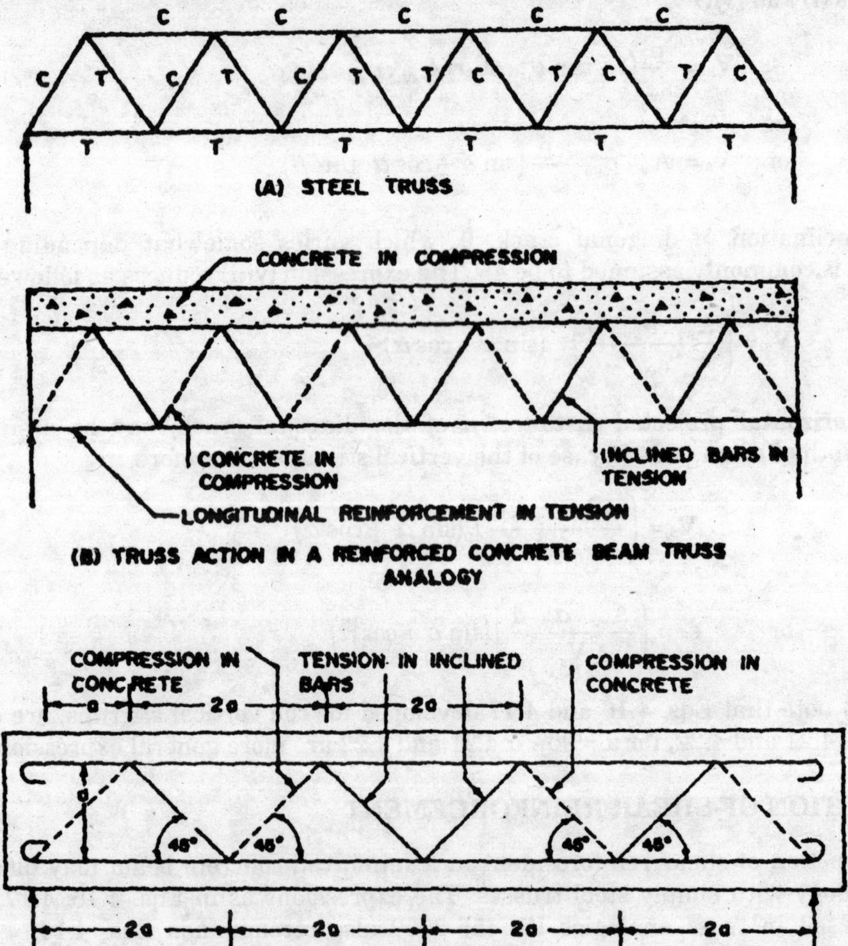

Fig. 4.12

The inclined stirrups inclined at 45° with the longitudinal reinforcement are used with spacing 2a as shown in Fig. 4.12(C). Force in the inclined bar near the support may be found from the triangle of forces. Therefore,

$$\left(\frac{\text{Force in inclined bar}}{\text{Force in horizontal bar}}\right) = \left(\frac{\sin 45^\circ}{\sin 90^\circ}\right)$$

$$\left(\frac{\text{Force in inclined bar}}{A_{sv} \cdot \sigma_{sv}}\right) = \left(\frac{\sin 45^\circ}{\sin 90^\circ}\right) \quad \text{(i)}$$

Therefore, the force in inclined bar

$$= (A_{sv} \cdot \sigma_{sv}). \sin 45^\circ = (0.707\, \sigma_{sv}\, A_{sv}) \quad \text{(ii)}$$

The portion of shear force resisted by the inclined bar in such a case shall be as follows :

$$V_s = (0.707 \sigma_{sv} \cdot A_{sv}). \sin 45^\circ \quad \text{(iii)}$$

IS : 456-2000 does not recommend this as the basis, since, the portion of shear force resisted by the inclined bar reduces by 0.707 as compared to the portion of shear force resisted and recommended as $V_s = \sigma_{sv}\, A_{sv} \sin 45^\circ$.

Truss analogy - 2

In this truss analogy, the angle of inclination of concrete in compression as web member is considered as $67\frac{1}{2}$ and that of inclined bars is taken as 45°, with the longitudinal reinforcement as shown in Fig. 4.13(A). The distance between holding bars at top and longitudinal reinforcement is assumed equal to lever arm, (a = jd). The spacing between the inclined bars from the geometry of any one triangle may be written as follows

$$S_v = \left[a \cot (45^\circ) + a \cot\left(67\frac{1^\circ}{2}\right)\right] = \left[s\left(\cot 45^\circ + \cot 67\frac{1^\circ}{2}\right)\right] = 1.414 \text{ a} \quad \text{(i)}$$

The tensile forces in longitudinal reinforcement and inclined bars are maintained equal. The tensile force in the inclined bars is equal to $A_{sv} \cdot \sigma_{sv}$. The portion of shear force resisted by the inclined bar

$$V_s = (A_{sv} \cdot \sigma_{sv}) . \sin 45^\circ \quad \text{(ii)}$$

The inclined bar used as shear reinforcement remains effective for distance equal to lever arm on either side measured along the neutral axis.

In case, the angle of inclination of inclined bars in tension is a with the longitudinal reinforcement, the angle of inclination for concrete in compression is assumed as $(90 - \alpha/2)$ as shown in Fig. 4.13 (B).

In Fig. 4.12 and Fig. 4.13(A) and (B) it is considered that a single inclined bar crosses the diagonal crack, (viz., n = 1). Such a system of arrangement is referred as *single system of lattice girder.*

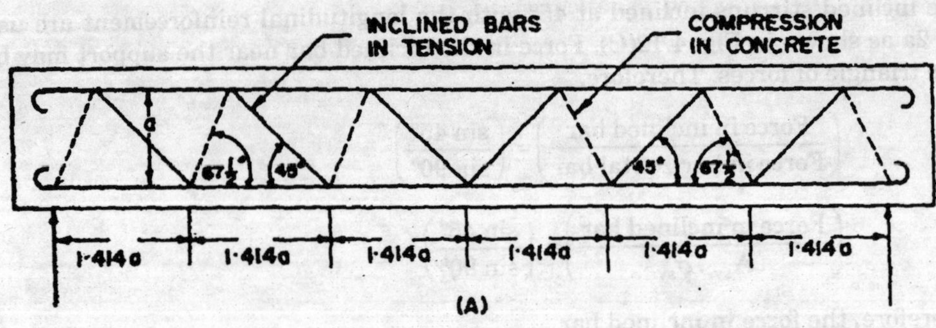

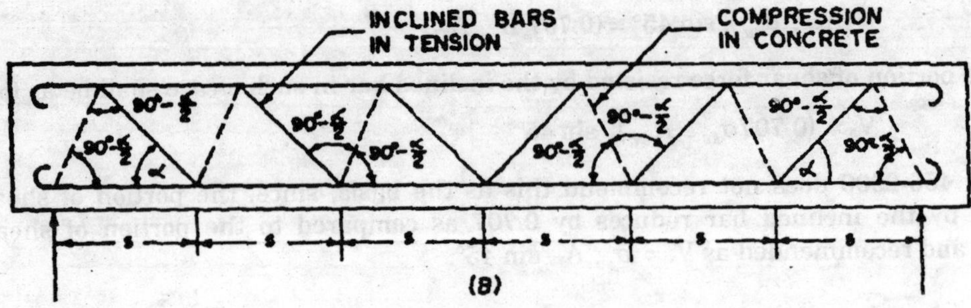

Fig. 4.13

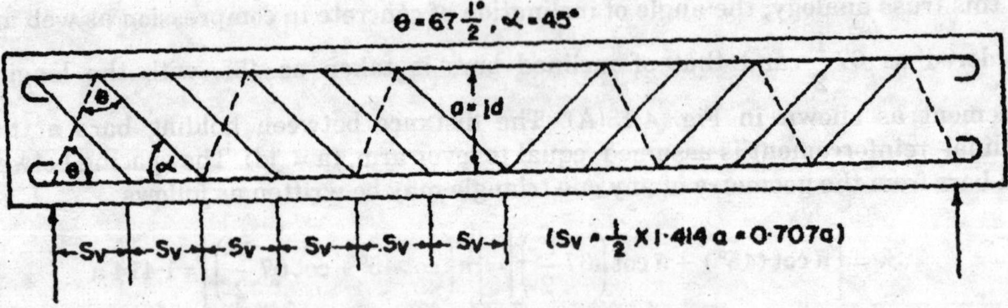

Fig. 4.14

In case, two inclined bars (inclined at an angle a = 45° with the longitudinal reinforcement) cross the diagonal crack (viz., n = 2) such a system is referred as double system of lattice girder. The spacing between two adjacent inclined bars becomes equal to half of that for a single system. Therefore,

$$s_v = (0.5 \times 1.141\ a) = 0.707\ a \qquad \text{(i)}$$

The portion of shear force resisted by the inclined bars shall be

$$V_s = [A_{sv} \,.\, \sigma_{sv} \,.\, (\sin\alpha + \cos\alpha) .\, d/s_v] \qquad \text{(ii)}$$

IS : 456-2000 recommends that angle α should not be less than 45°. For α = 45°, the value of $(\sin\alpha + \cos\alpha)$ is $\sqrt{2}$ maximum.

$$V_s = (2 \times 0.707\ A_{sv}\ \sigma_{sv}) \quad \text{(iii)}$$

For vertical stirrups, α = 90°, and therefore

$$V_s = \left(A_{sv} \cdot \sigma_{sv} \cdot d/s_v\right) \quad \text{(iv)}$$

Truss analogy 3

The action of vertical shear reinforcement in a reinforced concrete beam Fig. 4.15 (A) may be described analogues to the truss action as shown in Fig. 4.15(B). The vertical shear reinforcement resists the vertical component of diagonal tension, while the concrete along the diagonal as shown by dotted line resists the compression. The portion of shear force resisted by vertical reinforcement is $V_s = (A_{sv} \cdot \sigma_{sv} \cdot d/sv,)$ where s_v is the spacing between the two adjacent vertical stirrups.

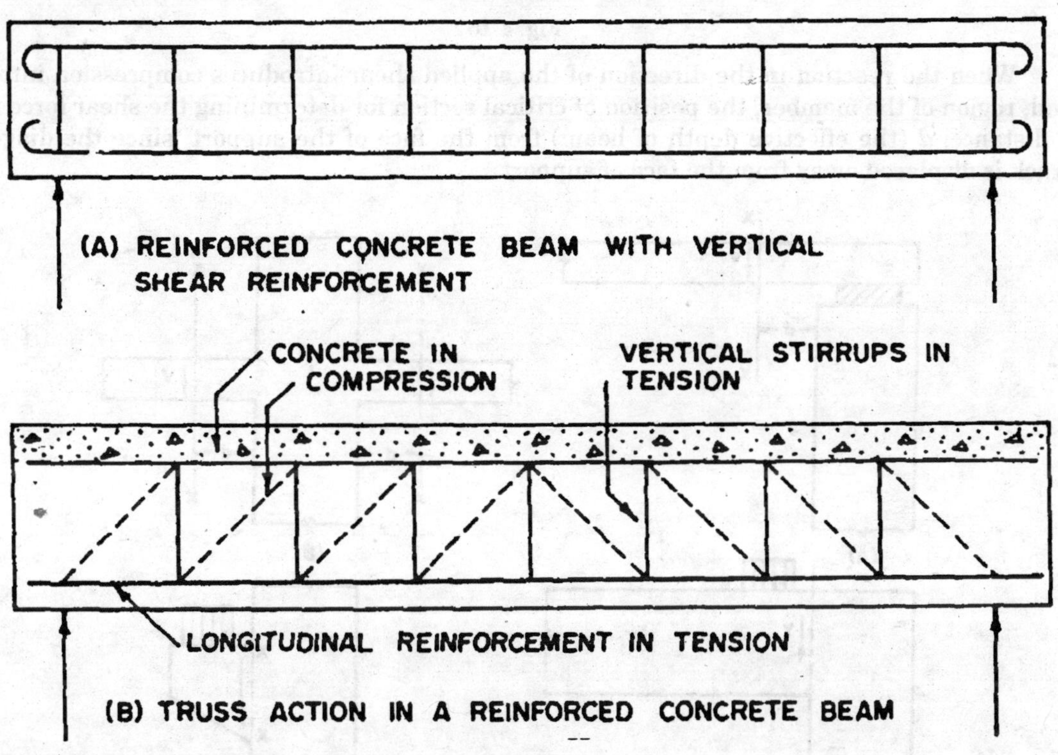

Fig. 4.15

4.12 CRITICAL SECTION FOR SHEAR FORCE

The position of critical section for calculating the shear force depends upon the types of supports. When the reaction in the direction of the applied shear introduces tension into the end region of the member, the position of critical section for calculating the shear force is at the face of the support as shown in Fig. 3.16 (A). In such a case, the support reaction does not introduce compression, as expected in rectangular water tank. A diagonal crack is likely to begin at the face.

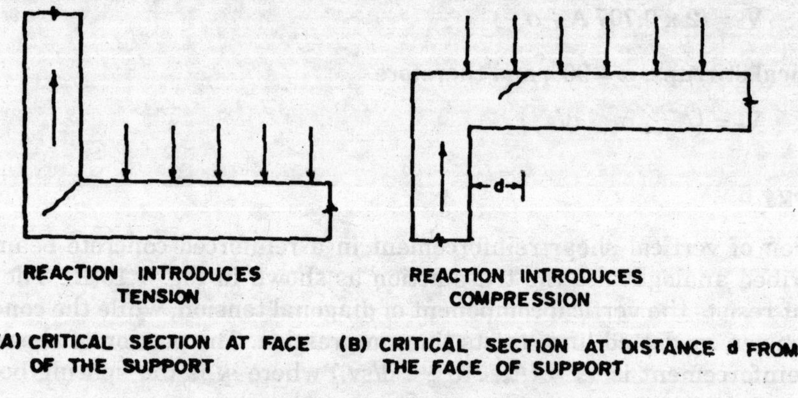

Fig. 4.16

When the reaction in the direction of the applied shear introduces compression into the end, region of the member, the position of critical section for determining the shear force is at *a* distance, *d* (the effective depth of beam) from the face of the support, since the diagonal crack is displaced away from the face of support.

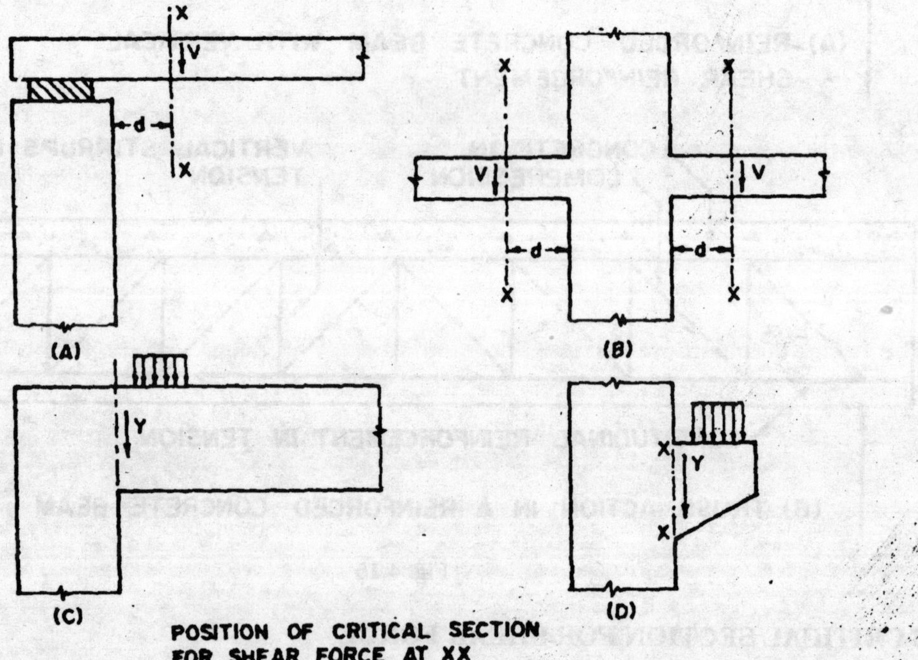

Fig. 4.17

The positions of critical sections for the various types of supports viz., beam is supported by bearing at the bottom of the beam, beam is integrally cast into another member, the beam carries concentrated load close to the support and bracket support are shown in Fig. 4.27 (A), (B), (C) and (D), respectively.

4.13 PERMISSIBLE SHEAR STRESSES IN CONCRETE

The values of permissible shear stresses in concrete in reinforced concrete beam without reinforcement as specified in IS : 456-2000 are as follows :

Table 4.1. PERMISSIBLE SHEAR STRESS IN CONCRETE [τ_c in N/mm² (MPa)]

$\left(\frac{100\,A_s}{b.d}\right)$	τ_c in N/mm²					
	M 15	M 20	M 25	M 30	M 35	M 40
0.25	0.22	0.22	0.23	0.23	0.23	0.23
0.50	0.29	0.30	0.31	0.31	0.31	0.32
0.75	0.34	0.35	0.36	0.37	0.37	0.38
1.00	0.37	0.39	0.40	0.41	0.42	0.42
1.25	0.40	0.42	0.44	0.45	0.45	0.46
1.50	0.42	0.45	0.46	0.48	0.49	0.49
1.75	0.44	0.47	0.49	0.50	0.52	0.52
2.00	0.44	0.49	0.51	0.53	0.54	0.55
2.25	0.44	0.51	0.53	0.55	0.56	0.57
2.50	0.44	0.51	0.55	0.57	0.58	0.60
2.75	0.44	0.51	0.56	0.58	0.60	0.62
3.00 and above	0.44	0.51	0.57	0.60	0.62	0.63

It is to note that A_s is that cross sectional area of longitudinal tension reinforcement which continues upto atleast one effective depth beyond the transverse section being considered. At supports, A_s is the cross-sectional area of complete tension reinforcement which exists there provided it is properly anchored.

Eq. 4.15 is the semi-emperical formula which gives the criterion for determining the value of permissible shear stress, τ_c.

The values of maximum permissible shear stress, $\tau_{c.max}$ for the reinforced concrete beam with shear reinforcement as specified in IS : 456-2000 is as follows :

Table 4.2 . MAXIMUM SHEAR STRESS, $\left(\tau_{c.max}\ in\ N/mm^2\right)$

Grade of concrete	M 15	M 20	M 25	M 30	M 35	M 40
$\tau_{c.max}\ (N/mm^2)$	1.6	1.8	1.9	2.2	2.3	2.5

So long as the nominal shear stress, in beam, τ_c remains less than $\tau_{c.max}$, the crushing of concrete in beam due to diagonal compressive stress shall not occur.

The value of $\tau_{c.max}$ as specified in IS : 456-1978 is calculated from the following expression :

$$\tau_{c.max} = 0.83\,(f_c)^{1/2} = 0.83\,(0.85\,f_{ck})^{1/2}\ N/mm^2 \tag{4.21}$$

Where f_c is the cylindrical strength of concrete. The cylindrical strength of concrete beam is converted to cube strength (f_c = 0.85 f_{ck}), and there after applying a factor of 0.85 being the partial safety factor for material strength (l/r_m).

4.14 DESIGN OF SHEAR REINFORCEMENT

Shear reinforcement in reinforced concrete beam is provided to resist the diagonal tension. At support, the nominal shear stress, τ_v is the measure of diagonal tension. The reinforcement in reinforced concrete beam is designed as follows :

Step 1. The maximum total shear force, V due to total transverse load on the beam occurring at the critical section is determined and the shear force diagram for the complete span of beam is drawn. The nominal shear stress, τ_c is found. The width and the effective depth b and d of the beam cross-section are known.

Step 2. This shear force V is resisted jointly by concrete and shear reinforcement. The value of permissible shear stress in concrete for the reinforced concrete beam without shear reinforcement, τ_c is found as described in Art. 4.13. The cross-sectional area of longitudinal tension reinforcement, A_s and the grade of concrete proposed to be used are already known. The value of maximum shear stress $\tau_{c.\max}$ for concrete in the reinforced concrete beam with shear reinforcement is also noted from IS : 456-2000. The value of nominal shear stress, τ_v calculated in step 1 is compared with $\tau_{c.\max}$ and τ_c .

The nominal shear stress, τ_v in beams should not exceed $\tau_{c.\max}$. In case, the value of τ_v exceeds $\tau_{c.\max}$, the effective depth of beam, d is increased and it is ascertained that τ_c is less than $\tau_{c.\max}$.

When the nominal shear stress, τ_v in beams is less than the τ_c found above, the minimum cross-sectional area for shear reinforcement as given by Eq. 4.19 is provided at a distance somewhat less than or at the most equal to the maximum spacing as described in Art. 4.8.

When the nominal shear stress, τ_v is more than τ_c and less than $\tau_{c.\max}$, the shear reinforcement is provided as designed in the following subsequent step 3 to resist the portion of shear force, V_s equal to $(V - \tau_c . b.d)$.

Step 3. Vertical stirrups, bent up bars along with vertical stirrups and inclined stirrups may be used as shear reinforcement.

3 (A) Vertical or inclined stirrups

In case vertical or inclined stirrups only are provided, it is usually convenient to select a trial shear reinforcement area, A_{sv}. The diameter of steel bars and shape of the stirrups (two-legged, four legged or multiple legged) are selected such that minimum spacing is not too small (preferably not less than 100 mm) and the total number of stirrups in a beam not too large.

The portion of shear force, V to be resisted, the cross-sectional area A_{sv} of the shear reinforcement, the permissible stress σ_{sv} in steel in tension and the effective depth, d of the beam being known, the spacing for the stirrups is determined by Eq. 4.15 for the vertical stirrups or by Eq. 4.22 for the inclined stirrups. The spacing s_v to be provided is kept somewhat smaller than its calculated value. However, s_v should not exceed the maximum spacing specified in Art. 4.8 for vertical stirrups. For the inclined stirrups at 45° the maximum spacing specified is effective depth, d or 450 mm whichever is less.

When the vertical stirrups are needed over a comparatively small distance, it is desirable to space these stirrups uniformly over the entire distance (the spacing being found for the maximum shear). If possible, the same spacing may even be provided in all similar beams in the entire structure in order to avoid errors and confusion. When the shear reinforcement is to be provided over a large distance and in case the shear force changes significantly throughout this distance, it is economical to calculate the spacing needed at more sections. The stirrups are placed accordingly in groups of varying spacing.

3 (B) Vertical stirrups and bent up bars

When the bent up bars are used along with the vertical stirrups, the portion of shear force, V_s is jointly resisted by the bent up bars and vertical stirrups. Therefore,

$$V_s = (V_{S.BB} + V_{S.VS}) \qquad \text{(i)}$$

The portion of shear force contributed towards shear resistance by the bent-up bars, $V_{S.BB}$ shall not be more than that of the total shear reinforcement (viz., $V_{S.BB} \not> 0.5\, V_S$). The strength of bent up bars bent at the same cross-section, $V_{S.BB}$ is equal to $\sigma_{sv}.\ A_{sv}.\ \sin\alpha$. The angle of inclination of bent up bars with the longitudinal reinforcement, a and diameter of bars (being same as that of longitudinal reinforcement) being known, $V_{S.BB}$ becomes known. Therefore, the remaining shear force $V_{S.VS}$ to be resisted by vertical stirrups also becomes known. The vertical stirrups and their spacings are found as described above.

In case, the cross-sectional area and form of shear reinforcement provide in a reinforced beam are specified along with their spacing and inclination, the strength of beam in shear may also be found, by using the appropriate expression Eq. 4.16, 4.20, 4.21. The total shear strength of a beam shall be as follows :

$$V = (V_C + V_s) = (\tau_c\,.\,bd) + V_s \qquad \text{(ii)}$$

Example 4.1 A reinforced concrete simply supported beam 400 mm wide and 550 mm effective depth is having 4 hysd steel bars of 20 mm diameter to resist the maximum bending moment at the centre. 2 hysd steel bars are bent at 45° with the horizontal and carried at the top near and support and 2 hysd steel bars are continued near the end. In addition to this, the vertical two- legged stirrups 8 mm diameter of hysd-steel are also provided at 110 mm spacing as shown in Fig. 3.28. Determine the total shear strength of beat at the transverse section crossing the bent up bars. The grade of concrete is M 20.

Solution : 2 hysd steel bars 20 mm in diameter continue as longitudinal tension reinforcement near the end of beam.

Step 1 : Area of steel reinforcement

$$A_s = 2 \times \left(\frac{\pi}{4}\right) \times 20^2 = 628.32\ mm^2 \qquad \text{(i)}$$

$$p = \left(\frac{100\ A_s}{b\ .\ d}\right) = \left(\frac{100 \times 628.32}{400 \times 550}\right) = 0.2856 \qquad \text{(ii)}$$

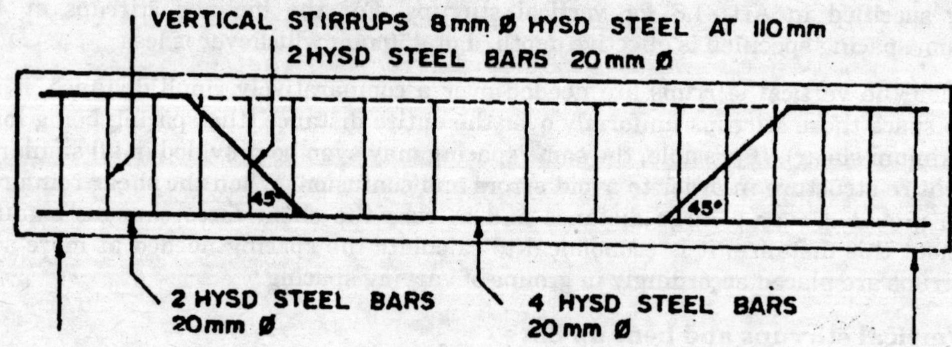

Fig. 4.18

Step 2 : Permissible shear stress in concrete

For M 20 grade of concrete, and 100 A_s/bd = 0.2856, the value of permissible shear stress in concrete from IS : 456-1978

$$\tau_c = \left[0.22 + \frac{(0.30 - 0.22)}{(0.50 - 0.25)} \times (0.2856 - 0.25)\right] N/mm^2$$

Step 3 : Strength of concrete in shear, V_c

$$V_c = \left(\frac{0.2314 \times 400 \times 550}{1000}\right) = 50.908\ kN \qquad \text{(iv)}$$

2 hysd steel bars of 20 mm diameter are bent at 45° at the same cross-section.

$$A_{sv} = 2 \times \left(\frac{\pi}{4} \times (20)^2\right) = 628.32\ mm^2 \qquad \text{(v)}$$

Step 4 : Strength of bent up bars in shear, $V_{S.BB}$

From IS : 456-1978 σ_{sv} is 230 N/mm²

$$V_{S.BB} = \sigma_{sv}\ A_{sv}\ \sin\alpha = \left(\frac{230 \times 628.32 \times 0.707}{1000}\right) = 102.17\ kN \qquad \text{(vi)}$$

Step 5 : Strength of vertical stirrups in shear, $V_{S.VS}$

Cross-sectional area of two legged vertical stirrups of 8 mm diameter

$$A_{sv} = 2 \times \left(\frac{\pi}{4}\right) \times 8^2 = 100.53\ mm^2 \quad and\ s_v = 110\ mm$$

$$V_{S.VS} = \left(\frac{\sigma_{sv} \cdot A_{sv} \cdot d}{s_v}\right) = \left(\frac{230 \times 100.53 \times 550}{110 \times 1000}\right) = 115.61 \ kN \quad \text{(vii)}$$

Step 6 : Total shear strength of beam at the transverse section crossing the bent up bars

$$V = (V_c + V_{S.BB} + V_{S.VS}) = (50.908 + 102.17 + 115.61) = 266.688 \text{ kN}.$$

Example 4.2. Design the beams for shear reinforcement.

Design : From Example 3.8, total intensity of uniformly distributed load per unit length acting over the beam is 27.2 kN/m and the effective span of beam, L is 5.58 m. The critical section for shear, XX Fig. 4.19 for the beam is at a distance d = 570 mm from the face of support.

Step 1 : Shear force, V at the critical section, XX

$$V = \left[\frac{1}{2} \times 27.2 \times 5.58 - 27.2 \times (5.70 + 0.300)\right] = 52.224 \ kN$$

Step 2 : Nominal shear stress

$$\tau_v = \left(\frac{V}{b \cdot d}\right) = \left(\frac{52.224 \times 1000}{400 \times 570}\right) = 0.229 \ N/mm^2$$

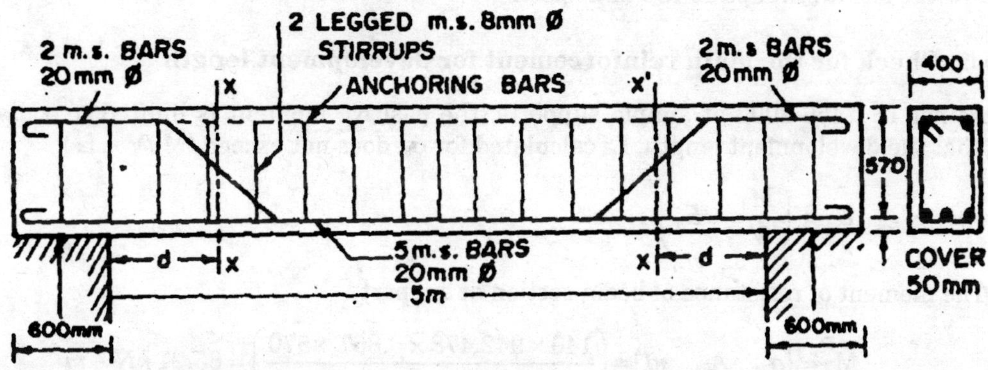

Fig. 4.19

Out of 5 mild steel bars of 20 mm ϕ, 2 bars have been bent- up near the support and 3 bars have been continued upto ends as shown in Fig. 4.19.

Step 3 : Area of steel reinforcement present at the critical section

$$A_{st} = \left(3 \times \frac{\pi}{4} \times 20^2\right) = 942.478 \ mm^2$$

$$\left(\frac{100 \cdot A_{st}}{b \cdot d}\right) = \left(\frac{100 \times 942.478}{400 \times 570}\right) = 0.4134$$

Step 4 : Permissible shear stress in concrete

For M 15 grade of concrete, the permissible shear stress in concrete without shear reinforcement from IS : 456-1978

$$\tau_c = \left[0.22 + (0.29 - 0.22) \times \frac{(0.4134 - 0.25)}{(0.50 - 0.25)}\right] = 0.266\ N/mm^2$$

Maximum shear stress, $\tau_{c.\max}$ from IS : 456-1978 for M 15 grade of concrete is 1.6 N/mm². The nominal shear stress τ_c is less than $\tau_{c.\max}$ and it is also less than τ_c . Hence, the beam needs only minimum shear reinforcement in the form of stirrups. Two legged shear stirrups 8 mm φ mild steel are provided throughout the length of beam as shown in Fig. 4.19.

$$A_{sv} = \left(2 \times \frac{\pi}{4} \times 8^2\right) = 100.531\ mm^2$$

From IS : 456-1978, $\left(\frac{A_{st}}{b\ .\ s_v}\right) \geq \left(\frac{0.4}{f_y}\right)$

$$\text{or } S_v = \left(\frac{A_{st}\ .\ f_y}{0.4 \times b}\right) = \left(\frac{100.531 \times 250}{0.4 \times 400}\right) = 157.08\ mm$$

Provide shear stirrups at 150 mm spacing.

Step 5 : Check for the main reinforcement for development length

As per IS : 456-2000 at simple supports, the positive moment is limited to a diameter such that the development length, L_d calculated for σ_{st} does not exceed ($M_1/V + L_o$).

$$L_d \not> \left(\frac{M_1}{V} + L_o\right)$$

The moment of resistance of beam section at support

$$M_1 = (\sigma_{st}\ .\ A_{st}\ .\ jd) = \left(\frac{140 \times 942.478 \times 0.867\ \times 570}{1000 \times 1000}\right) = 65.21\ kN - m$$

The shear force at support

$$V = \left(\frac{1}{2} \times 27.2 \times 5.58\right) = 75.888\ kN$$

From Eq. 3.53,

$$L_o = \left[\left(\frac{b_s}{2}\right) - x_1 + 13\ \phi\right]$$

End cover, x_1 is 25 mm and the width of support b_s is 600 mm. Therefore,

$$L_o = \left[\left(\frac{600}{2}\right) - 25 + 13 \times 20\right] = 535\ mm$$

$$\left(\frac{M_1}{V} + L_o\right) = \left(\frac{65.21}{75.888} + 0.535\right) = 1.394\ m$$

$$L_d = \left(\frac{58.3 \times 20}{1000}\right) = 1.166\ m$$

The development length, L_d does note exceed ($M_1/V + L_o$). The requirement of IS : 456-2000 is satisfied.

Example 4.3: Design the beam for shear reinforcement and check the longitudinal reinforcement for development length in Example 3.9 at supports.

Design : From Example 3.9, total intensity of uniformly distributed load per unit length acting over the beam is 22.86 kN/m and the effective span of be L is 5.25 m. The critical section for shear, XX Fig. 4.20 for the beam is at a distance d = 500 mm from the face of support.

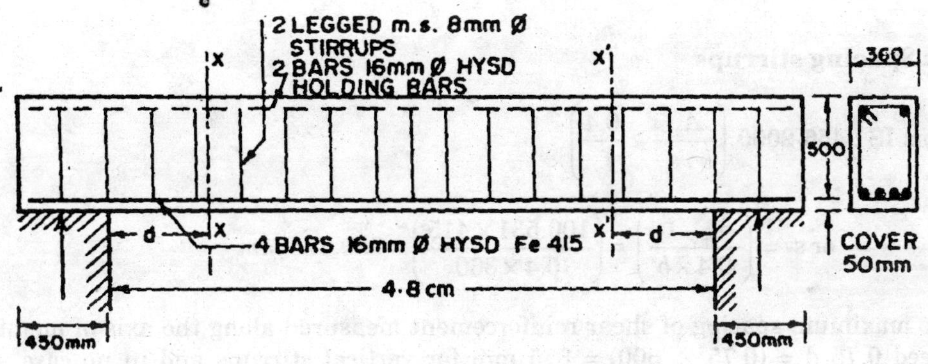

Fig. 4.20

Step 1 : Shear force, V at critical section, XX

$$V = \left[\frac{1}{2} \times 22.86 \times 5.25 - 22.86 \times \left(0.500 + \frac{0.450}{2}\right)\right] = 43.434\ kN$$

Step 2 : Nominal shear stress

$$\tau_v = \left(\frac{V}{b\,.\,d}\right) = \left(\frac{43.434 \times 1000}{360 \times 500}\right) = 0.2143\ N/mm^2$$

4 hysd-steel Fe 415 bars, continue upto ends as shown in Fig. 4.20. The area of steel reinforcement present at the critical section.

$$A_{st} = \left(4 \times \frac{\pi}{4} \times 16^2\right) = 804.248\ mm^2$$

Step 3 : Permissible shear stress in concrete

$$\left(\frac{100 \, . \, A_{st}}{b \, . \, d}\right) = \left(\frac{100 \times 804.248}{360 \times 500}\right) = 0.4468$$

For M 20 grade of concrete, the permissible shear stress in concrete without shear reinforcement from IS : 456-2000.

$$\tau_c = \left[0.22 + (0.29 - 0.22) \times \frac{(0.4468 - 0.25)}{(0.50 - 0.25)}\right] = 0.275 \, N/mm^2$$

Maximum shear stress, $\tau_{c.\max}$ from IS: 456-2000 for M 20 grade of concrete is 1.8 N/mm². The nominal shear stress, τ_v is less than $\tau_{c.\max}$ and it is also less than τ_c. Therefore, the beam requires only minimum shear reinforcement in the form of stirrups. Two legged shear stirrups 8 mm φ hysd-steel, Fe 415 are provided throughout the length of beam as shown in Fig. 4.20.

$$A_{sv} = \left(2 \times \frac{\pi}{4} \times 8^2\right) = 100.531 \, mm^2$$

Step 4 : Spacing stirrups

From IS : 456-2000 $\left(\frac{A_{st}}{b \, . \, s_v} \geq \frac{0.4}{f_y}\right)$

$$\text{or } S_v = \left(\frac{A_{sv} \cdot f_y}{0.4 \times b}\right) = \left(\frac{100.531 \times 415}{0.4 \times 360}\right) = 289.725 \, mm$$

The maximum spacing of shear reinforcement measured along the axis of member shall not exceed 0.75 d = (0.75 × 500) = 375 mm for vertical stirrups and in no case shall the spacing exceed 450 mm. Provide shear stirrups at 280 mm.

PROBLEMS

4.1 The floor of a class room 4 m × 7 m consisting of 150 mm thick reinforced concrete slab is supported by singly reinforced rectangular beams. Design the beams. Adopt the permissible stresses as recommended in IS: 456-1978. Also design the beam for shear reinforcement.

4.2 Design a singly reinforced rectangular beam simply supported at its two ends. The clear span of beam is 5.60 m. The intensity of uniformly distributed super-imposed dead and live loads are 8 kN/m and 6 kN/m. Use M 20 grade of concrete and hysd-steel, Fe 415. Adopt the permissible stresses from IS: 456-1978. Design the beam for shear reinforcement and check the longitudinal reinforcement. For the development length at supports.

4.3 A simply supported rectangular concrete beam 360 mm wide and 480 mm effective depth is having 4 hysd-steel bars of 18 mm diameter to resist the maximum bending moment at the centre, 2 hysd-steel bars are bent at 45° with the horizontal and carried at the top

near the support and 2 hysd-steel bars are continued near the end. In addition to this, two legged vertical stirrups 8 mm diameter of hysd-steel are also provided at 150 mm spacing. Determine the total shear strength of beam at the transverse section crossing the bent up bars. The grade of concrete is M 20.

5

BOND AND DEVELOPMENT LENGTH IN SINGLY REINFORCED RECTANGULAR BEAMS (SRRB - III)

5.1 INTRODUCTION

The bond in reinforced concrete beams is the adhesive force developed between concrete and the steel bars embedded in concrete, which resists any force that tends to push or pull the bars. It is collective expression for all forces such as adhesion, friction due to shrinkage and longitudinal shear in the concrete engage by the bar deformation that resist separation. The adhesive force (bond force) is the interface strength between concrete and steel which prevents slip to occur at the interface. The concrete after setting adheres to the surface of the steel bars placed in the concrete. It is most important requirement of reinforced concrete that a perfect bond exists between concrete and the steel bars used as reinforcement. As the concrete deforms, the steel bars also deforms equally. The strain in an embedded steel bars also deforms equally. The strain in an embedded steel bars, ε_s (unit extension or compression) is equal to the strain in the surrounding concrete, ε_c (viz., $\varepsilon_s = \varepsilon_c$).

The bond strength between concrete and steel bars is the additive effect of chemical adhesion, mechanical friction and frictional resistance (resistance to sliding). Chemical adhesion develops at the concrete-steel interface between cement paste and steel bars. It is attributed to the adhesive properties of the cement gel produced in the concrete during setting. Mechanical friction results due to shrinkage of the concrete and consequently gripping of the reinforcement. Resistance to sliding due to friction is provided by the natural roughness of the mill scale of hot rolled plain steel bars and by the closely spaced rib shaped surface deformations of hysd-steel bars. The relative contributions of these factors have not been separated so far. However, the resistance to slip of plain steel bars is about one- third to one-half of that with the deformed bars.

The beam reinforced with plain steel bars loose the bond strength between steel and concrete at large loads and the collapse of beam is likely to occur as the steel bars are likely to be pulled through the concrete. To prevent this, end anchorages are provided in the forms of hooks at both the ends. With adequate anchorages at both the ends of steel bars, such beam acts as tied arch and does not collapse, even if the bond breaks over the entire length.

The shoulders of the closed spaced projections of rib shaped deformations of hysd-bars bear on the surrounding concrete and provide a high degree of interlocking of two materials and result in greatly increased bond strength. Hysd-steel bars do not need any anchorage devices such as hooks. The crack widths and deflection in beams reinforced with hysd-steel bars also reduce.

The bond in a reinforced concrete beam at a transverse section is a three-dimensional phenomenon and it depends in addition to the cross-sectional properties of the steel bars, upon the anchorage lengths in both the directions therefrom.

5.2 BOND STRESSES

The bond stress is defined as the intensity of adhesive force per unit surface area. The bond stress in a reinforced concrete beam may be found by considering a portion of a beam between two vertical sections AD and BC as shown in Fig. 3.13. The beam section is considered as cracked section. The portion of concrete intension is neglected. Transverse sections AD and BC spaced small distance, δx apart carry moment M and (M + δM) and tensile forces T and δT due to moments, respectively. Then,

$$M = T \, . \, jd \tag{i}$$

and $$(M + \delta M) = (T + \delta T) \, . \, jd \tag{ii}$$

Therefore, $\delta M = \delta T \, . \, jd$

or $$\delta T = \delta M/(jd) \tag{iii}$$

The expression (iii) shows that the change in bending moment δM produces a change in tensile force, δT in the tension reinforcement. The steel bars are in equilibrium. Hence, this change in tensile force, δT is resisted at the contact surface between concrete and steel by an equal and opposite force developed by bond. Then

$$\tau_{bf} \, . \left(\sum O \, . \, \delta x \, . \, 1\right) = \delta T \tag{iii}$$

$$\tau_{bf} \, . \left(\sum O \, . \, 1\right) = \left(\frac{\delta T}{\delta x}\right) \tag{iv}$$

where τ_{bf} is the bond stress developed due to change in tensile force δT as a consequence of change in bending moment. Hence, it is called as flexural bond stress, $\sum O$ is the sum of the perimeters of the steel bars in tension. Substituting the value of δT from the expression (iii)

$$\tau_{bf} \, . \left(\sum O . 1\right) = \left(\frac{\delta M}{jd \, . \, \delta x}\right) \tag{v}$$

From mechanics, it is known that change in bending moment δM over a small length δx is equal to the external shear force, V. $\left(viz., \delta M / \delta x = V\right)$. Therefore,

$$\tau_{bf} = \left(\frac{V}{jd \, , \sum O}\right) \tag{5.1}$$

From Eq. 5.1 it is seen that, finally the flexural bond stress is a function of external shear force, V. It is uniformly distributed over the contact area.

In case, the steel bars are not at equal distance from the neutral axis, then, the general expression for flexural bond stress, τ_{bf} may be found by considering a steel bar in tension provided at a distance y from the neutral axis. The bending stress in steel bar in tension shall be

$$\sigma_{cc} = \left(\frac{M}{I} \,.\, y\right) \quad \text{and} \quad \sigma_{st} = (m \,.\, \sigma_{st})$$

$$\sigma_{st} = \left(m \,.\, \frac{M}{I} \,.\, y\right) \tag{vi}$$

The force in steel bar of diameter ϕ shall be T = (A_{st} . σ_{st}). Therefore

$$T = A_{st} \,.\, \sigma_{st} \tag{vii}$$

$$\text{and} \quad T = \left(\frac{\pi}{4}\phi^2\right).\left(m \,.\, \frac{M}{I} \,.\, y\right) \tag{viii}$$

The change in tensile force, δT

$$\delta T = \left(\frac{\pi}{4}\phi^2\right).\left(m \,.\, \frac{\delta M}{I} \,.\, y\right) \tag{ix}$$

From the expression (iii),

$$\delta T = \tau_{bf} \,.\, (\textstyle\sum O \,.\, \delta dx \,.\, 1)$$

For one steel bar $\sum O = (\pi\phi)$

$$\therefore\ \tau_{bf} \,.\, (\pi\phi . 1\, dx) = \left(\frac{\pi}{4}\phi^2 \,.\, m \,.\, \frac{\delta M}{I} \,.\, y\right)$$

$$\tau_{bf} = \left(\frac{m}{4I} \,.\, \frac{d\,M}{dx} \,.\, y \,.\, \phi\right) \tag{x}$$

$$\text{or} \quad \tau_{bf} = \left(\frac{V}{4I} \,.\, m \,.\, y \,.\, \phi\right) \tag{5.2}$$

where *m is the modular ratio* and *I* is the moment of inertia of equivalent area of beam about neutral axis.

The bond stresses between tensile steel and concreted used to be calculated at a particular transverse section and used to be checked till IS : 456-1978 was issued that these stresses do not exceed the value of safe or permissible stresses in the design of reinforced concrete construction. This check for bond stresses specified in the earlier code is now replaced by the concept of development length described in subsequent article.

5.3 DEVELOPMENT LENGTH OF REINFORCEMENT BARS IN TENSION

The development length of reinforcement bars is defined as the minimum length of the steel bars necessary on either side of a section to develop by bond, the tensile force $A_{st}\ \sigma_s$, where A_{st} is the cross sectional area of the bar and σ_s is the stress in the bar at the section being considered at the design load. The development length is denoted by L_d. It is also termed as embedment length. The expression for the development length may be derived as follows :

A steel bar embedded in concrete is shown in Fig. 5.1. The bar carries a tensile force T equal to $A_{st}.\sigma_{st}$. This force is transmitted to the concrete over the length, L_d by bond stresses over the surface area, $\pi\phi.L_d$, where ϕ is the diameter of bar. The bond force developed in the steel bar is equal to the tensile force and acts in the opposite direction. The steel bar is in equilibrium. Therefore,

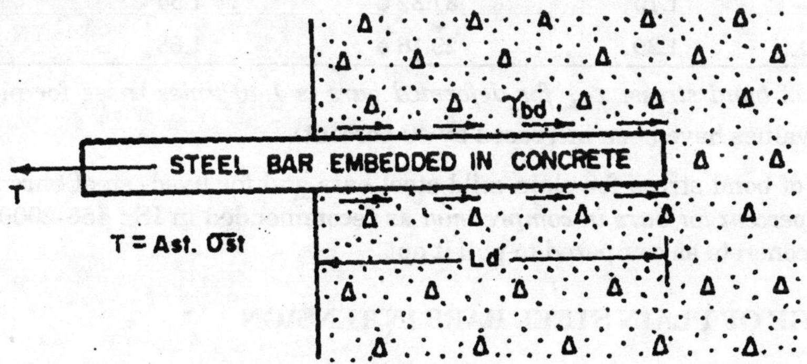

Fig. 5.1

$$\tau_{bd}\ .\ L_d\ .\ \pi\phi = \left[\left(\frac{\pi}{4}\right).\ \phi^2\right].\sigma_{st}$$

$$L_d = \left(\frac{\phi\ .\ \sigma_{st}}{4\ .\ \tau_{bd}}\right) \qquad (5.3)$$

where τ_{bd} is the design bond stress. Since, the bond force in steel bar develops due to anchorage, this bond stress is known as anchorage bond stress and it is different from flexural bond stress. In Eq. 5.3, the appropriate values of σ_{st} and τ_{bd} are to be used. In case the area of steel provided at a section is more than that necessary from the design consideration, the development length may be modified. The modified development length

$$L_{dm} = \left(\frac{A_{st}\ .\ needed}{A_{st}\ .\ provided}\right).\ L_d \qquad (5.4)$$

The values of permissible stresses in bond in tension and the development length, L_d for plain mild steel bars and hysd steel bars for different grades of concrete as specified in IS : 456-2000 and calculated by using Eq. 5.3, respectively are as follows :

Table 5.1. PERMISSIBLE STRESSES IN BOND, τ_{bd} AND DEVELOPMENT LENGTH, L_d IN TENSION

Grade of concrete	Plain mild steel bars $\phi \le 20\ mm$ $\sigma_{st} = 140\ N/mm^2$		Hysd-steel bars Fe 415 $\sigma_{st} = 230\ N/mm^2$	
	$\tau_{bd}\ (N/mm^2)$	L_d	$\tau_{bd}\ (N/mm^2)$	L_d
M 15	0.60	58.33 ϕ	0.84	68 .45 ϕ
M 20	0.80	43.75 ϕ	1.12	51.34 f
M25	0.90	38.88 ϕ	1.26	45.63 ϕ
M 30	1.00	35.00 ϕ	1.40	41.07 ϕ
M 35	1.10	31.82 ϕ	1.54	37.34 ϕ
M40	1.20	29.16 ϕ	1.68	34.23 ϕ

The values of *bond stress, τ_{bd} for deformed bars is 1.40 times* those for plain mild steel bars (*viz*., these values have been increased by 40 percent).

The values of bond stress for plain mild steel bars and for hysd- steel bars in tension are increased by *25 percent for bars in compression* as recommended in IS : 456-2000. It is difficult to push a bar in concrete as compared to pull it out.

5.4 ANCHORAGE OF PLAIN STEEL BARS IN TENSION

The *natural bond* strength in beams reinforced with plain mild steel bars used to be comparatively weak. It is likely that bond between concrete and steel bars may break and may collapse as the desired tension in steel bar may not develop, by bond alone. In order to avoid even the possibility of collapse of beam, special anchorage are provided at the ends of bar in tension, commonly by means of 90° bends or 180° hooks. There shall be no possibility of collapse of beam with adequate anchorage at the ends of steel bar even if the bond breaks over the complete length between the anchorage. In such a case, the beam shall be acting as a tied arch.

Hooks and bends

Hooks and bends are the *mechanical devices*, which are provided at the ends of steel bars as anchorages. The hooks and bends are capable of developing strength in the steel bars without any damage to concrete. These hooks and bends have standard dimensions as shown in Fig. 5.2, confirming to IS : 2502-1963.

Hooks are 180° bends viz., semi-circles plus an extension of at least 4 times the bar diameter at the free end of bar as shown in Fig. 5.2 (A).

90° bends are right angle bends plus an extension of at least 4 times the bar diameter at the free end of bar as shown in Fig. 5.2 (B).

The minimum value of k for mild steel is 2 and for cold worked steel is 4.

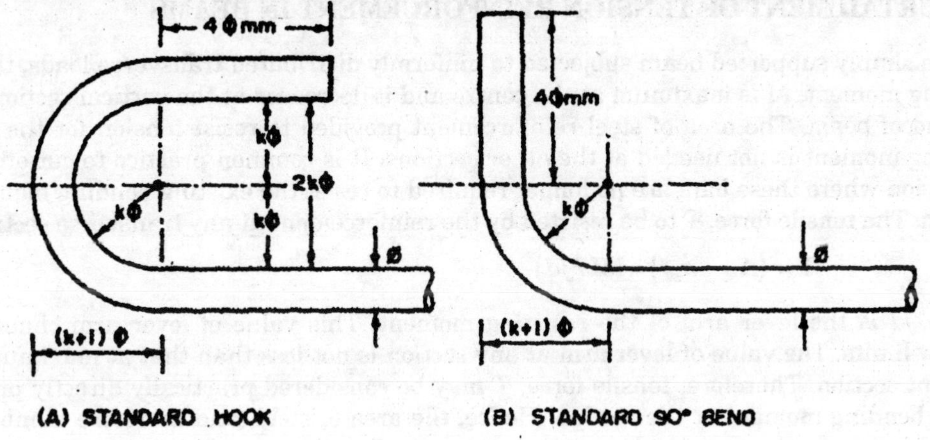

Fig. 5.2

Anchorage values of hooks and bends

Hooks and bends resist the tension in the steel bars by combined action of bond along the straight length of bar and anchorage provided by the hooks or bends at the ends. The tension T = A_{st}. σ_{st} acting in the straight portion of a steel bar (as shown in Fig. 5.3) of a right angled bend in resisted jointly by bearing resistance of vertical portion and resistance due to bond between concrete and steel bar over the length, L_d. The steel bar with right angle bend is equilibrium. Therefore,

$$\left(L_d \cdot \tau_{bd} \cdot \pi\phi + \sigma_{cbr} \cdot n\phi \cdot \phi\right) = (\pi/4)\phi^2\, \sigma_{st}$$

$$L_d + \left(\frac{n\ \sigma_{br} \cdot \phi}{\pi \cdot \tau_{bd}}\right) = \left(\frac{\phi \cdot \sigma_{st}}{4\ \tau_{bd}}\right) \qquad \text{(i)}$$

From Eq. 5.3, $L_d = L_d = [\phi\ \sigma_{st}/(4\tau_{bd})] = (L_d + k \cdot \phi)$ (ii)

where, *k. ϕ represents anchorage value of bend. The anchorage value of a bend is taken as 4 times the diameter of the bar for each 45° bend subject to a maximum of 16 times the diameter of the bar. The anchorage value of a standard U-type hook is 16 times the diameter of the bar. The anchorage value of 90° bend is 12 times the diameter* of the bar $(8\ \phi + 4\ \phi\ extension = 12\ \phi)$.

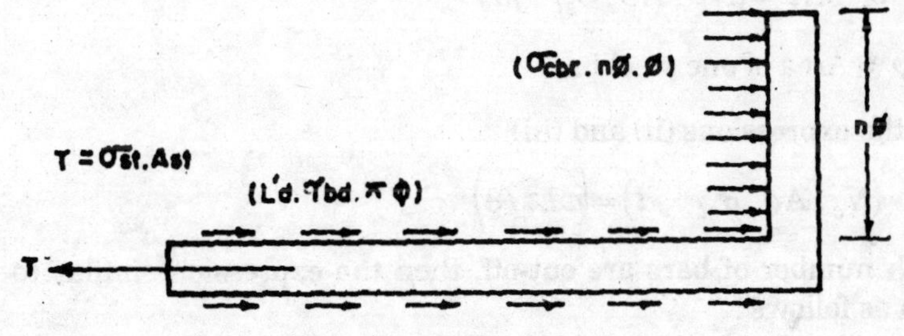

Fig. 5.3

5.5 CURTAILMENT OF TENSION REINFORCEMENT IN BEAMS

In simply supported beam subjected to uniformly distributed transverse loads, the value of bending moment, M is maximum at the centre and it decreases at the vertical sections towards the end of beam. The area of steel reinforcement provided to resist tension for the maximum bending moment is not needed at the other sections. It is common practice to cut-off steel bars in tension where these bars are no longer required to resist the existing bending moment at the section. The tensile force, T to be resisted by the reinforcement at any transverse section is

$$T = (A_{st} \cdot \sigma_{st}) = (M / jd) \qquad \text{(i)}$$

where, jd is the lever arm of the resisting moment. This value of lever arm changes within narrow limits. The value of lever arm at any section is not less than that at maximum bending moment section. Therefore, tensile force, T may be considered practically directly proportional to the bending moment at the section. Hence, the area of steel (and therefore number of bars) needed in tension is very nearly proportional to the bending moment. It is desirable that steel reinforcement in tension everywhere in the beam is as nearly fully stressed as possible. Therefore, the excess number of bars are cut-off.

Let N_c be the total number of steel bars in tension and N_x be the number steel bars cut-off at a distance x on either side measured from the centre. There

$$\left(\frac{N_x}{N_c}\right) = \left(\frac{4x^2}{L^2}\right) \qquad \text{(i)}$$

$$\text{or} \quad x = \frac{1}{2}\left(\frac{N_x}{N_c}\right)^{1/2} \qquad (5.5)$$

Alternatively, Eq. 5.5 may also be found by equating the bending moment at any section to the moment of resistance of the section calculated by considering only the continuing bars. The maximum bending moment for the uniformly distributed load w per unit length over the simply supported beam

$$M_{max} = (wL^2/8) \qquad \text{(ii)}$$

The moment of resistance of beam section resisting the maximum bending moment

$$MR = (T \cdot jd) = (A_{st} \cdot \sigma_{st} \cdot jd)$$

$$\text{or } MR = (N_c \cdot A\phi \cdot \sigma_{st} \cdot jd) \qquad \text{(iii)}$$

where, $A\phi$ = Area of one steel bar.

Equating the expressions (ii) and (iii)

$$(N_c \cdot A\phi \cdot \sigma_{st} \cdot jd) = (wL^2/8) \qquad \text{(iv)}$$

In case, N_x number of bars are cut-off, then the expression similar to the expression (iv) may be written as follows :

$$[(N_c - N_x)A\phi](\sigma_{st} \cdot jd) = \left(\frac{wL^2}{8} - \frac{w\,(2x)^2}{8}\right) \qquad \text{(v)}$$

By dividing the expression (v) by the expression (iv)

$$\left(\frac{(N_c - N_x)}{N_c}\right) = \left(\frac{wL^2/8 - w.4x^2/8}{wL^2/8}\right) \qquad \text{(vi)}$$

$$\text{or} \quad \left(1 - \frac{N_x}{N_c}\right) = \left(1 - \frac{4\,x^2}{L^2}\right) \qquad \text{(vii)}$$

$$\text{Therefore, } x = \frac{L}{2}\left(\frac{N_x}{N_c}\right)^{1/2} \qquad \text{(viii)}$$

The tension reinforcement is not curtailed exactly at the theoretically determined cut-off points. Prior to cracking of concrete, the tensile force in steel bars at any point is proportional to the moment at a transverse section passing through the point. After formation of a crack due to diagonal tension, an internal redistribution of forces occurs in a beam. The tensile force is then controlled by the moment at a section nearer the midspan, which one is more than that at the theoretical cut-off points. The actual bending moment diagram may differ from that used for the design of beam due to approximation of real loads, approximation in analysis or the superimposed of settlement or lateral loads. For these reasons, IS : 456-2000 recommends that a bar to be curtailed at a theoretical cut-off point is continued at least at a distance equal to the effective depth or 12 times the diameter of bar (whichever is greater) except at simple support or end of cantilever as shown in Fig. 5.4.

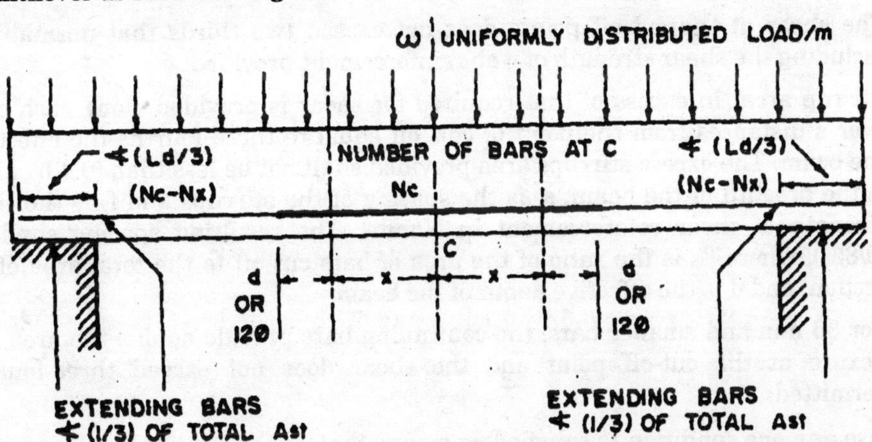

Fig. 5.4

In addition to this, it is also essential that the calculated tensile stress in reinforcement at each section be developed by adequate embedded length or by end anchorage or by combination of both. Hence, a development length, L_d must be provided beyond the critical sections at which the maximum stress (maximum bending moment), exists in the bars.

Further, IS: 456-2000, recommends that at least one third the positive moment reinforcement is extended along the same face of the member into the support, to a length equal to $L_d/3$. This takes care of possibility of change in location of maximum bending stress.

IS : 456-2000 also recommends that atleast one-third of the total reinforcement provided for the negative moment at the support shall extend at least L_d beyond the theoretical point of cut-off of bars and in addition must extend beyond the point of inflection for a distance not less than the effective depth of 12ϕ or one-sixteenth of the clear span, viz., $L_n/16$ whichever is greater.

The steel bars provided to resist the bending moment may be curtailed in tension zone only when one of the conditions specified in Art. 5.6 is satisfied.

5.6 CONDITIONS FOR CURTAILMENT OF FLEXURAL REINFORCEMENT IN TENSION ZONE

Curtailment of flexural reinforcement in tension zone results in a sharp discontinuity in the steel. In causes early formation and opening of flexural cracks, which may change into premature formation of diagonal tension cracks in the surrounding of the cut end. As a result of this, the shear capacity and overall ductility of the beam may decrease. In order to take special precautions, following conditions have been specified in IS: 456-2000. The flexural reinforcement may be curtailed in tension zone only when one of the following condition is satisfied :

1. The shear at the cut-off points does not exceed two thirds that normally permitted, including the shear strength of web reinforcement provided.
2. Stirrup area in excess of that required for shear is provided along each curtailed bar over a distances from the point of cut- off equal to three fourths the effective depth of the beam. The excess stirrups area provided shall not be less than $(0.4\ b\ .\ s_v/f_y)$, where b is the breadth of the beam, s_v is the spacing of the stirrups and f_y is the characteristic strength of shear reinforcement in N/mm². The resulting spacing shall not exceed $(d/8\beta_b)$, where β_b is the ratio of the area of bars cut-off to the total area of bars at the section, and d is the effective depth of the beam.
3. For 36 mm and smaller bars, the continuing bars provide double the area required for flexure at the cut-off point and the shear does not exceed three-fourths that is permitted.

In case any one condition is satisfied, it means that either the shear stresses are kept low (condition -1) or extra shear reinforcement is provided (condition-2) or excess flexural steel is available along with excess shear capacity (condition - 3). In the rare case, the condition 1 or 2 cannot be satisfied and bars are greater than 36 mm diameter.

Condition - 3 is applicable only when the diameter of longitudinal reinforcement bar is equal to or less than 36 mm. In case, the diameter of bar is more than 36 mm, then either condition - 1 or the condition - 2 has to be satisfied.

5.7 SPECIAL REQUIREMENTS NEAR POINTS OF ZERO MOMENT FOR CURTAILMENT OF TENSION REINFORCEMENT

The provision of full development length, L_d beyond the point where the bar is having maximum stress in the basic requirement. To insure safety against bond stress, this requirement may not be sufficient. Consider a simply supported beam carrying uniformly distributed load designed in such a way that design stress is developed at the central section, which is required to resist the maximum bending moment, $M = wL^2/8$. The shape of bending moment diagram is parabola. Equation of parabolic bending moment diagram with A (support) as the origin is given by

$$y = (4a/L^2) . x . (L - x) \tag{i}$$

At $x = L/2, y = M, a = M$

$$\therefore y = 4M. x . (L-x)/L^2 \tag{ii}$$

At $x = (L/4), y = 0.75 m$ (iii)

At L/4 from the support A the reinforcement is to resist 0.75 M where as the reinforcement would develop only half the design stress. Therefore, it is not possible to satisfy the requirement of code for development at quarter-span simply by satisfying this at mid span. Therefore, special requirement have been specified in IS : 456-2000 for the curtailment of tension reinforcement at supports and at point of zero moments. IS : 456-2000 recommends that simple supports and at points of zero moment, the diameter of positive moment tension reinforcement shall be limited such that the development length calculated for allowable stress, σ_{st} does not exceed $[(M_1/V) + L_o]$, viz.,

$$L_d \ngtr [(M_1/V)+L_o] \tag{5.6}$$

where M_1 is the moment of resistance of the section assuming all reinforcement at the section to be stressed to σ_{st}. V is the shear force at the section due to design loads and L_o is the sum of the anchorage beyond the centre of support and equivalent anchorage value of any hook or a mechanical anchorage at simple support; Fig. 5.5 (A); and at point of zero moment, L_o is limited to the effective depth of beam or 12ϕ, whichever is greater, ϕ is the diameter of bar.

The value of (M_1/V) in Eq. 5.6 may be increased by 30 percent when the ends of reinforcement are confined by a compressive reaction.

Consider an infinitesimal length of the beam very near the support. It is clear that if the development length requirement is satisfied at point P, the development length requirement specified in the code may always be satisfied in the portion to the left of point P. The point P may be located as follows :

From the expression (v). Art. 5.2, the bond force per unit along the tension requirement in a beam is given by

$$\tau_{bf} . (\sum O.1) = (dM / jd \; . d_x) \tag{iv}$$

Expression (iv) shows that the bond force per unit length of bar developed by bending is proportional to the slope of bending moment diagram. From the elementary mechanics, (dM/dx) at any point is equal to the shear force at that point, V. Draw a tangent to the bending moment diagram at the point A (point of zero moment) having a slope equal to the value of shear force, V, Fig. 5.5 (B).

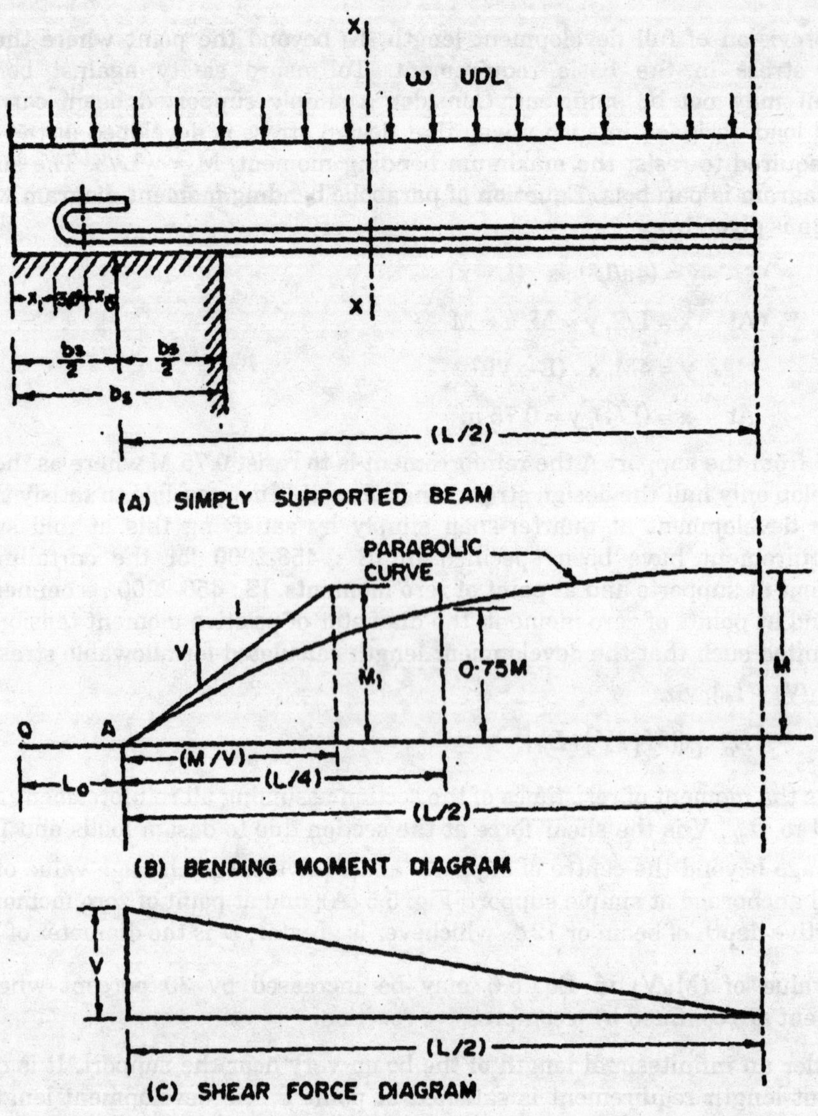

Fig. 5.5

In case M_1 is the moment of resistance provided by those bars that extend to the point of zero moment, and the bending moment diagram is conservatively assumed to vary linearly along the tangent, then,

$$(M_1/a_1) = V, \text{ and } a_1 = (M_1/V) \qquad \text{(v)}$$

Therefore, P is at a distance AP = a_1 = (M_1/V) from the support. The distance AP should be more than L_d, or L_d should not exceed (M_1/V), so that the pullout of bar would not occur.

However, the reinforcement bars are extended beyond the point A (the point of zero moment, towards the left Fig. 5.5 (B) to point Q. This extension of reinforcement is contribution towards satisfying the embedded length.

The length L_o = AQ may be calculated by knowing the width of support, b_s. Then, from Fig. 5.5 (A)

$$L_o = (x_o + 16\phi) \quad \text{(vi)}$$

$$\text{and } x_o = [(b_s/2) - x_1 - 3\phi] \quad \text{(vii)}$$

where, x_1 is the end cover for the reinforcement and 16ϕ is the anchorage value for the standard hook. Therefore, from expression (vi) and (vii)

$$L_o = [(b_s/2) - x_1 - 3\phi] + 16\phi \quad \text{(viii)}$$

$$L_o = [(b_s/2) - x_1 + 13\phi] \quad (5.7)$$

In case, the hook is not provided, then

$$L_o = [(b_s/2) - x_1] \quad (5.8)$$

In case, any inadequately occurs in Eq. 5.6, more bars may be brought within the support A. so that, M_1 is increased. Otherwise L_o may be increased by providing larger width for the support. But L_o is also limited to the effective depth of beam or 12ϕ, which- ever is greater. It is best to use smaller size steel bars, so that L_d is reduced.

5.8 BAR SPLICES

Small size steel bars are stocked by the supplier in lengths of 6 m or 12 m and the large size steel bars are stocked in 18 m lengths. Sometimes the length of bar required is more than the length of steel bars available. Sometimes, it is convenient to work with bars of small lengths. As such, it becomes necessary to splice the steel bars to be used as reinforcement in the structural members. The overlapping of the steel bars a sufficient length to transfer the stress by bond from one bar the other bar as shown in Fig. 5.6 is known as bar splicing.

Fig. 5.6

In order to keep the bars in position during the pouring of the concrete, the lapped bars placed in contact are usually lightly tied by wires. Alternatively, to provide full positive connection between the bars, the bars may be spliced by welding or by sleeves or by mechanical devices.

The splices in the reinforcing bars are provided at the points away from the sections of maximum stress and these are staggered.

IS: 456-2000 recommends that the bar splices in flexural members should not be located at sections where the bending moment is more than *fifty per cent* of the moment of resistance of the beam section. The number of bars spliced at a section should be less than half the total number of bars at that section.

Where the splices are made at points of maximum stress or where more than one-half of the bars are spliced at a section, special precautions shall be taken, such as increasing the length of lap and/or using spirals or closely spaced stirrups around the length of the splice. IS : 456-1978 specifies the recommendations for lap splices in bars in tension as follows :

Lap splices in tension

The lap splices shall not be used for bars larger than 36 mm. For larger diameters the bars may be welded. In cases, where welding is not practicable, the lapping of bars larger than 36 mm be allowed and in such case additional spiral should be provided around the lapped bars. The slap splices are considered as staggered if the centre to centre distance of the splices is not less than 1.3 times the lap length calculated as below :

The lap length including anchorage value of hooks in flexural tension shall be L_d or 30ϕ whichever is greater and for direction tension $2L_d$ or 30ϕ whichever is greater. The straight length of the lap shall not be less than 15ϕ or 200 mm.

The diameter of spiral bars used for enclosing shall not be less than 6 mm diameter with pitch not more than 100 mm. Hooks are provided at the ends of bars.

Where the bars of two different diameters are to be spliced, the lap length shall be calculated on the basis of diameter of the smaller bar.

In case of bundled bars, the lap splices shall be made by splicing one bat at a time. Such individuals splices within a bundle shall be staggered.

The design strength of a welded splice or a mechanical connection shall be taken as equal to eighty per cent. When the spliced area forms not more than twenty per cent of the total area at the section, 100 per cent design strength may be assumed.

When the splicing of welded wire fabrics is to be carried, lap splices of wires shall be made so that overlap measured between the extreme cross-wire shall be not less than the spacing of cross wires plus 100 mm.

Example 5.1. *In Example 4.2, check the main reinforcement for development length.*

From Example 4.2, As per IS: 456-1978, at simple supports, the positive moment is limited to a diameter such that the development length, L_d calculated for σ_{st} does not exceed (M_1/V + L_o), viz., from Eq. 5.6

$$L_d \ngtr \left(\frac{M_1}{V} + L_o\right)$$

The moment of resistance of beam section at support

$$M_1 = (\sigma_{st} \cdot A_{st} \cdot jd) = \left(\frac{230 \times 804.248 \times 0.9038 \times 500}{1000 \times 1000}\right) = 83.59\ kN - m$$

The shear force at support

$$V = \left(\frac{1}{2} \times 22.86 \times 5.25\right) = 60.01 kN$$

From Eq. 3.54 (since, hook is not provided in hysd-steel reinforcement)

$$L_o = \left(\frac{b_s}{2} - x_1\right)$$

End cover, x_1 is 25 mm and the width of support b_s is 450 mm. Therefore,

$$L_o = \left[\left(\frac{450}{2}\right) - 25\right] = 200 \; mm$$

$$\left(\frac{M_1}{V} + L_o\right) = \left(\frac{83.59}{60.01} + 0.2\right) = 1.593 \; m$$

$$L_o = \frac{68.45 \times 16}{1000} = 1.0952 \; m.$$

The development length, L_d does not exceed ($M_1/V + L_o$). The requirement of IS : 456-2000 is satisfied.

PROBLEMS

5.1. In Example 4.1, Check the main reinforcement for the development length.

5.2. In Example 4.2, Check the longitudinal reinforcement for the development length.

6

DEFLECTIONS AND THEIR LIMITATIONS IN SINGLY REINFORCED RECTANGULAR BEAMS (SRRB - IV)

6.1 INTRODUCTION

The reinforced concrete beams subjected to bending are designed to have adequate stiffness. In order to ensure serviceability certain control on deflection of beams has been imposed in code IS: 456-2000. Excessive, deflection in beam may lead to cracking of supported walls and partitions, ill-fitting doors and windows, poor roof drainage, misalignment of sensitive machinery and equipment or visually offensive safe-affecting the appearance of structure. Psychologically, the excessive deflection upsets the occupants. The deflection of beam shall not exceed the following limits :

The final deflection due to all loads (including the effects of temperature, creep and shrinkage) measured below the as cast level of floors, roofs and other horizontal members should not normally exceed (span/250). This limitation is based on limitation on crack width.

Part of the deflection (including the effects of temperature, creep and shrinkage) occurs after the erection of partitions and the application of finishes should not normally exceed span/350 or 20 mm whichever is less.

IS : 456-2000 code recommends that the above limits on vertical deflections of beam may be considered to be satisfied provided that the span to depth ratios specified therein are not exceeded. The basic values of span to effective depth ratios have been arrived by considering fully elastic simply supported rectangular beam carrying uniformly distributed load w per unit length. The maximum bending moment, M occurs at the centre of beam of effective span, L. Let the permissible stress in concrete in bending in compression be σ_{cbc}. The beam section may resist the bending moment M as under

$$M = \left(\sigma_{cbc} \, . \, Z\right) = \left(\sigma_{cbc} \, . \, bD^2/6\right) = \left(wL^2/8\right) \qquad \text{(i)}$$

$$\therefore \quad w = \left(8 \, . \sigma_{cbc} \, . \, b\right) . \left(D/L\right)^2/6 \qquad \text{(ii)}$$

The maximum deflection of beam occurring at the centre

$$y_{max} = \frac{5}{384}\left(\frac{wL^4}{El}\right) \quad \text{(iii)}$$

From the expressions (ii) and (iii)

$$\left(\frac{y_{max}}{L}\right) = \left(\frac{5}{24} \times \frac{\sigma_{cbc}}{E}\right) \times \left(\frac{L}{D}\right) \quad \text{(iv)}$$

$$\text{or} \quad \frac{L}{D} = K_1 \cdot \left(\frac{y_{max}}{L}\right). \ where\ K_1 = \frac{24}{5} \cdot \left(\frac{E}{\sigma_{cbc}}\right) \quad \text{(v)}$$

The deflection of beam remains limited to a given fraction of span, in case span/depth ratio is constant for a given elastic material. IS : 456-2000 specifies the basic values of span to effective depth ratios (instead of overall depth) for the following limits of span.

(i) For span upto 10 m

For cantilever beams ...7

For simply, supported beams ...20

For continuous beams ...26

(ii) For span above 10 m

Above specified values are multiplied by (10/span) 1 metres except for cantilever beams for which the actual calculations for deflection should be made.

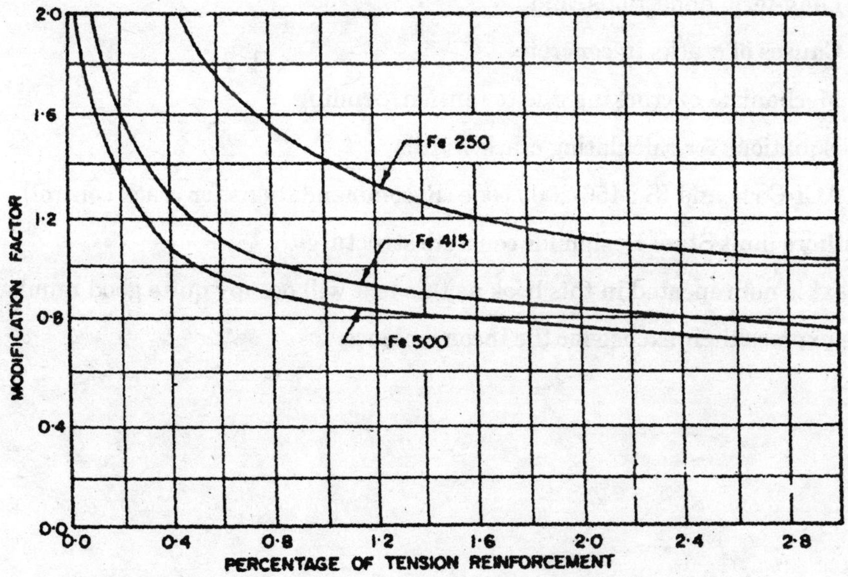

Fe 250 (MILD STEEL), Fe 415 HYSD STEEL, Fe 500 HYSD STEEL
MODIFICATION (MULTIPLICATION) FACTOR FOR TENSION REINFORCEMENT.

Fig. 6.1

The basic values of ratios of span of effective depth are modified depending on the area of steel reinforcement in tension as the deflection of beam is influenced by reinforcement and the service stress. The area of steel for tension reinforcement is taken at the centre of span of simply supported beam and at the support of span of cantilever beams. Above specified basic values are multiplied by multiplication factor, m (referred as modification factor in code). The value of multiplication factor, m is found for curves for different grades of steel as shown in Fig. 6.1 (reproduced from code IS : 456-2000).

6.2 SLENDERNESS LIMITS FOR BEAMS

In order to ensure lateral stability and to exclude failure of beams by sideways bending and bucklings, it is necessary to impose limits for the clear distance between lateral restraints (viz., lateral supports along the length of beam).

A simply supported or continuous beam shall be so proportioned that the clear distance between the lateral supports (as specified in IS : 456-1978) does not exceed 60.b or (250 b^2/ whichever is less, where b is the width of the compression face midway between the lateral restraints and d is the effective depth of the beam.

For a cantilever, the clear distance from the free end of the cantilever to the lateral support shall not exceed 25 b or (100b^2/d) whichever is less.

NOTE: The students, colleagues, and teachers of various Institutions and colleges are requested to study the text and illustrative examples concerning following

(i) Effects of deflections

(ii) Short-term deflections

(iii) Long-term deflections, and

(iv) Causes of cracks in concrete

(v) Mechanism of cracking due to tension, bending

(vi) Equations for calculating crack widths

(vii) ACI-Code and IS : 456-2000 code (Recommendations for crack control)

from authors limit State Design for concrete structures.

Above text is not repeated in this book as this text will occupy quite good number of pages.

Authors express their excuse for the inconvenience.

7

BENDING (FLEXURAL) STRENGTH AND DESIGN OF DOUBLY REINFORCED RECTANGULAR BEAMS (DRRB-I)

7.1 INTRODUCTION

When the singly reinforced beams need considered depth to resist large bending moment, then the beams are also reinforced in compression zone in addition to the reinforcement in tension as shown in Fig. 7.1 (A). Similarly, when the size of rectangular beam cross-section is limited (*viz.*, the dimensions of concrete section are fixed) because of architectural considerations and practical reasons and the bending moment is decidedly more than the moment of resistance of the beams section if ideally reinforced, the use of both tension and compression reinforcements becomes necessary. In such a case, it happens that the concrete alone cannot develop the compressive force and the compressive stresses in the extreme fibre required to resist the given bending moment, the beams are strengthened by providing reinforcement in compression zone. *The beams having reinforcements in tension and compression zones both are called as doubly reinforced beams.* When as beam is continuous over several supports, the section of beam at the support is usually designed as a doubly reinforced beam. In continuous beams, the moving the partial live loads may cause the points of infection to move closer to the supports under some conditions so that some tension reinforcement is needed near the supports. The doubly reinforced beams are also used when the beams are subjected to eccentric loading, shocks, impact or accidental lateral thrust or when the bending moment reverses due to loading. The lintels, braces, walls of storage reservoirs, brackets, etc. are the example of doubly reinforced beams. The doubly reinforced beams are considered as uneconomical beams, as the strength of compression reinforcement is not fully utilized.

The reinforcement in compression zone is also used for the reasons other than the strength. When the compressive reinforcement is added in the beams, the long-term deflection in beams reduce. In some cases, the steel bars are placed in compression zone for minimum-moment loading.

The strain in concrete in compression surrounding the steel bars placed in compression zone, ε_c and that in steel in steel bars used as compression reinforcement, ε_{sc} due to flexure are equal. Therefore, *the stress in steel bars, σ_{sc} is modular ratio, m_c times the stress in* *concrete*, α_{cbc_1}. It is seen that the actual stress in steel bars used as compression reinforcement always remains less than its respective permissible value.

$$\sigma_{sc} = (m_c \cdot \sigma_{cbc_1}) \quad (7.1)$$

From the bending stress distribution diagram, Fig. 7.1 (B)

$$\sigma_{cbc_1} = [(n - d_c)/n] \cdot \sigma_{cbc} \quad (7.2)$$

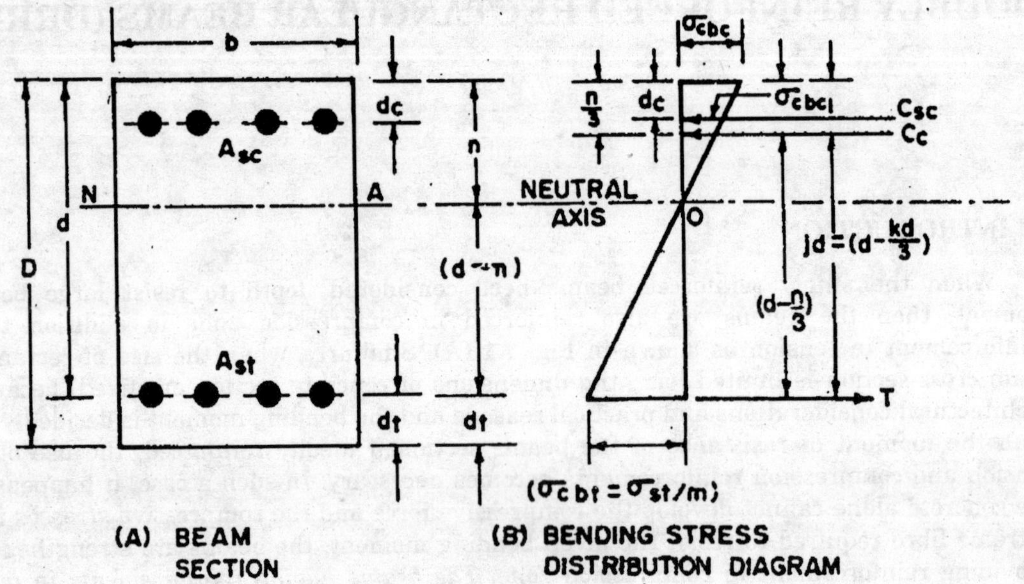

Fig. 7.1

The modular ratio of concrete in compression, m_c is *1.5 times* the modular ratio of concrete in tension. The assumptions made for theory of singly reinforced beams are also applicable for design of doubly reinforced beams by working stress method.

The permissible stresses in concrete in compression in bending and shear for different grades of concrete and those in steel reinforcement in tension are same as specified for the design of singly reinforced beams. However, the permissible stresses in compression in mild steel and high yield strength deformed reinforcements are 130 N/mm² and 190 N/mm², respectively.

In doubly reinforced beams, the steel and concrete in the compression zone of beam act together to resist compression. The steel in compression resists, C_{sc} and the concrete in compression resists, C_c, such that the sum of these two compressive forces is equal to total compressive force resisted by the beam. Therefore.

$$(C_c + C_{sc}) = C \quad (7.3)$$

where, C_{sc} = Compressive force developed in steel reinforcement in compression, as shown in Fig. 7.1 (B)

C_c = Compressive force developed in concrete in compression as shown in Fig. 7.1 (B)

7.2 LOCATION OF NEUTRAL AXIS

The *neutral axis* is the axis which passes the centroid of the beam section. At the neutral axis, the bending stress is zero, and the bending stress distribution diagram for the doubly reinforced beam is as shown in Fig. 7.1 (B).

Let n be the distance of neutral axis from the extreme fibre of concrete in compression, Fig. 7.1 (A). The neutral axis is, also expressed convertionally as kd, as a fraction of the effective depth d, where k is the **depth factor for the neutral axis**. The neutral axis is located by taking the moment of area of concrete in compression, the moment of equivalent area of steel in compression and the moment of equivalent area of steel in tension about the neutral axis. Therefore,

$$(b\,.\,n)\frac{n}{2} + [m_c\,.\,A_{sc}\,.\,(n - d_c) - A_{sc}\,(n - d_c)] - m\,A_{st}\,(d - n) = 0 \qquad \text{(i)}$$

$$\text{or} \quad \left(\frac{b\,.\,n^2}{2}\right) + [(m_c - 1)\,A_{sc}\,(n - d_c)] = m\,A_{st}\,(d - n) \qquad \text{(7.4A)}$$

$$\text{or} \quad \left(\frac{b\,.\,n^2}{2}\right) + [(1.5\,m - 1)A_{sc}\,(n - d_c)] = m A_{st}\,(d - n) \qquad \text{(7.4B)}$$

The term $A_{sc}\,.(n - d_c)$ in expression (i) has been deducted for the portion of concrete are being occupied by the area of steel in compression, A_{sc}. As usual, the strength of concrete in tension is very small and the area of concrete in tension is neglected. In Eq. 7.4,n is only unknown. By solving this quadratic equation, the depth of neutral axis. n may be determined.

Eq. 7.4(A) or 7.4 (B) may also be written as follows:

$$\left(b\,.\,kd\,.\,\frac{kd}{2}\right) + [(m_c - 1)\,A_{sc}\,.\,(n - d_c)] = m\,A_{st}\,(d - n) \qquad \text{(7.5A)}$$

$$\text{or} \quad \frac{b\,.\,(kd)^2}{2} + [(m_c - 1)\,A_{sc}\,.\,(kd - dc) = m\,.\,A_{st}\,(d - kd)] \qquad \text{(7.5B)}$$

In case, the actual stresses σ_{cbc}, the compressive stress in bending in concrete at the extreme fibre and σ_{st}, the tensile stress in bending in steel are known, the depth of neutral axis, n may be found form the bending in stress distribution diagram, Fig. 7.1 (B) The triangles above and below the neutral axis are similar. Therefore,

$$\left(\frac{\sigma_{abc}}{n}\right) = \left(\frac{\sigma_{st}/m}{(d - n)}\right) \qquad \text{(ii)}$$

$$\text{or} \quad \left(\frac{\sigma_{abc}}{kd}\right) = \left(\frac{\sigma_{st}/m}{(d - kd)}\right) \qquad \text{(iii)}$$

or $$k = \left(\frac{m\ \sigma_{cbc}}{m\ \sigma_{cbc} + \sigma_{st}} \right) \tag{7.6}$$

Eqs. 3.4 and 7.6 for determining the **depth factor for the neutral axis** for the singly reinforced rectangular beam and for the doubly reinforced rectangular beam, respectively are exactly same, In case of balanced section, σ_{cbc} and σ_{st} are the value of permissible stresses for compression in bending in concrete, and tension in steel, respectively, then it is seen form Eqs. 3.4 and 7.6 that the position of neutral axis remains same in duly reinforced bean section as it exists in the singly reinforced section. For the balanced section, the reinforcement in compression zone does not influence the position of neutral axis.

7.3 BENDING STRENGTH OF DOUBLY REINFORCED BEAM

The moment of resistance of the beam cross-section of a doubly reinforced beam gives the bending (flexural) strength of this beam.

From Fig. 7.1, it is seen that the doubly reinforced beam section carries a compressive force, C_c (force in concrete in compression), a compressive force, C_{sc} (force in steel in compression) and the tensile force, T (a force in steel in tension). The compressive forces, C_c in concrete in compression acts at a distance $(n/3)$ from the extreme fibre in compression (viz., form top in Fig.7.1(B), and its magnitude may be written as follows:

$$C_c = \left(\frac{1}{2} \cdot \sigma_{cbc} \cdot b \cdot n \right) \tag{i}$$

The compressive force C_{sc} in steel in compression acts at a distance, d_c from the extreme fibre in compression and its magnitude may be written as below:

$$C_{sc} = \left[(m_c - 1)\, A_{sc} \cdot \sigma_{cbc_1} \right] \tag{ii}$$

where, A_{sc} is the area of steel reinforcement in compression. From Eq. 7.2

$$\sigma_{cbc_1} = \left[(n - d_c) \cdot \sigma_{cbc} / n \right]$$

The tensile force, T in steel reinforcement in tension acts at the centre of steel bars in tension and it is at a distance, d from the extreme fibre of concrete in compression, (*viz.*, from top Fig.7.1). Its magnitude may be found as under

$$T = (A_{st} \cdot \sigma_{st}) \tag{iii}$$

where A_{st} is the area of steel reinforcement in tension. The moment of resistance of the doubly reinforced beam section. MR (Fig. 7.1) may be determined by taking the moment of the compressive force C_c about the centre of steel in tension. Therefore,

$$MR = C_c \cdot (d - n/3) + C_{sc} \cdot (d - d_c) \tag{iv}$$

or $$MR = \frac{1}{2} b \cdot n \cdot \sigma_{cbc} \cdot (d - n/3) + \left[(m_c - 1)\, A_{sc} \cdot \sigma_{cbc_1}\, (d - d_c) \right] \tag{v}$$

or $$MR = \frac{1}{2} b \cdot n \cdot \sigma_{cbc} \cdot (d - n/3) + (m_c - 1) \cdot A_{sc} \cdot \sigma_{cbc} \cdot \left(\frac{(n - d_c)}{n} \right) \cdot (d - d_c) \tag{7.7}$$

Since, $n = k \cdot d$, and $(d - n/3)$ = jd, Eq. 7.7 may be written as follows :

$$MR = \frac{1}{2} b \cdot kd \cdot \sigma_{cbc} \cdot jd + (m_c - 1) \cdot A_{sc} \cdot \sigma_{cbc} \cdot \left(\frac{(n - d_c)}{n}\right) \cdot (d - d_c) \qquad \text{(vi)}$$

or
$$MR = \left(\frac{1}{2} \cdot \sigma_{cbc} \cdot k \cdot j\right) bd^2 + (m_c - 1) \cdot A_{sc} \cdot \sigma_{cbc} \cdot \left(\frac{(n - d_c)}{n}\right) \cdot (d - d_c) \qquad \text{(vii)}$$

or
$$MR = R \cdot bd^2 + \left[(m_c - 1) \cdot A_{sc} \cdot \sigma_{cbc} \cdot \left(\frac{(n - d_c)}{n}\right)(d - d_c)\right] \qquad (7.8)$$

In order to determine the bending strength (viz., the moment of resistance of a doubly reinforced beam section (the size of beam section, b and d, the area of steel in compression, A_{sc} and the area of steel in tension, A_{st} are known), the depth of neutral axis m, n = n_B for the balanced section is first found from Eq. 7.4. Therefore,

$$k_B = \left(\frac{m\,\sigma_{cbc}}{(m\,\sigma_{cbc} + \sigma_{st})}\right) \qquad \text{(viii)}$$

$$n_B = k_B \cdot d \qquad \text{(ix)}$$

Actual depth of neutral axis, n is calculated from Eq. 7.4. Therefore

$$(bn^2/2) + [(m_c - 1)\,A_{sc} \cdot (n - d_c)] = m\,A_{st}\,(d - n) \qquad \text{(x)}$$

The values of n_B and n are compared. In case n_B and n are equal, the section of beam is **balanced**. In case, n is smaller than n_B, the beam section is **under-reinforced.** In which the value of tensile stress in steel in tension, σ_{st} is equal to its permissible stress. From the bending stress distribution diagram. Fig. 7.1 (B), from the basis of similar triangles, the stresses σ_{cbc} and σ_{cbc_1} may be computed. In case, n is more than n_B, the beam section is **over-reinforced.** In which case the value of σ_{cbc}, the compressive stress in bending in concrete is equal to its permissible stress for the given grade of concrete. Other stresses σ_{cbc_1} and σ_{st} may be found from the similar triangles of bending stress distribution, Fig. 7.1 (B).

The bending strength (*viz.*, **moment of resistance**) of the doubly reinforced beam section may then be calculated from either Eq. 7.7 or Eq. 7.8.

7.4 STRESSES IN CONCRETE AND STEEL

In case, a doubly reinforced beam of known size (*viz.*, the dimensions of beam cross-section, area of steel reinforcement in compression, A_{sc} and area of steel reinforcement in tension, A_{st}), is subjected to known value of the bending moment, M, the actual stresses developed in the various component materials of the beam due to bending may be determined as follows :

The actual depth of neutral axis of the beam cross-section n, and that for the balanced section $n_B = k_B.d$ are found from Eq. 7.5. The bending strength of beam (*viz.*, the moment of resistance) may be written by taking the moment of compressive forces, C_c in the concrete in compression, and C_{sc} in the steel reinforcement in compression, about the centre of steel reinforcement in tension, (Fig. 7.1 (B)). Thus, from Eq. 7.7

$$MR = \left[\frac{1}{2} b.\, kd\, .\, \sigma_{cbc}\, .\, (d - n/3)\right] + \left[(m_c - 1)\, A_{sc}\, .\, \sigma_{cbc}\, .\left(\frac{(n - d_c)}{n}\right).\, (d - d_c)\right] \quad \text{(i)}$$

The bending strength of beam MR is equated to the given bending moment, $M\, .\, (\because MR = M)$ Therefore,

$$M = \left[\frac{1}{2} b\, .\, n\, .\, \sigma_{cbc}\, .\, (d - n/3)\right] + \left[(m_c - 1)\, A_{sc}\, .\, \sigma_{cbc}\, .\left(\frac{(n - d_c)}{n}\right).\, (d - d_c)\right] \quad \text{(ii)}$$

The compressive stress in bending in concrete, σ_{cbc} at the extreme fibre of concrete in compression, is obtained by solving the expression (ii). The compressive stress in bending in concrete, surrounding the steel reinforcement in compression, σ_{cbc_1} is calculated from Fig. 7.1 (B). Since, two triangles of bending stress distribution diagram above the neutral axis are similar,

$$\sigma_{cbc_1} = \left(\frac{(n - d_c)}{n}\right).\, \sigma_{cbc} \quad \text{(iii)}$$

The stress in steel reinforcement in compression, σ_{sc} shall be

$$\sigma_{sc} = (m_c\, .\, \sigma_{cbc_1}) \quad \text{(iv)}$$

The stress in steel reinforcement in tension, σ_{st} is calculated from the similar triangles above the below the neutral axis, Fig. 7.1 (B). Therefore

$$\sigma_{st} = m\, .\left(\frac{(d - n)}{n}\right).\, \sigma_{cbc} \quad \text{(v)}$$

The actual stress in concrete in compression, σ_{cbc}, in steel reinforcement in compression, σ_{sc} and in steel reinforcement in tension, σ_{st} should be less than their respective values as specified in IS : 456-2000.

Example 7.1. A doubly reinforced rectangular beam is 400 mm wide. The effective depth of beam section is 550 mm and the effective cover to both tension and compression reinforcements from the outer surfaces is 50 mm. The tension and compression reinforcement consist of 4 bars of 20 mm diameter and 4 bars of 16 mm diameter, respectively. Locate the neutral axis and determine the moment of resistance of doubly reinforced beam. M15 grade of concrete and mild steel reinforcement are used.

Solution : The grade of concrete is M 15. From IS : 456-2000, the permissible stress in compression in bending in concrete σ_{cbc} is 5 N/mm².

Step 1 : Modular ratio of concrete

$$m = \left(\frac{280}{3\, .\, \sigma_{cbc}}\right) = \left(\frac{280}{3 \times 5}\right) = 18.67 \quad \text{(i)}$$

Step 2 : Area of steel reinforcement in tension

$$A_{st} = \left(4 \times \frac{\pi}{4} \times 20^2\right) = 1256.637\ mm^2$$

Step 3 : Area of steel reinforcement in compression

$$A_{sc} = \left(4 \times \frac{\pi}{4} \times (16)^2\right) = 804.248\ mm^2$$

Step 4 : Location of neutral axis

The doubly reinforced beam section is shown in Fig. 7.2. In order to locate the neutral axis, equate first moment of equivalent areas of concrete in compression and steel bars in tension and compression about the neutral axis. From Eq. 7.4 (A)

$$(b \,.\, n^2/2) + (m_c - 1)\, A_{sc} \,.\, (n - d_c) = m\, A_{st} \,.\, (d - n)$$

$$400 \times n^2/2 + (1.5 \times 18.67 - 1) \times 804.248\,(n - 50)$$

$$= 18.67 \times 1256.637\,(550 - n) \qquad \text{(ii)}$$

$$n = 174.59 \text{ mm}$$

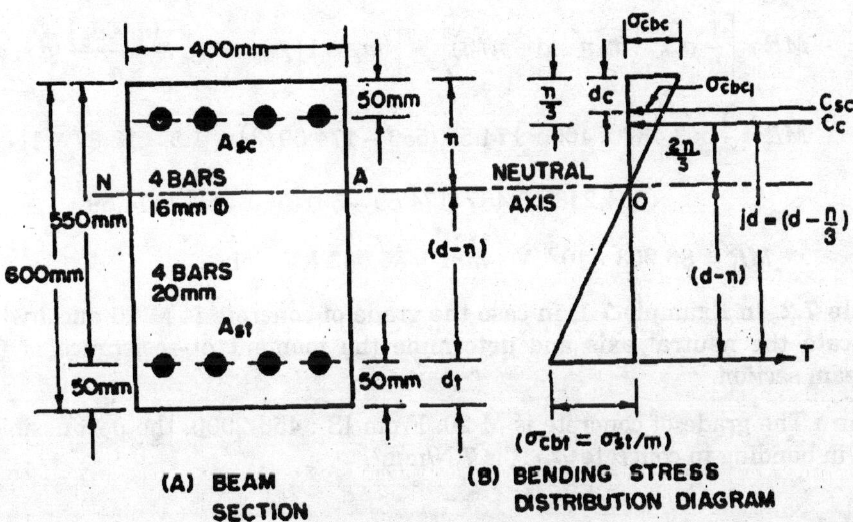

Fig. 7.2

Step 5 : Moment of resistance of beam section

In order to determine the moment of resistance of the doubly reinforced beam section, depth of neutral axis, for the balanced section is found. From IS : 456:2000.

$$\sigma_{cbc} = 5\ N/mm^2 \ and \ \sigma_{st} = 140\ N/mm^2$$

$$k_B = \left(\frac{m\ \sigma_{cbc}}{(m\ \sigma_{cbc} + \sigma_{sn})}\right) = \left(\frac{18.67 \times 5}{18.67 \times 5 + 140}\right) = 0.4$$

Depth of neutral axis for the balanced section

$$n_B = k_B\ .d = (0.4 \times 550) = 220\ mm$$

By comparing n and n_B, it is seen that the given beam is under-reinforced since n is less than n_B. Therefore, σ_{st} reaches its permissible value of stress (*viz.*, 140 N/mm²) first. From the bending stress distribution diagram Fig. 7.2 (B)

$$\sigma_{cbc} = \left(\frac{n}{(d-n)}\right) . \left(\frac{\sigma_{st}}{m}\right)$$

$$\sigma_{cbc} = \left(\frac{174.59}{(550 - 174.59)}\right) \times \frac{140}{18.67} = 3.487\ N/mm^2$$

The moment of resistance of the double reinforced given beam section is found by taking moment of the compressive forces C_c and C_{sc} about the centre of steel bars in tension. From Eq. 7.7 Fig. 7.2 (B)

$$MR = [C_c\ .\ (d - n/3)] + [C_{sc}\ .\ (d - d_c)]$$

$$MR = \left[\frac{1}{2}\sigma_{cbc}\ .b\ .n\ .(d - n/3)\right] + \left[(m_c - 1) A_{sc}\ .\sigma_{cbc} . \frac{(n - d_c)}{n} (d - d_c)\right]$$

$$MR = \frac{1}{2} \times 3.487 \times 400 \times 174.59\ (550 - 174.59/3) + (1.5 \times 18.67 - 1)$$

$$\times 804.248 \times 3.487\ (174.59 - 50)\ (550 - 50) / 174.59$$

$$\therefore\ MR = 86.903 \times 10^6\ N - mm = 86.903\ kN - m$$

Example 7.2. In Example 7.1, in case the grade of concrete is M 20 and hyd steel bars are used, locate the neutral axis and determine the moment of resistance of the doubly reinforced beam section.

Solution : The grade of concrete is M 20. From IS : 456-2000, the permissible stress in compression in bending in concrete σ_{cbc} is 7 N/mm².

Step 1 : Modular ratio of concrete

$$m = \left(\frac{280}{3 \times \sigma_{cbc}}\right) = \left(\frac{280}{3 \times 7}\right) = 13.33 \qquad \text{(i)}$$

Step 2 : Neutral axis of the beam section

From Fig. 7.2 and Eq. 7.4 (A)

$$\left(bn^2/2\right) + \left[(m_c - 1) A_{sc} \cdot (n - d_c)\right] = m \ A_{st} \ (d - n)$$

Substituting the values from Example 7.1

$$400 \times n^2/2 + (1.5 \times 13.33 - 1) \times 804.248 \ (n - 50)$$

$$= 13.33 \times 1256.637 \ (550 - n) \qquad \text{(ii)}$$

$$n = 157.21 \ mm$$

Step 3 : Moment of resistance of the beam section

From IS : 456-2000 $\sigma_{cbc} = 7 \ N/mm^2$ *and* $\sigma_{st} = 230 \ N/mm^2$

Factor for depth of neutral axis for the balanced section of beam

$$k_B = \left(\frac{m \ \sigma_{cbc}}{(m \ \sigma_{cbc} + \sigma_{sn})}\right) = \left(\frac{13.33 \times 7}{(13.33 \times 7 + 230)}\right) = 0.2886$$

$$n_B = k_B \cdot d = (0.2886 \times 550) = 158.735 \ mm$$

Since n is less than n_B, the beam is under reinforced. Therefore, the stress in steel reinforcement, σ_{st} reaches its permissible stress (*viz.,* 230 N/mm²), first From Fig. 7.2 (B) the bending stress distribution diagram

$$\sigma_{cbc} = \left(\frac{n}{(d - n)}\right) \times \left(\frac{\sigma_{st}}{m}\right)$$

$$\text{or} \quad \sigma_{cbc} = \left(\frac{157.21}{(550 - 157.21)}\right) \times \left(\frac{230}{13.33}\right) = 6.906 \ N/mm^2$$

Moment of resistance of the beam section from Eq. 7.7, Fig. 7.2 (B)

$$MR = \left[C_c \cdot (d - n/3)\right] + \left[C_{sc} \cdot (d - d_c)\right]$$

$$MR = \left[\frac{1}{2} \sigma_{cbc} \cdot b \cdot n \cdot (d - n/3)\right] + \left[(m_c -) A_{sc} \cdot \sigma_{cbc} \cdot \left(\frac{(n - d_c)}{n}\right) (d - d_c)\right]$$

$$MR = \left[\frac{1}{2} \times 6.906 \times 400 \times 157.21 \ (550 - 157.21/3)\right]$$

$$+ \left[(1.5 \times 13.33 - 1) \ 804.248 \times 6.906 \ (157.21 - 50) \ (550 - 50)/157.21\right]$$

$$MR = 144.033 \times 10^6 \ N - mm = 144.033 \ kN - m$$

Example 7.3. **The bending moment at a transverse section of a doubly reinforced beam, Example 7.1 is 80 kN-m. Determine the compressive stress in bending in concrete, σ_{cbc} at the extreme fibre and stresses in steel in compression and tension, σ_{sc} and σ_{st}, respectively.**

Solution **: It is necessary to find the depth of neutral axis in order to determine the stresses in the different materials of th beam.** ***From Example 7.1***

$$n = 174.59 \text{ mm} \qquad \text{(i)}$$

The external bending moment M = 80 kN-m acting at a transverse section of the beam is resisted by the strength of beam in bending, (*viz.*, the moment of resistance). Therefore,

$$MR = M = 80 \times 10^6 \ N-mm \qquad \text{(ii)}$$

Step 1 : Stress in compression in bending in concrete, σ_{cbc}

From Fig. 7.2 (B), Equ. 7.7

$$MR = [C_c \, . (d - n/3)] + [C_{sc} \ (d - d_c)]$$

$$MR = \left[\frac{1}{2}\,\sigma_{cbc} \,.\, b \,.\, n\,(d - n/3)\right] + \left[(m_c - 1)\, A_{sc} \,.\, \sigma_{cbc} \,.\frac{(n - d_c)}{n}(d - d_c)\right] \qquad \text{(iii)}$$

$$80 \times 10^6 = \frac{1}{2}\sigma_{cbc} \times 400 \times 174.59\,(500 - 174.59/3)$$

$$+[(1.5 \times 18.67 - 1) \times 804.248 \times \sigma_{cbc} \times (174.59 - 50)\,(550 - 50)/174.59\,]$$

$$\sigma_{cbc} = 3.452 \ N/mm^2$$

It is less than its permissible value for M 15 grade of concrete 5 N/mm². From bending stress distribution diagram, Fig. 7.2(B), stress in concrete surrounding the steel reinforcement in compression

$$\sigma_{cbc_1} = \left(\frac{(n - d_c)}{n}\right) \times \sigma_{cbc} \qquad \text{(iv)}$$

$$\text{or} \quad \sigma_{cbc_1} = \left(\frac{(174.59 - 50)}{174.59}\right) \times 3.452 = 2.463 \ N/mm^2$$

Step 2 : The stress in steel reinforcement in compression

$$\sigma_{sc} = (m_c \,.\, \sigma_{cbc}) = (1.5 \times 18.67 \times 2.463) = 68.976 \ N/mm^2$$

Step 3 : The stress in steel reinforcement in tension from, Fig. 7.2 (B)

$$\left(\frac{\sigma_{st}/m}{(d - n)}\right) = \left(\frac{\sigma_{cbc}}{n}\right)$$

$$\sigma_{st} = \left(\frac{m \,.\, (d - n)}{n}\,\sigma_{cbc}\right) = \left(\frac{18.67\ (550 - 174.59)}{174.59}\right) \times 3.452 \text{ N/mm}^2$$

$$= 138.58 \text{ N/mm}^2$$

It is less than its permissible value 140 N/mm² for mild steel reinforcement in tension (diameter of bars $\leq 20 \ mm$).

Example 7.4. The bending moment at a transverse section of a doubly reinforced beam, Example 7.2 is 120 kN-m. Determine the compressive stress in bending in concrete, σ_{cbc} at the extreme fibre and the stresses in steel in compression and tension, σ_{sc} and σ_{st}, respectively.

Solution : It is necessary to find the depth of neutral axis of the beam section to determine the stresses in the different materials of the beam.

From Example 7.2, n = 157.21 mm (i)

The external bending moment M = 120 kN-m acting a transverse section of the beam is resisted by the strength of beam in bending at (*viz.*, the moment of resistance). Therefore,

$$MR = M = 120 \times 10^6 \; N - mm \quad \text{(ii)}$$

Step 1 : Stress in compression in bending in concrete, σ_{cbc}

From Fig. 7.1, Eq. 7.7

$$MR = [C_c \, . \, (d - n/3)] + [C_{sc} \, . \, (d - d_c)]$$

$$MR = \left(\frac{1}{2}\,\sigma_{cbc} \, . \, b \, . \, n \, . \, (d - n/3)\right) + \left[(m_c - 1) A_{sc} \, . \, \sigma_{cbc} \, . \, \frac{(n - d_c)}{n} \, . \, (d - d_c)\right] \quad \text{(iii)}$$

$$120 \times 10^6 = \left[\frac{1}{2} \times \sigma_{cbc} \times 400 \times 157.21\,(550 - 157/3)\right]$$

$$+ \left[(1.5 \times 13.33 - 1)\, 804.248 \times \sigma_{cbc} \times (157.21 - 50)\,(550 - 50)/157.21\right]$$

$$\sigma_{cbc} = 5.7537 \; N/mm^2$$

It is less than its permissible value for M 20 grade of concrete 7 N/mm².

From, the bending stress distribution diagram, Fig. 7.2 (B), the stress in concrete surrounding the steel reinforcement in compression

$$\sigma_{cbc_1} = \left(\frac{(n - d_c)}{n}\,\sigma_{cbc}\right) \quad \text{(iv)}$$

$$\text{or} \quad \sigma_{cbc_1} = \left(\frac{(157.21 - 50)}{157.21}\right) \times 5.7537 = 3.924 \; N/mm^2$$

Step 2: Stress in hysd steel reinforcement in compression

$$\sigma_{sc} = (\mathrm{m_c} \times \sigma_{\mathrm{cbc_1}}) = (13.33 \times 3.924) = 52.307 \; \mathrm{N/mm^2}$$

It is less than its permissible value 190 N/mm²

Step 3 : Stress in hysd steel reinforcement in tension, from Fig. 7.2 (B)

$$\sigma_{st} = m \, . \left(\frac{(d - n)}{n}\right) . \, \sigma_{cbc}$$

$$\sigma_{st} = 13.33 \,.\left(\frac{(550 - 157.21)}{157.21}\right) \times 5.7537 = 191.627\ N/mm^2$$

It is less than its permissible value 230 N/mm^2 for hysd steel reinforcement in tension.

7.5 DESIGN OF DOUBLY REINFORCED BEAM

For the doubly reinforced beam subjected to known dead and live transverse loads, the maximum bending moment, M and the maximum shear force, V occurring at the critical sections are calculated. Usually, the dimensions of beam to be used as doubly reinforced beam, $b \times d$ are known. In the design of doubly reinforced beam, the total area of steel reinforcement, A_{st} in tension and the area of steel reinforcement in compression, A_{sc} to be provided in the beam cross-section as shown in Fig. 7.3 (A) are to be found. The design in carried out as follows : After deciding the grade of concrete and type of steel to be used as compression and tension reinforcement, the various design constants m, k_B, j_B, p_B and R_B, for the balanced beam section are calculated.

The maximum bending moment M is separated into its two components M_1 and M_2. M_1 is the bending moment to be resisted by the rectangular beam section of balanced designed with the steel reinforcement A_{st_1} in tension as in singly reinforced rectangular beam as shown in Fig. 7.3 (B). M_2 is the bending moment to be resisted by the steel in tension, A_{st_2} and a area of steel in compression, A_{sc} as shown in Fig. 7.3 (C). For the balanced design, it is understood that with an area of steel reinforced $A_{st_1} = p_B \,.\, b \,.\, d$, the maximum stresses both in concrete in compression, σ_{cbc} and steel in tension, σ_{st} are reached at the same time. The values of these stresses are equal to those specified in IS : 456-2000. The depth of neutral axis for the balanced section, n = n_B = k_B . d, where

$$k_B = \left(\frac{m \,.\, \sigma_{cbc}}{(m \,.\, \sigma_{cbc} + \sigma_{st})}\right) \qquad \text{(i)}$$

The moment of resistance of a simple beam of balanced design, M_1 is found as under :

$$M_1 = R_B \,.\, bd^2 = \left(\frac{1}{2}\, \sigma_{cbc} \cdot k_B \cdot j_B\right) bd^2 \qquad \text{(ii)}$$

The moment of resistance due to steel in tension $A_{st.2}$ and the steel in compression, M_2 is obtained as below :

$$M_2 = (M - M_1) \qquad \text{(iii)}$$

Total area of steel A_{st} = $(A_{st_1} + A_{st_2})$ in tension and steel in compression, A_{sc} are calculated as follows :

Step 1 : The area of steel reinforcement in tension, A_{st_1} for simple beam of balanced design is found as below :

$$(\sigma_{st} \cdot A_{st_1} \cdot j_B \cdot d) = M_1$$

$$\text{or } A_{st_1} = [M_1 / (\sigma_{st} \cdot j_B \cdot d)] \qquad (7.9)$$

Step 2 : The area of steel reinforcement in tension A_{st_2} to resist the additional moment $M_2 = (M - M_1)$ is obtained as under :

$$\sigma_{st} \cdot A_{st_2} \cdot (d - d_c) = M_2$$

$$\text{or} \quad A_{st_2} = [M_2 / (\sigma_{st}) . (d - d_c)] \tag{7.10}$$

Total area of steel reinforcement in tension shall be the sum of A_{st_1} and A_{st_2}, $(viz., A_{st} = A_{st_1} + A_{st_2})$.

Step 3 : The beam section has been designed to resist the moment, M_1 as the balanced beam section. Therefore, the position of its neutral axis found above should not shift. Therefore, the area of steel reinforcement in compression, A_{sc} is found by equating the moment of equivalent (transformed) areas A_{sc} and A_{st_2} about neutral axis. Therefore,

$$m\, A_{st_2}\, (d - n) = (m_c - 1)\, A_{sc}\, (n - d_c) \tag{7.11A}$$

$$\text{Therefore,} \quad A_{sc} = \left[\frac{m \cdot A_{st_2} \cdot (d - n)}{(m_c - 1)(n - d_c)}\right] \tag{7.11B}$$

This method as described above is exact. In case, the area of steel reinforcement in compression, A_{sc} happens to be equal to or larger than A_{st}, the area of steel A_{sc} is found using steel beam theory described in the subsequent article.

The design of transverse or shear or web reinforcement shall be described in articles described later.

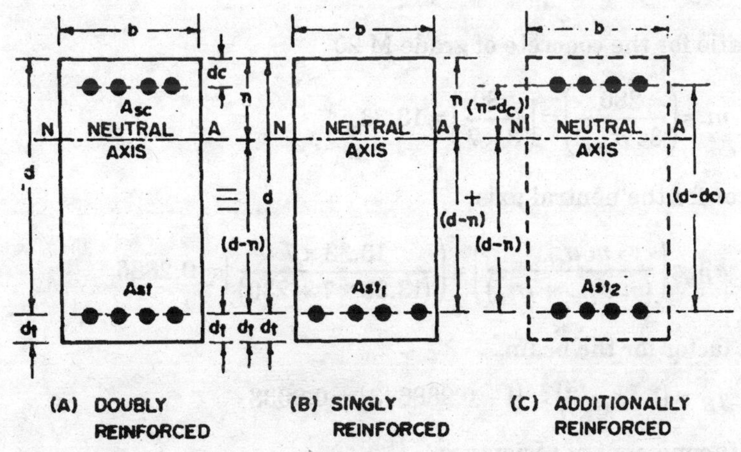

Fig. 7.3

Maximum compression reinforcement

The maximum are of compression reinforcement shall not exceed 0.04 bD. The compression reinforcement in beams shall be enclosed by stirrups for effective lateral restrained just as in columns.

Example 7.5. Design a doubly reinforced rectangular beam to carry a service live load of 16 kN/m and a calculated dead load of 12 kN/m on 5.40 m simple clear span. The cross-section of beam for architectural reasons is kept a 400 mm in width and 650 mm in overall depth. Use M 20 grade of concrete and hysd-steel Fe 415. Adopt the permissible stresses from IS : 456:2000.

Design : Let the width of support at each end be 450 mm.

Step 1 : Effective span, ES of simply supported beam

ES = centre to centre of supports

or ES = (5.40 + 0.450/2 + 0.4502/2) = 5.850 m (i)

The overall depth of beam, D is 650 mm. Let the cover upto the centre of steel reinforcement both in compression and tension be 50 mm. The effective depth, d of the beam shall be 600 mm. The effective span, ES of the beam

ES = (clear span + effective depth)

or ES = (5.400 + 0.600) = 6.000 m (ii)

The effective span of beam, L is minimum of the values in the expression (i) and (ii). As such, L = 5.85 m.

Step 2 : Design constants

For M 20 grade of concrete and hysd-steel Fe 415 are as follows : From IS : 456-2000.

$$\sigma_{cbc} = 7\ N/mm^2, and\ \sigma_{st} = 230\ N/mm^2$$

Modular ratio for the concrete of grade M 20

$$m = \left(\frac{280}{3\ .\ \sigma_{cbc}}\right) = \left(\frac{280}{3 \times 7}\right) = 13.33$$

Depth factor for the neutral axis

$$k_B = \left(\frac{m\ \sigma_{cbc}}{(m\ \sigma_{cbc} + \sigma_{st})}\right) = \left(\frac{13.33 \times 7}{(13.33 \times 7 + 230)}\right) = 0.2886$$

Lever arm factor for the beam

$$j_B = (1 - k_B/3) = (1 - 0.2886/3) - 0.9038$$

Factor for the moment of resistance

$$R_B = \left(\frac{1}{2}\sigma_{cbc}\ .\ k_B\ .\ j_B\right) = \left(\frac{1}{2} \times 7 \times 0.2886 \times 0.9038\right) = 0.9129$$

Depth of neutral axis for the balanced section

$$n = n_B = (0.2886 \times 500) = 173.16\ mm$$

Step 3 : Total load coming over the beam

Superimposed dead load = 12 kN/m

Self-weight of the beam

$$(25 \times 0.400 \times 0.650 \times 1) = 6.5 \ kN/m$$

Live load = 16 kN/m

Total load w = 34.5 kN/m (ii)

Step 4 : Maximum bending moment, M

For simply supported beam carrying uniformly distributed load occurs at the centre.

$$M = \left(\frac{w \cdot L^2}{8}\right) = \left(\frac{34.5 \times 5.85 \times 5.85}{8}\right) = 147.58 \ kN-m \quad \text{(iii)}$$

The moment of resistance of a simple beam of balanced design

$$M_1 = R_B \cdot bd^2$$

$$\therefore M_1 = \left(\frac{0.929 \times 400 \times 600 \times 600}{1000 \times 1000}\right) = 131.458 \ kN-m$$

$$\therefore M_2 = (M - M_1) = (147.58 - 131.458) = 16.122 \ kN-m$$

Step 5 : Area of steel reinforcement required in tension, A_{st_1}

For the simple beam of balanced design, from Eq. 7.9

$$A_{st_1} = \left(\frac{M_1}{(\sigma_{st} \cdot j_B \cdot d)}\right) = \left(\frac{131.458 \times 10^6}{230 \times 0.9129 \times 600}\right) = 1043.48 \ mm^2$$

The area of steel reinforcement required in tension, A_{st_2} for the beam to resist the additional moment, from Eq. 7.10

$$A_{st_2} = \left(\frac{M_2}{\sigma_{st} \cdot (d - d_c)}\right) = \left(\frac{16.122 \times 10^6}{230 \times (600 - 50)}\right) = 127.446 \ mm^2$$

Total area of steel reinforcement required in tension

$$A_{st} = (1043.48 + 127.446) = 1170.926 \ mm^2$$

Step 6 : Area of steel reinforcement required in compression

It is found by equating the moment of equivalent areas Asc and A_{st_2} about neutral axis. Therefore, from Eq. 7.11 (A)

$$\left[m \cdot A_{st_2}\,(d-n)\right] = \left[(m_c - 1)\,A_{sc} \cdot (n - d_c)\right]$$

$$A_{sc} = \left[\frac{m \cdot A_{st_2} \cdot (d-n)}{(m_c - 1) \cdot (n - d_c)}\right] = \left[\frac{13.33 \times 127.446 \times (600 - 173.16)}{(1.5 \times 13.33 - 1)(173.16 - 50)}\right]$$

$$= 309.883 \text{ mm}^2.$$

Step 7 : Number of steel bars in tension.

Let 20 mm hysd-steel bars be provided tension reinforcement. Area of cross-section of each steel bar

$$\left(\frac{\pi}{4} \times (20)^2\right) = 314.159\ mm^2$$

$$\text{Number of bars} = \left(\frac{1170.226}{314.159}\right) = 3.727\ (say\ 4\ bars)$$

4 hysd steel bars are provided as tension reinforcement. Therefore,

$$A_{st} \text{ provided} = (4 \times 314.159) = 1256.637\ mm^2.$$

Step 8 : Number of steel bars in compression

Let 4 hysd steel bars of 12 mm diameter be provided as compression reinforcement. Therefore,

$$A_{sc}\ provided = \left(4 \times \frac{\pi}{4} \times 12^2\right) = 452.389\ mm^2.$$

The design of beam for transverse reinforcement to resist shear force V and development length for steel bars have been done in *Example 8.1.* The details of complete design is shown in Fig. 8.2.

7.6 STEEL BEAM THEORY

A doubly reinforced beam may be considered as a steel beam. The area of concrete as a whole is neglected as far as bending is concerned.

The steel reinforcement in compression and tension may be treated as the flanges of an imaginary I-section, which provide the total strength of beam in bending. The area of concrete is considered as the web.

The bending strength of doubly reinforced beam as per steel theory may be written by taking the moment of tensile force in steel reinforcement about the centre of steel reinforcement in compression. Therefore,

$$\text{MR} = \sigma_{st} \cdot A_{st} \cdot (d - d_c) \qquad (7.12)$$

This method gives the approximate value of bending strength of the beam particularly when the area of steel reinforcement in compression, A_{sc} is equal to or larger than the area of steel reinforcement in tension. However, this method should not be used when the percentage of steel reinforcement in tension is more than 3 percent of the beam cross-section.

Example 7.6 : Design a cantilever rectangular beam of span 3.60 m. The beam is to carry a superimposed load of 40 kN/m. M 20 grade of concrete and hysd-steel bars of Fe 415 quality shall be used.

Design : For M 20 grade of concrete and hysd-steel of Fe 415 quality, the permissible stresses as per IS : 456-2000 are as follows :

$\tau_{cbc} = 7\ N/mm^2$ and $\sigma_{st} =$ 230 N/mm². The modular ratio of M 20 grade of concrete

$$m = \left(\frac{280}{3\times 7}\right) = 13.33$$

Step 1 : Design constants

For balanced section of beam, the design constants are as follows :

Depth factor for the neutral axis, k_B

$$k_B = \left(\frac{13.33\times 7}{13.33\times 7\ + 230}\right) = 0.2886$$

Lever arm factor, j_B

$$j_B = (1 - 0.2886/3) = 0.9038$$

Factor for the moment of resistance, R_B

$$R_B = (1/2\times 7\times 0.2886\times 0.9038) = 0.9129$$

The percentage of reinforcement for the balanced section, p_B

$$p_B = \left(\frac{1}{2}\times\frac{7\times 0.2886}{230}\right)\times 100 = 0.4392\ percent$$

Step 2 : Estimation of overall depth and width of beam

At support, the overall depth of cantilever beam may be assumed as one-seventh of the span

$$D = \left(\frac{1}{7}\times 3.60\times 1000\right) = 515.28\ mm\ \text{(approximately)}$$

D = 550 mm (say).

The width of beam, b may be assumed as *half* the overall depth.

$$b = (1/2\times 550) = 275\ mm.$$

Let the effective cover for the reinforcement be 50 mm. The effective depth of beam shall be 500 mm.

For a cantilever beam for lateral stability, as per IS : 456-2000, the clear distance from the free end of the cantilever to the lateral restraint shall not exceed 25 b or (100 b^2/d). Therefore,

$$25 \times b = (25 \times 275) = 6875\ mm$$

Actual clear distance is 3600 mm, which one is less than 6875 mm

$$\left(\frac{100\,b^2}{d}\right) = \left(\frac{100 \times 275 \times 275}{500}\right) = 15125\ mm$$

At the free end, the depth of beam is reduced to 200 mm, where the bending moment is zero.

Step 3 : Loads

Superimposed load is 40 kN/m. It includes superimposed dead load and live load. The self-weight of beam

$$\left(\frac{0.550 + 0.200}{2}\right) \times 0.25 \times 3.6 \times 25 = 8.44\,kN \quad \text{(approximately)}$$

The line of action of self-weight from the face of support

$$\left(\frac{0.550 + 2 \times 0.200}{0.550 + 0.200}\right) \times \frac{3.6}{3} = 1.52\ m$$

Step 4 : Maximum bending moment

Maximum bending moment due to superimposed load at the face of support

$$M_1 = \left(\frac{40 \times 3.6 \times 3.6}{2}\right) = 259.2\ kN - m$$

Maximum bending moment due to self-weight of the beam at the face of support

$$M_2 = (8.44 \times 1.52) = 12.83\ kN - m$$

Total BM, M = (259.2 + 12.83) = 272.03 kN-m

Step 5 : Maximum shear force

Due to superimposed load at the face of support

$$V_1 = (40 \times 3.6) = 144\ kN$$

Maximum shear force due to self-weight at the face of support

$$V_2 = 8.44\ kN$$

Total shear force

$$V = (144 + 8.44) = 152.44\ kN$$

The beam section may be designed as doubly reinforced. The bending moment resisted by the cantilever beam acting as balanced section

$$M'_1 = \left(R_B\, bd^2\right) = \left(\frac{0.9129 \times 275 \times 550 \times 550}{1000 \times 1000}\right) = 75.942\ kN - m$$

Additional moment to be resisted

$$M'_2 = (272.03 - 75.942) = 196.088\ kN - m$$

Step 6 : Tension Reinforcement

Tension reinforcement required to resist the bending moment M'₁

$$A_{st_1} = \left(\frac{75.942 \times 10^6}{230 \times 0.9038 \times 500}\right) = 730.65\ mm^2$$

Tension reinforcement required to resist the bending moment M'₂

$$A_{st_2} = \left(\frac{196.088 \times 10^6}{230 \times (500 - 50)}\right) = 1894.57\ mm^2$$

Total area of steel reinforcement required in tension

$$A_{st} = 2625.22\ mm^2$$

Step 7 : Compression reinforcement

The area of steel reinforcement required in compression is found by equating the moment of equivalent areas A_{sc} and A_{st2} about neutral axis. The depth of neutral axis for balanced section

$$n = k_B\ d = (0.2886 \times 500) = 144.30\ mm$$

Therefore,

$$m \,.\, A_{st2}\ (d - n) = (m_c - 1)\, A_{sc} \,.\, (n - d_c)$$

$$A_{sc} = \left(\frac{m \,.\, A_{st} \,.\, (d - n)}{(m_c - 1)(n - d_c)}\right)$$

$$\text{or}\quad A_{sc} = \left(\frac{13.33 \times 1894.57\ (500 - 144.30)}{(1.5 \times 13.33 - 1)(144.30 - 50)}\right) = 1050.93\ mm$$

Step 8 : Number of steel bars

Let 20 mm hysd-steel bars be provided as tension reinforcement. Area of cross-section of each steel bar

$$\left(\frac{\pi}{4}\times(20)^2\right) = 314.159\ mm^2$$

Number of steel bars in tension

$$\left(\frac{2625.22}{314.159}\right) = 8.355\ say 10.$$

Area of steel provided in tension at top

$$(10\times 314.159) = 3141.59\ mm^2$$

These bars be provided in two layer. Number of steel bars in compression.

$$\left(\frac{1050.93}{314.159}\right) = 3.345\ say\ 4.$$

Step 9 : Design of shear reinforcement

Total shear force occurring at the face of support

V = 152.44 kN

Area of steel reinforcement present at the critical section. As is 3141.59 mm²

$$\left(\frac{100\times 3141.59}{975\times 500}\right) = 2.285$$

And, M 20 grade of concrete shall be used permissible shear stress in concrete

$\tau_c = 0.51\ N/mm^2$

Nominal shear stress at the critical section

$$\tau_v = \left(\frac{152.44\times 10^3}{275\times 500}\right) = 1.108\ N/mm^2$$

$\tau_{c.max}$ from IS : 456-2000 for M 20 concrete

$$\tau_{c.max} = 1.8\ N/mm^2$$

τ_v is less than $\tau_{c.max}$ and it is more than τ_v. Hence, the shear reinforcement shall be provided, strength of concrete section in shear

$$\tau_c . b . d = \left(\frac{0.51\times 275\times 500}{1000}\right) = 70.125\ kN$$

Shear force to be resisted by vertical shear stirrups

$$V_s = (V - \tau_c\ . bd) = (152.44 - 70.125) = 82.315\ kN$$

Let 10 mm ϕ hysd-steel bars be provided for the vertical stirrups. Area of cross-section of two legged shear stirrups

$$A_{st_2} = \left(2 \times \frac{\pi}{4} \times 100\right) = 157.08\ mm^2$$

Spacing of the vertical stirrups

$$s_v = \left(\frac{230 \times 157.08 \times 500}{82.135 \times 1000}\right) = 219.45\ mm$$

say = 200 mm c/c near supports.

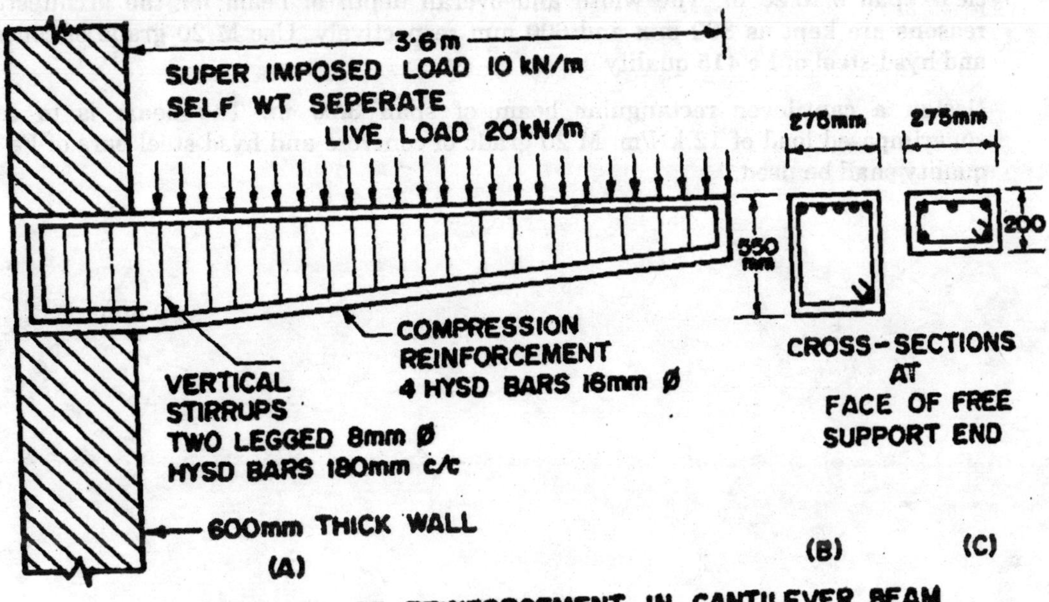

Fig. 7.4

PROBLEMS

7.1 A doubly reinforced rectangular beams is 360 mm wide. The effective depth of beam section is 450 mm and the effective cover to both tension and compression reinforcements from the outer surfaces is 50 mm. The tension and compression reinforcements consist of 4 bars of 18 mm diameter and 4 bars of 16 mm diameter, respectively. Locate the neutral axis and determine the moment of resistance of doubly reinforced beam. M 20 grade of concrete and hysd-steel reinforcement of Fe 415 quality are used. The two legged stirrups have been provided at appropriate spacing.

7.2 In *Example 7.1,* in case the grade of concrete is M 15 and mild-steel bars re used, locate the neutral axis and determine the moment of resistance of the doubly reinforced rectangular beam.

7.3 The bending moment at a transverse section of a doubly reinforced-rectangular beam, *Problem 7.1* is 90 kN-m. Determine the compressive stress in bending in concrete, σ_{cbc} at the extreme fibre and the stresses in steel in compression and tension, σ_{sc} and σ_{st} respectively.

7.4 The bending moment at a transverse section of a doubly reinforced rectangular beam, *Problem 7.2* is 112 kN-m. Determine the compressive stress in bending in concrete, σ_{cbc} at the extreme fibre and the stresses in steel in compression and tension, σ_{sc} and σ_{st} respectively.

7.5 Design a doubly reinforced rectangular beam simply supported at its both the ends to carry a service live load of 15 kN/m and superimposed dead load of 16 kN/m over a clear span of 6.20 m. The width and overall depth of beam for the architectural reasons are kept as 360 mm and 600 mm respectively. Use M 20 grade of concrete and hysd-steel of Fe 415 quality.

7.6 Design a cantilever rectangular beam of span 3.20 m. The beam is to carry superimposed load of 12 kN/m. M 20 grade of concrete and hysd steel bars of Fe 415 quality shall be used.

8

SHEAR STRENGTH AND DESIGN OF DOUBLY REINFORCED RECTANGULAR BEAMS (DRRB-II)

8.1 INTRODUCTION

A doubly reinforced rectangular beam is shown in Fig. 8.1 (A). The cracked equivalent beam section for this beam is shown in Fig. 8.1(B). The distribution of shear stress in the doubly reinforced beam is as shown in Fig. 8.1(C). The shear stress distribution diagram, Fig. 8.1(C) has been explained here by considering the cracked equivalent beam section.

The intensity of shear stress at ny plane at a height y above the neutral axis, in general, is given by (Eq. 8.2) in the layer CD

$$\tau = \left(\frac{V \cdot A\,\bar{y}}{I \cdot z}\right) \tag{i}$$

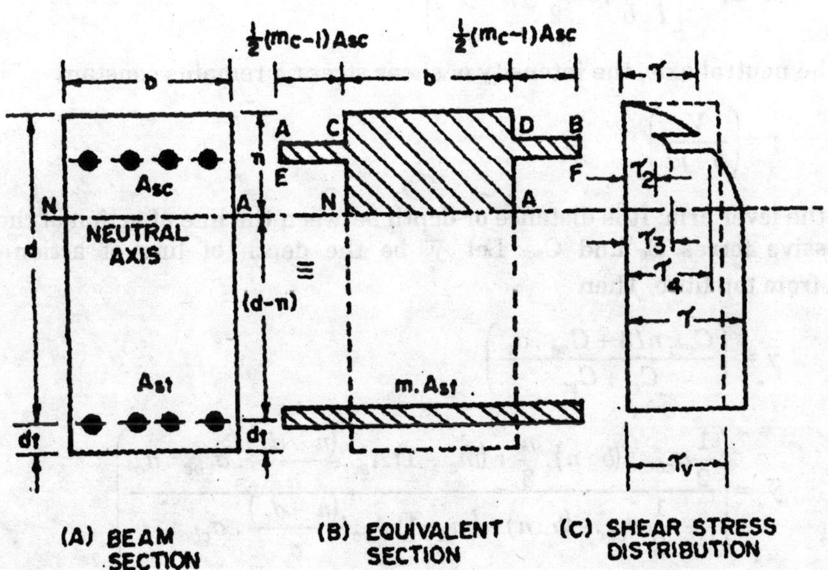

Fig. 8.1

where, z is the width of layer being considered. The intensity of shear stress just above steel reinforcement in compression shall be

$$\tau_1 = \left(\frac{V\,.\,A\,\bar{y}}{I\,.\,b}\right) \qquad \text{(ii)}$$

The intensity of shear stress in layer AB just below CD, τ_2 shall be as follows. It is seen from the cracked equivalent beam section, Fig. 8.2 (B) that the width of beam section, z increases suddenly from b to $[b + (m_c - 1).\, A_{sc}]$. Therefore, the intensity of shear stress decreases suddenly from the value τ_1 to τ_2.

$$\tau_2 = \left(\frac{V\,.\,A\,\bar{y}}{I\,.\,[b + (m_c - 1).\, A_{sc}]}\right) \qquad \text{(iii)}$$

The intensity of shear stress increases parabolically over the depth AE. It becomes τ_3 in the layer EF. Below EF, the intensity of shear stress again increases suddenly as the width of beam section reduces to b.

$$\tau_4 = \left(\frac{V\,.\,[A\,\bar{y} + (m_c - 1)\,A_{sc}\,(n - d_c)]}{I\,.\,b}\right) \qquad \text{(iv)}$$

The intensity of shear stress may be written from Eq. 8. as under for the layer (in between EF and the neutral axis) at aa height y above the neutral axis upto the layer EF. Therefore,

$$\tau_4 = \frac{V}{I\,.\,b}\int_y^{y_1} y\,.\,z\,dy \qquad \text{(v)}$$

$$\text{or } \tau_4 = \left[\frac{V}{I\,.\,b}, b\,.\,\frac{1}{2}\left(y_1^2 - y^2\right)\right] \qquad \text{(vi)}$$

Below the neutral axis, the intensity of shear stress, τ remains constant

$$\tau = \left(\frac{V}{b\,.\,jd}\right) \qquad \text{(vii)}$$

where, *jd* is the lever arm. It is distance or depth between the line of action of the resultant of two compressive forces C_c and C_{sc}. Let $\bar{y}$ be the depth of line of action of resultant compressive from top fibre. Then

$$\bar{y} = \left(\frac{C_c\,.\,n/3 + C_{sc}\,.\,d_c}{C_c + C_{sc}}\right) \qquad \text{(viii)}$$

$$\bar{y} = \left(\frac{\frac{1}{2}\sigma_{cbc}\,.\,(b\,.\,n).\,\frac{n}{3} + (m_c - 1)\,A_{sc}\,\frac{(n - d_c)}{n}.\,\sigma_{cbc}\,.\,d_c}{\frac{1}{2}\,\sigma_{cbc}\,.\,(b\,.\,n) + (m_c - 1)\,A_{sc}\,\frac{(n - d_c)}{n}.\,\sigma_{cbc}}\right) \qquad \text{(ix)}$$

Therefore, the lever arm depth *jd* is equal to (d - $\bar{y}$)

Alternatively

The lever arm depth, *jd* may be calculated from the following

$$(\sigma_{st} \, . \, A_{st} \, . \, j \, . \, d) = M$$

$$\text{or} \quad j \, . \, d = \left(\frac{M}{(\sigma_{st} \, . \, A_{st})} \right) \qquad \text{(x)}$$

Nominal shear stress

IS : 456-2000 introduced a concept of nominal shear stress, τ_v in beams of uniform depth. It is determined as follows :

$$\tau_v = \left(\frac{V}{b \, . \, d} \right) \qquad \text{(xi)}$$

The distribution of nominal shear stress, t_v along the full effective depth, *d* is treated as uniform as shown in by dotted line, Fig. 8.1 (C).

Example 8.1. Design the beam for shear reinforcement.

Design : From Example 7.5, total intensity of uniformly distributed load per unit length acting over the beam is 34.5 kN/m and the effective span of beam, L is 5.85 m. The critical section for shear XX, Fig. 8.2 for the beam is at a distance d = 600 mm from the face of support.

Step 1 : Shear force, V at the critical section, XX

$$V = \left[\left(\frac{1}{2} \times 34.5 \times 5.85 \right) - 34.5 \times \left(\frac{0.450}{2} + 0.6 \right) \right] = 72.45 \, kN$$

Step 2 : Nominal shear stress

$$\tau_c = \left(\frac{V}{b . d} \right) = \left(\frac{72.45 \times 1000}{400 \times 600} \right) = 0.302 \, N/mm^2 .$$

4 hysd steel bars 20 mm diameter provided as tension reinforcement continue upto ends as shown in Fig. 8.2.

Step 3 : Area of steel reinforcement present at the critical section

$$A_{st} = \left(4 \times \frac{\pi}{4} \times 20^2 \right) = 1256.637 \, mm^2$$

$$\left(\frac{100 \times A_{st}}{b \times d} \right) = \left(\frac{100 \times 1256.637}{400 \times 600} \right) = 0.524$$

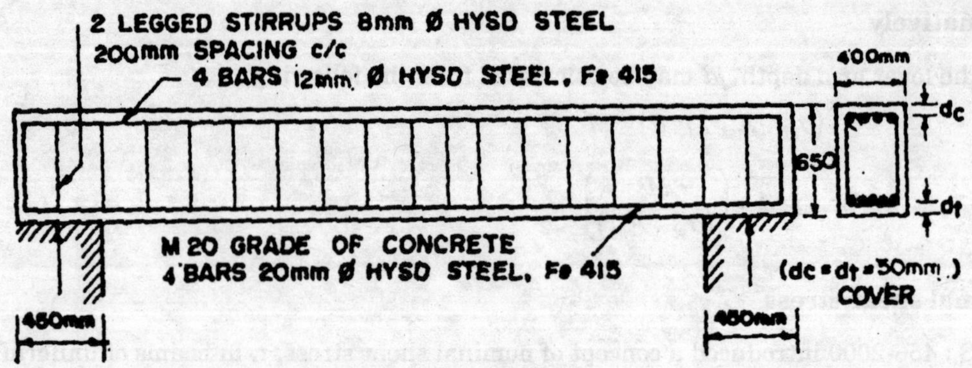

Fig. 8.2

Step 4 : Permissible shear stress in concrete

For M 20 grade of concrete, the permissible shear stress in concrete without shear reinforcement from IS : 456-2000

$$\tau_c = 0.30 + \left[\frac{(0.524 - 0.500)(0.35 - 0.30)}{(0.75 - 0.50)}\right] = 0.3048\ N/mm^2$$

Step 5 : Spacing of shear stirrups

Maximum shear stress, $\tau_{c.max}$ from IS : 456-2000 for M 20 grade of concrete is 1.8 N/mm². The nominal shear stress, τ_v is less than $\tau_{c.max}$ and it is also less than τ_c. Therefore, the beam needs only minimum shear reinforcement in the form of stirrups. Two legged shear stirrups 8 mm ϕ hysd-steel, Fe 415 are provided throughout the length of beam as shown in Fig. 7.5.

$$A_{sv} = \left(2 \times \frac{\pi}{4} \times 8^2\right) = 100.531\ mm^2$$

From IS : 456-2000, $\left(\frac{A_{sv}}{b \cdot s_v} \geq \frac{0.4}{f_y}\right)$

$$\text{or } s_v = \left(\frac{A_{sv} \cdot f_y}{0.4 \times b}\right) = \left(\frac{100.531 \times 415}{0.4 \times 400}\right) = 260.75\ mm.$$

The maximum spacing of shear reinforcement measured along the axis of member shall not exceed 0.75 d = (0.75 × 600) = 450 mm for the vertical stirrups and in no case, shall the spacing exceed 450 mm. Provide shear stirrups at 250 mm.

PROBLEMS

8.1 In Problem 7.1, design the beam for shear reinforcement.

8.2 In Problem 7.2, design the beam for shear reinforcement.

8.3 In Problem 7.3, design the beam for shear reinforcement.

8.4 In Problem 7.4, design the beam for shear reinforcement.

8.5 In Problem 7.5, design the beam for shear reinforcement.

8.6 In Problem 7.6, design the beam for shear reinforcement.

9

BOND AND DEVELOPMENT LENGTH IN DOUBLY REINFORCED RECTANGULAR BEAMS (DRRB-III)

9.1 INTRODUCTION

The permissible stress in bond for bars in compression is increased by 25 percent the values of permissible stress in bond for bars in tension.

As a part of the total force in compression is transferred by bond along the embedded length and a part is transferred by end bearing of the bars in concrete.

For mild steel plain bars in compression, the bond stress

$$\tau_{bd.compr} = (1.25 \times \tau_{bd}) \quad \text{(i)}$$

For hysd steel bars in compression, the bond stress

$$\tau_{bc.compr} = (1.25 \times 1.40 \times \tau_b)$$

where, τ_{bd} is the permissible stress in bond (average) for plain bars in tension. The values of τ_{bd} are given in Table 5.1.

9.2. DEVELOPMENT LENGTH OF REINFORCEMENT BARS IN COMPRESSION

In case of bars in compression, a part of total compressive force is transferred by bond along the embedded length and part is transferred by end bearing of the bars on the concrete. Therefore, the values of permissible stress of bond for bars in compression is increased by 25 percent of the values of bond stress for bars in tension. The surrounding concrete remains relatively free of cracks. Therefore, the reinforcement bars in compression need shorter basic development length as compared with that of tension

$$L_{d.compr} = \left(\frac{\phi\, \sigma_s}{(1.25 \times 4\, \tau_{bd})}\right) \quad (9.1)$$

where, τ_{bd} is the permissible bond stress (average) for plain mild steel bars in tension. The development length of reinforcement hysd steel bars in compression.

$$L_{d.compr} = \left(\frac{\phi \cdot \sigma_s}{(1.25 \times 1.40 \times 4\tau_{bd})}\right) \tag{9.2}$$

where, σ_s is the stress in bar at the section considered. The values of τ_{bd} are given in Table 5.1.

9.3 ANCHORAGE OF PLAIN BARS IN COMPRESSION

The anchorage length of straight bars in compression is taken equal to the development length of bars in compression.

The projected length of hooks, bends and straight lengths beyond bends if provided for a bar in compression shall be considered for the development length.

9.4 LAP SPLICES IN COMPRESSION

In addition to the requirements for the bar splices described in Art. 3.31 and for lap splices in tension, following additional requirement must be satisfied for lap splices for the bar in compression.

The lap length in compression shall be equal to the development length in compression as described in Art. 9.2, but not less than 24 times the diameter of the bar.

The compression bars may be spliced by lapping, by direct end bearing or by welding or mechanical devices that provide positive connections. The direct end bearing of the bars has been found by test and experience to be an effective means for transmitting compressive force. In such a case, the bars must be held in proper alignment by a suitable device and the bar ends must terminate in flat surface within specified tolerance (within 1.5' of a right angle).

Example 9.1. In Example 7.6, check the main reinforcement for development length as per IS : 456-2000.

Step 1 : Check for the main reinforcement for development length

The moment of resistance of beam section at support

$$M = \sigma_{st} \cdot A_{st} \cdot jd = \left(\frac{230 \times 3141.59 \times 0.9038 \times 530}{1000 \times 1000}\right) = 326.517\ kN-m$$

Shear force at the support

$$V = 152.44 \text{ kN and } L_o = (b_s/2 - x_1)$$

Let the width of support be 600 mm, x_1 is 25 mm. Therefore,

$$L_o = [(600/2) - 25] = 275\ mm$$

$$(M_1/V + L_o) = \left(\frac{326.527}{152.44} + 0.275\right) = 2.417\ m$$

For hysd steel

$$L_d = (68.45 \times 20/1000) = 1.369\ m$$

The development length, L_d does not exceed (M_1/V + L_o). The requirement of IS : 456-2000 is satisfied. The details of tension and shear reinforcements are shown in Fig. 9.1.

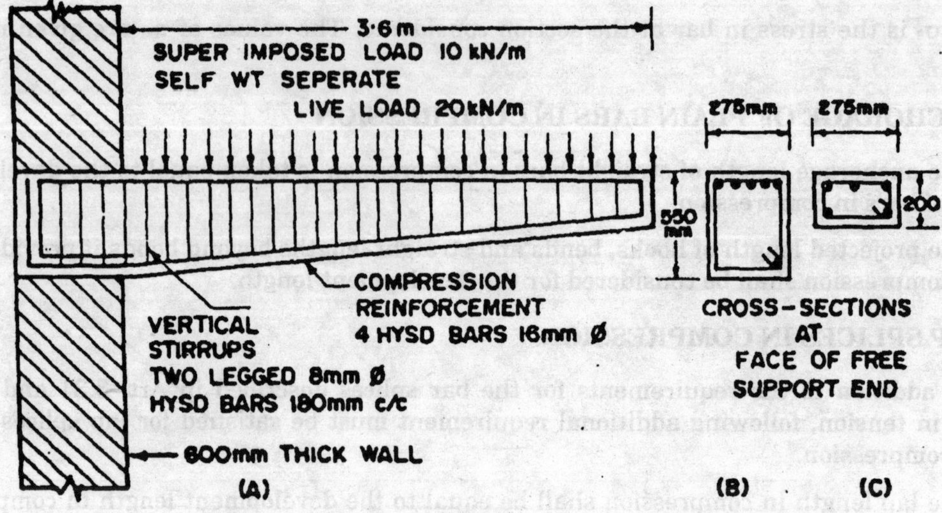

Fig. 9.1.

Example 9.2. In Example 8.1, check the main reinforcement for development length as per IS : 456-2000.

Step 1 : Check for the main reinforcement for development length

As per IS : 456-2000, at simple supports, the tension reinforcement limited to a diameter such that the development length, L_d calculated for s_{st} does not exceed (M_1/V + L_o), viz., from Eq. 3.52

$$L_d \not> \left(\frac{M_1}{V} + L_o\right)$$

The moment of resistance of beam section at support

$$M_1 = (\sigma_{st} \, . \, A_{st} \, . \, jd) = \left(\frac{230 \times 1256.637 \times 0.9038 \times 600}{1000 \times 1000}\right) = 156.733 \; kN - m$$

The shear force at support

$$V = \left(\frac{1}{2} \times 34.5 \times 5.85\right) = 100.9125 \; kN$$

From Eq. 3.54 (since, hook is not provided in hysd-steel reinforcement)

$$L_o = \left(\frac{b_s}{2} - x_1\right)$$

End cover, x_1 is 25 mm and the width of support b_s is 450 mm. Therefore

$$L_o = \left[\left(\frac{450}{2}\right) - 25\right] = 200\ mm$$

$$\left(\frac{M_1}{V} + L_o\right) = \left(\frac{156.733}{100.9125} + 0.2\right) = 1.753\ m$$

$$L_d = \left(\frac{68.45 \times 16}{1000}\right) = 1.0952\ m.$$

The development length, L_d does not exceed (M_1/V + L_o). The requirement of IS : 456-2000 is satisfied.

PROBLEMS

9.1 **In Problem 7.1, check the main reinforcement for the development length as per IS : 456-2000.**

9.2 **In Problem 7.2, check the main reinforcement for the development length.**

9.3 **In Problem 7.3, check the main reinforcement for the development length.**

9.4 **In Problem 7.4, check the main reinforcement for the development length.**

9.5 **In Problem 7.5, check the main reinforcement for the development length.**

9.6 **In Problem 7.6, check the main reinforcement for the development length.**

10

BENDING (FLEXURAL) STRENGTH AND DESIGN OF FLANGED (TEE) BEAMS (FB-I)

10.1 INTRODUCTION

In a reinforced concrete construction, the slab is supported over beams. The use of simple concrete slabs of moderate depth and weight is commonly limited to spans of 3 m to 5 m, where it is desired to use concrete for long spans without excessive weight and material, the slab is built monolithically with the reinforced concrete beams, the beams are considered as **flanged beams.** The slab is assumed to act as a flange of the beams as shown in Fig. 10.1. At the interior portions of floor, the slab extends on both the side of beams. Such beams as shown in Fig. 10.1 resembles to the shape of alphabet-T, therefore, these flanged beams re called as **Tee-beams** (or simply *T-beams)*, whereas at the end portions of the floor, the slab extends on one sides of beams. Such beams as shown in Fig. 10.1 resembles to the shape of inverted alphabet-L, therefore, these flanged beams are called as **L-beams**.

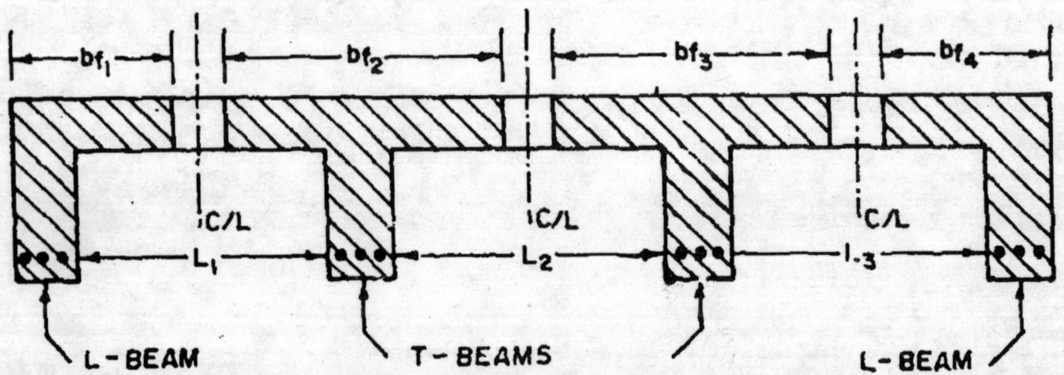

Fig. 10.1 Flanged beams

For reinforced concrete floors, roofs and decks, form works are built for beam soffits and sides and for the underside of the slabs. For the entire construction, the concrete is poured at once from the bottom of deepest beam upto the top of slab. The shear reinforcement of beam (*viz.,* stirrups) and bent bars extent into the slab. The complete construction is cast integrally. A part of the slab acts with the upper part of the beam and resists longitudinal bending compressive stresses. The portion of slab forms the flange for the beam. While the part of beam below the slab forms web for the beam.

In the rectangular beams, the area of concrete above neutral axis (in case of simply supported beams carrying transverse loads acting downward) is comparative small, and therefore, it carries a compressive force (0.5 s_{cbc}). (b.n), Fig. 3.4 (A). In case of T-beams, the width of beam at the top and thereby the area of concrete in compression have been increased, which carries a larger compressive force as compared to the rectangular beams. The portion of concrete below the neutral axis is assumed to be cracked section (portion of concrete in tension). It serves the purpose of embedding steel reinforcement in tension.

The assumptions made for theory of singly reinforced beams real so applicable for the design of T-beams by working stress method. The permissible stresses in concrete in compression in bending and shear for different grades of concrete and those for steel reinforcement in tension are same as specified for the design of singly reinforced beams and these values are adopted as specified in IS : 456-2000.

T-beams are also built as doubly reinforced under the necessity as doubly reinforced rectangular beams. In such cases, the steel reinforcement bars are also placed in compression zone in addition to that in tension zone.

10.2 EFFECTIVE WIDTH OF FLANGE

The width of flange, which is supposed to act as top flange of the beam cannot be found theoretically so far. However, it is clear that the elements of flange midway between the webs of two adjacent beams, Fig. 10.1 are less highly stressed in longitudinal compression than those elements directly over the webs of beams, as the deformation due to shear relieves the more distant elements of some compressive stress. The actual longitudinal compression varies due to this effect.

An effective width of flange, b_f is used in the design of flanged beam. It is treated to be uniformly stressed at the maximum value. It is smaller than the actual width of flange. It is seen that the effective width of flange primarily depends on the span of beam, breadth of the web, b_w and thickness of flange, D_f (depth of slab, D_f). IS : 456-2000 recommends the following effective width of flanges of T-beams and L-beams.

A. (i) For symmetrical T-beams

$$b_f = [(l_o / 6) + b_w + 6.D_f] \qquad (10.1)$$

(ii) For beams having a slab on one side only (L-beams)

$$b_f = [(l_o / 12) + b_w + 3.D_f] \qquad (10.2)$$

The effective flange width, b_f calculated as per Eq. 10.1 or Eq. 10.2, shall be not greater than the breadth of web plus half the sum of the clear distances to the adjacent beam on either side, viz., $[0.5\,(L_1 + L_2) + b_w]$ or $[0.5\,(L_2 + L_3) + b_w]$, Fig. 10.1.

Some of T-beams and L-beams are used in detached situation in which the flange is used only for the purpose of providing additional compressive area. These beams are referred as

isolated beams. *An isolated T-beam* is a single T-beam with a constant flange width (discontinuous at the ends). The effective width of flange for isolated beams as recommended in IS : 456-2000 are as follows. T-beams are treated as isolated beams in case it is specially mentioned.

B. (i) Isolated T-beams

$$b_f = \left(\frac{l_o}{(l_o/b)+4} + b_w\right) \quad (10.3)$$

(ii) Isolated L-beams

$$b_f = \left(\frac{0.5 \, . \, l_o}{(l_o/b)+4} + b_w\right) \quad (10.4)$$

$$b_f \not> \left[0.5\,(L_1 + L_2) + b_w\right]$$

$$\text{or} \quad b_f \not> \left[0.5\,(L_2 + L_3) + b_w\right]$$

where, l_o is the distance between points of zero moments in the beams and b is the actual width of flange. For continuous beams and frames, l_o is assumed as *0.7 times the effective span.*

10.3 LOCATIONS OF NEUTRAL AXIS

Depending upon the proportions (dimensions) of the cross-section, the area of steel reinforcement in tension, the strengths of the materials, the neutral axis of a T-beam in one case may lie in the flange (*viz.*, the depth of neutral axis, n is less than or equal to the thickness of flange or depth of slab, D_f) and in the second case, it may lie in the web, (*viz.*, the depth of neutral axis, n is more than the thickness of slab, D_f). The neutral axis of a T-beam in these two cases may be found s follows :

1. Neutral axis lies within the flange

Let b_f be the effective width of the flange, D_f be the thickness of slab, *d* be the effective depth and n be the depth of the neutral axis. The area of steel reinforcement in tension is A_{st}.

The neutral axis or effective centroidal axis of T-beam, Fig. 10.2 (A) is located by equating the first moment of concrete flange area in compression and that of equivalent concrete area of steel reinforcement in tension about the neutral axis. It is to note that the cross-section of T-beam is treated as cracked section and therefore, the area of concrete in tension below the neutral is neglected. Therefore,

$$(b_f \, . \, n \, . \, n/2) = m \, . \, A_{st}\,(d-n) \quad (10.5)$$

It is just possible that the neutral axis may lie at the junction of the slab and web, i.e., n = D_f. Then

$$(b_f \, . \, D_f \, . \, D_f/2) = m \, . \, A_{st}\,(d-n) \quad (10.6)$$

In both the above two cases, it is to note that the neutral axis may be located as if it were a rectangular beam of width b = bf.

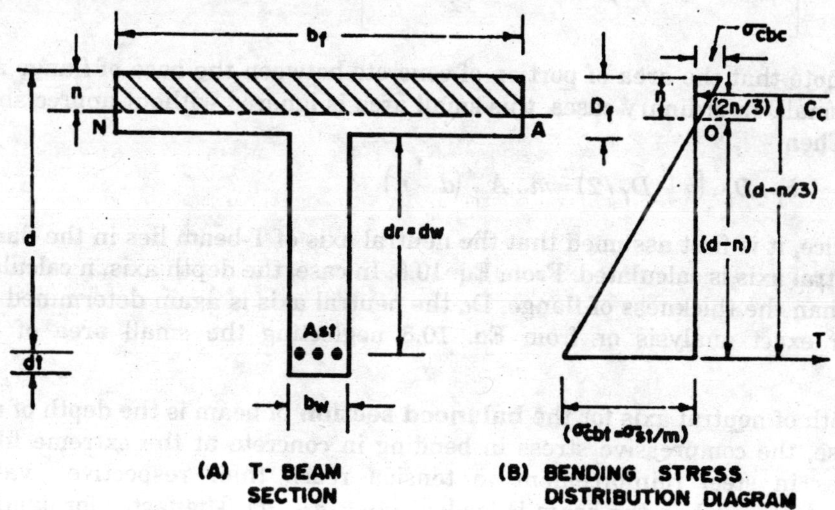

Fig. 10.2

2. Neutral axis lies below the flange (in the web)

When the neutral lies in the web, the area of concrete in compression shall be area of concrete in flange and small area of concrete in web (the area between base of flange and the neutral axis). The neutral axis of T-beam. Fig. 10.3 (A) is located by equating the first moment of area of concrete in compression and that of equivalent concrete area of steel in tension. Therefore,

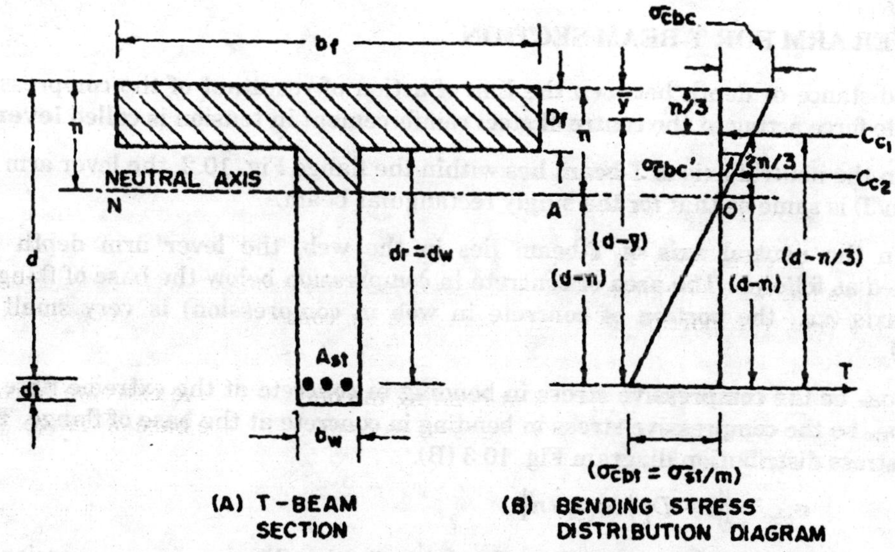

Fig. 10.3

$$\left[b_f . D_f . \left(n - D_f/2\right) + \frac{\left(b_w\right)\left(n - D_f\right)^2}{2}\right] = m . A_{st}\ (d - n) \qquad (10.7)$$

It is to note that the area of portion of concrete between the base of flange and neutral axis is very small. In ordinary cases, this small area is ignored without appreciably affecting the results. Then

$$b_f . D_f . \left(n - D_f/2\right) = m . A_{st}\ (d - n) \qquad (10.8)$$

In practice, it is first assumed that the neutral axis of T-beam lies in the flange and the depth of neutral axis is calculated. From Eq. 10.5. In case, the depth axis, n calculated is seen to be more than the thickness of flange, D_f, the neutral axis is again determined either from Eq. 10.7 for exact analysis or from Eq. 10.8 neglecting the small area of concrete in compression.

The depth of neutral axis for the **balanced section** of beam is the depth of neutral axis in which case, the compressive stress in bending in concrete at the extreme fibre and the tensile stress in steel reinforcement in tension reach their respective values of the permissible stresses when the beam is loaded, $(n_B = k_B . d)$. The factor for depth of neutral axis, k_B may be found by knowing the permissible stresses, σ_{cbc} and σ_{st}.

$$k_B = \left(\frac{m\ \sigma_{cbc}}{m\sigma_{cbc} + \sigma_{st}}\right) \qquad (i)$$

When the neutral depth of neutral axis, n is less than that for balanced or critical section, n_B, the T-beam shall be **under-reinforced.** When the actual depth of neutral axis, n is more than that for the balanced section, n_B, the beam shall be **over-reinforced.** Actua stresses for the under-reinforced and over-reinforced beam are found accordingly.

10.4 LEVER ARM FOR T-BEAM SECTION

The distance or depth between the line of action of resultant of the compressive force and tensile force acting at the centre of steel reinforcement in tension is called **lever arm.**

When the neutral axis of T-beam lies within the flange Fig. 10.2, the lever arm depth a = jd = (d – n/3) is same as that for the singly rectangular beam.

When the neutral axis of T-beam lies in the web, the lever arm depth a = jd i determined as follows : The area of concrete in compression below the base of flange and th neutral axis (i.e., the portion of concrete in web in compression) is very small and it i neglected.

Let σ_{cbc} be the compressive stress in bending in concrete at the extreme fibre, Fig. 10. (B) and σ_{cbc} be the compressive stress in bending in concrete at the base of flange. Then, fro bending stress distribution diagram Fig. 10.3 (B).

$$\sigma_{cbc} = \left[\left(n - D_f\right) . \sigma_{cbc} / n\right] \qquad (10.$$

The trapezoidal portion of bending stress distribution diagram in compression, Fig. 10 (B) may be divided into two triangles. Let $\bar{y}$ be the depth of centroid of this portion of stre

diagram from the top. The depth $\overline{y}$ may be determined by taking the moment of two areas of this portion of stress diagram. Therefore,

$$\overline{y} = \left[\frac{\left(\frac{1'}{2}\sigma_{cbc} \cdot D_t \cdot D_f/3\right) + \left(\frac{1}{2}\sigma_{cbc}' \cdot D_f \cdot 2D_f/3\right)}{\left(\frac{1}{2}\sigma_{cbc} \cdot D_f + \frac{1}{2}\sigma_{cbc}' D_f\right)}\right] \quad \text{(i)}$$

or $$\overline{y} = \frac{D_f}{3} \cdot \left(\frac{\sigma_{cbc} + 2\sigma_{cbc}'}{\sigma_{cbc} + \sigma_{cbc}'}\right) \quad \text{(ii)}$$

or $$\overline{y} = \frac{D_f}{3} \cdot \frac{\lfloor \sigma_{cbc} + 2 \cdot \sigma_{cbc} \cdot (n - D_f)/n \rfloor}{\lfloor \sigma_{cbc} + \sigma_{cbc} \cdot (n - D_f)/n \rfloor} \quad \text{(iii)}$$

or $$\overline{y} = \frac{D_f}{3} \cdot \left(\frac{3n - 2D_f}{2n - D_f}\right) \quad \text{(iv)}$$

For trapezoidal portion of the stress block. The depth for lever arm a = jd = (d − $\overline{y}$). Therefore,

$$(d - \overline{y}) = \left[d - \frac{D_f}{3} \cdot \left(\frac{3n - 2D_f}{2n - D_f}\right)\right] \quad \text{(v)}$$

or $$(d - \overline{y}) = d - \frac{D_f}{3}\left[\left(\frac{2n - D_f}{2n - D_f}\right) + \left(\frac{n - D_f}{2n - D_f}\right)\right] \quad \text{(vi)}$$

or $$(d - \overline{y}) = d - \frac{D_f}{3}\left[1 + \left\{\left(\frac{2n - D_f}{2n - D_f}\right) - \left(\frac{n}{2n - D_f}\right)\right\}\right] \quad \text{(vii)}$$

or $$(d - \overline{y}) = d - \frac{D_f}{3}\left[1 + \left\{1 - \frac{n}{2n(1 - D_f/2n)}\right\}\right] \quad \text{(viii)}$$

or $$(d - \overline{y}) = d - \frac{D_f}{3}\left[1 + \left\{1 - \frac{1}{2}\left(1 - \frac{D_f}{2n}\right)^{-1}\right\}\right] \quad \text{(ix)}$$

or $$(d - \overline{y}) = \left[d - \frac{D_f}{3}\left(\frac{3}{2} - \frac{D_f}{4n}\right)\right]$$

or $$(d - \overline{y}) = \lfloor d - (D_f/2) + (D_f^2/12n) \rfloor \quad \text{(10.10)}$$

10.5. BENDING STRENGTH OF T-BEAMS

The moment of resistance of the beam cross-section of a T-beam gives the bending strength of this beam.

1. When the neutral axis of T-beam lies in the flange

From Fig. 10.2, it is seen that the neutral axis of T-beam lies in the flange. The beam section carries a compressive force, C_c (force in concrete in compression) and the tensile force, T (a force in steel in tension). The compressive force, C_c in concrete in compression acts at a distance n/3 from the extreme fibre in compression (viz., from top in Fig. 10.2 (B)) and its value may be written as follows :

$$C_c = [(1/2)(\sigma_{cbc}).(b_f \cdot n)] \quad \text{(i)}$$

The moment of resistance of T-beam section, Fig. 10.2 may be calculated by taking th moment of the compressive force, C_c about the centre of steel reinforcement in tension. Therefore,

$$MR = C_c \,.\, (d - n/3) = [(1/2)(\sigma_{cbc} \cdot b_f \cdot n).(d - n/3)] \quad (10.11)$$

2. When the neutral axis of T-beam lies in web

From Fig. 10.3, it is seen that the neutral axis of T-beam lies in the web. The beam section carries a compressive force C_{c_1},(force in concrete in the flange in compression) and a compressive force C_{c_2}, (force in concrete in the web portion in compression). The values of these compressive forces may be written as follows :

$$C_{c_1} = (\textit{Average stress}).(\textit{area of flange})$$

$$\text{or} \quad C_{c_1} = \left(\frac{\sigma_{cbc_1} + \sigma_{cbc}}{2}\right).(b_f \cdot D_f) \quad \text{(i)}$$

where, from Eq. 10.9

$$\sigma_{cbc}' = [(n - D_f).\sigma_{cbc}/n] \quad \text{(ii)}$$

$$\text{And} \quad C_{c_2} = (\textit{Average stress})(\textit{area of web})$$

$$\text{or} \quad C_{c_2} = [(1/2)(\sigma_{cbc}).(b_w)(n - D_f)] \quad \text{(iii)}$$

The moment of resistance of T-beam section Fig. 10.3 may be determined by taking the moment of these two compressive forces C_{c_1} and C_{c_2} about the centre of steel reinforcement in tension. Therefore,

$$MR = C_{c_1}(d - \bar{y}) + C_{c_2}\left[(d - n) + \frac{2}{3}.(n - D_f)\right]$$

$$\text{or} \quad MR = [(\sigma_{cbc} + \sigma_{cbc}/2)].(b_f \, . \, D_f).(d - \bar{y})$$
$$+ (1/2\,\sigma_{cbc}) \, . \, b_w \, . \, (n - D_f) \, [(d - n/3 - 2\,D_f/3)] \tag{10.12}$$

where, the lever arm, $(d - \bar{y})$ may be written from Eq. 10.10.

Alternatively

The moment of resistance of T-beam section, Fig. 10.3 may also be found by first determining the resultant compressive force in concrete in compression, C_c for the two compressive forces C_{c_1} and C_{c_2} and then calculating the lever arm (the distance or depth between the compressive force C_c and the tensile force, T acting at the centre of steel bars in tension). The value of resultant compressive force, may be written as follows :

$$C_c = (C_{c_1} + C_{c_2}) \tag{iv}$$

$$\text{or} \quad C_c = \left[\frac{(\sigma_{cbc} + \sigma_{cbc})}{2} \, . \, (b_f \, . \, D_f) + (1/2)(\sigma_{cbc})\, b_w \, . \, (b - D_f)\right] \tag{v}$$

The line of action of resultant compressive, C_c may be obtained by taking the moment of two compressive forces C_{c_1} and C_{c_2} about top fibre (extreme fibre of concrete in compression). Therefore,

$$y_1 = \left\{\frac{\left(\frac{\sigma_{cbc} + \sigma_{cbc}}{2}\right)(b_f . D_f) . \frac{D_f}{3}\left[\frac{(\sigma_{cbc} + 2\,\sigma_{cbc})}{(\sigma_{cbc} + \sigma_{cbc})}\right] + \left(\frac{1}{2} . \sigma_{cbc}\right) . b_w \, . \, (n - D_f)\left[D_f + (n - D_f)/3\right]}{\left(\frac{\sigma_{cbc} + \sigma_{cbc}}{2}\right)(b_f \, . \, D_f) + (1/2\;\sigma_{cbc} \, . \, b_w)(n - D_f)}\right\} \tag{vi}$$

The lever arm distance, then, shall be $(d - \bar{y})$ the moment of resistance of T-beam

$$\text{MR} = \left[\left(\frac{\sigma_{cbc} + \sigma_{cbc}}{2}\right)(b_f . D_f) + (1/2\;\sigma_{cbc} \, . b_w)(n - D_f)\right] . (d - \bar{y}) \tag{10.13}$$

In case, the portion of concrete in compression in web and thereby the compressive force C_{c_2} are neglected, then, the moment of resistance of T-beam section, Fig. 10.3 may be found as follows :

$$M\,R = C_{c_1} \, . \, (d - \bar{y}) \tag{vii}$$

$$\text{or} \quad MR = \frac{1}{2}(\sigma_{cbc} + \sigma_{cbc}) \, . \, b_f \, . \, D_f \, . \, (d - \bar{y}) \tag{viii}$$

$$\text{or} \quad MR = \frac{1}{2}\,\sigma_{cbc}\left[1 + \frac{(n - D_f)}{n}\right] . \, b_f \, . \, D_f \, . \, (d - \bar{y}) \tag{ix}$$

The value of lever arm $(d - \bar{y})$ may be substituted from Eq. 10.10. Therefore,

$$MR = \frac{1}{2}\sigma_{cbc}\left[1+\left(\frac{n-D_f}{n}\right)\right]\left[b_f . D_f\right]\left[d-\frac{D_f}{2}+\frac{D_f^2}{12\,n}\right] \quad (10.14)$$

The moment of resistance of T-beam, Fig. 10.3 may also be calculated from tensile force in steel reinforcement. Therefore,

$$MR = T \;.\; lever\ arm$$

$$\text{or} \quad MR = \sigma_{st} \cdot A_{st} \cdot (d-\bar{y}) \quad (10.15A)$$

$$\text{or} \quad MR = \sigma_{st} \cdot A_{st} \cdot (d-\bar{y}) \quad (10.15B)$$

Eq. (10.15A) is used when the portion of compressive force in concrete in web is neglected. Eq. (10.15B) is used when this portion is considered.

Example 10.1 A single reinforced slab 120 mm thick is supported by T-beams spaced 3 m centre to centre. The effective depth and breadth of web are 600 mm and 450 mm, respectively. The effective cover to mild steel reinforcement 4 bars of 20 mm diameter is 50 mm. The effective span of simply supported beams is 6.30 m. The grade of concrete is M 15. Determine the depth of neutral axis of T-beam cross-section. Also determine the moment of resistance.

Solution : The grade of concrete is M15. From IS : 456-2000, the permissible stress in compression in bending in concrete s_{cbc} is 5 N/mm^2. The modular ratio of concrete

$$m = \left(\frac{280}{3\times\sigma_{cbc}}\right) = \left(\frac{280}{3\times5}\right) = 18.667 \quad (i)$$

Step 1 : Area of steel reinforcement in tension

$$A_{st} = \left(4\times\frac{\pi}{4}\times20^2\right) = 1256.637\ mm^2$$

Step 2 : Effective width of flange of T-beam

From IS : 456-2000

$$b_f = \left(\frac{l_o}{6}+b_w+6D_f\right) = \left(\frac{6300}{6}+450+6\times120\right) = 2200\ mm \quad (ii)$$

Check $b_f \not> [0.5\,(L_1+L_2)+b_w]$

or $b_f \not> [1275+1275+450] = 3000\ mm$ satisfied

Step 3 : Neutral axis of T-beam

T-beam cross-section is shown in Fig. 10.4 (A). In order to locate the neutral axis, equate moment of equivalent areas of concrete in compression and steel reinforcement in tension

about neutral axis. From Fig. 10.4 (A), Eq. 10.5 (it is first assumed that the neutral axis lies in the flange).

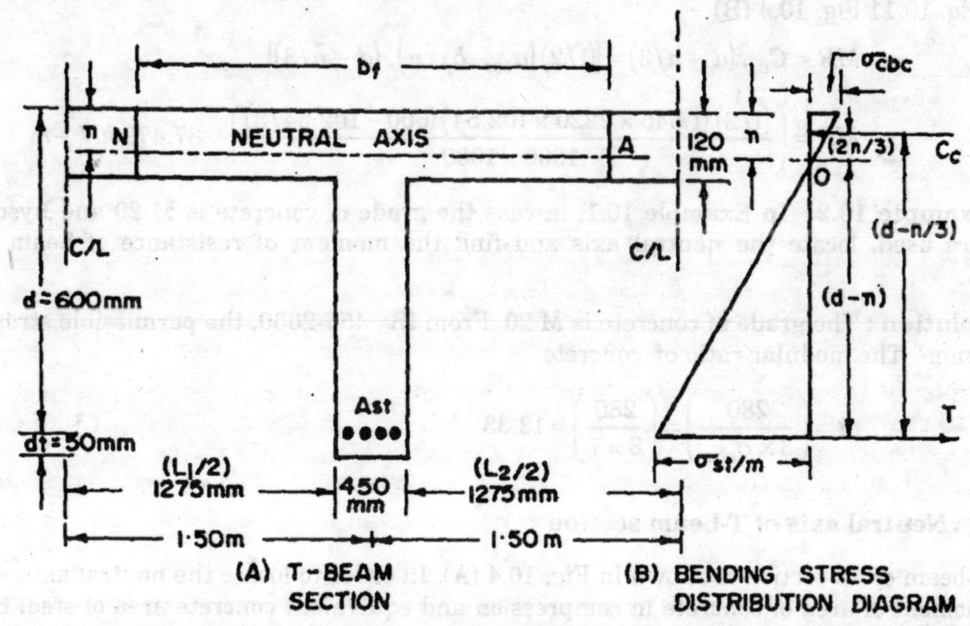

Fig. 10.4

$$\left(b_f \, . \, n \, . \, n/2\right) = m \, A_{st} \, (d - n)$$

$$\left(2220 \, n^2/2\right) = \left[18.67 \times 1256.637 \times (600 - n)\right]$$

n = 102.962 mm.

Step 4 : Moment of resistance of T-beam

In order to find the moment of resistance of T-beam section, depth of neutral axis, for the balanced section is calculated. From IS : 456-2000, σ_{cbc} = 5 N/mm² and σ_{st} = 140 N/mm².

$$k_B = \left(\frac{m \, \sigma_{cbc}}{m \, \sigma_{cbc} + \sigma_{st}}\right) = \left(\frac{18.67 \times 5}{18.67 \times 5 + 140}\right) = 0.4$$

Depth of neutral axis for the balanced section

$$n_B = \left(k_B \, . \, d\right) = \left(0.4 \times 600\right) = 240 \; mm \qquad \text{(iii)}$$

By comparing *n* and n_B, it is seen that the given beam is under-reinforced since, *n* is less than n_B. Therefore, steel reinforcement bars develop the stress equal to its permissible value (*viz.*, σ_{st} = 140 N/mm²). From Fig. 10.4 (B)

$$\sigma_{cbc} = \left(\frac{n}{d - n}\right) . \left(\frac{\sigma_{st}}{m}\right) = \left(\frac{102.54}{600 - 102.54}\right) \times \left(\frac{140}{18.67}\right) = 1.545 \; N/mm^2$$

The moment of resistance of T-beam section is calculated by taking moment of the compressive force C_c in concrete in compression about the centre of steel bars in tension. From Eq. 10.11 Fig. 10.4 (B)

$$MR = C_c \,.\, (d - n/3) = \left[(1/2)\left(\sigma_{cbc} \,.\, b_f \,.\, n\right).\, (d - n/3)\right]$$

$$MR = \left(\frac{(1/2)(1.545 \times 2220 \times 102.54)(600 - 102.54/3)}{1000 \times 1000}\right) = 87.479 \; kN - m$$

Example 10.2. In Example 10.1, in case the grade of concrete is M 20 and hysd-steel bars are used, locate the neutral axis and find the moment of resistance of beam cross-section.

Solution : The grade of concrete is M 20. From IS : 456-2000, the permissible stress, σ_{cbc} is 7 N/mm^2. The modular ratio of concrete

$$m = \left(\frac{280}{3 \times \sigma_{cbc}}\right) = \left(\frac{280}{3 \times 7}\right) = 13.33 \qquad \text{(i)}$$

Step 1 : Neutral axis of T-beam section

T-beam cross-section is shown in Fig. 10.4 (A). In order to locate the neutral axis, equate first moment of area of concrete in compression and equivalent concrete area of steel bars in tension about neutral axis. It is first assumed that the neutral axis lies in flange. From Fig. 10.4 (A), Eq. 10.5

$$\left(b_f \,.\, n \,.\, n/2\right) = m \,.\, A_{st} \,.\, (d - n)$$

Substituting the values of b_f, A_{st} and d from Example 10.1

$$\left(2200 \times n^2/2\right) = \left[13.33 \times 1256.637 \times (600 - n)\right]$$

n = **87.909 mm.**

Step 2 : Depth of neutral axis for balanced section

From IS : 456-2000, σ_{cbc} = 7 N/mm^2 and σ_{st} = 230 N/mm^2

Factor for depth of neutral axis for the balanced section

$$k_B = \left(\frac{m\,\sigma_{cbc}}{m\,\sigma_{cbc} + \sigma_{st}}\right) = \left(\frac{13.33 \times 7}{18.33 + 230}\right) = 0.2886$$

$$n_B = \left(k_B \,.\, d\right) = (0.2886 \times 600) = 173.16 \; mm$$

Since n is less than n_B, th beam is under-reinforced. Therefore, the stress in steel reinforcement, σ_{st} reaches its permissible stress value (*viz.,* 230 N/mm^2) first.

Step 3 : Actual compressive bending stress in concrete

From Fig. 10.4 (B), the bending stress distribution diagram

$$\sigma_{cbc} = \left(\frac{n}{d-n}\right) \cdot \left(\frac{\sigma_{st}}{m}\right) = \left(\frac{87.909}{600 - 87.909}\right) \times \left(\frac{230}{13.33}\right) = 2.962\ N/mm^2$$

Step 4 : Moment of resistance of beam section

$$MR = C_c \cdot (d - n/3) = (1/2)(\sigma_{cbc} \cdot b_f \cdot n)(d - n/3)$$

$$MR = \left(\frac{(1/2)(2.962 \times 2220 \times 87.909)(600 - 87.909/3)}{1000 \times 1000}\right) = 164.948\ kN-m$$

Example 10.3. A singly reinforced slab 120 mm thick is supported by T-beams spaced 3 m centre to centre. The effective depth and width of web are 580 mm and 450 mm respectively. Eight mild steel bars of 20 mm diameter have been provided in tension in two layer as shown in Fig. 10.5. The effective cover to steel bars in lower tier is 50 mm. The effective span of simply supported beam is 6.30 m. The grade of concrete is M 15. Determine the depth of neutral axis and the moment of resistance of T-beam section.

Solution : The grade of concrete is M 15. From IS : 456-2000, the permissible stress in compression in bending in concrete σ_{cbc} is 5 N/mm². The modular ratio of concrete

$$m = \left(\frac{280}{3 \cdot \sigma_{cbc}}\right) = \left(\frac{280}{3 \times 5}\right) = 18.67 \quad \text{(i)}$$

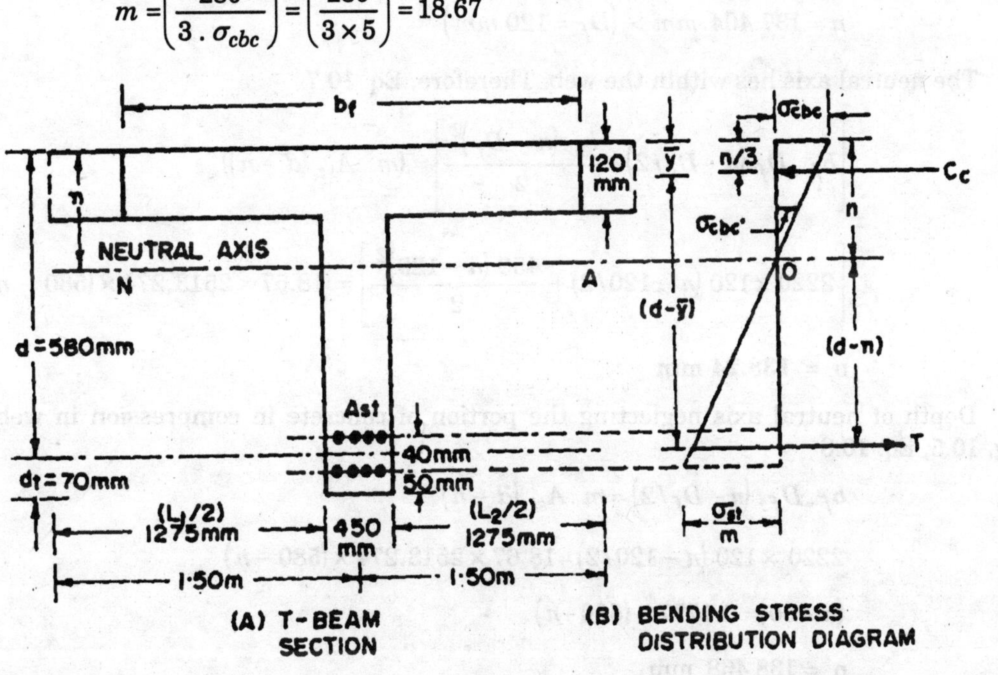

Fig. 10.5

Area of steel reinforcement in tension

$$A_{st} = \left(8 \times \frac{\pi}{4} \times 20^2\right) = 2513.274\ mm^2$$

Step 1 : Effective width of flange of T-beam

From IS : 456-2000, Fig. 10.5 (A)

$$b_f = \left(\frac{l_o}{6} + b_w + 6\, D_f\right) = \left(\frac{6300}{6} + 450 + 6 \times 120\right) = 2220\ mm \quad \text{(ii)}$$

Check $b_f \not> [0.5\,(L_1 + L_2)\,b_w]$

or $b_f \not> [(1275 + 1275 + 450) = 3000\ mm]$ satisfied

Step 2 : Neutral axis of T-beam

T-beam cross-section is shown in Fig. 10.5 (A). The neutral axis is located by taking first moment of area of concrete in compression and equivalent concrete area to steel reinforcement in tension. First, it is assumed that the neutral axis lies within the flange. Therefore, from Eq. 10.5

$$b_f\,.\,n\,.\,n/2 = m\,A_{st}\,(d - n)$$

$$2220\,b_f\,.\,n\,.\,n/2 = [18.67 \times 2513.274 \times (580 - n)]$$

$$n = 137.404\ mm > (D_f = 120\ mm) \quad \text{(iii)}$$

The neutral axis lies within the web. Therefore, Eq. 10.7

$$\left[b_f\,.\,D_f\,(n - D_f/2) + \frac{b_w\,(n - D_f)^2}{2}\right] = (m\,.\,A_{st}\,(d - n))$$

$$\left[2220 \times 120\,(n - 120/2) + \frac{450\,(n - 120)^2}{2}\right] = (18.67 \times 2513.274 \times (580 - n))$$

$$n = \mathbf{138.24\ mm} \quad \text{(iv)}$$

Depth of neutral axis neglecting the portion of concrete in compression in web. From Fig. 10.5, Eq. 10.8

$$b_f\,.\,D_f\,.\,(n - D_f/2) = m\,.\,A_{st}\,(d - n)$$

$$2220 \times 120\,(n - 120/2) = 18.67 \times 2513.274 \times (580 - n)$$

$$(n - 60) = 0.17774\ (580 - n)$$

$$n = 138.468\ \text{mm} \quad \text{(v)}$$

It is to note that the difference in two values of neutral-axis as above is very small.

Step 3 : Moment of resistance of T-beam

The neutral axis of T-beam lies in the web.

3 (A) Lever arm depth (d - $\bar{y}$)

Let $\bar{y}$ be the depth of line of action of the compression in concrete in flange. From the expression (iv), Art. 10.4

$$\bar{y} = \left(\frac{D_f}{3}\right) \cdot \left(\frac{3n - 2\ D_f}{2n - D_f}\right) = \frac{120}{3} \cdot \left(\frac{3 \times 138.24\ -\ 2 \times 120}{2 \times 138.24\ - 120}\right) = 44.6626\ mm$$

$$\therefore\ (d - \bar{y}) = (580\ -\ 44.6626) = 535.3374\ mm \qquad \text{(vi)}$$

3 (B) Depth of neutral axis for the balanced section

$$\sigma_{cbc} = 5\ N/mm^2, \sigma_{st} = 140\ N/mm^2, m = 18.667$$

3 (4) Depth factor for neutral axis

$$k_B = \left(\frac{m\ \sigma_{cbc}}{m\ \sigma_{cbc}\ +\ \sigma_{st}}\right) = \left(\frac{18.67 \times 5}{18.67 \times 5 + 140}\right) = 0.4$$

$$n_B = (k_B \times d) = (0.4 \times 580) = 232\ mm$$

Step 4 : Stresses in material

Actual depth of neutral axis n = 138.24 mm is less than the depth of neutral axis for the balanced design. The steel reinforcement in lower tier attain its stress equal to the permissible stress value in tension. Therefore, $\sigma_{st} = 140\ N/mm^2$

$$\sigma_{cbc} = \left(\frac{n}{600 - n} \times \frac{140}{18.67}\right) = \left(\frac{138.24}{(600\ -\ 138.24)} \times \frac{140}{18.67}\right) = 2.245\ N/mm^2$$

$$\sigma_{cbc}' = \left(\frac{n\ -\ D_f}{n}\right) . \sigma_{cbc} = \left(\frac{(138.24 - 124)}{138.24} \times 2.245\right) = 0.296\ N/mm^2$$

Step 5 : Moment of resistance

Moment of resistance of T-beam considering the portion of concrete in web also is found by taking moment of compressive forces C.c₁ and C.c₂ about the e.g. of steel reinforcement. Therefore,

$$MR = C.c_1\ (d - \bar{y}) + C.c_2\ [(d - n) + (2/3)(n - D_f)]$$

From Eq. 10.12

$$MR = \left(\frac{\sigma_{cbc}\ +\ \sigma_{cbc}'}{2}\right) . b_f\ . D_f\ (d - \bar{y}) + \frac{1}{2}\ \sigma_{cbc}' . b_w\ . (n - D_f)(d - n/3 - 2\ D_f/3)$$

$$MR = \left(\frac{2.245 + 0.296}{2}\right) \times \left(\frac{2220 \times 120 \times 535.3374}{1000 \times 1000}\right)$$

$$+ \frac{1}{2} \times 0.296 \times 456 \times (138.24 - 120) \times \left(\frac{(580 - 138.24/3 - 2 \times 120/3)}{1000 \times 1000}\right)$$

$$MR = 180.735 \text{ kN-m}$$

The moment of resistance of T-beam neglecting the portion of concrete is compression

$$MR = \left(\frac{2.245 + 0.296}{2}\right) \times \left(\frac{2220 \times 120 \times 535.3374}{1000 \times 1000}\right) = 179.52169\ kN - m$$

Example 10.4 In Example 10.3, in case the grade of concrete is M 20 and hysd-steel bars are used, locate the neutral axis, and find the moment of resistance of the beam.

Solution : The grade of concrete is M 20. From IS : 456-2000, the permissible stress is compression is bending is concrete σ_{cbc} is 7N/mm². The modular ration of concrete

$$m = \left(\frac{280}{3 \times \sigma_{cbc}}\right) = \left(\frac{283}{3 \times 7}\right) = 13.33 \quad \text{(i)}$$

Effective width of flange from *Example 10.3* is 2200 mm.

Step 1 : Neutral axis of the beam section

From Example 10.3, the neutral axis of T-beam section lies in the web

$$n = 138.24 \text{ mm} \quad \text{(ii)}$$

Depth of neutral axis neglecting the portion of concrete in compression in web

$$n = 138.468 \text{ mm} \quad \text{(iii)}$$

Step 2 : Moment of resistance of T-beam section

From IS : 456-2000, σ_{cbc} = 7 N/mm² and σ_{st} = 230 N/mm²

From *Example 10.3,* the layer arm depth

$$(d - \bar{y}) = 535.3347\ mm \quad \text{(iv)}$$

Step 3 : Depth of neutral axis for the balanced section

Depth factor for the neutral axis

$$k_B = \left(\frac{m\ \sigma_{cbc}}{m\ \sigma_{cbc} + \sigma_{st}}\right) = \left(\frac{13.33 \times 7}{13.33 \times 7 + 230}\right) = 0.2886$$

$$n_B = (k_B\ .\ d) = (0.2868 \times 580) = 167.388\ mm \quad \text{(v)}$$

The actual depth of neutral axis n = 138.24 mm is less than that for the balanced section n_B = 167.338 mm. The beam section is under reinforced. Therefore, the steel reinforcement

in lower tier attains its stress equal to the permissible stress value in tension. Therefore, σ_{st} = 230 N/mm².

Step 4 : Actual compressive bending stress in concrete

$$\sigma_{cbc} = \left(\frac{n}{600-n}\right) \times \frac{230}{13.33}\ N/mm^2 = \left(\frac{138.24}{600-138.24}\right) \times \frac{230}{13.33} = 5.166\ N/mm^2$$

and $$\sigma_{cbc} = \left(\frac{n-D_f}{n}\right) . \sigma_{cbc} = \left(\frac{38.24-120}{138.24}\right) \times 5.166 = 0.682\ N/mm^2$$

Step 5 : Moment of resistance of beam section

Moment of resistance of T-beam section considering the portion of concrete in web also, is obtained by taking the moment of compressive forces C_{c_1} and C_{c_2} about the e.g. of steel reinforcement. Therefore,

$$MR = C_{C_1} . (d - \bar{y}) + C_{c2}\ [(d-n) + (2/3)(n - D_f)]$$

From Eq. 10.12

$$MR = \left[\left(\frac{\sigma_{cbc} + \sigma_{cbc}}{2}\right) b_f . D_f . (d-\bar{y}) + (1/2)\sigma_{cbc}' . b_w . (n - D_f)(d - n/3 - 2\ D_f/3)\right]$$

or $$MR = \left[\left(\frac{5.166 + 0.682}{2}\right) \times \left(\frac{2220 \times 120 \times 535.3374}{1000 \times 1000}\right)\right]$$

$$+ (1/2) \times 0.682 \times 450\ (138.24 - 120) \times \left(\frac{580 - 138.24/3 - 2 \times 120/3}{1000 \times 1000}\right)$$

or MR = (413.246 + 2.797) = 416.043 kN.

10.6 STRESSES IN CONCRETE AND STEEL

In case, a T-beam of known dimensions, area of steel reinforcement in tension and the area of steel reinforcement in compression (*i.e.*, the beam is doubly reinforced) is subjected to given value of the bending moment, *M*, the actual stresses developed in the various component materials of the beam due to bending moment my be calculated as follows.

The actual depth of neutral axis of the beam cross-section, *n* and that for the balanced section are found. The bending strength of beam (viz., the moment of resistance) may be written by taking the moment of compressive forces C_{c_1} and C_{c_2} in the concrete in compression in the flange and web, respectively, (and C_s in the steel reinforcement in compression in any) about the centre of steel reinforcement in tension. Thus, from Fig. 10.2 (B), from Eq. 10.12

$$MR = C_{c1} . (d - \bar{y}) + C_{c2}\ [(d-n) + (2/3)(n - D_f)] \quad \text{(i)}$$

$$MR = \left(\frac{\sigma_{cbc} + \sigma_{cbc}'}{2}\right) b_f \,.\, D_f \,.\, (d - \bar{y}) + (1/2\, \sigma_{cbc}') \,.\, b_w \,.\, (n - D_f) \left[d - n/3 - 2D_f/3\right] \quad \text{(ii)}$$

The bending strength of beam MR is equated to the given bending moment, M. Therefore,

$$MR = ((\sigma_{cbc} + \sigma_{cbc}')/2)\, b_f \,.\, D_f \,.\, (d - \bar{y}) + (1/2)\, \sigma_{cbc}' \,.\, b_w \,.\, (n - D_f) \left[d - n/3 - 2D_f/3\right] \quad \text{(iii)}$$

The stress σ_{cbc}' at the bottom flange (slab) is expressed in terms of σ_{cbc}'. From Fig. 10.2 (B),

$$\sigma_{cbc}' = \left(\frac{n - D_f}{n}\right) . \sigma_{cbc} \quad \text{(iv)}$$

The compressive stress in bending in concrete, σ_{cbc} at the extreme fibre of concrete in compression is calculated by solving the expression (iii).

The stress in steel reinforcement in tension, σ_{st} is determined from the similar triangles above the below the neutral axis in bending stress distribution diagram. From Fig. 10.2 (B),

$$\sigma_{st} = m \,.\, \left(\frac{d - n}{n}\right) . \sigma_{cbc} \quad \text{(v)}$$

In case, T-beam is doubly reinforced, then the compressive stress in bending in concrete surrounding the steel reinforcement in compression is obtained from Fig. 10.2 (B),

$$\sigma_{cbc1} = \left(\frac{n - d_c}{n}\right) . \sigma_{cbc} \quad \text{(vi)}$$

The stress in steel reinforcement in compression, σ_{st} shall be

$$\sigma_{sc} = m_c \,.\, \sigma_{cbc1} \quad \text{(vii)}$$

The actual stress in different component materials of T-beam should be less than their respective permissible values as specified in IS : 456-2000.

Example 10.5. Determine the actual stresses developed in the different component materials of a T-beam in Example 10.3 if it is subjected to bending moment, M equal to 140 kN-m.

Solution : It is essential to determine the depth of neutral axis in order to calculate the stresses in the different component materials of T-beam. From *Example 10.3*

$$n = 138.24 \text{ mm} \quad \text{(i)}$$

And the effective width of flange of T-beam, from *Example 10.3*

$$b_f = 2220 \text{ mm} \quad \text{(ii)}$$

Step 1 : Moment of resistance of beam section

The external bending moment M = 140 kN-m acting at a transverse section of T-beam, Fig. 10.5 (*Example 10.3*) is resisted by the strength of beam in bending (*viz.*, the moment of resistance). Therefore,

$$MR = M = 140 \times 10^6\ N - mm \quad \text{(iii)}$$

From Example 10.3 lever arm $(d - \bar{y})$ is 535.3374 mm.

Step 2 : Stress in compression in bending in concrete, σ_{cbc}

From Fig. 10.5 (B), Eq. 10.12

$$MR = [(\sigma_{cbc} + \sigma_{cbc}')/2]\, b_f \,.\, D_f \,.\, (d - \bar{y}) + (\sigma_{cbc}'/2)\,.\, b_w \,.\, (n - D_f)\, [d - n/3 - 2D_f/3] \quad \text{(iv)}$$

$$\sigma_{cbc}' = [n - D_f / n] \times \sigma_{cbc} \quad \text{(v)}$$

Substituting this in expression (iv)

$$MR = \left[\left(\sigma_{cbc} + \frac{n - D_f}{n}\,.\,\sigma_{cbc}\right)/2\right] b_f \,.\, D_f \,.\, (d - \bar{y})$$

$$+ \frac{1}{2}\,.\left(\frac{n - D_f}{n}\right)\sigma_{cbc}\,.\, b_w\,(n - D_f)\left(d - \frac{n}{3} - \frac{2\,D_f}{3}\right)$$

$$140 \times 10^6 = \sigma_{cbc}\left[\left\{\left(1 + \frac{(138.24 - 120)}{138.24}\right) \times \frac{1}{2}\right\} 2220 \times 120 \times 535.3374\right.$$

$$\left. + \frac{1}{2}\,\frac{(138.24 - 120)}{138.24} \times 450 \times (138.24 - 120) \times \left(580 - \frac{138.24}{3} - \frac{2 \times 120}{3}\right)\right]$$

or σ_{cbc} **= 1.751** N/mm² < (5 N/mm²) (vi)

Step 3 : Stress in the steel bars in tension in lower-tier

$$\sigma_{st1} = m\,.\left(\frac{600 - n}{n}\right) \times \sigma_{cbc} \quad \text{(vii)}$$

$$\sigma_{st1} = 18.67 \times \left(\frac{600 - 138.24}{138.24}\right) \times 1.751\ \text{N/mm}^2$$

= **109.198** N/mm² < (140 N/mm²).

Step 4 : Stress in the steel bars in tension in upper-tier

$$\sigma_{st1} = m\,.\left(\frac{560 - n}{n}\right) \times \sigma_{cbc}$$

$$\sigma_{st1} = 18.67 \times \left(\frac{580 - 138.24}{138.24}\right) \times 1.751 = 104.468\ N/mm^2 < \left(140\ N/mm^2\right)$$

Example 10.6. Determine the actual stresses developed in the different component materials of a T-beam in Example 10.4 if it is subjected to bending moment M equal to 360 kN-m.

Solution : It is to necessary to calculate the depth of neutral axis in order to determine the actual stresses in the different component materials of T-beam. From *Example 10.4*

$$n = 138.24 \text{ mm} \quad \text{(i)}$$

Effective width of flange of T-beam from Example 10.4

$$b_f = 2220 \text{ mm} \quad \text{(ii)}$$

Lever arm depth $(d - \bar{y})$ is 535.3374 mm.

Step 1 : Moment of resistance

The external bending moment M = 360 kN-m acting at a transverse section of a T-beam Fig. 10.5 (Example 10.4) is resisted by the strength of beam in bending (*viz.*, the moment of resistance). Therefore,

$$MR = M = 360 \times 10^6 \text{ N-mm} \quad \text{(iii)}$$

Step 2 : Stress in compression in bending in concrete, σ_{cbc}

From Eq. 10.12, Fig. 10.5 (B)

$$MR = \left[\frac{(\sigma_{cbc} + \sigma_{cbc}')}{2}\right] b_f \,.\, D_f \,.\, (d - \bar{y}) + \left(\frac{\sigma_{cbc}'}{2}\right) . \, b_w \,.\, (n - D_f)\left[d - \frac{n}{3} - \frac{2\,D_f}{3}\right] \quad \text{(iv)}$$

$$\sigma_{cbc}' = \left[\frac{(n - D_f)}{n}\right] . \, \sigma_{cbc} \quad \text{(v)}$$

Substituting this in expression (iv)

$$MR = \left[\frac{\left\{\sigma_{cbc} + \sigma_{cbc}\left(\frac{n - D_f}{n}\right)\right\}}{2}\right] b_f \,.\, D_f \,.\, (d - \bar{y})$$

$$+ \left\{\sigma_{cbc}\,\frac{(n - D_f)}{n}\right\} . \, b_w \,.\, (n - D_f)\left(d - \frac{n}{3} - \frac{2\,D_f}{3}\right)$$

$$360 \times 10^6 = \sigma_{cbc}\left[\left[\left\{\frac{1 + \frac{(138.24 - 120)}{138.24}}{2}\right\} 2220 \times 120 \times 535.3374 + \right.\right.$$

$$\left.\frac{1}{2} \times \frac{(138.24 - 120)}{138.24} \times 450 \times (138.24 - 120) \times \left(580 - \frac{138.24}{3} - \frac{2 \times 120}{3}\right)\right]$$

$$\sigma_{cbc} = \mathbf{4.47\ N/mm^2} < 7 \text{ N/mm}^2$$

Step 3 : Stress in the steel bars in tension in lower tier

$$\sigma_{st1} = m \cdot \left(\frac{600 - n}{n}\right) \times \sigma_{cbc}$$

$$\sigma_{st1} = 13.33\left(\frac{66 - 138.24}{138.24}\right) \times 4.47 = 199\ N/mm^2\ \left(< 230 N/mm^2\right)$$

Step 4 : Stress in the steel bars in tension in upper tier

$$\sigma_{st2} = m \cdot \left(\frac{560 - n}{n}\right) \times \sigma_{cbc}$$

$$\sigma_{st2} = 13.33\left(\frac{560 - 138.24}{138.24}\right) \times 4.47 = 181.789\ N/mm^2\ \left(< 230\ N/mm^2\right)$$

10.7 EFFECTIVE DEPTH OF T-BEAM

The effective depth of T-beam, d is determined from the consideration of requirements of bending moments and also from the consideration of economy and these are referred as *minimum depth* and *economical depth,* respectively. The minimum depth and economical depth are calculated as follows :

1. Minimum depth of T-beam

The minimum depth of T-beam necessary to resist the maximum bending moment, M occurring any where at a transverse section is obtained by equating the bending moment, M and moment of resistance of an ideally reinforced T-beam section (*viz.,* balanced section of T-beam). The moment of resistance of a balanced section of a T-beam neglecting the compression in concrete in the portion of web, therefore.

$$\text{MR} = (\text{compressive force in concrete in the flange}) \times (\text{lever arm}) \tag{i}$$

Compressive force in concrete in the flange,

$$C = \left[\left(\frac{\sigma_{cbc} + \sigma_{cbc}'}{2}\right)\right] \cdot b_f \cdot D_f \tag{ii}$$

$$\sigma_{cbc}' = \left[\left(\frac{n - D_f}{n}\right)\right] \cdot \sigma_{cbc} \tag{iii}$$

And, lever arm depth

$$(d - \bar{y}) = \left[d - \left(\frac{D_f}{2}\right) + \left(\frac{D_f^2}{12\, n}\right)\right] \tag{iv}$$

Therefore, the compressive force in concrete in the flange of T-beam

$$C = \left[\left(\frac{\sigma_{cbc} + \left(\frac{n - D_f}{2}\right) . \sigma_{cbc}}{2}\right)\right] b_f . D_f$$

$$\text{or } C = \sigma_{cbc} \left[\frac{(2n - D_f)}{2n}\right] . b_f . D_f \qquad \text{(v)}$$

The moment of resistance of T-beam

$$MR = C . (d - \bar{y}) = \sigma_{cbc} . \left[\frac{(2n - D_f)}{2n}\right] . b_f . D_f (d - \bar{y}) \qquad \text{(vi)}$$

The moment of resistance of T-beam as given by the expression (vi) is equated to the maximum bending moment, M calculated. Since $n = n_B = k_B . d$ for the balanced section of T-beam

$$M = \sigma_{cbc} \left[\frac{(2k_B . d - D_f)}{2k_B . d}\right] b_f . D_f \times \left[d - \left(\frac{D_f}{2}\right) + \left(\frac{D_f^2}{12k_B . d}\right)\right] \qquad (10.16)$$

In Eq. 10.16, all terms except, d are known. Therefore, the minimum depth (effective depth required) d is determined by solving Eq. 10.16.

Alternatively. The lever arm for T-beam section is approximately assumed as 0.9 d (as its actual value varies from 0.88 d to 0.95). Then

M = C . lever arm

$$\text{or } M = \sigma_{cbc} . \left[\frac{(2k_B . d - D_f)}{2k_B . d}\right] b_f . D_f \times 0.9\, d \qquad (10.17)$$

The effective depth, d required is found approximately.

2. Economical depth of T-beam

The economical depth of T-beam is defined as the depth corresponding to which the cost of materials of T-beam used per unit length shall be minimum. The thickness of slab (flange) of T-beam remains fixed. Therefore, volume or the cost of concrete in the web and volume or the cost of steel reinforcement in tension per unit (1 mm) length of T-beam shall be considered.

Volume of concrete used in the web per 1 mm length of T-beam

$$V_c = b_w . (d - D_f + d_t) \times 1\, mm^3 = b_w . (D - D_f)\, mm^3 \qquad \text{(i)}$$

where, d_t is the effective cover for the steel reinforcement in tension.

The area of steel reinforcement in tension

$$A_{st} = \left(\frac{M}{\sigma_{st} \cdot j \cdot d} \right) mm^2 \tag{ii}$$

Therefore, the volume of steel reinforcement per 1 mm length of T-beam

$$V_s = (A_{st} \times 1)\, mm^3 \tag{iii}$$

Cost of component materials of T-beam per 1 mm length (without considering portion of flange)

$$\text{cost} = (R_c \cdot V_c + R_s \cdot V_s) = \left| R_c \cdot b_w \left(d - D_f + d_t \right) + R_s \cdot A_{st} \right|$$

$$\text{cost} = \left[R_c \cdot b_w \left(d - D_f + d_t \right) + R_s \cdot \left(\frac{M}{\sigma_{st} \cdot j \cdot d} \right) \right]$$

where, R_c is the cost of concrete per unit volume and R_s is the cost of steel per unit volume. Let r be the cost ratio of cost of steel to cost of concrete ($r = R_s/R_c$). Then, the expression (iv) may be written as follows

$$\text{cost} = \left[R_c \cdot b_w \left(d - D_f + d_t \right) + r \cdot R_c \cdot \frac{M}{(\sigma_{st} \cdot j \cdot d)} \right] \tag{v}$$

For the economical depth, the expression (v) is differential with respect to d and d (cost)/d_d is equated to zero. Therefore

$$\frac{d\,(\text{cost})}{dd} = \left[R_c \cdot b_w - r \cdot R_c \left(\frac{M}{\sigma_{st} \cdot j \cdot d^2} \right) \right] = 0 \tag{vi}$$

$$\text{or } d = \left[\frac{r \cdot M}{(\sigma_{st} \cdot j \cdot b_w)} \right]^{1/2} \tag{10.18}$$

The economical depth of T-beam may be found from Eq. 10.18. The effective depth of T-beam shall be larger of the values obtained from the above two consideration. It shall also be seen that effective depth, *d* obtained or adopted is sufficient when the shear in T-beam is considered.

10.8 DESIGN OF T-BEAMS

After deciding the grade of concrete and type of steel reinforcement, the various design constants, *m*, k_B, j_B and R_B for the balanced beam section are calculated.

The dead and live loads to be carried by the beams and the effective span of bums either simply supported at both the ends or cantilever beams are determined as described in Art. 10.6.

The slab spanning from beam to beam in each panel is designed as one way slab or slab spanning in one direction. (The design of slab has been described in Chapter). The thickness of slab *i.e.*, the depth of flange of T-beam, D_f thus becomes known. It is necessary to ensure the integrity of compression flange of T-beam (*viz.*, the slab) with web portion of T-beam. The

steel bars in the flange (slab) (viz., tension reinforcement) are provided in the direction transverse to this main. In general, the slab reinforcement serves this purpose.

In case, the main reinforcement of the slab is provided in the direction parallel to the min span of T-beams, the separate bars must be added in the transverse direction to the main span to permit the overhanging flanges to act as cantilever beams and to carry the loads directly applied. As per IS : 456-2000, such reinforcement shall not be less than 60 percent of the main reinforcement at mid-span of the slab.

Then, T-beams are designed as follows :

Step 1. In order to calculate the self-weight of T-beam, the various dimensions i.e., the proportions of T-beam are decided according to guide lines described below :

Dimensions of T-beams

The overall depth of beam, D, the breadth of web, b_w, the depth of rib, d_r and the effective width of flange, b_f are adopted as below :

1. Overall depth of beam, D

In order to estimate the self-weight of T-beam, the overall depth of beam, D is assumed as one-twelfth to one-fifteenth (1/12 to 1/15)th of simply supported span. An effective cover of 40 mm to 50 mm may be fixed.

2. Depth of rib, d_r

The vertical depth between the bottom of T-beam flange upto the centre of steel bars in tension gives the depth of rib (web), d_r. Therefore, the depth of rib (Fig. 10.2).

$$d_r = (D - \text{effective cover} - D_f) \qquad \text{(i)}$$

3. Width of web (breadth of rib), b_w

The width of web (*viz.,* the breadth of rib), b_w as shown in Fig. 10.2 is adopted as *one-third* to *one-half* of the depth of rib, d_r. The width of web, b_w is kept adequate to provide laterally stability.

4. Effective width of flange, b_f

The effective width of flange, b_f for T-beam is calculated from Eq. 10.1. For isolated T-beam, it is determined from Eq. 10.3. The effective width of flange, b_f shall be greater than the breadth of web *plus* half the sum of clear distances to the adjacent beams on either side viz. $[0.5\,(L_1 + L_2) + b_w]$ or $[0.5\,(L_2 + L_3) + b_w]$, fig. 10.1 s specified in IS : 456-2000.

Step 2 : The maximum bending moment, M and the shear force, V are determined at their respective critical transverse sections for the total loads acting over the beam and the effective span of beam.

Step 3 : The effective depth of T-beam, *d* is determined either that is necessary for the balanced beam section (*viz.,* by equating the moment of resistance of balanced section of T-beam to the maximum bending moment, M) or that is essential which gives the economical section of T-beam.

In actual practice, T-beam section is kept **under-reinforced**. For *under-reinforced* T-beam section, it is shown that the moment of resistance of ideally reinforced T-beam section (*viz.,* balanced section) is larger than the maximum bending moment, *M* due to externally applied total loads.

An effective cover to 40 mm or 50 mm is provided and the overall depth, D necessary is determined. The numerical value of D is round-off to a convenient value and the actual overall depth, D and effective depth d are fixed accordingly.

Step 4 : The area of steel reinforcement in tension, A_{st} is determined as below

$$A_{st} = \left(\frac{M}{\sigma_{st} \cdot (lever\ arm)}\right) \tag{ii}$$

The lever arm depth j.d may be assumed approximately for purpose of calculating A_{st}. The approximate value of j may be adopted as 0.9. The area of steel reinforcement necessary to resist the bending moment, M is found. The actual area of steel bars provided reinforcement in tension is calculated as the product of number of steel bars and the cross-sectional area of such (one) steel bar.

The actual area of steel reinforcement provided in tension is checked whether it is sufficient or not by calculating actual lever rm depth $(d - \bar{y})$ and substituting this is the expression (ii). Therefore,

$$A_{st} = \left(\frac{M}{\sigma_{st} \cdot (d - \bar{y})}\right) \tag{iii}$$

The steel bars provided as reinforcement in tension may be provided in *one or two (layers)* with proper spacing. The area of steel reinforcement provided should satisfy the requirement for minimum and maximum as specified in IS : 456-2000.

In addition to design beams for bending strength and main reinforcement, the shear strength and development length considerations are also made. The transverse (viz., shear) reinforcement is also designed for T-beams to resist shear V at the various sections of T-beams.

Sometimes T-beams are also made doubly reinforced (*viz.,* the steel bars are also provided in compression zone in addition to those in tension zone). In general, the size of T-beam (*viz.,* all dimensions and proportions) is known. The area of steel reinforcement in tension A_{st} and that in compression, A_{sc} are found to resist the bending moment developed due to externally applied total loads.

Example 10.7. Design T-beams for a class room 6 m x 12 m in size. The thickness of a singly reinforced slab is 120 mm. The top floor is also likely to be used as a class-room. Use M 15 grade of concrete and mild steel reinforcement.

Design : For M15 grade of concrete, the permissible stress in compression in bending in concrete, σ_{cbc} from IS : 456-2000 is 5 N/mm². Therefore, the modular ratio for the concrete

$$m = \left(\frac{280}{3 \times \sigma_{cbc}}\right) = \left(\frac{280}{3 \times 15}\right) = 18.67 \tag{i}$$

Step 1 : Design constants

From IS : 456-2000, σ_{cbc} is 5 N/mm^2 and stress in steel bars (upto 20 mm diameter) in tension, σ_{st} is 140 N/mm^2. The depth factor for neutral axis for the balanced section

$$k_B = \left(\frac{m\,\sigma_{cbc}}{m\,\sigma_{cbc} + \sigma_{st}}\right) = \left(\frac{18.67 \times 5}{18.67 \times 5 + 140}\right) = 0.4 \quad \text{(ii)}$$

Lever arm factor

$$j_B = \left(1 - \frac{k_B}{3}\right) = \left(1 - \frac{0.4}{3}\right) = 0.867 \quad \text{(iii)}$$

Moment of resistance factor

$$R_B = \left(\frac{1}{2}\right)\sigma_{cbc} \, . \, k_B \, . \, j_B = (0.5 \times 5 \times 0.4 \times 0.867) = 0.867 \quad \text{(iv)}$$

Step 2 : Effective span

The beams shall be simply supported at both the ends. Assume the wall thickness of class room as 600 mm. The effective span of T-beams

ES = Centre to centre of supports

$$\text{or} \quad ES = \left(600 + \frac{0.6}{2} + \frac{0.6}{2}\right) = 6.6\ m \quad \text{(v)}$$

Effective depth of T-beams may be assumed as 1/12 to 1/5th of span. Therefore,

$$d = \left(\frac{1}{12} \times 6.60\right) = 0.550\ m$$

$$ES = (C\,.\,S + d) = (6 + 0.550) = 6.550\ m \quad \text{(vi)}$$

The effective span of T-beams shall be whichever is less. Therefore, ES shall be 6.550 m. Let the effective cover be 50 mm. Then, the overall estimated depth of T-beams is (0.550 + 0.050) = 0.600 m.

Step 3 : Dimensions of T-beams

The thickness of slab, D_f is 120 mm and the overall depth of T-beams is 600 mm. The depth of rib of T-beams

$$d_{fr} = (D - D_f) = (600 - 120) = 480 \text{ mm} \quad \text{(vii)}$$

The width of web (rib)

$$b_w = 350 \text{ mm} \quad \text{(viii)}$$

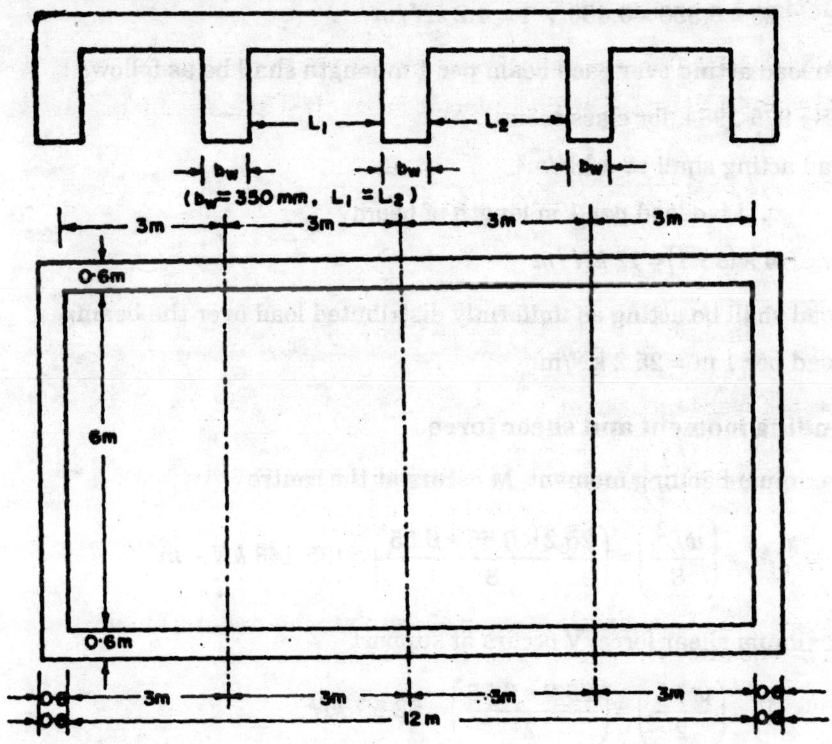

Fig. 10.6

Step 4 : Effective width of flange of T-beams

Effective width of flange of T-beams

$$b_f = \left[\left(\frac{l_o}{6}\right) + b_w + 6.\,D_f\right]$$

$$\text{or} \quad b_f = \left[\left(\frac{6.550}{6}\right) + 0.350 + 6 \times 0.120\right] = 2161.66\ mm \qquad \text{(ix)}$$

From Fig. 10.6, clear adjacent distances $L_1 = L_2 = 2650$ mm

$$b_f \not> \left[0.5\,(L_1 + L_2) + b_w\ \right] \not> \left[0.5\,(2650 + 2650) + 350\right] = 3000\ mm$$

Step 5 : Total load

The dead load acting over each beam per 1 m length shall be as follows :

Self weight of slab

$$25 \times (3 \times 0.120 \times 1) = 9\ kN/m$$

Self weight of rib

$$25 \times 0.350 \times 0.480 \times 1 = 4.2\ kN/m$$

The live load acting over each beam per 1 m length shall be as follows :

From IS : 875-1984, for class room.

Live load acting shall be 4 kN/m².

∴ Live load per 1 m length of beam

$$4 \times (3 \times 1) = 12\ kN/m$$

Total load shall be acting as uniformly distributed load over the beam.

Total load per 1 m = 25.2 kN/m.

Step 6 : Bending moment and shear force

The maximum bending moment, M occurs at the centre

$$M = \left(\frac{wL^2}{8}\right) = \left(\frac{25.2 \times 6.55 \times 6.55}{8}\right) = 135.143\ kN-m$$

The maximum shear force, V occurs at support

$$V = \left(\frac{w\,.\,L}{2}\right) = \left(\frac{25.2 \times 6.55}{2}\right) = 82.53\ kN.$$

Step 7 : Area of steel, A_{st} in tension

Area of steel reinforcement, A_{st} in tension may be found by taking moment about the line of action of resultant of compressive force in concrete. The approximate lever arm may be assumed as 0.9d. Therefore,

$$A_{st} = \left(\frac{M}{\sigma_{st} \times (0.9\,d)}\right) = \left(\frac{135.143 \times 10^6}{140 \times 0.9 \times 550}\right) = 1950.12\ mm^2$$

Provide 8 bars of 20 mm diameter is two layers as shown in Fig. 10.7.

Actual A_{st} provided

$$\left(8 \times \frac{\pi}{4} \times 20^2\right) = 2513.27\ mm^2.$$

Step 8 : Neutral axis of actual T-beam

It is assumed that the neutral axis of actual T-beam cross-section lies within the flange. From Eq. 10.5

$$\left(b_f\,.\,n\,.\,\frac{n}{2}\right) = m\,.\,A_{st}\,.\,(d-n)$$

$$\left(2161.66 \times \frac{n^2}{2}\right) = 18.67 \times 2513.27 \times (550 - n)$$

$$\therefore \text{n} = 133.531 \text{ mm} > 120 \text{ mm}$$

The neutral axis of beam lies in the web. Therefore, from Eq. 10.7,

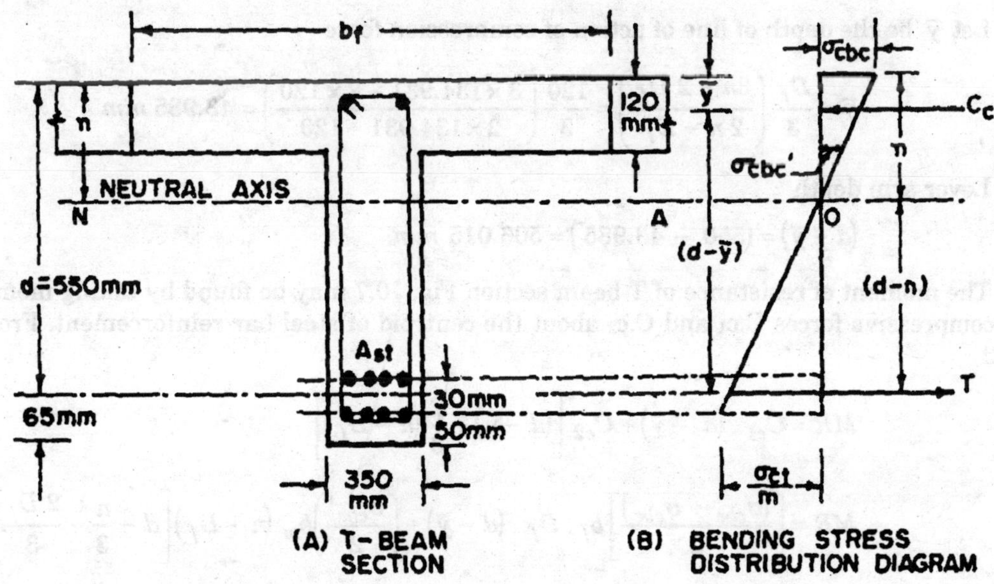

Fig. 10.7

$$\left[b_f \, . \, D_f \left(\frac{n - D_f}{2}\right) + b_w \, \frac{(n - D_f)^2}{2}\right] = m \, . \, A_{st} \, . \, (d - n)$$

$$2161.66 \times 120 \times (n - 60) + 350 \, \frac{(n - 120)^2}{2} = 18.67 \times 2513.27 \times (550 - n)$$

$$\therefore n = 134.931 \; mm < (n_B = 0.4 \times 550 = 220 \; mm)$$

Therefore, the beam is under-reinforced.

Step 9 : Moment of resistance of T-beam

In under-reinforced beam, the steel bars in tension in lower layer Fig. 10.7 (B) attain the stress equal to 140 N/mm² (the maximum permissible stress in steel bars of diameter ≤20 mm).

9. (A) Stress in compression in bending in concrete at the extreme (top) fibre

$$\sigma_{cbc} = \frac{\sigma_{st}}{m} \times \left(\frac{n}{d - n}\right) = \frac{140}{18.6} \times \left(\frac{134.931}{550 - 134.931}\right) = 2.438 \; N/mm^2$$

9. (B). Stress at the fibre of flange (slab)

$$\sigma_{cbc}' = \sigma_{cbc} \cdot \left(\frac{n - D_f}{n}\right) = 2.438 \times \left(\frac{134.931 - 120}{134.931}\right) = 0.2698\ N/mm^2$$

9 (C) Lever arm depth of T-beam

Let $\overline{y}$ be the depth of line of action of compression force

$$\overline{y} = \frac{D_f}{3}\left(\frac{3n - 2D_f}{2n - D_f}\right) = \frac{120}{3}\left(\frac{3 \times 134.931 - 2 \times 120}{2 \times 134.931 - 120}\right) = 43.985\ mm$$

Lever arm depth

$$(d - \overline{y}) = (550 - 43.985) = 506.015\ mm.$$

The moment of resistance of T-beam section Fig. 10.7 may be found by taking moment of two compressive forces C.c1 and C.c2 about the centroid of steel bar reinforcement. From Eq. 10.12

$$MR = C_{c1} \cdot (d - \overline{y}) + C_{c2}\left[(d - n) + \frac{2}{3}(n - D_f)\right]$$

$$MR = \left[\frac{(\sigma_{cbc} + \sigma_{cbc})}{2}\right] b_f \cdot D_f \cdot (d - \overline{y}) + \left(\frac{\sigma_{cbc}'}{2}\right) b_w (n - D_f)\left[d - \frac{n}{3} - \frac{2D_f}{3}\right]$$

$$MR = \left[\frac{\left(\frac{2.438 + 0.2698}{2}\right) \times 2161.66 \times 120 \times (506.015)}{1000 \times 1000} + \frac{0.2698 \times 350\,(134.931 - 120) \times \left(550 - \frac{134.931}{3} - \frac{2 \times 120}{3}\right)}{2 \times 1000 \times 1000}\right]$$

or MR = (177.713 + 0.2996) = **178.013 kN-m**

The moment of resistance of T-bam section MR = 178.013 kN-m is more than the maximum bending moment, M= 135.143 kN-m. Hence, the design of beam is safe for the bending strength.

The design of T-beams for transverse reinforcement to resist shear force, V and development length of steel bars have been done in Example 10.8. The details of complete design is shown in Fig. 10.7.

Example 10.8. A doubly reinforced T-beam is shown in Fig. 10.8. The effective depth of beam section is 600 mm and the effective cover to both tension and compression reinforcements from the outer surfaces is 50 mm. The tension and compression reinforcements consist of 4 bars of 20 mm φ and 4 bars of 18 mm φ, respectively. The effective width of flange of T-beam is 2200 mm. The depth of flange is 120 mm. The breadth of web is 450 mm. Determine the depth of neutral axis and the moment of resistance of doubly reinforced T-beam. M15 grade of concrete and mild steel reinforcements are used.

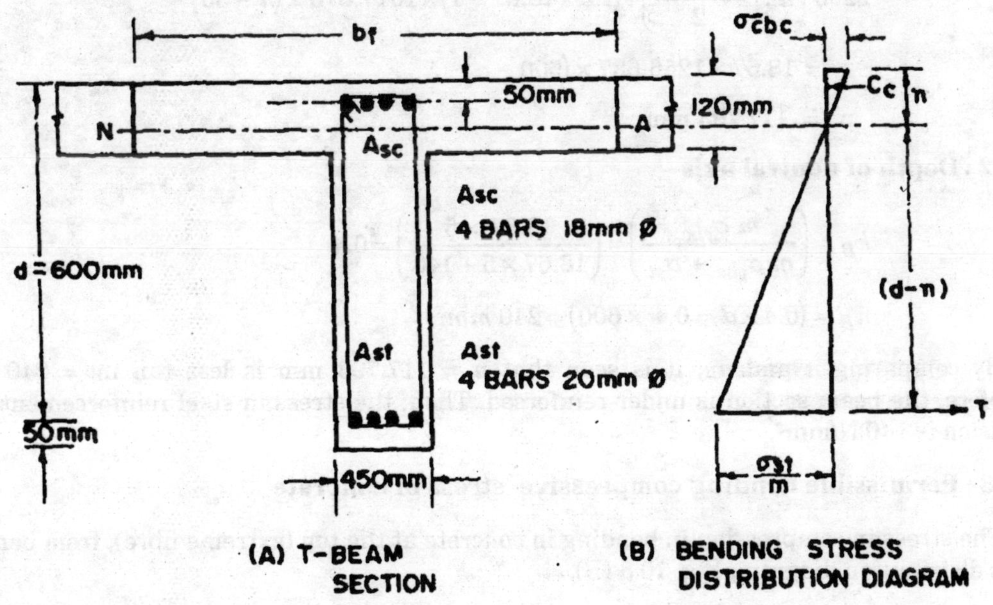

Fig. 10.8

Solution : The grade of concrete is M 15. From IS : 456-2000, permissible stress in compression in bending in concrete, σ_{cbc} is 5 N/mm². The modular ratio of concrete

$$m = \left(\frac{280}{3\times\sigma_{cbc}}\right) = \left(\frac{280}{3\times5}\right) = 18.67 \qquad \text{(i)}$$

Area of steel reinforcement in tension

$$A_{st} = \left(4\times\frac{\pi}{4}\times20^2\right) = 1256.637\ mm^2$$

Area of steel reinforcement in compression

$$A_{sc} = \left(4\times\frac{\pi}{4}\times18^2\right) = 1017.876\ mm^2$$

Step 1 : Location of neutral axis

The T-beam section is shown in Fig. 10.8. It is assumed that the neutral axis of beam section lies in the flange. In order to find the depth of neutral axis equate first moment of equivalent area of concrete in compression and steel bars in tension and compression about the neutral axis

$$\left[b_f \cdot n \cdot \left(\frac{n - D_f}{2}\right) + (m_c - 1) \cdot A_{sc} \cdot (n - d_c)\right] = m \cdot A_{st} \cdot (d - n)$$

$$2200 \cdot n \cdot \left(\frac{n - 120}{2}\right) + (1.5 \times 18.67 - 1) \times 1017.876 \times (n - 50)$$

$$= 18.67 \times 1256.637 \times (600 - n)$$

$$\text{n} = 117.793 \text{ mm} \qquad \text{(ii)}$$

Step 2 : Depth of neutral axis

$$k_B = \left(\frac{m\,\sigma_{cbc}}{m\,\sigma_{cbc} + \sigma_{st}}\right) = \left(\frac{18.67 \times 5}{18.67 \times 5 + 140}\right) = 0.4$$

$$n_B = (0.4 \times d = 0.4 \times 600) = 240\ mm.$$

By comparing n and n_B, it is seen that n = 117.793 mm is less tan n_B = 240 mm. Therefore, the beam section is under-reinforced. Then, the stress in steel reinforcement bars in tension is 140 N/mm².

Step 3 : Permissible bending compressive stress in concrete

The stress in compression in bending in concrete at the top (extreme fibre), from bending stress distribution diagram, Fig. 10.8 (B).

$$\sigma_{cbc} = n \cdot \left(\frac{\sigma_{st}/m}{d - n}\right) = \left(\frac{117.793 \times 140}{(600 - 117.793) \times 18.67}\right) = 1.832\ N/mm^2$$

The stress in compression in bending in concrete at the level of centre of steel reinforcement in compression, from bending stress distribution diagram.

$$\sigma_{cbc}' = \left(\frac{n - d_c}{n}\right) \cdot \sigma_{cbc} = \left(\frac{117.793 - 50}{117.793}\right) \times 1.832 = 1.054\ N/mm^2$$

Step 4 : Moment of resistance of beam section

In order to determine the moment of resistance of beam section, take moment of compressive forces in concrete, C_c and in steel, C_{sc} about the centre of steel reinforcement bars in tension. Therefore,

$$MR = [C_c \cdot (LA)_c + C_{sc} \cdot (LA)_s]$$

$$MR = \left[\frac{1}{2}\,\sigma_{cbc} \cdot b_f \cdot n\left(\frac{d - n}{3}\right) + (m_c - 1)\,A_{sc} \cdot \sigma_{cbc}' \cdot (d - d_c)\right]$$

$$\text{or} \quad MR = \left[\frac{\left(0.5 \times 1.832 \times 2220 \times 117.793 \times \left(600 - \frac{117.793}{3}\right)\right)}{1000 \times 1000}\right]$$

$$+\left(\frac{(1.5\times 18.67-1)\times 1017.876\times 1.054\times (600-50)}{1000\times 1000}\right)$$

$$\text{MR} = 149.04 \text{ kN-m}$$

The stress in steel reinforcement bars in compression

$$\sigma_{sc} = (m_c\ .\ \sigma_{cbc}) = (1.5\times 18.77\times 1.054) = 29.517\ N/mm^2.$$

PROBLEMS

10.1 A singly reinforced slab 100 mm thick is cast integrally with the web. The centre to centre spacing between T-beams is 2.8 m. The effective depth and breadth of web are 560 mm and 420 mm, respectively. The effective cover to mild-steel reinforcement 4 bars of 18 mm diameter is 40 mm. The effective span of simply supported beams is 5.80 m. The grade of concrete is M 15. Determine the depth of neutral axis of T-beam cross-section. Also determine the moment of resistance (flexural strength) of the beam.

10.2 In Problem 10.1, in case the grade of concrete is M 20 and hysd-steel bars Fe415 are used, locate the neutral axis and find the moment of resistance of T-beam cross-section.

10.3 A singly reinforced slab 100 mm thick is cast monolithically with the web. The centre to centre spacing between T-beams is 3.20 m. The effective depth and the width of web are 520 mm and 480 mm, respectively. Eight mild steel bars have been provided in tension in two layers. The centre to centre distance between the steel bars in two layers is 40 mm. The effective cover to steel bars in lower layer is 40 mm. The effective span of simply supported beams is 6.80 m. The grade of concrete is M 15. Determine the depth of neutral axis and the moment of resistance (flexural strength) of T-beam section. The diameter of reinforcing steel bars is 20 mm.

10.4 In Problem 10.3, in case the grade of concrete is M 20 and hysd steel bars Fe 415 are used, locate the neutral axis and find the moment of resistance (flexural strength) of T-beam section.

10.5 Determine the actual stresses developed at the extreme fibre of concrete in compression and in the steel bars in Problem 10.3, if it is subjected to bending moment 120 kN-m.

10.6 Determine the actual stresses developed in the extreme fibre of concrete in compression and in the steel bars, if it is subjected to bending moment 240 kN-m in Problem 10.4.

11

SHEAR STRENGTH AND DESIGN OF FLANGED (TEE) BEAMS (FB-II)

11.1 INTRODUCTION

A simply reinforced T-beam is shown in Fig. 11.1 (A). The distribution of shear stress in T-beam is also shown in Fig. 11.1 (B).

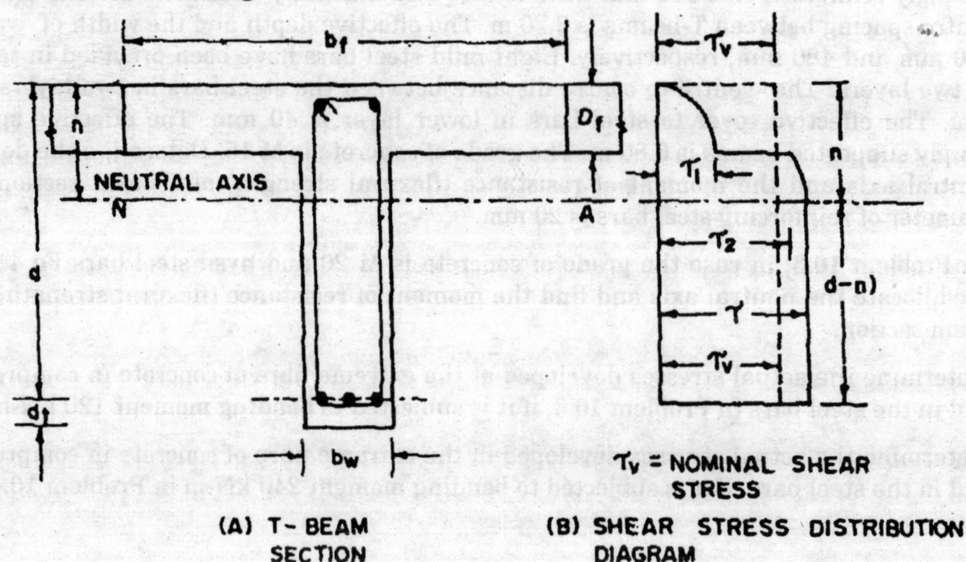

Fig. 11.1

The intensity of shear stress at any place at a height y above the neutral axis, in general is given by

$$\tau = \left(\frac{V \, . \, A \, \overline{y}}{I \, . \, z} \right) \qquad \text{(i)}$$

where, z is the width being considered. The intensity of shear stress within the flange (upto bottom of flange) increases parabolically.

$$\tau_1 = \left(\frac{V \cdot A\,\overline{y}}{I \cdot b_f}\right) \tag{ii}$$

Just below the flange, the width of beam decreases suddenly to b_w. Therefore, the shear stress increases suddenly.

$$\tau_2 = \left(\frac{V \cdot A\,\overline{y}}{I \cdot b_w}\right) \tag{iii}$$

The shear stress further increases parabolically and becomes maximum at the neutral axis.

$$\tau_{max} = \left(\frac{V}{b_w \cdot j \cdot d}\right) = \left(\frac{V}{b_w\,(d - \overline{y})}\right) \tag{iv}$$

Below the neutral axis, the shear stress remains constant. The concrete below the neutral axis in tension is neglected and the beam section is considered as cracked section upto neutral axis. The shear stress below the neutral axis

$$\tau = \left(\frac{V}{b_w\,(d - \overline{y})}\right) \tag{v}$$

As per IS : 456-2000, the distribution of shear across the complete depth is considered as uniform and it is called as **nominal shear stress,** symbolically represented by τ_v, where

$$\tau_v = \left(\frac{V}{b_w \cdot d}\right) \tag{11.1}$$

Example 11.1 Complete the design of T-beams, Example 10.7 for shear.

Design : From Example 10.7, the maximum shear force, V occurring at support is 82.53 kN. Let -bars be continued t support at the lower layer and 4-bars of upper layer be bent and carried at the top. Then, the reinforcement available in tension at the support, from Example 10.7.

$$A_s = (0.5 \times 2513.27) = 1256.63\ mm^2$$

Step 2 : Permissible shear stress in concrete

$$\left(\frac{100\,A_s}{b_w \cdot d}\right) = \left(\frac{100 \times 1256.63}{350 \times 550}\right) = 0.653$$

The grade of concrete is M 15. Therefore, the permissible shear stress and the maximum permissible shear stress in concrete from IS : 456-2000 are as under :

$$\tau_c = \left(0.29 + \frac{0.653 - 0.50}{0.50} \times 0.05\right) = 0.3053\ N/mm^2$$

$$\tau_{max} = 1.6\ N/mm^2$$

Step 2 : Nominal shear stress

$$\tau_v = \left(\frac{V}{b_w \cdot d}\right) = \left(\frac{82.53 \times 1000}{350 \times 550}\right) = 0.429\ N/mm^2$$

Nominal shear stress, τ_v is not greater than $\tau_{c.max}$. However, it is more than τ_c. Therefore, the beams need shear reinforcement.

Step 3 : Shear strength of concrete

$$V_c = (\tau_{sc} \cdot b \cdot d) = \left(\frac{0.3053 \times 350 \times 550}{1000}\right) = 58.77\ kN$$

Step 4 : Shear force to be resisted by the shear reinforcement

$$V_s = (V - V_s) = (82.53 - 58.77) = 23.76\ kN$$

Before bent-up bars, two legged vertical stirrups of 10 mm ϕ shall be provided

$$A_{sv} = \left(2 \times \frac{\pi}{4} \times 10^2\right) = 157.08\ mm^2$$

Step 5 : Spacing of vertical stirrups

$$S_v = \left(\frac{\sigma_{sv} \cdot A_{sv} \cdot d}{V_s}\right) = \left(\frac{140 \times 157.08 \times 550}{23.76 \times 1000}\right) = 509.05\ mm.$$

The maximum spacing of shear reinforcement for vertical stirrups should not exceed 0.75 d = (0.75 × 550) = 412.50 mm or 450 mm. Provide the vertical stirrups at support upto bent-up bars at 300 mm spacing.

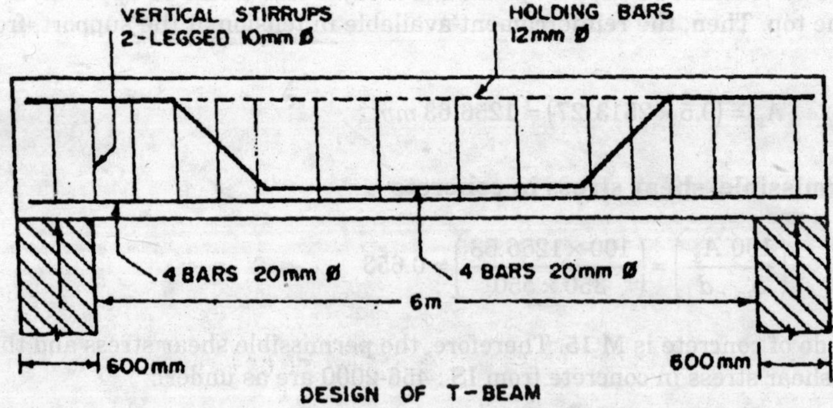

Fig. 11.2

At the bent-up bars, the bent-up bars also resist the shear force. Therefore, the shear force resisted by vertical stirrup shall be less than 23.76 kN and consequently the calculated

spacing shall be more than 509.05 mm. Therefore, a uniform spacing of 300 mm for the vertical stirrups is provided throughout the length of beam as shown in Fig. 11.2.

PROBLEMS

11.1 **Design T-beam for a class-room 5 m × 9 m in size. The thickness of singly reinforced slab is 100 mm. The top floor is also likely to be used as a class-room. A finishing surface of 20 mm thick shall be provided over the slab use M 15 grade of concrete and mild steel reinforcement. Also design the shear force reinforcement.**

11.2 **Design T-beams in Problem 10.7 in case, M 20 grade of concrete and hysd-steel bars of Fe 415 shall be used. Also design the shear force reinforcement.**

12

TORSIONAL STRENGTH AND DESIGN OF FLANGED (L) BEAMS (FB-III)

12.1 INTRODUCTION

When the slab is cast monolithically on one side (only) with a beam, a portion of slab acts as a flange of the beam. Such beams are referred as **L-beams**. The total uniformly distributed among load acting over the slab acts as an eccentric load from the vertical axis of the beam, as shown in Fig. 12.1 (A). As a result of this, the beam has to support a total load W and a twisting moment, M_t as shown in Fig. 12.1 (B). Due to load W, the beam is subjected to bending moment and transverse shear, and these internal forces are resisted by the rectangular portion of beam and its flange. In addition to bending moment and transverse shear, the twisting moment, M_t tends to twist the beam about its longitudinal axis and there is a tendency of angular rotation of the beam, and the beam is subjected to *torsion*. This torsion has to be resisted by the rectangular portion of the beam alone. The flange does not provide torsional resistance. The twisting moment is also called as **torsional** moment and *torsional stresses* develop in the beam as shown in Fig. 12.1 (C).

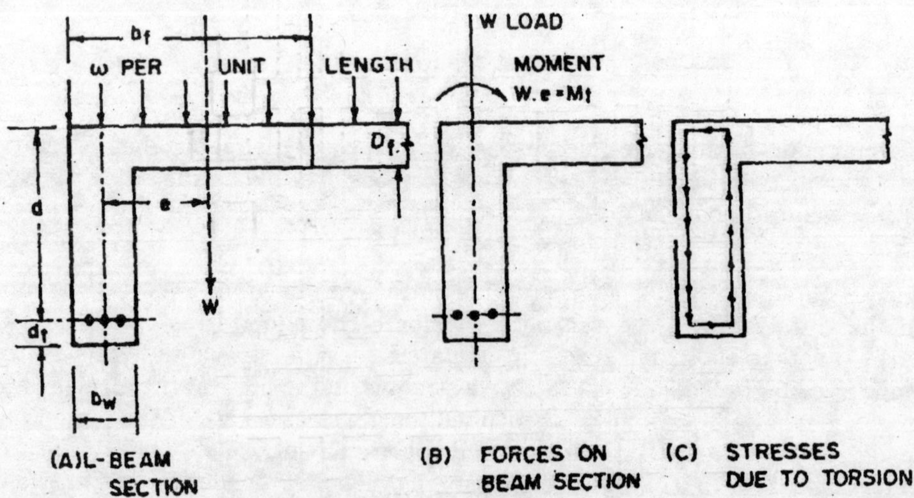

Fig. 12.1 L-beam

While considering torsion in reinforced concrete structures, it is useful to understand the difference between **primary torsion** and **secondary torsion.** *Primary torsion*, is sometimes called *equilibrium torsion* or *statically determinate torsion.* It exists when the external load is supported by torsion. The twisting moment, M_t acting along the length of the beam as above is the example of primary torsion. It is equilibrated by the resisted torque. The *secondary torsion* is also called *compatibility torsion* or *statically indeterminate torsion*. It arises from the requirement of continuity, i.e., compatibility of deformation between adjacent parts of structures. Based on static equilibrium alone, the torsional moments (secondary torsion) cannot be found.

The depth of neutral axis and the moment of resistance of L-beams are found as these are determined for T-beams.

In L-beams, it is assumed that the neutral axis is parallel to the compression flange and that the distribution of stresses will be the same as accepted for symmetrical T-beams.

Professor Bach by conducting tests for unsymmetrical beams (L-beams) proved that the neutral axis in L-beams slopes upward toward the side provided with the flange and may even intersect the flange. The maximum stresses in concrete occur at the rectangular corners. The stresses in steel are not influenced much by the unsymmetrical arrangement of the flange. For the various expressions developed for this condition reference may be made in the document by Karl Hager ``Vorlesungen iiber Theories des Eisenbetons".

12.2 TORSION IN PLAIN CONCRETE MEMBERS

A prismatic member of plain concrete subjected to equal and opposite twisting moment, M_t at the ends is shown in Fig. 12.2. The materials of the member is elastic. According to St. Venant's torsion theory, the torsional shear stresses are distributed over the cross-section as shown in Fig. 12.2 (C). The maximum shear stresses, τ_{max} occur at the middle of wide faces. The magnitude of τ_{mx} is equal to,

$$\tau_{max} = \left(\frac{M_t}{\alpha \, . \, b^2 \, d} \right) \tag{12.1}$$

where, α is the shape factor. Its value is approximately equal to 0.25 b and D are the shorter and the longer sides of the rectangle, respectively. In case, the material of prismatic member is inelastic, the distribution of torsional shear stresses shall be as shown by the dotted lines. However, the magnitude of the maximum torsional shear stress is still given by Eq. 12.1. But a larger value is assumed for the shape factor, α.

Consider an element at or near the wide surface as shown in Figs. 12.1 (A) and (B); the rear and the front faces of the prismatic member. The shear stresses act in pairs on the element. In the text books on strength of materials, it is shown that this state of stress corrcsponds to equal tensile and compressive stresses on the faces of the element at 45° to the direction of shear. The nature of these inclined tensile stresses are of the same type as those caused by transverse shear. But the torsional shear stresses are of opposite signs in the two halves of member. Fig. 12.2 (C). The corresponding diagonal tensile stresses in two halves are perpendicular to each other, Fig. 12.2 (A) and (B).

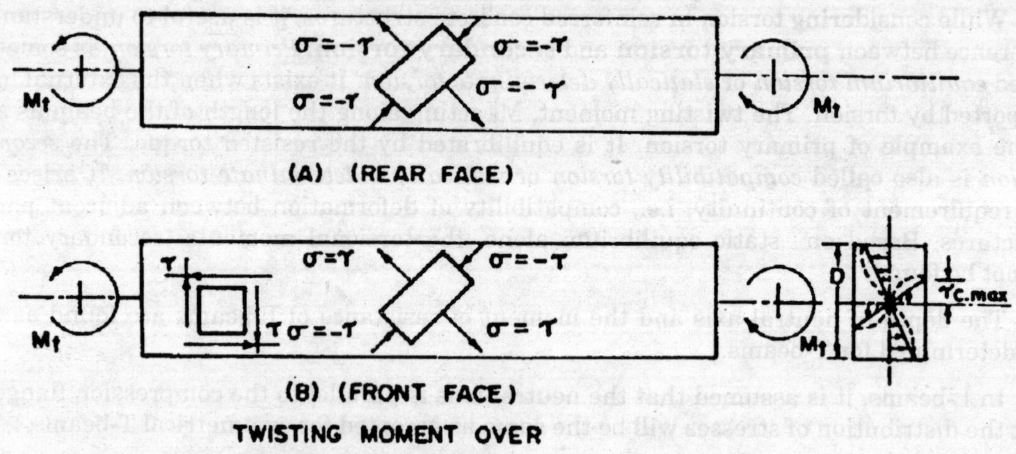

Fig. 12.2

A crack develops at some accidentally weaker section and spreads immediately across the beam, as shown in Fig. 12.3, when the diagonal tensile stresses exceed the tensile strength of concrete. It is seen that this tension crack (on the near face, Fig. 12.3) develops at practically 45°, i.e., in the direction perpendicular to the diagonal tensile stresses. When the diagonal tensile stresses are smaller, the cracks on the two narrow faces are of more indefinite inclination. The fracture line on the near face connects the crack at the short faces. An entire fracture surface across the prismatic member is formed and it results in the failure of the member.

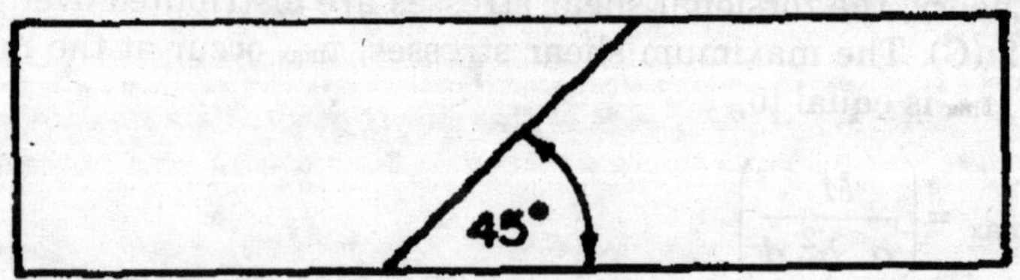

Fig. 12.3

This warped fracture surface may be replaced by a plane section inclined at 45° to the axis, s shown in Fig. 12.4 (A) for the purpose of analysis. Professor T.T.C. Hsu has shown in the paper titled as `torsion of structural concrete-plain concrete rectangular sections' in torsion of structural concrete published in ACI special publication SP-18, 1968 (pp 203-238) that on such a plane, the failure is more nearly by bending than by twisting. The applied twisting moment, M_t may be resolved into two components M_{tb} and M_{tt}. The component, M_{tb} causes bending about the axis a-a of the failure plane as shown in Fig. 12.4 (B).

Then, $M_{tb} = (M_t \, . \cos 45°)$ (i)

The section modulus of failure plane about a-a

$$Z = \left[\frac{1}{6}(D \, . \operatorname{cosec} 45°) \, . \, b^2\right] \qquad \text{(ii)}$$

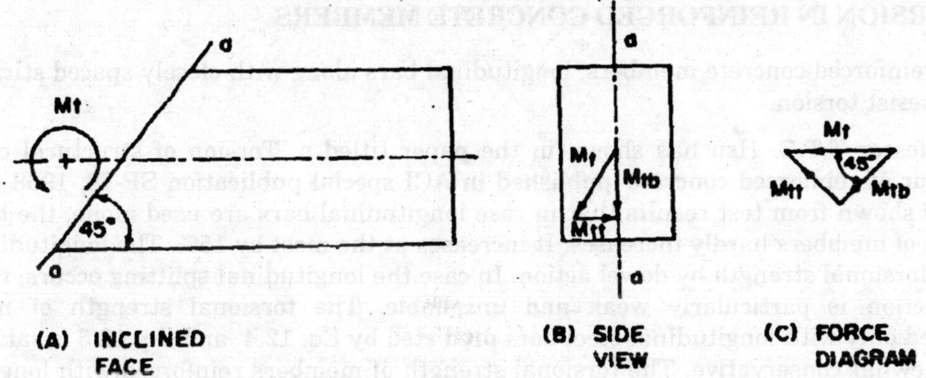

Fig. 12.4

The maximum tensile bending stress in the concrete

$$\sigma_{tb} = \left(\frac{M_{tb}}{Z}\right) = \left(\frac{M_t \, . \cos 45^\circ \, . \, 6 \sin 45^\circ}{D \, . \, b^2}\right) \tag{iii}$$

$$\text{or} \quad \sigma_{tb} = \left(\frac{3 \, . \, M_t}{b^2 \, . \, D}\right) \tag{12.2}$$

It is to note that the tensile stress so determined is identical with the St. Venant shear stress τ_{max} or with the corresponding diagonal tension stress σ for $\alpha = \frac{1}{3}$.

In case, σ_{tb} were the only stress acting, crack develops when σ_{tb} is equal to σ_r , where σ_r is the modulus of rupture. For normal density concrete, the modulus of rupture

$$\sigma_{cr} = 0.62 \, (f_c{'})^{1/2} \tag{12.3}$$

There exists a compressive stress in bending σ . For biaxil state of stress, the tensile strength of concrete reduces by about 15 per cent in the presence of equal perpendicular compressive bending stress. As a result of which, crack develops and the failure of prismatic member occurs at approximately $\sigma_{tb} = 0.85, \sigma_r = 0.50 \, (f_c{'})^{1/2}$. Let the value of σ_{tb} be designed as the cracking stress, σ_{cr}, where

$$\sigma_{cr} = \tau_{cr} = 0.50 \, (f_c{'})^{1/2} \tag{12.4}$$

By substituting σ_{cr} for σ_{tb} in Eq. 12.2, the magnitude of twisting moment, M_t which will crack and failure of plain rectangular concrete member occurs, may be determined.

$$M_t = M_{tcr} = \left(0.50 \, (f_c{'})^{1/2} \frac{b^2 D}{3}\right) \tag{12.5}$$

12.3 TORSION IN REINFORCED CONCRETE MEMBERS

In reinforced concrete members, longitudinal bars along with closely spaced stirrups are used to resist torsion.

Professor T.T.C. Hsu has shown in the paper titled s `Torsion of structural concrete-Behaviour of reinforced concrete' published in ACI special publication SP-18, 1968 (pp 261-306) has shown from test results that in case longitudinal bars are used alone, the torsional strength of members hardly increases. It increases at the most by 15%. The longitudinal bars provide torsional strength by dowel action. In case the longitudinal splitting occurs, then, the dowel action is particularly weak and unreliable. The torsional strength of members reinforced only with longitudinal steel bars predicted by Eq. 12.4 and Eq. 12.5 is satisfactory and somewhat conservative. The torsional strength of members reinforced with longitudinal bars long with transverse reinforcement is greatly increased.

Fig. 12.5 shown an adequately reinforced rectangular member. The crack develops in concrete at a torque equal to or only somewhat more than that necessary for reinforced member and given by Eq. 12.5. These cracks occurs in spiral pattern. Fig. 12.6 shows the pattern *abcd* for one single crack. Inclination of *ab* and *cd* is ϕ (in plan). larger number of such spiral cracks form at close spacing in a reinforced concrete member. The torsional resistance of the concrete decreases (after cracking) to about half of the uncracked member. The reinforcement resists the remaining torsion or twisting moment. Fig. 12.7 shows the redistribution of internal resistance. It is seen from torque-twist diagram that t cracking torque, the twist continues at constant torque. The concrete does not resist any torque. After cracking of the concrete, the reinforcement provides the resistance against torsion. At failure, the crushing of concrete occurs along a line as a-d, Fig. 12.6 (A). In a well designed member, the crushing of concrete should occur after the beginning of yielding of stirrups.

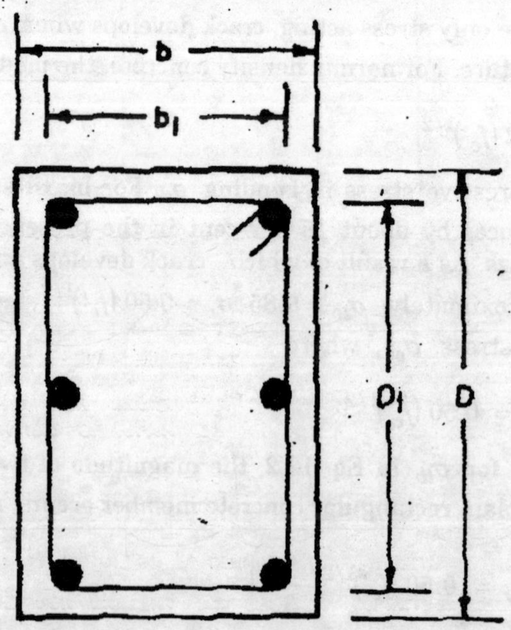

Fig. 12.5

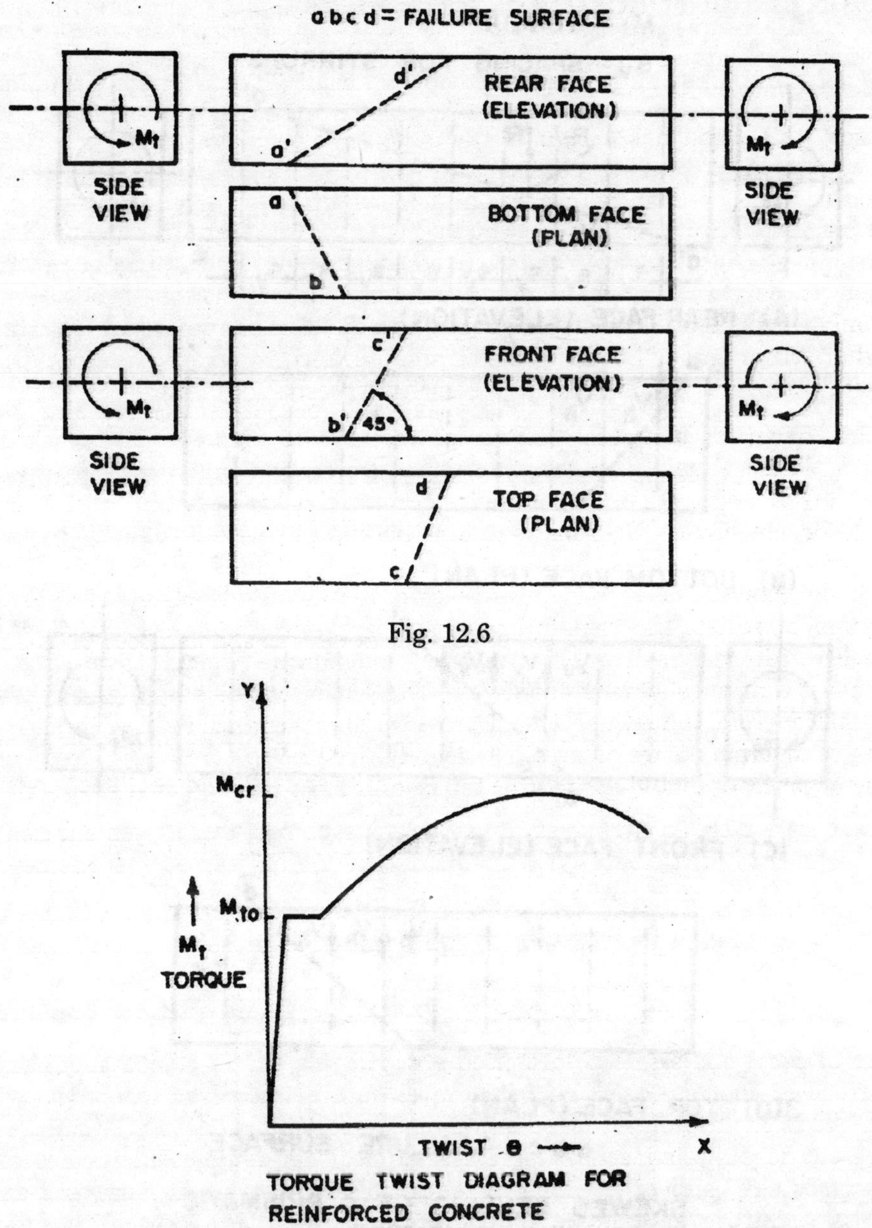

Fig. 12.6

Fig. 12.7

The torsional strength of a reinforced concrete member may be found by considering the equilibrium of the internal forces which are transmitted across the potential failure surface, abcd. (Fig. 12.6). This failure surface may be seen as bounded by a 45° tension crack across one wider face, by ϕ ° (an angle between 45° and 90° two tension cracks across the narrower faces, and the zone of concrete crushing along a-d. A concrete compression zone develops adjacent to a-d. Basically, the failure is flexural type, as it occurs for plain beams.

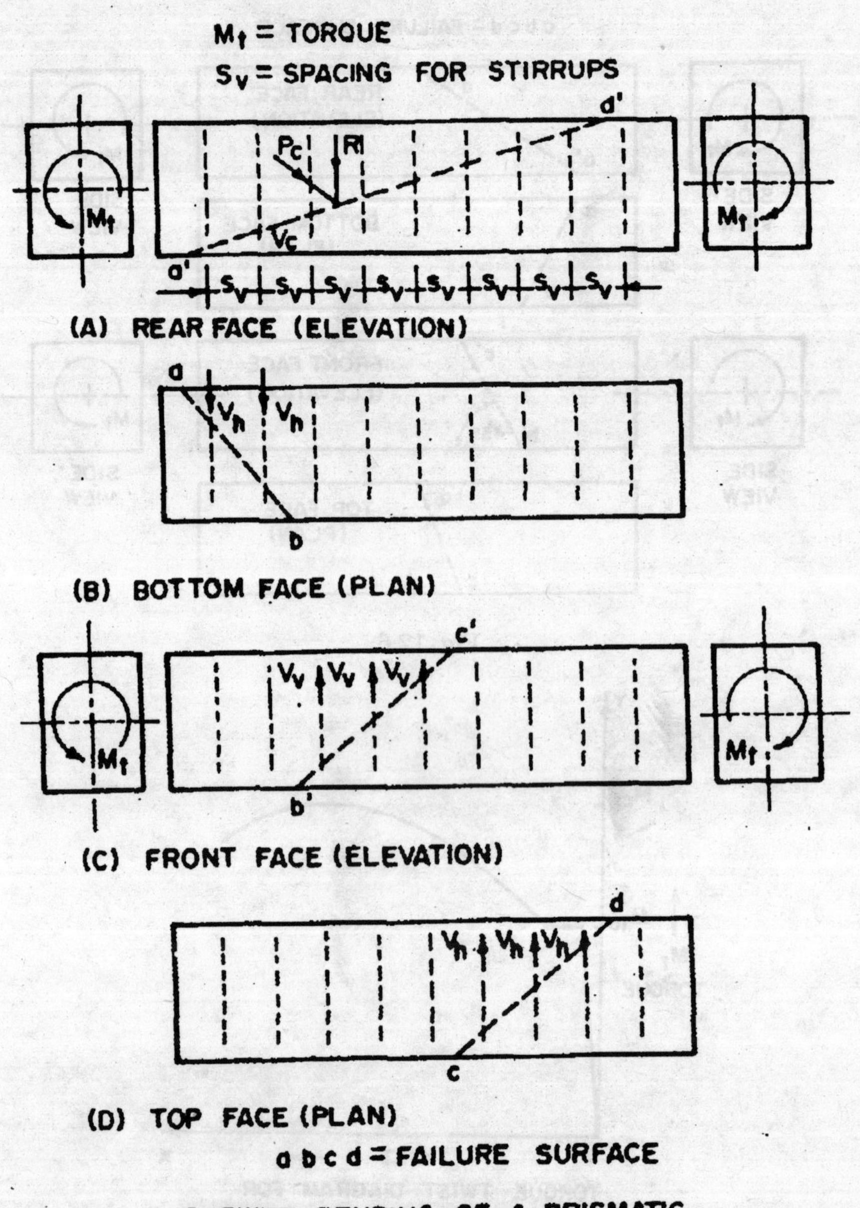

Fig. 12.8

A partially cracked failure surface abcd is shown in Fig. 12.8. A concrete compression zone develops adjacent to failure surface abcd. V_k and V_v are the horizontal and vertical forces, respectively in the shear stirrups (provided as transverse reinforcement at regular spacing, s_v). The number of horizontal legs of shear stirrups (top or bottom).

$$N_k = [(b_1 \,.\, \cot \phi)/s_v] \qquad \text{(i)}$$

The number of vertical legs of shear stirrups (fron or rear), opposite to compression zone

$$N_v = (D_l / s_v) \tag{ii}$$

From the tests, it was seen that at failure, the vertical legs of the shear stirrups yield and the horizontal legs are normally not stressed to yielding. The twisting couple developed by the horizontal forces in the stirrups.

$$M_{t.h} = (N_h \cdot V_h \cdot D_1) = \left(\frac{b_1 \cdot \cot\phi}{s_v}\right) \cdot \sigma_{sv} \cdot A\phi_v \cdot D_1 \tag{iii}$$

$$\text{or} \quad M_{t.h} = k_h \cdot \left(\frac{b_1 \cdot D_1}{s_v}\right) \cdot A_{sv} \cdot \sigma_y \tag{iv}$$

where, $A\phi_v$ = area of one leg of the stirrup

σ_{sv} = tensile stress in horizontal leg of the stirrup

σ_y = yield stress

$$k_h = \cot . \phi . (\sigma_{sv} / \sigma_y) \tag{v}$$

The torque produced by the vertical forces of the stirrups near the front face may be found as follows. It is to note that equilibrating forces near the face in the compression zone, are fairly indeterminate. The indeterminate forces consist, of a shear force V_c and a compressive force P_c in the concrete and the forces in the stirrups legs located in that zone.

Since the member is in equilibrium, all these forces have a resultant force, R. The resultant force R is equal to and acts in the direction opposite to the sum of vertical forces V_v in the shear stirrups. The torque produced by the vertical forces in the shear stirrups may be expressed as follows :

$$M_{t.v} = (N_v \cdot V_v \cdot x_v) = \left(\frac{D_1}{s_v}\right) \cdot A\phi_v \cdot \sigma_y \cdot k_v \cdot x_1 \tag{vi}$$

$$\text{or} \quad M_{t.v} = k_v \cdot \left(\frac{b_1 D_1}{s_v}\right) \cdot A\phi_v \cdot \sigma_y \tag{vii}$$

where, x_v is the lever arm of the internal forces V_v and R, and $k_v = (x_v/x_1)$. The total torque produced by the forces is therefore,

$$M_t = (M_{t.h} + M_{t.v}) \tag{viii}$$

Substituting the values of $M_{t.h}$ and $M_{t\text{-}v}$,

$$M_t = k_h \cdot \left(\frac{b_1 \cdot D_1}{s_v}\right) \cdot A\phi_v \cdot \sigma_y + k_v \cdot \left(\frac{b_1 \cdot D_1}{s_v}\right) A\phi_v \cdot \sigma_y \tag{ix}$$

$$\text{or} \quad M_t = (k_h + k_v) \cdot \left(\frac{b_1 \cdot D_1}{s_v}\right) \cdot A_{\phi v} \cdot \sigma_y \tag{x}$$

$$\text{or} \quad M_t = \alpha_t \cdot \left(\frac{b_1 \,.\, D_1}{s_v}\right) . A_{\phi v} \,.\, \sigma_y \tag{12.6}$$

where $\alpha_t = (k_h + k_v)$ (xi)

k_h and k_v constants have not been found analytically so far.

From the reference cited above, it is shown that α_t depends basically on the ratio of cross-sectional dimension.

$$\alpha_t = \left[0.66 + 0.33 \,.\left(\frac{D_1}{b_1}\right)\right] \le 1.50 \tag{12.7}$$

The torque due to concrete in compression zone, $M_{t.o}$ is about half of the critical torque, $M_{t.cr}$ (Eq. 12.5). Conservatively, $M_{t.o}$ may be taken 40 percent of $M_{t.cr}$. Then,

$$M_{t.o} = \left(0.40 \times 0.50\, (f_c')^{1/2}\, \frac{b^2 D}{3}\right) \tag{xii}$$

$$\text{or} \quad M_{t.o} = \left(0.20\, (f_c')^{1/2}\, \frac{b^2 D}{3}\right) \tag{12.8}$$

The total nominal torsional strength is the sum of $M_{t.o}$ and M_t. Therefore,

$$M_{t.n} = (M_{t.o} + M_t)$$

$$\text{or} \quad M_{t.n} = \left[0.20\, (f_c')^{1/2} \left(\frac{b^2 D}{3}\right) + \alpha_t \cdot \left(\frac{b_1 D_1}{s_v}\right) A_{\phi v} \,.\, \sigma_y\right] \tag{12.9}$$

It is to note that the nominal torsional torque develops when the stirrups are closely spaced sufficiently and a failure intersects a sufficient number of stirrups.

The longitudinal steel bars provided as reinforcement in the members anchor the stirrups, particularly at corners and enables them to develop their full yield strength. These longitudinal steel bars provide at least some resisting torque because of the dowel forces. These dowel forces develop where the bars cross torsional cracks. After cracking, the spiral crack widen and become more pronounced. As a result of this, the members subjected to torsion tend to lengthen. These longitudinal steel bars counteracts the lengthening tendency of members and control the crack width.

It is seen from the tests that for Eq. 12.9 to be valid, the volume of longitudinal reinforcement in a unit length of the member should be between 0.7 and 1.5 times the total volume of stirrups in that length. In the design of members fro torsional strength, these two volumes are kept equal. For this purpose, the total area of longitudinal reinforcement.

$$A_{st} = 2\, A_{\phi v} \cdot \left(\frac{b_1 + D_1}{s_v}\right) \tag{12.10}$$

12.4 MEMBERS WITHOUT SHEAR STIRRUPS SUBJECTED TO COMBINED TORSION AND SHEAR

In general, the beams are subjected to bending moments and shear forces. Sometimes, in addition to this, the beams also carry torsion. In uncracked member, the shear forces and torsion both produce shear stresses. Simultaneously acting bending shear and torsion interact in such a way that the strength of member is reduced as compared with that it would carry bending shear or torsion alone. So far, no satisfactory theories of this complex interaction have yet been developed.

Professor U. Erosy and P.M. Ferguson have done extensive experimental tests. They have published the paper titled as `concrete beams subjected to combined torsion and shear-Experimental trends in torsion of structural concrete' in ACI special publication SP - 18, 1981 (pp. 441-460). Let $M_{t.o}$ and V_o be cracking torque and shear of the member when subjected, respectively to torsion or flexural shear alone. The cracking torque for the members without shear stirrups is given by (Eq. 12.8).

$$M_{t.o} = \left(0.20\,(f_c')^{1/2}.\frac{b^2\,D}{3}\right) \qquad \text{(i)}$$

The cracking shear for the members without shear stirrups is given by

$$V_o = \left(0.17\,(f_c')^{1/2}.b.d\right) \qquad \text{(ii)}$$

Let M_{tc} and V_c be the torsional capacity and shear capacity of the member (without shear stirrups subjected to combined torsion and shear), respectively. The circular-interaction equation (given below) based on numerous test results may be satisfied.

$$\left[\left(\frac{V_c}{V_o}\right)^2 + \left(\frac{M_{tc}}{M_{t.o}}\right)^2\right] = 1 \qquad (12.11)$$

This interaction equation, Eq. 12.11 may be graphically represented as shown in Fig. 12.10. This interaction curve is quite favorable. These two modes do not interfere with each other very strongly. For example, in case a member carries a torque $M_{t.o}/2$ (i.e. one half of its pure torsion capacity), it is seen from the curve that the member may simultaneously carry about 0.85 V_c (i.e. only 15 percent less than it would carry, if no torsion were present at all).

The value of $M_{t.o}$ is noted from Eq. 12.8 instead of from Eq. 12.5 Eq. 12.5 represent the torsion strength of previously uncracked concrete. Eq. 12.7 gives the strength after prior cracking. The cracking of a member subjected to the combined action of bending, shear and torsion occurs long before torsional failure. This fact is not certain in all or even the majority of the cases.

12.5 MEMBERS WITH SHEAR STIRRUPS SUBJECTED TO COMBINED TORSION AND SHEAR

In Art. 12.3, it has been shown that the nominal torsional strength $M_{t.n}$ (Eq. 12.9) of members reinforced with the longitudinal steel bars and shear stirrups and subjected to torsion alone is given by the torsional strength of concrete in compression zone alone, $M_{t.o}$ (Eq. 12.8) and the torsional resistance of the steel bars, M_t (Eq. 12.6).

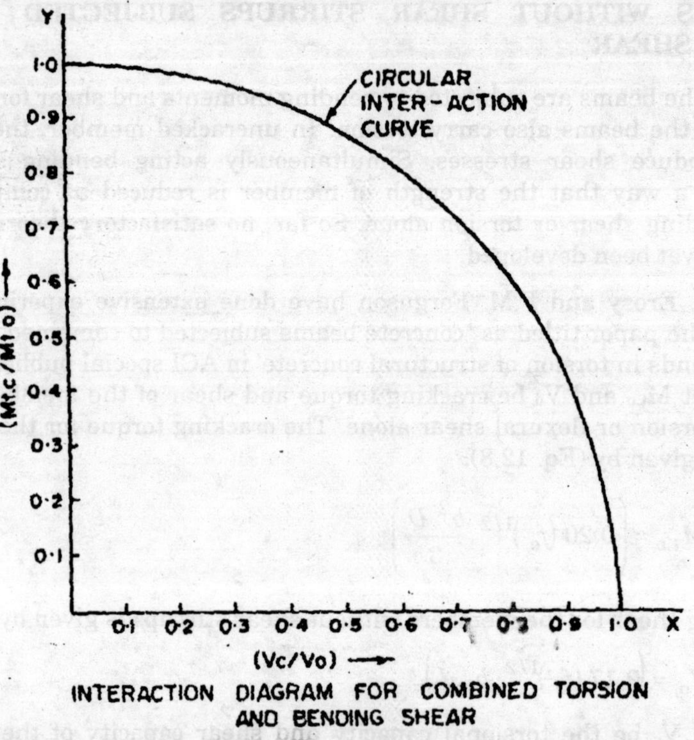

Fig. 12.9

Professor D.L. Osborn, B. Mayoglou and A.H. Mattock in their paper titled as `strength of reinforced concrete beams with web reinforcement in combined torsion, shear and bending' published in Journal of ACI, Vol. 66, No. 1, January, 1969 (pp. 31-41) have shown by experimental evidence that the strength of such reinforced members carrying simultaneously torsion and torsion is fairly thinly scattered, so far.

Based on the papers published by professors T.T.C. Hsu and E.L. Kemp titled as `Background and practical application of tentative design criteria for torsion in Journal of ACI, Vol. 66, No. 1, January 1969 (pp. 12-23) and Professors P. Lampert and B. Thurlimann titled as `Ultimate strength and design of reinforced concrete beams in torsion and bending' in Int. Assoc. Bridge and Structural Engineering Publication 31-I, Zurich, 1971 (pp. 107-131) and Tentative Recommendations for the design of reinforced concrete members to resist torsion' in ACI Committee 438, Journal of ACI, Vol. 66, No. 1, January, 1969 (pp. 1-8), following approach may be considered reasonable and it correlates satisfactorily and conservatively.

For member with a shear stirrups, the portion of the total torsion carried by the concrete is found by the same type of interaction equation (Eq. 12.11) as in members without stirrups.

The excess torque over the above that resisted by the concrete may be carried by the same amount of reinforcement provided in members subjected to torsion and shear as would be necessary for purely torsional members.

13

DEVELOPMENT LENGTH OF MAIN REINFORCEMENT BARS FOR FLANGED (TEE) BEAMS (FB-IV)

13.1 INTRODUCTION

Development length of main reinforcement bar in tension, which is necessary in T-beams is same as described for singly reinforced rectangular beams in Art. 3.26. The anchorage requirements for plain steel bars has been described in Art. 3.27.

Sometime T-beams are also made doubly reinforced. For such T-beams, development length of reinforcement in compression which is essential in these beams is same as described for doubly reinforced rectangular beams in Art. 4.9. The anchorage requirements for plain steel bars have been described in Art. 4.10.

The necessity of curtailment of tension reinforcement, the theoretical length of reinforcement curtailment and the actual length of reinforcement curtailment for T-beams are same as those described for the singly reinforced rectangular beams in Art. 3.28. The conditions for curtailment of the flexural reinforcement in tension zone and the special requirements near points of zero moments for curtailment in tension are also same as those for singly reinforced beams described in Arts. 3.29 and 3.30, respectively.

PROBLEMS

13.1 In Problem 11.1, check the main reinforcement bars t supports for the development length.

13.2 In Problem 11.2, check the main reinforcement bars at supports for thc dcvclopmcnt length.

14

BENDING (FLEXURAL) STRENGTH AND DESIGN OF CONTINUOUS BEAMS (CB-I)

14.1. INTRODUCTION

Beam supported by more than two supports and which cover more than one span are called **continuous beams.** In continuous beams, one support is usually kept as an immovable hinged support and the other supports are kept as hinged roller supports. The intermediate supports act like fixed supports except that these supports allow rotations on their both sides. The rotation at an intermediate support on its one side is equal and of opposite sign of that on the other side. The value of rotation at an intermediate support is influenced not only by the loads on the spans of either side of it but also by all other loads on all the spans of the continuous beam. The continuous beams are the indeterminate structures. The values of reactions at the supports cannot be determined simply by the application of laws of equilibrium. In order to find the bending moments and shear forces at all the points on the continuous beams, either the reactions or the bending moments are first calculated. The usual procedure is to obtain the bending moments at the supports. There are various methods of analysing the continuous beams described in the various text books of strength of materials and theory of structures.

The continuous beams are often more economical than the simply supported beams of same span. The maximum bending moment in a continuous beam is much less than that in a corresponding simply supported beam. The value of bending moment and the shear at any section of the beam depend in additions to loads, on the effective spans. After determination of moments and shears, the beam is proportioned so that no section is overstressed under the design loads. The design of beams of uniform section is controlled by the sections of maximum bending moments and the shears. In case of variable section, economy may be achieved by making the beams sufficiently strong for the larger part of the beam and reinforcing it at points of high moments or shears.

The continuous beams are used for both, the buildings and the bridges. There is difference between the continuous beams used in bridges and those used in building frames. In continuous span bridges, the continuous beam simply rests upon the intermediate supports (*viz.,* piers) and the beam is not monolithically connected with the supports. The continuous beam may be supported by small intersecting girders. In such cases of the continuous beam, full bending moment from one span is transferred directly to the adjoining span and the beams are sensitive to unequal loading. The positive bending moment are based upon unequal loading of the adjoining spans.

In *building frames*, the continuous beam remains connected with the intermediate columns. The bending moment from one span is not transferred directly in full to the adjoining span. A part of this bending moment is shared by the columns. The effect of the loading in one span upon the adjoining span is much smaller than for the continuous beams in the bridges. Similar provision is necessary for unequal loading.

14.2 EFFECTIVE SPAN

A continuous beam may have one end simply supported as shown in Fig. 14.1 (A) or one end fixed as shown in Fig. 14.1 (B). In case the width of the support is less than *one-twelfth of the* clear span, *the effective span of the member that is not built integrally with its supports shall be taken as clear span plus the effective depth of beam or centre to centre of supports, whichever is less.* Therefore

$$E.S = (C.S + d) \tag{14.1}$$

or $\quad E.S = \text{(centre to centre of supports)}$

$$\text{or} \quad E.S = \left(C.S + \frac{1}{2}b_s + \frac{1}{2}b_s\right) \tag{14.2}$$

where, E.S is the effective span, C.S is the clear span, L_c and b_s is the width of the support.

In case, the width of supports is wider than *one twelfth* of the clear span, L_c or 600 mm whichever is less, then, the *effective span* shall be taken as follows :

(i) For end span with one end simply supported (free) and the other continuous

The effective span shall be equal to the clear span plus half the effective depth of the beam or the clear span plus half the width of the discontinuous supports, whichever is less. Therefore,

$$E.S = \left(C.S + \frac{1}{2}.d\right) \tag{14.3}$$

$$\text{or} \quad E.S = \left(C.S + \frac{1}{2}b_s\right) \tag{14.4}$$

(ii) For end span with one end fixed and the other continuous

The effective span shall be the clear span between supports. Therefore,

$$E.S = C.S \tag{14.5}$$

(iii) For the intermediate supports

The effective span shall be clear span between supports. Therefore,

$$E.S = C.S \tag{14.6}$$

It is to note that in case of spans with roller or rocker bearings, the effective span shall always be the distance between centres of bearings.

In the analysis of continuous frame, considering the frame as a whole, the effective span of a member subjected to bending shall be the distance between the centre to centre lines of supported members. Even though the moments are calculated on the basis of effective span, in monolithic construction, it is permissible to design the section at the support on the basis of the moment at the face of the support.

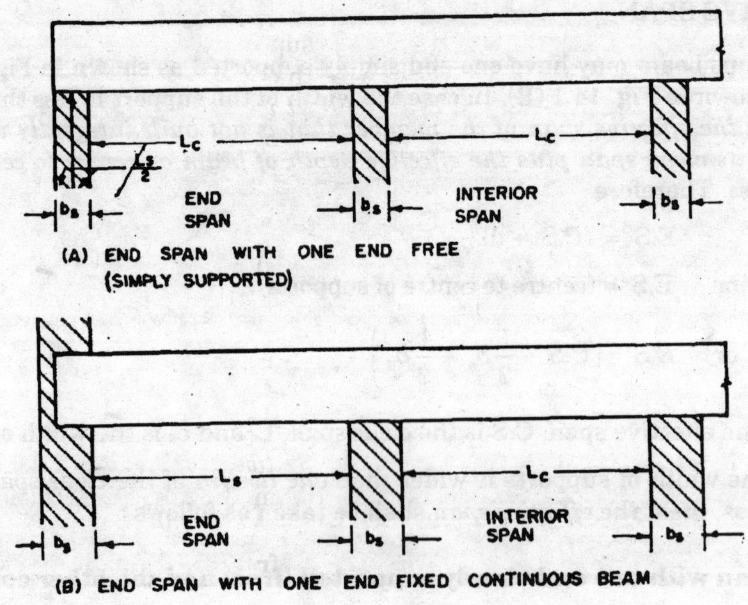

Fig. 14.1

14.3 CRITICAL SECTION FOR MOMENT

IS : 456-2000 permits the representation of a frame (structure having monolithic construction between beams and columns) by a simple line diagram, based dimensionally on the centre line distance between columns (and by extension, between floor beams). Actually the sectional dimensions of the beams and columns amount to sizable fraction of their respective lengths.

The bending moment diagram for continuous beams is generally quite steep in the region of support as shown in Fig. 14.2. Let this bending moment be M_1. Let the moment at support face be M_2. There is a substantial difference between M_1 and M_2. In case the bending moment M is used in proportioning the member, the section of member shall be unnecessary large. IS : 456-2000 recommends, that *for the monolithic construction, the moments computed at the face of the supports (columns) shall be used in the design of members at those sections.* As such, it accounts for the finite width of the supports.

However, for the columns, the curve for the bending moment is not very steep, so that the difference between the centre-line moment and the moment at the face (top or bottom) of the beam is small and it may be ignored.

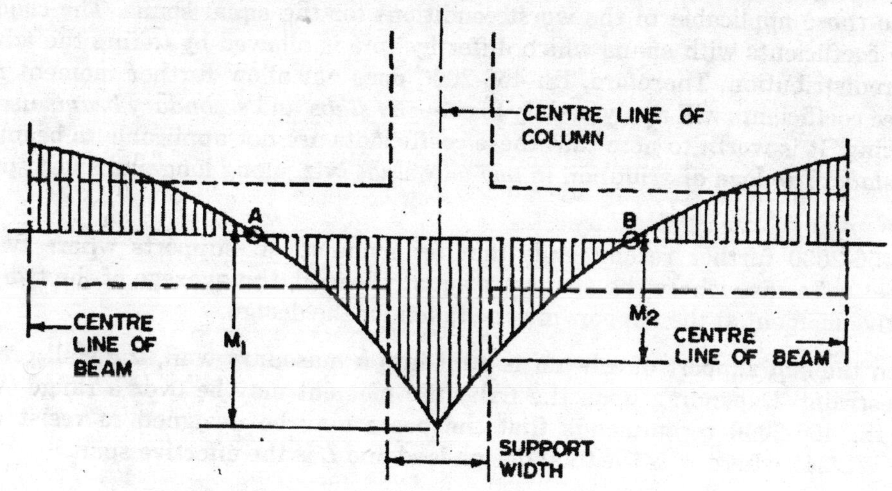

A, B = Points of contra flexure
M_1 = Actual BM as per analysis at the centre
M_2 = BM at the face of the support (Column) for Design
BM = Bending Moment

Fig. 14.2

For the non-monolithic construction, obviously, such reductions are not possible at the support. The design of section of beam members begins in the mid-span regions.

14.4 COEFFICIENTS FOR MOMENT FOR CONTINUOUS BEAMS

IS: 456-2000 specifies the coefficients for moments for continuous beams. These coefficient are also given in many of standard text books of theory of structures.

Unless more exact estimates are made, for beams of uniform cross-section which support substantially uniformly distributed loads over three or more spans which do not differ by more than 15 per cent of the longest span, the bending moments in design may be obtained using the coefficients given in Table 14.1.

TABLE 14.1. COEFFICIENTS FOR MOMENTS

Type of Loading	Span moments		Support moments	
	Near middle of end span	At middle of interior span	At support next to the end support	At other interior supports
Dead load and imposed load (fixed)	+ 1/12	+1/24	–1/10	–1/12
Imposed load (not fixed)	+ 1/10	+1/12	– 1/9	– 1/9

(Note : For obtaining the bending moments, these coefficients shall be multiplied by the total design load and effective span.)

These coefficients given in Table 14.1 are close to those which would be obtained from accurate analysis of an infinite number of equal spans on point supports and correspond roughly to those applicable to the worst conditions for the equal spans. The concession for using the coefficients with spans which differ by 15% is allowed by taking the advantage of moment redistribution. Therefore, IS: 456-2000 does not allow further moment redistribution. These coefficients will apply mainly to *one-way slabs* and *secondary beams* used at fairly close spacing. It is worth to note that these coefficients are not applicable to beams carrying *two-way slabs*. The load distribution in *two-way slabs* (viz. along long and short spans) is not uniform.

IS: 456-2000 further recommends that for moments at supports where two unequal spans meet or in case where the spans are equally loaded, the average of the two values for the negative moment at the support may be taken for the design.

When the end support of a beam is built into a masonary wall, the wall provides only partial restraint. Depending upon the fixity the moment may be over a range (WL/120) to (WL/12) IS: 456-2000 recommends that the beam may be designed to resist a negative moment (WL/24), where W is the total design load and L is the effective span.

14.5. CONCRETE BUILDING FRAMES

The reinforced concrete building frames have usually the monolithic construction. In the *monolithic construction,* the concrete in floor slabs, beams, girders and columns is placed continuously. A typical monolithic construction is shown in Fig. 14.3 and the slab, girders and the columns have been shown. The beams B_1, B_2 and B_3 are the *intermediate beams* and these beams are supported on the girders. These beams are built monothically with the girders as shown in Fig. 14.4 (A). The beams B_4, B_5 and B_6 and the beams situated on the column lines. These beams are supported by the columns. The beams G_1, G_2 and G_3 are beams of comparatively large dimensions and these beams are called *girders*. These girders go directly into the columns.

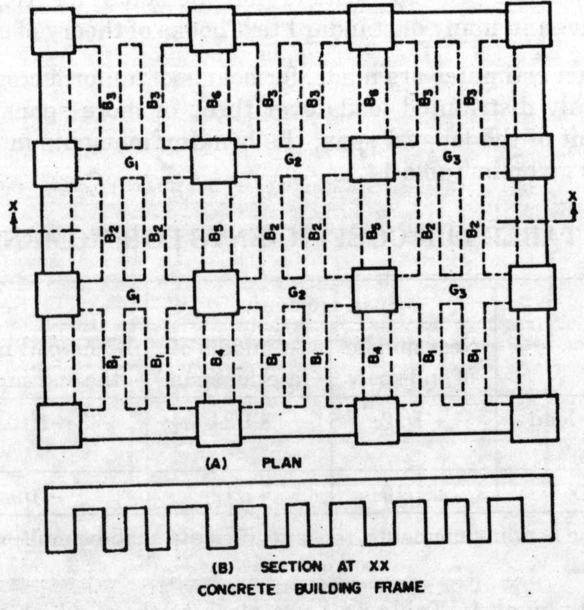

Fig. 14.3

The beams B_4, B_5 and B_6 are the continuous beams and these beams are built integrally with the upper and lower columns as shown in Fig.14.4(B). The girders G_1, G_2 and G_3 are also the continuous beams and these beams are also cast monolithically with the upper and lower columns as shown in Fig. 14.4(C). The continuous slab is supported on the beams.

For the accurate analysis, the buildings frames which may consist of number of storeys are analysed as *rigid frames* as two dimensional structures in longitudinal direction and against as two dimensional structures in transverse direction. For the approximate analysis and practical purposes, however, the beams and the girders and the adjacent columns may be isolated as *substitute frames* as shown in Fig. 14.5. The far ends of the columns are assumed as fixed. The stiffness factors for the beams, girders and columns are first assumed and later on, these are revised. The supporting beams and girders possess some torsional rigidity which may be supported by using equivalent supporting columns with fixed far ends.

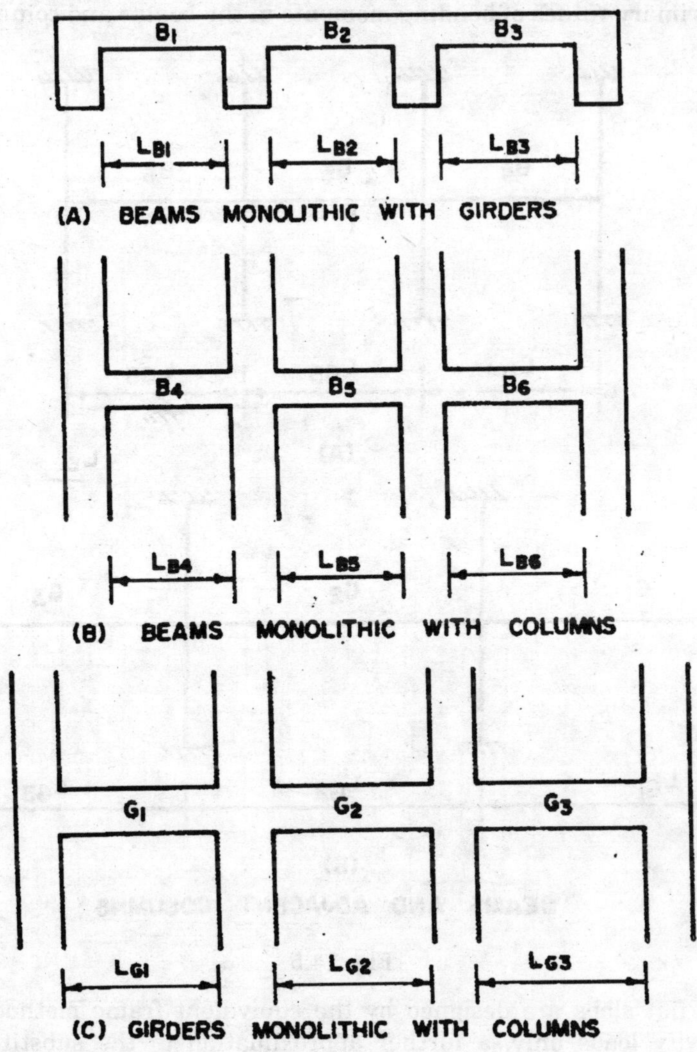

Fig. 14.4

These *substitute frames* are analysed for the gravity loads and the moments and shears at any floor or roof level are determined. A substitute frame for the floor level EF of a multistory building shown in Fig. 14.5(A) is shown in Fig. 14.5(B). It is permissible simplification of the frame for the analysis.

In IS : 456-2000, it was permissible to calculate the bending moments in the beams of a frame by assuming that the beams are continuous over supports and capable of free rotation at the supports. This assumption is approximately correct for analysing the effects of vertical loads, but IS : 456-2000 (third revised version of code does not permit this assumption.

The effects of deflections and axial deformations (which may arise) are generally negligible and these are not taken into consideration. Therefore, the method of analysis recommended by the code in case is *first order* method of analysis. However, in case of slender columns, the additional moments developed by the slenderness of the column must be added to the primary values of bending moments in the beams and column.

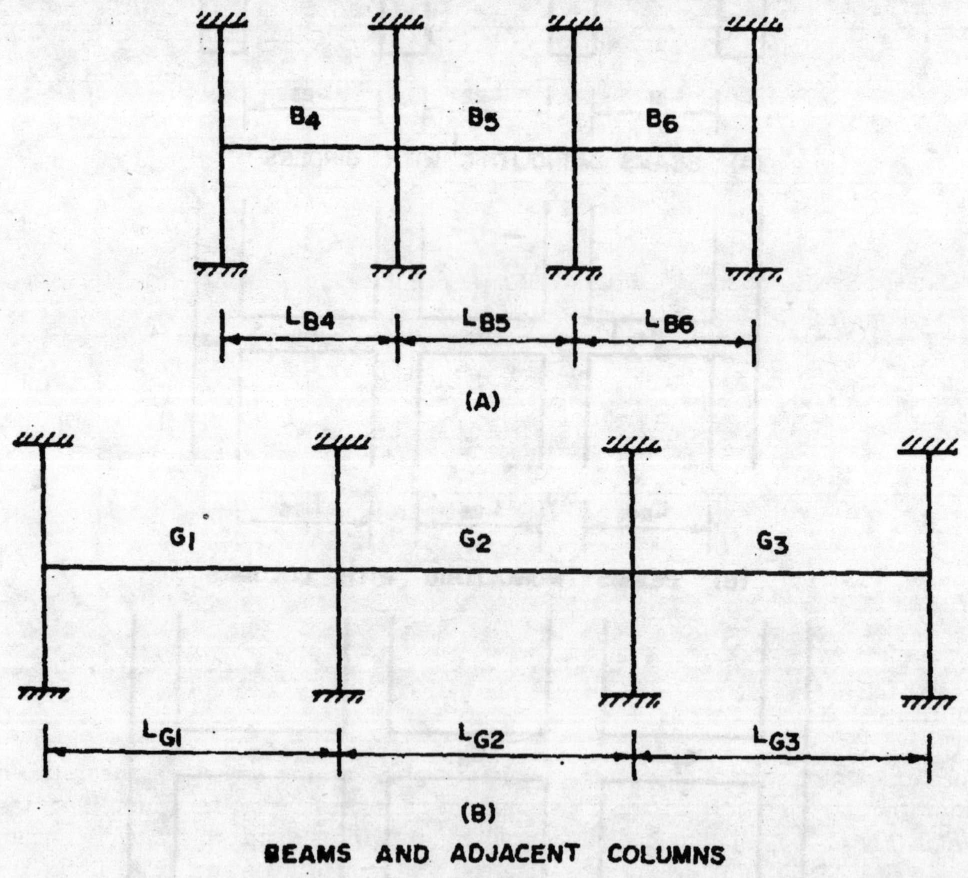

Fig. 14.5

In case, the flat slabs are designed by the equivalent frame method and if these are subjected to gravity loads only, a further approximation to the substitute frame will be possible.

For the lateral loads (winds and earthquake forces), IS : 456-2000, recommends that the simplified methods (such as portal method) are used to calculate moments, shear and reactions for the symmetrical structures. The portal method of analysis and the various assumptions made in the method are given in author's volume second (Design of Steel Structures). It is reasonable to use the approximate methods since considerable undertainty prevails regarding the determination of magnitude as well as the distribution of the wind earthquake forces.

For the unsymmetrical methods or very tall structures (structures having more than *twenty* storeys), more vigorous methods should be used.

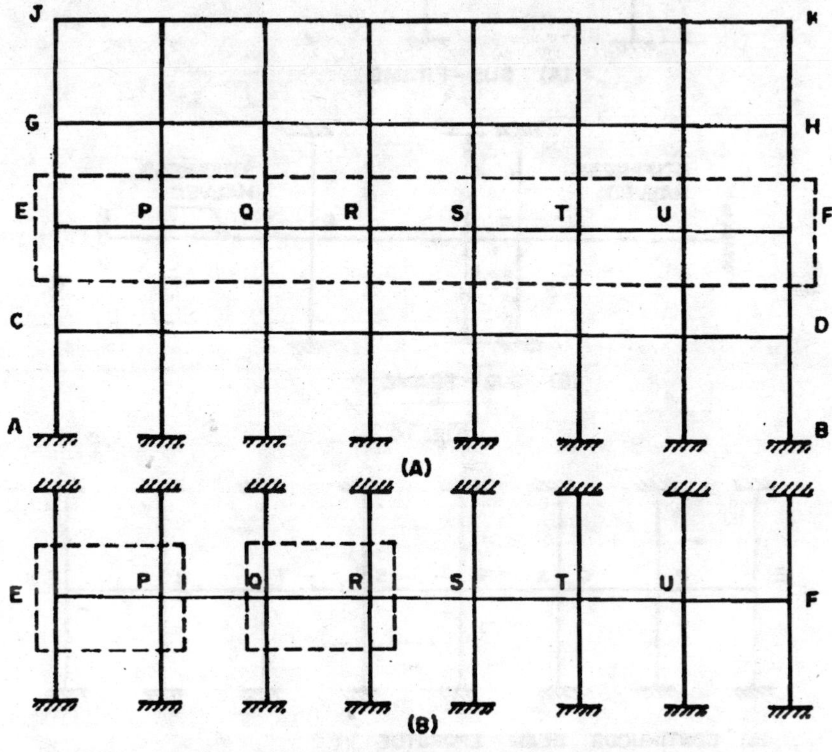

Fig. 14.6

14.6 ARRANGEMENT OF LIVE LOAD FOR BENDING MOMENT

The arrangement of live load to be used in the analysis of slabs, beams and girders to produce the most severe effect at any point in the frame structure theoretically needs consideration of the whole frame. This may not be necessary as the frame [for example, as shown in Fig. 14.6 (A)] may be considered by drawing substitute frame [for example, as shown in Fig. 14.6 (B)] at any floor/roof level. The substitute frame [Fig. 14.6 (B)] may be broken down into *sub-frames* as shown in Figs. 14.7 (A) and (B).

The position of live load to be used in the analysis for the continuous beam EPQRSTUF shown in Fig. 14.8 (A) may be decided by drawing the influence line diagrams, for bending moment in the central portion of span QR and at a section just to the left (infinitesimally distance) of support R as shown in Figs. 14.8 (B) and (C), respectively. From the influence line diagram, the position of uniformly distributed live load may be observed for the following :

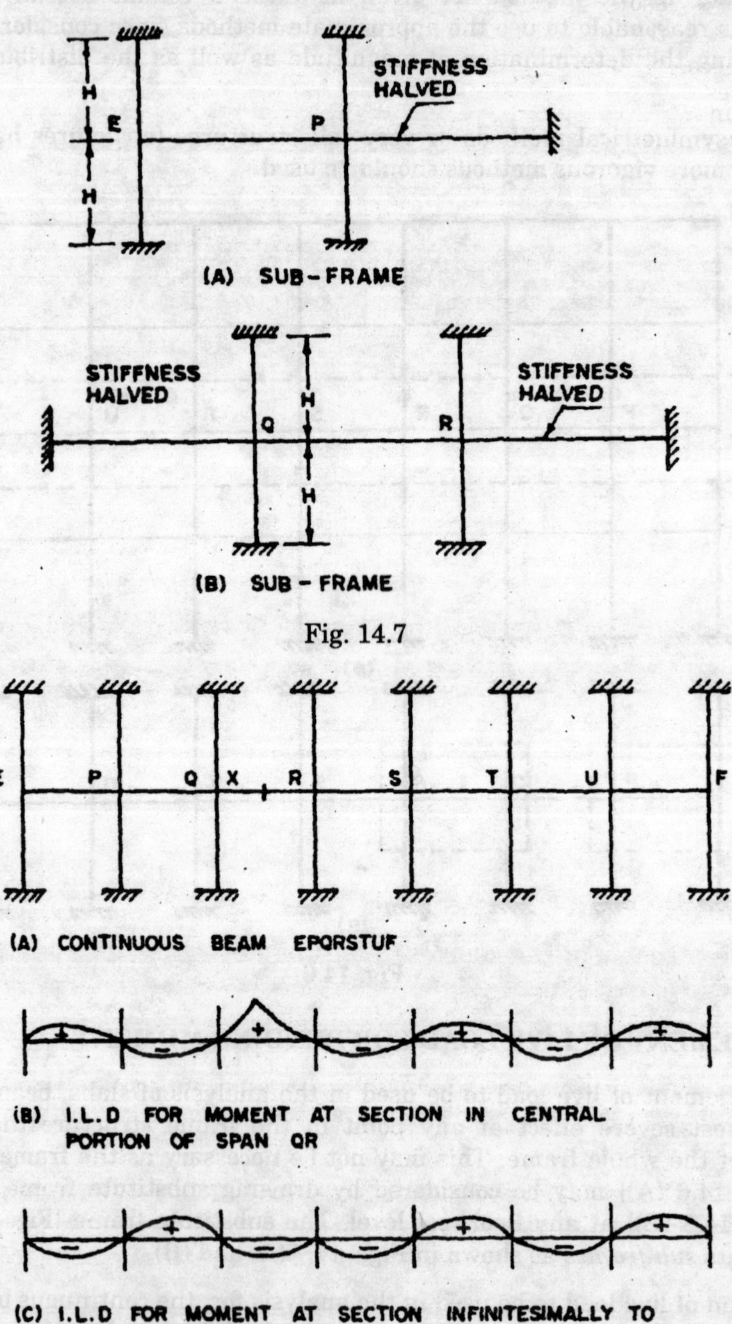

Fig. 14.7

Fig. 14.8

1. For maximum positive moment within a span (for example span, QR)

The uniformly distributed live load may be placed over than span (viz., span QR) and all other alternate spans (*viz.,* spans EP, ST and UF).

2. For maximum negative moment within a span (for example, span R)

The uniformly distributed live load may be placed over two adjacent spans to that span (viz., span RS) and all other alternate spans (viz., RS and PQ and TU).

3. For maximum negative moment at a support (for example, support R)

The uniformly distributed live load may be placed over the two adjacent spans to that support (viz., support R) and all other alternate spans (viz., QR, RS and EP TU).

4. For maximum positive moment at a support (viz., support R)

The uniformly distributed live load may be placed over the two spans beyond each of the two spans adjacent to the support (viz., support R and beyond adjacent spans QR and RS) and all other alternate spans (viz., PQ, ST and UF).

IS : 456-2000 recommends as follows as regards the arrangement of live load :

(a) The considerations may be limited to the combinations of :

(1) design dead load on all spans with full design live load on two adjacent spans; and

(2) design dead load on all spans with full design live load on alternate spans.

(b) When the design live load does not exceed *three-fourths* of the design dead load, the load arrangement may be design dead load and design live load on all the spans.

It is recommends in IS : 456-2000 in a note that for beams and slabs continuous over support, consideration as described in paragraph (a) above may be assumed.

14.7 COMPRESSION STEEL AT SUPPORT

In continuous beams at the support, the bottom of beam is subjected to compression. Frequently, additional compressive area is provided by reinforcement in the compressive area of the beam. Thus, the continuous beam is the beam reinforcement at top and bottom. This reinforcement in compression (compresses steel) takes its share of the stresses and strengthens the beam in its compression zone. In addition to this, this compressive steel increases the stiffness of the beam. There are *three methods* of providing this additional compressive reinforcement.

First method. In a reinforced concrete beam, part of the bottom steel is bent up and the rest, usually *fifty percent* of the steel area, is carried horizontally to the supports as shown in Fig. 14.9. This compressive steel reinforcement bars get their maximum compression at the edge of the support.

As such these bars are extended into the supports for a sufficient length beyond the edge of the support to provide requisite development length. This is the *simplest method* of providing reinforcement in compression and it is used wherever possible.

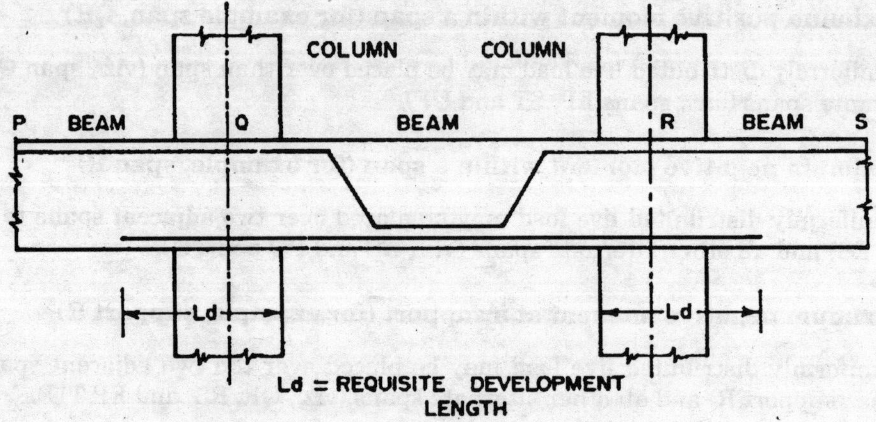

Fig. 14.9

Second method. In case the area of compression reinforcement provided as explained above by the straight bars from span QR is not sufficient, additional reinforcement in compression may be provided by extending into the support, Q under consideration the straight bars from the adjoining span, PR as shown in Fig. 14.10. The length of the extended bar, measured from the edge of the support depends upon the bending moment and provides the requisite development length. The compression reinforcement from both sides of the support consists of the sum of the horizontal bars from both the spans, PQ and QR.

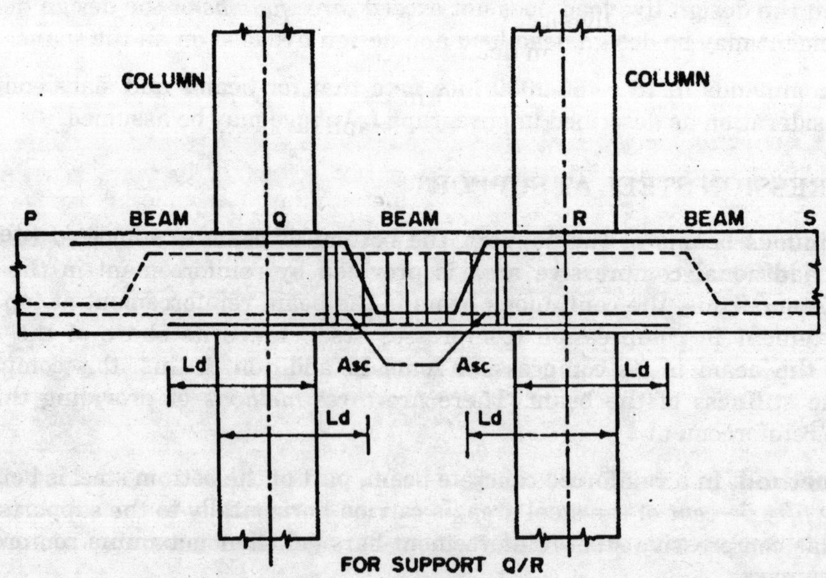

Fig. 14.10

Each horizontal bar at support Q is subjected to compressive stresses from the span PQ and also equal compressive stresses from the adjoining span PQ as shown in Fig. 14.11. From Fig. 14.11 it is clear that the compressive stresses from span QR act in the opposite direction to the compressive stresses in the adjoining span PQ and so also at support, R. Therefore, the compressive stresses in the bar are balanced and not added. The compressive stresses in the bar from one span acts as reaction to the compressive stress from the other span. The compressive stresses in one bar are equal to those, as if the bar had been extended from one span.

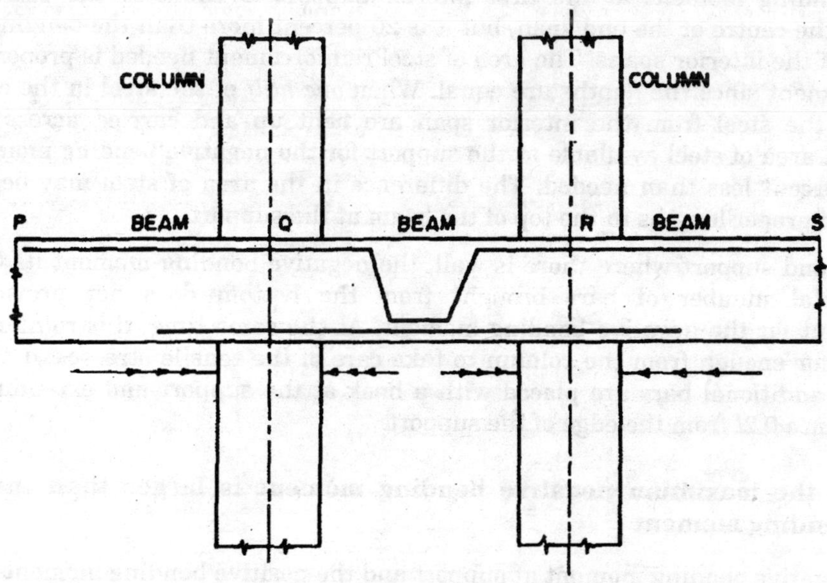

Fig. 14.11

Third method. Sometimes, the total area of compression reinforced provided as explained in the second method is also not sufficient. Then, the additional reinforcement is provided by short bars placed at the support at the bottom of the beam. The length of these bars depends upon the bending moment. The minimum length (projections beyond the column faces on each side) should be atleast equal to requisite development length for the expected compressive stresses.

14.8 TENSION STEEL AT SUPPORT

After calculating the area of steel reinforcement in tension in the centre of span and deciding the number and size of the bars, the requirement at the supports is studied, keeping in the view the values of maximum positive and negative bending moments at the centre of the span and at the support, respectively.

1. In case the maximum positive and negative bending moments are equal

When the continuous beam consists of a number of equal spans and the depth of beam is uniform, the arrangement of steel for the tension at the support is simple. The positive reinforcement is provided such that it may be divided into two equal portions. *One-half* of these bars are kept straight and the *other half* are bending up as shown in Fig. 14.10 and carried across the support, (say support Q) into the adjoining span PQ. These one-half bars bent up resist the tension at the support. An equal area of steel is brought across from adjoining the span (say, PQ) into the span (span QR) under consideration. Total number of the bars at the support becomes equal to that in the centre. The requirement as to the area of steel in tension at the support is satisfied. Then, the points of bending up the bars may be decided.

The bending moment at the first interior support is equal to the positive bending moment in the centre of the end span, but it is 20 percent more than the bending moment at the centre of the interior spans. The area of steel reinforcement needed is proportional to the bending moment since the depths are equal. When *one-half* of the steel in the end span and *one-half* of the steel from the interior span are bent up and carried across the interior support, the area of steel available at the support for the negative bending moment remains about 10 percent less than needed. The difference in the area of steel may be provided by extra bars of proper lengths in the top of the beam at the support.

At the end support where there is wall, the negative bending moment its ($wl^2/16$). One half the total number of bars brought from the bottom does not provide sufficient reinforcement for the negative bending moment. At the same time, this reinforcement does not extend for enough from the column to take care of the tensile stresses at the points. In such cases, additional bars are placed with a hook at the support and extending to a point upto a distance $0.2l$ from the edge of the support.

2. In case the maximum negative bending moment is larger than the maximum positive bending moment

The negative bending moment at support and the positive bending moment at the centre of span of a continuous beam running into the columns are ($wl^2/16$), respectively, *two-thirds* of the bottom steel may be bent-up and carried across the support to provide the necessary area of steel reinforcement in tension.

In such cases, the straight bars may not be sufficient to take care of the compressive stresses. Then, it is good practice to bend up smaller number of bars and extra bars are provided to complete the require area of steel in tension at the support.

3. In case, the bars are not bent-up

Sometimes, the reinforcement bars are kept straight both for the positive and negative bending moments. In short and deep beams, such arrangement is made. The bent-up bars do not reach the top at proper place to be effective. Some of the straight bars are kept straight and extend the full length of the beam while the remaining extend on both sides beyond the points of contra flexures.

The top bars at the support should extend on both the sides of an intermediate support beyond the points of contra flexures. At the extreme end support, where there is wall should be hooked and other bars should extend beyond the point of contra flexures.

In continuous beam in which the spans are not equal, the design must be done after study with more care.

14.9 TENSION STEEL AT CENTRE OF SPAN

Required cross-sectional area of reinforcement in the tension in the centre of each span is calculated independently. In beam of small widths, the number of bars should not be less than two. In beams of large width, the number of bars should not be less than four. The bars of largest diameter which one is suitable for the development length (viz., bond) should be selected. It reduces the number of bars and therefore the cost of handling. The number of bars provided should be such that which allow the needed proportion of bars to be bent up. In case the bending of bars of *one-half* is intended, an even number of bars may be provided.

As far as possible, the bars should be placed in one layer. The width of beam may be increased sufficiently to accommodate all the bars in one layer. When the bars are provided in two layers, the depth of beam increases. A strip of concrete equal to width of beam and thickness equal to diameter of bar plus 12 mm goes wastc.

14.10 HAUNCHES IN CONTINUOUS BEAMS

In continuous beams, at the support, the bottom surface carries compressive stresses. In one method, compression area is increased by providing compressive reinforcement as described in Art. 14.7. In another method, the beam at the support by a shallow haunch. This method is less desirable under ordinary conditions as compared to first method. In many cases, the haunch is objectionable from the architectural point of view. While there is saving in steel, the excess of cost of forms tends to overbalance this saving.

A haunch is useful, where it is desirable to avoid excessive compression reinforcement or where more lateral stiffness is needed for the structure. Trial procedure is used to find the required depth of haunch.

The depth of haunch may also be determined from the rectangular beam formula. The effect of compression reinforcement may be neglected, because for deep haunches, the bars are placed distance from the extreme fibre. In any case, the error is on the safe side.

Under ordinary conditions, the calculations may be done only for the maximum bending moment. In case, the continuous beams acts like an inverted T-beam, the point at which the slope of the haunch may be ended, may be readily determined as follows :

Let M_b be the negative bending moment next to the support and M_t be the moment of resistance of the inverted T-beam without the haunch controlled by the concrete for the uniformly distributed loaded beam. Let x be the length of haunch measured from the face of the column and L_c be the clear span of the beam. Then

$$x = \left(\frac{M_b - M_r}{M_b}\right) \cdot \frac{L_o}{5} \; (approximately) \qquad (14.7)$$

Eq. 14.7 is based upon the fact that the point of contraflexure is at approximately 1/5 of the span. It is assumed that the moment between the support and the 1/5th point varies by straight line. Therefore, the ratio of different between the bending moment, M_b and the moment of resistance, M_r to the bending moment is same as the ratio of the distance from the point where the haunch is required to the point of contraflexure is not approximately as 1/5th the span, this fraction may be changed accordingly.

14.11 EXTENSION OF REINFORCEMENT

The extension length of reinforcement depends upon the nature of bending moment whether positive or negative. IS : 456-2000 recommends as follows :

1. For positive bending moment

In the continuous beams, the *one-fourth* of positive moment reinforcement shall extend along the same face of the member into the support, to a length equal to $L_d/3$ as shown in Fig. 14.12.

When the flexural member is part of the primary lateral load resisting the system, the positive reinforcement required to be extended into the support as described above, shall be anchored to develop its design stress the tension at the face of the support.

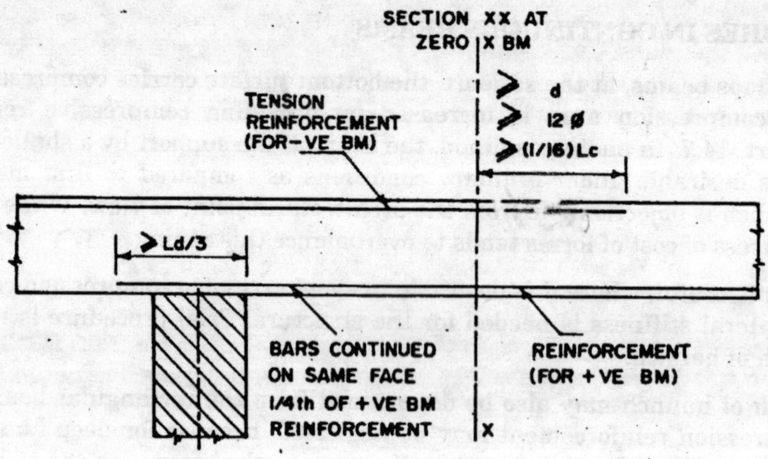

Fig. 14.12

2. For negative bending moment

At least *one-third* of the total reinforcement provided for negative moment at the support (*viz., tensile reinforcement)* shall extend the point of zero bending moment for a distance not less than the *effective depth of the member* or 12 times the diameter of bar or one-sixteenth of the clear span. L_c whichever is more as shown in Fig. 14.12.

14.12 DESIGN OF CONTINUOUS BEAMS

The continuous beams are designed as described in following step. The spacing of beams is fixed and the slab is designed. Then, the continuous beams are designed.

Step 1. The dead and live loads per unit length along the span of beam are calculated by multiplying the sum of dead and live loads by the spacing of the beams. The estimated dead load (*self-weight)* of the beam per unit length is added.

Step 2. The maximum bending moments of the various spans (span moments and support moments) and the shear forces at the various supports are calculated using the respective coefficients as given in IS : 456-2000 knowing the conditions of restraints at the ends.

Step 3. The depth of beam is then found. In a continuous beam of equal spans, it is desirable to keep uniform depth for all the spans, although the bending moments in the end spans are more than in the centre spans.

The depth of rectangular beam is found for the absolute maximum bending moment and designed for the largest shear for nominal shear stresses. T-beam sections are designed as described in Chapter 5. The depth of beam for continuous span must be satisfactory for all spans. As such the decision for the proper depth for a continuous span is somewhat complicate. It is desirable to keep same depth for all beams even if some beams are subjected to a disproportionate large bending moment at the support and a large amount of reinforcement is provided in compression for such beams. After determining the depth of beam and width of web, the design is continued further.

Step 4. The areas of reinforcement needed in the various spans of the beam for positive and negative bending moments are calculated. The number and size of bars for positive bending moment are decided as described in Art. 6.9.

Step 5. The areas of bars and their extension for compression and tension reinforcements at the supports are kept as described in Arts. 6.7, 6.8 and 6.14.

Step 6. The bending of bars is then decided as described in subsequent Article (Art. 14.7).

Step 7. The web reinforcement is then designed as described for the rectangular simply supported beams in Chapter 3 depending upon the shear forces at various locations.

14.13 BENDING OF BARS

The bars are bent with due consideration for the bending moments. The bars are bent far enough from the centre of span where the bending moment has reduced sufficiently to allow the reduction for the reinforcement. The bars are bent at different locations for the uniformly distributed loads and for the concentrated loads.

In small beams, the bars are bent at one location at each end. It reduces the types of bars and facilities the construction. However, it is better to bend the bars at different locations. It gives better distribution of reinforcement in the beam and increases the effectiveness as diagonal tension reinforcement.

In case the bars are bent at one location, the large number of bars are closely spaced and form an inclined plane which obstructs the flow of concrete. This inclined plane becomes a plane of weakness, specially when the bars are not placed exactly in position. The proper distribution of the locations of bending of bars avoids such conditions.

There is a natural tendency to bend the bars symmetrically in the cross-section. Though the symmetrical arrangement in a section is preferrable, it is not absolutely necessary. The bars may be bent at two locations with unsymmetrical arrangement in a cross-section.

In simply supported beams, and in continuous beams, the maximum bending moments reduce to zero at the points of support and at points of contraflexure. Therefore, for beams of constant cross-section, the required area of reinforcement at the intermediate positions reduces proportionately. As such, a part of the positive moment reinforcement may be bent up at proper locations and utilized it either as negative moment reinforcement or as diagonal tension reinforcement. The decrease in area of steel depends upon the decrease of bending moments, the bending moment diagrams may be used as a guide and the sketches for bending of the bars may be prepared. There are maximum and minimum distances of bending of the bars at top and at bottom of beam as explained here under :

1. Maximum distance for bending up bars

The maximum distance of a bent up bar from the edge of the support at the bottom of a beam should not exceed 0.25 L. As far as the positive bending moments concerned, the bars may be bent nearer towards the support and the distance of bent up bar from the support may be smaller than the maximum distance specified as above.

2. Minimum distance for bending up bars

The minimum distance of a bent up bar from the edge of the support at the top of a beam should not be less than 0.15 L. As far as the negative bending moment is concerned, the bars may be bent farther from the support at top and the distance of bent up bar from the support may be more than the minimum distance specified as above.

While deciding the actual bending of the bars, the diagonal tension should also be considered. It is advisable to consider the possibility of small in accuracies in bending of bars. Therefore the location at the bottom for bending of bar may be a few millimeters nearer the support and the position at the top a few millimeters farther from the support, than the theoretical values.

In practice, the position of bending of bar at the bottom is accordingly assumed and the bar is bent at an angle 45° or more. The position of bending of bat at the top is obtained automatically. In case, the distance from the support to this top position is equal to or more than the minimum specified, it is satisfactory. In case, the position at the top comes too nearer the support, a steeper angle may be used.

Example 14.1 Design a reinforced concrete rectangular beam continuous over three spans as shown in Fig. 14.15. The beam is to carry a superimposed live load of 40 kN/m and a superimposed dead load of 15 kN/m in addition to the self-weight of the beam. M 20 grade of concrete and hysd steel bars of Fe 415 quality shall be used.

Design : For M 20 grade of concrete and hysd steel bars of Fe 415 quality, the permissible stresses as per IS : 415-2000 are as follows :

$$\sigma_{cbc} = 7\ N/mm^2 \text{ and } \sigma_{st} = 230\ N/mm^2$$

The modular ratio of M 20 grade of concrete

$$m = \left(\frac{280}{3 \times 7}\right) = 13.33$$

Step 1 : Design Constants. For balanced section of beam, the design constants are as follows :

Depth factor for the neutral axis, k_B

$$k_B = \left(\frac{13.33 \times 7}{13.33 \times 7 + 230}\right) = 0.2886$$

Lever arm factor, j_B

$$j_B = \left(1 - \frac{0.2886}{3}\right) = 0.9038$$

Factor for the moment of resistance, R_B

$$R_B = \left(\frac{1}{2} \times 7 \times 0.2886 \times 0.9038\right) = 0.9129$$

The percentage of reinforcement for the balanced section, p_B,

$$p_B = \frac{1}{2} \times \left(\frac{7 \times 0.2886}{230}\right) \times 100 = 0.4392 \text{ percent.}$$

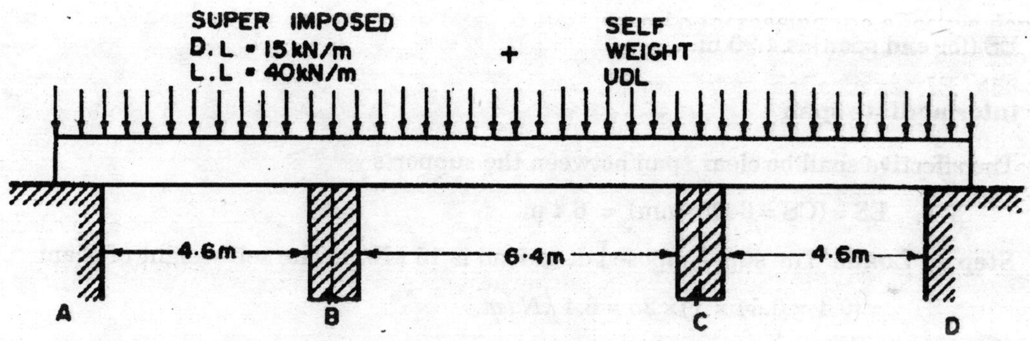

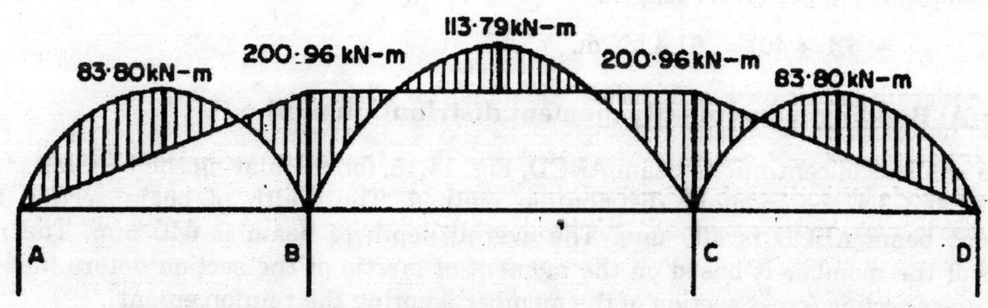

Fig. 14.13

Step 2 : Estimation of overall depth and width of beam. The overall depth of beam may be assumed as *one-tenth* of the span

$$D = \left(\frac{1}{10} \times 6.4 \times 1000\right) = 640\ mm$$

The width of beam, b may be assumed between *half to two-third* of the depth. Therefore,

b = 400 mm

Let the effective cover for the reinforcement be 40 mm. Therefore, the effective depth of beam shall be 600 mm.

Step 3 : Effective span. For the end span, the beam is simply supported on one side and continuous over the other side. The width of supporting wall be assumed as 600 mm. Therefore, **the effective span of end span** is as follows. The width of support 600 mm is wider than (4600/12) = 383.33 mm. Then

$$ES = \left(CS + \frac{1}{2}d\right) = \left(4600 \times \frac{1}{2} \times 600\right) = 4900\ mm$$

$$ES = \left(CS + \frac{1}{2}b_s\right) = \left(4600 \times \frac{1}{2} \times 600\right) = 4900\ mm$$

ES (for end span) is 4.90 m.

For intermediate span

The effective shall be clear span between the supports

ES = (CS = 6400 mm) = 6.4 m.

Step 4 : Loads. The superimposed dead load is 15 kN/m. The self-weight of beam

$$= (0.4 \times 0.64 \times 1) \times 25 = 6.4\ kN/m.$$

The superimposed live load acting over the beam is 40 kN/m. The intensity of uniformly distributed total load per meter length.

+ 6.4 + 40) = 61.4 kN/m.

Step 5 : (A) Bending moments (By moment distribution method).

The analysis of continuous beam ABCD, Fig. 14.15, for calculating the bending moments is carried out by the moment distribution method. The width of beam section for the continuous beam ABCD is 400 mm. The overall depth of beam is 640 mm. The relative stiffness of the member is based on the moment of inertia of the section determined on the basis of *gross-section* (cross section of the member ignoring the reinforcement).

Relative stiffness of the end spans

$$\left(\frac{I}{l}\right) = \left(\frac{1}{12} \times \frac{400 \times (640)^3}{4900}\right) = 17.83 \times 10^5\ mm^3$$

Relative stiffness of the intermediate span

$$\left(\frac{I}{l}\right)=\left(\frac{1}{12}\times\frac{400\times(640)^3}{6400}\right)=13.65\times10^5\,mm^3$$

End supports are hinged supports.

Therefore, the relative stiffness of the end members

$$\left(\frac{3}{4}\times\frac{I}{l}\right)=\left(\frac{3}{4}\times 17.83\ \times 10^5\right)=13.37\times10^5\ mm^3$$

Distribution factors

$$\text{For spans } \frac{BA}{CD}=\left(\frac{13.37\times10^5}{13.37+13.65}\times10^5\right)=0.495$$

$$\text{For span } BC=\left(\frac{13.65\times10^5}{13.37+13.65}\times10^5\right)=0.505$$

Bending moments at supports

$M_B = M_c$ = 200.96 kN-m (negative)

Bending moments at mid-span

$$\text{Span } \frac{AB}{CD}=\left(\frac{184.28\ -200.96}{2}\right)=83.80\ kN-m\ (positive)$$

Span BC = (314.37 – 200.58) = 113.79 kN-m (positive).

Step 5 : (B) Bending moments (By method of coefficients)

As per the procedure described in IS : 456-2000, the bending moment are found by utilising the coefficients given in tables (IS : 456-2000). These coefficients shall be multiplied by total design load and effective span. **Total design loads** for different spans are as under:

For end spans AB and CD

Total design dead load over complete span

$$W_{D.L}=(15+6.4)\times4.9\ =104.86\ kN$$

Total design live load over complete span

$$W_{L.L}=(40\times4.9)=196.00\ kN$$

Total load is = 300.86 kN

For intermediate span BC (effective span 6.4 m)

Total design dead load over complete span

$$W_{D.L} = (15 + 6.4) \times 6.4 = 136.96\ kN$$

Total design live load over complete span

$$W_{L.L} = (40 \times 6.4) = 256.00\ kN$$

Total load is = 392.96 kN

Step 6 : Support moments

6 (A) Bending moment due to dead load

At support next to the end support

$$M_{BA} = M_{CD} = -1 \times \left(\frac{1}{10} \times 104.86 \times 4.9\right) = -51.38\ kN-m$$

$$M_{BC} = M_{CB} = -1 \times \left(\frac{1}{10} \times 136.96 \times 6.4\right) = -87.04\ kN-m$$

6 (B). Bending moment due to imposed live load

At support next to the end support

$$M_{BA} = M_{CD} = \left(-1 \times \frac{1}{9} \times 196 \times 4.9\right) = -106.71\ kN-m$$

$$M_{BC} = M_{CB} = \left(-1 \times \frac{1}{9} \times 256 \times 6.4\right) = -182.04\ kN-m$$

6 (C) Total support moment

$$M_{BA} = M_{CD} = (-51.38 - 106.71) = -158.09\ kN-m$$

$$M_{BC} = M_{CD} = (-87.04 - 182.04) = -269.08\ kN-m$$

Step 7 : Span moment

Bending moment due to dead load

Near middle of end span

$$Span\frac{AB}{CD} = \left(\frac{1}{12} \times 104.96 \times 4.9\right) = 42.82\ kN-m$$

At middle of the interior span

$$\text{Span BC} = \left(\frac{1}{24} \times 136.96 \times 6.4\right) = 36.62\ kN-m$$

(A) Bending moment due to imposed live load

Near middle of the end span

$$Span\ \frac{AB}{CD} = \left(\frac{1}{10} \times 196 \times 4.9\right) = 96.04\ kN-m$$

7 (B) At middle of the interior span

$$Span\ BC = \left(\frac{1}{12} \times 256 \times 6.4\right) = 136.53\ kN-m$$

7 (C) Total span moments

$$Span\ \frac{AB}{CD} = (42.82 + 96.04) = 138.86\ kN-m$$

$$Span\ BC = (36.52 + 136.53) = 173.05\ kN-m$$

Step 8 : (A) Shear forces

Total uniformly distributed load over span AB/CD

(21.4 + 40) x 4.9 = 300.86 kN

Total uniformly distributed load over span BC

(i) Reactions at outer supports A and D shall be reaction for simply supported span + reaction for the moment at the intermediate support B. Therefore,

$$R_{AB} = R_{DC} = \left(\frac{300.86}{2} - \frac{200.96}{4.9}\right) = 109.42\ kN$$

(ii) Reactions at intermediate support B/C shall be reaction for simply supported spans AB and BC, + reaction for the moment at the intermediate support + reaction for the difference of moments.

(iii) Reaction at outer side

$$R_{CD} / R_{BA} = \left(\frac{300.86}{2} + \frac{200.96}{4.9}\right) = 191.44\ kN$$

(iv) Reaction at inner side

$$R_{BC} / R_{CB} = \left(\frac{392.96}{2} + 0\right) = 196.48\ kN$$

Step 8 : Moments at the face of supports

As per IS : 456-2000, for monolithic construction, the moments calculated at the face of supports are to be used in the design of members at those sections. The moments at the face of supports are calculated by approximate calculations. The moments at the face of supports are found as follows :

Note : The values of bending moments by moment distribution method and onwards the reactions determined have been used for further calculations :

At support B (outer side)

$$M_{BA} = \left(200.96 - \frac{191.44 \times 0.600}{2}\right) = 143.53\ kN$$

$$M_{BC} = \left(200.96 - \frac{196.48 \times 0.600}{2}\right) = 142.02\ kN$$

Step 9 : Check for overall depth of beam

The overall depth of beam s checked for the maximum bending moment 200.96 kN-m occurring at support B. The effective depth required for the slab to resist this bending moment

$$d = \left(\frac{200.96 \times 10^6}{0.9126 \times 500}\right)^{1/2} = 741.84\ mm$$

For the singly reinforced beam, the depth is large. The beam is made doubly reinforced.

The moment of resistance of singly reinforced beam of balanced section having 400 mm width and 640 mm depth.

$$M_1 = R_B.bd^2$$

or $$M_1 = \left(\frac{0.9129 \times 400 \times 600 \times 600}{1000 \times 1000}\right) = 131.46\ kN-m$$

$$\therefore\ M_2 = (M - M_1) = (200.96 - 131.46)\ \text{kN-m}$$

or $$M_2 = 69.50\ \text{kN-m}$$

Step 10 : Reinforcement for the beam

The bending moment 200.96 kN-m is the negative bending moment (viz., bending moment at supports B and C). Therefore, tension reinforcement, A_{st1} is provided at the top face for the bending moment M_1 = 131.46 kN-m. Additional tension reinforcement, A_{st2} is provided to resist the bending moment M_2 = 69.50 kN-m and the compression reinforcement, A is also necessary at the bottom face.

(i) Tension reinforcement, A_{st1} and A_{st2}

$$A_{st1} = \left(\frac{131.46 \times 10^6}{230 \times 0.9038 \times 600}\right) = 1054\ mm^2$$

$$A_{st2} = \left(\frac{69.50 \times 10^6}{230 \times 0.9038 \times (600 - 40)}\right) = 597.03\ mm^2$$

Total A_{st} = (1054 + 597.03) = 1651.03 mm².

Let 20 mm φ hysd steel bars of Fe 415 quality be used for the reinforcement. Area of cross-section of one bar is 314.159 mm². Therefore, the number of bars required

(1651.03/314.159) = 5.255 say 6 bars

(ii) Compression reinforcement, A_{sc}

The compression reinforcement, A_{sc} is found as under.

$$n_B = k_B \,.\, d = (0.2886 \times 600) = 173.16\ mm$$

$$(m_c - 1)\, A_{sc} \,.\, (n_B - d_c) = m \,.\, A_{st_2}\, (d - n_B)$$

$$(1.5 \times 13.33 - 1)\, A_{sc} \,.\, (173.16 - 40) = 13.33 \times 597.03 \times (600 - 173.16)$$

$$A_{sc} = \left(\frac{133.33 \times 597.03 \times 426.84}{18.995 \times 133.16}\right) = 1343.01\ mm^2$$

Number of bars needed for the compression reinforcement

(1343.01/314.159) = 4.275 say 5 bars

(iii) Tension reinforcement at bottom face for mid-span of AB and CD

$$A_{st} = \left(\frac{83.80 \times 10^6}{230 \times 0.9038 \times 600}\right) = 671.88\ mm^2$$

Number of bars required

(671.88/314.159) = 2.139 say 3 bars

(iv) Tension reinforcement at bottom for mid-span of BC

$$A_{st} = \left(\frac{113.79 \times 10^6}{230 \times 0.9038 \times 600}\right) = 912.33\ mm^2$$

Number of bars required

(912.33/314.159) = 2.904 say 3 bars

Note : The shear forces, and the shear force reinforcement have been calculated in Example 15.1 complete reinforces for positive and negative bending moments and the reinforcement stirrups have been shown in Fig. 15.2.

PROBLEMS

14.1 Design a reinforced concrete rectangular beams continuous over three spans AB = 3.6 m, BC = 4.8 m and CD = 3.6 (clear spans). The beam is to carry a superimposed dead load of 12 kN/m and a superimposed live load of 30 kN/m in addition to the self-weight of the beam. M 20 grade of concrete and hysd steel bars of Fe 415 quality shall be used.

15

SHEAR STRENGTH AND DESIGN OF CONTINUOUS BEAMS (CB-II)

15.1 INTRODUCTION

The design of shear reinforcement is dependent of the variation of shear forces along the span. The maximum shear force at any section due to the combination of dead and live loads should be calculated. In case the live load consists of concerned loads such as in the design of highway bridge spans, accurate calculations of such maximum shears at all the sections should be done.

The *critical section for shear* may be considered at the face of the support and the shear forces may be calculate and used in the design of the member at the section. The exceptions to this have been described in chapter 3 design of singly reinforced rectangular beams.

15.2 COEFFICIENTS FOR SHEAR FORCER FOR CONTINUOUS BEAMS

Unless more exact estimates are made, for beams of uniformly distributed loads over three or more spans which do not differ by more than 15 percent of the longest span, the shear forces used in design may be obtained by using the coefficient for shear as per IS : 456-2000 as given in Table 15.1.

TABLE 15.1. COEFFICIENT FOR SHEAR

Types of Loading	At end support	At support next to end support		At all other interior support
		Outer side	Inner side	
Dead load and imposed load (fixed)	0.40	0.60	0.55	0.50
Imposed load not fixed	0.45	0.60	0.60	0.60

(Note : For obtaining the shear force, above coefficient shall be multiplied by the total design load).

For beams, built into masonry wall which develop partial restraint the shear coefficient given in Table 15.1 at the end support may be increased by 0.05.

15.3 ARRANGEMENT OF LIVE LOAD FOR SHEAR FORCE

The influence line for end shear at Q of span QR in a substitute frame of a rigid frame is shown in Fig. 15.1 (B). It may be seen that, in the case of uniformly distributed live load, the position of load for maximum end shear is identical with that for maximum negative end moment. The position of load for maximum shear at one end shall be different from that for maximum shear at the other end. The position of load is decided by drawing the influence line diagrams for the support.

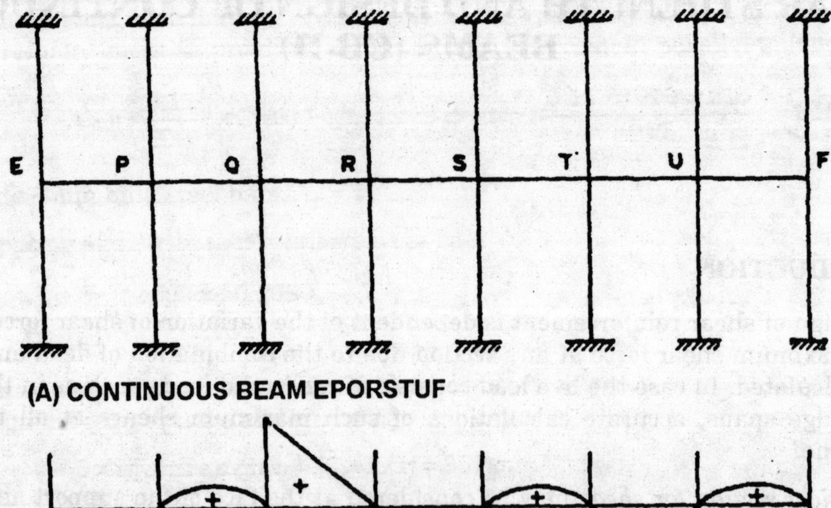

Fig. 15.1

Let 10 mm ϕ hysd steel bars be used for two logged vertical stirrups :

$$A_{sv} = \left(2 \times \frac{\pi}{4} \times 10^2\right) = 157.08\ mm^2$$

$$s_v = \left(\frac{230 \times 157.08 \times 600}{45.58 \times 1000}\right) = 475.58\ mm$$

Provide the vertical stirrups at 300 mm spacing centre to centre (less than maximum permissible 0.75 x 600 = 450 mm or 450 mm).

(ii) Shear reinforcement in span BC

Six bars are available for tension reinforcement. Therefore,

$$A_x = \left(6 \times \frac{\pi}{4} \times 10^2\right) = 1884.954\ mm^2$$

$$\left(\frac{100\ A_s}{bd}\right) = \left(\frac{1884.954 \times 100}{400 \times 600}\right) = 0.785$$

The permissible shear stress in shear concrete from IS : 456-2000.

$$\tau_c = \left[0.35 + \frac{0.04 \times (0.785 - 0.75)}{(1.00 - 0.75)}\right]$$

$$= 0.356 \text{ N/mm}^2$$

Nominal shear stress at support B/C

$$\tau_v = \left(\frac{196.48 \times 1000}{400 \times 600}\right) = 0.819\ N/mm^2$$

$\tau_{c.\max}$ **for M 20 is 1.8 N/mm².**

Shear force to be resisted by the shear force

$$V_s = \left(196.48 - \frac{0.356 \times 400 \times 600}{1000}\right)$$

$$= 111.04 \text{ kN}$$

Let 10 mm ϕ hysd steel bars be used for two legged vertical stirrups.

$$A_{sv} = \left(2 \times \frac{\pi}{4} \times 10^2\right) mm^2 = 157.08\ mm^2$$

Example 15.1 In Example 14.1, find the shear force. Calculate the shear reinforcement for the continuous beam shown in Fig. 14.13. Draw the sketch and show complete reinforcement. Also check the development length at supports.

Step 1 : Shear forces (By method of coefficients)

As per the procedure recommended in IS : 456-2000, the shear forces are obtained by noting the coefficients given in tables (IS : 456-2000) :

(i) Shear force due to dead load

At end support

(0.4 x 104.96) = 41.944 kN

At support next to the end support (out side)

$$(0.6 \times 136.96) = 75.328\ kN$$

(Inner side), $(0.55 \times 136.96) = 75.328\ kN$

(ii) Shear force due to imposed live load

$$\text{At end support,} = (0.45 \times 196) = 88.2\ kN$$

At support next to the end support (out side)

$$(0.6 \times 256) = 153.6\ kN$$

$$\text{(Inner side),} = (0.6 \times 256) = 153.6\ kN$$

(iii) Total shear force

At end support, = (41.944 + 88.2) = 130.144 kN

At support next to the end support

(Out side) = (82.176 + 153.6) = 235.776 kN

(Inner side) = (75.328 + 153.6) = 228.928 kN

$$s_v = \left(\frac{230 \times 157.08 \times 600}{111.04 \times 1000}\right) mm = 195.22\ mm$$

Provide the vertical stirrups at 150 mm spacing centre to centre.

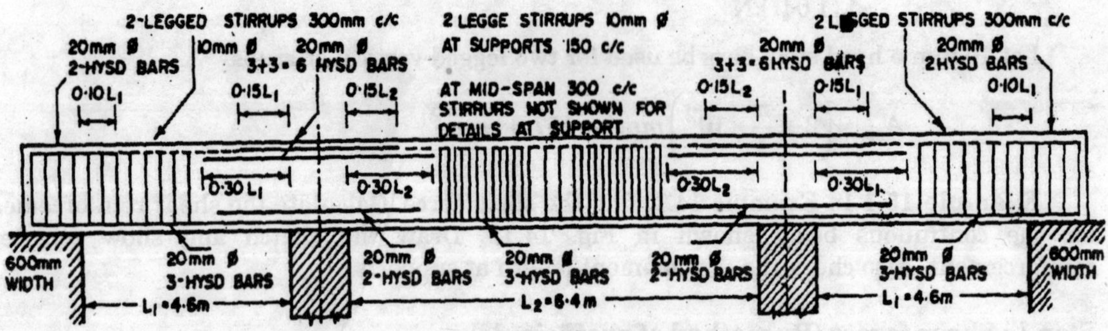

Fig. 15.2.

Step 2 : Check for the development length at supports A and D

A_s = 942.477 mm² (3-bars 10 mm ϕ)

$$M_1 = \left(\frac{230 \times 942.477 \times 0.9038 \times 600}{1000 \times 1000}\right) = 117.55\ kN - m$$

$$V = 109.42\ \text{kN}$$

$$\left(\frac{M}{V}\right) = \left(\frac{117.55}{109.42}\right) = 1.074\ m$$

$$L_o = \left(\frac{b_s}{2} - x_1\right) = \left(\frac{600}{2} - 20\right) = 275\ mm$$

$$\left(\frac{M_1}{V}+L_o\right)=(1.074+0.275)=19.34\ m$$

$$L_d=\left(\frac{68.45\times 20}{1000}\right)=1.369\ m$$

The development length, L_d and $[(M_1/V)+L_o]$ values are practically equal. The details of tension, compression and shear reinforcement are shown in Fig. 15.1.

PROBLEMS

15.1 In Problem 14.1, complete the design for the continuous beam for the shear forces. Also check the main reinforcement for the development length at the supports.

16

PART IV
DESIGN OF SLABS

DESIGN OF SLABS SPANNING IN ONE DIRECTION (SSOD)

16.1 INTRODUCTION

The reinforced concrete slabs may be supported on two opposite long edges only and free of any support along the two opposite short edges. The structural action of the slab is essentially *one-way.* The loads are carried by the slab in the direction perpendicular to the supporting beams (supporting edges) as shown in Fig. 16.1. Such slabs are called as **slabs spanning in one direction.** These are also called as **one-way slabs.**

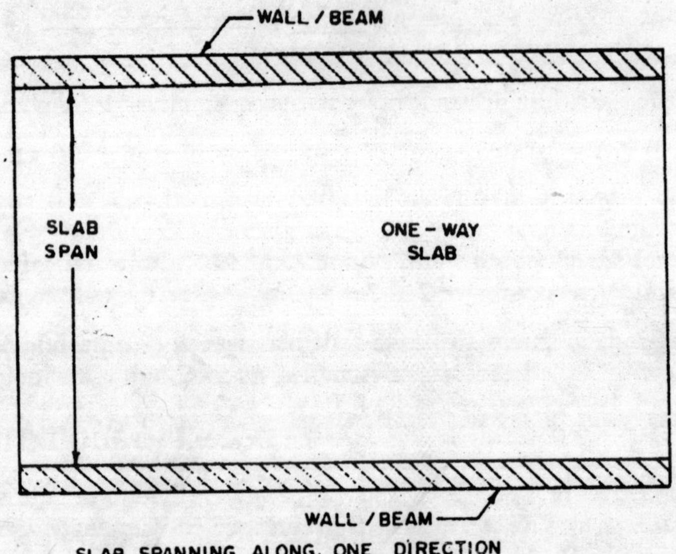

Fig. 16.1

The reinforced concrete slabs may also be supported on all its four sides as shown in Fig. 16.2. The structural action of the slab is *two-way.* The loads are carried by the slab along its short span, (*viz.,* the width of slab) and along its long span, (*viz.,* the length of slab) in different proportions depending upon the ratio of long span to short span. In case, the ratio of long span to short span is more than 2, the most of the load is carried along the short direction to the supporting sides and the structural action is practically *one-way*, even though the supports are provided on all the four sides. In such cases of the slabs, the slabs are designed as spanning in one direction (viz., one-way).

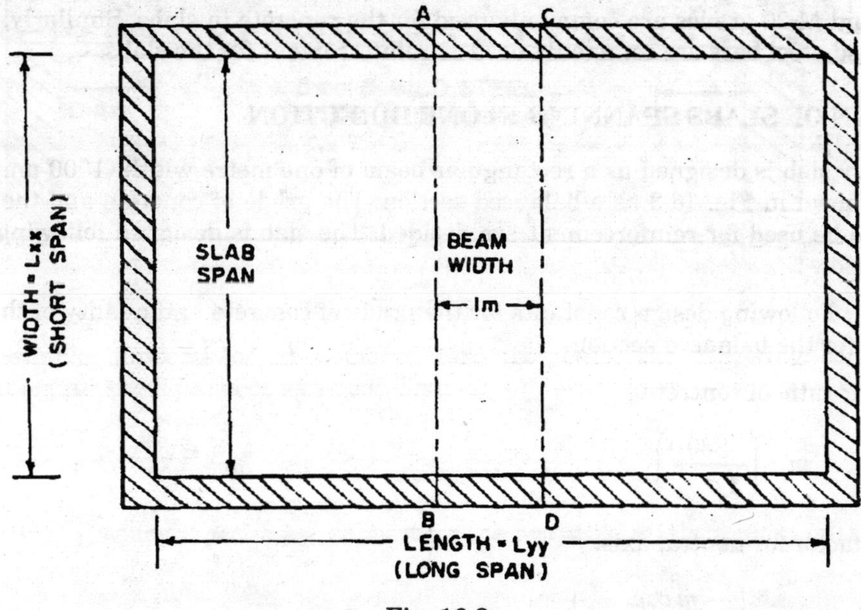

Fig. 16.2

The slab spanning in one-direction (viz., one way) are analysed and designed as rectangular beam of one metre (1000 mm), width in cross-section as shown in Fig. 16.3 and clear span as AB, Fig. 16.2 (viz., the distance between supporting sides). The curvature and consequently bending moments are same in all the strips spanning in the short direction between the supported sides.

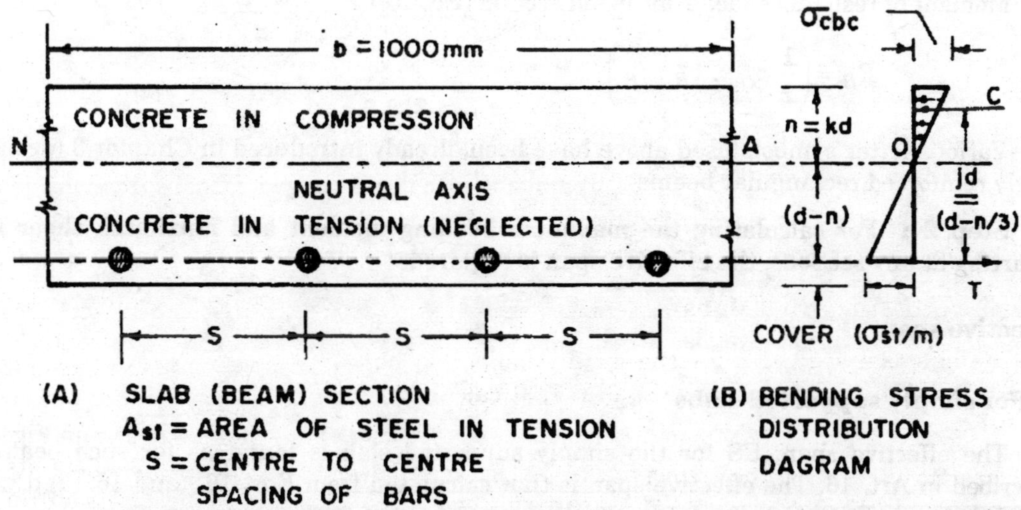

Fig. 16.3

The curvature and the bending moment along the long strips parallel to the supporting edges remain zero. As such, a one-way slab consists of a set of rectangular beams side by side.

M 15 and M 20 grades are commonly used for the concrete in slabs. Similarly, mild steel bars and hysd-steel bars are commonly used as reinforcement for the slabs.

16.2 DESIGN OF SLABS SPANNING IN ONE DIRECTION

One-way slab is designed as a rectangular beam of one metre width (1000 mm) in cross-section as shown in Fig. 16.3 as a balanced section. The grade of concrete and the quality of steel bars to be used for reinforcement are decided. The slab is designed following the steps described below :

Step 1 : Following design constants for the grade of concrete and quality of the steel are determined for the balanced section.

Modular ratio of concrete

$$m = \left(\frac{280}{3\,\sigma_{cbc}}\right) \qquad \text{(i)}$$

Depth factor for neutral axis.

$$k_B = \left(\frac{m\sigma_{cbc}}{m\sigma_{cbc} + \sigma_{st}}\right) \qquad \text{(ii)}$$

Lever arm factor for the beam section.

$$j_B = \left(1 - \frac{k_B}{3}\right) \qquad \text{(iii)}$$

Moment of resistance factor for beam section (Eq. 16.)

$$R_B = \left(\frac{1}{2}\,\sigma_{cbc} \,.\, k_B \,.\, j_B\right) \qquad \text{(iv)}$$

Various letter symbols used above have been already introduced in Chapter 3 (design of singly reinforced rectangular beams).

Step 2 : For calculating the maximum bending moment and maximum shear force occurring at any sections, the effective span is required.

Effective span

(i) For simply supported slabs

The effective span, ES for the simply supported slab is found as for such beams as described in Art. 16. The effective span is that calculated from Eqs. 16. and 16. and which ever is less.

(ii) For cantilever slabs

The effective span ES is calculated from Eq. 16.

(iii) For continuous slabs

The effective span ES is decided by considering the support width and restraint conditions for the end span as described in Art. 16.2.

Step 3. For calculating maximum bending moment and shear force, it is also necessary to know the types of loads to be supported by the slab.

Loads. The dead loads of materials attached and supported by the slab are obtained by knowing the unit weight of the materials. The self-weight of the slab is calculated by estimating the thickness of slab. The width of slab for design is taken as width of beam as 1000 mm.

Estimation of thickness of slab

The thickness of slab is estimated by considering the span depth ratio described in Art. 16. depending upon end restraint conditions. Exact depth necessary to resist the resulting maximum bending moment is obtained in the subsequent steps.

The self-weight of slab may be found by considering unit area of slab and thickness of the slab. The self-weight of slab shall be the unit-weight of reinforced concrete multiplied by volume of concrete as above. It is added to the dead loads attached.

The intensity of uniformly distributed live load per square metre is noted from IS : 875-1984 for the type of floor and its use (e.g., residential, class room hostel, restaurant or workshop).

Total load to be support by slab per unit length along the short span is obtained.

Step 4 : Required thickness of slab is determined after calculating the maximum bending moment, M resulting due to the load, the effective span of the slab and the support conditions.

(i) For simply supported over supports

The maximum bending moment, M for uniformly distributed load over the whole span occurs at the centre and M is calculated as (WL/8) where W is total and L is the effective span of the slab.

(ii) For slabs continuous over supports

The slabs continuous over supports are designed by calculating the bending moments for the continuous beams used the bridges and as described in Chapter 6.

(iii) For slabs monolithic with supports

IS : 456-2000 recommends that the bending moments in slabs constructed monolithically with the supports shall be calculated by taking such slabs either as continuous over supports and capable of free rotation, or as members of continuous framework with the supports, taking into account the stiffness of such supports. If such supports are formed due to beams which justify fixity at the support of slabs, then the effects on the supporting beam, such as, the bending of the web in the transverse direction of the beam and the torsion in the longitudinal direction of the beam, wherever applicable, shall also be considered in the design of the beam.

For the purpose of calculation of moments in slabs in a monolithic structure, it will generally be sufficiently accurate to assume that the members at the ends of such slabs are fixed in position and direction at the ends from their connections with the slabs.

The slabs carrying concentrated loads shall be discussed separately in the subsequent article.

One-way slab is designed as a beam of 1000 mm width as balanced section. As such the bending moment due to external loads, M is equated to moment of resistance of the beam section calculated from compressive force in concrete. Therefore,

$$M = R_B \cdot bd^2 \quad (\because b = 1000 \text{ mm})$$

$$\text{or} \quad d = \left(\frac{M}{R_B \times 1000}\right)^{1/2} \tag{16.1}$$

where R_B is the factor for moment of resistance.

Step 5 : A certain thickness of concrete cover outside of the outermost surface of the steel reinforcement bars for the reasons described in Art. 3.6 as recommended by IS : 456-2000 as follows :

Cover to reinforcement

The clear cover for the reinforcement (for tension, compression or shear or any other) shall be 15 mm minimum.

The *nominal maximum size* of coarse aggregate should usually be 5 mm less than the minimum cover to the reinforcement (viz., 15-5 = 10 mm).

Step 6 : Following reinforcements are provided in slabs.

6 (A) Main reinforcement

The area of steel reinforcement in tension, A_{st} is calculated from Eq. 3.23.

$$A_{st} = \left(\frac{M}{\sigma_{st} \cdot j_B \cdot d}\right) \qquad .v)$$

8 mm, 10 mm and 12 mm diameter bars are usually used for the main reinforcement bars. As per IS : 456-2000, the diameter of reinforcing bars shall not exceed one-eighth of the total thickness of the slab. Instead of deciding the number of bars used as tension reinforcement, in beas, the spacing of bars, s is found as follows by consider 1 m width of slab as shown in Fig. 16.3 (A).

The number of bars in 1 m (1000 mm) provided at a spacing, s from centre of adjacent bars

$$= (1000/s) \qquad \text{(vi)}$$

In case, A_ϕ the area of cross-section of the bar used, then the number of bars in 1 m may also be written as below :

$$= \left(A_{st} / A_\phi\right) \qquad \text{(vii)}$$

Equating the expressions (vi) and (vii),

$$\left(A_{st} / A_\phi\right) = (1000/s) \qquad \text{(viii)}$$

$$\text{or} \quad s = \left(\frac{1000 \, . \, A_\phi}{A_{st}}\right) \qquad (16.2)$$

The requirements for reinforcement for maximum reinforcement, spacing of reinforcement and minimum distance between individual bars have been described in Art. 3.6 for reinforcement in beams. The reasons for providing the limit for maximum reinforcement are partially for economy (as the saving in steel associated with increasing the effective depth is more than to compensate for the cost of the additional concrete) and partially because thin slabs with high steel ratios would be likely to allow large deflections. IS: 456-2000 recommends for the minimum reinforcement in slabs as under :

The reinforcement in either direction in slabs shall not be less than 0.15 percent of the total cross-sectional area. However this value can be reduced to 0.12 per cent when high yield strength deformed bars are used.

IS : 456-2000 has specified this requirement for the minimum value of the main reinforcement to resist the stresses developed due to temperature and shrinkage. The minimum area of reinforcement recommended is empirical. It is intended to take care of shrinkage and temperature stresses. This reinforcement is called *distribution reinforcement.*

6 (B) Distribution reinforcement

The distribution reinforcement, A_d in one-way slab is provided in the direction (parallel to long span) perpendicular to the main reinforcement (provided in the direction parallel to short span). The requirement for the distribution reinforcement has been specified in IS : 456-2000 as the requirement for the minimum reinforcement described in above paragraph. Usually 6 mm and 8 mm diameter bars are used for the distribution reinforcement. The spacing for the distribution reinforcement bars, s_d is calculated as below :

$$s_d = \left(\frac{1000 \times A_\phi}{A_d}\right) \qquad (16.3)$$

The necessity for providing the temperature and shrinkage reinforcement has been described in the subsequent articles separately.

6 (C) Spacing of reinforcement

The horizontal distance (viz., the spacing) between parallel main reinforcement bars as calculated from Eq. 16.2 shall not be more than three times the effective depth of a solid slab or 450 mm whichever is smaller.

The horizontal distance (viz., spacing) between parallel reinforcement bars provided against shrinkage and temperature shall not be more than *five times* the effective depth of a solid slab or 450 mm, whichever is smaller.

The shear force in slab seldom controls the design of *one-way* slabs, specially if small tensile steel percentage is used. It will be seen that the shear resisting capacity of the concrete. $\tau_c\ bd$, will almost without exception be well above the needed strength at the service loads.

All the tensile reinforcement bars resisting the positive moment are either kept straight and carried into the support or *fifty percent* are bent up into the compression zone. The bars are provided with sufficient anchorage. The location of the bent up bars is based on the requirement of area of steel only. The straight bars are usually extended into the support at least 150 mm. The location for bent up bars at the bottom face and at the top face from the support have been described in detail in Chapter-6, design of continuous beams.

The total thickness of slab is usually rounded to the next higher 6 mm for slab upto 150 mm thickness and to the next higher 12 mm for thicker slabs.

16.3 TEMPERATURE AND SHRINKAGE (DISTRIBUTION) REINFORCEMENT

Concrete structural components need extensive consideration for the **effects of temperature specially** for safety against fire in buildings. In the design of roof slabs spanning in one direction, concrete tanks which are used to store hot liquids, concrete chimneys, and expansion joints in concrete in roads and airport payments, the effects of temperature are seen with all importance. The tensile strength of concrete is very small. The concrete cracks even at low value of tensile strain 0.00015. The change in temperature develops tensile strains and results in cracking of concrete. The extent of cracks depends upon the coefficient of thermal expansion of the concrete and the variation of daily temperature.

1. Coefficient of thermal expansion of concrete, α

The change in unit length of a concrete structural component caused due to one degree change in temperature is called as coefficient of thermal expansion or contraction of concrete, α. It depends on the type of cement, aggregate used, the quantity of cement in concrete mix, the relative humidity and dimensions of structural component section. The value of coefficient of thermal expansion of concrete α is 0.000010 per *C. The values of thermal expansion for concrete with quartzite, sandstone, granite, basalt and limestone coarse aggregate are given in IS : 456-2000. These values range from 1.3×10^{-5} to 6×10^{-5} per *C. In case the structural components are rigidly connected to the immovable supports, then, the tensile stresses develop. It is worth to note that the coefficient of thermal expansion for steel is practically same as that for the concrete. Therefore, there is no differential temperature stresses between the steel bars provided as reinforcement and the surrounding concrete.

2. Variation in temperature

The variation in daily temperature depends upon season. This variation in winter is less and in summer is more. There may be maximum daily variation of about 22°C in summer.

For the variation of about 22°C in daily temperature, a strain equal to °C in daily temperature, a strain equal to $\alpha.\Delta T = 0.10 \times 10^{-4} \times 22 = 0.0022$ may develop and the concrete will crack. For M 15 grade of concrete, the value of coefficient of thermal expansion, α is 0.10 x 10^{-4}. The value of Young's modulus of Elasticity E_c is 5700 $\sqrt{f_{ck}} = 5700\sqrt{15} = 2.2074 \times 10^{+4}\ N/mm^2$. The stress developed for one degree change in temperature is

$$\left(\frac{1.\alpha \times E_c}{(1-\upsilon)}\right) = \left(\frac{1 \times 0.10 \times 10^{-4} \times 2.2074 \times 10^{+4}}{(1-0.2)}\right) = 0.2759\ N/mm^2$$

and the tensile stress developed for 22°C change in temperature is $\left(\frac{22 \times 0.10 \times 10^{+4} \times 2.2074 \times 10^{+4}}{(1-0.2)}\right) = 6.07\ N/mm^2$. The estimated tensile strength of M 15 grade of concrete is $0.7\sqrt{f_{ck}} = 0.7\sqrt{15} = 2.711\ N/mm^2$. The value of tensile stress 6.07 N/mm² is more than the tensile strength of the concrete. Therefore, the cracking of this concrete occurs. Even a variation of (2.711/0.2759) = 9.826 °C in temperature is sufficient to crack the concrete.

In the design of roof slabs, concrete arches and the presented concrete structures, it is essential to consider the **effects of shrinkage**. Total shrinkage of concrete depends upon the constitutions of concrete, size of the member and the environmental conditions. The concrete shrinks as the wet cement paste hardens. The shrinkage of concrete is approximately proportional to the quantity of cement paste in the concrete regardless of the composition of the paste. By using the smallest quantity of water and cement compatible with other requirements (such as strength and workability) and by thorough moist curing of sufficient duration, the shrinkage in concrete may be minimised. High relative humidity also reduces the extend of shrinkage. The saturation usually eliminates shrinkage or may cause slight expansion.

The linear coefficient of shrinkage varies from 0.00025 to 0.00050. The approximate value of the total shrinkage strain for design may be taken as 0.0003 (as per IS : 456-2000), which corresponds to tensile stress of 5.25 N/mm², tensile stresses will develop in a concrete member in case it is restrained and the decrease in length due to shrinkage is not allowed. For M 15 grade of concrete, the estimated tensile strength is 0.7 $\sqrt{15} = 2.2074\ N/mm^2$, which one is less than the tensile stress 5.25 N/mm² developed. As a result of which, the random cracks will develop in the roof slabs. This causes the weakening of strength of concrete and it may aggravate the disintegration. Besides taking all the precautions, the shrinkage of concrete in small extent is normally unavoidable.

The combined effect of temperature and shrinkage is to develop tensile stresses which develop cracks in concrete.

The temperature and shrinkage reinforcement (*viz.*, **distribution reinforcement**) is provided to counteract the expansion and contraction and resists the stresses developed due to these effects jointly or individually. The purpose of providing the distribution reinforcement is to knot the structure together and to avoid the localised cracks. The cracks are distributed in such a way that the numerous vary small cracks (which remain often invisible to eyes) are formed. These hair line cracks are not detrimental. As a result of distribution reinforcement provided, the few large cracks do not form. The distribution reinforcement is provided in the direction normal (parallel to long span) to the main reinforcement (provided parallel to short span) in floors and roof slabs where the main reinforcement extends in one direction only. The reinforcing bars of small diameter and in sufficient area are provided as close to the surfaces as practicable. Hysd-bars with irregular surfaces keep mechanical bond with the concrete and these bars are more effective in distributing the cracks than round mild steel bars which are having smooth surface.

The area of temperature and shrinkage reinforcement depends upon the conditions and degree to which these cracks are to be reduced. In water tight construction, more distribution reinforcement should be provided. The area of steel that is needed to distribute the cracks is not definite. IS : 456-2000 recommends that the **distribution (minimum) reinforcement** may be assumed not less than 0.15 percent of the total cross-section of concrete. It may be assumed as 0.12 percent when the high strength deformed bars are used.

16.4 CONCENTRATED LOADS OVER SLABS

Many times, the reinforced concrete slabs are supported on two opposite sides (edges/beams) only as shown in Fig. 16.1. In case of bridges culverts, the slabs rest on two opposite masonry retaining walls (which act as wide beams). The structural action of such slabs is essentially one way. The loads are carried by the slab in the direction perpendicular to the supporting beams/edges. When slab immediately below the load but the concentrated load is carried by a considerable width of the slab, called as the *effective width*. The ordinary formulate used for the slab assume uniform distribution of stresses along the whole width. These formulae cannot be used directly without some assumptions. In a wide slab under concentrated load, along the width of the slab, there is considerable variation in stress in the centre of the span. The stresses and deflections in the centre of spans are largest in the section of slab which are directly under the load. For sections away from the load, the stresses and deflections in the centre decrease rapidly at first and then more slowly, until for the wide slabs, these become insignificant the edges of the slab.

In order to be able to use ordinary slab formulae for the design of wide slab loaded by concentrated loads, it is necessary to determine the effective width of slab. **The effective width of slab** is the width of slab of same depth, which if stressed uniformly along its whole width, would have the same strength as the wide slab with its varying distribution of stresses, and for which the moment of resistance, based on an accepted maximum unit stress, would be equal to the moment of resistance of the whole width of slab with varying distribution of stress and with the largest stress equal to the maximum accepted stress. The effective width of slab is less than the total width of the slab; and the ratio of effective width to the total width depends upon the ratio of the total width of the span.

On the basis of extensive research on bridge slabs, the code of practice IS : 456-2000 recommends for (i) solid slabs supported on two opposite edges, and (ii) cantilever slabs, that if solid slabs carry concentrated loads, the maximum bending moment caused by concentrated loads shall be assumed to be resisted by *an effective width of slab* (measured parallel to the supported edges/edge), which may be found as follows :

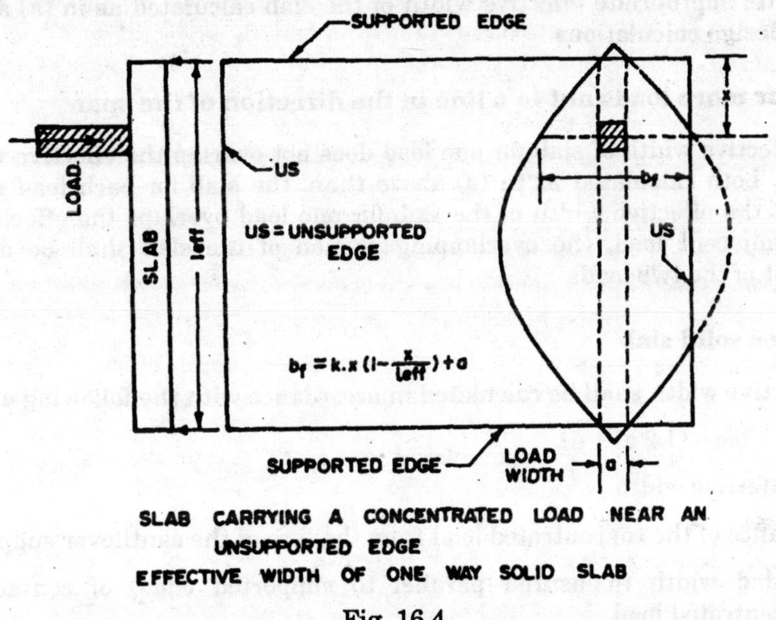

Fig. 16.4

(i) Solid slab supported on two opposite edges

(a) For a single concentrated load

The effective width of slab carrying a single concentrated load (as illustrated in Fig. 16.4) is calculated in accordance with the following expression provided that it shall not exceed the actual width of the slab

$$b_{eff} = \left[k \,.\, x \,.\left(1 - \frac{x}{l_{eff}}\right) + a\right] \tag{16.4}$$

where b_{eff} = effective width of the slab (measured parallel to the supporting edges/edge)

k = constant having the values as given in Table 16.1 depending upon the ratio of the width of slab, l to the effective span, l_{eff}.

x = distance of the centroid of the concentrated load from nearer support.

l_{eff} = effective span.

a = loaded width (measured parallel to the supported edge) of the contact area of the concentrated load and provided further that in case of a load near the unsupported edge of a slab, the effective width shall not exceed the above value nor half the above value plus the distance of the load from the unsupported edge.

(b) For two or more concentrated loads placed in a line in the direction of the span

The bending moment per metre width of slab shall be calculated separately for each load according to its appropriate effective width of the slab calculated as in (a) above and added together for design calculations.

(c) For two or more loads not in a line in the direction of the span

If the effective width of slab for one load does not overlap the effective width of slab for another load, both calculated as in (a) above than, the slab for each load can be designed separately. If the effective width of the slab for one load overlaps the effective width of the slab for an adjacent load, the overlapping portion of the slab shall be designed for the combine effect of the two loads.

(ii) Cantilever solid slab

The effective width shall be calculated in accordance with the following expression :

$$b_{eff} = (1.2\, a_1 + a) \qquad (16.5)$$

where b_{eff} = effective width

a_1 = distance of the concentrated load from the face of the cantilever support, and

a = loaded width (measured parallel to supported edge) of contact area, of the concentrated load.

Provided that the effective width of the cantilever slab shall not exceed *one-third* the length of the cantilever slab measured parallel to the fixed edge.

And provided further that when the concentrated load is placed near the extreme ends of the length of cantilever slab in the direction parallel to the fixed edge, the effective width shall not exceed the above value, nor shall it exceed half the above value plus the distance of the concentrated load from the extreme end measured in the direction parallel to the fixed edge.

TABLE 16.1. VALUES OF K FOR SIMPLY SUPPORTED AND CONTINUOUS SLAB (As per IS : 456 : 456-2000)

S.No.	Ratio $\frac{l'}{l_{eff}}$	Values of k	
		Simply supported slabs	Continuous slabs
1	0.1	0.40	0.40
2	0.2	0.80	0.80
3	0.3	1.16	1.16
4	0.4	1.48	1.44
5	0.5	1.72	1.68
6	0.6	1.96	1.84
7	0.7	2.12	1.96

8	0.8	2.24	2.08
9	0.9	2.36	2.16
10	1.0 and above	2.48	2.24

For slabs other than solid slabs, the effective width shall depend on the ratio of the transverse and longitudinal flexural rigidities of the slab. Where this ratio is unity, that is, where the transverse and longitudinal flexural rigidities are approximately equal, the value of the effective width as found for the solid slabs may be used. But as the ratio decreases, proportionately smaller values shall be taken.

Example 16.1 Design a simply supported slab over a class-room of size 5 m $\times 10\ m$. A finishing surface of cement concrete of 20 mm shall be provided over the slab. The slab shall be used as a class-room floor. M 15 grade of concrete and mild steel reinforcement shall be used.

Design : The class-room of given size is shown in Fig. 16.5.

The clear length and clear width of the class-room are 10 m and 5 m respectively.

The ratio of clear length of clear width (10/5 = 2.0) (i)

The slab is to be designed as slab spanning in one direction over the clear span of 5 m. The section of slab is designed as a beam ABCD of 1 m width as shown in Fig. 16.5 and as balanced section, following the steps described as below :

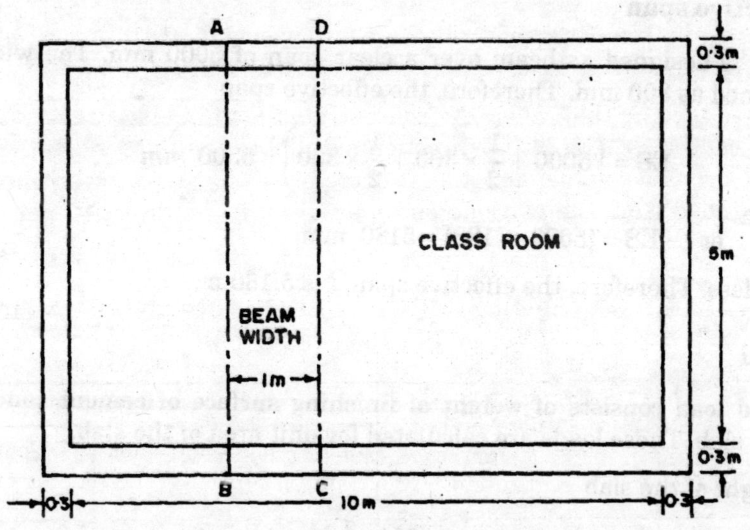

Fig. 16.5

Step 1 : Design constants

The design constants for M 15 grade of concrete and mild steel reinforcement for the balanced section are as follows :

The modular ratio of concrete

$$m = 18.67,\ \sigma_{cbc} = 5\ N/mm^2,\ \sigma_{st} = 140\ N/mm^2$$

$$k_B = 0.4,\ j_B = 0.867\ and\ R_B = 0.867.$$

Step 2 : Estimation of slab thickness

The thickness of slab is estimated from span to effective depth ratio specified for satisfactory control of the vertical deflection in IS : 456-2000. Therefore, the effective depth, d for the simply supported slab

$$d = \left(\frac{Span}{20 \times modification\ factor}\right) \quad \text{(ii)}$$

For M 15 grade of concrete and mild steel reinforcement, the percentage ratio of steel for the balance section, p_B is 0.7143 percent and therefore, the modification factor is 1.6. Therefore,

$$d = \left(\frac{5 \times 1000}{20 \times 1.6}\right) = 156.25\ mm \quad \text{(iii)}$$

The overall thickness of slab may be assumed as 200 mm and the effective cover may be considered as 20 mm. Therefore, the effective depth (estimated) may be taken as 180 mm.

Step 3 : Effective span

The slab is designed as beam over a clear span of 5000 mm. The width of supporting walls is assumed as 300 mm. Therefore, the effective span

$$ES = \left(5000 + \frac{1}{2} \times 300 + \frac{1}{2} \times 300\right) = 5300\ mm$$

or $$ES = (5000 + 180) = 5180\ mm,$$

whichever is less. Therefore, the effective span, l is 5.180 m.

Step 4 : Load

The dead load consists of weight of finishing surface of cement concrete and the self weight of the slab. These loads are calculated for unit area of the slab.

Self-weight of the slab

$$(1 \times 1 \times 0.20) \times 25 = 5.00\ kN/m^2$$

Weight of finishing surface

$$(1 \times 1 \times 0.020) \times 0.48\ kN/m^2$$

Live load for the floor for class-room

$$(1 \times 1) \times 24 = 4.00\ kN/m^2$$

Total load per unit area = 9.48 kN/m²

say = 9.5 kN/m²

The intensity of uniformly distributed load per 1 metre length along AB, Fig. 16.5, acting over the simply supported beam ABCD

$$w = 9.5 \text{ kN/m}^2 \tag{iii}$$

Step 5 : Maximum bending moment

The maximum bending moment, M for the simply supported beam occurs at the centre

$$M = \left(\frac{w \,.\, L^2}{8}\right) = \left(\frac{9.5 \times 5.18 \times 5.18}{8}\right) = 31.86 \; kN - m$$

Step 6 : Required effective depth, d and overall thickness

For the balanced section of 1 m wide beam, (*viz.*, b = 1000 mm)

$$R_B . bd^2 = M$$

$$d = \left(\frac{M}{R_B \times 1000}\right)^{1/2} = \left(\frac{31.86 \times 10^6}{0.867 \times 1000}\right)^{1/2} = 191.7 \; mm$$

Effective cover = 20 mm

Overall thickness = (191.7 + 20) = 211.7 mm say = 210 mm. (approx.)

Therefore, the effective depth, d is 190 mm

Step 7 : Area of steel reinforcement in tension

$$A_{st} = \left(\frac{M}{\sigma_{st} \times j.d}\right) = \left(\frac{31.86 \times 10^6}{140 \times 0.867 \times 190}\right) = 1381.48 \; mm^2 \tag{iv}$$

Area of cross-section for 10 mm ϕ bar

A_ϕ = 78.84 mm².

Step 8 : Spacing of main reinforcement

Centre to centre spacing between main steel bars

$$s = \left(\frac{1000 \times A_\phi}{A_{st}}\right) = \left(\frac{1000 \times 78.54}{1381.48}\right) = 56.85 \; mm$$

Provide 10 mm ϕ mild steel bars at 50 mm centre to centre spacing between the adjacent bars, parallel to short span, AB Fig. 16.5. The clear spacing (50 – 10 = 40 mm) is more than the diameter of bar and it is also more than (5 mm + size of the aggregate say 10 mm). The maximum allowable spacing 3 x effective depth (3 x 190 = 570 mm) or 450 mm.

Step 9 : Clear cover for main reinforcement

$$\left(20-\frac{\phi}{2}\right)=\left(20-\frac{10}{2}\right)=15\ mm$$

and it is not less than the diameter of bar, 10 mm.

Step 10 : Distribution reinforcement

The mild steel distribution reinforcement, A_d (formerly called as temperature and shrinkage reinforcement) is provided as minimum reinforcement parallel to long span

$$A_d=\left(\frac{0.15}{100}\right)\times Cross-\text{sectional}\ area=\left(\frac{0.15\times1000\times210}{100}\right)=315\ mm^2$$

Provide 8 mm mild steel bars, $A_\phi = 50.265\ mm^2$

The spacing of distribution reinforcement bars

$$s_d=\left(\frac{1000\times50.265}{315}\right)=159.57\ mm\ (say)=150\ mm,\ (centre\ to\ centre)$$

The maximum allowable horizontal clear distance between parallel reinforcement bars provided against temperature and shrinkage is

$$(5\times effective\ depth\ =(5\times190)=950\ mm)$$

or 450 mm whichever is less. Actual spacing is less than maximum allowable.

The details of main reinforcement in tension and the distribution of reinforcement are shown in Fig. 16.6.

Alternate main steel bars have been bent up near the supports. These bars may be bent 0.15 of span short of the support (*viz.*, 0.15 x 5.18 = 0.777 m). However, these bars have been bent at 0.4 m from the face of support.

Step 11 : Check for the development length at support

(a) The minimum length of bar embedded into the simple support is ($L_d/3$)

$$L_d=\left(\frac{\sigma_{st}\cdot\phi}{4\ \tau_{bd}}\right)=\left(\frac{140\times10}{4\times0.6}\right)=583.33\ mm$$

$$\left(\frac{L_d}{3}\right)=\left(\frac{583.33}{3}\right)=194.44\ mm$$

Actual length of the bar embedded into the support (viz., width of support-end cover = 300 – 15 = 285 mm).

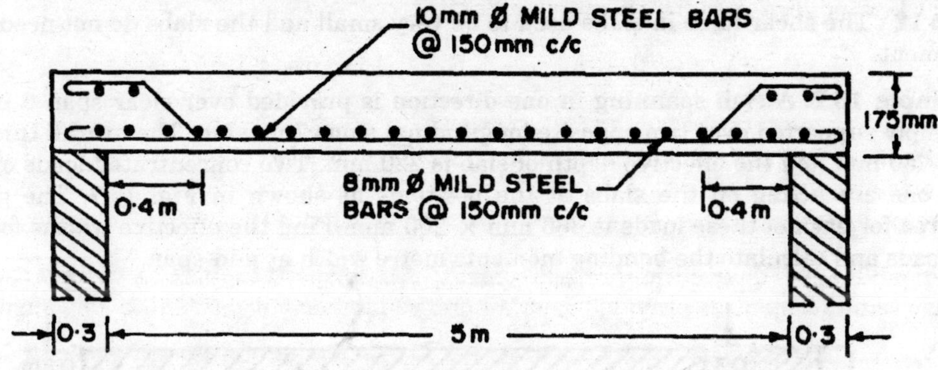

Fig. 16.6

(b) At simple supports or at point of zero moments, the following conditions of development length should also be satisfied. That is,

$$L_d \not> \left(\frac{M_1}{V} + L_o\right)$$

The moment of resistance of section at support for the available reinforcement in tension

$$M_1 = \sigma_{st} \cdot \left(\frac{A_{st}}{2}\right) . j . d$$

$$\text{or} \quad M_1 = \left(\frac{140 \times 1381.48 \times 0.867 \times 190}{2 \times 1000 \times 1000}\right) = 15.93\ kN - m$$

The shear force at support

$$V = \left(\frac{9.50 \times 5.18}{2}\right) = 24.605\ kN - m$$

$$\left(\frac{M_1}{V}\right) = \left(\frac{15.93}{24.605}\right) = 0.647\ m$$

The anchorage length beyond the centre of support

$$L_o = \left(\frac{b_s}{2} x' + 13\phi\right) = \left(\frac{\frac{300}{2} - 15 + 13 \times 10}{1000}\right) = 0.265\ m$$

$$\left(\frac{M_1}{V} + L_o\right) = (0.647 + 0.265) = 0.912\ m$$

L_d = 0.583 m does not exceed $\left(\frac{M_1}{V} + L_o\right) = 0.912\ m$. Hence, it is safe for the anchorage length.

Step 12 : The shear force in slabs used to be very small and the slabs do not need shear reinforcement.

Example 16.2. A slab spanning in one direction is provided over clear span 6 m. The slab is simply supported over two opposite longitudinal sides 7 m wide. The overall thickness of slab is 250 mm and the effective depth of slab is 220 mm. Two concentrated loads of 68 m kN each one are acting on the slabs at the positions as shown in Fig. 16.7. The ground contact area for each of these loads is 380 mm × 200 mm. Find the effective widths for each of these loads and calculate the bending moments/metre width at mid-span.

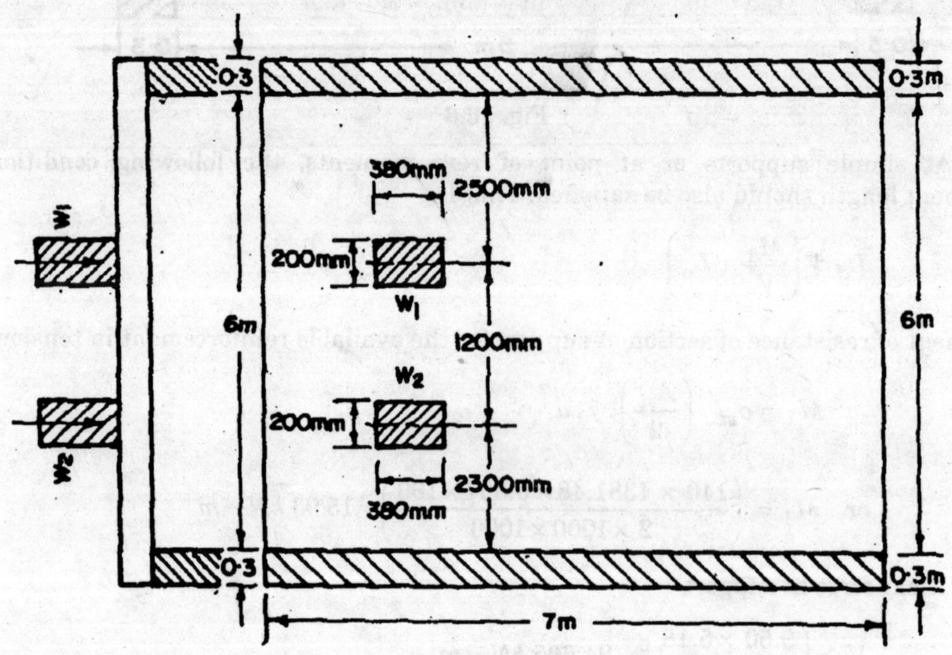

Fig. 16.7

Solution : The clear span for the slab is 6 m. It is worth to note that these given loads.

Step 1 : Effective span

The width of support may be assumed as 300 mm at each support. The effective span ES = l_{eff} may be found as follows. The slab is simply supported at both the supports.

$$ES = \left(clear\ span + \frac{1}{2} b_s + \frac{1}{2} b_s \right)$$

$$\text{or} \quad ES = \left(6 + \frac{1}{2} \times 0.3 + \frac{1}{2} \times 0.3 \right) = 6.3\ m \qquad \text{(i)}$$

where, b_s is the width of support. Also

ES = (clear span + effective depth)

or $ES = (6 + 0.220) = 6.22$ m (ii)

The effective span, l_{eff} is whichever is less. Therefore, l_{eff} is 6.22 m. The longitudinal side is supported edge and it is 7 m long. The ratio of length of longitudinal (supported) edge to the effective span

$$\left(\frac{l'}{l_{eff}}\right) = \left(\frac{7000}{6220}\right) = 1.1254 \quad \text{(iii)}$$

From IS : 456-2000, the value of k for simply supported slab, for above ratio

$$k = 2.48 \quad \text{(iv)}$$

Step 2 : Effective width of slab

The effective width of slab for two or more concentrated loads placed in a line in the direction of span is found from Eq. 16.4 for each load separately

$$b_{eff} = k \,.\, x \,.\left(1 - \frac{x}{l_{eff}}\right) + a \quad \text{(v)}$$

where, a is load width of the contact area measured parallel to supported edge and x is the distance of the centroid of concentrated load from nearer support.

2 (A). Effective width for concentrated, W_1

$x_1 = 2500$ mm and $a_1 = 380$ mm

$$b_{eff.1} = 2.48 \times 2500 \left(1 - \frac{2500}{6220}\right) + 380 \; mm$$

$$\text{or} \quad b_{eff.1} = \left(\frac{3708.04 + 380}{1000}\right) = 4.088 \; m \quad \text{(vi)}$$

2 (B). Effective width for concentrated load, W_2

$x_2 = 2300$ mm and $a_2 = 380$ mm

$$b_{eff.2} = 2.48 \times 2300 \left(1 - \frac{2300}{6220}\right) + 380 \; mm$$

$$\text{or} \quad b_{eff.2} = \left(\frac{3594.80 + 380}{1000}\right) = 3.595 \; m$$

Step 3 : Maximum bending moment in the slab at mid-span

The distance of mid-span from support

$$\left(\frac{l_{eff}}{2}\right) = \left(\frac{6220}{2}\right) = 3110 \; mm \qquad \text{(vii)}$$

3. (A) Maximum bending moment in the slab due to load W_1 per metre width

$$M_1 = \left(\frac{68 \times 1000}{6220} \times \frac{2500 \times 3110}{4.088} \times \frac{1}{10^6}\right) kN-m/m$$

or $M_1 = 20.793$ kN-m/metre (ix)

3 (B). Maximum bending moment of the slab due to load W_2 per metre width

$$M_2 = \left(\frac{68 \times 1000}{6220} \times \frac{2300 \times 3110}{3.595} \times \frac{1}{10^6}\right) kN-m/m$$

or $M_2 = 21.752$ kN/m/metre (x)

3 (C). Maximum bending moment due to self-weight the slab at mid-span, M

Self-weight of the slab

$$(1 \times 1 \times 0.250 \times 25) = 6.25 \; kN/m^2$$

$$M = \left(\frac{6.25 \times 6.22 \times 6.22}{8}\right) = 30.225 \; kN-m/m \qquad \text{(xi)}$$

Total maximum bending moment per metre width at the mid-span of slab

$$M_{total} = (M_1 + M_2 + M) = (20.793 + 21.752 + 30.225) = 72.770 \text{ kN-m/m}$$

Note : Now, the slab may be considered as if it consists of beams of one metre width. The effective depth of slab is 220 mm (given). The reinforcement may be calculated and repeated over the effective width, 4.088 m (found). The reinforcement for the self-weight for the remaining width of the slab (viz., 7 – 4.088 = 2.912 m) may be calculated.

In this method of design, it is assumed that each one metre of the slab spanning in one direction acts independently and Poisson's effect is neglected.

16.5 SLABS CONTINUOUS OVER WALLS

The slabs spanning in one direction (supported on all the four edges or supported on two opposite longitudinal edges may be continuous over more than two wall supports and which may cover more than one short spans as shown in Fig. 16.8 (A). The slab is continuous over short spans l_{x1}, l_{x2} and l_{x3}. The ratio of long span to short span for each panel is more than 2. Therefore, the slab is considered one way (slab spanning in one direction) continuous slab. This slab is designed as a continuous rectangular beam ABCD, Fig. 16.8. (A) having one metre width (1000 mm) in its cross-section as shown in Fig. 16.3 as a balanced section. The deflected shape of this continuous beam carrying uniformly distributed load is shown in Fig. 16.8 (B). The bending moment and shear force diagrams for the ABCD are shown in Figs. 16.8 (C) and (D), respectively. It is seen from Fig. 16.8 (C) that the bending moments are positive at mid-spans and negative at supports. The design of continuous beams has been described in Chapter.

The effective spans for various individual spans for different conditions of supports (for end span with end fixed and for end span with for end fee) are found as described in Art. and as recommended in IS : 456-2000.

The assumptions as regards relative stiffness and moment of inertia for the structural analysis (viz., for the determination of bending moments and shears) should be consistent for all members of the structures throughout the analysis.

The arrangement of live load over the various span for calculating the maximum values of bending moments and shears (at supports and at mid-spans) depends upon the influence lines for the beam. It has been described in detail in Art.

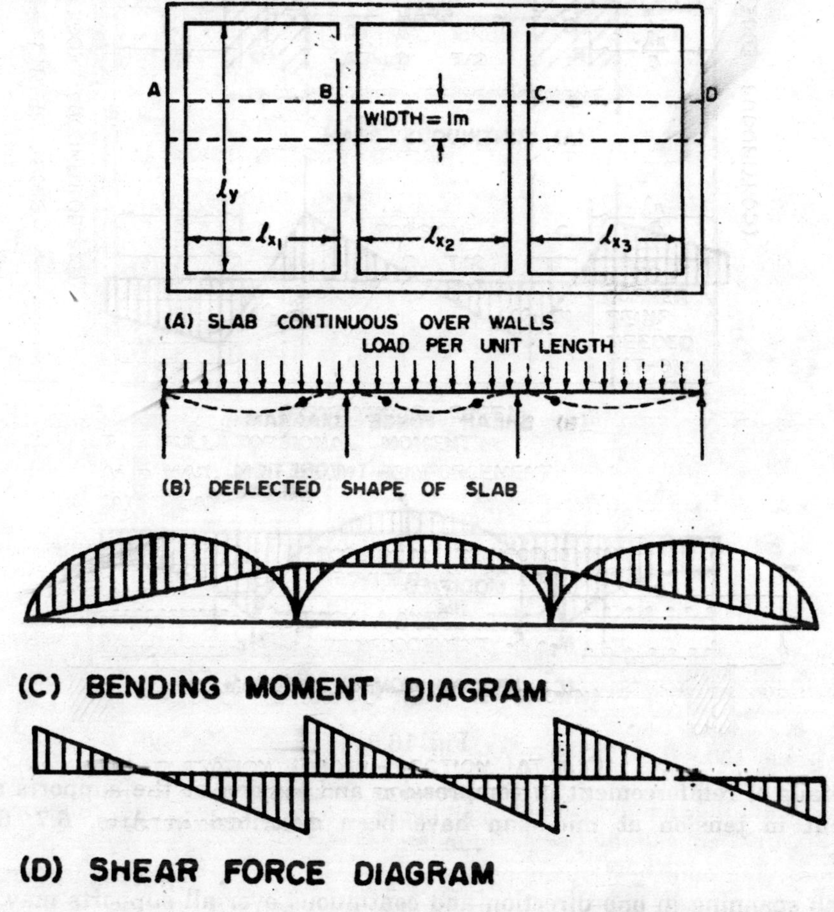

Fig. 16.8

The critical sections and coefficients for moments for the continuous beams as recommended by IS : 456-2000 have been described in Arts and, respectively. The span moments (near middle of end span and near middle of interior span) and at all other interior supports are found for the load acting cover the beam.

In actual construction, the support walls have definite width. In theoretical analysis, the support provides concentrated reactions along a line. The values of bending moment are modified due to the effect of support-width as shown in Fig. 16.9. It is seen that the values of bending moment at support reduce. The values of shear force remain maximum at the face of support. Therefore, it is not necessary to calculate the shear forces at the line of actions of support reactions.

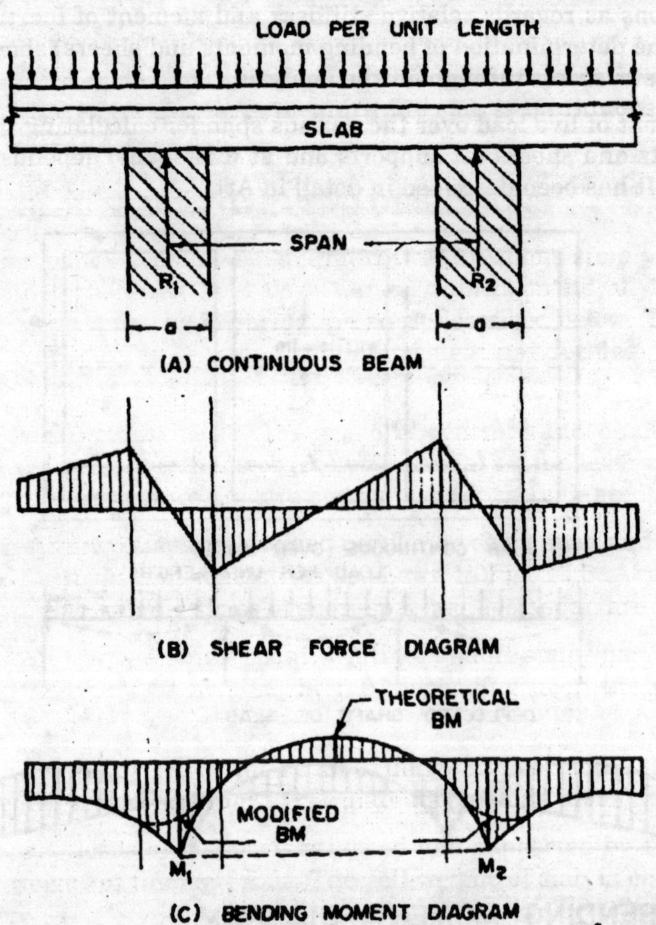

Fig. 16.9.

The details of reinforcement in compression and tension at the supports and those of reinforcement in tension at mid-span have been described in Arts. 6.7, 6.8 and 6.9, respectively.

The slab spanning in one direction and continuous over all supports may be designed following the various steps described above.

Example 16.3. Three garages are to be built side by side with masonry walls as shown in Fig. 16.8. The clear dimensions of each garage are 4 m and 10 m. A continuous reinforced concrete slab is to be provided to roof the garages. A finishing surface 20 mm thick shall be provided over the roof. The roof shall be used as the residential floor. Design the slab.

Design: The ratio of long span to short span is (10/4 = 2.5) for slab over each garage. As such the slab is designed as slab spanning in one direction (viz., one way slab) over the clear span of 4 m. The slab is designed as a continuous beam *ABCD* of 1 m width as shown in Fig. 16.8 and as balanced section following the steps described below:

Step 1: Design constant

The design constant for M 20 grade of concrete and hysd steel reinforcement for the balanced section of the beam are as under. From IS: 456-1978, for M 20 grade of concrete and hysd-steel

$$\sigma_{abc} = 7 N/mm^2 \quad and \quad \sigma_{st} = 230\ N/mm^2$$

Modular ratio for M 20 grade of concrete

$$m = \left(\frac{280}{3 \times 7}\right) = 13.33$$

Depth factor for the neutral axis

$$k_B = \left(\frac{13.33 \times 7}{13.33 \times 7 + 230}\right) = 0.2886$$

Lever arm factor for beam

$$j_B = \left(1 - \frac{0.2886}{3}\right) = 0.9038$$

Factor for the moment of resistance

$$R_B = \left(\frac{1}{2} \times 7 \times 0.2886 \times 0.9038\right) = 0.9129$$

$$R_B = \left(\frac{1}{2} \times \frac{7 \times 0.2866}{230}\right) \times 100 = 0.4392 \text{ per cent}$$

Step 2: Estimation of thickness of slab

Thickness of slab is estimated from span to effective depth ratio specified for the satisfactory control of the vertical deflection in IS: 456-2000. Therefore, the effective depth, *d* for the continuous slab

$$d = \left(\frac{\text{span}}{26 \times \text{modification factor}}\right)$$

For M 20 grade of concrete and hysd steel, the percentage ratio of steel for the balanced section of the beam, p_B is 0.4392 per cent and therefore, the modification factor, from IS : 456-2000 IS 1.2. Therefore,

$$d = \left(\frac{4000}{26 \times 1.2}\right) = 128.21\ mm$$

Providing effective cover of 25 mm, the overall thickness of the slab may be kept as (128.21 + 25) = 153.2 mm say 160 mm. Therefore, the effective depth of slab shall be (160–25) = 135 mm.

Step 3 : Effective span

The continuous slab is designed as continuous beam over each clear span of 4000 mm. The width of supporting walls is assumed as 450 mm. Therefore, the effective spans are as follows :

End span with one end free (simply supported)

The width of support is eider than

$$\left(\frac{span}{12}\right) = \left(\frac{4000}{12}\right) = 333.33\ mm$$

$$\text{ES} = \left(C.S + \frac{1}{2} \times d\right) = \left(4000 + \frac{1}{2} \times 135\right) = 4067.5\ mm$$

$$\text{or ES} = \left(CS + \frac{1}{2} \times 450\right) = 4225\ mm$$

$$\therefore \text{ES} = 4.0675\ \text{m (whichever is less)}$$

Intermediate span

The effective shall be clear span between the supports

$$\text{ES} = \text{CS} = 4.00\ \text{m}$$

Step 4 : Load

The dead load consists of weight of finishing surface and the self-weight of the slab. These loads are calculate for unit area of the slab.

Weight of finishing surface

$$(1 \times 1 \times 0.020) \times 24 = 0.48\ kN/m^2$$

Self-weight of the slab

$$(1 \times 1 \times 0.160) \times 25 = 4.00\ kN/m^2$$

Live load for the floor for residential building

$$(1 \times 1 \times 2) = 2.00\ kN/m^2$$

Total load per unit area

$$6.48 \ kN/m^2 \ say = 6.50 \ kN/m^2$$

The intensity of uniformly distributed load per metre length along the span is 6.5 kN/m as shown in Fig. 16.10.

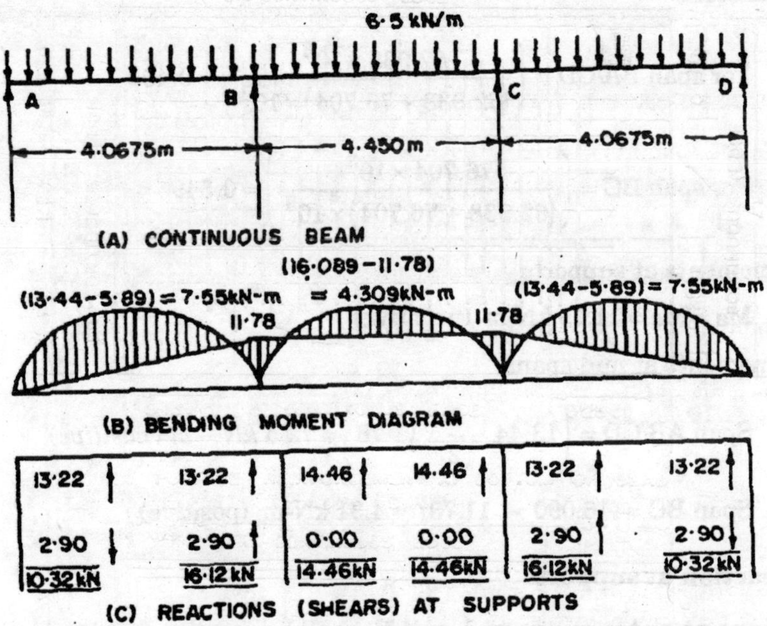

Fig. 16.10

Step 5 (A). Bending moments (By moment distribution method)

The analysis of continuous beam ABCD, Fig. 16.10 for calculating the bending moments is carried out by the method of moment distribution. The width of beam section for the continuous beam ABCD is 1000 mm. The relative stiffness of the member is based on the moment of inertia of the section determined on the basis of *gross section* (cross-section of the member ignoring the reinforcement).

Relative stiffness of end spans

$$\left(\frac{I}{l}\right) = \left(\frac{1}{12} \times \frac{1000 \times 160^3}{4067.5}\right) = 83.917 \times 10^3 \ mm^3$$

Relative stiffness of intermediate span. In this portion the effective span is taken as the distance between centre of supports

$$\left(\frac{I}{l}\right) = \left(\frac{1}{12} \times \frac{1000 \times 160^3}{4450}\right) = 76.704 \times 10^3 \ mm^3$$

End supports are hinged supports. Therefore, the relative stiffness of the end members

$$\left(\frac{3}{4}\times\frac{l}{l}\right)=\left(\frac{3}{4}\times 83.917\times 10^3\right)=62.938\times 10^3\ mm^3$$

Distribution factors

$$\text{For span BA/CD} = \left(\frac{62.938\times 10^3}{(62.938+76.704)\times 10^3}\right)=0.451$$

$$\text{For span BC} = \left(\frac{76.704\times 10^3}{(62.938+76.704)\times 10^3}\right)=0.549$$

Bending moments at supports

$M_B = M_C = 11.78$ kN-m (negative)

Bending moments at mid spans

$$\text{Span AB/CD} = \left(13.44-\frac{1}{2}\times 11.78\right)=7.55\ kN-m\ (positive)$$

Span BC = (16.090 − 11.78) = 4.31 kN-m (positive)

Step 5 (B). Reaction at supports

(i) **Reactions at outer support A and B** shall be reaction for simply supported span + reaction for the moment at the intermediate support B. Therefore,

$$R_A=R_D=\left(\frac{7.5\times 4.0675}{2}-\frac{11.78}{4.6075}\right)=10.32\ kN$$

Reactions at intermediate support B/C shall be reaction for simply supported spans AB and BC, + reaction for the moment at the intermediate support B - reaction for the difference of moments at supports B and C. Therefore,

$$R_B=R_C=\left(\frac{6.5\times 4.0675}{2}+\frac{6.5\times 4.450}{2}+\frac{11.78}{4.0675}+0\right)=30.58\ kN$$

R_{BA} = 16.12 kN and R_{BC} = 14.46 kN

Note : As per the procedure described in IS : 456-2000, the bending moments and shear forces (reactions at support) are found by utilising the coefficients given in tables.

For spans AB and CD : Total design dead loads

$$W_{D.L} = (4.48\ \times 4.0675) = 18.22\ kN$$

Total design live loads

$$W_{L.L}=(2\times 4.0675)=8.135\ kN$$

For span BC (Effective span = 4 m) : Total design dead load

$$W_{D.L} = (4.48 \times 4) = 17.92 \ kN$$

Total design live load

$$W_{L.L} = (2 \times 4) = 8 \ kN$$

Step 6 : Support moments

(i) Bending moment due to dead load

At support next to the end support

$$M_{BA} = M_{CD} = \left(-\frac{1}{10} \times 18.22 \times 4.0675 \right) = -7.411 \ kN-m$$

$$M_{BC} = M_{CB} = \left(-\frac{1}{10} \times 17.92 \times 4 \right) = -7.168 \ kN-m$$

(ii) Bending moment due to imposed live load

At support next to the end support

$$M_{BA} = M_{CD} = \left(-\frac{1}{9} \times 8.135 \times 4.0675 \right) = -3.676 \ kN-m$$

$$M_{BC} = M_{CB} = \left(-\frac{1}{9} \times 8 \times 4 \right) = -3.555 \ kN-m$$

(iii) Total support moment

$$M_{BA} = M_{CD} = (-7.411 - 3.686) = -11.087 \ kN-m$$

$$M_{BC} = M_{CB} = (-7.168 - 3.555) = -10.723 \ kN-m$$

Step 7 : Span moments

(i) Bending moment due to dead load

Near middle of end span

$$\text{Span AB/BC} = \left(\frac{1}{12} \times 18.22 \times 4.0675 \right) = +6.176 \ kN-m$$

At middle of the interior span

$$\text{Span BC} = \left(\frac{1}{24} \times 17.92 \times 4 \right) = +2.987 \ kN-m$$

(ii) Bending moment due to imposed live load

Near middle of the end span

$$\text{Span AB/BC} = \left(\frac{1}{10} \times 8.135 \times 4.0675\right) = +3.309\ kN-m$$

At middle of the interior span

$$\text{Sapn BC} = \left(\frac{1}{12} \times 8 \times 4\right) = +2.667\ kN-m$$

(iii) Total span moments

Span AB/BC = (6.176 + 3.309) = +9.485 kN-m

Span BC = (2.987 + 2.667) = + 5.654 kN-m

It is to note that the span moments found by using coefficients given in IS : 456-2000 are more than those found by moment distribution method.

Step 8 : Shear forces

(i) Shear force due to dead load

$$\text{At end support} = (0.4 \times 18.22) = 7.288\ kN$$

At support next to the end support

$$\text{Outer side} = (0.6 \times 17.92) = 10.752\ kN$$

$$\text{Inner side} = (0.55 \times 17.92) = 9.856\ kN$$

(ii) Shear force due to imposed live load

$$\text{At end support} = (0.45 \times 8.135) = 3.661\ kN$$

At support next to the end support

$$\textit{Outer side} = (0.6 \times 8) = 4.8\ kN$$

$$\textit{Inner side} = (0.6 \times 8) = 4.8\ kN$$

(iii) Total shear force

At end support = (7.288 + 3.661) = 10.949 kN

At support next to the end support

Outer side = (10.752 + 4.8) = 15.552 kN

Inner side = (9.856 + 4.8) = 14.656 kN

It is also to note that the shear forces found by using the coefficient given in IS : 456-2000 are generally more than those found by moment distribution method.

Step 9 : Moments at the face of supports

As per IS : 456-2000, for monolithic construction, the moments calculated at the face of supports are to be used in the design of members at those sections. The moments at the face of supports are calculated by approximate calculations. The moments at the face of supports are found as follows :

At support B (outer side)

$$M_{BA} = \left(\frac{11.78 - 16.12 \times 0.452}{2}\right) = 8.137\ kN - m\ (negative)$$

At support B (inner side)

$$M_{BC} = \left(\frac{11.78 - 14.46 \times 0.452}{2}\right) = 8.512\ kN - m\ (negative)$$

Step 10 : Check for overall thickness of slab

Absolute maximum bending moment is 11.78 kN-m. The effective depth required for the slab to resist this bending moment

$$d = \left(\frac{11.78 \times 10^6}{0.9129 \times 1000}\right)^{1/2} = 113.595\ mm$$

Effective cover assumed = 25 mm

Overall thickness 138.595 mm required is less than the estimated thickness 160 mm (adopted).

Step 11 : Reinforcement for the slab

(i) Tension reinforcement at bottom face

At the middle of end span for positive bending moment calculated using coefficient from IS : 456-2000.

$$A_{st} = \left(\frac{9.485 \times 10^6}{230 \times 0.9038 \times 135}\right) = 337.99\ mm^2$$

At the middle of interior span

$$A_{st} = \left(\frac{5.654 \times 10^6}{230 \times 0.9038 \times 135}\right) = 201.48\ mm^2$$

(ii) Tension reinforcement at top face

At the top of supports for negative bending moment calculated using co-efficients from IS : 456-2000 (moments at the face of supports)

$$A_{st} = \left(\frac{8.512 \times 10^6}{230 \times 0.9038 \times 135}\right) = 303.318\ mm^2$$

∴ Provide 10 mm diameter hysd-steel

$A_\phi = 78.54\ mm^2$

Spacing of the steel bars

$$s = \left(\frac{1000 \times 78.4}{337.99}\right) = 232.37\ mm$$

10 mm diameter bars are provided at 200 mm spacing centre to centre at the bottom surface (in the end span and in the interior span) and top surface (at the top of supports as shown in Fig. 16.11).

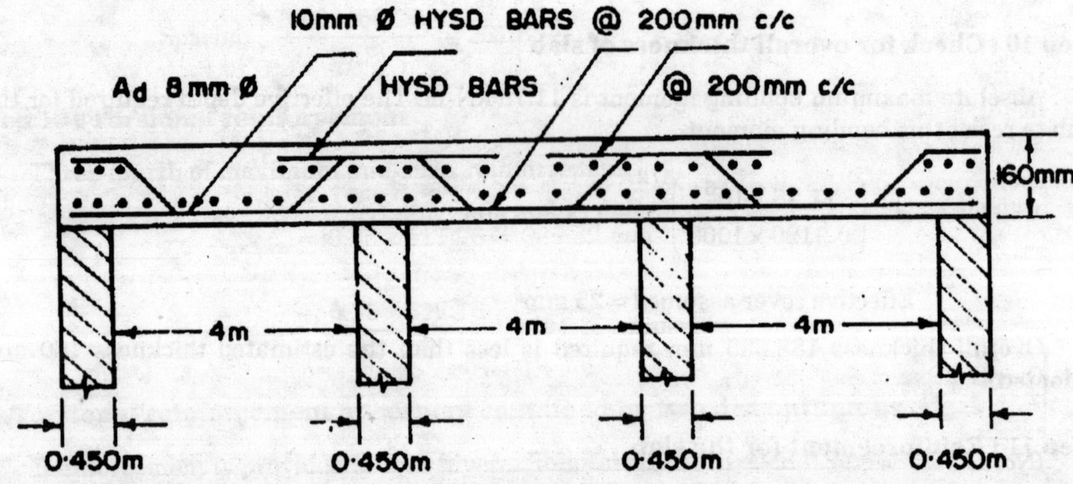

Fig. 16.11

Step 12 : Distribution (minimum) reinforcement in the slab

Distribution (temperature and shrinkage) reinforcement equal to 0.12 per cent of the total cross-sectional area is provided in the direction perpendicular to the main reinforcement (viz., parallel to the top long span of the slab). Therefore,

$$A_d = \left(\frac{0.12}{100} \times 1000 \times 160\right) = 192\ mm^2$$

Provide 8 mm diameter hysd-bars. Spacing for the distribution reinforcement

$$A_{\phi d} = 50.265\ mm^2$$

$$s_d = \left(\frac{1000 \times 50.265}{192}\right) = 261.799\ mm$$

8 mm diameter bars are provided at 200 mm c/c as shown in Fig. 16.11.

The nominal shear stress used to be very small and the slab does not need shear reinforcement.

16.6 SHEAR IN SLABS

The shear stress in slabs is not commonly very critical. The nominal shear stress, τ_v is calculated from Eq. 3.28 (B) as

$$\tau_v = \left(\frac{V}{b\,.\,d}\right) \tag{i}$$

where V is the shear force at the critical section, b is the width of slab under consideration, and d is the effective depth.

Professor Taylor, H.P.J. in his paper titled as `shear strength of large beams' published in Proc. ASCE, 98, No. ST-11, November, 1972 has shown by tests that shallow member fail at higher nominal shear stress. For slabs of overall thickness 300 mm or less, the allowable shear stress in solid slabs is obtained as $k\,.\,\tau_c$, where k is the modified shear stress coefficient, which varies linearly from k = 1.00 for a slab thickness of 300 mm or more to k = 1.30 for a slab thickness of 150 mm or less. (other values of k are given in Table 16.2).

TABLE 16.2. VALUES OF MODIFIED SHEAR STRESS COEFFICIENT k (as per IS : 456-2000)

Overall depth of slab (mm)	300 or more	275	250	225	200	175	150 or less
k	1.00	1.05	1.10	1.15	1.20	1.25	1.30

(Note : This does not only apply to flat slabs).

The values of t_c (shear strength of plain cement without shear reinforcement) may be noted from Table 16.2 or from IS : 456-2000.

The failure of slab in shear is likely to occur at the face of contact of the support. In case the resting face of the slab is different from the load face, the shear failure may occur at a distance, d from the face of contact supports.

The use of shear reinforcement in slabs less than 200 mm thickness is considered impracticable. The nominal shear stress in the slabs should not exceed half the allowable shear stress. This restriction is reasonable for slabs subjected to concentrated loads.

The depth of slab is commonly adjusted so that the requirement for shear is satisfied.

PROBLEMS

16.1 Design a simply supported slab over a class-room of size 3.5 × 8 *m*. A finishing surface of cement concrete of 20 mm shall be provided over the slab. The slab shall be used as a class room floor. M 20 grade of concrete and hysd-bars shall be used.

16.2 A one way slab is provided over a clear span 4.8 m. The slab is simply supported over two opposite longitudinal sides 7 m wide. The overall thickness of slab is 230 mm and the effective thickness of slab is 200 mm. Two concentrated loads of 68 kN each are acting on the slab in one line parallel to unsupported edge at a distance of 1500 mm the distances upto the centroid of the loads from the supported one edge is 2000 mm and 3200 mm, respectively. The ground contact area for each of these loads is 380 mm × 200 mm. Find the effective widths for each of these loads and calculate the bending moments per metre width at mid-span.

16.3 Three shops are to be constructed side by side with masonry walls. The clear dimensions of each shop are 3.20 m and 6.80 m. A continuous reinforced concrete slab is to be provided to roof the shops. A finishing surface 20 mm thick shall be provided over the roof. The roof shall be used as the residential floor. Design the slab.

17

DESIGN OF SLABS SPANNING IN TWO DIRECTIONS (SSTD)

17.1 INTRODUCTION

The reinforced concrete slabs supported on all its four sides on beams or walls having the ratio of long span to short span less than or equal to 2 (two) are called as **slabs spanning in two directions** or referred as *two way slabs.* The structural action in such slabs is two way. The loads are carried by the slab along short span and long span both. The deflected surface of such slabs has double curvature. These slab resist bending moments along both the spans. These slabs have reinforcement in two perpendicular directions parallel to the supporting edges. The reinforcement (steel bars) in each direction is considered as main reinforcement. The thickness of two-way designed slab shall be smaller than the thickness of slab designed as one-way designed slab for similar loading and similar support conditions. Therefore, these slabs are more economical. These slabs are rectangular slabs. These slabs are frequently used specially in large, continuous monolithic floor systems.

17.2 BEHAVIOUR OF SLABS SPANNING IN TWO DIRECTIONS

The rectangular slab spanning in two directions having supported edges enough stiff (i.e., which may be considered as unyielding) carrying uniformly distributed load deforms into a dished surface rather than a cylindrical surface as in case of one-way slab. Such slab bends in both principal directions. Therefore, the bending moments develop in both the directions. The values of bending moments remains proportional to curvatures.

These slabs are highly indeterminate. These slabs have complex behaviour a differential equation of four order arises in case these slabs are analysed by classical theory. This differential equation may be solved for vertical deflections and the stresses. However, the solution of such differential equation is very tedious. Therefore, approximate method (which gives satisfactory results when such slabs are analysed and designed) are used for the analysis of such slabs.

A slab spanning in two directions simply supported on its all the four edges is shown in Fig. 17.1 when this slab is loaded with uniformly distributed load and the load is increased till the failure is reached, and the cracks as shown by dotted lines develop at its bottom surface due to positive (*viz.,* sagging) moments. The bending of slab near the corner (e.g., A) occur along the line X_1Y_1 perpendicular to crack EA, whereas the bending of slab in the centre occurs along X_2Y_2 perpendicular to crack EF. In ultimate method of design, the method is based on the pattern of the cracks (*viz.*, yield lines).

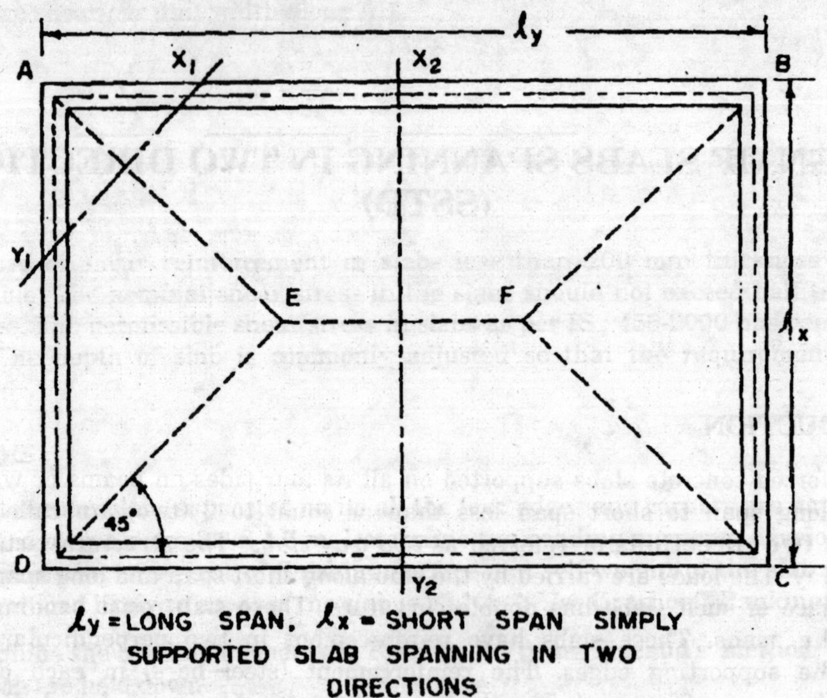

Fig. 17.1

17.3 GRASHOFF-RANKINE METHOD

Grashoff-Rankine proposed on elementary or simplified approach of analysis of a highly indeterminate system of two-way slab having simply supported edges (which may be considered unyielding). In this approximate method, the complete system is considered as equivalent simple determinate system of slab carrying uniformly distributed load w per unit area. The flexural performance of the slab may be visualized by considering the simply supported slab consisting of series of interconnected longitudinal and lateral strips (viz., two mutually perpendicular intersecting slab strips). Fig. 17.2 shows two central strips of unit width of the two way rectangular slab with l_x and l_y as short span and long span, respectively. These strips AB and CD act approximately as simple beam. The strip AB carries uniformly distributed load w_y per unit length along long direction and the strip CD carries uniformly distributed load w_x per unit length along short direction. The load w is shared by these two strips such that the deflections of two central strips at the point of intersection are equal. These imaginary strips are actual part of the same monolithic slab. The actions of adjacent strips to these strips are neglected. Therefore,

$$w = (w_x + w_y) \tag{i}$$

and $$y_{AB} = y_{CD} \tag{ii}$$

$$\therefore \frac{5}{384}\cdot\left(\frac{w_y \cdot l_y^4}{EI}\right)=\frac{5}{384}\cdot\left(\frac{w_x \cdot l_x^4}{EI}\right) \qquad \text{(iii)}$$

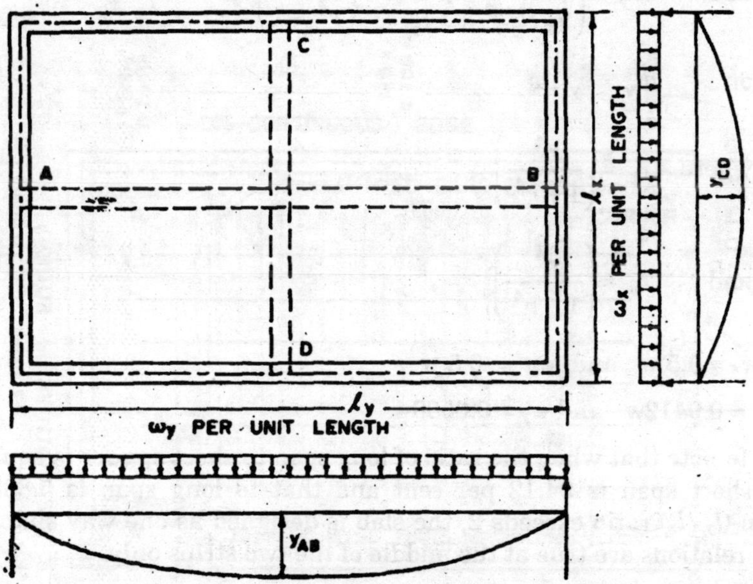

Fig. 17.2

It is assumed that the moments of inertias I_x and I_y are equal (*viz.*, $I_x = I_y = I$) and there is no torsion.

From the expression (iii)

$$\left(w_x \cdot l_x^4\right) = \left(w_y \cdot l_y^4\right)$$

$$\text{or} \quad w_x = w_y \cdot \left(l_y / l_x\right)^4 \qquad \text{(iv)}$$

From the expressions (i) and (iv)

$$w_x = \left(\frac{l_y^4}{l_y^4 + l_x^4}\right) \cdot w \qquad (17.1)$$

$$\text{and } w_y = \left(\frac{l_x^4}{l_y^4 + l_x^4}\right) \cdot w \qquad (17.2)$$

Let the ratio of long span to short span (l_x / l_y) be r. Then, from Eqs. 17.1 and 17.2

$$w_x = \left(\frac{1 \cdot w}{1 + \left(l_x / l_y\right)^4}\right) = \left(\frac{1}{1 + (1/r)^4}\right) \cdot w \qquad \text{(v)}$$

or $w_x = r_x \cdot w$ (17.3)

and $w_y = \left(\frac{1}{(l_x/l_y)^4+1}\right) \cdot w = \left(\frac{1}{r^4+1}\right) \cdot w$ (vi)

or $w_y = r_y \cdot w$ (17.4)

where $r_x = \left(\frac{r^4}{1+r^4}\right)$ (17.5)

and $r_y = \left(\frac{r^4}{1+r^4}\right)$ (17.6)

when, r = 1, $w_x = 0.5$ w and $w_y = 0.5$ w

when, r = 2, $w_x = 0.9412$w and $w_y = 0.0588$w

It is worth to note that when the ratio of long span to short span, *r* is equal to 2, the load transferred to short span is 94.12 per cent and that to long span is 5.88 per cent only. Therefore, when (l_y/l_x) ratio exceeds 2, the slab is designed as one way slab. It is to further note that above relations are true at the middle of the two strips only.

The maximum bending moment at the middle of the strip CD, Fig. 17.2 (*viz.*, along short span)

$$M_x = \left(\frac{w_x \cdot l_x^2}{8}\right) = \left(\frac{l_y^4}{l_y^4 + l_x^4}\right) \cdot \left(\frac{w \cdot l_x^2}{8}\right)$$

or $$M_x = \left(\frac{(l_y/l_x)^4}{(l_y/l_x)^4+1}\right) \cdot \left(\frac{w \cdot l_x^2}{8}\right)$$

or $M_x = \frac{1}{8}\left(\frac{r^4}{1+r^4}\right) w \cdot l_x^2$ (17.7A)

or $M_x = \alpha_x \cdot w \cdot l_x^2$ (17.7B)

Where $$\alpha_x = \frac{1}{8} \cdot \left(\frac{(l_y/l_x)^4}{(l_y/l_x)^4+1}\right) = \frac{1}{8}\left(\frac{r^4}{1+r^4}\right)$$

The maximum bending moment at the middle of the strip AB, Fig. 17.2 (*viz.*, along span)

$$M_y = \left(\frac{w_y^2 \cdot l_x^2}{8}\right) = \left(\frac{l_x^4}{l_y^4 + l_x^4}\right) \cdot \left(\frac{w \cdot l_y^2}{8}\right)$$

$$M_y = \frac{1}{[(l_y/l_x)^4 + 1]} \cdot \frac{w}{8} \cdot l_x^2 \cdot \left(\frac{l_y}{l_x}\right)^2$$

$$M_y = \frac{1}{8} \cdot \left(\frac{r^2}{1+r^4}\right) \cdot w \cdot l_x^2 \quad (17.8A)$$

$$M_y = \alpha_y \cdot w \cdot l_x^2 \quad (17.8B)$$

The values of α_x *and* α_y may be tabulated as in Table 17.1.

TABLE 17.1. COEFFICIENTS FOR BENDING MOMENTS FOR SLABS (SPANNING IN TWO DIRECTIONS AT RIGHT ANGLES) SIMPLY SUPPORTED ON FOUR SIDES

(l_y/l_x)	1.0	1.1	1.2	1.3	1.4
α_x	0.062	0.074	0.084	0.093	0.099
α_y	0.062	0.061	0.059	0.055	0.951
(l_y/l_x)	1.5	1.75	2.0	2.5	3.0
α_x	0.104	0.113	0.118	0.122	0.124
α_y	0.046	0.037	0.029	0.020	0.014

It is worth to note that the coefficients for bending moment α_x and α_y for slabs (spanning in two perpendicular directions at right angles) simply supported on four sides based on Grashoff and Rankine theory have been adopted and specified in IS : 456-2000 for design of such slabs. It is further to note that there is no provision for torsion.

17.4 RESTRAINED SLABS

The corners of a two-way slab (if these are not prevented from lifting) have a tendency to lift. In such slabs provisions are not kept to resist torsion at corners. When the corners of the slab are prevented from lifting, the slab is called as *restrained slab*. In the restrained slab, the maximum bending moment and deflection are smaller than those for the simply supported slabs where the corners are not held down. The behaviour of restrained slab carrying uniformly distributed load is identical to thin elastic plate. In restrained slab, it is essential to provide adequate provisions to resist torsion at corners, since torsion develops at the corners.

The slab is supported along all the four edges. The edges of slab may be *simply supported, fixed, partially fixed and continuous.* At least the edge is not kept free, since the free edge does not have any support (for example, free edge of a cantilever slab). The simply supported edges have wall supports. The simply supported edges do not resist any bending moments and permit free rotations without any bending restraint. Such slab edges are also referred as *discontinuous edges*. The fixed edges resist the bending moments developed and do not allow rotations. The continuous edges continue over adjacent spans. In the continuous edges, rotations occur and these edges resist bending moments. The slopes at the common support depend upon the loading on the adjacent panels. In the continuous beams, the slopes on two sides remain equal and the rotations also remain equal though they are of opposite sign. The fixed and continuous edges both are referred as *continuous edges.*

Two-way slabs are classified depending on continuity and discontinuity of long edges or short edges in nine standard cases as follows as shown in Fig. 17.3. The continuous and fixed edges are marked as continuous (letter C is used) by thick lines. The simply supported edges are marked as discontinuous (letters DC are used) by thin lines.

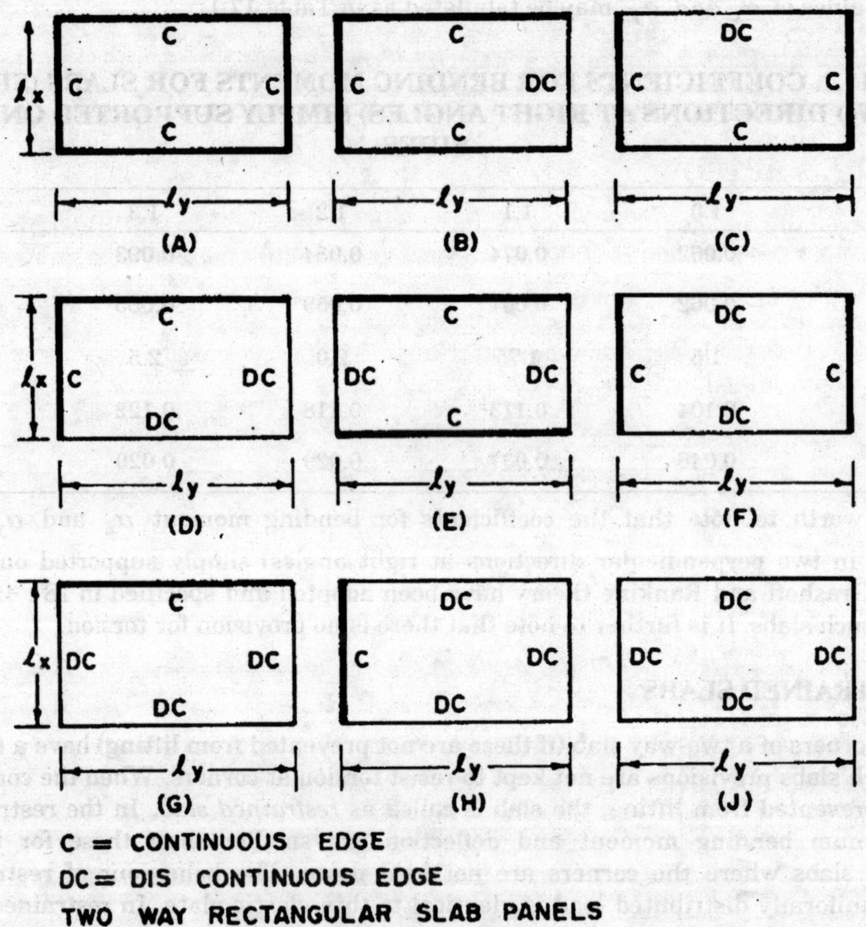

Fig. 17.3

1. All the four edges of two-way slab are fixed/continuous. Fig. 17.3 (A).

2. One short edge of the slab is discontinuous (three edges continuous), Fig. 17.3 (B).

3. One long edge of the slab is discontinuous (three edges continuous), Fig. 17.3 (C).

4. Two adjacent edges of the slab are discontinuous (two adjacent edges continuous), Fig. 17.3 (D).

5. Two short edges of the slab are discontinuous (two long edges continuous), Fig. 17.3 (E).

6. Two long edges of the slab are discontinuous (two short edges continuous), Fig. 17.3 (F).

7. Three edges of the slab are discontinuous (one long edge continuous), Fig. 17.3 (G).

8. Three edges of the slab are discontinuous (one short edge continuous), Fig. 17.3 (H).

9. All the four edges of the slab are continuous, Fig. 17.3 (J).

Two -way restrained slabs are analysed and designed by the following methods :

1. Marcus method. 2. Pigeaud's method. 3. IS : code method.

These methods have been described in the subsequent articles.

17.5 MARCUS METHOD

The bending moments calculated for two-way slab by Grashoff-Rankine method (in which torsion is not considered) are more than the bending moments in slab when determined by considering the torsion. Dr. H. Marcus of Germany proposed a simplified and approximate approach to find bending moments by considering the necessary *torsional effects.* The values of bending moments obtained by Marcus method vary by only 2 percent from those determined by rigorous elastic plate theory.

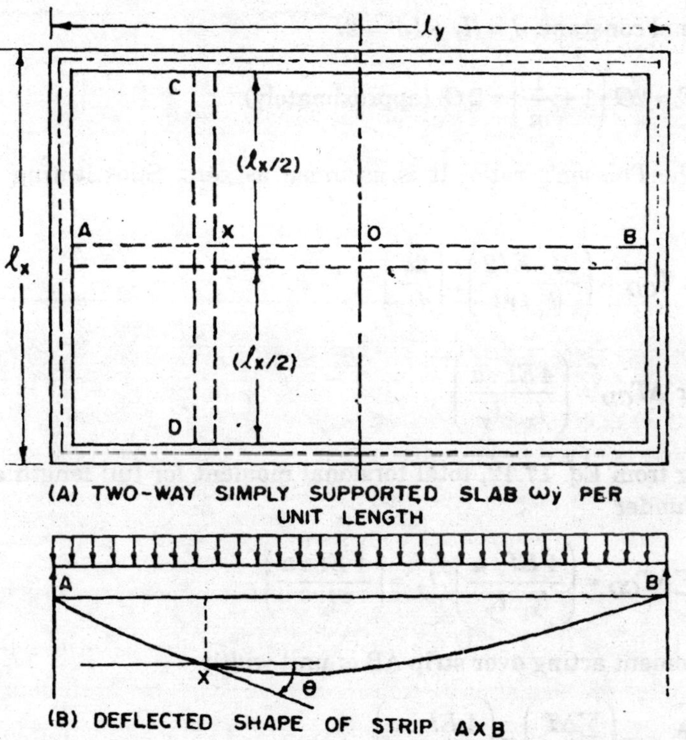

(A) TWO-WAY SIMPLY SUPPORTED SLAB ωy PER UNIT LENGTH

(B) DEFLECTED SHAPE OF STRIP AXB

Fig. 17.4.

A simply supported rectangular slab is shown in Fig. 17.4. Consider strip AXB of unit width, AXB, Fig. 17.4 (B) shows the deflected shape of the strip along long span. The deflected shape of strip AXB is assumed parabolic. Equation of this parabolic shape with A as the origin is follows.

$$y = \frac{4a}{l_y^2} . x . (l_y - x) \tag{17.9}$$

By differentiating Eq. 17.9 with respect to x

$$\left(\frac{dy}{dx}\right) = \frac{4a}{l_y^2} . (l_y - 2\,x) \tag{17.10}$$

The slope at distance $x = l_y/4$, θ may be obtained from Eq. 17.10. Therefore,

$$\theta = \frac{4a}{l_y^2}\left(l_y - 2 . \frac{l_y}{4}\right) = \left(\frac{2\,.\,a}{l_y}\right) \tag{17.11}$$

This slope θ at X of the strip AB at ($x = l_y/4$) twists the strip CD by an angle ($2\,.\,a/l_y$). From theory of torsion, the torsion in strip CD.

$$\left(\frac{\Delta\, T_{CD}}{J}\right) = \left(\frac{G.\theta}{(l_x/2)}\right) \tag{i}$$

where, J = torsional constant, $J = (I_x + I_y) = 2I$

$$E = 2G\left(1 + \frac{1}{m}\right) = 2\,G \text{ (approximately)} \tag{ii}$$

where (1/m) is the Poisson's ratio. It is assumed as zero. Substituting for J and G in the expression (i)

$$\Delta\, T_{CD} = \left(\frac{2I\,.\,E/2}{(l_x/2)}\right) . \left(\frac{2a}{l_y}\right)$$

$$\text{or } \Delta T_{CD} = \left(\frac{4EI\;.\;a}{l_x\;.\;l_y}\right) \tag{17.12}$$

By summing from Eq. 17.12, total torsional moment for full length along long span AB may be found as under

$$\Sigma \Delta T_{CD} = \left(\frac{4\,EI\;.\;a}{l_x\;.\;l_y}\right) . \, l_y = \left(\frac{4\,EI\;.\;a}{l_x}\right) \tag{iii}$$

Torsional moment acting over strip AB of unit width

$$T_{AB} = \left(\frac{\Sigma \Delta T}{l_y}\right) = \left(\frac{4\,EI\;.\;a}{l_x^2}\right) \tag{17.13}$$

From Grashoff-Rankine method, the uniformly distributed load acting along AB, from Eq. 17.2

$$w_y = \left(\frac{l_x^4}{l_y^4 + l_x^4}\right).w \quad \text{(iv)}$$

The maximum bending moment occurring at the middle of the strip AB

$$M_y'' = \left(\frac{w_y \cdot l_y^2}{8}\right) \quad \text{(v)}$$

The maximum deflection occurring at the middle of the strip AB

$$a = \frac{5}{384}\left(\frac{w \cdot l_y^4}{EI}\right) \quad \text{(vi)}$$

The torsion at middle of the strip AB may be found by substituting the values from the expressions (iv), (v) and (vi) in Eq. 17.13

$$T_m = \left(\frac{4EI}{l_x^2}\right) \cdot \left(\frac{5}{384}\right) \cdot \left(\frac{w_y \cdot l_y^2}{8}\right) \cdot \frac{8}{1} \cdot \frac{l_y^2}{EI} \quad \text{(vii)}$$

$$T_m = \frac{5}{6} \cdot \left(\frac{M_y''}{2}\right) \cdot \left(\frac{l_y^2}{l_x^2}\right) \quad \text{(17.14)}$$

Dr. H. Marcus arbitrarily increased the value of torsional moment, T_m at the middle of strip AB along long span by a ratio ($2w_y/w$) and the final value of torsional moment at the middle of the strips AB is obtained as below :

$$T_{mf.AB} = \frac{5}{6} \cdot \left(\frac{w_y}{w}\right) \cdot \left(\frac{l_y}{l_x}\right)^2 \cdot M_y'' \quad \text{(viii)}$$

The bending moments at the middle of the strips AB and CD are as under

$$M_y'' = \left(\frac{w_y \cdot l_y^2}{8}\right) \; and \; M_x'' = \left(\frac{w_x \cdot l_x^2}{8}\right) \quad \text{(ix)}$$

$$\text{Let } M_y' = \left(\frac{w \cdot l_y^2}{8}\right), \; and \; M_x' = \left(\frac{w \cdot l_x^2}{8}\right) \quad \text{(x)}$$

From the expressions (ix) and (xii)

$$\frac{M_y''}{M_x''} = \left(\frac{w_y}{w}\right) \cdot \left(\frac{l_y^2}{l_x^2}\right) \quad \text{(xi)}$$

Substituting for right side of the expression (xii) in the expression (viii),

$$T_{mf.AB} = \frac{5}{6} \cdot \frac{\left(M_y''\right)^2}{M_x'} \tag{17.15}$$

Similarly, the final value of torsional moment at the middle of the strip CD may be written as under :

$$T_{mf.CD} = \frac{5}{6} \cdot \frac{\left(M_x''\right)^2}{M_x'} \tag{17.16}$$

From Eq. 17.15, it is seen that the part of load on the strip AB will cause twisting of the transverse strip CD and the remaining part of the load on AB will cause bending. Therefore, the final value of bending moment on the strip AB is reduced by $T_{mf.AB}$. Similarly, from Eq. 17.16, it is observed that the part of load on the strip CD will cause twisting of the strip AB and the remaining part of the load on CD will cause bending. Therefore, the final value of bending moment on the strip CD is reduced by $T_{mf.CD}$. The net values of bending moments in the strips AB and CD are as under :

$$M_{yf.AB} = M_y'' \left(1 - \frac{5}{6} \cdot \frac{M_y''}{M_x'}\right) = C_y \,.\, M_y'' \tag{17.17}$$

$$\text{and } M_{xf.CD} = M_x'' \left(1 - \frac{5}{6} \cdot \frac{M_x''}{M_y'}\right) = C_x \,.\, M_x'' \tag{17.18}$$

where C_x and C_y are Marcus correction factors for the bending moment expressions given by Grashoff-Rankine, and

$$C_y = \left(1 - \frac{5}{6} \cdot \frac{M_y''}{M_x'}\right) \tag{17.19}$$

$$\text{and } C_x = \left(1 - \frac{5}{6} \cdot \frac{M_x''}{M_y'}\right) \tag{17.20}$$

Equ. 17.19 and 17.20 for the correction factors C_y and C_x respectively may be simplified as under : From the expressions (xiii) and Eq. 17.2.

$$\left(\frac{M_y''}{M_x'}\right) = \left(\frac{w_y}{w}\right) \cdot \left(\frac{l_y^2}{l_x^2}\right). \text{ Since } w_y = \left(\frac{l_x^4}{l_y^4 + l_x^4}\right) . w$$

$$\text{Therefore, } \left(\frac{M_y''}{M_x'}\right) = \left(\frac{l_x^4}{l_y^4 + l_x^4}\right) \cdot \left(\frac{l_y^2}{l_x^2}\right)$$

or $$\frac{M_y''}{M_x'} = \left(\frac{(l_y/l_x)^2}{(l_y/l_x)^4 + (l_x/l_x)^4}\right) = \left(\frac{r^2}{r^4+1}\right) \quad \text{(xiv)}$$

Similarly, $$\left(\frac{M_x''}{M_y'}\right) = \left(\frac{w_y}{w}\right) \cdot \frac{l_x^2}{l_y^2}$$

and $$w_x = \left(\frac{l_y^4}{l_y^4 + l_x^4}\right) \cdot w$$

Then $$\left(\frac{M_x''}{M_y'}\right) = \left(\frac{l_y^4}{l_y^4 + l_x^4}\right) \cdot \left(\frac{l_x^2}{l_y^2}\right)$$

or $$\frac{M_x''}{M_y'} = \left(\frac{(l_y/l_x)^2}{(l_y/l_x)^4 + (l_x/l_x)^4}\right) \times \frac{1}{(l_y'/l_x)^2}$$

$$\left(\frac{M_y''}{M_y'}\right) = \left(\frac{r^4}{r^4+1}\right) \cdot \left(\frac{1}{r^2}\right) = \left(\frac{r^2}{r^4+1}\right) \quad \text{(xv)}$$

The values from expressions (xiv) and (xv) may be substituted in Eq. 17.19 and 17.20, respectively. Therefore,

$$C_y = \left(1 - \frac{5}{6} \cdot \frac{r^2}{1+r^2}\right) \quad (17.21)$$

and $$C_x = \left(1 - \frac{5}{6} \cdot \frac{r^2}{1+r^4}\right) \quad (17.22)$$

As such, Marcus correction factors for both the directions for moment in their simplified form may be denoted as under :

$$C = \left(1 - \frac{5}{6} \cdot \frac{r^2}{1+r^4}\right) \quad (17.23)$$

Marcus correction factor may be calculated from Eq. 17.23 and it is applied to the values of bending moments as per Grashoff-Rankine theory (Eqs. 17.7 and 17.8). The final values of bending moments at the middle of strips AB and CD of unit width along long and short span be written as under :

$$M_{yf.AB} = \left(C \,.\, \alpha_x \,.\, w \,.\, l_x^2\right) \quad (17.24)$$

$$M_{xf.CD} = \left(C \,.\, \alpha_y \,.\, w \,.\, l_x^2\right) \quad (17.25)$$

when $(l_y/l_x) = r = 1$, the value of correction factor C is 0.583. The bending moment for a square slab with the corners held down reduces by 41.66 percent. For $(l_y/l_x) = r = 3$, the value of correction factor C is 0.9085. The bending moment will reduce by 0.146 per cent.

Thus, it is seen that a highly indeterminate slab may be reduced to an equivalent simple system. Marcus correction approach gives a simple and quick procedure for the analysis and design of a two-way slab.

Values of Marcus correction factors have been tabulated in Table 17.2.

TABLE 17.2. MARCUS CORRECTION (REDUCTION) FACTORS, C

Ratio r = (l_y/l_x)	Correction factors	
	C	C
1.00	0.861	0.583
1.05	0.862	0.585
1.10	0.864	0.595
1.15	0.866	0.599
1.20	0.870	0.610
1.25	0.874	0.622
1.30	0.878	0.635
1.40	0.888	0.663
1.50	0.897	0.691
1.60	0.906	0.718
1.75	0.918	0.754
2.00	0.935	0.804
2.50	0.957	0.870
3.00	0.970	0.909

Note : The values of Marcus correction factors, C' (above) are for slabs simply supported along all the four edges.

17.6 PIGEAUD'S METHOD

Pigeaud has also analysed the restrained slabs and suggested the modified coefficients slightly to accommodate the then experimental results. The bending moments of strips of unit width along short span and long span were found by Grashoff-Rankine from Eqs. 17.7 (A) and 17.7 (B), respectively as follows :

$$M_x = \left(\frac{r^4}{1+r^4}\right) \cdot \left(\frac{w \cdot l_x^2}{8}\right) \quad \text{(i)}$$

and

$$M_y = \left(\frac{r^1}{1+r^4}\right) \cdot \left(\frac{w \cdot l_y^2}{8}\right) \quad \text{(ii)}$$

TABLE 17.3. GRASHOFF-RANKINE AND PIGEAUD'S COEFFICIENTS FOR BENDING MOMENTS

Ratio r = (l_y/l_x)	Coefficients for bending moments			
	Grashoff-Rankine		Pigeaud	
	r_x	r_y	r_x'	r_y'
1.00	0.500	0.500	0.300	0.300
1.05	0.549	0.451	0.330	0.271
1.10	0.594	0.406	0.360	0.246
1.15	0.636	0.364	0.389	0.222
1.20	0.675	0.325	0.418	0.202
1.25	0.709	0.291	0.446	0.183
1.30	0.741	0.259	0.474	0.166
1.40	0.790	0.207	0.526	0.137
1.50	0.835	0.165	0.575	0.114
1.60	0.868	0.132	0.621	0.095
1.75	0.904	0.096	0.682	0.073
2.00	0.941	0.059	0.768	0.048
2.50	0.975	0.025	0.892	0.023
3.00	0.988	0.012	0.972	0.012

Note : Grashoff-Rankine coefficients r_x and r_y are for rectangular slabs (simply-supported) and corners not held down. And Rankine coefficients r_x' and r_y' are for simply supported/fixed edges with corners held down.

These expressions (i) and (ii) were also expressed by Grashoeff-Rankine as below :

$$M_x = r_x \cdot \left(\frac{w \cdot l_x^2}{8}\right) \quad \text{(iii)}$$

$$M_y = r_y \cdot \left(\frac{w \cdot l_y^2}{8}\right) \quad \text{(iv)}$$

Grashoeff-Rankine has also tabulated the values of r_x and r_y for different values of ratio, r as given in Table 17.3. Pigeaud has modified these expression as under :

$$M_x = r_x' \cdot \left(\frac{w \cdot l_x^2}{8}\right) \quad \text{(v)}$$

$$M_y = r_y' \cdot \left(\frac{w \cdot l_y^2}{8}\right) \quad \text{(vi)}$$

Pigeaud has also tabulated the values of modified coefficients r_x' and r_y' for different values of ratio, r and these are also given in Table 17.3.

17.7 WESTERGAARD'S METHOD

Westergaard H.M. and W.A. Slater published a paper titled as `moments and stresses in slabs' in proceedings of ACI, Vol. 17, 1921. Westergaard, H.M. published a separate paper titled as `Formulae for the design of rectangular floor slabs and the supporting girders' in proceedings ACI, Vol. 22, 1926. Depending upon experimental results. Westergaard modified the moment coefficients slightly in the method of analysis and design of restrained slabs. Westergaard's method remained predominant in many codes of practice for the last sixty years. This method is still prevailing in many codes of practice. In the past, IS : 456-2000, specified the modified moment coefficients developed by Westergaard.

Example 17.1 A rectangular slab (supported on its all the four edges and corners are not held down) is provided over a room 4.8 m x 6.2 m. Determine the bending moments developed. The dead load consisting of weight of finishing surface and self-weight of the slab is 5.48 kN/m². The live load likely to act over the slab is 4 kN/m². Use Grashoeff-Rankine method.

Solution : The slab is simply supported on both the edges. The width of supporting walls is assumed as 600 mm. Therefore, the effective short span is as follows. The effective depth of slab is assumed as 150 mm.

Step 1 : Effective span

1 (A) Effective short span. The effective depth of slab is assumed as 150 mm.

$$ES = (CS + d) = (4800 + 150) = 4950 \text{ mm} \quad \text{(i)}$$

$$ES = \left(CS + \frac{1}{2}.b_s + \frac{1}{2}.b_s\right) = \left(4800 + \frac{1}{2} \times 600 + \frac{1}{2} \times 600\right) = 5400\ mm \quad \text{(ii)}$$

$$\therefore\ ES = 4950 \text{ mm (whichever is small)} \quad \text{(iii)}$$

1 (B) Effective long span

$$ES = (6200 + 150) = 6350 \text{ mm} \quad \text{(iv)}$$

$$\text{or } ES = \left(6200 + \frac{1}{2} \times 600 + \frac{1}{2} \times 600\right) = 6800\ mm \quad \text{(v)}$$

$$\therefore\ ES = 6350 \text{ mm (whichever is small)} \quad \text{(vi)}$$

Step 2 : Total uniformly distributed load

Total dead load = 5.48 kN/m²

Live load = 4 kN/m²

Total load = 9.48 kN/m² (vii)

Ratio of long span to short span

$$r = \left(\frac{l_y}{l_x}\right) = \left(\frac{6350}{4950}\right) = 1.283 \quad \text{(viii)}$$

Step 3 : Bending moments for the slab

Bending moments for the middle strip of unit width along short span using Grashoeff-Rankine method.

$$M_x = \frac{1}{8} \cdot \left(\frac{r^4}{1+r^4}\right) \cdot w \cdot l_x^2$$

$$\text{or } M_x = r_x \cdot \left(\frac{w \cdot l_x^2}{8}\right) \quad \text{(ix)}$$

Bending moments for the middle strip of unit width along long span using Grashoeff-Rankine method

$$M_y = \frac{1}{8} \cdot \left(\frac{1}{1+r^4}\right) \cdot \frac{w \cdot l_y^2}{8}$$

$$\text{or } M_y = r_y \cdot \left(\frac{w \cdot l_y^2}{8}\right) \quad \text{(x)}$$

Grashoeff-Rankine coefficients r_x and r_y are noted from Table 17.3.

Ratio	(Long span/short span)	
1.25	1.30	1.283
$r_x = 0.709$	0.741	0.730
$r_y = 0.291$	0.259	0.270

The bending moments for the middle strips of unit width along long and short spans are as under :

$$M_x = 0.730 \times \left(\frac{9.48 \times 4.950^2}{8}\right) = 21.196\ kN/m$$

$$M_y = 0.270 \times \left(\frac{9.48 \times 6.350^2}{8}\right) = 12.901\ kN-m$$

Example 17.2. Determine the bending moments in Example 17.1 using Pigeaud's method. The corners of the slab are held down.

Solution : Bending moments for the middle strips of unit width along short and long spans are given by the expressions (v) and (vi), Art. 17.6 by Pigeaud's method.

$$M_x = r_x' \cdot \left(\frac{w \cdot l_x^2}{8}\right) \quad \text{(i)}$$

$$M_y = r_y' \cdot \left(\frac{w \cdot l_y^2}{8}\right) \quad \text{(ii)}$$

From Example 17.1, the ratio of long span to short span, r is 1.283. Pigeaud's coefficients r_x' and r_y' are noted from Table 17.3.

Ratio	(long span/short span)	
1.25	1.30	1.283
$r_x' = 0.446$	0.474	0.464
$r_y' = 0.183$	0.166	0.172

The bending moments are as follows :

$$M_x = 0.464 \times \left(\frac{9.48 \times 4.950^2}{8}\right) = \mathbf{13.472\ kN-m}$$

$$M_y = 0.172 \times \left(\frac{9.48 \times 6.350^2}{8}\right) = \mathbf{8.218\ kN-m}$$

17.8 IS : 456-2000 METHOD

The slabs spanning in two directions at right angles and carrying uniformly distributed load are designed by IS : 456-2000 method by using the coefficients for the bending moments for the rectangular panels supported on four sides given in Appendix-C, IS : 456-2000. The slabs spanning in two directions are classified in two categories, (i) simply supported slabs and (ii) restrained slabs. These two categories of the slabs are analysed and the bending moments are calculated as follows :

1. Simply supported slabs

The corners of simply supported slabs are not prevented from lifting. The corners remain free to lift. These slabs do not have adequate provisions to resist torsion at corners. The maximum moments per unit width for the short span and for the long span are calculated from the following expressions :

$$M_x = \alpha_x \cdot w \cdot l_x^2 \quad \text{(17.26A)}$$

$$M_y = \alpha_y \cdot w \cdot l_y^2 \quad \text{(17.27A)}$$

where, w = total design uniformly distributed load per unit area

l_x and l_y = lengths of the short span and long span.

M_x and M_y = bending moments on the strips of unit width spanning along short span, l_x and long span l_y, respectively.

α_x *and* α_y = coefficients for calculating bending moments (given in Table 17.1).

It is worth to note that the coefficients given in Table 17.1 are derived from Grashoeff-Rankine formulae which are based on the equality of deflections of two strips of the slabs, (one each along the directions parallel to the supporting edges) at the middle at the intersection.

2. Restrained slabs

The restrained slabs are the slabs, the corners of which are prevented from lifting. The slabs are subjected to uniformly distributed loads. The slab is supported on all the four edges. The edges of slab may be simply supported (viz., discontinuous edges), of fixed or continuous (viz., continuous edges). Depending on the various edge supports, the restrained slabs have been classified in nine categories as shown in Fig. 17.3 and described in Art. 17.4.

The bending moment M_x for strip of unit width spanning along short span, l_x (viz., along x-axis) and the bending moment M_y for strip of unit width spanning along long span, l_y (*viz.,* along y-axis) carry only one subscript, but each moment in reality is a function of both the axes x and y. Each moment varies on the plane of middle line with the maximum at the middle point and reduces in a sinusoidal manner. The critical moment on the critical strip, *for example,* the bending moment M_x depends on the boundary conditions at $x = 0$ to $x = l_x$. The bending moment M_x at $x = (L_x/2)$ for $y = 0$ to $y = l_y$ varies close to a sinusoidal variation. The depth of slab and the reinforcement depend on the value of critical moment at the central unit strip. Therefore, the spacing of reinforcement for M_x along y-axis varies with maximum at the edges (i.e., $y = 0$ to $y = l_y$), and the minimum at central unit strip. In order to minimize the problem of detailing of the reinforcement, IS : 456-2000 specified that the slab is divided into strips as follows :

The restrained slab are considered as divided parallel to short span and parallel to long span directions into *middle strip* and *edge strips* as shown in Figs. 8.5 (A), (B), respectively. The maximum bending moments are calculated for the middle strips. The maximum bending moments per unit width for the short span and for the long span are determined from the following expressions

$$M_x = \alpha_x \cdot w \cdot l_x^2 \tag{17.26B}$$

$$M_y = \alpha_y \cdot w \cdot l_x^2 \tag{17.27B}$$

where, w, l_x, l_y, M_x and M_y are same as defined for the simply supported spans described above, α_x and α_y are the coefficients for finding the bending moments for the middle strips and the redistribution is not made. These coefficients are based on the modern concept of yield line theory and adjusted to take into consideration the fact that, the slab is divided into middle and edge strips. Arbitrary parameters, such as the ratio of support moment to span moment (M support/M span), have been selected to give moments that correspond to those

which would be obtained from the elastic analysis. The values of coefficients for the nine cases of the restrained slab are given in Table 17.4 as given in IS : 456-2000.

It must specially be noted that l_x^2 appears in both Eq. 17.26 (B) and 17.27 (B) for the bending moments M_x and M_y.

TABLE 17.4. BENDING MOMENT COEFFICIENT FOR RECTANGULAR PANELS SUPPORTED ON FOUR SIDES

Case No.	Type of panel and moments considered	Short span coefficients α_x (Values of l_y/l_x)								Long span coefficients α_y for all values of l_y/l_x
		1.0	1.1	1.2	1.3	1.4	1.5	1.75	2.0	
(1)	(2)	(3)	(4)	(5)	(6)	(7)	(8)	(9)	(10)	(11)
1.	Interior Panels (Fig 17.3A)									
	Negative moment at continuous edge	0.032	0.037	0.047	0.043	0.051	0.053	0.060	0.065	0.032
	Positive moment at mid span	0.024	0.028	0.032	0.036	0.039	0.041	0.045	0.049	0.024
2.	One short edge discontinuous: (Fig. 17.3 B)									
	Negative moment at continuous edge	0.037	0.043	0.048	0.051	0.055	0.057	0.064	0.068	0.037
	Positive moment at mid span	0.028	0.032	0.036	0.039	0.041	0.044	0.048	0.052	0.028
3.	One long edge discontinuous : (Fig. 17.3 C)									
	Negative moment at continuous edge	0.037	0.044	0.052	0.057	0.063	0.067	0.077	0.085	0.037
	Positive moment at mid span	0.028	0.033	0.039	0.044	0.047	0.051	0.059	0.065	0.028
4.	Two adjacent edges discontinuous : (Fig. 17.3 D)									
	Negative moment at continuous edge	0.047	0.053	0.060	0.065	0.071	0.075	0.084	0.091	0.047
	Positive moment at mid-span	0.035	0.040	0.045	0.049	0.054	0.056	0.063	0.069	0.035
5.	Two short edges discontinuous (Fig. 17.3 E)									
	Negative moment at continuous edge	0.045	0.049	0.052	0.056	0.059	0.060	0.065	0.069	—
	Positive moment at mid-span	0.035	0.037	0.040	0.043	0.044	0.045	0.049	0.052	0.035
6.	Two long edges discontinuous; (Fig. 17.3 F)									
	Negative moment at continuous edge	–	–	–	–	–	–	–	–	0.045
	Positive moment at mid span	0.035	0.043	0.051	0.057	0.063	0.068	0.080	0.088	0.035
7.	Three edge discontinuous; (Fig. 17.3 G)									
	Negative moment at continuous edge	0.057	0.064	0.071	0.076	0.080	0.084	0.091	0.097	—

	Positive moment at mid-span	0.043	0.048	0.053	0.057	0.060	0.064	0.069	0.073	0.043
8.	Three edges discontinuous (One short edge discontinuous); (Fig. 17.3 H)									
	Negative moment at continuous edge	–	–	–	–	–	–	–	–	0.057
	Positive moment mid span	0.043	0.051	0.059	0.065	0.701	0.076	0.087	0.096	0.043
9.	Four edges discontinuous; (Fig. 17.3 J)									
	Positive moment mid span	0.056	0.064	0.072	0.079	0.085	0.089	0.100	0.107	0.056

(Note : The restrained slabs shall have provision for torsion at corner)

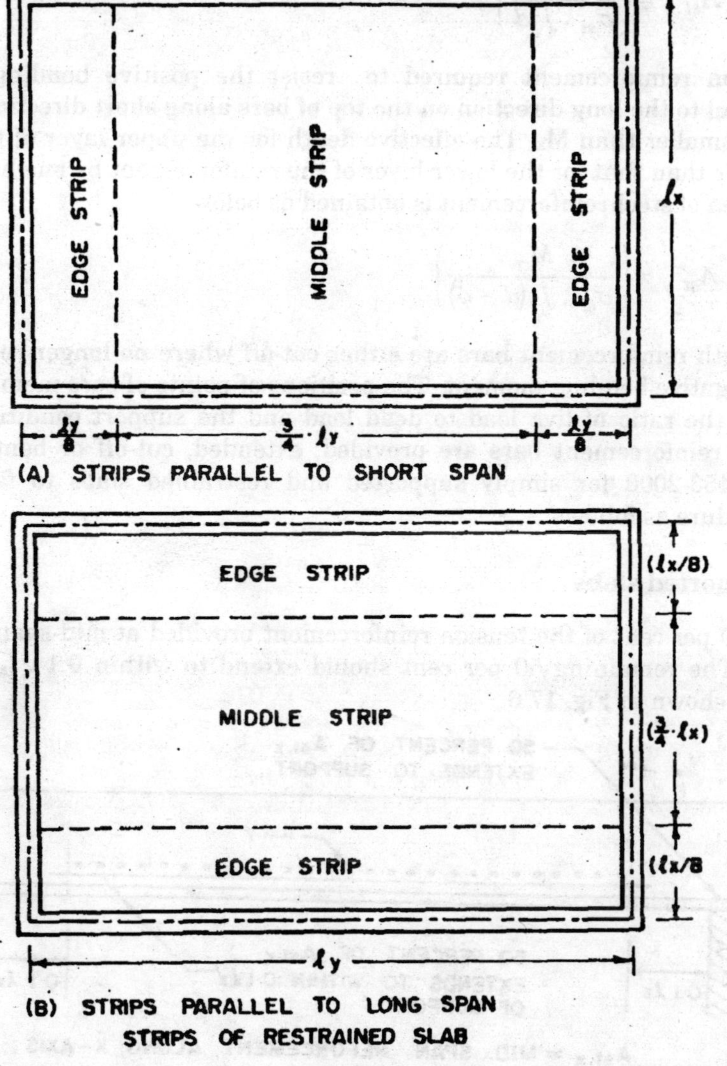

(A) STRIPS PARALLEL TO SHORT SPAN

(B) STRIPS PARALLEL TO LONG SPAN

STRIPS OF RESTRAINED SLAB

Fig. 17.5

It is to note that the IS : 456-2000 refers the moment per unit width in *three-fourth* of the total width of the slab and the middle strip, which is *three-fourth* of the total width is reinforced. Therefore the sum of coefficients for positive and negative bending moments recommended by IS : 456-2000 is approximately 1.33 times the corresponding values derived by Marcus theory.

17.9 TENSION REINFORCEMENT FOR TWO-WAY SUPPORTED SLABS

The tension reinforcement required to resist the positive bending moment, M_x is provided parallel to the short direction near the bottom face of the slab and calculated as follows :

$$A_{st.x} = \left(\frac{M_x}{\sigma_{st} \cdot j \cdot d} \right) \quad \text{(i)}$$

The tension reinforcement required to resist the positive bending moment, M_y is provided parallel to the long direction on the top of bars along short direction, as the bending moment M_y is smaller than M_x. The effective depth for the upper layer of the reinforcement becomes smaller than that for the lower layer of the reinforcement by one bar diameter, [viz., $(d - \phi)$]. The area of steel reinforcement is obtained as below :

$$A_{st.y} = \left(\frac{M_y}{\sigma_{st} \cdot j \cdot (d - \phi)} \right) \quad \text{(ii)}$$

The strength reinforcement bars are either cut-off where no longer required or bent-up to resist the negative bending moment. The positions of points of zero moments depend upon the side ratio, the ratio of live load to dead load and the support conditions at the edges. Therefore, the reinforcement bars are provided, extended, cut-off or bent-up as recommended in IS : 456-2000 for simply supported and restrained slabs to facilitate a simple detailing procedure as follows :

1. Simply supported slabs

At least 50 per cent of the tension reinforcement provided at mid-span should extend to the supports. The remaining 50 per cent should extend to within 0.1 . l_x or 0.1. l_y of the appropriate as shown in Fig. 17.6.

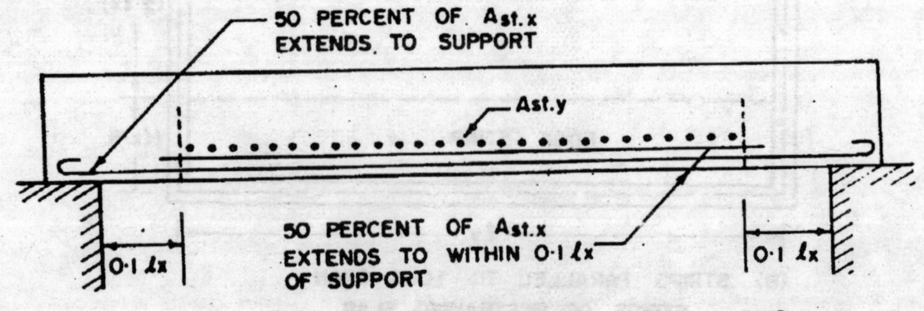

Fig. 17.6

2. Restrained slabs

The reinforcement in restrained slab is provided near the bottom face or top face depending upon whether the bending moment is *positive or negative.* The reinforcement extends to within the specified distance depending upon the edges of slab are continuous or discontinuous. IS : 456-2000 recommends the detailing procedure as follows :

(a) Reinforcement for positive bending moment for the middle strip

The tension reinforcement provided at mid-span in the middle strip for positive bending moment shall extend in the lower part of the slab (viz., near the bottom face of slab) within 0.25 . l of a continuous edge or 0.15 . l of a discontinuous edge as shown in Fig. 17.7.

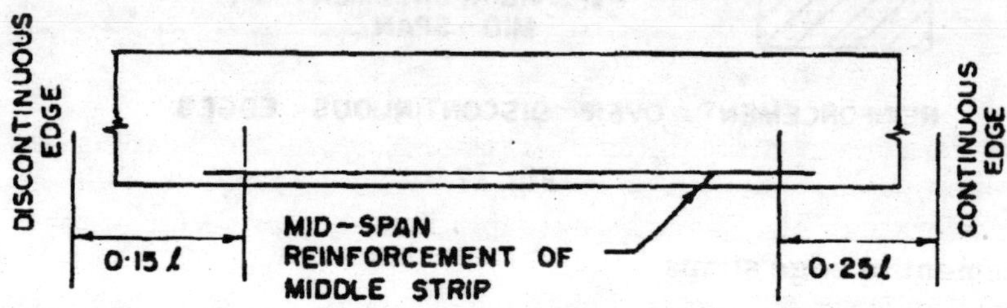

Fig. 17.7

(b) Reinforcement for negative bending moment over continuous edge

The negative bending moment develops over the continuous edge of a middle. For the negative bending moments, the reinforcement is provided in the upper part (viz., near the top face) of the slab. This reinforcement (100 per cent) extends a distance 0.15.l from the support on one side and 0.15.l from the support on the other side as shown in Fig. 17.8. 50 per cent of this very reinforcement extends a distance 0.30 l from the support on one side and 0.30.l from the support on the other side.

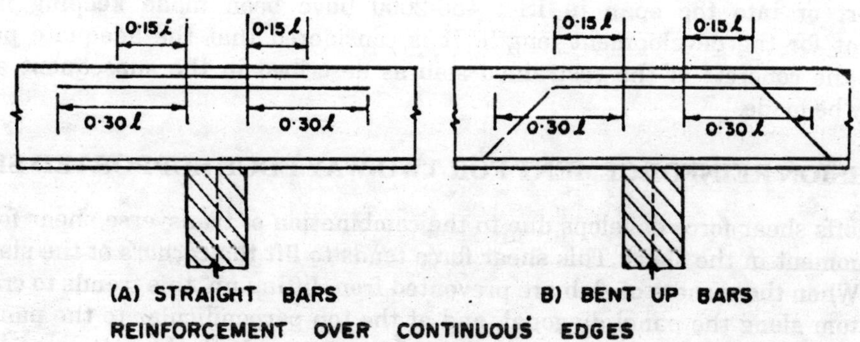

Fig. 17.8

(c) Reinforcement for negative bending moment over discontinuous edge

The negative bending moment may develop at a discontinuous edge of the slab depending upon the degree of fixity of the slab. In ordinary cases, this bending moment may be assumed *one-half* of the positive bending moment at the span. IS : 456-2000 recommends that, in general, tension reinforcement equal to 50 per cent of that provided at mid-span extending 0.1l into the span as shown in Fig. 17.9 will be sufficient.

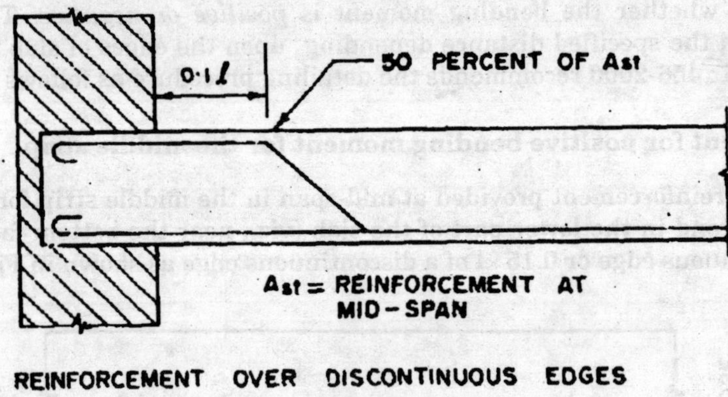

Fig. 17.9

(d) Reinforcement in edge strips

The area of reinforcement in edge strip, parallel to that edge shall be equal to the minimum reinforcement required for slab, (viz., the reinforcement shall not be less than 0.15 per cent of the cross-sectional area of the edge strip for mild steel and 0.12 per cent of the cross-sectional area of the edge strip for hysd-steel).

The minimum reinforcement in each direction for two-way slabs is that required for temperature and shrinkage crack control. The spacing of tension reinforcement resisting the bending moment in two-way slab at the Critical sections must not exceed twice the thickness of slab.

Above simplified recommendations of the extensions of steel bars (reinforcement) from the support or into the span in IS : 456-2000 have been made keeping in view the requirement for the development length. It is considered that the adequate provision for torsion at the concrete of the restrained slab as described in the subsequent article shall necessarily be made.

17.10 TORSION REINFORCEMENT FOR TWO-WAY EDGE SUPPORTED SLABS

Kirchoffs shear force develops due to the combination of transverse shear force and the twisting moment in the slabs. This shear force tends to lift the corners of the slabs from the supports. When the corners of slab are prevented from lifting up, this trends to crack the slab at the bottom along the panel diagonal, and at the top perpendicular to the panel diagonal. Therefore, the corners of slab need special reinforcement at both the bottom and top of slab. Formerly, the reinforcement used to be provided at the top of the slab in the direction parallel to the diagonal, while at the bottom of the slab in the direction perpendicular to the diagonal. Alternatively, either layer of steel may be placed in two bands parallel to the sides of the slab. IS : 456-2000 recommends this alternative way of providing the torsional reinforcement in the two-way slab.

(a) (i) IS : 456-2000 recommends that the torsional reinforcement shall be provided at any corner where the slab is simply supported on both edges meeting at that corner, (*viz., at the corners where the two adjacent edges are discontinuous*).

(ii) The torsional reinforcement shall be provided at the top and bottom of the slab, each with layers of bars placed parallel to the sides of the slab.

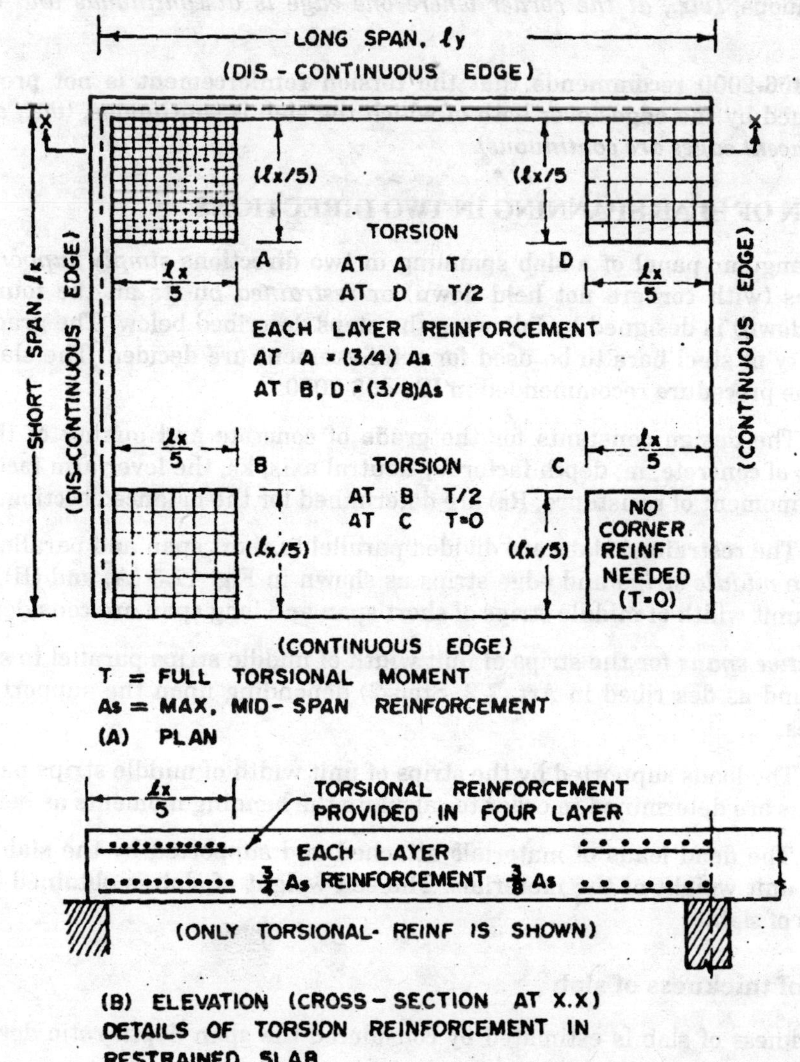

Fig. 17.10

(iii) This torsional reinforcement extends from the edges a minimum distance in each direction equal to one-fifth of the shorter span.

(iv) The area of reinforcement in each of these four layers shall be three-quarters of the area required for the maximum mid-span moment in the slab.

Let A_s be the area of reinforcement required for the maximum mid-span moment for the strip of unit width in the slab. Then, above recommendations of IS : 456-2000 may be illustrated as shown in Fig. 17.10.

(b) IS : 456-2000 recommends that torsional reinforcement equal to half that described above [a (i) to (iv)] shall be provided at a corner contained by edges over only one of which the slab is continuous, (*viz., at the corner where one edge is discontinuous and other edge is continuous).*

(c) IS : 456-2000 recommends that the torsion reinforcement is not provided at any corner contained by the edges over both of which the slab is continuous, (*viz., at the corners where the adjacent edges are continuous).*

17.11 DESIGN OF SLAB SPANNING IN TWO DIRECTIONS

The rectangular panel of a slab spanning in two directions *simply supported* on its all the four edges (with corners not held down) or *restrained* on its all the four edges (with corners held down) is designed by following the steps described below. The grade of concrete and the quality of steel bars to be used for reinforcement are decided. The slab is designed conforming the procedure recommended in IS : 456-2000.

Step 1: The design constants for the grade of concrete and quality of the steel (viz., modular ratio of concrete, m, depth factor for neutral axis, k_B, the lever arm factor, j_B and the factor for the moment of resistance, R_B) are determined for the balanced section.

Step 2 : The restrained slabs are divided parallel to short span and parallel to long span directions into *middle strips* and edge strips as shown in Fig. 17.5 (A) and (B), respectively. The strips of unit width of middle strips of short span and long span are considered.

The *effective spans* for the strips of unit width of middle strips parallel to short and long spans are found as described in Art. 7.2, Step 2) depending upon the support conditions of opposite edges.

Step 3 : The loads supported by the strips of unit width of middle strips parallel to short and long spans are determined in order to calculate the bending moments as below :

Loads : The dead loads of materials attached and supported by the slab are found by knowing the unit weight of the materials. The self-weight of slab is obtained by estimating the thickness of slab.

Estimation of thickness of slab

The thickness of slab is estimated by considered the span depth ratio described in Art. 3.7 depending upon end restraint conditions.

As per recommendation of IS : 456-2000, for slab spanning in two directions, *the shorter of two spans* should be used for calculating the span to effective depth ratios. For slabs of small spans (upto 3.5 m) with mild steel reinforcement, the *span to overall depth* ratios given below (Table 17.5) may generally be assumed to satisfy vertical deflection limits for loading class upto 3 kN/m². For high strength deformed bars of grade Fe 415, the values given for mild steel are multiplied by 0.8.

TABLE 17.5. SPAN TO OVERALL DEPTH RATIOS

End support	Span to overall depth ratios	
	mild steel	hysd steel
Simply supported	35	28
Continuous	40	32

Actual depth of slab required to resist the maximum bending moment is obtained in the subsequent steps.

The self-weight of slab may be calculated by considering the unit area of the slab and its thickness. It is added to the dead loads attached.

The intensity of uniformly distributed live load per square metre is noted from IS : 875-1984 for the type of floor and its use.

Total load per metre length to be supported by strips of unit width of middle strips along short and long spans are obtained.

Step 4 : Required thickness of slab is calculated after determining the maximum bending moments M_x and M_y for the strips of unit width of the middle strips of short and long spans, respectively.

Simply supported slab

The maximum bending moment per unit width for the short and the long spans of the simply supported slab are calculated from Eqs. 17.26 (A) and 17.27 (A) as below :

$$M_x = \alpha_x \,.\, w \,.\, l_x^2 \quad \text{(i)}$$

$$M_y = \alpha_y \,.\, w \,.\, l_x^2 \quad \text{(ii)}$$

The values of co-efficients α_x and α_y for the simply supported slab (with corners not held down) are noted from Table 17.1.

Restrained slab

The maximum bending moments per unit width for the short and the long spans of the restrained slab (viz., corners held down) are determined from Eqs. 17.26 (B) and 17.27 (B) as under :

$$M_x = \alpha_x \,.\, w \,.\, l_x^2 \quad \text{(iii)}$$

$$M_y = \alpha_y \,.\, w \,.\, l_x^2 \quad \text{(iv)}$$

The values of co-efficients, α_x and α_y for the ratio $(l_y / l_x) = r$ for the respective support conditions of the edges from Table 17.4. The absolute maximum bending moment M out of M_x and M_y is noted and the effective depth of the slab is determined as under :

$$M = R_B \,.\, bd^2 \qquad (\because b = 1000) \quad \text{(v)}$$

$$\text{or} \quad d = \left(\frac{M}{R_B \times 1000} \right)^{1/2} \qquad \text{(vi)}$$

The minimum clear cover for the reinforcement (for tension, compression or shear or any other) recommended in IS : 456-2000 is 15 mm. By adding the appropriate effective cover to effective depth and rounding off, the overall thinking of slab is decided.

Step 5 : Following tension reinforcements as described and detailed in Art. 8.9 are provided in the slab.

Main reinforcement

The area and spacings of steel reinforcements in tension, $A_{st.x}$ and $A_{st.y}$, and s_x and s_y for the strips of unit width of the middle strips along short and long spans, respectively are determined as follows :

$$A_{st.x} = \left(\frac{M_x}{\sigma_{st} \cdot j \cdot d} \right) \; and \; A_{st.y} = \left(\frac{M_y}{\sigma_{st} \cdot j \cdot (d - \phi)} \right) \qquad \text{(vii)}$$

$$s_x = \left(\frac{1000 \, A_{\phi.x}}{A_{st.x}} \right) \; and \; s_y = \left(\frac{1000 \, A_{\phi.y}}{A_{st.y}} \right) \qquad \text{(viii)}$$

where, $A_{\phi.x}$ and $A_{\phi.y}$ are the areas of one steel bar provided in the middle strips along short and long spans, respectively and as shown in Fig. 17.11 (A).

Reinforcement in edge strips are provided as described in Art. 17.9 section (d) as minimum reinforcement necessary (viz., temperature and shrinkage reinforcement) and as shown in Fig. 17.11 (B).

Step 6 : Torsional reinforcement

Torsional reinforcement to be provided at the corners depends upon the restrained conditions of the edges containing the corners whether simply supported (viz., discontinuous) or continuous as described and detailed in Art 17.10 in section (a), (b) and (c), and as illustrated in Fig. 17.10.

Example 17.3. Design a rectangular slab (supported on its all the four edges) over a class-room of size 4.8 x 6.2 m. Two adjacent edges of the slab are discontinuous and the remaining two edges are continuous. A finishing surface of cement concrete of 20 mm shall be provided over the slab. The slab shall be used as a class-room floor. M 20 grade of concrete and hysd-steel bars shall be used.

Design : The short and long spans of slab are 4.8 m and 6.2 m, respectively. The ratio of long span to short span

$$r = \left(\frac{l_y}{l_x} \right) = \left(\frac{6.2}{4.8} \right) = 1.292$$

This ratio is less than 2. The slab is to be designed as slab spanning in two directions (viz., two-way slab).

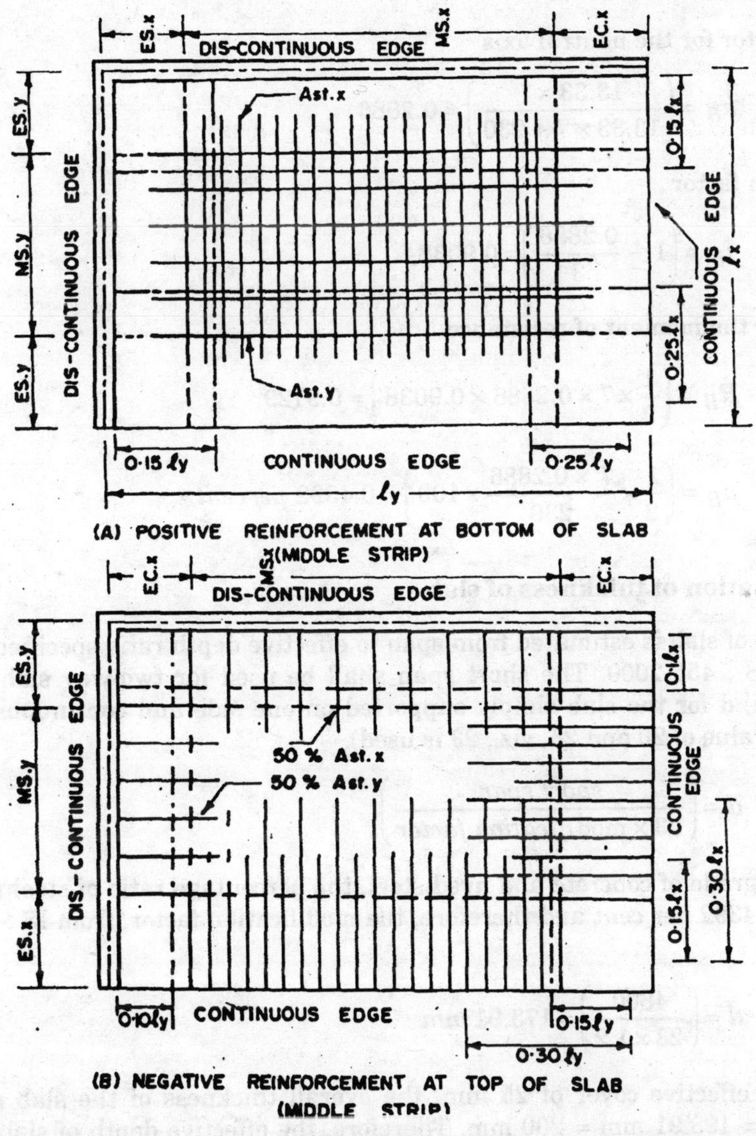

Fig. 17.11

Step 1 : Design constants

The design constants for M 20 grade of concrete and hysd-steel reinforcement for the balanced section are as under. From IS : 457-1978, for M 20 grade of concrete, and hysd-steel,

$$\sigma_{cbc} = 7\ N/mm^2 \ \ and \ \ \sigma_{st} = 230\ N/mm^2$$

Modular ratio for M 20 grade of concrete

$$m = \left(\frac{280}{3 \times 7}\right) = 13.33$$

Depth factor for the neutral axis

$$k_B = \left(\frac{13.33 \times 7}{13.33 \times 7 + 230}\right) = 0.2886$$

Lever arm factor

$$j_B = \left(1 - \frac{0.2886}{3}\right) = 0.9038$$

Factor for the moment of resistance

$$R_B = \left(\frac{1}{2} \times 7 \times 0.2886 \times 0.9038\right) = 0.9129$$

$$p_B = \left(\frac{1}{2} \times \frac{7 \times 0.2886}{230} \times 100\right) = 0.4392 \text{ percent}$$

Step 2 : Estimation of thickness of slab

Thickness of slab is estimated from span to effective depth ratio specified for the vertical deflection is IS : 456-2000. The short span shall be used for two-way slab. Therefore, the effective depth, d for the slab simply supported on one side and continuous over the other side (average value of 20 and 26, viz., 23 is used).

$$d = \left(\frac{\text{short span}}{23 \times \text{modification factor}}\right)$$

For M 20 grade of concrete and hysd-steel, the percentage ratio of steel for the balanced section p_B is 0.4392 per cent and therefore, the modification factor, from IS : 456-2000 is 1.2. Therefore,

$$d = \left(\frac{4800}{23 \times 1.2}\right) = 173.91 \text{ mm} \qquad \text{(ii)}$$

Providing effective cover of 25 mm, the overall thickness of the slab may be kept as (173.91 + 25) = 198.91 mm = 200 mm. Therefore, the effective depth of slab shall be (200 – 25) = 175 mm.

Step 3 : Effective span

The slab is discontinuous (*viz.,* simply supported) on one side and continuous on the other side. The width of supporting walls is assumed as 600 mm. Therefore, the *effective short span* is as follows :

The width of support 600 mm is wider than

$$\left(\frac{4800}{12}\right) = 400\ mm$$

$$\text{Then ES} = \left(CS + \frac{1}{2}d\right) = \left(4800 + \frac{1}{2}\times 150\right) = 4875\ mm$$

$$\text{and ES} = \left(CS + \frac{1}{2}.b_s\right) = \left(4800 + \frac{1}{2}\times 600\right) = 5100\ mm$$

$\therefore$ *ES (short) = 4.875 m (whichever is less)* (iii)

The effective long span shall be as under :

The width of support 600 mm is wider than

(6200/12) = 516.66 mm

$$\text{Then } ES = \left(6200 + \frac{1}{2}\times 150\right) = 6275\ mm$$

$$ES = \left(6200 + \frac{1}{2}\times 600\right) = 6500\ mm$$

ES (long) = 6.275 m (whichever is less) (iv)

Step 4 : Load

The dead load consists of weight of finishing surface and the self-weight of the slab. These loads are calculated for unit area of the slab.

Weight of finishing surface

$$(1\times 1\times 0.020)\times 24 = 0.48\ kN/m^2$$

Self-weight of the slab

$$(1\times 1\times 0.200)\times 25 = 5.00\ kN/m^2$$

Live load for the class-room floor

$$(1\times 1\times 4) = 4.00\ kN/m^2$$

Total load per unit area

= 9.48 kN/m^2 say = 9.50 kN/m^2

Ratio of effective long span to effective short span

r = (6275/4875) = 1.287 (v)

Step 5 : Bending moments

The slab shall be divided into middle and edge strips.

Width of strips of short span

$$\text{Middle strip} = \left(\frac{3}{4} \times 6275\right) = 4706.25\ mm$$

$$\text{Edge strip} = \left(\frac{1}{8} \times 6275\right) = 784.375\ mm$$

Width of strips of long span

$$\text{Middle strip} = \left(\frac{3}{4} \times 4875\right) = 3656.25\ mm$$

$$\text{Edge strip} = \left(\frac{1}{8} \times 4875\right) = 609.375\ mm$$

Bending moments per unit width strip of the middle strips

$$M_x = \alpha_x \cdot w \cdot l_x^2 \qquad \text{(vi)}$$

$$M_y = \alpha_y \cdot w \cdot l_x^2 \qquad \text{(vii)}$$

The coefficients a_x and α_y are noted from IS : 456-2000 for ratio

r = (l_y/l_x) = 1.287

	Ratio r	
1.20	1.30	1.287
α_x + ve = 0.045	0.049	0.0485
-ve = 0.060	0.065	0.06435
α_y + ve = –	-	0.035
- ve = -	-	0.047

$$M_x\ (+ve) = \left(0.0485 \times 9.50 \times 4.875^2\right) = 10.95\ kN-m$$

$$M_x\ (-ve) = \left(0.06435 \times 9.50 \times 4.875^2\right) = 14.53\ kN-m$$

$$M_y\ (+ve) = \left(0.035 \times 9.50 \times 4.875^2\right) = 7.902\ kN-m$$

$$M_y\ (-ve) = \left(0.047 \times 9.50 \times 4.875^2\right) = 10.61\ kN-m$$

Absolute maximum bending moment out of the four above values

M = 14.53 kN-m (viii)

Step 6 : Effective depth of the slab, d

$$d = \left(\frac{M}{R_B \times 1000}\right)^{1/2} = \left(\frac{14.53 \times 10^6}{0.9129 \times 1000}\right) = 126.16\ mm \qquad \text{(ix)}$$

Effective cover = 25 mm

Overall thickness of the slab

= (126.16 + 25) = 151.16 mm (x)

Let the overall thickness of slab be 160 mm and therefore, the effective depth of slab shall be

d = (160 – 25) = 135 mm

Step 7 : Tension reinforcement in middle strip

The tension reinforcement for the +ve and negative values of the bending moments required are as follows : 10 ϕ mm hysd-steel bars shall be used.

(i) For bending moment, M_x (+ve)

$$A_{st.x}(+ve) = \left(\frac{10.95 \times 10^6}{230 \times 0.9038 \times 135}\right) = 390.19\ mm^2$$

(ii) For bending moment, M_x (-ve)

$$A_{st.x}(-ve) = \left(\frac{14.53 \times 10^6}{230 \times 0.9038 \times 135}\right) = 517.76\ mm^2$$

(iii) For bending moment, M_y (+ve)

$$A_{st.y}(+ve) = \left(\frac{7.902 \times 10^6}{230 \times 0.9038 \times (135-10)}\right) = 304.11\ mm^2$$

(iv) For bending moment, M_y (-ve)

$$A_{st.y}(-ve) = \left(\frac{10.95 \times 10^6}{230 \times 0.9038 \times (135 \times 10)}\right) = 408.32\ mm^2$$

Area of cross-section of 10 mm ϕ bar A_ϕ = 78.5398 mm²

Step 8 : Spacing of bars

Spacing of the bars for $A_{st.x}$ (+ve)

$$s_x(+ve) = \left(\frac{1000 \times 78.5398}{390.19}\right) = 201.286\text{ mm} \quad \text{say 200 mm c/c}$$

Spacing of the bars for A_{st} (–ve)

$$s_x \cdot (-ve) = \left(\frac{1000 \times 78.5398}{517.76}\right) = 151.692\ mm,\ say = 150\ mm\ c/c$$

Spacing of the bars for $A_{st.y}$ (+ve)

$$s_y.(+ve) = \left(\frac{1000 \times 78.5398}{304.11}\right) = 258.261\ mm,\ say = 250\ mm\ c/c$$

Spacing of the bars for $A_{st.y}$ (-ve)

$$s_y.(-ve) = \left(\frac{1000 \times 78.5398}{408.32}\right) = 192.349\ mm,\ say = 190\ mm\ c/c.$$

Step 9 : Reinforcement in edge strips

Hysd-steel bars are used. Therefore, the reinforcement (for temperature and shrinkage) in edge strips

$$\left(\frac{0.12 \times 1000 \times 160}{100}\right) = 192\ mm^2$$

Provide 8 mm diameter bar $A_\phi = 50.26\ mm^2$

$$\text{Spacing}\left(\frac{1000 \times 50.26}{192}\right) = 261.80\ mm\ say = 250\ mm\ c/c$$

Step 10 : Torsional reinforcement

Three-fourth of maximum mid-span reinforcement

$$\frac{3}{4}.\ A_{s.max} = (0.75 \times 517.76) = 388.32\ mm^2$$

$$\frac{3}{8}.\ A_{s.max} = \left(\frac{0.75 \times 517.76}{2}\right) = 194.16\ mm^2$$

(i) Torsional reinforcement at corners contained by two discontinuous edges

Reinforcement is provided in four layers. Reinforcement in each layer

$$\left(\frac{3}{4}.\ A_{s.max}\right) = 388.32\ mm^2$$

8 mm ϕ bars are provided. Therefore spacing of bars in each layer

$$s = \left(\frac{1000 \times 50.26}{388.32}\right) = 129.429\ mm,\ say = 120\ mm.$$

(ii) Torsional reinforcement at corners contained by one discontinuous edge and one continuous edge

$$\left(\frac{3}{8}.\ A_{s.max}\right) = 194.16\ mm^2$$

8 mm φ bars are provided. Therefore spacing of bars n each layer

$$\left(\frac{1000 \times 50.26}{194.26}\right) = 258.858\ mm,\ say = 240\ mm.$$

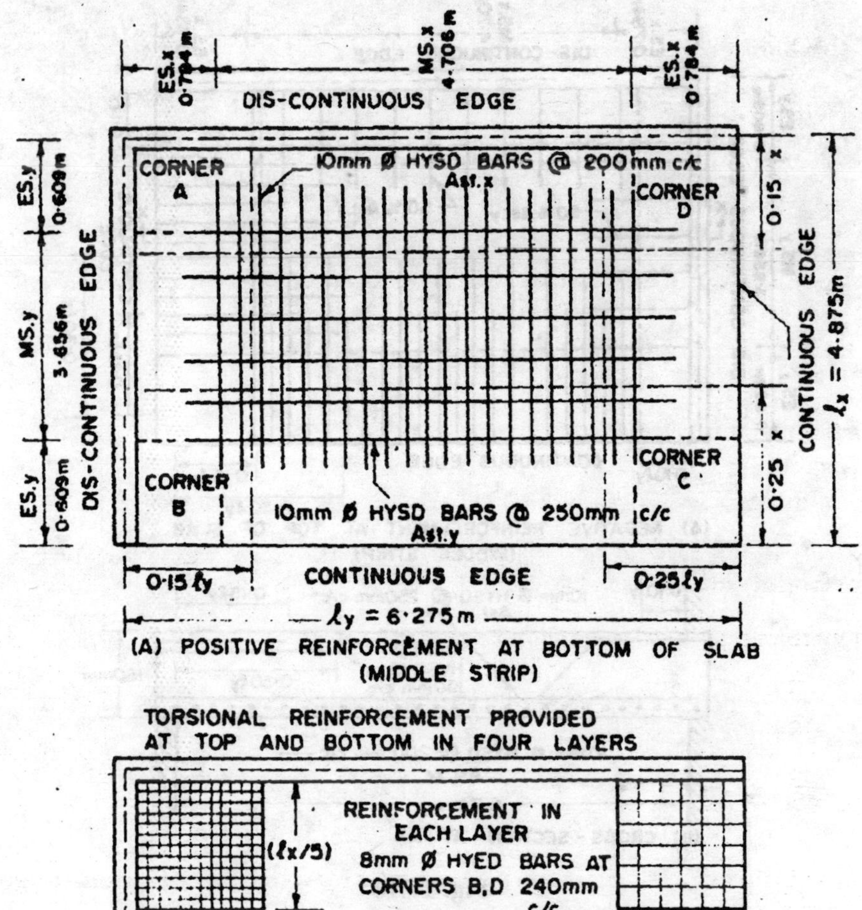

Fig. 17.12

(iii) Torsional reinforcement at corner contained by two continuous edges

Total reinforcement is not provided at such corners.

Dimensions for corners

$$\left(\frac{1}{5} \times short\ span\right) = \left(\frac{1}{5} \times 4875\right) = 975\ mm$$

The details of positive and negative reinforcement in tension for the middle strips, reinforcement for the edge strips of the slabs and the torsional reinforcement at the corners of the slab are shown in Fig. 17.12 and Fig. 17.13.

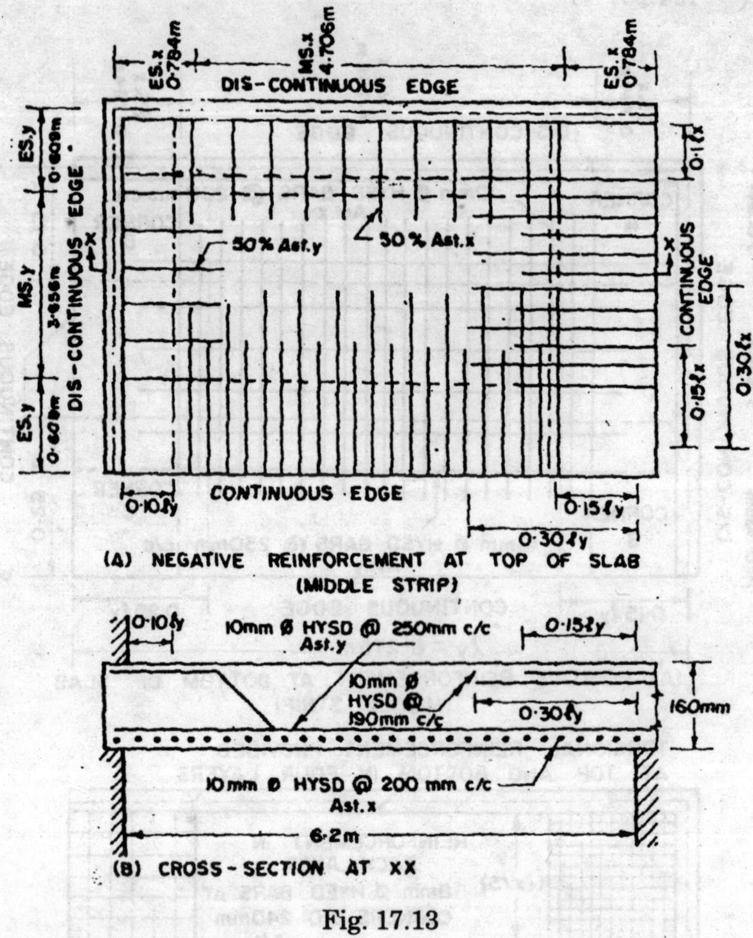

Fig. 17.13

17.12 SHEAR IN TWO-WAY SLAB

In two-way slab simply supported along its four edges, it is seen in Grashoeff-Rankine method that the uniformly distributed load w per unit area is shared along the short span as w_x and along the long span as w_L. From Eq. 17.,

$$w_x = \left(\frac{l_y^4}{l_y^4 + l_x^4}\right).w$$

$$\text{or} \quad w_x = \left(\frac{(l_x/l_x)^4}{(l_y/l_x)^4 + (l_x/l_x)^4}\right).w$$

$$\text{or} \quad w_x = \left(\frac{r^4}{1+r^4}\right).\, w \qquad \text{(i)}$$

From Fig. 17.,

$$w_y = \left(\frac{l_x^4}{l_y^4 + l_x^4}\right).\, w$$

$$\text{or} \quad w_y = \left(\frac{(l_x/l_x)^4}{(l_y/l_x)^4 + (l_x/l_x)^4}\right).\, w$$

$$\text{or} \quad w_y = \left(\frac{r^4}{1+r^4}\right).\, w \qquad \text{(ii)}$$

The values of bending moments found by using w_x and w_y values of loads are reasonably satisfactory. However, the values of reactions at edge support are quite different from these determined experimentally, as the slab spans along X_1Y_1 (Fig. 17.1) near the corner A, and in similar manner near other corners.

The shears in two-way slab may be obtained by assuming that the load from area AED (Fig. 17.1) is carried by the short edge DA and that from area AEFB is carried by the long edge AB. Total shear in slab across DA

$$= \textit{intensity of load} \times \textit{area AED (Fig.17.1)}$$

$$= w \times \frac{1}{2} \times l_x \times \left(\frac{l_x}{2}\right) = \left(\frac{w l_x^2}{4}\right) \qquad \text{(iii)}$$

Average shear per unit width along DA

$$= \frac{1}{l_x} \cdot \left(\frac{w\, l_x^2}{4}\right) = \left(\frac{w \,.\, l_x}{4}\right) \qquad \text{(iv)}$$

The maximum average shear per unit width occurs at the middle of DA. For all practical use, it may be assumed as

$$= \left(w \,.\, \frac{l_x}{3}\right) \qquad \text{(v)}$$

Total shear in slab across AB (Fig. 17.1)

$$= \textit{intensity of load} \times \textit{area AEFB} = w. \left(\frac{AB + EF}{2}\right) . \frac{l_x}{2}$$

$$= w \,.\, \frac{1}{2}\left[l_y + \left(l_y - 2 \,.\, \frac{l_x}{2}\right)\right] . \frac{l_x}{2}$$

$$= \left(\frac{w \cdot l_x \left(2l_y - l_x \right)}{4} \right) \qquad \text{(vi)}$$

Average shear per unit width along AB

$$\left(\frac{w \cdot l_x}{l_y} \right) \cdot \left(\frac{l_y \left(2 - \frac{l_x}{l_y} \right)}{4} \right) = \left(\frac{w \cdot l_x \left(2 - \frac{l_x}{l_y} \right)}{4} \right) \qquad \text{(vii)}$$

The use of shear reinforcement in slabs less than 200 mm thickness is considered impracticable. The nominal shear stress in the slabs should not exceed half the permissible shear stress. The permissible shear stress in slabs as per IS : 456-2000 has been discussed in Art. 7.6. The depth of slab is commonly adjusted so that the requirement for shear is satisfied.

PROBLEMS

17.1 A rectangular slab (supported on its all the four edges and corners are not held down) is provided over a room 3.6 x 4.6 m. Determine the bending moments developed. The dead load consisting of weight of finishing surface and self-weight of the slab is 4.60 kN/m². The live load likely to act over the slab is 2 kN/m². Use Grashoeff-Rankine method.

17.2 Determine the bending moments in Problem 8.1 using Pigeaud's method. The corners of the slab are held down.

17.3 Design a rectangular slab (supported on its all the four edges) over a residential room of size 3.8 m x 5.4 m. Two short edges of the slab are discontinuous. A finishing surface of cement concrete of 20 mm shall be provided over the slab. The slab shall be used as a residential floor. Use M 20 grade of concrete and hysd-steel bars.

18

DESIGN OF FLAT SLABS

18.1 INTRODUCTION

The reinforced concrete slabs with or without drops, supported generally without beams by columns with or without drops as shown in Fig. 18.1, Fig. 18.2 and Fig. 18.3 are called **flat slabs.** The flat slabs are also referred as **two-way column-supported slabs**. In that slabs, the slabs are cast integrally with the columns and beams are not provided. The flat slabs may be solid or may have recesses formed on the soffit which comprise of a series of ribs in two directions. The recesses may be formed by removable or permanent filler blocks. The flat slabs as shown in Fig. 18.1 (viz., flat slabs without drop and column without column head) are built where the span is small and the load is light. Such flat slabs are termed as flat plates. The flat slabs as shown in Fig. 18.2 (viz., flat slabs without drop and column with column with column head) are constructed where, large bending moments develop a round the columns. The top of column is enlarged in the shape of an inverted frustum, referred as the *column head* or *column capital*.

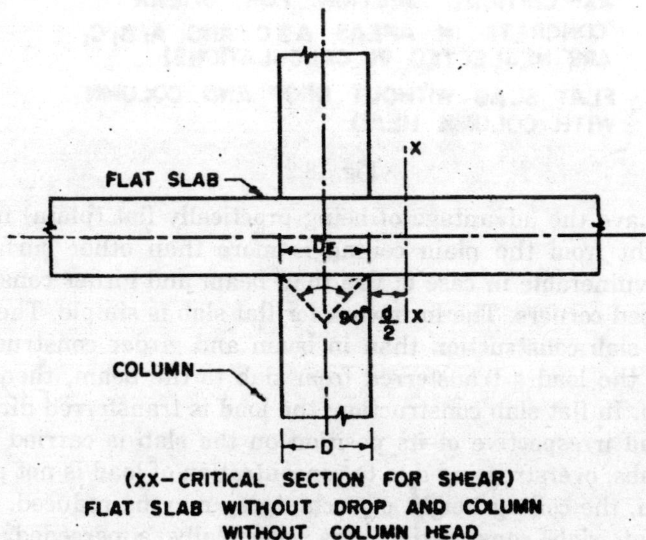

Fig. 18.1

The columns with column capital are referred as flared columns. The flat slabs as shown in Fig. 18.3, (*viz.* flat slabs with drop and column with column head) are made where the large shear forces also develop along with large bending moments. The thickness of slab around the column is increased to resist the shear (or diagonal tension). This thickened portion or increased thickness of slab is known as the *drop panel*, or simply *the drop*. Initially, the flat slabs were developed for heavy loads and large spans and typically using the flared column capital and often the thickened slab around the column (a drop panel). The flat slabs have long been recognized as the most economical construction for heavy service loads upto 5 kN/m².

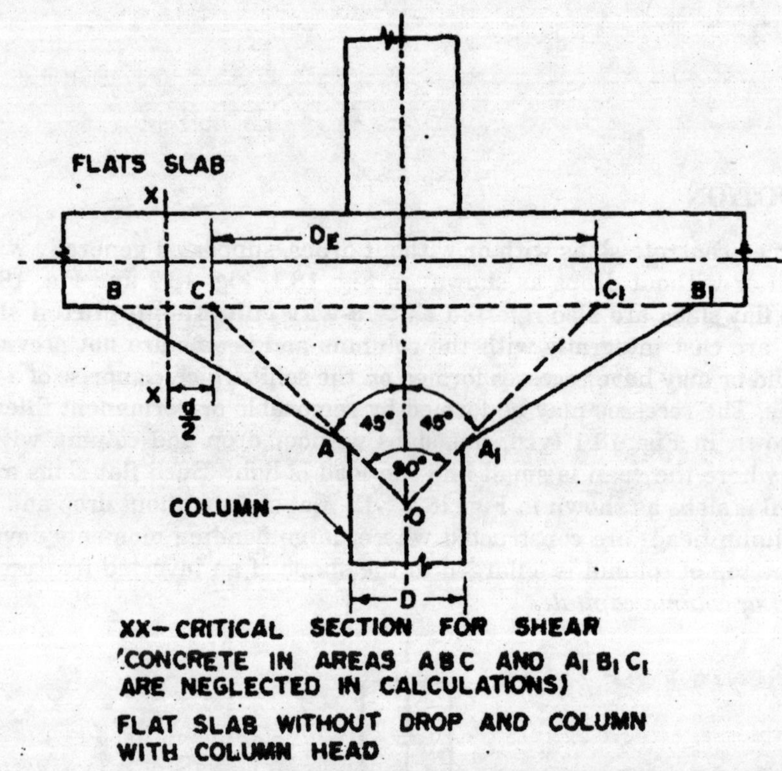

Fig. 18.2

The flat slabs have the advantage of being practically flat (plain) from the underside. The diffusion of light from the plain ceiling is more than other surfaces. The flat slab construction is less vulnerable in case of fire than beam and girder construction because of the absence of exposed corners. The formwork for flat slab is simple. The use of concrete is more logical in flat slab construction than in beam and girder construction. In beam and girder construction, the load s transferred from slab to the beam, then from beam to the girder to the column. In flat slab construction, the load is transferred directly by the slab to the column. The load irrespective of its position on the slab is carried by the whole slab. Therefore, in flat slabs, overstressing due to concentration of load is not possible. Because of absence of the beam, the ceiling height of each storey may be reduced. Owing to its many advantages, the flat slab construction has practically superseded beam and girder construction for spans upto 10 m and service loads upto 5 kN/m².

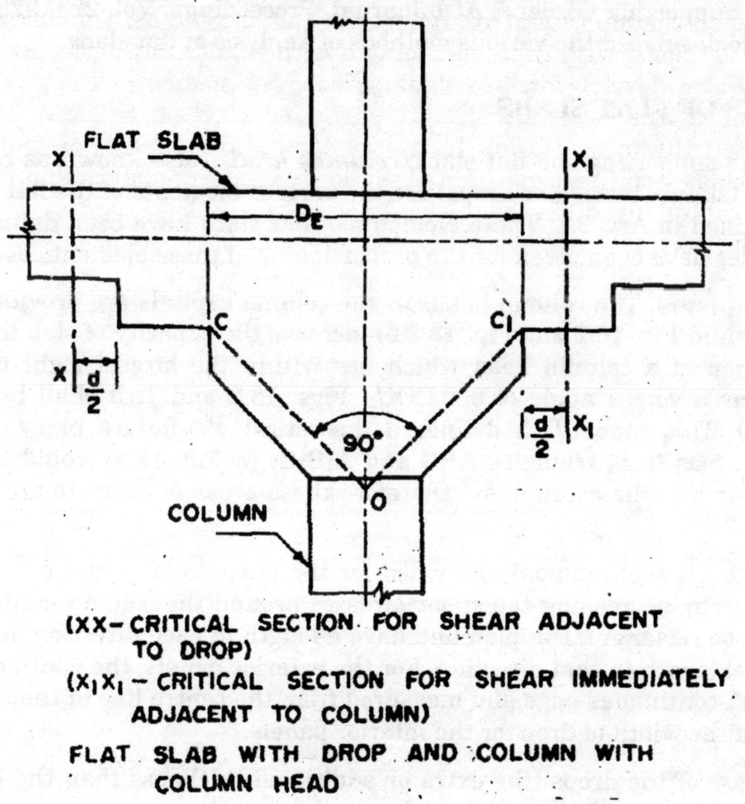

Fig. 18.3

The columns, column heads, slabs and drop panels are the various elements of the flat slab.

18.2 DEVELOPMENT OF FLAT SLABS

Originally Professor O.W. Norcross patented the flat slab floors in the United States on April 29, 1902. Since then, several systems of providing reinforcement (namely, the four-way system, two way system, three way system and the circumferential system developed and patented. Professor C.A.P. Turner and American was one of the early advocates of a flat-slab system (called as `*mushroom*' system. Professor Turner explained the functioning of slab by intuition and applied a proof load to the completed slab. From 1910 to 1920, numerous flat slab structures were load tested. These slabs performed well under test. In 1947, Professor, J. Neils Thompson and Professor Phil M. Ferguson carried out such a test on a flat slab constructed about in 1912. The flat slab although cracked badly but its strength was seen surprising.

Professor J.R. Nichols (Statical Limitation upon the steel Requirement in Reinforced concrete Flat Slab Floors' Trans. ASCE, Vol. 77, 1914), Professor H.M. Westergaard and Professor W.A. Slater (`Moments and stresses in Slabs', Proc. ACI. Vol. 17, 1921). Professor Joseph A. Wise (`Calculation of Flat Plates by the Elastic web Method', ACI Journal, Proceedings, Vol. 24, 1928), H.M. Westergaard (`Formulae for the design of Rectangular

Floor Slabs and Supporting Girders', ACI Journal, Proceedings, Vol. 22, 1926).. . are several investigators who developed the various methods of analysis of flat slabs.

18.3 ELEMENTS OF FLAT SLABS

The *columns* supporting the flat slabs, *columns heads* (also known as column capitals) and *drop panels* (simply termed as drops) are the various elements of the flat slabs and these have been explained in Art. 9.1. These elements of flat slabs have been defined in IS : 456-2000 and the rules have been given for the proportioning of these elements as under.

1. Column heads. The column heads or the column capitals are provided on the top of columns as shown in Fig. 18.2 and Fig. 18.3 to increase the capacity of slab to resist two-way shear. The portion of a column head which lies within the largest right circular cone or pyramid that has a vertex angle of 90° COC_1. Figs. 18.2 and 18.3 shall be considered for design purposes. This cone COC_1 defines a theoretical 45° failure plane. The portion of concrete in the areas (e.g., triangles ABC and $A_1B_1C_1$ in Fig. 18.2) would be ineffective in transferring shear into the column and therefore these areas of concrete are not included in the calculations.

2. Drops. Drops are commonly provided for the purpose of increasing the thickness of the slab and thereby decreasing the shear stresses around the column supports. The drops provided should be rectangular in plan and have a length in each direction not less than one-third of the panel length in that direction. For the exterior panels, the width of drops at right angles to the non-continuous edge and measured from the centre line of the columns shall be equal to one-half the width of drop for the interior panels.

The thickness of the drops (the extra or additional thickness than the overall depth of the slab) may be at least 0.25 times the thickness of the slab.

The drops when provided also decrease the area of reinforcement for negative bending moments at the supports.

3. Panels. The panels of flat slabs are the portions of slab bounded on each of its four sides by the centre-lines of the columns.

4. Thickness of flat slab. The thickness of flat slab shall be generally controlled by considerations of span to effective depth ratios. For calculating the modification factor for percentage of reinforcement in tension, the average percentage of steel across the whole width of the panel at mid-span should be used. The minimum thickness of slab shall be 125 mm.

For the slabs with drops satisfying the requirement for drops given above, span to effective depth ratios are applied directly.

For the slabs with drops which do not satisfy the requirements for drops given above, the span to effective depth ratios shall be multiplied by 0.9.

For the span to effective depth ratio, the longer span shall be considered.

For the purpose of design the flat slab is considered as divided in each direction into *column strip* and *middle strip* as shown in Fig. 18.4. The column strip and the middle strip are defined as below :

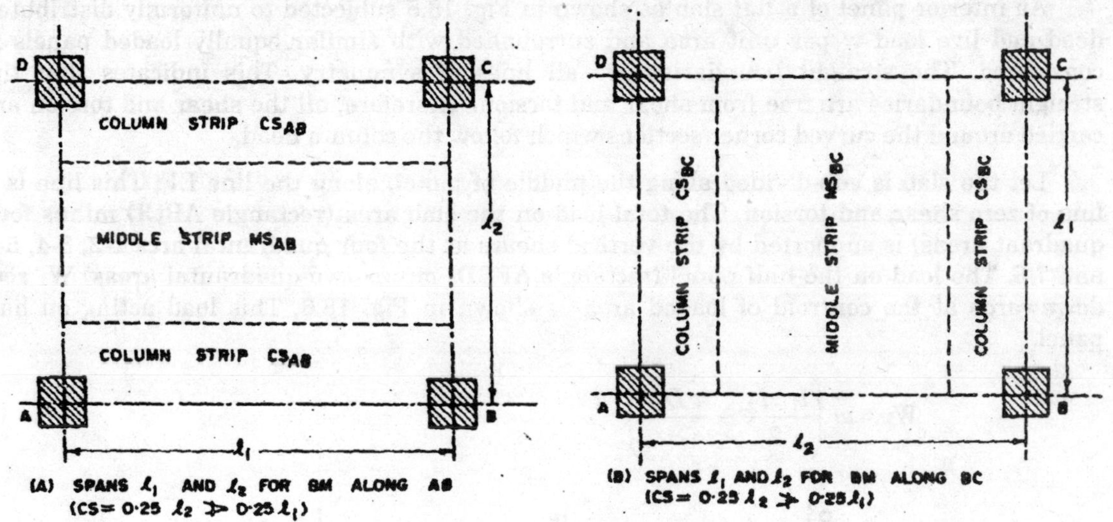

Fig. 18.4

4. Column strip, CS. The column strip of a flat slab is defined as a design strip having width of 0.25 l_2, but its width should not be greater than 0.25 l_1 on each side of the column centreline.

5. Middle strip, MS. The middle strip of a flat slab is defined as a design strip bounded on each of its opposite sides by the column strip.

The span l_1 is the distance measured from centre to centre of supports in the direction in which are total bending moment, M_o is calculated. The span l_2 is the distance measured from centre to centre of supports in the direction transverse to the span, l_1.

6. Clear span. The clear span is defined as the distance measured from face to face of column capitals, brackets or walls in the direction of span, l_1. The clear span should not be less than 0.65 l_1. This suggests that the maximum total width of column head is limited to 0.35 l_1 for the purpose of design.

IS : Code has used the symbol l_n to represent the clear span. It is suggested that this symbol may be modified as l_{1-n}, which indicates that it is to be measured in the direction of span, l_1.

18.4 STATICS OF A FLAT SLAB

Professor J.R. Nichols in 1914 has shown in his paper titled as `Statical Limitation upon the Steel Requirement in Reinforcement Concrete Flat Slab Floors' published in ASCE Trans Vol. 77, 1914 that as per requirement of statics, total design moment for a span, M_o is equal to the sum of the negative moment, M_n and the positive moment, M_p, of that span (viz., $M_o = M_n + M_p$). However, the present procedures are primarily based upon the papers published by Westergaard (H.M. Westergaard and W.A. Slater, `Moments and stresses in slabs' Proc. ACI, Vol. 17, 1921 and H.M. Westergaard, Formulae for the Design of Rectangular Floor Slabs and supporting Girders, ACI, Journal, Proceedings, Vol. 22, 1926).

An interior panel of a flat slab as shown in Fig. 18.5 subjected to uniformly distributed dead and live load w per unit area and surrounded with similar equally loaded panels is considered. The straight boundaries are all lines of symmetry. This indicates that the straight boundaries are free from shear and torsion. Therefore, all the shear and torsion are carried around the curved corner sections which follow the column head.

Let the slab is sub-divided along the middle of panel, along the line EF. This line is a line of zero shear and torsion. The total load on the slab area (rectangle ABCD minus four quadrant areas) is supported by the vertical shears at the four quadrantal arcs 1-2, 3-4, 5-6 and 7-8. The load on the half panel (rectangle AFBD) minus two quadrantal areas) W_1 acts downwards at the centroid of loaded area as shown in Fig. 18.6. This load acting on half panel.

$$W_1 = w\left(\frac{l_1 . l_2}{2} - \frac{\pi D_E^2}{8}\right) \qquad \text{(i)}$$

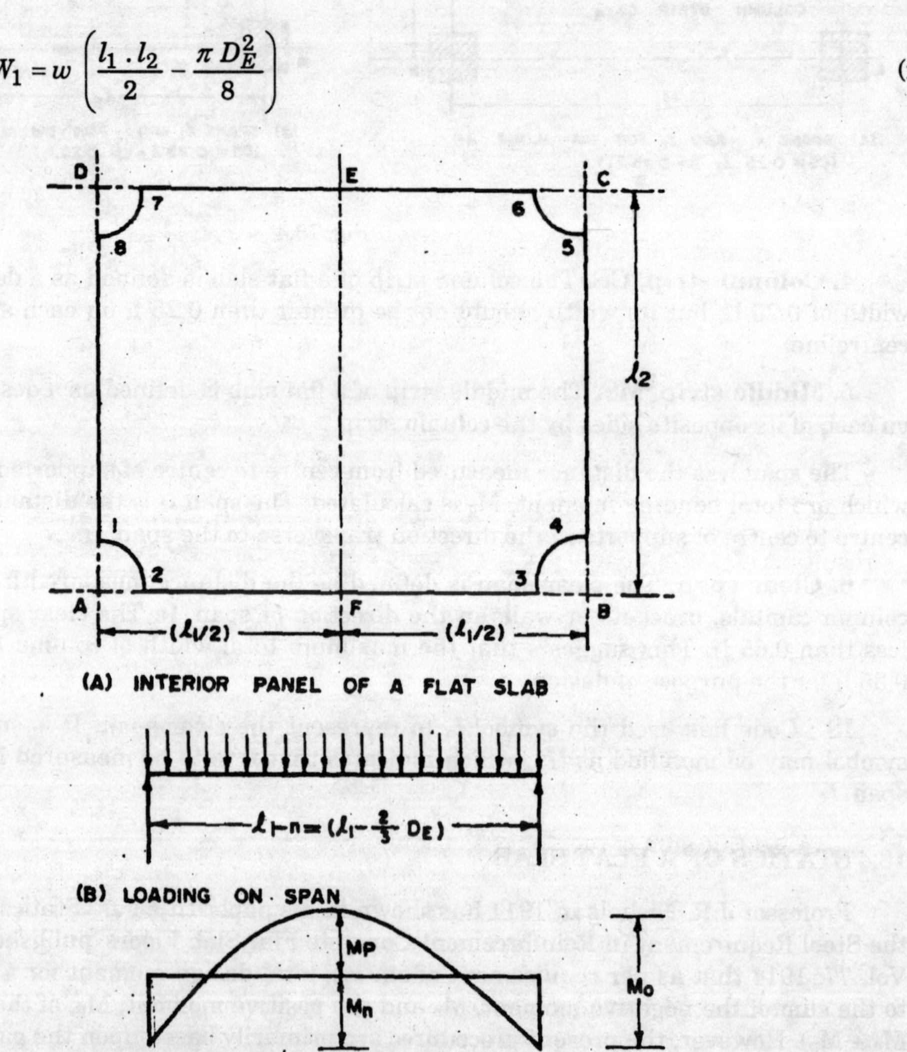

Fig. 18.5

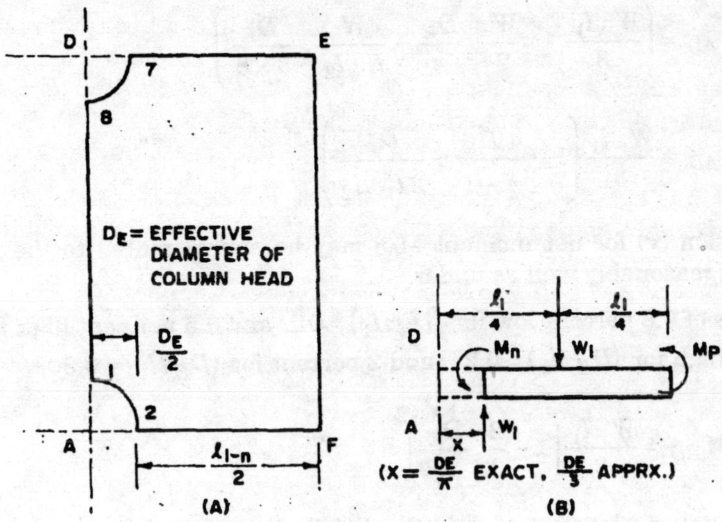

Fig. 18.6

The moment of this load about AD

$$M_{AD.W} = \left(\frac{w \,.\, l_1 \,.\, l_2}{2} \,.\, \frac{l_1}{4} - w\, \frac{\pi \,.\, D_E^2}{8} \,.\, \frac{2}{3} \, \frac{D_E}{\pi} \right)$$

$$\text{or } M_{AD.W} = \left(\frac{w \,.\, l_1 \,.\, l_2 \,.\, l_1}{8} - \frac{w \,.\, D_E^3}{12} \right) \quad \text{(ii)}$$

The upward shear W_1 is uniformly distributed around the quadrants 1-2 and 7-8. The resultant acts at a distance (D_E / π) from AD. The moment of this shear about AD

$$M_{AD.S} = w \left(\frac{l_1 \,.\, l_2}{2} - \frac{\pi \,.\, D_E^2}{8} \right) . \frac{D_E}{\pi}$$

$$\text{or } M_{AD.S} = \left(\frac{w \,.\, l_1 \,.\, l_2}{2} \,.\, \frac{D_E}{\pi} - \frac{w \,.\, D_E^3}{8} \right) \quad \text{(iii)}$$

Algebraic sum of moments $M_{AD.W}$ and $M_{AD.S}$

$$M_{AD} = \left(\frac{w \,.\, l_1 \,.\, l_2 \,.\, l_1}{8} - \frac{w \,.\, D_E^3}{12} - \frac{w \,.\, l_1 \,.\, l_2}{2} \,.\, \frac{D_E}{\pi} + \frac{w \,.\, D_E^3}{8} \right) \quad \text{(iv)}$$

Let $(W = w \,.\, l_1 \,.\, l_2)$, then

$$M_{AD} = \left(\frac{W \,.\, l_1}{8} - \frac{w \,.\, l_1 \,.\, l_2}{2} \,.\, \frac{D_E}{\pi} - \frac{w \,.\, D_E^3}{12} + \frac{w \,.\, D_E^3}{8} \right)$$

or $$M_{AD} = \left(\frac{W \,.\, l_1}{8} - \frac{W}{2} \cdot \frac{D_E}{\pi} + \frac{W}{l_1 \,.\, l_2} \cdot \frac{D_E^3}{3 \times 8} \right)$$

or $$M_{AD} = \frac{W \,.\, l_1}{8} \left(1 - \frac{4\, D_E}{\pi \,.\, l_1} + \frac{D_E^3}{3\, l_1^2 \,.\, l_2} \right) \tag{v}$$

The expression (v) for net moment M_{AD} may be approximated to the terms within the long parenthesis reasonably well as under.

(The error is of 0.5 percent low for $(D_E / l_1) = 0.1$ and 0.5 per cent high for $(D_E / l_1) = 0.2$. It is 1.3 percent high for $(D_E / l_1) = 0.25$ and 2 percent for $(D_E / l_1) = 0.30$.)

$$M_{AD} = \frac{W \,.\, l_1}{8} \left(1 - \frac{2}{3} \cdot \frac{D_E}{l_1} \right)^2 \tag{18.1}$$

This static analysis does not explain the distribution of total design moments into its two components positive moment and negative moment. It also does not explain the variation of these moments along the width of slab. The possible torsional moments around the column head likely to act due to non-uniform tangential bending are also neglected in this static analysis. The static analysis fails to give the total picture as regards the behaviour of large areas of this type of construction.

The static analysis and the experimental tests carried out on the flat slab were co-related. The interior panel of flat slabs acts in some form of an arch, and there is a horizontal thrust. Under favourable conditions, the arching action greatly strengthens interior panel. It adds some reserve strength to edge panels. Looking to the extent of the reserve strength of flat slabs in experimental tests carried in 1960's, ACI code recommended the use of

$$M_o = \left(\frac{w \,.\, l_2 \,.\, l_1^2 - n}{8} \right) = \left(\frac{W \,.\, l_1 - n}{8} \right) \tag{18.2}$$

where, W = (w . l_2 . l_1 – n)

This value of total design moment, M_o is based upon the clear spans l_1 – n between the square supports. In case the supports are not square or rectangular, these are replaced (for calculating the value of l_1 – n with square columns of equal area.

18.5 BEHAVIOUR OF FLAT SLABS IN BENDING

In case two-way slabs are supported by relatively shallow and flexible beams, a number of new considerations (as regards local behaviour of slabs surrounding the internal columns with respect to transfer of moment from the slab to the column, local behaviour of edge and corner columns at their intersections with slab, overall design and detailing.. . etc.) are introduced. the flat slabs are reinforced in two or more directions to transfer the loads to the columns. Therefore, the flat slabs themselves becomes grid of two-way shallow beams supported directly on the columns. The behaviour of two-way slabs along the column lines may be considered to act as beams. As these strips pass through the columns, these are called as **column strips**. The strip of flat slab between two adjacent column strips of slab is called as **middle strip.**

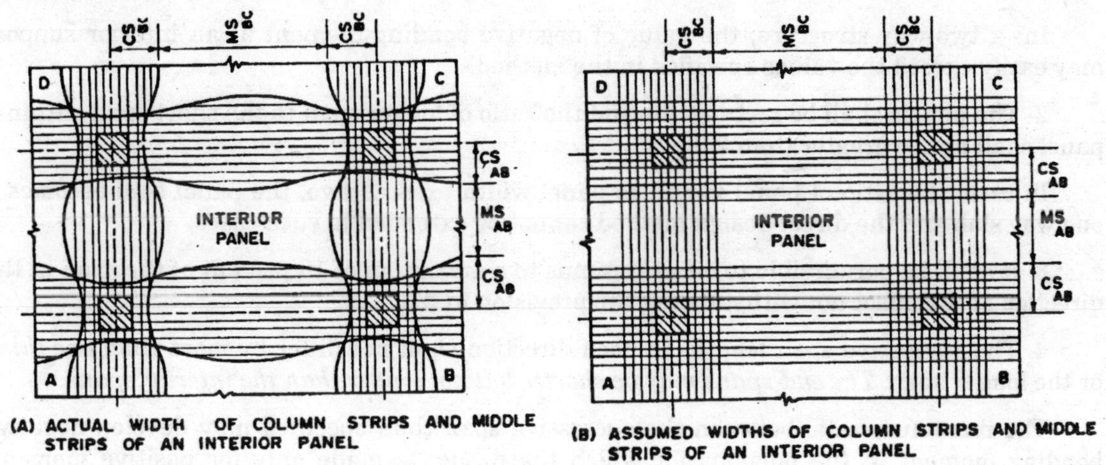

Fig. 18.7

Professors Regan, P.E. and C.W. Yu have discussed the various aspects of behaviour of flat slabs in `Limit State Design of Structural Concrete' Chato and Windus, England, 1973. The actual width of column strips (the portions of slab acting s beam between the columns) is shown in Fig. 18.7 (A). For simplification, the width of column strip has been idealised and assumed as uniform width and adopted in the various codes as shown in Fig. 18.7 (B).

The behaviour of column strip is like that of a continuous beam supported on columns. The deflected shapes of column strips remain identical to that of the continuous beam, (viz., deflections at column supports remain zero, and maximum at mid-spans). The middle strips may be seen as suspended between the column strips. The middle strip deflects like a plate supported on all its four sides by flexible beams. The resultant deflections of flat slabs are generally more than more rigid beam-slab construction. At the centre of the panel, the flat slab assumes saucer shape. Around the column, the flat slab remains convex upward, which gradually becomes concave at the centre of panel. There shall be point of inflection on such deflected shape. The locus of point of infection is a circle round the column. The column strips get negative moments near the column supports and the positive moments in the middle portion. The middle strip gets negative moments near the middle of column strips and positive moment at its centre.

The behaviour of flat slab in shear shall be discussed later.

18.6 DIRECT DESIGN METHOD

The direct design method was formerly called as *empirical method.* The direct design method is generally simpler to use than the other method of analysis and design of the flat slab. However, the direct design method is restricted to very regular layouts and normal live-load to dead-load ratios which shall be outlined in this section. The direct design method of analysis and design of flat slab may be adopted in case the following limitations specified in IS : 456-2000 are satisfied :

Limitations. The flat slabs shall fulfil the following conditions :

1. There shall be minimum of three continuous spans in each direction.

In a two-way structure, the value of negative bending moment at an interior support may easily exceed the values specified in the method.

2- The panels shall be rectangular and the ratio of longer span to the short span within a panel shall not be greater than 2.0.

In case, the ratio of panel length to panel width exceeds two, the panel slab becomes a one way slab and the direct design method cannot be extended to such cases.

3. It shall be permissible to offset columns to a maximum of 10 percent of the span in the direction of the offset not withstanding the provision in (2).

4. The successive span lengths in each direction shall not differ by more than *one-third* or the longer span. *The end spans may be shorter but not longer than the interior spans.*

The maximum ratio between the successive span than specified may develop negative bending moment in the portions for which the design is made only for positive moment. When the successive spans differ in the transverse direction, the average value should be considered for the determination of total design moment, M_o.

5. The design live load shall not exceed *three times* the design dead load.

This limitation on the ratio of live load to dead load is imposed in consideration of the effects of the pattern loading. If this ratio exceeds three, moments produced by the pattern loading would be more severe than those calculated according to the direct design method.

The flat slabs carrying uniformly distributed gravity loads may be analysed by the direct design method in case above five conditions described under limitations are fulfilled. The flat slabs are then analysed and total design moment in each direction shall be determined as under : (as step 1).

1. Total design moment

In the direct design method, the total design moment in each direction shall be obtained for a strip bounded laterally by the centre line of the panel on each side of the centre line of the supports as below :

Total design moment M_o for span AB, Fig. 18.5 as per Eq. 18.2 (B)

$$M_o = \left(\frac{W \,.\, l_1 - n}{8}\right) \qquad \text{(i)}$$

where, W = design load on the area $l_2 \,.\, l_1 - n$

or $\quad W = (w_d + w_l) \,.\, l_2 \,.\, l_1 - n$ (ii)

w_d = uniformly distributed dead load per unit area

w_l = uniformly distributed live load per unit area

$l_1 - n$ = clear span extending from face to face of columns, capitals, brackets walls, but not less than 0.65 l_1.

l_1 = length of span measured between the centre line to centre line of the two adjacent columns in the direction of M_o, and

l_2 = length of span measured between the centre line to centre line of the two adjacent columns in the direction transverse to span l_1.

Total design moment M_o is the absolute sum of positive and negative bending moments as shown in Fig. 18.5(B). The design requirements are for rectangular columns, since rectangular shapes are more widely used and more experimental data are available for the flat slabs supported by the rectangular columns. If the columns are circular or of any other shapes, equivalent rectangular (square) shape having equal areas should be considered.

Two separate values of negative design moment, M_n and positive design moment, M_p (for an interior span and for an end span) are then found (as step 2) as fractions of M_o as follows :

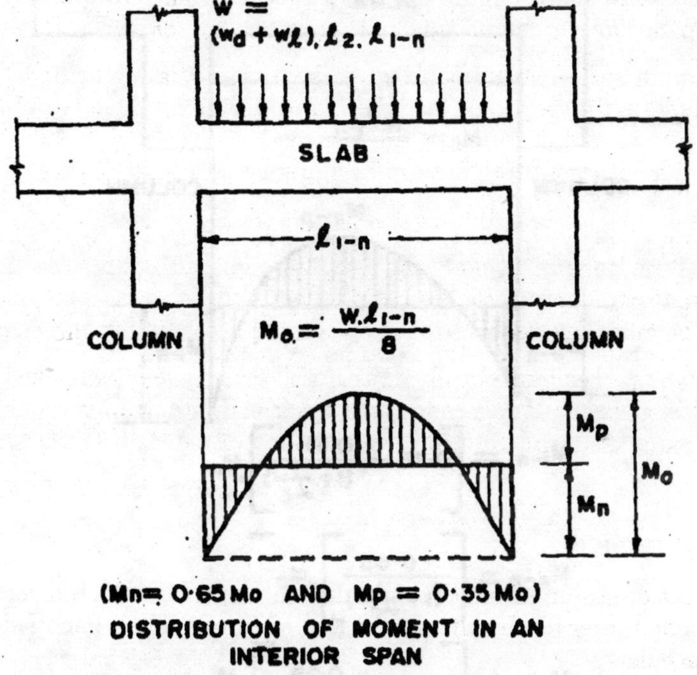

Fig. 18.8

2. Negative and positive design moments

The negative design moment shall be located at the face of rectangular supports (circular supports are considered as square supports having the equal area).

2 (A) An interior span

The total design moment, M_o shall be distributed as follows, as shown in Fig. 18.8.

(i) Negative design moment, $M_n = (0.65\ M_o)$ (18.3)

(ii) Positive design moment, $M_p = (0.35\ M_o)$ (18.4)

It is to note that these coefficients 0.65 and 0.35 for negative moment, and positive moment, respectively, may have to be modified for the effects of pattern loading described in Art. 18.7.

2 (B) An end span

The total design moment, M_o shall be distributed as follows as shown in Fig. 18.9.

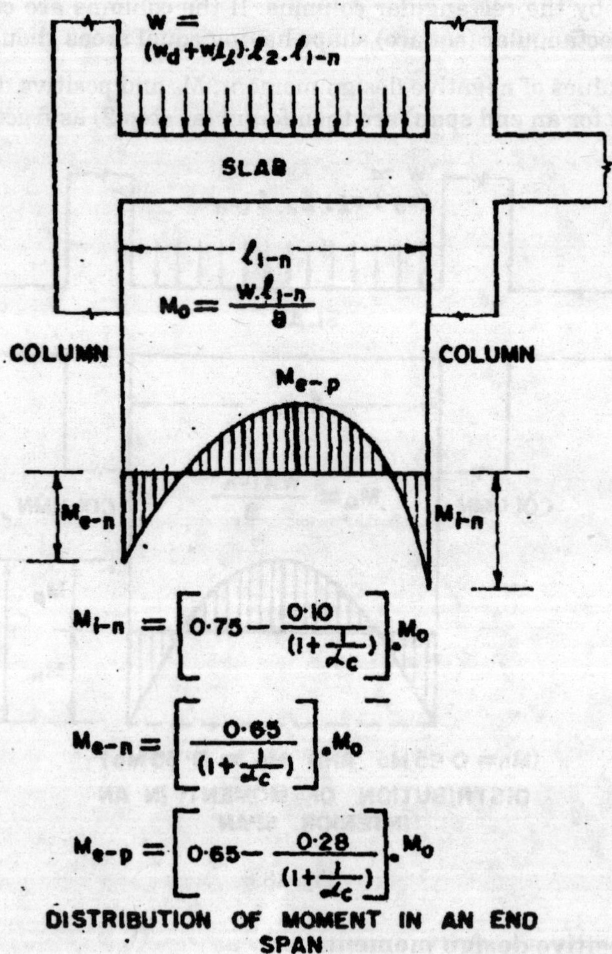

Fig. 18.9

(i) Interior negative design moment

$$M_{i-n} = \left[0.75 - \frac{0.10}{\left(1 + \dfrac{1}{\alpha_c}\right)}\right] M_o \qquad (18.5)$$

(ii) Exterior negative design moments

$$M_{e-n} = \left[\frac{0.65\ M_o}{\left(1+\dfrac{1}{\alpha_c}\right)}\right] \tag{18.6}$$

(iii) Positive design moment

$$M_{e-p} = \left[0.63 - \frac{0.28}{\left(1+\dfrac{1}{\alpha_c}\right)}\right] M_o \tag{18.7}$$

where α_c is the ratio of flexural stiffness of the exterior columns to the flexural stiffness of slab at a joint taken in the direction moments are being determined

$$\alpha_c = \left(\sum K_c / K_s\right) \tag{18.8}$$

where $\sum K_c$ = sum of the flexural stiffness of the columns meeting at the joint, and

K_s = flexural stiffness of the slab expressed as moment per unit rotation

It is to note that these coefficients of Eqs. 18.5 and 18.6 may have to be modified for the effects of pattern loading described in Art. 18.7.

The negative moment section shall be designed to resist the larger of two interior negative design moments determined for the spans framing into a common unless an analysis is made to distribute the unbalanced moment in accordance with the stiffness of the adjoining parts.

These negative and positive design moments found above (as in step 2) are adjusted in due proportions to column strip and middle strips (as step 3) as under :

3. Distribution of bending moments across the panel width

The bending moments at critical cross-section shall be distributed to the column strips and middle strips as specified in IS : 456-2000 as applicable.

3 (A) Negative moments in the column strips, M_{nc} at an interior support

At an interior support the column strip shall be designed to resist 75 per cent of the total negative moment in the panel at that support, viz.,

$$M_{nc} = 0.75\ M_n \tag{18.9}$$

$$\text{or}\quad M_{nc} = 0.75\,.\left(0.65\ M_o\right) \therefore M_{nc} = 0.4875\ M_o \tag{18.10}$$

There are two column strips of a span. Therefore, the moments, M_{nc} is distributed in two portions.

(The critical section for negative bending moment shall be at support faces of each span and that for positive bending moment is the mid-span).

3 (B) Negative moments in the column strips, M_{nc} at an exterior support

(i) At an exterior support, the column strip shall be designed to resist the total negative moment in the panel at that support, viz.,

$$M_{nc} = M_n \tag{18.11}$$

There are two column strips in a span. Therefore, the moment, M_{nc} is distributed in two portions.

(ii) Where the exterior support consists of a column or a wall extending for a distance equal to or greater than three-quarters of the value of l_2, the length of span transverse to the direction moments are being determined, the exterior negative moment shall be considered to be uniformly distributed across the length, l_2.

3 (C) Positive moment in the column strip M_{pc} for each span

For each span, the column shall be designed to resist 60 percent of the total positive moment in the panel, viz.,

$$M_{pc} = 0.60, M_p \tag{18.12}$$

For the interior span, $M_{pc} = 0.60 \times (0.35\, M) \;\therefore = 0.210\, M_o$ (18.13)

There are two column strips in a span. Therefore, the moment, M_{pc} is distributed in two portions.

3 (D) Moments in the middle strip, M_{pm} and M_{nm}

The middle strip shall be designed on the following bases :

(i) That portion of the design moment not resisted by the column strip shall be assigned to the adjacent middle strip.

Since, $M_{pc} = 0.60\, M_p$, $M_{pm} = 0.40\, M_p$ (18.14)

For the interior span

$$M_{pm} = (0.40 \times 0.35\, M_o), \therefore M_{pm} = 0.14\ M_c \tag{18.15}$$

Since $M_{nc} = 0.75\ M_n$ $M_{nm} = 0.25\, M_n$ (18.16)

$$M_{nm} = 0.25\ (0.65\ M_o) = 0.1625\, M_o \tag{18.17}$$

(ii) Each middle strip shall be proportioned to resist the sum of the moments assigned to its two half middle strips.

(iii) The middle strip adjacent and parallel to an edge supported by a wall shall be proportioned to resist *twice* the moment assigned to half the middle strip corresponding to the first row of interior column.

The coefficients specified in this step 3 from (A) to (D) are based on a linear elastic analysis of moments in slabs. Actual distribution of the moments across the panel at EF and at AB, Fig. 18.5 are as shown in Fig. 18.10 (A) and (B), respectively. The exact variation of moments is governed by presence or absence of drops and column heads, as well as the intensity of loads. For the design purpose, it is assumed as shown by dotted lines in Fig. 18.10 for simplification and convenience for design purpose. A panel is treated as composed of imaginatory column strips and middle strips across which the moments are assumed to have constant values.

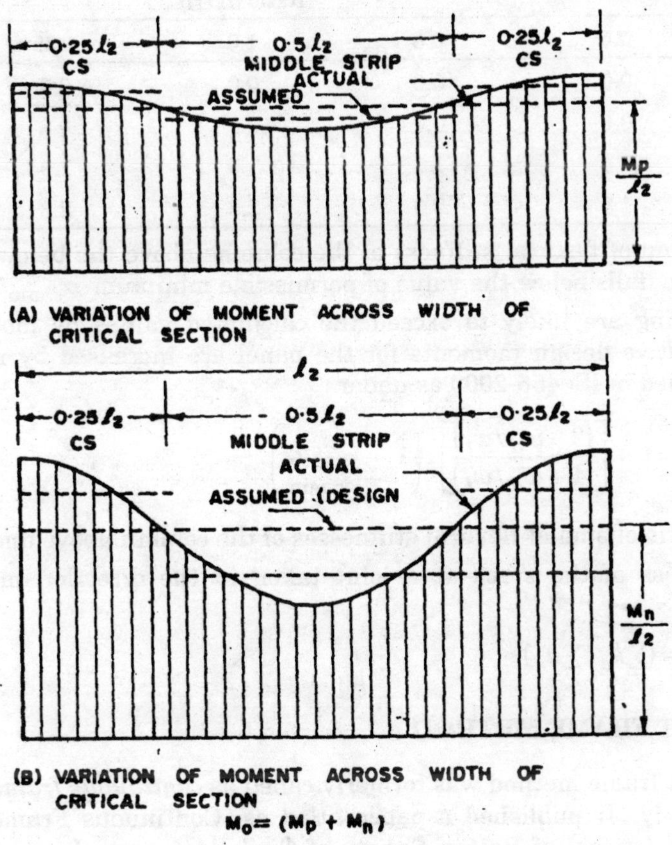

Fig. 18.10

18.7 EFFECTS OF PATTERN LOADING (Direct design method)

In the direct design method, the limitation on the ratio of live load to dead load is imposed in consideration of pattern loading. However, the direct design method does not specify any provision for the unbalanced live loads. The unbalanced live loads give the moments in columns arising from the effects of pattern loading. It is possible that the bending moments on the slab may exceed the prescribed values for loading over all spans by as much as *one hundred per cent.* The possible increases in the moments due to effects of pattern loading at service load level are limited by this procedure recommended in IS : 456-2000.

When the ratio of live load to dead load (w_l/w_d) exceeds 0.5 it is seen that the sum of the flexural stiffnesses of columns above and below the slab, $\sum k_c$ shall be such that the value of α_c is not less than appropriate minimum value, $\alpha_{c.\min}$ specified in Table 18.1.

TABLE 18.1. PERMISSIBLE VALUES OF $\alpha_{c.\min}$

Ratio $\left(\dfrac{\textit{Live load}}{\textit{Dead load}}\right)$	Permissible values of $\alpha_{c.\min}$				
	Ratio (l_2/l_1)				
	0.5	0.8	1.0	1.25	2.00
0.5	0.0	0.0	0.0	0.0	0.0
1.0	0.6	0.7	0.7	0.8	1.2
2.0	1.3	1.5	1.6	1.9	4.9
3.0	1.8	2.0	2.3	2.8	13.0

In case, the sum of flexural stiffness of the columns above the below the slab are such that the value of α_c falls below the value of permissible minimum $\alpha_{c.\min}$, the moments due to patterning loading are likely to exceed the calculated values by more than one-third. Therefore, the positive design moments for the panel are increased by multiplying by the coefficient β, specified in IS: 456-2000 as under :

$$\beta_s = 1 + \left[\frac{(2 - w_d / w_l)}{(4 + w_d / w_l)}\right] \cdot \left(1 - \frac{\alpha_c}{\alpha_{c.\min}}\right) \tag{18.18}$$

where α_c is the ratio of sum of flexural stiffnesses of the column above and below the slab to the flexural stiffness of the slabs at a joint taken in the direction moments are being determined.

$$\alpha_t = \left(\sum k_c / \sum k_s\right) \tag{18.19}$$

18.8 EQUIVALENT FRAME METHOD

The equivalent frame method was formerly called as *continuous frame method.* In 1948, Professor D. Peabody, Jr published a paper titled as `Continuous Frame Analysis of Flat Slabs' published in Journal of Boston Society of Civil Engineers, January, 1948. Professor Peabody proposed continuous frame method of analysis of that slabs. It appeared in refined form in ACI Code as the equivalent frame method. the complete documentation of this method may be seen in papers published by W.G. Corley, M.A. Sozen and C.P. Siess (``The Equivalent Frame Method for Reinforced Concrete Slabs', University of Illinois, USA, Department of Civil Engineering, Structures, Res. Series 218, June 1961) and W.G. Corley and J.O. Jirsa (`Equivalent Frame Analysis for Slab Design', Journal of ACI Vol. 67, No. 11, November, 1970).

The equivalent frame method is a general method. However this method is intended for analysis when the direct design method of analysis of flat slab cannot be used as each of six limitations on geometry and load are not satisfied for the proposed structure. The equivalent frame method is compulsorily adopted when the proposed structure is required to resist the lateral loads. The shear forces and bending moments may be calculated by analysing the structure as a continuous frame by moment distribution method. Kani's method or stiffness method. But, the equivalent frame method becomes quick and superior to other methods when the calculations may be carried out by electronic digital computer. The bending moments and shear forces may be determined by an *analysis of the structure as a continuous frame* by having the following assumptions :

Assumptions

(a) The reinforced concrete flat slabs supported by the columns (a three-dimensional slab and column structure) shall be considered to be made up of equivalent frames on column lines taken longitudinally and transversely through the building (a series of two dimensional frames) as shown in Fig. 18.11, and which are analysed for loads acting in the plane of the frames. Each frame consists of row of equivalent columns or supports and a broad continuous beam. The beam (*viz.,* slab beam) includes the portion of the slab bounded by panel centrelines on either side of the columns, together with column-line beams or drop panels if used.

(b) Each such frame may be analysed in its entirely (when the lateral loads are to be resisted by the frame), or for vertical loading, each floor thereof and the roof may be analysed separately with its columns being assumed fixed at their remote ends. Where slabs are thus analysed separately, it may be assumed in determining the bending moment at a given support that the slab is fixed at any two support panels distant thereform provided the slab continuous beyond that point.

Simplified method such as portal methods, are permitted for lateral loads on the normal types of structural frames. These simplified methods should not be applied to the equivalent frames constituting the flat slabs. The equivalent frames should be analysed in its entirely by Hardy Cross-method or other suitable method.

When the flat slabs carry the vertical (gravity) loads only, IS : 456-2000 permits the use of substitute frames with the further simplification that the slab is fixed at a support two panels away. The spans used for analysis should be the distance between the centre-lines of the supports and not the clear span.

(c) For the purpose of determining relative stiffness of members, the moment of inertial of any slab or column may be assumed to be that of the gross cross-section of the concrete alone.

(d) Variation of moment of inertia along axis of the slab an account of provision of drops shall be taken into account. In case of recessed or coffered slab which is made solid in the region of the columns, the stiffening effect may be ignored provided the solid part of the slab does not extend more than 0.15 I_{eff}, into the span measured from the centre line of the columns. The stiffening effect off flared column head may be ignored.

The relative stiffness of slabs and columns referred above in assumption (c) and the equivalent column referred above in assumption (a) have been discussed in detail in subsequent articles.

18.9 RELATIVE STIFFNESS OF SLABS

For the purpose of determine the relative stiffness, the moment of inertial of any slab may be based on the gross-sections of the concrete alone (neglecting the reinforcement) and the variation in the moment of inertia along the axis of the slab-beam between supports is taken into account.

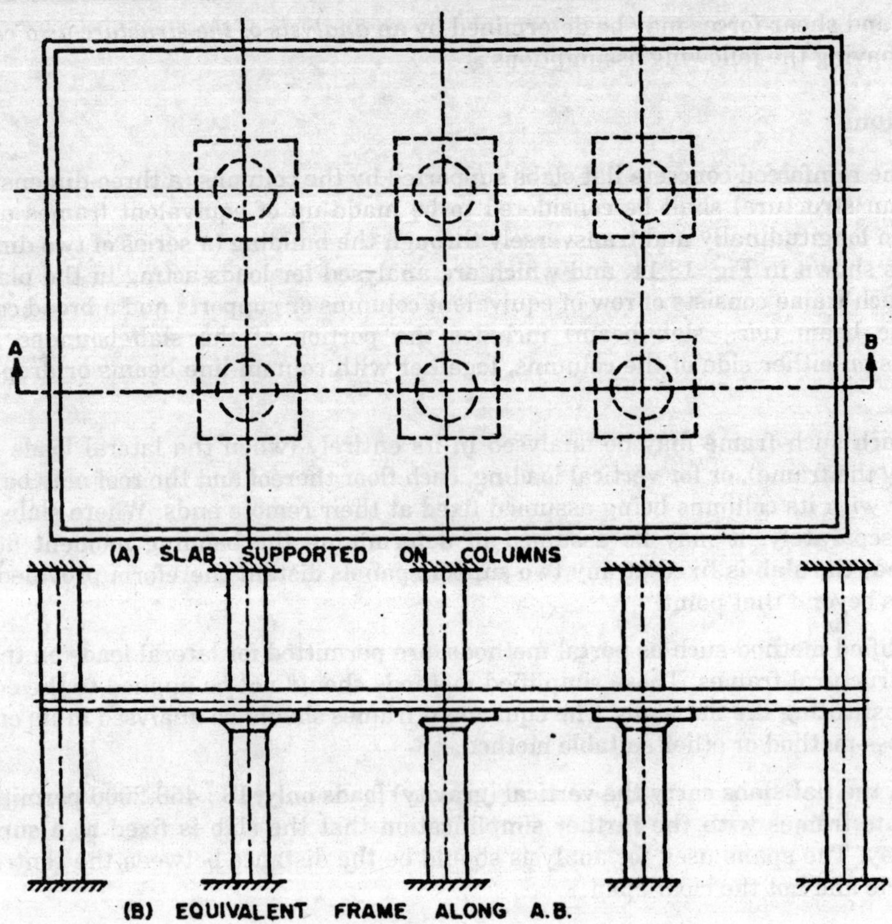

Fig. 18.11

The absolute flexured stiffness, K_s for the slab of *uniform section* is calculated from the following expression

$$K_s = \left(\frac{4\,E_c\,.\,I}{l_1}\right), \quad \textit{where} \tag{i}$$

I = moment of inertia of concrete slab based on the gross-section of concrete alone (neglecting the reinforcement)

l_1 = Appropriate length of the slab (length of the slab along the direction moment is considered)

$$I = \left(\frac{1}{12}\,l_2\,.\,D_f^3\right) \tag{ii}$$

l_2 = Width of the slab (span in the direction perpendicular to l_1)

D_f = Overall thickness of concrete slab.

In case the slab is having variable thickness along the length (*viz.*, the slab is having drop panel) then, the absolute flexural stiffness, K_s is calculated from the following expression

$$K_s = \left(\frac{k \cdot E_c \cdot I}{l_1}\right) \qquad \text{(iii)}$$

where, k is the stiffness factor for the slab (member). Its value shall be more than 4. The value of k for the slab without drop and the slab with drop may be noted from Table 18.2 and 18.3, respectively.

TABLE 18.2. STIFFNESS FACTORS FOR SLABS WITHOUT DROP PANELS (Fig. 18.12 (A) as per SP : 24-1983)

Column dimension (C_1/l_1)	Stiffness factor K
0.00	4.00
0.05	4.05
0.10	4.18
0.15	4.40
0.20	4.72
0.25	5.14
0.30	5.69
0.35	6.42

(It is to note that these values are applicable when $C_1/l_1 = C_2/l_2$. For other relationships between these ratios, these constants shall be slightly approximate).

TABLE 18.3. STIFFNESS FACTORS FOR SLABS WITH DROP PANELS (Fig. 18.3 (A) as per SP : 24-1983)

Column dimension (C_1/l_1)	Stiffness factor K
0.00	4.78
0.05	4.84
0.10	4.98
0.15	5.22
0.20	5.55
0.25	5.98
0.30	6.64

(It is to note that these values are applicable when ($C_1/l_1 = C_2/l_2$). For other relationships between these ratios, these constants shall be slightly approximate).

For the slab, the first change in the moment of inertia from the mid-span normally occurs at the edge of drop panels (if they exist). The second change occurs at the edge of the column or column capital. The idealization for column type of slab elements are shown in Fig. 18.12, Fig. 18.13 and Fig. 18.14.

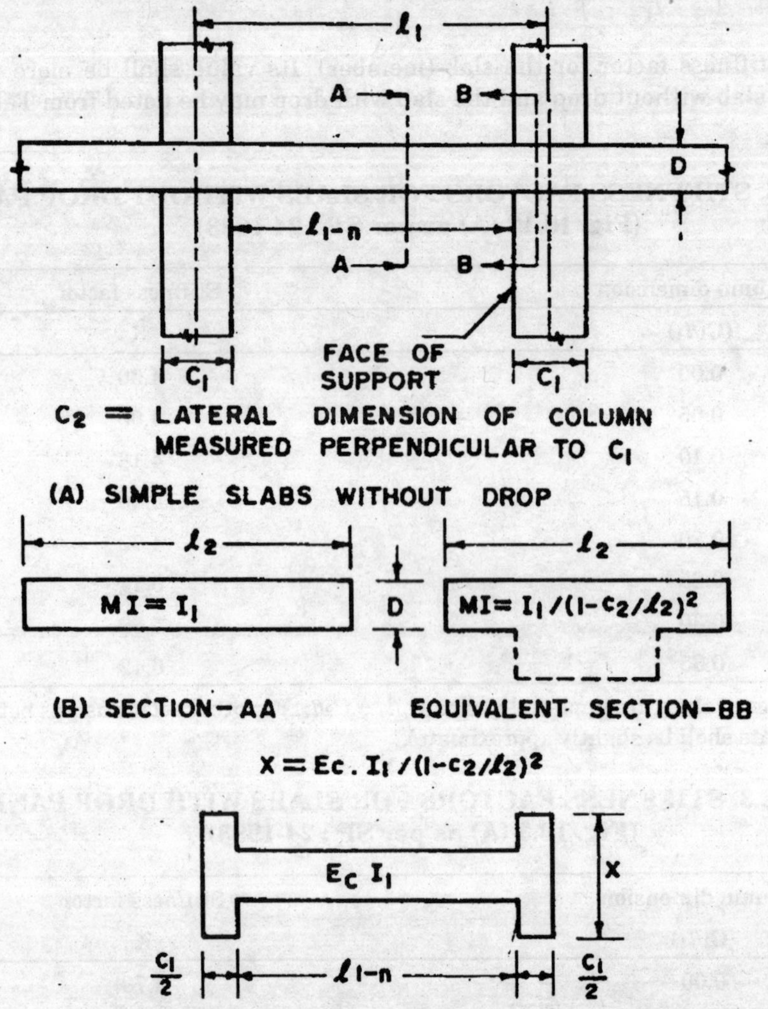

Fig. 18.12

The moment of inertia of the slab-beam from the face of support to the centre-line of support is assumed equal to the moment of inertia of the slab-beam at the face of support divided by $(1 - C_2/l_2)^2$, where C_2 is the width of rectangular support and l_2 is the span, both referring to the direction perpendicular to that in which the analysis is made.

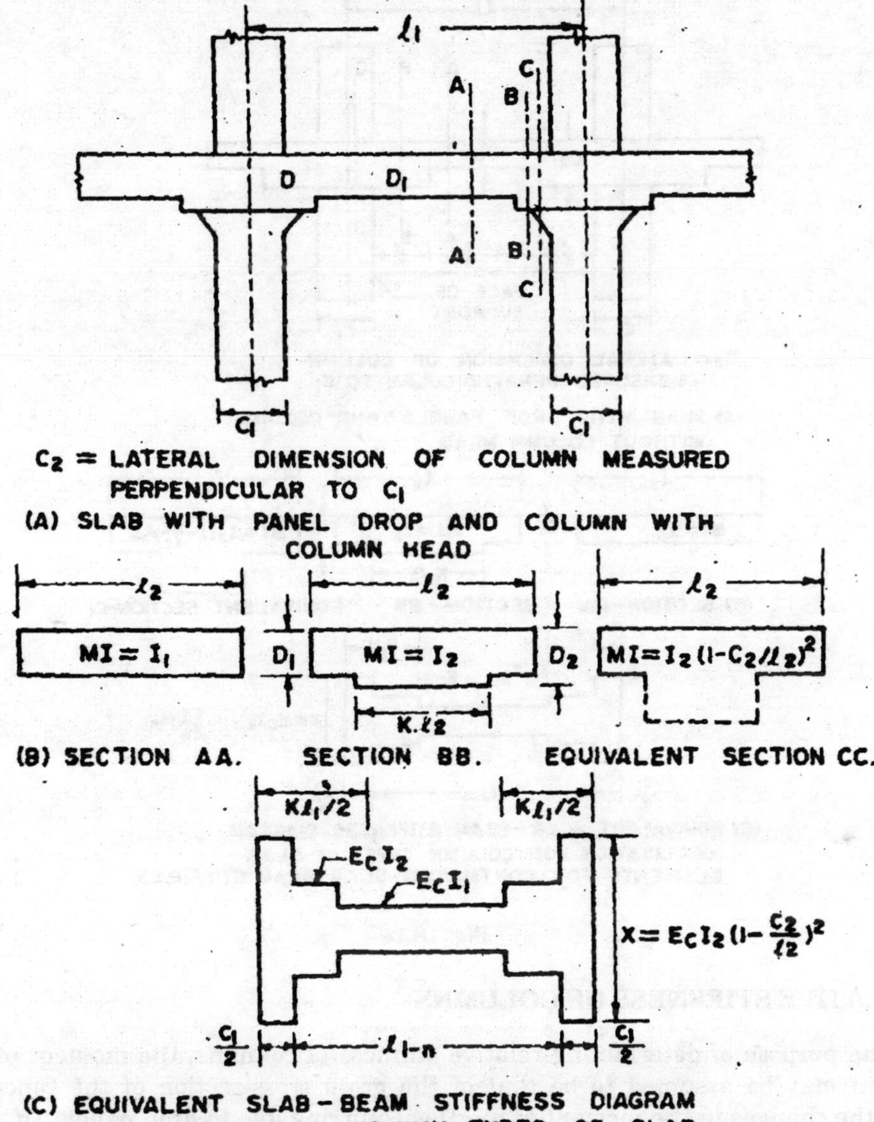

Fig. 18.13

The coefficients for *fixed end moment, stiffness,* K and the carryover factors for a slab without drop panels, for a slab with drop panels (depth equal to 1.25 times the slab depth and of length equal to *one-third* the span length) and for slab with column capital are given in SP 24-1983 (Explanatory handbook on IS : 456-2000). It is to note that the conventional coefficients for fixed-end moment, stiffness and carryover factors derived for the prismatic members will not be appropriate when the moment of inertia varies along the axis of the member.

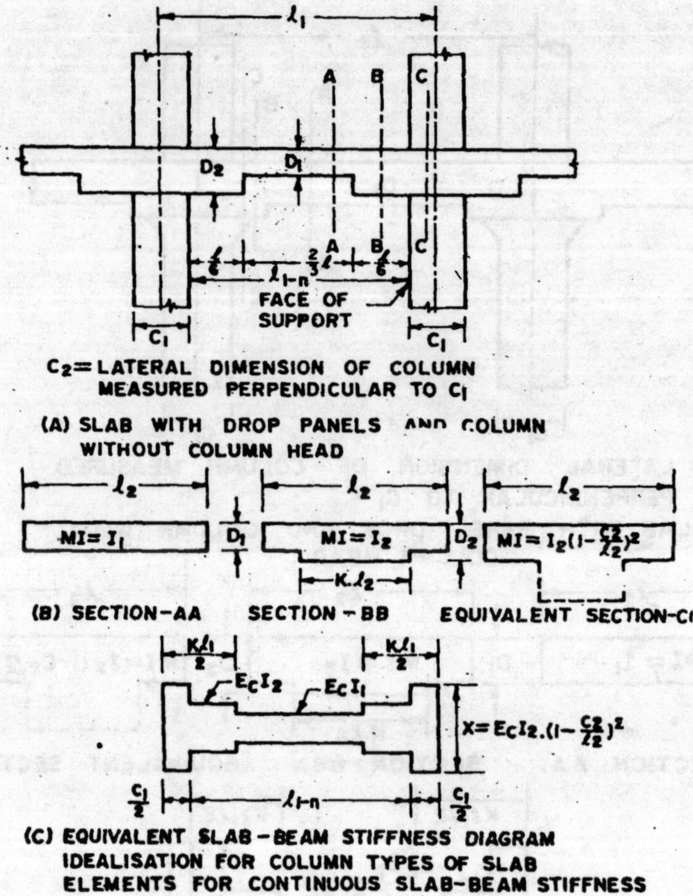

Fig. 18.14

18.10 RELATIVE STIFFNESS OF COLUMNS

For the purpose of determining relative stiffness of columns, the moment of inertia of any column may be assumed to be that of the gross cross-section of the concrete alone. However, the changes in the moment of inertias occurring due to drop panels, (if they exist), or due to column capital has to be taken into account. The idealisation for different types of column elements are shown in Fig. 18.15, Fig. 18.16 and Fig. 18.17.

For columns of uniform circular size, the flexural stiffness of the column shall be calculated from the following expression

$$K_c = \left(\frac{4 \, . E_c \, . I_c}{l_c}\right) \tag{i}$$

where, l_c is the length of column measured from the line of slab upto the centre line of adjacent slab as shown in Fig. 18.15.

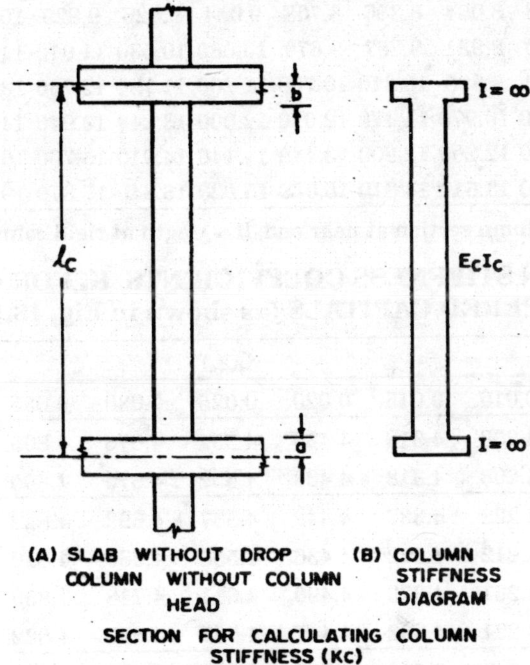

Fig. 18.15

In general, the flexural stiffness of the column may be expressed as under :

$$K_c = \left(\frac{k_c \, . \, E_c \, . \, l_c}{4l_c}\right), \quad where \quad I_c = \left(\frac{1}{12} \, . \, D\right) \tag{ii}$$

The values of column stiffness coefficients, k_c for the columns without tapered capitals (Reference : Simmonds, S.H. and Misic, J, `Design factors for Equivalent Frame Method' proceedings ACI, Vol. 68 (II), November, 1971) may be noted from Table 18.4.

The values of column stiffness coefficients, k_c for columns with 45° tapered capitals may be noted from Table 18.5.

TABLE 18.4. COLUMN STIFFNESS COEFFICIENTS, K_c FOR COLUMNS WITHOUT TAPERED CAPITALS (as shown in Fig. 18.15)

$\left(\frac{a}{l_c}\right)$	(b/lc)												
	0.00	0.02	0.04	0.06	0.08	0.10	0.12	0.14	0.16	0.18	0.020	0.22	0.24
0.00	4.000	4.082	4.167	4.255	4.348	4.444	4.545	4.651	4.762	4.878	5.000	5.128	5.623
0.02	4.337	4.433	4.533	4.648	4.747	4.862	4.983	5.110	5.242	5.384	5.533	5.690	5.856
0.04	4.709	4.882	4.940	5.063	5.193	5.330	5.475	6.627	5.787	5.958	6.138	6.329	6.533
0.06	5.122	5.252	5.393	5.539	5.693	5.855	6.027	6.209	6.403	6.608	6.827	7.060	7.310
0.08	5.581	5.735	5.898	6.070	6.252	6.445	6.650	6.868	7.100	7.348	7.613	7.897	8.203
0.10	6.091	6.271	6.462	6.665	6.880	7.109	7.353	7.614	7.893	8.192	8.513	8.859	9.203
0.12	6.659	6.870	7.094	7.333	7.587	7.859	8.150	8.461	8.796	9.157	9.546	9.967	10.430

0.14	7.292	7.540	7.803	8.084	8.385	8.708	9.054	9.426	9.829	10.260	10.740	11.250	11.810
0.16	8.001	8.291	8.600	8.931	9.287	9.670	10.080	10.530	11.010	11.540	12.110	12.740	13.420
0.18	8.796	9.134	9.498	9.888	10.310	10.760	11.260	11.790	12.730	13.010	13.700	14.470	15.310
0.20	9.687	10.080	10.510	10.970	11.470	12.010	12.600	13.240	13.940	14.710	15.560	16.490	17.530
0.22	10.960	11.160	11.660	12.200	12.800	13.440	14.140	14.910	15.760	16.690	17.210	18.870	20.150
0.24	11.820	12.730	12.960	13.610	14.310	15.080	15.920	16.840	17.870	19.000	20.260	21.650	23.260

Note : a = length of rigid column section at near end, b = length at rigid column section at far end.

TABLE 18.5. COLUMN STIFFNESS COEFFICIENTS, K_c FOR COLUMNS WITH 45° TAPERED CAPITALS (as shown in Fig. 18.17)

(b'/l_c)	(a/l_c)										
	0.00	0.005	0.010	0.015	0.020	0.025	0.030	0.035	0.040	0.045	0.050
0.0	4.000	4.102	4.208	4.318	4.433	4.552	4.676	4.805	4.940	5.080	5.226
	4.000	4.102	4.208	4.318	4.433	4.552	4.676	4.805	4.940	5.080	5.226
0.02	4.013	4.115	4.222	4.332	4.448	4.567	4.692	4.822	4.957	5.098	5.245
	4.003	4.105	4.212	4.322	4.436	4.558	4.680	4.810	4.944	5.085	5.232
0.04	4.050	4.154	4.261	4.373	4.490	4.611	4.738	4.869	5.006	5.149	5.298
	4.012	4.115	4.221	4.332	4.447	4.567	4.692	4.822	4.958	5.099	2.46
0.06	4.108	4.214	4.323	4.438	4.557	4.680	4.809	4.943	5.083	5.229	5.382
	4.025	4.129	4.236	4.347	4.463	4.584	4.170	4.841	4.978	5.120	5.269
0.08	4.185	4.293	4.406	4.523	4.645	4.772	4.905	5.042	4.186	5.336	5.492
	4.042	4.147	4.222	4.367	4.484	4.606	4.733	4.866	5.003	5.147	5.297
0.10	4.280	4.392	4.508	4.629	4.755	4.886	5.022	5.164	5.313	5.467	5.629
	4.063	4.168	4.277	4.391	4.590	4.633	4.741	4.895	5.034	5.180	5.331
0.12	4.393	4.509	4.628	4.754	4.884	5.020	5.161	5.308	5.462	5.623	5.790
	4.086	4.192	4.303	4.418	4.538	4.662	4.792	4.928	5.069	5.216	5.370
0.14	4.522	4.642	4.767	4.897	4.032	5.174	5.321	5.474	5.634	5.801	5.975
	4.112	4.219	4.331	4.448	4.569	4.695	4.827	4.964	5.107	5.257	5.413
0.16	4.667	4.793	4.923	5.059	5.200	5.347	5.301	5.661	5.828	6.002	6.184
	4.139	4.248	4.362	4.480	4.603	4.731	4.864	5.004	5.149	5.301	5.459
0.18	4.830	4.961	5.097	5.239	5.387	5.541	5.701	5.869	6.044	6.266	6.417
	4.169	4.279	4.394	4.514	4.639	4.769	4.904	5.046	5.193	5.347	5.508
0.20	5.009	5.146	5.289	5.438	4.493	5.754	5.923	6.098	6.282	6.473	6.674
	4.200	4.312	4.429	4.550	4.677	4.809	4.947	5.090	5.240	5.397	5.560
0.22	5.205	5.349	5.499	5.656	5.818	5.988	6.615	6.360	6.543	6.744	6.955
	4.233	4.346	4.465	4.588	4.717	4.851	4.991	5.137	5.289	5.449	5.615
0.24	5.419	5.571	5.729	5.893	6.604	6.243	6.430	6.624	6.287	7.040	7.262
	4.266	4.382	4.504	4.628	4.758	4.805	5.037	5.185	5.340	5.502	5.672

Note : Upper value is for capital end lower value is for base end of column.
a = length of column section considered rigid, b' = depth of 45 degree capital below soffit of slab

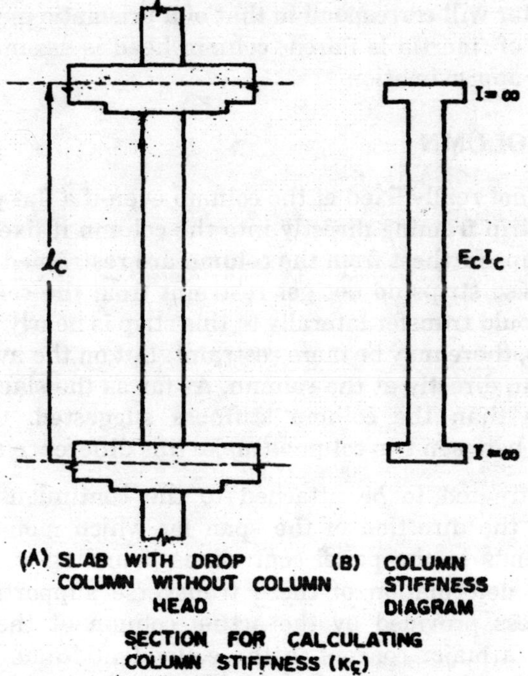

Fig. 18.16

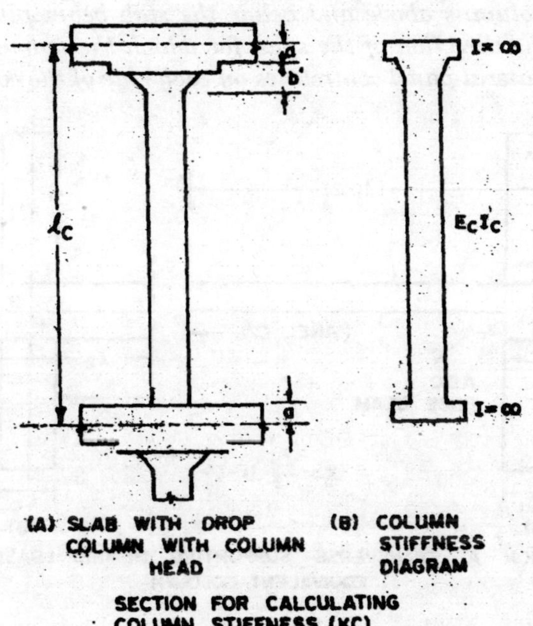

Fig. 18.17

In case, the stiffening effect due to flares is neglected, (as permitted in IS : 456-2000), Fig. 18.15 and Fig. 18.16 will be relevant and the factors for fixed and moments, stiffness k

and the carry power factor will correspond to that of a prismatic member. It is to note that in Fig. 18.17, the moment of inertia is flared column head is assumed to vary linearly along the axis which one is an approximation.

18.11 EQUIVALENT COLUMN

The slab spans are not really fixed at the column even if a flat plate is supported on very stiff column. The slab strip framing directly into the column if fixed over the column width, but the parallel slab strips farthest from the column are restrained only by the adjacent slab span and practically these strips do not get restraint from the column itself. The torsional restraint that are slab could transfer laterally to this strip is nearly negligible. The slabs with heavy transverse beams, there may be more restraint. But on the average across the slab, the restraint is still less than directly at the column. As far as the slab is concentrated, the slab restraint averages less than the column stiffness suggested. Only a transverse beam infinitely stiff in torsion between two columns make this difference negligible.

The columns are treated to be attached to the continuous slab-beam by torsional members transverse to the direction of the span for which moment are being found, the torsional members extends to the panel centrelines bounding each side of the slab beam under study. Torsional deformation of these transverse supporting members reduce the effective flexural stiffness provided by the actual column at the support. Their *reduced column effectiveness* is a major concept in the equivalent frame method of analysis. This effect is considered for in the analysis by using the term *an equivalent column* having the stiffness less than that of the actual column. **An equivalent column** *may be assumed to comprise of the actual columns above and below the slab beam plus an attached torsional members transverse to the direction of the span for which the moments are being calculated and extending to bound lateral panel centrelines on each side of the column.*

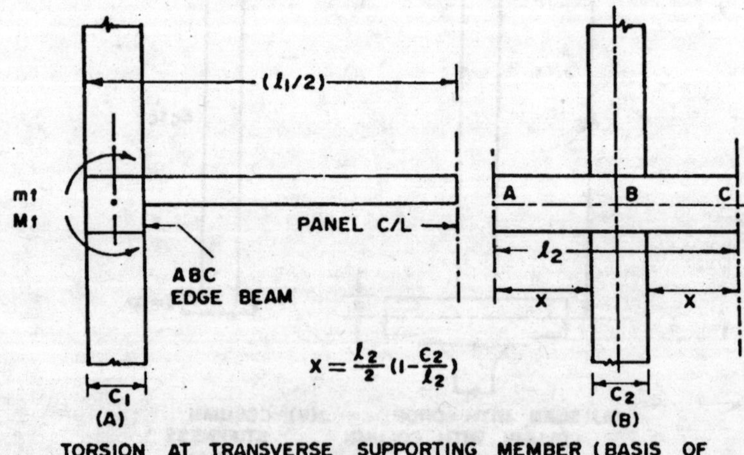

Fig. 18.18

A column and a transverse beam at the exterior support of a continuous slab beam strip are shown in Fig. 18.18 to illustrate the action of a column and the transverse torsional member. The slab spans in the direction l_1. The flexural stiffness of the column and the torsional stiffness of the edge beam AC. Let m_t be the distributed torque applied by the slab and M_t be the resisting torque provided by the column. The rotations at A and C of the edge-beam will be more than at the section at B due to torsional deformation of the edge beam. The actual columns and edge beam are replaced by *an equivalent column* to allow for this unequal effect of rotations.

The equivalent column is so defined that the torsional flexibility (inverse or stiffness) is the sum of the flexibilities of the actual columns and edge beam. Therefore,

$$\left(\frac{1}{K_{ec}}\right) = \left(\frac{1}{\sum K_c} + \frac{1}{K_t}\right) \tag{18.20}$$

where K_{ec} = flexural stiffness of equivalent column

$\sum K_c$ = sum of the flexural stiffnesses of columns above and below the joint, and

K_t = torsional stiffness of the slab or beam on each side of the column.

The moment of inertia of actual column at any cross-section outside the joints or column capitable may be based on the gross-sectional area of concrete. The variation in moments of inertia along the axis of the columns should be taken into consideration. The moments of inertia of columns should be assumed to be infinite from the top of the slab to the bottom of the slab-beam. Eq. 18.20 may be expressed in terms of stiffness rather than flexibility as under :

$$K_{ec} = \left(\frac{1}{\left(\frac{1}{1/\sum K_c}\right) + \left(\frac{1}{K_t}\right)}\right) \tag{18.21}$$

According to the equivalent frame method, the concept of the equivalent column illustrated with an exterior column is used at all supported column for each continuous slab-beam.

The stiffness K_t of an attached torsional member is calculated as described in the subsequent article.

18.12 TORSIONAL MEMBER

A torsional member attached with the slab is assumed to have a constant cross-section throughout their length consisting of the larger of the following (as prescribed by ACI 318-77 Code).

(a) a portion of the slab having width equal to that of the column, bracket or capital in the direction of span for which the moments are being determined ;

(b) for monolithic or fully composite construction, the portion of slab in (a) plus that part of the transverse beam above and below the slab; and

(c) the transverse beam including the portion of slab on either side of beam.

The beam frames into the column. The action of beam is assumed as that of flanged beam (T-beam or L-beam). The flanges extend on each side of beam a distance equal to the projection of the beam above or below the slab but not greater than *four-times* the thickness of slab. It is also assumed that the torsional rotation does not occur in the beam over the width of the support. The governing criteria to identify the torsional members as per paragraphs (a), (b) and (c) above are shown in Fig. 18.19.

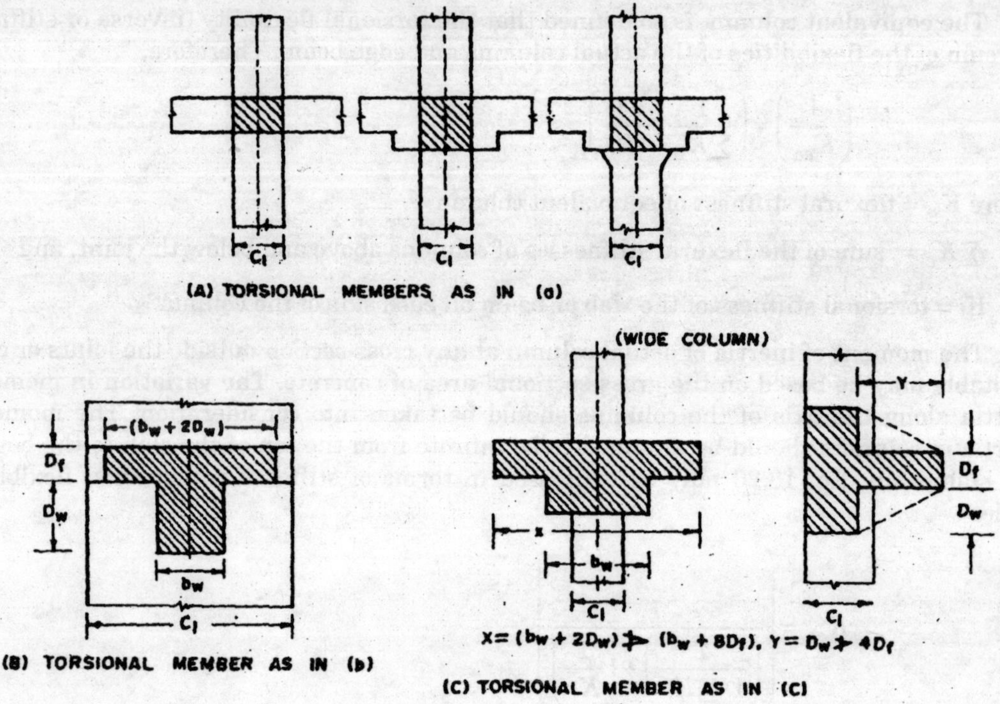

Fig. 18.19

The torsional stiffness, of edge beam, K_t is evaluated by having simple (rather crude) assumptions and basic theory with final adjustment by an empirical factor to fit test results. A transverse beam (or slab) loaded in torque by unbalanced slab moments that vary from zero at mid-span to a maximum at the centreline of the column is shown in Figs. 18.20 (A) and (B) respectively.

The torque (area of the load curve) is 0.5 for each span and the shape of the torque curve is a second power parabola with an ordinate at the face of the column

$$T_u = \frac{1}{2}\left(1 - \frac{c_2}{l_2}\right)^2 \qquad \text{(i)}$$

The unit rotation angle at any point is the torque divided by both G, the shear modulus, and C, the torque stiffness of the cross-section. The difference between column rotation and mid-span beam rotation may be obtained by summation of area under this curve

$$\theta_t = \left(\frac{1}{3}\right)\left[\left(\frac{l_2}{2}\right)\left(1 - \frac{c_2}{l_2}\right) 0.5\left(1 - \frac{c_2}{l_2}\right)^2\right] \times \frac{1}{C \cdot G}$$

$$\text{or} \quad \theta_t = \left(\frac{1}{12}\right)(l_2)\left(1-\frac{c_2}{l_2}\right)^3 . C . G \qquad (\because E = 2G)$$

$$\text{or} \quad \theta_t = \left(\frac{1}{6}\right).(l_2).\left(1-\frac{c_2}{l_2}\right)^3 . C . E \qquad \text{(ii)}$$

The average value of θ for the slab as $\theta_t/3$ or a factor of (1/3) may be used to fit the equation to test results. It is necessary to represent a non-linear torque load. The angle is twist, θ_t is therefore,

$$\theta_t = \left(\frac{1}{18}\right).(l_2).\left(1-\frac{c_2}{l_2}\right)^3 . E . C \qquad \text{(iii)}$$

The stiffness (for arm) is the ratio of torque (used as 0.5 per mm) to this, θ_t

$$\therefore K_t = \left(\frac{0.5}{\theta_t}\right) = \left(\frac{9.E.C}{l_2\left(1-\frac{c_2}{l_2}\right)^3}\right) \qquad \text{(iv)}$$

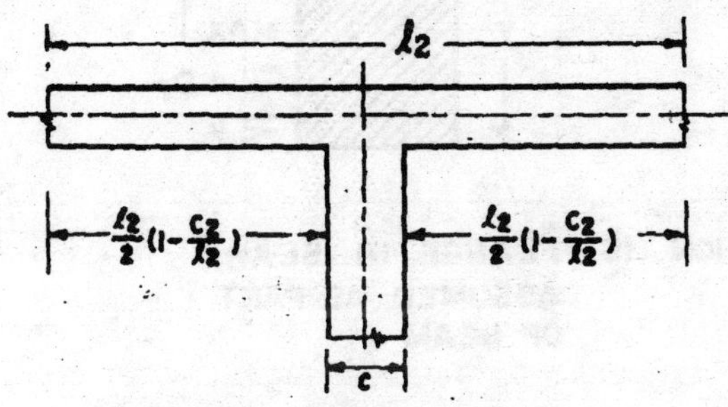

(A) BEAM-COLUMN COMBINATION

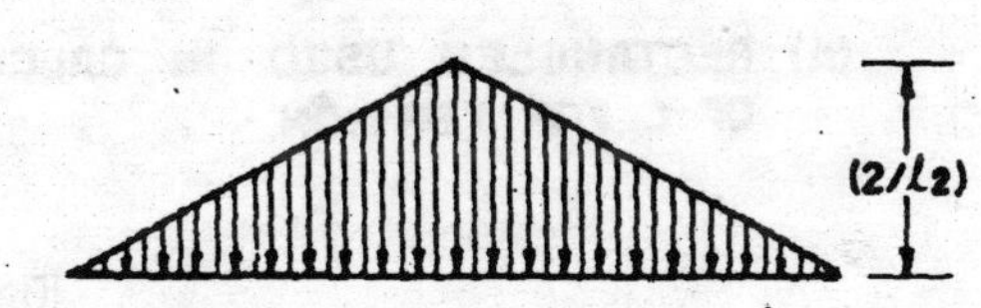

(B) DISTRIBUTION OF TORQUE OR TWIST LOAD

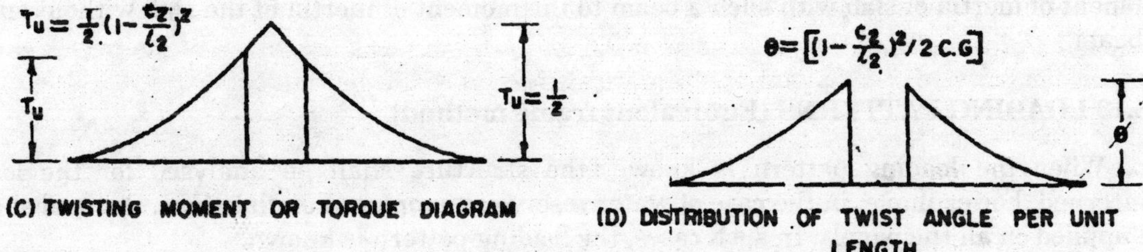

(C) TWISTING MOMENT OR TORQUE DIAGRAM

(D) DISTRIBUTION OF TWIST ANGLE PER UNIT LENGTH

Fig. 18.20

The stiffness of torsional member (which one is twice for two arms in the symmetrical) may be written as

$$K_t = \left(\frac{\sum 9\,E\,.C}{l_2 \left(1 - \frac{c_2}{l_2}\right)^3} \right) \tag{18.22}$$

The value of C, polar moment of inertia in the above equation may be obtained for the cross-section by dividing it into various rectangles

$$C = \sum \left(1 - \frac{0.63\,x}{y}\right).\left(x^3 \,.\, \frac{y}{3}\right) \tag{18.23}$$

where x and y are small and large dimensions of the various rectangles making up the cross-section as shown in Fig. 18.21 (A). Where no beam stem is present, the beam as a width of slab equal to the column width. Where a beam stem is present, the beam includes the stem plus the adjoining the slab on each side of width equal to 4 times D_f if smaller as shown in Fig. 18.21 (B).

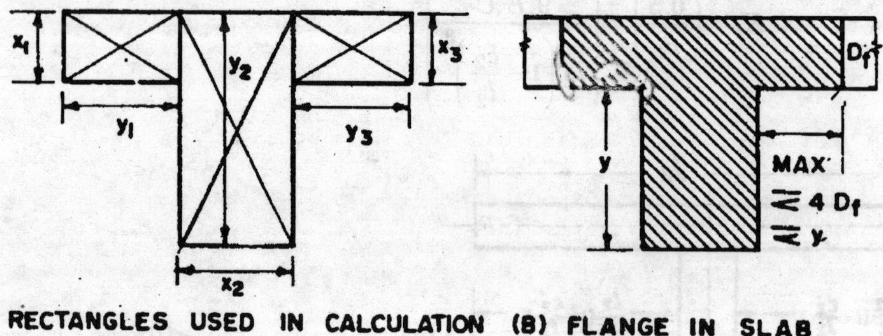

(A) RECTANGLES USED IN CALCULATION OF C FOR TORSION

(B) FLANGE IN SLAB ASSUMED AS PART OF BEAM

Fig. 18.21

Where the beam frames into the columns in the direction of span for which moments are being determined, the value of K_t as calculated above should be multiplied by the ratio of moment of inertia of slab with such a beam to the moment of inertia of the slab without such a beam.

18.13 LOADING PATTERNS (Equivalent frame method)

When the loading pattern is known, the structure shall be analysed for the load concerned. For example, in the case of water reservoirs supported on flat slabs, the load shall be applied on all the panels. In such cases, the loading pattern is known.

When the live load is variable but does not exceed *three-quarters* of the dead load, (viz. $w_l > 0.75\ w_d$), or the nature of the live load is such that all the panels will be loaded simultaneously, the analysis of frame is made when full design live load is on the entire slab system. The dead load (of course) remains on all the spans.

When the live load exceeds *three-quarters* of the dead load, (viz., $w_l > 0.75\ w_d$), it is acceptable to design only for three-quarters of full live load for the following positions of live load occupying the spans :

(A) For maximum positive moment

The alternate spans as shown in Fig. 18.22 (A) are considered load for the maximum positive moment in the panel, BC near mid-span.

(B) For maximum negative moment

Two adjacent spans as shown in Fig. 18.22 (B) are considered loaded for the maximum negative moment at a support, B.

(C) The design positive and negative moments in no case shall be taken less than those occurring with full live loads as shown in Fig. 18.22 (C) are considered loaded.

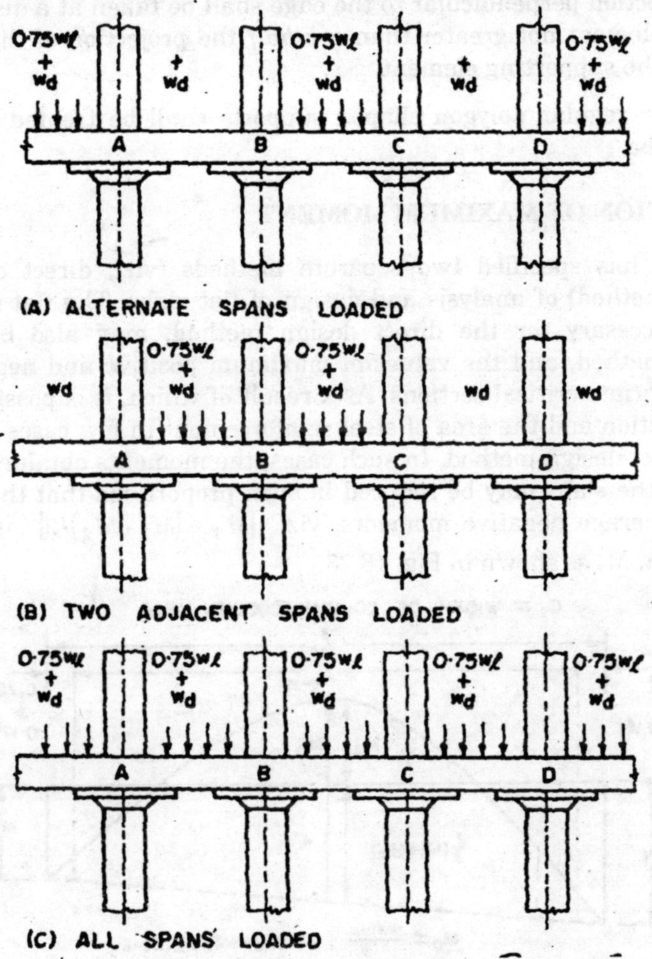

Fig. 18.22

18.14 CRITICAL SECTIONS FOR NEGATIVE DESIGN MOMENTS

The critical sections for negative design moments for the interior supports and for the exterior supports are as under :

(A) Interior supports

The critical section for negative moment at interior supports, in both, the column strip and middle strip shall be taken at the face of the rectilinear supports, but in no case, at a distance greater than 0.175 l_1 from the centre line of the column where l_1 is the length of the span in the direction moments are being calculated. It is measured from centreline to centreline of supports.

(B) Exterior supports

The critical section for negative moment (at exterior supports provided with brackets or capitals) in the direction perpendicular to the edge shall be taken at a distance from the face of the supporting element not greater than *one-half* the projection of the bracket or capital beyond the face of the supporting element.

The circular or regular polygon shaped supports shall be treated as square supports having the same area.

18.15 MODIFICATION OF MAXIMUM MOMENT

IS : 456-2000 has specified two separate methods (viz., direct design method and equivalent frame method) of analysis and design of flat slabs. The flat slabs, which satisfy the limitations necessary for the direct design method, may also be analysed by the equivalent frame method, and the values of maximum positive and negative moments are found at the appropriate critical sections. As a result of which, it is possible that the overall thickness of the section and the area of steel reinforcement in few cases may be larger than those found by direct design method. In such cases, the moments obtained by the equivalent frame method, for the slabs may be reduced in such proportions that the numerical sum of the positive and average negative moments, viz., $[M_3 + (M_1 + M_2)/2]$ is not less than the value of total design, M_o as shown in Fig. 18.23.

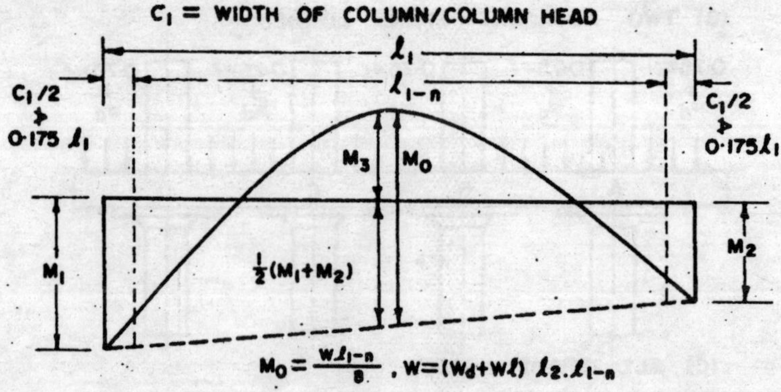

Fig. 18.23

The permissible reduction for moments M_1, M_2 and M_3 shall be

$$\left[\frac{wl_2 : l_o^2}{8}\right] \cdot \left[M_3 \frac{(M_1 + M_2)}{2}\right]$$

18.16 BENDING MOMENTS IN PANELS WITH MARGINAL BEAMS OR WALLS

When the slabs are supported by a marginal beam (edge beam) with a depth greater than 1.5 times the thickness of the slab or by a wall then the total load to be carried by the beam or wall shall comprise those loads directly on the wall or beam plus a uniformly distributed load equal to one-quarter of the total load on the slab.

The bending moment on the half-column strip adjacent to the beam or wall shall be one-quarter of the bending moments for the first interior column strip.

18.17 TRANSFER OF BENDING MOMENTS TO COLUMNS

The flat slabs (which contains no beams) transfer the bending moments (in addition to shear) to the columns directly in the following cases :

(a) interior columns in the flat slabs when the adjacent spans are different in length or the adjacent spans carry unbalanced gravity load.

(b) exterior columns of flat slabs.

(c) flat slab systems which resist lateral loads (e.g., wind, earthquake or other lateral loads).

The bending stresses in the slabs are found by using a fraction, α of the moment given by Eq. 18.24 as under :

$$\alpha = \left(\frac{1}{1 + \frac{2}{3}\left(\frac{a_1}{a_2}\right)^{1/2}}\right) \tag{18.24}$$

where a_1 = overall dimension of the critical section for shear in the direction in which the moment acts

a_2 = overall dimension of the critical section for shear transverse to the direction in which the moment acts.

A slab width between lines that are *one and one-half* slab thickness (or drop panel thickness), 1.5 times D on each side of the column (or column capital) may be considered effective where, D being the size of the column. The concentration of reinforcement over column head by closer spacing or additional reinforcement may be used to resist the moment on this section.

From the research on flat slab system without beams supported by square columns (Hanson, N.W. and Hanson, J.M. `shear and moment transfer between concrete slabs and columns' Journal, PCA Research and Development Laboratories Vol. No. 1, January, 1968) it is seen that 60 percent of the unbalanced moment can be assumed to be transferred to the column by bending of the slab and 40 percent by eccentricity of the shear force about the centroid of the critical section as shown in Fig. 18.24.

The dimensions for critical sections for shear will be

$$a_1 = (c_1 + d) \text{ and } a_2 = (c_2 + d),$$

respectively, where d is the effective depth of the slab (or effective depth of drop). The additional steel required for the transfer of moment through flexure may be distributed over a width of ($C + 3D_f$) and not ($C + D_f$) alone.

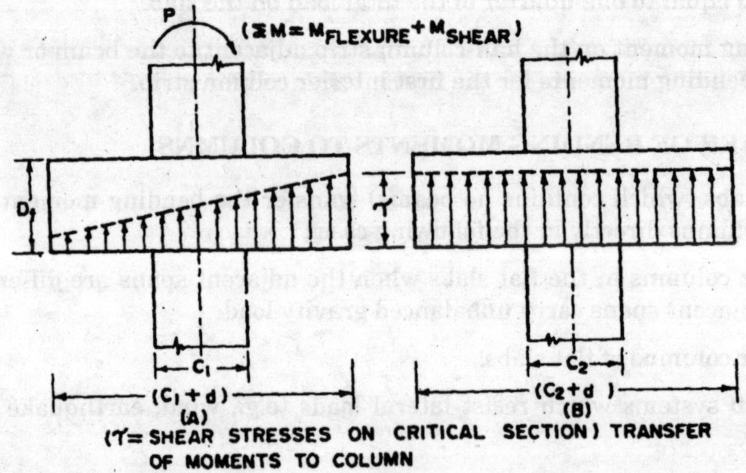

Fig. 18.24

The total unbalanced moment is to be distributed to the columns above and below the joint in proportion to their stiffness. The appropriate moment capacity must be provided in the columns.

It is to note that for the equivalent frame method, the unbalanced moments are calculated with respect to the critical sections for negative moments.

18.18 MOMENTS IN COLUMNS

In flat slabs, the columns are built integrally with the slab system. The columns are designed to resist moment arising from loads on the slab.

At an interior support, slab negative moments are obtained, assuming that load and full live load act. For the column design, a more severe loading results from the partial removal of the live load. In case of two adjacent unequal spans, the longer span carries full dead load plus half live load (viz., $w_d + 0.5\ w_l$) and the shorter span carries the dead load only (viz., w_l), thereby the effects of pattern loading may be accounted for as under.

The columns above and below the slab shall be designed to resist the moment, M calculated from Eq. 18.25, distributed to upper and lower columns in direct proportions to their stiffnesses unless a general analysis is made

$$M = \frac{0.08\left[(w_d + 0.5\,w_l)\,l_2\,.\,l_{1-n}^2 - w_d'\,.\,l_2'\,(l_{1-n}')^2\right]}{\left(1+\dfrac{1}{\alpha_c}\right)} \tag{18.25}$$

where w_d, w_l = design dead load and live load, respectively, per unit area

l_2 = length of span transverse to the direct of M

$l_{1\text{-}n}$ = length of the clear span in the direction of M measured from face to face of supports

w_d', l_2' and $l_{1\text{-}n}'$ refer to shorter span

$$\alpha_c = \left(\sum K_c / \sum K_s\right)$$

$\sum K_c$ = sum of the flexural stiffnesses of the columns above and below slab (meeting at the joint)

$\sum K_s$ = sum of the flexural stiffnesses of the slab panels meeting at the joint.

At an exterior support, the design moment, M is calculated from Eq. 18.26

$$M = \left(\frac{0.65}{1+\dfrac{1}{\alpha_c}}\right)\left(\frac{(w_d + w_l)\,l_2\,.\,l_{1-n}^2}{8}\right) \tag{18.26}$$

18.19 SLAB REINFORCEMENT

In flat slab systems, the flexural reinforcement is placed in an orthogonal grid, with bars parallel to the sides of the panels. The straight bars are commonly provided throughout. Although in some cases, the positive steel is bent up where no longer required, in order to provide for part or all of the negative moment.

When the bars are placed in perpendicular layers, these steel bars result in stacking. The inner steel bafs will have an effective depth 1 bar diameter les than the outer steel bars. The stacking problem is connected with the middle strip positive steel bars and the column strip positive and negative bars. In the flat slabs with beams on the columns lines, the stocking occurs for the middle strip positive steel, and in the column strips in important mainly for the column line beams because slab moments are usually very small in the region where the column strips intersect.

For the complete panel, including both middle strips and column strips in each direction, the moments in the long direction shall be more than those for the short direction, as it is seen by determining the static moment, M_o in each direction for a rectangular panel. It may be noted while adjusting in due proportion of M_o first to negative and positive moment sections, and then laterally to column and middle strips, it is done by applying exactly the same ratios in each direction to the corresponding section. Therefore, it is clear that the middle strip positive moments (for example) are more in the long direction than the short direction, exactly the opposite of the situation for the slab with stiff edge beams. In the column strip, positive and negative moments are more in the long direction than in the short direction. Therefore, the steel bars for negative and positive moments in both middle and column strips

are placed closer to the top or bottom surface of the slab, respectively, with the larger effective depth.

Bar cut off points or bend points could be determined from moment envelopes if available. However, when the direct design method is used, the moment envelopes and lines of inflection are not found explicity. In such a case and often when the equivalent frame method is used as well, standard bar cutoff or bend points from Fig. 18.25 and Fig. 18.26 are used as recommended by IS : 456-2000.

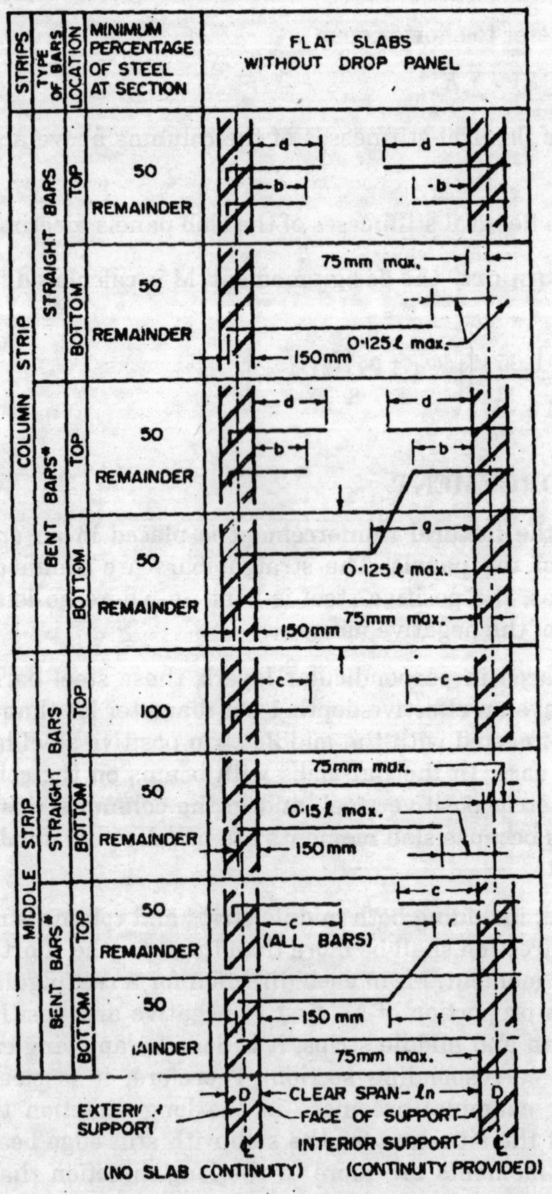

Fig. 18.25

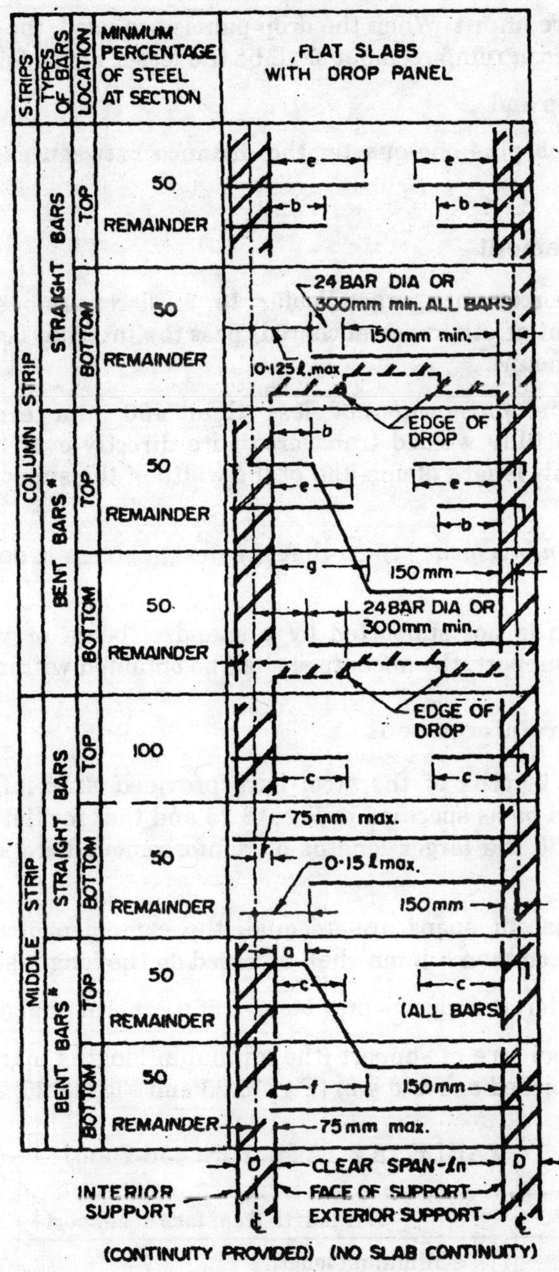

Fig. 18.26

The specifications specified in IS : 456-2000 as given below are followed for the slab reinforcement for flat slabs.

1. Spacing. The spacing of bars in flat slab shall not exceed 2 times the slab thickness, except where a slab is of cellular or ribbed construction. This takes care for the local concentrated load as well as it ensures that tensile cracks are narrow and well distributed.

2. Area of reinforcement. When the drop panels are used, the thickness of drop panel for determination of area of reinforcement shall be the lesser of the following :

(a) thickness of drop and

(b) thickness of slab plus one-quarter the distance between edge of drop and edge of capital.

3. Anchoring reinforcement

(a) All slab reinforcement perpendicular to a discontinuous edge shall have an anchorage (straight, bent or otherwise anchored) past the internal face of the spandrel beam wall or column, of an amount.

(1) *For positive reinforcement* not less than 150 mm except that with fabric reinforcement having a fully welded transverse wire directly over the support, it shall be permissible to reduce this length of one-half of the width of the support of 50 mm, whichever is greater; and

(2) *For negative reinforcement* : Such that the design stress is developed at the internal face.

(b) Where the slab is not supported by a spandrel beam or wall or where the slab cantilevers beyond the support, the anchorage shall be obtained within the slab.

4. Minimum length of reinforcement

(a) The minimum lengths of the steel bars provided as reinforcement in flat slabs without drop panel shall be as specified in Fig. 18.25 and that for flat slabs with drop panels as specified in Fig. 18.26 the larger lengths of reinforcement bars shall be provided when required for analysis.

(b) Where the adjacent spans are unequal, the extension of negative reinforcement beyond each face of the common column shall be based on the longer span.

The bent bars at exterior supports may be used if a general analysis is made.

The bar lengths from face of support (the minimum lengths marked as a, b, c, d, and e and maximum length marked as f and g in (Fig. 18.25 and Fig. 18.26) are given in Table 18.2.

TABLE 18.2. (As per IS: 456-2000)

	Bar length from face of support						
	Minimum length					Maximum length	
Mark	a	b	c	d	e	f	g
Length	$0.14l_n$	$0.20\ l_n$	0.221_n	$0.30l_n$	$0.33l_n$	$0.20l_n$	$0.24ln_n$

$(l_n = l_{1\text{-}n})$

18.20 BEHAVIOUR OF FLAT SLAB IN SHEAR

The flat slabs are the slabs which are supported directly by the columns. In footing, the slabs carry concentrated loads. In flat slabs and footings, the shear near the columns is of critical importance. From the test of flat plates and flat slabs, it is seen that in most practical cases, the capacity in controlled by shear. (Reference) : `The Shear Strength of Reinforced Concrete Members-slabs', ASCE-ACI Task Committee 426, Journal of Structural Division, ASCE, Vol. 100, No. ST 8, August 1974 and W.G. Corley and N.M. Howkins, `Shearhead Reinforcement for Slabs', Journal of ACI, Vo.. 65, No. 10, October, 1978). In the design of flat plates and flat slabs, following two types of shear may be critical.

1. Beam-type shear (Diagonal tension failure). The beam-type shear is also called as one-way shear. In long narrow slabs and footings, it is considered in the analysis that the slab acts like a wide beam spanning between the supports provided by the perpendicular column strips. A measure diagonal crack extends in a plane across the complete width l_2 of the slab. As the name implies, beam shear in slabs is considered in the same manner as shear in beams. The critical section is taken a distance d from the face of the column or capital. The nominal shear stress is calculated as $\tau_v = (V/b_w.d)$ where ($b_w = l_2$).

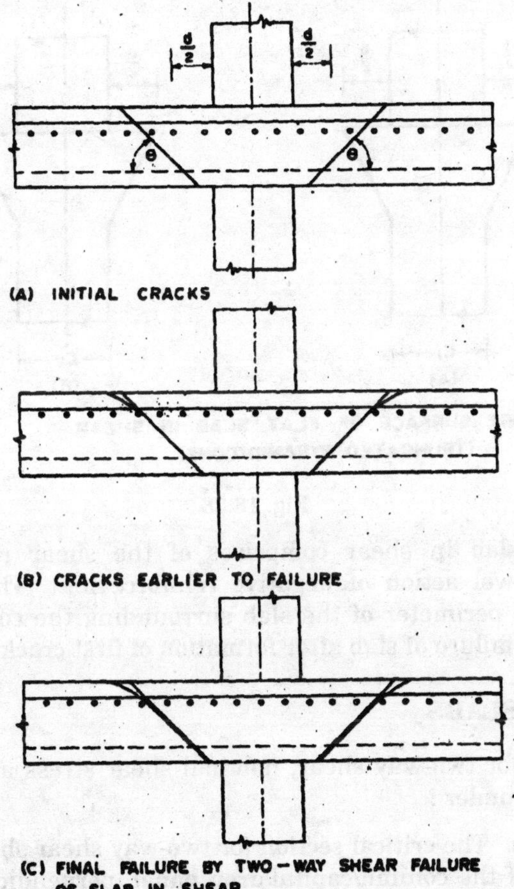

Fig. 18.27

2. Two-way shear. In the flat slabs, the middle strips are supported by the column strips. These middle strips transfer the loads to the column strips. The common strips further transfer the loads to the columns. The portion of slab adjacent to the column carries excessive shear stress. It results in initiation of shear cracks at a distance, d from the face of column, where d is the effective depth of slab. These cracks do not remain visible. These cracks propagate towards the top from the bottom of slab. These crack exhibit adequate reserve strength before failure. Just prior to failure, the projected horizontal lengths of these cracks seem to be approximately 1.5 times the effective depth of the slab or even more. The failure occurs at the bottom compressed edge of the slab surrounding the column through two way shear as shown in Fig. 18.27.

The angle of inclination with the horizontal, θ depends upon type and area of reinforcement in the slab. It may range between 20° and 45°. The critical section for two-way is taken perpendicular to the plane of the slab and to have a perimeter, b_o that is a minimum, but approaches no closer than d/2 to the periphery of the support or concentrated load as shown in Fig. 18.27 (A). These cracks follow the surface of a truncated pyramid or truncated cone, as shown in Fig. 18.28.

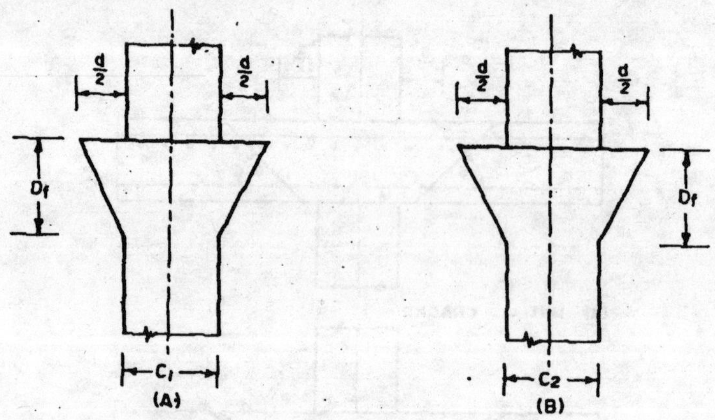

Fig. 18.28

The resistance of slab in shear comprises of the shear resistance of concrete in compression and the dowel action of negative reinforcement. The dowel action of shear resistance occurs a large perimeter of the slab surrounding the column. As a result of this dowel action, the sudden failure of slab after formation of first crack, does not occur.

18.21 SHEAR IN FLAT SLABS

The critical section for two-way shear, nominal shear stress and the permissible shear stress in flat slabs are as under :

1. Critical sections. The critical section for two-way shear shall be taken at a distance d/2 from the periphery of the column/capital/drop panel, perpendicular to the plane of slab where d is the effective depth of the section as shown in Fig. 18.1 and Fig. 18.2.

When the flat slabs are provided with drops, two critical section for two way shear (one within the drop and the other at a distance d/2) will have to be considered as shown in Fig. 18.3.

The shape in plan is geometrically similar to the support immediately below the slab as shown in Fig. 18.29 (A) and (B). For column sections with reentrant angles, the critical sections shall be taken as indicated in Fig. 18.29 (C) and (D).

In the case of columns near the free edge of a slab, the critical section shall be taken as shown in Fig. 18.30.

2. Calculation of shear stress. The shear stress, τ_v shall be the sum of the values calculated as under :

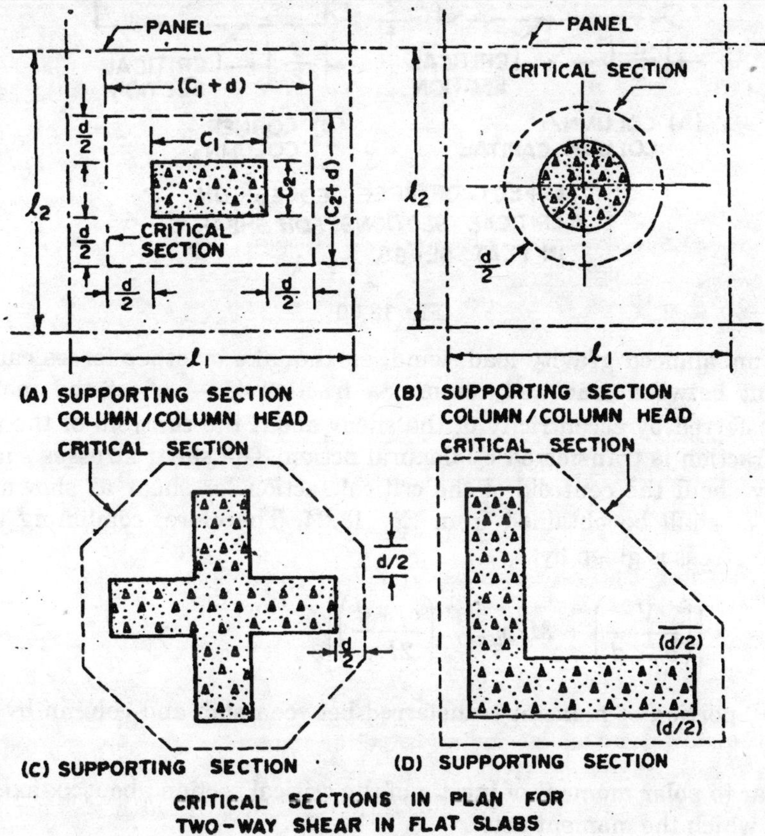

Fig. 18.29

The nominal shear stress in flat slabs shall be taken as in Eq. 18.27, below :

$$\tau_v = \left(\frac{V}{b_o \cdot d}\right), \text{ where} \tag{18.27}$$

V = design shear force due to design load

$V = (w_d + w_l)\,[(l_1 \cdot l_2 - (c_1 + d)\,(c_2 + d)]$

b_o = Perimeter of the critical section

$b_o = 2\ [(c_1 + d) + (c_2 + d)]$

d = effective depth of the section.

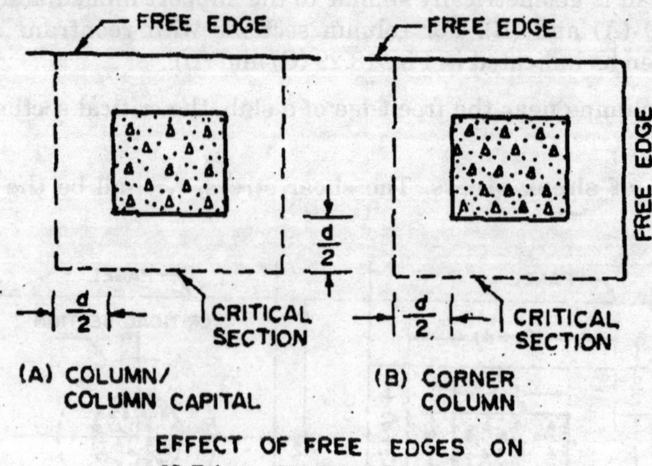

Fig. 18.30

When the unbalanced gravity load, wind, earthquake or other forces cause transfer of bending moment between slab and column, a fraction $(1 - \alpha)$ of the moment shall be considered transferred by eccentricity of the shear about the centroid of the critical section and the other fraction is transferred by flexural action. The shear stresses shall be taken as varying linearly about the centroid of the critical section for shear as shown in Fig. 18.24. The value of α shall be obtained from Eq. 18.24. Therefore, combining those two, the maximum shear stress is given by

$$\tau = \left[\left(\frac{V}{b_o \cdot d}\right) + M_{(shear)} \cdot \left(\frac{c_1 + d}{2I_{xx}}\right)\right] \tag{18.28}$$

where $M_{(shear)}$ = portion of moment transferred between slab and column by eccentricity of shear

J_{xx} = similar to polar moment of inertia of the critical section about xx-axis (viz., the axis about which the moment acts).

3. Permissible shear stress. The shear stress, horizontal flexural compressive stresses in two directions at the bottom side of the slab and the compressive stresses from the column in vertical direction exist at the sections adjacent to the column. The simultaneous presence of compressive stresses in two horizontal directions and one vertical direction create confining effect of the concrete and increase the resistance to shear. The limiting value of shear stress in the case of flat slabs is calculated from the consideration of two way shear and it will be considerably larger than the value of shear stress in concrete recommended in the case of bending action in beams.

When the shear reinforcement is not provided, the calculated shear stress at the critical section shall not exceed $k_s \cdot \tau_c$,

where $k_s = (0.5 + \beta_c)$ but not greater than 1

β_c = the ratio of short side to long side of the column/capital

For working stress method of design :

$$\tau_c = 0.16\, (f_{ck})^{1/2} \tag{18.29}$$

When the shear stress at the critical section exceeds the value $k_s.\tau_c$, but it is less than 1.5 τ_c, shear reinforcement shall be provided. If the shear stress exceeds 1.5 τ_c, the flat slab shall be redesigned. The shear stresses shall be determined at successive sections more distant from the support and shear reinforcement shall be provided upto a section where the shear stress does not exceed 0.5 τ_c. While designing the shear reinforcement, the shear stress carried by the concrete shall be assumed to be 0.5 τ_c and the reinforcement shall carry the remaining shear.

Example 18.1 Design an interior panel of a flat slab 4.8 x 5.8 m in size. The slab is supported on columns 500 mm in diameter. The height of columns above and below the slab is 3.8 m. A finishing surface of 20 mm thickness is provided over the slab. The floor of the slab is likely to be used as a class-room. Use M 20 grade of concrete and Hysd steel bars of grade Fe 415.

Use direct design method as recommended in IS : 456-2000.

Design : M 20 grade of concrete and Hysd steel bars of grade Fe 415 will be used for the flat slabs.

Step 1: Design constants. The permissible stress in concrete in bending in compression and that for Hysd steel bars of grade Fe 415 in bending in tension are $\sigma_{cbc} = 7\ N/mm^2$ and $\sigma_{st} = 230\ N/mm^2$.

Modular ratio for concrete

$$m = \left(\frac{280}{3\ \sigma_{cbc}}\right) = \left(\frac{280}{3 \times 7}\right) = 13.33$$

The design factors for balanced section are as under :

Factor for depth of neutral axis

$$k_B = \left(\frac{m\ \sigma_{cbc}}{m\ \sigma_{cbc} + \sigma_{st}}\right) = \left(\frac{13.33 \times 7}{13.33 \times 7 + 230}\right) = 0.2886$$

Factor for lever arm

$$j_B = \left(1 - \frac{k_B}{3}\right) = \left(1 - \frac{0.2886}{3}\right) = 0.9038$$

Factor for moment of resistance

$$R_B = \left(\frac{1}{2}\sigma_{cbc} \cdot k_B \cdot j_B\right) = \left(\frac{1}{2} \times 7 \times 0.2886 \times 0.9038\right) = 0.9130$$

Percentage of steel

$$p_B = \left(\frac{1}{2}\frac{\sigma_{cbc}}{\sigma_{st}} \cdot k_B\right) = \left(\frac{1}{2} \times \frac{7}{230} \times 0.2886 \times 100\right) = 0.4392$$

Step 2 : Clear spans

The flat slab shall be designed without drop panel and the supporting columns shall be have columns heads of 400 mm height with 45° Inclination and additional vertical height of 40 mm as shown in Fig. 18.31.

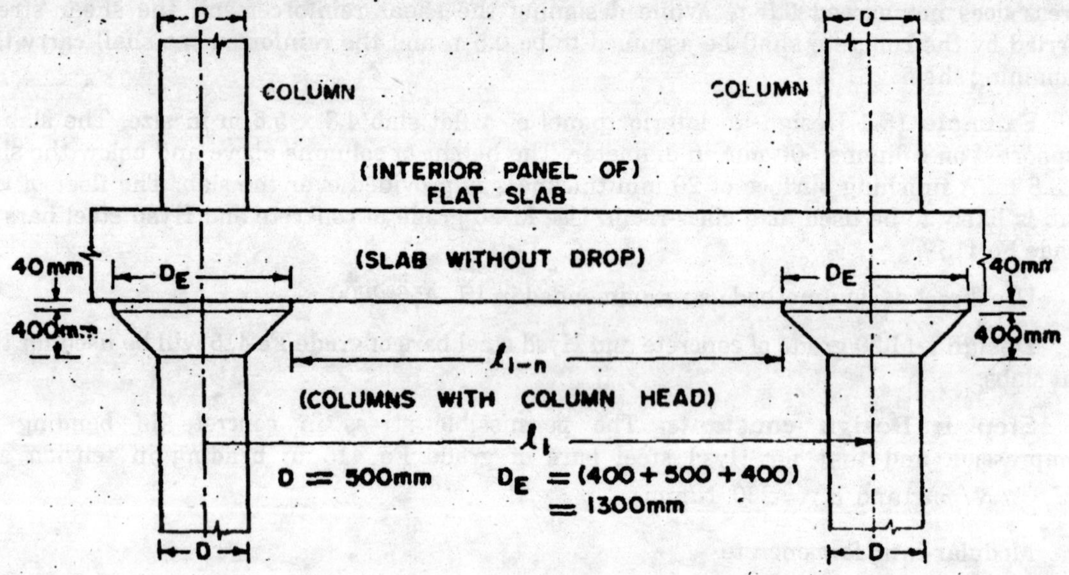

Fig. 18.31

As per IS : 456-2000, the columns of circular sections shall be treated as square sections having the same cross-sectional area. Therefore, let a be the size of square column

$$a^2 = \left(\frac{\pi}{4} \times 1500^2\right),\ a = 115\ mm = 1.15\ m$$

Clear span along AB (long span), Fig. 18.4 (B)

$$l_{1-n} = \left(5.8 - \frac{1.15}{2} - \frac{1.15}{2}\right) = 4.65\ m$$

Clear span along BC (short span), Fig. 18.4 (B)

$$l_{1-n} = \left(4.8 - \frac{1.15}{2} - \frac{1.15}{2}\right) - 3.65\ m$$

Step 3 : Estimated thickness of the flat slab

The flat slab remain continuous over minimum of three span (Direct design method). For slabs without drops :

$$\left(\frac{\textit{Longer span}}{\textit{Effective depth}}\right) = (0.9 \times 26 \times m')$$

where m' is the modification factor. For Hysd steel of Fe 415 grade

p_B = 0.4392 per cent, m' = 1.4

Estimated effective depth

$$\left(\frac{5.8 \times 1000}{0.9 \times 26 \times 1.4}\right) = 177.05\ mm$$

Let the overall estimated thickness be 200 mm (More than 125 mm minimum thickness specified).

Step 4 : Estimated loads

(A) Dead load

It consists of the weight of finishing surface plus the self-weight of the slab.

Weight of finishing surface

$$\left(\frac{20}{1000} \times 1 \times 1 \times 24\right) = 0.480\ kN/m^2$$

Self-weight of the slab

$$\left(\frac{200}{1000} \times 1 \times 1 \times 25\right) = 5.000\ kN/m^2$$

Total w_d = 5.48 kN/m².

(B) Live load

The floor of slab is likely to be used for class-room

w_l = 4 kN/m²

Total load

$(w_d + w_l) = (5.48 + 4) = 9.48$ kN/m²

Step 5 : Width of the strips for interior panel

(A) Width of strip along AB (long span), Fig. 18.4 (A)

Span AB, l_1 = 5.8 m, span BC, l_2 = 4.8 m

Width of column strip

$0.25\,l_2 = (0.25 \times 4.8) = 1.2$ m (Adopted)

$\not> (0.25\,l_1 = (0.25 \times 5.8) = 1.45\,m)$

Width of middle strip

$(4.8 - 1.2 - 1.2) = 2.4$ m.

(B) Width of strip along BC (short span), Fig. 18.4 (B)

Span BC, $l_1 = 4.8$ m, span AB, $l_2 = 5.8$ m

Width of column strip

$0.25 l_2 = (0.25 \times 5.8) = 1.45\,m$

$\not> (0.25\,l_1 = 0.25 \times 4.8 = 1.2\,m)$ Adopted

Width of middle strip

$(5.8 - 1.2 - 1.2) = 3.4$ m.

Step 6 : Total design moment for a span

(A) Total design moment, M_o along AB, (long span), Fig. 18.4 (A)

$l_1 = 5.8$ m, $l_{1\text{-}n} = 4.65$ m, $l_2 = 4.8$ m

Total design load over an area $l_2 - l_{1\text{-}n}$

$$W = (w_d + w_1) . l_2 . l_{1-n} = (9.48 \times 4.8 \times 4.65) = 211.59\ kN$$

Total design moment along the direction AB

$$M_{o.AB} = \left(\frac{W . l_{1-n}}{8}\right) = \left(\frac{211.59 \times 4.65}{8}\right) = 122.9\ kN-m$$

(B) Total design moment, M_o along BC (short span), Fig. 18.4 (B)

$l_1 = 4.8$ m, $l_{1\text{-}n} = 3.65$ m, $l_2 = 5.8$ m

Total design load over an area $l_2 . l_{1\text{-}n}$

$$W = (w_d + w_1) \times l_2 \times l_{1-n} = (9.48 \times 5.8 \times 3.65) = 200.69\ kN$$

Total design moment along the direction BC

$$M_{o.BC} = \left(\frac{W . l_{1-n}}{8}\right) = \left(\frac{200.69 \times 3.65}{8}\right) = 91.56\ kN-m$$

Step 7 : Negative and positive design moments

In an interior span, the total design moment M_o is distributed in the following proportions in each direction.

(A) Distribution of total design moment $M_{o.AB}$, Fig. 18.4 (A)

Negative design moment (at the faces of left and right rectangular supports)

$$M_{n.AE} = 0.65\ M_{o.AE} = (0.65 \times 122.99) = 79.943\ kN - m$$

Positive design moment (at the centre)

$$M_{p.AB} = 0.35\ M_{o.AB} = (0.35 \times 122.99) = 43.047\ kN - m$$

(B) Distribution of total design moment $M_{o.BC}$, Fig. 18.4 (B)

Negative design moment (at the faces of left and right rectangular supports)

$$M_{n.BC} = 0.65\ M_{o.BC} = (0.65 \times 91.56) = 59.514\ kN - m$$

Positive design moment (at the centre)

$$M_{p.BC} = 0.35\ M_{o.BC} = (0.35 \times 91.56) = 32.046\ kN - m$$

Step 8 : Effects of pattern of loading

The effects of pattern of loading for negative and positive bending moments may be seen before *distribution of these moments across the pael width.*

Ratio of live load to dead load

$$\left(\frac{w_l}{w_d}\right) = \left(\frac{4}{5.48}\right) = 0.73.$$

The ratio of live load to dead load exceeds 0.5.

(A) For the direction along AB, Fig. 18.4 (A)

The ratio of flexural stiffness of the columns above and below the slab to the flexural stiffness of the slabs at a joint in the direction along AB, Fig. 18.4 (A)

$$\alpha_c = \left(\sum K_c / \sum K_s\right)$$

(i) Stiffness co-efficient for slab

Size of equivalent square column (diameter of column is 500 mm)

$$C_1 (=a) = \left(\frac{\pi}{4} \times 500^2\right)^{1/2} = 443\ mm$$

Centre to centre distance between columns

Along AB l_1 = 5.8 m and Along BC l_2 = 4.8 m

$$\left(\frac{c_1}{l_1}\right) = \left(\frac{0.443}{5.8}\right) = 0.076 = 0.1\ and \left(\frac{c_2}{l_2}\right) = \frac{0.443}{4.8} = 0.092 = 0.1$$

From Table 18.2, stiffness factor for the slab without drop panel (as per SP : 24-1983), k = 4.8

For the slab of uniform thickness

$$K_s = \left(\frac{4.18\, E_c l}{l_1}\right) = \left(\frac{4.18 \times E_c}{5.8} \times \frac{1 \times 4.8 \times 2.2200^3}{12}\right) = 2.306\, E_c \times 10^{-3}\ kN-m$$

(ii) Stiffness co-efficient for column

Consider Fig. 18.17 and Fig. 18.31. Length of column section (considered rigid)

a = 100 mm = 0.1 m

Depth of 45 degree capital (below the soffit of the slab)

b' = 440 mm = 0.44 m

l_c = (3.8 + 0.1 + 0.1) = 4.0 m

$$\left(\frac{a}{l_c}\right) = \left(\frac{0.1}{4}\right) = 0.025 \text{ and } \left(\frac{b'}{l_c}\right) = \frac{0.44}{4} = 0.110$$

From Table 18.5, column stiffness coefficients for columns with 45° tapered capital for above ratios

For upper (capital) end, $k_c = \left(\frac{4.886 + 5.02}{2}\right) = 4.953$

For lower (base) end, $k_c = \left(\frac{4.590 + 4.538}{2}\right) = 4.564$

Stiffness of column for upper (capital) end $\left(\because I_c = \frac{1}{12} D^4\right)$

$$k_c = \left(\frac{4.953\, E_c I_c}{l_c}\right) = \left(\frac{4.953 \times E_c}{4} \times \frac{1}{12} \times 0.5^4\right) = \left(6.45\ E_c \times 10^{-3}\right) kN-m$$

Stiffness of column for lower (base) end

$$k_c = \left(\frac{4.564 \times E_c}{4} \times \frac{1}{12} \times 0.5^4\right) = \left(5.943\ E_c \times 10^{-3}\right) kN-m$$

$$\alpha_c = \left(\frac{(6.45 + 5.943)\, E_c \times 10^{-3}}{2 \times 2.306\, E_c \times 10^{-3}}\right) = 2.687$$

$$\left(\frac{l_2}{l_1}\right) = \left(\frac{4.8}{5.8}\right) = 0.827 \text{ and } \left(\frac{w_1}{w_d}\right) = \left(\frac{4}{5.48}\right) = 0.73$$

From Table 18.1 (as per IS : 456-2000) by interpolation

$\alpha_{c.\min} = 0.322$ which one is less than $(\alpha_c = 2.687)$

Therefore, the corrections for pattern of loading in the direction AB, Fig. 18.4 (A) is not necessary.

(B) For the direction along BC, Fig. 18.4 (B)

The ratio of flexural stiffness of the columns above and below the slab to the flexural stiffness of the slab at a joint in the direction along BC, Fig. 18.4 (B)

$$\alpha_c = \left(\sum K_c / \sum K_s\right)$$

(i) Stiffness coefficient for slab

Size of equivalent square column (diameter of column is 500 mm).

$$c_1 \; (=a) = \left(\frac{\pi}{4} \times 500^2\right)^{1/2} = 443 \; mm$$

Centre to centre distance between columns

Along BC, l_1 = 4.8 m and Along AB, l_2 = 5.8 m

$$\left(\frac{c_1}{l_1}\right) = \left(\frac{0.443}{4.8}\right) = 0.092 = 0.1 \; and \left(\frac{c_2}{l_2}\right) = \left(\frac{0.443}{5.8}\right) = 0.076 = 0.1$$

From Table 18.2, stiffness factor for the slab without drop panel (as per SP : 24-1983)

k = 4.18

$$k_s = \left(\frac{4.18 \times E_c}{4.8} \times \frac{1}{12} \times 5.8 \times 0.200^3\right) = 3.367 \; E_c \times 10^3 \; kN-m$$

(ii) Stiffness coefficients for column

Consider Fig. 18.17 and Fig. 18.31 length of column section (considered rigit)

a = 100 mm = 0.1 m

Depth of 45 degree capital (below the soffit of slab)

b' = 440 mm = 0.44 m

l_2 = (3.8 + 0.1 + 0.1) = 4.0 m

$$\left(\frac{a}{l_c}\right) = \left(\frac{0.1}{4}\right) = 0.025 \; and \left(\frac{b'}{l_c}\right) = \left(\frac{0.44}{4}\right) = 0.11$$

From Table 18.5, column stiffness coefficients for column with 45° tapered capital.

For upper (capital) end for above ratios, $k_c = \left(\frac{4.886 + 5.02}{2}\right) = 4.953$

For lower (base) end for ratios, $k_c = \left(\frac{4.590 + 4.538}{2}\right) = 4.564$

Stiffness of column for upper (capital) end,

$$k_c = \left(\frac{4.953 \times E_c}{4} \times \frac{1}{12} \times 0.5^4\right) = 6.45\, E_c \times 10^{-3}\ kN-m$$

Stiffness of column for lower (base) end,

$$k_c = \left(\frac{4.564 \times E_c}{4} \times \frac{1}{12} \times 0.5^4\right) = 5.943\, E_c \times 10^{-3}\ kN-m$$

$$\alpha_c = \left(\frac{(6.45 + 5.943)\, E_c \times 10^{-3}}{2 \times 3.367 \times E_c \times 10^{-3}}\right) = 1.840$$

$$\left(\frac{l_2}{l_1}\right) = \left(\frac{5.8}{4.8}\right) = 1.208\ and\ \left(\frac{w_l}{w_d}\right) = \left(\frac{4}{5.48}\right) = 0.73$$

From Table 18.1 (as per IS : 456-2000) by interpolation.

$\alpha_{c.\min} = 0.361$ **which one is less than** $(\alpha_c = 1.840)$.

Therefore, the correction for pattern of loading in the direction BC, Fig. 18.4 (B) is not needed.

Step 9 : Distribution of bending moments across the panel width

(A) Across the strips of span AB, Fig. 18.9 (A)

(i) Negative moment at an interior support

(a) Column strips : 74 per cent of the total negative moment in the panel at this support is resisted by the column strips. There are two column strips. Therefore, the moment resisted by each column strip $(\because M_{n.AB} = 79.943\ kN-m)$

$$\left(\frac{1}{2} m_{nc}\right) = \left(\frac{1}{2} \times 0.75\, M_{n.AB}\right)\left(\frac{1}{2} \times 0.75 \times 79.943\right) = 29.978\ kN-m$$

(b) Middle strips : The negative moment resisted by the middle strip

$$m_{nm} = 0.25\, M_{n.AB} = (0.25 \times 79.943) = 19.986\ kN-m$$

Check : (29.978 + 19.986 + 29.978) = 79.943 kN-m

(ii) Positive moment at an interior support

(a) Column strips : 60 per cent of the total positive moment in the panel is resisted by the column strips. There are two column strips. Therefore, the moment resisted by each column strip $(\because M_{p.AB} = 43.047\ kN-m)$

$$\left(\frac{1}{2} m_{pc}\right) = \left(\frac{1}{2}\, 0.60 \times M_{p.AB}\right) = \left(\frac{1}{2} \times 0.60 \times 43.047\right) = 12.914\ kN-m$$

(b) **Middle strips :** The positive moment resisted by the middle strip

$$m_{p.m} = 0.40\, M_{P.AB} = (0.40 \times 43.047) = 17.22\ kN-m$$

Check : (12.194 + 17.22 + 12.914) = 43.047 kN-m

(B) Across the strips of span BC, Fig. 18.9 (B)

(i) Negative moment at an interior support

(a) Column strips : 75 per cent of the total negative moment in the panel at this support is resisted by the column strips. There are two column strips. Therefore, the moment resisted by each column strip $(\because M_{p.BC} = 59.514\ kN-m)$

$$\left(\frac{1}{2} m_{nc}\right) = \left(\frac{1}{2} \times 0.75 \times M_{p.BC}\right) = \left(\frac{1}{2} \times 0.75 \times 59.514\right) = 22.318\ kN-m$$

(b) Middle strip : The negative moment resisted by the middle strip

$$m_{nm} = 0.25\ M_{n.BC} = (02.5 \times 59.514) = 14.878\ kN-m$$

Check ; (22.318 + 14.878 + 22.318) = 59.514 kN-m

(ii) Positive moment at an interior support

(a) **Column strips :** 60 per cent of the total positive moment in the panel is resisted by the column strips. There are two column strips. Therefore the moment resisted by each column strip $(\because M_{p.BC} = 32.046\ kN-m)$

$$\left(\frac{1}{2} m_{p.c}\right) = \left(\frac{1}{2} \times 0.60 \times M_{p.BC}\right) = \left(\frac{1}{2} \times 0.60 \times 32.046\right) = 9.614\ kN-m$$

(b) Middle strip : The negative moment resisted by the middle strip

$$m_{p.m} = (0.40\ M_{p.AB}) = (0.40 \times 32.046) = 12.818\ kN-m$$

Check : (9.614 + 12.818 + 9.614) = 32.046 kN-m

Step 10 : Strip moments per metre width

(A) Strips of span AB, Fig. 18.9 (A)

(i) Negative moment

(a) Column strips : The width of column strip of the interior panel is 1.2 m. Therefore, the moment per metre width

$$\left(\frac{1}{2} m_{nc}\ /metre\right) = \left(\frac{29.978}{1.2}\right) = 24.982\ kN-m/m$$

(b) **Middle strips** : The width of middle strip of the interior panel is 2.4 m. Therefore, the moment per metre width

$$\left(\frac{1}{2} m_{nm}\ /metre\right) = \left(\frac{19.986}{2.4}\right) = 8.328\ kN-m/m$$

(i) Positive moment

(a) Column strips : The width of column strip of the interior panel is 1.2 m. Therefore, the moment per metre width

$$\left(\frac{1}{2} m_{pc} / metre\right) = \left(\frac{12.914}{1.2}\right) = 10.762\ kN-m/m$$

(b) Middle strip : The width of middle strip of the interior panel is 2.4 m. Therefore, the moment per metre width

$$\left(\frac{1}{2} m_{pm} / metre\right) = \left(\frac{17.22}{2.4}\right) = 7.175\ kN-m/m$$

(B) Strips of span BC, Fig. 18.9 (B)

(i) Negative moment

(a) Column strips : The width of column strip of the interior panel is also 1.2 m. Therefore, the moment per metre width

$$\left(\frac{1}{2} m_{nc} / metre\right) = \left(\frac{22.318}{1.2}\right) = 18.598\ kN-m/m$$

(b) Middle strip : The width of middle strip of the interior panel is 3.4 m. Therefore, the moment per metre width

$$\left(\frac{1}{2} m_{nm} / metre\right) = \left(\frac{14.878}{3.4}\right) = 4.376\ kN-m/m$$

(ii) Positive moment

(a) Column strips : The width of column strip of the interior panel is also 1.2 m. Therefore, the moment per metre width

$$\left(\frac{1}{2} m_{pc} / metre\right) = \left(\frac{9.614}{1.2}\right) = 8.012\ kN-m/m$$

(b) Middle strips : The width of middle strip of the interior panel is 3.4 m. Therefore, the moment per metre width

$$\left(\frac{1}{2} m_{pn} / metre\right) = \left(\frac{12.818}{3.4}\right) = 3.770\ kN-m/m$$

Step 11 : Effective depth of the flat slab

The absolute maximum bending moment out of eight values of values of step 9 is 24.982 kN-m/m. The effective depth of the flat slab.

$$d = \left(\frac{M}{R \times 1000}\right)^{1/2} = \left(\frac{24.982 \times 10^6}{0.9130 \times 1000}\right)^{1/2} = 165.54\ mm$$

Let the effective cover be 25 mm. Then, the overall thickness of the flat shall be (165.54 + 25) = 190.54 mm, say 200 mm. The effective depth of slab, d shall be (200-25) = 175 mm.

Step 12 : Slab reinforcement

The spacing of bars in flat slab shall not exceed 2 times the slab thickness (viz., 2 x 200 = 400 mm). 12 mm diameter Hysd steel bars shall be provided.

$$A_\phi = \left(\frac{\pi}{4} \times 12^2\right) = 113.097\ mm^2$$

(A) Reinforcement along the span AB, Fig. 18.9 (A)

(i) Negative moment reinforcement

(a) Column strips

$$A_{st} = \left(\frac{24.982 \times 10^6}{230 \times 0.9038 \times 175}\right) = 685.71\ mm^2$$

$$\text{Spacing} = \left(\frac{1000 \times 113.097}{685.71}\right) = 164.928\ mm$$

Provide bars at 160 mm spacing at top.

(b) Middle strips

$$A_{st} = \left(\frac{8.328 \times 10^6}{230 \times 0.9038 \times 175}\right) = 228.93\ mm^2$$

$$\text{Spacing} = \left(\frac{1000 \times 113.097}{228.93}\right) = 494.02\ mm$$

Provide bars at 400 mm spacing at top.

(ii) Positive moment reinforcement

(a) Column strips

$$A_{st} = \left(\frac{10.762 \times 10^6}{230 \times 0.9038 \times 175}\right) = 295.84\ mm^2$$

$$\text{Spacing} = \left(\frac{1000 \times 113.097}{295.84}\right) = 382.30\ mm$$

Provide bars at 380 mm spacing at bottom.

(b) Middle strips

$$A_{st} = \left(\frac{7.175 \times 10^6}{230 \times 0.9038 \times 175}\right) = 197.23\ mm^2$$

$$\text{Spacing} = \left(\frac{1000\times 113.097}{197.23}\right) = 573.43\ mm$$

Provide bars at 380 mm spacing at bottom.

(B) Reinforcement along the span BC, Fig. 18.9 (B)

(i) Negative moment reinforcement

(a) Column strip :

$$A_{st} = \left(\frac{18.958\times 10^6}{230\times 0.9038\times(175-6)}\right) = 529.39\ mm^2$$

$$\text{Spacing} = \left(\frac{1000\times 113.097}{529.39}\right) = 213.64\ mm$$

Provide bars at 200 mm spacing at top.

(b) Middle strips :

$$A_{st} = \left(\frac{4.376\times 10^6}{230\times 0.9038\times(175-6)}\right) = 124.56\ mm^2$$

$$\text{Spacing} = \left(\frac{1000\times 113.097}{124.56}\right) = 907.96\ mm$$

Provide bars at 400 mm spacing at top.

(ii) Positive moment reinforcement

(a) Column strips

$$A_{st} = \left(\frac{8.012\times 10^6}{230\times 0.9038\times(175-6)}\right) = 228.06\ mm^2$$

$$\text{Spacing} = \left(\frac{1000\times 113.097}{228.06}\right) = 495.90\ mm$$

Provide bars at 400 mm spacing at bottom.

(b) Middle strips

$$A_{st} = \left(\frac{3.770\times 10^6}{230\times 0.9038\times(175-6)}\right) = 107.31\ mm^2$$

$$\text{Spacing} = \left(\frac{1000\times 113.097}{107.31}\right) = 1053.88\ mm$$

Provide bars at 400 mm spacing at bottom.

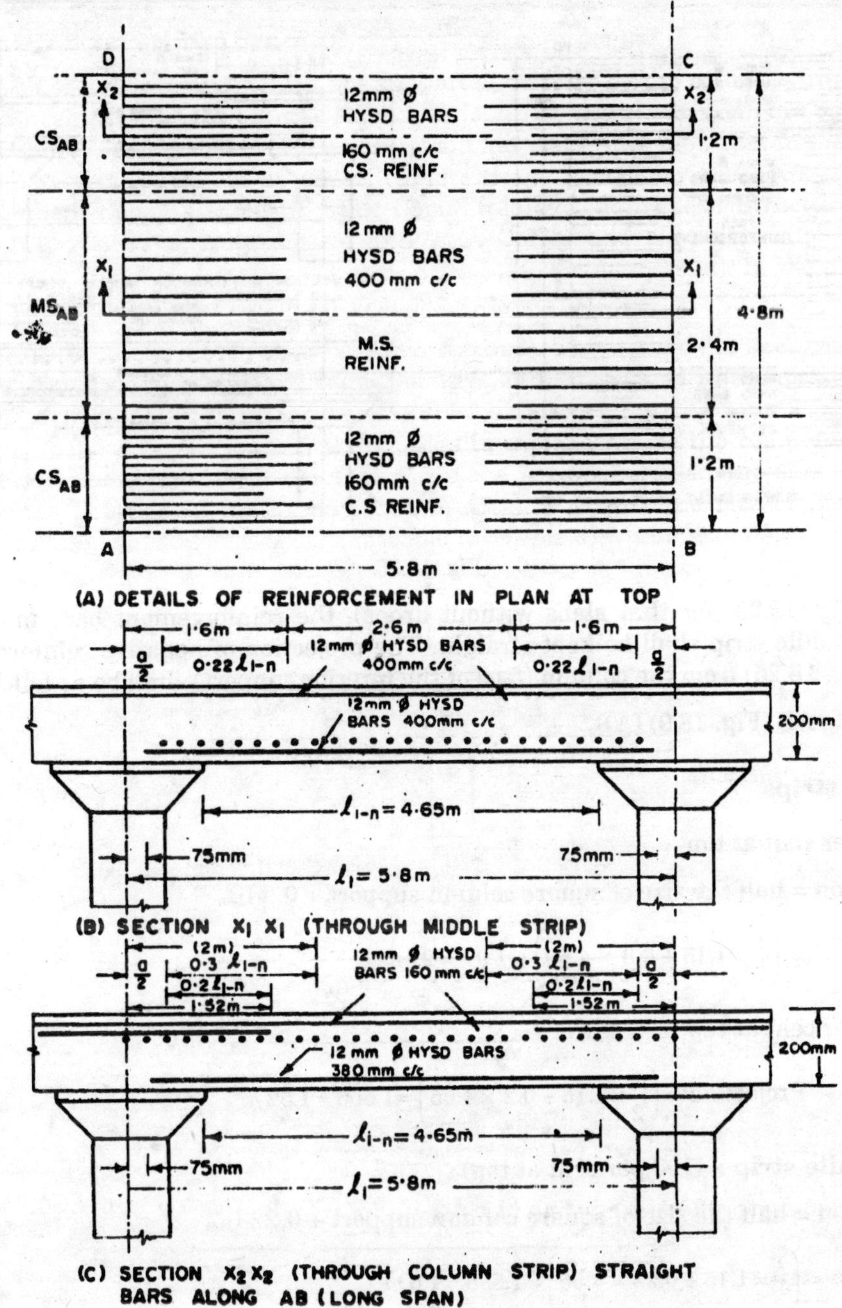

Fig. 18.32

The details of reinforcement for column strips and middle strips for negative and positive bending moments along long and short spans are shown in Fig. 18.32 and Fig. 18.33, respectively.

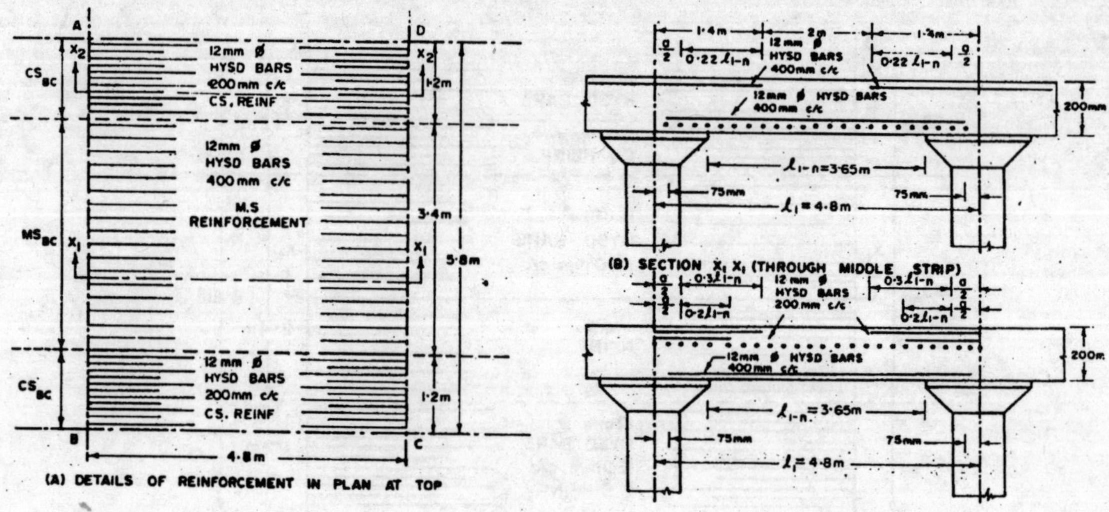

Fig. 18.33

From Fig. 18.25 (for that slabs without drops), the reinforcement bars in the column strips and middle strip shall be kept straight. The projection of negative reinforcement into the span (Fig. 18.25) from the column (face of the interior support) shall be as follows :

(A) Along AB (Fig. 18.9) (A))

(a) Column strips

(a) 50 per cent at top

Projection = half the size of square column support + 0.3 l_{1-n}

$$= \left(\frac{1}{2} \times 1.15 + 0.3 \times 4.65\right) = 1.97 = 2\ m$$

(b) 50 per cent at top

$$\text{Projection} = \left(\frac{1}{2} \times 1.15 + 0.2 \times 4.65\right) = 1.505 = 1.52\ m$$

(ii) **Middle strip :** (100 per cent at top)

Projection = half the size of square column support + 0.22 l_{1-n}

$$\left(\frac{1}{2} \times 1.15 + 0.22 \times 4.65\right) = 1.598 = 1.6\ m$$

(B) Along BC (Fig. 18.9 (B))

(i) Column strips

(a) 50 per cent at top

Projection = half the size of square column support + 0.3 l_{1-n}

$$\left(\frac{1}{2}\times 1.15 + 0.3\times 3.65\right) = 1.67 = 1.70\ m$$

(b) 50 per cent at top

$$\text{Projection} = \left(\frac{1}{2}\times 1.15 + 0.2\times 3.65\right) = 1.305 = 1.32\ m$$

(ii) **Middle strip :** (100 per cent at top)

Projection = half the size of square column support + 0.22 l_{1-n}

$$\left(\frac{1}{2}\times 1.15 + 0.22\times 3.65\right) = 1.378 = 1.4\ m$$

The projections of reinforcement from supports are also shown in Fig. 18.32 and Fig. 18.33.

Step 13 : Shear in flat slabs

The diameter of column head is 1300 mm and the effective depth of flat slab, d is 175 mm. The diameter of critical section for two-way round the column head, Fig. 18.29 (B)

$$\left(D_E + \frac{d}{2} + \frac{d}{2}\right) = \left(1300 + \frac{175}{2} + \frac{175}{2}\right) = 1475\ mm$$

Peripheral distance

$$b_o = (\pi \times 1.475) = 4.634\ m$$

Shear force at the critical section, Fig. 18.29 (B)

$$V = (w_d + w_l)\left[l_1 . l_2 - \frac{\pi}{4}(D_E + d)^2\right]$$

$$\text{or}\quad V = (5.48 + 4)\left[5.8\times 4.8 - \frac{\pi}{4}(1.475)^2\right] = 247.72\ kN$$

Nominal shear stress

$$\tau_v = \left(\frac{V}{b_o . d}\right) = \left(\frac{247.72\times 1000}{4634\times 175}\right) = 0.305\ N/mm^2$$

Permissible shear stress

The grade of concrete is M 20. The permissible shear stress in flat slab in concrete (working stress method)

$$\tau_c = 0.16\,(f_{ck})^{1/2} = 0.16\,(20)^{1/2} = 0.72\ N/mm^2$$

When the shear reinforcement is not provided, the nominal shear stress, τ_v calculated at the critical section shall not exceed $k_s.\tau_c$.

$$k_s = (0.5 + \beta_c) \text{ and } \not> 1.0$$

$$\beta_c = 1.00 \text{ for circular column and } \therefore k_s = 1.00$$

$$k_s \,.\, \tau_c = 0.72 \; N/mm^2$$

$$\left(\tau_v = 0.305 \; N/mm^2\right) \not> \left(k_s \,.\tau_c = 0.72 \; N/mm^2\right) \text{ satisfied.}$$

18.22 OPENINGS IN FLAT SLABS

Openings of any size may be provided in the flat slab. It is shown by analysis that the requirements of strength and serviceability are satisfied. However, for openings conforming to the following, no special analysis is needed.

(a) Openings of any size may be placed within the middle half of the span in each direction, provided the total quantity of reinforcement needed for the panel without the opening is maintained.

(b) In the area common to two column strips, not more than *one-eighth* of strip in either span shall be interrupted by the openings. The equivalent of reinforcement interrupted shall be added on all the sides of the openings.

(c) In the area common to one column strip and one middle strip, not more than *one-quarter* of the reinforcement in either strip shall be interrupted and shall be added on all the sides of the openings.

(d) The requirements of shear are satisfied.

PROBLEMS

18.1 Design an interior panel of a flat slab 5.2 m x 6.2 m in size. The slab is supported on columns of 600 mm in diameter. The height of columns above and below the slab is 4 m. A finishing surface of 20 mm thickness is provided over the slab. The floor of the slab is likely to be used as a class-room. Use M 20 grade of concrete and Hysd steel bars of grade Fe 415.

19

DESIGN OF CIRCULAR SLABS

19.1 INTRODUCTION

The roof covering, the base slabs and circular raft foundations in the circular water tanks are the examples of *circular slabs.* The roof coverings of circular auditoriums are also the **circular slabs.** The pump houses built-over the tube wells have circular slabs for roof covering. The circular slabs are also provided over traffic control posts at road crossings. The elastic analysis for isotropic circular panels carrying symmetrical loads is dealt in standard text books of theory of plates and shells. For circular panels (in plan) simply supported or fully fixed along the circumference and carrying symmetrically distributed loads, the expressions for total bending moments which should be provided for across each of two diameters mutually at right-angles have been derived by S. Timoshenko and S. Woinowsky-Krieger (Theory of Plates and Shells, New York, Mc Graw-Hill, second edition, 1959).

The circular plate carrying loads symmetrically distributed about the axis perpendicular to the plate through its centre, the middle plane of the plate deflects symmetrically. The deflections at all the points equally distant from the centre of plate will be equal. It is sufficient to consider the deflections in one diametrical section through the axis of symmetry. The circular slabs carrying uniformly distributed load deflects in the shape of a saucer. The radial and circumferential stresses develop in the slabs. The radial and circumferential cracks are always seen near the centre of slabs. The convex and concave surfaces of the saucer surface carries tensile and compressive stresses, respectively. The reinforcement is provided in the radial and circumferential directions as shown in Fig. 19.1 (A). Instead of this, the reinforcement is also provided in two directions mutually at right-angles as shown in Fig. 19.1 (B). Normally, the reinforcement for the positive bending moments is provided in two directions mutually perpendicular near the centre and the reinforcement for the negative bending moments near the edge should be provided by radial and circumferial bars.

In general, the radial and tangential moments vary according to the position being considered. A circular slab may therefore be designed by one of the following elastic methods.

(i) The circular slab may be designed for the maximum positive bending moment at the centre of slabs. The quantity of reinforcement or the thickness of slab may be reduced towards the circumference. In case the circular slab is not truly simply supported, the reinforcement is provided for the negative bending moment around the circumference.

(ii) The circular slab may be designed for average positive bending moment across a diameter and the same thickness and the quantity of reinforcement is provided throughout the entire slab. However, in case the slab is not truly simply supported around the circumference, the reinforcement is provided for the appropriate negative bending moment.

When the ratio of diameter of the slab and the thickness of slab is more than 10, the circular slabs are considered as thin plates. Such reinforced circular concrete slabs as analysed for the purpose of determining the bending moments as thin plates. The poisson's ratio of the concrete in the slabs is considered as zero.

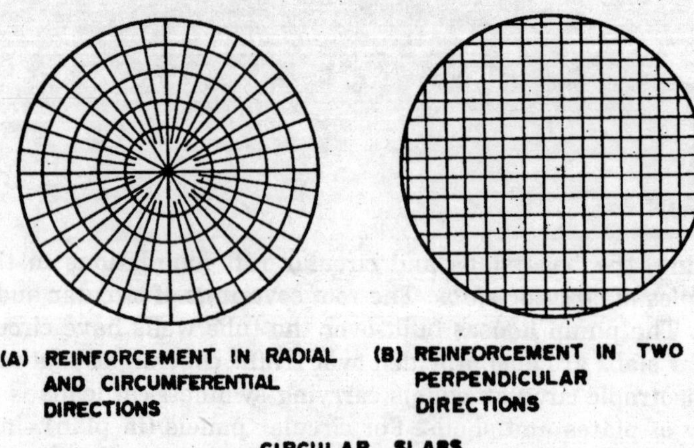

Fig. 19.1

The circular slabs may be simply supported at their edges. These slabs may have fixed support at the edges. The isolated circular slabs (*for example,* circular foundation slabs) are considered as free. The various cases of circular slabs with different support conditions and carrying different symmetrical loads have been discussed in the subsequent articles.

19.2 CIRCULAR SLAB (SIMPLY SUPPORTED AT EDGE) CARRYING UNIFORMLY DISTRIBUTED LOAD

A circular slab of radius R, simply supported at the edge, carrying uniformly distributed load w kN/m^2 over its complete surface is shown in Fig. 19.2 (A). The moments and shear in the radial and circumferential direction are as under :

1. Radial direction. The distribution of bending moment along the radial direction is as shown in Fig. 19.2 (B) and it is *parabolic*. This bending moment, M_r at any point at a radial distance r is obtained from the following expression

$$M_r = \frac{1}{16}(3+v)\left(R^2 - r^2\right). w \textit{ per unit width} \tag{19.1}$$

where, ν is the Poisson's ratio of material of the slab. For concrete, the Poisson's ratio is taken as zero. Then,

$$M_r = \frac{3}{16}. \left(R^2 - r^2\right) w \textit{ per unit width} \tag{19.2}$$

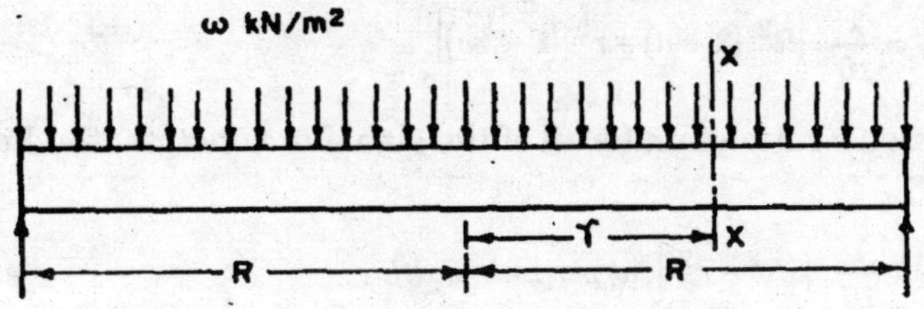

(A) CIRCULAR SLAB (SIMPLY SUPPORTED)

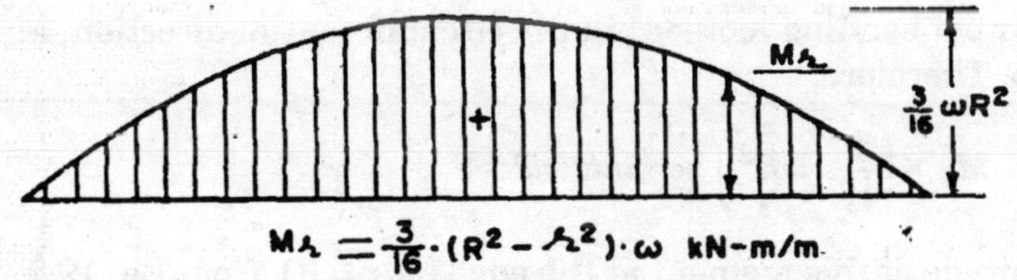

(B) MOMENT IN RADIAL DIRECTION (Mr + Ve)

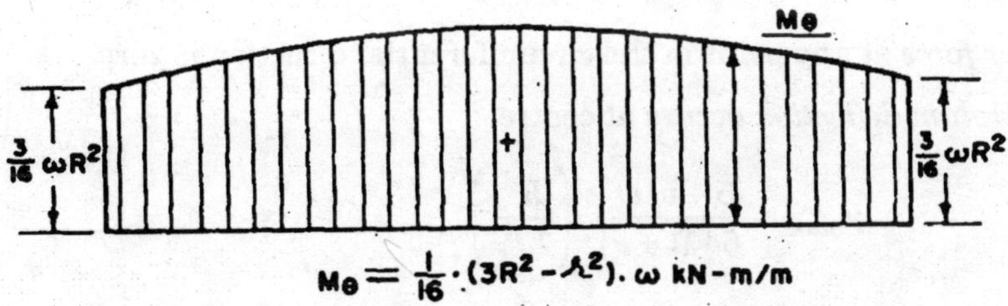

(C) MOMENT IN CIRCUMFERENTIAL DIRECTION (Mθ +VE)

Fig.19.2

The maximum bending moment in the radial direction occurs at the centre of the slab, at which r is zero. Therefore,

$$M_r = \frac{3}{16}\, wR^2 \;\; w \;\; per \; unit \; width \tag{19.3}$$

It is to note from Eq. 19.2 that M_r is zero at r = R.

The value of *shear force,* F at a distance *r* from the centre of slab

$$F_r = \left(\frac{1}{2} w \,.\, r\right) \; per \; unit \; width \tag{19.4}$$

$$\text{At} \qquad r = R, \;\; F_r = \left(\frac{1}{2} w \,.\, R\right) \; per \; unit \; width \tag{19.5}$$

2. Circumferential direction. The distribution of bending moment along the circumference is shown in Fig. 19.2 (C). This bending moment, M_θ is determined from the equation given below

$$M_\theta = \frac{1}{16} \cdot \left[R^2 (3 + v) - r^2 (1 + 3v)\right] . w \tag{19.6}$$

where, ν is the poisson's ratio of material of the slab. For concrete, the Poisson's ratio is taken as zero. Then

$$M_\theta = \frac{1}{16} \cdot \left[3R^2 - r^2\right] . w \text{ per unit width} \tag{19.7}$$

It is seen from Eq. 19.7 that this moment varies parabolically.

The maximum bending moment in the circumferential direction, M_θ occurs at the centre, where r is zero. Therefore,

$$M_\theta = \left(\frac{3}{16} . wR^2\right) \text{ per unit width} \tag{19.8}$$

The magnitude of this moment at the edge (at r = R), from Eq. 19.7

$$M_\theta = \left(\frac{2}{16} \; w . R^2\right) \text{ per unit width} \tag{19.9}$$

The *shear force* at any point in the circumferential direction is zero.

The *maximum deflection* occurs at centre.

$$y_{\max} = \frac{(5 + v) w .}{64 (1 + v)} \cdot \left(\frac{R^4}{EI_e}\right) \tag{19.10A}$$

$$\text{For} \quad \nu = 0, \; y_{\max} = \left(\frac{5w . R^4}{64 . El_e}\right) \tag{19.10B}$$

At support, the deflection is zero.

19.3 CIRCULAR SLAB (FIXED AT EDGES) CARRYING UNIFORMLY DISTRIBUTED LOAD

A circular slab of radius, R fixed at the edges carrying uniformly distributed load, w kN/m² over the entire surface is shown in Fig. 19.3 (A). The moments and shears in the radial and circumferential directions are as below :

1. **Radial direction.** The distribution of bending moment along the radial direction is as shown in Fig. 19.3. This bending moment, M_r at any point at a radial distance, r is found from the expression as under

$$M_r = \frac{1}{16} \left(R^2 (1 + v) - r^2 (3 + v)\right) w \tag{19.11}$$

where, ν is the Poisson's ratio of material of the slab. For concrete, the Poisson's ratio is taken as zero. Then,

$$M_r = \frac{1}{16} \cdot \left(R^2 - 3r^2\right) w \text{ per unit width} \tag{19.12}$$

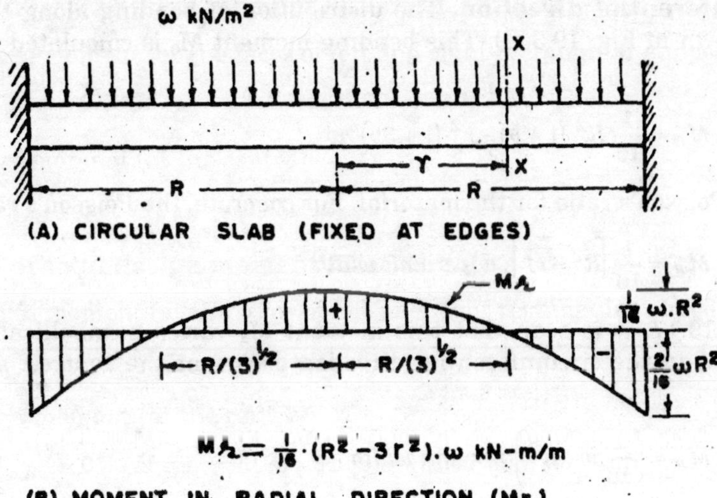

(A) CIRCULAR SLAB (FIXED AT EDGES)

$M_r = \frac{1}{16} \cdot (R^2 - 3r^2) . \omega$ kN-m/m

(B) MOMENT IN RADIAL DIRECTION (Mr)

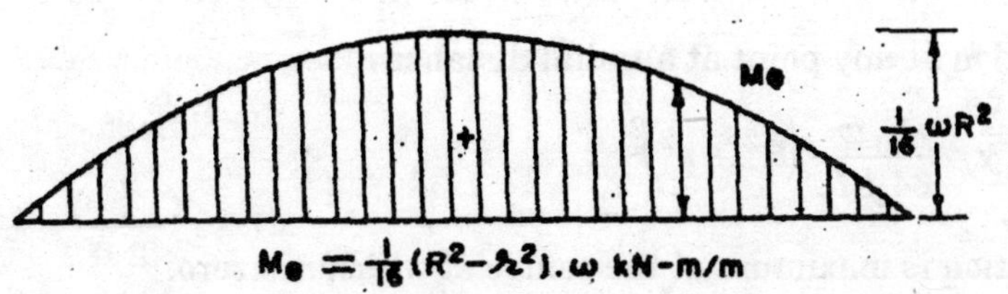

$M_\theta = \frac{1}{16} (R^2 - r^2) . \omega$ kN-m/m

(C) MOMENT IN CIRCUMFERENTIAL DIRECTION (Mθ +VE)

Fig. 19.3

It is seen from Eq. 19.10 that this moment varies parabolically. The maximum *positive* moment occurs at the centre, at which r is zero. Therefore,

$$M_r = \left(\frac{1}{16} wR^2\right) \textit{ per unit width} \tag{19.13}$$

The maximum *negative* bending moment occurs at the edge, at which r is equal to R. Therefore,

$$M_r = \left(\frac{-2}{16} wR^2\right) \textit{ per unit width} \tag{19.14}$$

The points of contraflexure (viz., $M_r = 0$) occurs at a distance $r = R/(3)^{1/2}$.

The shear force, F_r at any distance *r* from the centre of slab

$$F_r = \left(\frac{1}{2} w . r\right) \tag{19.15}$$

2. Circumferential direction. The distribution of bending along the circumferential direction is shown in Fig. 19.3(C). This bending moment M_θ is calculated from the following equation

$$M_\theta = \frac{1}{16} R^2 (1 + v) - r^2 (1 + 3v) . w \tag{19.16}$$

where ν is the Poisson's ratio for the material. For concrete, the Poisson's ratio is zero. The,

$$M_\theta = \frac{1}{16} \left[R^2 - r^2\right] . w \textit{ per unit width} \tag{19.17}$$

From Eq. 19.17, it is seen that this moment M_θ varies parabollically. The maximum bending moment in the circumferential direction occurs at the centre, at which r is zero. Therefore,

$$M_\theta = \left(\frac{1}{16} w . R^2\right) \textit{ per unit width} \tag{19.18}$$

From Eq. 19.17, it may be observed that M_θ is zero at R = r.

The shear force in the circumferential direction at any point is zero.

The deflection at any point at a radial distance *r*

$$y = \frac{1 . w}{64\, EI_e} \left(R^2 - r^2\right)^2 \tag{19.19}$$

This deflection is maximum at the centre at which *r* is zero.

$$y_{\max} = \frac{1}{64} \left(\frac{w . R^4}{EI_e}\right) \tag{19.20}$$

When the circular slab is partially at the edge

A circular slab partially fixed at the edge and carrying uniformly distributed load may be treated in between the circular slabs simply supported and fixed at the edges depending upon the degree of fixidity. In case, *fifty per cent* fixidity is taken, then, the values of moments may be determined as average of the values for these two cases at the respective points.

1. **Radial direction.** The moments in the radial direction at any point at distance r from the cnetre for simply supported edge and fixed edge are as under : (From Eq. 19.2 and Eq. 19.12 respectively).

For simply supported edge (*Positive* B.M.)

$$M_r = \frac{3}{16} \left(R^2 - r^2\right) . w \tag{i}$$

For fixed edge (*Positive* B.M.)

$$M_r = \frac{1}{16} \left(R^2 - r^2\right) . w \tag{ii}$$

For the edges partially fixed *(with fifty per cent)*

$$M_r = \frac{2}{16}\left(R^2 - r^2\right).\,w \text{ per unit width} \tag{19.21}$$

At centre, M_r is maximum

$$M_r = \left(\frac{2}{16} w \,.\, R^2\right) \tag{19.22}$$

Negative bending moment

$$M_r = \left(\frac{1}{16} wR^2\right) \tag{19.23}$$

For simply supported (*negative* B.M.)

$$M_r = 0 \tag{iii}$$

For fixed edge (*negative* B.M.)

$$M_r = \left(\frac{-2}{16} w \,.\, R^2\right) \text{ per unit width} \tag{iv}$$

For the edges partially fixed (*negative* B.M.)

$$M_r = \left(\frac{-1}{16} w\, R^2\right) \text{ per unit width} \tag{19.24}$$

The moment in the radial direction is zero at a distance $r = R\,(2/3)^{1/2}$. From the centre.

2. **Circumferential direction.** As per discussion above, the moment in circumferential direction in the circular slab at the centre.

For simply supported edge, from Eq. 19.8

$$M_\theta = \left(\frac{+3}{16} w \,.\, R^2\right) \tag{v}$$

For fixed edge, from Eq. 19.18

$$M_\theta = \left(\frac{+1}{16} w \,.\, R^2\right) \tag{vi}$$

For the edge partially fixed at centre (*positive* B.M.)

$$M_\theta = \left(\frac{+2}{16} w \,.\, R^2\right) \text{ per unit width} \tag{19.25}$$

For the simply supported edge at support, from Eq. 19.9

$$M_\theta = \left(\frac{+2}{16} w \,.\, R^2\right) \tag{vii}$$

For the fixed edge at support

$M_\theta = 0$ (viii)

For the edge partially fixed at support

$$M_\theta = \left(\frac{+1}{16} w\ R^2\right) \tag{19.26}$$

For negative bending moment, at least *one-third* of the total reinforcement for negative moment at the support shall extend beyond the point of inflection a distance not less than the effective depth of the member of 12ϕ or *on-sixteen* of the clear span which ever is more.

19.4 CIRCULAR SLAB (SIMPLY SUPPORTED AT EDGE) CARRYING UNIFORMLY DISTRIBUTED TOTAL LOAD W ALONG CIRCUMFERENCE OF A CONCENTRIC CIRCLE

A circular slab of radius R, simply supported at the edges carrying total uniformly distributed load W along the circumferential of a concentric circle of radius, b is shown in Fig. 19.4. The bending moments and shears in the radial and circumferential directions are as under :

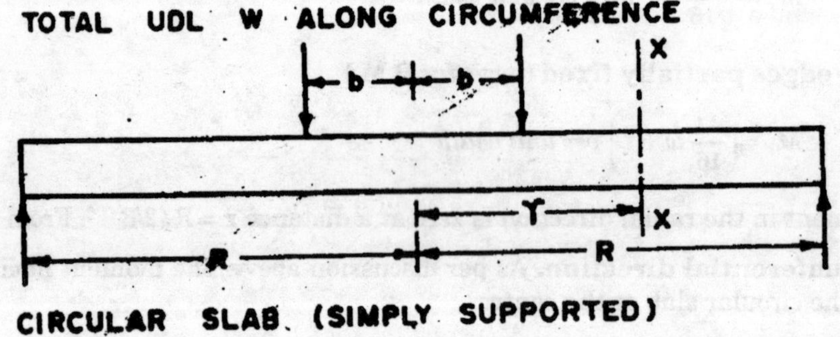

Fig. 19.4

1. Radial direction

(a) For r less than b

$$M_r = \frac{W}{8.\pi}\left[2\log_e\left(\frac{R}{b}\right)+1-\left(\frac{b}{R}\right)^2\right] \tag{19.27}$$

$$F_r = 0. \tag{19.28}$$

(B) *For r more than b*

$$M_r = \frac{W}{8\ .\ \pi}\left[2\ \log_e\left(\frac{R}{r}\right)-\left(\frac{b}{R}\right)^2+\left(\frac{b}{r}\right)^2\right] \tag{19.29}$$

$$F_r = \left(\frac{W}{2\pi\ .\ r}\right) \tag{19.30}$$

2. Circumferential direction

(a) For r less than b

$$M_r = \frac{W}{8.\pi}\left[2\log_e\left(\frac{R}{b}\right)+1-\left(\frac{b}{R}\right)^2\right] \quad (19.31)$$

(B) For r more than b

$$M_r = \frac{W}{8\,\pi}\left[2\log_e\left(\frac{R}{r}\right)-\left(\frac{b}{R}\right)^2+2-\left(\frac{b}{r}\right)^2\right] \quad (19.32)$$

(Note $\log_e x = 2.3026 \log_{10} x$).

19.5 CIRCULAR SLAB (SIMPLY SUPPORTED AT EDGE) CARRYING UNIFORMLY DISTRIBUTED LOAD INSIDE A CIRCLE

A circular slab of radius, R simply supported at the edge carrying uniformly distributed load w kN/m² inside a circle of radius, b is shown in Fig. 19.5. The bending moments end shear along radial and circumferential directions are as follows :

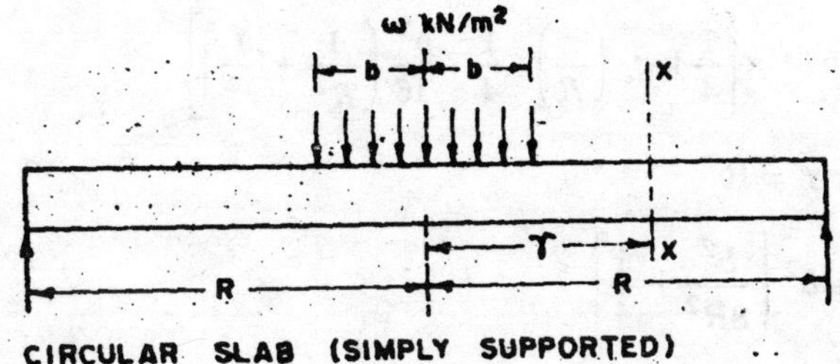

Fig. 19.5

1. Radial direction

$$M_r = -\frac{3}{16}w\,.\,r^2+\frac{1}{4}w.b^2\times\left[1-\log_e\left(\frac{b}{R}\right)-\left(\frac{b^2}{4R^2}\right)\right] \quad (19.33)$$

per unit width. At r = 0,

$$M_r = \frac{1}{4}wb^2\times\left[1-\log_e\left(\frac{b}{R}\right)-\left(\frac{b^2}{4R^2}\right)\right] \quad (19.33A)$$

At r = 0, and b = R

$$M_r = \left(+\frac{3}{16}wR^2\right) \quad (19.33B)$$

$$F_r = \left(\frac{1}{2}wr\right) \quad (19.34)$$

(b) *For r more than b*

$$M_r = -wb^2 \times \left[\frac{1}{4}\log_e\left(\frac{r}{R}\right) + \frac{b^2}{16}\left(\frac{1}{R^2} - \frac{1}{r^2}\right)\right] \tag{19.35}$$

per unit width

$$F_r = \left(\frac{1}{2}\frac{w.b^2}{r}\right) \tag{19.36}$$

2. Circumferential direction

(a) *For r less than b*

$$M_\theta = -\frac{1}{16}w.r^2 + \frac{1}{4}w.b^2 \times \left[1 - \log_e\left(\frac{b}{R}\right) - \left(\frac{b^2}{4R^2}\right)\right] \tag{19.37}$$

per unit width.

(b) *For r more than b*

$$M_\theta = -wb^2 \times \left[\frac{1}{4}\log_e\left(\frac{r}{R}\right) - \frac{1}{4} + \frac{b^2}{16}\left(\frac{1}{R^2} + \frac{1}{r^2}\right)\right] \tag{19.38}$$

per unit width. For r = R

$$M_\theta = -wb^2\left[\frac{b^2}{8R^2} - \frac{1}{4}\right] \tag{19.38A}$$

19.6 CIRCULAR SLAB (SIMPLY SUPPORTED AT EDGE WITH A CENTRAL HOLE) CARRYING TOTAL UNIFORMLY DISTRIBUTED LOAD W ALONG CIRCUMFERENCE OF A CONCENTRIC CIRCLE

A circular slab of radius, R simply supported at the edges carrying total uniformly distributed load W kN along the circumference of a concentric circle of radius, c is shown in Fig. 19.6. There is a circular hole at the centre of the slab of radius, b. The bending moment and shear in the radial and circumferential directions are obtained from the following expressions :

1. Radial direction

(a) For r less than c

$$M_r = \frac{W}{4\pi}\cdot\frac{R^2}{(R^2 - r^2)}\cdot\left(1 - \frac{b^2}{r^2}\right)\times\left[\log_e\left(\frac{R}{c}\right) + \frac{1}{2} - \left(\frac{c^2}{2R^2}\right)\right] \tag{19.39}$$

per unit width, $F_r = 0$ (19.40)

(B) For r more than c

$$M_r = \frac{W}{4 \,.\, \pi}\left[\log_e\left(\frac{c}{r}\right) - \frac{1}{2} + \left(\frac{r^2}{R^2 - r^2}\right)\left(\frac{r^2 - b^2}{r^2}\right) \times \left\{\log_e\left(\frac{R}{c}\right) + \frac{1}{2} - \left(\frac{c^2}{2R^2}\right)\right\}\right] \quad (19.41)$$

$$F_r = \left(\frac{W}{2\pi \,.\, r}\right) \quad (19.42)$$

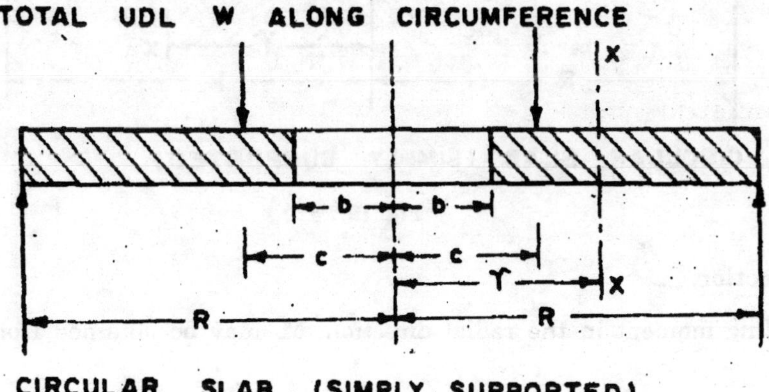

Fig. 19.6

2. Circumferential direction

(a) For r less than c

$$M_\theta = \frac{W}{4 \,.\, \pi}\frac{R^2}{\left(R^2 - r^2\right)}\left(1 + \frac{b^2}{r^2}\right) \times \left[\log_e\left(\frac{R}{c}\right) + \frac{1}{2} - \left(\frac{c^2}{2R^2}\right)\right] \quad (19.43)$$

per unit width

(b) *For r more than c*

$$M_\theta = \frac{W}{4 \,.\, \pi}\left[\log_e\frac{c}{r} + \frac{1}{2} + \left(\frac{R^2}{R^2 - r^2}\right)\left(\frac{r^2 + b^2}{r^2}\right) \times \left\{\log_e\left(\frac{R}{c}\right) + \frac{1}{2} - \left(\frac{c^2}{2R^2}\right)\right\} - \left(\frac{c^2}{2r^2}\right)\right] \quad (19.44)$$

per unit width

19.7 CIRCULAR SLAB (SIMPLY SUPPORTED AT EDGE WITH CENTRAL HOLE) CARRYING UNIFORMLY DISTRIBUTED LOAD

A circular slab of radius, R simply supported at edge carrying uniformly distributed load w kN/m² over the complete surface is shown in Fig. 19.7. There is a hole at the centre of the slab of radius, b. The bending moments and shears in the radial and circumferential directions are calculated as under :

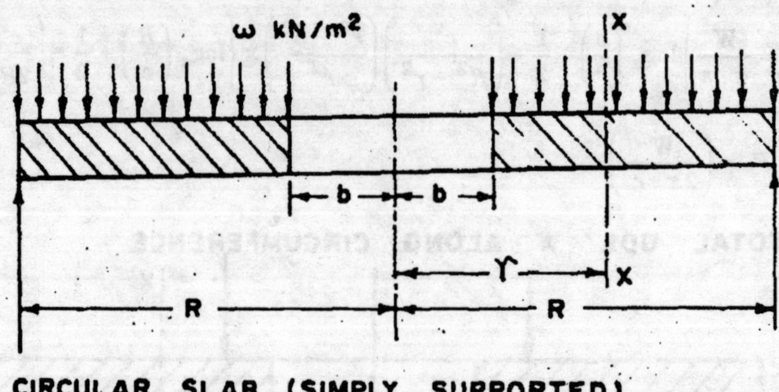

Fig. 19.7

1. Radial direction

The bending moment in the radial direction, M_r may be obtained from the following equation

$$M_r = -\frac{3}{16} wr^2 + \frac{wb^2}{4}\left[\log_e\left(\frac{r}{R}\right) + \frac{3}{4}\left\{\left(1 + \frac{R^2}{b^2}\right) - \left(\frac{R^2}{r^2}\right)\right\} + \left(\frac{R^2 - r^2}{r^2}\right) \cdot \left(\frac{b^2}{R^2 - b^2}\right) . \log_e\left(\frac{R}{b}\right)\right] \quad (19.45)$$

per unit width

$$\text{Shear force, } F_r = \left(\frac{1}{2} w\ r - \frac{1}{2}\frac{w . b^2}{4}\right) \quad (19.46)$$

2. Circumferential direction

The bending moment in the circumferential direction M_θ may be determined from the following equation :

$$M_\theta = -\frac{1}{16} w . r^2 + \frac{wb^2}{4} \times \left[\log_e\left(\frac{r}{R}\right) + \frac{3}{4}\left\{-\frac{1}{3} + \left(\frac{R^2}{b^2}\right) + \left(\frac{R^2}{r^2}\right)\right\} - \left(\frac{R^2 + r^2}{r^2}\right) \cdot \left(\frac{b^2}{R^2 - b^2}\right) . \log_e\left(\frac{R^2}{b^2}\right)\right] \quad (19.47)$$

per unit width.

19.8 CIRCULAR SLAB CARRYING A CONCENTRATED LOAD AT CENTRE

A circular slab of radius R carrying a concentrated load at the centre is shown in Fig. 19.8. The bending moments in the radial and circumferential direction in the circular slab when the slab is simply supported and when the slab is having fixed edges are as under :

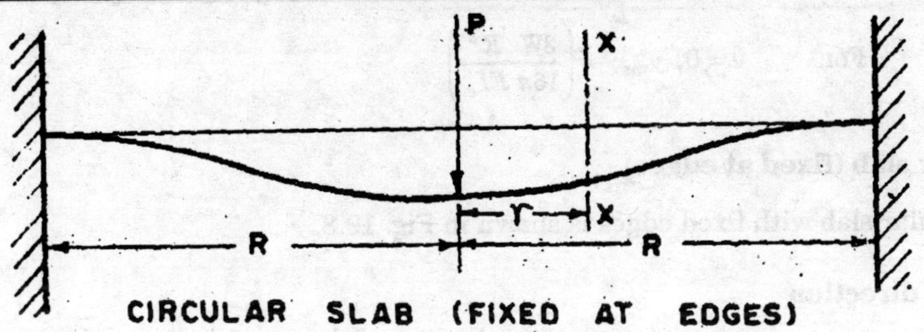

Fig. 19.8

1. Circular slab (simply supported at edges)

(A) Radial direction

The bending moment in the radial direction for the points not very close to the load

$$M_r = \frac{W}{4 . \pi}(1+v)\log_e\left(\frac{R}{r}\right) \quad (19.8)$$

For $v = 0, M_r = \frac{W}{4 . \pi} . \log_e\left(\frac{R}{r}\right)$ (19.49)

(B) Circumferential direction

The bending moments in the circumferential direction for the points not very close to the load

$$M_\theta = \frac{W}{4 . \pi}\left[(1+v)\log_e\left(\frac{R}{r}\right) + (1-v)\right] \quad (19.50)$$

For $v = 0, M_\theta = \frac{W}{4 . \pi}\left[\log_e\left(\frac{R}{r}\right) + 1\right]$ (19.51)

The deflection at any point of the slab at a distance r from the centre

$$y = \frac{W}{16\pi . EI_e} . \left[\left(\frac{3+v}{1+v}\right)\left(R^2 - r^2\right) + 2r^2 \log_e\left(\frac{r}{R}\right)\right] \quad (19.52)$$

For ν = 0

$$y = \frac{W}{16 . \pi . EI_e} = \left[3\left(R^2 - r^2\right) + 2r \log_e\left(\frac{r}{R}\right)\right] \quad (19.53)$$

The maximum deflection at the centre of simply supported slab

$$y_{max} = \left(\frac{(3+v)W . R^2}{16\pi(1+v)EI_e}\right) \quad (19.54)$$

For $\quad \nu = 0, \; y_{max} = \left(\dfrac{3W \, . R^2}{16\pi \, EI_e}\right)$ (19.55)

2. Circular slab (fixed at edges)

A circular slab with fixed edges is shown in Fig. 19.8.

(A) Radial direction

The bending moment at any point not very close to the load in the radial direction

$$M_r = \left[\frac{W}{4 \, . \, \pi}(1 + v) \log_e \left(\frac{R}{r}\right) - 1\right] \tag{19.56}$$

For $\quad \nu = 0, \; M_r = \dfrac{W}{4 \, . \, \pi}(1 + v)\left[\log_e \left(\dfrac{R}{r}\right) - 1\right]$ (19.57)

(B) Circumferential direction

The bending moment at any point not very close to the load

$$M_\theta = \frac{W}{4 \, . \, \pi}\left[(1 + v) \log_e \left(\frac{R}{r}\right) - v\right] \tag{19.58}$$

For $\quad \nu = 0, \; M_\theta = \dfrac{W}{4 \, . \, \pi} \, . \log_e \left(\dfrac{R}{r}\right)$ (19.59)

The deflection of a circular slab with fixed edges carrying a concentrated load at the centre as shown in Fig. 19.8

$$y = \left(\frac{W \, . \, r^2}{8 \, \pi \, . \, EI_e}\right) \log_e \left(\frac{r}{R}\right) \frac{W}{16\pi \, . \, EI} \left(R^2 - r^2\right) \tag{19.60}$$

Example 19.1. Design a circular slab for a room 6 m in effective diameter. The circular slab is simply supported at edges and carries a lived load of 4 kN/m². The slab shall have a finishing coat of 20 mm cement concrete. Provide M 20 grade of concrete and Hysd steel bars of quality Fe 415.

Design : The design constants for 20 grade of concrete and Hysd steel bars of quality Fe 415 are as follows :

Step 1 : Design constants

The permissible stress in bending in compression for M 20 grade of concrete

$$\sigma_{cbc} = 7 \; N/mm^2$$

The permissible for Hysd steel bars in being in tension

$$\sigma_{st} = 230 \; N/mm^2$$

Modular ratio for M 20 grade of concrete

$$m = \left(\frac{280}{3 \times 7}\right) = 13.33$$

The design factor for balanced section are as under :

Factor for depth of neutral axis

$$k_B = \left(\frac{13.33 \times 7}{13.33 \times 7 + 230}\right) = 0.2886$$

Factor for lever arm

$$j_B = \left(1 - \frac{0.2886}{3}\right) = 0.9038$$

Factor for moment of resistance

$$R_B = \left(\frac{1}{2} \times 7 \times 0.2886 \times 0.9038\right) = 0.9130$$

Percentage of steel

$$p_B = \left(\frac{1}{2} \times \frac{7}{230} \times 0.2886 \times 100\right) = 0.4392 \textit{ per cent}$$

Step 2: Design loads

The dead load and live load are as under :

Superimposed load on the slab shall be weight of finishing coat

$$(0.020 \times 1 \times 1 \times 24) = 0.48\ kN/m^2$$

Let the overall thickness of slab be 150 mm. The self-weight of slab

$$(0.150 \times 1 \times 1 \times 25) = 3.75\ kN/m^2$$

Live load = 4.00 kN/m²

Total uniformly distributed load

w = 8.23 kN/m².

Step 3 : Bending moments per unit width

The effective (span) diameter of the slab is 6 m (R = 3 m).

The circular slab simply supported at edges carrying uniformly distributed load is shown in Fig. 19.2 (A).

The maximum bending moment in the radial direction occurs at the centre of slab. Therefore, from Eq. 19.3

$$M_r = \frac{3}{16} wR^2 \text{ (Positive)} = \left(\frac{3}{16} \times 8.23 \times 3^2\right) = 13.888\ kN - m/m$$

The moment in the radial direction at the support (at r = R) is zero.

The maximum bending moment in the circumferential direction also occurs at the centre. From Eq. 19.8

$$M_\theta = \frac{3}{16} wR^2 \text{ (Positive)} = \left(\frac{3}{16} \times 8.23 \times 3^2\right) = 13.888\ kN - m/m$$

The bending moment in the circumferential direction at support, from Eq. 19.9

$$M_\theta = \left(\frac{2}{16} \times wR^2\right) = \left(\frac{2}{16} \times 8.23 \times 3^2\right) = 9.259\ kN - m/m$$

Step 4 : Effective depth of slab

The effective depth of slab is as under :

$$d = \left(\frac{M}{R \times 1000}\right)^{1/2} = \left(\frac{13.888 \times 10^6}{0.9130 \times 1000}\right)^{1/2} = 123.34\ mm$$

Let the effective core be 25 mm. Then, the overall thickness of slab

(123.34 + 25) = 148.34 mm = 150 mm

Therefore, the effective depth of slab is 125 mm.

Step 5 : Steel reinforcement

Hysd steel bars are provided for reinforcement. The reinforcement necessary per unit width in the radial and circumferential direction near centre

$(\because M_r = M_\theta)$

$$A_{st} = \left(\frac{M}{\sigma_{st} \cdot j.d}\right) = \left(\frac{13.888 \times 10^6}{230 \times 0.9038 \times 125}\right) = 534.48\ mm^2/m$$

Minimum area of steel needed

$$A_{st.\min} = \left(\frac{0.12}{1000} \times 1000 \times 150\right) = 180\ mm^2/m$$

10 mm diameter Hysd-steel bars shall be provided.

$$A_\phi = \left(\frac{\pi}{4} \times 10^2\right) = 78.54\ mm^2$$

Spacing of bars centre to centre

$$s = \left(\frac{1000 \times 78.54}{534.48}\right) = 146.95\ mm$$

10 mm diameter Hysd steel bars shall be provided in two perpendicular directions as shown in Fig. 19.1 (B) at 130 mm spacing centre to centre at the *bottom* of the slab.

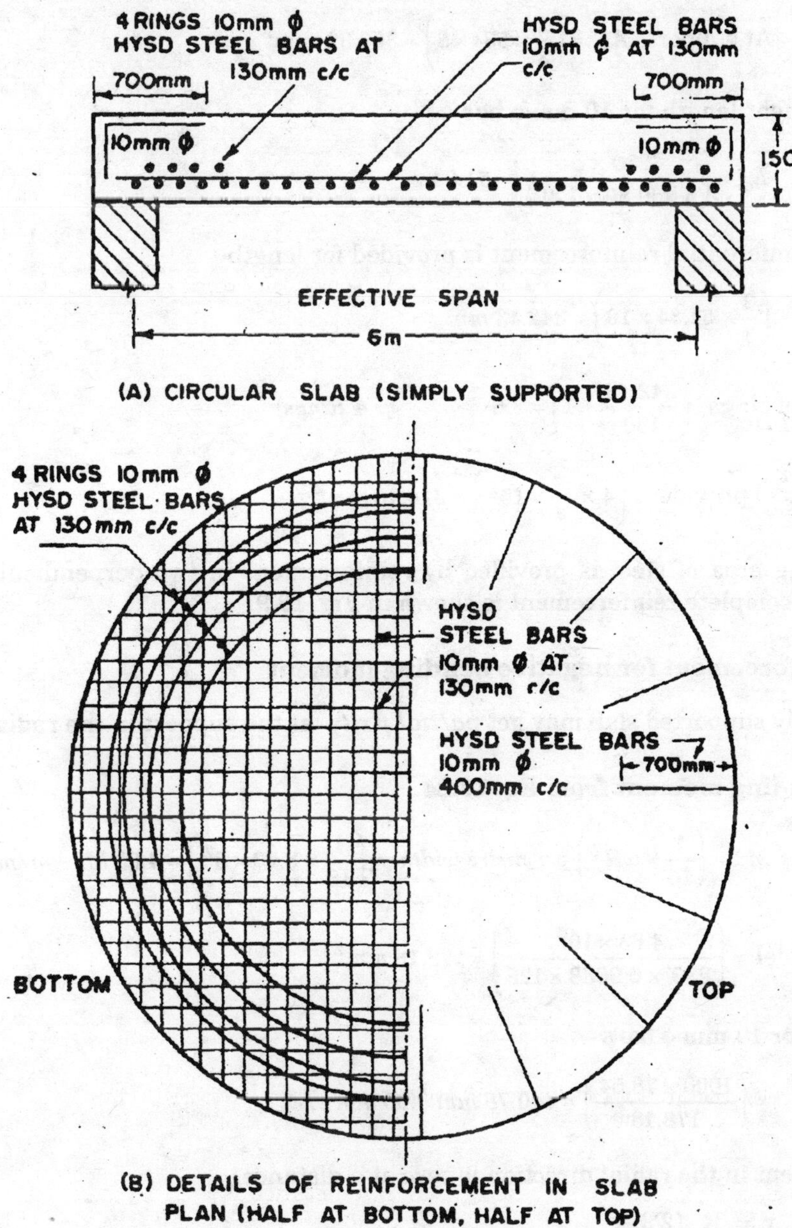

Fig. 19.9

The slab is simply supported at the edges. Above reinforcement remains free at the support and do not have sufficient development length. As such, these bars may slip.

Total reinforcement need at the support in the circumferential direction

$$A_{st}\,(\text{support}) = \frac{2}{3}\,.\,A_{st}\,.\,(\text{centre}).\ \textit{Since}\left(M_\theta\ .\text{support} = \frac{2}{3}\,.\,M_\theta\ .\,\text{centre}\right)$$

At support, $A_{st} = \left(\frac{2}{3} \times 534.48\right) = 356.32\ mm^2$

Development length for 10 mm ϕ bar

$$L_d = \left(\frac{230 \times \phi}{4 \times (0.80 \times 1.40)}\right) = 51.34\ \phi$$

The circumferential reinforcement is provided for length

$$\left(\frac{2}{3} \times 51.34 \times 10\right) = 342.43\ mm$$

Number of rings, $\left(\frac{342.43}{130} + 1\right) = 3.6 \qquad (\because\ 4\ \textit{Rings})$

Area of steel provided, $\left(4 \times \frac{\pi}{4} \times 10^2\right) = 314.15\ mm^2$

Remaining area of steel is provided by reinforcement in two perpendicular directions. The details of complete reinforcement is shown in Fig. 19.9.

Step 6 : Reinforcement for negative bending moment

The simply supported slab may get *partial fixity* at the support in the radial direction.

Negative bending moment from Eq. 19.24

$$M_r = \left(\frac{1}{16} \times wR^2\right)\ \textit{per metre width} = \left(\frac{1}{16} \times 8.23 \times 3^2\right) = 4.63\ kN-m/m$$

Area of steel $= \left(\frac{4.63 \times 10^6}{230 \times 0.9038 \times 125}\right) = 178.18\ mm^2$

Spacing for 10 mm ϕ bars

$$= \left(\frac{1000 \times 78.54}{178.18}\right) = 440.78\ mm\ (400\ mm\ c/c)$$

The moment in the radial direction is zero at a distance

$$r = R\,.\,(2/3)^{1/2}$$

From the centre of support

$$r = 3 \times (2/3)^{1/2} = 2.450\ m.$$

From the support, the point of inflection

(3000 – 2450) = 550 mm

The reinforcement in the radial direction is provided for a distance whichever is more from the following :

(a) (500 + d) = (550 + 125) = 675 mm

(b) $(550 + 12\phi) = (550 + 120) = 670\ mm$

(c) $L_d = \left(\frac{230 \times 10}{4 \times (1.4 \times 0.8)}\right) = 513.39\ mm$

The reinforcement in the radial direction is provided upto 700 mm distance from the support.

Step 7 : Check for shear force

The shear force at support in the radial direction from Eq. 19.5 per metre width

$$F_r = \left(\frac{1}{2} wR\right) = \left(\frac{1}{2} \times 8.23 \times 3\right) = 12.345\ kN/m$$

Nominal shear stress

$$\tau_v = \left(\frac{12.345 \times 1000}{1000 \times 125}\right) = 0.099\ N/mm^2$$

Steel in tension at support

$$\left(\frac{100\ A_s}{bd}\right) = \left(\frac{100 \times 78.54}{1000 \times 125}\right) \times \left(\frac{1000}{130}\right) = 0.48$$

For M 20 grade of concrete

$$\tau_c = \left(0.22 + \frac{0.23}{0.25} \times 0.08\right) = 0.294\ N/mm^2$$

The nominal shear stress in much less than shear strength of concrete. As such, the slab is safe in shear.

Example 19.2. A circular base slab of a circular tank used for water supply is 10 m in diameter. The slab is supported by 8 m circular ring beam as shown in Fig. 19.10. The height of tank is 4.20 m above the top of base slab. The thickness of side walls of the tank is 150 mm. Use M 20 grade of concrete and Hysd steel bars of quality Fe 415 for the reinforcement. Design the circular slab.

Design : The design constants for M 20 grade of concrete and Hysd steel bars of quantity Fe 415 are as under.

Step 1 : Design constants. The permissible stress in bending in compression for M 20 grade of concrete and the permissible in bending in tension of Hysd steel bars in tension are 7 N/mm^2 and 230 N/mm^2, respectively. The design constants for the balanced section are as below :

$$m = \left(\frac{280}{3 \times 7}\right) = 13.33$$

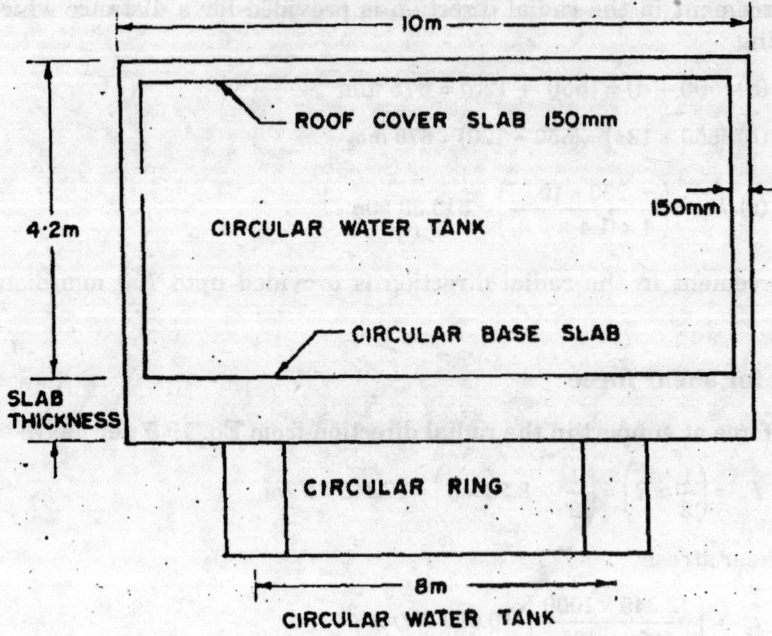

Fig. 19.10

Factor for depth of neutral axis

$$k_B = \left(\frac{13.33 \times 7}{13.33 \times 7 + 230}\right) = 0.2886$$

Factor for lever arm

$$j_B = \left(1 - \frac{0.2886}{3}\right) = 0.9038$$

Factor for moment of resistance

$$R_B = \left(\frac{1}{2} \times 7 \times 0.2886 \times 0.9038\right) = 0.9130$$

Percentage of steel

$$p_B = \left(\frac{1}{2} \times \frac{7}{230} \times 0.2886 \times 100\right) = 0.4392 \textit{ per cent}$$

Step 2 : Loads. The unit weight of water is 10 kN/m³. The weight of water column above slab acts as uniformly distributed load. The weight of water per 1 sq. metre

$$(4.2 - 0.150) \times 1 \times 1 \times 10 = 40.5\ kN/m^2$$

Let the overall thickness of slab be 480 mm. The self-weight of slab per 1 m²

$$(0.480 \times 1 \times 1 \times 25) = 12\ kN/m^2$$

Total uniformly distributed load acting over the slab

$$(40.5 + 12) = 52.5 \text{ kN/m}^2$$

The weight of roof covering slab

$$\left(\frac{\pi}{4} \times 10^2 \times 0.15 \times 25\right) = 294.52 \ kN$$

This total load shall be acting distributed load over the circumference

$$\left(\frac{294.52}{\pi\,(10 - 0.150)}\right) = 9.52 \ kN/m$$

The weight of wall of the water tank (also acts as distributed over the circumference)

$$(4.20 - 0.150) \times 1 \times 0.150 \times 25 = 15.19 \ kN/m$$

Total distributed load over the circumference per unit load

$$(9.52 + 15.19) = 24.71 \text{ kN/m}$$

Total vertical load acting downward

$$52.5 \times \left(\frac{\pi}{4}\right) \times (10 - 0.3)^2 + (24.71) \times \pi \times (10 - 0.150)$$

$$= (3879.65 + 764.64) = 4644.29 \text{ kN}$$

Let w be the distributed load acting over the circumference per 1 m length of the circular ring acting upward. Then,

$$(\pi \times 8 \times w) = 4644.29$$

$$\therefore \ w = 184.79 \text{ kN/m}$$

Let the circular slab be fictitiously simply supported at its edge round the circumference. The various loads acting over the fictitiously simply supported at its edge acting downward and upward are as shown in Fig. 19.11.

It is to note that the moment at the edge due to integral connection with the wall to the tank is small and therefore it is not considered.

Step 3 : Bending moments. The radius of slab is 5 m.

(a) **Bending moments due to uniformly distributed load 52.5 kN/m².**

The bending moment in the radial direction per unit width from Eq. 19.2

$$M_r = \frac{3}{16}\left(3R^2 - r^2\right).w \qquad \text{(i)}$$

The bending moment in the circumferential per unit width from Eq. 19.7

$$M_\theta = \frac{1}{16}\left(3R^2 - r^2\right).w \qquad \text{(ii)}$$

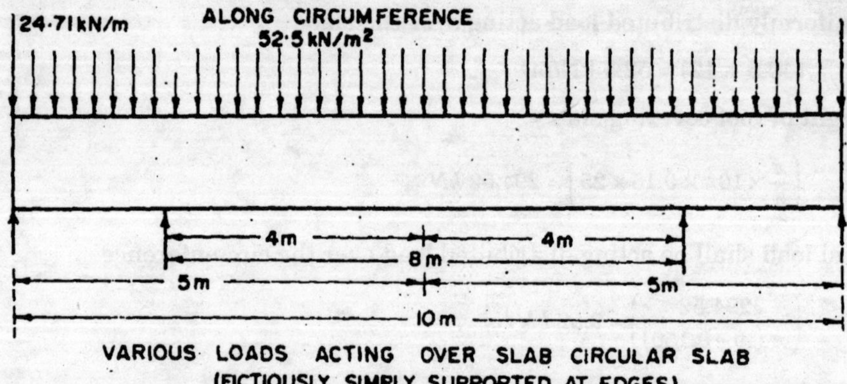

Fig. 19.11

The shear force in the slab, from Eq. 19.4.

$$F_r = \left(\frac{1}{2} w \cdot r\right) \textit{ per unit width}$$

The values of M_r, M_θ and F_r have been calculated and listed in Table 19.1.

Table 19.1

Distance r (m)	B.M. M_r (kN-m/m)	B.M. M_θ (kN-m/m)	F_r (kN/m)
0	246.09	246.09	0.00
1	236.25	242.81	26.50
2	206.72	232.97	52.50
3	157.50	216.56	78.75
4	88.59	193.59	105.00
5	0.00	164.06	131.2

(b) Bending moment due to total load 4644.29 kN distributed along the concentric circle of radius 4 m

Radius of the slab, R = 5 m

Radius of the concentric circle, b = 4 m

The bending moments in the radial and circumferential directions and the shear force in the slab are as below :

(i) For the distance r less than b

From Eqs. 19.27 and 19.31

$$M_r = M_\theta = \frac{W}{8\pi}\left[2 \log_e \left(\frac{R}{r}\right) + 1 - \left(\frac{b}{R}\right)^2\right] kN - m/m$$

From Eq. 19.28, $F_r = 0$

(ii) For the distance r more than b

From Eqs. 19.29 and 10.32

$$M_r = -1 \times \frac{W}{8\pi}\left[2\log_e\left(\frac{R}{r}\right) - \left(\frac{b}{R}\right)^2 + \left(\frac{b}{r}\right)^2\right]$$

$$M_\theta = -1 \times \frac{W}{8\pi}\left[2\log_e\left(\frac{R}{r}\right) - \left(\frac{b}{r}\right)^2 + 2 - \left(\frac{b}{R}\right)^2\right]$$

From Eq. 19.30, $F_r = \left(\frac{W}{2\pi \cdot r}\right)$

The values of M_r, M_θ and F_r have been calculated and listed in Table 19.2.

TABLE 19.2

Distance r (m)	B.M. M_r (kN-m/m)	B.M. M_θ (kN-m/m)	F_r (kN/m)
0	–148.99	–148.99	0
1	–148.99	–148.99	0
2	–148.99	–148.99	0
3	–148.99	–148.99	0
4	–148.99	–148.99	–184.79
5	–82.47	–135.05	–147.83

The net values of bending moments in radial direction, M_r and in circumferential direction, M_q and the shear force, F_r at various value of distance measured from the centre are algebraic sum of respective values listed in Tables 19.1 and 19.2, respectively. The net values have been calculated and listed in Table 19.3.

TABLE 19.3

Distance r (m)	B.M. M_r (kN-m/m)	B.M. M_θ (kN-m/m)	F_r (kN/m)
0.0	97.10	97.10	0.00
1	87.26	93.82	26.50
2	57.73	83.98	52.50
3	8.51	67.57	78.75
4	–60.40	44.60	–79.79
5	–82.47	31.01	–16.63

The maximum bending moment is 97.10 kN-m/m.

Step 4 : Effective depth of the base slab

$$d = \left(\frac{M}{R \times 1000}\right)^{1/2} = \left(\frac{97.10 \times 10^6}{0.9130 \times 1000}\right)^{1/2} = 326.12\ mm$$

Let the effective cover be 25 mm. Then, the overall thickness of slab

$$(326.12 + 25) = 351.12 \text{ mm say} = 360 \text{ mm}$$

The effective depth of slab is 335 mm.

Step 5 : Steel reinforcement bars :

(a) The reinforcement bars for positive radial moment

$$A_{st} = \left(\frac{M}{\sigma_{st} \cdot j \cdot d}\right) = \left(\frac{97.10 \times 10^6}{230 \times 0.9038 \times 335}\right) = 1394.36\ mm^2$$

Hysd-steel bars of 10 mm ϕ shall be provided. The cross-sectional area of one bar

$$A_\phi = \left(\frac{\pi}{4} \times 10^2\right) = 78.54\ mm$$

Spacing of the bars

$$s = \left(\frac{1000 \times 78.54}{1394.36}\right) = 56.33\ mm$$

The steel bars shall be provided in two perpendicular direction at 50 mm spacing centre to centre near the centre of the slab.

(b) The reinforcement bars for negative radial moment

$$A_{st} = \left(\frac{82.47 \times 10^6}{230 \times 0.9038 \times 335}\right) = 1184.27\ mm^2$$

Spacing for 10 mm Hysd steel bars

$$s = \left(\frac{1000 \times 78.54}{1184.27}\right) = 66.32\ mm$$

These bars may also be provided at 50 mm centre to centre.

(c) The reinforcement bars for positive circumferential moment

Since the maximum bending moments in radial and circumferential directions are equal, the reinforcement for positive circumferential bending moment is same. The details of reinforcements in the circular base slab of a circular tank are shown in Fig. 19.12.

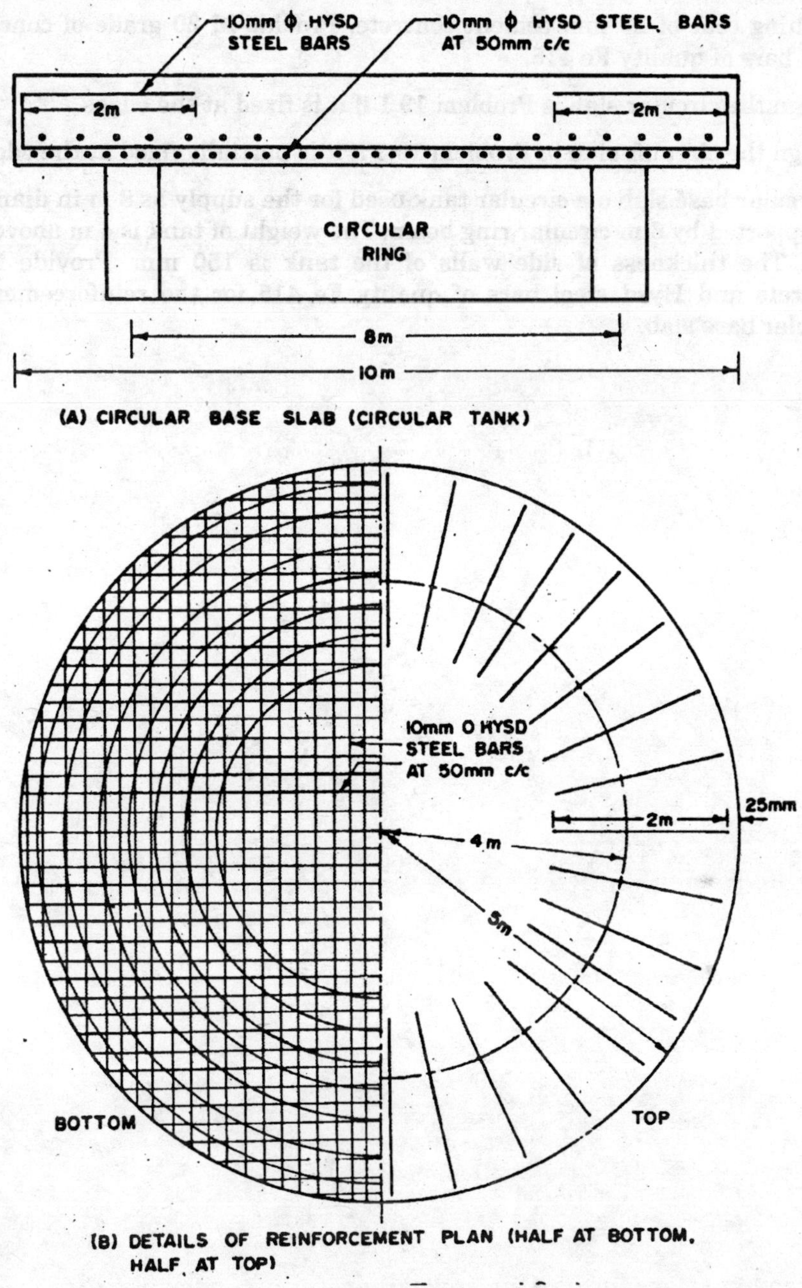

Fig. 19.12

PROBLEMS

19.1 Design a circular slab for a room 4.8 m in effective diameter. The circular slab is simply supported at edges and carries a live load of 4 kN/m². The slab shall have a finishing coat of 20 mm cement concrete. Provide M 20 grade of concrete and Hysd steel bars of quality Fe 415.

19.2 Design the circular slab is Problem 19.1 if it is fixed at the edges.

19.3 Design the circular slab in Problem 19.1 if it is partially fixed at the edges.

19.4 A circular base slab o a circular tank used for the supply as 8 m in diameter. The slab is supported by 6 m circular ring beam. The weight of tank is 4 m above the top of bas slab. The thickness of side walls of the tank is 150 mm. Provide M 20 grade of concrete and Hysd steel bars of quality Fe 415 for the reinforcement. Design the circular base slab.

PART V

DESIGN OF COLUMNS

20

DESIGN OF AXIALLY LOADED COLUMNS AND COMPRESSION MEMBERS

20.1 INTRODUCTION

A *column* is defined as a structural member subjected to compressive force in a direction parallel to its longitudinal axis. As per IS: 456-2000, the columns or struts are defined as the compression members, the effective length of which exceeds *three-times* the least lateral dimension. When the effective length of a compression member *does not exceed three times* the least lateral dimensions, then such compression member is called as *pedestal.* In axially *loaded columns and compression members,* the loads are applied at the centroid of the section in a direction parallel to the longitudinal axis of the members. The terms *centrally loaded* and *concentrically loaded* are also used for the axially loaded columns and the compression members.

When the compression members are overloaded, then their failure may take place in direct compression (viz., in crushing), or in excessive bending combined with twisting. The failure of columns depends upon their slenderness ratios. the load required to cause these types of failures, decreases as the length of compression members increases, the cross-sectional areas of the members remaining constant. Therefore, the columns and the compression members are classified as *short* and *slender* (viz. long) members. The compression members are considered as *short* when both the slenderness ratios (l_{ex}/D) and (l_{ey}/b) are less than 12, where l_{ex} and l_{ey} are the effective length with respect to major and minor axes, respectively and b is the width of member and D is the depth with respect to the major axes. In the short reinforced concrete columns, the lateral deflections are not significant. The columns and the compression members are treated as *slender* (viz. long) when either (l_{ex}/b) or (l_{ey}/b) or both the slenderness ratios are 12 or more than 12. In the slender reinforced concrete columns, the lateral deflections have an important effect on the strength of columns.

In case, *plain concrete columns* are used, as shown in Fig. 1.2 (A) (Chapter - 1), it is seen that their load carrying capacity is small, though the concrete carries compression only. In case, the steel bars are embedded in the columns of same size as shown in Fig. 1.2(B)

(Chapter - 1), since the steel bars have high compressive strength, the original (existing) load carrying capacity of plain cement concrete column is very much increased by providing additional strength by using steel bars. The columns become *reinforced concrete columns.* In the concrete and the part of the load is carried by steel (as the steel bars have much greater strength in compression also. The steel bars are placed nearer to the outer periphery of the section leaving the clear concrete cover as needed. The cross-sectional dimensions of reinforced columns and compression members may be reduced, if desired. Very few members are truly axially loaded, the steel bars are also provided for the reason that the steel bars are *essential* to resist the bending that may exist. The steel bars provided as reinforcement in the plain cement concrete columns are laterally tied at appropriate spacing in order to avoid the buckling of steel bars.

The columns may be square, rectangular, circular or polygonal in cross-section. The minimum number of steel bars provided as reinforcement in columns depends upon its cross-sectional shape.

20.2 AXIALLY LOADED COLUMNS

An axially loaded column as defined by the structural engineers transmits a compressive force without an explicit design requirements to carry lateral loads or end moments. There exist accidental end eccentricity, initial crookedness and initial curvature in the columns. By having appropriate modifications in the analytical expressions and using suitable factor of safety, the design formula makes the account for these factors. Due to these factors, the strength of an actual columns used to be less than that of a perfectly axially loaded column.

An ideal column is assumed to be perfectly straight and centrally loaded. A slender ideal column vertically fixed at the base and free at the upper end and carrying an axial load P is shown in Fig. 20.1. The column is assumed to be perfectly elastic.

When the axial load P is less than critical load and the stress is within the limit of proportionality, the column remains straight. The columns remains in **stable equilibrium** (*viz.,* if a small lacteal load is applied at the free end, the column deflects. On withdrawal of the lateral load, the column resumes its vertical position and deflection vanishes). When the magnitude of axial load P is gradually increased, a stage will be reached when the vertical position of the column becomes in **unstable equilibrium** (*i.e.,* if a small lateral load is applied a deflection will be produced, which will not disappear on the withdrawal of the lateral load). The axial load which is sufficient to keep the column in such a slight deflected shape is called as **critical load.** The *critical load* is also called as **buckling load** or **crippling load.** The *buckling load is defined as the load at which a member or a structure as a whole collapses in service (or buckles in a test load).* The buckling is defined as the sudden bending, warping, curling or crumbling of the elements or members under compressive stresses. The direction of buckling of a column depends upon the flexural rigidity, EI of the column. The columns buckle in a direction perpendicular to the axis about which the moment of inertia of the column section is minimum.

In about 1759 Professor Leonhard Euler (a Swis mathematician) gave the basic information on the behaviour of straight, concentrically loaded slender column. The formula for critical load for the column as shown in Fig. 20.1 was derived as under :

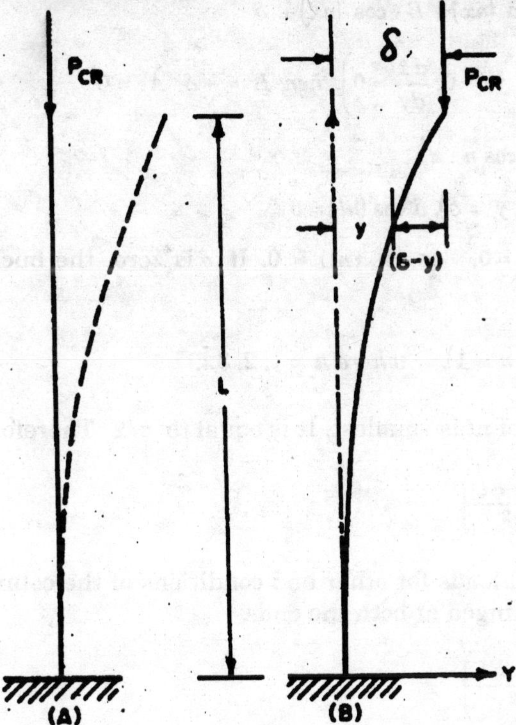

Fig. 20.1

The differential equation of the deflected shape of the column is as below :

$$\left(EI\frac{d^2y}{dx^2}\right) = +M \text{ (hogging)} \qquad \text{(i)}$$

The bending moment at any point on the deflected shape of the column

$$M = +P(\delta - y) \qquad \text{(ii)}$$

Therefore, $$\left(EI\frac{d^2y}{dx^2}\right) = +P(\delta - y) \qquad \text{(iii)}$$

$$\left(\frac{d^2y}{dx^2} - \frac{P}{EI}(\delta - y)\right) = 0 \qquad \text{(iv)}$$

Let n be equal to $\left(\frac{P}{EI}\right)^{1/2}$. **Then, the differential equation becomes**

$$\left(\frac{d^2y}{dx^2} + n^2 . y - n^2 . \delta\right) = 0 \qquad \text{(v)}$$

The general solution of this equation is

$$y = A \ . \sin\ (nx) + B \ . \cos\ (nx) + \delta \tag{vi}$$

$$At \left(x = 0, \ y = \ 0, \frac{dx}{dy} = 0 \right), then \ B \ = -\ \delta, \ A \ = 0$$

Therefore, $y = \delta\ (1 - \cos n \ . \ x)$

$$\text{At} \quad (x = l, y \ = \delta), \delta \cos\ (nl) = 0$$

From this, either $\delta = 0$, or cos (nl) = 0. If δ is zero, the buckling of column does not occur. If cos nl = 0, then

$$nl = \frac{\pi}{2}\ (2\ n - 1), \quad where\ n = 1, 2, 3....$$

For n = 1, the value of n*l* is smallest. It is equal to $\pi/2$. Therefore,

$$P_{cr} = \left(\frac{\pi^2 \ . \ EI}{4l^2} \right) \tag{20.1}$$

The values of critical loads for other end conditions of the columns may be derived from this case. For a column, hinged at both the ends

$$P_{cr} = \left(\frac{\pi^2 \ . \ EI}{l^2} \right) \tag{20.2}$$

where, P_{cr} is the critical load for the column. Since, each half of the column is in the same position (a half sine wave) as the whole of the column, the column hinged at both ends is called as **fundamental case or standard case** of buckling of the columns. In generalized form, the expression of Eq. 20.1 or Eq. 20.2 may be written as

$$P_{cr} = \left(\frac{\pi^2 \ . \ EI}{(kl^2)} \right) \tag{20.3}$$

At the critical load given by Eq. 20.1, the originally straight member buckles into a half sine wave as in Fig. 20.1 (B). In this bent configuration, the bending moments $P \ . \ (\delta - y)$ act at that section and the deflection at that section y continuous to increase. The deflections at the top increases more rapidly as compared to y. The bending stress caused by the increasing moment, together with the original compressive stress over stresses and fails the member.

20.3 EFFECTIVE LENGTH OF COMPRESSION MEMBERS

The effective length of a compression member depend upon end restraint conditions. The end restraint conditions are of two types, viz. (i) **position restraint** and (ii) **direction restraint.**

(i) Position restraint. In position restraint, the end of a column is not free to change its position but rotation about the end of a column, may take place, e.g. hinged end of a column as shown in Fig. 20.2 (A).

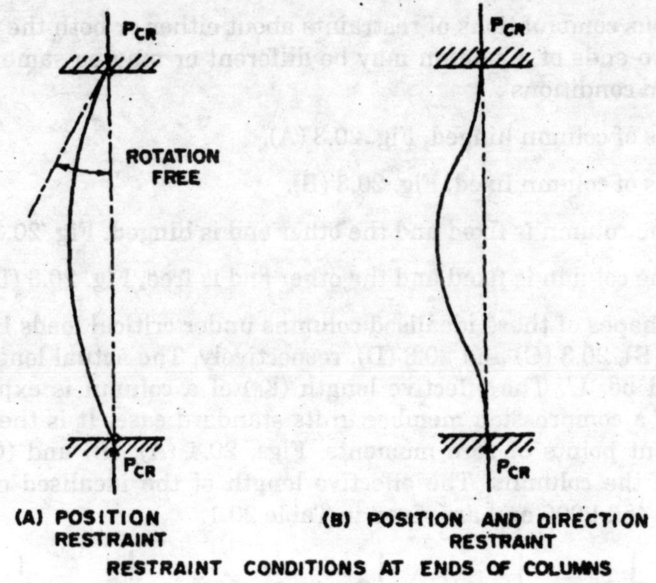

Fig. 20.2

PCR PCR PCR PCR
L L L L
PCR PCR PCR PCR
(A) (B) (C) (D)
IDEAL CASES OF COLUMNS WITH DIFFERENT END CONDITIONS

Fig. 20.3

(ii) **Direction restraint.** In direction restraint, end of the column is free to change its position but rotation at the end of the column cannot take place.

When any end of a column is having restraint in position and direction both, then the end is not free to change its position, and the rotation about the end of column also cannot take place as shown in Fig. 20.2 (B).

There are various combinations of restraints about either or both the axes. The restraint conditions at the two ends of a column may be different or may be same. Following are the ideal cases of the end conditions :

1. Both the ends of column hinged, Fig. 20.3 (A).
2. Both the ends of column fixed, Fig. 20.3 (B).
3. One end of the column is fixed and the other end is hinged, Fig. 20.3 (C).
4. One end of the column is fixed and the other end is free, Fig. 20.3 (D).

The deflected shapes of these idealised columns under critical loads have been shown in Figs. 20.3 (A), 20.3 (B), 20.3 (C) and 20.3 (D), respectively. The actual lengths of the columns have been indicated be `L'. The effective length (l_{eff}) of a column is expressed in terms of *equivalent length* of a compression member in its standard case. It is the length of columns between two adjacent points of zero moments. Figs. 20.4 (A), (B) and (C) also show three other ideal cases of the columns. The effective length of the idealised columns have been recommended in IS: 456-2000 and are given in Table 20.1.

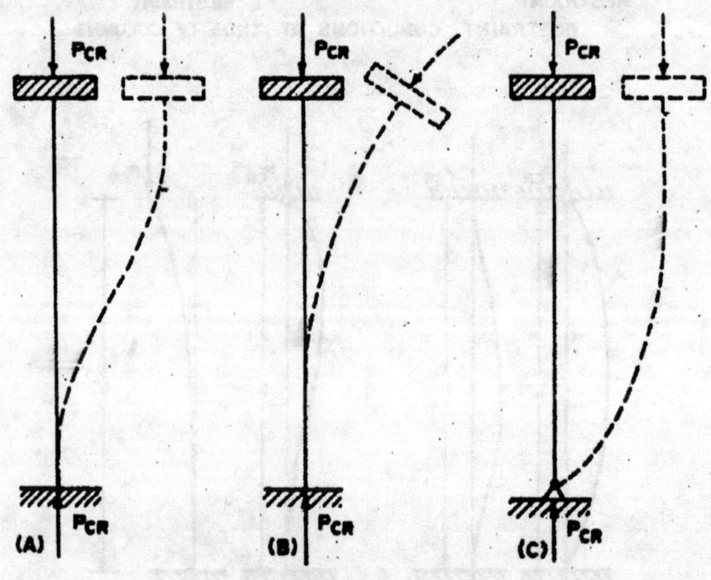

Fig. 20.4

TABLE 20.1. EFFECTIVE LENGTH OF COMPRESSION MEMBERS (As per IS : 456-2000)

S.No.	Degree of end restraint of compression members	Effective length	
		Theoretical	Recommended
1.	Effectively held in position at both ends, but not restrained against rotation. [Fig. 20.3 (A)].	1.00 l	1.00 l
2.	Effectively held in position and restrained against rotation at both ends. [Fig. 20.3 (B)].	0.50 l	0.65 l

3.	Effectively held in position at both ends, restrained against rotation at one end. [Fig. 20.3 (C)].	0.70 l	0.80 l
4.	Effectively held in position and restrained against rotation at one end but not held in position nor restrained against rotation at the other end. [Fig. 20.3 (D)].	2.00 l	2.00 l
5.	Effectively held in position and restrained against rotation at one end and at the other end restrained against rotation but not held in position. [Fig. 20.4 (A)].	1.00 l	1.20 l
6.	Effectively held in position and restrained against rotation at one end, and at the other end partially restrained against rotation but not held in position. [Fig. 20.4 (B)].	—	1.50 l
7.	Effectively held in position at one end but not restraint against rotation, and at the other end restraint against rotation but not held in position. [Fig. 20.4 (C)].	2.00 l	2.00 l

(**Note :** l is unsupported length of the compression member).

In reinforced concrete structures, one is generally concerned with rigid frames rather than with the single members. The buckling behaviour of frames is shown in Fig. 20.5. The loads are applied to the columns concentrically. In case the sideway is prevented as shown diagramatically by the brace in Fig. 20.5 (A), the buckling behaviour shall be as indicated. The effective length, kl shall be between 1.00 l and 0.50 l. The exact value depends on the degree of restraint i.e., on the ratio of *(EI/l)* of the column to the sum of rigidities *(EI/l)* of the restraining members at both the ends. It is seen that one end was prevented from moving laterally relative to the other end by horizontal bracing or otherwise. In such case, the effective length kl is always smaller than (or at the most it is equal to) the actual length l.

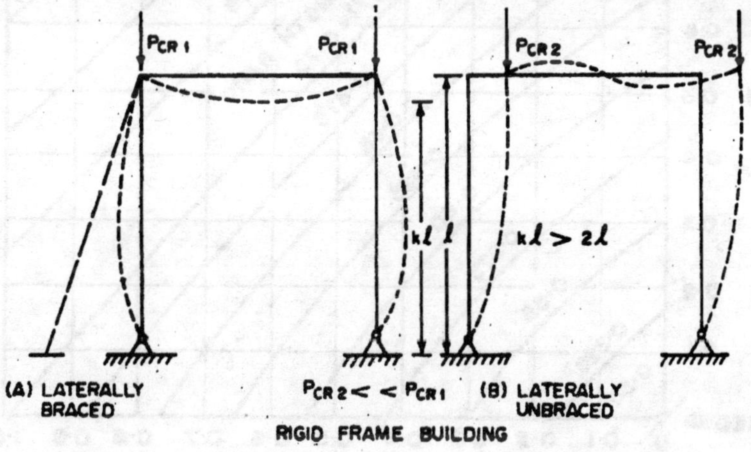

Fig. 20.5

In case no sideway bracing is provided to an otherwise identical frame, the buckling of frame occurs as shown in Fig. 20.5 (B). The unbraced frame will buckle at a radically smaller load than the braced frame. In general, the compression members free to buckle in a sides way mode are always considerably weaker than when braced against sidesway. If the column is rotationally fixed at both the ends but one end can move laterally with respect to the other, the effective length *kl* is equal to *l*. In case, one compares this column fixed at both the ends and free to sidesway with a fixed column, the effective length of former is *twice* that of the latter. When the columns are hinged at the support and free for sidesway, the effective length of such columns exceeds twice the unsupported length.

When the exact frame analysis is not done, the effective length of columns in frame structures may be found from the ratio of effective length to unsupported length (l_{eff}/l). When the relative displacement of the ends of the columns is prevented (*viz.,* the frames with no sway), the effective length may be noted from Fig. 20.6.

When the relative displacement of the ends of column are not prevented (i.e., the frames without restraint against sway), the effective length may be noted from, Fig. 20.7, IS : 456-2000 recommends that in these cases, the effective length ratio (l_{eff}/l) may not be taken to be less than 1.2.

It is to note that in Figs. 20.6 and 20.7

$$\beta_1 = \left(\frac{\sum k_c}{\sum k_c + \sum k_b}\right) \tag{20.4}$$

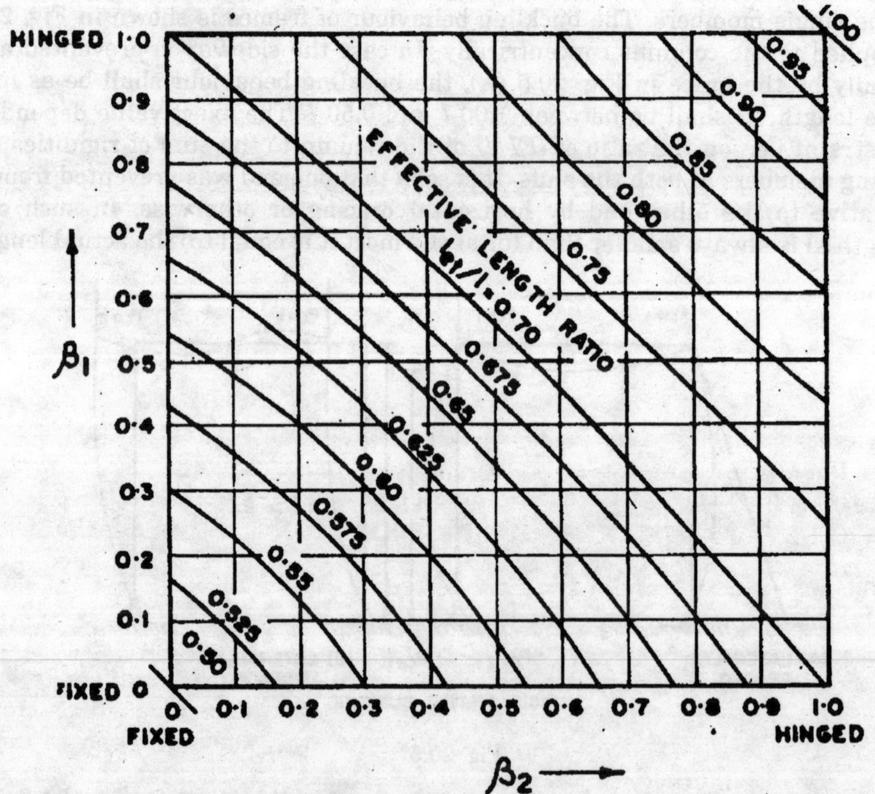

EFFCTIVE LENGTH RATIOS FOR A COLUMN IN A FRAME WITH NO SWAY

Fig. 20.6

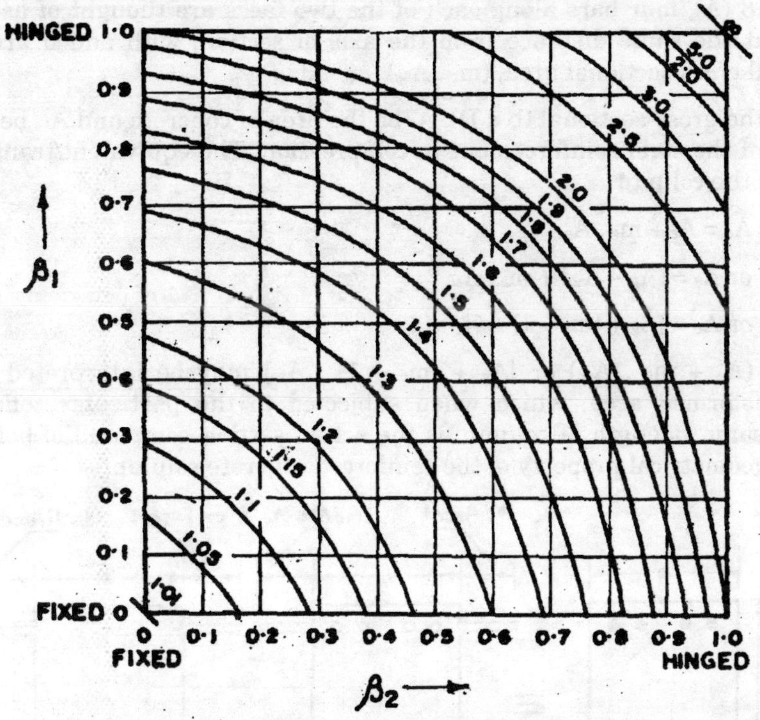

Fig. 20.7

This summation is to be done for the members framing into a joint at the top.

$$\beta_2 = \left(\frac{\sum k_c}{\sum k_c + \sum k_b}\right) \tag{20.5}$$

This summation is to be done for the members framing into a joint at the bottom. Where, k_c and k_b are the flexural stiffnesses of column and beam, respectively.

20.4 EQUIVALENT (TRANSFORMED) CROSS-SECTIONAL AREA OF COLUMNS

A reinforced concrete column [(e.g., a rectangular column as shown in Fig. 20.8(A)] consists of two materials concrete and steel bars, (*viz.,* it consists of composite section). The actual steel and concrete cross-section is replaced with a fictitious section though of consisting of concrete only. This fictitious section is called as **equivalent** or **transformed sectional** area as shown in Fig. 20.8(B) or in Fig. 20.8(C). This equivalent concrete area is seen to consist of actual concrete area A_c plus m_c (the modular ratio of concrete in compression, m_i = 1.5 m) *times* the area of reinforcement. That is, the actual area of reinforcement, A_{sc} is replaced with an equivalent concrete area equal to ($m_c \cdot A_{sc}$) located at the level of steel. The equivalent/transformed section pertaining to the rectangular section is shown in Fig. 20.8(C).

In Fig. 20.8 (A), four bars along each of the two faces are thought of as being removed and replaced at the same distance from the axis of section, with added areas of fictitious concrete of total cross-sectional area, ($m_c \,.\, A_{sc}$).

Let A_g be the gross-sectional (b x D), A_c be the area of concrete and A_{sc} be the total cross-sectional area of the steel reinforcement in compression. The equivalent/transformed area of cross-section of the column

$$A_e = A_c + m_c \,.\, A_{sc} \tag{20.6}$$

$$\text{or } A_e = (A_g - A_{sc}) + m_c \,.\, A_{sc} \tag{i}$$

$$\text{or } A_e = [A_g + (m_c - 1)\, A_{sc}] \tag{20.7}$$

The term ($A_e + m_c \,.\, A_{sc}$) or [$A_g + (m_c - 1) \,.\, A_{sc}$] may be interpreted as the area of equivalent/transformed area, which when subjected to the particular concrete stress, σ_c results in the same strength of column as the actual section composed of both concrete and steel. It is the geometrical property of the reinforced concrete column.

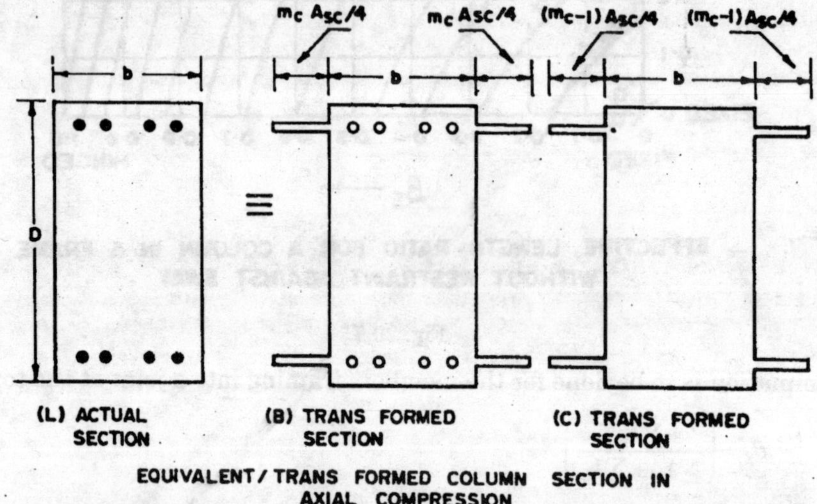

Fig. 20.8

20.5 RADIUS OF GYRATION OF A COLUMN SECTION

The radius of gyration of a reinforced concrete column section is a geometrical property of the section and it is determined as

$$r = (I_e/A_e)^{1/2} \tag{20.8}$$

where A_e = Equivalent sectional area of the reinforced concrete column section

I_e = Equivalent moment of inertia of the reinforced concrete column section.

The moment of inertia of the reinforced concrete column section as per IS : 456-2000 may be calculated on the basis of any one of the following :

(a) **Gross-section.** The gross cross-sectional of the member ignoring the reinforcement.

(b) **Transformed section**. The area of cross-section *plus* the area of reinforcement transformed on the basis of modular ratio, m_c (Eq. 20.6 or Eq. 20.7) in case of axially loaded columns.

(c) **Cracked section**. The area of concrete in compression *plus* the area of reinforcement transformed on the basis of modular ratio. The area of concrete in tension is neglected. It is to note that when the column section is subjected to axial load and bending moment, the column section may crack, then cracked section is to be considered.

20.6 SLENDERNESS RATIO OF COMPRESSION MEMBERS

The slenderness ratio of a compression member is defined as the ratio of *unsupported length* of the compression member, to the *appropriate radius of gyration* of the compression member, r

$$\lambda = (l/r) \tag{20.9}$$

The square and circular shapes are two chief shapes of the reinforced columns. When the column is free to bend about any axis, then, the value of radius of gyration of square or circular shape of the column about any cross-sectional axis remains the same. The value of radius of gyration of the column for other shape is smallest about the minor axis and maximum about the major principal axis. The smallest or minimum value of radius of gyration r_{min} is used for calculating the slenderness ratio of a column free to bend about any axis. Such a column bends about minor axis and this axis is considered as the weak axis.

Many times, external restraints (e.g., construction of walls) are attacked so that the bending of column about weak axis is not allowed. Then, the column remains free to bend about the strong axis (the value of radius of gyration about strong axis used to be largest and therefore the value of slenderness ratio is kept smallest). The appropriate radius of gyration, r is calculated looking to the external restraint or the axis about which the bending of column is possible.

When a short reinforced column (viz., the ratio (l_{ex}/D) and (l_{ey}/b) both are less than 12) is loaded with gradually increasing axial load in compression, the column fails by crushing of concrete and yielding of the reinforcement because the concrete and steel are stressed to their maximum carrying capacity simultaneously.

When a slender reinforced column (having same cross-sectional dimensions as that of short column and the same support condition at the ends) is loaded with gradually increasing axial load in compression, then, the column with large slenderness ratio will fail in buckling (i.e., by lateral bending of the member with consequent of overstressing of steel and concrete by flexural stresses which superpose over the axial compression stresses) at much smaller compression load than the failure load for the short column.

From Euler's formula in its generalised form, Eq. 20.3, it is seen that the

$$P_{cr} = \left(\frac{\pi^2 E_i . I}{(kl)^2} \right) \tag{i}$$

critical load decreases rapidly with increasing *slenderness ratio, (k.l / r)*. For the simplest case of a column hinged at both the ends and made of elastic material, E_i simply becomes Young's modulus of elasticity and *k.l* is equal to the actual length. The stress-strain curve for a short reinforced concrete column and buckling stress (P_{cr}/A_e) below the limit of proportionality, σ_p, E_t is equal to E. In case, (P_{cr}/A_e) is more than σ_p, the buckling of column occurs in the inelastic range, in which case, E_t is the slope of the tangent to the stress-strain curve and E_t is called as **tangent modulus** of the elasticity as shown in Fig. 20.9 (A). The value of E_t decreases as the buckling increases. The critical load-slenderness ratio is shown in Fig. 20.9 (B). It shows the effect of slenderness ratio on the strength of axially loaded column. The buckling strength of column depends on the slenderness ratio of the column. The buckling strength of the column decreases as the slenderness ratio of the column increases.

When the slenderness ratio of a column is less than limiting slenderness ratio, the failure of the columns occurs in crushing and when the slenderness ratio of a column is more than limiting slenderness ratio (kl/r) lim, the failure of the column occurs in buckling.

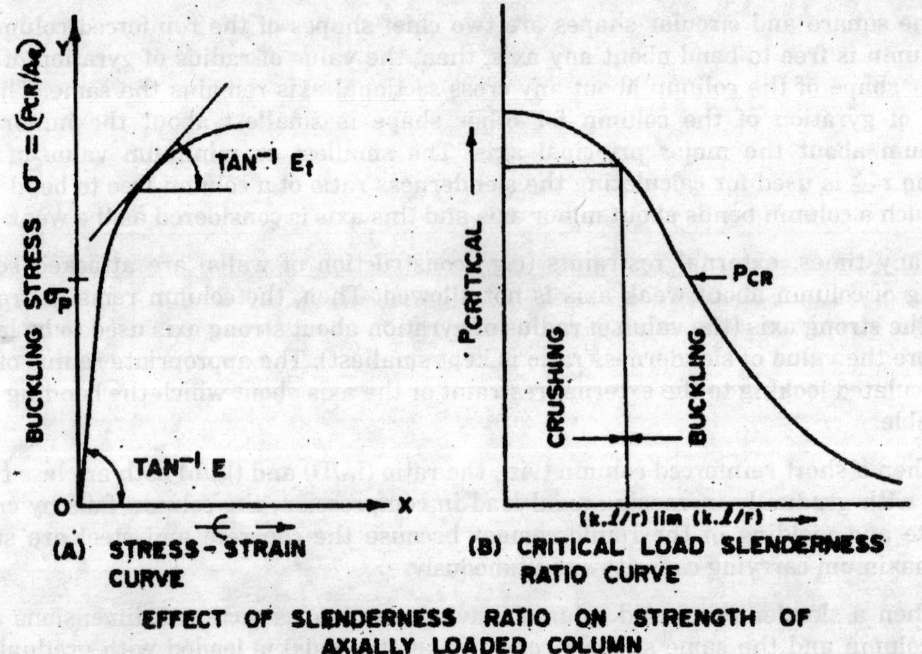

Fig. 20.9

Slenderness limits for column

IS : 456-2000 specifies that the unsupported length between end restraint shall not exceed 60 *times* the least lateral dimension of a column.

If in any given plane, one end of a column is unrestrained, its unsupported length, *l* shall not exceed ($100\ b^2/D$), where b is width of the column cross-section and D is the depth of cross-section measured in the plane under consideration.

20.7 UNSUPPORTED LENGTH OF COMPRESSION MEMBER

IS : 456-2000 defines the *unsupported length*, l of a compression member. It shall be taken as the clear distance between end restraint except the following :

IS : 456-2000 defines the unsupported length for flat slab construction, beam and slab construction, columns restrained laterally by struts and for columns restrained laterally by struts or beams respectively.

(a) Flat slab construction

The unsupported length, l shall be clear distance between the floor and the lower extremity of the capital, the drop panel or slab (whichever is less) as shown in Fig. 20.11 (A), (B) and (C), respectively.

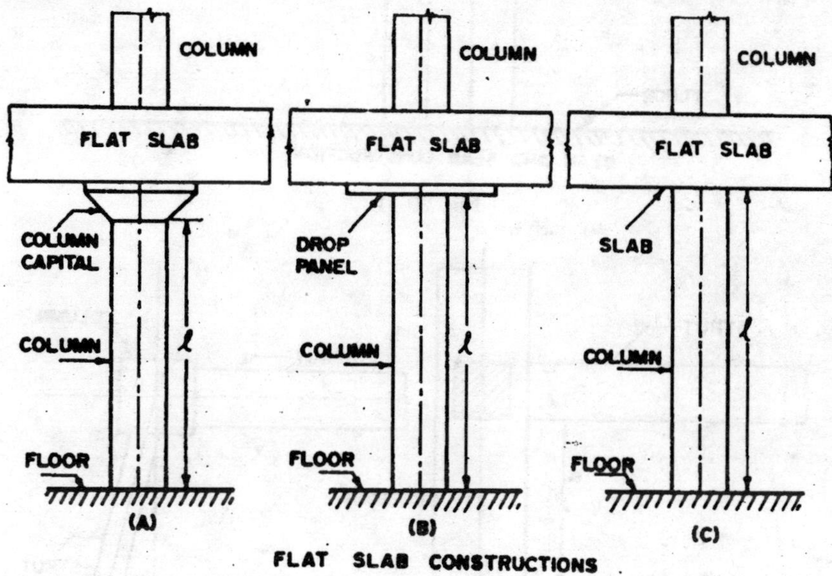

Fig. 20.10

(b) Beam and slab construction

The unsupported length, l shall be the clear distance between the floor and the underside of the shallower beam framing into the columns in each direction at the next higher floor level as shown in Fig. 20.11.

(c) Columns restrained laterally by struts

The unsupported length shall be the clear distance between consecutive struts in each vertical plane, provided that be an adequate support, two such struts shall meet the columns at approximately the same level and the angle between vertical planes through the struts shall not vary more than 30° from a right angle as shown in Fig. 20.12. Such struts shall be of adequate dimensions and shall have sufficient anchorage to restrain the member against lateral deflection.

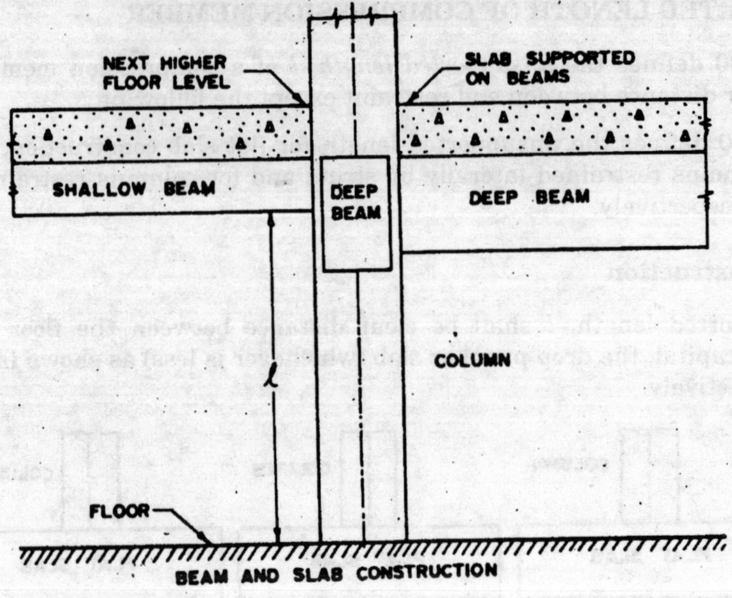

Fig. 20.11

STRUT
COLUMN
STRUT
l
COLUMN
STRUT
STRUT
(Ø ≯ 30°)
Ø
(A) A VIEW IN ELEVATION
(B) A VIEW IN PLAN
COLUMNS RESTRAINED LATERALLY BY STRUTS

Fig. 20.12

(d) Columns restrained laterally by struts or beams with brackets used at the junction

The unsupported length, l shall be the clear distance between the floor and the lower edge of the bracket, provided that the bracket width equals that of the beam or strut and is at least half that of the column as shown in Fig. 20.13.

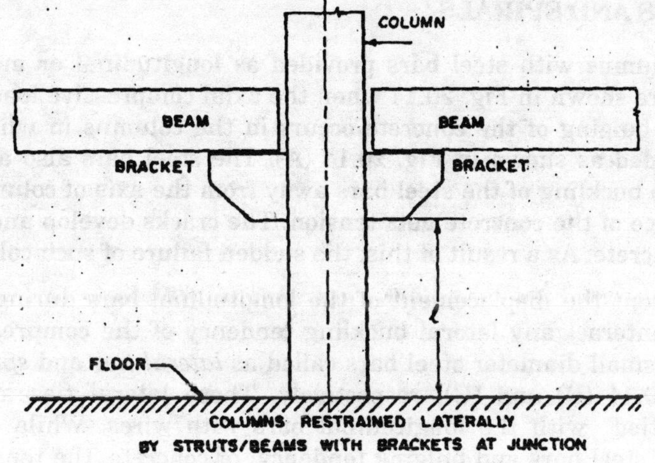

Fig. 20.13

P
P
P
LONGITUDINAL REINFORCEMENT WITHOUT LATERAL TIES
PITCH (SPACING)
LONGITUDINAL REINFORCEMENT
LATERAL TIES
PITCH (SPACING)
LONGITUDINAL REINFORCEMENT
SPIRALS (HELIX)
LATERAL TIES
SPIRALS (HELIX)
D
D_{CORE}
* (A)
* (B)
* (C)

* LONGITUDINAL (MAIN) REINFORCEMENT REINFORCED CONCRETE COLUMNS

Fig. 20.14

20.8 LATERAL TIES AND SPIRALS

The concrete columns with steel bars provided as longitudinal or main reinforcement parallel to the axis are shown in Fig. 20.14 when the axial compressive load is applied to the columns, the lateral bulging of the concrete occurs in the columns in which lateral ties or spirals are not provided as shown in Fig. 20.14 (A). The steel bars also act individually as slender columns. The buckling of the steel bars away from the axis of column also takes place. The outside surface of the concrete gets tension. The cracks develop and open in the thin outer cover of the concrete. As a result of this, the sudden failure of such columns occurs.

In order to prevent the displacement of the longitudinal bars during the construction operation and to counteract any lateral buckling tendency of the compression loaded steel bars, the transverse small diameter steel bars called as *lateral ties* and *spirals* are provided as shown in Fig. 20.14 (B) and (C), respectively. These lateral ties and spirals (small diameter bars) are tied with the longitudinal bars with wires. While counteracting the buckling tendency of steel bars and bulging tendency of concrete, the tensile force develops normal to the straight bars (main reinforcing bars) and the lateral ties or spirals remain effective in restraining these actions. These *lateral ties* are used in square and rectangular reinforced columns. The spirals are commonly used in reinforced concrete circular columns.

The arrangement of transverse reinforcement divides the columns into two general classes viz., **the columns with lateral ties and spirally reinforced** columns. In **tied columns,** the hoops or *ties* or rings are provided with longitudinal bars at appropriate spacing. In **spirally reinforced columns,** the *steel spirals* are provided enclosing the longitudinal bars at suitable pitch.

When the column tests were carried out and the axial load was increased from the service load to ultimate load for the columns, it was seen that the steel bars reached its elastic limit first and the increased deformation then occurring built up the stress in the concrete until its ultimate strength was reached. In case, the concrete approached its ultimate strength before the steel reached its elastic limit, the increased deformation of the concrete near its maximum stress forced the steel stress to build up more rapidly. As such, a simple and effective *addition law* was introduced in the axially loaded columns. The axially loaded columns reach their yield point only when the load became equal to 85 per cent of the ultimate strength measured by standard cylinder tests plus the yield point strength of the longitudinal steel bars provided as main reinforcement. The tied columns and spiral columns act almost identically upto the column yield point as shown by the load-deformation curve, Fig. 20.15.

In the **axially loaded tied columns,** the concrete immediately fails by crushing and shearing outward along inclined planes as soon as the yield point load is reached and the steel bars fail in buckling outward between ties. Thus the yield point and the ultimate strength of a tied column are same. The failure of the tied column is abrupt and complete (*viz.,* sudden and brittle).

In **an axially loaded spiral column,** the cracking or complete destruction of thin outer cover (outside the spiral) occurs at the yield point load. With the large increased deformation, the spirals come into effective action. the spiral provides radial compressive forces on the concrete within the core of the concrete and it prevents the outward buckling. The core concrete carries significant more load due to these confining stresses.

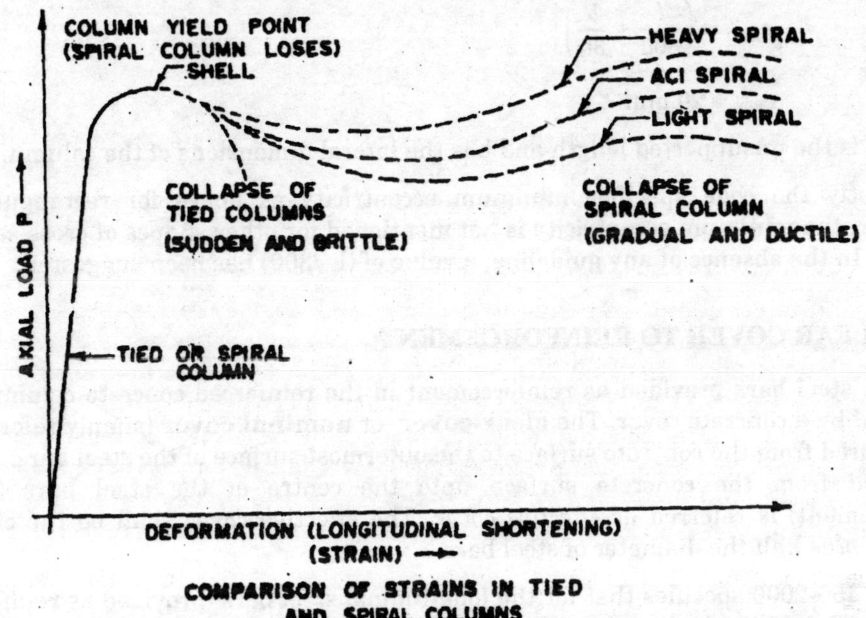

Fig. 20.15

In case **too light spirals** are used the column will continue to carry some load beyond the column yield point, but not as much as that which caused the spalling of the thin concrete cover. ACI building code specifies that the minimum quantity of spirals steel that will just replace (an estimated 10 per cent in excess of the shall strength is used just to be sure the strength after spalling is not less than before) the strength lost when the shell concrete spalls. The instantaneous crushing of concrete and buckling for steel are prevented. The failure becomes more gradual and ductile (i.e., a tougher column). In case of heavy spirals, the column carries more strength than that lost in the spalling or failure of the shell.

The initial cracking of the shell gives some warning of overload prior to failure. The spirals also add a considerable element of toughness to the column. The toughness is valuable in resisting explosions or earthquakes. Thoughness measures the energy that can be absorbed. It was seen that two columns (with heavy spirals) from the same storey of a building severely damaged by a strong earthquake prevented the total collapse of that storey.

20.9 MINIMUM ECCENTRICITY

All the axially loaded reinforced concrete columns are considered as axially loaded. In face, these columns are not perfectly axially loaded, There remains always small eccentricity in the loading due to inaccuracies in construction, lateral deflection of column and/or inaccuracies of loading. The eccentricity, e remains small and the moment of load, P, also remains small. Therefore, code IS : 4560-2000 recommends that all the axially columns shall be designed for minimum eccentricity e_{min} equal to unsupported length of column/500 *plus* plateral dimension/30, subject to a minimum of 20 mm., viz.,

$$e_{min} = \left(\frac{l}{500} + \frac{b}{30}\right) \quad \text{(i)}$$

$$e_{min} = 20 \text{ mm} \quad \text{(ii)}$$

where, l is the unsupported length and b is the lateral dimensions of the column.

Mostly the code specifies minimum eccentricity as above for rectangular columns. However, the minimum eccentricity is not mentioned for other shapes of cross-sections of the column. In the absence of any guideline, a value of (l_{eff}/300) has been suggested.

20.10 CLEAR COVER TO REINFORCEMENT

The steel bars provided as reinforcement in the reinforced concrete columns should be protected by a concrete cover. The **clear-cover** or **nominal cover** (simply referred as *cover*) is measured from the concrete surface to the outermost surface of the steel bars. The distance measured from the concrete surface upto the centre of the steel bars (longitudinal reinforcement) is referred as *effective cover.* The effective cover shall be the clear cover of concrete *plus* half the diameter of steel bars.

IS : 456-2000 specifies that for the longitudinal steel bars (provided as reinforcement) in a column should not be less than 40 mm, no less than the diameter of such bars. In the case of columns of minimum dimensions of 200 mm or under, whose reinforcing bars do not exceed 12 mm, a cover of 25 mm may be used.

The thickness of cover may be increased when the surfaces of concrete members are exposed to be action of harmful chemicals (as in the case of concrete in contact with earth, faces contaminated with such chemicals), acid vapour, saline atmosphere, sulphurous smoke (as in case of steam operated railways), etc. and such increase of cover may be between 15 mm and 50 mm beyond that specified above.

For reinforced concrete members totally immersed in sea water, the cover shall be 40 mm more than that specified above. For the reinforced concrete members, periodically immersed in sea water or subject to sea spray, the cover of concrete shall be 50 mm more than specified. For concrete of grade M 25, M 30, M 35 and M 40, the additional thickness of cover may be reduced to half. In all such cases the cover should not exceed 75 mm.

The concrete cover is also provided for resistance against fire. For protection against fire, three aspects (*namely,* retention of structural strength, resistance to penetration of flames and resistance to heat transmission) are usually considered. The minimum requirements of adequate protection against fire, and for dimensions of concrete to give resistance in hour, IS : 1642-1960 (code of practice for fire safety of buildings, general), materials and details of construction and the explanatory handbook to IS : 456-2000 may be referred.

20.11 PERMISSIBLE STRESSES IN CONCRETE AND STEEL BARS

The permissible stresses in direct compression in concrete of different grades as per IS : 456-2000 are adopted from Table 20.2 given as under :

Table 20.2. PERMISSIBLE STRESSES IN DIRECT COMPRESSION IN CONCRETE, σ_{cc} N/mm²

Grades of concrete	Permissible stress in direct compression σ_{sc} (N/mm²)
M 15	4.00
M 20	5.00
M 25	6.00
M 30	8.00
M 35	9.00
M 40	10.00

The *permissible stress in the steel bars* provided as longitudinal reinforcement in columns and compression members as per IS : 456-2000 for mild steel bars conforming to grade of IS : 432 (Part I) 1966 or deformed mild steel bars conforming to IS : 1139-1966 is 130 N/mm² and for medium tensile steel conforming to IS : 432 (Part I) 2966 or deformed medium tensile steel bars conforming to IS : 1139-1966 is 130 N/mm² and that for high yield strength deformed bars (briefly referred as Hysd-steel bars) conforming to IS : 1139-1965 or IS : 1786-1979 (grade Fe 415) is 190 N/mm².

20.12 STRENGTH OF AXIALLY LOADED COLUMNS AND COMPRESSION MEMBERS

The strength of axially loaded columns and compression members is the load carrying capacity (viz., permissible loads) for such members. The strengths of short, long and composite columns and compression members are as below :

(A) Strength of short columns with lateral ties

The strength of short columns with lateral ties in the sum of the concrete contribution and the steel contribution. The concrete and the steel work together and the overall capacity of the column is the sum of strength of the individual material capacities. The axial load, P permissible on *pedestal* or short column reinforced with longitudinal bars and lateral ties shall be that given by the following expression

$$P = (\sigma_{cc} \, . \, A_c + \sigma_{sc} \, . \, A_{sc}), \quad where \tag{20.10}$$

σ_{cc} = permissible stress in concrete in direct compression

A_c = cross-sectional area of concrete excluding any finishing and reinforcing steel

$A_c = (A_g - A_{sc})$

A_g = Gross-cross-sectional area of the column excluding any finishing material

σ_{sc} = permissible compressive stress for column bars, and

A_{sc} = cross-sectional area of the longitudinal steel.

It is to note that the minimum eccentricity described in Art. 20.9 may be deemed to be incorporated in Eq. 20.10.

(B) Strength of short column with helical reinforcement

The permissible load for columns with helical (viz., spiral) reinforcement shall be 1.05 times the permissible load for similar short column with lateral ties or rings provided the following condition as expressed below is satisfied

$$\left(\frac{V_h}{V_{core}}\right) \not< 0.36\left[\left(\frac{A_g}{A_c}\right)-1\right]\left(\frac{f_{ck}}{f_y}\right) \text{ where} \tag{20.11}$$

V_h = volume of helical reinforcement per 1 mm length of column

V_{core} = volume of the core of concrete per 1 mm length

$$V_h = \left(\frac{\text{Circumferencial length} \times \text{Cross-sectional area of spiral}}{\text{Pitch of spiral wire (spacing)}}\right)$$

$$\therefore V_h = \left(\frac{\pi \cdot d_c \cdot A_{sp}}{p}\right) \tag{i}$$

$$V_{core} = \left(\frac{\pi \cdot (D_{core})^2}{4}\right).1, \quad \text{where} \tag{ii}$$

D_{core} = diameter of core measured outside to outside of the spiral (helix)

d_c = centre-line diameter of spiral

p = pitch of the spiral

The volumetric ratio (viz., the ratio of volume of helix/spiral steel to the volume of core concrete)

$$p_s = \left(\frac{V_h}{V_{core}}\right) = \left(\frac{\pi\, d_c \cdot A_{sp}}{p}\right) \times \left(\frac{4}{\pi \cdot (D_{core})^2}\right) \tag{iii}$$

Professors F.E. Richhart, A. Brandzaeg and R.L. Brown in their papers titled as `A study of the failure of concrete under combined stresses' University of III. Eng. Exp. Stn. Bull. 185, November 1928 and `the failure of plain and spirally reinforced concrete in compression' University of III. Eng. Exp. Stn. Bull. 190, April 1929 and Professors S. Martinez, A.H. Nilson and F.E. Slate. In their paper titled as spirally reinforced high strength concrete columns', Journal of ACI, Vol. 81, No. 5 September/October 1984 found that the increase in compressive strength of the core of concrete in column providing by confining effect of spiral steel is closely represented by the expression.

$$(\sigma_c{}^* - 0.85 f_{ck}) = 4.0\, \sigma_2' \tag{iv}$$

where $\sigma_c{}^*$ = compressive strength of spirally confined core concrete

$0.85 f_{ck}$ = compressive strength of concrete if unconfined

σ_c' = lateral confinement stress in core concrete produced by spiral

It is assumed that the spiral steel reaches its yield stress f_y when the column eventually fails. By having analysis of hoop tension of an idealised model of a sort segment of column confined by one turn of lateral steel, the confinement stress σ_c' is calculated. Therefore,

$$\sigma_c' = \left(\frac{2\ A_{sp}\ .\ f_y}{d_c\ .\ p} \right) \tag{v}$$

From the expression (iii)

$$A_{sp} = \left(\frac{\rho_s\ .\ d_c\ .\ p}{4} \right) \tag{vi}$$

$$\sigma_2' = \left(\frac{\rho_s\ .\ f_y}{2} \right) \tag{vii}$$

In order to determine the right quantity of spiral steel, the strength contribution of the shell is calculated. It will be equal to

$$\text{where} = 0.85\ f_{ck}\ .\ (A_g - A_c) \tag{viii}$$

A_g = gross area of the section of column = $\left(\frac{\pi}{4}\ .\ D^2 \right)$

A_c = area of concrete of helically reinforced column = $\left(\frac{\pi}{4}\ .\ D^2{}_{core} \right)$

The gain in strength provided by the spiral should be atleast be equal to that lost when the shell (thin outside cover of concrete) spalls. The strength provided by the spiral is equal to

$$= 2\ \rho_s f_y\ .\ A_c \tag{viii}$$

By combining (vii) and (viii)

$$0.85\ f_{ck}\ .\ (A_y - A_c) = 2\ \rho_s\ .\ f_y\ .\ A_c \tag{ix}$$

$$\text{From which, } \rho_s = 0.425 \left(\frac{A_g}{A_c} - 1 \right) \frac{f_{ck}}{f_y} \tag{x}$$

As per ACI Code, this result is rounded upward slightly and ACI Code recommends that ps shall be not les than

$$\rho_s \nless 0.45 \left(\frac{A_g}{A_c} - 1 \right) \frac{f_{ck}}{f_y} \tag{xi}$$

IS : 456-2000 specifies inequality as in Eq. 20.11

(C) Strength of slender (long) columns

The strength of *slender (long) columns* may be determined by considering the effects of slenderness of the column. There are two methods of taking into consideration the effects of slenderness :

(i) By reduction of the axial load on the short columns

The load carrying capacity of short columns may be reduced to such a value that it corresponds to the safe load on the slender columns. This method has been recommended by IS: 456-2000 for working stress method. IS: 456-2000 specifies that the maximum permissible stress in a reinforced concrete column or part thereof having a ratio of effective length of the column to the lateral dimension more than 12 shall not exceed that which results from the multiplication of the appropriate permissible stresses specified for the appropriate grade of the concrete and respective equality of the steel bars used as the reinforcement.

However, in practice, the strength of a short column (viz., the load carrying capacity, P is determined and a reduction-factor is applied to this load. The effect of slenderness of column is thus accounted for IS : 456-2000 specifies the *reduction factor,* C_r as under :

$$C_r = \left(1.25 - \frac{l_{eff}}{48 \, . \, b}\right) \tag{20.12}$$

where, l_{eff} is the effective length of the column and b is the least lateral dimension. *For long (slender) columns with spiral (helical) reinforcement,* the core diameter is used instead of b. Therefore, the *reduction factor,* C_r.

$$C_r = \left(1.25 - \frac{l_{eff}}{48 \, . \, D_{core}}\right) \tag{20.13}$$

For more exact calculations, the maximum permissible stresses in a reinforced concrete column or part thereof having a ratio of effective length of the column to the least lateral radius of gyration more than 40 shall not exceed those which result from the multiplication of the appropriate maximum permissible stresses for the respective grade of concrete and the steel bars used as reinforcement. The *reduction factor* C_r may be calculated as follows :

$$C_r = \left(1.25 - \frac{l_{eff}}{160 \, . \, r_{min}}\right) \tag{20.14}$$

where, r_{min} is the least radius of gyration in case the external restraint is not attached by any construction or otherwise.

In case, P is the load carrying capacity of a short column (identical in shape and dimensions, main reinforcement and the lateral reinforcement, the strength of slender (long) columns shall be P.C_r).

(ii) By considering additional eccentricity

In the slender columns, the *additional eccentricity,* e_{add} develops due to the lateral deflection. The effect of slenderness of the column may be accounted for considering this additional eccentricity. In the limit state method of design of slender columns, this method has been introduced.

(D) Strength of composite columns

IS : 456-2000 recommends allowable load, metal core and reinforcement, splices and connections of metal cores and allowable load on metal core only as follows :

(i) Allowable load

The allowable axial load P on a *composite* column consisting of structural steel or cast-iron column thoroughly encased in concrete reinforced with both longitudinal and spiral reinforcement, shall be equal to the following :

$$P = (\sigma_{cc} \cdot A_c + \sigma_{sc} \cdot A_{sc} + \sigma_{mc} \cdot A_m), \quad \textit{where} \qquad (20.15)$$

σ_{cc} = permissible stress in concrete in direct compression

A_c = net area of concrete section which is equal to the gross area of concrete section, *minus*, the area of steel reinforcement, *minus* the area of metal core. (viz., $A_g - A_{sc} - A_m$)

σ_{sc} = permissible compressive stress for steel bars in column

A_{sc} = cross-sectional area of the longitudinal steel bars w (reinforcement)

σ_{mc} = allowable unit stress in the metal core, not to exceed 125 N/mm² for steel core or 70 N/mm² for cast iron core

A_m = cross-sectional area of steel or cast-iron core.

(ii) Metal core and reinforcement

The cross-sectional area of the metal core shall not exceed 20 per cent of the gross area of the column.

If a *hollow metal core* is used, it shall be filled with concrete.

The cross-sectional area of *longitudinal and spiral reinforcement* and the *spacing of steel bars,* details of splices and thickness of protective shell (cover) outside the spiral shall conform to the design requirements for the axially loaded reinforced concrete columns (described later in the subsequent sections).

A clearance of atleast 75 mm shall be maintained between the spiral and the metal core at all points, except that when the core consists of a structural steel H-column, then, the minimum clearance may be reduced to 50 mm.

(iii) Splices and connections of metal core

The metal cores in composite columns shall be accurately milled at splices and one core above another is properly aligned. At the base, the load shall be transferred to the footing such that the stresses at the base remain less than the permissible stresses.

The base of metal core shall be designed to transfer the load from the entire composite column to the footing, or it may be designed to transfer the load from the metal section only, provided it is placed in the pier or pedestal as to leave ample section of concrete above the base for the transfer of load from the reinforced concrete section of the column by means of bond on the vertical reinforcement and by the direct compression on the concrete.

The transfer of loads to the metal core shall be provided for by the use of bearing members, such as billets, brackets or other positive connections; these shall be provided at the top of metal core and at intermediate floor levels where required.

(iv) Allowable load on metal core only

The metal core of composite columns shall be designed to carry any construction or other loads to be placed upon them prior to their encasement in the concrete.

Example 20.1 A square reinforced concrete axially loaded column is 450 mm x 450 mm in size. The effective length of column is 4.5 m. 8 steel bars of 18 mm diameter are provided at appropriate spacing. Determine the strength of column. The grade of concrete is M 20 and the quality of steel is Hysd Fe 415.

Solution

From IS : 456-2000, the permissible stress in direct compression in concrete of grade M 20 is 5 N/mm² and that for Hysd steel bars of Fe 415 to 190 N/mm². The effective length of column is 4.5 m. The size of square column is 450 mm x 450 mm.

Step 1 : Area of concrete and steel bars

Gross-area of concrete in column

$$A_g = (450 \times 450) = 202500 \ mm^2 \quad \text{(i)}$$

Cross-sectional area of steel bars provided as longitudinal reinforcement

$$A_{sc} = \left(8 \times \frac{\pi}{4} \times 18^2\right) = 2035.75 \ mm^2 \quad \text{(ii)}$$

Area of concrete in column

$$A_{cc} = (A_g - A_{sc}) = (202500 - 2035.75) = 200464.25 \text{ mm}^2 \quad \text{(iii)}$$

Step 2 : Slenderness ratio

Ratio of the effective length of column to least lateral dimension

$$\left(\frac{l_{eff}}{b}\right) = \left(\frac{4.5 \times 1000}{450}\right) = 10 < 12 \quad \text{(iv)}$$

Step 3 : Safe load carrying capacity

The column is a short column. The safe load carrying capacity of the column, (viz., strength) from Eq. 20.10,

$$P = (A_c \cdot \sigma_{cc} + \sigma_{sc} \cdot A_{sc})$$

$$P = \left(\frac{200464.25 \times 5 + 190 \times 2035.75}{1000}\right) = \mathbf{1389.11 \ kN}$$

Example 20.2 (a) In case the column is circular having 450 mm diameter in Example 20.1, determine the strength of the column.

(b) In case, the effective length of column is 6 m, determine the strength of the column.

Solution : (a) Gross area of circular column

$$A_g = \left(\frac{\pi}{4} \times D^2\right) = \left(\frac{\pi}{4} \times 450^2\right) = 159043.13\ mm^2 \qquad \text{(i)}$$

Step 1 : Area of concrete and steel bars

From Example 20.1, cross-sectional area of steel bars provided as longitudinal reinforcement

$$A_{sc} = 2035.75 \text{ mm}^2 \qquad \text{(ii)}$$

Area of concrete in column

$$A_c = (159043.13 - 2035.75) = 157007\text{-}38 \text{ mm}^2 \qquad \text{(iii)}$$

Step 2 : Safe load capacity

From *Example 20.1,* it is seen that the column is a *short column*. The transverse reinforcement is provided as lateral ties with appropriate spacing. The safe load carrying capacity of the column (*viz.,* strength) from Eq. 20.10.

$$P = (A_c \,.\sigma_{cc} + \sigma_{sc} \,.\, A_{sc})$$

$$P = \left(\frac{157007.38 \times 5 + 190 \times 2035.75}{1000}\right) = \mathbf{1171.83\ kN}$$

(b) The effective length of column is 6 m. The column is having lateral ties (viz., circular rings) for transverse reinforcement. The diameter of column (least lateral dimension) is 450 mm. The ratio of effective length of the column to its least lateral dimension

$$\left(\frac{l_{eff}}{b}\right) = \left(\frac{6000}{450}\right) = 13.33 < 12 \qquad \text{(iv)}$$

The column is a *long column*. The safe load carrying capacity of column shall be P. C_r, where C_r is the reduction-factor. From Eq. 20.12.

$$C_r = \left(1.15 - \frac{l_{eff}}{48\,.\,b}\right) = \left(1.25 - \frac{6000}{43 \times 450}\right) = 0.972 \qquad \text{(v)}$$

Step 3 : Strength of long column

$$P.C_r = (1171.83 \times 0.972) = \mathbf{1139.02\ kN}$$

Example 20.3. A circular reinforced concrete column is 450 mm in diameter. The effective length of column is 4.50 m. 8 steel bars of 18 mm diameter are provided for the longitudinal reinforcement. The spiral bars of 8 mm diameter are provided at 60 mm pitch (spacing). Determine the safe load carrying capacity of the column. The grade of concrete is M 20 and the quality of steel is Hysd Fe 415. The clear cover of concrete over main reinforcement is 50 mm.

Solution : From IS : 456-2000, the permissible stress in direct compression in concrete of grade M 20 is 5 N/mm^2 and that for Hysd steel bars Fe 415 is 190 N/mm^2. The effective length of column is 4.50 m. The diameter of circular column is 450 mm.

Step 1 : Area of concrete and steel bars

$$A_g = \left(\frac{\pi}{4} \times 450^2\right) = 159043.13\ mm^2 \quad \text{(i)}$$

Cross-sectional area of steel bars provided as longitudinal reinforcement

$$A_{sc} = \left(8 \times \frac{\pi}{4} \times 18^2\right) = 2035.75\ mm^2 \quad \text{(ii)}$$

Area of concrete in the column

$$A_c = (A_g - A_c) = (159043.13 - 2035.75) = 157007.38\ \text{mm}^2 \quad \text{(iii)}$$

Step 2 : Slenderness ratio

Ratio of the effective length of column of column to its least lateral dimension

$$\left(\frac{l_{eff}}{b}\right) = \left(\frac{4.5 \times 1000}{450}\right) = 10 < 12$$

The column is a short column.

Step 3 : Safe load carrying capacity of short column (viz., strength)

From Eq. 20.10 $P = (\sigma_{cc} \cdot A_c + \sigma_{sc} \cdot A_{sc})$

$$P = \left(\frac{5 \times 157007.38 + 190 \times 2035.75}{1000}\right) = \mathbf{1171.83\ kN}$$

Step 4 : Check for volumetric ratio

The spirals of 8 mm diameter are used as the transverse reinforcement at 60 mm pitch (spacing). The clear cover of concrete over main reinforcement is 50 mm core diameter of column, Fig. 20.14 (C) (distance measured outside to outside of spiral).

$$D_{.core} = (Diameter\ \ of\ \ column\ - 2 \times clear\ \ cover\ \ of\ \ concrete\ \ + 2 \times diameter\ \ of\ \ spiral\ bars)$$

$$\therefore\ D_{.core} = (450 - 2 \times 50 + 2 \times 8) = 366\ \ mm$$

Gross area of concrete of the column

$$A_g = 159043.13\ \text{mm}^2$$

Area of core of helically reinforced column

$$A_c = \left(\frac{\pi}{4} \times 366^2\right) = 105208.8\ mm^2$$

$$d_c = (366 - 2 \times 4) = 358\ mm$$

Volume of helically reinforcement per 1 mm length of column

$$V_h = \left(\frac{\pi\ d_c\ .\ \frac{\pi}{4} \times \phi^2{}_{helix}}{P} \right)$$

$$\text{or}\ \ V_h = \left(\frac{\pi \times 358}{60} \times \frac{\pi}{4} \times 8^2 \right) = 942.22\ mm^3\ /mm$$

Volume of core of the concrete of column per 1 mm length

$$V_{.core} = \left(\frac{\pi}{4} \times 366^2 \times 1 \right) = 105208.8\ \ mm^3\ /\text{min}$$

From Eq. 20.11, the volumetric ratio

$$\left(\frac{V_h}{V_{.core}} \right) \nless 0.36 \left(\frac{A_g}{A_c} - 1 \right) . \frac{f_{ck}}{f_y}$$

$$\text{or}\ \ \left(\frac{942.22}{105208.8} \right) \nless 0.36 \left(\frac{159043.13}{105208.8} - 1 \right) \times \frac{20}{415}$$

$$\text{or}\ \ (0.008954 \nless 0.008883)$$

The left side is not less than right side. Therefore, the requirement of inequality stated above is satisfied. The load carrying capacity of column (viz., strength is increased by 5 per cent). As such

$$P = (1.05 \times 1171.83) = \mathbf{1230.42\ kN}$$

Example 20.4. A 400 mm x 200 mm rectangular concrete column is reinforced with 20 mm diameter 8 steel bars. The effective length of column is 2.20 m. The lateral ties of 8 mm diameter have been provided as transverse reinforcement at appropriate spacing. Determine the safe load carrying capacity of the column. (ii) In case the effective length of column is 4.40 m, what shall be permissible load for the column. M 20 grade of concrete and Hysd-steel bars of grade Fe 415 shall be used.

Solution : From IS : 456-2000, the permissible stress in direct compression in concrete of grade M 20 is 5 N/mm² and that for Hysd steel bars of grade Fe 415 is 190 N/mm².

Step 1 : Area of concrete and steel bars

Gross area of concrete of the column

$$A_g = (400 \times 200) = 80000\ mm^2 \qquad \text{(i)}$$

Cross-sectional area of steel bars provided for the longitudinal reinforcement

$$A_s = \left(8 \times \frac{\pi}{4} \times 20^2 \right) = 2513.27\ mm^2 \qquad \text{(ii)}$$

Area of the concrete of the column

$$A_c = (A_g - A_{sc}) = (80000 - 2513.27) = 77486.73 \text{ mm}^2 \quad \text{(iii)}$$

Step 2 : Slenderness ratio

(i) The effective length of column is 2.20 m ratio

$$\left(\frac{l_{eff}}{b}\right) = \left(\frac{2.2 \times 1000}{200}\right) = 11 < 12$$

$$\left(\frac{l_{eff}}{D}\right) = \left(\frac{2.2 \times 1000}{400}\right) = 5.5 < 12$$

Step 3 : Safe load carrying capacity

The column is a *short-column*. The safe load carrying of short column with lateral-ties at appropriate spacing from Eq. 20.10.

$$P = (\sigma_{cc} \cdot A_c - \sigma_{sc} \cdot A_{sc}) = \left(\frac{5 \times 77486.73 + 190 \times 2513.27}{1000}\right) = 864.95 \; kN$$

(ii) In case, the effective length of column is 4.4 m, then the ratio

$$\left(\frac{l_{eff}}{b}\right) = \left(\frac{4.4 \times 1000}{200}\right) = 22 < 12$$

$$\left(\frac{l_{eff}}{D}\right) = \frac{4.4 \times 1000}{400} = 11 \le 12$$

The column is a long-column. The reduction-factor for the column, from Eq. 20.11.

$$C_r = \left[1.25 - \left(\frac{l_{eff}}{48 \times b}\right)\right]$$

$$\text{or } C_r = \left(1.25 - \frac{4.4 \times 100}{48 \times 200}\right) = 0.792$$

Permissible load for the long column

$$P \cdot C_r = (864.95 \times 0.792) = \mathbf{685.04 \text{ kN}}$$

Example 20.5. A circular reinforced concrete column 700 mm in diameter is reinforced with 6 steel bars of 18 mm diameter. The lateral ties of 8 mm diameter have been provided at 60 mm spacing centre to centre. An ISHB 400, @ 822 N/metre has been embedded as a metallic core inside the column. The grade of concrete is M 20 and the quality of steel is Hysd Fe 415. Determine the safe permissible load for the composite column.

Solution : From IS : 456-2000, the permissible stress in direct compression in concrete of grade M 20 is 5 N/mm² and that for Hysd steel bars of grade Fe 415 used as longitudinal

reinforcement is 190 N/mm². The permissible stress in direct compression in metallic core is 125 N/mm².

Step 1 : Area of concrete and steel bars

Gross area of concrete of the column

$$A_g = \left(\frac{\pi}{4} \times 700^2\right) = 384845.10\ mm^2 \qquad \text{(i)}$$

Cross--sectional area of 6 steel bars of 18 mm diameter

$$A_{sc} = \left(6 \times \frac{\pi}{4} \times 18^2\right) = 1526.81\ mm^2 \qquad \text{(ii)}$$

Area of cement concrete in the column

$$A_c = (A_g - A_s) = (384845.10 - 1526.81)\ A_c = (A_{gr} - A_{sc} - A_m) = 383318.29\ \text{mm}^2 - A_m \qquad \text{(iii)}$$

Area of cross-section of ISHB 400, @ 822 N/m

$$A_m = 10466\ \text{mm}^2 \qquad \text{(iv)}$$

$$(\therefore\ 383318.29 - 10466) = 372852.29$$

Permissible load for the given composite column, from Eq. 20.15

$$P = (\sigma_{cc}\ .\ A_c + \sigma_{sc}\ .\ A_{sc} + \sigma_{mc}\ .\ A_m) \qquad \text{(v)}$$

$$P = \left(\frac{5 + 190 \times 1526.81 + 125 \times 10466}{1000}\right) = \mathbf{3514.94}\ kN$$

20.13 DESIGN OF AXIALLY LOADED REINFORCED CONCRETE COLUMNS AND COMPRESSION MEMBER

An axially loaded reinforced concrete column or a compression member is designed for the given load, actual length of the member and its restraint conditions at its two ends. The shape of column or compression member (square, rectangular, circular, polygonal or any other) is decided from the architectural requirements. Usually m 15 and M 20 grades of the concrete are used. the steel bars used as longitudinal reinforcement and transverse reinforcement (*mild steel bars*) conform to grade I of IS : 432 (Part I) 1966 or deformed mild steel bars conform to IS : 1139 - 1966, or *medium tensile steel* conform to IS : 432 (Part I) 1966 or deformed medium tensile steel bars conforming to IS : 1139-1966 or *high yield strength deformed bars,* conforming to IS : 1139-1966 or IS : 1786-1679 are used.

The permissible stresses for the respective grade of concrete and the quality of steel as recommended by IS : 456-2000 and described in Art. 20.11 are used. The modular ratio, m for the adopted grade of concrete is calculated by noting the stress in concrete in bending in compression, σ_{cbc} from IS : 456-2000 and the modular ratio of concrete in compression, m_c is determined as below :

$$m = \left(\frac{280}{3\ .\ \sigma_{cbc}}\right) \qquad \text{(i)}$$

and $m_c = 1.5 . (m)$ (ii)

Whether the column shall be short or long (viz., slender) is decided after knowing the gross cross-sectional area of the column and its respective dimensions. The gross cross-sectional area A_g, the area of concrete, A_c and the cross-sectional area of longitudinal reinforcement are determined by assuming that the column shall be short as follows :

The given axial load P shall be carried by the area of concrete, A_c and the area of steel reinforcement, A_{sc}. Therefore,

$$P = (\sigma_{cc} \, . A_c + \sigma_{sc} \, . A_{sc}) \quad \text{(iii)}$$

The compressive stress in the steel bars, σ_{sc} shall be equal to 1.5 *times* the modular ratio of concrete, in compression times the direct compressive stress in concrete. ($\because \; m_c = 1.5 \, m$)

$$m \; (viz., \sigma_{sc} = m_c \, . \, \sigma_{cc} = 1.5 \, m \, . \, \sigma_{cc})$$

Then, $P = (\sigma_{cc} \, A_c + 1.5 \, m \, . \, \sigma_{cc} \, . \, A_{sc})$ (iv)

and $A_c = (A_s - A_{sc})$ (V)

$$\therefore \; P = \left| \sigma_{cc} \, (A_g - A_{sc}) + 1.5 \, m \, \sigma_{cc} \, . \, A_{sc} \right| \quad \text{(vi)}$$

The area of steel A_{sc} may be assumed as the percentage of the gross area of the column section. (A_{sc} = p . A_g). In practice, the percentage steel *p* may be adopted as *one* or *two* per cent.

$$p = (p / 100)$$

$$p = \left| \sigma_{cc} \, . \, (A_g - p \, . \, A_g) + 1.5 \, m \, \sigma_{cc} \, . \, p \, . \, A_g \right| \quad \text{(vii)}$$

The gross cross-sectional area of concrete, A_g is calculated as below :

$$A_g = \left(\frac{P}{\sigma_{cc} \, . \, [(1-p) + 1.5 \, m \, . \, p]} \right) \quad \text{(viii)}$$

Depending upon the shape of the reinforced concrete column, the dimensions of column are decided.

(i) For the square column of size, a

$$a = (A_g)^{1/2} \quad \text{(ix)}$$

(ii) For the rectangular column of size b x D

The breadth of column, b may be expressed as some fraction of overall depth, D of the column. Then,

$$(b \times D) = A_g \quad \text{(x)}$$

(iii) For the circular column of diameter, D

The gross cross-sectional area of circular reinforced concrete column

$$\left(\frac{\pi}{4} . D^2\right) = A_g \qquad \text{(xi)}$$

The diameter of column, D is determined.

When the shape and its dimensions of the reinforced concrete column are known, then, the slenderness ratios (l_{ex}/D or l_{ey}/b), both are calculated. When, both the slenderness ratios are less than 12, then, the designed cross-sectional dimensions of the column will be those for the short column. In case one or both the slenderness ratios are more than 12, the column shall be long (viz., the slender) columns. The slenderness limits for the column recommended in IS : 456-2000 and described in Art. 20.6 may be checked.

In case, it is seen that the column is long (slender), then the reduction factor, C_r (as specified by IS : 456-2000 and described in Art. 20.12, Eq. 20.12, Eq. 20.13 or Eq. 20.14 is calculated. The axial load for the long column

$$P_{long} = C_r . P_{short} \qquad \text{(xii)}$$

The revised load for short column becomes

$$P_{short} = (P_{long}/C_r) \qquad \text{(xiii)}$$

The revised gross area of the column is found from the expression (viii) and the dimensions of the columns are also revised. The safe load carrying capacity for the respective column (short or long) is calculated, which should be slightly more than the given axial load. Alternatively, either the grade of concrete or the percentage of the steel or both may be improved.

The percentage of steel used as the fraction of gross-area of the column section should be within the minimum and maximum percentage specify by IS : 456-2000 and described in Art. 20.14. The cross-sectional area of the steel bars may than be obtained.

$$A_{sc} = p . A_g \text{ and} \qquad \text{(xiv)}$$

$$A_c = (A_g - A_{sc}) \qquad \text{(xv)}$$

The longitudinal (main) steel bars, and their number and the transverse reinforcement (lateral ties or spirals) are provided as described in Art. 20.14 and Art. 20.15, respectively.

20.14 REQUIREMENT OF LONGITUDINAL (MAIN) REINFORCEMENT FOR COLUMNS

IS : 456-2000 specified the requirements of longitudinal (main) steel bars to be used for columns and compression members regarding the percentage of reinforcement, number of steel bars, diameter of the bars, the spacing of the bars and the concrete cover for the steel bars as under :

1. Percentage of the longitudinal reinforcement

(i) The **minimum percentage** of cross-sectional area of the longitudinal reinforcement shall be not less than 0.8 per cent of the gross cross-sectional area of the column.

(ii) In any column that has a large cross-sectional area than that required to support the load, the *minimum percentage of steel* shall be based upon the area of concrete required to resist the direct stress and not upon the actual area.

(iii) The **maximum percentage** of cross-sectional area of the longitudinal reinforcement shall be not more than 6 per cent of the gross cross-sectional area of the column.

It is to note that the use of 6 per cent reinforcement may involve practical difficulties in placing and compacting the concrete; hence lower percentage is recommended. Where the steel bars from column below have to be lapped with those in column under consideration. The percentage of steel shall usually not exceed 4 per cent.

The minimum percentage of reinforcement is specified, as the reinforcement is needed to resist the bending moment, which may exist irrespectives of whether the design calculations show that the bending moment exists. This also decreases the effect of creep and shrinkage under sustained loading. The shrinkage and creep redistribute load from the concrete to reinforcement, and the stress in steel may reach the yield level even under service loads. The minimum percentage of steel protects columns in structural frames against failure in tension when, for example, the surrounding floors, near the column are unloaded above but heavily loaded below or when the structural frame is subjected to unequal foundation settlement.

2. Number of steel bars

(i) In the *rectangular columns,* the minimum number of bars provided in a column shall be four.

(ii) In the *circular columns,* the minimum number of bars in a column shall be *six.*

For shapes other than the rectangular or circular, at least one bar should be provided at each apex or corner.

3. Diameter of steel bars

The bars shall be not less than 12 mm diameter.

It is based on the requirements of stiffness and it is independent of the strength or type of steel.

4. Spacing of longitudinal bars

The spacing of longitudinal steel bars measured along the periphery of the column shall not exceed 300 mm.

5. Columns with helical reinforcement

(i) A reinforced concrete column having *helical reinforcement* shall have at least six bars of longitudinal reinforcement within the helical reinforcement.

(ii) In a *helically reinforcement column,* the longitudinal bars shall be in contact with the helical reinforcement and equidistance around its inner circumference.

In case of **pedestals** in which the longitudinal reinforcement is not taken into consideration in strength calculations. The nominal longitudinal reinforcement not less 0.15 per cent of the cross-sectional area shall be provided.

The **thickness of cover** to be outmost surface of the longitudinal steel bars has been recommended by IS : 456-2000 and it is kept as described in Art. 20.10.

Example 20.6. The size of column is 450 mm x 450 mm (i.e., the column is a square column). The axial load carried by the column, P is 600 kN. The grade of concrete is M20. Calculate the area of reinforcement.

Step 1 : Gross area of the column

$$A_g = (450 \times 450) = 202500 \; mm^2$$

The gross-sectional area of concrete A_g required. For M 20 grade of concrete

$$\sigma_{cc} = 5 \text{ N/mm}^2$$

$$A_g = \left(\frac{P}{\sigma_{cc} \cdot [(1-p) + 1.5\, m \,.\, p]}\right)$$

$$A_{g1} = \left(\frac{600 \times 1000}{5\,[(1 - 0.008) + 1.5 \times 13.33 \times 0.008]}\right)$$

In case, $p_{min} =$ 0.008 percent

$$m = \left(\frac{280}{3 \times 7}\right) = 13.33$$

$$A_{g1} = \left(\frac{600 \times 1000}{5\,[0.992 + 0.012 \times 13.33]}\right) \text{mm}^2$$

$$A_{g1} = \left(\frac{600 \times 1000}{5 \times 1.15196}\right) = 104170.28 \; mm^2$$

Size of square column required = 322.754 mm < 450 mm. So also $A_{g1} < A_g = 202500$ mm²)
In that case, the minimum area of reinforcement A_{sc} is found on the basis of A_{g1}

$$A_{sc} = \left(\frac{8}{1000} \times 104170.28\right) = 833.36 \; mm^2$$

Note : In such a case, A_{sc} is not found as $\left(\frac{8}{1000} \times 202500\right)$.

20.15 REQUIREMENTS OF TRANSVERSE (LATERAL) REINFORCEMENT

The general description of transverse reinforcement (lateral) ties and spirals (helixs), the requirements for arrangement of transverse reinforcement, for pitch and diameter of lateral ties and spiral reinforcements have been stated in IS : 456-2000 as follows :

1. General description

A reinforced concrete compression member shall have the transverse (viz., lateral ties or spiral/helical) reinforcement. Every longitudinal steel bar placed nearest to the face of compression member will get effective lateral support against buckling. The effective lateral support is given by the transverse reinforcement either in the form of circular rings capable of taking up reinforcement tension or by polygonal links (lateral ties) with the internal angles not more than 135°. The ends of transverse reinforcement shall be properly anchored.

2. Arrangement of transverse reinforcement

IS : 456-2000 recommends following four arrangements for the transverse reinforcements (viz., lateral ties or stirrups).

(i) If the longitudinal bars are not spaced more than 75 mm on either side, the transverse reinforcement need only to go round the corner and alternate bars for the purpose of providing effective lateral support as shown in Fig. 20.16(A).

(ii) If the longitudinal bars spaced at a distance of not exceeding 48 times the diameter of the ties (viz., $48.\phi_{tr}$) are effectively tied in two directions, additional longitudinal bars in between these bars need to be tied in one direction by *open ties* as shown in Fig. 20.16(B).

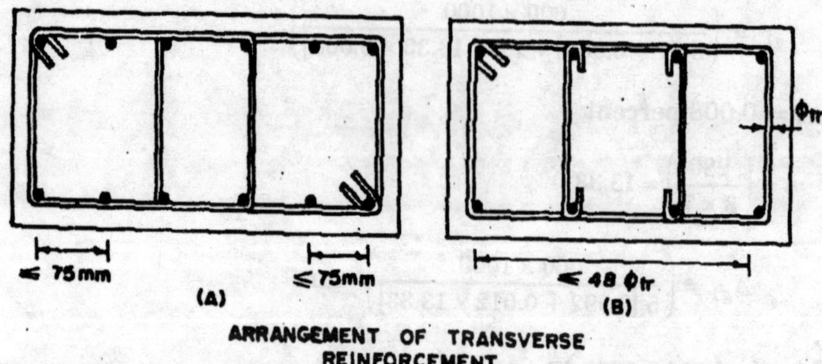

ARRANGEMENT OF TRANSVERSE REINFORCEMENT

Fig. 20.16

(iii) Where the longitudinal reinforcing bars in compression members are placed in more than one row, effective lateral support to the longitudinal bars in the inner rows may be assumed to have been provided if :

1. Transverse reinforcement is provided for the outer-most row, and
2. No bar of the inner row in closer to the nearest compression face than *three times* the diameter of the largest longitudinal bar (viz., $3\phi_{max}$) in the inner row as shown in Fig. 20.17 (A).

(iv) Where the longitudinal bars in compression member are grouped (not in contact) and each group adequately tied with transverse reinforcement the transverse reinforcement for compression member as a whole may be provided on the assumption that each group is a single longitudinal bar for the purpose of determining the pitch and diameter of the transverse reinforcement. The diameter of such transverse reinforcement need not, however, exceed 20 mm as shown in Fig. 20.17 (B).

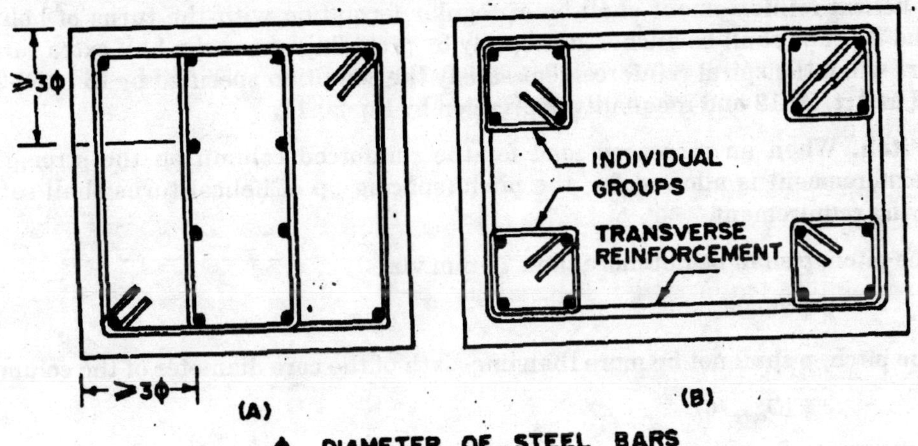

Fig. 20.17

IS : 456-2000 recommends above four arrangements to ensure that the concrete in core is adequately confined. In case the columns are overloaded, it results in dectile behaviour.

3. Lateral ties

IS : 456-2000 specifies pitch and diameter of the steel bars provided as transverse reinforcement as under :

(i) **Pitch.** The pitch (spacing) of transverse reinforcement shall be not more than the least of the following distances :

1. The least lateral dimension of the compression member, viz.,

$$p \not> b \quad \text{(i)}$$

2. Sixteen times the smallest diameter of the longitudinal reinforcement bar to be tied, viz.,

$$p \not> 16 \cdot \phi_{min} \quad \text{(ii)}$$

3. Forty-eight times the diameter of transverse reinforcement

$$p \not> 48 \cdot \phi_{tr} \quad \text{(iii)}$$

(ii) **Diameter.** The diameter of the polygonal links or lateral ties shall be not less than one-fourth of the diameter of largest longitudinal bar (viz., $\phi_{tr} \not< 0.25 \cdot \phi_{max}$) and in no case less than 5 mm.

4. Spirals (Helix)

IS : 456-2000 specifies pitch and diameter of steel bars provided as spirals (Helixs) as below :

The helical reinforcement shall be of regular formation with the turns of helix spaced evenly and its ends shall be anchored properly by providing one and a half extra turns of the spiral bars when the spiral reinforcement satisfy the condition specified by IS : 456-2000 and described in Art. 20.12 and inequality expression by Eq. 20.11.

(i) **Pitch.** When an increased load for the reinforced column on the strength of the helical reinforcement is allowed for, the pitch (spacing), p of helical turns shall satisfy also the following requirement.

1. The pitch, p shall be not more than 75 mm viz.,

$$p \ngtr 75mm \quad \text{(i)}$$

2. The pitch, p shall not be more than one-sixth of the core diameter of the column, viz.,

$$p \ngtr (D_{core}/6) \quad \text{(ii)}$$

3. The pitch, p shall not be less than 25 mm, viz.,

$$p \nless 25mm \quad \text{(iii)}$$

4. The pitch shall not be less than three times the diameter of the steel bar forming the helix, viz.,

$$p \nless (3 \cdot \phi_{helical}) \quad \text{(iv)}$$

(ii) **Diameter.** The diameter of helical reinforcement shall not be less than one-fourth of the diameter of the largest longitudinal bar viz.,

$$\phi_{helical} \nless (0.25\ \phi_{max})$$

and in no way less than 5 mm.

The purpose of links is mainly to avoid the outward buckling of the longitudinal steel bars. The diameter and pitch (spacing) of the transverse reinforcement are therefore related to the diameter of the longitudinal bars. The minimum diameter is specified as one-quarter of that of the largest longitudinal bar, but the minimum size of steel bar is IS : 456-2000 is 5 mm. The smaller size of links fails to hold the main longitudinal bars securely, and may themselves be pushed out of shape in the process of concreting.

Example 20.7. Design a square reinforcement concrete axially loaded column to carry an axial load of 1200 kN including dead load, live load and self-weight. The column shall remain continuous through reinforced concrete beam and slab floors at both the ends. Provide M 20 grade of concrete and Hysd steel bars of grade Fe 415. The clear height between the floor and soffit of shallow beam is 6.0 m.

Design : M 20 grade of concrete is to be used. The permissible stress in concrete in bending in compression from IS : 456-2000 is 7 N/mm^2. The modular ratio of M 20 grade of concrete in compression, $m_c = (1.5 \times m)$

$$\therefore\ m_c = \left(\frac{1.5 \times 280}{3 \times 7}\right) = 20 \quad \text{(i)}$$

Step 1 : Size of the column

Let the percentage of steel in compression for longitudinal reinforcement be *one per cent.* Then, p = 0.01. The gross cross-sectional area of concrete, A_g needed (from the expression (viii), Art. 20.13).

$$A_g = \left(\frac{P}{\sigma_{cc} \, . \, [(1-p) + 1.5\, m \, . \, p]}\right) \quad \text{(ii)}$$

The axial load P over the column is 1200 kN and the permissible stress in direct compression, σ_{cc} (from IS : 456-2000) for M 20 is 5 N/mm².

$$A_g = \left(\frac{1200 \times 1000}{5. \, [(1-0.01) + 20 \times 0.01]}\right) = 201680.67 \; mm^2.$$

Let a be the size of square column

$$a = (201680.67)^{1/2} = 449.09 \text{ mm (say 450 mm)}$$

Since, the column remains continuous through beam and slab floors at both the ends, the column may be considered as restrained in position and direction both. The unsupported length of column, l is the clear height between the floor and soffit of the shallow beam (viz., l = 6.0 m). Therefore, the effective length of column.

$$l_{eff} = (0.65 \times l) = (0.65 \times 6000) = 3900\, mm.$$

Ratio of the effective length to least lateral dimension

(3900/450) = 8.67 < 12. (The column is a short column).

Step 2 : Longitudinal reinforcement

The cross-sectional area of longitudinal reinforcement has been assumed as *one per cent.* Therefore,

$$A_{sc} = \left(\frac{1}{100} \times 201680.67\right) = 2016.8067 \; mm^2.$$

Let the diameter of steel bars Hysd Fe 415 be 18 mm. The cross-sectional area of one bar

$$A_\phi = \left(\frac{\pi}{4} \times 18^2\right) = 254.469 \; mm^2$$

Number of bars needed

(2016.8067/254.469) = 7.925 (say 8 bars)

8 steel bars of 18 mm diameter of quality Hysd grade Fe 415 shall be provided for the longitudinal reinforcement with clear concrete cover as 40 mm as shown in Fig. 20.18, centre to centre distance between the main bars at corners (let ϕ_{tr} be 8 mm).

$$(450 - 2 \times 40 - 2 \times 90) = 353 \; mm < [48 \times \phi_{tr} = (48 \times 8) = 384 \; mm]$$

Step 3 : Transverse reinforcement

Let the lateral ties of 8 mm diameter shall be provided as transverse reinforcement. The transverse reinforcement shall be arranged as shown in Fig. 20.18 (B).

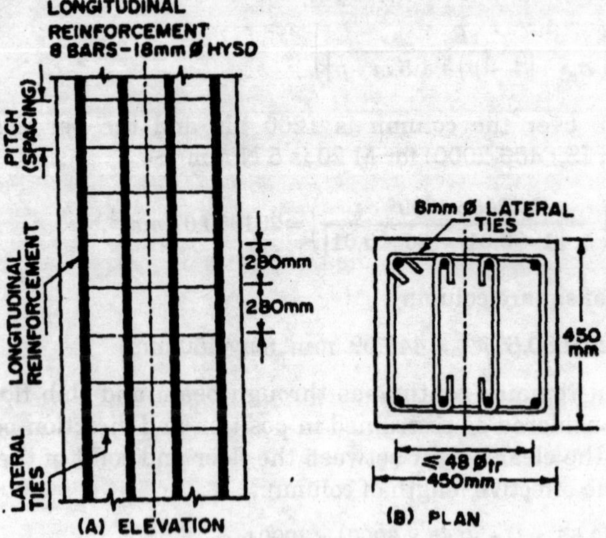

Fig. 20.18

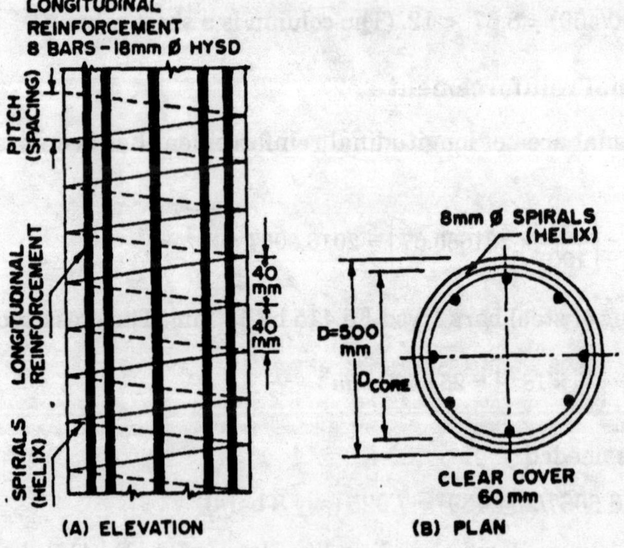

Fig. 20.19

Step 4 : Pitch of lateral ties

Least lateral dimension of the column

b = 450 mm

16 *times* diameter of longitudinal steel bar

$(16 \times 18) = 288\ mm$

48 times diameter of lateral ties

$(48 \times 8) = 384\ mm.$

The pitch of transverse reinforcement shall not be more than the least of three above. Therefore, pitch of lateral ties shall be 280 mm centre to centre along the height of column.

Example 20.8. Design a circular reinforced concrete axially load column in Example 20.6. Provide spiral reinforcement.

Design : From Example 20.6, the modular ratio of concrete M 20 in compression, m_c = 20

Step 1 : Size of the column

Let the percentage of steel in compression for longitudinal reinforcement be *one percent.* Then, p = 0.01. The gross cross-sectional area of concrete, A_g needed [from the expression (viii), Art. 20.13].

$$A_g = \left(\frac{P}{\sigma_{cc} \cdot [(1-p) + 1.5\ m.p]}\right) \quad \text{(ii)}$$

The axial load over the column with spiral reinforcement is 1200 kN. It is assumed that the ratio of volume of helical reinforcement for 1 mm length of the column to the volume of core of the concrete of column per 1 mm length is not less than

$$0.36\left[\left(\frac{A_g}{A_v} - 1\right)\right] \cdot \left(\frac{f_{ck}}{f_y}\right)$$

Then, the axial load for short column

$$P = \left(\frac{1200}{1.05}\right) = 1142.86\ kN$$

The permissible stress in direct compression, σ_{cc} (from IS : 456-2000) for M 20 is 5 N/mm².

$$A_g = \left(\frac{1142.86 \times 1000}{5\ [(1-0.01) + 20 \times 0.01]}\right) = 192077.31\ mm^2$$

Let D be the diameter of circular column.

$$\left(\frac{\pi}{4} \times D^2\right) = 192077.31\ mm^2,\ D = 494.53\ (say\ 500\ \ mm)$$

Step 2 : Longitudinal reinforcement

The cross-sectional area of longitudinal reinforcement has been assumed as *one percent.* Therefore,

$$A_{sc} = \left(\frac{1}{100} \times 192077.31\right) = 1220.7331\ mm^2$$

Let the diameter of steel bars Hysd Fe 415 be 18 mm. The cross-sectional area of one bar

$$A_{\phi} = \left(\frac{\pi}{4} \times 18^2\right) = 254.469\ mm^2$$

Number of bars

(1920.7731/254.469) = 7.55 (say 8 bars)

Step 3 : Transverse reinforcement

The spirals of 8 mm diameter are used as the transverse reinforcement at 40 mm pitch (spacing). The clear cover of concrete over main reinforcement is 60 mm. Core diameter of column, Fig. 20.14 (C).

$$D_{core} = \left(500 - 2 \times 60 + 2 \times 8\right) = 396\ mm^2$$

Pitch of spirals. The pitch of spirals should be less than equal to the following :

$$p \leq 75\ mm,\ p \leq \frac{1}{6} \times core\ diameter \leq \left(\frac{1}{6} \times 396 = 66 mm\right)$$

So also, the pitch of spirals should be more than or equal to the following :

$$p \geq 25\ mm, \geq \left(3 \times \phi_{helix}\right) \geq \left(3 \times 8 = 24\ mm\right)$$

Hence, the pitch of spiral is adopted as 40 mm, which satisfies all the four above specified requirements for the pitch.

Step 4 : Check for volumetric ratio

Gross area of concrete of the column

$$A_g = \left(\frac{\pi}{4} \times 500^2\right) = 196349.54\ mm^2$$

$$A_{sc} = \left(8 \times \frac{\pi}{4} \times 18^2\right) = 2035.75\ mm^2$$

Area of core of helically reinforced column

$$A_c = \left(\frac{\pi}{4} \times 396^2\right) = 123163.0\ mm^2$$

$$d_c = \left(396 - 2 \times 4\right) = 388\ mm$$

Volume of helical reinforcement per 1 mm length of column

$$V_h = \left(\frac{\pi\, d_c \,.\, \frac{\pi}{4} \,.\, \phi_{helix}}{p}\right) = \left(\frac{\pi \times 388}{40} \times \frac{\pi}{4} \times 8^2\right) = 1531.76\ mm^2\ /mm$$

Volume of core of the concrete of column per 1 mm length

$$V_{.core} = \left(\frac{\pi}{4} \times 396^2\right) = 123163.0\ mm^3 / mm$$

From Eq. 20.11, the volumetric ratio

$$\left(\frac{V_h}{V_{core}}\right) \nless 0.36 \left[\frac{A_g}{A_c} - 1\right] . \frac{f_{ck}}{f_y}$$

$$\left(\frac{1531.76}{123163.0}\right), \nless 0.36 \left[\frac{196349.54}{123163.0} - 1\right] . \frac{20}{415}$$

$$0.0124 \nless 0.36\ [0.594] \times 0.0482$$

$$0.0124 \nless 0.0103.\ (As\ assumed)$$

The details of longitudinal and spiral reinforcement are shown in Fig. 20.19.

Example 20.9. Design a rectangular reinforced concrete column to carry on axial load of 1000 kN. The actual length of column is 8.00 m. The column is restrained in position and direction at its both the ends. M 20 grade of concrete and Hysd steel bars of grade Fe 415 are to be used. Adopt the permissible stresses in direct compression for concrete and steel as specified in IS : 456-2000.

Design : M 20 grade of concrete is to be used. The permissible stress in concrete in bending in compression from IS : 456-2000 is 7 N/mm². The modular ratio of M 20 grade of concrete in compression, $\therefore m_c = (1.5 \times m)$

$$\text{or } m_c = \left(\frac{1.5 \times 280}{3 \times 7}\right) = 20 \qquad \text{(i)}$$

Step 1 : Size of the column

The axial load to be carried by the column is 1000 kN. In case, the column is likely to be long column, the *reduction factor*, C_r may be assumed as 0.85. The axial load to be carried by the short column

$$P = \left(\frac{1000}{0.85}\right) = 1176.45\ kN$$

Let the percentage of steel in compression for longitudinal reinforcement be *one per cent*. Then, p = 0.01. The gross cross-sectional area of concrete, A_g needed (from the expression (viii), Art. 20.13).

$$A_g = \left(\frac{P}{\sigma_{cc} . [(1-p) + 1.5\ m\ .\ p]}\right) \qquad \text{(ii)}$$

The permissible stress in direct compression, σ_{cc} (from IS : 456-2000) for M 20 grade of concrete is 5 N/mm².

$$A_g = \left(\frac{1176.47 \times 1000}{5 . [(1 - 0.01) + 20 \times 0.01]}\right) = 197726.05\ mm^2 \qquad \text{(iii)}$$

Let b x D be the size of rectangular column and the width of column b be assumed half of the overall depth of column. Therefore,

$$\frac{1}{2} D \, . \, D = (197726.05)\, mm^2 \, , \, D = 628.85 \, mm$$

Step 2 : Slenderness ratio

Let the overall depth of column be 650 mm. The width of column would be 325 mm. The actual length of column is 8.00. The column is restrained in position and direction at both the ends. The effective length of column

$$l_{eff} = (0.65 \times 8.00) = 5.20 \, m$$

$$\text{Ratio } \frac{l_{eff}}{b} = \left(\frac{5.20 \times 1000}{325}\right) - 16 > 12$$

$$\frac{l_{eff}}{D} = \left(\frac{5.20 \times 1000}{650}\right) - 8 > 12$$

The column is a long column. The actual reduction-factor for the column

$$C_r = \left(1.25 - \frac{l_{eff}}{48 \times b}\right) = \left(1.25 - \frac{520}{48 \times 325}\right) = 0.9167$$

Step 3 : Check for the axial load for the short column

$$P = \left(\frac{1000}{0.9167}\right) = 1090.19 \, kN$$

It is more than the assumed axial load for the short column, 1176.47 kN. Hence, the overall dimensions of rectangular column 325 mm x 650 mm are satisfactory.

Step 4 : Longitudinal reinforcement

Hysd steel bars of grade Fe 415 are to be used for the main reinforcement for the column. The cross-sectional area of longitudinal reinforcement has been assumed as *one percent* of the gross area.

$$A_{sc} = \left(\frac{1}{100} \times 197726.05\right) = 1977.2605 \, mm^2$$

Let the diameter of steel bars Hysd Fe 415 be 20 mm. The cross-sectional area of one bar

$$A_{\phi} = \left(\frac{\pi}{4} \times 20^2\right) = 314 \, mm^2$$

Number of bars needed

(1977.2605/314) = 6.297 (say 8 bars)

8 steel bars of 20 mm diameter of quality Hysd grade Fe 415 shall be provided for the longitudinal reinforcement with clear concrete cover as 40 mm as shown in Fig. 20.20.

Step 5 : Transverse reinforcement

Let the lateral ties of 8 mm diameter Hysd steel bars of grade Fe 415 shall be provided as transverse reinforcement as shown in Fig. 20.20.

Pitch of lateral ties

Least lateral dimension of the column

$$b = 325 \text{ mm}$$

16 times diameter of the longitudinal steel bar

$$(16 \times 20) = 320 \; mm$$

48 times diameter of lateral ties

$$(48 \times 8) = 384 \; mm$$

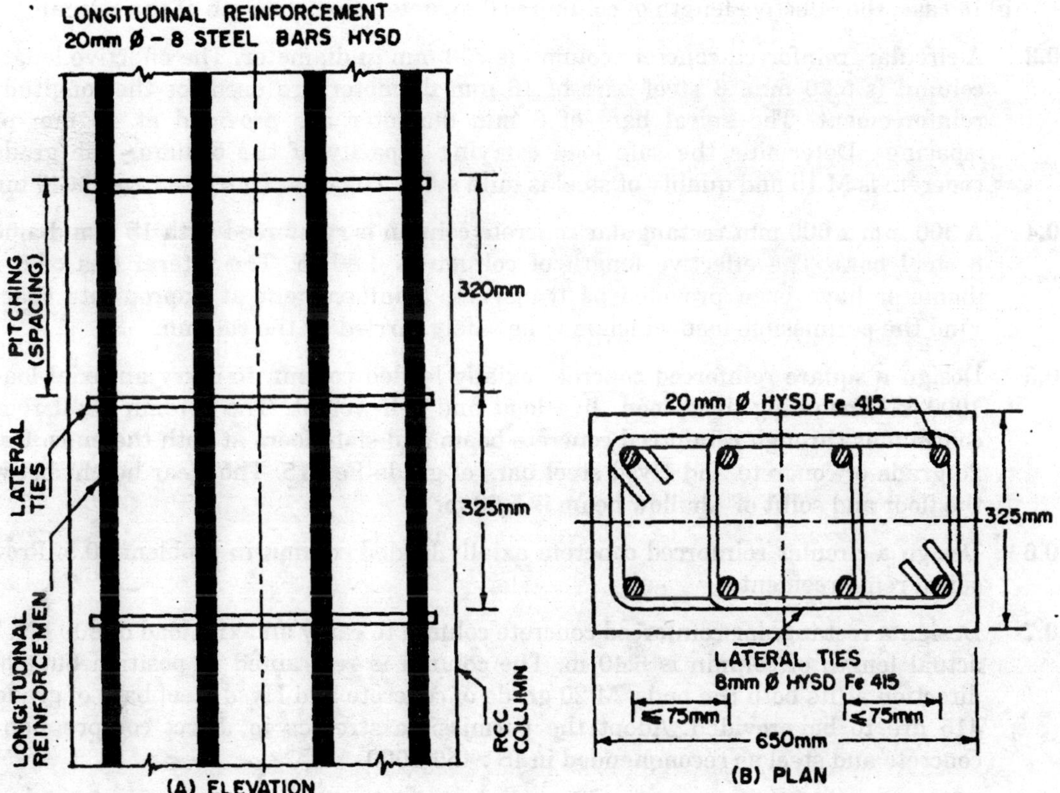

Fig. 20.20

The pitch of transverse reinforcement shall not be more than least calculated above. Therefore, pitch of lateral ties shall be 320 mm centre to centre along the height of column. The centre to centre distance between corner bars

$$(650 - 2 \times 40 - 2 \times 10) = 550\ mm$$

$$48\ \phi_{tr} = (48 \times 8) = 384\ mm$$

The lateral ties have been arranged as shown in Fig. 20.20 (B).

PROBLEMS

20.1 A square reinforced concrete axially loaded column is 400 mm x 400 mm in size. the effective length of column is 5.20 m. 8 steel bars of 16 mm diameter are provided for the longitudinal reinforcement. The lateral ties 6 mm have been provided at appropriate spacing. Determine the strength of column. The grade of concrete is M 15 and the quality of steel is mild steel Fe 250.

20.2 (a) In case the column is circular having 400 mm diameter in Prob. 20.1, determine the strength of the column.

(b) In case, the effective length of column is 6 m, determine strength of the column.

20.3 A circular reinforced concrete column is 360 mm in diameter. The effective length of column is 5.20 mm. 6 steel bars of 16 mm diameter are used for the longitudinal reinforcement. The spiral bars of 6 mm diameter are provided at 40 mm pitch (spacing). Determine the safe load carrying capacity of the column. The grade of concrete is M 15 and quality of steel is mild steel. The clear concrete cover is 40 mm.

20.4 A 300 mm x 600 mm rectangular concrete column is reinforced with 18 mm diameter 8 steel bars. The effective length of column is 4.80 m. The lateral ties of 6 mm diameter have been provided as transverse reinforcement at appropriate spacing. Find the permissible load which may be safely carried by the column.

20.5 Design a square reinforced concrete axially loaded column to carry an axial load of 1000 kN including dead load, live load and self-weight. The column shall remain continuous through reinforced concrete beam and slab floors at both the ends. Use M 20 grade of concrete and Hysd steel bars of grade Fe 415. The clear height between the floor and soffit of shallow beam is 5.60 m.

20.6 Design a circular reinforced concrete axially loaded column in problem 20.5. Provide spiral reinforcement.

20.7 Design a rectangular reinforced concrete column to carry an axial load of 800 kN. The actual length of column is 6.40 m. The column is restrained in position but not in direction at its both the ends. M 20 grade of concrete and Hysd steel bars of grade Fe 415 are to be provided. Adopt the permissible stresses in direct compression for concrete and steel as recommended in IS : 456-2000.

21

DESIGN OF COLUMNS AND COMPRESSION MEMBERS SUBJECTED TO COMBINED AXIAL LOADS AND MOMENTS

21.1 INTRODUCTION

In actual construction, it is not possible to fabricate perfectly straight columns and compression members with perfectly symmetrical reinforcement and to load them axially (i.e., concentrically) through the effective centroid. These columns and compression members mainly carry loads in compression and simultaneously almost always accompanied by the bending moments. In fact, the building columns are parts of monolithic frames in which the moments develop in the joints because of frame action under both applied load and deformations induced by temperature, shrinkage and settlement. The support moments of beams are resisted by the abutting columns, by transverse loads (such as wind loads), by loads carried eccentrically on column brackets, or in arches when the arch axis does not coincide with the pressure line. Even if, the design calculations indicate that a member is to be loaded purely axially, the inevitable imperfections of construction will introduce eccentricities and consequently, the bending moments shall develop. When the columns and the compression members are subjected to combined axial loads and the moments, the columns and the compression members are often called as eccentrically loaded.

The action of an axial load P and a moment M is equivalent to the action of a load P (the line of action of which is parallel to the longitudinal axis of the column) acting at an eccentricity, e = (M/P) as shown in Fig. 21.1. These two loadings are statically equivalent. All the columns may then be categorized in terms of the equivalent eccentricity. Those columns which have *small eccentricity*, e are characterized by compression over the entire concrete sections. These columns fail by crushing of the concrete (when overloaded) accompanied by yielding of the steel bars in compression or more heavily loaded side. The **small eccentricity**, e (viz., the distance of line of action of load measured from the centroidal axis of concrete section of the column) shall be less than *one-fourth* the overall depth of the section. These columns which have **large eccentricity**, e have tension over at least a part of the column section. In case, these columns are overloaded, these columns fail due to tensile yielding of the steel on the side farthest from the load. The **large eccentricity,** e shall be more than *1.5 times* the overall depth of the section. When the *eccentricity*, *e* is more than *one-fourth* the overall depth of section and less than *1.5 times* the overall depth of the section, it is referred as **intermediately eccentricity.**

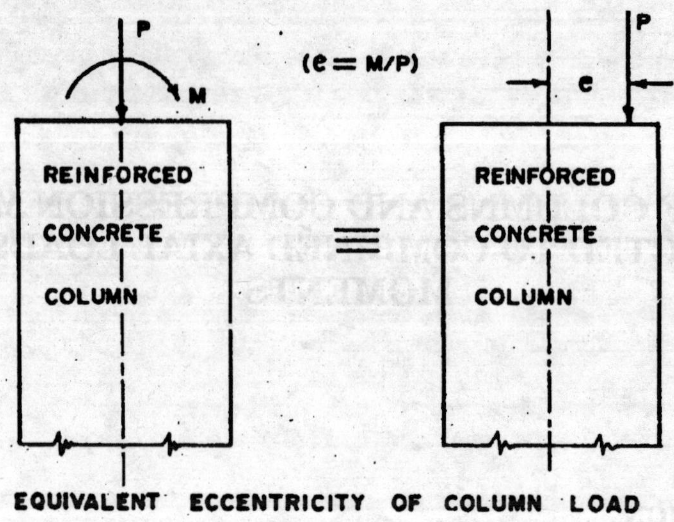

Fig. 21.1

21.2 ECCENTRICALLY LOADED COLUMNS AND COMPRESSION MEMBERS

An eccentrically loaded column as shown in Fig. 21.1 is subjected to bending moment, M besides the axial compression, P. the column section has the bending stresses and the direct compressive stress. In case, the column is subjected to axial load P only, then, the cross-sectional area needed is given by

$$A_a = (P / \sigma_{cc}) \qquad \text{(i)}$$

where, σ_{cc} is the permissible axial compressive stress in concrete. When the column is subjected to bending moment, P. e the cross-sectional area required is obtained by

$$A_b = \left[P.\, e\, .\, y_{\max} / \left(\sigma_{cbc}\, .\, r^2\right)\right], \textit{ where} \qquad \text{(ii)}$$

σ_{cbc} = permissible bending compressive stress in concrete

y_{max} = Distance to the extreme fibre of the section measured from the neutral axis

$A_b.r^2$ = Moment of inertia of the section about the appropriate axis

r = radius of gyration of the section about the appropriate axis

P.e = moment due to eccentricity, e of the load P

e = Distance of line of action of load measured from the centroidal axis.

From theory of simple bending,

$$\left(\frac{P.e}{I}\right) = \left(\frac{\sigma_{cbc}}{y_{\max}}\right) \qquad \text{(iii)}$$

and $$\sigma_{cbc.cal} = \left(\frac{P.e.y_{max}}{I}\right) = \left(\frac{P.e.y_{max}}{A_b.r^2}\right) \quad \text{(iv)}$$

when the column is subjected to axial load P and the bending moment, M both, then, the total area needed shall be

$$A_e = (A_a + A_b) \quad \text{(v)}$$

or $$A_e = \left(\frac{P}{\sigma_{cc}}\right) + \left(\frac{P.e.y_{max}}{\sigma_{cbc}.r^2}\right) \quad \text{(vi)}$$

Both the sides of expression (vi) may be divided by A_e as below :

$$\left(\frac{P/A_e}{\sigma_{cc}} + \frac{P.e.y_{max}/A_e.r^2}{\sigma_{cbc}}\right) = 1$$

or $$\left(\frac{\sigma_{cc.cal}}{\sigma_{cc}} + \frac{\sigma_{bc.cal}}{\sigma_{cbc}}\right) = 1, \text{ where} \quad (21.1)$$

$$\sigma_{cc.cal} = \left(\frac{P}{A_e}\right) = \text{calculated average axial compressive stress}$$

$$\sigma_{cbc.cal} = \left(\frac{P.e.y_{max}}{A_e.r^2}\right) = \text{calculated bending stress in the extreme fibre}$$

It is to note that a member subjected to axial load and bending (due to eccentricity of load, monolithic construction, lateral forces, etc.) shall be considered safe provided Eq. 21.1 is satisfied.

Eq. 21.1 gives the fundamental interaction condition for the columns and the compression members subjected to axial load and bending. It is a straight line when plotted graphically, $(\sigma_{cc.cal}/\sigma_{cc})$ along y-axis and $(\sigma_{cbc.cal}/\sigma_{cbc})$ along x-axis. IS : 456-2000 adopted this interaction condition for the working stress method of design of eccentrically loaded columns. The columns subjected to both axial compression and bending shall be so proportioned that the quantity $(\sigma_{cc.cal}/\sigma_{cc}) + (\sigma_{cbc.cal}/\sigma_{cbc})$ should not exceed unity.

When the bending of the column occurs about both the axes of the member, then, the interaction condition may be written as under,

$$\left(\frac{\sigma_{cc.cal}}{\sigma_{cc}} + \frac{\sigma_{cb.cx.cal}}{\sigma_{cbc.x}} + \frac{\sigma_{cbc.y.cal}}{\sigma_{cbc.y}}\right) \leq 1.0 \quad (21.2)$$

The value of $\sigma_{cbc.x}$ and $\sigma_{cbc.y}$ to be used in the above condition shall be lesser of the values of the maximum permissible stress σ_{cbc} for bending about the appropriate axis.

The calculated direct compressive stress in concrete, $\sigma_{cc.cal}$ may be determined as under:

$$\sigma_{cc.cal} = \left(\frac{P}{A_e}\right) = \left(\frac{P}{[A_{cc} + 1.5\,m\,.\,A_{sc}]}\right), \text{ where} \quad (20.3)$$

A_e = Equivalent area of the column, $[A_{cc} + 1.5\ m.\ A_{sc}]$

A_{cc} = Cross-sectional area of concrete in direct compression

$A_{cc} = [(A_g - A_{sc})]$

A_g = Gross cross-sectional area of the column section

A_{sc} = Cross-sectional area of longitudinal steel bars provided as main reinforcement

$m_c = 1.5\ m$

$m = (280/3\ s_{cbc})$, modular ratio of concrete

The calculated bending compressive stress in concrete, $\sigma_{cbc.cal}$ is determined as below :

$$\sigma_{cbc.cal} = (M/Z)$$

(where, M equals moment and Z equals modulus of section)

$$Z = (I_e/y_{max})$$

I_e = Equivalent moment of inertia of the column section.

In case of the column section is subjected to moments in two directions, the stresses are calculated separately and added algebraically.

21.3 RESULTANT COMPRESSIVE AND TENSILE STRESSES IN COLUMNS

In Art. 21.1, it has been stated that an eccentrically loaded column carrying a load P parallel to the longitudinal axis of the column and acting at an eccentric distance, e from the centroid of column is equivalent to the column subjected to an axial load P and the moment M = (P . e) as shown in Fig. 21.2.

The distribution of stress due to axial load P may be assumed uniform [as shown in Fig. 21.2 (A)]. It is calculated as below :

$$\sigma_{cc.cal} = (P/A_e) \tag{i}$$

where A_e is the equivalent cross-sectional area of the column section. The stresses developed due to bending moment M = (P.e) are shown in Fig. 21.2 (B), These are determined as under

$$\sigma_{cbc.cal} = \left(\frac{M}{Z}\right) = \left(\frac{P\ .e\ .\ y_{max}}{A_b\ .\ r^2}\right) \tag{ii}$$

or $$\sigma_{cbc.cal} = (M\ .\ y_{max}\ /\ I_e) \tag{iii}$$

and $$\sigma_{cbc.cal} = \sigma_{cbc.cal} \tag{iv}$$

Depending upon the direction of moment, the stress due to moment M acts upward on one side (viz., compressive) and downward on the other side (viz., tensile). The combine (viz., resultant) stress depends upon the magnitudes of the direct compressive stress $\sigma_{cc.cal}$ calculated and the bending compressive/tensile stress $(\sigma_{cbc.cal}\ /\ \sigma_{cbt.cal})$. The variation of combined stress is assumed linear in working stress method. When the intensity of direct compressive stress in concrete, $\sigma_{cc.cal}$ due to axial load is more than the stress developed due

to bending moment (i.e., $\sigma_{cc.cal} > \sigma_{cbt.cal}$), the resultant stress on one side of the concrete section shall be compressive.

$$\sigma_{r1} = (\sigma_{cc.cal} + \sigma_{cbc.cal}) \quad \text{(v)}$$

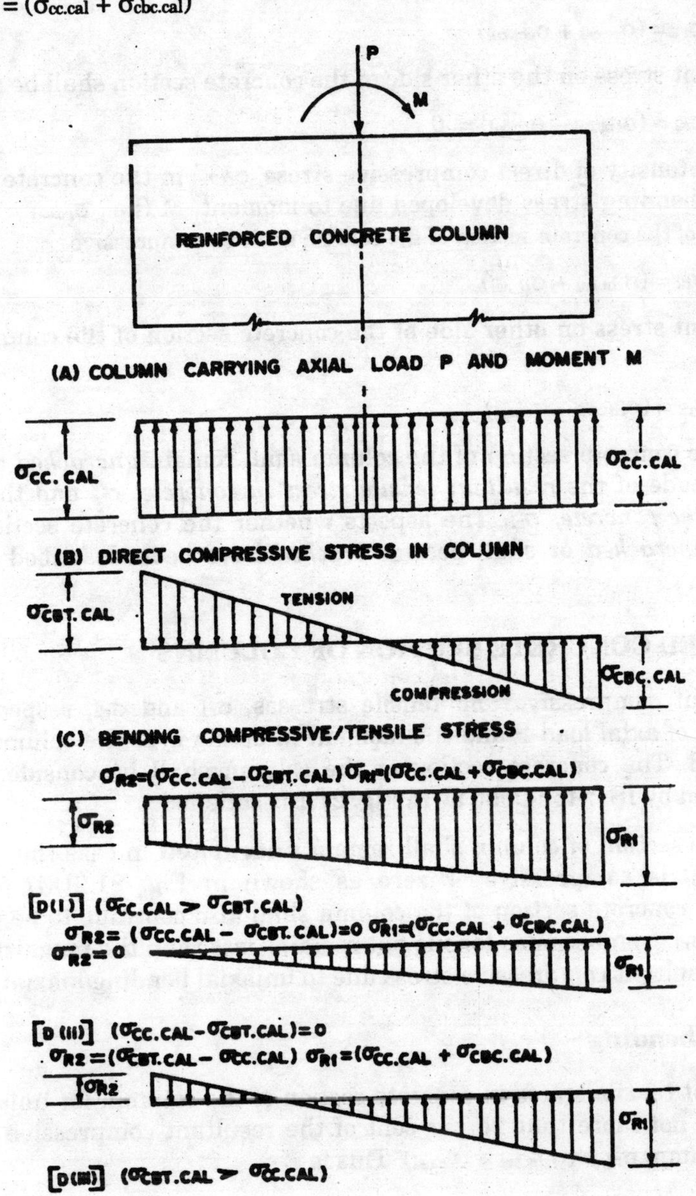

Fig. 21.2

The resultant stress on the other side of the concrete section shall also be compressive [Fig. 21.2 D (i)]

$$\sigma_{r2} = (\sigma_{cc.cal} - \sigma_{cbt.cal}) \quad \text{(vi)}$$

When the intensity of direct compressive stress in concrete, $\sigma_{cc.cal}$ due to axial load is equal to the stress developed due to bending moment (i.e., $\sigma_{cc.cal} = \sigma_{cbt.cal}$), the resultant stress on one side of the concrete section of the column shall be

$$\sigma_{r1} = (\sigma_{cc.cal} + \sigma_{cbt.cal}) \qquad \text{(vii)}$$

The resultant stress on the other side of the concrete section shall be zero [Fig. 21.2 (ii)],

$$\sigma_{r2} = (\sigma_{cbt.cal} - \sigma_{cc.cal}) = 0 \qquad \text{(viii)}$$

When the intensity of direct compressive stress, $\sigma_{cc.cal}$ in the concrete due to axial load P is less than the bending stress developed due to moment, M (i.e., $\sigma_{cc.cal} < \sigma_{cbt.cal}$), the resultant stress on one side of the concrete section of the column shall be compressive

$$\sigma_{r2} = (\sigma_{cbt.cal} + \sigma_{cc.cal}) \qquad \text{(ix)}$$

The resultant stress on other side of the concrete section of the column shall be tensile, [Fig. 21.2 D (iii)]

$$\sigma_{r2} = (\sigma_{cbt.cal} - \sigma_{cc.cal}) \qquad \text{(x)}$$

Whether the concrete section of the column shall remain *uncracked* or *cracked* depends upon the magnitude of the *resultant tensile stress in concrete, σ_{r2}* and the *resultant compressive stress in the concrete, σ_{r1}*. The aspects whether the concrete section of column shall continue to be *uncracked* or shall become *cracked* have been described in the subsequent sections.

20.4 UNCRACKED CONCRETE SECTION OF COLUMNS

The resultant compressive and tensile stresses, σ_{r1} and σ_{r2}, respectively, due to the combined action of axial load P and the moment M acting over the column have been described in Art. 21.3. The concrete section of the column shall be considered *safe* when the condition specified by IS : 456-2000 as in Fig. 21.1 is satisfied.

The concrete section of column shall remain **uncracked** in case the resultant stress $\sigma_{r2} = (\sigma_{cc.cal} - \sigma_{cbc.cal})$ is compressive or zero as shown in Fig. 21.2D(i) and Fig. 21.2D(ii), respectively. The concrete section of the column shall also continue to be **uncracked,** if the resultant stress σ_{r2} even becomes tensile but remains less than the magnitude specified by IS : 456-2000, for resultant compressive stress due to uniaxial bending/biaxial bending.

(A) for uniaxial bending

The resultant tensile stress in concrete section of the column, for uniaxial bending $\sigma_{r2} = (\sigma_{cbt.cal} - \sigma_{cc.cal})$ is not more that 25 per cent of the resultant compressive stress in concrete section of the column, $\sigma_{r1} = (\sigma_{cbc.cal} + \sigma_{cc.cal})$. This is

$$\sigma_{r2} \not> 0.25\,(\sigma_{cbc.cal} + \sigma_{cc.cal}) \qquad \text{(i)}$$

In case, the resultant tensile stress in concrete section of the column section σ_{r2} shall not be more than *three-fourths* the 7 day's modulus of rupture of concrete than also, the section shall remain **uncracked.**

(B) For bi-axial bending

The resultant tensile stress for biaxial bending in concrete section of the column, σ_{r2} is not more than 35 per cent of the resultant compressive stress in concrete section of the column, viz.,

$$\sigma_{r2} \not> 0.35\ \sigma_{r1} \quad \text{(ii)}$$

The condition specified for modulus of rupture as above for uniaxial bending is also applicable for biaxial bending.

For **uncracked section of the column, the equivalent area of the rectangular column of size** b.D

$$A_e = [b.D + \sum (m_c - 1) A_{sc1}] \quad (21.3)$$

For **uncracked section of the column,** the equivalent moment of inertia of the rectangular column of size b.D about neutral axis. Fig. 21.3.

$$I_e = \left[\frac{1}{12} b.D^3 + b.D.\bar{y}^2 + \sum (m_c - 1) A_{sc1} . a_{sc1}^2\right] \quad (21.4\text{ A})$$

For symmetrical placed steel bars

$$I_e = \left[\frac{1}{12} b.D^3 + \sum (m_c - 1) A_{sc1} . a_{sc1}^2\right] \quad (21.4\text{B})$$

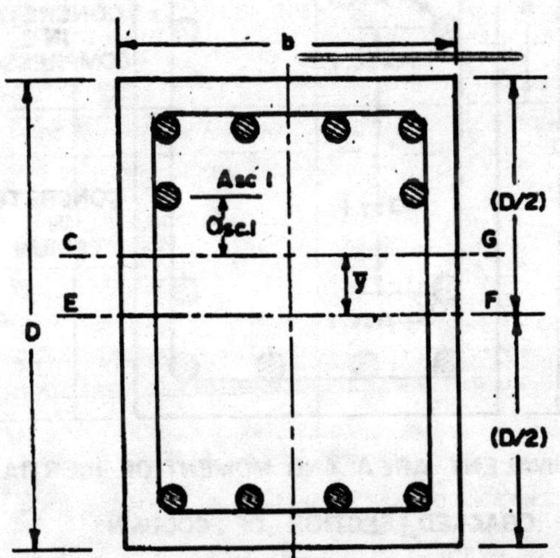

EQUIVALENT AREA AND MOMENT OF INERTIA

UNCRACKED CONCRETE SECTION OF COLUMN

Fig. 21.3

where $\bar{y}$ = Distance between the centroidal axis CG and the middle line EF of concrete section of the column.

A_{sc1} = Cross-sectional area of one steel bar provided as longitudinal reinforcement.

a_{sc1} = Distance between the centroid of one bar from the centroidal axis of the column section.

The distance $\bar{y}$ between the centroidal axis CG and middle line of the column section may be found by moment of areas of the column section about the top surface of Fig. 21.3. Therefore,

$$\left(\frac{D}{2}-\bar{y}\right)=\left(\frac{b.D.D./2+\sum(m_c-1)\,A_{sc1}\,a_{sc.1}}{b.D+\sum(m_c-1).\,A_{sc1}}\right) \quad (21.5)$$

21.5 CRACKED CONCRETE SECTION OF COLUMNS

The resultant compressive and tensile stresses, σ_{r1} and σ_{r2}, respectively, due to the combined action of the axial load P and the moment M acting over the column section have been described in Art. 21.3. The concrete section of the column shall be considered *safe* when the condition specified by IS: 456-2000 as in Eq. 21.1 is satisfied.

The concrete section of the column shall become cracked in case the resultant tensile stress exceeds the magnitude specified by IS : 456-2000 for the resultant compressive stress due to unaxial bending/biaxial bending.

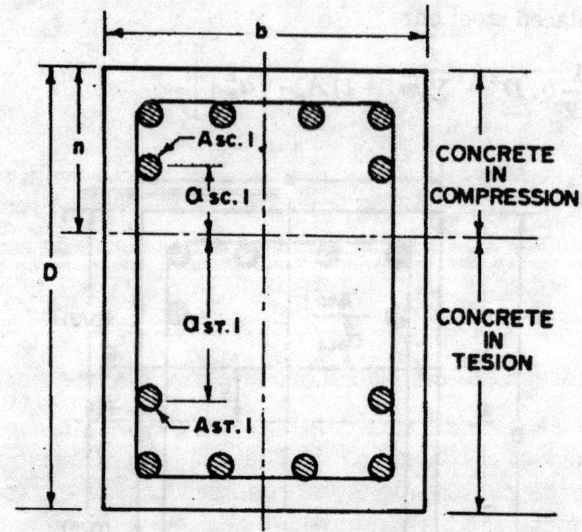

EQUIVALENT AREA AND MOMENT OF INERTIA

CRACKED SECTION OF COLUMN

Fig. 21.4

(A) Uniaxial bending

The resultants tensile stress for uniaxial bending in concrete section of the column $\sigma_{r2} = (\sigma_{cbc.cal} - \sigma_{cc.cal})$ exceeds 25 per cent of the resultant compressive stress in concrete section of the column $\sigma_{r1} = (\sigma_{cbc.cal} + \sigma_{cc.cal})$. That is

$$\sigma_{r2} > 0.25\ (\sigma_{cbc.cal} + \sigma_{cc.cal}) \quad \text{(i)}$$

In case, the resultant tensile stress in concrete section of the column, σ_{r2} exceeds *three-fourths* the 7 days modulus of the rupture, then also, the concrete section of column also become *cracked.*

(B) Bi-axial bending

The resultant tensile stress for biaxial bending of the concrete section of the column s_{r2} exceeds 35 per cent of the resultant compressive stress in concrete section of the column, viz.,

$$(\sigma_{r2} > 0.35\ \sigma_{r1}) \qquad \text{(ii)}$$

The condition specified for modulus of rupture as above for uniaxial bending is also applicable for bi-axial bending.

It is to note that for the **cracked section** of the concrete, the tensile resistance of concrete is ignored. For the cracked section of the column, the equivalent area of rectangular column of size, b.D

$$A_e = \left[b.n + \sum(m_c - 1)\,A_{sc1} + \sum m\,.\,A_{st1}\right] \qquad (21.5)$$

For the cracked section of column, the equivalent moment of inertia of the rectangular column of size b. D, about neutral axis. Fig. 21.4.

$$I_e = \left[\frac{1}{3}\,bn^3 + \sum(m_c - 1)\,A_{sc}\,.\,a_{sc1}^2 + \sum mA_{st1}\,.\,a_{stl}^2\right], \textit{where} \qquad (21.6)$$

n = depth of neutral axis

A_{sc1} = Cross-sectional area of one steel bar in compression

a_{sc1} = Distance between the centroid of one steel bar in compression from the neutral axis

A_{st1} = Cross sectional area of one steel bar in tension.

a_{st1} = Distance between the centroid of one steel bar in tension from neutral axis.

For the **cracked section of the column,** the stresses in concrete and steel shall be calculated by the theory of cracked section. The maximum stress in concrete and steel are found by determining the *no-stress* line (viz., neutral axis) of the section. The *no-stress line* should satisfy the following requirements.

(a) The direct load should be equal to the algebraic sum of forces on concrete and steel.

(b) The moment of external loads about any reference line (e.g., centre-line of steel bars in tension) should be equal to the algebraic sum of moment of forces in concrete (ignoring) the tensile force in concrete) and steel about the same line.

(c) The moment of external loads about any other reference line should be equal to the algebraic sum of the moment of the forces in concrete (ignoring the tensile force in concrete) and steel about the same line.

Example 21.1. A reinforced concrete square column of size 500 mm x 500 mm carries a vertical load of 480 kN at an eccentricity, e_1 equal to 60 mm as shown in Fig. 21.5. 4 mild steel bars of 22 mm diameter have been provided as longitudinal reinforcement. The grade of

concrete is M 15. Find whether the column is safe or not. The effective cover for the steel bars is 50 mm. Also find whether the column section acts as uncracked or cracked.

Solution : The grade of concrete is M 15. The permissible stress for this grade of concrete in bending in compression from IS : 456-2000 is 5 N/mm².

Step 1 : Modular ratio of concrete

$$m = \left(\frac{280}{3 \times 5}\right) = 18.67 \text{ and } m_c = (1.5 \times 18.67) = 28.005 \quad \text{(i)}$$

The permissible stress in direct compression for M 15 grade of concrete is 4 N/mm². (ii)

Step 2 : Cross-sectional area of main steel reinforcement

$$A_{sc} = \left(4 \times \frac{\pi}{4} \times 22^2\right) = 1520.53\ mm^2.$$

Step 3 : Equivalent area of the concrete section of the column

$$A_e = A_g + (m_c - 1)\, A_{sc} = [(500 \times 500) + (28.005 - 1) \times 1520.53]\ mm^2 = 291061.94\ mm^2 \quad \text{(iii)}$$

Step 4 : Calculated direct compressive stress in concrete

$$\sigma_{cc.cal} = \left(\frac{480 \times 1000}{291061.94}\right) = 1.649\ N/mm^2 \quad \text{(iv)}$$

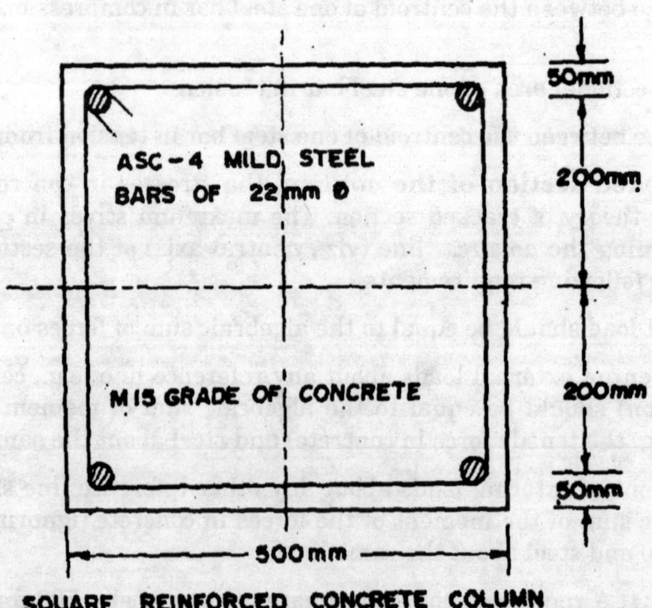

Fig. 21.5

Step 5 : Moment due to load

$$P\,.\,e = (480 \times 60) = 28800 \; kN-mm$$

In the initial stage, the concrete section of the column is assumed as uncracked. Equivalent moment of inertia of the concrete section of column (for symmetrical placed reinforcement).

$$I_e = \left[\frac{1}{12} bD^3 + \Sigma(m_c - 1) A_{sc1} \,.\, a_{sc1}^2\right]$$

$$I_e = \left[\frac{1}{12} \times 500 \times 5600^3 + \times (28.005 - 1) \times 380.133 \times 200^2\right] mm^4$$

$$I_e = (52.083 + 16.425) \times 10^8 = 68.508 \times 10^8 \; mm^4$$

Step 6 : Calculated stress in concrete in bending in compression

$$\sigma_{cbc.cal} = \left(\frac{28800 \times 10^3}{68.508 \times 10^8}\right) \times 250 = 1.051 \; N/mm^2$$

Step 7 : Check for interaction expression

From the condition Eq. 21.1

$$\left(\frac{\sigma_{cc.cal}}{\sigma_{cc}} + \frac{\sigma_{cbc.cal}}{\sigma_{cbc}}\right) = \left(\frac{1.649}{4} + \frac{1.051}{5}\right) = 0.622 < 1$$

As this condition is satisfied, the given concrete section is safe under given eccentric load.

Step 8 : Resultant stresses

Resultant compressive stress in concrete

$$\sigma_{r1} = (\sigma_{cc.cal} + \sigma_{cbc.cal}) = (1.649 + 1.051) = 2.7 \; N/mm^2 \;\; and \;\because\; (\sigma_{cbt.cal} = \sigma_{cbc.cal})$$

Resultant tensile stress in concrete

$$\sigma_{r2} = (\sigma_{cbt.cal} - \sigma_{cc.cal}) = (1.051 - 1.649) = 0.598 \; N/mm^2$$

(viz., it is also compressive).

The concrete section of column shall remain **uncracked.**

Example 21.2. In Example 21.1, find the value of maximum eccentricity, e_{max} for the given vertical load such that the resultant tensile stress is zero and the column section is safe.

Solution: The resultant tensile stress σ_{r2} shall be zero, when the $(\sigma_{cbt.cal} - \sigma_{cc.cal}) = 0$. Therefore, $\sigma_{cc.cal}$ and $\sigma_{cbt.cal}$ are equal.

$$\left(\frac{P}{A_e} - \frac{P \cdot e_{max} \cdot y_{max}}{I_e}\right) = 0$$

$$\text{or} \quad e_{max} = \left(\frac{I_e}{A_e} \times \frac{1}{y_{max}}\right) \qquad \text{(i)}$$

Substituting the respective values from Example 21.1

$$= \left(\frac{68.508 \times 108}{291061.94} \times \frac{1}{250}\right) = 94.149\ mm$$

From Example 21.1, $(\sigma_{cc.cal} = \sigma_{cbt.cal}) = 1.649\ N/mm^2$

From the condition, Eq. 21.1

$$\left(\frac{\sigma_{cc.cal}}{\sigma_{cc}} + \frac{\sigma_{cbc.cal}}{\sigma_{cbc}}\right) = \left(\frac{1.649}{4} + \frac{1.649}{5}\right) = 0.742 < 1$$

As this condition is satisfied, the column section remains safe.

Example 21.3. **In Example 21.1, find the value of the maximum eccentricity, e_{max}, so that the concrete section of the column carries just allowable resultant tensile stress and it acts as uncracked section and still the column remain safe.**

Solution : **The concrete section of column shall act as uncracked section till, the resultant tensile stress σ_{r2} is not more than 25 per cent of resultant compressive stress for uni-axial bending.**

Therefore,

$$(\sigma_{cbt.cal} - \sigma_{cc.cal}) \ngtr 0.25\,(\sigma_{cc.cal} + \sigma_{cbc.cal})$$

$$\left(\frac{P.\,e.\,y_{max}}{I_e} - \frac{P}{A_e}\right) \ngtr \left(\frac{P\,.\,e\,.\,y_{max}}{I_e} - \frac{P}{A_e}\right)$$

$$\text{or} \quad \left(0.75\,\frac{e.y_{max}}{I_e} \ngtr 1.25\,\frac{1}{A_e}\right)$$

$$\text{or} \quad e \ngtr \left(\frac{1.25}{0.75} \times \frac{I_e}{A_e} \times \frac{1}{y_{max}}\right) \qquad \text{(i)}$$

Substituting the values of I_e, A_e and y_{max} from Example 21.1

$$e \ngtr \left(\frac{1.25}{0.75} \times \frac{68.508{\times}10^8}{291061.94} \times \frac{1}{250}\right) = \mathbf{156.91}\ mm$$

$$\sigma_{cc.cal} = 1.649\ N/mm^2 \qquad \text{(ii)}$$

$$\sigma_{cbc.cal} = \left(\frac{P.e.\,y_{max}}{I_e}\right) = \left(\frac{480 \times 1000 \times 156.91 \times 250}{68.508 \times 10^8}\right) = 2.748\ N/mm^2 \qquad \text{(iii)}$$

$$\left(\frac{\sigma_{cc.cal}}{\sigma_{cc}}+\frac{\sigma_{cbc.cal}}{\sigma_{cbc}}\right)=\left(\frac{1.649}{4}+\frac{2.748}{5}\right)=0.962<1.00$$

Hence, the column section is still safe.

Example 21.4. **In Example 21.1, find out the maximum value of vertical load and the value of maximum eccentricity such that the tension is zero and the column section remains safe.**

Solution : The concrete section of column is having zero tension. Therefore,

$$(\sigma_{cbt.cal}-\sigma_{cc.cal})=0$$

$$\text{or}\quad \left(\frac{P\,.\,e\,.\,y_{max}}{I_e}\right)=\left(\frac{P}{A_e}\right) \qquad \text{(i)}$$

The concrete section of column also remains safe. Therefore,

$$\left(\frac{\sigma_{cbt.cal}}{\sigma_{cc}}+\frac{\sigma_{cbc.cal}}{\sigma_{cbc}}\right)=1$$

$$\text{or}\quad \left[\frac{(P/A_e)}{4}+\frac{(P\,.\,e\,.\,y_{max}/I_e)}{5}\right]=1 \qquad \text{(ii)}$$

From the expression (i) above

$$\left(\frac{P.\,e.\,y_{max}}{I_e}\right)=\left(\frac{P}{A_e}\right)$$

Therefore, the expression (ii) becomes as under

$$\left[\frac{(P/A_e)}{4}+\frac{(P/A_e)}{5}\right]=1$$

$$0.25\left(\frac{P}{A_e}\right)+0.2\left(\frac{P}{A_e}\right)=1$$

$$\text{or}\quad 0.45\,.\left(\frac{P}{A_e}\right)=1$$

$$\text{Therefore,}\quad P=\left(\frac{1}{0.45}\times A_e\right) \qquad \text{(iii)}$$

Substituting the value of A_e from Example 21.1

$$P=\left(\frac{1}{0.45}\times\frac{291061.94}{1000}\right)=\mathbf{646.80}\ kN$$

Example 21.5. **In Example 21.1, the resultant tensile stress in the concrete section remains maximum and the column section still remains safe and uncracked, find the maximum value of vertical load which may act over the column.**

Solution : The resultant tensile stress in concrete section of the column is equal to maximum allowable and the column remains uncracked. Therefore,

$$(\sigma_{cbc.cal} - \sigma_{cc.cal}) \ngtr 0.25\,(\sigma_{cbc.cal} + \sigma_{cc.cal})$$

$$\left(\frac{P.e.y_{max}}{I_e} - \frac{P}{A_e}\right) \ngtr 0.25\left(\frac{P.e.y_{max}}{I_e} + \frac{P}{A_e}\right)$$

$$\left(0.75\,.\,\frac{P.e.y_{max}}{I_e}\right) \ngtr \left(1.25\,\frac{P}{A_e}\right). \textit{That is, } e \ngtr \left(\frac{1.25}{0.75} \times \frac{I_e}{A_e} \times \frac{1}{y_{max}}\right)$$

From Example 21.1

$$e \ngtr \left(\frac{1.25}{0.75} \times \frac{68.508 \times 10^8}{291061.94} \times \frac{1}{250}\right) \ngtr 156.91\ mm$$

The concrete section of the column remains safe. Therefore

$$\left(\frac{\sigma_{cc.cal}}{\sigma_{cc}} + \frac{\sigma_{cbc.cal}}{\sigma_{cbc}}\right) = 1. \textit{ That is, } \frac{(P/A_e)}{4} + \frac{(P.e.y_{max}/I_e)}{5} = 1$$

$$\text{or} \quad \left(0.25\,\frac{P}{A_e} + \frac{P.e.y_{max}}{I_e}\right) = 1$$

Substituting A_e, I_e and y_{max} from Example 21.1

$$\left(\frac{0.25 \times P}{291061.94}\right) + \left(\frac{0.20 \times P \times 156.91 \times 250}{68.508 \times 10^8}\right) = 1$$

P = **500.632 kN**

Example 21.6. In Example 21.1, the concrete section of column carries vertical load of 480 kN and the moment 30 kN-m. Find out whether the column section shall remain safe and uncracked.

Solution : From Example 21.1

y_{max} = 250 mm, A_e = 291061.94 mm², and I_e = 68.500 x 10⁸ mm⁴.

The calculated compressive stress in column,

$$\sigma_{cc.cal} = \left(\frac{480 \times 10^3}{291061.94}\right) = 1.649\ N/mm^2 \qquad \text{(i)}$$

The calculated stress in concrete in bending in compression,

$$\sigma_{cbc.cal} = \left(\frac{M.y_{max}}{I_e}\right) = \left(\frac{30 \times 10^6 \times 250}{68.508 \times 10^8}\right) = 1.095\ N/mm^2 \qquad \text{(ii)}$$

For the column section remaining safe

$$\left(\frac{\sigma_{cc.cal}}{\sigma_{cc}} + \frac{\sigma_{cbc.cal}}{\sigma_{cbc}}\right) \not> 1. \textit{ That is,} \left(\frac{1.649}{4} + \frac{1.095}{5}\right) = 0.631 \not> 1.00$$

Hence, the column section is safe. That is, $\sigma_{cc.cal} > (\sigma_{cbc.cal} = \sigma_{cbt.cal})$

Therefore, there is no tension in the concrete section of the column. Hence, the concrete section of column remains **uncracked.**

21.6 DISTRIBUTION OF STRESS OVER COLUMN SECTION (ECCENTRICITY OF LOAD LESS THAN 0.25 D)

The compressive load P acting at a distance e_1 from the middle line of a reinforced concrete rectangular column section of size b x D is shown in Fig. 21.6. The distance $\bar{y}$ between the line and the centroidal axis may be determined from Eq. 21.5, viz.,

$$\left(\frac{D}{2} - \bar{y}\right) = \left(\frac{b.D.\, D/2 + \sum(m_c - 1)\, A_{sc1} \cdot a_{sc1}}{b.D + \sum(m_c - 1).\, a_{sc1}}\right)$$

The eccentricity of load P from the centroidal axis

$$e = (e_1 - \bar{y}) \tag{ii}$$

The *eccentricity* e is considered small in case it is less than 0.25 *times* the overall depth of the section. These columns having *small eccentricity,* e are characterised by compression over the entire concrete sections. These columns fail by *crushing* of the concrete (when over loaded) accompanied by *yielding* of the steel bars in compression on more heavily loaded side.

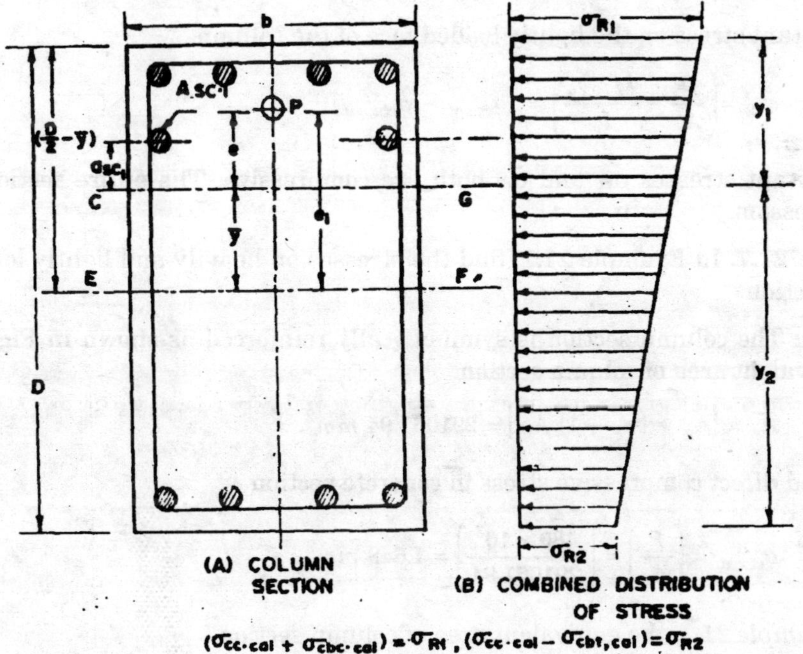

Fig. 21.6

The distribution of stress over the column section is shown in Fig. 21.6(B). The resultant stress on the heavily loaded face of the column.

$$\sigma_{r1} = \left(\frac{P}{A_e} + \frac{M \cdot y_1}{I_e}\right) = (\sigma_{cc.cal} + \sigma_{cbc.cal}) \tag{21.7}$$

The column section acts as *uncracked section.* The equivalent area of the uncracked section may be obtained from Eq. 21.3, viz.,

$$A_e = [b \cdot D + \sum(m_c - 1) A_{sc1}] \tag{iii}$$

The equivalent moment of inertial of the uncracked section may be found from Eq. 21.4 (A), viz.,

$$I_e = \left(\frac{1}{12} b.D^3 + b.D \cdot \bar{y}^2 + \sum(m_c - 1) A_{sc1} \cdot a_{sc1}^2\right) \tag{iv}$$

When the column section is symmetrically reinforced, then, the centroidal line and the middle of column section coincide. Then, the equivalent moment of inertia of the uncracked section is calculated from Eq. 21.4(B), viz.,

$$I_e = \left(\frac{1}{12} b.D^3 + \sum(m_c - 1) A_{sc1} \cdot a_{sc1}^2\right) \tag{v}$$

equivalent moment of inertia of the uncracked section is calculated from Eq. 21.4 (B), viz.,

$$I_e = \left(\frac{1}{12} b.D^3 + \sum(m_c - 1) A_{sc1} \cdot a_{sc1}^2\right) \tag{vi}$$

The resultant stress on the lightly loaded face of the column

$$\sigma_{r2} = \left(\frac{P}{A_e} - \frac{M \cdot y_2}{I_e}\right) = (\sigma_{cc.cal} - \sigma_{cbt.cal}) \tag{21.8}$$

The resultant stresses σ_{r1} and σ_{r2} both are compressive. The entire section of column carries compression.

Example 21.7. In Example 21.1, find the stresses on heavily and lightly loaded faces of the column section.

Solution: The column section is symmetrically reinforced as shown in Fig. 21.5. From 21.2, the equivalent area of column section.

$$A_e = [A_g + (m_c - 1) A_{sc}] = 291061.94 \ mm^2$$

Calculated direct compressive stress in concrete section

$$\sigma_{cc.cal} = \left(\frac{P}{A_e}\right) = \left(\frac{480 \times 10^3}{291061.94}\right) = 1.649 \ mm^2$$

From *Example 21.1,* the equivalent area of column section

$$I_e = 68.508 \times 10^8 \ mm^4$$

Calculated stress in concrete in bending in compression

$$\sigma_{cbc.cal} = \left(\frac{M \ . \ y_1}{I_e}\right) = \left(\frac{P \ . \ e. \ y_1}{I_e}\right)$$

$$\text{or} \quad \sigma_{cbc.cal} = \left(\frac{480 \times 10^3 \times 60 \times 250}{68.508 \times 10^8}\right) = 1.051 \ N/mm^2$$

The stress on heavily loaded face of the column, from Eq. 21.7

$$\sigma_{r1} = \left(\frac{P}{A_e} + \frac{M \ . \ y_1}{I_e}\right) = (\sigma_{cc.cal} + \sigma_{cbc.cal})$$

$$\text{or} \quad \sigma_{r1} = (1.641 + 1.051) = 2.7 \ N/mm^2 \ (compressive)$$

The stress on lightly loaded face of the column from Eq. 21.8 (B).

$$\sigma_{r2} = \left(\frac{P}{A_e} - \frac{M \ . \ y_2}{I_e}\right) = (\sigma_{cc.cal} - \sigma_{cbt.cal})$$

or $\sigma_{r2} = (1.649 - 1.051) =$ **0.598 N/mm²** (compressive)

21.7 DISTRIBUTION OF STRESSES OVER COLUMN SECTION (ECCENTRICITY OF LOAD MORE THAN 1.5 D)

The compressive load P acting at a distance e_1 from the middle axis of a reinforced concrete rectangular column section of size b x D is shown in Fig. 21.7(A). The distance $\bar{y}$ between the middle line and centroidal axis may be determined as under :

$$\left(\frac{D}{2} - \bar{y}\right) = \left(\frac{b.D \ . \ \frac{D}{2} + \sum(m_c - 1)A_{sc1} \ . \ a_{st1} + \sum A_{st1} \ . \ a_{st1}}{b.D + \sum(m_c - 1) \ A_{sc1} + \sum m \ . \ A_{st1}}\right)$$

The eccentricity, of load P from the central axis **(i)**

$$e = (e_1 - \bar{y}) \qquad \textbf{(ii)}$$

The eccentricity e_1 is considered as large in case is more than 1.5 times the over depth of the section. These columns which have large eccentricity, e_1 have tension over atleast a part of the column section. In case these columns are overloaded, these columns fail due to tensile yielding of the steel on the side farthest from the load. The concrete section of the column cracks below the centroidal axis (viz., neutral axis). The concrete section of column acts as *cracked section.* The tensile resistance of concrete below the neutral axis is neglected. The tensile resistance of concrete below the neutral axis is neglected. The tensile stresses are taken up by the steel bars provided as tension reinforcement.

The distribution of direction compressive stress due to load P (viz., $\sigma_{cc.cal} = P/A_e$) is as shown in Fig. 21.7(B). The distribution of stress in concrete in bending in compression due to bending moment $M = P.e$ (viz, $\sigma_{cbc.cal} = P.e.y_1/I_e$) is as shown in Fig. 21.7 (C). The neutral axis of the concrete section of column for bending alone passes through the point O. The

combined stress distribution diagram shall be as shown in Fig. 21.7 (D), in which, the neutral axis passes through O_1. Since, eccentricity e_1 is large (viz., $e_1 > 1.5$ D), the bending moment, M = P . e is also large. The large bending moment predominates and the distance (viz. shift of neutral axis) OO_1 between the two positions of neutral axis in Fig. 21.7 (C) and in Fig. 21.7 (D) is very small. Therefore, *as an approximation* the shift in the position of neutral axes may be neglected. The analysis of concrete section of the column may then be carried on the basis of bending stress distribution diagram only viz., as shown in Fig. 21.7 (C)].

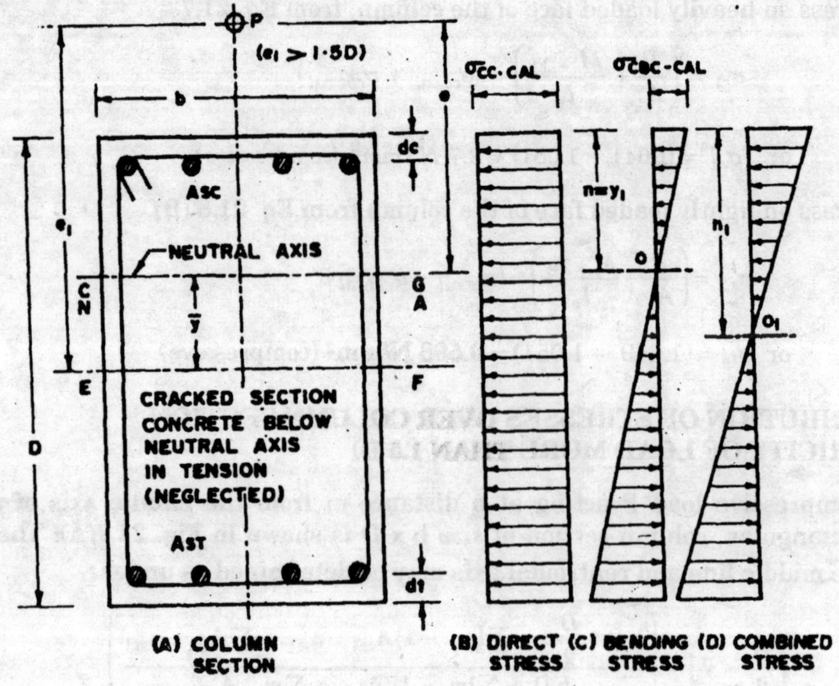

Fig. 21.7

The depth of neutral axis, n may then the obtained by equating first moment of equivalent areas of concrete and the steel bars provided for compression and tension reinforcement. Therefore,

$$\left[\frac{b \cdot n^2}{2} + (m_c - 1)\, A_{sc}\, (n - d_c)\right] = m\, A_{st}\, (D - n - d_t)\; where, \qquad (21.9)$$

A_{sc} = cross-sectional area of steel bars in compression

A_{st} = cross-sectional area of steel bars in tension

n = depth of neutral axis

m_c = modular ratio of concrete in compression (m_c = 1.5 m)

d_c = effective cover for the steel bars in compression

d_t = effective cover for the steel bars in tension.

The compressive stress at the top of the concrete section $\sigma_{cbc.cal}$, that at the bottom of the concrete section, $\sigma_{cbt.cal}$, the compressive stress in the steel bars in compression, σ_{sc} and tensile stress in the steel bars in tension, σ_{st} may be determined by taking the moment of internal forces and the external forces about the centre of steel in tension. Therefore,

$$\left[C_c \,.\, (LA)_c + C_{sc}\,.\,(LA)_{.sc}\right] = P.\,(LA)_{.p} \tag{iii}$$

The compressive forces in concrete in compression

$$C_c = \left[b.n.\frac{1}{2}\left(\sigma_{cbc.cal}\right)\right] \tag{iv}$$

The lever arm distance $(LA)_{.c}$ measured from the centroid of triangular stress distribution diagram upto the centre of steel bars in tension.

$$(LA)_{.c} = \left(D - d_t - \frac{n}{3}\right) \tag{v}$$

The compressive force in steel bars in compression

$$C_{sc} = (m_c - 1)\, A_{sc}\,.\,\sigma_{cbc1} \tag{vi}$$

$$\sigma_{cbc1} = \left(\frac{n - d_c}{n}\right) \times \sigma_{cbc.cal} \tag{vii}$$

The lever arm distance (LA) σ_{sc} measured from the centre of steel bars in compression upto the centre of steel bars in tension.

$$(LA)\,\sigma_{sc} = (D - d_c - d_t) \tag{viii}$$

The lever arm distance $(LA)_p$ for the load measured from centre of steel bars in tension

$$(LA)_{.p} = \left(e_1 + \frac{D}{2} - d_t\right) \tag{ix}$$

Substituting these respective values in the expression (iii), then

$$\left[b.n.\left(\frac{\sigma_{cbc.cal}}{2}\right) + (m_c - 1)\, A_{sc}.\left(\frac{n - d_c}{n}\right)\sigma_{cbc.cal.}\ (D - d_c - d_t)\right]$$

$$= P.\left(e_1 + \frac{D}{2} - d_t\right) \tag{21.10}$$

The depth of neutral axis, n has already been calculated from Eq. 21.9. The solution of Eq. 21.10 gives the stress $\sigma_{cbc.cal}$. The stress in steel bars in compression

$$\sigma_{sc} = m_c\left(\frac{n - d_c}{n}\right).\,\sigma_{bcc.cal} \tag{x}$$

The stress in steel bars in tension

$$\sigma_{st} = m\,.\left(\frac{D - d_t - n}{n}\right).\,\sigma_{cbc.cal} \tag{xi}$$

The equivalent area of concrete section of the column

$$A_e = [b \cdot n + (m_c - 1)A_{sc} + m \cdot A_{st}] \tag{xii}$$

The calculated direct stress in compression over the concrete section

$$\sigma_{cc.cal} = (P / A_e) \tag{xiv}$$

The resultant stress in steel bars in compression

$$(\sigma_{sc} + m_c \cdot \sigma_{cc}) = \left[m_c \cdot \left(\frac{n - b_c}{n} \right) \sigma_{cbc.cal} + m_c \cdot \sigma_{cc.cal} \right] \tag{xv}$$

The distribution of stress over column section (when the eccentricity e_1 of the load is more than 1.5 D) is shown in Fig. 21.17 (D). Above analysis may be studied and done by accurate method (without any approximation) as described in the subsequent article.

21.8 DISTRIBUTION OF STRESS OVER COLUMN SECTION (ECCENTRICITY OF LOAD MORE THAN 0.25 D AND LESS THAN 1.5 D)

The compressive load P acting at a distance e_1 from the middle axis of a reinforced concrete rectangular column section of size b x D is as shown in Fig. 21.8(A).

The distance $\bar{y}$ between the middle line and the centroidal axis may be obtained as below :

$$\left(\frac{D}{2} - \bar{y} \right) = \left(\frac{b.D. \dfrac{D}{2} + \sum (m_c - 1) A_{sc1} \cdot a_{st1} + \sum m A_{st1} \cdot a_{st1}}{b.D + \sum (m_c - 1) A_{sc1} + \sum m. A_{st1}} \right) \tag{i}$$

The eccentricity of load P from the centroidal axis

$$e = (e_1 - \bar{y}) \tag{ii}$$

When the eccentricity e_1 is more than 0.25 D and less than 1.5 D, it is considered as *moderate.* The column section is subjected to an axial load P and a moment M = P. e_1. The concrete section of column cracks below the centroidal axis (viz., neutral axis). The concrete section of column acts as a *crocked section.* The tensile stress of the concrete below the neutral axis is neglected. The tensile stresses are taken up by the steel bars provided as tension reinforcement.

The distribution of direct compressive stress to the load P (viz., $\sigma_{cc.cal} = P/A_e$) is as shown in Fig. 21.8 (B). The distribution of stress in concrete in bending in compression due to bending moment M = P.e [(viz., $\sigma_{cbc.cal} = P.e.y_1/I_e$) is as shown in Fig. 21.8 (C). The neutral axis of the column section for bending alone passes through O. The combined stress distribution diagram shall be as shown in Fig. 21.8 (D), in which the neutral axis passes through O_1.

The column section is in equilibrium. therefore, total sum of internal forces developed in the column section is equal to the external force acting over the column. Therefore,

$$(C_c + C_{sc} - T) = P \tag{i}$$

where, C_c is the compressive force in the concrete in the portion of column section under compression, C_{sc} is the compressive force in the steel bars provided as reinforcement in compression and the T is the tensile force in the steel bars as reinforcement in tension. The compression force C_c, in the concrete in the portion of column section under compression.

$$C_c = \left(\frac{b.n_1.\sigma_{cbc.cal}}{2}\right) \tag{ii}$$

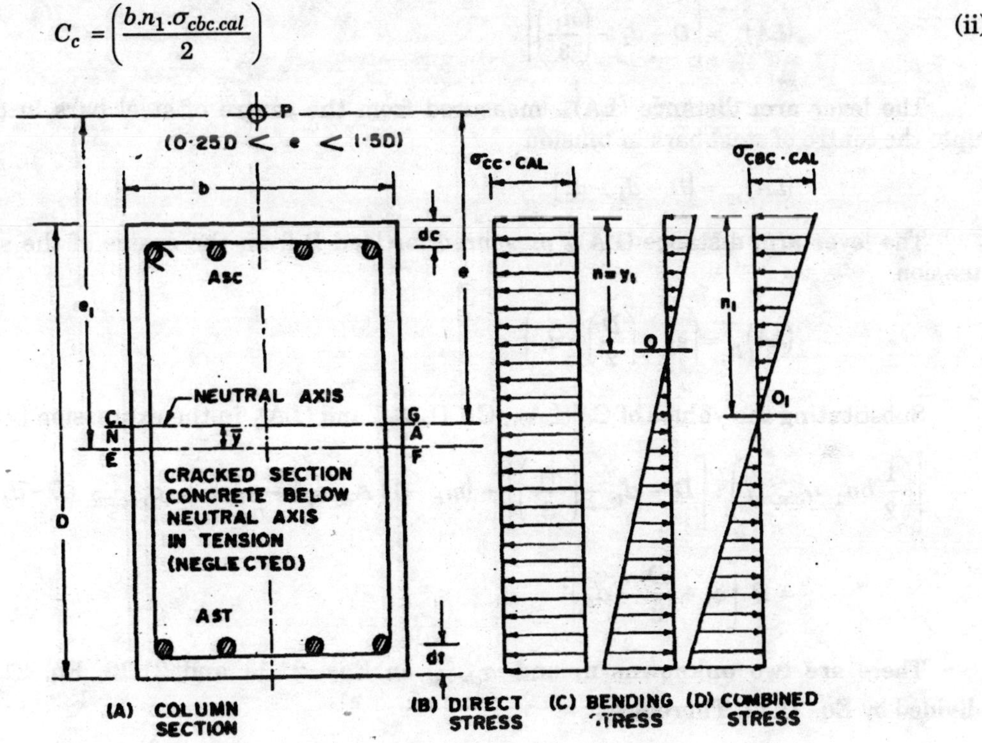

Fig. 21.8

The compressive force C_{sc} in the steel bars in compression

$$C_{sc} = (m_c - 1)A_{sc}.\left(\frac{n_1 - d_c}{n_1}\right).\sigma_{cbc.cal} \tag{iii}$$

The tensile force T in the steel bars in tension

$$T = m.A_{st}.\left(\frac{D - dt - n_1}{n_1}\right).\sigma_{cbc.cal} \tag{iv}$$

Substituting these values in the expression (i)

$$\left[\frac{1}{2}b.n_1.\sigma_{cbc.cal} + (m_c - 1)A_{sc}.\left(\frac{n_1 - d_c}{n_1}\right).\sigma_{cbc.cal} - m.A_{st}.\left(\frac{D - d_t - n_1}{n_1}\right).\sigma_{cbc.cal}\right] = P \tag{21.11}$$

Since the column section is in equilibrium, the moment of all the internal forces developed and the external force acting over the column about any point shall be zero. Hence, take the moment of all the internal forces developed in the column section and the external force acting cover the column section about the centre line of steel bars in tension. Therefore,

$$C_c \cdot (LA)_{.c} + C_{sc} \cdot (LA)_{.sc} = P \cdot (LA)_{.p} \quad \text{(v)}$$

The lever arm distance $(LA)_{.c}$ measured from the centroid of triangular stress distribution diagram. Fig. 21.8 (D) upto the centre of steel bars in tension

$$(LA)_{.c} = \left[D - d_t - \left(\frac{n_1}{3}\right)\right] \quad \text{(vi)}$$

The lever arm distance $(LA)_{.sc}$ measured from the centre of steel bars in compression upto the centre of steel bars in tension

$$(LA)_{.sc} = [D - d_t - d_c] \quad \text{(vii)}$$

The lever arm distance $(LA)_{.p}$ measured for load P from the centre of the steel bars in tension

$$(LA)_{.p} = \left(e_1 + \left(\frac{D}{2}\right) - d_t\right) \quad \text{(viii)}$$

Substituting the values of C_c, $(LA)_c$, C_{sc}, $(LA)_{sc}$, and $(LA)_p$ in the expression (v)

$$\left[\left(\frac{1}{2} bn_1 \; \sigma_{cbc.cal}\right) \cdot \left[D - d_t - \left(\frac{n_1}{3}\right)\right] + (m_c - 1) A_{sc} \cdot \left(\frac{n_1 - d_c}{n_1}\right) \cdot \sigma_{cbc.cal} \; (D - d_t - d_c)\right]$$

$$= P \cdot \left[e_1 + \frac{D}{2} - d_t\right] \quad \text{(21.12)}$$

There are two unknowns n_1 and $\sigma_{cbc.cal}$ in Eqs. 21.11 and 21.12, Eq. 21.12 may be divided by Eq. 21.11. Therefore,

$$\left(\frac{\frac{1}{2} b.n_1 \cdot \left[D - d_t - \left(\frac{n_1}{3}\right)\right] + (m_c - 1) \cdot A_{sc} \cdot \left(\frac{n_1 - d_c}{n_1}\right) \cdot [D - d_t - d_c]}{\frac{1}{2} b.n_1 + (m_c - 1) A_{sc} \cdot \left(\frac{n_1 - d_c}{n}\right) - m \, A_{st} \cdot \left(\frac{D - d_t - n_1}{n_1}\right)}\right)$$

$$= \left[e_1 + \left(\frac{D}{2}\right) - d_t\right] \quad \text{(21.13)}$$

Eq. 21.13 is a cubic equation having n_1 only as one unknown. The value of n_1 may be found by solving Eq. 21.13. The value of stress $\sigma_{cbc.cal}$ may be calculated by solving either Eq. 21.11 or Eq. 21.12. The stress in steel bars provided as reinforcement in compression

$$\sigma_{sc} = m_c \cdot \left(\frac{n_1 - d_c}{n}\right) \cdot \sigma_{cbc.cal} \quad \text{(ix)}$$

The stress in steel bars provided as reinforcement in tension

$$\sigma_{st} = m \cdot \left(\frac{D + d_t - n_1}{n_1}\right) \cdot \sigma_{cbc.cal} \quad \text{(x)}$$

This method is accurate and there is no approximation in analysis.

Alternative method

The depth of neutral n_1 and the bending stress in concrete in compression $\sigma_{cbc.cal}$ at the top of column section may alternatively be found by calculating equivalent cross-sectional area, A_e and the equivalent moment of inertia, I_e of the cracked column section as follows :

A vertical load P is acting over the column at an eccentric distance e_1 from the middle line of the rectangular column of size b x D as shown in Fig. 21.9. The eccentricity e_1 is more than 0.25 D and less than 1.5 D, where D is the overall depth of column section. Let ΔA_e be an elementary area at a distance y from the centroidal axis (viz., neutral axis of the column section).

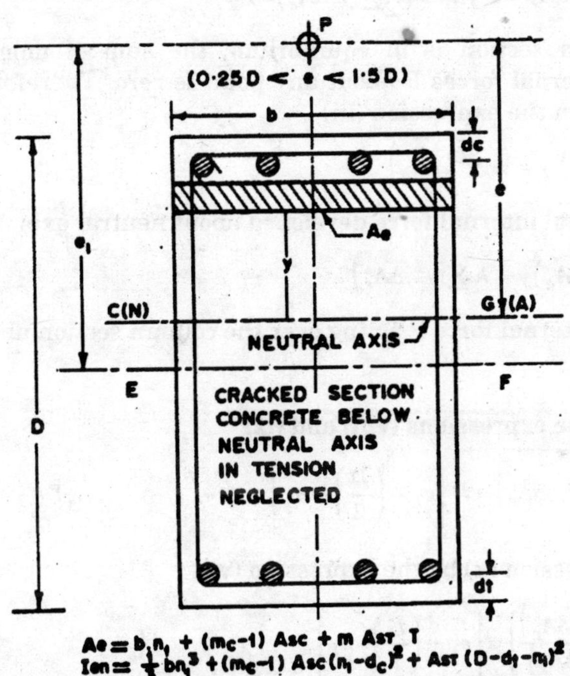

Fig. 21.9

From the theory of bending, it is known that the bending stress at any layer is linearly proportal to ts distance from the neutral axis. Therefore, the bending stress in the elementary layer of area ΔA_e.

$$\sigma_b \;\alpha\; y \qquad \text{(i)}$$

$$\sigma_c = k.y \qquad \text{(ii)}$$

The force in the elementary layer of area ΔA_e

$$\left(\sigma_b \,.\, \Delta A_e = k.y.\,\Delta A_e\right) \qquad \text{(iii)}$$

where k is a constant of proportionality.

Total internal force developed in the total area of column section

$$\left(\sum \sigma_b \,.\, \Delta A_e = \sum k \,.\, y.\, \Delta A_e\right) \tag{iv}$$

The column section is in equilibrium.

Therefore, the elementary force developed in the column section is equal to the external load P acting over the column. Therefore,

$$\left(\sum \sigma_b \; \Delta A_e = P\right) \tag{v}$$

From the expression (iv), then

$$\left(\sum k.y \,.\, \Delta A_e = P\right) and \left(k.\sum (y.\Delta A_e) = P\right) \tag{vi}$$

Since, the column section is in equilibrium, the sum of moment of internal forces developed and the external forces P about any point is zero. Therefore, the moment of force about neutral axis, from the expression (iii)

$$(\sigma_b.\Delta A_e).y = \left(k.y^2.\Delta A_e\right) \tag{vii}$$

The moment of total internal force developed about neutral axis

$$\sum(\sigma_b, \Delta A_e).y = k.\sum\left(y^2.\Delta A_e\right) \tag{viii}$$

The moment of external force P acting over the column section about neutral axis

$$= P.\,e \tag{ix}$$

Therefore, from the expressions (viii) and (ix)

$$k \,.\, \sum\left(y^2 \,.\Delta A_e\right) = P\left[e_1 - \left(\frac{D}{2}\right) + n_1\right] \tag{x}$$

Dividing the expression (x) by the expression (vi)

$$\left(\frac{\sum\left(y^2.\Delta A_e\right)}{\sum(y.\Delta A_e)}\right) = \left[e_1 - \left(\frac{D}{2}\right) + n_1\right] \tag{21.14}$$

Eq. 21.14 may be written as below

$$\left[e_1 - \left(\frac{D}{2}\right) + n_1\right] = \left(\frac{I_{en} \text{ about neutral axis}}{\text{Firat moment of the area about neutral axis}}\right)$$

The moment of inertia I_{en} of a cracked section of a column about the neutral axis

$$I_{.en} = \left[\frac{1}{2}bn_1^3 + (m_c - 1)\, A_{sc}.\,(n_1 + d_c)^2 + m.A_{st}\;(D - d_t - n_1)^2\right]$$

First moment of the equivalent area about the neutral axis

$$\sum y.\Delta A_e = \left[\frac{1}{2}\,bn_1^2 + (m_c - 1).\left(A_{sc}.\,(n_1 + d_c)^2 - m\,.\,A_{st}\;(D - d_t - n)\right)\right]$$

Therefore,

$$e = \left(\frac{I_{.en}}{\sum(y.\Delta A_e)}\right) = \left(\frac{\frac{1}{3}bn_1^3 + (m_c - 1).A_{sc}.(n_1 + d_c)^2 + m.A_{st}.(D - d_t - n_1)^2}{\frac{1}{2}bn_1^2 + (m_c - 1)A_{sc}.(n_1 + d_c) - m.A_{st}(D - d_t - n_1)}\right)$$

where, $I_{en} = \sum(y^2 . \Delta A_e)$ = Second moment of the area (viz., moment of inertia of equivalent area) about neutral axis

$\sum(y.\Delta A_e)$ = *First moment of equivalent area about neutral axis.*

Equ. 21.14 is a cubic equation and it determines the depth of neutral axis, n_1.

The bending stress σ_b may be found by considering expression (ii), from which

$$k = (\sigma_b/y) \qquad \text{(xi)}$$

Substituting the value of constant of proportionately, k from the expression (xi) in the expression (x),

$$\left(\frac{\sigma_b}{y}.\sum(y^2 . \Delta A_e)\right) = P.\left[e_1 - \left(\frac{D}{2}\right) + n_1\right]$$

$$\text{and} \left(\frac{\sigma_b}{y}.I_{en}\right) = P\left[-\left(\frac{D}{2}\right) + n_1\right] \qquad \text{(xii)}$$

Since $P.\left[e_1 - \left(\frac{D}{2}\right) + n_1\right]$ is the moment of external load P about the neutral axis, it may be put equal to M. Therefore, from the expression (xii).

$$(\sigma_b / y).I_{en} = M$$

$$\text{or } \sigma_b = (M.y/I_{.en}) \qquad (21.15)$$

Eq. 21.15 gives the bending stress at any layer at a distance y from the neutral axis. The bending stress at the top layer of column section at $y = y_{max} = n_1$, which have already been calculated in Eq. 21.14

$$\sigma_{cbc.cal} = (M.n_1 / I_{en}) \qquad (21.16)$$

Eq. 21.16 is a cubic equation. Trial and error method is used to solve this equation.

Example 21.8 A rectangular reinforced concrete column of size 450 mm x 600 mm carries a vertical load of 90 kN at an eccentric distance 920 mm from the middle line of the section as shown in Fig. 21.10. 4 bars of 20 mm diameter as compression reinforcement have been used with 50 mm effective cover from the outer forces. M 20 grade of concrete and HYSD steel bars Fe 415 have been provided. Find the stresses in concrete and steel.

Solution : M 20 grade of concrete has been used. The permissible stress in concrete in bending in compression from IS : 456-2000 is 7 N/mm².

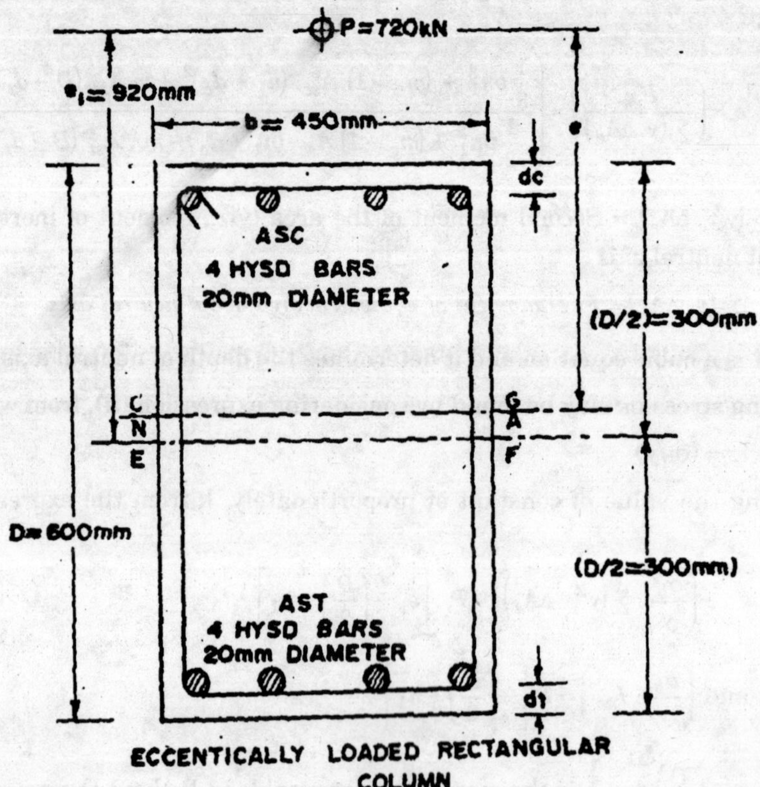

Fig. 21.10

Step 1 : Modular ratio of M 20 grade of concrete

$$m = \left(\frac{280}{3 \times 7}\right) = 13.33, \; and \; m_c = (1.5 \times 13.33) = 20 \tag{i}$$

Step 2 : Area of steel bars

Cross-sectional area of steel bars in compression

$$A_{sc} = \left(4 \times \frac{\pi}{4} \times 20^2\right) = 1256.54 \; mm^2$$

Cross-sectional area of steel bars in tension

$$A_{sc} = \left(4 \times \frac{\pi}{4} \times 20^2\right) = 1256.64 \; mm^2 \tag{ii}$$

Step 3 : Neutral axis for column section

The eccentricity of load P from the middle line e_1 = 920 mm is more than 1.5 D = (1.5 x 604) = 900 mm. The depth of neutral axis from the compression face of concrete may be determined by *appropriate method (viz.,* as if the column section is subjected to bending moment alone). The first moment of equivalent areas of concrete and compressive steel and tensile steel bars about the neutral axis are equal. The column section acts as cracked section. Therefore, the resistance of concrete in tension is neglected. Therefore,

$$\left[\frac{b.n.n}{2} + (m_c - 1)\,A_{sc}.\,(n - d_c)\right] = m\,A_{sc}\,(D - d_t - n)$$

$$\left[\frac{450n^2}{2} + (20-1)\,(1256.64).\,(n-50)\right] = 13.33 \times 1256.64 \times (600 - 50 - n)$$

$$\left(n^2 + 180.57\,n - 46252.73\right) = 0 \qquad \text{(iii)}$$

Solving this, n = 142.96 mm.

Step 4 : Eccentric distance of vertical load P from the neutral axis

$$e = \left[e_1 - \left(\frac{D}{2}\right) + n\right] = (920 - 300 + 142.96) = 762.96\ mm \qquad \text{(iv)}$$

Step 5 : Actual compressive bending stress in concrete

The column section is in equilibrium. Therefore, the moment of internal forces developed and the external force acting over the column about any point is zero. Therefore, take the moment of compressive forces in concrete and steel and the vertical load P about the centre-line of steel bars in tension

$$\left[C_c\,.\,(LA)_{.c} + C_{sc.}\,(AL)_{.sc}\right] = P\,.(LA)_{.p}$$

$$\left(\frac{b\,.\,n\,.\,\sigma_{cbc.cal}}{2}\right)\left(D - d_t - \frac{n}{3}\right) + (m_c - 1).\,A_{sc}$$

$$\times\left(\frac{n - d_c}{n}\right).\,\sigma_{cbc.cal}\,.(D - d_c - d_t) = P.\left[e_1 + \frac{D}{2} - d_t\right] \qquad \text{(vi)}$$

Substituting the numerical values

$$\left[\left(\frac{450 \times 142.29 \times \sigma_{cbc.cal}}{2}\right) \times \left(600 - 50 - \frac{142.96}{3}\right) + (20-1)\right.$$

$$\left. 1256.64 \times \left(\frac{142.96 - 50}{142.96}\right).\sigma_{cbc.cal}\,.\,(600 - 50 - 50)\right]$$

$$= \left[90 \times 10^3\ (920 + 300 - 50) + \left(16158.483 \times 10^3 + 7762.758 \times 10^3\right)\sigma_{cbc.cal} = 105300 \times 10^3\right]$$

$$\therefore \sigma_{cbc.cal} = \left(\frac{105300 \times 10^3}{23921.24 \times 10^3}\right) = 4.402\ N/mm^2 \qquad \text{(vii)}$$

Step 6 : Average compressive in column

Equivalent cross-sectional area of the column section

$$A_e = [b.D + (m_c - 1) . A_{sc} + m . A_{st}]$$

$$\text{or} \quad A_e = [450 \times 600 + (20 - 1) 1256.64 + 13.33 \times 1256.64] mm^2$$

$$\therefore A_e = [(270 \times 10^3 + 40.627 \times 10^3)] = 310.627 \times 10^3 \ mm^2 \qquad \text{(viii)}$$

Step 7 : Total compressive stress at the top face of concrete column

$$(4.402 + 0.289) = 4.462 \text{ N/mm}^2$$

Compressive stress in steel bars in compression

$$\sigma_{.sc} = m_c . \sigma_{cc.cal} + m_c \left(\frac{n-d}{n}\right) \sigma_{cbc.cal}$$

$$= \left[20 \times 0.579 + 20 \times \left(\frac{142.96 - 50}{142.96}\right) \times 4.402\right] = \mathbf{68.828} \ N/mm^2$$

Step 8 : Tensile stress in steel bars in tension

$$\sigma_{.st} = \left[m \ \sigma_{cbc.cal} \left(\frac{D - d_t - n}{n}\right) - m \ \sigma_{cc.cal}\right]$$

$$\left[13.33 \times 4.402 \times \left(\frac{600 - 50 - 142.96}{142.96}\right) - 13.33 \times 0.289\right] = 159.354 \ N/mm^2$$

Example 21.9. Find the stresses in concrete at the top and steel bars in compression and in tension for the column in Example 21.8 by accurate method.

Solution: In order to find the stresses in the various components of the column by accurate method, following geometrical properties of the column section are calculated.

Step 1 : First moment of equivalent areas of the column about the neutral axis of the column section

$$\Sigma(y \ \Delta A_e) = \left[\frac{bn_1^2}{2} + (m_x - 1) . A_{sc} . (n_1 - d_c) - m . A_{st} \ (D - d_t - n_1)\right] \qquad \text{(i)}$$

where n is the depth of neutral axis substituting the numerical values from Example 21.8 in the expression (i).

The column section acts as a cracked section. Therefore, portion of concrete in tension is neglected.

$$\Sigma(y \ \Delta A_e) = \left(\frac{450 \times n_1^2}{2}\right) + (20 - 1).1256.64 \ (n_1 - 50) - 13.33 \times 1256.64 \ (600 - 50 - n_1)$$

$$\text{or } \Sigma(y.\Delta A_e) = 225\, n_1^2 + 40627.17\, n_1 - 10406.864 \times 10^3 \quad \text{(ii)}$$

Step 2 : Moment of inertia of equivalent column section about the neutral axis, Fig. 21.9

(The column section acts as a cracked section).

$$I_{en} = \frac{1}{3} b n_1^3\, (m_c - 1)\, A_{sc} \,.\, (n_1 - d_c)^2 + m\, A_{st} \,.\, (D - d_t - n_1 - d_c)^2 \quad \text{(iii)}$$

Substituting the numerical values from *Example 21.8* in the expression (iii). Therefore,

$$I_{.en} = \frac{1}{3} \times 450 \times n_1^3 + (20 - 1)\, 1256.64\, (600 - 50 - n_1)^2$$

$$+ 13.33 \times 1256.64\, (600 - 50 - n)^2$$

$$I_{.en} = 150\, n_1^3 + 23876.16\, (n_1 - 50)^2 + 16751.01\, (550 - n_1)^2$$

$$I_{.en} = 150\, n_1^3 + 40.627 \times 10^3 \times n_1^2 - 20.813 \times 10^6 \times n_1 + 5126.87 \times 10^6 \quad \text{(iv)}$$

Step 3 : Distance of vertical load from the neutral axis, Fig. 21.9

$$e = \left(e_1 - \frac{D}{2} + n_1\right) = (920 - 300 + n_1) \quad \text{(v)}$$

The distance of vertical load from the neutral axis is given by Eq. 21.14

$$(620 + n_1) = \left(\frac{I_{en}}{\Sigma(y.\Delta A_e)}\right),$$

$$\text{or} \quad (620 + n_1) = 150.n_1^3 + 40.627 \times 10^3 \times n_1^2$$

$$+ \left(\frac{-20.813 \times 10^6 \times n_1 + 5126.87 \times 10^6}{225 n_1^2 + 406127.17 n_1 - 10406.864 \times 10^3}\right)$$

$$\text{or} \quad n_1 = 182.29 \text{ mm} \quad \text{(vi)}$$

Moment of inertia of equivalent column section about neutral axis by substituting value of n_1 expression (iv)

$$I_{.en} = 150 \times (182.29)^3 + 40.627 \times 10^3 \times (182.29)^2 - 20.813 \times 10^6$$

$$\times (182.29) + 5126.87 \times 10^6 = 3591.50 \times 10^6\ mm^4 \quad \text{(vii)}$$

The distance of vertical load P from the neutral axis

$$e = (e_1 - D/2 + n_1) = (920 - 300 + 182.29) = 802.29\ mm \quad \text{(viii)}$$

Moment of the vertical load P about the neutral axis

$$M = (P.e = 90 \times 0.80229) = 72.206\ kN - m \quad \text{(ix)}$$

Step 5 : Compressive stress in concrete at the top of concrete column (Fig. 21.10, from Eq. 21.16)

$$\sigma_{cbc.cal} = \left(\frac{72.206 \times 10^6 \times 182.29}{3591.50 \times 10^6}\right) = 3.665\ N/mm^2.$$

Step 6 : Compressive stress in the steel bars provided as reinforcement in compression

$$\sigma_{sc} = m_c \left(\frac{n - d_c}{n}\right) . \sigma_{cbc.cal} = 20 . \left(\frac{182.29 - 50}{182.29}\right) \times 3.665 = \mathbf{53.193}\ N/mm^2$$

Step 7 : Tensile stress in the steel bars provided as reinforcement in tension

$$\sigma_{st} = m \left(\frac{D - d_t - n_1}{n_1}\right) \times \sigma_{cbc.cal} = 13.33 \left(\frac{600 - 50 - 182.29}{182.29}\right) \times 3.665\ \text{N/mm}^2$$

$$= 98.548\ \text{N/mm}^2.$$

These stresses differ slightly than those calculated by approximate methods.

Example 21.10. In case the distance to the vertical load measured from the middle line of column in Example 21.8 is 720 mm, find the stresses at the top of the column section, steel reinforcement in compression and that in tension.

Solution : The column section in *Example 21.8* is shown in Fig. 21.10. Distance to the vertical load P, now given from the middle line

$e_1 = 720$ mm (i)

Overall depth of column section is D = 600 mm (ii)

0.25 D = 150 mm, and 1.5 D = 900 mm (iii)

$[0.25\ D < e_1 < 1.5\ D]$.

Step 1 : Depth of neutral axis

Therefore, it is the case of *moderate eccentricity* and may be analysed by accurate method only. The column section, is in equilibrium. Total sum of internal forces developed in the column section are equal to the load acting over the column. Therefore,

$$(C_c + C_{sc} - T = P) \quad \text{(iv)}$$

$$\left[\frac{b.n_1 . \sigma_{cbc.cal}}{2} + (m_c - 1)A_{sc} . \left(\frac{n_1 - d_c}{n_1}\right) . \sigma_{cb.cal} - mA_{st}\left(\frac{D - d_t - n}{n_1}\right)\sigma_{cbc.cal}\right] = P \quad \text{(v)}$$

The column section is in equilibrium. The sum of moment of internal forces developed about the centre line of tension reinforcement is equal to the moment of the vertical load P about the line. Therefore,

$$[C_c . (LA)_c + C_{sc} . (LA)_{sc}] = P . (LA)_{p.}$$

$$\frac{1}{2} b.n_1 \ \sigma_{cbc.cal} . [D - d_t - n_1/3] + (m_c - 1) \ A_{sc} . \left(\frac{n_1 - d_c}{n_1} \right),$$

$$\sigma_{cbc.cal} . [D - d_t - d_c] = P . [e_1 + D/2 - d_t] \quad \text{(vi)}$$

Dividing the expression (vi) by the expression (v)

$$\left(\frac{\frac{1}{2} b.n_1 \left[D - d_t - \frac{n_1}{3} \right] + (m_c - 1) A_{sc} . \left(\frac{n_1 - d_c}{n} \right) [D - d_t - d_c]}{\frac{1}{2} b.n_1 + (m_c - 1) A_{sc} . \left(\frac{n_1 - d_c}{n} \right) - m A_{st} . \left(\frac{D - d_t - n_1}{n_1} \right)} \right)$$

$$= \left[e_1 + \left(\frac{D}{2} \right) - dt \right] \quad \text{(vii)}$$

Substituting the numerical values from Example 21.8 in the expression (vii)

$$\left(\frac{\frac{1}{2} n_1 \left[600 - 50 - \frac{n_1}{3} \right] + (20 - 1)(1256.64) \left(\frac{n_1 - 50}{n_1} \right) [600 - 50 - 50]}{\frac{1}{2} \times 450 \times n_1 + (20 - 1) \ 1256.64 \left(\frac{n_1 - 50}{n_1} \right) - 13.33 \times 1256.64 \times \left(\frac{600 - 50 - n_1}{n_1} \right)} \right)$$

$$= [720 + 300 - 50] \quad \text{(viii)}$$

$$\therefore n_1 = 194.85 \ mm$$

Step 2 : Compressive stress at top of column in bending

Substituting the value of depth of neutral axis n1 in the expression (v)

$$\sigma_{cbc.cal} \left[\frac{1}{2} \times 450 \times 194.85 + 19 \times 1256.54 \times \left(\frac{194.85 - 50}{194.85} \right) \right.$$

$$\left. - 13.33 \times 1256.64 \times \left(\frac{600 - 50 - 194.85}{194.85} \right) \right] = 90 \times 10^3$$

Therefore, the compressive stress at top of column section

$$\sigma_{cbc.cal} = \mathbf{2.898} \ N/mm^2$$

Step 3 : Compressive stress in the steel bars provided as reinforcement in compression

$$\sigma_{sc} = 20 \times \left(\frac{194.85 - 50}{194.85} \right) \times 2.898 = \mathbf{43.09} \ N/mm^2$$

Step 4 : Tensile stress in the steel bars provided as reinforcement in tension

$$\sigma_{sc} = 13.33 \times \left(\frac{600 - 50 - 194.85}{194.85} \right) \times 2.898 = \mathbf{70.41} \ N/mm^2$$

Example 21.11. Find the stresses in concrete at the top of column section, steel bars provided as compressive reinforcement and steel bars provided as tensile reinforcement by calculating the moment of inertia of equivalent area of the column section and the first moment of equivalent area about the neutral axis in Example 21.10.

Solution : The column section of Example 21.10 is shown in Fig. 21.10.

Step 1 : First moment of equivalent area of the column section about the neutral axis

$$\Sigma(y \,.\, \Delta A_e) = \left[\frac{bn_1^2}{2} + (m_c - 1) \,.\, A_{sc} \,.\, (n_1 - d_c) - m \,.\, A_{st} \ (D - d_t - n_1) \right] \quad \text{(i)}$$

where, n_1 is the depth of neutral axis

Substituting the numerical values from Example 21.10

$$\Sigma(y \,.\, \Delta A_e) = \left[\left(\frac{450 \times n_1^2}{2} \right) + (20 - 1) \times 1256.64 \ (n_1 - 50) - 13.33 \times 1256.64 \ (600 - 50 - n_1) \right]$$

$$\text{or} \ \ \Sigma(y \,.\, \Delta A_e) = 225 \, n_1^2 + 40627.17 \, n_1 - 10406.864 \times 10^3 \quad \text{(ii)}$$

Step 2 : Moment of inertia of equivalent column section about the neutral axis

Fig. 21.10. (The column section acts as a cracked section).

$$I_{en} = \frac{1}{3} bn_1^3 + (m_c - 1) \, A_{st} \,.\, (n_1 - d_c)^2 + m \, A_{st} \ (D - d_t - n_1)^2 \quad \text{(iii)}$$

Substituting the numerical values from Example 21.10

$$I_{en} = \frac{1}{3} \times 450 \times n_1^3 + (20 - 1) 1256.64 . (n_1 - 50)^2 + 13.33 \times 1256.64 \ (600 - 50 - n_1)^2$$

$$I_{en} = 150 \, n_1^3 + 40.627 \times 10^3 \times n_1^2 - 20.813 \times 10^6 \times n_1 + 5126.87 \times 10^6 \quad \text{(iv)}$$

Step 3 : Distance of the vertical load from the neutral axis

$$e = \left(e_1 - \frac{D}{2} + n_1 \right) = (720 - 300 + n_1) = (420 + n_1) \quad \text{(v)}$$

The distance of the vertical load from the neutral axis is given by Fig. 21.14

$$(420 + n_1) = \left(\frac{I_{en}}{\Sigma y \,.\, \Delta A_e} \right) \quad \text{(vi)}$$

$$\text{or} \quad (420 + n_1) = \left(\frac{150.n_1^3 + 40.627 \times 10^3 \times n_1^2 - 20.813 \times 10^6 \times n_1 + 5126.87 \times 10^6}{225\, n_1^2 + 40627.17\, n_1 - 10406.864 \times 10^3} \right)$$

or $n_1 = 194.85$ (vii)

Substituting the value of depth of neutral axis n_1 in the expression (iii)

$$I_{en} = 150 \times (194.85)^3 + 40.627 \times 10^3 \times (194.85)^2 - 20.816 \times 10^6$$

$$\times (194.85) + 5126.87 \times 10^6 = 3723.50 \times 10^6 \ mm^4 \quad \text{(viii)}$$

The distance of the vertical load P from the neutral axis

$$e = (e_1 - D/2 + n_1) = (720 - 300 + 194.85) = 614.85 \ mm \quad \text{(ix)}$$

Moment of the vertical load P about the neutral axis

$$M = P.e = (90 \times 0.61485) = 55.3365 \ kN - m \quad \text{(x)}$$

Step 4 : Compressive stress in the concrete at the top of concrete column (Fig. 21.10 from Eq. 21.16)

$$\sigma_{cbc.cal} = \left(\frac{55.3365 \times 10^6 \times 194.85}{3723 \times 10^6} \right) = \mathbf{2.896} \ N/mm^2$$

Step 5 : Compressive stress in the steel bars provided as reinforcement in compressive

$$\sigma_{sc} = 20 \times \left(\frac{194.85 - 50}{194.85} \right) \times 2.896 = \mathbf{43.06} \ N/mm^2$$

Step 6 : Tensile stress in the steel bars provided as reinforcement in Tension

$$\sigma_{st} = 13.33 \left(\frac{600 - 50 - 194.85}{194.85} \right) \times 2.896 = 70.36 \ N/mm^2$$

It is to note that the stress calculated in *Example 21.10* and *Example 21.11* are almost equal as both the methods are alternative to each other method.

21.9 ECCENTRICALLY LOADED CIRCULAR COLUMNS (SPIRALLY REINFORCED)

Spirally reinforced circular columns are highly ductile. These columns have the advantage whether axially loaded or uniaxially eccentrically loaded so long as biaxial bending does not occur. The calculations involved in the analysis and design of columns are burden-some but easily manageable. The main reinforcement is dealt with the concept of *equivalent ring* to discuss the elastic behaviour of the columns. The circular columns are widely used in the buildings and bridges as these are efficient in the use of formwork and have elegant appearance. The minimum number of steel bars provided as main reinforcement is 6 (six). However, 8, 10, 12, 16 and more number of steel bars are also commonly used. The use of 8 steel bars in actual cross-section of the circular column is shown in Fig. 21.11(A). These steel bars are assumed as consisting of a thin circular shell of equivalent thickness as shown in Fig. 21.11 (B).

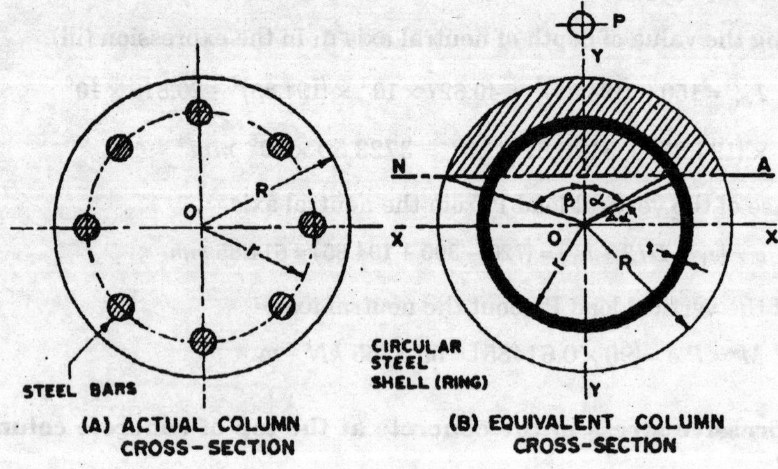

Fig. 21.11

The radius of circular column is R and the radius to the centre of reinforcement circle is *r*. The angle subtended by the reinforcement ring is α and the angle subtended by the concrete at the neutral axis is β. The equivalent thickness of circular shell is *t*. Consider as elementary layer of thickness d_y at a distance *y* from the neutral axis as shown in Fig. 21.12 (A). The generalised angle for the elementary layer being considered is θ. The *force-method* is used for the analysis of spirally reinforced circular column subjected to the vertical load P at an eccentric distance, *e* from the neutral axis long yy-axis. Fig. 21.12 (B) shows the combined stress distribution diagram. The area of n steel bars of diameter ϕ provided as main reinforcement

$$= \left(n \cdot \frac{\pi}{4} \phi^2\right) \quad \text{(i)}$$

The area of circular steel shell of equivalent thickness, *t*

$$= (2\,\pi r.t) \quad \text{(ii)}$$

$$(2\,\pi\, r.t) = \left(n \cdot \frac{\pi}{4} \cdot \phi^2\right) \quad \text{(iii)}$$

$$\text{Percentage of steel} = \left(\frac{2\pi\, r\, t}{\pi\, R^2}\right) \times 100 \quad \text{(iv)}$$

The column section is in equilibrium. Therefore, the internal forces developed in the column section are equal to the external force P acting over the column section. The concrete above neutral axis is in compression and that below the neutral axis is in tension. The column section acts as *cracked section.* The resistance of concrete in tension is neglected. Then

$$(C_c + C_{sc} - T) = P \tag{v}$$

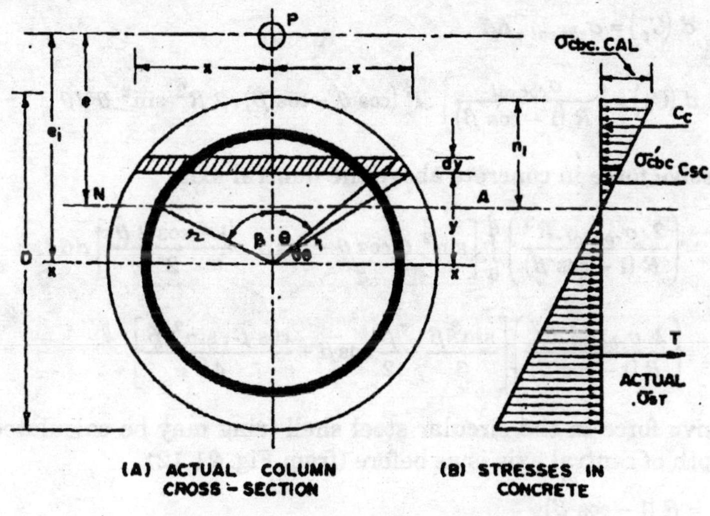

Fig. 21.12

where C_c = compressive force in the concrete section above neutral axis

C_{sc} = compressive force in the steel reinforcement in compression

T = tensile force in the reinforcement in tension.

The width of elementary layer of thickness dy at a distance y from the neutral axis, Fig. 21.12 (B).

$$2\,x = 2\,R \sin\theta \tag{vi}$$

The distance of elementary layer from the centre-line.

$$y = R\,.\cos\theta \tag{vii}$$

The thickness of elementary layer

$$dy = d\,(R\cos\theta)$$

$$\text{or} \quad dy = R\,(\sin\theta\,.\,d\theta) \tag{viii}$$

The negative sign is neglected. The area of elementary layer

$$\Delta A = (width \times thickness) = 2R^2 \sin^2\theta\, d\theta \tag{ix}$$

Let the intensity of compressive stress at the top be $\sigma_{cbc.cal}$ and that at the level of elementary layer be $\sigma'_{cbc.cal}$.

$$\sigma'_{cbc.cal} = \sigma_{cbc.cal}\left(\frac{y - R\cos\beta}{n_1}\right) = \sigma'_{cbc.cal}\cdot\left(\frac{R\,(\cos\theta - \cos\beta)}{R\,(1 - \cos\theta)}\right) \tag{x}$$

The compressive force on this elementary layer

$$d\left(C_c\right) = \sigma'_{cbc.cal} \cdot \Delta A \quad \text{(xi)}$$

$$\text{or} \quad d\left(C_c\right) = \left(\frac{\sigma_{cbc.cal}}{R\left(1-\cos\beta\right)}\right) . R\left(\cos\theta - \cos\beta\right) . 2R^2 \sin^2\theta \, d\theta$$

Total compressive force in concrete above the neutral axis

$$C_c = \left(\frac{2.\,\sigma_{cbc.cal} . R^3}{R\left(1-\cos\beta\right)}\right)\int_0^\beta \left[\sin^2\theta\cos\theta - \cos\beta . \frac{\left(1-\cos 2\theta\right)}{2}\right] d\theta \quad \text{(xii)}$$

$$C_c = \left(\frac{2.\,\sigma_{cbc.cal} . R^2}{R\left(1-\cos\beta\right)}\right)\left[\frac{\sin^3\beta}{3} - \frac{\beta}{2}\cos\beta + \frac{\cos\beta . \sin^2\beta}{4}\right] \quad \text{(21.16)}$$

The compressive force in the circular steel shell (ring may be calculated by considering Fig. 21.13. The depth of central axis is as before (from Fig. 21.12).

$$n_1 = R\left(1-\cos\beta\right) \quad \text{(i)}$$

From Fig. 21.12 (A) and Fig. 21.13 (A)

$$R\cos\beta = r\cos\alpha \quad \text{(ii)}$$

$$n_1 = R\left(1-\cos\beta\right) = \left(R - \cos\alpha\right)$$

The area of small element of circular steel shell (ring)

$$\left(\Delta A_s = r\, d\theta . t\right) \quad \text{(iii)}$$

The compressive stress at the level of this element of steel shell (ring)

$$\sigma_{sc} = \left(m_c - 1\right)\left[\frac{\left(r\cos\theta - r\cos\alpha\right)}{\left(R - r\cos\alpha\right)}\right]\sigma_{cbc.cal} \quad \text{(iv)}$$

Compressive force in this element of steel shell (ring)

$$d\left(C_c\right) = \sigma_{sc} . \Delta A_s$$

$$\text{or} \quad d\left(C_{sc}\right) = \left(\frac{r^2\, d\theta . t . \sigma_{cbc.cal}}{\left(R - r\cos\alpha\right)}\right)\left(\cos\theta - \cos\alpha\right)\left(m_c - 1\right)$$

Therefore, total compressive force in steel shell (ring)

$$C_{sc} = 2\,\sigma_{cbc.cal}\int_0^\alpha \frac{r^2 . t}{\left(R - r\cos\alpha\right)}\left(\cos\theta - \cos\alpha\right)\left(m_c - 1\right) d\theta$$

$$\text{or} \quad C_{sc} = \left(\frac{2\,\sigma_{cbc.cal}\left(m_c - 1\right)}{\left(R - r\cos\alpha\right)}\right)\left(\sin\alpha - \alpha\cos\alpha\right) r^2 . t \quad \text{(21.17)}$$

Tensile force in the steel shell (ring) below the neutral axis

$$T = \left(\frac{2\sigma_{cbc.cal} \cdot m.r^2 t}{(R - r \cos \alpha)} \right) [\sin \alpha + (\pi - \alpha) \cos \alpha] \quad (21.18)$$

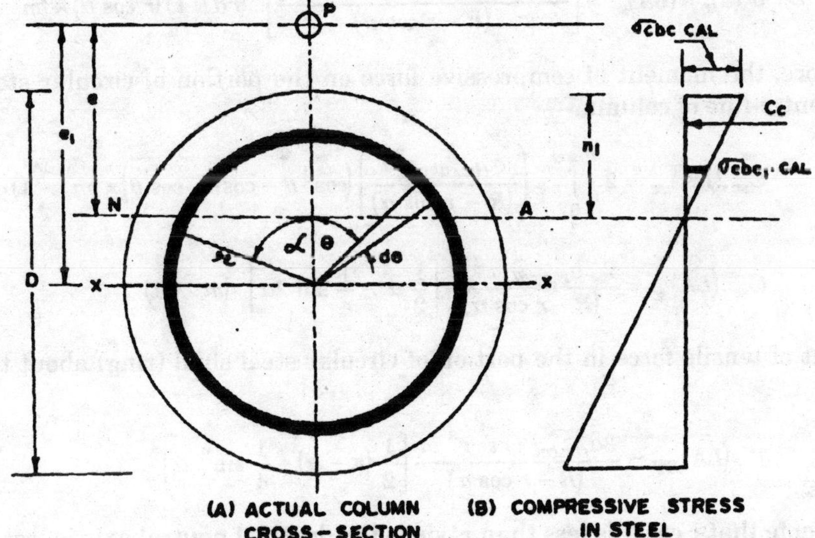

Fig. 21.13

Since the column section is in equilibrium, the sum of moment of internal forces developed in the column and external force P about any line is zero. The moment of internal forces and the external load about xx-axis.

$$C.(LA)_c + C_{.sc} + T.(LA)_{st} = P.e \quad (21.19)$$

where, e1 is the distance of vertical load from xx-axis. The moment of compressive force about the centre-line of column.

$$C_c.(LA)_c = \int_0^{\beta} Comp.\ \ force\ in\ ele.\ layer \times R \cos \theta$$

$$= \frac{2\,\sigma_{cbc.cal} \cdot R^3}{(1 - \cos \beta)} \int_0^{\beta} \left(\cos^2 \beta \sin^2 \theta - \cos \beta \,.\, \sin^2 \theta \,.\, \cos \theta\right) d\theta$$

$$= \frac{2\,\sigma_{cbc.cal} \cdot R^3}{(1 - \cos \beta)} \int_0^{\beta} \left[\frac{1}{4} (\sin 2\theta) - \cos \beta \cos \theta \sin^2 \theta\right] d\theta$$

$$= \frac{2\,\sigma_{cbc.cal} \cdot R^3}{(1 - \cos \beta)} \int_0^{\beta} \left[\frac{(1 - \cos 4\,\theta)}{8} - \cos \beta \,.\, \cos \theta \,.\, \sin^2 \theta\right] d\theta$$

$$C_c.(LA)_c = \frac{2\,\sigma_{cbc.cal} \cdot R^3}{(1 - \cos \beta)} \left[\frac{1}{8}\beta - \frac{1}{32} \sin 4\,\beta - \frac{1}{3} \cos \beta \sin^3 \beta\right] \quad (21.20)$$

Moment of the compressive force on the element of circular steel shell (ring) Fig. 21.13 about the centre-line of column.

$$d\,(C_{sc}).\,(LA)_{sc} = \left[\frac{2\sigma_{cbc.cal}.\,(r\cos\theta - r\cos\alpha)}{(R - r\cos\alpha)}\right].\,(r\,d\theta.\,t)\,(r\cos\theta)\times(m_c - 1)$$

Therefore, the moment of compressive force on the portion of circular steel shell (ring) about the centre-line of column.

$$C_{sc}.(LA)_{.sc} = 2\int_0^{\theta=\alpha}\left(\frac{\sigma_{cbc.cal}.\,r^3.t}{(R - r\cos\alpha)}\right)(\cos^2\theta - \cos\alpha.\cos\theta)\times(m_c - 1)\,d\theta$$

$$C_{sc}.(LA)_{.sc} = \frac{2.\,\sigma_{cbc.cal}.\,r^3.t}{(R - r\cos\alpha)}\left[\frac{1}{2}\alpha - \frac{1}{4}\sin 2\alpha\right].\,(m_c - 1) \qquad (21.21)$$

Moment of tensile force in the portion of circular steel shell (ring) about the centre-line of column

$$T.\,(LA)_{ST} = \frac{2\alpha_{cbc.cal}.\,m.\,r^3.\,t}{(R - r\cos a)}\left[\frac{1}{2}(\pi - \alpha) + \frac{1}{4}\sin^2\alpha\right] \qquad (21.22)$$

It is to note that r cos α is less than r, since the depth of neutral axis is less than r. Other positions of neutral axis may also be considered. Constantly it is checked that the behaviour of column is controlled by concrete or steel.

Example 21.12 A reinforced circular column of 600 mm diameter is reinforced with 8 Hysd steel bars of 22 mm at an effective cover of 50 mm equally distributed along the circumference. The column reinforcement is tied with 10 mm diameter spirals. The grade of concrete is M 20. The column is subjected to an axist load of 300 kN and a bending moment of 55.50 kN-m. Find the stress in concrete at the top and in steel.

Solution : The permissible stress in M 20 grade of concrete in bending in compression from IS : 456-2000 is 7 N/mm².

Step 1 : Modular ratio for concrete

$$m = \left(\frac{280}{3\times 7}\right) = 13.33 \text{ and } m_c = (1.5\,m) = 20 \qquad \text{(i)}$$

Step 2 : Thickness of equivalent steel shell ring

Let the thickness of equivalent steel shall be t. Then,

$$2\pi.\,r.\,t = 8\times\frac{\pi}{4}\times\phi^2,\ \therefore\ t = \left(\frac{8\times 22^2}{2\times 4\times 250}\right) = 1.936\text{ mm}$$

Step 3 : Eccentricity of load

where, the radius of steel shell (ring) r = 250 mm. The eccentricity of load P = 300 kN measured from centre line of column.

$$e_1 = \left(\frac{108 \times 1000}{300}\right) = 360\ mm$$

The reinforced circular column with 8 - Hysd steel bars is shown in Fig. 21.14 with equivalent steel shell (ring) of thickness *t*. Consider an elementary layer of thickness dy at a distance y from the neutral axis. The generalised angle for the elementary layer being considered as θ. The force-method is used for the analysis of spirally reinforced column subjected to the vertical load P = 300 kN at an eccentric distance, e from the neutral axis. The column section is in equilibrium. Therefore, the internal forces developed in the column section are equal to the external load P. The concrete above the neutral axis in compression and that below the neutral axis is in tension. The concrete section acts as a *cracked section*. The resistance of concrete in tension is neglected. Therefore,

$$(C_c + C_{sc} - T) = P \qquad \text{(iii)}$$

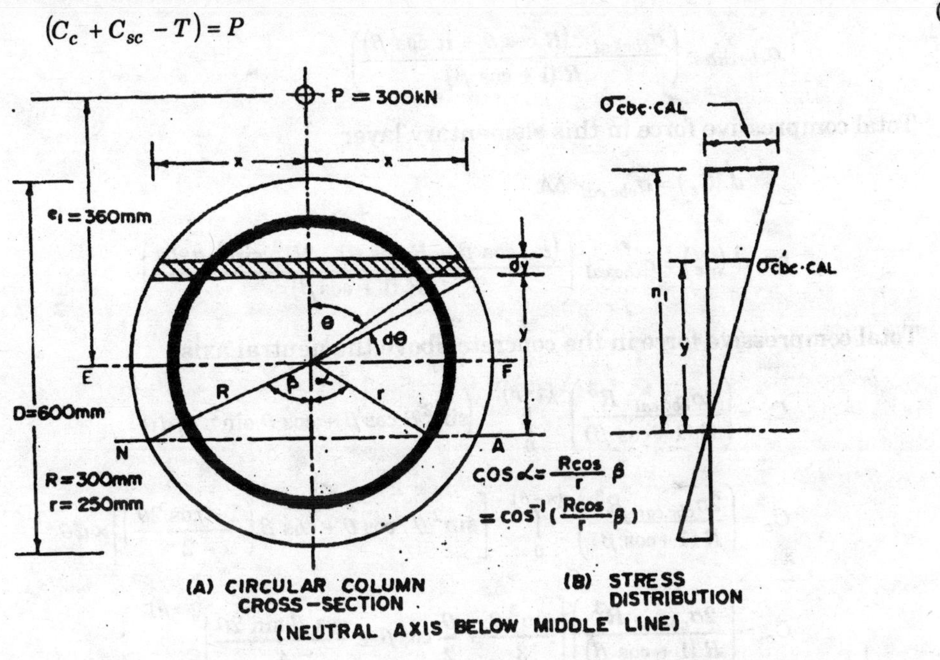

Fig. 21.14

where, C_c is the compressive force in the concrete section above the neutral axis, C_{sc} is the compressive force in the steel bars in compression and T is the tensile force in the reinforcement in tension.

The width of elementary layer of thickness dy at a distance y from the neutral axis Fig. 21.14 (A).

$$2x = 2\,R \sin \theta \qquad \text{(iv)}$$

The distance of elementary layer from the centre-line is = R cos θ

Thickness of elmentary layer

$$dy = d\,(R \cos \theta)\, dy = R \sin \theta\, d\theta \qquad \text{(v)}$$

The negative sign is neglected.

The area of elementary layer

$$\Delta A = (width \times thickness) = \left(2R^2 \sin^2 \theta d\theta\right) \quad \text{(vi)}$$

The distance of elementary layer from the neutral axis

$$y = (R \cos \beta + R \cos \theta) \quad \text{(vii)}$$

The depth of neutral axis

$$n_1 = (R + R \cos \beta) \quad \text{(viii)}$$

Let the intensity of compressive stress at the top be $\sigma_{cbc.cal}$ and that at the level of elementary layer be $\sigma'_{cbc.cal}$.

$$\sigma'_{cbc.cal} = \left(\frac{\sigma_{cbc.cal} \cdot (R \cos \theta + R \cos \beta)}{R(1 + \cos \beta)}\right) \quad \text{(ix)}$$

Total compressive force in this elementary layer

$$d(C_c) = \sigma'_{cbc.cal} \; \Delta A$$

$$\text{or } d(C_c) = \sigma'_{cbc.cal} \left(\frac{(R \cdot \cos \beta + R \cos \theta),. \; 2R^2 \sin^2 \theta \, d\theta}{R(1 + \cos \beta)}\right)$$

Total compressive force in the concrete above the neutral axis

$$C_c = \left(\frac{2\sigma_{cbc.cal} \cdot R^3}{R(1 + \cos \beta)}\right) \int_0^{(\pi-\beta)} \left(\sin^2 \theta \cos \beta + \cos \theta \sin^2 \theta\right) d\theta$$

$$C_c = \left(\frac{2\sigma_{cbc.cal} \cdot R^3}{R(1 + \cos \beta)}\right) \int_0^{(\pi-\beta)} \left[\sin^2 \theta \cdot \cos \theta + \cos \beta \left(\frac{1 - \cos 2\theta}{2}\right)\right] \times d\theta$$

$$C_c = \left(\frac{2\sigma_{cbc.cal} \cdot R^3}{R(1 + \cos \beta)}\right) \left[\frac{\sin^3 \theta}{3} + \frac{\theta}{2} \cos \beta - \frac{\cos \beta \sin 2\theta}{4}\right]_0^{(\pi-\beta)}$$

$$C_c = \left(\frac{2R^2 \sigma_{cbc.cal}}{(1 + \cos \beta)}\right) \left[\frac{\sin^3(\pi - \beta)}{3} + \frac{(\pi - \beta)}{2} \cos \beta - \frac{\cos \beta}{4} \cdot \sin(2\pi - 2\beta)\right]$$

$$C_c = \left(\frac{2R^2 \sigma_{cbc.cal}}{(1 + \cos \beta)}\right) \left[\frac{\sin^3 \beta}{3} + \frac{(\pi - \beta)}{2} \cos \beta - \frac{\cos \beta}{4} \sin 2\beta\right] \quad \text{(21.23)}$$

Let the depth of neutral axis be 0.6 D. Then

$$n_1 = (0.6 \times 600) = 360 \; mm$$

Radius of column section and steel shell ring

R = 300 mm, r = 250 mm

$$n_1 = (R + R\cos\beta) = R(1 + \cos\beta)$$

$$\left(\frac{n_1}{R} - 1\right) = \cos\beta, \textit{Therefore, } \beta = \cos^{-1}\left(\frac{n_1}{R} - 1\right)$$

$$\beta = \cos^{-1}\left(\frac{360}{300} - 1\right) = \cos^{-1}(0.2)$$

$$\beta = 78.46° = \left(\frac{78.46 \times \pi}{180}\right) = 1.369 \textit{ radians.}$$

$\sin\beta = 0.9798, \cos\beta = 0.2$

$\sin 2\beta = 0.3919, \sin 4\beta = 0.7112$

$$\cos\alpha = \left(\frac{R\cos\beta}{r}\right), \cos\alpha = \left(\frac{R}{r}\cos 78.46°\right)$$

$$\cos^{-1}\alpha = \left(\frac{300}{250} \times 0.2\right) = \left(76.11° \times \frac{\pi}{180}\right) = 1.328 \textit{ radians}$$

$\sin\alpha = 0.9708, \cos\alpha = 0.2399, \sin 2\alpha = 0.4661$

Substituting these values in Eq. 21.13

$$C_c = \left(\frac{2 \times 300^2 \times \sigma_{cbc.cal}}{(1 + 0.2)}\right)\left[\frac{(0.9798)^3}{3} + \frac{(\pi - 1.369)}{2} \times 0.2 + \frac{0.2}{4} \times 0.3919\right]$$

$$C_c = \left(\frac{2 \times 300^2 \times 0.5104}{1.2}\right).\sigma_{cbc.cal} = 76558.93\ \sigma_{cbc.cal} \qquad \text{(i)}$$

The compressive force in the circular steel shell ring may be calculated by considering Fig. 21.15. The depth of neutral axis as before from Fig. 21.14.

$$n_1 = (R + R\cos\beta) \qquad \text{(ii)}$$

From Fig. 21.15(A)

$$R\cos\beta = r\cos\alpha \qquad \text{(iii)}$$

$$\therefore n_1 = (R + R\cos\beta) = (R + r\cos\alpha) \qquad \text{(iv)}$$

The area of small element of circular steel shell (ring)

$$\Delta A_s = r\,d\theta . t \qquad \text{(v)}$$

The compressive stress at the level of this element of steel shell (ring).

$$\sigma_{sc} = (m_c - 1)\left(\frac{(r\cos\theta + r\cos\alpha)}{(R + r\cos\alpha)}\right).\ \sigma_{cbc.cal} \qquad \text{(vi)}$$

The compressive force in this element of steel shell ring.

$$d(C_{.sc}) = \sigma_{sc}\ .\ \Delta A_s$$

$$\text{or } d\left(C_{sc}\right)=\left(\frac{r^2\, d\theta\, .\, t\ \ \sigma_{cbc.cal}}{\left(R+r\cos\alpha\right)}\right)\left(\cos\theta+\cos\alpha\right)\left(m_c-1\right)$$

Therefore the total compressive force in steel shell (ring) also

$$C_c=\left(\frac{2r^2t\left(m_c-1\right)\sigma_{cbc.cal}}{\left(R+r\cos\alpha\right)}\right)\int_0^{(\pi-\alpha)}\left(\cos\theta+\cos\alpha\right)d\theta$$

$$C_c=\left(\frac{2r^2t\left(m_c-1\right)\sigma_{cbc.cal}}{\left(R+r\cos\alpha\right)}\right)\left[\sin\theta+\theta\cos\alpha\right]_0^{(\pi-\alpha)}$$

$$C_c=\left(\frac{2r^2t\left(m_c-1\right)\sigma_{cbc.cal}}{\left(R+r\cos\alpha\right)}\right)\left[\sin\alpha+\left(\pi-\alpha\right)\cos\alpha\right] \quad (21.24)$$

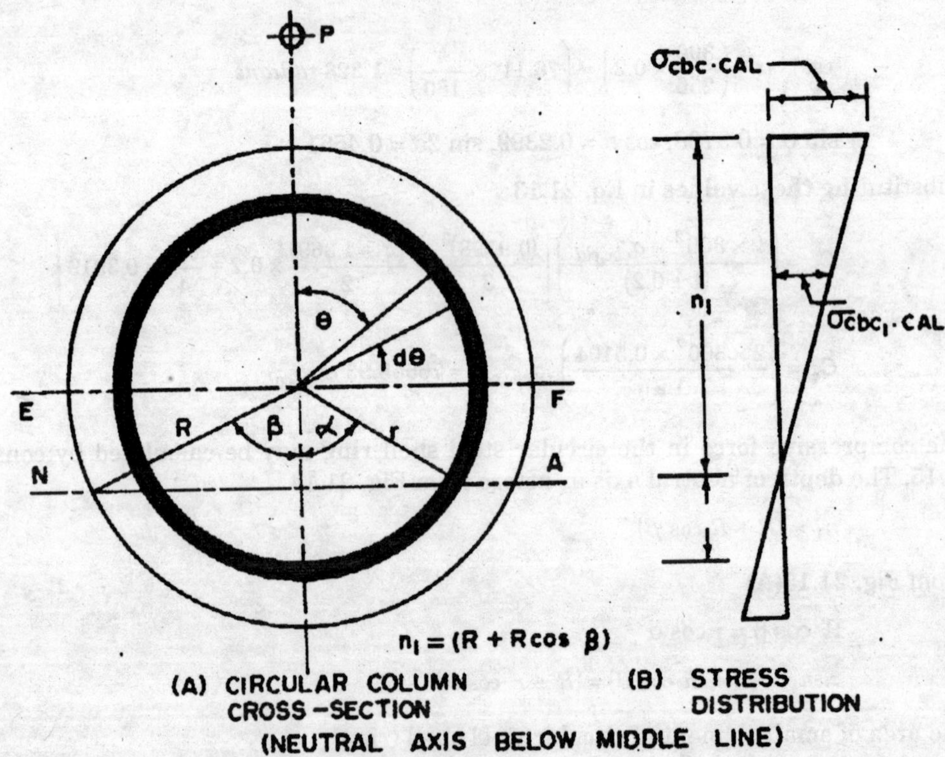

Fig. 21.15

Substituting the numerical values in Eq. 21.24

$$C_{sc}=\left(\frac{2\times 250^2\times 1.936\,(20-1)\,\sigma_{cbc.cal}}{\left(300+280\times 0.2399\right)}\right)\left[0.9708+\left(\pi-1.328\right)\times 0.2399\right]$$

$$C_{sc}=\left(\frac{4598000}{347.98}\times 1.4059\right)\sigma_{cbc.cal}$$

$$C_{sc} = (13213.4 \times 1.4059) \times \sigma_{cbc.cal} = (18576.466)\ \sigma_{cbc.cal} \quad (2)$$

Total tension T in steel shell (ring) below the neutral axis

$$T = \left(\frac{2mr^2 t.\sigma_{cbc.cal}}{(R + r\cos\alpha)}\right) \int_0^{\alpha} (\cos\theta + \cos\alpha)\, d\theta$$

$$T = \left(\frac{2mr^2 t.\sigma_{cbc.cal}}{(R + r\cos\alpha)}\right) [\sin\theta - \theta\cos\alpha]_0^{\alpha}$$

$$T = \left(\frac{2mr^2 t.\sigma_{cbc.cal}}{(R + r\cos\alpha)}\right) [\sin\alpha - \alpha\cos\alpha]_0^{\alpha} \quad (21.25)$$

Substituting the numerical values in Eq. 21.25

$$T = \left(\frac{2 \times 13.33 \times 250^2 \times 1.936 \times \sigma_{cbc.cal}}{(300 + 250 \times 0.2399)}\right) [0.9708 - 1.328 \times 0.2399]$$

$$T = \left(\frac{3225860}{347.98} \times 0.6522\right) \times \sigma_{cbc.cal}$$

$$T = (9270.24 \times 0.6522) \times \sigma_{cbc.cal} = (6047.17)\ \sigma_{cbc.cal} \quad (3)$$

The column section is in equilibrium. Therefore,

$$(C_{.c} + C_{sc} - T) = P$$

$$\text{or } (76558.93 + 18576.466 - 6046.17)\ \sigma_{cbc.cal} = 300 \times 1000$$

$$89.09\ \sigma_{cbc.cal} = 300. \textit{ Therefore, } \sigma_{cbc.cal} = 3.367\ N/mm^2$$

Since, the column section is in equilibrium, the sum of moment of internal forces developed in the column and the external force P about any line is zero. The moment of internal forces and the external forces about the centre line (xx-axis).

$$C_c.(LA)_c + C_{sc}.(LA)_{.sc} + T.(LA)_{st} = P.e_1 \quad (21.26)$$

where, e_1 is the distance of vertical load from xx-axis. The moment of compressive force about the central line of column.

$$C_c.(LA)_c = \int_0^{(\pi-\beta)} \textit{compressive force in ele. layer} \times R\cos\theta$$

$$C_c.(LA)_c = \left(\frac{2R^3.\sigma_{cbc.cal}}{(1+\cos\beta)}\right) \int_0^{(\pi-\beta)} (\cos^2\theta\sin^2\theta + \cos\beta\sin^2\theta\cos\theta)\, d\theta$$

$$C_c.(LA)_c = \left(\frac{2R^3.\sigma_{cbc.cal}}{(1+\cos\beta)}\right) \left[\frac{1}{8}\left(\theta + \frac{\sin 4\theta}{4}\right) + \frac{\cos\beta.\sin^3\theta}{3}\right]_0^{(\pi-\beta)}$$

$$C_c.(LA)_c = \left(\frac{2R^3.\sigma_{cbc.cal}}{(1+\cos\beta)}\right)\left[\frac{1}{8}\left\{(\pi-\beta)+\frac{\sin^4(\pi-\beta)}{4}\right\}+\frac{1}{3}\cos\beta.\sin^3(\pi-\beta)\right]$$

$$C_c.(LA)_c = \left(\frac{2R^3.\sigma_{cbc.cal}}{(1+\cos\beta)}\right)\left[\frac{(\pi-\beta)}{8}+\frac{\sin^4\beta}{32}+\frac{1}{3}\cos\beta.\sin^3\right] \quad (21.27)$$

Substituting the numerical values in Eq. 21.27

$$C_c.(LA)_c = \left(\frac{2\times 300^3\times\sigma_{cbc.cal}}{(1+0.2)}\right)\left[\frac{(\pi-1.369)}{8}+\frac{(-0.7212)}{32}+\frac{1}{3}\times 0.2\times(0.9798)^2\right]$$

$$C_c.(LA)_c = \left(\frac{54\times 10^6.\sigma_{cbc.cal}}{1.2}\right)\times 0.2617 = \left(11.777\times 10^6\right)\sigma_{cbc.cal} \quad (4)$$

The moment of total compressive force in steel shell (ring) about the central line (xx-axis).

$$C_{sc}.(LA)_{sc} = \int_0^{(\pi=\alpha)} \frac{[(m_c-1)\,r\,d\theta\,.\,t\,.\,\sigma_{cbc.cal}]}{(R+r\cos\alpha)}\times(r\cos\theta+\cos\alpha).r\cos\theta$$

$$C_{sc}.(LA)_{sc} = \left(\frac{2\,.\,r^3\,.\,(m_c-1)\,.\,t\,.\,\sigma_{cbc.cal}}{R+r\cos\alpha}\right)\int_0^{(\pi-\alpha)}\left(\cos^2\theta+\cos\theta\,.\cos\alpha\right)d\theta$$

$$C_{sc}.(LA)_{sc} = \left(\frac{2\,.\,r^3\,.(m_c-1).\,t\,.\sigma_{cbc.cal}}{(R+r\cos\alpha)}\right)\int_0^{(\pi-\alpha)}\left[\frac{(1+\theta\cos 2\theta)}{2}+\cos\theta\cos\alpha\right]d\theta$$

$$C_{sc}.(LA)_{sc} = \left(\frac{2.r^3.(m_c-1).t.\sigma_{cbc.cal}}{(R+r\cos\alpha)}\right)\left[\frac{\theta}{2}+\frac{\sin 2\theta}{4}+\cos\alpha.\sin\theta\right]_0^{(\pi-\alpha)}$$

$$C_{sc}.(LA)_{sc} = \left(\frac{2.r^3.(m_c-1).t.\sigma_{cbc.cal}}{(R+r\cos\alpha)}\right)\left[\left(\frac{\pi-\alpha}{2}\right)+\frac{\sin 2(\pi-\alpha)}{4}+\cos\alpha.\sin(\pi-\alpha)\right]$$

$$C_{sc}.(LA)_{sc} = \left(\frac{2.r^3.(m_c-1).t.\sigma_{cbc.cal}}{(R+r\cos\alpha)}\right)\left[\frac{(\pi-\alpha)}{2}+\frac{\sin 2\alpha}{4}\right] \quad (21.28)$$

Substituting the numerical value in Eq. 21.28

$$C_{sc}.(LA)_{sc} = \left(\frac{2\times 250^2\times 1.936\times(20-1)\,\sigma_{cbc.cal}}{(300+250\times 0.2399)}\right)\times\left[\frac{(\pi-1.328)}{2}+\frac{0.4661}{4}\right]$$

$$C_{sc}.(LA)_{sc} = \left(\frac{1149.5\times 10^6}{347.98}\times 1.0233\right)\sigma_{cbc.cal} = 3.3804\times 10^6\,\sigma_{cbc.cal} \quad (5)$$

The moment of tensile force in the portion of circular steel shell (ring) about the centre line of column

$$T.(LA)_{ST} = \left(\frac{2\sigma_{cbc.cal}\ .\ m\ .\ r^3\ .\ t}{(R + r\ .\ \cos\alpha)}\right)\int_0^{\alpha}(\cos\theta - \cos\alpha)\ \cos\theta\ d\theta$$

$$T.(LA)_{ST} = \left(\frac{2m\ .\ r^3\ .\ t\ \sigma_{cbc.cal}}{(R + r\ .\ \cos\alpha)}\right)\int_0^{\alpha}\left(\cos^2\theta - \cos\theta\cos\alpha\right)d\theta$$

$$T.(LA)_{ST} = \left(\frac{2m\ .\ r^3\ .\ t\ \sigma_{cbc.cal}}{(R + r\ .\ \cos\alpha)}\right)\int_0^{\alpha}\left[\left(\frac{1 + \cos 2\theta}{2}\right) - \cos\theta\cos\alpha\right]d\theta$$

$$T.(LA)_{ST} = \left(\frac{2\,m\ .\ r^3\ .\ t\ \sigma_{cbc.cal}}{(R + r\ .\ \cos\alpha)}\right)\left[\frac{\theta}{2} + \frac{\sin^2\theta}{2} - \sin\theta\cos\alpha\right]_0^{\alpha}$$

$$T.(LA)_{ST} = \left(\frac{2r^3\ .\ m\ .\ t\ \sigma_{cbc.cal}}{(R + r\ .\ \cos\alpha)}\right)\left[\frac{\alpha}{2} - \frac{\sin 2\alpha}{4}\right] \tag{21.29}$$

Substituting the numerical values in Eq. 21.29

$$T.(LA)_{ST} = \left(\frac{2\times250^3\times13.33\times1.936\ \sigma_{cbc.cal}}{300+250\times023.99}\right)\times\left(\frac{1.328}{2} - \frac{.4661}{4}\right)$$

$$T.(LA)_{ST} = \left(\frac{806.465\times10^6}{347.98}\sigma_{cbc.cal}\times0.5475\right) = 1.269\ \sigma_{cbc.cal} \tag{6}$$

From Eq. 21.26

$$\left(C_c\ .\ (LA)_c + C_{sc}\ .\ (LA)_{sc} + T\ .\ (LA)_{ST} = P.e_1\right)$$

$$\left[11.777\times10^6 + 3.3804\times10^6 + 1.269\times10^6\right]\sigma_{cbc.cal} = 55.50\times10^6$$

$$16.426\ \sigma_{cbc.cal} = 55.50,\ \sigma_{cbc.cal} = 3.379\ N/mm^2$$

Two values of $\sigma_{cbc.cal}$ calculated from the equilibrium of forces and the equilibrium of moments are practically equal. Otherwise, other trial would have carried by assuming the depth of neutral axis different.

Compressive stress in concrete at the top of column

$$\sigma_{cbc.cal} = \left(\frac{3.367 + 3.379}{2}\right) = \mathbf{3.373}\ N/mm^2$$

From Fig. 21.15 (B)

$$\sigma_{st} = \left(\frac{m\times\sigma_{cbc.cal}}{n_1}\right)(d - n_1)$$

$$\sigma_{st} = \left(\frac{13.33\times3.373\ (600 - 50 - 360)}{360}\right) = \mathbf{23.73}\ N/mm^2$$

21.10 DISTRIBUTION OF STRESS OVER MEMBER SECTION [ECCENTRICITY OF TENSILE LOAD (LESS THAN 0.25 D)]

In reinforced concrete members used in axial tension, the concrete contributes very small to its strength, as the tensile strength of concrete is only a small friction of its compressive strength. As such, the reinforced concrete is not suitable if it is used as tension members. However, there are, situations where the reinforced concrete is stressed in axial tension (*for example,* in the tie rods as arches).

Both, concrete and steel bars behave elastically so long as the axial tensile load over the member is small enough for stress in the concrete to be considerably below its *tensile strength.* When the tensile load is increased, the concrete cracks across the entire cross-section at a stress and strain of the order of *one-tenth* of what it could sustain in compression. Then the concrete does not resist any part of the applied tensile load. At any load more tan that which caused concrete to crack, the steel bars resist the total tensile load. At this stage

$$\sigma_{st} \cdot A_{st} = T \qquad \text{(i)}$$

In circular water tanks, the water tightness is achieved and the hoop tension caused by the water pressure is prevented from causing the concrete to crack. In this case, the axial tensile strength is found from the following expression

$$T = \sigma_{st} \cdot (A_{st} + m\, A_{st}) \qquad \text{(ii)}$$

where, σ_{st} is the tensile stress, in concrete an appropriate fraction of tensile strength of concrete i.e., of that stress which would cause the concrete to crack.

The rectangular water tanks, bunkers and the silos are the examples where the reinforced concrete cross-section are subjected to *combined tension* and *bending*. Such cross-sections may be analysed in a manner similar to that of combined compression and bending but, the entire portion of concrete remains in tension and it is neglected. The combined tensile load and bending is resisted by the steel bars provided in two layers as A_{st1} and A_{st2}.

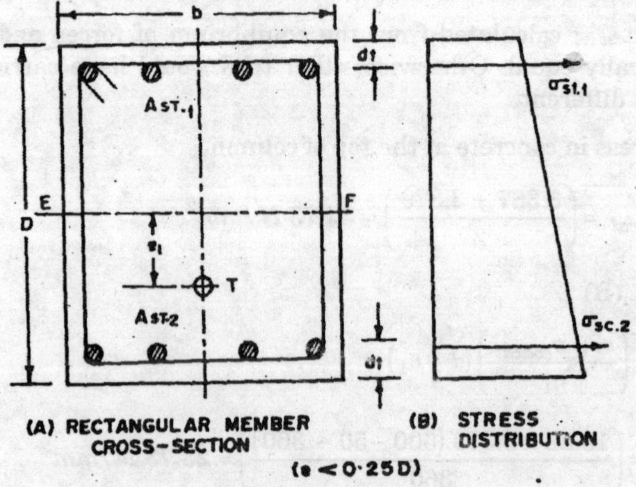

Fig. 21.16

A tensile load T acting at an eccentricity e_1 from the middle of a reinforced concrete rectangular section of size b x D is shown in Fig. 21.16. The eccentricity e_1 is considered small, as it is less than 0.25 D, the areas of the steel bars resisting smaller tension and larger tension are A_{st1} and A_{st2} respectively. The entire portion of concrete in tension is neglected. The cross-section of such a member is designed by trial and error. The dimensions of the member b and D and A_{st1} and A_{st2} are known. Then,

$$T_1 = \sigma_{st.1} \cdot A_{st.1} \quad \text{(i)}$$

$$\text{or } T_2 = \sigma_{st.2} \cdot A_{st.2} \quad \text{(ii)}$$

The member cross-section is in equilibrium. Therefore, for equilibrium of forces

$$(T_1 + T_2 = T) \quad \text{(iii)}$$

The tensile force T_1 may be found by taking moment about the line of action of tensile force, T_2. Then,

$$T_1.(D - 2d_t) = T \,.\, [(D/2) - e - D_t]$$

$$\therefore\ T_1 = T \,..\left(\frac{[(D/2) - e - d_t]}{(D - 2\,d_t^{'})}\right) \quad (21.29)$$

$$\sigma_{st.1} = (T_1 / A_{st.1}) \quad (21.30)$$

The tensile force T_2 may be found by taking moment about the line of action of tensile force, T_1. Then,

$$T_2 \,.\, (D - 2\,d_t) = T \,.\, [e + (D/2) - d_t]$$

$$T_2 = T.\ [e + D/2) - d_t] \quad (21.31)$$

$$\text{and} \quad \sigma_{st.2} = (T_2 / A_{st.2}) \quad (21.32)$$

The stress in steel bars provided as tension reinforcement near two faces are as shown in Fig. 21.13 (B).

Example 21.13. A reinforced concrete truss member 320 mm x 320 mm in cross-section carried a tensile force of 120 kN at an eccentric distance of 60 mm from the middle line. The effective cover at both the facres is 40 mm diameter Hysd steel have been provided at each face. Find the tensile stresses developed in the steel bars. The grade of concrete is M 20.

Solution: The overall dimensions of truss member is 320 mm x 320 mm. Therefore, the overall depth of section

$$D = 320 \text{ mm} \quad \text{(i)}$$

The effective c over at each face

$$d_t = 40 \text{ mm} \quad \text{(ii)}$$

Step 1 : Area of steel bars at each face

$$A_{st} = \left(4 \times \frac{\pi}{4} \times 20^2\right) = 1256.64\ mm^2 \quad \text{(iii)}$$

Step 2 : Tensile forces in member

Truss member is in equilibrium. Therefore,

$$(T_1 + T_2 = T) \tag{iv}$$

$$(\sigma_{st1} . A_{st1} + \sigma_{st2} A_{st2}) = 120 \times 10^3 \tag{v}$$

Take the moment about the centre of steel bars near top face

$$T_2 . (D - 2 d_t) = T [e + D/2 - d_t]$$

$$\therefore T_2 = \left(\frac{120 . [60 + 160 - 40]}{(320 - 2 \times 40)}\right) = 90 \; kN$$

$$T_1 = (120 - 90) = 30 \text{ kN}$$

Step 3 : Stress in steel bars

$$\sigma_{st1} = \left(\frac{30 \times 10^3}{1256.64}\right) = \mathbf{23.87} \; N/mm^2$$

and $$\sigma_{st2} = \left(\frac{90 \times 10^3}{1256.64}\right) = \mathbf{71.62} \; N/mm^2.$$

21.11 DISTRIBUTION OF STRESS OVER MEMBER SECTION (ECCENTRICITY OF TENSILE LOAD IS MORE THAN 1.5 D)

When the eccentricity of tensile load e_1 measured from the middle line is more than 1.5 D, it is considered large eccentricity. A portion of member cross-section remains in compression. The member cross-sectional may be analysed in the usual conventional manner.

The tensile load T acting at a distance e_1 from the middle of reinforced concrete rectangular member section of size b x D is shown in Fig. 21.17(A). A part of member section carries compression. The distribution of direct tensile stress σ_{at} due to tensile force (viz., $\sigma_{at.cal}$ = T/A_e) is as shown in Fig. 21.17 (B).

The distribution of stress in concrete in bending in compression due to bending moment M = T.e (viz., $\sigma_{cbc.cal}$ = T.e. y_2/I_e) is shown in Fig. 21.17 (C). The neutral axis of the concrete section of column for bending moment alone passes through point O. The combined stress distribution diagram shall be as shown in Fig. 21.17 (D), in which the neutral axis passes through O_1. Since, the eccentricity e_1 is large (viz., e_1 > 1.5 D), the bending moment, M = T.e is also large. This large bending moment predominates and the distance (viz., the shift of neutral axis) OO_1 between the two positions of neutral axis in Fig. 21.17 (C) and Fig. 21.17 (D) is very small. Therefore, *as an approximation*, the shift in position of neutral axis may be neglected. The analysis of member section may then be carried on the basis of bending stress distribution diagram only [viz., as shown in Fig. 21.17 (C)].

The depth of neutral axis n may then be obtained by equating first moment of equivalent areas of concrete and the steel bars provided for compression and tension reinforcement. Therefore,

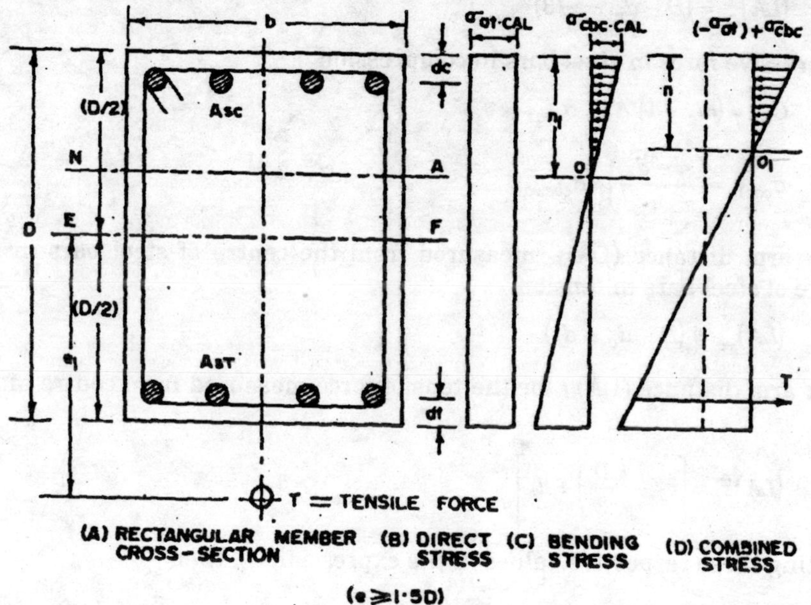

Fig. 21.17

$$\left[\frac{b.n^2}{2} + (m_c - 1)\, A_{st}\right] = m\ .\ A_{st}\ (d - n), \textit{ where} \tag{21.33}$$

A_{sc} = cross-sectional area of steel bars in compression

A_{st} = cross-sectional area of steel bars in tension

n = depth of neutral axis

m_c = modular ratio of concrete in compression (m_c = 1.5 m)

d_c = effective cover for the steel bars in compression

d_t = effective cover for the steel bars in tension.

The compressive stress at the top of concrete section, $\sigma_{cbc.cal}$ that at the bottom of concrete section $\sigma_{cbc.cal}$, the compressive stress in the steel bars in compression σ_{sc} and the tensile stress in the steel bars in tensile, σ_{st} may be determined by taking the moment of internal forces and the external forces about the centre of steel in tension. Therefore,

$$\left[C_c\ .\ (LA)_c + C_{.sc}\ .\ (LA)_{sc}\right] = T\ .\ (LA)_{st} \tag{i}$$

The compressive force in concrete in compression

$$C_c = \left[b\ .\ n\ .\ \frac{1}{2}\,(\sigma_{cbc.cal})\right] \tag{ii}$$

The lever arm distance $(LA)_{.c}$ measured from the centroid of triangular stress distribution upto the entire of steel bars in tension

$$(LA)_{.c} = (D - d_c - n/3) \quad \text{(iii)}$$

The compressive force in steel bars in compression

$$C_{sc} = (m_c - 1) A_{sc} . \sigma_{cbc1} \quad \text{(iv)}$$

$$\sigma_{cbc1} = \left(\frac{n - d_c}{n}\right) \times \sigma_{cbc.cal} \quad \text{(v)}$$

The lever arm distance $(LA)_{.sc}$ measured from the centre of steel bars in compression upto the centre of steel bars in tension

$$(LA)_{.sc} = (D - d_c - d_t) \quad \text{(vi)}$$

The lever arm distance $(LA)_{.T}$ for the tensile force measured from centre of steel bars in tension

$$(LA)_{.}T = \left[e_1 - \left(\frac{D}{2}\right) + d_t\right] \quad \text{(vii)}$$

Substituting these respective values in the expression (i), then

$$\left[b.n.\frac{\sigma_{cbc.cal}}{2} + (m_c - 1) A_{sc} \cdot \left(\frac{n - d_c}{n}\right) \sigma_{cbc.cal} \times (D - d_t - d_c)\right] = T.\left[e_1 - \left(\frac{D}{2}\right) + d_t\right] \quad (21.34)$$

The depth of neutral axis, n has already been calculated from Eq. 21.39. The solution of Eq. 21.34 gives the stress $\sigma_{cbc.cal}$. The stress in steel bars in compression

$$\sigma_{sc} = m_c \cdot \left(\frac{n - d_c}{n}\right) \sigma_{cbc.cal} \quad \text{(viii)}$$

The stress in steel bars in tension

$$\sigma_{st} = m \cdot \left(\frac{D - d_t - n}{n}\right) \cdot \sigma_{cbc.cal} \quad \text{(ix)}$$

The equivalent area of concrete section of the member

$$A_e = [b.n + (m_c - 1) . A_{sc} + m . A_{st}] \quad \text{(x)}$$

The calculated direct stress in tension

$$\sigma_{at.cal} = (T / A_1) \quad \text{(xi)}$$

The resultant stress at the top of concrete section, Fig. 21.17 (D)

$$= (\sigma_{cbc.cal} - \sigma_{at.cal}) \quad \text{(xii)}$$

The resultant stress in steel bars in compression

$$= (m_c \; \sigma_{cbc.cal} - \sigma_{at.cal}) \quad \text{(xiii)}$$

The distribution of stress over member section (when the eccentricity e_1 of the tensile force is more than 1.5 (D) is shown in Fig. 21.17 (D). Above analysis may be studied and done by accurate method (without any approximation) as described in the subsequent article.

21.12 DISTRIBUTION OF STRESS OVER MEMBER SECTION (ECCENTRICITY OF TENSILE FORCE MORE THAN 0.25 AND LESS THAN 1.5 D)

The tensile force, T acting at a distance e_1 from the middle axis of a reinforced concrete rectangular member section of size b x D is as shown in Fig. 21.18 (A).

When the eccentricity e_1 is more than 0.25 D and less than 1.5 D, it is considered as *moderate.* The member section is subjected to an axial tensile for T and moment M = T. e_1. A portion of the member cross-section remains in compression and remaining portion of the cross-section remains in Tension. The concrete section of member acts as *cracked section.* The tensile portion of concrete is neglected.

The distribution of direct tensile stress due to the tensile force T (viz., $\sigma_{at.cal} = T/A_e$) is as shown in Fig. 21.18 (B). The distribution of stress in concrete in bending stress in concrete due to bending moment M = T.e [(viz., $\sigma_{cbc.cal}$ = T.e n_1/I_e) is as shown in Fig. 21.18(C)]. The neutral axis of the member section for bending alone passes through O. The combined stress distribution diagram shall be as shown in Fig. 21.18(D), in which the neutral axis passes through O_1. The shift in neutral axis between two diagram Fig. 21.18(C) and (D) is appreciable and it cannot be neglected.

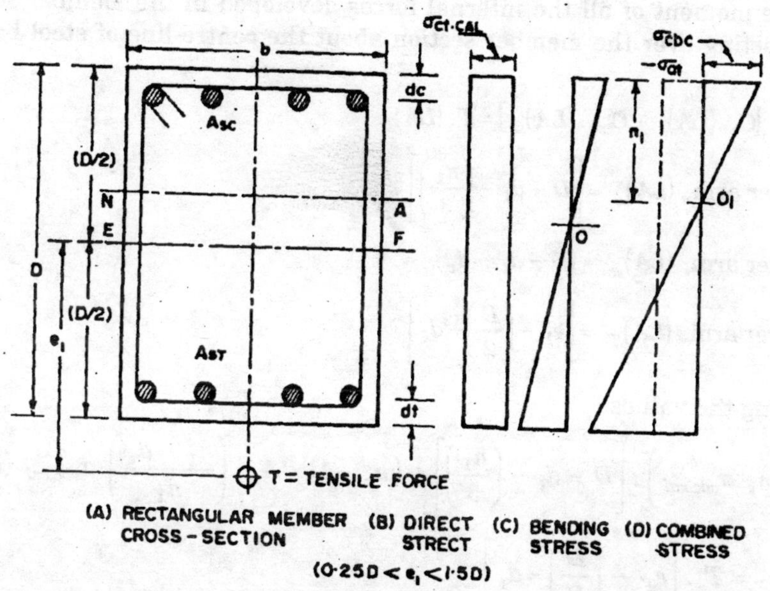

Fig. 21.18

The number section is in equilibrium. Therefore, total sum of internal forces developed in the member section is equal to the external tensile force, T acting over the member. Therefore,

$$(-C_c - C_{sc} + T) = T \qquad \text{(i)}$$

where, C_c is the compressive force in concrete in the portion of member section in compression, C_{sc} is the compressive force in the steel bars provided as reinforcement in compression and T is the tensile force in the steel bars provided as reinforcement in tension. The compressive force, C_c, in the portion of member section

$$C_c = \left(\frac{b.n_1.\sigma_{cbc.cal}}{2}\right) \quad \text{(ii)}$$

The compressive force $C_{.sc}$ in the steel bars in compression

$$C_{.sc} = (m_c - 1).\, A_{sc}.\left(\frac{n_1 - d_c}{n_1}\right)\sigma_{cbc.cal} \quad \text{(iii)}$$

The tensile force T in the steel bars in tension

$$T' = m\,.\, A_{st}.\left(\frac{D - d_t - n_1}{n_1}\right).\,\sigma_{cbc.cal} \quad \text{(iv)}$$

Substituting these values in the expression (i)

$$\left[-\frac{1}{2}b.n_1\,.\,\sigma_{cbc.cal}\,(m_b - 1).\, A_{sc}.\left(\frac{n_1 - d_c}{n_1}\right)\times\sigma_{cbc.cal} + m\,A_{st}\left(\frac{D - d_t - n_1}{n_1}\right)\right] = T \quad (21.35)$$

Since the column section is in equilibrium, the moment of all the internal forces developed and the external force acting over the member about any point shall be zero. Hence, take the moment of all the internal forces developed in the member section and the external force acting over the member section about the centre-line of steel bars in tension. Therefore,

$$\left[C_c.\,(LA)_c + C_{.sc}.\,(LA)_{sc}\right] = T.\,(LA)_T \quad \text{(v)}$$

$$\text{The lever arm, } (LA)_{.c} = \left[D - d_t - \left(\frac{n_1}{3}\right)\right] \quad \text{(vi)}$$

$$\text{The lever arm, } (LA)_{sc} = (D - d_t - d_c) \quad \text{(vii)}$$

$$\text{The lever arm, } (LA)_{.T} = \left[e_1 - \left(\frac{D}{2}\right) + d_t\right] \quad \text{(viii)}$$

Substituting the values

$$\left[\left(\frac{1}{2}b.\, n_1\;\sigma_{cbc.cal}\right).\left[D - d_t - \left(\frac{n_1}{3}\right)\right] + (m_c - 1)\times A_{sc}.\left(\frac{n_1 - d_c}{n_1}\right).\sigma_{cbc.cal}\,(D - d_t - d_c)\right]$$

$$= T\,.\left[e_1 - \left(\frac{D}{2}\right) - d_t\right] \quad (21.36)$$

There are two unknown n_1 and $\sigma_{cbc.cal}$ in Eqs. 21.35 and 21.36. Eq. 21.36 may be divided by Eq. 21.35. Therefore,

$$\left(\frac{\frac{1}{2}bn_1.\left[D - d_t - \left(\frac{n_1}{3}\right)\right] + (m_c - 1).\,A_{sc}.\left(\frac{n_1 - d_c}{n_1}\right).[D - d_t - d_c]}{-\frac{1}{2}bn_1 - (m_c - 1).A_{sc}.\left(\frac{n_1 - d_c}{n_1}\right) + m.A_{st}.\left(\frac{D - d_t - n_1}{n_1}\right)}\right) = \left[e_1 - \left(\frac{D}{2}\right) + d_t\right] \quad (21.37)$$

Eq. 21.37 is a cubic equation having n_1 only as one unknown. The value of n_1 may be found by solving Eq. 21.37.

The value of stress $\sigma_{cbc.cal}$ may be calculated by solving either Eq. 21.35 or Eq. 21.36. The stress in steel bars provided as reinforcement in compression

$$\sigma_{sc.} = m_c.\left(\frac{n_1 - d_c}{n}\right).\sigma_{cbc.cal} \quad \text{(ix)}$$

The stress in steel bars provided as reinforcement in tension

$$\sigma_{st} = m.\left(\frac{D - d_c - n_1}{n_1}\right).\sigma_{cbc.cal} \quad \text{(x)}$$

This method is accurate and there is no approximation in analysis.

Example 21.14. A rectangular reinforced concrete member of size 450 mm x 600 mm in size carries a tensile force of 90 kN at an eccentric distance 920 mm (below) from the middle line of the section as shown in Fig. 21.19. 4 bars of 20 mm diameter as compression reinforcement and 4 bars of 20 mm diameter as tension reinforcement have been provided with 50 mm effective cover from the outer faces. M 20 grade of concrete and HYSD steel bars Fe 415 have been used. Find the stresses in concrete and steel.

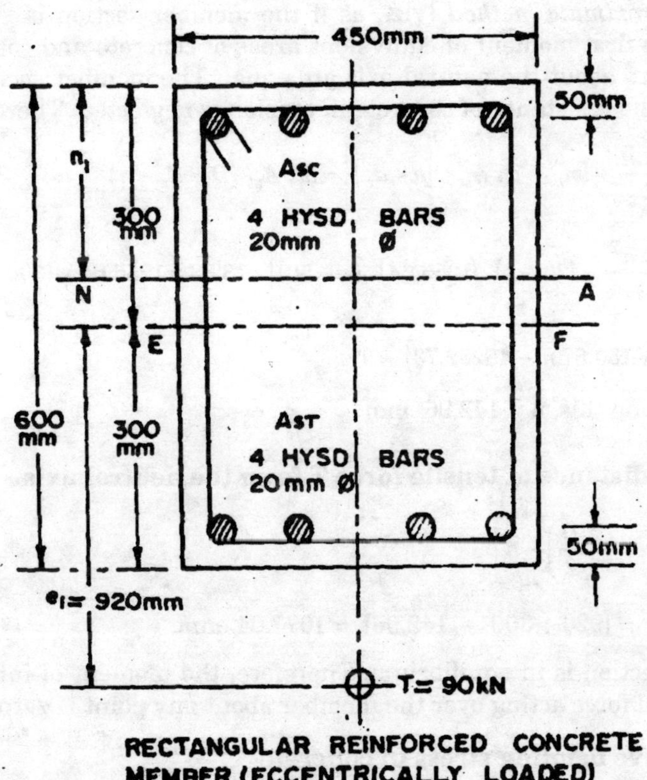

RECTANGULAR REINFORCED CONCRETE MEMBER (ECCENTRICALLY LOADED)

Fig. 21.19

Solution : M 20 grade of concrete has been used. The permissible stress in concrete in bending in compression from IS : 456-2000 is 7 N/mm². The modular ratio of M 20 grade of concrete

$$m = \left(\frac{280}{3 \times 7}\right) = 13.33 \text{ and } m_c = (1.5 \times 13.33) = 20$$

Step 1 : Cross-sectional area of steel bars in tension

$$A_{sc} = \left(4 \times \frac{\pi}{4} \times 20^2\right) = 1256.64 \ mm^2$$

Step 2 : Cross-sectional area of steel bars in tension

$$A_{st} = \left(4 \times \frac{\pi}{4} \times 20^2\right) = 1256.64 \ mm^2.$$

Step 3 : Neutral axis for member cross-section

The eccentricity of tensile force T from the middle e_1 = 920 mm; is more than 1.5 D = (1.5 x 600) = 900 mm. The depth of neutral axis from the compression face of concrete may be determined by *approximate method* (viz., as if the member section is subjected to bending moment alone). The first moment of equivalent areas of concrete and compressive steel bars and tensile steel bars about the neutral axis are equal. The member section acts as *cracked-section.* Therefore, the resistance of concrete in tension is neglected. Therefore,

$$\left[\frac{b.n.n}{2} + (m_c - 1).\ A_{sc}\ .\ (n - d_c)\right] = m.\ A_{st}\ (D - d_c - n)$$

$$\left[\frac{450\ n^2}{2} + (20 - 1).\ (1256.64).\ (n - 50)\right] = 13.33 \times 1256.64 \times (600 - 50 - n)$$

$$\left(n^2 + 180.57\ n - 46252.73\right) = 0 \qquad \text{(iii)}$$

Solving this, n = 142.96 mm

Step 4 : Eccentric distance of tensile force T from the neutral axis

$$e = \left[e_1 + \left(\frac{D}{2}\right) - n\right]$$

or e = [920 + 300 − 142.96] = 1077.04 mm.

The member section is in equilibrium. Therefore, the moment of internal forces developed and the external force acting over the member about any point is zero.

Step 5 : Compressive bending stress in concrete

Therefore, take the moment of compressive force in concrete and steel and the tensile force T about the centre-line of steel bars in tension

$$[C_c \,.\, (LA)_c + C_{sc} \,.\, (LA)_{sc} = T \,.\, (LA)_T]$$

$$\left[\frac{b.n.\ \sigma_{cbc.cal}}{2} \,.\left(D - d_t - \frac{n}{3}\right) + (m_c - 1) \,.\, A_{sc} \,.\left(\frac{n - d_c}{n}\right).\sigma_{cbc} \,.\, (D - d_c - d_t)\right]$$

$$= T \,.\left[e_1 - \left(\frac{D}{2}\right) + d_t\right] \quad \text{(vi)}$$

Substituting the numerical values

$$\left(\frac{450 \times 142.96 \times \sigma_{cbc.cal}}{2}\right) \times \left(600 - 50 - \frac{142.96}{3}\right) + (20 - 1)\ (1256.64)\,.\,.$$

$$\left(\frac{142.96 - 50}{142.96}\right).\ \sigma_{cbc.cal} \times (600 - 50 - 50) = 90 \times 10^3\ (920 - 300 + 50)$$

$$(16158.483 \times 10^3 + 7762.758 \times 10^3)\,\sigma_{cbc.cal} = 90 \times 10^3 \times 670$$

$$\sigma_{cbc.cal} = \left(\frac{90 \times 10^3 \times 670}{23921.24 \times 10^3}\right) = 2.521\ N/mm^2 \quad \text{(vii)}$$

Step 6 : Equivalent cross-section area of the member cross-section

$$A_c = b.D + (m_c - 1).\ A_{sc} + m.A_{st}$$

$$A_c = [450 \times 600 + (20 - 1)\,1256.64 + 13.33 \times 1256.64]$$

$$\therefore \quad A_c = 310.627 \times 10^3\ mm^2 \quad \text{(viii)}$$

Step 7 : Average tensile stress due to the tensile force acting over the member

$$\sigma_{at.cal} = \left(\frac{T}{A_c}\right) = \left(\frac{90 \times 10^3}{310.627 \times 10^3}\right) = 0.289\ N/mm^2$$

Total compressive stress at the top face of concrete member

$(2.521 - 0.289) = 2.232$ N/mm^2

Step 8 : Compressive stress in steel bars in compression

$$\sigma_{sc} = m \,.\, \sigma_{at.cal} + \left(m_c \left(\frac{n - d_c}{n}\right) \sigma_{cbc.cal}\right)$$

$$= (-13.33 \times 2.232) + 20 \times \left(\frac{142.96 - 50}{142.96}\right) \times 2.521$$

$$= (-30.968 + 32.785) = 1.816\ N/mm^2$$

Step 9 : Tensile stress in steel bars in tension

$$\sigma_{st} = m\ \sigma_{at.cal} + m\ \sigma_{cbc.cal}\left(\frac{D - d_t - n}{n}\right)$$

$$\sigma_{st} = \left[30.968 + (13.33 \times 4.402) \times \left(\frac{600 - 50 - 142.96}{142.96}\right)\right]$$

$= 198.04$ N/mm^2.

Example 21.15. In case the distance to the tensile force measured from the middle line in Example 21.14 is 720 mm (below), find the stresses at the top of member, steel reinforcement in compression and that in tension.

Solution : The member section in Example 21.14 is shown in Fig. 21.14. Distance to the tensile force T, now given from the middle line

$$e_1 = 720 \text{ mm} \tag{i}$$

Overall depth of member section

$$D = 600 \text{ mm} \tag{ii}$$

$$(0.25 \times D) = 150\ mm, and\ 1.5\ D = 900\ mm \tag{iii}$$

Step 1 : Depth of neutral axis

$[0.25\ D < e_1 < 1.5D]$ Therefore, it is the case of *moderate eccentricity* and may be analysed by accurate method only. The member section is in equilibrium. Total sum of internal forces developed in the member section are equal to the tensile action over the member. Therefore,

$$(C_c - C_{.sc} + T) = T \tag{iv}$$

$$\left[-\frac{b\ .\ n_1 . \sigma_{cbc.cal}}{2} - (m_c - 1)\ A_{sc} \cdot \left(\frac{n_1 - d_c}{n_1}\right) . \ \sigma_{cbc.cal} + m\ A_{st} \cdot \left(\frac{D - d_c - n_1}{n_1}\right) \sigma_{cbc.cal}\right] = T \tag{v}$$

The member section is in equilibrium. The sum of moment of internal forces developed about the centre-line of tension reinforcement is equal to the moment of tensile force T about the line. Therefore,

$$[C_c \cdot (LA)_{.c} + C_{.sc}\ (LA)_{.sc}] = T.\ (LA)_T$$

$$\left[\frac{1}{2} b.n_1\ \sigma_{cbc.cal} \cdot \left[D - d_t - \frac{n_1}{3}\right] + (m_c - 1)\ A_{sc} \cdot \left(\frac{n_1 - d_c}{n_1}\right) \sigma_{cbc.cal} \cdot (D - d_t - d_c)\right]$$

$$= T \cdot \left[e_1 - \frac{D}{2} + d_t\right] \tag{vi}$$

Dividing the expression (vi) by the expression (v)

$$\left(\frac{\frac{1}{2}.bn_1.\left[D-d_t-\left(\frac{n_1}{3}\right)\right]+(m_c-1).A_{sc}.\left(\frac{n_1-d_c}{n_1}\right).[D-d_t-d_c]}{-\frac{1}{2}b.n_1-(m_c-1)A_{sc}.\left(\frac{n_1-d_c}{n_1}\right)+m.A_{st}.\left(\frac{D-d_t-n_1}{n_1}\right)}\right)$$

$$=\left[e_1-\left(\frac{D}{2}\right)+d_t\right] \qquad \text{(vii)}$$

Substituting the numcrical values from *Example 21.14* in the expression (vii)

$$\left(\frac{\frac{1}{2}\times 450\times n_1\left[600-50-\frac{n_1}{3}\right]+(20-1)(1256.64)\left(\frac{n_1-50}{n_1}\right).[600-50-50]}{-\frac{1}{2}\times 450\times n_1-(20-n_1)1256.64\left(\frac{n_1-50}{1}\right)+13.33}\right.$$

$$\left.\times 1256.64\left[\frac{600-50-n_1}{n_1}\right]\right)=[720-300+50]$$

$= [720 - 300 + 50]$

$n_1 =$ 225.22 mm

Step 2 : Compressive bending stress in concrete

Substituting the value of depth of neutral axis in the expression (v)

$$\left[\frac{-450\times 225.22\times \sigma_{cbc.cal}}{2}-19\times 1256.64\left(\frac{225.22-50}{225.22}\right)\right.$$

$$\left.\times\,\sigma_{cbc.cal}+13.33\times 1256.64\times\left(\frac{50-22.22}{225.22}\right)\sigma_{cbc.cal}\right]=90\times 20^3$$

$$\sigma_{cbc.cal}=1.996\ N/mm^2\ (compression)$$

Step 3 : Stress in steel bars provided as reinforcement in compression

$$\sigma_{sc}=20\times 1.996\times\left(\frac{225.22-50}{225.22}\right)=31.058\ N/mm^2$$

Step 4 : Stress in steel bars provided as reinforcement in tension

$$\sigma_{st}=13.33\times 1.996\times\left(\frac{550-225.22}{225.22}\right)=38.368\ N/mm^2.$$

PROBLEMS

21.1 A reinforced concrete square column of size 400 mm x 400 mm carries a vertical load of 360 kN at an eccentricity e_1 equal to 80 mm as shown in Fig. 21.20. 4 Hysd steel bars of 20 mm diameter have been provided as longitudinal reinforcement. The grade of concrete is M 20. Find whether the column is safe or not. The effective cover for the steel the column is safe or not. The effective cover for the steel bars is 40 mm. Also find whether the column section acts as uncracked or cracked.

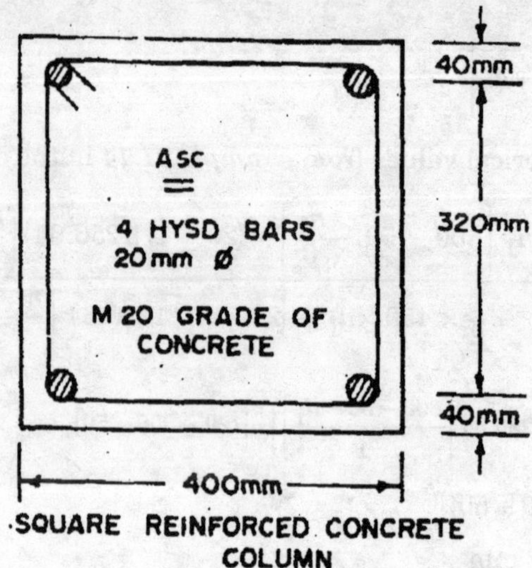

Fig. 21.20

21.2 In Problem 21.1, find the value of maximum eccentricity, e_{max} for the given vertical load such that the resultant tensile stress is zero and the column section is safe.

21.3 In Problem 21.1, find the value of the maximum eccentricity, e_{max}, so that the concrete section of the column carries just allowable resultant tensile stress and it acts as uncracked section still the column remains safe.

21.4. In Problem 21.1, find out the maximum value of the vertical load and the value of maximum eccentricity such that the tension is zero and the column section remains safe.

21.5. In Problem 21.1, the resultant tensile stress in the concrete section remains maximum and the column section still remains safe and uncracked, find the maximum value of vertical load which may act over the column.

21.6 In Problem 21.1, the concrete section of column carries vertical load 360 kN and the moment 25 kN-m. Find out whether the column section shall remain safe and uncracked.

21.7 In Problem 21.1, find the stresses on heavily and lightly loaded faces of the column section.

21.8. A rectangular reinforced concrete column of size 420 mm x 580 mm carries a vertical load of 72 kN at an eccentricity 880 mm from the middle line of the section as shown in Fig. 21.21. 4 bars of 22 mm diameter as compression reinforcement and 4 bars of 22 mm diameter as tension reinforcement have been used with 60 mm effective cover from the outer faces. M 20 grade of concrete and HYSD steel bars Fe 415 have been provided. Find the stresses in concrete and steel.

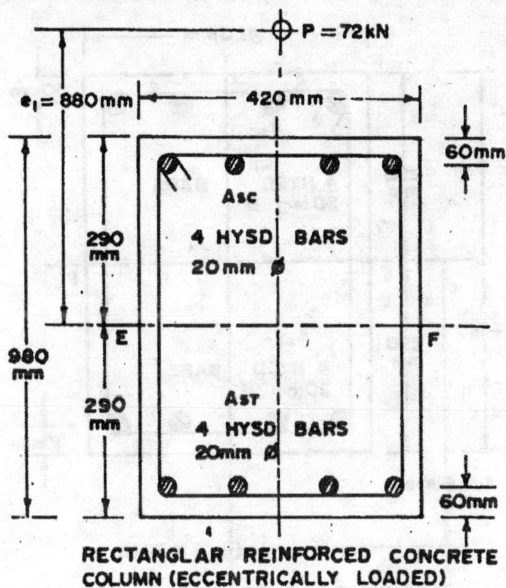

Fig. 21.21

21.9 Find the stresses in concrete at the top and steel bars in compression and tension for the column in Problem 21.8 by accurate method.

21.10 In case the vertical load measured from the middle line of column in Problem 21.8 is 670 mm, find the stresses at the top of the column section, steel reinforcement in compression and that in tension.

21.11 Find the stresses in concrete at the top of column section, steel bars provided as compressive reinforcement and the steel bars provided as tensile reinforcement by calculating the moment of inertia equivalent areas of the column section and the first moment of equivalent area about the neutral axis in Problem 21.10.

21.12 A reinforcement circular column of 480 mm diameter is reinforced with 8 Hysd-steel bars of 18 mm at an effective cover of 60 mm equally distributed along the circumference. The column reinforcement is tied with 8 mm diameter spirals. The grade of concrete is M 20. The column is subjected to an axial load of 240 kN and a bending moment of 36 kN-m. Find the stresses in concrete at the top and in steel.

21.13 A reinforced concrete truss member 280 mm x 280 mm in cross-section carries at tensile force of 80 kN at an eccentricity of 60 mm from the middle line. The effective cover at both faces is 40 mm. Four bar of 18 mm diameter Hysd steel bars have been provided at each face. Find the tensile stresses developed in the steel bars. The grade of concrete is M 20.

21.14 A rectangular reinforced concrete member of size 420 mm x 580 mm in size carries a tensile force of 72 kN at an eccentric distance 800 mm (below) from the middle line of the section as shown in Fig. 21.22. 4 bars of 20 mm diameter as compression reinforcement and 4 bars of 20 mm diameter as tension reinforcement have been provided with 50 mm effective cover from the outer faces. M 20 grade of the concrete and Hysd steel bars Fe 415 have been used. Find stresses in concrete and steel.

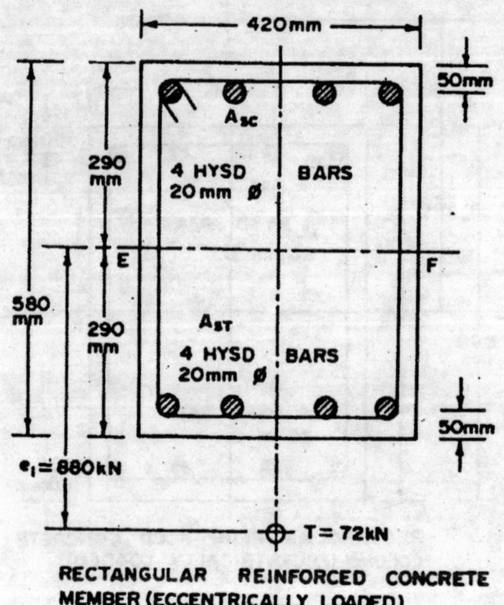

Fig. 21.22

21.15 **In case the distance to the tensile force measured from the middle line in Problem 21.14 is 670 mm (below), find the stresses at the top of member, steel reinforcement in compression and that in tension.**

PART VI

DESIGN OF FOOTINGS AND FOUNDATIONS

22

DESIGN OF INDEPENDENT FOOTINGS

22.1 INTRODUCTION

The portion a structure above the plinth is called as **super-structure** and the portion of a structure below the plinth is know as **sub-structure.** The *sub-structure* is also defined as that part of the structure which is usually placed below the surface of the ground. *The **footing or foundation** is defined as that part of sub-structure which remains in contact with the soil or rock.* The footing or foundation transmits safely the load [the dead load of the super-structure, the live load (as in dwellings, warehouses, industrial buildings) or fill (as in silos, bunkers, tank supporting structures and the like), the ground effects of various lateral forces (as in silos, bunkers, tank supporting structures and the like), the ground effects of various lateral forces (such as wind, blast or earthquake) to the soil stratum of sufficient strength and distributes the load over a sufficiently large area of that stratum so that the bearing pressure developed in the soil and it is ensured that any settlement which may occur shall be nearly uniform as possible and the differential settlement of the various parts of the structures shall be eliminated as nearly as possible. In case the satisfactory soil directly underlies the structure, it is simply essential to spread the load by the footing. In case the adequate soil is not seen immediately below the structure, it becomes essential to provide deep foundations (such as piles or caissons) to transfer the load to deeper and firm layers. In case the forces to be transferred to the soil are of dynamic character, an allowance for impact should be included in the estimate of their magnitude unless a more exact evaluation is needed. The footings also include the foundation elements that are provided to resist the uplift, overturning and the sliding.

There are many ways and means to transmit these forces into the soil. These depend primarily on the magnitude of load, the stiffness and the structural behaviors of the super-structure, the bearing capacity of the soil in general and that of a certain stratum in particular, the depth below the ground surface where the soil stratum of sufficient strength is encountered. The design of foundation for a structure comprises of three stages. In the first stage, the nature of the ground is seen and the safe bearing capacity of the soil is evaluated. In the second stage, the type of foundation is decided and the suitability of one or more types is compared. In the third stage, the selected foundation to transmit and distribute the loads from the structure to the ground is designed. Professor Terzaghi gave the criterion and defined *the shallow foundation* in case its depth is equal to, or less than its width and *the deep foundation,* in case its depth is more than its width. The design of independent or isolated footings has been described in this chapter while the design of combined footings have been described in chapter 23.

22.2 TYPES OF INDEPENDENT FOOTINGS

The **independent footings** are also known as **isolated footings.** *The independent footings are defined as the footings which support the individual columns.* The independent footings distribute and spread the load over a sufficiently large area of the soil stratum to minimize the bearing pressure and therefore, these types of footings are referred as *spread footings.* The independent spread footings are of two types, namely, *axially loaded independent spread footings, and eccentrically loaded independent spread footings.* The axially loaded spread footings are used to resist column loads that are concentric (axial) with the centroid of footing. Theoretically, any shape may be used for such footings. The *wall footing* (which is a continuous strip of concrete that support a bearing wall) as shown in Fig. 22.1(A), the square footing and the rectangular footing as shown in Fig. 22.1(B) the circular footings as shown in Fig. 22.1(C) are the footings which are commonly used. However, round and polygonal footings are provided under certain conditions. These footings may be solid or without a central portion (viz., like a ring). In these footings, the bearings pressure distribution is usually assumed uniform over the complete area. Many times, a pedestal is provided between the column and the footing slab. The pedestal provides for a more favourable transfer of the load. In many case, the pedestal is needed in order to provide the necessary development length. The spread of footing is done by providing steps as shown in the wall footing and the footing is referred as *stepped footing.* In stepped footing, the concrete must be poured in a single pour in order to provide monolithic action. The sloped footings need less concrete as compared to the stepped footing. However, the additional labour required to produce the slopping surface (form work) usually makes the stepped footings convenient and more economical.

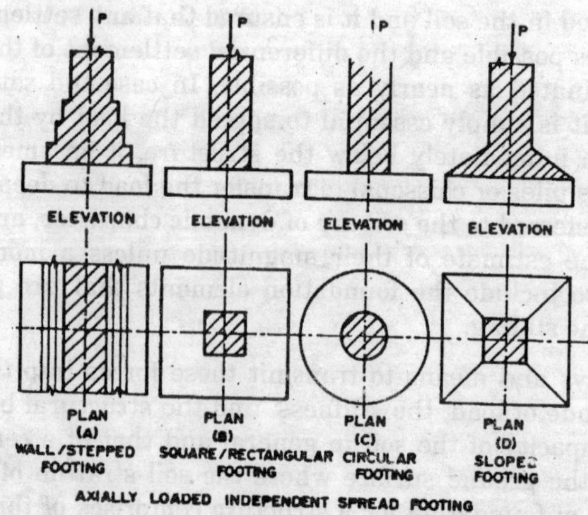

Fig. 22.1

The eccentrically loaded independent spread footings or independent spread footings subjected to axial load and moment at the base shall be described later in this chapter.

22.3 BEARING PRESSURE DISTRIBUTION IN SOIL

The load from a wall or a column is transferred vertically and axially to the footing. All the soil compress noticeably when loaded, as a result of which the bearing pressure develops in the soil, which acts in the upward direction and supports the load. The vertical axial load acting downward is symmetrical with respect to the bearing area. The actual bearing pressure distribution under the footing resting on the coarse-grained soil is larger at the centre of the footing and reduces toward the perimeter (edges) as shown in Fig. 22.2(A). The individual grains in the granular soils are somewhat mobiles. Such soils near the edge shift very slightly in the outward direction of smaller soil pressure. Whereas, the actual bearing pressure distribution under the footings resting on the clay soils is higher nearer the edges than at the centre of the footing as shown in Fig. 22.2 (B). In clay soil (cohesive soils), a shear resistance around the perimeter develops, which adds to the upward pressure. The numerical values of the bearing pressure in both these types of soils are uncertain and highly variable. It is customary to disregard the non-uniformities. The bearing pressure distribution is assumed as uniformly distributed as shown in Fig. 22.2 (C) over the entire bearing area for reasons of simplicity and it is common design practice. This assumption is on the safe side as the most of the common soil types develop uniform bearing pressure distribution.

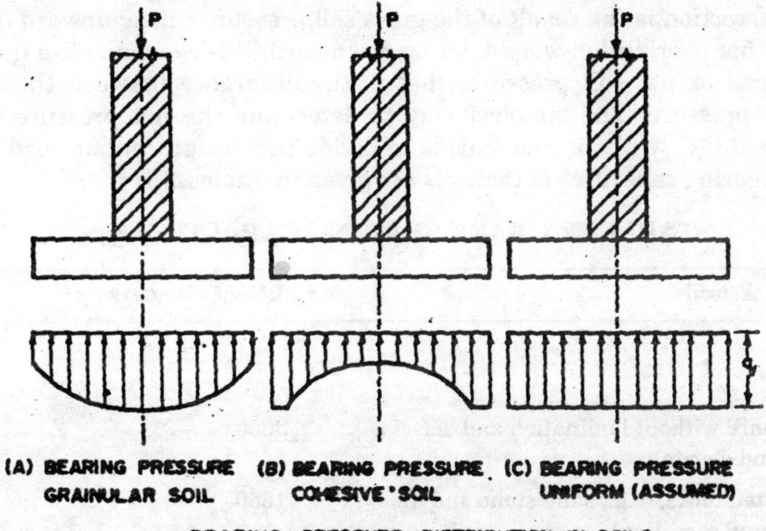

Fig. 22.2

22.4 BEARING CAPACITY OF SOIL

The maximum bearing capacity of the soil is evaluated from the considerations of soil mechanics. The bearing capacity of soil is the *maximum allowable bearing pressure.* It is a combined function of both the site conditions and the characteristics of the super-structure proposed to be constructed on the ground. It is determined from the *ultimate bearing capacity of soil* (defined as the intensity of loading at the base of a foundation which will result shear failure of the soil support) or type of tolerable settlement expected and the capability of the proposed structure to take up this settlement. The *safe bearing capacity* (viz., the maximum intensity of loading that the soil will carry safely without any possibility of the shear failure irrespective of any settlement that may occur) is found on the basis of plate load tests and other experimental determinations. Professors R.B. Peck, W.E. Hanson, and T.H. Thornburn `Foundation Engineering' second edition, published by John Wiley and Sons, Inc., New York, 1974 and Professors K. Terzaghi and R.B. Peck `Soil Mechanics in Engineering Practice' second edition, John Wiley and Sons, Inc., New York. 1967 references may be seen for this purpose. The allowable bearing pressures, q_a under service loads are obtained by assuming appropriate factor of safety (2.5 to 3.0) against exceeding the ultimate bearing capacity of the particular soil and the settlements are kept within the tolerable limits. Many building codes specify the allowable bearing pressures for the various types of soils and soil conditions found in the particular locality. The **safe bearing pressure** (defined as the maximum intensity of loading that the soil will carry without exceeding the settlement prescribed for the structure. The bearing capacity of soil is equal to safe bearing capacity or safe bearing pressure whichever is less.

The area of footing base must be adequate to care for the column load, self weight of the footing, and weight of any cover burden (if any) under the footing. Sometimes this overall pressure is called *gross soil pressure* to distinguish it from the *net soil pressure*. The weight of footing and overburden is usually nearly uniform over the footing area. The design moment or shear on any section is the result of the gross soil pressure acting upward *less* the design of footing and other overburden weight acting downward. It is convenient to think in terms of the resultant load or net soil pressure, that is the difference between these upward and downward unit pressure. The simplest way to determine the net pressure caused by the vertical axial load (i.e., without moment) is to divide the design column load by the footing area. The safe bearing capacities of the soils are given in Table 22.1.

TABLE 22.1. SAFE BEARING CAPACITY, σ_{br}

S.No.	Types of rocksisoil	σ_{br} kN/m²	Remarks
1	2	3	4
1	**Rocks**		
	(a) Rocks-hard without lamination and defects, e.g., trap and diorite)	3300	–
	(b) Laminated rocks, (e.g., sand stone and lime stone in sound condition)	1650	–
	(c) Residual deposits of shattered and broken bed rock and hard shale, cemented material	900	–
	(d) Soft rock.	450	–
2	**Non-cohesive soils**		
	(a) Gravel, sand and gravel, compact and offering high resistance to penetration when excavated by tools	450	(See NB : 2)

	(b) Coarse sand, compact and dry*	450	*Depth of water table $\nless$ width of foundation below the base of the foundation
	(c) Medium sand, compact and dry*	250	
	(d) Fine sand, silt (dry lumps easily pulverized by the fingers)	150	–
	(e) Loose gravel or sand gravel mixture, loose coarse to medium sand, dry	150	–
	(f) Fine sand, loose and dry	100	–
3	**Cohesive soils**		
	(a) Soft shale, hard or stuff clay in deep bed dry	450	Susceptible to long term consolidation settlement
	(b) Medium clay, readily indented with a thumb-nail	250	
	(c) Moist clay and sand clay mixture which can be indented with strong thumb pressure	150	–
	(d) Soft clay indented with moderate thumb pressure	100	–
	(e) Very soft clay which can be penetrated several centimetres with the thumb	50	(See NB : 3)
	(f) Black cotton soil or other shrinkage or expansive clay in dry condition (50 percent saturation)	–	To be found by investigation
4	**Peat**		
	Peat	–	(See NB : 3 and 4)
5	**Made-up ground**		
	Fills or made up ground	–	(See NB : 3 and 4)

Note : 1. Values listed in Table 22.1 above are from the shear considerations only

2. Values listed above are very much rough due to the following reasons :

(i) Effect of characteristics of foundations (i.e., effect of depth, width, shape, roughness, etc.) has not been considered.

(ii) Effect of range of soil properties (i.e., angle of frictional resistance, cohesion, water table, density, etc. has not been considered.

(iii) Effect of eccentricity and indication of loads has not been considered.

3. For non-cohesive soils, the values listed in Table 22.1 shall be reduced by 50 percent if the water table is above or near the base of footing.

4. Compression or looseness of non-cohesive soils may be determined by driving the cone of 65 mm diameter and 60° apex angle by a hammer of 0.650 kN falling from 750 mm. If corrected number of blows (N) for 300 mm penetration are less than 10, the soil is called *loose,* if N lies between 10 and 30, it is called *medium,* if more than 30, the soil is called dense.

(Reference : IS 6403-1971 : Code of practice for determination of allowable bearing pressure on shallow foundations).

22.5 DEPTH OF FOUNDATION

The depth of foundation is decided by keeping the following factors into consideration.

(i) The safe bearing capacity of soil is secured.

(ii) In case of clay, the foundation is to be penetrated below the zone where shrinkage and swelling of soils due to seasonal weather changes are likely to cause significant movement.

(iii) In case of fine sands and silts, the foundation is to be penetrated below the zone in which the trouble may be expected from the frost.

(iv) The foundation is carried sufficiently below the maximum depth of scour (if likely to occur).

IS : 1080-1972 recommends that a **minimum depth of foundation** of atleast 500 mm below natural ground level should be provided in all types of soils. In case of rock or such other weather-resisting natural ground, the top soil may be removed. The surface is cleaned. If necessary, the surface is stepped or otherwise prepared to provide a suitable bearing. The slipping or any other movement is prevented. The minimum depth of foundation is determined by Rankine formula

$$h_{\min} = \frac{p}{r}\left(\frac{1-\sin\phi}{1+\sin\phi}\right)^2, \textit{ where} \tag{22.1}$$

h_{min} = minimum depth of foundation

p = gross bearing capacity of the soil

r = density of soil

ϕ = angle of repose of the soil

In case there had been excavation, ditch, pond, water course and the earth is filled up (viz., in filled up ground or similar conditions adjoining to the subsoil on which the structure is proposed to be constructed) which is likely to damage the stability of a structure, either the foundation of such structure shall be placed to a depth beyond the damage influence of these conditions or earth retaining structures (viz., retaining walls) are built for the purpose of shielding from the damaging effects.

22.6 THICKNESS AT EDGE OF FOOTING

IS : 456-2000 recommends that in the reinforced and plain concrete footings, the thickness at edge shall be not less than 150 mm for footings or soils, nor less than 300 mm above the tops of piles for footings on piles.

The *sloped* or *stepped footings,* are designed as a unit shall be constructed to assure action as a unit. The effective cross-section in compression shall be limited by the area above the neutral plane, and the angle of slope or depth and location of steps shall be such that the design requirements are satisfied at every section.

In case of plain pedestals, the angle α between the plain passing through the bottom edge of the pedestal and the corresponding junction edge of the column with pedestal and the horizontal plane as shown in Fig. 22.3. It shall be controlled by Eq. 22.2

$$\tan\alpha \nless 0.9\left(\frac{100\, q_o}{f_{ck}}+1\right)^{1/2}, \textit{where} \tag{22.2}$$

q_o = calculated maximum bearing pressure at the base of the pedestal in N/mm² and

f_{ck} = characteristic strength of concrete at 28 days in N/mm²

Deciding the width of projection BE/DF, the depth of pedestal AE/CF may be calculated as under :

$$\tan(90-\alpha) = (BE/AE) \tag{i}$$

Therefore, $AE = BE \,.\, \tan\alpha$ (ii)

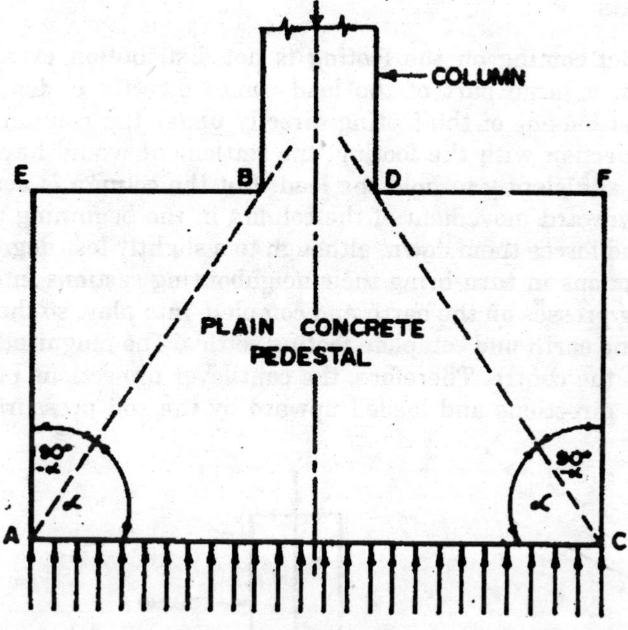

Fig. 22.3

22.7 MOMENTS AND BENDING MOMENTS IN FOOTINGS

IS : 456-2000 specifies that in case of footings on piles, computation for moments may be based on the assumption that the reaction from any pile is concentrated at the centre of the pile.

For the purpose of calculating stresses in footings which support a round or hexagonal, octagonal concrete column or pedestal, the face of the column or pedestal shall be taken as the side of a square inscribed within the perimeter of the round or octagonal column or pedestal as shown in Fig. 22.4.

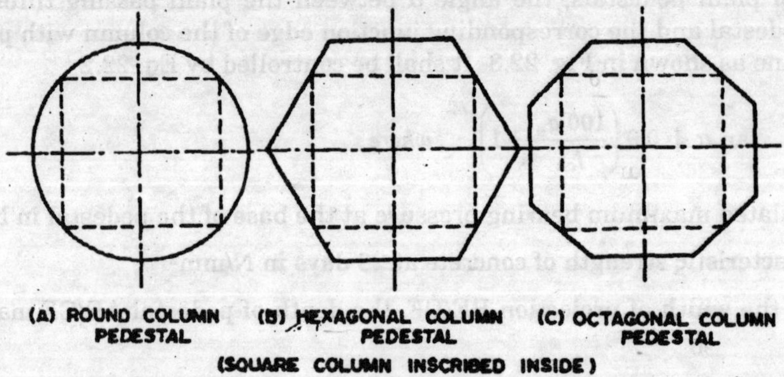

Fig. 22.4

1. Bending moments

The column load coming on the footing is not distribution evenly over the complete foundation. At first, a large part of the load comes directly under the column and this produces a small settlement of the footing directly under the column. In case, the column would have no connection with the footing, the settlement would have continued until the ground compressed sufficiently to hold the load. But the column is actually connected with the footing, the downward movement of the column in the beginning produces shear in the adjacent sections and forces them down, although to a slightly less degree because of bending produced. These sections in turn bring their neighbouring sections into play, so that finally the complete footing presses on the earth and complete into play, so that finally the complete footing presses on the earth and complete footing settles. The magnitude of settlement is less at the edge than at the centre. Therefore, the cantilever projections projecting out from the column in both the directions and loaded upward by the soil pressure bends in a curve as shown in Fig. 22.5.

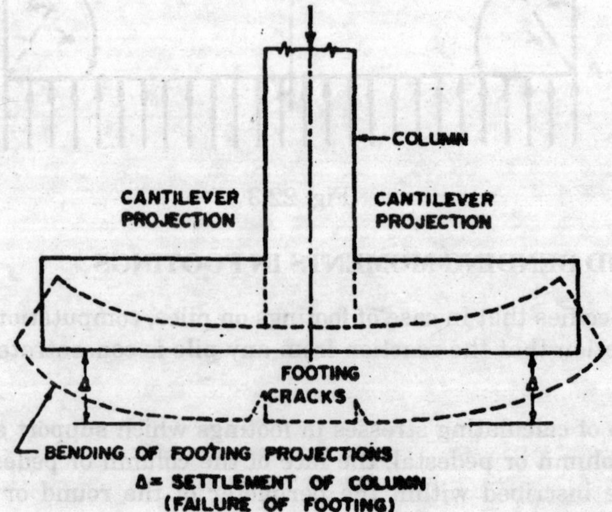

Fig. 22.5

Due to bending of the cantilever projections as shown in Fig. 22.5, the bending tensile stresses develop in both these projections at the bottom surface. Therefore, the steel bars are provided as reinforcement in tension in two layers perpendicular to each other and parallel to the edges.

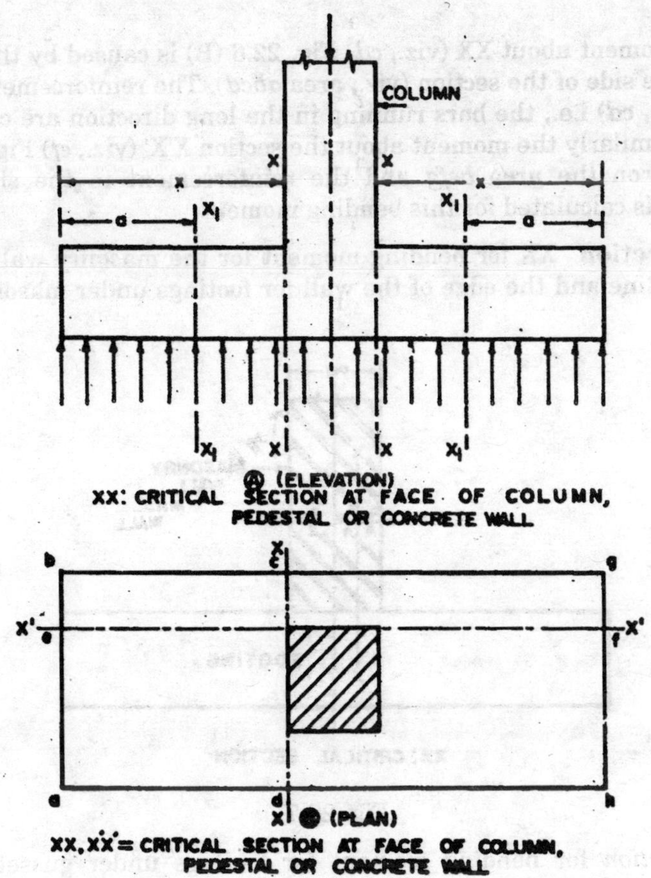

Fig. 22.6

The bearing pressure of soil q_o at the base of footing as shown in Fig. 22.6 (as discussed in Art. 22.3) is assumed uniform. The bending moment at any section X_1X_1 at a distance from free edge shall be determined by passing through the section a vertical plane which extends completely across the footing and calculating the moment of forces acting over due entire area of footing on one side of the said plane.

$$M_{X1X1} = \left(q_o \, .a^2/2\right) \tag{22.2}$$

The maximum bending moment to be used in the design of an isolated concrete footing which supports a column, pedestal or concrete wall, shall be the moment calculated at the **critical section** XX as shown in Fig. 22.6(A) at any section in Fig. 22.6(A) at any section XX at a distance x from the free edge

$$M_{XX} = \left(q_o \, . \, x^2/2\right) \tag{22.4}$$

2. Critical sections for bending moment

The critical section XX for bending moment shall be at the face of the column, pedestal or concrete wall, for footing supporting a concrete column, pedestal or wall as shown in Fig. 22.6 (B) in plan.

The bending moment about XX (viz., *cd*) Fig. 22.6 (B) is caused by the upward pressure q_o on the area to one side of the section (viz., area *abcd*). The reinforcement perpendicular to this section XX (viz., *cd*) i.e., the bars running in the long direction are calculated from this bending moment. Similarly the moment about the section X'X' (viz., *ef*) Fig. 22.6 (B) is caused by the pressure q_o on the area *befg* and the reinforcement in the short direction (i.e., perpendicular to *ef*) is calculated for this bending moment.

The **critical section** XX for bending moment for the masonry walls shall be halfway between the centre line and the edge of the wall for footings under masonry walls as shown in Fig. 22.7.

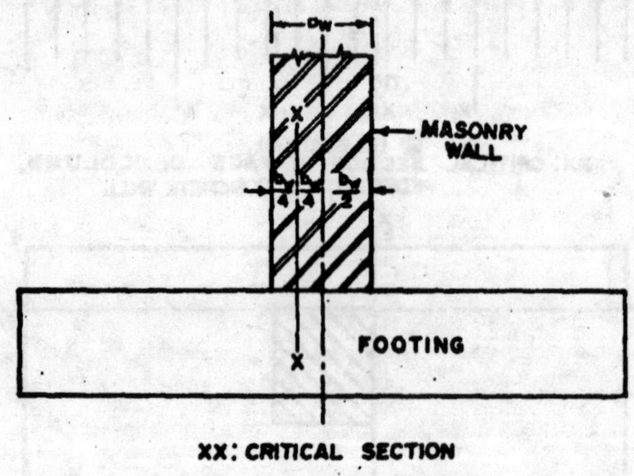

Fig. 22.7

The *critical section* for bending moment for footings under gusseted bases shall be halfway between the face of column or pedestal and the edge of gusseted base.

3. Critical section for development length

The critical sections for checking the development length in a footing shall be assumed at the same planes as those described for bending moments (as above) and at all other vertical planes where the abrupt changes of footing section occur. In case the reinforcement is curtailed, the anchorage requirement shall also be checked as described for the design of beams.

In a single column footing, the available development length of steel bar is from the face of column to the edge of the footing less the end cover.

4. Failure of footing

When a footing fails it does not mean that reactions of the soil move the footing upward, but that the projecting footing is not able to hold the column itself from settlement. The column settles, while the projections of the footing, instead of following the column crack as shown in Fig. 22.5 and break away from it.

22.8 SHEAR FORCES IN FOOTINGS

IS : 456-2000 recommends that in case of footings on piles, the calculations for shears may be based on the assumption that the reaction from any pile is concentrated at the centre of the pile.

For the purpose of calculating stresses in footings which support a round or hexagonal or octagonal concrete column or pedestal, the face of the column or pedestal shall be taken as the side of square inscribed within the perimeter of the round or hexagonal or octagonal column or pedestal as shown in Fig. 22.4.

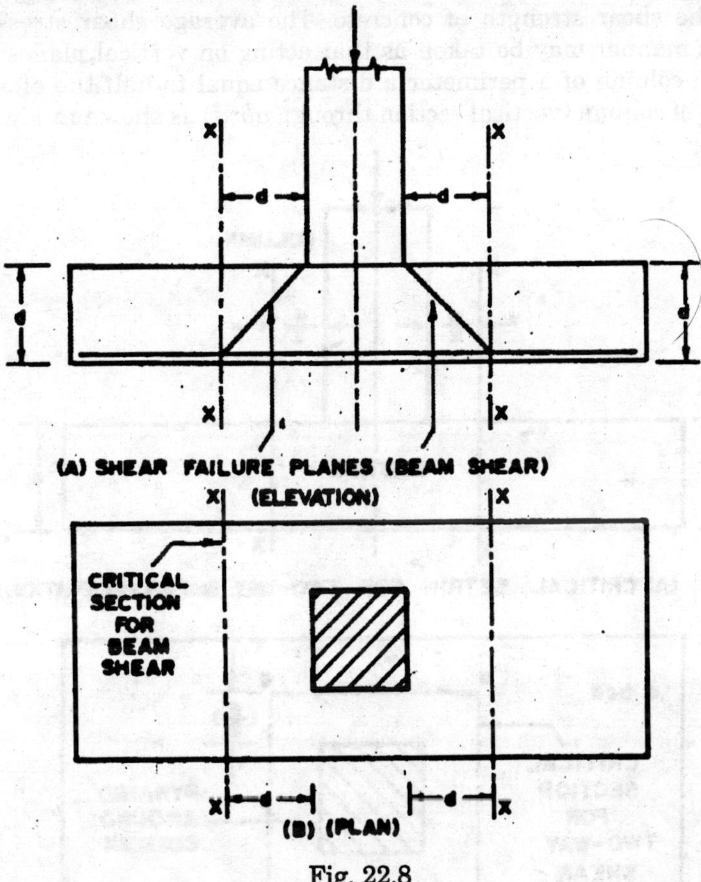

Fig. 22.8

Two different types of shears, namely **beam shear (one-way shear)** and **shear as in flat slabs (two way shear)** are distinguished in footings.

In beam shear (viz., **one-way shear**), it is considered that the footing essentially acts as a wide beam and a potential diagonal tension crack extends on an approximately 45° plane. The shear crossing this crack is produced by the upward load to the left of X, that is, the load beyond a plane XX a distance equal to the *effective depth*, *d* from the face of column footing or wall footing as shown in Fig. 22.8. The sections XX are the *critical sections* for beam shear.

In case of footings on piles, *the critical section* shall be at a distance equal to half the effective depth of footing.

In flat slab type shear (i.e., two way shear), when failure occurs, the fracture takes the form of truncated cone (for a round column) or truncated pyramid (for a square or rectangular column) with sides sloping outward on all the faces. The problems reason for this type of failure is the heavy compressive stresses between the diagonal cracks as the column load spread out into the surrounding footing, with this compression accentuated by the upward soil reaction. The concrete subjected to shear stress is also in vertical compression from the stresses spreading out from the column, and in horizontal compression in both major directions because of the bi-axial bending moments in the footing. This triaxially of stress increases the shear strength of concrete. The average shear stress in the concrete which fails in this manner may be taken as that acting on vertical planes laid through the footing around the column of a perimeter a distance equal to half the effective depth (viz., d/2) from the faces of column (vertical section through *abcd*) as shown in Fig. 22.9.

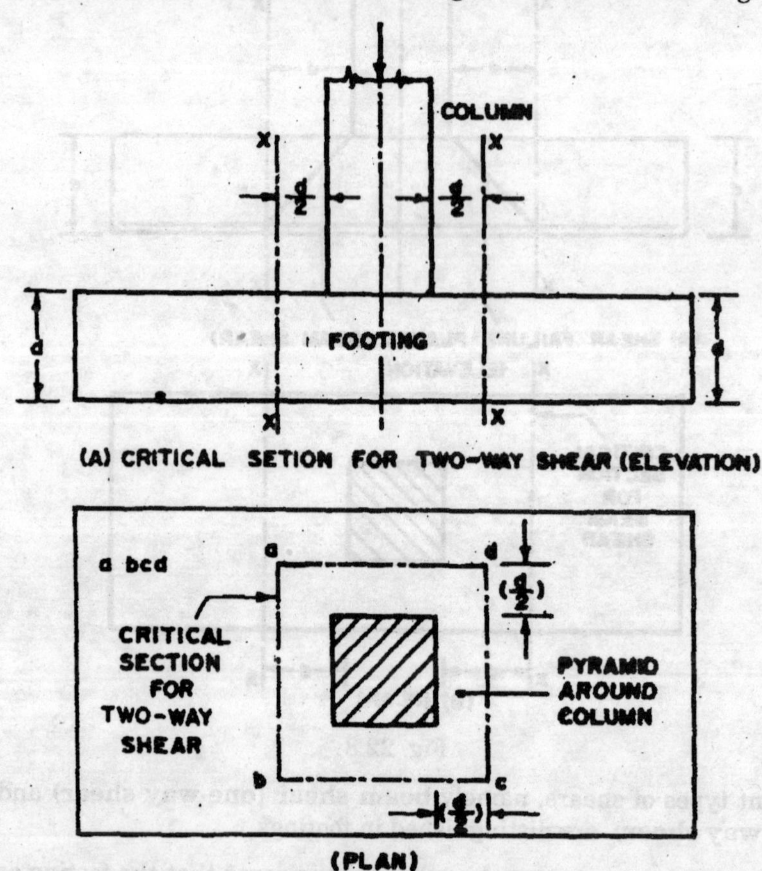

Fig. 22.9

1. Nominal shear stress in footing,

$$\tau_v = \left(\frac{V}{b_o . d}\right), \textit{ where} \tag{22.5}$$

V = shear force due to the design load

b_o = perimeter of the critical section (*abcd* in Fig. 22.9 (B))

d = effective depth.

2. Permissible shear stress

When the shear reinforcement is not provided, the calculated shear stress at the critical section shall not exceed the permissible shear stress equal to $k_s \cdot \tau_c$., where

$k_s = (0.5 + b_c)$, but not greater than unity

b_c = ratio of short to long side of the footing

$$\tau_c = 0.16\,(f_{ck})^{1/2} \tag{22.6}$$

It is generally not economical and convenient to use shear reinforcement in footings. For this reason only, the footings are designed such that all shear is carried by the concrete.

22.9 TRANSFER OF LOAD AT BASE OF COLUMN (BEARING UNDER COLUMNS)

A column rests on a footing or a pedestal. It transfers its load to only a part of the total area of the supporting member. The adjacent concrete of footing provides lateral support to the directly loaded part of the concrete. This causes triaxial compression stresses that increases the strength of the concrete which is loaded directly under the column.

The bearing from the column on the footing is allowed to be more than the value of the same concrete in the column. That is, lower strength concrete is allowed in the footing without lowering the load carrying capacity of the column or necessarily increasing the dowels provided.

The compressive stress in concrete at the base of a column or pedestal shall be considered as being transferred by bearing to the top of the supporting pedestal or footing. IS : 456-2000 (originally based on tests and provisions of ACI Code) that when the supporting area is wider than the loaded area on all the sides, the design bearing stress on the loaded area of footing shall not exceed the permissible bearing stress in direct compression multiplied by a value equal to $(A_1/A_2)^{1/2}$ but not greater than 2 (two), viz.,

$$q_o \ngtr \sigma_{cbr} \cdot (A_1 / A_2)^{1/2} \tag{22.7A}$$

The permissible bearing stress on full area of concrete shall be taken as $0.25 \cdot f_{ck}$, viz.,

$$\sigma_{cbr} = (0.25\ f_{ck}), \tag{22.7B}$$

Eq. 22.7, then, becomes

$$q_o \ngtr 0.25\ f_{ck} \cdot (A_1 / A_2)^{1/2} \tag{22.8}$$

The advantage of confinement of the bearing area is because of triaxial state of compressive stresses, (as described in paragraph one above), in the immediate vicinity of the loaded area of the footing. The concrete around the loaded area in the footing provides considerable support through the confinement.

In Eqs. 22.7 and 22.8, A_2 is the loaded bearing area under the column base and A_1 is the maximum area of the portion of the supporting surface. The area A_1 is the area of the lower base of the largest frustrum of a pyramid, cone, or tapered wedge contained wholly within the support and having for its upper base the loaded area. The area A_1 is geometrically similar to concentric with the loaded area as shown in Fig. 22.10.

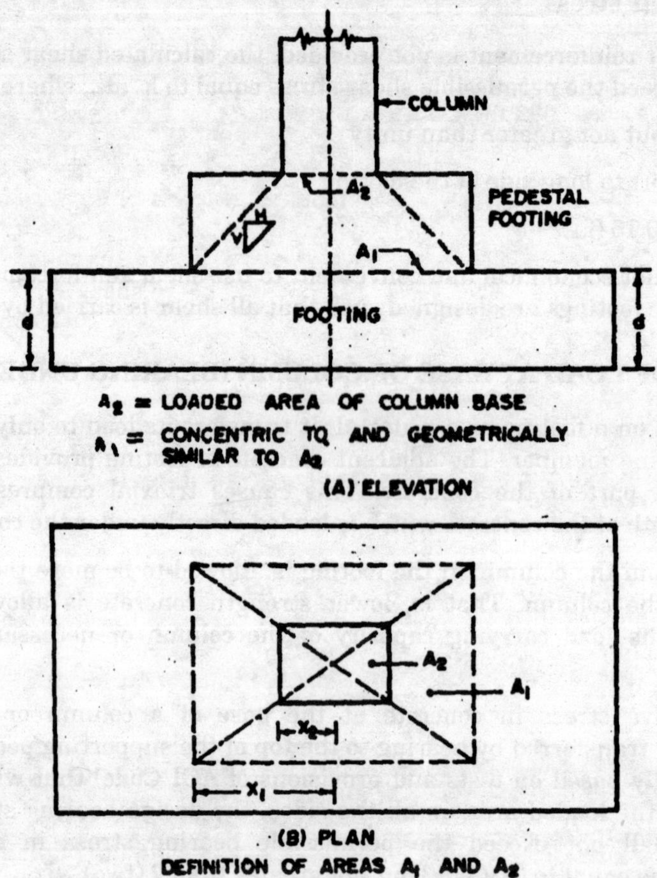

Fig. 22.10

It is established by going down and out from the loaded area at the slope of V (vertically) and H (horizontally), [V : H : 1 : 2] until one intersects the boundary closest to the column. The boundary fixes one side of A_1 and it is simplest to draw the complete area on the plan view similar to A_2. From the similar areas.

$$\left(\frac{A_1}{A_2}\right) = \left(\frac{x_1}{x_2}\right)^2 \quad \text{(i)}$$

$$\left(\frac{A_1}{A_2}\right)^{1/2} = \left(\frac{x_1}{x_2}\right) \leq 2 \quad \text{(ii)}$$

It shows how limit for $(A_1/A_2)^{1/2}$ has been specified as 2 (two).

All axial forces and bending moments which act at the bottom section of a column may be transferred to the footing at the bearing surface. The compressive forces are transferred by bearing by compression in the concrete. In case, the design values of bearing stresses are exceeded, the reinforcement or dowels should be detailed so as to take up the remaining forces and should be extended across the interface into the column as well as footing so that sufficient development lengths are provided on side.

The tensile forces (uplift) are transferred by reinforcements properly anchored into column as well as footing.

The lateral forces are transferred by shear friction, shear keys etc.

Dowels into footing

The steel bars which extend into two adjoining portions of concrete construction, so as to connect the two portions and transfer shear loads are referred as **dowels.**

The longitudinal reinforcement is extended into the column footing or by providing dowels, which are embedded in the footing and project above it. In the latter case, the column bars merely rest on the footing and in most cases are tied to dowels. This results in a simpler construction procedure than extending the column bars into the footing. IS : 456-2000 (so also ACI Code) specifies that the minimum area of reinforcement that crosses the bearing surface (dowels or column bars) bar 0.005 *times* (viz., 0.5 per cent) the gross cross-sectional area of the supported column or pedestal and a minimum of four bars shall be provided. Where the dowels are used, their diameter shall not exceed the diameter of column bars by more than 3 mm. The column bars of diameter larger than 36 mm, in compression can only be dowelled at the footings with the bars of smaller size of necessary area. The dowel shall extend into the column, a distance equal to the development length of the column bar and into the footing, a distance equal to the development length of the dowel.

Under normal circumstances, the lap splicing of bars is prohibited because tests have shown that welded splices or other positive connections are necessary to develop these heavy bars fully. The long-standing successful use was relied and the code allowed the splicing of heavy bars to dowels of lesser diameter provided the dowels have a development length into the column corresponding to that of the column bar and into the footing as specified for the particular dowel.

22.10 BASIC REQUIREMENTS FOR COLUMNS

For **isolated footing** and so also for **combined footing** if these are held in static equilibrium solely by bearing pressure acting against its base, the footing has to satisfy the following four basic requirements :

1. The resultant of all bearing pressures, acting against the footing base *(reaction)* must be equal intensity and opposite direction to the resultant of all loads and/or vertical effects due to moments and lateral forces, acting on the footing element *(action)*.

2. The line of action of the resultant vector of the reaction must coincide with the line of action of the resultant vector of the action applied at the footing base.

3. The maximum intensity of the bearing pressure under the most severe combination of service loads must be smaller than, or equal to, the maximum bearing pressure allowed for this kind of loading and type of soil, as determined by the principles of soil mechanics.

4. The resultant vector of the least favourable combination of vertical loads, horizontal shears and bending moments that may occur under service load conditions, including wind or earthquake, must intersect the footing base within a maximum eccentricity that will provide safety against overturning.

The method most commonly used for the design of footings and related elements for ordinary building construction, is the one in which the static equilibrium is obtained by bearing pressures against the footing base only. This method is also the standard method. ACI Code 318-83 `Building code requirements for reinforcement concrete' has also included this method.

When the columns are having axial loads P (viz., the eccentricity is zero), the bearing pressures will be uniformly distributed over the entire base area of the footing, A_F.

22.11 WALL FOOTINGS

A footing provided under a wall used to be *continuous* along the length of wall. The nature of load transferred by the wall to be footing is axial and uniform. Therefore, for the design of a footing under a wall, a design strip of *one metre* length of the wall is considered. The simple principles of beam action applies to wall footing with only minor modifications. The position of critical section for bending moment depends upon the types of material used for the construction of the wall. A wall may be a *masonry wall* or *a concrete wall.*

(A) Analysis of footing under a masonry wall

A continuous footing under a masonry wall with the forces acting on it is shown in Fig. 22.11.

The self-weight of wall is assumed as ten per cent of the column load. The area of footing, A_F is calculated as under :

$$A_F = \left(\frac{1.10P}{q_o}\right) \quad (\textit{where } A_F = B \times 1)$$

where B is the width of footing in metres.

The critical section XX for bending moment is halfway between the centre-line and edge of the wall as shown in Fig. 22.11A [as described in Art. 22.7, Fig. 22.7]. The maximum bending moment in footing at section XX for 1 metre design strip may be calculated as under:

$$M = \frac{1}{2} q_o \cdot \left[\left(\frac{B - b_w}{2}\right) + \frac{b_w}{4}\right]^2 - \frac{1}{2} q \cdot \left(\frac{b_w}{4}\right)^2 \qquad \text{(i)}$$

$$M = \frac{1}{2}\left(\frac{P}{B \times 1}\right) \cdot \left[\left(\frac{B - b_w}{2}\right) + \frac{b_w}{4}\right]^2 - \frac{1}{2}\left(\frac{P}{b_w \cdot 1}\right) \cdot \left(\frac{b_w}{4}\right)^2 \qquad (22.9)$$

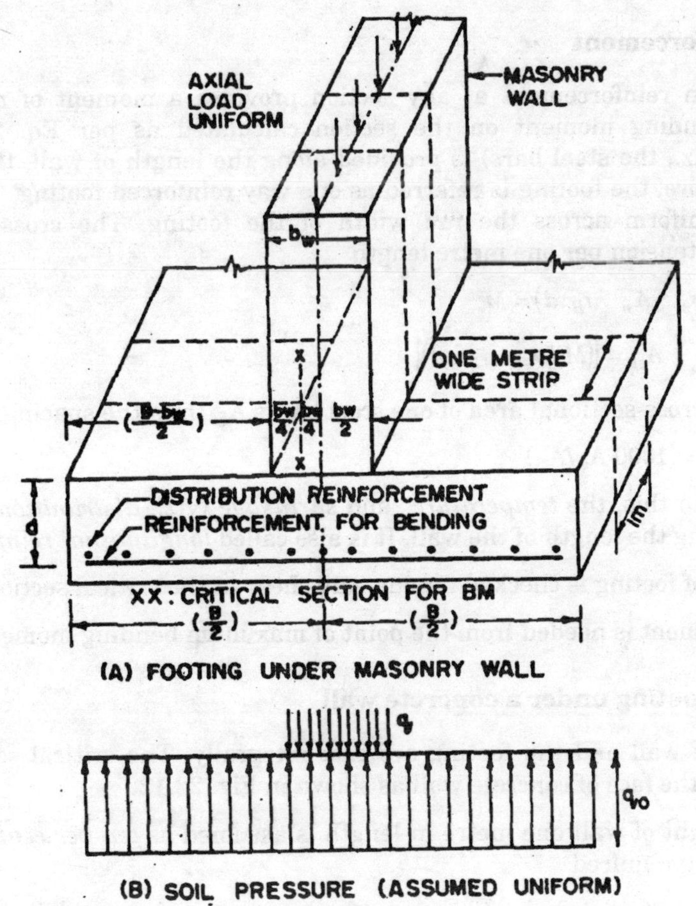

Fig. 22.11

The maximum shear force in footing at section XX for 1 metre design strip may be calculated as below :

$$F = q_o \cdot \left[\left(\frac{B - b_w}{2}\right) + \frac{b_w}{4}\right] - q \cdot \left(\frac{b_w}{4}\right) \qquad \text{(ii)}$$

$$F = \left(\frac{P}{B \times 1}\right)\left[\left(\frac{B - b_w}{2}\right) + \frac{b_w}{4}\right] - \left(\frac{P}{b_w \cdot 1}\right)\left(\frac{b_w}{4}\right) \qquad (22.10)$$

1. Effective depth

The design strip of one metre length of the footing is designed as a balanced section. Therefore,

$$M = R_B \cdot (1000) \cdot d^2$$

$$d = [M/(1000 \cdot R_B)]^{1/2} \tag{22.11}$$

2. Tension reinforcement

Total tension reinforcement at any section provides a moment of resistance at least equal to the bending moment on the section calculated as per Eq. 22.9. The tension reinforcement (viz., the steel bars) is provided along the length of wall, that is, only in one direction. Therefore, the footing is referred as one-way reinforced footing. The reinforcement is distributed uniform across the full width of the footing. The cross-sectional area of reinforcement in tension per one metre length

$$(\sigma_{st} \cdot A_{st} \cdot j_B \cdot d) = M$$

$$\therefore \quad A_{st} = [(M/\sigma_{st} \cdot j_B \cdot d)] \tag{22.12}$$

In case, the cross-sectional area of one steel bar is A_ϕ, than the spacing of steel bars

$$s = (1000\, A_\phi / A_{st}) \tag{22.13}$$

In addition to this, the *temperature* and *shrinkage* (viz., *distribution) reinforcement* is also provided along the length of the wall. It is also called *longitudinal reinforcement.*

The section of footing is checked for one-way shear at the critical section.

Full development is needed from the point of maximum bending moment.

(B) Analysis of footing under a concrete wall

The concrete wall and the footing are cast integrally. The critical section for bending moment XX is at the face of concrete wall as shown in Fig. 22.12.

The self-weight of wall one metre in length is assumed as *ten per cent* of the wall load. The area of footing required

$$A_F = \left(\frac{1.10\,P}{q_o}\right) \qquad .(A_F = B \times 1)$$

where, B is the width of footing for 1 m length.

The maximum bending occurring at the critical section XX

$$M = \frac{1}{2} \cdot q_o \cdot \left(\frac{B - b_w}{2}\right)^2 = \frac{1}{2} \cdot \left(\frac{P}{B \times 1}\right) \cdot \left(\frac{B - b_w}{2}\right)^2 \tag{22.14}$$

$$F = q_o \cdot \left(\frac{B - b_w}{2}\right) = \left(\frac{P}{B \times 1}\right) \cdot \left(\frac{B - b_w}{2}\right) \tag{22.15}$$

The effective depth of footing, main tension reinforcement, the distribution reinforcement and the development length remain same as described for the footing under masonry wall above.

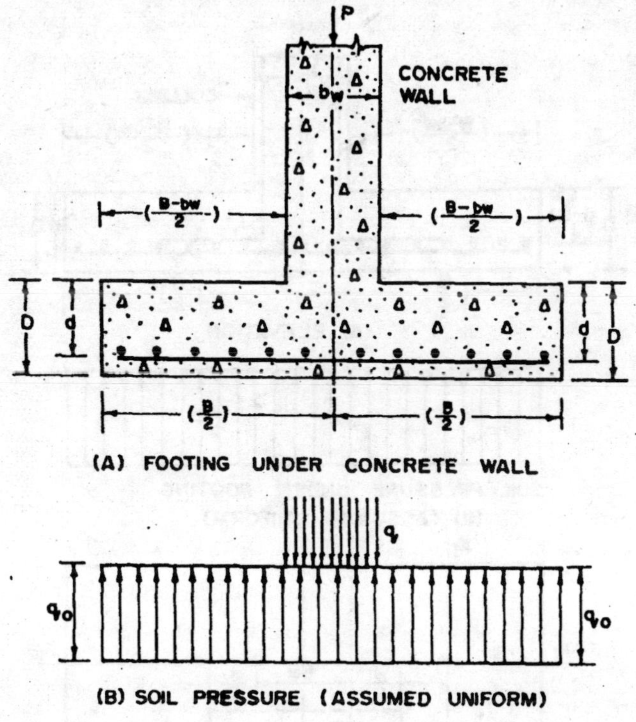

Fig. 22.12

The section of footing is checked for one-way shear force at the critical section.

22.12 SQUARE FOOTING

The square footings are usually provided under single-columns. In the square footing, the cantilevers project out from the column in both the directions. The column transfers the axial (viz., concentric) load acting downward to the footing. the footing is loaded upward by the soil pressure. The tension stresses develop in both these directions at the bottom surface. Such footings are therefore, reinforced by two layers of steel bars, perpendicular to each other and parallel to the edge.

A square footing under a single column with the forces acting on it, is shown in Fig. 22.13. The required bearing area of footing is found by dividing the total load including the self-weight of the footing (the column load P + ten per cent of column load) by the bearing capacity of the soil. The required bearing area of footing

$$A_F = \left(\frac{1.10\ P}{q_o}\right) \qquad \left(\because\ A_F = (B \times B)\,.\,m^2\right) \qquad \text{(i)}$$

where B = width of the square footing measured in metre.

The critical section XX for bending moment is at the face of column as shown in Fig. 22.13. The maximum (design) bending moment in the footing at the section XX may be calculated as under :

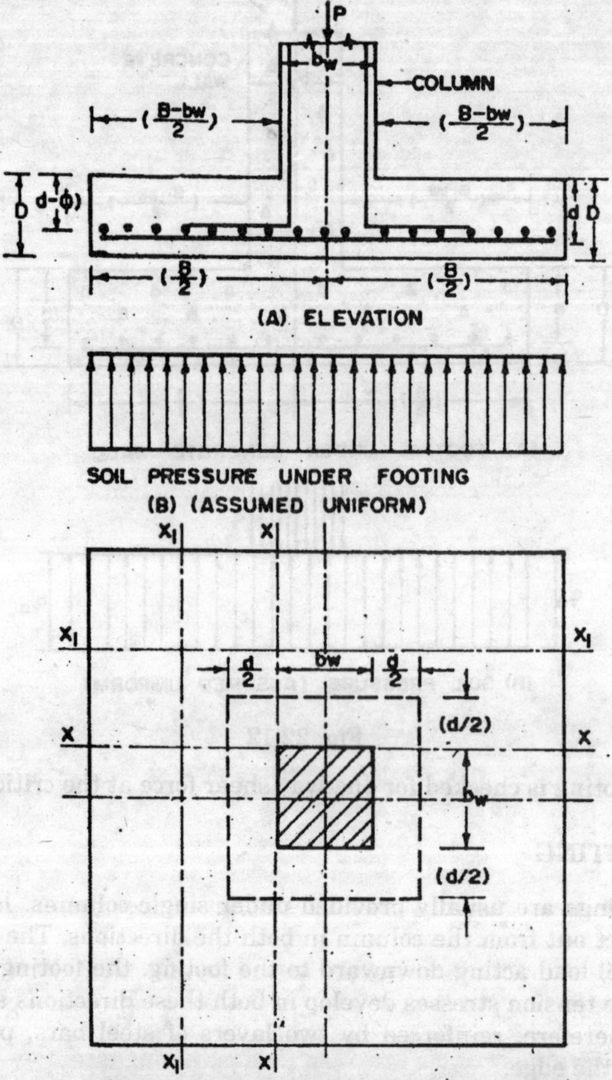

Fig. 22.13

$$M = \frac{1}{2} .q_o . \left(\frac{B - b_w}{2}\right) . \left(\frac{B - b_w}{2}\right)$$

or $$M = \frac{1}{2} . \left(\frac{P}{(B \times B)}\right) . \left[\left(\frac{B - b_w}{2}\right) . \left(\frac{B - b_w}{2}\right)\right] \quad (22.16)$$

The shear force at the critical section XX.

$$V = \left(\frac{P}{B \times B}\right) \cdot \left[\left(\frac{B - b_w}{2}\right)\right] \tag{22.17}$$

Effective depth of footing, d

$$d = [M / (R_B \, . \, B)]^{1/2} \tag{22.18}$$

Tension reinforcement

The total tension reinforcement at any section provides a moment of resistance at least equal to the bending moment at the section. The tension reinforcement extending in each direction is distributed uniformly across the full width of the footing. In the square footing, the steel bars are provided in two directions. Therefore, it is called as two-way reinforcement. The effective depths d for the upper layer is less by *one diameter* of the bar than that of the steel bars in lower layer, consequently, the area of steel reinforcement is larger for the upper layer. Instead of using two different areas of reinforcements, the area of steel bars for upper layer is calculated and reinforcement in two layers are kept equal. The area of steel reinforcement in each direction,

$$A_{st} = \left(\frac{M}{\sigma_{st} \cdot j_B \, (d - \phi)}\right) \tag{22.19}$$

The steel bars are provided uniformly in the full width of the footing. In case, A_ϕ is the cross-sectional area of one steel bar, then, the spacing of steel bars shall be as under :

$$s = \left(\frac{1000 \, . \, A_\phi}{A_{st}}\right) \tag{22.20}$$

The depth of square footing shall be checked for one-way (beam shear) and for two-way shear as described in Art. 13.8. The steel bars shall be provided for the full development length from the critical section for the bending moment.

22.13 RECTANGULAR FOOTING

The rectangular footings are also provided under the single column in case there is space restriction or if the supported columns are of elongated rectangular cross-section. In the simplest form, the rectangular footings consist of single slab. In the rectangular footing, the cantilevers project out from the column in both the directions. The column transfers the axial (concentric) load acting downward to the footing. The footing is loaded upward by the soil pressure. The tension stresses develop in both these directions at the bottom face. Such footings are therefore, reinforced by two layers of steel bars perpendicular to each other and parallel to the edge.

A rectangular footing under a single column with the forces acting on it, is shown in Fig. 22.14. The required bearing of footing is obtained by dividing the total load including the self-weight of the footing (the column load P + ten per cent of the column load) by the bearing capacity of the soil. The required bearing area of the footing.

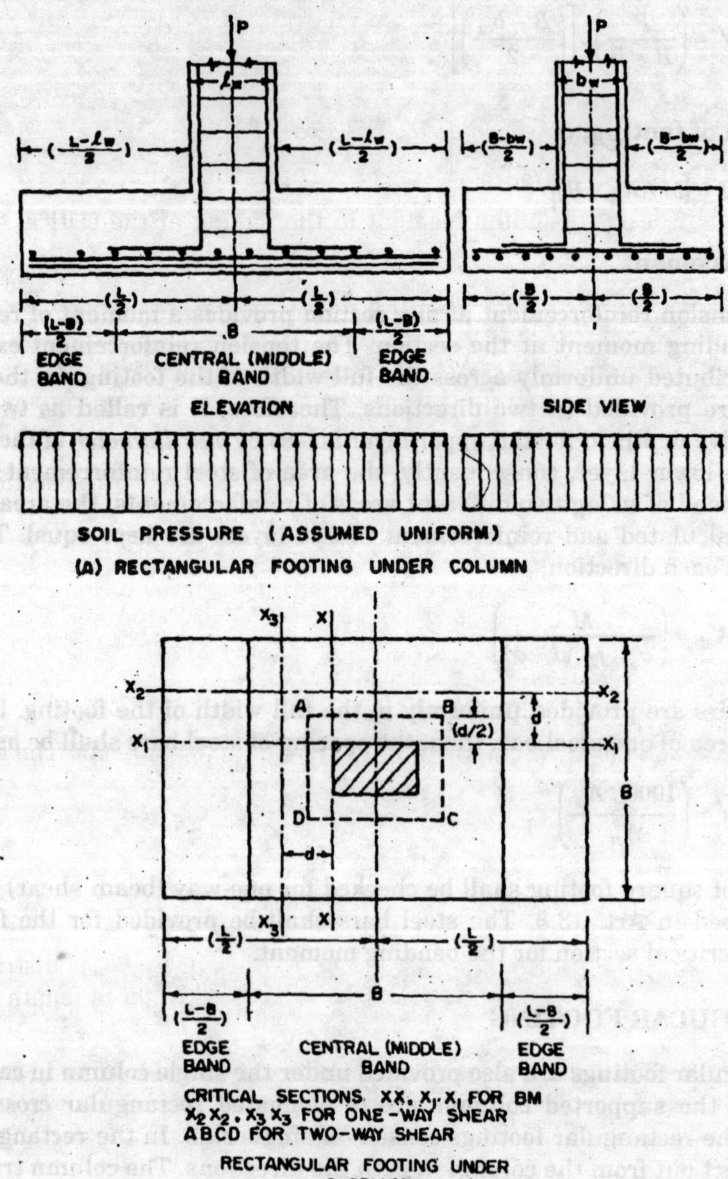

Fig. 22.14

$$A_F = \left(\frac{1.10\ P}{q_o}\right) \qquad \text{(i)}$$

$$A_F = (L \times B) \qquad \text{(ii)}$$

The critical sections XX and X_1X_1 for bending moments are at the faces of column as shown in Fig. 22.14. The maximum (design) bending moment in the footing at the section XX may be calculated as under :

$$M_L = \frac{1}{2} q_o \cdot \left[\left(\frac{L - l_w}{2}\right) \cdot \left(\frac{L - l_w}{2}\right)\right]$$

$$\text{or} \quad M_L = \frac{1}{2} \times \left(\frac{P}{L \times B}\right) \cdot \left[\left(\frac{L - l_w}{2}\right) \cdot \left(\frac{L - l_w}{2}\right)\right] \qquad (22.21)$$

The maximum (design) bending moment in the footing at the critical section X_1X_1 may be obtained as below :

$$M_B = \frac{1}{2} q_o \cdot \left[\left(\frac{B - b_w}{2}\right) \cdot \left(\frac{B - b_w}{2}\right)\right]$$

$$\text{or} \quad M_B = \frac{1}{2} \times \frac{P}{(L \times B)} \cdot \left[\left(\frac{B - b_w}{2}\right) \cdot \left(\frac{B - b_w}{2}\right)\right] \qquad (22.22)$$

Similarly the (design) shear forces are determined at the critical sections XX and X_1X_1 are as under :

$$V_L = q_o \cdot \left(\frac{L - l_w}{2}\right) = \left(\frac{P}{(L \times B)}\right) \cdot \left(\frac{L - l_w}{2}\right) \qquad (22.23)$$

$$V_B = q_o \cdot \left(\frac{B - b_w}{2}\right) = \left(\frac{P}{L \times B}\right) \cdot \left(\frac{B - b_w}{2}\right) \qquad (22.24)$$

The bending moment, M_L is larger than the bending moment M_B. The effective depth of footing is determined as under :

$$d = [M_L / R_B . B]^{1/2} \qquad (22.25)$$

Tension reinforcement

In rectangular footing, the reinforcement in the long direction is distributed uniformly across the full width viz., the shorter dimensions, that is, parallel to length of the footing. It is calculated as below :

$$A_{st.L} = \left(\frac{M_L}{\sigma_{st} \cdot j_B \cdot d}\right) \qquad (22.26)$$

The spacing of the bars across the width

$$s_L = \left(\frac{1000 \cdot A_\phi}{A_{st.L}}\right) \qquad (22.27)$$

For reinforcement in the short direction (viz., across the length, that is parallel to width), the length of footing is divided into two bands, namely *middle band* (or central band) and *edge bands*. The *central band shall have width equal to the width of the footing. The width of each edge band shall be equal to 0.5 (L – B)*. The middle portion of rectangular footing (viz., *central band),* gets concentrated support from the column. Therefore, the curvature of footing is sharpest, that is, the moment per metre is largest, immediately under

the column. Hence, the central portion needs a larger steel area than the edge bands (the far ends of footing). The area of steel bars required

$$A_{st.B} = \left(\frac{M_B}{\sigma_{st} \cdot j_B \cdot (d - \phi)}\right) \quad (22.28)$$

The reinforcement in the *central band* is increased by an amount equal to $[2/(\beta + 1)]$, where β is the ratio of long side to the short side of the footing. That is,

$$\left(\frac{\text{Reinforcement in central band}}{\text{Total reinforcement in short direction}}\right) = \left(\frac{2}{\beta + 1}\right)$$

$$\text{That is, } (A_{st})_{central\ band} = \left(\frac{2}{\beta + 1}\right) \cdot A_{st} \cdot B \quad (22.29A)$$

This reinforcement is uniformly distributed across the central band.

$$(S)_{central\ band} = \left(\frac{1000 \cdot A_\phi}{(A_{st})_{central\ band}}\right) \quad (22.30A)$$

The remaining area of steel reinforcement (viz., A_{st} B – A_{st} central band) shall be uniformly distributed in the two outer portions (edge bands) of the footing.

The minimum reinforcement specified by the code to prevent precipitous fracture applies in footings separately to each of the two directions unless the reinforcement provided is *one-third* more than needed.

The footing may be checked for one-way and two-way shears as described in Art. 22.8.

The critical sections for the bending moments are also the critical sections for the development length. The development length is also checked at all vertical planes where the changes of cross-section of footing or the reinforcement occur, as at the edge of pedestals or where the part of reinforcement is terminated.

22.14 STEPPED FOOTINGS

In isolated spread footings, different depths are needed at the different critical sections. These depths so calculated are not required at the edge of the footings. The depth of footing provided near the centre is reduced by providing steps (as shown in Fig. 22.15) towards the edges without exceeding the permissible stresses. By providing the steps in the cross-section of footing, the concrete is saved in the areas where, it is not structurally needed. These steps are provided where the thickness (or depth) of single slab of the footing is becoming excessively thick (usually more than *one metre*). The size and thickness are designed in such a way that each step (viz., change in cross-section), is safe to resist the bending and shear.

The area of footing at the base required is calculated for the total load, that is, the column load and the self-weight of the footing (viz., the column load, P plus *ten per cent* of the column load as the self-weight of the footing). Therefore,

$$A_F = \left(\frac{1.10\ P}{q_o}\right) \quad (i)$$

where, q_o is the safe bearing capacity of the soil.

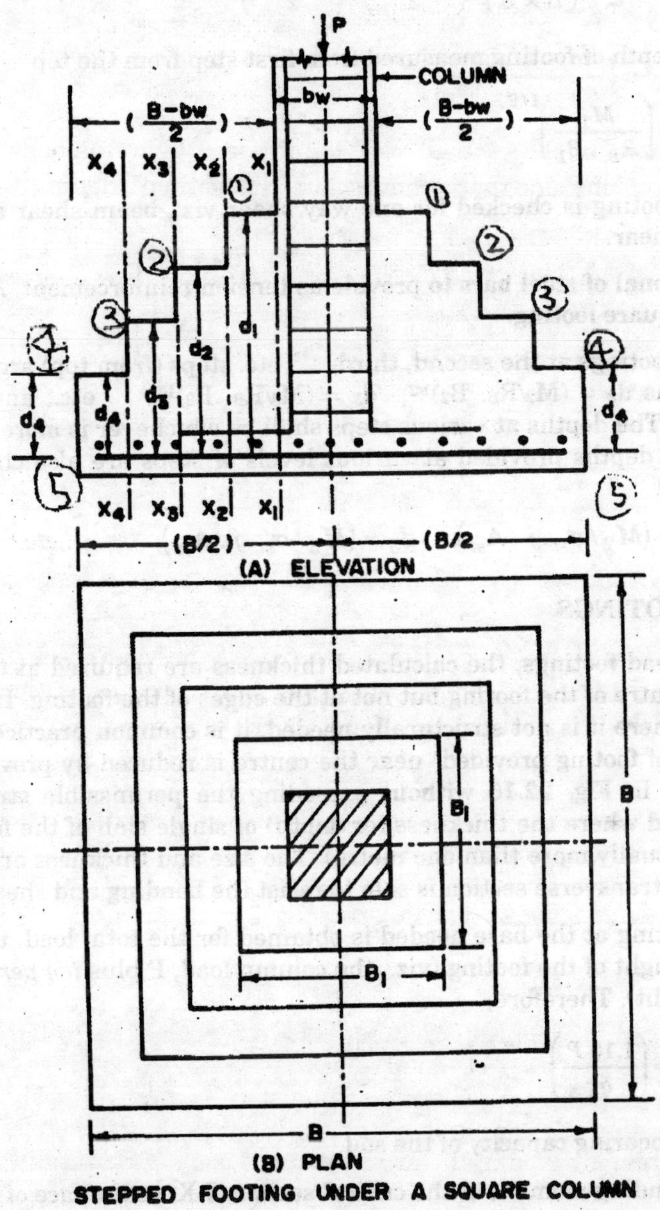

Fig. 22.15

The bending moments and shear forces at the various critical sections (X_1X_1, X_2X_2. . . etc.) are determined. The bending moment at critical section X_1X_1 at the face of column

$$M_1 = \frac{1}{2} q_u \left(\frac{B - b_w}{2}\right) \cdot \left(\frac{B - b_w}{2}\right) \tag{22.29B}$$

$$\text{or}\quad M_1 = \frac{1}{2}\left(\frac{P}{B \times B}\right) \cdot \left(\frac{B - b_w}{2}\right) \cdot \left(\frac{B - b_w}{2}\right) \tag{22.30B}$$

The effective depth of footing measured from first step from the top

$$d_1 = \left(\frac{M_1}{R_B \,.\, B_1}\right)^{1/2} \tag{22.31}$$

The depth of footing is checked for one way shear viz., beam shear and two-way shear viz., flat slab type shear.

The cross sectional of steel bars to provide as tension reinforcement, A_{st}, is calculated as it is found for the square footing.

The depths of footings at the second, third, . . . etc. steps (from top) are obtained from the bending moments as $d_2 = (M_2/R_B.\ B_2)^{1/2}$, $d_3 = (M_3/R_B\ .B_3)^{1/2}$. . . etc., and so also from the reinforcement side. The depths at various steps shall be whichever is more obtained from two considerations. The depths provided at various levels of steps are also checked for one-way and two-way shears.

$$d_2 = \left(M_2 / \sigma_{st} \cdot j \,.\, A_{st}\right), \quad d_3 = \left(M_3 / \sigma_{st} \cdot j \,.\, A_{st}\right), \quad \text{......etc.}$$

22.15 SLOPED FOOTINGS

In isolated spread footings, the calculated thickness are required at the various critical sections near the centre of the footing but not at the edges of the footing. In order to save the concrete in areas where it is not structurally needed, it is common practice to provide sloped footing. The depth of footing provided near the centre is reduced by providing a taper upto the edge as shown in Fig. 22.16 without exceeding the permissible stresses. The sloped footings are provided where the thickness (or depth) of single slab of the footing is becoming excessively thick (usually more than one metre). The size and thickness are designed in such a manner that each transverse section is safe to resist the bending and shear.

The area of footing at the base needed is obtained for the total load, that is, the column load and the self-weight of the footing (viz., the column load, P plus *ten percent* of the column load as the self-weight). Therefore,

$$A_F = \left(\frac{1.10\, P}{q_o}\right) \tag{i}$$

where q_o is the safe bearing capacity of the soil.

The factored bending moment at the critical section X_1X_2 at the face of column

$$M_1 = \frac{1}{2} q_o \cdot \left(\frac{B - b_w}{2}\right) \cdot\cdot \left(\frac{B - b_w}{2}\right)$$

$$\text{or}\quad M_1 = \frac{1}{2} \cdot \left(\frac{P}{B \times B}\right) \cdot \left(\frac{B - b_w}{2}\right) \cdot \left(\frac{B - b_w}{2}\right) \tag{22.32}$$

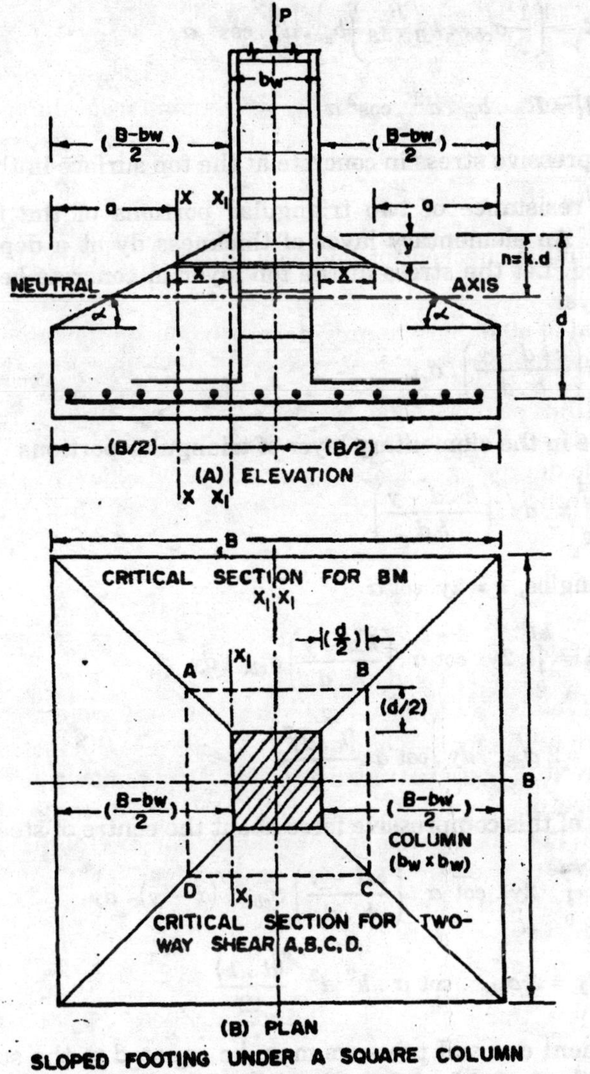

Fig. 22.16

The cross-sectional shape of the footing, Fig. 22.16 (A), resisting this bending moment is *trapezoidal*. The moment of resistance of this trapezoidal cross-section shall comprise of the moment of resistance of the middle rectangular portion and that of the two triangular portions at the sides. The horizontal component of total compressive force (which shall be acting horizontally)

$$C_H = \left(\frac{1}{2}\,\sigma_{cbc}\,.\,b_w\,.\,k_B\,d\,.\cos^2\alpha\right) \qquad \text{(ii)}$$

The moment of resistance of middle rectangular portion

$$MR_1 = C_H\; j_B\;.\,d \qquad \text{(iii)}$$

$$\text{or } MR_1 = \left(\frac{1}{2}\sigma_{cbc} \cdot k_B \cdot j_B\right) b_w \cdot d^2 \cdot \cos^2 \alpha \quad \text{(iv)}$$

$$\text{or } MR_1 = R_B \cdot b_w \cdot d^2 \cdot \cos^2 \alpha \quad (22.34)$$

where, σ_{cbc} is the compressive stress in concrete at the top surface in the middle portion.

The moment of resistance of two triangular portions of the footing at the sides is determined as under. An elementary layer of thickness dy at a depth y from the top Fig. 22.16 (A) is considered. Let the stress at the top layer in concrete be σ_{cbc}. The stress at the level of elementary layer

$$\sigma_{cbc}' = \left(\frac{k \cdot d - y}{k \cdot d}\right) \cdot \sigma_{cbc} \quad \text{(v)}$$

Compressive force in the elementary layer of triangular portions

$$C_T = \int_0^{kd} x \cdot dy \cdot \left(\frac{k \cdot d - y}{k.d}\right) \quad \text{(vi)}$$

From similar triangles, $x = 2y \cdot \cot \alpha$

$$\text{or } C_{T2} = \int_0^{kd} 2y \cdot \cot \alpha \cdot \left(\frac{k.d - y}{k \cdot d}\right) \sigma_{cbc} \cdot dy \quad \text{(vii)}$$

$$\text{or } C_{T2} = 2\,\sigma_{cbc} \cdot dy \cdot \cot \alpha \cdot \frac{(k \cdot d)^2}{6} \quad (22.35)$$

Take the moment of this compressive force about the centre of steel reinforcement.

$$MR_2 = \int_0^{kd} 2y \cdot \cot \alpha \cdot \left(\frac{k.d - y}{k \cdot d}\right) \sigma_{cbc} \cdot (d - y) \cdot dy \quad \text{(viii)}$$

$$\text{or } MR_2 = 2\,\sigma_{cbc} \cdot \cot \alpha \cdot k^2\, d^3\, \frac{(d \cdot k)}{12} \quad (22.36)$$

The bending moment due soil pressure may be equated to the sum of MR_1 and MR_2 to find the effective depth. Instead of this, an equivalent rectangular cross-section of width $[b_w + (B - b_w)/8]$ and depth, d may be used as an approximation for the trapezoidal cross-section and the depth of footing may be determined.

The footing may be checked for one-way and two-way shears and also for the development lengths. The area of steel bars in each direction may be calculated and the reinforcement is provided as usual for the square footing.

Example 22.1 **A 450 mm masonry wall supports a dead load of 180 kN/m and a live load of 160 kN/m. The allowable bearing pressure on soil is 150 kN/m². Design a reinforced concrete footing under the masonry wall. Use M 15 grade of concrete and mild steel bars less than or equal to 20 mm. The footing shall be 1.24 m below the ground surface. The height of wall is 3 m.**

Design : The permissible stresses for M 15 grade of concrete and mild steel bars of less than or equal to 20 mm diameter are 5 N/mm² and 140 N/mm², respectively. The modular ratio of M 15 grade of concrete.

$$m = \left(\frac{280}{3 \times 5}\right) = 18.67$$

Step 1 : Design constants

The section of footing shall be designed as balanced section. The design constants are as follows :

Factor for depth of neutral axis

$$k_B = \left(\frac{18.67 \times 5}{18.67 \times 5 + 140}\right) = 0.4$$

$$j_B = \left(1 - \frac{0.4}{3}\right) = 0.867$$

$$R_B = \left(\frac{1}{2} \times 5 \times 0.4 \times 0.867\right) = 0.867$$

$$p_B = \left(\frac{1}{2} \times \frac{5}{40} \times 0.4 \times 100\right) = 0.7143 \textit{ per cent.}$$

Step 2 : Load supported by footing

Consider *1* m length of masonry wall. Weight of the masonry wall

$$\left(\frac{450}{1000} \times 3 \times 1 \times 20\right) = 27 \; kN/m$$

Dead load supported by the wall, = 180 kN/m

Live load supported by wall, = 160 kN/m

Total load = 367 kN/m

The self-weight of footing is assumed as 10 per cent of total load to be transferred

= 3.67 kN/m

Total load in order to calculate the width of footing

= (367 + 3.67) = 370.67 kN/m

The level of footing is 1.240 m below the ground surface. The overall depth of footing may be assumed as 350 mm. The weight of earth per square metre

$$= (1.24 - 0.35) \times 1 \times 1 \times 20 = 17.8 \; kN/m$$

Consequently, the portion of allowable bearing pressure which shall be available or effective for carrying the wall load

$$q = (150 - 17.8) = 132.2 \; kN/m^2$$

Step 3 : Width of footing required

$$B = \left(\frac{370.67}{132.2}\right) = 2.804 \; mm$$

Provide a width of 3 m under the masonry wall as shown in Fig. 22.17.

Step 4 : Bending moment and shear force

Total load to be transferred to the footing is 367 kN/m. Therefore, the actual soil pressure

$$q_o = \left(\frac{367}{3}\right) = 122.33 \; kN/m^2$$

For *1* m width, q_o = 122.33 kN/m

$$\left(\frac{B - b_w}{2}\right) = \left(\frac{3 - 0.450}{2}\right) = 1.275 \; mm$$

For the masonry wall, the critical section for moment lies at the quarter point of the wall thickness. The critical section XX for the masonry wall is shown in Fig. 22.17. The distance of section XX from the edge

$$\left(\frac{B - b_w}{2}\right) + \frac{b_w}{4} = \left(1.275 + \frac{0.450}{4}\right) = 1.3875 \; m$$

Step 5 : Bending moment about the critical section XX, (Fig. 22.17)

$$M = \left(\frac{1}{2} \times 122.33 \times 1.3875^2\right) = 117.75 \; kN - m$$

The position of critical section XX for bending is explained by Fig. 22.17 (B). The shear force diagram is shown in Fig. 22.17(B). This diagram shows that the maximum bending moment occurs at the middle of the wall. However, in case the cantilever projection is extended back to the quarter point of the wall thickness, as shown by the dotted line, the area under the shear force diagram is nearly the same.

Step 6. Depth of footing

Let the effective depth of footing be d. Then

$$d = \left(\frac{M}{R_B \, . \, B}\right)^{1/2} = \left(\frac{117.75 \times 10^6}{0.867 \times 3000}\right)^{1/2} = 212.77 \; mm$$

Provide effective cover of 60 mm. Over all depth of footing

$$= (212.77 + 60) = 282.77 \; mm$$

Provide 300 mm thick footing. The effective depth of footing shall be

$$d = (300 - 60) = 240 \text{ mm}$$

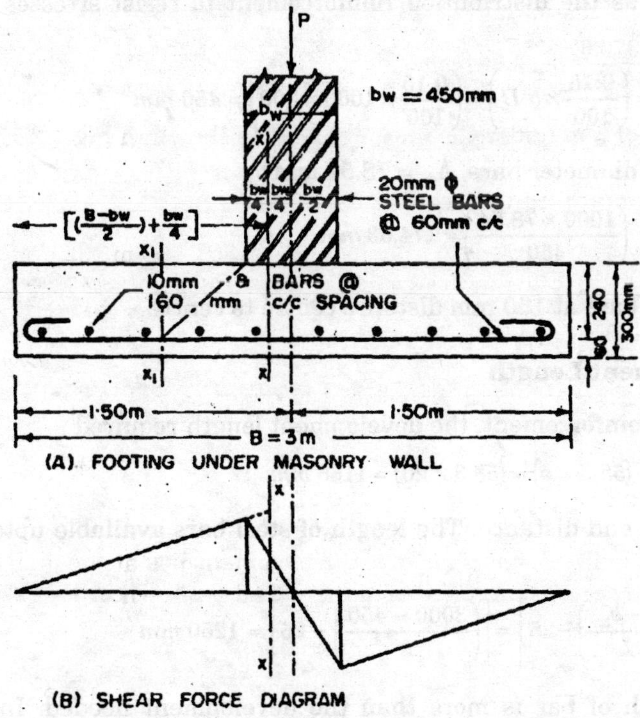

Fig. 22.17

Step 7. Shear force about the critical section

X_1X_1 (Fig. 22.17), at a distance d from the wall

$$V = 122.33 \times (1.3875 - 0.240) = 140.37\ kN$$

Step 8 : Reinforcement for bending

Main reinforcement: The cross-sectional area of the steel bars to be provided as reinforcement in tension

$$A_{st} = \left(\frac{M}{\sigma_{st} \cdot j_B \cdot d}\right) = \left(\frac{117.75 \times 10^6}{140 \times 0.867 \times 240}\right) = 4042.058\ mm^2$$

Provide 20 mm diameter bars. The cross-sectional area of one bar

$$A_\phi = 314.16\ mm^2$$

$$\text{Spacing of the bars, } s = \left(\frac{1000 \times 314.16}{4042.058}\right) = 77.72\ mm.$$

Provide the steel bars at 60 mm spacing centre to centre.

Step 9 : Distribution reinforcement : 0.15 percent of the cross-sectional area of the footing is provided as the distribution reinforcement to resist stresses due to tension and shrinkage. (b = 1 m)

$$A_d = \left(\frac{0.15}{100} \times b\ D\right) = \left(\frac{0.15}{100} \times 1000 \times 300\right) = 450\ mm^2$$

Provide 10 mm diameter bars, A_ϕ = 78.54 mm²

$$s_d = \left(\frac{1000 \times 78.54}{450}\right) = 174.53\ mm$$

Provide 10 mm bars at 160 mm distance centre to centre

Step 10 : Development Length

For mild steel reinforcement, the development length required

$$L_d = (58.3 \times \phi) = (58.3 \times 20) = 1166\ mm$$

Provide 25 mm end distance. The length of steel bars available upto the critical section for bending moment

$$\left[\left(\frac{B - b_w}{2}\right) - 25\right] = \left[\left(\frac{3000 - 450}{2}\right) - 25\right] = 1250\ mm$$

Available length of bar is more than the development needed. In addition to this, a standard hook is also provided at the end of the bar.

Note : The design of footing under a concrete wall is similar to this except that the critical section for bending moment shall be at the face of concrete wall.

Example 22.2 A column 450 mm x 450 mm in size with 8 steel bars of 18 mm diameter transfer a dead load of 620 kN and a live load of 860 kN to the footing. The bearing capacity of soil is 120 kN/m². M 20 grade of concrete and Hysd-steel bars of Fe 415 shall be used. Design a square footing to support the column.

Design : The permissible stresses for M 20 grade of concrete and Hysd Steel bars of Fe 415 grade for bending are 7 N/mm² and 230 N/mm², respectively. The modular ratio of M 20 grade of the concrete.

$$m = \left(\frac{280}{3 \times 7}\right) = 13.33 \text{ and } m_c = (1.5 \times 13.33) = 20$$

Step 1 : Design constants

The section of footing shall be designed as balanced section. The design constants are as follows : Factor for depth of neutral axis

$$k_B = \left(\frac{13.33 \times 7}{13.33 \times 7 + 230}\right) = 0.2886$$

$$j_B = \left(1 - \frac{k_B}{3}\right) = \left(1 - \frac{0.2886}{3}\right) = 0.9038$$

$$R_B = \left(\frac{1}{2} \times 7 \times 0.2886 \times 0.9038\right) = 0.9129$$

$$p_B = \left(\frac{1}{2} \times \frac{7}{230} \times 0.2886 \times 100\right) = 0.4392 \textit{ per cent}$$

Step 2 : Size of footing to maintain allowable soil pressure

Dead load and live load to be transferred by the column

P = (620 + 860) = 1480 kN

Self-weight of footing may be assumed *ten percent* of the column load

= 148 kN

Total load = 1628 kN

Area of the footing required in plan

$$A_F = \left(\frac{1628}{120}\right) = 13.506 \; m^2$$

A square footing shall be provided to support the load from the square column. Size of footing

$$B = (13.506)^{1/2} = 3.683 \; m$$

A square footing of size 3.70 m x 3.70 m shall be provided as shown in Fig. 22.18. Net soil pressure acting upward

$$q_o = \left(\frac{1480}{3.700 \times 3.700}\right) = 108.11 \; kN/m^2$$

< bearing capacity of the soil. Hence safe.

Step 3 : Bending moment

Bending moment at the critical section XX at the face of the column for 1 m strip of the footing

$$M = \frac{1}{2} q_o \left(\frac{B - b_w}{2}\right)^2 \times 1 \; kN-m/m$$

or $$M = \frac{1}{2} \times 108.11 \times \left(\frac{3.70 - 0.450}{2}\right)^2 \times 1 = 142.739 \; kN-m/m$$

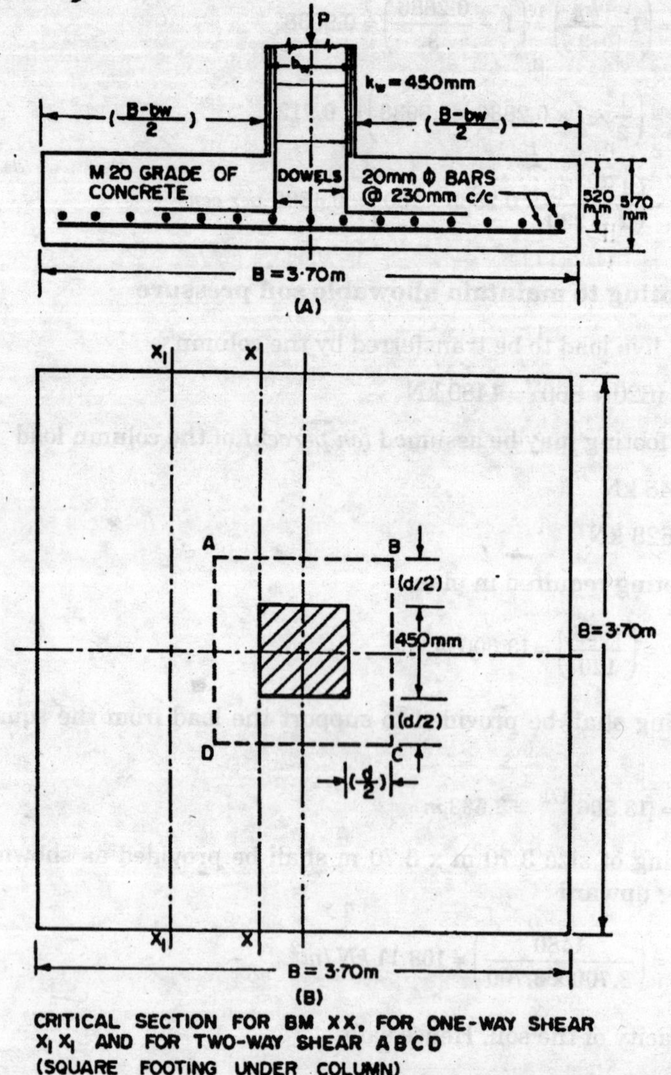

Fig. 22.18

Step 4. Depth of footing

The effective depth of footing, since 1 m strip is considered

$$d = \left(\frac{M}{R_B\ .1000}\right)^{1/2} = \left(\frac{142.739 \times 10^6}{0.9129 \times 1000}\right)^{1/2} = 395.42\ mm$$

Let the effective of footing be 50 mm. The overall depth of footing

D = (395.42 + 50) = 445.42 mm, say D = 450 mm,

Therefore, d = 400 mm.

Step 5 : Check for single shear

Shear force at the critical section X_1X_1 at a distance d = effective depth from the face of the footing.

$$V = q_o\left[\left(\frac{B - b_w}{2}\right) - d\right] \times 1$$

or $$V = 108.11\left[\left(\frac{3.70 - 0.450}{2}\right) - 0.4\right] \times 1\ kN/m$$

or $$V = (108.11 \times 1.225) = 132.435\ kN/m$$

Nominal shear stress across the depth of footing

$$\tau_v = \left(\frac{132.435 \times 1000}{1000 \times 400}\right) = 0.331\ N/mm^2$$

For the balanced section for footing, percentage area of the steel reinforcement

$$p_B = \left(\frac{100\ A_s}{bd}\right) = 0.4392\ percent$$

From IS : 456-2000, the permissible shear stress in concrete (without shear reinforcement)

$$\tau_c = \left(0.22 + \frac{0.08 + 0.1892}{0.25}\right) = 0.28\ N/mm^2$$

Therefore, the depth of footing needed for one way shear

$$d = \left(\frac{132.435 \times 1000}{1000 \times 0.28}\right) = 472.98\ mm$$

Effective cover EC = 50 mm

Overall thickness of concrete

D = (472.98 + 50) = 522.98 mm

However, D may be adopted as 525 mm. Revised effective depth shall be 485 mm.

$$\tau_v = \left(\frac{132.435 \times 1000}{1000 \times 485}\right) = 0.273\ N/mm^2$$

It is less than the permissible stress, τ_c. Hence, safe.

Step 6 : Check for two-way shear

The critical section for two-way shear shall be BCDA, Fig. 22.18(B). The peripherical perimeter

$$b_o = 4 \times \left(450 + \frac{485}{2} + \frac{485}{2}\right) = 3740 \; mm$$

$$\therefore AB = 935 \; mm = 0.935 \; m$$

Total shear force

$$V = q_o \left[\left(B^2 - AB^2\right)\right] = 108.11 \left[\left(3.7^2 - 0.935^2\right)\right] = 1385.513 \; kN$$

$$= 108.11 \times 12.816 = 1385.513 \; kN$$

Nominal shear stress

$$\tau_v = \left(\frac{V_{Fd}}{b_o . d}\right) = \left(\frac{1385.513 \times 1000}{3740 \times 485}\right) = 0.7638 \; N/mm^2$$

Permissible shear stress in concrete for two-way shear for M 20 grade

$$\tau_c' = k_s \, . \, \tau_c$$

$$k_s = (0.5 + \beta_c) \ngtr 1, \quad \left(\beta_c = \frac{B}{B} = 1.0\right) \; and \; k_s = 1.00$$

$$\tau_c = 0.16 \, (f_{ck})^{1/2}$$

$$\text{or } \tau_c' = 0.16 \times (20)^{1/2} = 0.7155 \; N/mm^2$$

The effective depth of footing needed to resist two-way shear

$$d = \left(\frac{1385.513 \times 1000}{3740 \times 0.7155}\right) = 517.73 \; mm$$

Add the effective cover of 50 mm. The overall depth of footing required

D = (517.73 + 50) = 567.73 mm

say D = 570 mm, EC = 50, d = 520 mm

Fig. 22.18 (B)

$$AB = \left(450 + \frac{520}{2} + \frac{520}{2}\right) = 970 mm$$

$$b_o = (4 \times 970) = 3880 \; mm$$

$$\tau_v = \left(\frac{1385.513 \times 1000}{3880 \times 520}\right) = 0.6867 \; N/mm^2$$

It is less than $\tau_c = 0.7155 \; N/mm^2$. Hence, the footing is safe in two-way shear also.

Step 7. Reinforcement for bending

In the square footing, the area of steel bars to be provided in the upper layer shall be more than the area of steel bars to be provided in the lower layer. Therefore, the area of steel bars needed, for upper layer is also provided for that in the lower layer. The area of steel bars required for upper layer for *1* m strip of the footing (Let ϕ = 20 mm)

$$A_{st} = \left(\frac{M}{\alpha_{st} \times j_B . (d - \Phi)}\right)$$

$$A_{st} = \left(\frac{142.739 \times 10^6}{230 \times 0.9038\ (520 - 20)}\right) mm^2$$

or $A_{st} = 1373.32\ mm^2, and\ A_{\phi} = 314.159\ mm^2$

spacing of the steel bars in the upper layer and lower layer both are equal

$$s = \left(\frac{1000 \times 314.159}{1373.32}\right) = 228.78\ mm$$

Provide 20 mm diameter Hysd bars of grade Fe 415 at 220 mm spacing centre to centre in both the layers.

Step 8 : Check for the transfer of load at the base of the column

The concrete of footing is having the same strength as the concrete of column, the bearing at the top of the footing is adequate [See step 9, Example 22.3].

Example 22.3 **A rectangular column 450 mm x 600 mm transfer a dead load of 880 kN and a live load of 1420 kN without any moment and there is no overburden. The safe bearing capacity of soil is 140 kN/m². M 20 grade of concrete and Hysd-steel bars of Fe 415 grade shall be used. Design the rectangular footing to support the column.**

Design : **The permissible stresses for M 20 grade of concrete and Hysd steel bars of Fe 415 grade for bending are 7 N/mm² and 230 N/mm², respectively. The modular ratio of M 20 grade of concrete and Hysd steel bars of grade 415 shall be used.**

$$m = \left(\frac{280}{3 \times 7}\right) = 13.33\ and\ m_c = (1.5 \times 13.33) = 20$$

Step 1 : Design constants

The section of footing shall be designed as a balanced section. The design constants are as follows :

$$k_B = \left(\frac{13.33 \times 7}{13.33 \times 7 + 230}\right) = 0.2886$$

$$j_B = \left(1 - \frac{2886}{3}\right) = 0.9038$$

$$R_B = \left(\frac{1}{2} \times 7 \times 0.2886 \times 0.9038\right) = 0.9129$$

$$p_B = \left(\frac{1}{2} \times \frac{7}{230} \times 0.2886 \times 100\right) = 0.4392\ per\ cent.$$

Step 2. Size of footing to maintain allowable soil pressure

Dead load and live load to be transferred by the column

$$P = (880 + 1420) = 2300\ kN$$

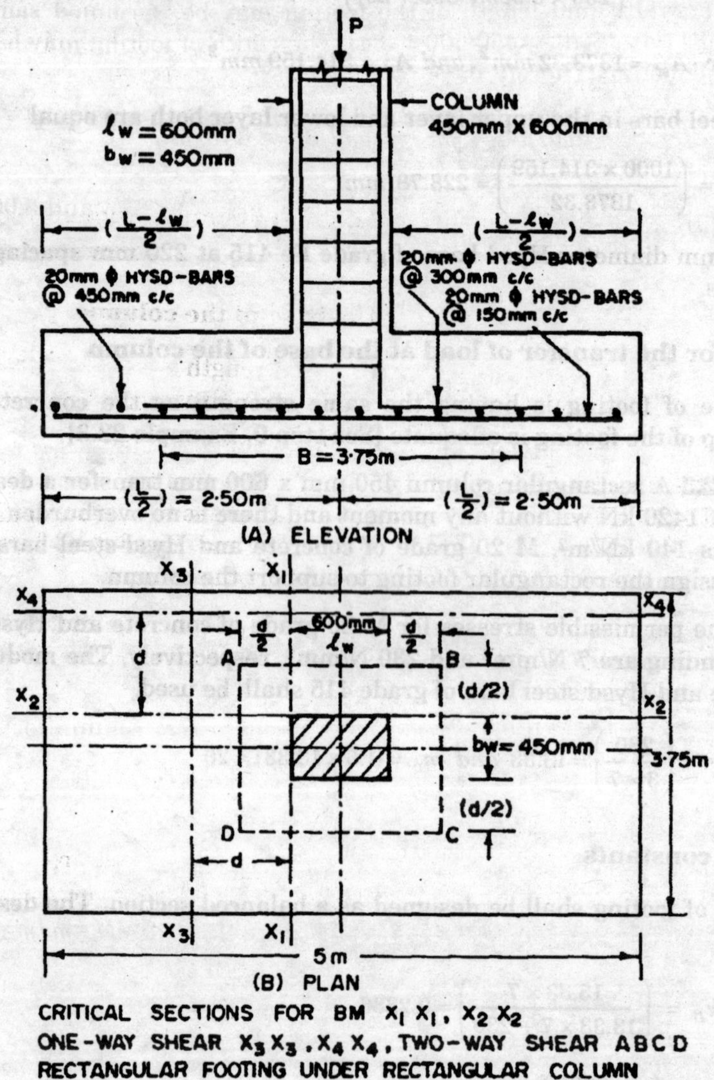

Fig. 22.19

Self-weight of the footing may be assumed *ten percent* of the column load

= 230 kN

Total load = 2530 kN

Area of the footing required in plan

$$A_F = \left(\frac{2530}{140}\right) = 18.07\ m^2$$

A rectangular footing shall be provided to support the load from the rectangular column. The proportions of width and length of the footing may be assumed same as that of the column dimensions, (viz., b_w/l_w = 450/600 = 3/4). The width of footing may be taken as (3/4)th of the length of the footing. Therefore,

$$\left(\frac{3}{4} L \times L\right) = 18.07, L = 4.9087\ m \ \ and\ B = 3.6815\ m$$

The width and length of the footing may be adopted as 3.75 m and 5.00 m, respectively as shown in Fig. 22.19, Net soil pressure acting upward

$$q_o = \left(\frac{880 + 1420}{3.75 \times 5.0}\right) = 122.67 < 140\ kN/m^2$$

Step 3. Bending moment

Bending moment at the critical section X_1X_1 at the face of column for 1 m strip along the length.

$$M_L = \frac{1}{2} \cdot q_o \cdot \left(\frac{L - l_w}{2}\right)^2 .1\ kN-m/m$$

$$M_L = \frac{1}{2} \times 122.67 \times \left(\frac{5.00 - 0.60}{2}\right)^2 \times 1 = 296.86\ kN-m/m$$

Bending moment at the critical section X_2X_2 at the face of column for 1 strip along the width

$$M_B = \frac{1}{2} \times q_o \left(\frac{B - b_w}{2}\right)^2 \times 1\ kN-m/m$$

or $$M_B = \frac{1}{2} \times 122.67 \left(\frac{3.75 - 0.450}{2}\right)^2 \times 1 = 166.985\ \ kN-m/m$$

($M_B < M_L$)

Step 4 : Depth of footing

The effective depth of footing, (since 1 m strip is considered)

$$d = \left(\frac{M}{R_B \times 1000}\right)^{1/2} = \left(\frac{296.86 \times 10^6}{0.9129 \times 1000}\right)^{1/2} = 570.25\ mm$$

Let the effective cover be 60 mm. Then, the overall depth of footing

D = (570.25 + 60) = 630.25 mm

Provide overall depth footing as 650 mm then, the effective depth shall be 590 mm.

Step 5 : Check for single shear

Shear force at the critical section X_3X_3 at a distance d = effective depth from the face of the footing for 1 m strip along the length

$$V_L = q_o \cdot \left[\left(\frac{L - l_w}{2}\right) - d\right] \times 1\ kN/m$$

$$V_L = 122.67 \times \left[\left(\frac{5 - 0.60}{2}\right) - 0.59\right] \times 1 = 197.50\ kN/m.$$

Shear force at the critical section X_4X_4 at a distance d = effective depth from the face of the footing for 1 strip along the width

$$V_B = q_o \cdot \left[\left(\frac{B - b_w}{2}\right) - d\right] \times 1\ kN/m$$

$$V_B = 122.67 \times \left[\left(\frac{3.75 - 0.450}{2}\right) - 0.590\right] \times 1 = 130.03\ kN/m$$

$(V_B < V_L)$

Nominal shear stress across the depth of footing

$$\tau_v = \left(\frac{V_L}{1000 \times d}\right) = \left(\frac{197.50 \times 100}{1000 \times 590}\right) = 0.335\ N/mm^2$$

For the balanced section for footing percentage area of the steel reinforcement

$$p_B = \left(\frac{100\ A}{bd}\right) = 0.4392\ percent$$

From IS : 456-2000, the permissible shear stress is concrete (without shear reinforcement)

$$\tau_{sc} = \left(0.22 + 0.08 \times \frac{0.1892}{0.25}\right) = 0.28\ N/mm^2$$

Therefore, the depth of footing required for one way shear

$$d = \left(\frac{197.50 \times 1000}{1000 \times 0.28}\right) = 705.36\ mm$$

Effective cover has been 60 mm. Therefore, the overall depth

$$D = (705.36 + 60) = 765.36\ mm$$

Keep overall depth, D of the footing as 770 mm. Then, the effective depth of footing shall be 710 mm. Therefore,

$$\tau_v = \left(\frac{197.50 \times 1000}{1000 \times 710}\right) = 0.278\ N/mm^2$$

It is less than the permissible stress, τ_c. Hence, safe.

Step 6 : Check for two-way shear

The critical section of two-way shear shall be ABCD, Fig. 13.19 (B), the peripherical perimeter

$$b_o = 2 \times \left[\left(600 + \frac{710}{2} + \frac{710}{2}\right) + \left(450 + \frac{710}{2} + \frac{710}{2}\right)\right] mm$$

$$\therefore b_o = 2 \times [(1310) + 1160] = 4940 \, mm$$

Total shear force

$$V = q_o \, [(5.0 \times 3.75) - (0.6 \times 0.450)] \, kN$$

$$\text{or} \quad V = 122.67 \times [(18.75 - 0.27)] = 2266.94 \, kN$$

Nominal shear stress

$$\tau_v = \left(\frac{V}{b_o \, . \, d}\right) = \left(\frac{2266.94 \times 1000}{4940 \times 710}\right) = 0.646 \, N/mm^2$$

Permissible shear stress in concrete for two-way shear for M 20 grade

$$\tau_c = k_s \, . \, \tau_c$$

$$k_s = (0.5 + \beta_c) \not> 1, \left(\beta_c = \frac{3.75}{5} = 0.75\right), \therefore k_s = 1$$

$$\tau_c' = 0.16 \, (f_{ck})^{1/2} = 0.16 \times (20)^{1/2} = 0.7155 \, N/mm^2$$

The nominal shear stress τ_v is less than the permissible stress, τ_c. Hence, the depth is satisfactory.

Step 7. Reinforcement for bending moment

The cross-sectional area of the steel bars needed *along the length* for 1 m strip.

Let $\quad \phi_L = 20 \, mm$

$$A_{st.L} = \left(\frac{M_L}{\sigma_{st} \cdot j_B \, . \, d}\right) = \left(\frac{296.86 \times 10^6}{2.30 \times 0.9038 \times 710}\right) = 2011.28 \, mm^2$$

$$A_\phi = 314.159 \, mm^2$$

Spacing of the steel bars alone the length (distributed uniformly)

$$s_L = \left(\frac{1000 \times 314.159}{2011.38}\right) = 156.19 \, mm$$

Provide these bars at 150 mm spacing. The cross-sectional area of the steel bars required *along the width* for 1 m strip.

Let $\phi_B = 20\ mm.\ (\phi_B = \phi_c = \phi)$

$$A_{st.B} = \left(\frac{M_B}{\sigma_{st} \cdot j_B \cdot (d - \phi)}\right) = \left(\frac{166.985 \times 10^6}{2.30 \times 0.9038 \times (710 - 20)}\right) = 1164.20\ mm^2$$

Reinforcement in central (middle) band

$$A_{st.CB} = \left(\frac{2}{\beta + 1}\right) \times \textit{Total reinforcement in short direction}$$

$$\beta = \left(\frac{\textit{Long side of footing}}{\textit{Short side of footing}}\right) = \left(\frac{5.00}{3.75}\right) = 1.333$$

Therefore,

$$A_{st.CB} = \left(\frac{2}{1.333 + 1}\right) \times 1164.20 = 997.886\ mm^2$$

Reinforcement in *two side (edge)* bands

$$A_{st.EB} = (1164.20 - 997.886) = 166.31\ mm^2$$

$$\left(\frac{1}{2} A_{st.EB}\right) = \left(\frac{1}{2} \times 166.31\right) = 83.155\ mm^2$$

Spacing of bars in central (middle) band

$$s._{CB} = \left(\frac{1000 \times 314.159}{997.886}\right) = 314.825\ mm$$

s.CB (say) = 300 mm

Spacing of bars in side (edge) band

$$s._{EB} = \left(\frac{1000 \times 314.159}{83.155}\right) = 37.7797\ mm \not> 450\ mm$$

s.EB (say) = 450 mm.

Step 8 : Check for development length

Development required for Hysd steel bars

$$L_d = \left(\frac{\sigma_{st} \cdot \phi}{4 \times 1.40\ (\tau_{bd})}\right) = \left(\frac{230 \times \phi}{4 \times 1.40 \times 0.8}\right) = 51.34\ \phi$$

$$= (51.34 \times 20) = 1026.8\ mm$$

Development length available along the length (from the face of column)

$$\left[\left(\frac{L-l_w}{2}\right)-end\ distance\right]=\left[\left(\frac{5000-600}{2}\right)-25\right]$$

= 2175 mm (satisfactory)

Development length available along the width (from the face of column)

$$\left[\left(\frac{B-b_w}{2}\right)-end\ distance\right]=\left[\left(\frac{3750-450}{2}\right)-25\right]$$

= 1625 mm (satisfactory)

Step 9. Check for transfer of load at the base of column

Loaded area at the column base

$$A_2=(450\times 600)=27\times 10^4\ mm^2$$

Supporting area for bearing of footing which one is sloping at the rate of 2 : 1.

Spread along the length upto the effective depth

$$(2d+l_w+2d)=(2\times 710+450+2\times 710)=3440\ mm$$

Spread along the width upto the effective depth

$$(2d+b_w+2d)=(2\times 710+450+2\times 710)=3290\ mm$$

Supporting area for bearing

$$A_1=(3440\times 3290)=1131.76\times 10^4\ mm^2$$

$$\left(\frac{A_1}{A_2}\right)^{1/2}=\left(\frac{1131.76\times 10^4}{27\times 10^4}\right)^{1/2}=6.47>2.0$$

Permissible bearing stress on full area of concrete

$$\sigma_{cbr}=0.25\,.\,f_{ck}=(0.25\times 20)=5\ N/mm^2$$

$$\sigma_{cbc}\cdot\left(\frac{A_1}{A_2}\right)^{1/2}=(5\times 2)=10\ N/mm^2$$

Compressive stress in concrete at the base of a column

$$q=\left(\frac{880+1420}{600\times 450}\right)=8.593\ N/mm^2$$

It does not exceed the permissible stress.

It is to note that because of the slope of spread of footing is 2 : 1, *where the effective depth of footing is more than half the width of column* the permissible stress q will not exceed the permissible stress as calculated above.

It is further to note that when the concrete in footing is having same strength as the concrete [which shall be enhanced by $(A_1/A_2)^{1/2}$ shall be clearly capable of carrying that part of the column load transmitted by the column concrete.

Example 22.5. Design a stepped square footing to support the column, in Example 22.2.

Design : From Example 22.2, the modular ratio of M 20 grade of concrete.

m = (13.33) and m_c = 20

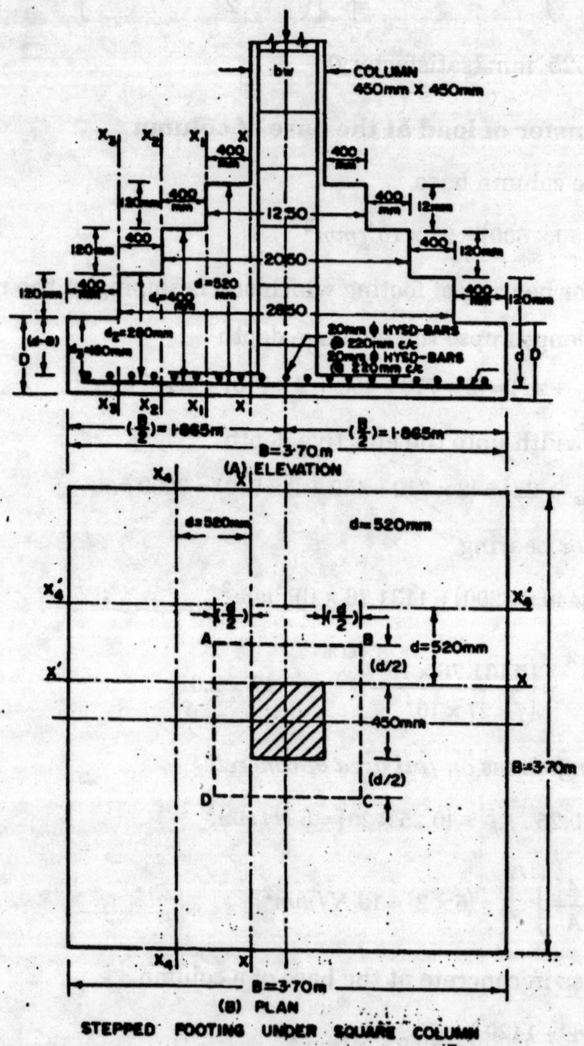

Fig. 22.20

Step 1. Design constants

The section of footing shall be designed as the balanced section. From *Example 22.2,* the design constants are as under; for m 20 grade of concrete and Hysd-steel bars of Fe 415 grade.

$k_B = 0.2886$, and $j_B = 0.9038$

$R_B = 0.9129$ and $p_B = 0.4392$ per cent.

Step 2. Size of footing to maintain allowable soil pressure

Dead load and live load to be transmitted by the column

P = (620 + 860) = 1480 kN

Self-weight of the footing may be assumed *ten percent* of the column load

= 148 kN

Total load = 1628 kN

Area of footing required in plan

$$A_F = \left(\frac{1628}{120}\right) = 13.566\ m^2$$

A square footing shall be provided to support the load from the square column. Size of footing

$$B = (13.566)^{1/2} = 3.683\ m$$

A square footing of size 3.70 m x 3.70 m shall be provided as shown in Fig. 22.20. Net soil pressure acting upward

$$q_o = \left(\frac{1480}{3.70 \times 3.70}\right) = 108.11\ kN/m^2$$

It is less than the ultimate bearing capacity of soil. Hence, safe.

Step 3 : Bending moment

Overall depth of footing (From Example 22.2, overall depth of footing needed for two-way shear) is assumed as 570 mm. An effective cover of 50 mm shall be kept. Therefore, the effective depth shall be 520 mm. It will now be checked for bending moments, one way shear and two way shear. The (design) bending moment at the critical section XX, X'X' for 1 m strip of footing considered

$$M = \frac{1}{2} . q_o . \left(\frac{B - b_w}{2}\right)^2 \times 1 = \frac{1}{2} \times 108.11 \times \left(\frac{370 - 0.450}{2}\right)^2 \times 1 = 142.739 \text{ kN-m/m}$$

Step 4. Depth of footing

The effective depth of footing required to resist the bending moment. (Only 1 m strip is considered)

$$d = \left(\frac{M}{R_B \times 1000}\right)^{1/2} = \left(\frac{142.739 \times 10^6}{0.9129 \times 1000}\right)^{1/2} = 395.42\ mm$$

An effective depth of footing 520 mm assumed is more than this effective depth. It is safe.

Bending moment at the critical section X_1X_1

$$M_1 = \frac{1}{2} \cdot q_o \cdot \left[\left(\frac{B - b_w}{2}\right) - 0.400\right]^2 \times 1$$

$$M_1 = \frac{1}{2} \times 108.11 \left[\left(\frac{3.70 - 0.450}{2}\right) - 0.4\right]^2 \times 1 = 81.116\ kN-m/m$$

Effective depth of the footing after one step. (Only 1 m strip is considered)

$$d_1 = \left(\frac{81.116 \times 10^6}{0.9129 \times 1000}\right)^{1/2} = 298.086\ mm$$

Effective depth of footing provided is (520-120) = 400 mm is more than needed.

Bending moment at the critical section X_2X_2

$$M_2 = \frac{1}{2} \cdot q_o \cdot \left[\left(\frac{B - b_w}{2}\right) - 2 \times 0.4\right]^2 \times 1$$

$$M_2 = \frac{1}{2} \times 108.11 \left[\left(\frac{3.70 - 0.450}{2}\right) - 0.8\right]^2 \times 1 = 36.791\ kN-m/m$$

Effective depth of the footing after two steps. (Only 1 m strip considered)

$$d_2 = \left(\frac{36.791 \times 10^6}{0.9129 \times 1000}\right)^{1/2} = 200.752\ mm$$

Effective depth of footing a provided is (400 – 120) = 380 mm is more than needed.

Bending moment at the critical section XX

$$M_3 = \frac{1}{2} \cdot q_o \cdot \left[\left(\frac{B - b_w}{2}\right) - 3 \times 0.4\right]^2 \times 1$$

$$M_3 = \frac{1}{2} \times 108.11 \left[\left(\frac{3.70 - 0.450}{2}\right) - 1.2\right]^2 \times 1 = 22.973\ kN-m/m$$

Effective depth of footing required after three steps

$$d_3 = \left(\frac{22.973 \times 10^6}{0.9129 \times 1000}\right)^{1/2} = 158.635\ mm$$

Effective depth of footing provided is (380–120) = 260 mm, which one is more than required. Hence safe.

The thickness of the footing at the edge actually provided (570 – 3 x 120) = 210 mm is more than 150 mm specified by IS : 456-2000.

Step 5. Check for single shear

The shear force at the critical sections X_4X_4 and $X_4'X_4'$, at a distance d = effective depth of footing for 1 m strip of the footing

$$V = q_o \cdot \left[\left(\frac{B - b_w}{2}\right) - d\right]$$

$$V = 108.11\left[\left(\frac{3.70 - 0.450}{2}\right) - 0.520\right] = 119.462\ kN - m/m$$

Nominal shear stress across the depth of footing

$$\tau_v = \left(\frac{119.462 \times 1000}{1000 \times 520}\right) = 0.2297\ N/mm^2$$

For the balanced section for footing percentage of area of steel reinforcement

$$p_B = \left(\frac{100\ A_s}{bd}\right) = 0.4392\ percent$$

From IS: 456-2000, the permissible shear stress in concrete (without shear reinforcement)

$$\tau_c = \left(0.56 + 0.08 \times \frac{0.1892}{0.25}\right) = 0.28\ N/mm^2$$

Nominal shear stress, τ_v is less than the permissible shear stress, τ_c. Hence, it is safe. Similarly, the depth of footing after one, two and three steps may also be checked for one-way shear.

Step 6 : Check for two-way shear

The critical section for two-way shear shall be ABCD, Fig. 22.20(B). The peripherial perimeter

$$b_o = 4 \times \left[450 + \frac{520}{2} + \frac{520}{2}\right] = 3880\ mm.$$

Total shear force

$$V = q_o \times \left|\left(B^2 - AB^2\right)\right| = 108.11 \times \left|\left(3.70^2 - 0.970^2\right)\right| = 1378.305\ kN/m$$

Nominal shear stress

$$\tau_v = \left(\frac{1378.305 \times 1000}{3880 \times 520}\right) = 0.683\ N/mm^2$$

Permissible shear stress in concrete for two-way shear for M 20 grade

$$\tau_c' = k_s \cdot \tau_c$$

$$k_s = (0.5 + \beta_c) \not> 1, \left(\beta_c = \frac{B}{B} = 1.0\right), \text{ and } k_s = 1.00$$

$$\tau_c = 0.16 \times (f_{ck})^{1/2} = 0.16 \times (20)^{1/2} = 0.7155 \ N/mm^2$$

The nominal shear stress, τ_v calculated above is less than the permissible shear stress, τ_c'. Hence safe.

Step 7. Check for transfer of load at the base of column

The concrete of footing is having the same strength as the concrete of column, the bearing at the top of footings is adequate [See step 9, Example 22.3].

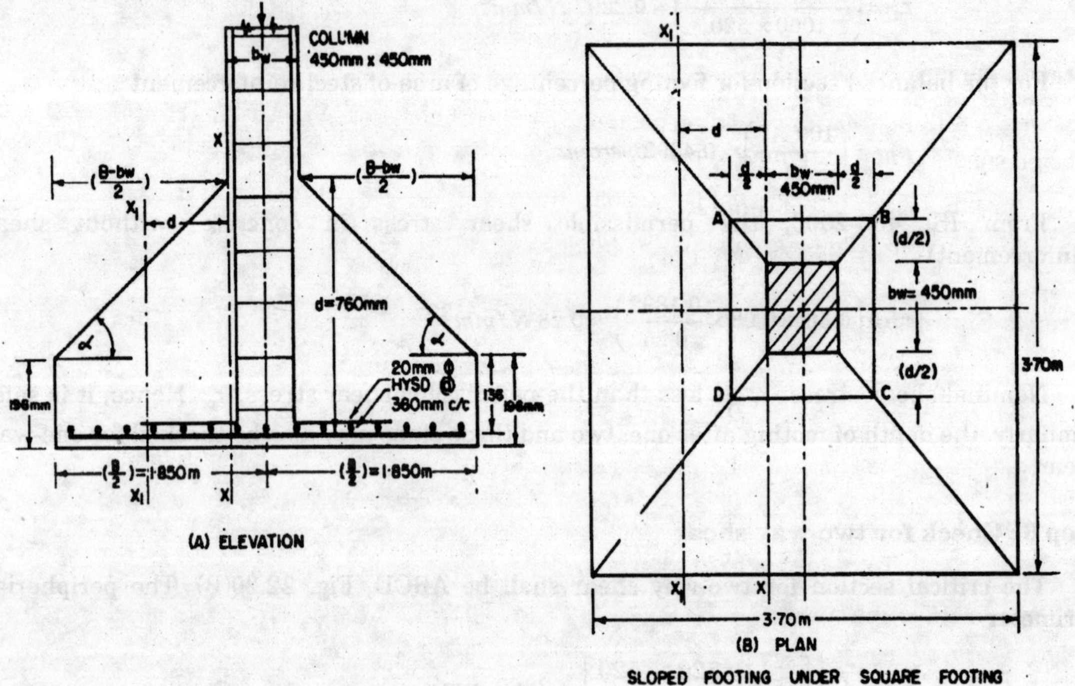

Fig. 22.21

Step 8. Reinforcement for bending moment

In the square footing, the area of steel bars to be provided in the upper layer shall be more than the area of steel bars to be provided in the lower layer. Therefore, the area of steel bars required for upper layer is also provided for that in the lower layer.

The area of steel bars needed for the upper layer for 1 m strip of the footing. Let ϕ = 20 mm.

$$A_{st} = \left(\frac{M}{\sigma_{st} \cdot j_B \cdot (d - \phi)}\right) = \left(\frac{142.739 \times 10^6}{230 \times 0.9038 \times (520 - 20)}\right) mm^2$$

$$A_{st} = 1373.32\ mm^2 \ and\ A_\phi = 314.159\ mm^2$$

Spacing of the steel bars in the upper layer and the lower layer both are equal

$$s = \left(\frac{1000 \times 314.159}{1373.32}\right) = 228.78\ mm$$

Provide 20 mm diameter Hysd bars of grade Fe 415 at 220 mm spacing centre to centre in both the layers.

Example 22.5. **Design a sloped square footing to support the column in Example 22.2.**

Design : From Example 22.2, the modular ratio of M 20 grade of concrete

$$m = 13.33, \text{ and } m_c = 20$$

Step 1. Design constants

The section of footing shall be designed as the balanced section. From Example 22.2, the design constants are as under for M 20 grade of concrete and Hysd-steel bars of Fe 415 grade

$$k_B = 0.2886, \text{ and } j_B = 0.9038$$

$$R_B = 0.9129, \text{ and } p_B = 0.4392$$

Step 2. Size of footing to maintain allowable soil pressure

Dead load and live load to be transferred by the column

$$P = (620 + 860) = 1480\ kN$$

Self-weight of the footing shall be assumed as *ten percent* of the column load

$$= 148\ mm$$

$$\text{Total load} = 1628$$

Area of footing needed

$$A_F = \left(\frac{1628}{120}\right) = 13.566\ m^2$$

A square footing shall be provided to support the load fiom the square column. Size of footing

$$B = (13.566)^{1/2} = 3.683\ m.$$

A square footing of size 3.70 m x 3.70 m shall be provided as shown in Fig. 22.21. Net soil pressure acting upward

$$q_o = \left(\frac{1480}{3.70 \times 3.70}\right) = 108.11\ kN/m^2$$

It is less than the bearing capacity of the soil. Hence, safe.

Step 3 : Bending moment

Overall depth of footing (From *Example 22.2,* overall depth of footing needed for two-way shear) is assumed as 570 mm. An effective cover of 50 mm shall be used. Therefore, the effective depth of footing shall be 520 mm. It will now be checked for bending moment, one-way shear and two-way shear.

The bending moment at the critical section XX, X'X' for 1 m strip of footing is considered.

$$M = \frac{1}{2} \times q_o \times \left(\frac{B - b_w}{2}\right)^2 \times 1\, kN - m/m$$

$$M = \frac{1}{2} \times 108.11 \times \left(\frac{3.70 - 0.450}{2}\right)^2 \times 1 = 142.739\, kN - m/m.$$

Total bending moment for complete width

$$M = (142.739 \times 3.70) = 528.134\, kN - m$$

Step 4 : Depth of footing

The cross-sectional shape of the footing. Fig. 22.21 (A), resisting this bending moment is *trapezoidal shape.* The moment of resistance of this trapezoidal section shall comprise of the moment of resistance of the middle rectangular portion and that of the two triangular portions at the sides. The horizontal component of total compressive force (which shall be acting horizontally).

$$C_H = \left(\frac{1}{2}\sigma_{cbc} \,.\, b_w \,.\, k_B \,.\, d \,.\, \cos^2 \alpha\right)$$

The moment of resistance of the middle rectangular portion

$$MR_1 = C_H \,.\, j_B \,.\, d$$

$$MR_1 = \left(\frac{1}{2}\sigma_{cbc} \cdot k_B \cdot j_B\right) b_w \,.\, d^2 \,.\, \cos^3 \alpha = R_B \,.\, b_w \,.d^2 \,.\, \cos^2 \alpha$$

where σ_{cbc} is the compressive stress in the concrete at top surface in the middle portion. The moment of resistance of two triangular portions

$$MR_2 = 2\sigma_{cbc} \,.\, \cot \alpha\, {k_B}^2\, d^3 \left(\frac{2 - k_B}{12}\right)$$

$$\cot \alpha = \left[\left(\frac{B - b_w}{2}\right)/(d - 0.2)\right] = \left[\left(\frac{3700 - 450}{2}\right)/(0.8d)\right] = \left(\frac{2031.25}{d}\right)$$

The moment of resistance of middle rectangular portion, from Eq. 22.34,

$$MR_1 = (R_B \,.\, b_w \,.\, d^2 \,.\, \cos^2 \alpha)$$

The value of $\cos^2 \alpha$ may be assumed as 0.85 for practical use.

$$MR_1 = \left(0.9129 \times 450 \times d^2 \times 0.85\right) = 349.184\ d^2$$

The moment of resistance of two triangular portions, from Eq. 22.36,

$$MR_2 = \left[2\ \sigma_{cbc}\ .\ \cot \alpha\ k_B^2\ .\ d^3 \left(\frac{2 - k_B}{12}\right)\right]$$

$$2 \times 7 \times \frac{2031.25}{d} \times 0.2886^2 \times d^2 \times \left(\frac{2 - 0.2886}{12}\right) = 337.797\ d^2$$

Total moment of resistance of the section

$$(MR_1 + MR_2) = (349.184\ d^2 + 337.797\ d^2) = 729.357\ d^2$$

Compressive force for rectangular portion from Eq. 22.33,

$$C_{H1} = \left(\frac{1}{2}\ \sigma_{cbc}\ .\ b_w\ .\ k_B\ .\ d\ .\ \cos^2 \alpha\right)$$

$$= \left(\frac{1}{2} \times 7 \times 450 \times 0.2886 \times d \times 0.85\right) = 386.363\ d$$

Compressive force for two triangular portions from Eq. 22.35,

$$C_{H2} = \left(2\ \sigma_{cbc.}\ \cot \alpha . \frac{k_B{}^2\ d^2}{6}\right)$$

$$= \left(2 \times 7 \times \frac{2031.25}{d} \times \frac{0.2886^2 \times d^2}{6}\right) = 394.760\ d$$

Total compressive force

$$(C_{H1} + C_{H2}) = (386.363\ d + 394.760\ d) = 781.123\ d$$

Lever arm factor

$$\left(\frac{729.357\ d^2}{781.123\ d}\right) = 0.934\ d, \textit{and}\ \therefore j = 0.934$$

Alternatively. **The trapezoidal section may be considered as the equivalent rectangular section of width [$b_w + (B - b_w)/8$] and depth, d may be used as an approximation.**

Width of equivalent rectangular section

$$[b_w + (B - b_w)/8] = [450 + (3700 - 450)/8] = 856.25 \text{ mm}$$

Therefore, the moment of resistance of the section as an approximation

$$R_B\ .\ b_e\ .\ d^2 = \left(0.9129 \times 856.25 \times d^2\right) = 781.67\ d^2\ \left(\textit{against}\ 729.357\ d^2\right)$$

Equating the moment of resistance to bending moment

$$681.67\ d^2 = \left(528.134 \times 10^6\right)$$

$$d = \left(\frac{528.134 \times 10^6}{781.67}\right)^{1/2} = 675.65\ mm$$

Let the effective cover be 60 mm. Then, the overall depth of footing shall be (675.65 + 60) = 735.648 mm. The overall depth of footing may be adopted as 740 mm. Then, the effective depth of footing shall be 680 mm. Then effective depth of footing at the edge shall be

$$0.2\,d = (0.2 \times 680) = 136\ mm$$

The overall thickness at the edge shall be

(136 + 60) = 196 mm > 150 mm minimum (Specified in IS : 456-2000).

The cross-section of the footing resisting one-way shear in shown in Fig. 22.22.

Step 5. Check for one-way (beam) shear

Shear force at the critical section X_1X_1 Fig. 22.22, at a distance d from the face of the column, for complete width of footing

$$V = q_o\,.\,B\,.\left[\left(\frac{B - b_w}{2}\right) - d\right]$$

$$= \frac{108.11 \times 3700}{1000} \times \left[\left(\frac{3700 - 450}{2}\right) - 0.680\right] = 378.01\ kN$$

Effective depth at the critical section X_1X_1, Fig. 22.22

$$= 136 + \left(\frac{680 \times 945}{\frac{3700 - 450}{2}}\right) = 531.446\ mm$$

Width at the critical section X_1X_1, Fig. 22.22

$$450 + 2 \times 1625 \times \left[\frac{(680 - 136)\ - (531.446 - 136)}{(680 - 136)}\right]$$

$$[450 + (2 \times 1625\ \times 148.554/544)] = (450\ + 2 \times 443.75) = 1337.50\ mm$$

Depth of neutral axis for this section resisting the shear

$$k_B \times 531.446 = (0.2886 \times 531.446) = 153.375\ mm$$

Width at the level of neutral axis

$$450 + \left[2 \times 1625 \times \left(\frac{148.554\ + 153.375}{544}\right)\right] = 2253.80\ mm$$

Nominal shear stress at the section X_1X_1

$$\tau_v = \left(\frac{378.01 \times 1000}{2253.80 \times 531.446}\right) = 0.3156\ N/mm^2$$

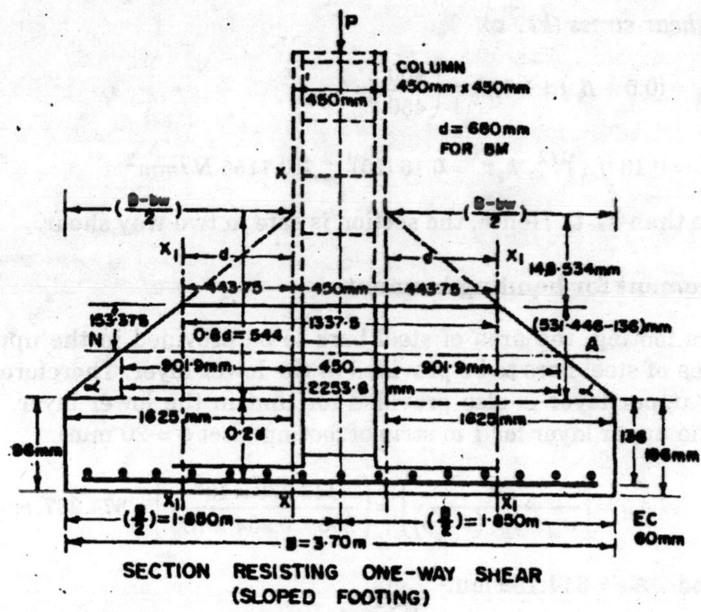

Fig. 22.22

For balanced section p_B = 0.4392 for M 20 concrete

$$\tau_c = \left[0.22 + \left(\frac{0.4392 - 0.25}{0.50 - 0.25}\right) \times 0.08\right] = 0.2805\ N/mm^2$$

The effective depth of complete section is increased in the ratio of (τ_v/τ_c). The revised depth of complete section

$$\left(600 \times \frac{0.3125}{0.2805}\right) = 757.456\ mm = 760$$

Overall depth of complete section

(760 + 60) = 820 mm.

Step 6. Check for two-way (Flat slab type) shear

The critical section for two-way shear round the column shall be ABCD. The peripherial length of the perimeter

$$b_o = 4 \times (450 + 760) = 4840\ mm$$

Shear force V

$$108.11\ [3.7 - 0.450 \times 0.450] = (108.11 \times 13.48.75) = 1458.134\ kN$$

Nominal shear stress for two-way shear

$$\tau_v = \left(\frac{V}{b_o \cdot d}\right) = \left(\frac{1458.134 \times 1000}{4840 \times 760}\right) = 0.3964\ N/mm^2$$

Permissible shear stress ($k_s \cdot \tau_c$)

$$k_s = (0.5 + \beta_c) \not> 1.0, \beta_c = \left(\frac{450}{450}\right) = 1$$

$$\tau_c = 0.16\,(f_{ck})^{1/2},\ k_s\tau_c = 0.16\,(20)^{1/2} = 0.7155\ N/mm^2$$

τ_v is not more than $k_s . t_c$. Hence, the section is safe in two-way shear.

Step 7. Reinforcement for bending moment

In the square footing, the area of steel bars to be provided in the upper layer shall be more than the area of steel bars to be provided in the lower layer. Therefore, the area of steel bars required for upper layer is also provided for that in the lower layer. The area of steel bars needed for the upper layer for 1 m strip of footing. (Let ϕ = 20 mm)

$$A_{st} = \left(\frac{M}{\sigma_{st} \cdot j_B \cdot (d - \phi)}\right) = \left(\frac{142.739 \times 10^6}{230 \times 0.934 \times 760}\right) = 874.287\ mm^2$$

and $A_{st} = 314.159\ mm^2$

Spacing of the steel bars along the length (distributed uniformly)

$$s = \left(\frac{1000 \times 314.159}{874.287}\right) = 359.33\ mm\ Say = 350\ mm.$$

Step 8. Check for transfer of load at base of column

Loaded area at the column base

$$A_2 = (450 \times 450) = 20.25 \times 10^4\ mm^2$$

Supporting area for bearing of footing which one is sloping at the rate of 2 : 1. Spread of the footing upto the effective depth

$$(2d + b_w + 2d) = (2 \times 760 + 450 + 2 \times 760) = 3490\ mm$$

Supporting area of the bearing

$$A_1 = (3490 \times 3490) = 1218.01 \times 10^4\ mm^2$$

$$\left(\frac{A_1}{A_2}\right)^{1/2} = \left(\frac{1218.01 \times 10^4}{22.25 \times 10^4}\right)^{1/2} = 7.938 > 2.0$$

Permissible bearing stress on full area of concrete

$$\sigma_{cbr} = 0.25\ f_{ck} = (0.25 \times 20) = 5\ N/mm^2$$

$$\sigma_{cbr} \cdot \left(\frac{A_1}{A_2}\right)^{1/2} = (5 \times 2) = 10\ N/mm^2$$

Compressive stress in concrete at the base of the column

$$q = \left(\frac{(620 + 860) \times 1000}{450 \times 450}\right) = 7.308\ N/mm^2$$

It does not exceed the permissible stress. Hence safe.

It is to note that the dowel bars are provided in the column and footing in case the permissible stress becomes less than the actual bearing stress.

Step 9. Check for the development length

Development length required for Hysd steel bars

$$L_d = \left(\frac{\sigma_{st} \cdot \phi}{4 \times 1.40 \cdot \tau_{bd}}\right) = \left(\frac{230 \times \phi}{4 \times 1.40 \times 0.8}\right) = (51.34 \times 20) = 1026.8\ mm$$

Development length available from the critical section XX for the bending moment that is, at the face of column

$$\left[\left(\frac{B - b_w}{2}\right) - end\ distance\right] = \left[\left(\frac{3700 - 450}{2}\right) - 25\right] = 1600\ mm$$

Actual length available is more than the development length needed. Hence, satisfactory. Design of sloped footing is shown in Fig. 22.21.

22.16 CIRCULAR FOOTING

The circular footings are usually provided under single circular columns. In the circular footing, the cantilever project out from the column in all the directions. The column transfers the axial (viz., concentric) load acting downward to the footing. The footing is loaded upward by the soil pressure. The tensile stresses develop in the directions at the bottom surface. Such footings are therefore reinforced by two layers of steel bars perpendicular to each other. A circular footing under a single circular column with the forces acting on it, is shown in Fig. 22.23. The required bearing area of footing is determined by dividing the total load including the self-weight of the footing (the column load P + ten per cent of column load) by the bearing capacity of the soil.

$$A_F = \left(\frac{1.10P}{q_o}\right) \tag{i}$$

Let R be the radius of circular footing and r be the radius of circular column. Then

$$\pi R^2 = A_F,\ R = \left(\frac{A_F}{\pi}\right)^{1/2}$$

For the purpose of calculating stresses in footings which support a circular concrete column or pedestal, the face of the column or pedestal shall be taken as the side of a square

inscribed within the perimeter of the circular column or the pedestal as shown in Fig. 22.23. The perpendicular distance from the centre of the column upto the side of square inscribed within the perimeter of column is ($r/\sqrt{2}$). The critical section XX for calculating the maximum bending moment shall at the face of the square as shown in Fig. 22.23 and Fig. 22.24.

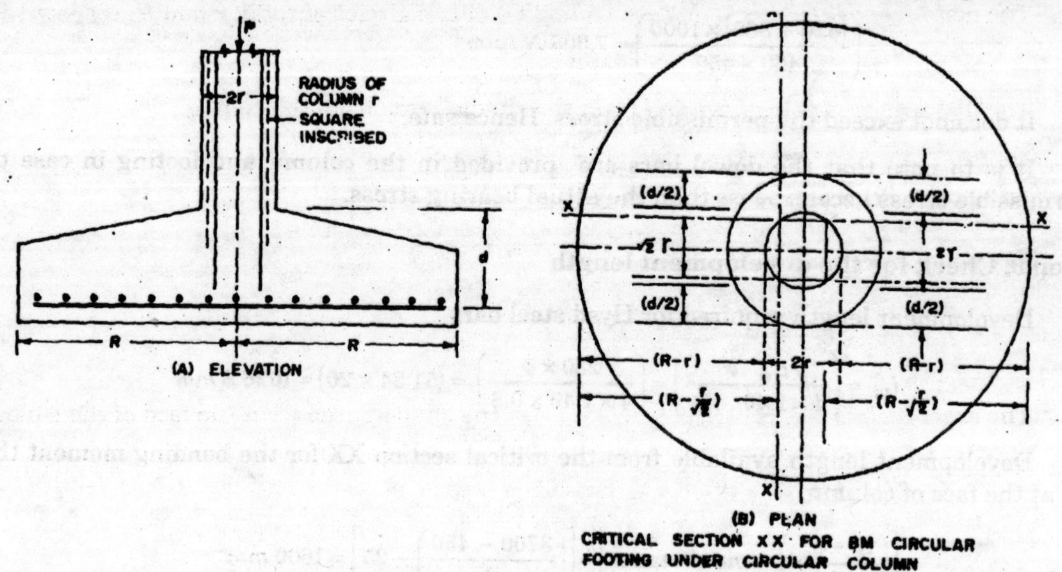

Fig. 22.23

The maximum bending moment in the circular footing supporting a circular column may be determined by considering a quadrant of a circle OPQ as shown in Fig. 22.24.

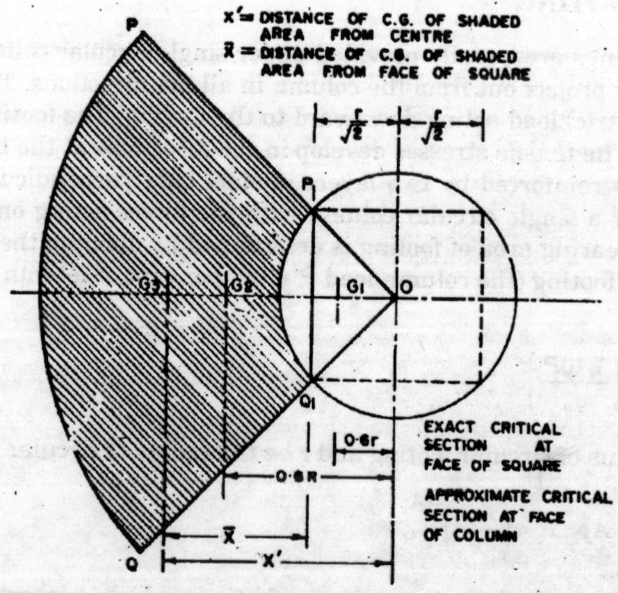

Fig. 22.24

The areas of quadrants of two circles of radii, r and R, respectively.

$$A_1 = \left(\frac{\pi}{4}.r^2\right) \text{ and } A_2 = \left(\frac{\pi}{4}.R^2\right) \qquad \text{(i)}$$

The distances of centres of gravity G_1 and G_2 of two circles of radii, r and R, respectively.

$$OG_1 = 0.6\,r \quad \text{and } OG_2 = 0.6R \qquad \text{(ii)}$$

The distance of centre of gravity G_3 of the shaded area from the centre

$$X' = \frac{\left[\frac{1}{2}R^2 \times (0.6R) - \frac{\pi}{4} \times r^2 \times (0.6r)\right]}{\left(\frac{\pi}{4}.R^2 - \frac{\pi}{4}.r^2\right)} = 0.6\left(\frac{R^3 - r^3}{R^2 - r^2}\right) \qquad \text{(iii)}$$

$$\text{or} \quad X' = 0.6.\left(\frac{R^2 + R.r + r^2}{R + r}\right) \qquad \text{(iv)}$$

The *exact distance* of centre of gravity, G_3 of the shaded area from the face of the square

$$\bar{x} = \left[0.6\left(\frac{R^2 + R.r + r^2}{R + r}\right) - \frac{r}{\sqrt{2}}\right] \qquad \text{(v)}$$

$$\bar{x} = \left(\frac{(0.6R^2 + 0.6R.r + 0.6r^2 - 0.707R.r - 0.707r^2)}{(R + r)}\right) \qquad \text{(vi)}$$

$$\bar{x} = 0.107\left[\frac{5.6075\,R^2 - R.r - r^2}{(R + r)}\right] \qquad \text{(22.37)}$$

The *approximate distance* of centre of gravity, G_3 of the shaded area from the face of circular column instead of face of the square

$$\bar{x} = \left[0.6\left(\frac{R^2 + Rr + r^2}{R + r}\right) - r\right]$$

$$= \left(\frac{0.6R^2 + 0.6Rr + 0.6r^2 - Rr - r^2}{(R + r)}\right) \qquad \text{(vii)}$$

$$\text{or } \bar{x} = \frac{0.2}{(R + r)}\left[3R^2 - 2\,Rr - 2.r^2\right] \qquad \text{(22.38)}$$

The soil pressure acting upward on the shaded area

$$= \frac{\pi}{4}\left(R^2 - r^2\right).q_o \qquad \text{(viii)}$$

Bending moment at the face of circular column

$$M = \frac{\pi}{4}\left(R^2 - r^2\right).q_o \times \frac{0.2}{(R+r)} \times \left[3R^2 - 2.Rr - 2.r^2\right] \quad \text{(ix)}$$

$$= \frac{1}{20}\pi.(R-r)q_o.\left[3R^2 - 2Rr - 2r^2\right] \quad (22.39)$$

Bending moment at the face of square inscribed within the column

$$M = \frac{\pi}{4}\left(R^2 - r^2\right).q_o \times \frac{0.107}{(R+r)} \times \left[5.6R^2 - R.r - r^2\right] \quad \text{(x)}$$

or $M = 0.084\ q_o (R - r).\ [5.6R^2 - R.r - r^2]$ **(22.40)**

Width of the quadrant P_1Q_1, Fig. 22.24 shall be

$$\left(\frac{2\pi r}{4}\right) = \left(\frac{1}{2}\pi.r\right) \quad \text{(xi)}$$

Width of the face of square, P_1Q inscribed within the perimeter of column

$$\left(\frac{2.r_1}{\sqrt{2}}\right) = \sqrt{2} \quad \text{(xii)}$$

The effective depth of circular footing at the critical section

$$d = \left(\frac{M}{R_B\ .width}\right)^{1/2} \quad (22.41)$$

where, R_B moment resistance factor for the balanced section of the footing.

The cross-sectional area of steel bars in one direction in lower layer.

$$A_{st1} = \left(\frac{M}{\sigma_{st}\ .\ j_B\ .d}\right) \quad \text{(xiii)}$$

The cross-sectional area of steel bars in the other (viz., perpendicular) direction in upper layer

$$A_{st2} = \left(\frac{M}{\sigma_{st}\ .\ j_B\ .\ (d - \phi)}\right) \quad \text{(xiv)}$$

It is to note that A_{st2}, shall be more than A_{st1}, therefore, A_{st2} area of steel shall be provided in lower layer and so also in the upper layer in uniform spacing in both the directions.

Check for two way shear

The critical section for two-way shear as shown in Fig. 22.23 shall be at a distance (d/2) from the periphery of the column. The shear force

$$V = q_o\ .\ \pi\left[R^2 - \left(r + \frac{d}{2}\right)^2\right] kN \quad \text{(xix)}$$

The peripherial distance of critical section,

$$b_o = 2\pi\left(r + \frac{d}{2}\right) \quad \text{(xx)}$$

The nominal shear stress

$$\tau_v = \left(\frac{V}{b_o \cdot d}\right) \quad (22.42)$$

The nominal shear stress shall not exceed the permissible stress for two-way shear.

Check for the development length

The available length of steel bars from the face of square upto the edge of footing less end distance shall be more than the development needed for the type of steel bars used.

Note : In case a square footing is provided under a circular column, then also, the square shall be inscribed within the perimeter of the circle. Then, it shall be the case of square footing supporting the square column, and which shall be designed as usual.

Example 22.6. A circular column of 480 diameter transfers an axial dead load of 650 kN and an axial live load of 500 kN. The column is having 8 steel bars of 18 mm diameter. The safe bearing capacity of the soil is 140 kN/m² use M 20 grade of concrete and Hysd steel bars of Fe 415 grade. Design a circular footing to support the circular column.

Design: The modular ratio of M 20 grade of concrete m is 13.33 and m_c is 20.

Step 1: Design constants

$k_B = 0.2886$, and $j_B = 0.9038$

$R_B = 0.9129$, and $p_B = 0.4392$

Step 2 : Size of the footing to maintain allowable pressure

Dead load and live load to be transferred by the column

$P = (650 + 500) = 1150$ kN

Self-weight of the footing shall be assumed as *ten percent* of the column load

= 115 kN

Total load = 1265 kN

Area of the footing required

$$A = \left(\frac{1265}{140}\right) = 9.0357\ m^2$$

The radius of circular footing

$$R = \left(\frac{A_F}{\pi}\right)^{1/2} = \left(\frac{9.0357}{\pi}\right)^{1/2} = 1.696\ m$$

The radius of circular footing may be adopted as 1.80 m and then, its diameter shall be 3.60 m. The net soil pressure (R = 1.80 m)

$$q_o = \left(\frac{1150}{\pi \,.\, (1.80)^2}\right) = 112.98\ kN/m^2$$

Step 3. Bending moment

The critical section for bending moment, XX shall be at the face of square inscribed within the column. The distance of face of square from the centre of the column, Fig. 22.23 (B). (The radius of column is 240 mm).

$$\left(\frac{r}{\sqrt{2}}\right) = \left(\frac{240}{\sqrt{2}}\right) = 169.7\ mm$$

$$\left(2 \,.\, \frac{r}{\sqrt{2}}\right) = (2 \times 169.7) = 339.41\ mm$$

The distance of C.G. of the shaded area from the face of square, Fig. 22.24, Eq. 22.37 [R = 1.80 m, r = 0.240 m].

$$\bar{x} = 0.0107\left[\frac{5.6075\,R^2 - Rr - r^2}{(R+r)}\right]$$

$$\bar{x} = 0.107\left[\frac{5.6075 \times 1.80^2 - 1.80 \times 0.240 - 0.240^2}{(1.80 + 0.240)}\right] = 0.9273\ m$$

Soil pressure acting upward on the shaded area. Fig. 22.24.

$$\left(112.98 \times \frac{\pi}{4} \times \left(R^2 - r^2\right)\right) = \left[112.98 \times \frac{\pi}{4} \times \left(1.80^2 - 0.240^2\right)\right] = 282.388\ kN$$

Maximum bending moment at the critical section XX, located at the face of square, Fig. 22.23

$$(282.388 \times 0.9273) = 261.858\ kN-m$$

Step 4. Effective depth of footing

The effective depth of footing

$$d = \left(\frac{M}{R_B \times width\ of\ sq.\ face}\right)^{1/2} = \left(\frac{261.858 \times 10^6}{0.9129 \times 339.41}\right)^{1/2} = 919.304\ mm$$

Let the effective cover be 60 mm. Then the overall depth of footing shall be

D = (919.304 + 60) = 979.304 mm

say D = 980 mm, and d = 920 mm

Step 5. Check for two-way shear

The critical section for two-way shear shall be the circumference of a circle of radius (r + d/2) or at a distance (d/2) of the column.

Shear force for two-way shear

$$V = q_o \,.\, \pi \left[R^2 - \left(r + \frac{1}{2} d \right)^2 \right]$$

$$V = 112.98 \times \pi \left[1.80^2 - \left(0.240 + \frac{1}{2} \times 0.920 \right)^2 \right] = 976.077 \; kN$$

The peripherial distance or circumference of the critical section for two-way shear

$$b_o = 2\pi \,.\, \left(r + \frac{1}{2} d \right) = \left[2\,\pi \left(240 + \frac{1}{2} \times 920 \right) \right] = 4398.23 \; mm$$

The effective depth at the outer edge is assumed as 184 mm.

The effective depth of footing under the critical section for two-way shear shall be as follows : (as shown in Fig. 22.25).

$$d = 920 - \left(\frac{(920 - 184) \times \dfrac{920}{2}}{(1800 - 240)} \right) = 702.974 \; mm$$

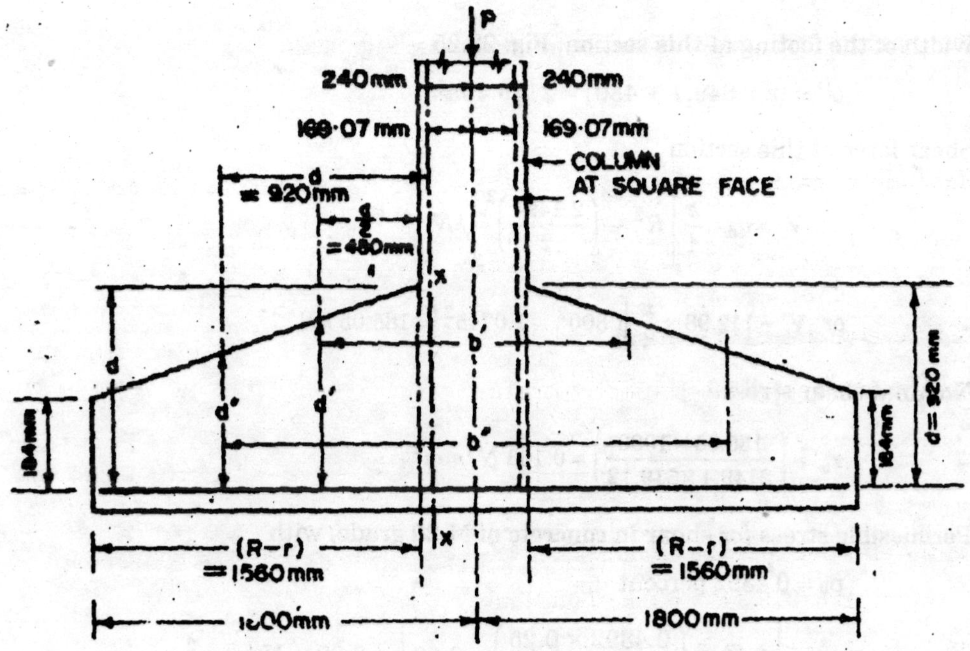

Fig. 22.25

Nominal shear stress

$$\tau_v = \left(\frac{V}{b_o.d'}\right) = \left(\frac{976.077 \times 1000}{4398.23 \times 702.974}\right) = 0.316\ N/mm^2$$

Permissible shear stress for two way shear under triaxial compression

$$\tau_c = 0.16\ (f_{ck})^{1/2} = 0.16\ (20)^{1/2} = 0.7155\ \text{N/mm}^2$$

The nominal shear stress is less than the permissible shear stress. Hence, the depth of footing satisfactory.

Step 6. Check for one-way (beam) shear

The critical section for one-way (beam) shear X_1X_1 shall be at a distance d from the face of square inscribed within the column. The distance of critical section X_1X_1 from the centre shall be

$$\left(d = \frac{r}{\sqrt{2}}\right) = (920 + 169.7) = 1089.7\ mm$$

The distance of critical section X_1X_1 from the face of column, Fig. 22.25

$$(1089.7 - 240) = 849.7\ \text{mm}$$

The effective depth of footing at the critical section X_1X_1, Fig. 22.25

$$d'' = \left(920 - \frac{[(920 - 184) \times 849.7]}{(1800 - 240)}\right) = 519.12\ mm$$

Width of the footing at this section, Fig. 22.25

$$b'' = (2 \times 849.7 + 480) = 2149.4\ mm$$

Shear force at this section

$$V' = q_o\ ..\frac{\pi}{4}\left[R^2 - \left(\frac{2.1494}{2}\right)^2\right]\ kN$$

$$\text{or}\quad V' = 112.98 \times \frac{\pi}{4}\left[1.800^2 - 1.0745^2\right] = 185.05\ kN$$

Nominal shear stress

$$\tau_v' = \left(\frac{185.05 \times 1000}{2149.4 \times 519.12}\right) = 0.166\ N/mm^2$$

Permissible stress for shear in concrete of M 20 grade, with

$$p_B = 0.4392\ \text{percent}$$

$$\tau_v' = \left[0.22 + \left(\frac{0.4392 \times 0.25}{0.50 - 0.25}\right) \times 0.08\right] = 0.281\ \ N/mm^2$$

Hence safe.

Step 7. Reinforcement for bending moment

The steel bars shall be provided in two layers as provided in the square footing. The area of steel bars provided in the lower layer shall be kept equal that provided in upper layer. The area of steel bars. (Let the diameter of bars be 20 mm).

$$A_{st} = \left(\frac{M}{\sigma_{st} \cdot j_B \cdot (d - \phi)}\right) = \left(\frac{261.858 \times 10^6}{230 \times 0.9038 \times (920 - 20)}\right) = 1399.66\ mm^2$$

Area of steel bars required for 1 m width

$$\left(\frac{1399.66}{3.6}\right) = 388.795\ mm^2$$

Area of one steel bars is $A_\phi = 314.159\ mm^2$

Spacing of the steel bars

$$s = \left(\frac{1000 \times 314.159}{388.795}\right) = 808.033\ mm$$

Spacing of steel bars shall not be more than 450 mm. Therefore 20 mm diameter bars shall be provided at 450 mm centre to centre as shown in Fig. 22.25.

Step 8. Check for development length

Development length required for Hysd steel bars for M 20 grade of concrete

$$L_d = \left(\frac{\sigma_{st} \times \phi}{4 \times 1.4 \times \tau_{bd}}\right) = \left(\frac{230 \times \phi}{4 \times 1.4 \times 0.8}\right) = 51.34\phi$$

$$L_d = (51.34 \times 20) = 1026.8\ mm$$

Development length available from the critical section for bending, that is, from the face of square

$$(R - 169.07 - 25) = (1800 - 169.07 - 25) = 1605.93\ mm$$

Available development length of the steel bars is more than the development length required.

Step 9. Check for transfer of load at column base

Loaded area at the column base

$$A_s = \left(\frac{\pi}{4} \times 0.480^2\right) = 0.181\ m^2$$

Supporting area of bending of footing which one is sloping at the rate of 2 : 1. Spread of the footing upto the effective depth

$$(2d + r + 2d) = (2 \times 0.920 + 0.240 + 2 \times 0.920) = 3.92\ m$$

Supporting area of the bearing

$$A_1 = \left(\frac{\pi}{4} \times 3.92^2\right) = 12.069\ m^2$$

$$\left(\frac{A_1}{A_2}\right)^{1/2} = \left(\frac{12.069}{0.181}\right)^{1/2} = 8.166 > 2$$

Permissible bearing stress for full area

$$\sigma_{cbr} = 0.25 f_{ck} = 0.25 \times 20 = 5\ N/mm^2$$

$$\sigma_{cbr} = \left(\frac{A_1}{A_2}\right)^{1/2} = (5 \times 2) = 10\ N/mm^2$$

Compressive stress at the base of column

$$q = \left(\frac{1150 \times 1000}{\frac{\pi}{4} \times 480 \times 480}\right) = 6.355\ N/mm^2$$

It does not exceed the permissible bearing stress.

Step 10. Check for dowels into the footings

Note : In case the actual bearing stress exceeds the permissible bearing stress, the load is transferred by dowels. Code of practice IS: 456-2000 also permits the use of dowels into the footing. In case the permissible stress would have worked out as 4.8 N/mm², then, the dowels have to be provided to transfer the excess of the load.

In case a minimum of 4 dowels bars of 16 mm diameter are provided, then, the capacity of concrete plus the dowels to transfer the load

$$4.8\left(\frac{\pi}{4} . (480)^2 - 8 \times \frac{\pi}{4}(18)^2\right) + 4 \times \frac{\pi}{4} \times 16^2 \times 230$$

$$= \left(\frac{4.8(178919.98) + (184976.98)}{1000}\right) = (858.816 + 184.97698) = 1043.793\ kN$$

The column load to be transferred

$$P = (650 + 550) = 1150\ kN$$

Excess load to be transferred

$$= (1150 - 1043.793) = 106.21\ kN$$

Instead of 4 dowel bars, 8 dowel bars are provided. Additional load transferring capacity

$$= \left(4 \times \frac{\pi}{4} \times \frac{16^2 \times 2}{1000}\right) = 184.797\ kN$$

Hence, the load transferring capacity

(1043.793 + 184.797) = 1228.59 kN

It is more than the load to be transferred. Hence, it would be have been satisfactory.

22.17 FOOTINGS ON ROCK

Very commonly, only a nominal footing is needed on rock because rock is usually at least as strong as concrete. Moreover, the rock is loaded on only a small part of its surface.

A heavy steel column needs a significant footing. In such case, instead of calculated unit stresses, the deformation and relative stiffness become significant. A reinforced concrete footing functions only on elongation of the steel bars provided as reinforcement and on deflection of the cantilever projections. The rock underlying the footing used to be more stiffer against the compressive deformation than the reinforced concrete footing is, again bending deflection. As such the footing does not distribute the load laterally to any significant extent. Therefore, a reinforced concrete footing on rock has a little value.

The transfer of stress through shear with small deformations is more reliable. Therefore, it needs deep footing. A plain concrete footing on rock is just as effective as the reinforced footing on rock.

The frictional (interlocking) forces in the usual *rough rock surface* is sufficient to prevent any significant stretching of the bottom surface of the footing. A large depth (viz., the depth more than the length of projection) of the footing on rock gives shear stiffness. The quality of rock, below the footing, its surface roughness (interlock) and its lack of volume change are the factors which influence the development of tension on the bottom surface. For a footing on a poor rock, the concrete is designed to resist the total moment as for a footing on soil. However, under the ideal rock conditions, the tension from bending may be neglected.

22.18 BALANCING FOOTING PRESSURES

Formerly, it was the accepted concept that the footings would undergo essentially uniform settlements in case these footings develop the same soil loading per unit area under the usual loads on the stresses. These footings used to have *balanced footing areas.* The usefulness of this concept is greatly restricted in the modern soil mechanism. In most practical cases, it is established that the uniform soil pressure does not lead to uniform settlement. For outside wall footings, the balanced footing pressures need a large size. In a heavy structure on a thick stratum of material that consolidates slowly under additional load, such an increase in exterior footing size actually *increases* the final differential settlement. In case, the footing is on *expansive soils,* and on application of downward loads, the foundation may move upward. The concept of *balanced footing areas* should be used after ascertaining that special foundation situation justifies this approach. Very commonly, the differential settlements increase rather than decreases.

The term *usual load* is applied to the *average load* or the load frequently on the structure, for a balanced footing design.

In case, all footings are designed for the *usual load,* at complete permissible soil pressure, the footing that cared for the maximum percentage of live load (compared to dead load) would overstress the soil when the total live is acting. This result suffests the clue for balanced footing area design. That, the footing with the maximum ratio of live load to dead load is designed first by the usual procedures using largest load and the total permissible soil pressure. The usual soil pressure under this footing is established later on for the usual load, very commonly something between the dead load alone and dead *plus* half the live load. The bearing areas of all other footings are then decided to give this same usual soil pressure under the usual footing loads. (This action automatically gives them less than the maximum permissible soil pressure under the maximum live load). The isolated footings are then designed for thickness and reinforcement using these balanced areas and the total design live load.

22.19 ECCENTRICALLY LOADED INDEPENDENT SPREAD FOOTING

The independent footings supporting the columns carrying axial (viz., concentric) loads provide the uniform soil pressure. The footings are designed and built such that the centre of gravity of the footings coincide with the theoretical point of application of the downward load. The area of footing is kept symmetrical and its centre of gravity coincides with the centre of gravity of the column.

An eccentrically loaded footing may support a column which carries an eccentric load or an eccentric load in addition to an axial load or an eccentric load and a lateral load as shown in Fig. 22.26(A). The load, P acts at an eccentric distance, e from the longitudinal (centroidal) axis and the lateral load P_H acts at a height, h from the base of the footing. The footing may be considered as loaded with a direct load P, lateral load P_H at the base and the moment as shown in Fig. 22.26 (B). The magnitude of the moment in such case shall be

$$M = (P \cdot e + P_H \cdot h) \tag{i}$$

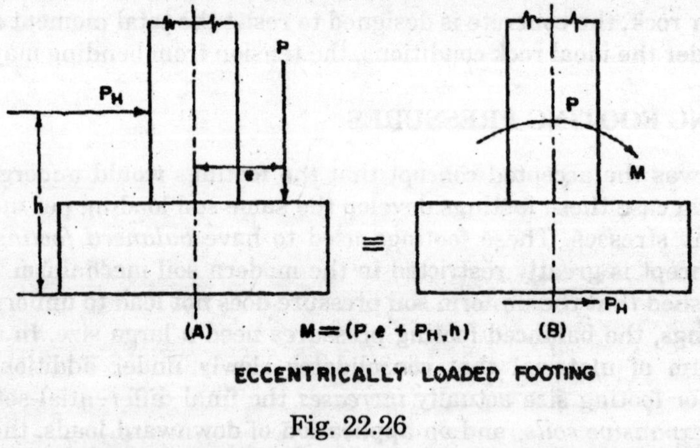

(A) M = (P.e + P_H.h) (B)

ECCENTRICALLY LOADED FOOTING

Fig. 22.26

The equivalent eccentricity of load

$$e_{equv.} = \left(\frac{M}{P}\right) = \left(\frac{P.e + P_H.h}{P}\right) \tag{ii}$$

In case, the lateral load does not exist.

$$M = (P.e) \text{ and } e_{equv} = \left(\frac{M}{P}\right) = \left(\frac{P.e}{P}\right) \tag{iii}$$

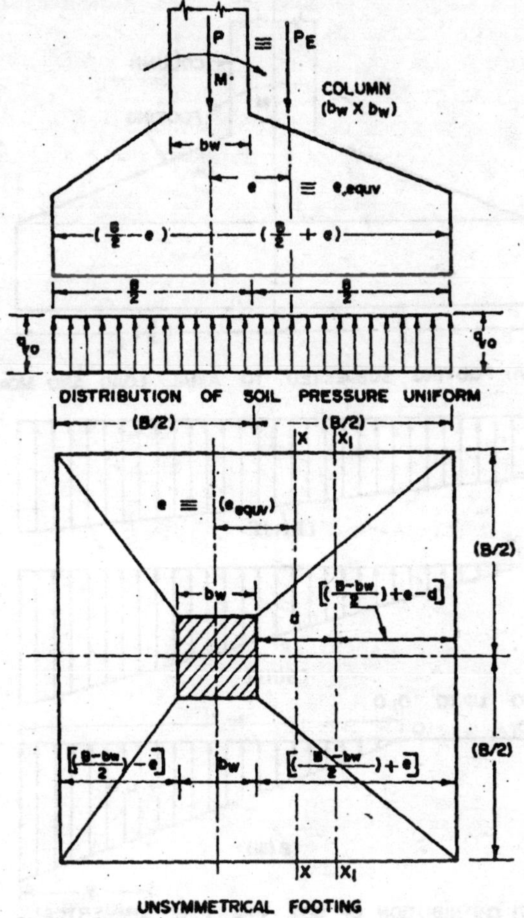

Fig. 22.27

If the magnitude of moment remains constant, it is desirable to keep the centre of gravity of the footing under this eccentric load. In practice, the base of the footing is extended in the plane of applied overturning moment as shown in Fig. 22.27, and the distribution of soil pressure remains uniform (viz., rectangular).

In case of uniaxial eccentric load, the pressure exterted by the edge of the footing is linearly transmitted to the soil in the direction of moment, as shown in Fig. 22.28.

There are three possibilities of resultant (viz., combined distribution of stresses over the soil as shown in Fig. 22.28 [B (i), [B (ii)] and [B (iii)]. In Fig. 22.28 [B (i)], the values of combined stresses q_1 and q_2 are as under :

$$q_1 = \frac{P}{B \cdot L} \cdot \left(1 + \frac{6 \cdot e}{L}\right) \text{ and} \tag{i}$$

$$q_2 = \frac{P}{B \cdot L} \cdot \left(1 - \frac{6 \cdot e}{L}\right) \tag{ii}$$

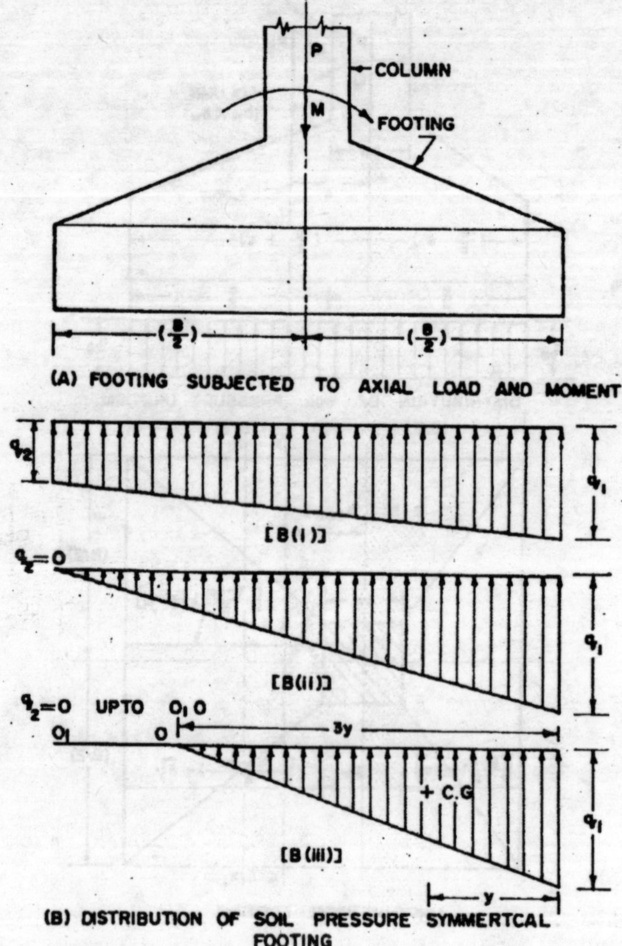

Fig. 22.28

where, q_1 and q_2 are the compressive stresses at the edges of the footing. L and B are the length and the bread of the footing, (in case the footing is rectangular in plan). P is the vertical load action at an equivalent eccentricity, e_{equv} = e. And, in case P is the vertical axial load, the moment M is equal to P. e. In Fig. 22.28 [B(ii)],

$$q_1 = \frac{P}{B \cdot L} \cdot \left(1 + \frac{6 \cdot e}{L}\right) \quad \text{(iii)}$$

But $q_2 = 0$. (zero) (iv)

However, in case the intensity of soil pressure due to axial load alone is less than the stress developed due to moment [viz., (P/B.L) is less than (6 P. e/BL^2)], then the intensity of soil pressure between the base of footing and soil is maximum at one edge and it is zero at point O within the base. The portion OO_1 will not remain in contact with the soil and the intensity of soil pressure will therefore be zero for all points between O and O_1. The resultant passes at a distance more than (L/6) from the centre of footing (i.e., outside the middle third portion. It acts at the centroid of the triangle at a distance, y from the edge. The resultant soil pressure acts upward and it equal to and coincides with the line of action of the total vertical load acting downward. The length of base of triangle of soil pressure is 3 y. The ordinate of triangle q_1 represents maximum intensity of soil pressure. This should not exceed allowable soil pressure. The area of a base giving upward pressure

$$= (3y \times B) \quad \text{(v)}$$

Total upward pressure (viz., resultant)

$$R = \left(\frac{1}{2} \times q_1 \times 3y \times B\right) \quad (But\ R = P) \quad \text{(vi)}$$

$$P = \left(\frac{1}{2}\ q_1 \times 3y \times B\right) \quad \text{(22.43)}$$

And, taking moment about centre of footing

$$R \cdot \left(\frac{L}{2} - y\right) = M \quad (M = P \cdot e) \quad \text{(vii)}$$

Therefore,

$$\left(\frac{L}{2} - y\right) = \left(\frac{M}{R}\right) \quad (\because R = P)$$

$$y = \left(\frac{L}{2} - \frac{M}{P}\right) \quad (\because M = P.e)$$

$$\therefore y = \left(\frac{L}{2} - e\right) \quad \text{(22.44)}$$

The length of footing base is determined by *trial and error* and the distance y may be calculated from Eq. 22.44. The width of footing may be determined from Eq. 22.43. The effective depth of footing is obtained by equating the moment of resistance of the footing with the moment due to intensity of soil pressure.

The magnitude of moment, M (=P.e) varies due to variation of live load. It is not possible to avoid all the eccentricities on the footing. The distribution of soil pressure modifies the magnitude of the design moment and the shear, but not the design procedure.

An eccentrically loaded footing is designed in the same manner as an axial loaded footing. Hence, the pressure on the soil is assumed to be exterted by design forces minus the self-weight of the footing.

22.20 KERN DISTANCE

When a footing is subjected to an axial load and a moment, the magnitude of stresses developed at the two edges of the footing are given by

$$q_1 = \left(\frac{P}{A_F} + \frac{M \cdot y_{max}}{I}\right) and \quad \text{(i)}$$

$$q_2 = \left(\frac{P}{A_F} - \frac{M \cdot y_{max}}{I}\right) \quad \text{(ii)}$$

where P is the axial load and M is the moment (M = P. e, in case the load P is acting at an eccentric distance, e from the axis). For a rectangular footing of length, L and breadth, B, the area of footing, A_F = L x B, the moment of inertia is (1/12) (L . B^3). The distance to the extreme fibre, y_{max} is equal to (1/2) (L), as shown in Fig. 22.29.

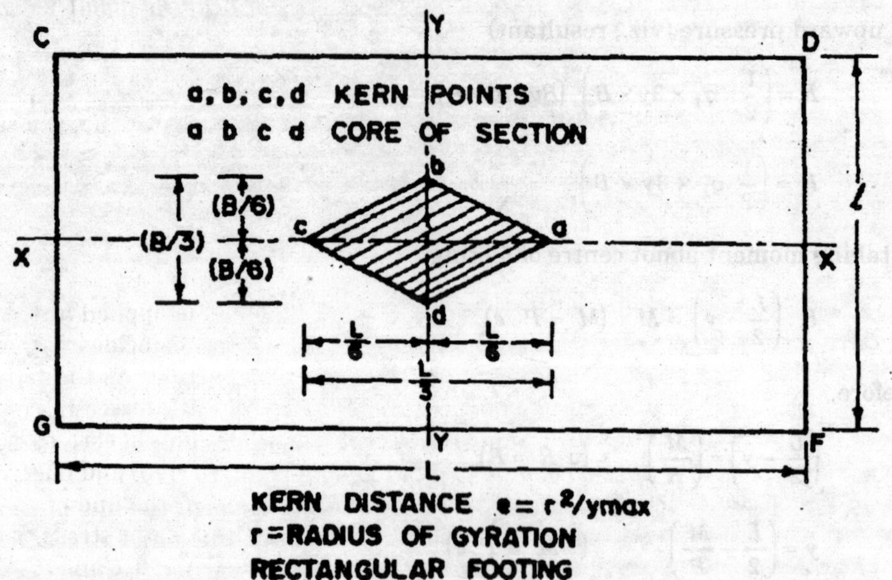

Fig. 22.29

From the expression (i) and (ii)

$$q_1 = \left(\frac{P}{A_F} + \frac{P \cdot e \cdot y_{max}}{A_F \cdot r^2}\right) \quad \text{(iii)}$$

$$\text{and } q_2 = \left(\frac{P}{A_F} + \frac{P \cdot e \cdot y_{max}}{A_F \cdot r^3}\right) \quad \text{(iv)}$$

$$q_1 = \frac{P}{A_F} \cdot \left(1 + \frac{e \cdot y_{max}}{A_F \cdot r^2}\right) \quad \text{(v)}$$

$$q_2 = \frac{P}{A_F} \cdot \left(1 - \frac{e \cdot y_{max}}{A_F \cdot r^2}\right) \quad \text{(vi)}$$

where, *r* is the radus of gyratien of the section of footing. From the expression (v), it is seen that the edge of footing in the direction of moment is subjected to maximum compressive stress and the other edge also carries the small compressive stress so long as the eccentricity, *e* is small, as shown in Fig. 22.28 [B(i)]. The maximum eccentricity at which the tensile stress would be zero (viz., q_2 = 0).

$$o = \left(1 - \frac{e \cdot y_{max}}{r^2}\right) \tag{vii}$$

or $$e = \left(\frac{r^2}{y_{max}}\right) \tag{22.45}$$

Eq. 22.45 defines the *core, or kern* of the footing section. *For bending about YY-axis, and eccentricity,* e_{xx} along YY-axis.

$$r^2 = \frac{B^2}{12}, and\ y_{max} = \left(\frac{B}{2}\right) and\ e_{xx} = \left(\frac{B}{6}\right) \tag{22.46}$$

For bending about XX-axis and eccentricity e_{yy} along YY-axis,

$$r^2 = \frac{1}{12}.L^2, and\ y_{max} = \left(\frac{L}{2}\right) and\ e_{yy} = \left(\frac{L}{6}\right) \tag{22.47}$$

The line of zero stress coincides with the side CG when the load is applied at the point a at distance (L/6) from the centroid. Similarly, the line of zero stress coincides with the side GF when the load is at the point *b* at a distance (B/6) from the centroid. As the load moves along the line ab, the neutral axis rotates about the point G without intersecting the cross-section of footing. Hence, 0B is one of the side of the core. The other sides of core follows from the symmetry. The core is therefore a *rhombus* with diagonal equal to (L/3) and (B/3). As long as the point of application of the load remains with this *rhombus abcd,* the line of zero stress does not intersect the cross-section and there will no reversal in the sig of stress. *Rhombus abcd* is called as the core of the section. The limiting points of the kern a, b, c and d are called as the kern points.

When the eccentricity of load is equal to, or smaller than the kern distance, e, the extreme soil pressure (q_{max} and q_{min}) may be calculated by superposition of bending bearing pressure over the axial bearing pressures.

When the eccentricity becomes more than the kern distance, the principle of superposition cannot be applied because, in such cases, it would result in tensile stresses between the soil and the footing near the lifted edge. However, the equilibrium may be established by resisting the resultant load by the resultant of the bearing pressure of equal magnitude and the location.

This condition will prevail until the eccentricity, e of the load, P reaches the edge of the footing base. Any larger eccentricity will result in overturning. Such a condition, however, may only occur on rock or on very hard stiff soils. For most practical cases edge-yielding may make a footing unusable and produce a condition that is equivalent to overturning. Edge-yielding will occur when the extreme bearing pressure at the pressed edge will cause failure in the bearing capacity of the sub-soil. The eccentricity causing this condition will therefore, limit the maximum useful eccentricity. Unless actual test results are available, the failure condition in the bearing capacity of the soil, q_F may be assumed with about 2.5 *times* the allowable bearing capacity of the soil, q_o. The minimum safety factor against over turning is usually specified as 1.50, although, somewhat greater safety factors are some-times desirable.

22.21 SYMMETRICAL FOOTINGS SUBJECTED TO AXIAL LOAD AND MOMENT

A footing subjected to an axial load P, and a moment, M as shown in Fig. 22.28 may be designed as *symmetrical footing*. The distribution of soil pressure may be as shown in Fig. 22.28 [B(i)]. The maximum bearing pressure compressive in nature may be equated to the permissible safe bearing capacity of the soil, q_o, and size of the square footing, (B x B) may be calculated as under :

$$\left(\frac{1.10P}{A_F} + \frac{M \cdot y_{\max}}{I}\right) = q_o \qquad \text{(i)}$$

$$\text{or} \quad \left(\frac{1.10P}{(B \times B)} + \frac{M \cdot (B/21)}{\frac{1}{12}\left(B \times B^3\right)}\right) = q_o \qquad \text{(ii)}$$

By trial and error, the size of footing, B may be obtained from the expression (i). The self weight of the footing is being assumed as the *ten per cent* of the axial load, P. The effective soil pressure upward, q_o is found as below :

$$q_o = \left(\frac{P}{(B \times B)} + \frac{M \cdot (B/2)}{\frac{1}{12}\left(B \times B^3\right)}\right) \qquad \text{(iii)}$$

The intensities of soil pressures at the two edges of the footing as shown in Fig. 22.28 [B (i)] are as follows :

$$q_1 = \left(\frac{P}{(B \times B)} + \frac{M \cdot (B/2)}{\left(\frac{1}{12}\right) B \times B^3}\right) \qquad \text{(iv)}$$

The cantilever projections on all the four sides of the footing from the face of the column are equal. But the maximum bending shall occur at the edge near the maximum soil pressure. The trapezoidal soil pressure diagram. Fig. 22.28 [B (i)] may be considered as consisting of a rectangular and a triangular distribution diagram. Then, the total upward force from the right of the edge of column shall be

$$F = \left[q_2 \cdot B \cdot \left(\frac{B - b_w}{2}\right) + \frac{1}{2}(q_1 - q_2) \cdot B \cdot \left(\frac{B - b_w}{2}\right)\right]$$

$$\text{or} \quad F = \left(\frac{q_1 + q_2}{2}\right) \cdot B \cdot \left(\frac{B - b_w}{2}\right) \qquad (22.48)$$

The maximum bending moment at the face of the column

$$M = \left[\frac{1}{2} q_2 \cdot B \left(\frac{B - b_w}{2}\right)^2 + \frac{1}{2}(q_1 - q_2)\left(\frac{B - b_w}{2}\right) \cdot \frac{2}{3}\left(\frac{B - b_w}{2}\right)\right] \qquad (22.49)$$

The effective depth of footing may be determined by equating the maximum bending moment and the moment of resistance of the section. The cross-sectional area of steel bars to be provided as reinforcement may be calculated as usual and checked for necessary development length. The design has been explained by working out *an illustrating example.*

22.22 UNSYMMETRICAL FOOTINGS SUBJECTED TO ECCENTRIC LOAD

A footing may be subjected to an eccentric load, P_E acting vertically at an eccentric distance, e from the centroidal or the longitudinal axis of the column as shown in Fig. 22.27. Such a loading shall be equivalent to an axial load P and a moment. M = P. e. Then, the equivalent eccentricity, e_{equv} = e may be determined as under :

$$e_{equv} = e = \left(\frac{M}{P}\right) \quad \text{(i)}$$

The distribution of soil pressure shall be uniform, (viz., rectangular) as shown in Fig. 22.27. The area of square footing (A_F = B x B) may obtained by

$$A_F = \left(\frac{1.10 \times P}{q_o}\right) \quad \text{(ii)}$$

Then, B = $(A_F)^{1/2}$, where (iii)

A_F = area of square footing, (B x B)

q_o = permissible (safe) bearing capacity of the soil

It is to note that the self-weight of the footing has been assumed as *ten per cent* of the vertical load, P. The effective soil pressure acting upward is given by

$$q = \left(\frac{P}{(B \times B)}\right) \quad \text{(iv)}$$

The cantilever projection of the footing on one side is larger than the other side of the column. The maximum bending moment occurs about the face of column towards large projection (Fig. 22.27). The bending moment about this face of column, (critical section XX, Fig. 22.27).

$$M = \frac{1}{2} . q . B . \left[\left(\frac{B - b_w}{2}\right) + e\right]^2 \quad (22.50)$$

The critical section X_1X_1 for one-way shear shall be at a distance d equal to the effective depth of the footing. The section of footing by having the critical section XX shall be trapezoidal. Its moment of resistance may be calculated as explained earlier or by the alternative method by replacing the trapezoidal section by an equivalent rectangular section of width [b_w + (B – b_w)/8]. The maximum bending moment, M and the moment of resistance of the section of footing may be equated and the effective depth of footing, d may be evaluated. The footing may be checked for *one way shear*

$$V = q . B . \left[\left(\frac{B - b_w}{2}\right) + e - d\right] \quad (22.51)$$

The footing may also be checked for two way shear and the effective depth of footing may be revised, if necessary. The footing may be checked for two-way shear. The cross-sectional area of the steel bars to be provided as reinforcement may be calculated as usual. The bars may be checked for development length. The design has been explained by working out an illustrative example.

Example 22.7. A reinforced concrete column of size 520 mm x 520 mm transfer a load of 654.545 kN and a moment of 24 kN-m. The safe bearing capacity of soil is 110 kN/m². Design a symmetrical square footing. Use M 20 grade of concrete and Hysd steel bars of grade Fe 415.

Design. The modular ratio of M 20 grade of concrete m is 13.33 and m_c is 20.

Step 1 : Design constants

The section of the footing shall be designed as a balanced section. The design constants are as below :

$k_B = 0.2886$, and $j_B = 0.9038$

$R_B = 0.9129$, and $p_B = 0.4392$.

Step 2 : Size of the footing to maintain allowable soil pressure

Reinforced concrete column transfers an axial load of 654-545 kN. Assume self-weight of the footing as *ten per cent* of the column load

= 65.545 kN

Total load, 1.10 P = 720 kN

Moment transferred by the column = 24 kN-m

Safe bearing capacity of the soil = 110 kN/m²

The soil pressure at one edge of the footing loaded heavily is equated to the bearing capacity of soil. Let the size of footing be (B x B). Then

$$\left(\frac{1.10P}{(B \times B)} + \frac{M \times (B/2)}{\left(\frac{1}{12} B \times B^3\right)}\right) = q_o \qquad \text{(i)}$$

$$\text{or} \quad \left(\frac{720}{B^2} + \frac{24 \times 6}{B^3}\right) = 110 \qquad \text{(ii)}$$

Multiplying the expression (ii) and B^3 on both the sides, then

$$720\,B + 144 = 110\,B^3 \qquad \text{(iii)}$$

$$\text{or} \quad 110\,B^3 - 720\,B - 144 = 0 \qquad \text{(iv)}$$

By trial and error the value of B works out to be 2.6531 m. Let the size of the square footing be 2.70 m. The soil pressure (effective) under heavily loaded edge of the footing

$$q_1 = \left[\frac{P}{(B \times B)} + \frac{M \times (B/2)}{\left(\frac{1}{12} . B \times B^3\right)}\right]$$

$$q_1 = \left(\frac{654.545}{2.7 \times 2.7} + \frac{24 \times (2.7/2)}{\frac{1}{12} \times 2.7 \times 2.7 \times 2.7}\right) (89.787 + 7.316) = 97.103\ kN/m^2$$

The soil pressure (effective) under the other edge of the footing

$$q_2 = (89.787 - 7.316) = 82.471\ kN/m^2$$

The soil pressure (effective) under the middle of the column

$$q = \frac{1}{2} \times (97.103 + 82.471) = 87.787\ kN/m^2$$

Soil pressure under the column face

$$= \left[87.787 + \frac{(97.103 - 87.787)}{1350} \times 200\right] = 89.581\ kN/m^3.$$

Step 3 : Bending moment about the face of column

Critical section XX

$$M' = \left[\left(\frac{1}{2} \times 89.581 \times 1.090^2\right) + \left(\frac{1}{2} \times 1.090\right) \times (97.103 - 89.581) \times \frac{2}{3} \times 1.090\right] = kN-m$$

$$(53.216 + 2.979) = 56.195\ kN-m/m$$

Total Bending moment $(56.195 \times 2.7) = 151.727\ kN-m$

Width of equivalent section of the trapezoidal section of the footing

$$B_{equv} = \left[b_w + \frac{1}{8} \times (B - b_w)\right] = \left[0.52 + \frac{1}{8} \times (2.7 - 0.52)\right] = 0.7925\ m.$$

Step 4. Effective depth of the footing

$$d = \left(\frac{151.727 \times 10^6}{0.9129 \times 792.5}\right)^{1/2} = 457.95\ mm.$$

Let the effective cover for the lower layer of the steel bars be 60 mm. The overall depth of the footing shall be 513.95 mm (say, 520 mm). Then, the effective depth of footing shall be (520–60) = 460 mm. The overall depth of footing at the edge may be assumed as 180 mm (more than 150 mm specified by IS : 456-2000). The effective depth of footing at edge shall be 120 mm.

Step 5. Check for one-way (beam) shear

The critical section X_1X_1 for *one way shear* shall be at a distance d = 460 mm from the face of column. The length of cantilever projection from X_1X_1 upto the edge of footing

$$= (1090 - 460) = 630 \text{ mm}$$

Soil pressure under the column face

$$\left[89.581 + \frac{(97.103 - 89.581)}{1090} \times 630\right] = 93.929 \; kN/m^2$$

Total shear force at the critical section X_1X_1

$$\left[(93.929 \times 0.630) + \frac{1}{2} \times 0.630\,(97.103 - 93.929)\right] \times 2.7$$

$$V_1 = (59.175 + 0.9999) \times 2.7 = 162.472\,kN$$

Width of footing between critical section X_1X_1

$$b'' = (2700 - 2 \times 630) = 1440 \; mm.$$

Effective depth of footing at the critical section X_1X_1

$$d'' = \left[120 + \frac{630}{1090} \times (460 - 120)\right] = 316.514 \; mm.$$

Depth of neutral axis for the balanced section

$$(k_B \,.d'') = 0.2886 \times 316.514 = 91.346 \; mm$$

Width of the section at neutral axis

$$\left[520 + 2 \times \left(\frac{316.514 - 120}{460 - 120}\right) \times 1090\right] = 1780 \; mm$$

Nominal shear stress

$$\tau_v = \left(\frac{162.472 \times 1000}{1780 \times 316.514}\right) = 0.288 \; N/mm^2.$$

Permissible shear stress in concrete for

$$p_B = 0.4392 \; percent \; and \; \tau_c = 0.2957 \; N/mm^2$$

Permissible shear stress for M 20 grade of concrete is more than the nominal shear stress. Hence, safe.

Step 6. Check for two-way shear

The critical section ABCD as shown in Fig. 22.30 (C), for two-way shear shall be at a distance d/2 = 230 mm from each face of the column.

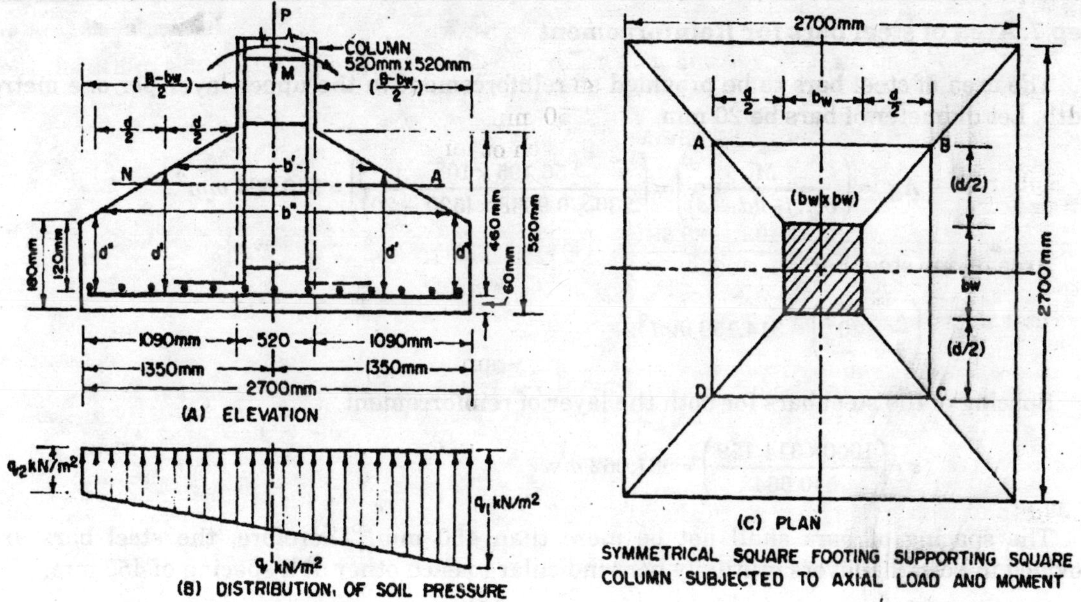

Fig. 22.30

Depth of the footing at a distance 230 mm from the face of column, Fig. 22.30 (A)

$$d' = \left[120 + \frac{(460 - 120) \times (1090 - 230)}{1090}\right] = 388.257\ mm$$

Width of the one side of critical section ABCD

$$\text{AB} = (520 + 230 + 230) = 980 \text{ mm}$$

Perimeter of the section ABCD

$$b_o = (4 \times 980) = 3920\ mm$$

Shear force for two-way shear

$$V = [P - q \cdot (b_w \times b_w)] = |654.545 - 87.787 \times 0.520^2| = 630.81\ kN$$

Nominal shear stress for two-way shear

$$\tau_v = \left(\frac{630.81 \times 10^3}{3920 \times 388.257}\right) = 0.414\ N/mm^2$$

Permissible shear stress for tri-axial compression

$$\tau_c = (k_s \times \tau_c) = 1 \times 0.16\ (f_{ck})^{1/2} = 0.7155\ N/mm^2.$$

Permissible shear stress is more than the nominal shear stress. Hence, the design is satisfactory.

Step 7. Area of steel bars for Reinforcement

The area of steel bars to be provided as reinforcement in the upper layer per one metre width. Let diameter of bars be 20 mm

$$A_{st} = \left(\frac{M}{\sigma_{st} \cdot j_B (d - \phi)}\right) = \left(\frac{56.195 \times 10^6}{230 \times 0.9038 \times (520 - 20)}\right) = 540.664\ mm^2.$$

Area of one steel bar

$$\left(\frac{\pi}{4} \times 20^2\right) = 314.159\ mm^2.$$

Spacing of the steel bars for both the layer of reinforcement

$$s = \left(\frac{1000 \times 314.159}{540.664}\right) = 581.062\ mm$$

The spacing of bars shall not be more than 450 mm. Therefore, the steel bars are provided in both the layers mutually perpendicular to each other at a spacing of 450 mm.

Step 8. Check for development length

$$L_d = \left(\frac{\sigma_{st} \cdot \phi}{4 \times 1.4 \times 0.8}\right) = \left(\frac{230 \times 20}{4 \times 1.4 \times 0.8}\right) = 1026.786\ mm.$$

The length of steel bars available from the face of column upto the edge of (1090–25) = 1065. Hence safe footing.

Example 22.8. In Example 22.7, design unsymmetrical footing.

Design : The modular ratio of M 20 grade of concrete m is 13.33 and m_c is 20.

Step 1 : Design Constants

The section of footing shall be designed as a balanced section. The design constants are as below:

$k_B = 0.2886$, and $j_B = 0.9038$

$R_B = 0.9129$, and $p_B = 0.4392$

Step 2 : Size of the footing to maintain allowable soil pressure

Reinforced concrete footing transfers an axial load of 654.545 kN and a moment of 24 kN-m

Assume self-weight of the footing as *ten-per cent* of the column load

= 65.459 kN

Total load 1.10 P = 720 kN

Safe bearing capacity of the soil is 110 kN/m²

Let the size of the footing be B x B. Then

$$A_F = \left(\frac{720}{110}\right) = 6.545\ m^2. \quad \therefore B = 2.558\ m.$$

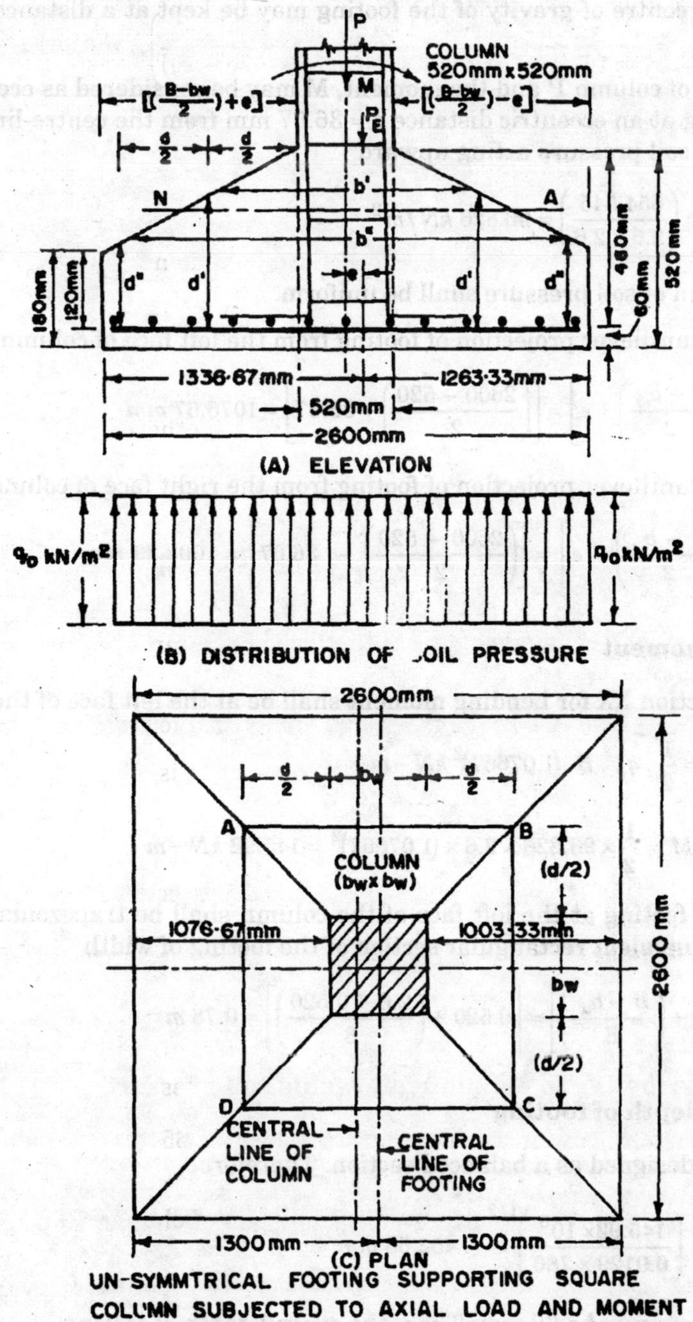

UN-SYMMTRICAL FOOTING SUPPORTING SQUARE COLUMN SUBJECTED TO AXIAL LOAD AND MOMENT

Fig. 22.31

Let the size of footing be 2.6 m x 2.6 m, Equivalent eccentricity

$$e_{equv} = e = \left(\frac{M}{P}\right) = \left(\frac{24 \times 1000}{654.545}\right) = 36.67\ mm$$

Therefore, the centre of gravity of the footing may be kept at a distance 36.67 mm from centre.

The axial load of column P and the moment, M may be considered as eccentric load as P_E = 654.545 kN acting at an eccentric distance e = 36.67 mm from the centre-line of the column. The effective (nett) soil pressure acting upward

$$q_o = \left(\frac{654.545}{2.6 \times 2.6}\right) = 96.826\ kN/m^2$$

The distribution of soil pressure shall be uniform.

The length of cantilever projection of footing from the left face of column

$$\left[\left(\frac{B - b_w}{2}\right) + e\right] = \left[\left(\frac{2600 - 520}{2}\right) + 36.67\right] = 1076.67\ mm$$

The length of cantilever projection of footing from the right face of column

$$\left[\left(\frac{B - b_w}{2}\right) - e\right] = \left[\left(\frac{2600 - 520}{2}\right) - 36.67\right] = 1003.33\ mm$$

Step 3. Bending moment

The critical section XX for bending moment shall be at the left face of the column. Then,

$$M = \frac{1}{2} \cdot q_o \cdot B \cdot (1.07667)^2\ kN-m$$

$$\text{or}\quad M = \frac{1}{2} \times 96.826 \times 2.6 \times (1.07667)^2 = 145.92\ kN-m$$

The section of footing at the left face of the column shall be trapezoidal section. It may be considered as equivalent rectangular section of the footing of width

$$\left[b_w + \left(\frac{B - b_w}{8}\right)\right] = \left[0.520 + \left(\frac{2.6 - 0.520}{8}\right)\right] = 0.78\ m$$

Step 4. Effective depth of footing

The footing is designed as a balanced section. Therefore

$$d = \left(\frac{145.92 \times 10^6}{0.9129 \times 780}\right)^{1/2} = 452.68\ mm$$

Let the effective cover be 60 mm. Then, the overall depth of footing

D = (452.68 + 60) = 512.68 mm (say, 520 mm)

The actual effective depth of footing shall be

d = (520 – 60) = 460 mm

Step 5. Check for one-way shear

The critical section for one-way (beam) shear shall be at X_1X_1 [viz., at a distance (1076.67 – 460) = 616.67 mm from the edge of footing]. The depth of footing at the critical section

$$d' = 120 + \left(\frac{1076.67 - 460}{1076.67}\right) \times (460 - 120) = 314.737\ mm$$

Shear force

$$V = q_o \cdot B.\ \text{cantilever length}$$

$$= 96.826 \times 2.6 \times (1.076 - 0.460) = 155.076\ kN$$

Depth of neutral axis

$$= k_B \cdot d = (0.2886 \times 314.737) = 90.83\ mm$$

Width of the footing at the level of neutral axis

$$b' = 520 + 2 \times \left[\frac{(460 - 120) - (314.737 - 120 - 90.83)}{(460 - 120)}\right] \times 1076.67 = 2015.26\ mm$$

Nominal shear stress

$$\tau_v = \left(\frac{155.076 \times 1000}{2015.26 \times 314.737}\right) = 0.244\ N/mm^2$$

Permissible shear stress in concrete without shear reinforcement for p_B = 0.4392 percent.

$$\tau_c = \left[0.22 + \left(\frac{0.4392 - 0.25}{0.25}\right) \times 0.08\right] = 0.281\ N/mm^2$$

Permissible shear stress in concrete is more than the nominal shear stress. Hence, safe.

Step 6. Check for two-way shear

The critical section ABCD for two-way shear shall be at a distance (d/2) = 230 mm from each face of the column. The peripheral distance

$$b_o = 4 \times [520 + 230 + 230] = 3920\ mm$$

The depth of footing below the critical section, ABCD

$$d'' = \left[120 + \left(\frac{1076.67 - 230}{1076.67}\right) \times (460 - 120)\right] = 387.369\ mm$$

Shear force for two-way shear

$$V = q_o \cdot \left[(B^2) - \left(\frac{3.920}{4}\right)^2 \right] = 96.826 \left[2.6^2 - \left(\frac{3.920}{4}\right)^2 \right] = 561.552\ kN$$

Nominal shear stress

$$\tau_v = \left(\frac{561.552 \times 1000}{3920 \times 387.369}\right) = 0.3698\ N/mm^2$$

Permissible shear stress in concrete in triaxial compression

$$k_s \cdot \tau_c = \left[1 \times 0.16 \times (20)^{1/2}\right] = 0.7155\ N/mm^2$$

Permissible shear stress in concrete is more than the nominal shear stress in concrete. Hence, the footing is safe.

Step 7. Steel bars for reinforcement

The maximum bending moment per metre width of the footing

$$M_1 = \left(\frac{145.92}{2.6}\right) = 56.123\ kN - m/m$$

Area of steel bars for reinforcement per one metre

$$A_{st} = \left(\frac{M_1}{\sigma_{st} \cdot j_B \cdot d}\right) = \left(\frac{56.123 \times 10^6}{230 \times 0.9038 \times 460}\right) = 586.926\ mm^2$$

Area of 20 mm diameter bar

$$A_\phi = \left(\frac{\pi}{4} \times (20)^2\right) = 314.159\ mm^2$$

Spacing of the steel bars

$$s = \left(\frac{1000 \times 314.159}{586.926}\right) = 535.262\ mm$$

The steel bars of 20 mm diameter of Hysd of Fe 415 grade are provided as reinforcement at 450 mm spacing (centre to centre).

Step 8. Check for development length

The development length of steel bars needed for M 20 grade of concrete and Hysd steel bars of 20 mm ϕ

$$L_d = \left(\frac{\sigma_{st} \cdot \phi}{4 \times (1.4 \times \tau_{bd})}\right) = \left(\frac{230 \times 20}{4 \times (1.4 \times 0.8)}\right) = 1026.786\ mm$$

The length of steel bars available from the face of column upto the edge of the footing

= (1076.67 – 25) = 1051.66 mm

It is more than the length required. Hence it is safe.

PROBLEMS

22.1 A 360 mm masonry wall supports a dead load of 160 kN and a live load of 140 kN. The allowable bearing pressure on soil is 130 kN/m². Design reinforced concrete footing under the masonry wall. Use M 20 grade of concrete and Hysd steel bars less than or equal to 20 mm. The footing shall be 1.20 m below the ground.

22.2 In Problem 22.1, the design footing in case the wall is a concrete wall.

22.3 A column 540 mm x 540 mm in size with 8 steel bars of 20 mm diameter transfers a dead load of 560 kN and a live load of 880 kN to the footing. The bearing capacity of soil is 130 kN/m². M 20 grade of concrete and Hysd steel bars of Fe 415 grade shall be used. Design a square footing to support the column.

22.4 A rectangular column 380 mm x 520 mm transfers a dead load of 760 kN and a live load of 1350 kN without any moment and there is no overburden. The safe bearing capacity of soil is 124 kN/m². M 20 grade of concrete and Hysd steel bars of Fe 415 grade shall be used. Design the rectangular footing to support the column.

22.5 Design a stepped footing to support the column in Problem 22.3.

22.6 Design a sloped square footing to support the column in *Problem 22.3.*

22.7. A circular column 440 mm diameter transfers an axial dead load of 520 kN and an axial live load of 450 kN. The column is having 8 steel bars of 12 mm diameter. The safe bearing capacity of the soil is 110 kN/m². Provide M 20 grade of concrete and Hysd steel bars of Fe 415 grade. Design a circular footing to support the circular column.

22.8 A reinforced concrete column of size 460 mm x 460 mm transfers a load of 548 kN and a moment of 36 kN-m. The safe bearing capacity of the soil is 100 kN/m². Use M 20 grade of concrete and Hysd steel bars of grade Fe 415.

Design (a) a symmetrical squaring footing, (b) an unsymmetrical square footing.

23

DESIGN OF COMBINED FOOTING

23.1 INTRODUCTION

The **combined footings** are the *spread footings* which support more than one column or a wall. In case, two interior columns carrying loads P_1 and P_2 are so closely spaced and the bearing capacity of the soil q is small, the isolated (single) footing areas, A_{F1} and A_{F2} are large and overlap as shown in Fig. 23.1, then, the combined footing is used.

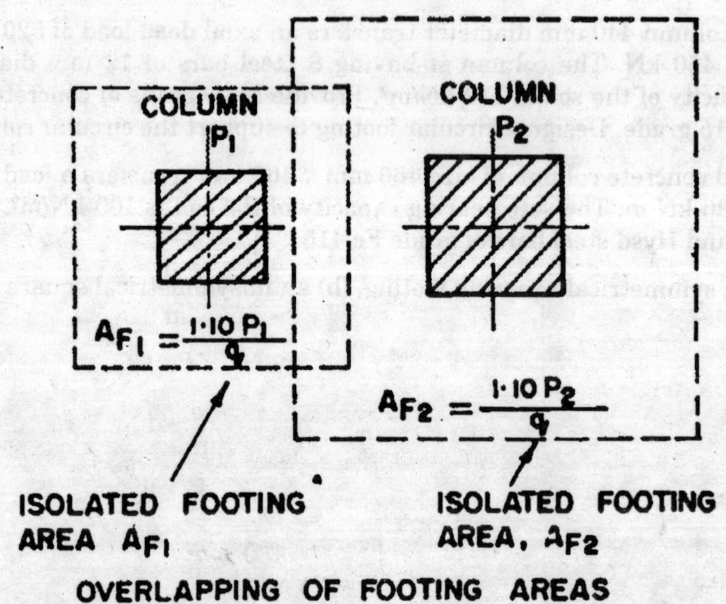

Fig. 23.1

The combined footing is also used, in case, the projection of footing of an exterior column is likely to project into the other's property line. In such cases, a combined rectangular footing or a combined trapezoid footing in plan as shown in Fig. 23.2, is used so that the projection from the footing remain within the own property line.

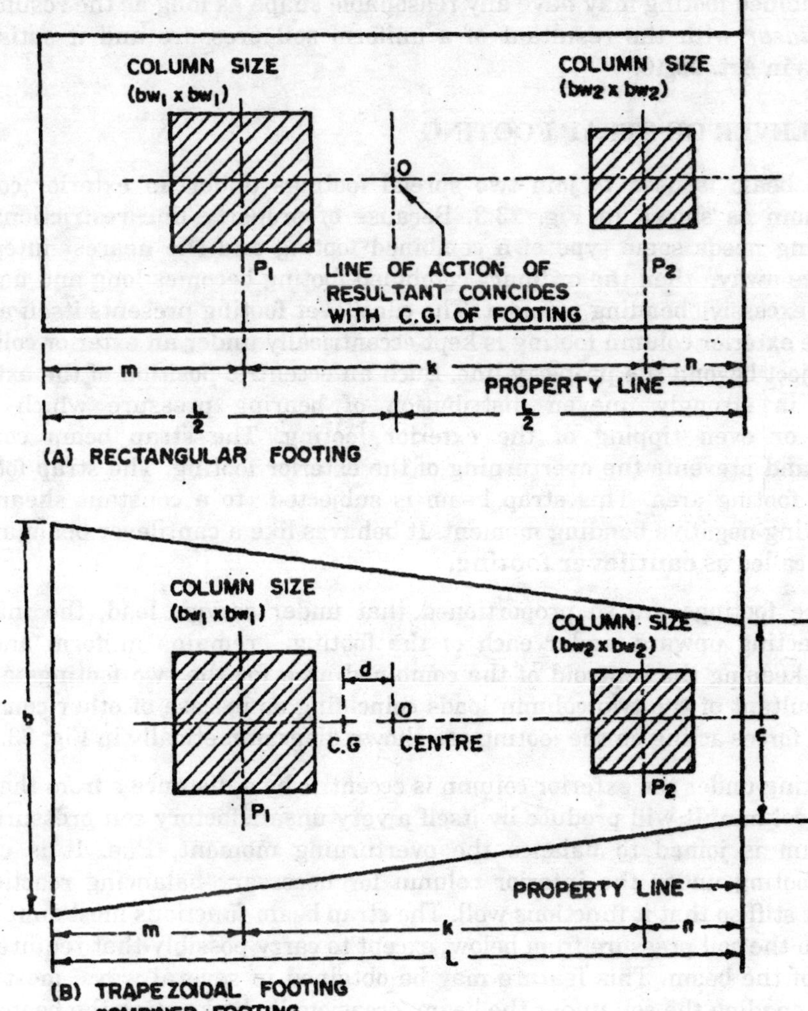

Fig. 23.2

In each case the centroid of the combined footing must coincide with the resultant of the two column loads marked as O.

In addition to above two essential circumstances, the combined footing is also used in case (i) insufficient bearing capacity of sub-soil requiring large areas, or (ii) sensitivity of the superstructure to differential settlements, or (iii) advantage in construction procedure such as trench excavation or similar.

The combined footings may be divided into two categories, viz., those which support two columns and those which support more than two (commonly large numbers of) columns. The ***continuous strip footings, the grid foundations, the mat foundations, the strip foundations, the raft foundations, ...*** **etc. are the combined footings and are used under different conditions. These have been described in subsequent sub-sections.**

The combined footing may have any reasonable shape as long as the resultant of column loads is *collinear* with the resultant of a uniform soil pressure and it satisfies the basic requirements in Art. 22.10.

23.2 CANTILEVER OR STRAP FOOTING

A strap beam is used to join two spread footings under an exterior column and an interior column as shown in Fig. 23.3. Because of property line restrictions, an exterior column footing needs some type of a combined footing and the nearest interior footing is some distance away, then the ordinary combined footing becomes long and narrow and it is subjected to excessive bending moment. The cantilever footing presents itself as an *alternate solution.* The exterior column footing is kept eccentrically under an exterior column so that it does not project beyond the property line. Such an eccentric position of the exterior column-may result in strongly uneven distribution of bearing pressure which may lead to overturning or even tipping of the exterior footing. The strap beam counteracts the eccentricity and prevents the overturning of the exterior footing. The strap footing does not provide any footing area. This strap beam is subjected to a constant shearing force and linearly varying negative bending moment. It behaves like a cantilever beam and the footing is therefore, called as **cantilever footing.**

Both the footings are so proportioned that under service load, the intensity of soil pressure q acting upward under each of the footing, remains uniform and equal. It is achieved by keeping the centroid of the combined area for the two footings and the line of action of resultant of the two column loads coinciding as in case of other combined footing. The various forces acting on the footing are shown diagrammetically in Fig. 23.3.

The footing under the exterior column is eccentric by a distance e from the centre line of the exterior column. It will produce by itself a very unsatisfactory soil pressure distribution. A strap beam is joined to balance the overturning moment, $P_1.e$. It is carried to the symmetric footing under the interior column for necessary balancing reaction. The strap beam is kept stiff so that it functions well. The strap beam functions most simply as it is kept relieved from the soil pressure from below, except to carry possibly that required to carry the self-weight of the beam. This feature may be obtained in several ways, most commonly by loosening or spading the soil under the beam, occasionally by keeping the beam free from the soil.

It is assumed that the self-weight of the strap beam is carried by the soil. Then, the cantilever footing may be analysed quite easily. The strap beam becomes effectively weightless. The resultant reactions under the exterior column footing R_1 and the interior column footing R_2 are also shown in Fig. 23.3. The column load P_2 and the reaction R_2 are collinear. The column load P_1 and the reaction under the exterior column footing R_1 form a resulting couple and balances the effect of the eccentricity of the column relative to the centre of footing. Consider the footing under the exterior column and the strap beam as shown in Fig. 23.3(C). The overturning couple and the balancing couple are equal. Therefore,

$$\Delta R \,.\, (L - e) = P_1 \,.\, e$$

$$R_1 = (P_1 + \Delta R) \; and \; R_2 = (P_2 - \Delta R)$$

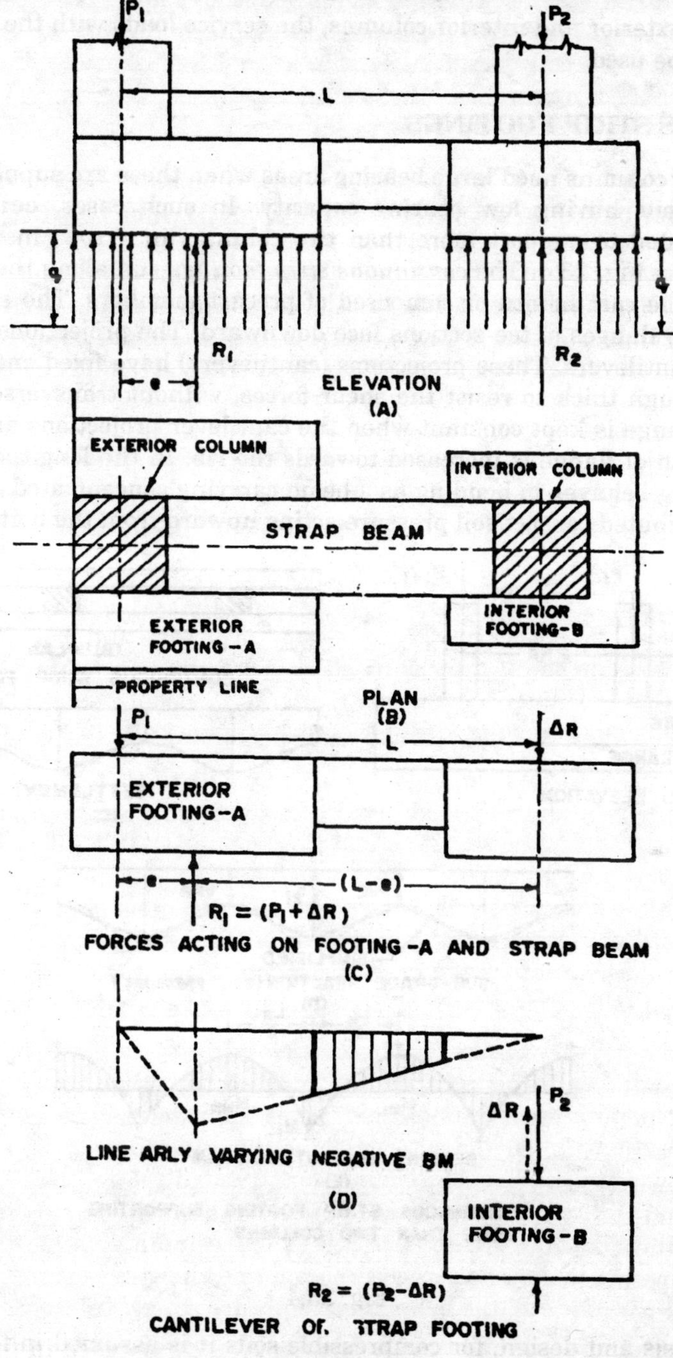

Fig. 23.3

The footing under exterior column is kept slightly larger than required for P_1 and alone and the footing under the interior column be slightly reduced than that for P_2 alone. The strap beam carries a constant shear force ΔR and a moment, $\Delta R\,.\,(L-e)$. In proportioning

the footings under exterior and interior columns, the service loads with the usual permissible soil pressures may be used.

23.3 CONTINUOUS STRIP FOOTINGS

Heavily loaded columns need large bearing areas when these are supported on relatively weak or uneven soils having low bearing capacity. In such cases, **continuously strip footings** are provided to support more than two columns in a row, instead of individual footings, as shown in Fig. 23.4. The *continuous strip footings* run along the column row. The strip footings may be cast in-situ or composed of precast members. The strip footings have tee-sections and the flanges of tee-sections face downward. The projections of the tee-section flange behave as cantilevers. These projections (cantilevers) have fixed ends in the ribs. The flange is made enough thick to resist the shear forces, without transverse or bent-up bars. The thickness of flange is kept constant when the cantilever projections are of small length. Otherwise the depth of flange is increased towards the rib. In the longitudinal direction, an isolated strip footing behaves in bending as a beam carrying concentrated column loads from above and the distributed contact soil pressure acting upward from the bottom side.

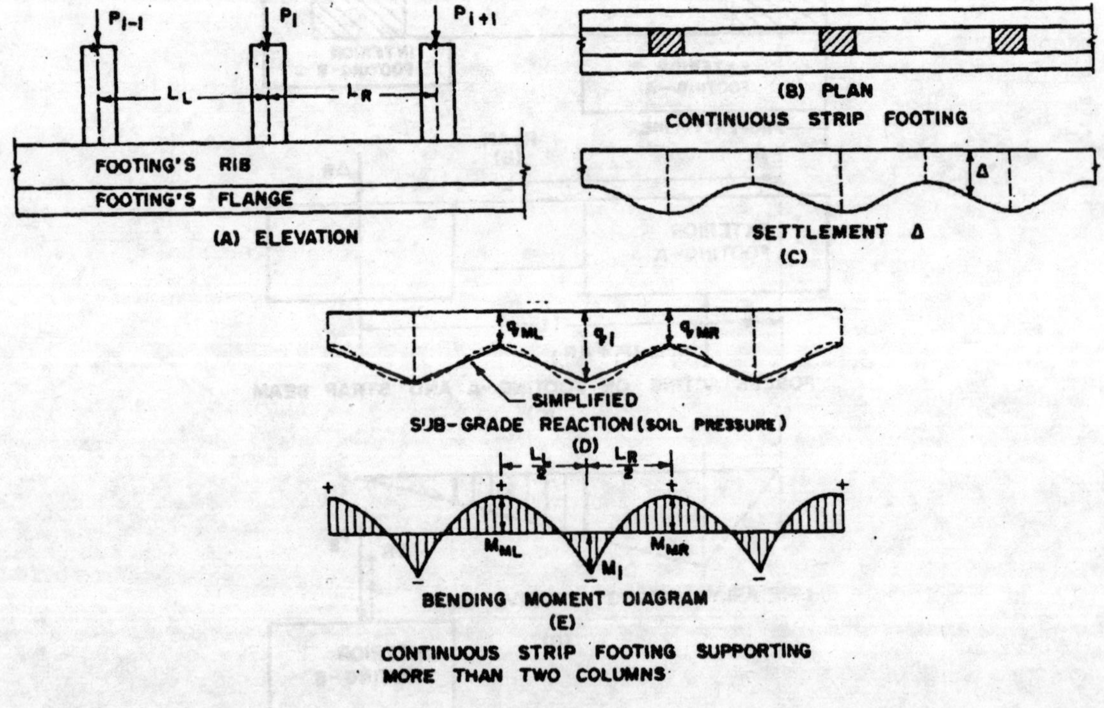

Fig. 23.4

For the analysis and design, for compressible soils it is assumed in first approximation that the deformation or settlement of the soil at a given location and the bearing pressure at that location are proportional to each other. In case, the columns are spaced at moderate distances and the strip footing is very rigid, the settlements in all proportions of the footing will be substantially the same. It means that the bearing pressure will also be equal in case the centroid of footing coincides with the line of application of the resultant of the column loads. In case these two do not coincide, then for such rigid footings, the sub-grade reactions (viz., *bearing pressures are called as sub-grade reaction*) may be assumed as linear and found from the statics, in the sameway as described for independent footings. In This case, the column loads acting downward and the soil pressure acting upward are known. Therefore, the moments and shear forces may be obtained by statics alone. Once these are known, the design of strip footing is similar to that of inverted continuous beams.

In case, the footing is relatively flexible and the spacing between columns is large, then, the settlement will not remain linear or uniform. The settlements under the heavily loaded columns will be large and the sub-grade reactions will also be large. The strip footing between the columns deflects upward. Therefore, the settlement of the soil and the subgrade reactions will be small as compared to those directly under the columns as shown in Fig. 23.4 (B) and (C), respectively. The analysis of strip footing may then be done by theory of beams on elastic foundation [Reference may be made in M. Heteneyi, `Beams on Elastic Foundation' University of Michigan Press, Ann Arbor, 1946].

In the report of ACI Committee 436, Journal of ACI, Vol. 63 No. 10, October 1966 (pp 1041-1057), the suggested Design Procedures for combined footings and Mats have been described.

As per above reference a simplified procedure for most frequent situations may be adopted. This simplified procedure defines the conditions according to which a footing may be regarded as rigid such that uniform or overall linear distribution of subgrade reaction may be assumed. The average of two adjacent span lengths in a continuous strip footing is given by

$$L_{ave} = \frac{1}{2}(L_L + L_R), \text{ where} \quad \text{(i)}$$

L_L = left span between the centre to centre of the columns being considered

L_R = right span between the centre to centre of the columns being considered.

In case, L_{ave} does not exceed $(1.75/\lambda)$, provided also that the adjacent span and column loads do not differ by more than 20 percent of the larger value, then,

$$\lambda = \left(\frac{k_s \, . \, b}{3 \, . \, E_c \, . \, I}\right)^{1/4} m^{-1}, \text{ where} \quad (23.1)$$

$k_s = S \, . \, k_s'$

k_s' = coefficient of subgrade reaction as defined in soil mechanics basically force per uni area needed to produce unit settlement kN/m³

b = width of footing, metre

E_c = modulus of elasticity of concrete, kN/m²

I = moment of inertia of the footing, m⁴

S = shape factor,

For granular soils such as sands, $S = [(b + 1)/2b]^2$ (23.2)

For cohesive soil such as clays, $S = [(n + 0.5)/1.5\,n]$ where (23.3) n = ratio of longer to shorter side of the strip footing.

The foundation shall be regarded as *flexible* in case L_{ave} exceeds (1.75/λ). In case, the adjcent spans and column loads do not differ by more than 20 per cent, the complex cantilever distribution of subgrade reaction may be substituted by a set of equivalent trapezoidal distribution reactions, as shown in Fig. 23.4(C).

Reference may also be made to Kramrisch, F. and Rogers P., `Simplified Design of combined Footings' Proceedings ASCE, Paper No. 2959, 87 (SM 5), October, 1961 and Kramrisch, Fritz `Footings' in Handbook of Concrete Engineering, second edition, edited by Mark Fintel.

The reinforcement in the rib is provided in the manner similar to that in multi-span beams. The cross-sectional area of longitudinal reinforcement is taken on the basis of normal section analysis for bending moments and that for transverse shear reinforcements (stirrups and bent up bars) is obtained by the inclined section shear analysis. The rigidity of footing is increased by choosing the cross-sectional area adequate with low percentage of reinforcement, but it should not be less than the minimum safe area for the bending members.

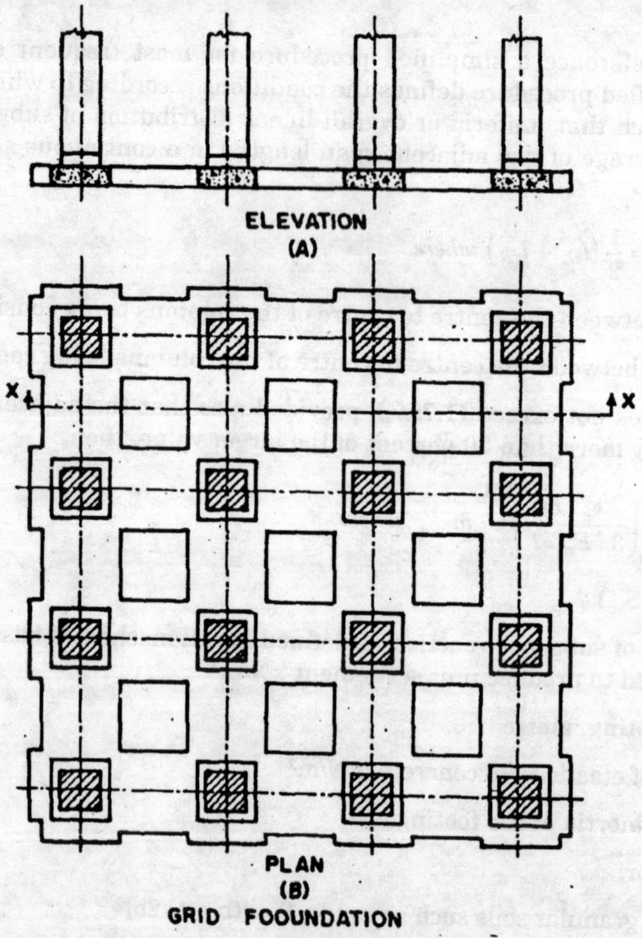

Fig. 23.5

23.4 GRID AND MAT (RAFT) FOUNDATIONS

Heavily loaded column require large bearing areas when these are supported on relatively weak or uneven soils having low bearing capacity. When the columns are located in rows mutually in perpendicular directions, then either sets of continuous strip footings are provided intersecting at right angles and forming one continuous grid foundation or **grid foundation** is provided as shown in Fig. 23.5. The grid foundation provides much larger bearing area more economically as compared to single footings because, the individual strips of the grid foundation represents continuous beams whose moments are much smaller than the cantilever moments in large single footings which project fair out from the column in all the four directions.

When the bearing capacity of soil is still smaller than as discussed above, the strip footings are merged resulting in mat **(raft) foundation** as shown in Fig. 23.6. In mat foundation, a solid reinforced concrete slab is provided below the complete building. The structural behaviour of mat foundation is similar to flat slab or a flat plate, upside down, that is, loaded upward by the soil pressure and the concentrated column reactions acting downward. The line of action of the resultant of al column loads coincide with the centre of gravity of the mat (raft) foundation. The mat foundation provides the maximum available bearing below the building. The mat foundation provides continuity and rigidity which help in reducing the differential settlement of individual columns relative to each other.

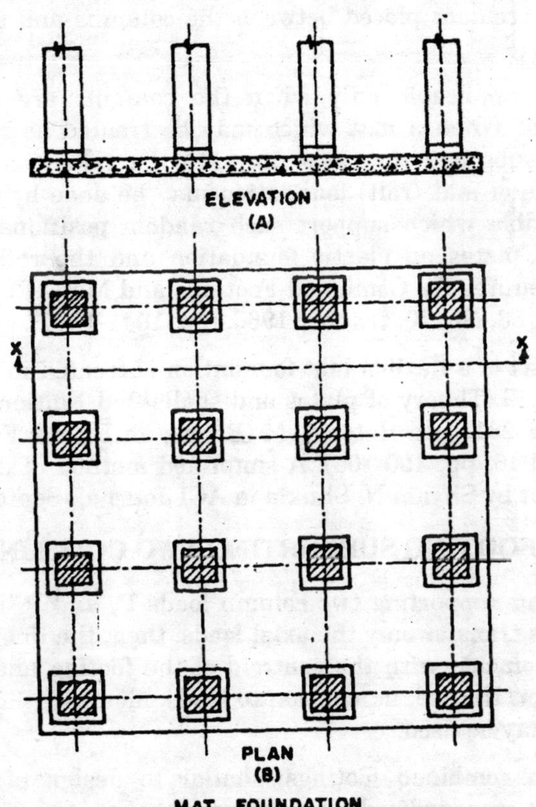

Fig. 23.6

It is economical to use mat foundations when the loads are so large that the footings would occupy more than 50 percent of the projected area of the building.

Mat (raft) foundation likewise, need different approaches for the analysis, depending on whether these may be classified as rigid or *flexible.* Similar to strip footings, in case the spacing between the adjacent column is less than $1/\lambda$ (λ as determined by Eq. 23.1), the mat (raft) foundation may be considered as *rigid*. For example, very thick and massive mat foundation under power plant structures or within the superstructures with its superimposed loads is rigid (as in case of silos and tanks). And the soil pressure may be assumed as uniformly or linearly distributed.

While the **mat (raft) foundation** shall be treated *flexible* as defined above, and if the variation of adjacent column loads and span is not more than 20 per cent, the simplified procedure which one used for the strip and grid footings may be used for the **mat (raft foundation).** In a mat foundation on an elastic subgrade, the soil pressure and deflection vary from point to point. The parameters governing the pressures and deflections are the loads, their distances from the points under consideration, rigidity of the mat, and the modulus of subgrade reaction. The mat foundation is divided into two sets of strip footings mutually at right angles to each other, of width equal to the distance between mid-spans, and the distribution of bearing pressures and bending moment is determined for each strip as explained earlier. Once, the moments are obtained, the mat in essence is regarded as flat slab or plate, with the reinforcement placed between the columns and middle strips as in these slab structures.

Above approach is applicable only when the columns are positioned in a regular rectangular grid pattern. When a mat which may be treated as rigid support columns at random positions, the subgrade reactions may still be taken as uniform or as linearly distributed. The analysis of mat (raft) foundation may be done by statics. In case, the mat (raft) foundation is flexible which support such random positions columns, the design is based on the theory of plates on elastic foundation and the reference may be made in `Suggested Design Procedures for Combined Footings and Mats' Reports of ACI Committee, 336, Journal of ACI, Vol. 63, No. 10, October 1966, (pp. 1041-1057).

The classical analysis of a flexible mat foundation is treated in detail by Timoshenko, S and Woinowsky-Krieger, S (Theory of plates and shells 2nd Edition, McGraw-Hill Book Co, New York, 1959, pp. 259-281) and Hetenyi, M (Beams an Elastic Foundation, University of Michigan, Ann Arbor, 1946, pp. 100-106). A simplified method of design of mats on elastic foundation has been dealt by Shyam N. Shukla in ACI Journal, September-October, 1984).

23.5 RECTANGULAR FOOTING SUPPORTING TWO COLUMNS LOADS

A rectangular footing supporting two column loads P_1 and P_2 is shown in Fig. 23.7 (in plan). When the columns transfer only the axial loads, then, the line of application of column loads may be made to coincide with the centroid of the footing and the distribution of soil pressure shall be uniform. In case, in addition to loads, moment is also transferred then two alternative procedures may be used.

The design of such combined footings, similar to design of many other reinforced concrete structures is not standardized. The structural behaviour of such combined footing is not yet certain. Two different approaches are used for the design of such footings. These approaches are as below :

The footing may be designed only as a slab or a beam and slab.

1. One approach

It is assumed that the structural behaviour of such footing consists of transverse slab strips below the column that act to distribute the load laterally and longitudinal slab strips that act like slabs supported on these cross-beams. Such an analysis is appropriate for long and narrow footings.

The slab as shown in Fig. 23.7(A) is regarded as spanning longitudinally and it carries upward soil pressure. Two columns support a strip ABCD only. The two portions of slabs ABFE and DCGH have no supports. In order to have longitudinal action some arrangement is needed. It is obtained by distributing the column loads with the help of transverse beams, JKLM and PQRS as shown in Fig. 23.7(C). These transverse beams span in the transverse direction. These transverse beams are not separate structural beams but these beams act as separate strips within the depth of the slab. These are reinforced in an appropriate manner so that the moments and shears are resisted by two cantilevers JKLM and PQRS, respectively.

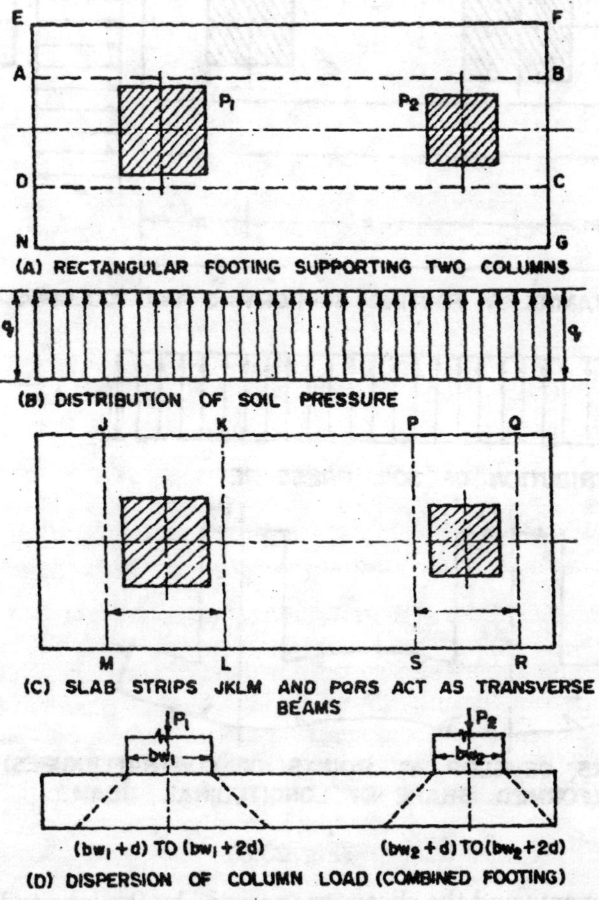

Fig. 23.7

The transverse beam do not exist as isolated strips below the columns different from the adjoining portion of the footing slab. The dispersion of column load P_1 occurs over a width $(b_{w1} + d)$ or $(b_{w1} + 2d)$ and similarly for column load P_2, it occurs over a width $(b_{w2} + d)$ or $(b_{w2} + 2d)$. The dispersion over a width $(b_{w1} + 2d)$ and $(b_{w2} + 2d)$ be used in design as the dispersion occurs with 45° with the horizontal. b_{w1} and b_{w2} are the widths of columns at the top surface.

Second approach. In this approach, the combined footing may be designed as a longitudinal beam along the length of the slab as shown in Fig. 23.8, and this beam connects the two columns. The remaining portion of the slab acts as slabs cantilevering out transversely. The structural behaviour in this approach is simple. The longitudinal beam may project below along the depth in case the sagging bending moment is more. Otherwise, if hogging moment is more, the beam may project upwards above the slab in depth.

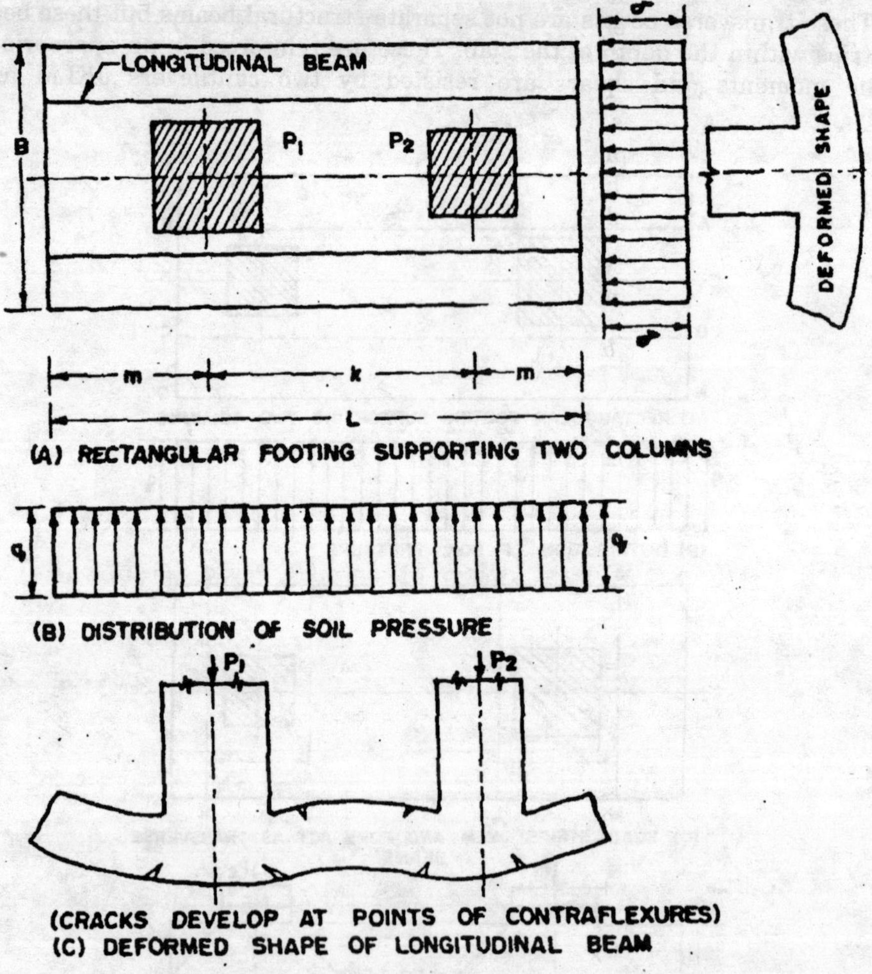

Fig. 23.8

The bending moments and the shears are resisted by the longitudinal beam. The area of steel reinforcement for the beam and the cantilever projection of slab are calculated as usual.

So, also, the area of steel to be provided as transverse reinforcement is also determined. The main reinforcement is checked for requisite development length.

Both these approaches have been further explained by working out illustrating examples.

23.6 TRAPEZOIDAL FOOTING SUPPORTING TWO COLUMNS LOADS

A trapezoidal footing supporting two columns is shown in Fig. 23.2 (B).

The line of action of the resultant of the column loads is made to coincide with the centre of gravity of the footing. Let the centre to centre distance between the two columns be K and it is known. The projections of footing are m and n. One of the projection, n is known and it is fixed because of the property line. The other projection, m is assumed such that, as far as possible, the bending moments under the columns and between two columns (viz., hogging and sagging) are approximately equal. However, it is by trial to fix up these projections most satisfactory. In case, there is any restriction for the projection, the other projection is fixed accordingly, as stated above. The length of footing, L is known, knowing the centre to centre distance between the columns K and fixing the projections of the footing. The area of base of the footing, A_F required to spread the two column loads is known, knowing the permissible bearing capacity of the soil. The width of two sides at the ends of the footing b and c as shown in Fig. 23.2 (B) are determined as follows :

Let d be the distance between centre of the footing and centroid (viz., centre of gravity, C.G.) of the footing. The distance of centre of gravity from the width b of footing is given by

$$\bar{x} = \frac{L}{3}\left(\frac{b+2c}{b+c}\right) \quad or\ \bar{x} = \frac{L}{3}\left(1+\frac{c}{b+c}\right),\ then \qquad \text{(i)}$$

$$d = \left[\frac{L}{2} - \frac{L}{3}\left(1+\frac{c}{b+c}\right)\right] \quad and\ 6d = \left[3L - 2L\left(1-\frac{c}{b+c}\right)\right] \qquad \text{(ii)}$$

or $$\frac{6d}{L} = \left[3 - 2 - \frac{2c}{b+c}\right] \quad or\ \frac{6d}{L} = \left[1 - \left(\frac{2c}{b+c}\right)\right] \qquad \text{(iii)}$$

or $$\frac{6d}{L} = \left[\frac{(b+c-2c)}{b+c}\right] \quad or\ \frac{6d}{L} = \left[\left(\frac{b-c}{b+c}\right)\right] \qquad \text{(iv)}$$

Multiply both the sides of the expression (iv) by (A_F/L). Then

$$\left(\frac{6d}{L} \times \frac{A_F}{L}\right) = \left(\frac{b-c}{b+c}\right) . \left(\frac{A_F}{L}\right) \qquad \text{(v)}$$

But, $$[(b + c) . L] = (2 . A_F) \qquad \text{(vi)}$$

Therefore, $$\left(\frac{6d}{L} \times \frac{A_F}{L}\right) = \left(\frac{(b-c) . A_F}{2 . A_F}\right)$$

or $$\left(\frac{6\,A_F . d}{L^2}\right) = \left(\frac{b-c}{2}\right) \qquad \text{(vii)}$$

And from the expression (vi), $\left(\frac{A_F}{L}\right) = \left(\frac{b+c}{2}\right)$ (viii)

Then, from the expressions (vii) and (viii)

$$b = \left(\frac{A_F}{L} + \frac{6 \,.\, A_F \,.d}{L^2}\right) \tag{23.4}$$

$$\text{and } c = \left(\frac{A_F}{L} - \frac{6 \,.\, A_F \,.d}{L^2}\right) \tag{23.5}$$

$$\text{or b and c are given by, } = \frac{A_F}{L}\left(1 \pm \frac{6\,.\,d}{L}\right) \tag{23.6}$$

where plus sign gives width b, and minus sign gives width c.

The effective soil pressure acting upward shall be uniform below the footing. It is obtained as

$$q_o = \frac{(P_1 + P_2)}{A_F}, \text{ where} \tag{ix}$$

$$A_F = L \,.\, \frac{(b+c)}{2} \tag{x}$$

Let x be the distance from left side width (i.e., width, b) of the trapezoidal footing upto a section at which the bending moment is to be calculated. Then, the bending moment due to effective soil pressure, q_o acting upward is

$$M = \frac{q_o \,.\, x^2}{2} \,.\left[b - \frac{x}{3}\left(\frac{b-c}{L}\right)\right] \tag{xi}$$

The bending moment, under the heavily loaded column

$$M_1 = \frac{q_o \,.\, x^2}{2} \,.\left[b - \frac{x}{3}\left(\frac{b-c}{L}\right)\right] - \frac{P_1}{b_{w1}}\frac{1}{2}\left[x - \left(m - \frac{b_{w1}}{2}\right)\right]^2 \tag{23.7}$$

The bending moment between the two columns

$$M_2 = \frac{q_o \,.\, x^2}{2} \,.\left[b - \frac{x}{3}\left(\frac{b-c}{L}\right)\right] - P_1 \,.\, (x - m) \tag{23.8}$$

The bending moment under the other column load

$$M_3 = \frac{q_o \,.\, x^2}{2} \,.\left[b - \frac{x}{3}\left(\frac{b-c}{L}\right)\right] - P_1 \,.\, (x - m) - \frac{P_2}{b_{w2}} \,.\, \frac{1}{2}\left[x - \left(m + k - \frac{b_{w2}}{2}\right)\right]^2 \tag{23.9}$$

The bending moments M_1, M_2 and M_3 are differentiated with respect to x, and are equated to zero individually. The values of x are obtained for the maximum bending moments in the above expressions. The maximum bending moments are calculated from above expressions by substituting the separate values of x. The maximum of maximum bending moments (viz., absolute maximum bending moment) is obtained and it is used for the design of depth of the footing.

23.7 APPROXIMATE METHOD OF DESIGN OF FOOTING SUPPORTING TWO COLUMNS

The line of action of resultant of two column loads is made to coincides with the centre of gravity of the footing as usual.

The sizes of columns (b_{w1} x b_{w1}) and $\left(b_{w2} \times b_{w2}\right)$are known. The footing acts as inverted slab loaded with uniform soil pressure acting upward supported at two columns. The footing has cantilever projections at both the sides. The footing is subjected to the cantilever moment at each end and the span moment in between the two columns. The cantilever moments and the span moments are of opposite sign.

The cantilever moments are assumed to be maximum at the centre line of the columns. The maximum span moment is assumed to occur at the point where the shear force changes sign. The maximum cantilever moments and the maximum span moments are thus, determined. The footing is designed for absolute maximum bending moment.

The shear force and the bending moment diagrams are drawn as shown in Fig. 23.9.

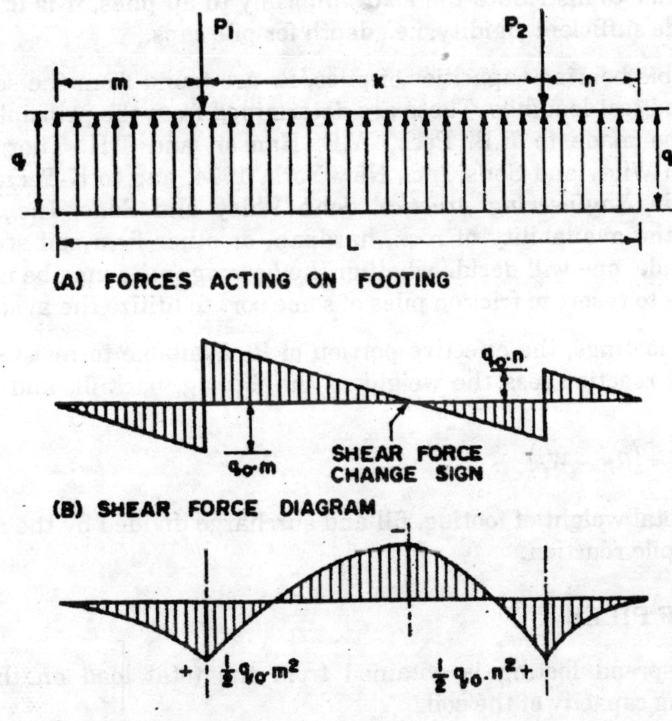

Fig. 23.9

23.8 FOOTINGS ON PILES

In case, the bearing capacity of the soil in the upper layers is in sufficient for a spread footing, and a firm stratum is available at greater depth, piles are used to transfer the loads from the super-structure down to the firm soil stratum, where the required resistance is available. These piles are generally arranged in groups or clusters one under each column. These piles act as a series of relatively concentrated reactions for the footing. These piles develop the resistance either by *end bearing* on the firm stratum (such piles are called *bearing piles*) or by *skin friction* (in which case, these piles are called *friction piles*).

The footings on piles distribute the column load to the piles in each group. These groups of piles are capped by the spread footings by *Piles caps* (the pile cap distributes the load to all the piles in a group). The loads from the *piles* is further transmitted to the subsoil.

The pile caps are in many ways very similar to the footings on soils, except for two distinctions. For one, the reactions on caps act as concentrated loads at the isolated piles rather than as distributed pressure. For another, in case the total of all pile reactions in a group is divided by the area of the footing to determine *an equivalent uniform pressure* (for the purpose of comparison only), it is seen that this equivalent pressure is significantly more in pile caps than for the spread footings. It means that the moments and particularly shears are also more, which needs more footing depths than for spread footing of similar horizontal dimensions. In order to distribute the load uniformly to all piles, it is in any circumstances, advisible to provide sufficient rigidity, i.e., depth for pile caps.

The permissible bearing capacities of piles, R_P are found from the soil exploration, pile-driving energy, and test loadings. These are determined from the principles of soil *mechanics* (Reference may be made to R.B. Peck, W.E. Hanson and T.H. Thornburn, `*Foundation Engineering*' John Wiley and Sons, Inc., New York, 1974, and to K-Terzaghi and R.B. Peck, `*Soil Mechanics* in *Engineering practice*' John Wiley and Sons, Inc., New York, 1967). Depending upon the availability of rock, hardpan, or other firm soil stratum and on their distance below grade, one will decide whether the *bearing piles* may be used economically or otherwise, one has to resort to friction piles of some sort to utilize the available soil condition.

As in spread footings, the effective portion of R_P available to resist the column loads is the allowable pile reaction less the weight of the footing, backfill, and surcharge per pile. That is,

$$R_c = (R_P - W_f) \tag{23.10}$$

where, W_f is the total weight of footing, fill and surcharge divided by the number of piles and R_c is the effective pile reaction.

23.9 NUMBER OF PILES

The size of spread footing is obtained from the total load on the footing and the permissible bearing capacity of the soil.

In case of pile cap, the number of piles in a group is found from the total load and the allowable load bearing capacity of each individual pile. The number of piles in a concentrically loaded group is the integer next larger than

$$N = \left(\frac{Total\ load}{R_e}\right) \tag{23.11}$$

where, the total load consists of dead load plus the live load and R_e is the effective reaction from individual pile as obtained from Eq. 23.11. As far as the effects of wind, earthquake moments at the foot of columns and the safety against overturning are considered, the design considerations are similar to those for the spread footings.

These effects generally result in an eccentrically loaded pile group in which different piles carry different loads. The number of piles and their location in a such a group is calculated by successive approximation from the condition that the load on the most heavily loaded pile shall not exceed the permissible pile reaction, R_P. The distribution of pile loads due to bending may be assumed linear.

It is common practice to use for stability, a minimum of three piles in a free standing pile group, a minimum of two piles if a foundation beam or similar provides lateral support, and single pile only if the lateral support may be provided in two directions. These minimum requirements have to be satisfied even if the capacity provided by the pile group far exceeds the value of the load to be supported. It is good practice to design the pile caps in any case for the full allowable (permissible) capacity of the group and the pile capacity is fully utilized.

In bearing piles, every pile in a group may be considered to act as an independent pier down to the bearing stratum and to share the load equally. In friction piles, the number of piles in a group influence their carrying capacity, especially that of the interior piles. Although this deficiency is usually averaged over the entire pile group, as far as the capacity of the group is concerned, the variation in the capacity of each individual pile needs, sometimes, consideration in the design of the pile cap.

It is always possible that some piles in the group may develop a smaller (or greater) resistance than the other, a pile cap ought to be stiff enough to equalize this condition. It is therefore, advisable not to keep the effective depth of a pile cap down to the minimum required, but no increase it somewhat wherever possible.

23.10 EVALUATION OF PILE REACTION

The reaction in a pile group is evaluated separated for concentric load conditions or eccentric load conditions (or concentric load with moment as follows :

1. Concentric load condition

The *permissible or allowable pile reaction, R_F* (as used in Eq. 23.10) is evaluated by principles of soil mechanics. The *effective pile reaction, R_e* is also determined from the principles of soil mechanics as discussed in Art 23.9 and given by Eq. 23.11 and number of piles in a pile group is found.

2. Eccentric load, condition (or concentric load with moment)

The maximum pile reaction, R_{max} for the eccentric load condition (or concentric load with moment) is evaluated as below

$$R_{max} = \left(\frac{P}{N} + \frac{M}{(I_{PG}/C)}\right), \text{ where} \tag{23.12}$$

P = maximum load (including weight of the cap, backfill, etc.).

M = moment to be resisted by the pile group, both referred to the bottom of the cap.

I_{PG} = moment of inertia of entire pile group about the centroidal axis about which bending moment is taken.

c = distance from the axis to the extreme pile.

In order to calculate the moment of inertia of a pile group, the centroid of the pile group is first located (that is, determined). In case, there are N number of piles in a pile group, the moment of inertia,

$$I_{PG} = \sum_{I}^{N} \left(1 \times y_N^2\right) \tag{23.13}$$

where y is the distance of each pile in the group from the centroidal axis. Each pile is considered as one unit and it is located at a distance y_N from the centroidal axis.

Where a pile group consists of m equal, parallel rows of piles, the moment of enertia of the entire group

$$I_{PG} = \left(\frac{m \,.\, I_p}{R_{ow}}\right) = m \,.\, \frac{N_{PR}\left(N_{PR}^2 - 1\right)}{12} C_{p^2} \tag{23.14}$$

And, the section modulus of the extreme piles in the group

$$Z_{PG} = m \,.\, N_{PR} \,.\left(\frac{N_{PR} + 1}{6}\right).\, C_p \tag{23.15}$$

However, if the parallel rows are not of the same configuration, sum up the moments of inertia for the various rows and find the section modulus of the extreme pile by dividing the moment of inertia of the entire group by the distance of the extreme pile from the centroid.

$$S_{PG} = (I_{PG}/C) \tag{23.16}$$

Example 23.1 Two reinforced concrete columns 800 mm x 800 mm and 600 mm x 600 mm in size carry axial loads of 1500 kN and 1000 kN, respectively. These columns are placed 4 m apart centre to centre. The safe bearing capacity of the soil is 200 kN/m². M 20 grade of concrete and Hysd steel bars be used for the footing. Design the combined slab type rectangular footing. The cantilever projection available from centre line of column 1 upto property line is 1.72 m.

Design. The modular ratio for M 20 grade of concrete

$$m = \left(\frac{280}{3 \times 7}\right) = 13.333.$$

Step 1. Design constants

The design constants for M 20 grade of concrete and Hysd steel bars as follows :

Factor for depth of neutral axis

$$k_B = \left(\frac{13.333 \times 7}{13.333 \times 7 + 230}\right) = 0.2886$$

Factor for lever arm

$$j_B = (1 - k_B/3) = (1 - 0.2886/3) = 0.9038$$

Factor for moment of resistance

$$R_B = \left(\frac{1}{2} \times 7 \times 0.2886 \times 0.9038\right) = 0.9129$$

Percentage of reinforcement for the balanced section

$$p_B = \left(\frac{1}{2} \times \frac{7}{230} \times 0.2886 \times 100\right) = 0.4392$$

Step 2. Area of combined rectangular footing

Axial load from column 1, $P_1 = 1500$ kN

Axial load from column 2, $P_2 = 1000$ kN

The self-weight of footing may be assumed as *ten per cent* of total load transferred through the columns.

$$0.10\,(P_1 + P_2) = \frac{10}{100}(1500 + 1000) = 250\ kN$$

Total load = 2750 kN

Area of combined rectangular footing required

$$A_F = \left(\frac{2750}{200}\right) = 13.75\ m^2$$

Length of the rectangular footing

The cantilever projection available from the centre-line upto the property line is 1.72 m as shown in Fig. 23.10.

The distance of line of action of the resultant of two column loads from 1500 kN load

$$\bar{x}_1 = \left(\frac{1500 \times 0 + 1000 \times 4}{(1500 + 1000)}\right) = 1.60\ m$$

The centre of gravity of footing is kept coinciding with the line of action of resultant of columns (that is, the footing has to be symmetrical about 0). Therefore, the length of the footing, L_F

$$2(1.72 + 1.60) = 2 \times 3.32 = 6.64\ m$$

Width of the rectangular footing

$$B_F = \left(\frac{A_F}{L_F}\right) = \left(\frac{13.75}{6.64}\right) = 2.07\ m\ say\ B_F = 2.20\ m.$$

Effective soil pressure acting upward

$$q_o = \left(\frac{1500 + 1000}{6.64 \times 2.20}\right) = 171.139\ kN/m^2$$

Intensity of upward load per metre length of the footing and for complete width of footing

$$= (171.139 \times 2.2) = 376.506\ kN/m$$

Step 3. Shear force in footing slab

The column load P_1 = 1500 kN acts over the size 0.8 m along the length. Therefore, the distributed load per metre length acting downward

(1500/0.8) = 1875 kN/m.

The column load P_2 = 1000 kN acts over the size 0.6 m along the length. Therefore, the distributed load per metre length acting downward

(1000/0.6) = 1666.667 kN/m

The loads acting downward and the intensity of soil pressure acting on the footing upward are shown in Fig. 23.8 (B). The shear force at B_1

$$(376.506 \times 1.32) = 496.988\ kN$$

Shear force at D_1, $[376.506 \times (1.32 + 0.8) - 1500] = 701.807\ kN$

Shear force at F_1, $[376.506 \times (1.32 + 0.8 + 3.3) - 1500] = 540.663\ kN$

Shear force at H_1, $[376.506 \times (1.32 + 0.8 + 3.3 + 0.6) - 2500] = 233.434\ kN$

In the portion B_1D_1, the shear force changes sign at C_1 under column 1. From the similar triangles

$$\left(\frac{B_1C_1}{496.988}\right) = \left(\frac{C_1D_1}{701.807}\right)\ (B_1C_1 = x_1)$$

$$x_1 = \left(\frac{496.988}{701.807}\right) \times (0.8 - x_1) = 0.332\ m.$$

In the portion D_1F_1, the shear force changes sign at E_1 between two columns. From the similar triangles

$$\left(\frac{D_1E_1}{701.807}\right) = \left(\frac{E_1F_1}{540.663}\right)\ (D_1E_1 = x_2)$$

$$x_2 = \left(\frac{701.807}{540.663}\right) \times (3.3 - x_2) = 1.864\ m.$$

In the portion F_1H_1, the shear force changes sign at G_1, under the column 2. From the similar triangles

$$\left(\frac{F_1G_1}{540.663}\right) = \left(\frac{G_1H_1}{233.434}\right) \quad (F_1G_1 = x_3)$$

$$x_3 = \left(\frac{540.663}{233.434}\right) \times (0.6 - x_3) = 0.419\ m.$$

(Shear force at point of contra flexures are found, after locating these points in the step 4 below). Shear force at x = 0.392 m right of point D, Fig. 23.10,

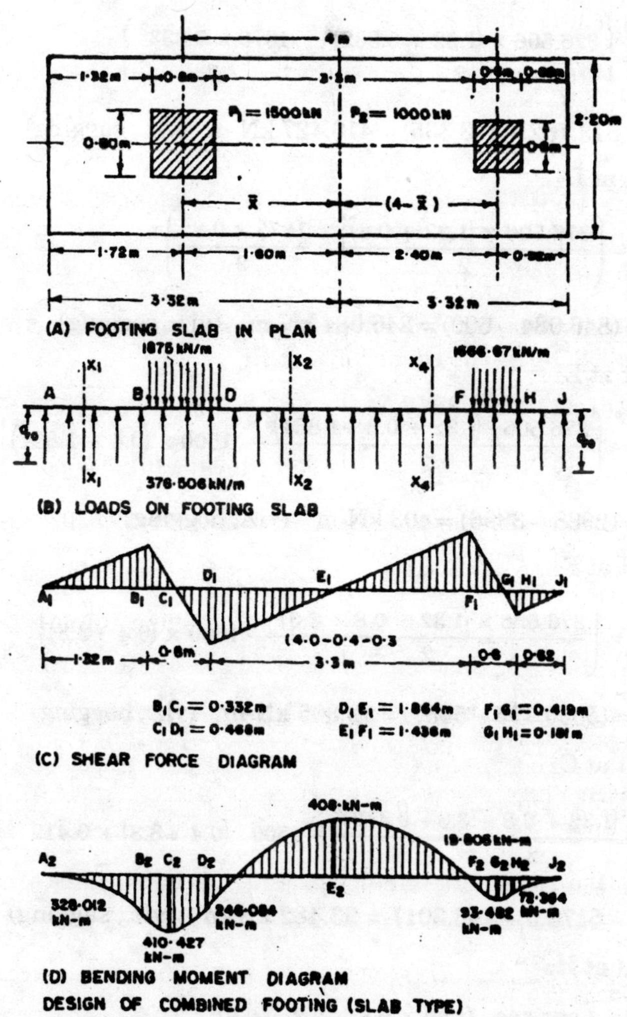

Fig. 23.10

$= [367.506\,(1.32 + 0.8 + 3.92) - 1500] = 554.217$ kN

Shear force at 0.191 m below the second column

$$[376.506\ (1.32 + 0.8 + 3.3 + 0.191) - 1500 - 1666.67 \times 0.191] = -294.24\ kN$$

Step 4. Bending moment in footing slab

The bending moments are calculated from left side.

Bending moment at B_2

$$M_{B_2} = \left(\frac{376.506 \times 1.32^2}{2}\right) = 328.012\ kN-m \quad \text{(viz., sagging)}$$

Bending moment at C_2

$$M_{c_2} = \left(\frac{376.506 \times (1.32 + 0.332)^2}{2} - \frac{1875 \times 0.333^2}{2}\right)$$

$= (513.762 - 103.335) = 410.427$ kN-m (viz., sagging)

Bending moment at D_2

$$M_{D_2} = \left(\frac{376.506 \times (1.32 + 0.8)^2}{2} - \frac{1875 \times 0.8^2}{2}\right)$$

$= (846.084 - 600) = 246.084$ kN-m (viz., sagging)

Bending moment at E_2

$$M_{E_2} = \left(\frac{376.506\ \ (1.32 + 0.8 + 1.864)^2}{2} - 1500 \times (0.4 + 1.864)\right)$$

$= (2988 - 3396) = 408$ kN-m (viz., hogging)

Bending moment at F_2

$$M_{F_2} = \left(\frac{376.506 \times (1.32 + 0.8 + 3.3)^2}{2} - 1500 \times (0.4 + 3.3)\right)$$

$= (5530.195 - 5550) = 19.805$ kN-m (viz., hogging)

Bending moment at C_2

$$M_{G_2} = \left(\frac{376.506 \times (1.32 + 0.8 + 3.3 + 0.419)^2}{2} - 1500\ \ (0.4 + 3.3) + 0.419 - \frac{1666.67 \times 0.419^2}{2}\right)$$

$= (6418.283 - 6178.5 - 146.301) = 93.482$ kN-m (viz., sagging)

Bending moment at H_2

$$M_{H_2} = \left(\frac{376.506\ \ (1.32 + 0.8 + 3.3 + 0.6)^2}{2} - 1500 \times (0.4 + 3.3 + 0.6) - 1000 \times 0.3\right)$$

$= (6822.364 - 6450 - 300) = 72.364$ kN-m (viz. sagging)

Let the point of zero moment (viz., point of contraflexure) be at a distance x from the point D, Fig. 23.10

$$M_x = \left[\frac{376.506\,(1.32 + 0.8 + x)^2}{2} - 1500\,(0.4 + x)\right] = 0$$

$$\text{or } M_x = \left[\frac{376.506\,(2.12 + x)^2}{2} - 1500\,(0.4 + x)\right] = 0$$

Therefore, x = 0.392 m and x = 3.336 m (**Note :** this value has to be found again by considering second load also). As such

$$M_x = \left[\frac{376.506\,(2.12 + x)^2}{2} - 1500\,(0.4 + x) - \frac{1666.67\,(x - 3.3)^2}{2}\right] = 0$$

Therefore, x = 34310 m [i.e., (3.4910 – 3.3) = 0.191 m] below the second column.

Step 5. Effective depth of footing slab

The absolute maximum bending occurs at C_2, M_{c2} = 410.427 kN-m for the complete width of the footing. The effective depth of footing

$$d = \left(\frac{410.427 \times 10^6}{0.9129 \times 2200}\right)^{1/2} = 452.058\,mm$$

Let the effective cover be 60 mm. Then, the overall depth of footing slab shall be (425.058 + 60) = 512.058 mm, say 520 mm, Then, the effective depth of slab shall be (520 – 60) = 460 mm.

Step 6. Check for two-way (flat slab type) shear

(i) The critical section for two-way shear shall be periphery at a distance d/2 from the periphery of the column (carrying 1500 kN column load)

$$b_o = 4 \times [800 + (460/2) + 460/2] = (4 \times 1260) = 5040\,mm$$

Effective soil pressure acting upward is 171.139 kN/m².

Shear force = column load – upward pressure on the area within the periphery

$$= (1500 - 171.139 \times 1.260 \times 1.260) = (1500 - 271.70) = 1228.30\,kN$$

Nominal shear stress

$$\tau_v = \left(\frac{1228.3 \times 1000}{5040 \times 460}\right) = 0.5928\,N/mm^2$$

Permissible shear stress for tri-axial compression is $k_s.\tau_c$ $(k_s = 1)$

$$k_s \times 0.16(f_{ck})^{1/2} = 1 \times 0.16 \times (20)^{1/2} = 0.716\ N/mm^2$$

The permissible stress, τ_c is more than the nominal shear stress. Hence, it is safe.

(i) The critical section for two-way shear shall be periphery at a distance d/2 from the periphery of the column (carrying 1000 kN column load)

$$b_o = 4 \times [600 + 460/2 + (460/2)] = 4 \times 1060 = 4240\ mm$$

Shear force = (column load – upward pressure) on the area within the periphery

$$= (1000 - 171.139 \times 1.06 \times 1.06) = (1000 - 192.292) = 807.708\ kN$$

Nominal shear stress

$$\tau_v = \left(\frac{807.708 \times 1000}{4240 \times 460}\right) = 0.414\ N/mm^2$$

Permissible shear stress for tri-axial compression

$$k_s \cdot \tau_c\ (k_s = 1) = 1 \times 0.16 \times (20)^{1/2} = 0.716\ N/mm^2$$

The permissible shear stress is more than the nominal shear stress. Hence, it is safe.

Step 7. Check for one-way (beam) shear

The critical section for one-way (beam) shear shall be taken at a distance, d (the effective depth of the slab) from the face of column carrying 1500 kN load (viz. (1.32 – 0.460) = 0.85 m) on the left side. The shear force at the critical section X_1X_1, Fig. 23.8 (B)

$$V_{X_1X_1} = (376.506 \times 0.86) = 323.80\ kN$$

$$\textbf{Nominal shear stress, } \tau_v = \left(\frac{323.80 \times 1000}{2.20 \times 1000 \times 460}\right) = 0.32\ N/mm^2$$

Permissible shear stress in concrete without shear reinforcement, τ_c, for percentage reinforcement for balanced section

$$p_B = (100\ A_s/bd) = 0.4392$$

$$\tau_c = \left(0.22 + \frac{0.08 \times 0.1892}{0.25}\right) = 0.28\ N/mm^2$$

It is less than the nominal shear stress. Therefore, the effective depth of footing slab shall be increased to

$$d = \left(460 \times \frac{0.32}{0.28}\right) = 525.714\ mm$$

The effective cover is 60 mm. The overall depth shall be 585.714 mm, say 590 mm. Therefore, the effective depth shall be 530 mm.

(ii) The critical section for one-way (beam) shear shall be taken at a distance, d (the effective depth of slab) from the face of the column carrying 1500 kN load (viz., 1.32 + 0.80 + 0.530 = 2.65 m)

on the right side

$$V_{X2X_2} = (376.506 \times 2.65 - 1500) = -502.26\ kN$$

Nominal shear stress, $\tau_v = \left(\dfrac{502.26 \times 1000}{2200 \times 530}\right) = 0.430\ N/mm^2$

Permissible shear stress (k_s = 1)

τ_c = 0.28 N/mm² (as above)

It is less than the nominal shear stress. As such, the effective depth of the footing slab is again increased

$$d = \left(530 \times \frac{0.43}{0.28}\right) = 813.93\,mm$$

Effective cover = 60 mm.

Overall depth shall be 873.93 mm, say 880 mm. The effective depth shall be 820 mm.

(iii) The critical section one-way (beam) shear shall be taken at a distance, d (the effective depth of slab) from the face of column carrying 1000 kN (viz., 0.62 – 0.82 m, i.e., out the side), on the right side. The shear force is zero.

(iv) The critical section for one-way (beam) shear shall be taken at a distance, d from the face of column carrying 1000 kN, (viz., 0.62 + 0.82) = 1.44 m) on the left side.

The shear force at X_4X_4 (from right side)

$$V_{X_4X_4} = (376.506 \times 1.44 - 1000) = -457.83\ kN$$

It is less than 502.26 kN, shear force.

Step 8. Reinforcement along the length of footing

(i) Reinforcement for tension M_{c2}

The maximum bending moment under the column load 1500 kN (on left side of the footing) occurs at C_2. The reinforcement for tension shall be calculated for M_{c2} (which one is (sagging). It shall be provided along the length of footing near the bottom surface under the column load 1500 kN. Area of steel bars to be provided as tension reinforcement

$$A_{st.c2} = \left(\frac{M_{c2}}{\sigma_{st} \cdot j_B \cdot d}\right) = \left(\frac{410.427 \times 10^6}{230 \times 0.9038 \times 820}\right) = 2407.81\,mm^2$$

20 mm ϕ Hysd-steel bars shall be provided is A_ϕ = 314.159 mm²

Number of bars $\left(\dfrac{2407.81}{314.159}\right) = 7.664$ (*say 8 bars*)

8 bars of 20 mm ϕ shall be provided as uniformly distributed in 2.20 m width of the footing.

(ii) Reinforcement for tension for M_{E2}

The maximum bending moment in between the two columns occurs at E_2. The reinforcement for tension shall be calculated for M_{E2} (which one is hogging). It shall be provided along the length of footing near the top surface in between the two columns. Area of steel bars to be provided as tension reinforcement

$$A_{st.E_2} = \left(\frac{M_{E_2}}{\sigma_{st} \cdot j_B \cdot d}\right) = \left(\frac{408 \times 10^6}{230 \times 0.9038 \times 820}\right) = 2393.57 \ mm^2$$

For 20 mm ϕ steel bars is $A_\phi = 314.159 \text{ mm}^2$

$$\text{Number of the steel bars, } \left(\frac{2393.57}{314.159}\right) = 7.62 \ (say\ 8\ bars)$$

8 bars shall be provided as uniformly distributed in 2.20 m width of the footing slab.

(iii) Reinforcement for tension at M_{G2}

The maximum bending moment under the column load 1000 kN (on right side) occurs at G_2. The reinforcement for the tension shall be calculated for M_{G2} (which one is sagging). It shall be provided along the length of footing near the bottom surface under the column load 1000 kN. Area of the steel bars to be provided as tension reinforcement

$$A_{st.G_2} = \left(\frac{M_{G_2}}{\sigma_{st} \cdot j_B \cdot d}\right) = \left(\frac{93.482 \times 10^6}{230 \times 0.9038 \times 820}\right) = 548.42 \ mm^2$$

For 20 mm ϕ steel bars is $A_\phi = 314.159 \text{ mm}^2$

Number of the steel bars

(548.4/314.159) = 1.745 (say 6)

6 bars of 20 mm ϕ shall be provided as uniformly distributed in 2.20 m width of the slab.

Step 9. Reinforcement along the width of footing

(i) Under 1500 kN column load

The projection of footing along the width, $\frac{1}{2} \times (2200 - 800) = 700 mm$

The width of transverse beam within the footing slab under 1500 kN column load

$$= (800 + 2 \times d) = [800 + (2 \times 460)] = 1720 \ mm$$

Width available on the left side of column face is 1320 mm (more than 460 mm). Hence, it is alright.

The effective soil pressure acting under this transverse beam

$$\left(\frac{1500}{2.200 \times 1.720}\right) = 396.41\ kN/m^2$$

Maximum bending at the face of column *per one metre* strip

$$M_{t1} = \left(\frac{396.41 \times 0.7^2}{2}\right) = 97.12\ kN-m$$

Effective depth required for the beam

$$d = \left(\frac{97.12 \times 10^6}{0.9129 \times 1000}\right)^{1/2} = 326.168\ mm$$

Actual effective depth of the footing slab 820 mm is more than 326.168 mm.

Area of the steel bars to the provided as tension reinforcement (using 20 mm ϕ bars)

$$A_{st.s1} = \left(\frac{97.12 \times 10^6}{230 \times 0.9038 \times (820 - 20)}\right) = 584.00\ mm^2$$

$$\text{Spacing of bars, } S_{t.1} = \left(\frac{1000 \times 314.159}{584.00}\right) = 537.94\ mm$$

Provide bars at 450 mm spacing.

(ii) Under 1000 kN column load. The section of footing along the width

$$\frac{1}{2}(2200 - 600) = 800\ mm$$

The width of transverse beam within the footing slab under 1000 kN column load

$$(600 + 2 \times 460) = 1520\ mm$$

Width available on the right side of column face is 920 mm (more than 460 mm). Hence, it is alright. The effective soil pressure acting under this transverse beam

$$\left(\frac{1000}{2.2 \times 1.520}\right) = 299.043\ kN/m^2$$

Maximum bending moment at the face of the column per one metre strip

$$M_{t.2} = \left(\frac{299.043 \times 0.8^2}{2}\right) = 95.694\ kN-m$$

Effective depth required for this beam

$$\left(\frac{95.694 \times 10^6}{0.9129 \times 1000}\right)^{1/2} = 323.765\ mm$$

Actual effective depth of the footing slab 820 mm is more than 323.765 mm. Area of the steel bars to be provided bars as tension reinforcement (using 20 mm ϕ)

$$A_{st2} = \left(\frac{95.694 \times 10^6}{230 \times 0.9038 \times (820 - 20)}\right) = 575.432\ mm^2$$

Spacing of the bars

$$s_{t.2} = \left(\frac{1000 \times 314.159}{575.432}\right) - 545.95\ mm$$

Provide bars at 450 mm spacing.

Step 10. Distribution reinforcement

The distribution reinforcement equal to 0.12 percent of the cross-sectional area for Hysd-steel bars) shall be provided at locations, when the transverse reinforcement is not provided. This reinforcement is also provided in transverse direction (viz., in the perpendicular direction to longitudinal reinforcement)

$$A_d = \left(\frac{0.12}{100} \times 1000 \times 880\right) = 1056\ mm^2$$

For 10 mm ϕ steel bars, $A_\phi = \left(\frac{\pi}{4} \times 10^2\right) = 78.54\ mm^2$

Spacing of the distribution reinforcement steel bars

$$S_d = \left(\frac{1000 \times 78.54}{1056}\right) = 74.375\ mm$$

10 mm ϕ steel bars shall be provided at 70 mm spacing centre to centre. Let 20 mm ϕ Hysd steel bars be provided as distribution reinforcement at the top surface

$$A_{\phi_1} = \left(\frac{\pi}{4} \times 20^2\right) = 314.159\ mm^2$$

Spacing of the distribution reinforcement steel bars

$$s_{d_1} = \left(\frac{1000 \times 314.519}{1056}\right) = 297.50\ mm$$

20 mm ϕ steel bars shall be provided at 280 mm spacing centre to centre near the top surface as distribution reinforcement.

Step 11. Check for development length

(i) (A) point of zero moment near D2

The point of zero moment near the point D is at a distance x = 0.392 m towards right of D. Let the number of steel bars needed at the point of zero moment be to satisfy the following condition specified in IS : 456-2000

$$[L_d \not> (M_1/V) + L_o]$$

The moment of resistance of section

$$M_1 = N \times \left(\frac{314.159 \times 230 \times 0.9038 \times 820}{1000 \times 1000}\right) = 53.50\ N\ kN-m$$

The shear force at point of zero moment from step 3

V = 554.217 kN

At point of zero moment, as per IS : 456-2000

L_o = d or 12 ϕ whichever is greater

= 820 mm or (12 x 20 = 240 mm) = 820 mm

$$L_d = \left(\frac{230 \times 20}{4 \times 1.40 \times 0.80}\right) = 1026.786\ mm$$

Therefore,

$$1026.786 = \left(\frac{53.50\ N \times 10^6}{554.50 \times 10^3} + 820\right), \quad \therefore N = 2.143$$

8 bars have already been provided at the bottom face and these bars shall be extended upto either edge of the footing.

(B) The number of bars near the top face should also satisfy the following condition specified is IS : 456-2000, $[L_d \not> (M_1/V) + L_o]$

From above step, M = 53.50 N kN-m

The shear force at the point of zero moment

V = 554.217 kN, L_o = 820 mm, L_d = 1026.786 mm

$$\text{Therefore, } 1026.786 = \left(\frac{53.50\ N \times 10^6}{554.50 \times 10^3} + 820\right)\ and\ \ V = 2.143$$

8 bars have already been provided near the top surface. These bars shall extend 820 mm beyond the point of contraflexure.

(ii) (A) At point of zero moment F_2

The point of zero moment near the point F_2 is at a distance x = 3.491 m from the point D_2 i.c., at a distance

(3.491 – 3.3) = 0.191 m

From F_2 (towards right). The shear force at the point of zero moment

V = 294.24 kN

Let the number of steel bars needed at the point of zero moment be N_1 to satisfy the following condition specified IS: 456-2000, $[L_d \not> (M_1/V) + L_o]$

The moment of resistance of section

$$M_1 = \left(\frac{N \times 314.159 \times 230 \times 0.9038 \times 820}{1000 \times 1000} \right) = 53.50\ N_1\ kN - m$$

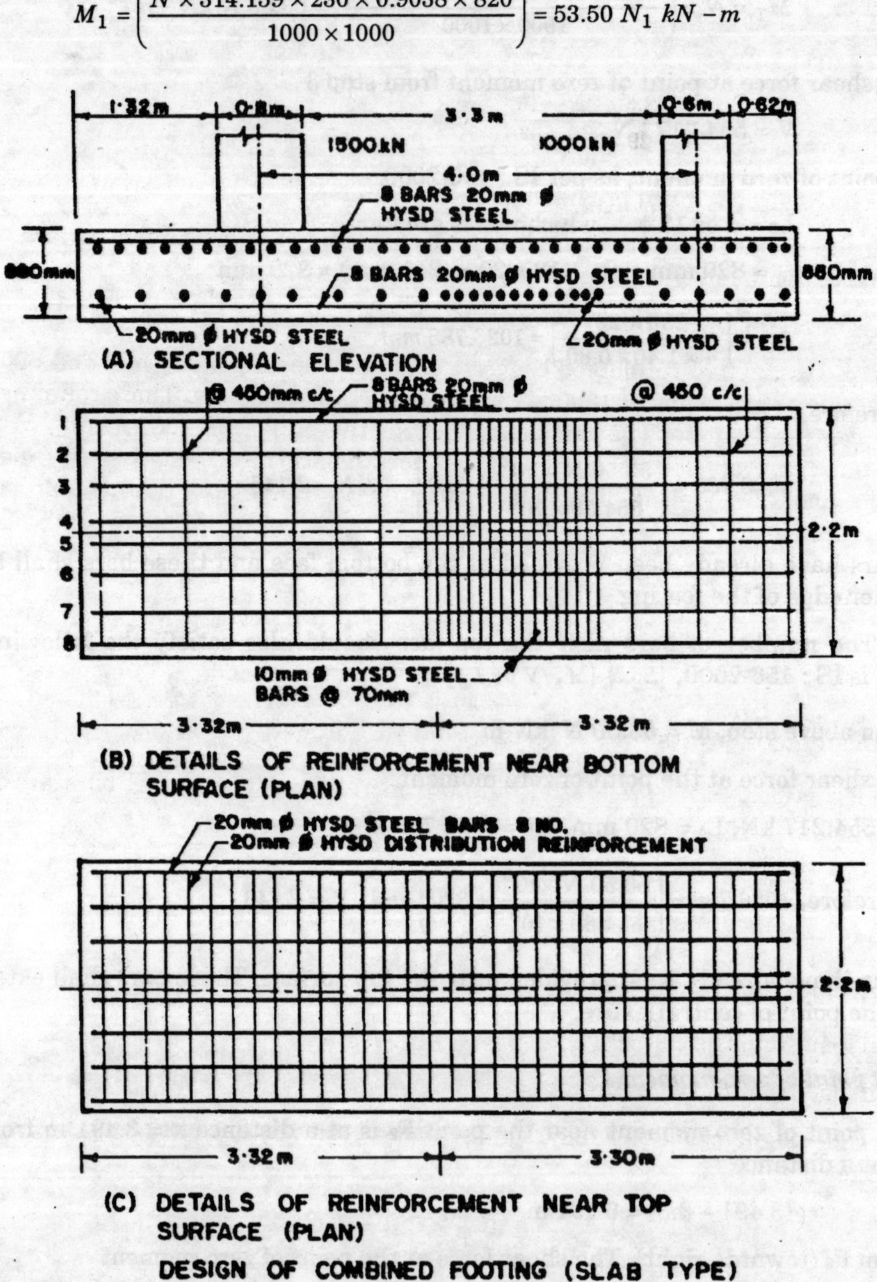

Fig. 23.11

At the point of zero moment, as per IS : 456-200,

L_o = d or 12 ϕ whichever is greater

= 820 mm or (12 x 20 = 240 mm) = 820 mm.

$$L_d = \left(\frac{230 \times 20}{4 \times 1.40 \times 0.80}\right) = 1026.786\ mm$$

$$\text{Therefore, } 1026.786 = \left(\frac{53.50\ N_1 \times 10^6}{294.24 \times 10^3} + 820\right) \text{ and } N_1 = 1.137$$

6 bars have already been provided at the bottom face.

(B) At top surface, 8 bars have already been provided near the top surface. These bars shall extend 829 mm beyond the point of zero moment.

Details of design of combined rectangular footing slab type are shown in Fig. 23.11.

Example 23.2. Two reinforced concrete columns 800 mm x 800 mm and 600 mm x 600 mm in size carry axial loads of 1500 kN and 1000 kN, respectively. These columns are placed 4 m apart centre to centre. The safe bearing capacity of the soil is 200 kN/m². M 20 grade of concrete and Hysd steel bars shall be used for the footing. Design beam and slab type rectangular combined footing. The cantilever projection available from the property line of column 1 is 1.72 m.

Design. The modular ratio of M 20 grade of concrete

$$m = \left(\frac{280}{3 \times 7}\right) = 13.33$$

Step 1 : Design constants

The design constants for M 20 grade of concrete and Hysd steel bars are as follows : (From Example 23.1)

$k_B = 0.2886, \quad j_B = 0.9038$

$R_B = 0.9129, \quad p_B = 0.4392$

Step 2 : Area of combined footing

Axial load from the column 1, $P_1 = 1500$ kN

Axial load from the column 2, $P_2 = 1000$ kN

The self-weight of footing may be assumed as ten percent of the total load transferred through the columns

$$0.10\left(P_1 + P_2\right) = \frac{10}{100}(150 + 1000) = 250\ kN$$

Total load = 2750 kN

Area of combined rectangular footing required

$$A_F = \left(\frac{2750}{200}\right) = 13.75\ m^2$$

From Example 23.1, the length and width of footing are 6.64 m and 2.20 m, respectively.

Effective soil pressure acting upward

$$q_o = \left(\frac{1500 + 1000}{6.63 \times 2.20}\right) = 171.139\ kN/m^2$$

Effective soil pressure acting upward per metre length of the footing and for complete width of the footing

$$q_o = (171.139 \times 2.2) = 376.506\ kN/m$$

Step 3. Shear forces

From Example 23.1, the loads acting downward and the intensity of soil pressure acting upward on the footing are shown in Fig. 23.10 (B). Therefore, the shear forces at different locations will remain same. As such, from Example 23.1,

S.F. at B_1 = 496.988 kN and S.F. at D_1 = 701.807 kN

S.F. at F_1 = 540.563 kN and S.F. at H_1 = 233.434 kN

The shear force diagram is shown in Fig. 23.10 (C). In Example 23.1, the points of change in shear force have been located. From Example 23.1 their positions are as below :

B_1C_1 = 0.332 m, D_1E_1 = 1.864 m, F_1G_1 = 0.419 m

C_1D_1 = 0.68 m, E_1F_1 = 1.436 m, G_1H_1 = 0.181 m

The shear forces at points of zero bending moments are found after location these points in the step 4 below.

From Example 23.1, the shear force at x = 0.329 m

right of point D, Fig. 23.10 is 554.217 kN

And, the shear force at x = 0.191 m

below the second column is 294.24 kN

Step 4. Bending moments

From Example 23.1, the bending moments at different locations, Fig. 23.10 (D) are as follows :

B.M. at B_2 = 328.012 kN-m(viz., sagging)

B.M. at C_2 = 410.427 kN-m (viz., sagging)

B.M. at D_2 = 246.084 kN-m (viz., sagging)

B.M. at E_2 = 408 kN-m (viz., hogging)

B.M. at F_2 = 19.805 kN-m (viz., hogging)

B.M. at G_2 = 93.482 kN-m (viz., sagging)

B.M. at H_2 = 72.364 kN-m (viz., sagging)

The points of zero moments have been located in *Example 23.1.* The points of zero moments are at following positions. From Example 23.1.

At a distance x = 0.392 m towards right from the point D, Fig. 23.10 (D),

At a distance x = 0.191 m towards right from the point F. The bending moment diagram is shown in Fig. 23.10(D).

Step 5 : Effective depth of beam

The combined footing shall be designed as the beam and slab. The width of beam shall be kept equal to the size of column, whichever is more, (i.e., b = 80 mm). The effective depth of beam shall be calculated for absolutely maximum bending moment. From above, M = 410.427 kN-m. The effective depth for the balanced section of rectangular beam, 800 mm wide

$$d = \left(\frac{|M|}{R_B \times b}\right)^{1/2} = \left(\frac{410.427 \times 10^6}{0.9129 \times 800}\right)^{1/2} = 749.65\ mm$$

Let the effective cover be 60 mm. Then, the overall depth of beam shall be (749.65 + 60) = 809.65 mm, say 820 mm. Then, the effective depth of beam shall be (820 – 60 = 760 m).

Step 6. Reinforcement along length of beam

(i) Reinforcement for tension for M_{C2}

The maximum sagging bending at C_2 is M_{C2} = 410.427 kN-m. Reinforcement for tension

$$A_{st.C_2} = \left(\frac{M_{C2}}{\sigma_{st} \cdot j_B \cdot d}\right) = \left(\frac{410.427 \times 10^6}{230 \times 0.9038 \times 760}\right) = 2597.89\ mm^2$$

Area of one 20 mm ϕ Hysd steel bar, A_ϕ = 314.159 mm²

$$\text{Number of bars} \left(\frac{2597.89}{314.159}\right) = 8.269$$

Provided 10 bars of 20 mm ϕ near the bottom face.

(ii) Reinforcement for tension for M_{F2}

The maximum bending moment at E_2 is M_{E2} = 408 kN-m (hogging)

Reinforcement for tension

$$A_{st.F_2} = \left(\frac{480 \times 10^6}{230 \times 0.9038 \times 760}\right) = 2582.54\ mm^2$$

$$\text{Number of bars} = \left(\frac{2582.54}{314.519}\right) = 8.22$$

Provide 10 bars of 20 mm diameter near the top face also.

(iii) Reinforcement for tension for M_{G2}

The maximum bending moment at G2 under the column load 1000 kN, M_{G2} is 93.482 kN-m. Reinforcement for tension

$$A_{st.G_2} = \left(\frac{93.482\times 10^6}{230 \times 0.9038 \times 760}\right) = 591.72\ mm^2$$

$$\text{Number of the steel bars} = \left(\frac{591.72}{314.159}\right) = 1.88$$

Provide 4 bars of 20 mm Hysd steel of grade Fe 415.

Step 7. Shear reinforcement in beam

The shear force is maximum at D, and it is equal to 701.807 kN. For p_B = 0.4392 percent and M 20 grade of concrete, from IS : 456-2000

$$\tau_c = \left(0.22 + 0.8 \times \frac{0.1892}{0.25}\right) = 0.28\ N/mm^2$$

Nominal shear stress

$$\tau_v = \left(\frac{701.807 \times 1000}{800 \times 760}\right) = 1.154\ N/mm^2$$

From IS : 456-2000, for M 20 grade of concrete

$$\tau_{c.max} = 1.8\ N/mm^2,\ \tau_v \not> \tau_{c\ max}.$$

The shear reinforcement shall be provided for

$$V_s = (V - V_c) = \left[701.807 - \left(\frac{0.28 \times 800 \times 760}{1000}\right)\right] = 531.567\ kN$$

Provide 4 legged stirrups of 12 mm ϕ Hysd steel bars

$$A_v = \left(4 \times \frac{\pi}{4} \times 12^2\right) = 452.39\ mm^2$$

The spacing of four legged stirrups

$$S_V = \left(\frac{A_{sv}\,.\sigma_{sv}\,.d}{V_s}\right) = \left(\frac{452.39 \times 230 \times 760}{531.567 \times 1000}\right) = 148.76\ mm.$$

Provide 4-legged stirrups of 12 mm ϕ Hysd steel of grade Fe 415 at 140 mm centre to centre throughout the length of beam.

Step 8. Effective depth of slab

The slab projects out as cantilever on each side of 800 mm wise beam and it is subjected to uniform upward soil pressure 171.139 kN/mm². The length of cantilever projection on each side

$$\frac{1}{2}(2200 - 800) = 700\ mm.$$

Consider 1 m strip of the slab. The bending moment to be resisted by the slab, (maximum bending moment occurs at the junction with the beam)

$$M = \left(\frac{1}{2} \times 171.139 \times 0.700^2\right) = 41.93\ kN - m/m$$

Effective depth of slab

$$d_s = \left(\frac{41.93 \times 10^6}{0.9129 \times 1000}\right)^{1/2} = 214.31\ mm$$

Let the effective cover for the slab reinforcement be 25 mm. Then, the overall depth of slab shall be 239.31 mm, say 250 mm. The effective depth shall be 225 mm.

Step 9. Tension reinforcement for slab

The tension reinforcement for slab

$$A_{st.s} = \left(\frac{41.93 \times 10^6}{230 \times 0.9039 \times 225}\right) = 896.48\ mm^2$$

Let 16 mm ϕ Hysd steel bars be provided for tension reinforcement near the bottom surface

$$A_\phi = \left(\frac{\pi}{4} \times 16^2\right) = 201.062\ mm^2$$

Spacing of 16 mm ϕ bars

$$s = \left(\frac{1000 \times 201.062}{896.48}\right) = 224.278\ mm$$

16 mm ϕ Hysd steel bars shall be provided at 220 mm spacing centre to centre.

Step 10. Distribution reinforcement for slab

The distribution reinforcement for slab for 1 m and overall depth

$$A_d = \left(\frac{0.12}{100} \times 1000 \times 250\right) = 300\ mm^2$$

Cross-sectional area of one bar 10 mm ϕ

$$A_{\phi_1} = \left(\frac{\pi}{4} \times 10^2\right) = 78.54\ mm^2$$

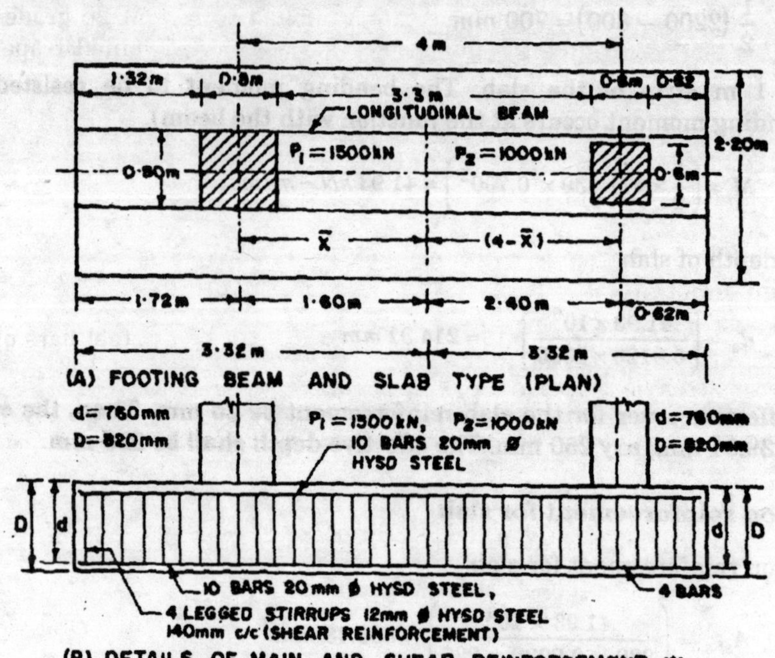

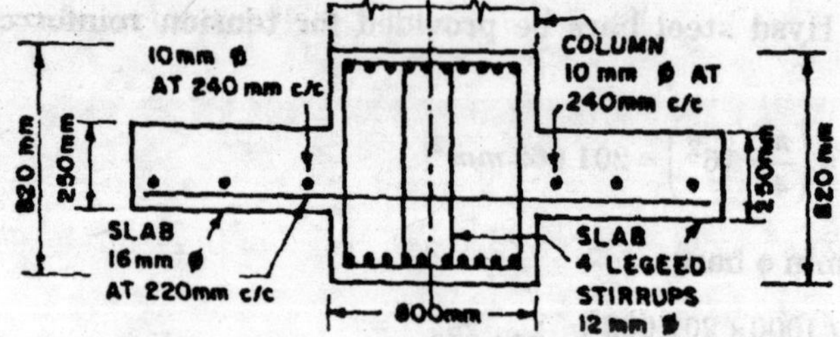

(C) CROSS - SECTIONAL ELEVATION OF BEAM

DESIGN OF BEAM AND SLAB TYPE FOOTING

Fig. 23.12

Spacing of the distribution reinforcement

$$s_d = \left(\frac{1000 \times 78.54}{300}\right) = 261.80 \; mm$$

Provide 10 mm ϕ Hysd steel bars at 240 mm c.c.

The details of reinforcement for the combined footing beam and slab type are shown in Fig. 23.12.

Example 23.3. Two reinforced columns 800 mm x 800 mm and 600 mm x 600 mm in size carry axial loads of 2250 kN and 1800 kN respectively. The columns is only 1.4 m from the centre line. The safe bearing capacity of soil is 200 kN/m². M 20 grade of concrete and Hysd steel bars of Fe 415 quality shall be used. Design a combined trapezoidal slab type footing.

Design. The modular ratio for M 20 grade of concrete

$$m = \left(\frac{280}{3 \times 7}\right) = 13.33$$

Step 1. Design constants

The design constants for M 20 grade of concrete and Hysd steel bars of Fe 415 quality are as follows : (from Example 23.2)

$k_B = 0.2886, \quad R_B = 0.9129$

$j_B = 0.9038, \; p_B = 0.4392$

Step 2. Area of combined trapezoidal footing

The combined footing supports two column loads.

Column load $P_1 = 2250$ kN

Column load $P_2 = 1800$ kN

Self-weight

$0.10\,(P_1 + P_2) = 0.10\,(2250 + 1800) = 405$ kN

Total load = 4450 kN

Safe bearing capacity of soil is 200 kN/m². Area of the footing required

$$A_F = \left(\frac{4450}{200}\right) = 22.275\ m^2$$

Provide 22.50 m² area for the footing.

Length of the trapezoidal footing

Distance between centre to centre of columns is K = 3 m

Available space by the side of 1800 kN column load is 1.40 m from the centre upto the property line towards right, as shown in Fig. 23.13 is n = 1.4 m

The distance of line of action of the resultant of two column loads from 2250 kN loads

$$\bar{x}_1 = \left(\frac{2250 \times 0 + 1800 \times 3}{(2250 + 1800)}\right) = 13.33\ m$$

$$(3 - \bar{x}_1) = 1.667\ m$$

Provide a cantilever projection on the left side of column load

$P_1 = 2250$ kN equal to m = 1.10 m

Length of the combined footing

$$L = (m + K + n) = (1.10 + 3.0 + 1.4) = 5.5 \text{ m}$$

Distance of C.G. of combined footing from the larger width of trapezium (*viz.*, from left side)

$$(1.1 + 1.333) = 2.433 \text{ m}$$

Distance of centre of combined footing

$$\left(\frac{L}{2}\right) = \left(\frac{1}{2} \times 5.5\right) = 2.75\ m$$

Distance between centre and the C.G. of the combined footing

$$d = (2.75 - 2.433) = 0.317 \text{ m}$$

Width of trapezoidal combined footing

Larger width of the trapezoidal footing

$$b = \frac{A_F}{L}\left(1 + \frac{6d}{L}\right) = \frac{22.5}{5.5} \times \left(1 + \frac{6 \times 0.317}{5.5}\right) = 5.50\ m$$

Smaller width of the trapezoidal footing

$$c = \frac{22.5}{5.5} \times \left(1 - \frac{6 \times 0.317}{5.5}\right) = 2.68\ m$$

Let y_1 be the width of trapezoidal footing at the centre of column load $P_1 = 2250$ kN

$$y_1 = \left(c + (b - c) \times \frac{K + n}{L}\right) = 2.68 + (5.50 - 2.68) \times \left(\frac{3 + 1.4}{5.5}\right) = 4.936\ m$$

Let y_2 be the width of trapezoidal footing at the centre of column load $P_2 = 1800$ kN

$$y_2 = \left(c + (b - c) \times \frac{n}{L}\right) = \left(2.68 + (5.50 - 2.68) \times \frac{1.4}{5.5}\right) = 3.398\ m$$

Area of the footing provided

$$= \left(\frac{b + c}{2}\right) \times L$$

$$\left(\frac{5.50 + 2.68}{2}\right) \times 5.50 = 22.495\ m^2 = 22.50\ m^2\ \left(\textit{more than } 22.275\ m^2\right)$$

Effective soil pressure acting upward

$$q = \left(\frac{2250 + 1800}{22.5}\right) = 180\ kN/m^2$$

The combined trapezoidal footing in plan is shown in Fig. 23.13(A).

Step 3. Bending moments

The bending moment due to effective soil pressure acting upward alone is given by

$$M = q\,.\frac{x^2}{2}\left[b - \frac{x}{3}\left(\frac{b-c}{L}\right)\right] \quad \text{(i)}$$

The bending moment at the face of column carrying column load P_1 = 2250 kN

$$M_{B_2} = \frac{180x^2}{2}\left[b - \frac{x}{3}\left(\frac{b-c}{L}\right)\right] \quad \text{(ii)}$$

Distance upto the face of column

x = (1.10 – 0.4) = 0.7 m

$$M_{B_2} = \frac{180 \times 0.7^2}{2}\left[5.50 - \frac{0.7}{3}\left(\frac{5.5-2.68}{5.5}\right)\right] = 237.274\ kN-m\ (viz., Sagging)$$

The maximum bending moment under the column carrying load P_1 = 2250 kN

$$M_{C_2} = \frac{180 \times x^2}{2}\left[5.50 - \frac{x}{3}\left(\frac{5.5-2.68}{5.5}\right)\right] - \frac{2250}{0.8}\left(\frac{x-0.7}{2}\right)^2 \quad \text{(iii)}$$

Differentiating M_{c2} w.r.t x and equating

$(d\,M_{C_2}/dx) = 0$, x = 1 m.

Substituting this in the expression (iii)

M_{C_2} = 255.74 kN-m (sagging)

The bending moment M_{D_2} at x = 1.5 m

$$M_{D_2} = \frac{180 \times x^2}{2}\left[5.50 - \frac{1.5}{3}\left(\frac{5.5-2.68}{5.5}\right)\right] - \frac{2250}{0.8}\left(\frac{1.5-0.7}{2}\right)^2$$

or $M_{D_2} = 161.836\ kN-m\ (sagging)$

The maximum bending moment between the two columns

$$M_{D_2} = \frac{180\ x^2}{2}\left[5.50 - \frac{x}{3}\left(\frac{5.50-2.68}{5.50}\right)\right] - 2250\,(x-1.10) \quad \text{(iv)}$$

Differentiating M_{E_2} w.r.t. x and equating

$(dM_{E_2}/dx) = 0,\ x = 2.60\ m.$

Substituting this in the expression (iv)

$M_{E_2} = -\ 257.143\ kN-m\ (viz., hogging)$

The bending moment shall be zero on the right side of D_2

$$M_1 = \frac{180\ x^2}{2}\left[5.50 - \frac{x}{3}\left(\frac{5.5-2.68}{5.5}\right)\right] - 2250\ (x - 1.10) = 0$$

At x = 1.71 m, and x = 3.50m. The distance x is measured from extreme left edge.

The bending moment at the face of the column carrying load P_2 = 1800 kN at x 3.8 m

$$M_{F_2} = \frac{180\times 3.8^2}{2}\left[5.50 - \frac{3.8}{3}\left(\frac{5.50-2.68}{5.50}\right)\right] - 2250\ (3.8 - 1.10)$$

$$\therefore M_{F_2} = 228.764\ kN-m\ (viz., sagging)$$

The maximum bending moments under the column carrying load P_2 = 1800 kN

$$M_{G_2} = \frac{180\times x^2}{2}\left[5.5 - \frac{x}{3}\left(\frac{b-c}{L}\right)\right] - 2250\ (x - 1.10) - \frac{1800}{0.6}\left(\frac{(x-3.80)^2}{2}\right)^2 \quad \text{(v)}$$

Differentiating M_{G_2} w.r.t.x and equating

$$(d\,M_{G_2}/dx) = 0,\ at\ x = 4.395\ m$$

Substituting this in the expression (v)

$$M_{G_2} = \frac{180\times 4.395^2}{2}\left[5.5 - \frac{4.395}{3}\left(\frac{5.5-2.68}{5.5}\right)\right]$$

$$- 2250\ (3.395 - 1.10) - \frac{1800}{0.6}\ (4.395 - 3.80)^2\ kN-m$$

$$M_{G_2} = 310.82\ kN-m\ (viz., sagging)$$

The bending moment at the face of column P_2 = 1800 kN (left face) i.e., at x = 4.4 m

$$M_{H_2} = \frac{180\times 4.4^2}{2}\left[5.50 - \frac{4.4}{3}\left(\frac{5.50-2.68}{5.5}\right)\right]$$

$$- 2250\ (4.4 - 1.10) - 1800\ (4.4 - 4.1) = 307.915\ kN-m$$

The bending moment diagram for the combined trapezoidal footing is shown in Fig. 23.13 (B).

Step 4. Shear forces

The shear force at a section is the algebraic sum of the forces acting on one side of the section. The upward force due to effective soil pressure is given by the intensity of soil pressure acting upward multiplied by the trapezoidal area of the footing.

The width of trapezium at a distance x from the larger edge

$$b_x = \left[c + (L-x)\left(\frac{b-c}{L}\right)\right] = \left[c + L\,.\,\frac{(b-c)}{L} - x\left(\frac{b-c}{L}\right)\right] = \left[b - x\left(\frac{b-c}{L}\right)\right] \quad \text{(i)}$$

Area of trapezoidal portion of the footing upto a distance x

$$A_x = \frac{1}{2}\left[b + b - x\left(\frac{b-c}{L}\right)\right].x = \frac{x}{2}\left[2b - x\left(\frac{b-x}{L}\right)\right] \quad \text{(ii)}$$

Shear force due to soil pressure alone

$$F_x = \frac{qx}{2}\left[2b - x\left(\frac{b-c}{L}\right)\right] \quad \text{(iii)}$$

Shear force at B_1 at x = 0.7 m

$$F_{B_1} = \frac{180 \times 0.7}{2}\left[2 \times 5.5 - 0.7\left(\frac{5.5 - 2.68}{5.5}\right)\right] = 670.388\ kN \quad \text{(iv)}$$

Shear force at D_1, at x = 1.5 m

$$F_{D_1} = \frac{180 \times 1.5}{2}\left[2 \times 5.5 - 1.5\left(\frac{5.5 - 2.68}{5.5}\right)\right] - 2250\ \ kN$$

$$= (1381.172 - 2250) = -868.827\ kN \quad \text{(v)}$$

Shear force at F_1 at x = 3.8 m

$$F_{F_1} = \frac{180 \times 3.8}{2}\left[2 \times 5.5 - 3.8\left(\frac{5.5 - 2.68}{5.5}\right)\right] - 2250\ kN$$

$$= (3095.305 - 2250) = 845.305\ kN \quad \text{(vi)}$$

Shear force at H_1 at x = 4.4 m

$$F_{H_1} = \frac{180 \times 4.4}{2}\left[2 \times 5.5 - 4.4 \times \left(\frac{5.5 - 2.68}{5.5}\right)\right] - 2250 - 1800\ kN$$

$$F_H = (3462.149 - 4050) = -587.851\ \text{kN}$$

The shear force diagram is shown in Fig. 23.13 (C).

In the portion B_1D_1, the shear force changes sign at C_1

$$\left(\frac{B_1C_1}{670.388}\right) = \left(\frac{C_1D_1}{868.827}\right),\ \therefore B_1C_1 = \left(\frac{670.388}{868.827}\right) \times (0.8 - B_1C_1)$$

$$B_1C_1 = \left(\frac{0.6173}{1.772}\right) = 0.348\ m\ and\ C_1D_1 = (0.8 - 0.398) = 0.452\ m$$

In the portion D_1F_1, the shear force changes sign at E_1

$$\left(\frac{D_1E_1}{868.827}\right) = \left(\frac{E_1F_1}{845.305}\right),\ \therefore D_1E_1 = \left(\frac{868.827}{845.305}\right)\ (2.3 - D_1E_1)$$

$$D_1E_1 = \left(\frac{1.0272 \times 2.3}{2.0272}\right) = 1.165\ m\ and\ \therefore\ E_1F_1 = (2.3 - 1.165) = 1.135\ m$$

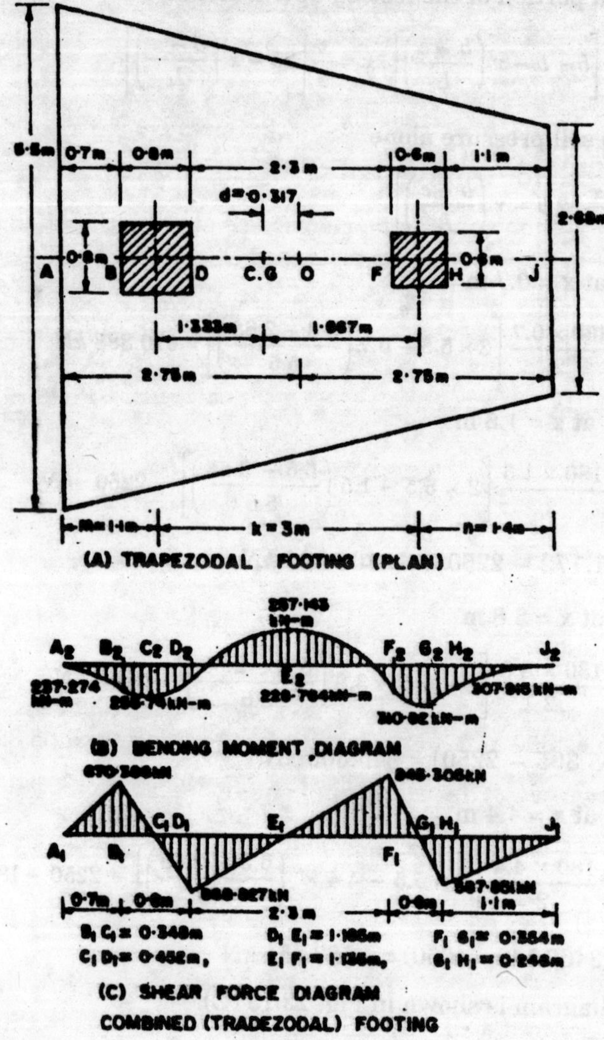

Fig. 23.13

In the portion F_1H_1, the shear force changes sign at G_1

$$\left(\frac{F_1G_1}{845.305}\right) = \left(\frac{F_1C_1}{587.851}\right) \; and \quad F_1G_1 = \left(\frac{845.305}{587.851}\right) \times (0.5 - F_1G_1)$$

$$F_1G_1 = \left(\frac{1.438 \times 0.6}{2.438}\right) = 0.354 \; m \; and \; G_1H_1 = (0.6 - 0.354) = 0.246 \; m$$

Shear force at a distance 1.71 m from A_1 at point of zero moment (viz., 1.71 – 1.50 = 0.21 from D_1). From Fig. 23.13 (C), D_1E_1 = 1.165 m,

$$F = 868.837 \times \left(\frac{1.165 - 0.21}{1.165}\right) = 712.21 \; kN$$

Shear force at a distance 3.5 m from A_1 at second point of zero moment (viz., 3.8 – 3.5 = 0.3 m F_1 towards left), from Fig. 23.13(C), E_1F_1 = 1.135 m

$$F = 845.305 \times \left(\frac{1.135 - 0.3}{1.135}\right) = 621.876\,kN$$

Step 5. Effective depth for footing slab

The absolute maximum bending moment per metre width (when calculated); it occurs at G_2, M_{G_2} = 310.82 kN-m for the width of footing at G_2. The width of footing at G_2

$$\left[b - x \,.\left(\frac{b - c}{L}\right)\right] = \left[5.5 - 4.395\left(\frac{5.5 - 2.68}{5.5}\right)\right] = 3.245\ m$$

Bending moment at G_2 per metre width

$$M_{G_2} = \left(\frac{310.82}{3.245}\right) = 95.784\ kN - m/m$$

Effective depth of footing slab

$$d = \left(\frac{95.784 \times 16^6}{0.9129 \times 1000}\right)^{1/2} = 323.92\ mm$$

Let the effective cover be 25 mm. The overall depth of slab shall be 348.92 mm. Let the overall depth of slab be 350 mm. The effective depth of slab shall be 325 mm.

Step 6. Check for two-way (flat slab type) shear

(i) The critical section for two-way shear shall be periphery at a distance d/2 from the periphery of column (carrying 2250 kN column load)

$$b_o = 4 \times [800 + (325/2) + (325/2)] = (4 \times 1125) = 4500\ mm$$

Effective soil pressure acting upward is 180 kN/m²

Shear force = (Column load – upward pressure on the area within the periphery)

$$= (2250 - 180 \times 1.125 \times 1.125) = 2022.188\ kN$$

Nominal shear stress

$$\tau_v = \left(\frac{2022.188 \times 1000}{4500 \times 325}\right) = 1.383\,N/mm^2$$

Permissible shear stress for tri-axial compression is k_s . τ_c. (k_s = 1)

$$\tau_c = k_s \times 0.16 \times (20)^{1/2} \quad or = 1 \times 0.16 \times (20)^{1/2} = 0.715\ N/mm^2$$

The effective depth of slab shall be increased

$$d = \left(\frac{1.383}{0.716} \times 325\right) = 627.76\ mm\ and\ EC = 25\ mm$$

Overall depth of the footing slab

$$(627.76 + 25) = 652.76 \text{ mm say} = 660 \text{ mm, then d} = 635 \text{ mm}$$

(ii) The critical section for two-way shear shall be periphery at a distance d/2 from the periphery of column (carrying 1800 kN column load)

$$b_o = 4 \times [600 + (635/2) + (635/2)] = (4 \times 1235) = 4940 \; mm$$

Shear force = (column load – upward pressure on the area within the periphery)

$$= (1800 - 180 \times 0.935 \times 0.935) = (1800 - 157.36) = 1642.64 \; kN$$

Nominal shear stress

$$\tau_v = \left(\frac{1642.64 \times 1000}{4950 \times 635}\right) = 0.523 \, N/mm^2$$

The permissible shear stress in tri-axial compression is $k_s \, . \, \tau_c$. ($k_s = 1$)

$$1 \times 0.16 \times (20)^{1/2} = 0.716 \, N/mm^2$$

The permissible shear stress is more than the nominal shear stress. Hence, the depth of footing is safe for two-way shear.

Step 7. Check for one-way (beam) shear

(i) The critical section for one-way (beam) shear shall be taken at a distance, d from the face of column carrying 2250 kN load (*viz.*, (0.7–0.635) = 0.065 m), on the left side. The width of trapezium at a distance x from the larger edge

$$b_x = \left[b - x\left(\frac{b-c}{L}\right)\right] = \left[5.5 - 0.065 \times \left(\frac{5.5 - 2.68}{5.5}\right)\right] = 5.467 m$$

Shear force is due to soil pressure alone at the critical section

$$F = \frac{180 \times 3.065}{2}\left[2 \times 5.5 - 0.065\left(\frac{5.5 - 2.68}{5.5}\right)\right] = 64.159 \; kN$$

Nominal shear stress

$$\tau_v = \left(\frac{64.159 \times 1000}{5467 \times 635}\right) = 0.018 \, N/mm^2$$

Permissible shear stress in concrete without shear reinforcement for the balanced section $p_B = (100 \, A_s/bd) = 0.4392$

$$\tau_c = \left(0.22 + \frac{0.08 \times 0.1892}{0.25}\right) = 0.28 \, N/mm^2$$

It is more than the nominal shear stress at the critical section. The effective depth of footing slab at that section is safe.

(ii) The critical section for one-way shear shall be taken at a distance d from the face of column carrying 2250 kN load (viz., (0.7 + 0.8 + 0.635) = 2.135 m) on the right side. The width of footing at a distance x from the larger side

$$b_x = \left[b - x \cdot \left(\frac{b-c}{L}\right)\right] = \left[5.5 - 2.135\left(\frac{5.5-2.68}{5.5}\right)\right] = 4.305\, m$$

Shear force at the critical section

$$F = \left(\frac{180 \times 2.135}{2}\right)\left[2 \times 5.5 - 2.135\left(\frac{5.5 \times 2.68}{5.5}\right)\right] - 2250 = -346.80\, kN$$

Nominal shear stress

$$\tau_v = \left(\frac{346.80 \times 1000}{4.405 \times 635}\right) = 0.124\, N/mm^2$$

Permissible stress in concrete from above

$\tau_c = 0.28$ N/mm²

It is more than the nominal shear stress at the critical section. The effective depth of footing slab at the section is again safe.

(iii) The critical section for one-way shear is taken at a distance d from the critical section from the column carrying column load of 1800 kN (viz., (3.8 – 0.635) = 3.165 m), on the left side of column. The width of trapezoidal footing from the large edge

$$b_x = \left[b - x\left(\frac{b-c}{L}\right)\right] = \left[5.5 - 3.165\left(\frac{3.5-2.68}{5.5}\right)\right] = 3.876\, m$$

Shear force, $\frac{qx}{2}\left[2b - x\left(\frac{b-c}{L}\right)\right] - 2250$

$$\frac{180 \times 3.165}{2}\left[2 \times 5.5 - 3.165\left(\frac{5.5-2.88}{5.5}\right)\right] - 2250 = 420.855\, kN$$

Nominal shear stress

$$\tau_v = \left(\frac{420.855 \times 1000}{3876 \times 635}\right) = 0.171\, N/mm^2$$

Permissible shear stress from above $\tau_c = 0.28$ N/mm²

It is more than the nominal shear stress. Therefore, the depth of footing slab is safe at that section.

(iv) The critical section for one-way shear is taken at a distance d = 0.635 m on the right side of the face of column, whereas the available cantilever projection is 1.1 m. The distance from the large edge

x = (3.8 + 0.6 + 0.635) = 5.035 m

width of the slab

$$b_x = \left[b - x.\left(\frac{b-c}{L}\right)\right] = \left[5.5 - 5.035 \times \left(\frac{5.5 - 2.68}{5.5}\right)\right] = 2.197\ m$$

$$\text{Shear force, } \frac{q_o\ .x}{2}\left[2b - x\left(\frac{b-c}{L}\right)\right] - 2250 - 1800$$

$$\frac{180 \times 5.035}{2}\left[2 \times 5.5 - 5.035\left(\frac{5.5 - 2.88}{5.5}\right)\right] - 2250 - 1800 = -235.816\ kN$$

Nominal shear stress

$$\left(\frac{235.816 \times 1000}{5035 \times 635}\right) = 0.074\ N/mm^2$$

Permissible shear stress from above τ_c = 0.28 N/mm².

It is more than the nominal shear stress. Therefore, the effective depth of the footing slab at that section is safe.

Step 8. Reinforcement along the length of footing

(i) Reinforcement for tension for M_{C_2}

The maximum bending moment under the column load 2250 kN (or the left side of the footing) occurs at C_2. The reinforcement for tension shall be calculated for M_{C2} (which one is sagging). It shall be provided along the length of footing near the bottom face. Area of the steel bars to be provided as tension reinforcement.

$$A_{st.C_2} = \left(\frac{M_{C_2}}{\sigma_{st} \cdot j_B \cdot d}\right) = \left(\frac{255.74 \times 10^6}{230 \times 0.9038 \times 635}\right) = 1937.42\ mm^2$$

20 mm ϕ Hysd steel bars shall be provided is A_ϕ = 314.159 mm²

$$\text{Number of bars } \left(\frac{1937.42}{314.159}\right) = 6.167\ (say\ 8\ bars)$$

8 bars of 20 mm ϕ shall be provided as uniformly distributed in the width of footing, near bottom surface.

(ii) Reinforcement for tension for M_{E_2}

The maximum bending moment in between the two column occurs at E_2. The reinforcement for tension shall be calculated for M_{E_2} (which one is hogging). It shall be provided along the length of footing near the top surface in between the two columns. Area of the steel bars to be provided as tension reinforcement

$$A_{st.E_2} = \left(\frac{M_{C_2}}{\sigma_{st} \cdot j_B \cdot d}\right) = \left(\frac{257.143 \times 10^6}{230 \times 0.9038 \times 635}\right) = 1948.05\ mm^2$$

20 mm ϕ Hysd steel bars shall be provided is $A_\phi = 314.159\ mm^2$

$$\text{Number of steel bars } \left(\frac{1948.05}{314.159}\right) = 6.2 \ (say\ 8\ bars)$$

8 bars of 20 mm ϕ shall be provided as uniformly distributed in the width of footing near top surface.

(iii) Reinforcement for tension for M_{G2}

The maximum bending moment under the column load, 1800 kN (on right side) occurs at G_2. The reinforcement for tension shall be calculated for M_{G2}. It shall be provided along the length of footing near the bottom surface under the column load 1800 kN. Area of the steel bars to be provided as tension reinforcement

$$A_{st.G2} = \left(\frac{M_{G2}}{\sigma_{st} \cdot j_B \cdot d}\right) = \left(\frac{310.82 \times 10^6}{230 \times 0.9038 \times 635}\right) = 2354.69\ mm^2$$

$A_\phi = 314.159\ mm^2$

$$\text{Number of bars } \left(\frac{2354.69}{314.159}\right) = 7.495 \ (say\ 8\ bars)$$

8 bars of 20 mm ϕ shall be provided as uniformly distributed in the width of footing near bottom surface.

Step 9. Reinforcement along the width of footing :

(i) Under 2250 kN column load

The centre of the column is at a distance x = (0.7 + 0.4) = 1.1 m from the larger face. The width of footing through the centre of column

$$b_x = \left[b - x\left(\frac{b-c}{L}\right)\right] = \left[5.5 - 1.1\left(\frac{5.5 - 2.68}{5.5}\right)\right] = 4.9357\ m$$

The length of cantilever projection from the face of column on each side

$$= \frac{1}{2} \cdot (4.935 - 0.8) = 2.068\ m$$

A pedestal of 1 m x 1 m size extra on each side be provided for the column. A beam shall be provided within the slab. The width of beam, (b + 2 d)

$$= (800 + 635 + 635) = 2070\ mm$$

Area of distribution of 2250 kN load, = 2.068 m x 2.070 m

$$\text{Effective pressure, } \left(\frac{2250}{2.068 \times 2.070}\right) = 525.6076\ kN/m^2$$

The moment at the face of pedestal

$$M = \left(\frac{528.6076 \times (2.068 - 1)^2}{2}\right) = 299.76\ kN - m/m$$

Effective depth, $d = \left(\frac{299.76 \times 10^6}{0.9129 \times 1000}\right)^{1/2} = 573.03\ m$

The effective depth already provided is more than this depth. The area of reinforcement

$$A_{st} = \left(\frac{299.76 \times 10^6}{230 \times 0.9038 \times (635 - 20)}\right) = 2344.76\ mm^2$$

Number of bars, $\left(\frac{2344.76}{314.159}\right) = 7.46$ (*say 8 bars*)

(ii) Under 1800 kN column load

The centre of the column load is at a distance

$$x = (3.8 + 0.3) = 4.1\ m$$

from the larger face. The width of footing through the centre of column

$$b_x = \left[b - x.\left(\frac{b-c}{L}\right)\right] = \left[5.5 - 4.1 \times \left(\frac{5.5 - 2.68}{5.5}\right)\right] = 3.40\ m$$

The length of the cantilever projection from the face of column on each side

$$= \frac{1}{2} \times (3.40 - 0.6) = 1.4\ m$$

A pedestal of 1 m x 1 m side extra on each side be provided for the column. A beam shall be provided within the slab. The width of beam (b + 2d)

$$= (600 + 635 + 635) = 1870\ mm$$

Area of distribution of 1800 kN load

$$= (1.4 \times 1.870)\ m^2$$

Effective pressure, $= \left(\frac{1800}{1.4 \times 1.870}\right) = 687.548\ kN/m^2$

The moment at the face of pedestal

$$M = \left(\frac{687.548 \times (1.4 - 1)^2}{2}\right) = 55\ kN - m/m$$

Effective depth

$$d = \left(\frac{55 \times 10^6}{0.9129 \times 1000}\right)^{1/2} = 245.46\ mm$$

The effective depth available is already more than this depth. The area of reinforcement

$$A_{st} = \left(\frac{55 \times 10^6}{230 \times 0.9038 \times (635 - 20)}\right) = 430.22\ mm^2$$

Number of bars , $\left(\frac{430.22}{314.159}\right) = 1.37$ (8 *bars*)

Step 10. Distribution reinforcement

The distribution reinforcement equal to 0.12 percent of the cross-sectional area (for Hysd steel bars) shall be provided at locations, where the transverse reinforcement is not provided. This reinforcement is also provided in the transverse direction (viz., in the perpendicular direction to longitudinal reinforcement)

$$A_d = \left(\frac{0.12}{100} \times 1000 \times 600\right) = 792\ mm^2$$

For 10 mm ϕ steel bars, $A_\phi = \left(\frac{\pi}{4} \times 10^2\right) = 78.54\ mm^2$

Spacing of the distribution reinforcement steel bars

$$s_d = \left(\frac{1000 \times 78.54}{792}\right) = 99.167\ mm$$

10 mm steel bars shall be provided at 90 mm spacing centre to centre. Let 20 mm diameter Hysd bars be provided as distribution reinforcement at top surface

$$A_\phi = \left(\frac{\pi}{4} \times 20^2\right) = 314.159\ mm^2$$

Spacing of the distribution reinforcement steel bars

$$s_{d_1} = \left(\frac{1000 \times 314.159}{792}\right) = 396.66\ mm$$

20 mm ϕ steel bars shall be provided at 380 mm spacing centre to centre near the top surface as distribution reinforcement.

Step 11. Check for development length

(i) (A). At point of zero moment near D_2

The point of zero moment near the point D is a distance (1.71 – 1.50) = 0.21 m towards right of D. Let the number of steel bars needed at the point of zero moment be N to satisfy the following condition specified in IS : 456-2000

$$[L_d \not> (M_1/V) + L_o]$$

The moment of resistance of section

$$M_1 = \left(\frac{N \times 314.159 \times 240 \times 0.9038 \times 635}{1000 \times 1000}\right) = 41.43\ kN-m$$

The shear force at point of zero moment from step 3

V = 712.21 kN

At point of zero moment, as per IS : 456-2000

L_o = d or 12 ϕ whichever is greater

= 635 mm or (12 x 20 = 240 mm)

L_o = 635 mm

$$L_d = \left(\frac{230 \times 20}{4 \times 1.40 \times 0.80}\right) = 1026.786\ mm$$

Therefore, $1026.786 = \left(\frac{41.43\ N \times 10^6}{712.21 \times 10^3} + 635\right)$ $and\ \therefore N = 6.735$

8 bars have already been provided at the bottom face and these bars shall be extended upto either edge of the footing.

(B) The number of bars near the top face should also satisfy the following condition specified in IS : 456-2000

$$[L_d \not> (M_1/V) + L_o]$$

From above step

M_1 = 41.43 N kN-m

The shear force at the point of zero moment

V = 712.21 kN, L_o = 635 mm, L_d = 1026.786 mm

Therefore, $1026.786 = \left(\frac{41.43N \times 10^6}{712.21 \times 10^3} + 635\right)$ $and\ \therefore N = 6.735$

8 bars have already been provided near the top surface. These bars shall extend 635 mm beyond the point of contraflexure.

(ii) (A) At point of zero moment near F_2

The point of zero moment near the point F_2 is at a distance x = 3.5 m from the point A (i.e., at a distance (3.8 – 3.5) = 0.3 m from F_2 towards left side). The shear force at the point of zero moment

V = 621.876 kN

Let the number of steel bars needed at the point of zero moment be N_1 to satisfy the following condition specified in IS : 456-2000

$$[L_d \not> (M_1/V) + L_o]$$

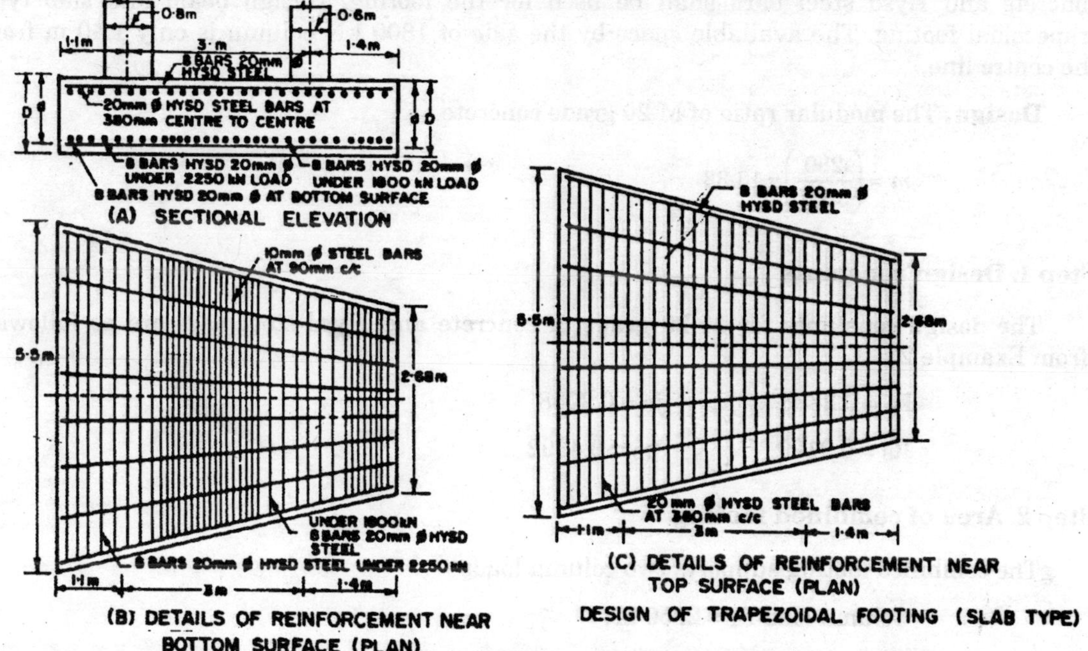

Fig. 23.14

The moment of resistance of the section

$$M_1 = \left(\frac{N_1 \times 314.159 \times 0.9038 \times 635}{1000 \times 1000}\right) = 41.43\, N_1\; kN - m$$

At the point of zero moment, as per IS : 456-2000

L_o = d or 12ϕ whichever is greater

= 635 mm or (12 x 20 = 240 mm) = 635 mm

$$L_d = \left(\frac{230 \times 20}{4 \times 1.40 \times 0.80}\right) = 1026.786\; mm$$

$$\text{Therefore, } 1026.786 - \left(\frac{41.43\, N_1 \times 10^6}{621.876 \times 10^3} + 635\right) \;\; and \;\; \therefore N_1 = 5.88$$

8 bars have already been provided at the bottom face

(B) At the top surface, 8 bars have already been provided near the top surface. These bars shall extend 635 mm beyond the point of zero moment.

Details of design of combined trapezoidal footing slab type are shown in Fig. 23.14.

Example 23.4. Two reinforced concrete columns 800 mm x 800 mm and 600 mm x 600 mm in size carry axial loads of 2250 kN and 1800 kN respectively. these columns are placed 3 m apart centre to centre. The safe bearing capacity of the soil is 200 kN/m². M 20 grade of concrete and Hysd steel bars shall be used for the footing. Design beam and slab type trapezoidal footing. The available space by the side of 1800 kN column is only 1.40 m from the centre line.

Design. The modular ratio of M 20 grade concrete

$$m = \left(\frac{280}{3 \times 7}\right) = 13.33$$

Step 1. Design constants

The design constants for M 20 grade of concrete and Hysd-steel bars are as follows : (from Example 23.3)

$k_B = 0.2886$ $\quad$ $j_B = 0.9038$

$R_B = 0.9129$ $\quad$ $p_B = 0.4392$

Step 2. Area of combined footing

The combined footing supports two column loads

Column load $P_1 = 2250$ kN

Column load $P_2 = 1800$ kN

Self-weight $= 0.10\ (P_1 + P_2) = 405$ kN

Total load = 4450 kN

Safe bearing capacity of the soil is 200 kN/m².

$$\text{Area of the footing required, } A_F = \left(\frac{4450}{200}\right) = 22.275\ m^2$$

Provide 22.50 m² area for the footing from Example 23.3, the length of footing

L = 5.5 m,

and the widths of trapezoidal footing, the larger width

b = 5.50 m,

and the smaller width

c = 2.68 m

The effective soil pressure acting upward

$$q_o = \left(\frac{2250 + 1800}{22.50}\right) = 180\ kN/m^2$$

Step 3. Bending moments

From Example 23.3, the bending moments at different locations are as follows :

B.M. at B_2 = 237.274 kN-m (*viz.*, sagging)

B.M. at C_2 = 255.74 kN-m (*viz.*, sagging)

B.M. at D = 161.836 kN-m (*viz.*, sagging)

B.M. at E_2 = 257.143 kN-m (*viz.*, hogging)

B.M. at F_2 = 228.764 kN-m (*viz.*, sagging)

B.M. at G_2 = 310.82 kN-m (*viz.*, sagging)

B.M. at H_2 = 307.915 kN-m (*viz.*, sagging)

The bending moments are zero at the points x = 1.71 m and x = 3.50 m from the larger width. The bending moment diagram is shown in Fig. 23.13 (B).

Step 4. Shear forces

From *Example 23.3*, the shear forces at different locations will remain same. As such, their values have been noted from Example 23.3.

S.F. at B_1 = 670.388 kN and S.F. at D_1 = 868.827 kN

S.F. at F_1 = 845.305 kN and S.F. at H_1 = 587.851 kN

The shear force diagram is shown in Fig. 23.13 (C).

Shear force at x = 1.71 m at point of zero moment is 712.21 kN.

Shear force at x = 3.5 at point 4 zero moment is 621.876 kN.

Step 5. Effective depth of beam

The combined footing is designed as the beam and slab. The width of beam shall be kept equal to the size of column, whichever is more, (i.e., b = 800 mm). The effective depth of beam shall be calculated for absolutely maximum bending moment from above, M = 310.82 kN-m. The effective depth for the balanced section of rectangular beam, 800 mm wide

$$d = \left(\frac{|M_1|}{R_B \times b}\right)^{1/2} = \left(\frac{310.82 \times 10^6}{0.9129 \times 800}\right)^{1/2} = 652.376\ mm$$

Let the effective cover be 60 mm. Then, the overall depth of shall be 652.376 + 60) = 712.376 mm, say 720 mm. Then, the effective depth of beam shall be 660 mm.

Step 6. Reinforcement along the length of beam

(i) Reinforcement for tension for M_{C_2}

The maximum sagging bending moment at C_2 is $M_{C_2} = 255.74\ kN-m$. Reinforcement for tension

$$A_{st.C_2} = \left(\frac{255.74 \times 10^6}{230 \times 0.9038 \times 660}\right) = 1864\ mm^2$$

20 mm ϕ Hysd-steel bars shall be provided

$$\left(A_\phi = 314.159\ mm^2\right)$$

$$\text{Number of bars}\left(\frac{1864}{314.159}\right) = 5.933\ (say\ 8\ bars)$$

Provide 8 bars of 20 mm ϕ near the bottom face.

(ii) Reinforcement for tension for M_{E_2}

The maximum bending moment at E2 is M_{E_2} = 257.143 kN-m. Reinforcement in tension

$$A_{st.E_2} = \left(\frac{257.143 \times 10^6}{230 \times 0.9038 \times 660}\right) = 1874.23\ mm^2$$

20 mm ϕ Hysd-steel bars shall be provided, (A_ϕ = 314.159 mm²)

$$\text{Number of bars}\left(\frac{1874.23}{314.159}\right) = 5.966\ (say\ 6\ bars)$$

Provide 8 bars of 20 mm ϕ near the top face.

(iii) Reinforcement for tension for M_{G_2}

The maximum bending moment at G2 is

$$M_{G_2} = \left(\frac{310.82 \times 10^6}{230 \times 0.9038 \times 660}\right) = 2265.46\ mm^2$$

20 mm ϕ Hysd steel bars shall be provided,

$$(A_\phi = 314.159\ mm^2)$$

$$\text{Number of bars}\left(\frac{2265.46}{314.159}\right) = 7.2\ (say\ 8\ bars)$$

Provide 8 bars of 20 mm ϕ Hysd-steel at bottom face.

Step 7 : Shear reinforcement in beam

The maximum shear force is at D2. It is V_{D_1} = 868.827 *kN*. For p_B = 0.4392 percent and M 20 grade of concrete, from IS : 456-2000

$$\tau_v = \left(0.22 + 0.08 \times \frac{0.1892}{0.25}\right) = 0.28\ N/mm^2$$

Nominal shear stress

$$\tau_v = \left(\frac{868.827 \times 10^3}{800 \times 660}\right) = 1.646\ N/mm^2$$

From IS : 456-2000, for M 20 grade of concrete

$$\tau_{c.\max} = 1.8\ N/mm^2,\ \tau_v \not> \tau_{c.\max}.$$

The shear reinforcement shall be provided for

$$V_s = (V - V_c) = \left(868.827 \times \frac{0.28 \times 800 \times 600}{1000}\right) = 720.99\ kN$$

Provide 4 legged stirrups of 12 mm ϕ Hysd-steel bars

$$A_{sv} = \left(4 \times \frac{\pi}{4} \times 12^2\right) = 452.39\ mm^2$$

The spacing of four legged stirrups

$$s_v = \left(\frac{A_{sv} \cdot \sigma_{sv} \cdot d}{V_s}\right) = \left(\frac{452.39 \times 230 \times 660}{720.99 \times 1000}\right) = 95.248\ mm$$

Provide 4 legged stirrups of 12 mm ϕ Hysd-steel bars of grade Fe 415 at 90 mm spacing centre to centre throughout the length of beam.

Step 8 : Effective depth of slab

The slab projects out as cantilever on each side of 800 mm wide beam and it is subjected to uniform upward soil pressure of 180 kN/m². The length of cantilever projection

$$\frac{1}{2}(4.935 - 0.8) = 2.068\ m$$

It is to note that width of trapezium through the centre of 2250 kN, from *Example 23.3* (reinforcement along the width of footing, step 9) is 4.935 m. The moment at the face of column

$$M = \left(\frac{180 \times 2.068^2}{2} \times 1\right) = 384.896\ kN - m^2$$

Effective depth

$$d_s = \left(\frac{384.896 \times 10^6}{0.9129 \times 1000}\right)^{1/2} = 649.322\ mm$$

Let the effective cover for the slab reinforcement be 25 mm. Then, the overall depth of slab shall be 674.322 mm, (say 680 mm). The effective depth shall be 655 mm.

Step 9 : Tension reinforcement for slab

The tension reinforcement for slab

$$A_{st.s} = \left(\frac{384.896 \times 10^6}{230 \times 0.9038 \times 655}\right) = 2826.84\ mm^2$$

Let 20 mm ϕ Hysd-steel bars be provided for tension reinforcement, $\left(A_\phi = 314.159\ mm^2\right)$. The spacing of bars

$$s_d = \left(\frac{1000 \times 314.159}{2826.84}\right) = 111.134\ mm$$

20 mm ϕ Hysd-steel bars shall be provided at 100 mm spacing centre to centre.

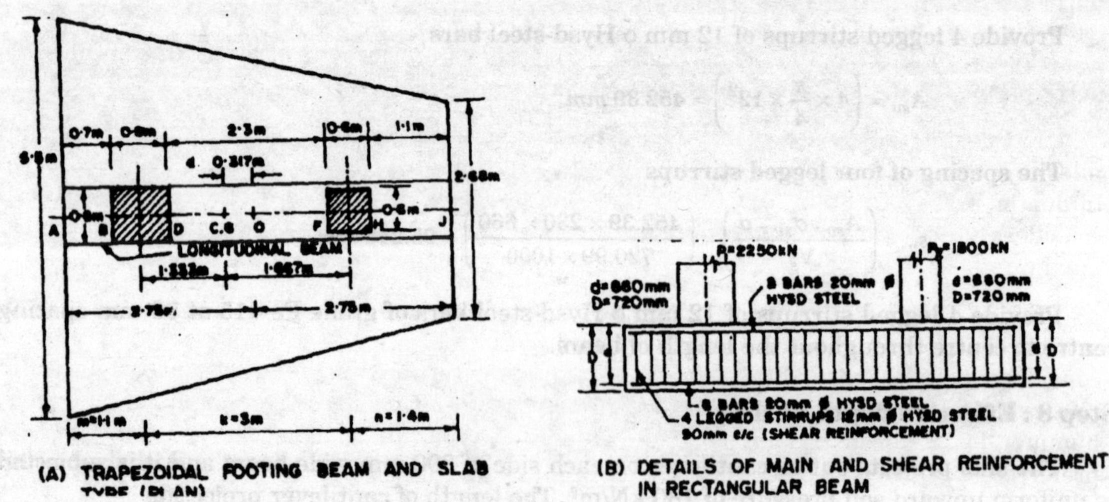

(A) TRAPEZOIDAL FOOTING BEAM AND SLAB TYPE (PLAN)

(B) DETAILS OF MAIN AND SHEAR REINFORCEMENT IN RECTANGULAR BEAM

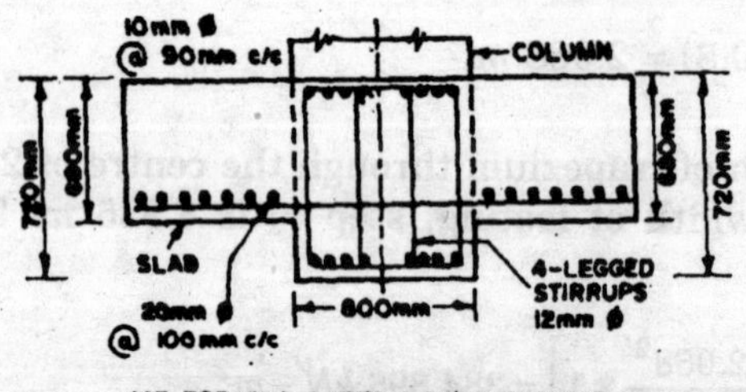

(AT TOP AND BOTTOM EACH SURFACE
8 BARS 20mm ϕ HYSD STEEL Fe 415)
(ENLARGED SKETCH)

(C) CROSS SECTIONAL ELEVATION OF BEAM

DESIGN OF BEAM AND SLAB TYPE FOOTING

Fig. 23.15

Step 10 : Distribution reinforcement for slab

The distribution reinforcement for slab for 1 m and overall depth 680 mm

$$\left(\frac{0.12}{100} \times 1000 \times 680\right) = 816\ mm^2$$

Let 10 mm ϕ Hysd-steel bars be provided as distribution reinforcement $\left(A_\phi = 78.54\ mm^2\right)$ The spacing of distribution reinforcement

$$s_d = \left(\frac{1000 \times 78.54}{816}\right) = 96.25\ mm$$

Provide 10 mm ϕ Hysd steel bars at 90 mm spacing centre to centre.

The details of reinforcement for the combined footing (beam and slab type) are shown in Fig. 23.15.

23.11 REINFORCED CONCRETE PILES

Precast piles, piles cast in place and caisson piles are three general groups of piles. The description of these piles of each group is as under.

1. Present concrete piles. The precast concrete piles are cast in the yard and these piles are transported at the construction site and then driven. These piles have been driven in depths exceeding 30 m. But in case, the length of pile is more than 20 m, it needs special consideration for the design of pile, for lifting and for driving. These files are used for 800 kN maximum load. These piles of less than 4.50 m length do not remain economical

For ordinary work, the precast piles are generally square, round or octagonal in cross-section. The length and size of the cross-section of the pile are governed only by practical considerations. The size of cross-section should be such that the handling and driving of the pile remain easy. The long piles of large cross-section need costly equipment and heavy hammer. Therefore, the use of large number of piles of smaller cross-section are often preferred.

Sometimes, these piles are kept tapered so that these may be easily be driven. The tapered piles provide more friction. However, these piles are not tapered when the soil is in compressible, or plastic clay. In such soil, there is a tendency of heaving up. These piles also heavy up when the adjacent piles are driven. The shoes are provided at the forward end to facilitate pile driving and to protect it from breaking (in case it strikes a boulder).

Several patented precast piles are also available in market. The method of driving, the design of top of the pile (so that splitting during driving is prevented and the design of shoe are the features of their patents. The choice of selection and use of piles between patented and non-patented depends mainly on economical considerations, *Chenoweth pile* (or circular cross-section), *Gilberth pile* (of octagonal cross-section), *Cumming piles* (having special circumferential and spiral reinforcement at the head), *Bignell pile* (having double jetting system), *Giant pile* (having special method of driving) etc., are the various petented precast piles mentioned in the chronological order to their appearance.

In precast piles, the hardening of the concrete is unaffected by deleterious ground water. The pile may be inspected before driven into the ground. The size of pile is not affected by the under ground. The precast piles may be driven into the ground that is below the water.

2. Cast-in place piles. The *cast-in place piles* are also called as *cast-in-situ* piles. These piles are built in the positions which these are intended to remain in the completed structures. The excavations are made at the desired positions of piles and the concrete is then cast at these positions. **Driven piles** (cased or uncased) and **bored piles** are two main group of these piles. **Pressure piles, pedestal piles** and **under-reamed** piles are the three main sub-groups of bored piles. The pedestal piles have enlarged (bulbed like) portion and the end resting on the soil. The bearing area of piles is increased. The under reamed piles are short bored cast-in-situ concrete piles. These piles also have bulbed shape enlargement near the base. These piles are commonly used in expansive soils. (Such as black cotton soil), filled up ground and other types of soils having poor bearing capacity. The enlarged pile base provide anchorage to the structure at a deep depth in additioned to increased bearing.

Remound pile (a composite pile of wood and concrete), **simplex pile** (having plain concrete shaft), **Peerless pile** (consisting of a cast-steel or cast-concrete point and a twin concrete shall of uniform cross-section resting on the shoe and solidly filled with concrete) are the various parented cast-in place piles.

The cast in situ piles have their own **advantages and disadvantage.** The advantages obtained with cast-in-situ piles (although all are not applicable to only one system) are summarised here. The cast-in-situ piles may be built without delaying the construction. These piles do not need any storage space and expensive handling equipment. The length of each pile conforms to the depth of the bearing. Neither any pile is too long nor too-short. The cutting-off surplus lengths or lengthening in-situ is not needed. The top of a pile may be at any level below the ground, or in some systems at any level above the ground. The enlarged foot gives large bearing area with some piles. Boring shows the class of soil through which pile is to cast and the nature of the stratum. The bored piles provide little frictional resistance and no ill-effect on the adjacent piles. Boring piles is less noisy and these are vibration less. These piles need small head room. The under-reamed piles are structurally safe and economical than the traditional type of foundation. The simple equipments are used in the construction. The highly skilled labour is not needed.

The cast-in-situ piles do have disadvantages. The shape of these piles cannot be inspected after erections. The progress of driving has to be properly regulated, in the absence of which the setting of concrete may be influenced by the vibration caused by driving adjacent piles.

3. Caisson piles. Professor Charles R. Gow introduced these piles in the eastern part of the United States. As such these are also named as **Gow piles.** A caisson pile consists of a cylindrical shaft of plain concrete, strong enough to carry the column load. These piles reach from the bottom of the basement or ground floor slab to the bearing soil. These piles are enlarged in the shape of frustum of a cone to provide the required bearing area. The slopes of sides remain 60° with the horizontal.

23.12. LOAD CARRYING CAPACITY OF PILES

The ultimate load carrying capacity of a pile is the maximum load which may be placed over a pile prior to which the pile continuous to settle with even infinitesimal increase in load. The safe load on a pile depends on the load that the pile may safely carry as a column and the load that produces settlement or further penetration of the pile into the ground. There are many factors which influence the load causing settlement for any particular pile that the calculated loads are not very reliable unless associated with loading tests on driven piles. Such tests are often inconvenient and expensive. Frequently one has to rely on calculated loads and a large factor of safety. The ultimate bearing resistance, appropriate factor of safety, allowable settlement, overall stability of pile foundation, the weight of hammer, the type of hammer, the height of fall etc., are the various factors which affect the allowable load which may be placed over a pile.

Numerous formulae were devised for calculating the safe load carrying capacity of the pile on the principle that the energy applied to pile for driving is equal to the work done (i.e., the product of weight of hammer and height of drop of hammer). Some of the commonly adopted **dynamic formulae** to determine the load carrying capacity of the pile are as under.

1. Engineering news formula. Professor Wellington, A.M. proposed `Engineering news Formula' in 1988. Allowable load for a column is given by

$$Q_{allows.} = \frac{1}{F.S}\left(\frac{W.H}{(S+C)}\right) kN, \text{ where} \tag{23.17}$$

W = weight of hammer in kN

H = height of fall of hammer in mm

S = average settlement of pile in mm per blow

C = empirical constant

F.S = appropriate factor of safety, adopted as 6.

(i) **For drop hammer,** the allowable load carrying capacity of pile

$$Q_{allows.} = \frac{1}{6}\cdot\left(\frac{W.H}{(S+2.5)}\right) kN \tag{23.18}$$

(ii) **For single acting steam hammer,** the allowable load carrying capacity of the pile

$$Q_{allows.} = \frac{1}{6}\cdot\left(\frac{W.H}{(S+2.5)}\right) kN \tag{23.19}$$

(iii) **For double acting steam hammer,** the allowable load carrying capacity of the pile

$$Q_{allows.} = \frac{1}{6}\left(\frac{W.a.p}{S+2.5}\right) H\ kN \tag{23.20}$$

where, a is the effective cross-sectional area of the piston in mm^2 and p is the mean effective pressure kN/mm^2.

2. Hiley's formula. Professor Hiley published a paper `A rational pile driving formula and its application in pile practice explained' in journal Engineering. May 29 and June 12, 1925, and the following formula was presented for ultimated bearing capacity of the pile. It was also recommended is IS : 291 : (part second). 1964

$$Q_{ult} = \left(\frac{n_h.W.H.n_b}{(S+0.5\,C)}\right) kN, \text{ where} \tag{23.21}$$

W = weight of hammer in kN

H = height of fall of hammer in mm

n_h = efficiency of the hammer

n_b = efficiency of the hammer blow

S = average settlement of pile in mm per blow.

C = sum of elastic compression pile, cap and ground = $[C_1 + C_2 + C_3]$

C_1, C_2 and C_3 are all proportional to the compressive stress in the pile induced by the resistance to driving. Let R be the resistance in kN and A_1 be the cross-sectional area of the pile in mm². The value of C_1 is as below :

(i) For reinforced concrete pile

$$C_1 = \left(72 \times \frac{R \cdot L}{A_P}\right) mm \tag{23.22A}$$

(ii) For timber piles

$$C_1 = \left(96 \times \frac{R \cdot L}{A_P}\right) mm \tag{23.22B}$$

where, L is the length of pile in metre.

The value of C_2 is as follows :

(i) For reinforced concrete pile with cushion (say 25 mm thick but no dolly or helmet)

$$C_2 = \left(177 \times \frac{R}{A_P}\right) mm \tag{23.22C}$$

(ii) For reinforced concrete pile with short dolly (upto 600 mm) and helmet with cushion upto 75 mm in thickness)

$$C_2 = \left(901 \times \frac{R}{A_P}\right) mm \tag{23.22D}$$

And $$C_3 = \left(354 \times \frac{R}{A_P}\right) mm \tag{23.22E}$$

The efficiency of hammer blow depends upon the weight of hammer (kN), the weight of pile, anvil, halmet, follower etc., (kN) and the coefficient of restitution, e.

(i) For (W/P) more than e

$$n_b = \left(\frac{W + e^2 . P}{W + P}\right) \tag{23.23A}$$

(ii) For (w/p) less than e

$$n_b = \left[\left(\frac{W - e^2 . p}{W + P}\right) - \left(\frac{w + e . p}{w + p}\right)\right] \tag{23.23B}$$

The product of η_h.H is known as **effective fall of hammer.** The values of η_h are as follows for different type of hammer.

(i) For a drop hammer with trigger release, $\eta_h = 1.00$

(ii) For a drop hammer falling against friction $n_h = 0.80$

(iii) For a single acting steam hammer $n_h = 0.90$

(iv) For a double acting steam hammer, W.n_h. H is to be replaced by (W.S. + A_P.s.p), where

W = weight of the piston in kN

S = stroke of piston in mm

A_P = area of the piston mm²

p = mean effective pressure in kN/mm².

Above dynamic formulae used for calculating the ultimate bearing capacity of a pile are largely inapplicable when the modern pile-driving equipment is used. Instead, so called **static formula** established on theory of soil mechanics are now developed to cover all types of piles, all methods of driving and for all the soil conditions.

Static formula. The static formula determine the ultimate, load bearing capacity of a pile by adding the end resistance of the tip of pile, R_e and frictional resistance at the skin surface of pile, R_f (both at the ultimate load). Therefore

$$Q_{ult} = \left(R_e + R_f\right), \quad where \tag{23.24}$$

$$R_f = l.\left[\frac{w\,.\,H^2}{2}\cdot\left(\frac{1+\sin\phi}{1-\sin\phi}\right)\right].\,P \tag{23.25}$$

b = coefficient of friction between soil and the pile (0.3 to 0.6)

w = unit weight of soil

p = perimeter of the pile

H = minimum length of pile needed

ϕ = angle of repose.

Group of piles. The practicability of driving and the effectiveness for carrying the loads by the piles are kept in view, while preparing plan of group of piles. The centre of gravity of all piles in a group should coincide with the line of action of resultant of applied loads so that each pile in group shall carry an equal portion of the applied load. The clear distance between two adjacent piles should generally be not less than 760 mm. As far as possible, the pile are arranged in straight lines in both the directions. The arrangement of piles in a group should be such that while driving, any displacement of earth due to consolidation in the piled area shall be free to take place in any direction away from the piles already drawn.

The load bearing capacity of a group of piles depends upon whether the piles are **load bearing piles or friction piles** or both the types of piles.

Load bearing piles. In case, the number of load bearing piles are used in a group of piles, the load bearing capacity of a group of pile

$$Q_{G.P} = (N \times Q_{BI}) \tag{23.26}$$

where, N is the number of end bearing piles in a group and Q_{BI} is the load bearing capacity of the individual pile. And there is no reduction in load bearing capacity of *Friction pile.* In case, the number of friction piles are used in a group of piles, then, the load bearing capacity of the group of piles is given by the following

$$Q_{G.P} = (\eta_g \cdot N \cdot Q_{IF}), \quad where \tag{23.27}$$

N = number of piles

$Q_{I.F}$ = load bearing capacity of an individual frictional pile

n_g = reduction factor. It is also called as the efficiency of the group of piles.

It is seen that there is some reduction due to overlapping of the zone of the stresses in soil supporting the frictional piles. This reduction depends upon the size and shape of the group of the piles as well as the size, spacing and the length of individual frictional pile.

23.13 STRESSES IN PRECAST PILES

The precast piles are designed to resist the stresses developed in the pile during handling, during driving and after the pile has been driven.

1. Stresses during handling. The precast piles are cast at different site from the positions at which these are driven. The concrete is usually poured during casting in the horizontal position. After curing and gaining the strength, the piles are lifted with a derrick and then transported to the site of driving. During lifting and transportation, the piles remain suspended either at *one point, or at two points or at three points* as shown in Fig. 23.16 and the piles are subjected to bending moment and *shear force* due to self-weight of pile. The points of suspension of piles are so provided so that the stresses developed during the handling of piles remain minimum. The stresses during handling of piles are determined for different positions of suspension as follows :

(i) **Pile suspended at one point.** The piles upto 12.20 m length are suspended at their centre of gravity (i.e., from one point) during handling as shown in Fig. 23.16 (A) such that two portions of piles AC and CB acts as cantilevers. The maximum bending moment occurs at the point of suspension

$$M = \frac{1}{2} w \left(\frac{L}{2}\right)\left(\frac{L}{2}\right) = \frac{w.L^2}{8} \tag{23.28}$$

The tensile and compressive stresses shall also be maximum at the point of suspension.

(ii) Pile suspended at two points. When the length of a pile is more than 12.20 m and in case this is suspended from one point, the stresses developed shall be large. Therefore, when the length of piles are more than 12.20 m these are suspended at two point C and D, such that the maximum bending moment at any point remains minimum. It is achieved by keeping hogging moment at C and D equal sagging moment at the middle of CD. The hogging bending moment C or D

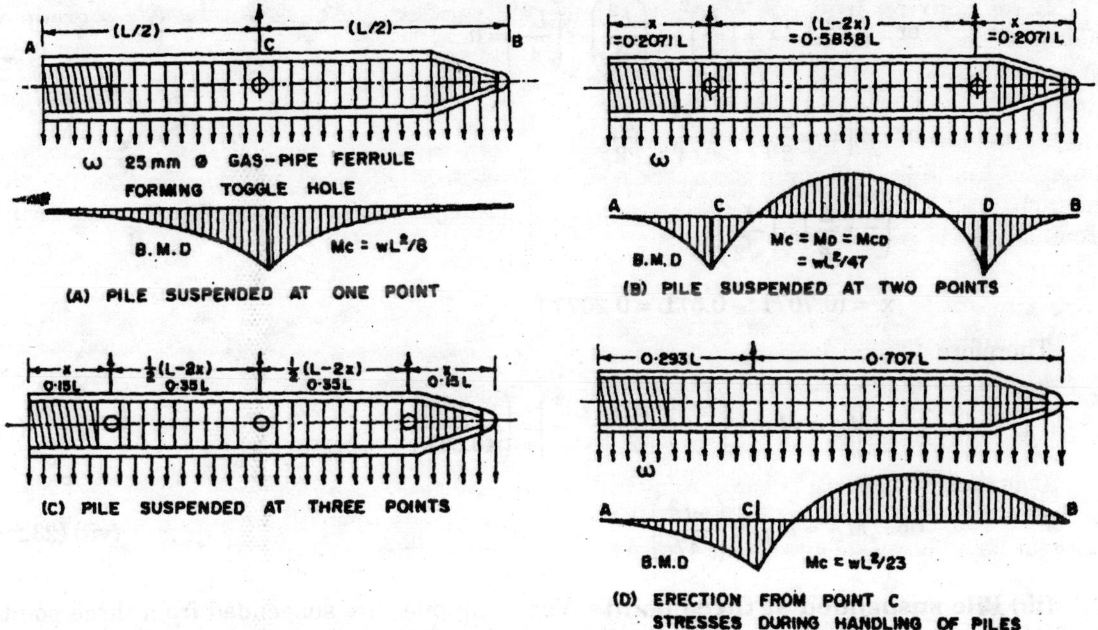

Fig. 23.16

$$M_C = M_D = \left(\frac{w \cdot x^2}{2}\right) \quad \text{(ii)}$$

Reaction at point of suspension

$$R_C = R_D = \left(\frac{w.L}{2}\right) \quad \text{(iii)}$$

The bending moment at the middle of CD

$$M_{CD} = \left(R_C \cdot \frac{1}{2} \cdot (L - 2x) - \frac{wL^2}{8}\right) \quad \text{(iv)}$$

For $M_C = M_D = M_{CD}$

$$\frac{wx^2}{2} = R_C \cdot \frac{1}{2}(L - 2x) - \left(\frac{wL^2}{8}\right)$$

or $$\frac{wx^2}{2} = \left(\frac{wL}{2} \cdot \frac{1}{2}(L - 2x) - \frac{wL^2}{8}\right)$$

or $$\frac{x^2}{2} = \left(\frac{L^2}{4} - \frac{2Lx}{4} - \frac{L^2}{8}\right)$$

or $$x^2 + L.x - (L^2/4) = 0$$

$$\text{or} \quad x^2 + Lx + \left(\frac{L}{4}\right)^2 - \left(\frac{L^2}{4}\right) - \left(\frac{L^2}{4}\right) = 0$$

$$\text{or} \quad \left(x + \frac{L}{2}\right)^2 = 2.\frac{L^2}{4} = \frac{L^2}{2}$$

$$\left(x + \frac{L}{2}\right) = \left(\frac{L}{\sqrt{2}}\right)$$

$$x = (0.7071 - 0.5)\,L = 0.2071\,L$$

Therefore

$$M_C = M_D = \left(\frac{w\,.\,(0.2071\,L)^2}{2}\right) = \left(\frac{wL^2}{46.627}\right) \quad \text{(vi)}$$

$$\text{or} \quad M_C = M_D = \left(\frac{wL^2}{47}\right) \quad \text{(vii) (23.29)}$$

(iii) **Pile suspended at three points.** Very long piles are suspended from three points as shown in Fig. 23.16(C). These toggles are provided in the length of pile. Two toggle holes are kept at a distance 0.15 L from each end and third toggle hole is kept at the middle of these two holes. The maximum bending moment

$$M_C = M_E = \left(wL^2/90\right) \quad \text{(viii) (23.30)}$$

(iv) **Erection of pile.** When the pile is erected from its horizontal position and brought to its vertical position, then, one forward end is kept on the ground and the pile is lifted from a point C, x = 0.293 L from the head as shown in Fig. 23.16(D). The maximum hogging and sagging bending moment in such a case shall be

$$M = (wL^2/23) \quad \text{(ix) (23.31)}$$

For the design of piles for the handling stresses, a much smaller factor of safety may be used the permissible stresses may be increased by 25 percent more than these specified by code of practice.

2. Stresses during driving. When the hammer blows are applied at the head of piles and the pile driving is uneven, then, the stresses induce in the pile during driving. The concrete is brittle material when the hammer blows are directly applied to the head, during heavy driving, it is likely that the head of pile may break and split, unless piles is protected by cushion or the head is reinforced by special transverse reinforcement. The cushions will reduce the effectiveness of the hammer blow and the progress of the pile driving becomes slow. The transverse (viz., lateral) reinforcement consisting of closed spaced spirals of bands of steel are preferred.

When the pile strikes a boulder, in such cases also, such types of stresses induce in the pile. A metal cone is provided at the forward or driving end of the pile. In addition to the main reinforcement, the closely spaced lateral reinforcement is also provided throughout the length of pile.

3. Stresses in piles after driven. The piles after driven in position and during supporting the load from the super structure, act like reinforced concrete columns of an unsupported length depending upon the nature of soil through which the piles are driven. Soil offers sufficient resistance against lateral movement, and the piles may be considered as short columns. Fill, silt, peat, mud or soil having water do not offer any lateral support and the piles act like long column. Its un-supported length shall be equal to depth of such material. The permissible stresses for the pile in all cases are taken as those specified for reinforced concrete columns and with the same slenderness ratio.

23.14 PILE CAP

When a single pile is provided under a column, the column is placed axially on the pier. When two or more piles are provided under a column or a bridge pier, it is essential to provide a pile cap. he function of a pile cap is to distribute the load from a column or pier of stanchion base to individual piles. A **pile cap** is a reinforced concrete rigid deep slab of uniform thickness which acts integrally with the group of piles. The size and space of the pile cap is decided by calculating the number of piles required to support the super-imposed load. The criterion for arrangement of piles in a pile group has been discussed in Art. 23.12. A pile cap extends beyond the outer piles by 100 mm to 150 mm. The piles are kept about 150 mm within the piles caps. The reinforcement in a pile cap remains about 100 mm above the head of a pile. A framework of rectangular beams (along which main reinforcement is provided) is provided over a complete area of the piles. These rectangular beams have width equal to the piles' width.

The **pile caps are designed** mainly for two-way shear around the heads of piles and around the column base and for moment or force due to transferring the load from the column to the piles. The pile caps are also designed to resist the normal shear forces, as in the case of beams carrying concentrated loads. The thickness of the cap must also be sufficient to provide adequate bond length for the bars projecting from the pile and for the dowel bars for the column. In general, the design of pile cap follows the same rules and regulations as that of a spread footing except that calculation for moments and shears are based on the assumption that the reaction from any pile is concentrated at the centre of the pile. While calculating the external shear on any section through a footing supported on the piles, the entire reaction from any pile whose centre is located $d_p/2$ (where d_p is the diameter of the pile at the upper end) or more outside the section shall be assumed as producing shear on the section. The reaction of any pile whose centre is located $(d_p/2)$ or more inside, the section shall be assumed as producing no shear for the intermediate positions of the centre of pile, the portion of the pile reaction to be assumed as producing shear as the section shall be based on the straight line interpolation between full value at $(d_p/2)$ outside the section and zero value at $(d_p/2)$ inside the section.

Superimposed loads are taken to be dispersed through a pile cap at 45° to the main tension steel. The critical section for shear should be that section where the lines of pressure intersect the tensile steel. It is practice in pile caps that the shear reinforcement is not provided, the shear resistance of the concrete at the critical section must be equal to or more than the maximum allowable shear strength at the critical section. In case, the thickness is such that the column load can all be transmitted to the piles by dispersion, no bending moments required to be considered. But generally when two or more piles are placed under one column, it is essential to reinforce the pile cap for the moments or forces produced.

There are two basic methods of analysis pile caps in common use. Firstly, a pile cap may be assumed to behave as a short deep beam, transferring the load from columns to piles by bending action. The piles of large bearing capacities are used. The number of piles in each group will be small when the ratio of over all depth/clear span of a simply supported beam exceeds 0.8, the member has to be designed for bending and shear as a deep flexural member. This method seems most appropriate for a two-pile cap. Alternatively, the pile cap may be considered to act as a space frame, the inclined lines of force linking the underside of the column to the tops of the piles being assumed to form compression members and the pile heads being linked together by reinforcement acting as horizontal tension members. This assumption appears particularly appropriate for analysing more, three dimensional caps, such as those needed for three or more piles.

For three or more pile groups, no provisions, however exist in the code so far dealing with two-way shear action in deep flexural members.

23.15 ANALYSIS OF PILE CAPS

The pile caps are very commonly used for three or more piles in a pile group. Professors Whittle, R.T. and Beattie, D. published a paper titled as `standard pile caps' in general concrete, Vol. 6, No. 1, January 1972 and Vol. 6, No.2, February, 1972 and described the beam method of analysis of the pile caps. The analysis of pile caps is carried out as follows. Each pile shares the column load equally.

1. Pile cap for two piles. A pile cap with two piles supporting a column carrying an axial load of P is shown in Fig. 23.17. A deep beam shall be provided over the two piles within the pile cap. The maximum bending moment at the centre shall be

$$M = \left(\frac{P \, . \, L}{4}\right) \tag{i}$$

Let the tensile force to be resisted by the reinforcement be T. Taking the lever arm factor j as unity.

$$T.d = M = \left(\frac{PL}{4}\right)$$

$$\text{Then, } T = \left(\frac{P \, . \, L}{4d}\right) \tag{23.32}$$

In case, size of column (a x b) is taken into consideration, then, (Reference : C.E. Reynolds and J.C. Steedman)

$$T = \left(\frac{P}{12 \, . \, L \, . \, d}\right)\left[3L^2 - a^2\right] \tag{23.33}$$

where, d is the effective depth of the pile. The dimensions of pile cap are shown in Fig. 23.17 (A).

2. Pile cap for three piles. A pile cap with three piles supporting a column-carrying an axial load of P is shown in Fig. 23.17 (B). The tensile force resisted by the reinforcement shall be (Reference : C.E. Reynolds and J.C. Steedman)

$$T = \left(\frac{P \cdot L}{9 \cdot d}\right) \tag{23.34}$$

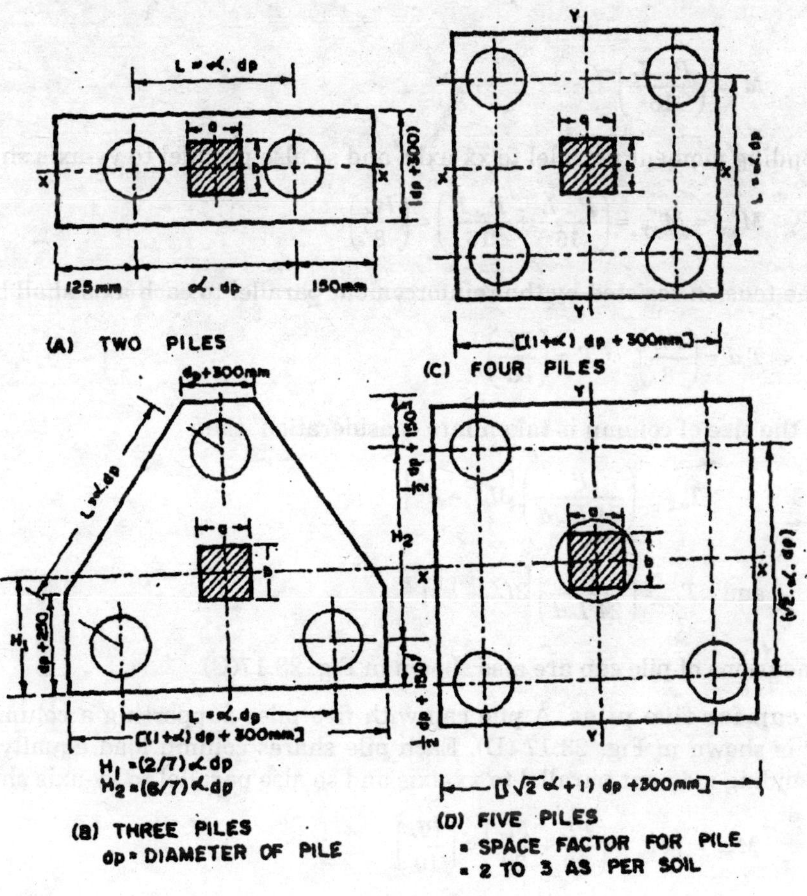

Fig. 23.17

In case, size of column (a x b) is taken into consideration, then, the tensile force parallel to xx-axis shall be

$$T_{xx} = \left(\frac{P}{36 . L . d}\right) . \left[4L^2 + b^2 - 3a^2\right] \tag{23.35}$$

And, the tensile force parallel to yy-axis shall be

$$T_{yy} = \left(\frac{P}{18\, L.d}\right)\left[2L^2 - b^2\right] \tag{23.36}$$

The dimensions of pile cap are also shown in Fig. 23.17(B).

3. Pile cap for four piles. **A pile cap with four piles supporting a column carrying an axial load F is shown in Fig. 23.17 (C). Two rigid beams shall be provided under two pairs of**

piles parallel to x-axis. Similarly two rigid beams shall be provided under two parts of piles parallel to yy-axis. Each pile shares column load equally, (i.e., equal to P/4). The bending moment at the middle of two piles parallel to xx-axis, and then so also parallel to yy-axis, shall be

$$M = \left(\frac{P \,.\, L}{16}\right) \tag{i}$$

Total bending moment parallel to xx-axis, and so also parallel to yy-axis shall be

$$M_{xx} = M_{yy} = \left(\frac{P \,.\, L}{16} + \frac{P \,.\, L}{16}\right) = \left(\frac{PL}{8}\right) \tag{ii}$$

Then, the tension resisted by the reinforcement parallel to each axis shall be

$$T.d = \left(\frac{PL}{8}\right), \because T = \left(\frac{PL}{8d}\right) \tag{23.37}$$

In case, the size of column is taken into consideration, then

$$T_{xx} = \left(\frac{P}{24\ L\,.\,d}\right)\left[3L^2 - a^2\right] \tag{23.38}$$

$$\text{and} \quad T_{yy} = \left(\frac{P}{24\ L.d}\right)\left[3L^2 - b^2\right] \tag{23.39}$$

The dimensions of pile cap are also shown in Fig. 23.17(C).

4. Pile cap for five piles. A pile cap with five piles supporting a column carrying an axial load P is shown in Fig. 23.17 (D). Each pile shares column load equally (i.e., equal to P/5). Total bending moment parallel to xx-axis and so also parallel to yy-axis shall be

$$M_{zx} = M_{yy} = \left(\frac{PL}{20} + \frac{PL}{20}\right) = \left(\frac{PL}{10}\right) \tag{i}$$

Then, the tension resisted by the reinforcement parallel to each axis

$$T.d = \left(\frac{PL}{10}\right), \quad \therefore T = \left(\frac{PL}{10d}\right) \tag{23.40}$$

In case, the size of column is taken into consideration, then

$$T_{xx} = \left(\frac{P}{30\ L.d}\right)\left[3L^2 - a^2\right] \tag{23.41}$$

$$\text{and } T_{yy} = \left(\frac{P}{30\,L.d}\right)\left[3L^2 - b^2\right] \tag{23.42}$$

The dimensions of pile cap are also shown in Fig. 23.17 (D).

Professors Taylor, H.P.J. and Clark, J.L. published a paper titled as `some detailing problems in concrete frame structures' in the Journal. The structural Engineer Vol. 54. No. 1, January 1976 reported the space frame method of analysis.

In addition to main reinforcement, *secondary reinforcement* is also provided essentially. It is placed near the bottom face of the pile cap. It runs around the main longitudinal reinforcement projecting from the piles into the pile cap. It provides a change in the direction at the head of every pile. The area of this second reinforcement should not be less than 20 percent.

23.16 DESIGN OF PILES

Reinforcement concrete piles are designed by considering the column load P applied on the piles, during handling including erection stress and stresses developed due to driving of the piles. Following factors are decided while the piles are designed :

1. Length of pile. Let w be the bearing capacity of the soil, ϕ be the angle of repose, μ be the coefficient of the friction and q be the bearing capacity of the soil. In case, Q_{allow} is the load bearing capacity of pile and h is the depth of penetration, then from **Rankine's theory**

$$Q_{allow} = (q \times area\ of\ cross-section\ of\ pile) + frictional\ resistance\ of\ pile$$

$$Q_{allow} = \left(q \,.\, A + C_p \,.\, \frac{w.l^2}{2} \,.\, \mu \times A_p\right), \text{ where} \qquad (23.43)$$

A = cross-sectional area of pile

C_p = coefficient of pressure

$$C_P = \left(\frac{1+\sin\phi}{1-\sin\phi}\right)$$

A_P = perimeter of the pile.

The angle of repose ϕ is assumed as 15° to 22° and the coefficient of friction may be taken as 0.30 to 0.35.

The depth of penetration of pile h may be obtained for the known values of ϕ, h, q, w, A, A_p and Q_{allow}. The effective length of pile may be decided keeping in view the condition of supply.

2. Main reinforcement. The area of main longitudinal reinforcement in a pile may be decided keeping in view the following specifications :

(a) 1.25 percent of the gross-cross-sectional area of the pile for piles of length 30 times their least width.

(b) 1.5 percent of the gross cross-sectional of the pile for piles of length between 30 to 40 times their least width.

(c) 2.0 percent of the gross cross-sectional area of the pile for piles of lengths more than 40 *times* their least width.

3. Lateral reinforcement. Following specifications decide the lateral reinforcement :

(a) The lateral links are provided in the piles. The diameter of link bars should not be less than 5 mm.

(b) The volume of lateral reinforcement provided should not be less than 0.2 percent of the gross volume of the pile.

(c) The centre to centre spacing of lateral reinforcement should not be more than half the least width of pile.

(d) At each end of the pile for a distance equal to 3 times the least width, the volume of the lateral reinforcement should not be less than 0.6 percent of the gross volume.

In case, the piles are required to be penetrated through hard stratum, the spiral reinforcement is provided for a length equal to 3 times the width. 12 mm to 16 mm diameter *spacers* are provided to keep the longitudinal reinforcement bars apart at intervals not more than 1.50 m. 6 mm ϕ links are provided with the spacer. A cast iron shoe is provided at the driving end of the pile.

4. Grade of concrete. M 20 grade of concrete is commonly used for the piles.

5. Cover. A minimum cover of 40 mm is provided. Where, the piles are likely to remain in contact with sea water, this cover may be increased to 55 mm.

The arrangement of piles in a pile group has already been discussed previously. The analysis and dimensions of the pile caps have been discussed in Art. 23.15.

Example 23.5. A drop hammer weighing 60 kN is used to drive a precast pile. The height of free fall of the hammer is 1.30 m. The average penetration noted in the last few blows is 6 mm per blow. Determine the safe load carrying capacity of the precast pile. Use a factor of safety of 6.

Solution. The safe load carrying capacity of a precast pile may be found from Engineering news formula, Eq. 23.17.

$$Q_{allow} = \frac{1}{F.S}\left(\frac{W.H}{(S+C)}\right) kN$$

Weight of drop hammer, W = 60 kN

Height of free fall of drop hammer H = (1.30 × 1000) = 1300 mm

Factor of safety (given), F.S = 6

Average penetration noted per blow, S = 6 mm

Value of constant, C = 25

Therefore, safe load carrying capacity of pile

$$Q_{allow} = \frac{1}{6} \times \left(\frac{60 \times 1330}{6 + 25}\right) = 419.35\ kN$$

Example 23.6. A steam hammer weighing 30 kN and having 0.60 m stroke length is used to drive a precast reinforced cement concrete pile. The average penetration noted in the last few blows 5 mm per blow. Determine the safe load carrying of the precast pile.

Solution. The safe load carrying capacity of a precast pile may be obtained from Engineering news formula, Eq. 23.19

$$Q_{allow} = \frac{1}{F.S.} \cdot \left(\frac{W.H}{(S+C)}\right) \qquad \text{(i)}$$

Weight of the steam hammer, W = 30 kN

Stroke length of the steam hammer, H = (0.60 × 1000) = 600 mm

Factor of safety (assumed), F.S = 6

Average penetration noted per blow, S = 5 m

Value of constant, C = 2.5

Therefore, safe load carrying capacity of precast pile

$$Q_{allow} = \frac{1}{6} \times \left(\frac{30 \times 600}{(5 + 2.5)}\right) = 400\,kN$$

Example 23.7 A square precast pile 400 mm × 400 mm in cross-section is 12 m long. A single acting steam hammer weighing 40 kN is used for pile driving. The free fall of hammer 0.8 m. The weight of the helmet and dolly is 6 kN. The ultimate resistance received while driving is 900 kN. Determine the average penetration per blow. Adopt coefficient is restitution, as 0.25 and the efficiency of hammer n_h as 0.85.

Solution. Resistance received during is 900 kN. The cross-sectional area of pile, A_p is (400 × 400) = 160000 mm². The length of pile, L is 12 m. The elastic compression of the pile, cap and the ground

$$C = (c_1 + c_2 + c_3) \qquad \text{(i)}$$

From Eq. 23.22(A)

$$c_1 = \left(72 \times \frac{R.L}{A_P}\right) = \left(\frac{72 \times 900 \times 12}{160000}\right) = 4.86\,mm \qquad \text{(ii)}$$

For reinforced concrete pile, from Eq. 23.22 (D)

$$c_2 = \left(\frac{901 \times R}{A_P}\right) = \left(\frac{901 \times 900}{160000}\right) = 5.068\,mm \qquad \text{(iii)}$$

From Eq. 23.22 (E)

$$c_3 = \left(\frac{354 \times R}{A_P}\right) = \left(\frac{354 \times 900}{160000}\right) = 1.99\,mm. \qquad \text{(iv)}$$

Therefore, the elastic compression of the pile, cap and the ground

$$C = (c_1 + c_2 + c_3)$$

Substituting from the compressions (ii), (iii) and (iv)

$$C = (4.86 + 5.068 + 1.99) = 11.918 \text{ mm}$$

P = (weight of pile + weight of helmet and dolly)

$$P = [(0.4 \times 0.4 \times 12) \times 25 + 6] = (48 + 6) = 54\,kN \qquad \text{(v)}$$

Weight of the hammer, W = 40 kN (vi)

$$\left(\frac{W}{P}\right) = \left(\frac{40}{54}\right) = 0.741 \quad \text{(vii)}$$

Since, $\left(\frac{W}{P} > e\right)$ (*viz.*, 0.741 > 0.25)

∴ Efficiency of the hammer blow, from Eq. 23.23 (A)

$$n_b = \left(\frac{W + p.e^2}{W + p}\right) = \left(\frac{40 + 54 \times 0.252}{(40 + 54)}\right) = 0.46 \quad \text{(viii)}$$

H = (0.80 × 1000) = 800 mm

From Eq. 23.21, from Hiley's formula, the ultimate capacity of the pile

$$W_{ult} = \left(\frac{n_h \,.\, W \,.\, H \,.\, n_b}{(S + 0.5\,C)}\right) = \left(\frac{0.85 \times 40 \times 800 \times 0.46}{(S + 0.5 \times 11.918)}\right) \quad \text{(ix)}$$

The ultimate capacity of pile is equal to the ultimate resistance received. Therefore, Q_{ult} = 900 kN

$$900 = \left(\frac{0.85 \times 40 \times 800 \times 0.46}{(S + 0.5 \times 11.918)}\right)$$

$$\therefore (S + 0.5 \times 11.918) = 13.90 \; and \; S = 7.943 \; mm$$

Average penetration per blow is 7.943 mm.

Example 23.8. A reinforced concrete column transmits an axial load of 3000 kN to a group of four piles. The piles are to be driven to a hard stratum available at a depth 7.50 m. Design a pile, Use M 20 grade of concrete and Hysd-steel bars of grade Fe 415.

Design. M 20 grade of concrete is to be provided. The permissible stress in concrete in bending in compression σ_{cbc} is 7 N/mm². Therefore, the modular ratio of concrete

$$m = \left(\frac{280}{3 \times 7}\right) = 13.333 \quad and \; m_c = (1.5 \times 13.333) = 20$$

Step 1 : Design constants

For M 20 grade of concrete and Hysd-steel bars, the *design constants* are as under :

$k_B = 0.2886$, $j_B = 0.9038$

$R_B = 0.9129$, $R_B = 0.4392$

The cross-section of the pile shall be designed as the balanced section bending action and the pile shall be designed as a column for the direct compression.

Step 2 : Design of a pile in direct compression

Axial load transmitted by a column, = 3000 kN (i)

There are four piles in the pile group. The load supported by each pile

$$P_1 = \left(\frac{3000}{4}\right) = 750\ kN \tag{ii}$$

The pile is likely to act as a long column. Assume the reduction coefficient for the pile

$$C_r = 0.85 \text{ (assumed)} \tag{iii}$$

$$\text{Axial load for the short column, } P = \left(\frac{750}{0.95}\right) = 822.353\ kN \tag{iv}$$

The load carrying capacity of a short column

$$P = (A_{cc} \,.\, \sigma_{ce} + A_{sc} .\, \sigma_{sc}) \tag{v}$$

Let the cross-sectional area, A_{sc} of the steel reinforcement in compression be 2% (assumed) of the gross-cross sectional area; A. Then, $A_{cc} = (A - 0.02A)$ and

$$A_{sc} = 0.02\,A \tag{vi}$$

From IS: 456-2000, the direct permissible stress in M 20 grade of concrete.

$$\sigma_{cc} = 5\ N/mm^2 \tag{vii}$$

And the direct permissible stress in Hysd-steel bars in compression.

$$\sigma_{sc} = 190\ N/mm^2 \tag{viii}$$

Then, $P = (A - 0.02\,A)\,\sigma_{cc} + m_c \,.\, \sigma_{cc} \,.\, (0.22\,A)$

or $P = [0.98\,A.\sigma_{cc} + 20\,\sigma_{cc} \,.\, (0.02A)]$

or $P = (0.98\,A + 0.4A)\,\sigma_{cc}$

or $A = \left(\frac{P}{1.38 \times 5}\right) = \left(\frac{882.353 \times 10^3}{1.38 \times 5}\right) = 1278.7710\ mm^2$

Let the cross-section area of the pile be a square. Then, the size of the pile, *s*.

$$s = 357.6\ mm \tag{ix}$$

Let the size of the pile be 380 mm × 380 mm. Let the length of the portion of pile above the ground be 0.8 m. Then, the total length of pile.

$$l = (7.50 + 0.8) = 8.3\ m \tag{x}$$

Ratio of length of pile to its least lateral dimension, *D*.

$$(l/D) = \left(\frac{8.8 \times 1000}{380}\right) = 21.84 \tag{xi}$$

Since, (l/D) ratio is more than 12, the pile shall be treated as the long column. The reduction coefficient for the column, from IS: 456-1978.

$$C_r = \left(1.25 - \frac{l}{48 \times b}\right) = \left(1.25 - \frac{8.3 \times 1000}{48 \times 380}\right) = 0.795 \qquad \text{(xii)}$$

Axial load for the short column shall be

$$\left(\frac{750}{0.795}\right) = 943.40\,kN$$

Check for the load carrying capacity of the column.

$$P = (A_{cc} \cdot \sigma_{cc} + A_{sc} \cdot \sigma_{sc}) = [(A - 0.02A).\ \sigma_{cc} - m_c \cdot \sigma_{cc}\ (0.02A)]$$

$$P = [0.98\ A.\sigma_{cc} + (20\ \sigma_{cc} \times 0.02A)] = 1.38\ A.\sigma_{cc}$$

$$\text{or}\ \ P = \left(\frac{1.38 \times 380 \times 380 \times 5}{1000}\right) = 996.36\ kN$$

It is more than 943.40 kN. Hence the pile cross-section with 2 percent reinforcement is safe.

Step 3: Main Reinforcement in column

Length of pile is 8.3 m. Least lateral dimension of pile is 380 mm. *Thirty times the least lateral dimension.*

$$\left(\frac{30 \times 380}{1000}\right) = 11.4\ m.$$

Actual length of column 8.3 is less than thirty times the least lateral dimension. The main reinforcement should not be less than 1.25 percent. Actual reinforcement is kept as 2 percent more than 1.25 percent. Cross-sectional area of the reinforcement.

$$A_{sc} = \left(\frac{2}{100} \times 380 \times 380\right) = 2888\ mm^2$$

Provide 8 bars of 22 mm ϕ : The cross sectional area of main reinforcement provided

$$\left(8 \times \frac{\pi}{4} \times (22)^2\right) = 3041.06\ mm^2$$

It is more than 2888 mm² required.

Step 4: Lateral reinforcement

The lateral reinforcement in the form of ties shall be provided in the body of the pile. The volume of the lateral ties shall be 0.2 percent volume of the pile per mm. Volume of the lateral ties required

$$\left(\frac{0.2}{100} \times 380 \times 380 \times 1\right) = 288.8\ mm^3$$

Provide effective cover for the pile as 50 mm. Length of each side of the tie (let 10 mm ϕ tie be provided)

$$(380 - 2 \times 50 + 10) = 290 \text{ mm}.$$

Volume of each tie

$$\left(4 \times 290 \times \frac{\pi}{4} \times (10)^2\right) = 91106.19\ mm^2$$

Pitch of tie

$$\left(\frac{91106.19}{288.8}\right) = 315.46mm$$

The pitch of transverse reinforcement should not be more than the following :

(i) Least lateral dimension is D = 380 mm

(ii) 16 times the diameter of bar is = (16 × 22) = 352 mm

(iii) 48 times the diameter of ties is = (48 × 10) = 480 mm

$$\text{Maximum permissible pitch is} = \left(\frac{1}{2} \times 380\right) = 190\ mm$$

Provide lateral ties in the middle portion (body) of the pile 10 mm in diameter Hysd-steel bars at 180 mm centre to centre.

Step 5 : Spiral Reinforcement

Spiral reinforcement is provided in the portion of pile (= *three times the least lateral dimension)* near the head of the column. The volume of the spiral reinforcement shall be 0.6 percent of the gross volume of the pile per 1 mm length

$$\left(\frac{0.6}{100} \times 380 \times 380 \times 1\right) = 866.4\ mm^3$$

Provide 10 mm ϕ Hysd steel bars for the spiral. Cross-section area of spiral bars

$$A_\phi = \left(\frac{\pi}{4} \times 10^2\right) = 78.54\ mm^2$$

Volume of spiral

$$((4 \times 290) \times 78.54) = 911106.4\ mm^3$$

Pitch of the spiral

$$\left(\frac{91106.4}{866.4}\right) = 105.16\ mm$$

Provide 10 mm ϕ spiral reinforcement at 100 mm pitch.

Step 6 : Lateral reinforcement near toe

This lateral reinforcement is provided near the driving (viz., forward) end of the pile. It is kept 0.6 percent of the gross volume of the pile

$$\left(\frac{0.6}{109} \times 308 \times 308 \times 1\right) = 866.4\ mm^3$$

Volume of 10 mm ϕ Hysd steel bars is = 91106.4 mm³

Spacing of the lateral ties is = $\left(\frac{91106.4}{866.4}\right) = 105.155\ mm$

Provide the lateral ties near the forward end at 100 mm spacing centre to centre.

In addition to the above 12 mm of spacer forks with 8 mm diameter ties shall be provided to avoid inward buckling at intervals of atleast 1.50 m.

Step 7 : Handling stresses

Three holes shall be provided for suspended the pile along its length. One hole for suspending the pile shall be provided at the middle for the pile length. Two holes for suspending the pile shall be made at a distance 0.15 L from each end.

The maximum hogging and the sagging bending moment shall be ($wL^2/90$), where w is the weight of the pile per metre length

$$w = (0.38 \times 0.38 \times 1 \times 25) = 3.61\ kN/m$$

$$\left(\frac{wL^2}{90}\right) = \left(\frac{3.61 \times 8.3 \times 8.3}{90}\right) = 2.763\ kN-m$$

Step 8 : Erection stresses

An additional hole shall be provided near 0.293 L from the head of the pile, viz.,

$$(0.293 \times 8.3) = 2.432\ m.$$

The maximum hogging and sagging bending moment shall be ($wL^2/23$).

$$M = \left(\frac{3.61 \times 8.3 \times 8.3}{23}\right) = 10.813\ kN-m$$

Effective depth of pile section

$$d = (380 - 50) - 330\ mm.$$

The cross-section of pile is doubly reinforced. Actual depth of neutral axis may be located by equating the first moment of equivalent areas about neutral axis. Therefore,

$$\left[\left(b.n.\frac{n}{2} + (m_c - 1)\,A_{sc}\,.\,(n - d_c) = mA_{st}\,(d - n)\right)\right]$$

$$\left[\frac{380 \times n^2}{2} + 19 \times 4 \times \frac{\pi}{4} \times (22)^2 \times (n-50)\right] = 13.33 \times 4 \times \frac{\pi}{4} \times 22^2 \times (330 - n)$$

$\therefore n = 114.65\ mm.$

The moment of resistance of the pile section may be obtained by taking the moment of compressive forces about the centre of steel in tension. Therefore,

$$b.n.\frac{\sigma_{cbc}}{2}\left(d - \frac{n}{3}\right) + (m_c - 1)\, A_{sc} \cdot \sigma_{cbc} \cdot (d - d_c) = M$$

$$\text{or} \left(\frac{380 \times 114.65}{2}\right) \times \sigma_{cbc}\left(330 - \frac{114.65}{3}\right) + 19 \times 4 \times \frac{\pi}{2} \times 22^2 \times \sigma_{cbc}$$

$$\times \left(\frac{114.65 - 50}{114.65}\right) \times (330 - 50) = 10.813 \times 10^6$$

$$\sigma_{cbc} = 0.99 N/mm^2$$

Actual stress is much less than the permissible stress. Therefore, the pile section shall be safe. The details of reinforcement are shown in Fig. 23.18.

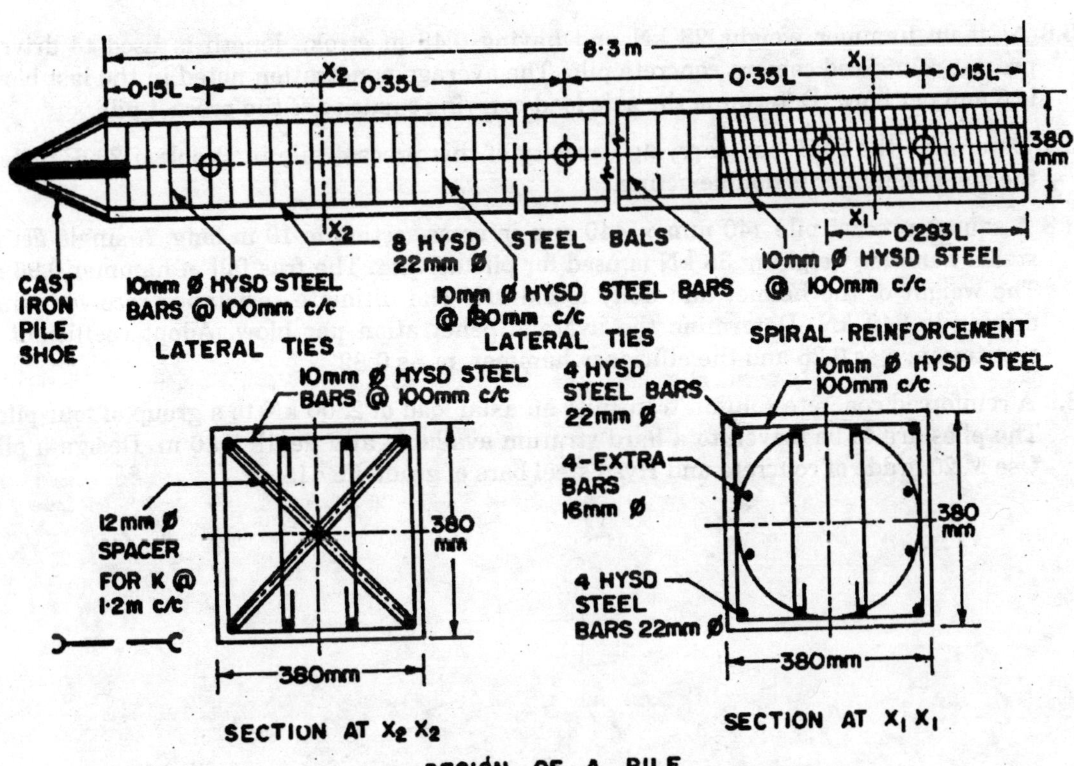

Fig. 23.18

PROBLEMS

23.1 Two reinforced concrete columns 600 mm × 600 mm and 500 mm × 500 mm in size carry axial loads of 1260 kN and 840 kN, respectively. These columns are placed 3.6 m apart centre to centre. The safe bearing capacity of the soil is 180 kN/m². M 20 grade of concrete and Hysd steel bars shall be used for the footing. Design combined slab type rectangular footing the cantilever projection available from centre line of column 1 upto the property line is 1.58 m.

23.2 In Problem 23.1, design combined beam and slab type footing.

23.3 Two reinforced columns 800 mm × 800 mm and 600 mm × 600 mm in size carrying axial loads of 1760 kN and 1340 kN, respectively. The columns are placed 3.2 m apart. The available space by the side of 1340 kN column is only 1.42 m from the centre line. The safe bearing capacity of soil is 160 kN/m². M 20 grade of concrete and Hysd steel bars of Fe 415 quality shall be used. Design a combined trapezoidal slab type footing.

23.4 In Problem 23.3, design combined beam and slab type trapezoidal footing.

23.5 A drop hammer weighing 56 kN is used to drive a precast pile. The height of free fall of the hammer is 1.28 m. The average penetration noted in the last few blows is 4 mm per blow. Determine the safe load carrying capacity of the precast pile. Use a factor of safety of 6.

23.6 A steam hammer weight 28 kN and having 0.48 m stroke length is used to drive a precast reinforced cement concrete pile. The average penetration noted in the last blows is 6 mm per blow. Determine the safe load carrying capacity of the precast pile.

23.7 Determine the safe load carrying capacity of this precast pile in Problem 23.6, in case the steam hammer is double acting.

23.8 A square precast pile 440 mm × 440 mm in cross-section is 10 m long. A single acting steam hammer weighing 38 kN is used for pile driving. The free fall of hammer 0.96 m. The weight of the helmet and dolly is 4.8 kN. The ultimate resistance received while driving is 840 kN. Determine the average penetration per blow. Adopt coefficient of restitution, e as 0.25 and the efficiency hammer, n_h as 0.82.

23.9 A reinforced concrete column transmits an axial load of 2800 kN to a group of four piles. The piles are to be driven to a hard stratum available at a depth 6.80 m. Design a pile. Use M 20 grade of concrete and Hysd steel bars of grade Fe 415.

PART VII

DESIGN OF OTHER CONCRETE STRUCTURES

24

DESIGN OF STAIRS

24.1 INTRODUCTION

Stairs are needed for ascending and descending from floor to floor. The stairs are the structures which consist of a series of steps with or without landing and give access from floor to floor. A flight between landing to landing is also called as **stair.** The stairs provide vertical circulation from one level to another level. The space within which a stair is built is known as staircase. The staircase provides an enclosure for a stair. A stair consists of number of steps. There may be 3 to 12 steps in one flight (i.e., between two adjacent landings). Each step has **one tread (going) and one rise.** The tread and rise are proportioned while designing a stair so that the stair provides a convenient access. The relationship between a tread and a rise is determined in one of two ways (i) *twice the rise plus tread equals 600 mm to 640 mm and (ii) the product of rise and tread equals 40,000 mm^2 to 42,600 mm^2 (*the rise and tread) both the measured in *millimetres.*

In a residential building, a star is located in a place easily accessible to all the members. In a public building, a stair is located near main entrance. The stairs are well planned so that these remain easily approachable from al the rooms. The rise of a step is kept easy (about 150 mm). It increases by 12 mm to 13 mm and depending upon necessity, the rise is also kept as 200 mm. A step is kept sufficiently wide. In a hospital building, the tread may be kept about 300 mm. It decreases by 25 mm and it may be kept about 200 mm. The width of a step of a stair is kept about 0.90 m in a residential building. It is kept about 1.80 m in a public building. Two persons may cross each other conveniently while passing in a stair.

The stair may also be defined as *an inclined structural form.* BIS recommends that the **pitch of a stair** should not be less than 25° and it should not be more than 40°. The clear distance between tread and the soffit of a flight immediately above (called as *head room),* should not be less than 2.14 m. A landing is usually provided after 12 to 15 steps. The width of landing should not be less than the width of stair.

Only the reinforced cement concrete stairs have been described in this chapter.

24.2 TYPES OF STAIRS

The differential types of reinforced cement concrete stairs are as under :

1. Straight stair. In a straight stair as shown in Fig. 24.1 (A), all the steps rise in one direction. The straight stair may have one single flight (i.e., it may not be having any landing) or two more flights having loadings in between the two adjacent flights. A straight stair is provided when the available space is straight, narrow and long and it is not possible to provide other types of the stairs.

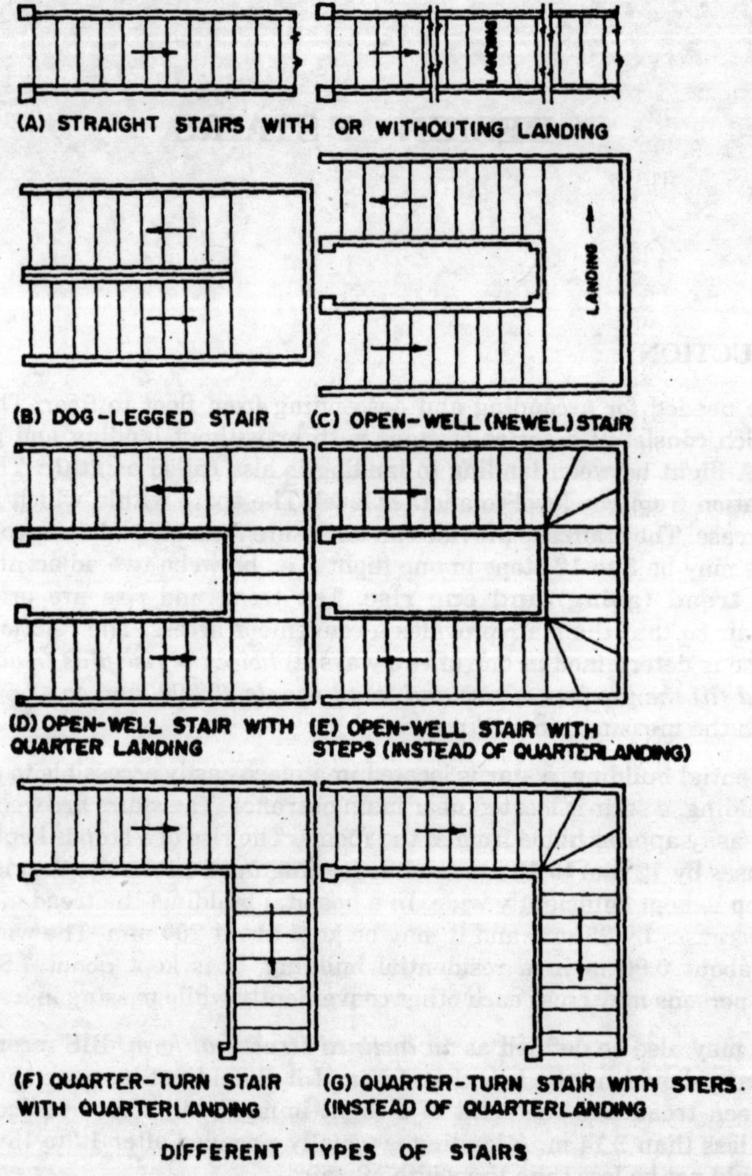

Fig. 24.1

2. Dog-legged stairs. In a dog-legged stair, two straight flights are provided in opposite directions as shown in Fig. 24.1(B) and there is no space between the flights in plan. A level landing is provided across the two flights. The dog-legged stair is provided when the available space width is just equal to *twice* the width of steps.

3. Open-well (open newel) stair. In an open-well stair, two straight flights are provided in opposite directions between two adjacent floors as shown in Fig. 24.1(C) and there remains a space between the two flights in plan. A level landing is provided across the two flights. The open-well stair is provided when the space width is more than *twice* the width of steps.

Sometimes, it is not possible to cover complete height between two floors, then, a third flight having 3 to 6 steps are provided instead of landing as shown in Fig. 24.1(D) and quarter landing is left between two adjacent perpendicular flights.

Sometimes, even instead of quarter landing, the steps are also provided in the space of quarter landing as shown in Fig. 24.1(E).

4. Quarter-turn stair. A quarter-turn stair having two adjacent flights mutually perpendicular to each other (viz., a stair turning through 90°) with a quarter landing as shown in Fig. 24.1(F) is adopted.

Sometimes, the steps are also provided as shown in Fig. 24.1(G) to cover the complete height between two floors.

In addition to the above types of stairs, the stairs of some geometrical shapes, as described below are also built.

1. Geometrical stair. The geometrical stair is similar to the open well stair except that the steps are also provided in a curved shape instead of landing [as shown in Fig. 24.2(A)] between forward and rear flights. The winders (steps in curved portion) are provided to change the direction. The curved portion of the stair may have any geometrical shape.

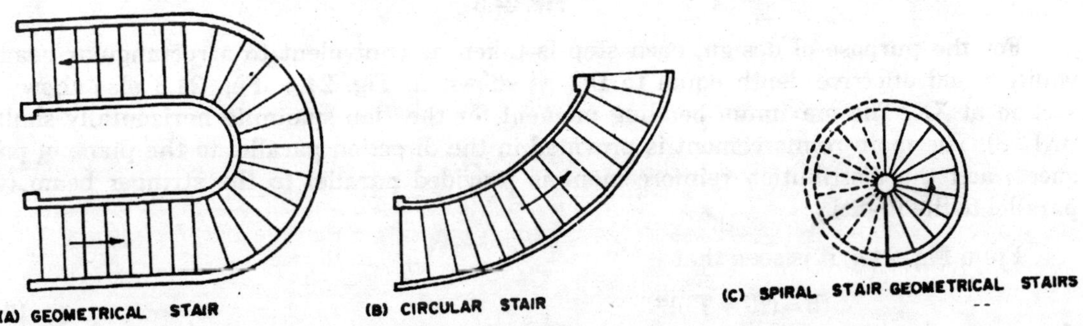

Fig. 24.2

2. Circular stair. In a circular stair as shown in Fig. 24.2 (B), all the steps radiate from a common point (viz., centre of a circle). These steps are of winder type.

3. Spiral stair. A spiral stair as shown in Fig. 24.3 (C) is also seen as circular stair in plan. In its elevational view, the stair has a spiral shape. All steps also radiate from newel post or well hole.

24.3 EFFECTIVE SPAN OF STAIRS

The stair slabs may span either in the horizontal direction or in the longitudinal direction. The effective span for each type is adopted as described below :

1. Stair slab spanning horizontally. When the stair slab spans in the direction of width of the steps, it is called as *stair slab spanning horizontally*. The stair slab is supported either on each side by side walls or by stringer beams or by a beam on one side and a wall on the other side as shown in Fig. 24.3. The stair slab remains *simply supported* parallel to the steps. The distance between centre to centre of supports AB is adopted as the effective span, *L*.

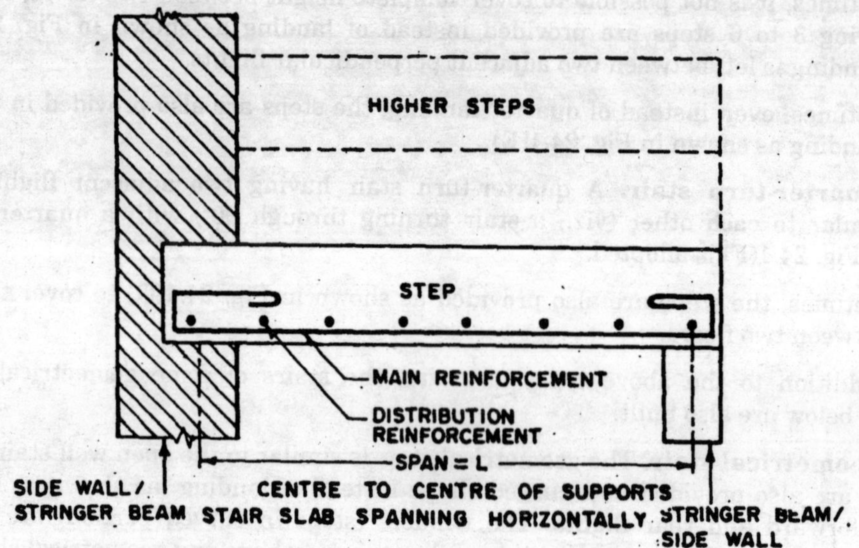

Fig. 24.3

For the purpose of design, each step is taken as equivalent to a rectangular beam of width b and effective depth equal to D/2 as shown in Fig. 24.4. Fig. 24.3 also shows the section at XX. The maximum bending moment for the slab spanning horizontally shall be $(w.L^2/8)$. The main reinforcement is provided in the direction parallel to the plane of paper sheet) and the distribution reinforcement is provided parallel to the stringer beam (viz., parallel to the flight).

From Fig. 24.4, it is seen that

$$b = (R^2 + T^2)^{1/2} \tag{24.1}$$

$$\text{Depth, } D = (D + y) \tag{i}$$

From similar triangles

$$\left(\frac{y}{AB}\right) = \left(\frac{AC}{BC}\right)$$

(Since AB = T, AC = R and BC = b)

$$y = (T.R/b) \tag{ii}$$

$$\therefore D = D_1 + (T.R/b) \tag{24.2}$$

The thickness of waist slab, D_1 may be assumed about 80 mm per metre span.

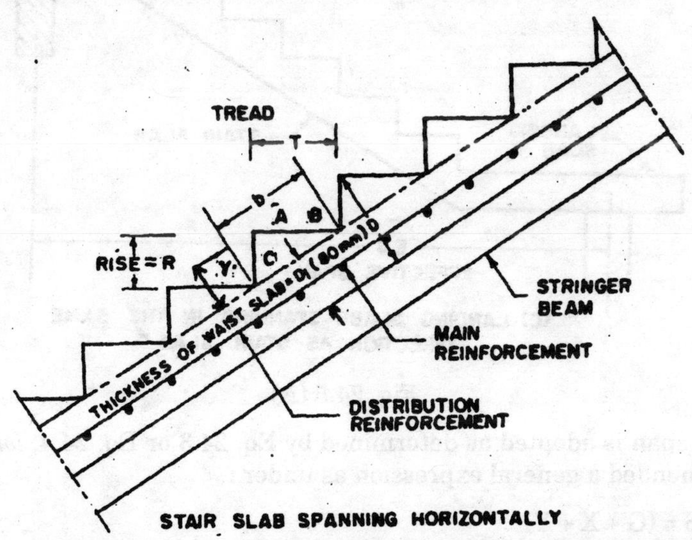

Fig. 24.4

2. Stair slab spanning longitudinally. When the stair slab spans in the direction of going, it is called as *stair slab spanning longitudinally*. The stair slab is supported at two ends (viz., at the top and at the bottom of the flight). There is a continuous support at each end of the flight, (i.e., the support is continuous along the width of flight). There is no stringer beam along the flight (neither at both the ends nor at the middle of the width of stair slab). The stair slab again remains *simply supported* at both the ends. The effective span of stair slab is adopted as following horizontal distance as recommended in IS : 456-2001.

(a) Where a stair slab is supported at the bottom riser and so also at the top riser by beams spanning parallel with the risers, the effective span of the stair slab shall be the distance between centre to centre of beams, as shown in Fig. 24.5(A).

(b) Where a stair slab is spanning on to the edge of a landing, which spans parallel with the risers, the effective span, ES of the stair slab is adopted as below: [as shown in Fig. 24.5 (B)]

(i) a distance equal to the going of the stairs, G *plus* half the width of landing, b_L at each end

$$ES = \left(G + \frac{1}{2}b_L + \frac{1}{2}b_L\right) \tag{24.3}$$

or (ii) a distance equal to the going of the stairs, G *plus one metre* at each end

$$ES = (G + 1\,m + 1\,m) \tag{24.4}$$

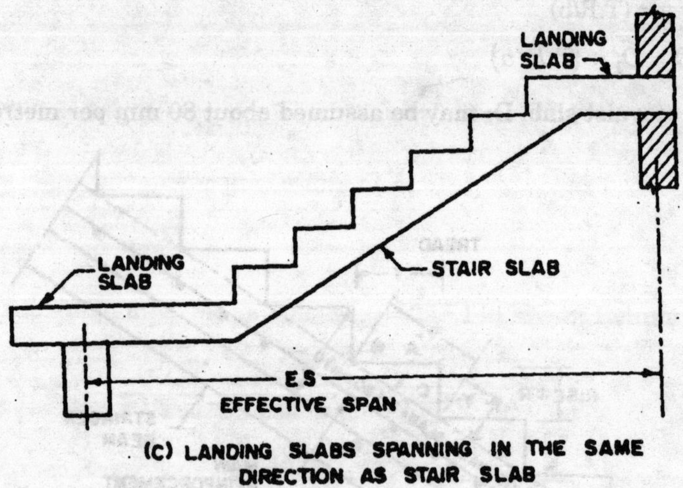

Fig. 24.5 (A)

The effective span is adopted as determined by Eq. 24.3 or Eq. 24.4, *whichever is less* IS : 456-2000 recommended a general expression as under :

$$ES = (G + X + Y) \qquad (24.5)$$

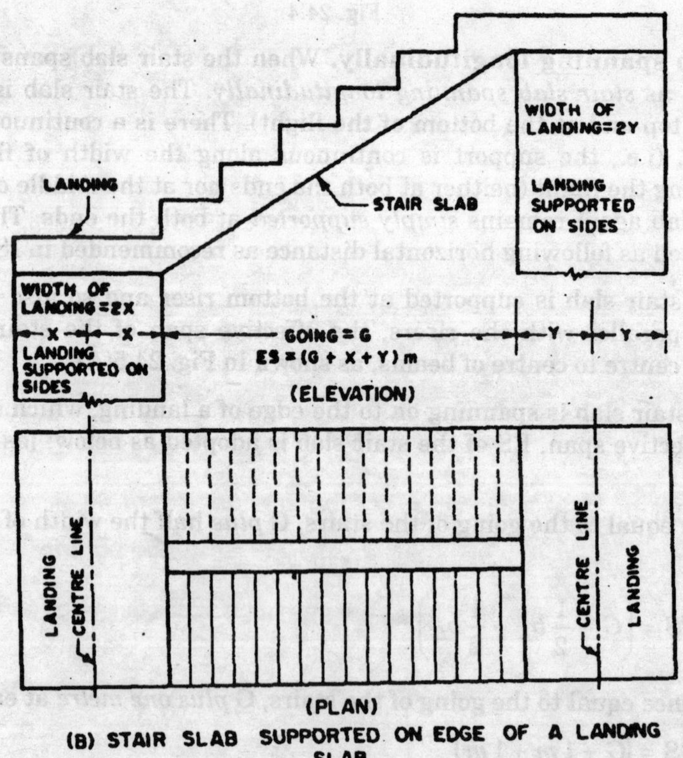

Fig. 24.5 (B)

The values of X and Y are adopted depending upon whether X and Y, whether each one is less than or equal to or more than 1 m. It is indicated in Table 24.1 given below :

TABLE 24.1. EFFECTIVE SPAN OF STAIR SLAB
(Spanning in longitudinal direction)

X	Y	Span in metres
< 1m	< 1m	ES = (G + X + Y)
< 1 m	≥ 1*m*	ES = (G + X + 1)
≥ 1*m*	< 1 m	ES = (G + X + 1)
≥ 1*m*	≥ 1*m*	ES = (G + 1 + 1)

(c) When the stair slab is spanning longitudinally and the landing slabs in the same direction as the stair, it is considered that the slabs and landing act together to form a single slab. The effective span is determined as the distance centre to centre of the supporting beams or walls as shown in Fig. 24.5 (C). The going is measured horizontally as usually.

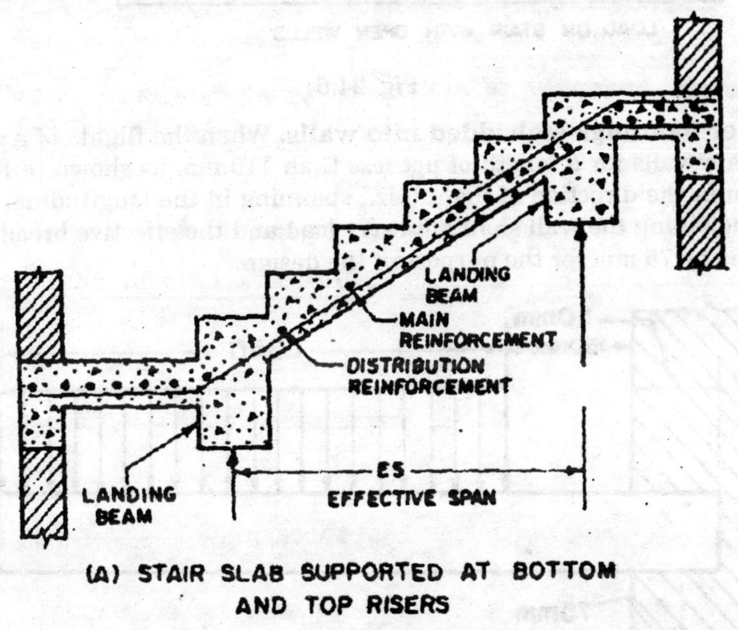

Fig. 24.5C

24.4 DISTRIBUTION OF LOADS ON STAIRS

The distribution of loads on stairs depends upon types of stairs and its construction. The loads are distributed on stairs as below :

1. Stairs with open wells (spans) partly crossing at right angles. In such types of stairs, the load on the areas common to any two spans (crossing at right angles) may be taken as *one-half* in each direction as shown in Fig. 24.6.

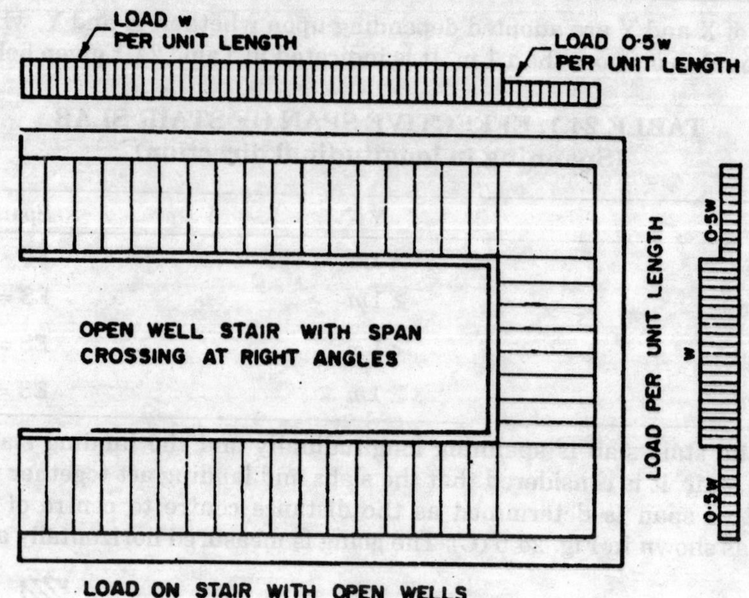

Fig. 24.6

2. Flights or bandings embedded into walls. When the flights of a stairs or landing are embedded into walls for a length of not less than 110 mm as shown in Fig. 24.7 and are designed to span in the direction of flight (viz., spanning in the longitudinal direction) a 150 mm wide strip adjoining the wall shall not carry load and the effective breadth of the section shall be increased by 75 mm for the purpose of the design.

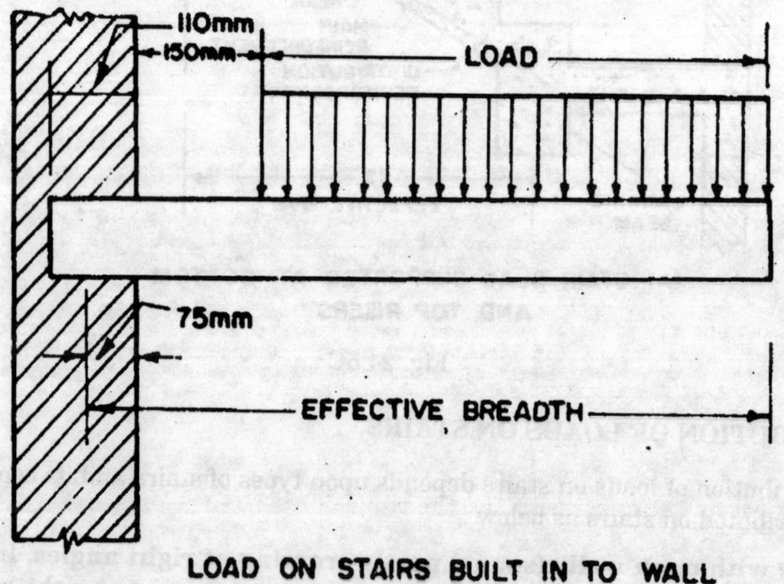

Fig. 24.7

24.5 LOADS ON STAIRS

The loads on stair are dead and live load. The loads are determined as under :

1. Dead load. The dead loads are self-weight of the waist slab and the self-weight of the steps. These self-weights are estimated as follows :

(i) Self-weight of the waist of the slab. The self-weight of the waist slab is first determined in the direction perpendicular to the slope of 1 m length along the flight and for 1 m along the width of steps. Let D_1 be the thickness of the waist slab. Then, the self-weight normal to the slab as shown in Fig. 24.8.

w_n = (Volume of waist slab) × unit weight

$$w_n = \left(\frac{D_1}{1000} \times 1\,m \times 1\,m\right) \times 24 = 24\,D_1\ kN/m^2 \qquad \text{(i)}$$

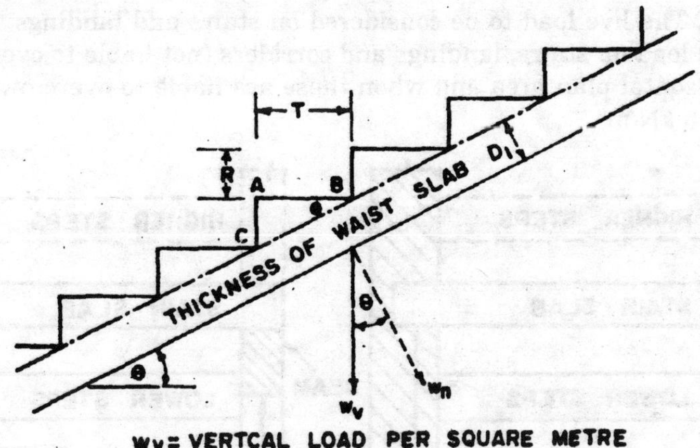

Fig. 24.8

Let w_v be the vertical load acting per square metre of horizontal plan area. Then

$$w_v = \cos\theta = w_n \qquad \text{(ii)}$$

and $w_v = (w_n / \cos\theta)$ (iii)

From the triangle ABC,

$$\cos\theta = (AB_1 / BC)$$

or $\cos\theta = T\,(T^2 + R^2)^{1/2}$ (iv)

Substituting from the expression (iv) the value of cos θ in the expression (iii)

$$w_v = w_n \cdot \left(\frac{(T^2 + R^2)^{1/2}}{T}\right) \qquad (24.5)$$

$$\text{or } w_v = 24D_1 \cdot \left(\frac{(T^2 + R^2)^{1/2}}{T}\right) kN/m^2 \tag{24.6}$$

where, D_1 is the thickness of waist slab. It is usually kept 80 mm. The self-weight of finishing and floor attached with the stair is calculated and treated as also acting vertically.

(ii) **Self-weight of steps.** The step is treated as equivalent rectangular beam of width b and depth D/2 as already mentioned in Art. 24.4.

From Eq. 24.1, and Fig. 24.4,

$$b = (T^2 + R^2)^{1/2} \tag{iv}$$

From Eq. 24.2 and Fig. 24.4,

$$D = D_1 + (T.R/b) \tag{v}$$

2. Live load. The live load to be considered on stairs and landings are specified in IS : 875-1984. The live load on stairs, landings and corridors (not liable to overcrowding) is taken as 3 kN/m² of horizontal plan area and when these are liable to overcrowding, then, the live load is adopted as 5 kN/m².

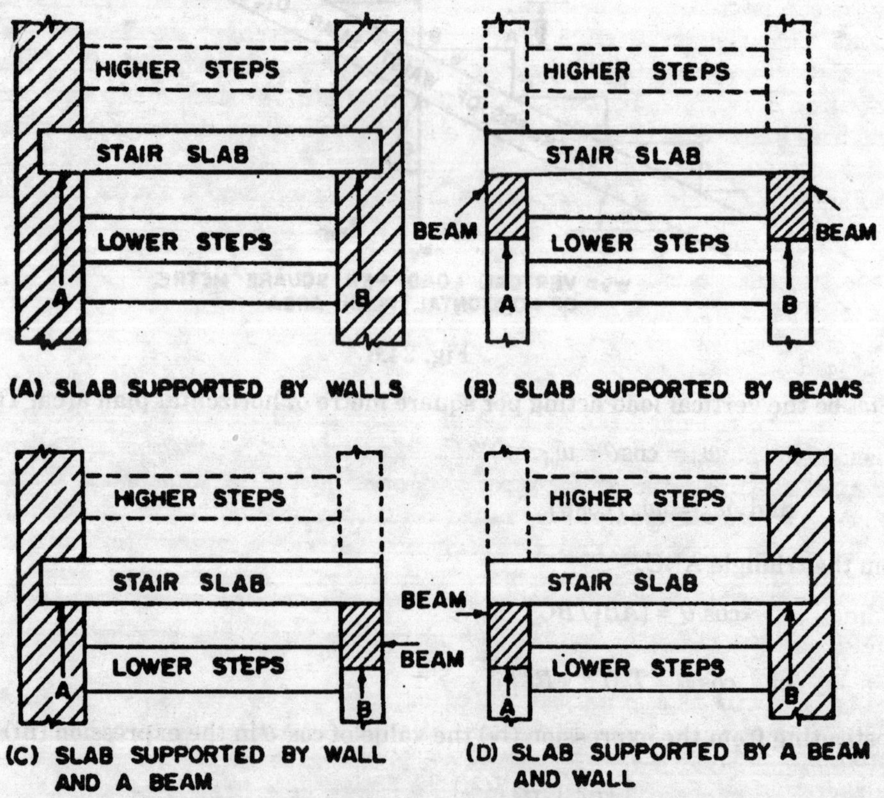

Fig. 24.9

Above specified live load is subjected to a minimum of 1.30 kN concentrated load at the unsupported end of each step for stairs construction out of the structurally independent cantilever steps.

24.6 STRUCTURAL SYSTEM OF STAIR SLAB

In Art. 24.3, the effective span of stair slab has been described. The stair slab may span horizontally or longitudinally. The structural system of these types of stair slabs is as follows:

1. Stair slab spanning horizontally. Structurally, a stair slab spanning horizontally may be simply supported or cantilevered as below :

(a) Simply supported waist of the slab. A stair slab spanning horizontally (i.e., in the direction of steps) may have beam/wall supports at both the ends or a wall at one support and a beam at the other support. In both these cases, the stair slab acts as simply supported at both the ends.

(b) **Cantilever waist slab.** A stair slab spanning horizontally (viz., in the direction of steps) may be built in a wall at one end only acts like a cantilevered slab as shown in Fig. 24.10.

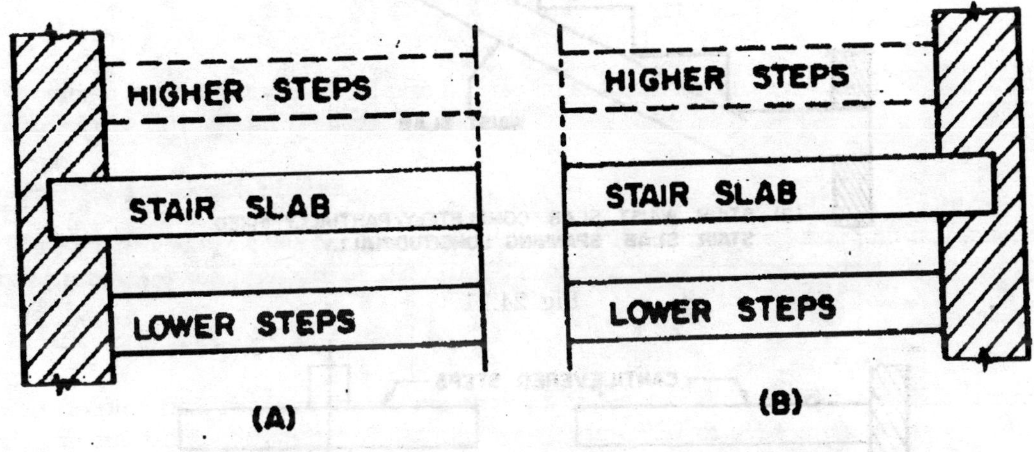

CANTILEVERED STAIR (WAIST) SLAB STAIR SLAB SPANNING HORIZONTALLY

Fig. 24.10

2. Stair slab spanning longitudinally. The waist slab of a stair slab spanning longitudinally (viz., *parallel to flight*) may act simply supported or completely fixed or partially fixed at the supported as shown in Fig. 24.11, depending upon the fixity from the supports.

Sometimes, instead of a waist slab, only steps are kept cantilevered from a wall edge beam from one side as shown in Fig. 24.12 (A). Many times, the steps are cantilevered on both sides from a middle stringer beam as shown in Fig. 24.12 (B). Such systems of stairs are referred as stair way.

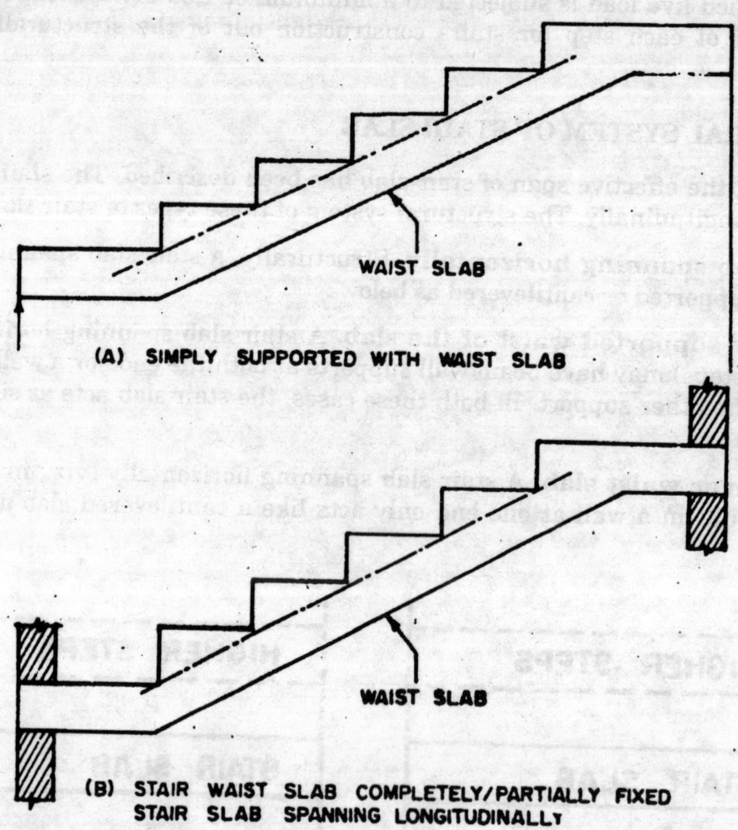

Fig. 24.11

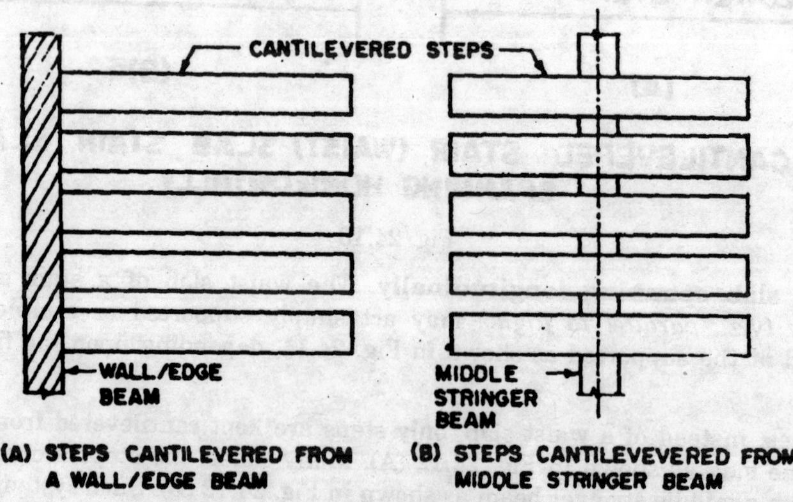

Fig. 24.12

Example 24.1. The steps of a stair of a residential building having rise of 160 mm and tread of 250 mm are supported at their ends by a wall on one side and a stringer beam on the other side. The distance between centre to centre of supports is 1.40 m. Design the stair slab. Provide M 20 grade of concrete and Hysd-steel bars of grade Fe 415.

Design. For M 20 grade of concrete and Hysd-steel bars, Fe 415 the *design constants* are as under : The stair slab shall be designed as balanced section.

$k_B = 0.2886, \quad j_B = 0.9038$

$R_B = 0.9129, \quad p_B = 0.4392$

Step 1. Effective span. The stair slab spans parallel to the steps, (viz., horizontally). The steps are supported by a wall on one side and a string beam on other side. The distance between centre to centre of supports is 1.40 m.

$\therefore$ Effective span = 1.40 m.

Step 2. Loads on stairs. The loads on stairs shall be dead load and live load as under.

(i) **Dead load.** The steps of the stair may be considered as equivalent rectangles of width, b and the depth, (D/2) as shown in Fig. 24.4. Let the thickness of a waist slab D_1 be 80 mm. Then, from Fig. 24.4, Eq. 24.1

$$b = (R^2 + T^2)^{1/2} \qquad \text{(i)}$$

The rise of each step R is equal to 160 mm and the tread T is equal to 250 mm. Therefore,

$$b = (160^2 + 250^2)^{1/2} = 296.82 \text{ mm}$$

$$\text{From Eq. 24.2, } D = \left(D_1 + \frac{T.R}{b}\right) = 80 + \left(\frac{250 \times 160}{296.82}\right) = 214.76 \, mm$$

$$\therefore (D/2) = \left(\frac{1}{2} \times D\right) = \left(\frac{214.76}{2}\right) = 107.38 \, mm$$

Self-weight of the steps per metre (parallel to steps running horizontally)

$$= \left(\frac{296.80 \times 107.38 \times 1}{1000 \times 1000} \times 24\right) = 765 \, kN/m$$

Self-weight of the waist slab, from Eq. 24.6

$$w_v = 24 \times D_1 \left(\frac{T^2 + R^2}{T}\right)^{1/2}$$

$$\text{or } w_v = \left[24 \times \frac{80}{100} \times \left(\frac{250 + 160}{250}\right)^{1/2}\right]$$

$$\text{or } w_v = \left(24 \times \frac{80}{100} \times \frac{296.82}{250} \times 1\right) = 2.280 \, kN/m$$

Let the weight of finishing be 0.075 kN/m. Total dead load

= 3.12 kN/m

(ii) Live load. The live load on stair for residential building may be adopted as 3 kN/m² horizontal plan area

$$(3 \times 0.250 \times 1) = 0.750\ kN/m$$

Total dead load plus live load, w = (3.12 + 0.750) = 3.870 kN/m

Step 3. Effective depth of section. The stair slab is spanning horizontally and it acts as simply supported beam. The maximum bending moment

$$M = \left(\frac{w.L^2}{8}\right) = \left(\frac{3.870 \times 1.40^2}{8}\right) = 0.948\ kN-m$$

$$\text{Effective depth, } d = \left(\frac{M}{R_B \times b}\right)^{1/2} = \left(\frac{0.948 \times 10^6}{0.9129 \times 296.82}\right)^{1/2} = 59.15\ mm$$

The available depth (D/2) is equal to 107.38 mm.

Step 4. Steel reinforcement. The area of steel bars provided as reinforcement in tension

$$A_{st} = \left(\frac{0.948 \times 10^6}{230 \times\ 0.9048 \times 107.38}\right) = 42.47\ mm^2$$

A 8 mm diameter Hysd-steel bar (A_f = 50.26 mm²) is provided in each step as main reinforcement. The distribution reinforcement is provided as under

$$A_d = \left(\frac{0.12}{1000} \times 296.82\ \times 214.76\right) = 76.94\ mm^2$$

Spacing of the distribution bars of 6 mm

$$s_d = \left(\frac{1000 \times 28.27}{76.494}\right) = 369.93\ mm$$

These bars are provided at 350 mm c/c.

Example 24.2. The clear dimensions of a staircase hall is 2.4 m x 4.750 m. The floor to floor height is 3.52 m. A two flight stair is to be provided for the two floors. Design the overall geometry of the stair.

Design. The clear width of staircase hall is 2.4 m. A dog-legged stair shall be provided. The clear width of each step

$$= \left(\frac{1}{2} \times 2.4\right) = 1.20\ m$$

The rise of each step is kept 150 mm to 160 mm, and the tread of each step is kept 250 mm to 300 mm. Let the rise of each step be 160 mm.

Then,

Step 1. Number of risers $= \left(\dfrac{3.52 \times 1000}{160}\right) = 22$

The number of risers in each flight $= \left(\dfrac{22}{2}\right) = 11$

Step 2 : Number of treads $= (11 - 1) = 10$

Let the tread of each step be 250 mm. Going of each flight

$= (10 \times 250) = 2500\ mm.$

The loadings are provided on both the sides. The clear width of each loading

$\left(\dfrac{4.750 \times 1000 - 2500}{2}\right) = 1125\ mm$

Step 3. Overall geometry of stair

The overall geometry of two flight stair is shown in Fig. 24.13.

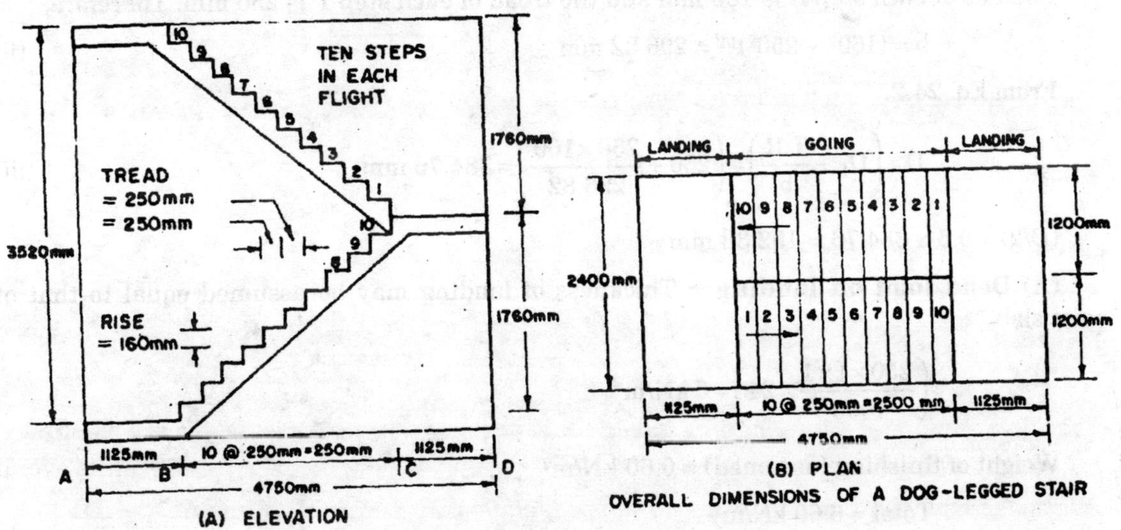

Fig. 24.13

Example 24.3. The overall dimensions of a dog-legged stair are shown in Fig. 24.13. The landing slabs span in the same direction as the stair and are supported by the walls at the ends. Design the stair slab. Provide M 20 grade of concrete and Hysd-steel bars of grade Fe415. The stair is used inside a residential building.

Design: For M 20 grade of concrete and Hysd-steel bars, the *design constants* are as under. The stair slab shall be designed as a balanced section.

$k_B = 0.2886$, $j_B = 0.9038$

$R_B = 0.9129$, $p_B = 0.4392$

Step 1. Effective span. The stair slab spans parallel to the steps, (viz., parallel to flight). The landing slabs span in the same direction as the stair and are supported by the walls at the ends. The effective span, ES shall be the distance centre-to-centre of the supporting walls. Let the thickness or width of supporting wall be 400 mm. Then, From Fig. 24.13.

$$\text{E.S.} = \left(\frac{1}{2}\times 300 + 1125 + 2500 + 1125 + \frac{1}{2}\times 300\right) \text{ mm}$$

$$= 5050 \text{ mm} = 5.05 \text{ m}$$

Step 2: Loads on the stair: The loads on the stair will consist of dead load and live load as below:

(*i*) **Dead load.** The step of the stair may be considered as equivalent rectangle of width, *b* and the depth (D/2). Let the thickness of waist slab, D_1 be assumed as 250 mm (*viz.,* @ 50 mm per meter span). From Eq. 24.1.

$$b = (R^2 + T^2)^{1/2} \qquad \text{(i)}$$

The rise of each step *R* is 160 mm and the tread of each step *T* is 250 mm. Therefore,

$$b = (160^2 + 250^2)^{1/2} = 296.82 \text{ mm} \qquad \text{(ii)}$$

From Eq. 24.2.

$$D = \left(D_1 + \frac{T.R}{b}\right) = \left(250 + \frac{250\times 160}{296.82}\right) = 384.76 \text{ mm} \qquad \text{(iii)}$$

$(D/2) = 0.5 \times 384.76 = 192.38$ mm

(A) **Dead load on landing =** Thickness of landing may be assumed equal to that of waist slab.

$$\left(\frac{250\times 1\times 1}{1000}\times 24\right) = 6 \text{ kN/m}^2$$

Weight of finishing (assumed) = 0.60 kN/m²

Total = 6.60 kN/m²

For 1.2 m width of step (stair) (6.60 x 1.2) = 7.92 kN/m

(B) **Dead load of going.** The vertical component of this dead load is as under

$$w_v = (0.250 \times 1 \times\times 24)\times\left(\frac{\left(R^2 + T^2\right)^{1/2}}{T}\right) = \left(6\times\frac{296.82}{250}\right) = 7.12 \text{ kN/m}^2$$

Weight of finishing (assumed as above) = 0.60 kN/m²

Weight of one step. The step may be considered as equivalent rectangle of width, b and depth $(D/2)$

$$\text{Self weight} = \left(\frac{296.82}{1000} \times \frac{192.38}{1000} \times 1 \times 24\right) = 1.37 \text{ kN/m}^2$$

Number of teps of tread 250 mm per m = (1000/250) = 4

Self weight of step = (1.37 x 4) = 5.48 kN/m²

Total weight = (7.12 + 0.6 + 5.48) = 13.2 kN/m²

Total weight for complete width of the step (stair) = (13.2 x 1.2) = 15.84 kN/m

(ii) Live load: The live load on the stair for the residential building may be adopted as 3 kN/m² horizontal plan area. For complete width of the stair, the live load.

= (3 x 12) = 3.6 kN/m

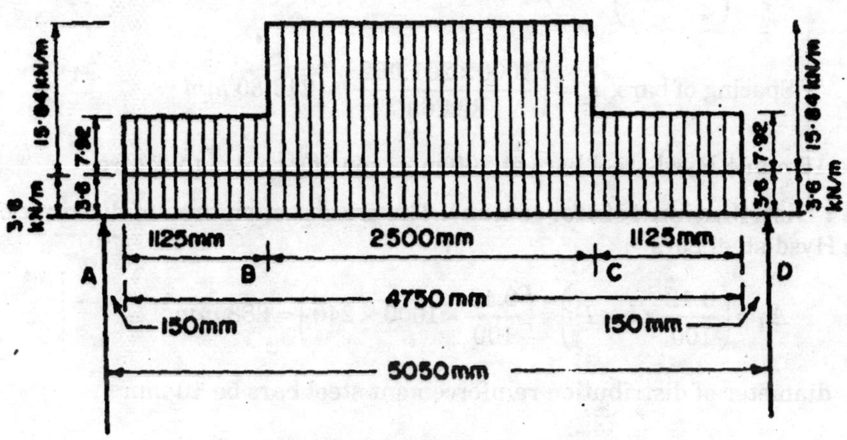

LOADS ON STAIR SLAB
LANDINGS SUPPORTED ON WALLS

Fig. 24.14

The loads acting on the stair shall be as shown in Fig. 24.14.

Step 3: Bending moment: The end reaction at the support.

$$R_A = \frac{1}{2}[3.60 \times 4.750 + 7.92 \times 2 + 15.84 \times 2.5] = 36.27 \text{ kN}$$

The maximum bending moment occurs at the centre

$$M = \left[36.75 \times 5.05 - (3.6 + 7.92) \times 1.25 \times \left(\frac{1.125}{2} + 1.25\right)\right.$$

$$\left. - (3.6 + 15.84) \times \frac{2.5 \times 1.25}{2}\right] = 37.75 \text{ kN - m}$$

Step 4: Effective depth: The cross-section of the stair slab shall be designed as balance section. Therefore, the required effective depth of slab.

$$d = \left(\frac{M}{R_B \times 1000}\right)^{1/2} = \left(\frac{37.715 \times 10^6}{0.9129 \times 1000}\right)^{1/2} = 203.26$$

Let the effective cover be 20 mm. The overall depth of the slab shall be 223.26 mm. Let the overall thickness of the slab be 240 mm. Then, the effective depth of slab will be 220 mm.

Step 5: Reinforcement. Let the cross-sectional area of steel bars be A_{st}.

$$A_{st} = \left(\frac{M}{\sigma_{st} \cdot j_B \cdot d}\right) = \left(\frac{37.715 \times 10^6}{230 \times 0.9038 \times 220}\right) = 824.69 \text{ mm}^2$$

Provide 16 mm ϕ steel bars

$$A_\phi = \left(\frac{\pi}{4} \times 16^2\right) = 201.062 \, mm^2$$

$$\text{Spacing of bars, } s_v = \left(\frac{1000 \times 201.062}{824.69}\right) = 243.80 \, mm$$

Provide 16 mm ϕ Hysd-steel bars at 200 mm, spacing centre to centre.

Step 6 : Distribution reinforcement. The cross-section area of distribution reinforcement for A_d Hysd-steel bars

$$A_d = \left(\frac{0.12}{100} \times b \times D\right) = \left(\frac{0.12}{100} \times 1000 \times 240\right) = 288 \, mm^2$$

Let the diameter of distribution reinforcement steel bars be 10 mm

$$A_{\phi d} = \left(\frac{\pi}{4} \times 10^2\right) = 78.54 \, mm^2$$

Spacing of the distribution reinforcement

$$s_d = \left(\frac{1000 \times 78.54}{288}\right) = 272.71 \, mm$$

Provide 10 mm f Hysd-steel bars at 250 mm spacing centre to centre. The details of reinforcement in one flight is shown in Fig. 24.15. The reinforcement is also provided in the same manner in the other flight.

Example 24.4. The overall dimensions of a dog-legged stair are shown in Fig. 24.13. The landing slabs are supported on the sides. Design the stair slab. Provide M 20 grade of the concrete and Hysd-steel bars of grade Fe 415. The stair is used inside a residential building.

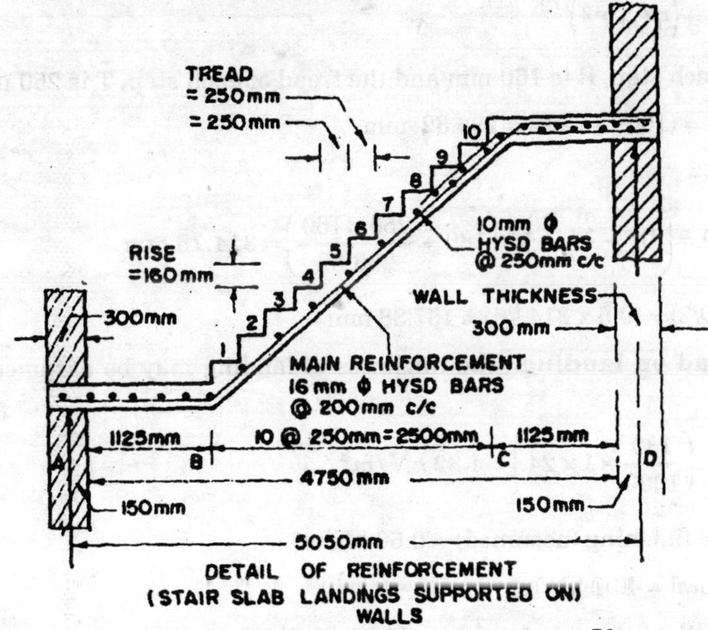

Fig. 24.15

Design : For M 20 grade of concrete and Hysd-steel bars, the *design constants* are as under. The stair slab shall be designed as a balanced section.

$k_B = 0.2886, j_B = 0.9038$

$R_B = 0.9129, p_B = 0.4392$

Step 1 : Effective span. The stair slab spans in the longitudinal direction (viz., parallel to the flight). The landing slabs are supported on the side walls.

Half the width of landing on one side

$$X = (0.15 \times 1125) = 562.5\ mm < 1\ m$$

Half the width of landing on other side

$$Y = (0.15 \times 1125) = 562.5\ mm < 1\ m$$

Going for the stair slab, G = 2500 mm.

Effective span, ES of the stair slab from IS : 456-2000

$$ES = (G + X + Y) = (2500 + 562.5 + 562.5) = 3625\ mm$$

Step 2 : Loads on the stair. The loads on the stair will consist of dead load and live load as below :

(i) Dead load. The steps of the stair may be considered as equivalent rectangles of width, b and the depth, D/2. Let the thickness of the waist slab, D_1 be assumed as 180 mm (viz., @ 50 mm per metre run). From Eq. 24.1.

$$b = \left(R^2 + T^2\right)^{1/2}$$

The rise of each step, R is 160 mm and the tread of each step, T is 250 mm. Therefore,

b = (160^2 + 250^2) = 296.82 mm

From Eq. 24.2

$$D = \left(D_1 + \frac{T.R}{b}\right) = \left(180 + \frac{250 \times 160}{296.82}\right) = 314.76\ mm$$

(D/2) = (0.5 x 314.76) = 157.38 mm

(A) **Dead load on landing.** The thickness of landing may be assumed equal to that of waist slab

$$= \left(\frac{180}{1000} \times 1 \times 24\right) = 4.32\ kN/m^2$$

Weight of the finishing (assumed) = 0.60 kN/m²

Total = 4.92 kN/m²

For 1.2 m width of the step (stair), $= (4.92 \times 1.2) = 5.90\ kN/m$

(B) Dead load on going. The vertical component of this dead load is as under

$$w_v = 4.32 \times \left(\frac{\left(160^2 + 250^2\right)^{1/2}}{250}\right) = 5.13\ kN/m^2$$

Weight of the finishing (assumed as above) = 0.60 kN/m²

$$\text{Weight of one step,} = \left(\frac{296.82}{1000} \times \frac{157.38}{1000} \times 1 \times 24\right) = 1.12\ kN/m^2]$$

Number of steps of tread 250 mm per m $(1000/250) = 4$

Self-weight of the steps, $(1.12 \times 4) = 4.48\ kN/m^2$

Total weight = (5.13 + 0.6 + 4.48) = 10.21 kN/m²

Total weight for complete width of the step (stair)

$= (10.21 \times 2.2) = 12.52\ kN/m$

(ii) **Live load.** The live load on the stair for the residential building may be adopted as 4 kN/m² horizontal plan area. For complete width of the stair, the live load per metre

$= (3 \times 1.2) = 3.6\ kN/m$

The loads acting on the stair slab shall be as shown in Fig. 24.16.

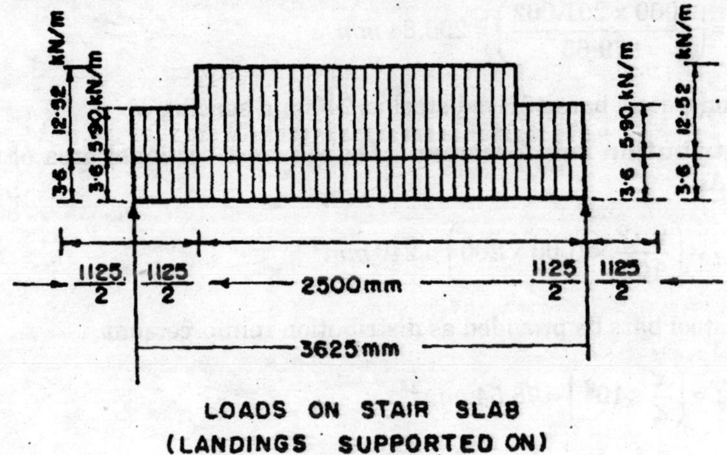

Fig. 24.16

Step 3 : Bending moment. The reaction at the end

$$= \frac{1}{2}[5.90 \times 0.5625 + 12.25 \times 2.5 + 3.6 \times 3.625] = 25.494\ kN$$

The maximum bending moment occurs at the centre

$$M = \left(25.494 \times \frac{3.625}{2} - (3.6 + 5.90) \times 0.5625 \times \left(\frac{0.5625}{2} + 1.25\right) - (3.6 + 12.52) \times 1.25 \times \frac{1.25}{2}\right) = 25.43\ kN - m$$

4. Effective depth. The cross-section of the stair slab shall be designed as balanced section. Therefore, the required effective depth of the slab

$$d = \left(\frac{M}{R_B \times 1000}\right)^{1/2} = \left(\frac{25.43 \times 10^6}{0.9129 \times 1000}\right)^{1/2} = 166.90\ mm$$

Let the effective cover be 20 mm. Then, the overall depth of slab shall be 186.60 mm. Let the overall thickness of the slab be 200 mm. The effective depth of the slab shall be 180 mm.

Step 5. Reinforcement. Let the cross-sectional area of the steel bars be A_{st}

$$A_{st} = \left(\frac{M}{\sigma_{st} \cdot j_B .d}\right) = \left(\frac{25.43 \times 10^6}{230 \times 0.9038 \times 180}\right) = 679.63\ mm^2$$

Provide 16 mm ϕ steel bars of Hysd-steel

$$A_{\phi} = \left(\frac{\pi}{6} \times 16^2\right) = 201.062\ mm^2$$

Spacing of the steel bars

$$s = \left(\frac{1000 \times 201.062}{679.63}\right) = 296.84\ mm$$

Provide 16 mm ϕ steel bars of Hysd-steel at 240 mm spacing c/c.

Step 6 : Distribution reinforcement. Let the cross-sectional area of the distribution reinforcement be A_d.

$$A_d = \left(\frac{0.12}{100} \times 1000 \times 200\right) = 240\ mm^2$$

Let 10 mm ϕ steel bars be provided as distribution reinforcement

$$A_\phi = \left(\frac{\pi}{4} \times 10^2\right) = 78.54\ mm^2$$

The spacing of the distribution reinforcement steel bars

$$s_d = \left(\frac{1000 \times 78.54}{240}\right) = 327.25\ mm$$

Provide 10 mm f Hysd-steel bars at 300 mm spacing centre to centre. The details of the reinforcement is shown in Fig. 24.17.

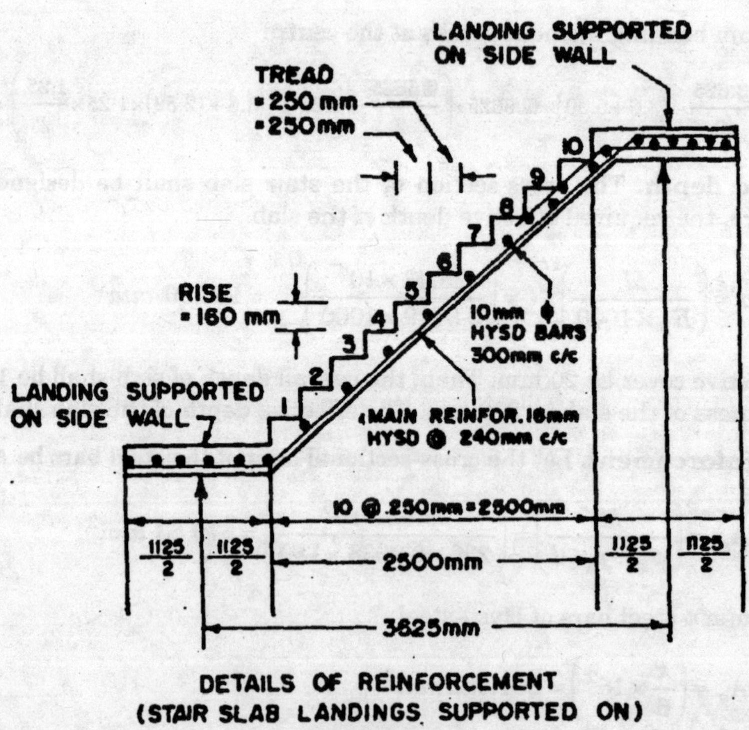

Fig. 24.17

Example 24.5 The overall dimensions of a dog-legged stair are shown in Fig. 24.13. The stair slab is supported by the beams spanning parallel with the risers. These beams are provided below the top and the bottom risers. Provide M 20 grade of the concrete and Hysd-steel bars of grade Fe 415. The stair is used in a residential building. Design the stair slab.

Design: For M 20 grade of concrete and Hysd-steel bars, the *design constants* are as under. The stair slab shall be designed as a balanced section.

$k_B = 0.2886, \quad j_B = 0.9038$

$R_B = 0.9129, \quad p_B = 0.4392$

Step 1 : Effective span. The stair slab is supported by the beams spanning parallel with the risers. These beams are provided below the top and the bottom risers. The effective span shall be the distance centre to centre of the beams. Therefore, the effective span, from Fig. 24.13.

$$ES = (2500 - 125 + 125) = 2500 \text{ mm}$$

Step 2 : Loads on the stair. The loads on the stair will consist of dead load and live load as below :

(i) **Dead load.** The steps of the stair may be considered as equivalent rectangles of width, b and the depth, (D/2). Let the thickness of the waist slab, D_1 be assumed as 125 mm (viz., @ 50 mm per metre run). From Eq. 24.1

$$b = (R^2 + T^2)^{1/2}$$

The rise of each step, R is 160 mm and the trend of each step, T is 250 mm. Therefore,

$$b = (160^2 + 250^2)^{1/2} = 296.82 \text{ mm}$$

From Eq. 24.2

$$D = \left(D_1 + \frac{T.R}{b}\right) = \left(125 + \frac{250 \times 160}{296.82}\right) = 259.76\ mm$$

$$(D/2) = (0.5 \times 259.76) = 129.88 \text{ mm}$$

Dead load on going. The vertical component of this dead load is as under :

$$w_v = \left[\left(\frac{125}{1000 \times 1 \times 1 \times 24}\right) \times \left(\frac{\left(160^2 + 250^2\right)^{1/2}}{250}\right)\right] = 3.12\ kN/m^2$$

Weight of the load finishing (assumed as below) = 0.60 kN/m²

$$\text{Weight of one step} = \left(\frac{259.76}{1000} \times \frac{129.88}{1000} \times 1 \times 24\right) = 0.81 \text{ kN/m}^2$$

Number of steps of tread 250 mm per m, (1000/250) = 4

$$\text{Self-weight of the steps,} = (0.81 \times 4) = 3.24\ kN/m^2$$

Total weight for complete width of the step (stair)

$$= (3.24 + 0.60 + 3.12) \times 1.2 = 8.352\ kN/m^2$$

(ii) **Live load.** The live load on the residential building may be adopted as 3 kN/m² horizontal plan area. For complete width of the stair, the live load per metre

$$= (3 \times 1.2) = 3.6\ kN/m$$

The loads acting on the stair slab is shown in Fig. 24.18.

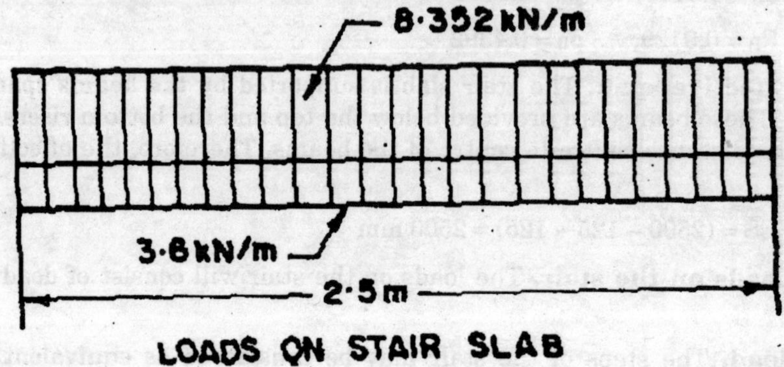

Fig. 24.18

Beams spanning parallel to the risers are provided under top and bottom risers.

Step 3 : Bending moment. Total uniformly distributed load (dead load *plus* live load) per metre

$$w = (8.352 + 3.6) = 11.952 \text{ kN/m}$$

Maximum bending moment occurs at the centre

$$M = \left(\frac{w.L^2}{8}\right) = \left(\frac{1.952 \times 2.5 \times 2.5}{8}\right) = 9.3375\ kN - m$$

Step 4 : Effective depth. The cross-section of the stair slab shall be designed as the balanced section. Therefore, the required depth

$$d = \left(\frac{M}{R_B \times 1000}\right)^{1/2} = \left(\frac{9.3375 \times 10^6}{0.9129 \times 1000}\right)^{1/2} = 101.14\ mm$$

Let the effective cover be 20 mm. Then, the overall depth of slab shall be 121.14 mm. Let the overall thickness of the stair slab be 125 mm. The effective depth of the slab shall be 105 mm.

Step 5 : Reinforcement. Let the cross-sectional area of the steel bars be A_{st}.

$$A_{st} = \left(\frac{M}{\sigma_{st}.j_B.d}\right) = \left(\frac{9.3375 \times 10^6}{230 \times 0.9038 \times 105}\right) = 427.8\ mm^2$$

Provide 12 mm ϕ steel bars of Hysd-steel, $A_\phi = \left(\frac{\pi}{4} \times 10^2\right) = 78.54\ mm^2$

Spacing of the steel bars, $s = \left(\frac{1000 \times 78.54}{427.8}\right) = 183.59\ mm$

Provide 10 mm ϕ steel bars of Hysd-steel bars at 180 mm spacing centre to centre.

Step 6 : Distribution reinforcement. Let the cross-sectional area of the distribution reinforcement be A_d

$$A_d = \left(\frac{0.12}{1000} \times 1000 \times 125\right) = 150\ mm^2$$

Provide 10 mm ϕ steel bars, $A_\phi = 78.54\ mm^2$

Spacing of the distribution steel bars

$$s_d = \left(\frac{1000 \times 78.54}{150}\right) = 523.6\ mm.$$

Provide 10 mm f Hysd-steel bars at 450 mm spacing centre to centre. The details of the reinforcement is shown in Fig. 24.19.

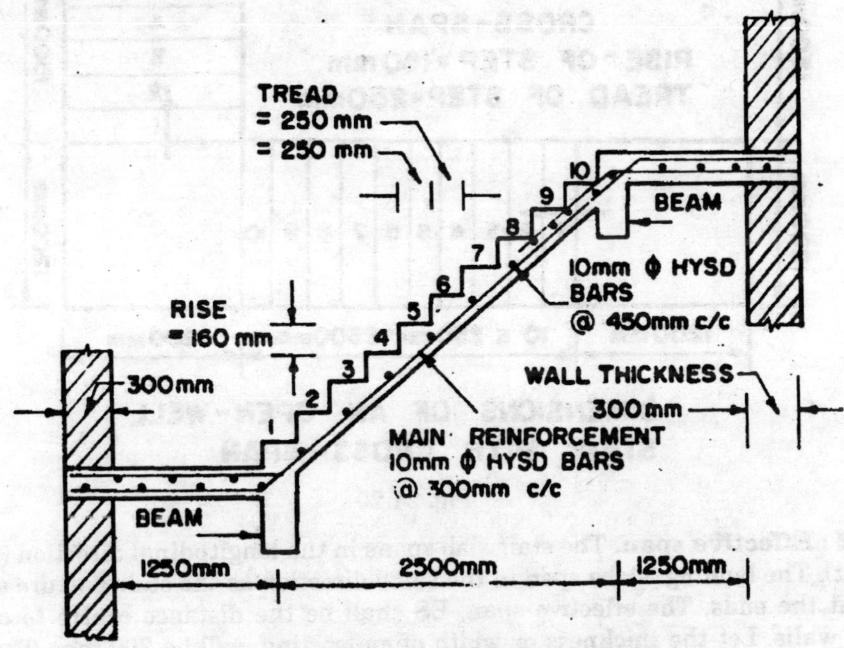

Fig. 24.19

Example 24.6. A stair with an open will consists of two flight and a span partly crossing at right angles. There are ten steps of rise 160 mm and tread 250 mm in each flight and six such steps in the cross span. The width of landings and stair is 120 m. The landing are supported on the walls at the ends. Design the stair slab. Provide M 20 grade of the concrete and Hysd-steel bars of grade Fe 415.

Design : For M 20 grade of concrete and Hysd-steel bars, the design constants are as under. The stair slab shall be designed as a balanced section

$k_B = 0.2886, \quad j_B = 0.9038$

$R_B = 0.9129, \quad p_B = 0.4392$

Step 1 : Overall dimensions. The overall dimensions of the stair are shown in Fig. 24.20.

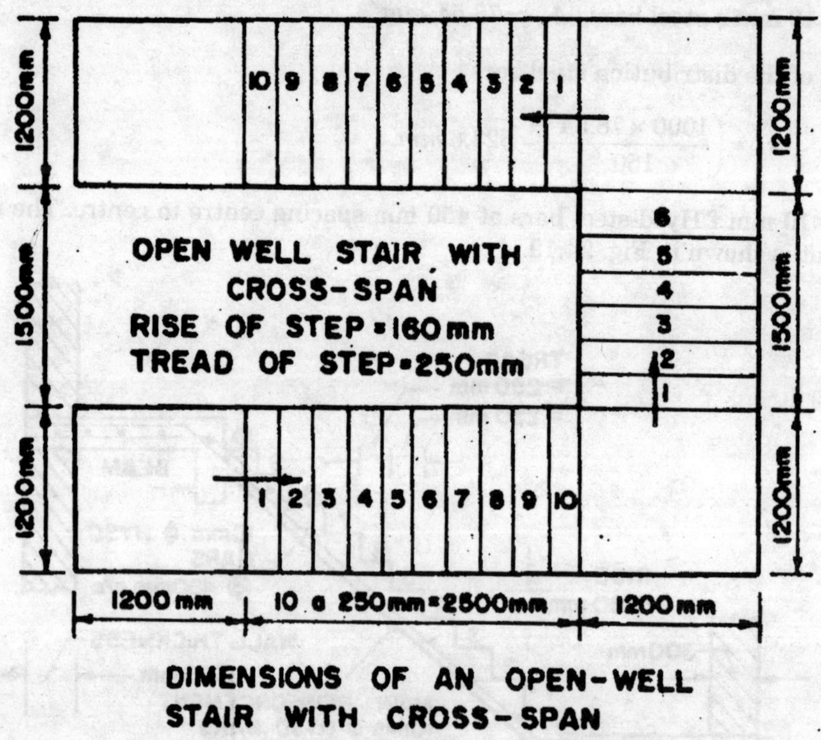

Fig. 24.20

Step 2 : Effective span. The stair slab spans in the longitudinal direction (viz., parallel to the flight). The landing slabs span in the same direction as the stair and are supported by the walls at the ends. The effective span, ES shall be the distance centre to centre of the supporting walls. Let the thickness or width of supporting wall be 300 mm. Then from Fig. 24.20.

$$ES = \left(\frac{1}{2} \times 300 + 1200 + 2500 + 1200 + \frac{1}{2} \times 300\right) = 5200 \; mm$$

Step 3 : Loads on the stair. The loads on the stair will consist of dead load and live load as below :

(i) **Dead load.** The steps of the stair may be considered as equivalent rectangle of the width, b and the depth (D/2). Let the thickness of waist slab D_1 be assumed as 250 mm (viz., @ 50 mm per metre span). From Eq. 24.1.

$$b = (R^2 + T^2)^{1/2} \quad \text{(i)}$$

The rise and tread are 160 mm and 250 mm respectively. Therefore,

$$b = (160^2 + 250^2)^{1/2} = 296.82 \text{ mm}$$

From Eq. 24.2

$$D = \left(D_1 + \frac{T.R}{b}\right) = \left(250 + \frac{250 \times 160}{296.82}\right) = 384.76\ mm$$

$$(D/2) = (0.5 \times 384.76) = 192.38\ mm$$

(A) Dead load on landing. Thickness of landing may be assumed equal to that of the waist slab

$$\left(\frac{250 \times 1 \times 1}{1000} \times 24\right) = 6\ kN/m^2$$

Weight of the finishing (assumed) = 0.60 kN/m²

Total = 6.60 kN/m²

For 1.2 m width of step (stair)

$$= 6.60 \times 1.2 = 7.92\ kN/m$$

(B) Dead load of going. The vertical component of this dead load is as under

$$w_v = \left[(0.250 \times 1 \times 1 \times 24)\left(\frac{(160^2 + 250^2)^{1/2}}{250}\right)\right] kN/m^2$$

$$\text{or } w_v = \left(\frac{6 \times 296.82}{250}\right) = 7.12\ kN/m^2$$

Weight of finishing (assumed as above) = 0.60 kN/m²

Weight of one step. The step may be considered as equivalent rectangle of width, b and depth (D/2)

$$\left(\frac{296.82}{1000} \times \frac{192.38}{1000} \times 1 \times 24\right) = 1.37\ kN/m^2$$

Number of steps of tread 250 mm per m, (1000/250) = 4

Self-weight of the steps, $(1.37 \times 4) = 5.48\ kN/m^2$

Total weight

$$(7.12 + 0.6 + 5.48) = 13.2 \text{ kN/m}^2$$

Total weight for complete width of the steel (stair)

$$(13.2 \times 1.2) = 15.84 \; kN/m$$

(ii) **Live load.** The live load on the stair for the residential building may be adopted as 3 kN/m² horizontal plane area. For complete width of the stair live load

$$(3 \times 1.2) = 3.6 \; kN/m$$

Step 4 : Distribution of loads on the stair. In case of stair with open-well wheel a span partly cross at right angles occurs, the load on areas common to any such spans may be taken as *one-half in each direction* as shown in Fig. 24.21.

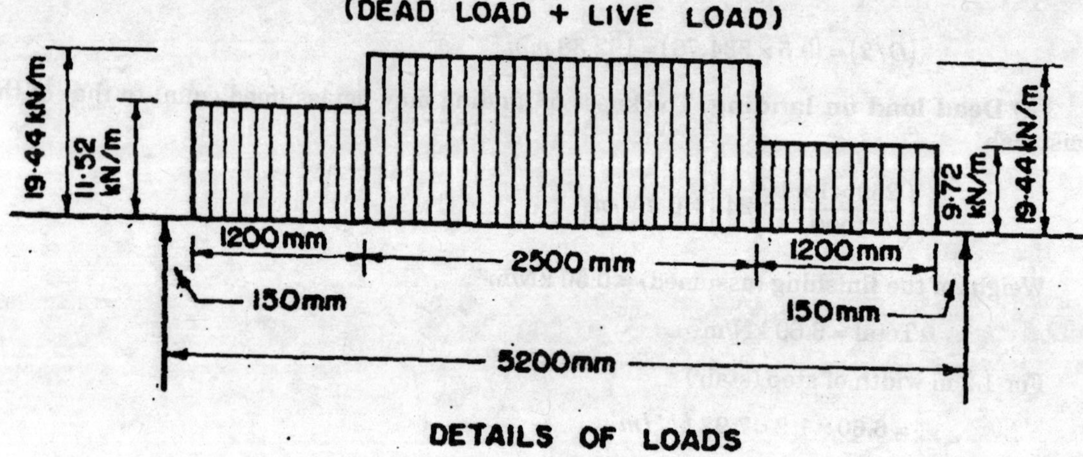

Fig. 24.21

Bending moment. The end reaction at left support, Fig. 24.21

$$R_A = \left[(11.52 \times 1200) \times \frac{4.450}{5.200} + \left(\frac{19.44 \times 2.500}{2}\right) + (9.72 \times 1.200) \times \frac{0.750}{5.250}\right] kN$$

$$\therefore \; R_A = (1.830 + 24.3 + 1.666) = 37.80 \; kN/m$$

Maximum bending moment occurs approximately near the centre

$$M = \left[37.80 \times 2.6 - (11.52 \times 12.00)(0.6 + 1.25) - (19.44 \times 1.250) \times \frac{1.250}{2}\right]$$

$$= [(98.27 - 25.57 - 15.19)] = 57.52 \; kN-m$$

Step 5 : Effective depth. The cross-section of the stair slab shall be designed as the balanced section. Therefore, the required effective depth

$$d=\left(\frac{M}{R_B \times 1000}\right)^{1/2}=\left(\frac{57.52\times10^6}{0.9129\times 1000}\right)^{1/2} = 250.08\ mm$$

Let the effective cover be 20 mm. Then, the overall depth of slab shall be 270.48 mm. Let the overall thickness of the stair slab be 280 mm. The effective depth of the slab shall be 260 mm.

Step 6 : Reinforcement. Let the cross-sectional area of the steel bars be A_{st}.

$$A_{st} = \left(\frac{M}{\sigma_{st} \cdot j_B \cdot d}\right) = \left(\frac{57.52\times10^6}{230 \times 0.9038\times 260}\right) = 1064.25\ mm^2$$

Provide 16 mm ϕ steel bars of Hysd-steel

$$A_\phi = \left(\frac{\pi}{4}\times 16^2\right) = 201.062\ mm^2$$

Spacing of the steel bars

$$s = \left(\frac{1000\times 201.62}{1064.25}\right) = 188.92\ mm$$

Provide 16 mm ϕ steel bars of Hysd at 180 mm spacing c/c.

Step 7 : Distribution reinforcement. Let the cross-sectional area of the distribution reinforcement be A_d

$$A_d = \left(\frac{0.12}{100}\times 1000\times 280\right) = 336\ mm^2$$

Let 10 mm ϕ steel bars be provided as distribution reinforcement

$$A_\phi = \left(\frac{\pi}{4}\times 10^2\right) = 78.54\ mm^2$$

The spacing of the distribution reinforcement steel bars

$$s_d = \left(\frac{1000\times 78.54}{336}\right) = 233.75\ mm.$$

Provide 10 ϕ Hysd-steel bars at 220 spacing centre to centre.

The reinforcement in the cross-span required shall be less. However, same reinforcement provided at the main flight (stair slab) may be provided in the cross-span slab.

The details of the reinforcement is shown in Fig. 24.22.

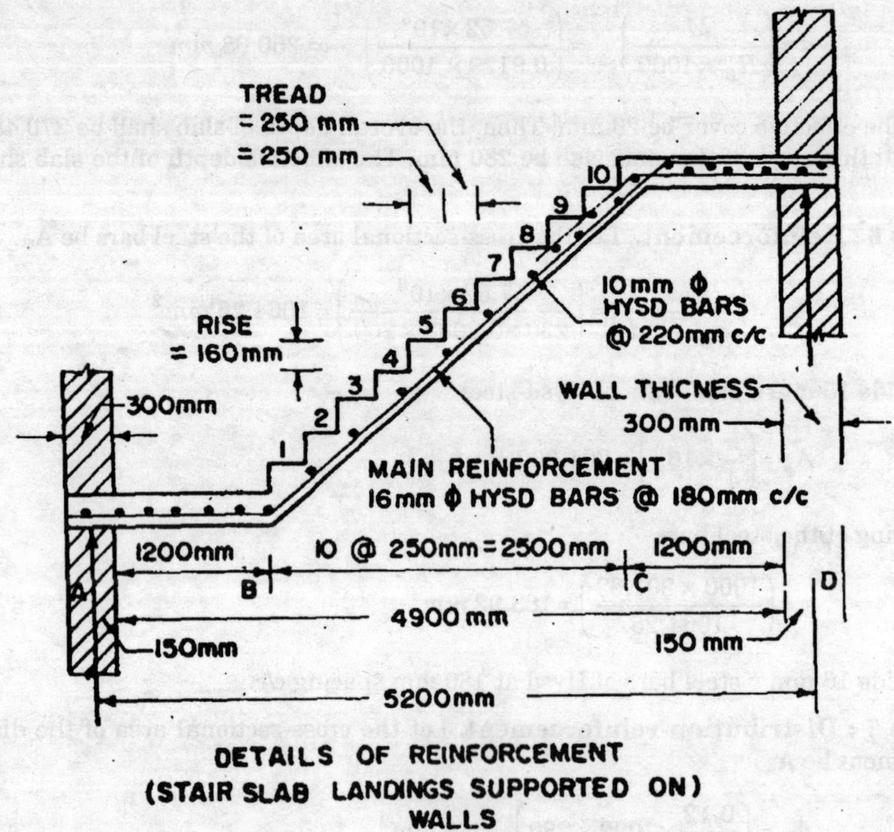

Fig. 24.22

PROBLEMS

24.1 The steps of a stair of a residential building having rise of 150 mm and tread of 300 mm are supported at end by wall on one side and a stringer beam on the other side. The distance between centre to centre of supports is 1.30 m. Design the stair slab. Provide M 20 grade of concrete and Hysd-steel bars of grade Fe 415.

24.2 The clear dimensions of a staircase hall is 3 m x 6 m. The floor to floor height is 3.60 m. A two flight stair is to be provided for the two floors. Design the overall geometry of the stair.

24.3 The overall dimensions of a dog-legged stair may be adopted as in Problem 24.2. Provide M 20 grade of concrete and Hysd steel bars of grade Fe 415. The stair is used inside a residential building. The landing slabs span in the same direction as the star and are supported by the walls at the ends.

24.4. Design the stair-slab in Problem 24.3, in case the landing slabs are to be supported on the sides.

24.5 **Design the stair slab in Problem 24.3, in case the stair is supported by the beams spanning parallel with the risers. These beams are provided below the top and the bottom risers.**

24.6 **A stair with an open well consists of two flights and a span partly crossing at right angles. These are twelve steps of rise 150 mm and tread 30 mm in each flight and six steps in the cross-span. The width of landings and stairs is 1.150 m. The landings are to be supported on the walls at the ends. Design the stair slab. Provide M 20 grade of the concrete and Hysd steel bars of grade Fe 415.**

25

Design of Retaining Walls

25.1 INTRODUCTION

Retaining walls are the structures used to retain the earth on either side of it at two different levels. The retaining walls hold back the masses of earth and loose material. The retaining walls prevent a bank from slipping down. The retaining walls provide soil stability at a vertical change in ground elevation. The retaining walls also prevent the soil to assume the natural angle of repose at locations where an abrupt change in ground elevation occur. In the railway and highway construction, the width of the right of way is fixed, and the cut or embankment is to be contained within that width. The basement walls of buildings are to be kept within the property. The soil surrounding the basement is to be retained. The wing walls and the abutments in the bridges also retain the earth. The loose material (viz., earth, soil., etc.,) retained is called as back *fill*. The retained material exerts a push on the structure and tends to overturn and or slide it. In order to achieve and maintain the stability, the weight of the retaining wall structures is of considerable importance. The retaining walls are the free-standing structures and do not form the part of the other structures.

25.2 TYPES OF RETAINING WALLS

The various type of retaining walls are shown in Fig. 25.1. The type of a retaining wall depends on the mode of resisting the earth pressure and their shape. The most common types of the retaining walls are as follow:

1. Gravity wall. The gravity retaining walls as shown in Fig. 25.1 (A) retain the earth entirely by its own weight to provide the necessary stability. Plain concrete and stone masonry gravity walls are most commonly used for walls up to about 3 m height. It is designed chiefly with keeping the thrust line within the middle third of the cross-section, and the tension does not develop anywhere in the cross-section.

2. Cantilever retaining walls. The reinforced concrete cantilever retaining walls are most commonly used for walls of height ranging from 3 to 8 m. It consists of the vertical arm which retains the earth and is held in position by the footing or base slab. In addition to the self-weight, the weight of backfill on the top of heel is utilized to obtain the stability of the structure. The stem, heel and toe of such a wall each acts as a cantilever slab, as shown in Fig. 25.1(B). L-shaped and the reversed L-shaped as shown in Fig. 25.1(C) & (D) are also cantilever retaining walls. The bending moment and therefore, the wall thickness of stem (viz., arm which represents the vertical cantilever) increase rapidly with the increasing height.

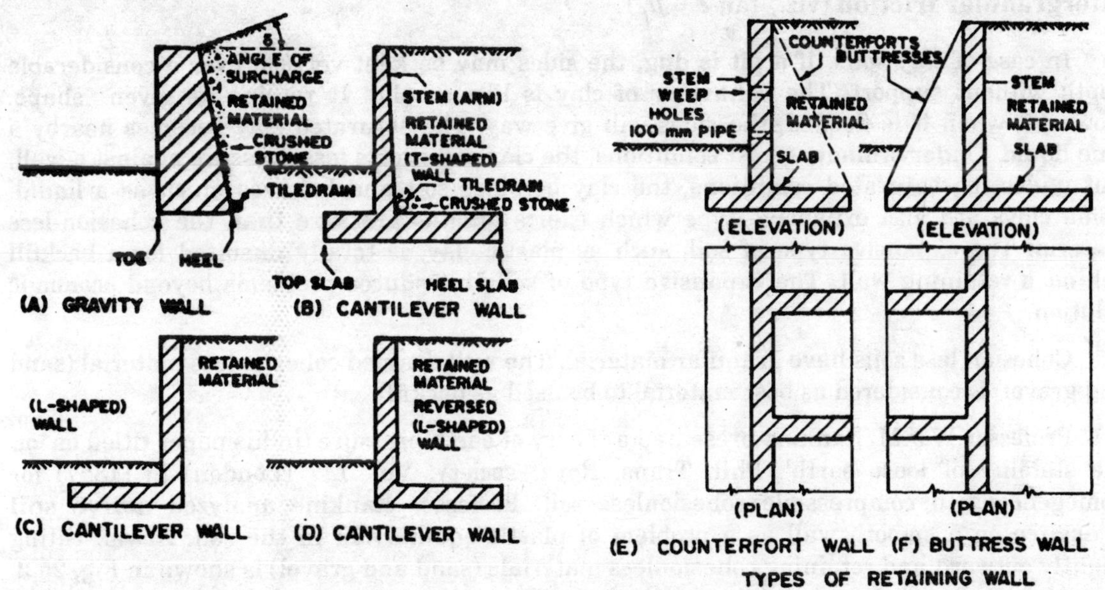

Fig. 25.1

3. Counter fort retaining walls. In order to reduce the bending moments and wall thickness, the triangular walls are attached with the vertical wall and the base slab both at horizontal spacing equal to or slightly more than one-half of the height as shown in Fig. Fig. 25.1 (E), towards the backfill side. The counterfort act as tension ties and provide horizontal force opposite to that due to earth pressure. The counterforts change the support conditions of the stem and heel slab. The stem becomes a slab spanning horizontally between the counterforts. The heel slab gets support from three sides. This type of retaining walls are more economical than the cantilever retaining walls. These are often economical for height over about 8.

4. Buttress retaining walls. The buttress retaining walls are similar to counterfort walls. The buttresses are also triangular walls (attached with the stem and the slab) placed on the side of stem opposite to the retained material as shown in Fig. 25.1(F). The buttresses act as compression struts. The buttresses are more efficient being compression elements than the tension ties (counterforts) and are economical in the same height range.

A counterfort retaining wall is more widely used than a buttress wall. The counterforts remain hidden beneath the retained material. The buttresses remain exposed and deemed unsightly. The buttresses occupy usable space in front of the wall.

25.3 ACTIVE EARTH PRESSURE

Soils and other granular masses occupy a position intermediate between liquids and solids as regards their physical behaviour. In case, sand is poured from a dump truck, it flows, but it will not assume a horizontal surface like frictionless liquid. The soil maintains in a stable heap. The sides of heap subtends an angle ϕ with the horizontal. This angle is called as **angle of repose.** The tangent of angel of repose is roughly equal to the **coefficient of intergranular friction** (viz., $\tan \phi = \mu_f$).

In case of *clay soils,* if a pit is dug, the sides may be kept vertical over a considerable depth without support. The behaviour of clay is like a solid. It retains its given shape. However, when it is flooded, the sides will give way. The saturated clay becomes nearby a true liquid. Under ordinary moist conditions, the clay soil exerts less pressure against a wall. But under the saturated conditions, the clay becomes soft and behaves much as a liquid. Some clays are also expansive type which exerts greater pressure than the cohesion-less material The expansive type of soil, such as plastic clay, is totally unsuited for a backfill behind a retaining wall. The expansive type of soils introduces problems beyond economic solution.

Cohesion less soils have granular material. The well drained cohesionless material (sand and gravel) is considered as best material to be used as backfill.

Professor W.J.M. Rankine presented a theory of earth pressure (in his paper titled as `on the stability of loose earth'. Phil. Trans. Royal society, Vol. 147 (London) in (1857) for homogeneous in compressible cohesionless soil. Professor Rankine analyzed **active soil pressure** on a smooth wall as a problem of plastic equilibrium in the soil. A wall tilting slightly outward and retaining cohesionless materials (sand and gravel) is shown in Fig. 25.2. The surface of wall is smooth (*viz.,* frictionless). The sliding occurs on two sets of planes in a wedge behind the wall. These planes make an angle of $(45 + \phi/2)^o$ with the horizontal, where ϕ is *the angle of friction* for soil on soil. These planes produce the maximum pressure that can follow through against a wall. (In case, there would not have any tilting at all, the larger pressures may exist).

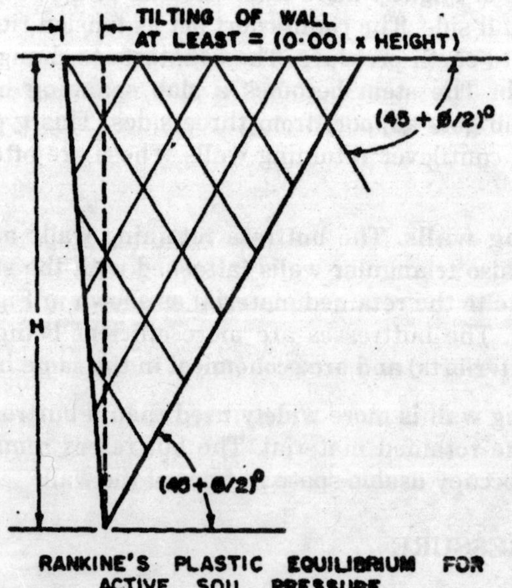

Fig. 25.2

Consider a wall AB as shown in Fig. 25.3 (A). The wall AB retains sand behind it. The sand tends to slide on some plane BC_1. A free body of wedge of sand ABC, is in equilibrium under three forces, *namely*, the weight of the soil, W_1, the reaction from the soil below BC_1, R_1 and the holding force from the wall p_{H1}. The reaction from the soil R_1 may be considered as a normal component R_{N1} normal to BC_1 and a frictional force R_{F1} resisting sliding along BC_1. For any given angle of slope, Q the force p_{H1} (and R_1 if desired) is obtained from the triangle of forces, Fig. 25.3(B). Similarly, the triangle of forces for other potential sliding planes BC_2 and BC_3 leading to other holding forces P_{H3} and P_{H3} and so on, may be drawn. The maximum possible value of p_H shall be obtained by number of such trials. A plot of ordinate P_{H1} over C_1, P_{H2} over C_2, and so on with curve drawn through the points so located. The sliding plane for the maximum value of P_H is located, say BC as shown in Fig. 25.3 (C).

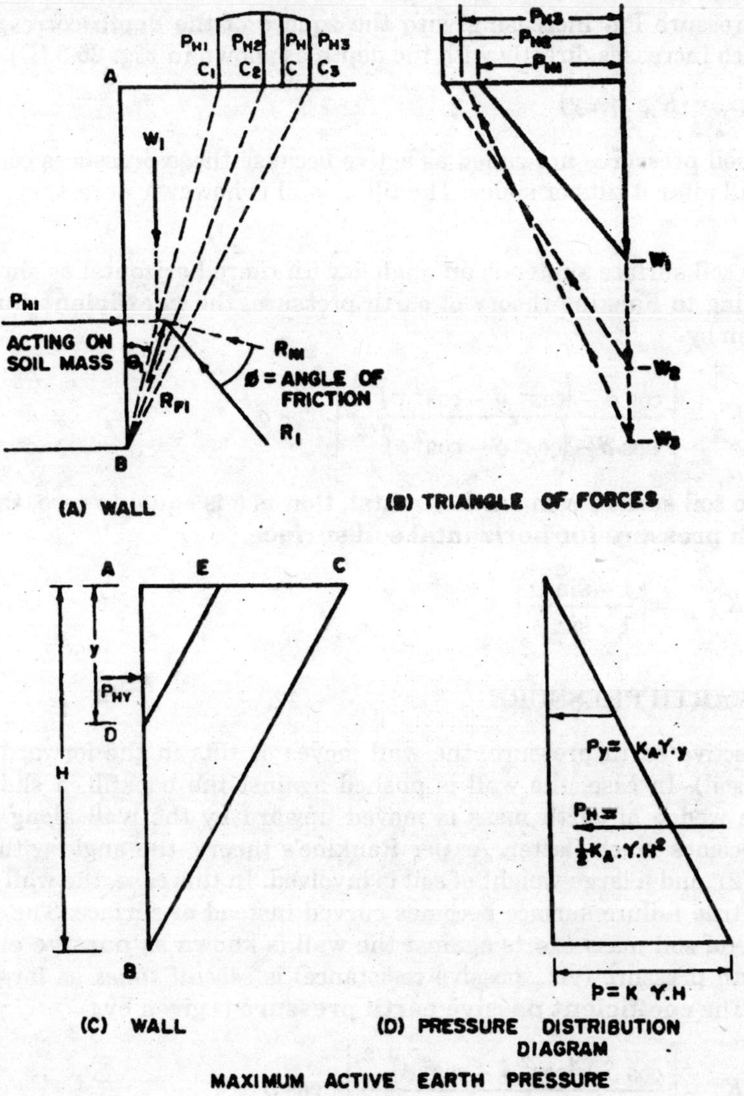

Fig. 25.3

A similar analysis of the soil above point D (a point between A and B, Fig. 25.3 (A) gives a critical sliding plane through D, parallel to BC. The holding force p_{Hy} for the depth AD is proportional to the weight of sand in the wedge ADE. It means, P_{Hy} is proportional to square of y and the unit weight of soil, γ.

$$\therefore \quad P_{H_y} \propto \gamma \cdot y^2 \tag{i}$$

$$\text{or} \quad P_{H_y} = \left(\frac{1}{2} K_A \cdot \gamma \cdot y^2\right) \tag{25.1}$$

where (K_A/2) is the constant of proportionality. The coefficient K_A is called as the *coefficient of active earth pressure.* It depends on the angle of sliding friction. It varies from 0.27 to 0.34.

The total pressure P_{Hy} increasing with the square of the depth corresponds to a unit pressure, P_y which increases directly with the depth as shown in Fig. 25.3 (D).

$$p_y = (K_A \cdot \gamma \cdot y) \tag{25.2}$$

The active soil pressures are called as active because these pressures continue to act on the retaining wall after it tilts or slides. The tilt of wall is however necessary, it may be 0.001 times its height.

In case, the soil surface subtends an angle δ with there horizontal as shown in Fig. 25.1 (A), then according to Rankine theory of earth pressure, the **coefficient for active earth pressure** is given by

$$K_A = \left[\frac{\cos\delta - \left(\cos^2\delta - \cos^2\phi\right)^{1/2}}{\cos\delta + \left(\cos^2\delta - \cos^2\phi\right)^{1/2}}\right] . \cos\delta \tag{25.3}$$

Usually, the soil surface remains horizontal, that is δ is equal to zero, then **coefficient for active earth pressure for horizontal soil surface**

$$K_{A.\,H} = \left(\frac{1-\sin\phi}{1+\sin\phi}\right) \tag{25.4}$$

25.4 PASSIVE EARTH PRESSURE

In case of active earth pressure, the wall moves or tilts in the forward direction (viz., away from the soil). In case, the wall is pushed against the backfill, a sliding plane is so formed that the wedge of earth mass is moved upward by the wall along the plane. The sliding plane becomes much flatter. As per Rankine's theory, the angle with the horizontal becomes (45 – ϕ/2)° and a large weight of soil is involved. In this case, the wall friction is more important. The true failure surface becomes curved instead of surface. The pressure which this large wedge of soil mass exerts against the wall is known as **passive earth pressure**. The passive earth pressure (viz., passive resistance) is *several times* as large as the active earth pressure. The **coefficient passive earth pressure** is given by

$$K_p = \left[\frac{\cos\delta + \left(\cos^2\delta - \cos^2\phi\right)^{1/2}}{\cos\delta - \left(\cos^2\delta - \cos^2\phi\right)^{1/2}}\right] . \cos\delta \tag{25.5}$$

when δ = 0, the earth surface is horizontal, then

$$K_{PH} = \left(\frac{1+\sin\phi}{1-\sin\phi}\right) \tag{25.6}$$

25.5 EFFECT OF SURCHARGE ON EARTH PRESSURE

Usually, the surface of backfill behind the retaining wall remains horizontal. Sometimes, this surface of backfill may slope upward (i.e., forming the angle of surcharge d positive) or downward (i.e., the angle surcharge d negative) from the top of retaining wall. Sometimes, the horizontal surface of backfill may carry uniformly distributed additional load (*for example,* goods in a storage yard or traffic in road). In latter two cases, their effect is to increase the horizontal pressure. The additional load on the surface of ground over a possible sliding plane is called as *surcharge.*

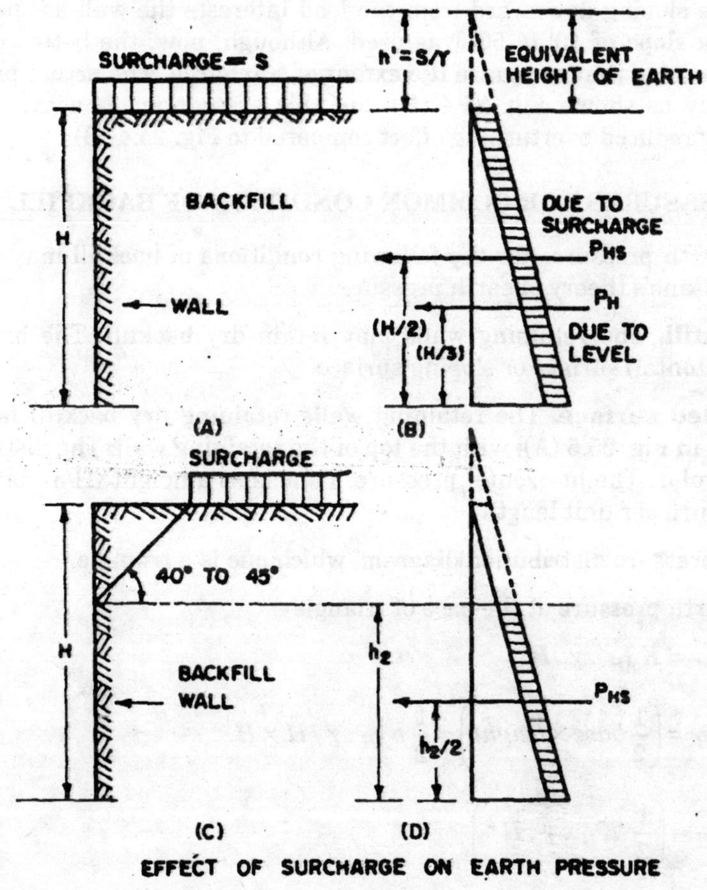

Fig. 25.4

The effect of surcharged is considered by taking an additional height of soil having an equivalent weight. Such a surcharge adds a uniform pressure to the triangular soil pressure. For the uniformly distributed load on the horizontal surface of backfill as shown in Fig. 25.4, the increase in pressure caused by uniform surcharge, s is calculated by converting its load into an equivalent, imaginary height of earth, h' above the top of the wall.

$$h = (s / \gamma) \tag{25.7}$$

where, γ = unit weight of the soil. Its effect is as if the height of backfill upto horizontal surface is replaced by (H + h').

$$\text{where } P_{HS} = K_{AH} \,.\, (\gamma h') \,.\, H \tag{25.8}$$

A load just to the right of C, Fig. 25.3(C) may influence the sliding on a slightly flatter plane and may make the plane critical.

A surcharge far to the right of C, Fig. 25.3(C) does not influence the sliding plane BC or the pressure, P_H.

It is commonly assumed that a surcharge does not influence the pressure above the point where a line sloping downward from the load interests the wall as shown in Fig. 25.4 (C). In the past, a slope of 40' to 50' was used. Although, now, the better methods in soils mechanics are available that recognize the extent of surcharge. The actual pressure does not change as abruptly as shown Fig. 25.4 (D), but this assumption is within the reason and indicates a greatly reduced overturning effect compared to Fig. 25.4 (B).

25.6 EARTH PRESSURES FOR COMMON CONDITION OF BACKFILL

The active earth pressures for the following conditions of backfill may be calculated as follows as per Rankine's theory of earth pressure.

1. Dry backfill. The retaining walls may retain dry backfill. The backfill may have levelled (viz., horizontal) surface or sloping surface.

1. (A) **Levelled surface.** The retaining walls retaining dry backfill having a levelled surface [as shown in Fig. 25.5 (A)] with the top of the retaining wall. The distribution of earth pressure is triangular. The horizontal pressure, P_H acts at a height (H/3) form the base. The active earth pressure per unit length

p_H = area of pressure distribution diagram, which one is a triangle.

The active earth pressure at the base of triangle

$$p_H = K_{AH} \,.\, \gamma \,.\, H \tag{i}$$

$$p_H = \left(\frac{1}{2} \, base \times height\right) = \frac{1}{2} \, k_{AH} \,.\, \gamma \,.\, H \times H$$

$$p_H = \left(\frac{1}{2} . K_H \,.\, \gamma \,.\, H^2\right) \tag{ii}$$

where γ is the unit weight of backfill and H is the height of retaining wall. The coefficient of active earth pressure is given by

$$K_{AH} = \left(\frac{1 - \sin\phi}{1 + \sin\phi}\right) \tag{iii}$$

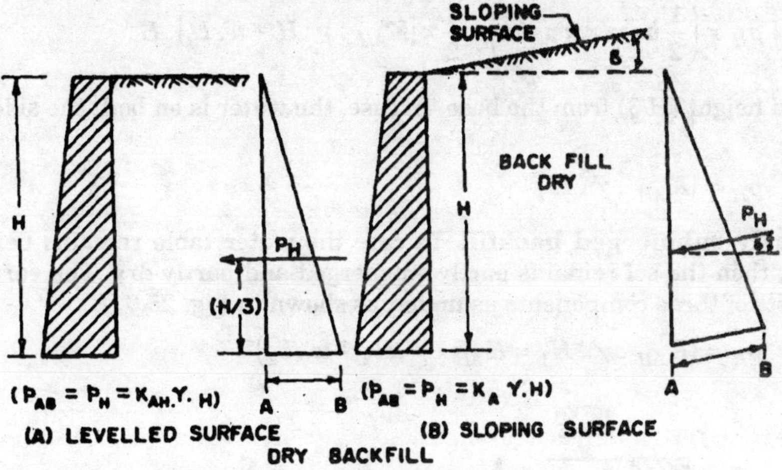

Fig. 25.5

1. (B) **Sloping surface.** In case the surface of the backfill is sloping up by an angle δ, [as shown in Fig. 25.5 (B)] with the horizontal, the distribution of active earth pressure still remains triangular. The earth pressure at the base of triangle.

$$p_H = K_A \,.\, \gamma \,.\, H \tag{i}$$

The earth pressure acts in the direction parallel to the sloping surface. It acts at a height (H/3) form the base. It is calculated as the area of pressure distribution diagram, which one is a triangle. Therefore,

$$P = \left(\frac{1}{2}\, base \times height\right) = \left(\frac{1}{2}\, K_A \,.\, \gamma \,.\, H \,.\, H\right) = \frac{1}{2}\, K_A \,.\, \gamma . H^2 \tag{ii}$$

The coefficient of active earth pressure is determined as under.

$$K_A = \left[\frac{\cos\delta - \left(\cos^2\delta - \cos^2\phi\right)^{1/2}}{\cos\delta + \left(\cos^2\delta - \cos^2\phi\right)^{1/2}}\right] . \cos\delta \tag{iii}$$

where, ϕ is the angle of repose for the soil.

2. **Submerged backfill**. The backfill material retained by the retaining material may be completely submerged or partially submerged.

2. (A) **Completely submerged backfill.** When the backfill material is completely submerged, the water table is at the level of top of the retaining wall. The submerged backfill becomes saturated. The lateral pressure consists of that due to submerged weight of soil and weight of water. Therefore, the lateral pressure at the base of triangle as shown in Fig. 25.6 (A) is given by

$$p_H = \left(K_{AH} \cdot \gamma \,.\, H + w \,.\, H\right) \tag{i}$$

where, γ is the unit weight of submerged soil and w is the unit weight of water. The lateral force acts horizontally. It is given by

$$p_H = \left(\frac{1}{2} base \times height\right) = \frac{1}{2} \times [K_{AH} \,.\, \gamma \,.\, H + w.\, H].\, H \qquad \text{(ii)}$$

It acts at a height (H/3) from the base. In case, the water is on both the sides of retaining wall, then,

$$p_H = (K_{AH} \,.\, \gamma' \,.\, H) \qquad \text{(iii)}$$

2. (B) **Partly submerged backfill.** In case the water table remains below the top of retaining wall, then the sol remains partly submerged and partly dry. The earth pressure at the base consists of three components as under, as shown in Fig. 25.6 (B)

$$p_H = (K_{AH} \,.\, \gamma' \,.\, H_1 = K_{AH} \,.\, \gamma' \,.H_2 + w \,.\, b_2) \qquad \text{(i)}$$

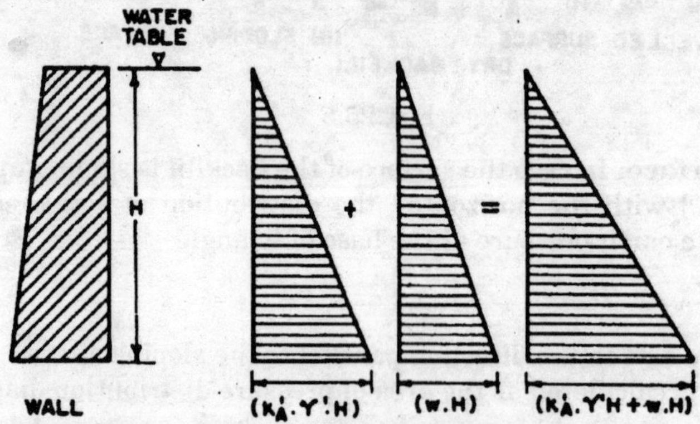

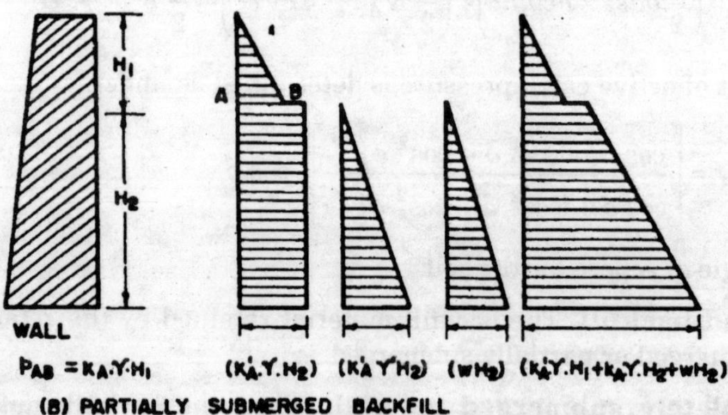

Fig. 25.6

3. Levelled backfill carrying uniformly distributed surcharge

In case, the dry backfill levelled with the top of the retaining wall carries uniformly distributed surcharged s per unit length, the surcharge is replaced by an imaginary equivalent soil of equal weight and height h'. The soil pressure at he base of the retaining wall consists of two components as shown in Fig. 26.7,

$$p_H = (K_A \,.\, \gamma \,.\, H + K_A \,.\, \gamma \,.\, h') \qquad \text{(i)}$$

The horizontal pressure may be determined as below

$$p_H = \left(\frac{1}{2} K_{AH} \,.\, \gamma \,.\, H^2 + K_{AH} \,.\, \gamma \,.\, h' \,.\, H\right) \qquad \text{(ii)}$$

The lines of action of both the components are shown in Fig. 25.7.

$$\text{or } p_H = \frac{1}{2} K_{AH} \,.\gamma.\, H\,(H + 2h')$$

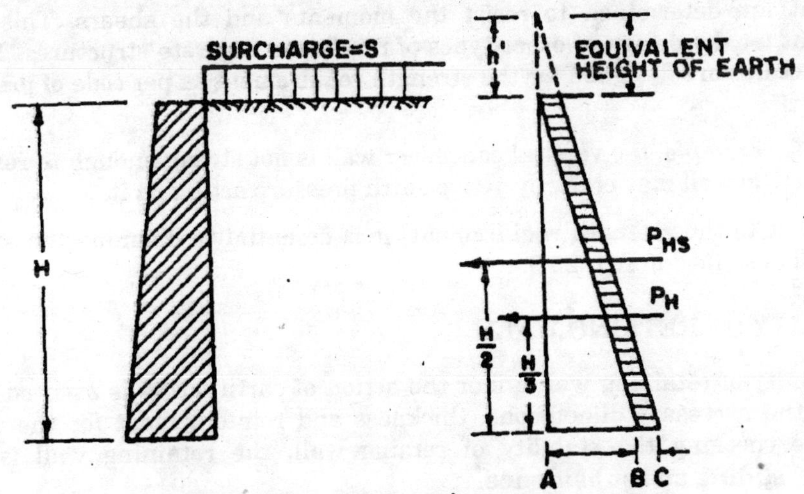

Fig. 25.7

25.7 FORCES ACTING ON RETAINING WALL

The various forces acting on the retaining wall are the vertical loads and lateral forces.

The vertical loads are the gravity loads. The self-weight of the stem W_1, the self weight of the base, W_2 and the weight of the earth W_3 over the heel. The resultant of all the gravity loads, W is found. The weight of the retaining wall is of considerable importance in achieving and maintaining the stability, (i.e., to resist overturning from the lateral earth pressure and to resist horizontal sliding from the same forces. The weight of the soil (viz., backfill) on the top of the heel, in addition to the weight of the retaining wall, contributes to the stability.

The backfill retained by the retaining wall causes lateral forces. The surcharge on the surface of back causes additional lateral forces. The lateral forces are also called as active earth pressures, which have been described in Art. 25.3. The effect of surcharge on earth pressure has been described in Art. 25.3.

In addition to the active earth pressure, sometimes, the passive earth pressure described in Art. 25.4, may act on the retaining wall. The passive earth pressure is again the lateral force and acts in the opposite direction to that of the active earth pressure. The passive earth pressure increases the stability of the retaining wall.

It is to note that the water must be prevented form accumulating in the backfill material. The retaining walls are rarely designed for or assumed to retain saturated material. Therefore, proper drainage must be provided.

25.8 STRENGTH OF RETAING WALL

The individual component parts of the retaining wall are made strong enough to resist the various forces acting on these components. The necessary dimensions, thickness and reinforcement are determined to resist the moments and the shears. This procedure is similar to that for the design of other types of reinforced concrete structures. The reinforced concrete structures are designed for the strength requirement as per code of practice IS : 456-2000.

In case, *for example*, the vertical cantilever wall is not strong enough to resist the acting forces, the vertical wall may crack by active earth pressure acting on it.

In addition to the strength requirement, it is essential to examine the stability of the retaining wall described in Art. 25.9.

25.9 STABILITY OF RETAING WALL

The stability of retaining wall under the action of earth forces is assured even prior to determining the necessary dimensions, thickness and reinforcement for the moments and shears. While checking the stability of retaing wall, the retaining wall is checked for **overturning, sliding** and **subsidence.**

A cantilever retaining wall is shown in Fig. 25.8 (A). Let W_1 be the weight of vertical wall of the retaining wall (viz., stem), W_2 be the weight of base slab and W_3 be the weight of earth acting over the heel slab. These weights W_1, W_2 and W_3 acting at the different locations are also shown in Fig. 25.8 (A). Let the x_1, x_2 and x_3 be the respective distances of the lines of action of the vertical loads W_1, W_2 and W_3, respectively, measured horizontally from the toe of the retaining wall. Let W be the resultant vertical load acting downward.

The lateral force P_H due to active earth pressure acts horizontally (in case, the backfill material retained by the retaining wall is having levelled horizontally surface passing through the top). It acts at a height (H/3) from the base, where H is the total height of the retaining wall as shown in Fig. 25.8. The distribution of active earth pressure is shown in Fig. 25.8 (B).

The retaining wall is checked for the following effects as under :

1. Overturning. The lateral force, P_H due to earth pressure tends to overturn the retaining wall about the toe of the retaining wall. The overturning moment.

$$M_o = \left(P_H \cdot \frac{H}{3}\right) = \left(\frac{1}{2} K_{AH} \cdot \gamma \cdot H^2 \cdot \frac{H}{3}\right) \quad \text{(i)}$$

or $$M_o = \left(\frac{K_{AH} \cdot \gamma \cdot H^2}{6} \right) \qquad (25.9)$$

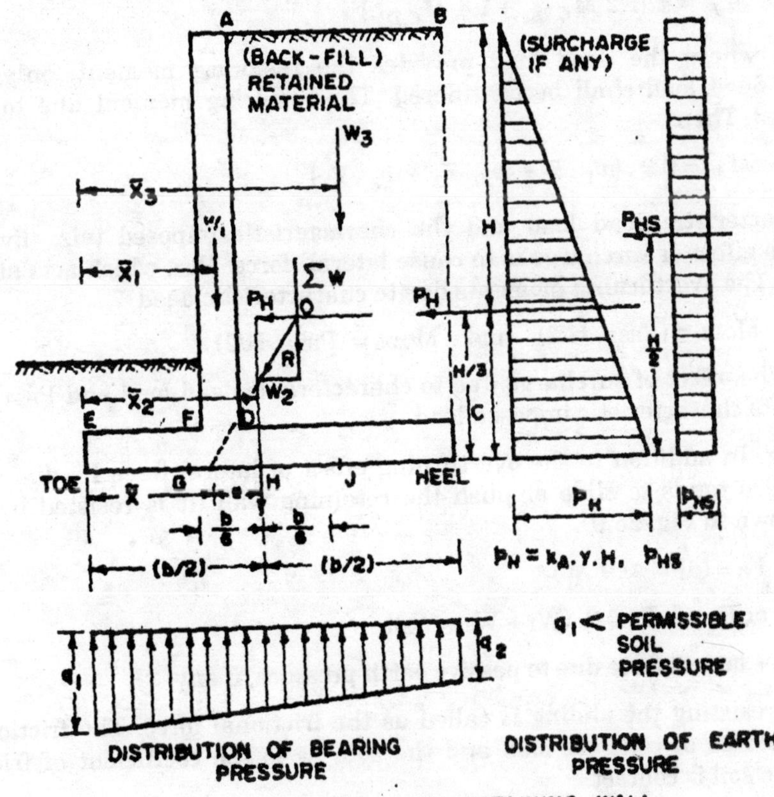

Fig. 25.8

The various weights W_1, W_2 and W_3 tend to restore the position of the retaining wall. These weights resist the overturning. The weight of earth above toe and passive earth pressure (if any) are neglected. The restoring moment about the toe of retaining wall

$$M_R = (w_1 \cdot \bar{x}_1 + w_2 \cdot \bar{x}_2 + w_3 \cdot \bar{x}_3) \qquad (25.10)$$

It is to note that the live load is also not taken while calculating the restoring moment. The live load is not transferred to the heel slab because of arching effect of soil. Factor of safety against overturning may be calculated by dividing the restoring moment M_R by the overturning moment, M_o. Therefore,

$$F.S\,(O.T) = \left(\frac{\textit{Moment resisting the overturning}}{\textit{Moment causing the overturning}} \right) = \left(\frac{M_R}{M_O} \right) \qquad (25.11)$$

IS : 456-2001 recommends that the stability of a structure as a whole against *overturning* shall be ensured so that the restoring maximum overturning moment M_R shall be not less than the sum of 1.2 times the maximum overturning moment due to the characteristic dead load, $M_{C.DL}$ and 1.4 times the maximum overturning moment due to the characteristic imposed loads, $M_{C.DL}$ that is,

$$M_R = \nless (1.2\, M_{C.DL} + 1.4\, M_{C.DL}) \tag{25.12}$$

In cases, where the dead load provides the restoring moment, only 0.9 times the characteristic dead load shall be considered. The restoring moment due to imposed loads shall be ignored. Then

$$M_R = 0.9\,.\,(w_1\,.\,\bar{x}_1 + w_2\,.\,\bar{x}_2 + w_3\,.\,\bar{x}_3) \tag{25.13}$$

The characteristic dead load and the characteristic imposed (viz., live) load act as surcharge. The effect of surcharge is to cause lateral force P_{HS}, which acts at a height (D/2) form the base. The overturning moments due to characteristic dead

$$M_{C.DL} = (P_{HS1}\,.\,H/2) \quad \text{and} \quad M_{C.DL} = (P_{HS2}\,.\,H/2)$$

where P_{H1} is the effect of surcharge due to characteristic dead load and P_{HS2} is the effect of surcharge due to characteristic imposed load.

2. Sliding. In addition to the overturning effect of lateral force P_H due to active earth pressure, P_H also tends to slide or push the retaining wall. It is resisted by the frictional force, F_R as shown in Fig. 25.9

$$F_R = (\mu F\ .\ 0.9\ W)$$

$$\text{or } F_R = [\mu F\,.\,0.9\ (W_1 + W_2 + W_3)] \tag{i}$$

$$+ \text{ lateral force due to passive earth pressure, if any} \tag{ii}$$

The force resisting the sliding is called as the frictional force. The frictional force acts between the surface of concrete slab and the soil, μ_F is the coefficient of friction between concrete and the soil is contact.

$$\text{Sliding force, } F_S = (P_H + P_{HS}) \tag{iii}$$

Factor of safety against sliding may be determined by dividing the force resisting the sliding by the force causing the sliding. Therefore,

$$F.S\ (S.L) = \left(\frac{\textit{Force resisting the sliding}}{\textit{Force causing the sliding}}\right)$$

$$F.S\ (S.L) = (F_R / F_s) \nless 1.4 \tag{25.14}$$

IS : 456-2001 specifies that the factor of safety against sliding shall not be less than 1.4 under the most adverse combination of the applied characteristic forces (loads).

It is note that in the expression (i) above 0.9 W has been taken as code recommends that only 0.9 times the characteristic dead load shall be taken into account.

The values of coefficient of friction, μ_F, angle of repose, ϕ and the unit weight of different types of backfill material (i.e., the density), γ are given in Table 25.1 below :

Sometimes a *shear key* as shown in Fig. 25.9 (vertical wall projecting from the slab into the soil) is built under the vertical wall of the retaining wall to check the sliding. In that case, the passive earth pressure acting on this built-up key should also considered.

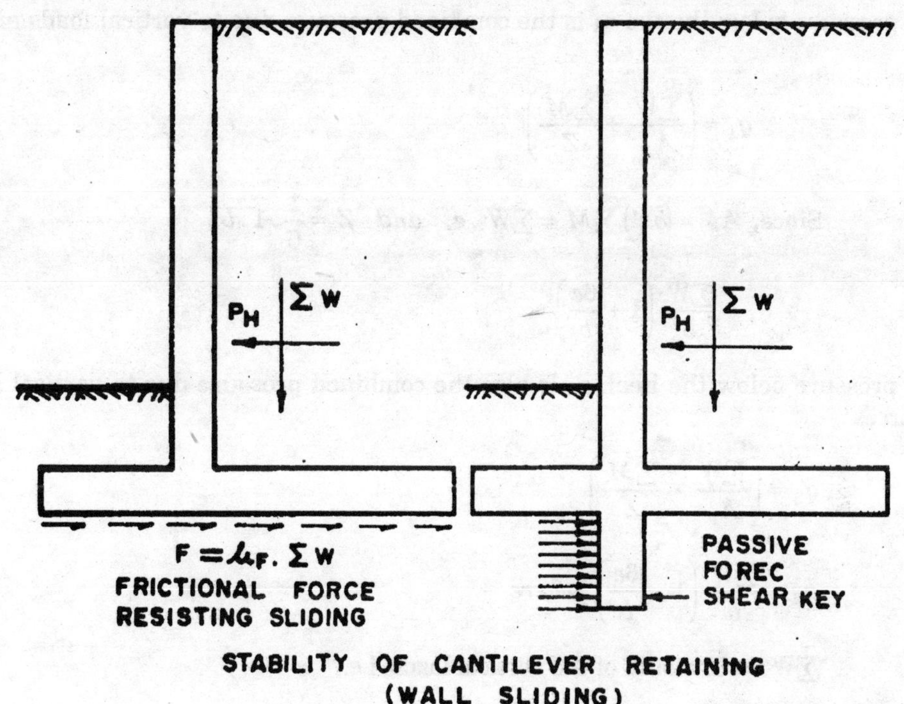

Fig. 25.9

TABLE 25.1. UNIT WEIGHTS, EFFECTIVE ANGLES OF INTERNAL FRICTION (VIZ., ANGLE OF REPOSE AND COEFFICIENTS OF FRICTION WITH CONCRETE)

S. No.		Unit weight kN/m^3	ϕ	μ_F
1	Sand or gravel without fine particles, highly permeable	17.28–18.85	33–40	0.5–0.6
2	Sand or gravel with silt mixture, low permeability	18.85–20.42	25–35	0.4–0.5
3	Silty sand, sand and gravel with high clay content	17.28–18.85	23–30	0.3–0.4
4	Medium or stiff clay	15.71–18.85	25–35	0.2–0.4
5	Soft clay, silt	14.14–17.28	20–25	0.2–0.3

(**Note** : For saturated conditions, ϕ for clays and silts may be close to zero).

3. Subsidence. It is necessary to ensure that (i) the *pressure* below the slab does not exceed the *permissible bearing pressure,* (ii) the *resultant* of the gravity loads and the lateral forces lies within the middle third width of the slab and (iii) there should not be any tension between the slab and the soil. Therefore, following checks are applied.

(i) Check for pressure. The distribution of the pressure due to the combined effect of vertical load and the lateral forces (i.e., due to the resultant force) is shown in Fig. 25.8 (C). The pressure below the toe of the horizontal slab is more and that below the heel is less.

The pressure below the toe q_1 is the combined pressure due to vertical loads and lateral forces.

$$q_1 = \left(\frac{\sum W}{A_F} + \frac{\sum M}{Z}\right) \tag{25.15A}$$

Since, $A_F = (bx'), \sum M = \sum W \cdot e$, *and* $Z = \frac{1}{6} \cdot 1 \cdot b^2$

$$q_1 = \frac{\sum W}{b}\left(1 + \frac{6e}{b}\right) \tag{25.15 B}$$

The pressure below the heel, q_2 is also the combined pressure due to vertical loads and lateral forces

$$q_2 = \left(\frac{\sum W}{A_F} - \frac{\sum M}{Z}\right) \tag{25.16 A}$$

$$q_2 = \frac{\sum W}{b}\left(1 - \frac{6e}{b}\right), \text{ where} \tag{26.16 B}$$

$\sum W$ = The sum of the vertical loads, i.e.,

$$\sum W = (W_1 + W_2 + W_3) \tag{i}$$

$\sum M$ = Algebraic sum of moments of the vertical loads and the lateral forces about the toe i.e.,

$$\sum M = (w_1 \bar{x}_1 + w_2 \bar{x}_2 + w_3 \bar{x}_3) - \left(P_H \cdot \frac{H}{3} + P_{HS} \frac{H}{2}, \text{if any}\right) \tag{ii}$$

A_F = area of the footing (viz., the slab)

$$A_F = (b \times 1) \tag{iii}$$

The retaining wall and therefore, the horizontal slab of 1 m length is considered for the analysis. b is the width of slab.

Z = section modulus of the footing (viz., slab)

$$Z = \left(\frac{1}{6} \cdot 1 \cdot b^2 = \frac{1}{6} \cdot b^2\right) \tag{iv}$$

The bearing pressure q_1 should not be more than the maximum permissible bearing pressure for the soil.

(ii) Check for location of resultant. Let $\bar{x}$ be the distance upto the point at which the resultant of the vertical loads and the lateral forces intersects the bottom surface of the horizontal slab of the retaining wall, measured form the toe. Then,

$$\bar{x} = (\sum M / \sum W) \tag{25.17}$$

Therefore, the eccentricity, e measured form the middle line of horizontal slab shall be

$$e = \left(\frac{b}{2} - \bar{x}\right) \tag{25.18}$$

The point of intersection of the resultant with the bottom surface of the horizontal slab should lie within the *middle-third* width of the slab that is, $(e \leq b/6)$. It shows that the distribution of pressure below the horizontal slab shall be compressive throughout the width of slab.

It is good practice, in general, to get the resultant located within the *middle-third* within. It reduces the magnitude of the maximum bearing pressure and prevents too large-non-uniformity of the pressure.

In case the soil is highly compressible, such as certain clays, there shall be much large settlement of the toe than the heel, which will result in the tilt of the vertical wall. In such a case, the resultant should intersect at or very near the centre of the footing. In case the soil is highly incompressible, for *example* well compacted gravel or rock, the resultant may be allowed to lie outside the *middle-third*.

(iii) Check for tension. In case, the resultant intersects the bottom surface of the horizontal slab outside middle-third width of slab, it indicates that there shall be tension and which cannot develop between the soil and the concrete slab. The concrete slab simply rests on the soil. It implies that the concrete slab will lift slightly at the rear point of the slab (viz., at the heel). The point of contact of the concrete slab with the soil will shift away form the and towards the centre. The footing has to be then, analysed accordingly. It is preferred that there shall be no tension between concrete slab and the soil.

25.10 BEHAVIOUR OF CANTILEVER RETAINING WALL

A cantilever retaining wall is shown in Fig. 25.10 (A). The cantilever retaining wall retains the backfill material. It is constructed on the soil. The *vertical wall* (viz., the stem) and the *horizontal slab* are the two main components of the wall. However, the horizontal slab again is having two different slabs, which are called as the *toe slab* and the *heel slab*. The stem, heel slab and the toe slab are three basic elements of the cantilever retaining wall. The behaviour of these three elements is as under :

1. Vertical wall. The vertical wall (viz., the stem) is subjected to active earth pressure acting laterally. The vertical wall acts a cantilever. It is subjected to the lateral force due to active earth pressure having triangular distribution. It bends/deflects away from the backfill side. The backfill surface of the vertical wall is subjected to the bending tensile stress and it is referred as tension side. The free surface of the vertical wall is subjected to the bending compressive side. The deflected shape of all the three elements of the retaining wall are shown in Fig. 25.10 (B). The vertical wall acting as a cantilever is subjected to maximum bending moment at the junction of vertical wall and the heel slab. Therefore, it is likely that a crack may develop at this junction and the vertical wall may fail in bending.

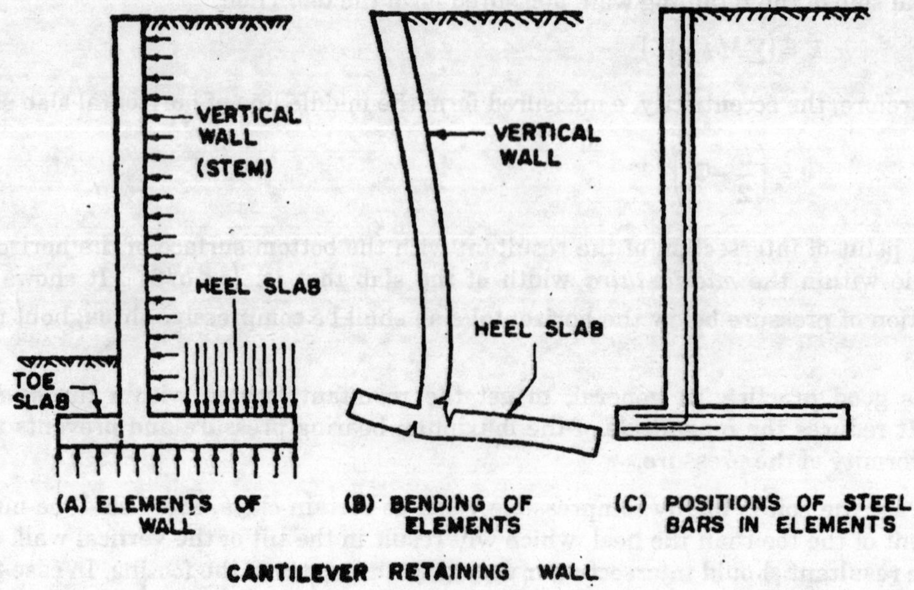

Fig. 25.10

2. Heel slab. The heel slab carries the self-weight, the weight of vertical wall and the weight the backfill material acting downward. It is subjected to soil pressure acting upward from the bottom surface. The heel slab also acts as a cantilever and deflects/bends downward as shown in Fig. 25.10 (B). The top surface of heel slab remains in tension and the bottom surface remains in compression. The heel slab is subjected to the maximum bending moment at its junction with the vertical. It is also possible that a crack may develop at the top surface of heel slab near the junction and it may fail.

3. Toe slab. During construction, the earth is removed and the top surface remains free from any burden. However, the toe slab is subjected to soil pressure acting upward. It deflects/bends upward as shown in Fig. 25.10 (B) acting as a cantilever. The bottom surface of the toe slab remains in tension and top surface remains in compression. A crack may developed in the toe slab just below the front surface of the vertical wall and the toe slab may fail.

It is interesting to note that *all the three elements of the cantilever wall act as cantilever and the retaining wall is called as* **cantilever retaining wall.**

4. Main reinforcement

The steel bars are provided as reinforcement on the tension side of structural member to resist the tension developed due to bending. Therefore, the steel bars are provided towards the backfill surface of the vertical wall as shown in Fig. 25.10 (C). The steel bars are provided as main reinforcement near the top surface of the heel slab and bottom surface of the toe slab as shown in Fig. 25.10 (C).

25.11 PRELIMINARY PROPORTIONS OF CANTILEVER RETAINING WALL

The preliminary proportions of different elements of the cantilever retaining wall *for example*, width and thickness of the base slab and the relative position of the vertical wall with respect to the base slab are needed to begin the design. The stability of retaining wall is then checked. In case, the stability requirements are satisfied, the different elements of the cantilever retaining wall shall be structurally designed.

The preliminary proportions (viz., the dimensions) of above mentioned elements are estimated with an approximate use of statics as belows :

1. Width of the base slab. The width of the base slab and the relative position of the vertical wall with respect to the base slab (i.e., the width of toe slab and the width of the heel slab) are estimated by following an approximate method. The width of base slab may be estimated from the consideration of overturning and so also from the consideration of sliding as below.

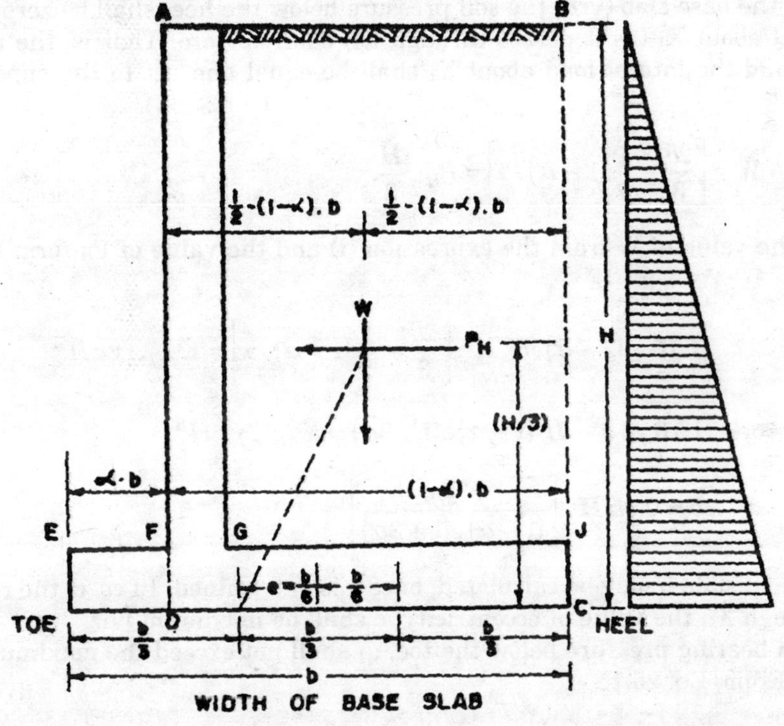

Fig. 25.11

1. (A) Overturning. A cantilever retaining wall retaining the backfill material having the levelled surface is shown in Fig. 25.11 (A) and the distribution of active earth pressure is shown in Fig. 25.11 (B). The resultant vertical load. W and the lateral force, P_H due to active earth pressure acting on the retaining wall are shown. The weight of toe slab (the slab upto the outer surface of the vertical wall) is neglected. Let γ be the unit weight of the backfill material. The average unit weight of concrete and the backfill material both lying within ABCD is assumed 1.10 γ. One metre length of base slab shall be considered. Then

W = (CD . BC . 1 × Average unit weight)

or W = [(1 – α) b . H . 1 . (1.10 γ)] (i)

where, α is the ratio of width of the heel slab to the complete width of the base slab, (i.e., α = FJ/EJ). The line of action of load W is at a distance 0.5 (1 – α) b from (i.e., midway between the width CD. The lateral force due to active earth pressure, P_H acts at height (H/3) form the base as shown in Fig. 25.11

$$P_H = \left(\frac{1}{2} K_{AH} \cdot \gamma . H^2\right) \quad \text{(ii)}$$

where K_{AH} is the **coefficient of active earth pressure** for the backfill material having levelled surface AB.

In case the resultant of the vertical load W and the lateral force, P_H passes through middle-third points of the base width X_1 and X_2 as shown in Fig. 25.11 (A), there shall be no tension below the base slab (viz., the soil pressure below the heel shall be zero). The moment of resultant, R about X_1 (as it passes through X_1) shall be zero. That is, the moment of the vertical load and the lateral load about X_1 shall be equal and act in the opposite direction. Therefore,

$$W = \left[\frac{2b}{3} - \frac{1}{2}(1-\alpha).b\right] = P_H . \frac{H}{3} \quad \text{(iii)}$$

substituting the value of W from the expression (i) and the value of P_H form the expression (ii)

$$1.10\,\gamma\,(1-\alpha).b.H.\left[\frac{2b}{3} - \frac{1}{2}(1-\alpha).b\right] = K_{AH} \cdot \gamma . H^3$$

$$\text{or} \quad 1.10\,\gamma . b^2 . H\,(1-\alpha).(1+3\alpha) = K_{AH} \cdot \gamma . H^3$$

$$\therefore \quad b = 0.95\,H . \left[\frac{K_{AH}}{(1-\alpha).(1+3\alpha)}\right] \quad (25.19)$$

The width of slab, b may be calculated, once a is determined. In case, the resultant force, R passes through X_1, the value of eccentricity, e shall be maximum [viz., (b/2) – (b/3) = b.6]. The maximum bearing pressure below the toe, q_1 shall not exceed the maximum permissible soil pressure. From, Eq. 25.15

$$q_1 = \left(\frac{\Sigma W}{A_f} + \frac{\Sigma M}{Z}\right) \quad \text{(iv)}$$

Here, $\Sigma W, = W, \Sigma M = W . e,$ *and*

$$A_F = b \times 1, Z = \left(\frac{1 \times 1}{6}\right) \times b^2$$

And e = b/6. Therefore,

$$q_1 = \left(\frac{W}{b \times 1} + \frac{6 . W . b}{b^2 \times 6}\right) = \frac{W}{b}(1+1)$$

$$\text{or} \quad q_1 = \left[\frac{2}{b} \times 1.10\, \gamma\, (1-\alpha) b . H\right] \tag{v}$$

Let q_1 be equal to q_o, maximum permissible soil pressure The $(1 - \alpha) = q_o/(2.2 \, . \, \gamma \, . \, H)$

$$\alpha = \left[1 - \left(\frac{q_o}{2.2 . \gamma . H}\right)\right] \tag{25.20}$$

The values of width of base slab, b and the ratio α may be found from Eqs. 25.19 and 25.20. These values provide approximate guide lines. When the backfill material is having an angle of slope, δ, with the horizontal, the values of α and b may be obtained form the following

$$\alpha = \left[1 - \left(\frac{q_o}{2.7 \gamma . H}\right)\right] \tag{25.21}$$

$$\text{And,} \quad b = H . \left[\frac{K_A . \cos \delta}{(1-\alpha).(1+3\,\alpha)}\right] \tag{25.22}$$

The approximate value of width of base slab, b may be adopted from Table 25.2 for the different conditions of backfill and the different types of retaining wall.

Table 25.2. APPROXIMATE VALUES OF b = WIDTH OF BASE SLAB

S.No.	Type of cantilever retaining wall	To surface of backfill	
		Levelled	sloping
1	T-shaped, Fig. 25.1 (B)	b ≈ 0.46 H	b ≈ 0.65 H
2	L-Shaped, Fig. 25.1 (C)	b≈ 0.55 H	b ≈ 0.74 H
3	Reversed L-shaped, Fig. 25.1 (D)	b ≈ 0.65 H	b ≈ 0.8 H

1. **(B) Sliding.** The width of base slab may also be estimated from the consideration of sliding. The factor of safety against sliding for a retaining wall should not be less than 1.4 under the most adverse combination of the applied characteristic forces as per IS : 456-2000. In this case only 0.9 times the characteristic dead load shall be taken into consideration. From Eq. 25.14

$$F.S\,(S.L) = \left(\frac{F_R}{F_s}\right) \nless 1.4 \tag{i}$$

$$\text{or} \quad \left(\frac{\mu_F . 0.9\, W}{P_H}\right) = 1.4 \tag{ii}$$

As $W = [1.10 \,.\, \gamma \,.\, b \,.\, H.\, (1-\alpha)]$ *and* $P_H = \left(\frac{1}{2} K_{AH} \,.\, \gamma \,.\, H^2\right)$

The expression (ii) may be written as under

$$[\mu_F \,.\, 0.9\times 1.10\, \gamma \,.\, b \,.\, H\,(1-\alpha)] = \left[\frac{1}{2} K_{AH} \,.\, \gamma \,.\, H^2 \times 1.4\right]$$

$$b = \left(\frac{0.707 \,.\, H}{(1-\alpha)}\right) . \left(\frac{K_{AH}}{\mu_F}\right) \tag{25.23}$$

The value of α is calculated form Eq. 25.21 and the value of width of the base slab is found from Eq. 25.23. Knowing the values of α and b, the individual widths of toe slab and heel slab shall be α . b and (1 – α) . b, respectively.

The thickness of base slab is roughly 7 to 10 per cent of the over all height with a minimum of 300 mm. The actual thickness needed shall be calculated later on from the consideration of bending and the shear force.

2. **Vertical wall (stem).** The position of vertical wall is located as soon as the value of a is determined. The height of vertical wall and its thickness are decided as under :

2. (A) **Height of vertical wall.** The height of the vertical wall, H shall be equal to the different between the elevations of the ground level on both the sides of the wall *plus* the depth below the ground level. That is

$$H = [(R.\, L_1 - R \,.\, L_2) + D], \textit{where}$$

R . L_1 = level of top surface of the backfill material to be remained.

R. L_2 = level of the ground towards outer surface of the vertical wall.

D = depth of the retaining wall below the ground level (R . L_2).

Minimum depth of foundation below the ground level may be calculated as per Rankine's formula

$$D_{\min} = \frac{q_o}{\gamma} . \left(\frac{1-\sin\phi}{1+\sin\phi}\right) \tag{25.24}$$

where, ϕ is the angle of repose, γ is the unit weight of soil and q_o is the safe bearing capacity of the soil. The minimum depth of foundation below the ground level should be about one metre.

2. (B) Thickness of vertical wall. The thickness of the vertical wall may be kept about 150 mm to 300 mm depending upon the overall height of the wall. The thickness at the base of vertical wall may be kept about 8 to 15 percent of the height of wall.

The stability of the retaining wall is checked, once the preliminary proportions (dimensions) of all the three components (viz., vertical wall, heel slab and toe slab) are decided. These estimated dimensions may be revised, if necessary. When the requirements of the stability are satisfied, the design of retaining wall may be done.

25.12 DESIGN OF CANTILEVER RETAINING WALL

The cantilever retaining wall is designed after estimating the preliminary dimensions of three elements of the wall as described in Art. 25.12 and checking the stability as described in Art. 25.9. The cantilever retaining wall is designed when the requirements of the stability are satisfied in the following sequence:

1. Design of heel slab 2. Design of toe slab 3. Design of vertical wall (stem)

1. **Design of heel slab:** The heel slab is subjected to the weight of earth fill, and its own weight acting downward, and the reaction of soil acting upward. The downward load predominates over the upward reaction. The distribution of upward reaction from the soil on the heel slab is triangular. In the design, the upward reaction from the soil is neglected. Actually, the distribution of reaction is parabolic and if it becomes more parabolic, the upward reaction disappears. It is not necessary that the upward triangular distribution may exist. Further, the effect of upward pressure is to reduce the bending moment and shear but to a small extent only. The heel slab acts as a cantilever. It projects from the back face of the vetial wall (stem). The supporting upward reaction comes from the tension in the stem heel.

The heel slab is often made just long enough to cause the resultant of the forces on the wall to strike the ground under the stem.

The heel length is chosen such that adequate friction is provided. A shear key is built below the base to prevent failure from the lesser frictional resistance of concrete on soil.

The increased length of heel slab increases the vertical load (weight of earth fill as well as self-weight), which will increase the frictional resistance also. However, the cost of retaining wall and so also the cost of excavation will also increase.

2. **Design of toe slab.** The toe slab is subjected to self-weight acting downward. The weight of earthfill over the toe is neglected. It is possible that the earthfill over the toe slab is removed or it get eroded.

The toe slab acts as a cantilever projecting outward form the face of vertical wall (stem). The maximum bending moment in the toe slab occurs where the shear force changes sign. The shear force changes sign inside the face of the stem because of large compression from the stem. However, the toe slab is considered as supported from the face. It bends upwards due to upward soil pressure. The bottom surface of toe slab remains in tension due to bending. Therefore, the tension reinforcement is provided near the bottom.

The shear stress controls the overall depth (viz., thickness of toe slab). However, the thickness of heel slab and toe slab are kept equal as these are two elements of one base below the vertical wall.

The diagonal tension remains critical at a distance equal to the effective depth of toe slab from the face of vertical wall.

The heel slab and toe slab are inter-related. However, their lengths are established is sequence.

3. **Design of vertical wall.** The vertical wall (stem) is designed after the design of heel slab and design of toe slab. As a matter of fact, the dimensions of the vertical wall (viz., the height, the thickness at the top and the thickness at the bottom) are revised.

The overall thickness of base slab influence the overall height of the vertical wall (stem). Therefore, the vertical wall is designed, after completing the design of heel slab and the design of toe slab.

The vertical wall (stem) bends due to the horizontal component of earth pressure. The maximum bending moment occurs at the junction with the top surface of heel/toe slab. The thickness required for the maximum bending moment and the reinforcement in tension are calculated for the maximum bending moment.

The bending moment in the vertical wall decreases rapidly as the distance from the bottom increases. Therefore, only part of main reinforcement is needed at higher elevations. the alternate bars may be discontinued where no longer required. The bending moments are calculated at two or more intermediate levels from the top. The bending moment diagram for the vertical wall (stem) may be drawn. The points of curtailment of main reinforcement may be found as usual. The available length of bars may be checked for the development length.

4. Distribution reinforcement. In addition to the main reinforcement in the vertical wall, the *distribution reinforcement* is provided to control the shrinkage and temperature cracking on the inner face. However, the distribution reinforcement is also provided in the vertical and horizontal directions both on the outer face.

5. Drainage. The retaining walls are not designed for water pressure. Complete drainage is provided for the backfill. A porous backfill, such as gravel is provided directly behind every wall to allow the water to reach the drains.

Example 25.1. A backfill material consisting of sand and gravel mixture with a rather moderate quantity of fine silty particles of 3.5 m height to be retained by a gravity retaining wall. A live load surcharge of 19.20 kN/m² shall be acting over the horizontal surface of the backfill. Design the gravity retaining wall.

Design : Gravity retaining wall is design as under :

Step 1 : Material properties of backfill material. The material properties of the given backfill material are as follows. Unit weight the backfil

$$\gamma = 18.85 \text{ kN/m}^3 \quad \text{(i)}$$

Angle of repose (effective angle of internal friction), $\phi = 30^\circ$ (ii)

Coefficient of friction between concrete surface and the soil is $\mu_F = 0.5$ (iii)

Coefficient of active earth pressure for the given backfill, form Eq. 25.4

$$K_{AH} = \left(\frac{1 - \sin\phi}{1 + \sin\phi}\right) = 0.333 \quad \text{(iv)}$$

Safe bearing capacity of soil (assumed), $q_o = 250$ kN/m² (v)

Unit weight or reinforced cement concrete, is 25 kN/m³

$$K_{PH} = \left(\frac{1 + \sin\phi}{1 - \sin\phi}\right) = 3.0 \quad \text{(vi)}$$

Horizontal surface of the backfill is subjected to a live load surcharge.

$= 19.20\ kN/m^2$ (vii)

Equivalent height of surcharge (h' = s/γ)

$$h' = \left(\frac{19.20}{18.85}\right) = 1.02\ m, \qquad \text{(viii)}$$

Step 2. Preliminary proportions (dimensions) or gravity retaining wall. The preliminary dimensions of gravity retaining wall are assumed as shown in Fig. 25.12. These dimensions may be revised if necessary.

Step 3. Gravity loads. The various components of weights W_1, W_3, ... etc. are shown in Fig. 25.12 (A) and their respective distance from the toe are $\bar{x}_1, \bar{x}_2$, ..., etc. These weights and restoring moments due to them have been calculated as under :

Weight of Components	W (kN)	$\bar{x}$ (m)	$M_R = W.\bar{x}$ (kN − m)
Base Height			
$W_1 = (3.20 \times 0.62 \times 1 \times 24)$	47.616	1.600	76.185
$W_2 = (0.46 \times 3.98 \times 1 \times 12)$	43.939	0.450	19.773
$W_3 = [(2.3/2) \times 3.98 \times 1 \times 24]$	109.848	1.447	158.950
$W_4 = [(2.3/2) \times 3.98 \times 1 \times 18.85]$	86.276	1.533	132.261
$W_5 = [0.22 \times 3.98 \times 1 \times 18.85]$	16.505	3.090	51.000
Total $\sum W' = 304.184\ kN$	M = 438.169 kN-m		

Step 4. Horizontal forces. Horizontal force due to active earth pressure is $\left[P_H = \frac{1}{2}(K_{AH}\ \gamma . H). H\right]$

$$\text{or } P_H = \frac{1}{2}[0.333 \times 18.85 \times 4.6 \times 4.6] = 66.411\ kN \qquad \text{(xi)}$$

This horizontal force, P_H acts at a height (H/3) = 1.533 m from the base. Horizontal force due to surcharge

$$P_{HS} = (K_{AH} . \gamma . h'). H = (0.333 \times 18.85 \times 1.02 \times 4.6) = 29.452\ kN$$

This acts at a height (H/2) = 2.3 m from the base.

Step 5. Check for stability

5. (A) Check for overturning. Resisting moment, M_R = 438.169 kN-m

Overturning moment

$$M_O = [P_H . (H/3) + P_{HS} . (H/2)]$$

$$M_O = [66.411 \times 1.533 + 29.452 \times 2.3] = 169.547\ kN - m$$

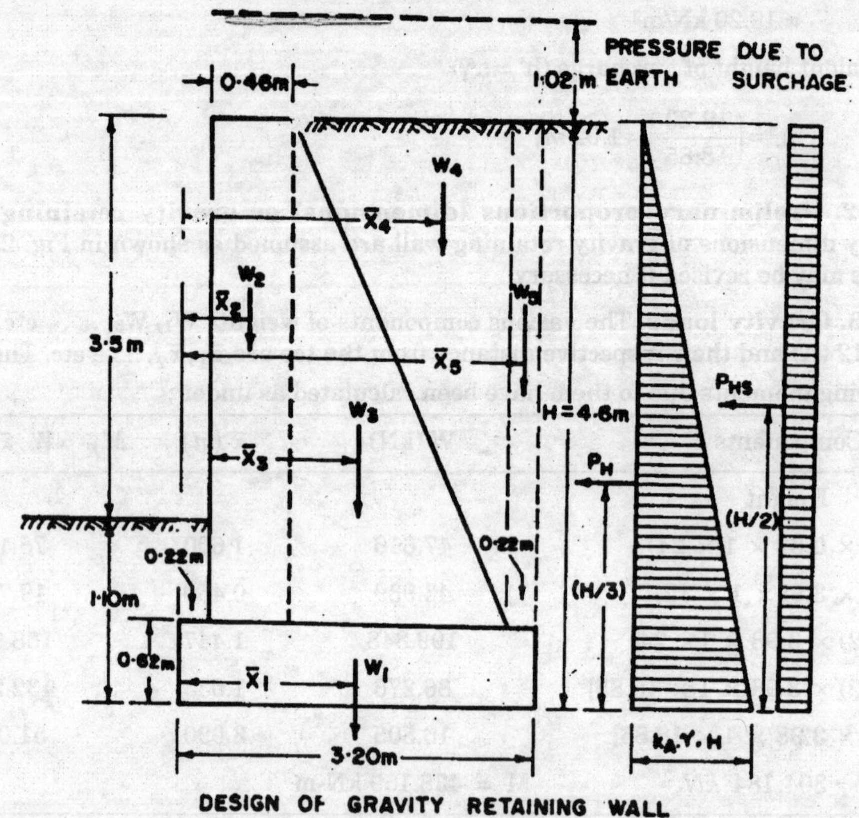

Fig. 25.12

The distance of resultant from toe, form Eq. 25.17

$$\therefore \quad \bar{x} = (\sum M / \sum W) = \left(\frac{438.169 - 169.547}{304.184} \right) = 0.8831\, m$$

It is just out side the middle third width of the base b is 3.20 m. Factor of safety against overturning

$$F\,S_{(O.T)} = \left(\frac{M_R}{M_O} \right) = \left(\frac{438.169}{169.547} \right) = 2.584.$$

As per IS : 456-2001, restoring moment, M_R = 438.169 kN-m

Overturning moment due to dead load

$$M_{O.DL} = \left(P_H \times \frac{H}{3} \right) = \left(\frac{66.411 \times 4.6}{3} \right) = 101.808\, kN - m$$

Since live load is not considered, the over turning moment due to live load

$M_{O.IL}$ = Zero

$$M_R \nless (1.2\,M_{O.DL} + 1.4\,M_{O.IL})$$

$$438.196\ kN-m \nless [(1.2\times 101.808) = 122.17\ kN-m]$$

Hence, it is safe against overturning.

5. (B) Check for sliding. The sliding force is as below :

$$F_S = (P_H + P_{HS}) = (66.411 + 29.452) = 95.863\ kN$$

Resisting force, $F_S = \mu_F \times 0.9\,(W_1 + W_2 + W_3 + W_4 + W_5)\,kN$

or $F_R = 0.5\times 0.9\times (304.184) = 136.883\ kN$

Factor of safety against sliding

$$F.S_{SL} = \left(\frac{F_R}{F_S}\right) = \left(\frac{136.883}{95.863}\right) = 1.428 \nless 1.4.$$

Hence, it is safe.

Step 6. Check for bearing stresses. The resultant is just out side the *middle-third* width of the base. Therefore,

$$e = \left(\frac{b}{2} - \bar{x}\right) = \left(\frac{3.2}{2} - 0.8831\right) = 0.7169\ m$$

The ultimate pressure below the toe of retaining wall, form Eq. 25.15 (B)

$$q_1 = \frac{\Sigma W}{b}\left(1 + \frac{6.e}{b}\right) = \frac{304.184}{3.2}\left(1 + \frac{6\times 0.7169}{3.2}\right) = 222.83\ kN/m^2$$

$q_1 < q_0$ = safe bearing capacity of soil.

The pressure below the heal, from Eq. 25.16 (B)

$$q_2 = \frac{304.184}{3.2}\left(1 - \frac{6\times 0.7169}{3.2}\right) = 32.72\ kN-m^2\ (Tension)$$

The point of zero pressure is slightly inside (viz., 0.41 m). The calculation for the pressures may be done accordingly. However, the gravity retaining wall shown in Fig. 25.12 is safe. It is not necessary to raise the dimcnsions.

Example 25.2. Design a cantilever retaining wall to retaining backfill material consisting of sand and gravel mixture with a rather moderate quantity of fine silty particles of 3.5 m. A live load surcharge of 19.20 kN/m² shall be acting over the horizontal surface of the backfill. M 20 grade of concrete and Hysd steel bars, Fe 415 shall be used. The maximum bearing capacity of soil is 140 kN/m².

Design. The cantilever retaining wall is designed as under :

Step 1. Material properties of backfill. The material properties of backfill material are as follows :

Let unit weight of the backfill be, $\gamma = 18.82$ kN/m² (i)

Angle of repose (effective angle of internal friction), $\phi = 30°$ (ii)

Coefficient of friction between concrete surface and the soil, $\mu_F = 0.50$ (iii)

Coefficient of active earth pressure

$$\therefore K_{AH} = \left(\frac{1-\sin\phi}{1+\sin\phi}\right) = 0.333 \quad \text{(iv)}$$

Safe bearing capacity of soil (assumed), $q_o = 140$ kN/m² (v)

Unit weight of reinforced cement concrete is 25 kN/m².

Coefficient of passive earth pressure

$$K_{PH} = \left(\frac{1+\sin\phi}{1-\sin\phi}\right) = 3.0 \quad \text{(vi)}$$

Horizontal surface of the backfill is subjected to a live load surcharge

= 19.20 kN/m² (vii)

Equivalent height of surcharge

$$h' = \left(\frac{s}{\gamma}\right) = \left(\frac{19.20}{18.85}\right) = 1.02\ m \quad \text{(viii)}$$

Height of retaining wall shall be (3.5 + 1.10) m depth of foundation. (H = 4.6 m)

Step 2. Preliminary proportions (dimensions) of cantilever retaining wall. The preliminary proportions (dimensions) of heel slab, toe slab, thickness at the top and that at the bottom of vertical wall (stem) are estimated from the guide line rules derived from approximate statics : Therefore,

(i) **Ratio of width of toe slab to the width of complete slab.**

From Eq. 25.20,

$$\alpha = \left[1 - \left(\frac{q_o}{2.2\,\gamma\,.\,H}\right)\right] = \left[1 - \frac{140}{2.2 \times 18.85 \times 4.6}\right] = 0.266$$

(ii) **Complete width of the base slab**

(A) From (consideration of overturning moment). Eq. 25.19

$$b = 0.95\,H\left[\frac{K_{AH}}{(1-\alpha)(1+3\alpha)}\right]^{1/2}$$

$$b = 0.95 \times 4.6\left[\frac{0.333}{(1-0.266)(1+3\times 0.266)}\right]^{1/2} = 2.195\ m$$

(B) From (consideration of sliding) Eq. 25.23,

$$b = \left(\frac{0.707\,H}{(1-\alpha)}\right)\cdot\left(\frac{K_{AH}}{\mu_F}\right) = \left(\frac{0.707\times 4.6}{(1-0.266)}\right)\times\left(\frac{0.333}{0.5}\right) = 2.951\,m$$

Average width, $b = \frac{1}{2}(2.195 + 2.951) = 2.573\,m$

Let the complete width of base slab be 2.76 m

(iii) **Width of toe slab,** $\alpha b = (0.266 \times 2.76) = 0.774$ m

Provide 0.70 m width for the toe slab

(iv) Width of heel slab $(1-\alpha).b = (1-0.266)\times 2.76 = 2.026\,m$

Provide (2.76 – 0.70) = 2.06 m, Let the thickness of heel slab be 0.4 m.

[Thickness of heel slab is kept equal to the thickness of vertical wall at the base, (8 to 15 percent of height of vertical wall)].

(v) **Vertical wall (*stem*)**

(A) **Thickness.** The thickness at the top of the vertical wall is taken as 150 mm to 300 mm. Let the thickness at the top of vertical wall be 0.240 m.

Thickness at the bottom of vertical wall

$$= \left[\left(\frac{8}{100}\times 4.6\right) to \left(\frac{15}{100}\times 4.6\right)\right] = (0.368\,m\ to\ 0.690\,m)$$

Let the thickness at the base of vertical wall be 0.40 m. If required, it may be revised.

(B) **Height of wall.** The height of earth retained plus (minimum) depth of foundation = (3.50 + 1.5) = 4.60. The preliminary proportions (dimensions) of the different elements of the retaing wall are shown in Fig. 25.13.

Step 3. Gravity loads. The various gravity loads of different components of the cantilever retaining wall, W_1, W_2.. . . . etc. and their respective distances $\bar{x}_1$, $\bar{x}_2$,......, etc. from the toe are shown in Fig. 25.13. These have been calculated and listed as below :

Consider one metre length of the retaining wall.

Weight of Components	W (kN)	$\bar{x}$ (m)	$M_R = W.\bar{x}$ (kN – m)
Base Height			
$W_1 = 0.24 \times 4.2 \times 1 \times 24$	25.20	0.820	20.664
$W_2 = \frac{1}{2} \times 0.16 \times 4.2 \times 1 \times 25$	8.40	0.873	7.333
$W_3 = 2.76 \times 0.40 \times 1 \times 25$	27.60	1.380	38.088
$W_4 = \frac{1}{2} \times 0.16 \times 4.2 \times 1 \times 18.85$	6.334	1.047	6.632

$W_5 = 1.66 \times 4.2 \times 1 \times 18.85$	131.42	1.930	253.406
Total gravity load $\sum W = 98.954\ kN$			326.123

Total vertical load is equal to = 198.954 kN

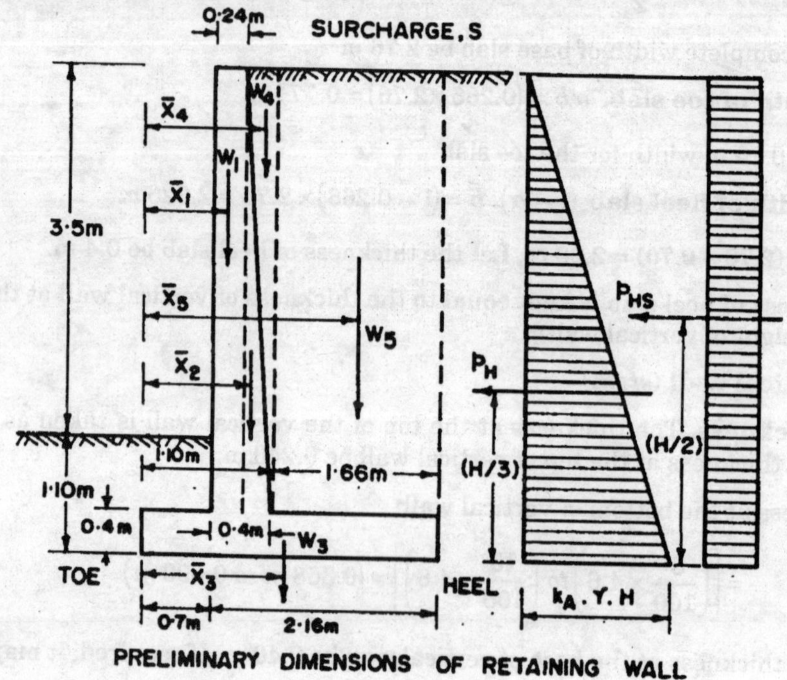

Fig. 25.13

Step 4. Lateral forces. The lateral forces are due to active earth pressure and surcharge :

Lateral forces due to active earth pressure

$$P_H = \left(\frac{1}{2} K_{AH} \cdot \gamma \cdot H . H\right) = \left(\frac{1}{2} \times 0.333 \times 18.85 \times 4.6 \times 4.6\right) = 66.41\ kN$$

This acts at a height, (H/3), Equivalent height of surcharge

$$h' = \left(\frac{Live\ load}{unit\ wt.\ of\ backfill}\right) = \left(\frac{19.20}{18.85}\right) = 1.02\ m$$

Horizontal force due to surcharge. This acts at height, (H/2) $(0.333 \times 18.85 \times 1.02 \times 4.6) = 29.452\ kN$

Step 5. Check for stability. The cantilever retaining wall is checked for stability, viz., against overturning and sliding as under :

5. (A). Overturning. Resisting moment, M_R = 326.123 kN-m

Overturning moment, $M_O = (66.411 \times 1.533 + 29.452) = 169.547\ kN - m$

The distance of resultant from toe

$$\bar{x} = \Sigma M / \Sigma W) = \left(\frac{326.123 - 169.547}{198.954}\right) = 0.787 \text{ m}$$

$$\left(\frac{b}{3}\right) = \left(\frac{2.76}{3}\right) = 0.92\, m$$

Since $\bar{x}$ is more than (b/3), the resultant intersects within the *middle-third* width of the base. There shall be no tension between bottom surface of the base slab and the soil. Overturning moment due to dead load

$$M_{O.DL} = P_H \,.\, (H/3) = \left(\frac{66.411 \times 4.6}{3}\right) = 101.808\ kN - m$$

The live load is not considered. As such the overturning moment due to live load $M_{O.IL}$ = zero

$$M_R \nless (1.2\, M_{O.DL} + 1.4\, M_{O.DL})$$

$$326.123 \text{ kN-m} \nless (1.2 \times 101.808 = 122.17\ kN - m)$$

Hence, it is safe against overturning.

5. (B) Sliding. The sliding force is as follows :

$$F_S = (P_H + P_{HS}) = (66.411 + 29.452) = 95.863\ kN$$

Resisting force, $F_R = (\mu \times 0.9\, \Sigma W') = (0.5 \times 0.9 \times 198.954) = 89.529\ kN$

Factor of safety against sliding

$$F.S_{SL} = \left(\frac{F_R}{F_S}\right) = \left(\frac{89.529}{95.863}\right) = 0.934 < 1.4$$

As per IS : 456-2001, the factor of safety against sliding should not be less than 1.4. A shear key shall be provided below the base slab i.e., below the vertical wall), such that the factor of safety against sliding shall be satisfied. Therefore,

F_R = (89.529 + Passive earth pressure)

or $$F_R = \left(89.529 + \frac{1}{2} k_{PH} \,.\, \gamma \,.\, h^2\right) = \left(89.529 + \frac{1}{2} \times 3 \times 18.85 \times h^2\right)$$

Now $$\left(\frac{F_R}{F_S}\right) = \left(\frac{89.529 + 28.275h^2}{95.863}\right) = 1.4$$

$$\therefore h = \left(\frac{44.68}{28.275}\right) = 1.2574 m.$$

The height of shear key shall be kept 1.26 m. In case, the erosion of soil over the toe slab does not occur, than, it is necessary to provide shear key of small depth. The depth of foundation 1.10 m shall provide passive earth pressure. Then, the depth of shear key (1.26 – 1.10) = 0.16 m = 160 mm.

Step 6. Check for subsidence. The pressure below the toe and that below the heel are checked for compression and tension, respectively, as under :

6 (A) **Check for pressure under toe.** The distance of resultant from the toe $\bar{x} = 0.787 m$

Eccentricity, e from the central line of the slab base

$$e = \left(\frac{b}{2} - \bar{x}\right) = \left(\frac{2.76}{2} - 0.787\right) = 0.593\ m$$

The eccentricity (e = 0.294 m) is less than [(b/6) = 0.43 m]. Actual ultimate pressure below the toe, from Eq. 25.15 (B)

$$q_1 = \frac{\Sigma W}{b}\left[1 + \left(\frac{6 \times e}{b}\right)\right]$$

$$q_1 = \frac{198.954}{2.76}\left[1 + \left(\frac{6 \times 0.593}{2.76}\right)\right] = 165.011\ kN/m^2$$

It is less than the maximum ultimate bearing pressure (assumed as *twice* of safe bearing capacity) of soil 2 x 140 kN/m². Hence, it is safe.

6. (B) **Check for pressure below heel.** Actual bearing pressure below the heel from Fig. 25.16 (B)

$$q_2 = \frac{\Sigma W}{b}\left(1 - \frac{6.\,e}{b}\right) = \frac{198.954}{2.76}\left(1 - \frac{6 \times 0.593}{2.76}\right) = 20.842\ kN/m^2$$

The distribution of actual bearing pressure below the base slab is shown in Fig. 25.14. Pressure below F

$$\left[20.842 + \frac{2.06}{2.76} \times (165.011 - 20.842)\right] = 128.446\ kN/m^2$$

Pressure below, G

$$\left[20.842 + \frac{(2.06 - 0.4)}{2.76} \times (165.011 - 20.842)\right] = 107.552\ kN/m^2$$

The pressure q_2 is also compressive. Hence, the retaining wall satisfies the requirements of the stability and so also for the pressure. The elements of retaining wall may now be designed.

Step 7. Design of heel slab. Following loads and forces act on the heel slab : Weight of backfill material acting downward

$$W_S = (1.66 \times 4.2 \times 1 \times 18.85) = 131.42\ kN$$

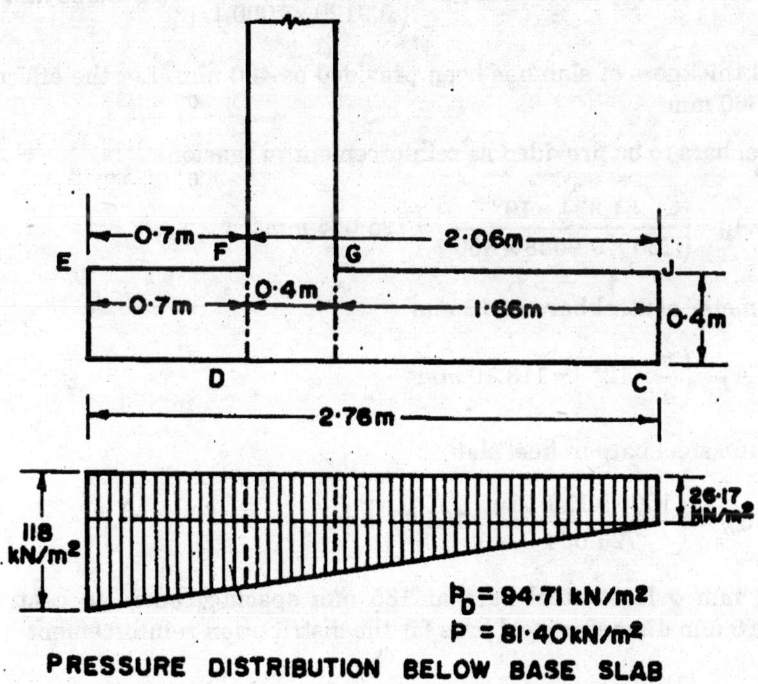

Fig. 25.14

The distance of line of action of W_s from G is (1.66/2) = 0.83 m. Self-weight of heel slab acting downward

$$W_{SH} = (1.66 \times 0.4 \times 1 \times 25) = 16.60\ kN$$

The distance of line of action of W_{SH} from G is (1.66/2) = 0.83 m. Soil reaction acting upward

$$\left(\frac{107.552 + 20.842}{2}\right) \times 1.66 = 106.567\ kN$$

The distance of line of action of soil reaction from G

$$\frac{1.66}{3} \times \left(\frac{107.552 + 2 \times 20.842}{107.552 + 20.842}\right) = 0.643\ m.$$

Downward loads predominates over upward soil reaction. Therefore, the heel slab bends downward. It acts as a cantilever slab. The maximum moment occurs at G. From Fig. 25.14.

$$M_G = (131.42 \times 0.83 + 16.60 \times 0.83 - 106.567 \times 0.643) = 54.334\ kN - m$$

Maximum shear force occurs at G

$$F_G = (131.42 + 16.60 - 106.567) = 41.453\ kN$$

Effective depth required, $d = \left(\dfrac{54.334 \times 10^6}{0.9129 \times 1000}\right)^{1/2} = 243.963\ mm$

The overall thickness of slab has been provided as 400 mm. Let the effective cover be 40 mm. Then, d = 360 mm.

Area of steel bars to be provided as reinforcement in tension

$$A_{st} = \left(\frac{54.334 \times 10^6}{230 \times 0.9038 \times 360}\right) = 726.054\ mm^2$$

Let the diameter of steel bars be 12 mm.

$$A_\phi = \left(\frac{\pi}{4} \times 12^2\right) = 113.10\ mm^2$$

Spacing of the steel bars in heel slab

$$S_H = \left(\frac{1000 \times 113.10}{726.054}\right) = 155.77\ mm$$

Provide 12 mm ϕ Hysd-steel bars at 130 mm spacing centre to centre near the top surface. Provide 8 mm diameter steel bars for the distribution reinforcement

$$A_\phi = \left(\frac{0.12}{100} \times 1000 \times 400\right) = 480\ mm^2$$

$$A_\phi = \left(\frac{\pi}{4} \times 8^2\right) = 50.26\ mm^2$$

Spacing of the distribution reinforcement

$$s_d = \left(\frac{1000 \times 50.26}{480}\right) = 104.72\ mm$$

Provide distribution reinforcement at 100 mm spacing c/c.

Nominal shear stress in heel at G

$$\tau_v = \left(\frac{58.64 \times 1000}{1000 \times 360}\right) = 0.163\ N/mm^2$$

It is less than the permissible shear stress for M 20 grade of concrete without shear reinforcement. ($\tau_c = 0.61\ N/mm^2$)

Step 8. Design of toe slab. The weight of soil acting over the toe slab is not considered. It is likely that the erosion of this soil may take place. Following load and force act on the toe slab :

Self-weight of toe slab, $W_{ST} = (1.10 \times 0.40 \times 1 \times 25) = 11\ kN$

The distance of line of action of this load from F, Fig. 25.14 is (1.10/2) = 0.55 mm.

Total soil reaction acting upward, $\left(\dfrac{94.71+118}{2}\right)\times 1.10 = 116.99\ kN$

The distance of line of action of the soil reaction from, G

$$\frac{1.10}{3}\times\left(\frac{94.71+2\times 118}{94.71+118}\right) = 0.57\ m$$

The soil reaction acting upward predominates over downward load. Therefore, the toe slab bends upward. It acts as a cantilever slab.

Maximum bending is considered to occur at F.

$$M_F = (116.99\times 0.57 - 11\times 0.55) = 60.03\ kN-m$$

Shear force at G, F_G = (116.99 –11) = 105.99 kN

Effective depth required for toe slab

$$d = \sqrt{\left(\frac{60.63\times 10^6}{0.9129\times 1000}\right)} = 257.71\ mm$$

Overall thickness of toe slab is 400 mm. Let the effective cover provided be 20 mm. Then, the effective depth, d shall be 380 mm.

Area of steel provided as tension reinforcement

$$A_{st} = \left(\frac{60.63\times 10^6}{230\times 0.9038\times 380}\right) = 767.54\ mm^2$$

Let the diameter of the steel bars be 12 mm, $A_\phi = \left(\dfrac{\pi}{4}\times 12^2\right) = 113.10\ mm^2$

Spacing of the steel bars, $S_T = \left(\dfrac{1000\times 113.10}{767.54}\right) = 147.35\ mm$

Provide 12 mm φ steel bars at 130 mm spacing centre to centre near the bottom surface.

Distribution reinforcement

$$A_d = \left(\frac{0.12}{100}\times 1000\times 400\right) = 480\ mm^2$$

Provide 8 mm φ steel bars, $A_{\phi d} = \left(\dfrac{\pi}{4}\times 8^2\right) = 50.26\ mm^2$

Spacing of the distribution reinforcement bars, $s_d = \left(\dfrac{1000\times 50.26}{480}\right) = 104.72\ mm$

Provide 8 mm diameter bars at 100 mm spacing centre to centre.

Nominal shear stress, $\tau_v = \left(\dfrac{105.99 \times 1000}{1000 \times 380}\right) = 0.278\ N/mm^2$

Step 9. Design of vertical wall (stem)

The vertical wall (stem) acts as a cantilever slab. The maximum bending moment occurs at its junction with the heel slab.

Lateral earth force from step 4, P_H = 66.41 kN

P_H acts at a distance (H/3) from the base. 4.6/3 = 1.533 m.

Lateral force due to surcharge from step 4, P_{HS} = 29.452 kN

P_{HS} acts at a distance 4.6/2 = 2.3 m from the base. Take moment of lateral force about the junction

$$M = [66.41(1.533 - 0.4) + 29.452 \times (2.3 - 0.4)] = 131.22\ kN - m$$

Effective depth of slab required the junction, $d = \left(\dfrac{131.22 \times 10^6}{0.9129 \times 1000}\right)^{1/2} = 379.14\ mm$

Let the effective cover be 20 mm. Then, the thickness slab be 399.14 mm. The thickness of vertical wall at the junction is 400 mm. The effective depth slab be 380 mm². Area of steel reinforcement

$$A_{st} = \left(\frac{131.22 \times 16^6}{230 \times 0.9038 \times 380}\right) = 407.56\ mm^2$$

Provide 12 mm diameter Hysd-steel bars (A_ϕ = 113.10 mm²). The spacing of the steel bars

$$s = \left(\frac{1000 \times 113.20}{407.56}\right) = 277.49\ mm.$$

These bars may be provided at spacing 225 mm centre to centre from the junction with base slab.

As the height of vertical wall decreases, the bending moment at any section reduces very rapidly. Therefore, the area of steel reinforcement provided at the bottom may be curtailed towards top. The area of steel reinforcement height H_1 from top, and Similarly, the area of steel reinforcement for height H_2 from top and similarly, the area of steel reinforcement for height H_2 from top.

$$A_{st_1} = \left(\frac{M_1}{\sigma_{st} \cdot j_B \cdot d_1}\right) \quad and \quad A_{st_2} = \left(\frac{M_2}{\sigma_{st} \cdot j_B \cdot d_2}\right) \qquad \text{(i)}$$

$$\text{Then} \left(\frac{A_{st_2}}{A_{st_1}}\right) = \left(\frac{M_2}{M_1}\right)\left(\frac{d_1}{d_2}\right) \qquad \text{(ii)}$$

The bending moment, M_1 due to active earth pressure for height H_1.

$$M_1 = K_A \cdot \gamma \cdot H_1^3/6 \quad \text{(iv)}$$

$$M_2 = K_A \cdot \gamma \cdot H_2^3/6 \quad \text{(v)}$$

$$\text{Then } \left(\frac{M_2}{M_1}\right) = \left(\frac{H_2}{H_1}\right)^3 \quad \text{(vi)}$$

Substituting this in the expression (iii)

$$\left(\frac{A_{st2}}{A_{st1}}\right) = \left(\frac{H_2}{H_1}\right) \cdot \left(\frac{d_1}{d_2}\right) \quad \text{(vii)}$$

The area of steel for the vertical wall of height, H_2 may be curtailed to half of that for H_1.

$$\text{Then, } A_{st2} = 0.5\, A_{st1}, \text{ and } 0.5\left(\frac{d_2}{d_1}\right) = \left(\frac{H_2}{H_1}\right)^3 \quad \text{(viii)}$$

As the top of retaining wall, the thickness is 240 mm. For the effective cover of 20 mm, the effective thickness is 220 mm. Then, the effective depth at height, H_2.

$$d_2 = (220 + x) \text{ mm} \quad \text{(ix)}$$

The effective depth at height, H_1

$$d_1 = (220 + 380) \text{ mm} \quad \text{(x)}$$

From similar triangles,

$$\left(\frac{H_2}{x}\right) = \left(\frac{H_1}{380}\right), \quad H_2 = \left(\frac{x}{380}\right) H_1 \quad \text{(xi)}$$

$$\text{Then, } d_2 = \left(220 + 380 \cdot \frac{H_2}{H_1}\right) \quad \text{(xii)}$$

Substituting this in expression (viii)

$$\frac{0.5}{d_1} \cdot \left(220 + 380 \cdot \frac{H_2}{H_1}\right) = \left(\frac{H_2}{H_1}\right)^3$$

At the height H_1, d_1 is equal to 380 mm. Therefore,

$$\frac{1}{2} \times \frac{1}{380} \times (220 + 380\,y) = y^3 \quad \text{(xiii)}$$

where y = ratio of (H_2/H_1)

The cubic equation may be solved by trial and error

$$\text{or } y^3 - \left(\frac{1}{2} \times \frac{220}{380}\right) - \frac{1}{6} \cdot \frac{380}{380}\, y = 0 \quad \text{(xiv)}$$

or $y^3 - 0.5\,y - 0.2895 = 0$

But trial and error, y = 0.9055

Then, $H_2 = 0.9055\ H_1$, $\therefore\ H_2 = (0.9055 \times 3.8) = 3.169$ m

For the vertical wall,

Thickness at top = 240 mm, and thickness at bottom = 400 mm

Average thickness = 320 mm, Distribution reinforcement needed

$$A_d = \left(\frac{0.12}{100} \times 1000 \times 320\right) = 384\ mm^2$$

Let 8 mm diameter of steel bars be provided. Then

$$A_\phi = \left(\frac{\pi}{4} \times 8^2\right) = 50.26\ mm^2$$

Spacing of the distribution reinforcement

$$s_d = \left(\frac{1000 \times 50.26}{383}\right) = 130.9\ mm$$

Provide 8 mm ϕ Hysd steel bars are 100 mm spacing c/c.

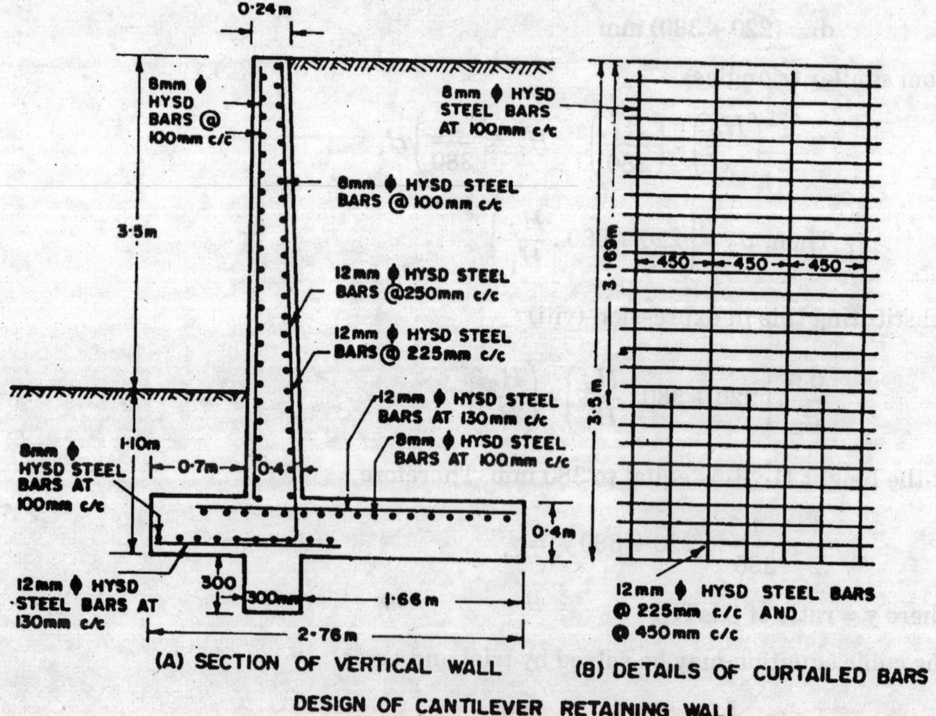

Fig. 25.15

This distribution reinforcement is also provided on the front face in both the directions.

A shear key of 300 mm x 300 mm in size has been provided.

The details of reinforcement for cantilever retaining wall is shown in Fig. 25.15.

25.13 COUNTERFORT RETAINING WALLS

A counterfort retaining wall is used when the height of backfill to be retained exceeds 8 m. The vertical wall and the base slab both become tied together at regular intervals by the counterforts, as shown in Fig. 25.16 (A). Sometimes, buttresses are also attached over the toe slab and the vertical wall below the ground level as shown in Fig. 25.16 (B).

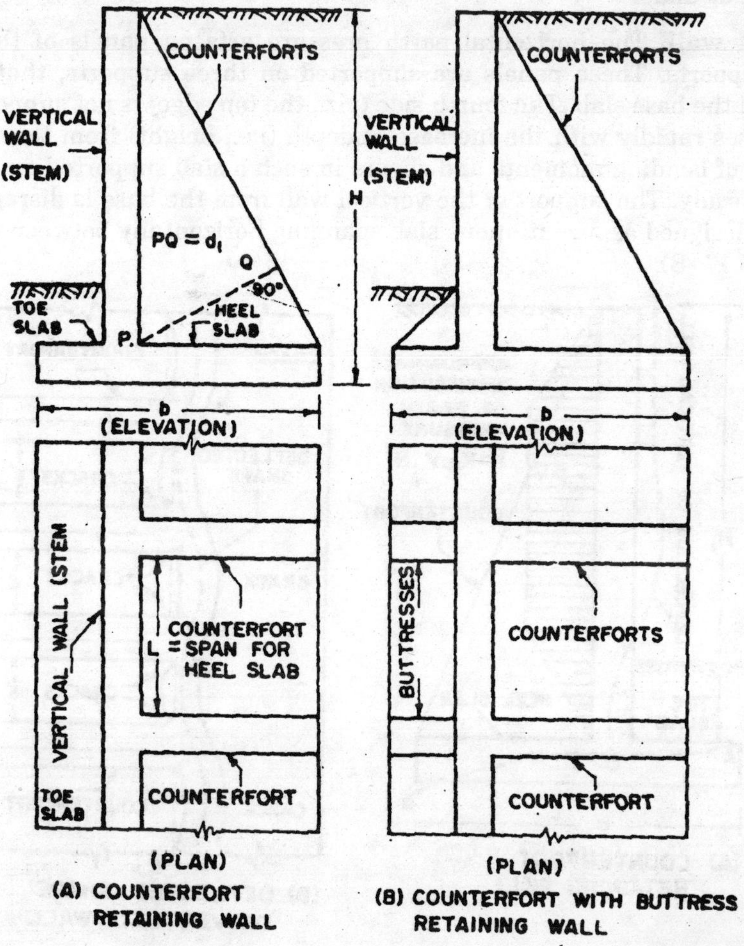

Fig. 25.16

The counterforts which are built on the same side of the backfill material act as tension ties while the buttresses, which are constructed or built on the opposite side of the vertical wall, act in compression.

The general proportions (dimensions) of the counterfort retaining wall base slab and other elements are kept same as those of the cantilever retaining wall. The thickness of heel slab may be kept about *one-twentieth* of the height. The maximum spacing between counterforts may be taken as 3 m for wall in 8 m to 10 m range to about 2.40 m for walls 16 m to 15 m range. These dimensions are checked for the stability in the same manner as that for the cantilever retaining wall. When the requirements of stability are satisfied, then, the counterforts walls are designed.

25.14 BEHAVIOUR OF COUNTERFORT RETAINING WALLS

The counterfort totally change the support conditions for the vertical wall (stem) and the heel slab. The vertical wall becomes a slab spanning horizontally between counterforts. The heel slab becomes a slab supported on three sides. The behaviour of vertical wall and that of the heel slab are as under :

1. Vertical wall. The horizontal earth pressure acts on panels of the vertical wall between two supports. These panels are supported on three supports, that is, at the two counterforts and the base slab. The fourth side (viz., the top edge) is not supported. The earth pressure increases rapidly with the increase of depth (i.e., height) from the top. In involves the calculations of bending moments and shears in such a slab supported on three sides and loaded non-uniformly. The support of the vertical wall from the base is disregarded, and the vertical wall is designed as a continuous slab spanning horizontally between counterforts, as shown in Fig. 25.17 (B).

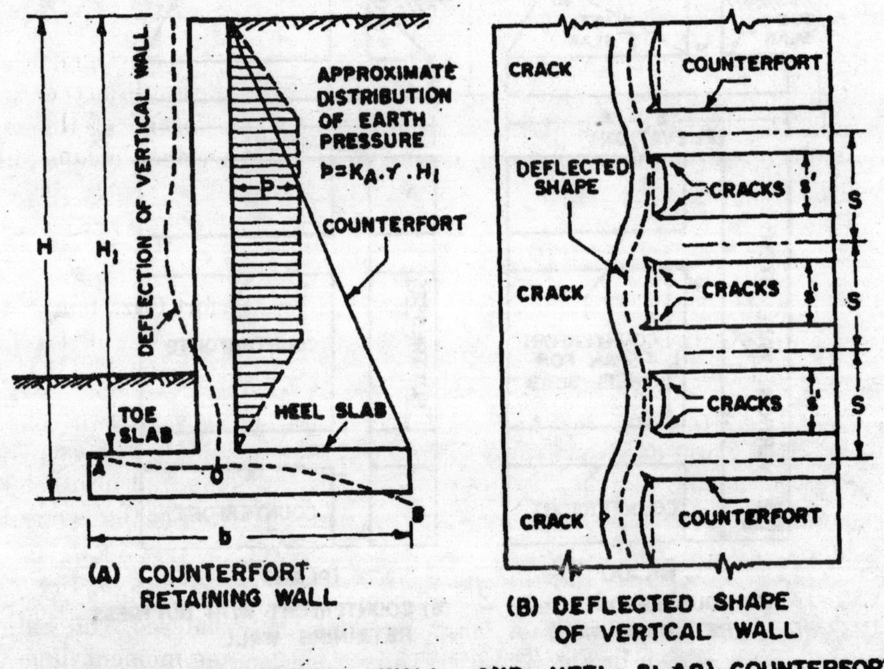

Fig. 25.17

Structurally the vertically wall deforms as shown in Fig. 25.17. The deflection of vertical wall near the top edge is small. However, it will be maximum at some height above the base slab as shown in Fig. 25.17 (A). The vertical wall acts as a slab bending in two directions with its edges fixed against rotation at the counterforts and at the footing. The vertical wall needs reinforcement in vertical and horizontal directions, both. A point of zero bending moment for vertical bending occurs near the bottom of wall, which needs reinforcement in the back side below that region. The points of zero bending for horizontal bending occurs at about 0.3 times the spacing between the counterforts, which need horizontal surface reinforcement across the counterforts.

The cracks in vertical wall may develop due to earth pressure because of insufficient tension reinforcement.

2. Heel slab. The heel slab is also supported similar to the vertical wall. It is loaded downward by the weight of the backfill resting on the heel slab, its self-weight and surcharge, (if any). The down load is partially counteracted by the soil reaction acting upward. The influence of support from the vertical wall is neglected for the simplified analysis.

The heel slab is also subjected to two-way bending. Midway between the counterforts, and in the direction parallel to the counterfort, the heel slab deflects as AOB as shown in Fig. 25.17 (A). The heel slab deflects as a continuous beam (supported at the counterforts) in the direction along the vertical wall.

The vertical wall may crack and fail at the locations of maximum *positive* and *negative* bending moments (viz., the points having maximum tension due to bending) as shown in Fig. 25.17 (B) because of insufficient tension reinforcement.

In addition to this, the crack may develop between the counterfort and the heel slab (because of insufficient tension reinforcement at the outer face of counterfort) or between the counterfort and the vertical wall (because of insufficient anchorage of the wall to the counterfort). The seperation of vertical wall from counterfort, therefore, means failure of the retaining wall.

25.15 DESIGN OF COUNTERFORT RETAINING WALL

The vertical wall (stem), heel slab, toe slab and the counterfort (viz., four elements of a counterfort retaining wall) are designed after satisfying the requirement of stability (*that is*, overturning, sliding and bearing pressure) as follows :

1. Vertical wall (stem). A panel of the vertical wall between two counterforts is subjected to the horizontal earth pressure as shown in Fig. 25.18 (D). The vertical wall acts as a slab in two-way bending (viz., bending in the vertical plane and also bending in the horizontal plane). The vertical bending and the horizontal bending of the vertical wall both may be considered separately as below :

1. (A) Vertical bending. The vertical wall bends in the vertical plane as shown in Fig. 25.18 (A). For design, a vertical strip of *one* metre width is considered. The actual bending moment diagram is as shown in Fig. 25.18(B). The actual bending moment diagram may be approximated as shown in Fig. 25.18(C). The bending moment at the bottom of the vertical slab is given by Eq. 25.14, as under. It is a conservative emperical expression.

$$M_1 = 0.03 \,.\, p \,.\, H_1^2 \,.(S/H_1), where$$

p = intensity of active earth pressure

$$p = K.\gamma.h'_1 \quad \text{(i)}$$

H_1 = height of the vertical wall measured from top upto the top surface of base slab.

S' = clear spacing of the counterforts as shown in Fig. 25.17(B).

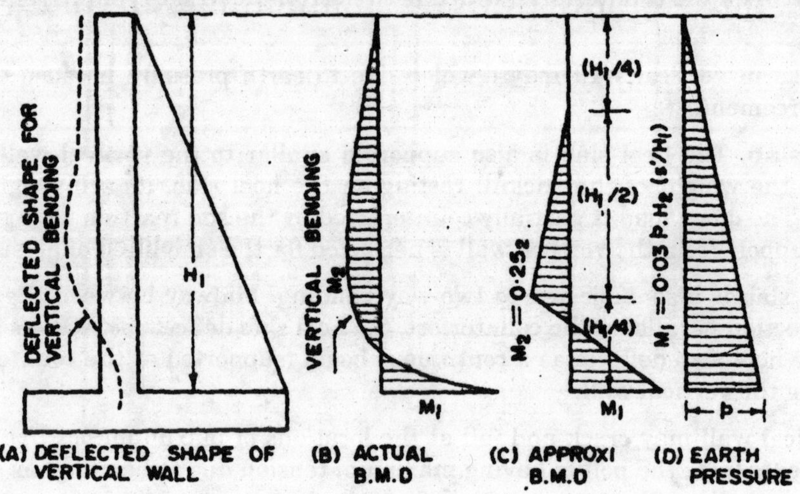

Fig. 25.18

The expression in Eq. 25.24 varies from nearly exact for fixed edges at (S'/H_1) = 0.75 to about 30 percent high for (S'/H_1) = 0.50. It becomes about *three times* the theoretical values for (S'/H_1) = 0.25.

The maximum positive bending moment occurs at about the *lower quarter point* (in the range (S'/H_1) less than *unity*). It may be reasonably be taken as *one-fourth* of M_1. Therefore,

$$M_2 = 0.0075\ p\,.\,H_1^2\,(S'/H_1) \quad (25.25)$$

The vertical bending moment becomes zero at the counterforts and approximately varies parabolically between them. Professor Huntington (in his book `Earth pressures and retaining walls, New York, John Wiley and Sons Inc. 1957) suggests taking the vertical bending moment constant longitudinally over the middle third of the distance the counterforts and linearly varying from the maximum to zero over the end *one-thirds.* The vertical moment diagram may be taken linearly between the quarter points as shown in Fig. 25.18 (C). Frequently, a 0.001 minimum percentage of reinforcement is provided vertically in the front face.

1. (B) **Horizontal bending.** For horizontal bending a strip of *one* metre height is considered. This strip is treated as a continuous beam as shown in Fig. 25.19. The loading on each strip will be uniform and equal to the intensity of active earth pressure.

$$((p = K_A\,.\,\gamma\,.\,h).$$

where, h is the height of vertical wall measured from top upto the middle of the strip being considered.

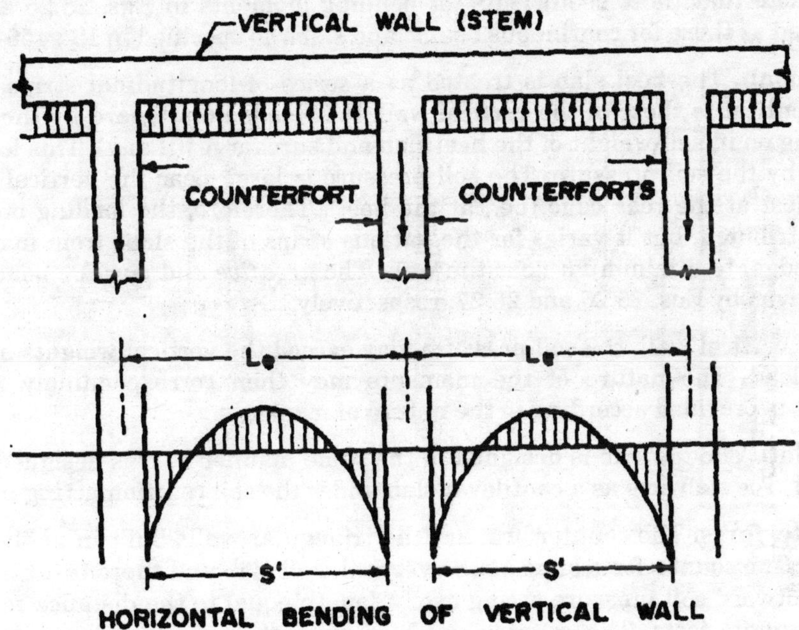

Fig. 25.19

The *negative* bending moment over the counterfort supports

$$M_n = \left(\frac{w \,.\, L_e^2}{12}\right), \quad \text{where} \tag{25.26}$$

$$w = (p \times 1 \times 1) = (K_A \,.\, \gamma \,.\, h)\; kN/m$$

S' = clear spacing between the counterforts

L_e = clear span + effective depth

The *positive* bending moment at mid-span

$$M_p = \left(\frac{w \,.\, L_e^2}{24}\right) \tag{25.27A}$$

However, it may be adopted as

$$M_p = \left(\frac{w \,.\, L_e^2}{16}\right) \tag{25.27B}$$

The moment values for the bottom strips may be reduced to account for the fact that the base slab provides additional support.

Three or four equally spaced additional strips at higher level be considered. The earth pressure on the different strips decreases with increasing elevations (i.e., decreasing height from top).

It is to note that these coefficients for bending moments in Eqs. 25.25 and 25.27 may now be adopted as those for continuous beams and slabs as specified in IS : 456-2000.

2. Heel slab. The heel slab is treated as a series of longitudinal strips, with support conditions identical to those of the vertical wall. It is loaded downward by the weight of the backfill resting on it, self-weight of the heel slab and surcharge, (if any). This load is partially counteracted by the soil pressure. The soil pressure is large near the vertical wall and it is usually smallest at the rear edge (i.e., at the heel). Therefore, the loading on each strip is uniformly distributed, but it varies for the various strips of the slabs from maximum at the heel (i.e., at edge) to a minimum near the wall. The *negative* and *positive* bending moments are same as given by Eqs. 25.26 and 25.27, respectively.

Near the vertical wall, the soil pressure may exceed the vertical weights and result in a net upward load. The nature of the moments may then correspondingly reversed. The reinforcement is provided according to the nature of moments.

3. Toe slab. The toe slab is designed in the same manner as it is designed in cantilever retaining wall. Toe slab acts as a cantilever slab under the soil reaction acting upward.

4. Counterforts. The counterforts are the triangular walls built in at the bottom with the base slab. The counterforts support the vertical wall slab and therefore, these are loaded by the total outward soil pressure acting over a length equal to the distance centre to centre between the counterforts. The counterforts also support the net pressure (which acts, in general, downward). Therefore, the angle between the vertical wall and the heel slab (i.e., 90°) tends to increase because of outward soil pressure acting horizontally and net downward pressure on the heel slab. The counterfort ties both the vertical wall and the heel slab with itself. Therefore, the counterforts serve as tension members between the vertical wall and the heel slab. Therefore, the horizontal steel bars are provided as reinforcement to tie the counterfort to the vertical wall and the vertical steel bars are provided as reinforcement to tie the counterfort to the heel slab. the tension in the counterfort is carried by the main steel bars located near the back of the counterfort. The tension may be calculated at any section by taking the moments, about a point located at the centre of slab, of the forces acting above a horizontal section.

The counterfort acts in combination with the vertical wall as a T-beam as shown in Fig. 25.20. A portion of the vertical wall acts as the flange and the counterfort acts as the web. The maximum bending moment due to total earth pressure about the bottom of slab (critical location for moment) may be calculated, as follows.

Total earth pressure acting horizontally

$$P_H = \left(\frac{1}{2} \cdot K_A \cdot \gamma \cdot H^2 \cdot s \cdot \cos\delta\right), \qquad \textit{where} \qquad \text{(i)}$$

s = spacing between the counterforts

δ = angle of inclination of earth pressure.

$$M = P_H \cdot (H/3) = \left(\frac{1}{6} \times K_A \cdot \gamma \cdot H^3 \cdot s \cdot \cos\delta\right) \qquad \text{(ii)}$$

For the levelled ground, the angle δ is zero and consequently, cos δ is unity. Therefore,

$$P_H = \left(\frac{1}{2} K_{AH} . \gamma . H^2 . s\right) and \tag{iii}$$

$$M = P_H . (H/3) = \frac{1}{6} K_{AH} . \gamma . H^3 \tag{iv}$$

This moment is held in equilibrium by the tensile force in the steel bars (tension reinforcement as usual, provided near the inclined edge of the counterfort). Therefore, the effective depth of T-beam shall be the perpendicular drawn from the centre of bottom section of the vertical wall to the inclined surface of the counterfort).

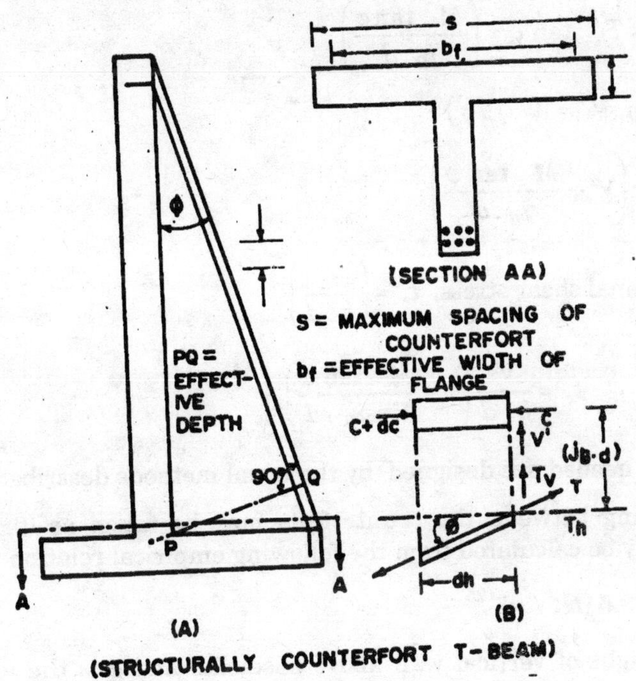

Fig. 25.20

The area of steel to resist the bending moment may be calculated in the same manner as for the ordinary T-beam. That is, from Fig. 25.20 (B)

$$M = (T_h . j_B . d)$$

$$\text{or } M = (T . \cos\phi, j_B . d) = C . j_B . d \tag{v}$$

The compressive area, $b_f . t_f$ is relatively large. Therefore, the design is controlled from the area of steel. The area of steel

$$A_{st} = \left(\frac{M}{\sigma_{st} . j_B . d \cos\phi}\right), \quad where \tag{25.28}$$

ϕ = angle of the inclined face of the counterfort with the vertical as shown in Fig. 25.20 (A).

The bending moment decreases rapidly in the upper parts of the counterfort, the area of steel bars my be reduced by bending some bars diagonally across the counterfort toward the face slab and anchoring them in the compression zone.

The shear in the counterfort may be determined from

$$V = \left(V_B' + T_V\right), \quad where \tag{vi}$$

V' = shear resisted by the concrete, including any carried by the stirrups

Tv = portion of shear carried by the steel bars in tension

$$T_V = T_h \cdot \tan\phi = \left(\frac{M \cdot \tan\phi}{j_B \cdot d}\right) \tag{25.29}$$

Then, $V' = (V - T_V)$

$$V' = \left(V - \frac{M \cdot \tan\phi}{j_B \cdot d}\right) \tag{25.30}$$

$$\text{Nominal shear stress, } \tau_c = \left(\frac{V'}{b \cdot d}\right) \tag{vii}$$

$$\therefore \quad \tau_c = \frac{V}{b \cdot d}\left(1 - \frac{M \cdot \tan\phi}{V \cdot j_B \cdot d}\right) \tag{25.31}$$

Any stirrups, if needed are designed by the usual methods described for beams.

The clear spacing between the counterforts may be taken as (0.33 to 0.5) *times* the height of wall. It may be calculated from the following empirical relation

$$S' = 3.5\,(H_1 / \gamma)^{0.25} \tag{25.32}$$

where, H_1 is the height of vertical wall above base slab and g is the unit weight of soil in kN/m³.

Example 25.3. Design a counterfort retaining wall to retain a backfill material consisting of sand and gravel mixture with a rather moderate quantity of the fine silty particles of 8 m height. A live load of surcharge of 19.20 kN/m² shall be acting over the horizontal surface of the backfill. M 20 grade of concrete and Hysd steel bars shall be used. The safe bearing capacity of soil is 250 kN/m². Take foundation 1.20 m deep.

Design : For M 20 grade of concrete, and Hysd steel,

$$\sigma_{abc} = 7\ N/mm^2,\ \sigma_{st} = 230\ N/mm^2$$

$$\text{Modular ratio of concrete, } m = \left(\frac{280}{3 \times 7}\right) = 13.33.$$

Step 1. Design constants. The design constants are as follows :

$k_B = 0.2886$, $j_B = 0.9038$, $R_B = 0.9129$, $p_B = 0.4398$.

Step 2. Material properties of backfill. The material properties of the backfill material are as below.

Unit weight of the backfill, $\gamma = 18.85$ kN/m². (i)

Angle of repose (effective r angle of internal friction), $\phi = 30°$ (ii)

Coefficient of friction between concrete surface and the soil, $\mu_F = 0.50$ (iii)

$$\text{Coefficient of active earth pressure, } K_{AH} = \left(\frac{1-\sin\phi}{1+\sin\phi}\right) = 0.333 \quad \text{(iv)}$$

Safe bearing capacity of the soil, $q_o = 250$ kN/m² (v)

Unit weight of reinforced cement concrete is 25 kN/m².

$$\text{Coefficient of passive pressure, } K_{PH} = \left(\frac{1+\sin\phi}{1-\sin\phi}\right) = 3.0 \quad \text{(vi)}$$

Horizontal surface of the backfill is subjected to a live load surcharge

= 19.20 kN/m² (vii)

Equivalent height of surcharge

$$h' = \left(\frac{s}{\gamma}\right) = \left(\frac{19.20}{18.85}\right) = 1.02\, m \quad \text{(viii)}$$

Height of retaining wall shall be

H = (8.00 + 1.20) = 9.20 m

Step 3. Preliminary proportions (dimensions) of counterfort retaining wall.

(i) **Ratio of width of toe slab to the width of complete slab.**

From Eq. 25.20

$$\alpha = \left[1-\left(\frac{q_o}{2.2\,\gamma\, H}\right)\right] = \left[1-\left(\frac{250}{2.2\times 18.85\times 9.2}\right)\right] = 0.345$$

(ii) Complete width of the base slab.

(A) From (consideration of overturning) Eq. 25.19

$$b = 0.95\,.\,H\left[\frac{K_{AH}}{(1-\alpha)(1+3\alpha)}\right]^{1/2}$$

$$b = 0.95\times 9.2\left[\frac{0.333}{(1-0.345)(1+3\times 0.345)}\right]^{1/2} = 4.368\, m$$

(B) From (consideration of sliding), Eq. 25.23

$$b = \left(\frac{0.707\,H}{(1-\alpha)}\right).\left(\frac{K_{AH}}{\mu_F}\right) = \left(\frac{0.707 \times 9.2}{(1-0.345)}\right) \times \left(\frac{0.333}{0.5}\right) = 6.614\ m$$

This value is too high.

The width of base slab may be kept as = $(0.6 \times 9.2) = 5.52\ m$

(iii) Width of toe slab, $\alpha b = (0.345 \times 5.52) = 1.904\ m$

Width of toe slab is kept 2.00 m

(iv) Width of heel slab, (5.52 –2.00) = 3.52 m

Let the thickness of heel slab be 0.4 m. Thickness of heel slab is taken as the thickness of vertical wall (stem) at the base equal to 8 to 15 percent of height.

(v) Vertical wall (stem)

(A) **Thickness of wall.** The thickness at the top of the vertical wall is taken as 150 mm to 300 mm. Let the thickness at the vertical wall be 0.240 m.

Thickness at the bottom shall be equal to that of heel slab, i.e., 0.4 m.

(B) Height of wall. The height of wall

H = (8 + 1.2) = 9.2 m

Step 4. Gravity loads. One metre length of wall is considered. The various gravity loads of different components of the counterfort retaining wall. W_1, W_2. . . . etc., and their respective distances $\bar{x}_1, \bar{x}_2, \ldots$ . . . etc. from the toe are shown in Fig. 25.21. These have been calculated and listed as below.

Weight of Components	W (kN)	$\bar{x}$ (m)	$M_R = W . \bar{x}\ (kN - m)$
Base Height			
$W_1 = (0.24 \times 8.8 \times 1 \times 25)$	52.80	1.870	98.726
$W_2 = \left(\frac{1}{2} \times 0.16 \times 8.8 \times 1 \times 25\right)$	17.66	2.043	35.957
$W_3 = (5.02 \times 0.4 \times 1 \times 25)$	55.20	2.760	152.352
$W_4 = \left(\frac{1}{2} \times 0.16 \times 8.8 \times 1 \times 18.85\right)$	13.27	2.097	27.827
$W_5 = (3.12 \times 8.8 \times 1 \times 18.85)$	517.55	3.96	2049.498
Total gravity loads $\sum W = 656.42\ kN$			2364.37

Total vertical load = 656.42 kN.

Step 5. Lateral forces. The lateral forces are due to active earth pressure and the surcharge

Lateral earth pressure due to active earth pressure

$$P_H = \left(\frac{1}{2} K_{AH} \cdot \gamma \cdot H \cdot H\right)$$

$$\text{or } P_H = \left(\frac{1}{2} \times 0.333 \times 18.85 \times 9.2^2\right) = 265.64\ kN$$

This acts at a height (H/3) 3.067 m.

Equivalent height of surcharge

$$h' = \left(\frac{\textit{Live load}}{\textit{Unit wt. of backfill}}\right) = \left(\frac{19.20}{18.85}\right) = 1.02\ m.$$

Horizontal force due to surcharge

$$P_{HS} = (0.333 \times 18.85 \times 1.02 \times 9.2) = 58.90\ kN.$$

This acts at a height (H/2) = 4.6 m.

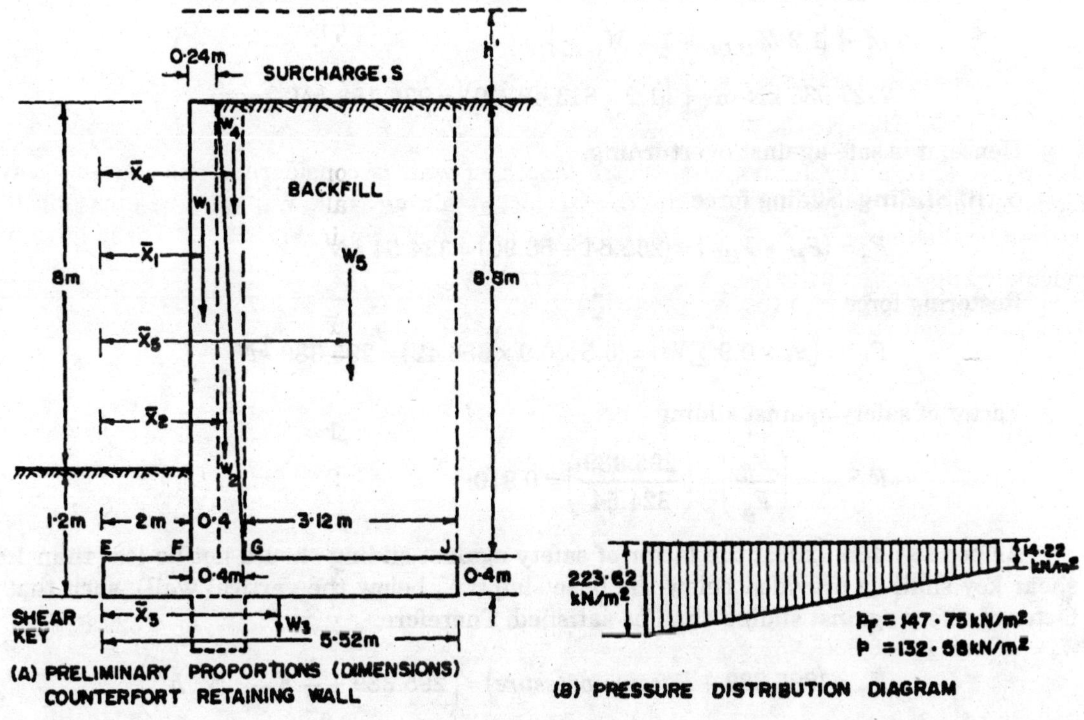

Fig. 25.21

Step 6. Check for stability. The counterfort retaining wall shall be checked for stability (viz., against *overturning* and sliding) and pressures below the base slab as below :

6. (A) **Overturning.** Resisting moment, M_R = 2364.37 kN-m

Overturning moment

$$M_O = (265.24 \times 3.067 + 58.90 \times 4.6) = 1084.43\ kN - m$$

The distance of the resultant from the toe, $\bar{x} = (\sum M / \sum W)$

$$\bar{x} = \left(\frac{2364.37 - 1084.43}{656.42}\right) = 1.95\ m$$

$$\left(\frac{b}{3}\right) = \left(\frac{4.6}{3}\right) = 1.533\ m$$

Overturning moment due to dead load

$$M_{O.D} = (265.24 \times 3.067) = 813.49\ kN - m$$

The live load is not considered. As such the overturning moment due to live load, $M_{I.IL}$ = zero.

Restoring moment is due to dead loads only. Therefore, restoring moment

$$M_R = 0.9 \sum W\ \bar{x} = (0.9 \times 2364.37) = 2127.933\ kN - m$$

$$M \nless [1.2\ M_{O.DL} + 1.4\ M_{O.IL}]$$

$$2127.933 \text{ kN-m} \nless [(1.2 \times 813.49 + 0)] = 976.186\ kN - m$$

Hence, it is safe against overturning.

6. (B) **Sliding.** Sliding force

$$F_s = (P_H + P_{HS}) = (265.64 + 58.90) = 324.54\ kN$$

Restoring force

$$F_R = (\mu_F \times 0.9 \sum W) = (0.5 \times 0.9 \times 656.42) = 295.389\ kN$$

Factor of safety against sliding

$$F.S_{SL} = \left(\frac{F_R}{F_S}\right) = \left(\frac{295.389}{324.54}\right) = 0.910$$

As per IS : 456 : 2001, the factor of safety against sliding should not be less than 1.4. A shear key shall be provided below the base slab (i.e., below the vertical wall), such that the factor of safety against sliding shall be satisfied. Therefore,

$$F_R = (295.389 + Passive\ pressure) = \left(295.389 + \frac{1}{2} K_{PH} \cdot \gamma \cdot h^2\right)$$

$$\text{or } F_R = \left(295.389 + \frac{1}{2} \times 3 \times 18.85 \times h^2\right)$$

$$Now \left(\frac{F_R}{F_S}\right) = \left(\frac{295.389 + \frac{1}{2} \times 3 \times 18.85 \times h^2}{324.54}\right) = 1.4$$

h' = 2.636 m.

Depth of foundation is 1.20 m. Then, the depth of shear key below the base slab shall be (2.626 – 1.20) = 1.426 m.

Passive earth pressure

$$\left(\frac{1}{2}\times 3\times 18.85\times 2.626^2\right) = 194.98\ kN.$$

$$\therefore\ \ F\,S_{SL} = \left(\frac{295.389 + 194.98}{324.54}\right) = 1.511 > 1.4.$$

6(C). **Pressures below base slab.** The distance of resultant from the toe of the base slab :

$$\bar{x} = (\Sigma M/\Sigma W) = \left(\frac{2467.37 - 1084.93}{656.42}\right) = 1.95\ m$$

$$\left(\frac{b}{3}\right) = \left(\frac{4.6}{3}\right) = 1.533\ m.$$

The resultant acts at 1.95 m from the toe, i.e., the middle third width of base slab. The eccentricity, e of the resultant (distance of resultant from the centre-line of base slab)

$$e = \left(\frac{b}{2} - \bar{x}\right) = \left(\frac{5.52}{2} - 1.95\right) = 0.81\ m$$

The pressure below the base slab

$$q_1 = \frac{\Sigma W}{b}\left[1 + \frac{6.e}{b}\right] = \frac{656.42}{5.52}\left[1 + \frac{6\times 0.81}{5.52}\right] = 223.62\ kN/m^2 \qquad \left(< 250\ kN/m^2\right)$$

$$q_2 = \frac{656.42}{5.52}\left[1 - \frac{6\times 0.81}{55.2}\right] = 14.22\ kN/m^2$$

Pressure at F

$$= \left[14.22 + (223.62 - 14.22)\left(\frac{3.52}{5.52}\right)\right] = 147.75\ kN/m^2$$

Pressure at G

$$= \left[14.22 + (223.62 - 14.22)\left(\frac{3.12}{5.52}\right)\right] = 132.58\ kN/m^2$$

The pressure distribution diagram is shown in Fig. 25.21 (B).

Step 7. Design of vertical wall (stem). Height of vertical wall above heel slab

$H_1 = (9.2 - 0.4) = 8.8$ m.

Consider a strip of 1 m height. This strip acts as a continuous beam. Clear spacing of counterforts, from Eq. 25.32

$$S' = 3.5\left(\frac{H_1}{\gamma}\right)^{0.25} = 3.5\left(\frac{8.8}{18.85}\right)^{0.25} = 2.893\ m$$

Let the thickness of counterforts be 0.4 m (assumed). Effective span of the above strip

S = ES = [2.893 + (0.2 + 0.2) = 3.293 m.

Intensity of active earth pressure (average of top and bottom of strip

$$p = K_{AH}\cdot\gamma\cdot\left(\frac{8.8+7.8}{2}\right) = (0.333\times 18.85\times 8.3) = 52.10\ kN/m^2.$$

Bending moment over the counterforts (*negative*)

$$M_{CF} = \left(-\frac{52.10\times 3.293^2}{12}\right) = -47.08\ kN-m$$

Effective depth of vertical slab

$$d = \left(\frac{47.08\times 106}{0.9129\times 1000}\right)^{1/2} = 227.095\ mm$$

Let the effective cover be 30 mm. Then, the overall thickness shall be 257.095 mm. The thickness of slab (assumed) was 400 mm at the base. Hence it is alright. The effective depth of slab shall be 370 mm.

Step 7 (A) Steel reinforcement for negative BM, Area of steel bars required per 1 m height

$$A_{st} = \left(\frac{M}{\sigma_{st}\cdot j_B\cdot d}\right) = \left(\frac{47.08\times 10^5}{230\times 0.9038\times 370}\right) = 612.12\ mm^2.$$

Provide 12 mm ϕ Hysd steel bars. Area of one steel bar

$$A_\phi = \left(\frac{\pi}{2}\times 12^2\right) = 113.10\ mm^2.$$

$$\text{Spacing of bars, } s = \left(\frac{1000\times 113.10}{612.12}\right) = 184.76\ mm$$

Provide 12 mm ϕ steel bars horizontally near counterforts face at 180 mm vertical spacing towards counterforts face shear force near the counterforts

$$V = \frac{1}{2}(3.293 - 0.2)\times 52.10 = 80.57\ kN$$

$$\text{Nominal shear stress, } \tau_v = \left(\frac{V}{b\cdot d}\right) = \left(\frac{80.57\times 10^3}{1000\times 370}\right) = 0.218\ N/mm^2$$

For balanced action, (p_B = 0.4398), $\left(\dfrac{100\,A_S}{b.d}\right) = 0.4398$

and for M 20 grade of concrete, the permissible shear stress in concrete without shear reinforcement, from IS : 456-2000

$$\tau_c = \left(0.22 + \frac{0.1893}{0.25} \times 0.08\right) = 0.2806\ N/mm^2.$$

τ_c is more than τ_v, the vertical walls slab is safe in shear. It does not need shear reinforcement.

Step 7 (B) Steel reinforcement for positive BM. Bending moment mid-way between the counterforts

$$M_{MCF} = \left(\frac{52.10 \times 3.293^2}{17}\right) = 35.31\ kN-m$$

Area of steel bars required for 1 m weight

$$A_{st} = \left(\frac{35.31 \times 10^6}{230 \times 0.9038 \times 370}\right) = 459.10\ mm^2$$

Provide 12 mm ϕ Hysd steel bars, A_ϕ =113.10 mm^2.

$$\text{Spacing of the steel bars, } s = \left(\frac{1000 \times 113.10}{459.10}\right) = 246.36\ mm$$

Provide 12 mm ϕ Hysd steel bars horizontally at 240 mm vertical spacing on the outer face.

Step 7 (C) Distribution reinforcement. Area of steel for distribution reinforcement

$$A_d = \left(\frac{0.12}{100} \times 1000 \times 400\right) = 480\ mm^2.$$

Provide 8 mm ϕ Hysd steel bars, $A_{\phi d}$ = 50.26 mm^2.

$$\text{Spacing, } s_d = \left(\frac{100 \times 50.26}{480}\right) = 104.71\ mm.$$

Provide 8 mm ϕ Hysd steel bars at 100 mm spacing on both the faces.

The main reinforcement may be checked for the development length near the point of zero moments.

The bending moments may be calculated at higher elevations and the spacing of bars may be calculated.

Step 8. Design of heel slab. Clear spacing between the counterforts (from above)

= 2.893 mm.

Thickness of counterforts assumed as 0.4 m = 400 mm.

Effective span, (2.893 + 0.2 + 0.2) = 3.293 m

Consider a strip of 1 m width. This strip acts as a continuous beam supported by the counterforts at the ends.

Downward load due to weight of backfill

$$= (1 \times 1 \times 8.8 \times 18.85) = 165.88\ kN/m^2$$

Downward load due to self-weight $= (1 \times 1 \times 0.4 \times 25) = 10\ kN/m^2$

Upward soil pressure at G, from Fig. 25.21 (B) = 132.58 kN/m²

Upward soil pressure at 2.12 m from J (Heel)

$$\left[14.22 + (223.62 - 14.22) \times \frac{2.12}{5.52}\right] = 94.14\ kN/m^2$$

Average upward pressure due to soil, $\left(\frac{132.58 + 94.14}{2}\right) = 113.36\ kN/m^2$

Net downward load, (165.88 + 10 – 113.36) = 62.52 kN/m²

Maximum bending moment over the counterforts (negative)

$$\left(-\frac{62.52 \times 3.293^2}{12}\right) = -56.50\ kN-m$$

$$\text{Effective depth required, } d = \left(\frac{56.50 \times 10^6}{0.9129 \times 1000}\right)^{1/2} = 248.78\ mm$$

Let the effective cover be 30 mm. Then overall thickness shall be 205.258 mm. The thickness of counterfort slab has been kept as 400 mm. Hence, it is alright.

$$\text{Shear force, } V = \left(\frac{62.50 \times 3.293}{2}\right) = 102.91\ kN$$

$$\text{Nominal shear stress, } \tau_v = \left(\frac{102.91 \times 10^3}{1000 \times 370}\right) = 0.278\ N/mm^2$$

Permissible shear stress in M 20 grade of concrete without shear reinforcement for (p_B = 100 A_s/bd) = 0.4398 percent

$$\tau_c = \left(0.22 + \frac{0.1898}{0.25} \times 0.08\right) = 0.2806\ N/mm^2$$

τ_c is more than τ_v, the heel slab is safe is in shear. It does not need shear reinforcement.

Step 9 (A) Steel reinforcement for negative BM. Area of Hysd steel bars required for 1 m width

$$A_{st} = \left(\frac{M}{\sigma_{st} \cdot j_B \cdot d}\right) = \left(\frac{56.50 \times 10^6}{230 \times 0.9038 \times 370}\right) = 734.59\ mm^2$$

Provide 12 mm ϕ Hysd steel bars, A_ϕ = 113.10 mm².

Spacing of the steel bars for negative bending moment (near the counter forts)

$$s = \left(\frac{1000 \times 113.10}{734.59}\right) = 153.96\ mm$$

Provide 12 mm ϕ Hysd steel bars at 140 mm spacing centre to centre.

Step 9 (B). Steel reinforcement for positive BM. Bending moment mid-way between the counterforts heel slab strip shall be positive

$$\left(\frac{62.52 \times 3.293^2}{16}\right) = 42.37\ kN - m$$

Area of steel reinforcement

$$A_{st} = \left(\frac{42.37 \times 10^6}{230 \times 0.9038 \times 370}\right) = 550.91\ mm^2$$

Provide 12 mm ϕ Hysd steel bars, A_ϕ = 113.10 mm²

Spacing of the steel reinforcement bars for positive BM

$$s = \left(\frac{1000 \times 113.10}{550.91}\right) = 205.30\ mm$$

Provide 12 mm ϕ Hysd steel at 200 mm spacing centre to centre in heel slab near the bottom face.

Step 9 (C) Distribution reinforcement. Heel slab is not exposed to temperature. However shrinkage may take place. Area of distribution reinforcement

$$A_d = \left(\frac{0.12}{100} \times 1000 \times 400\right) = 480\ mm^2.$$

Provide 8 mm ϕ Hysd steel bars, $A_{\phi d}$ = 50.26 mm²

$$\text{Spacing, } s_d = \left(\frac{1000 \times 50.26}{480}\right) = 104.71\ mm.$$

Provide 8 mm ϕ Hysd steel bars at 100 mm spacing on the sides.

The main reinforcement steel bars may be checked for development for point of zero moment.

Two or three more strips of 1 m width may be considered and calculations may be carried for bending moment, reinforcement and the spacing of the steel bars. The bending moment due to down ward loads shall be increasing as the soil pressure shall be reducing towards heel point.

Step 10. Design of toe slab. Toe slab acts as a cantilever due to soil pressure as shown in Fig. 25.21 (B) acting upward.

Total upward reaction due to soil action over toe slab

$$\left(\frac{263.62 + 147.75}{2}\right) \times 2 = 371.37\ kN$$

Distance of line of reaction from the face of the vertical wall (stem) viz., *point* F, Fig. 25.21 (B)

$$\frac{2}{3} \times \left(\frac{147.75 + 2 \times 220.62}{147.75 + 223.62}\right) = 1.068\ m.$$

Bending is maximum about the face

$$(371.37 \times 1.068) = 396.06\ kN - m$$

The slab shall be designed as the balanced section. The moment of resistance of the toe slab with 370 mm effective depth and 1000 mm strip

$$(R_B \, . \, 1000 \times 370 \times 370) = \left(\frac{0.9129 \times 1000 \times 370 \times 370}{1000 \times 1000}\right) = 124.98\ kN - m$$

Toe slab shall be designed as doubly reinforcement beam, M_1 = 124.98 kN-m

M_2 = (396.04 − 124.98) = 271.08 kN-m

Step 10 (A) Tension reinforcement for B.M., M_1

$$A_{st_1} = \left(\frac{124.98 \times 10^6}{230 \times 0.9038 \times 370}\right) = 1625\ mm^2$$

Step 10 (B) Area of steel reinforcement for B.M., M_2

$$A_{st_2} = \left(\frac{271.08 \times 10^6}{230 \times 0.9038 \times (370 - 12)}\right) = 3641.55\ mm^2$$

Step 10. (C) Total area of steel reinforcement in tension

$$(A_{st_1} + A_{st_2}) = (1625 + 3645.75) = 5266.55\ mm^2$$

Provide 20 mm φ steel bars, $A_\phi = \left(\frac{\pi}{4} \times 20^2\right) = 314.159\ mm^2$

$$\text{Spacing of the steel bars, } s = \left(\frac{1000 \times 314.159}{5266.55}\right) = 59.62\ mm$$

Provide 20 mm ϕ Hysd steel bars at bottom surface at 50 mm spacing centre to centre.

Step 10 (D) Area of steel on compression side (i.e., near the top surface)

$$(m_c - 1)\, A_{sc} \,.\, (n_B - d_c) = mA_{st_2}\,(d - n_B)$$

$$n_B = (0.2886 \times 370) = 106.782\ mm \quad and\ d_c = 30\ mm$$

$$(20 - 1)\, A_{sc} \,.\, (106.782 - 30) = 13.33 \times 3641.55 \times (370 - 106.782)$$

$$A_{sc} = \left(\frac{13.33}{19}\right) \times \left(\frac{370 - 106.782}{106.782 - 30}\right) \times 3641.55 = 8758.28\ mm^2$$

$$\text{Spacing, } \left(\frac{1000 \times 314.159}{8758.28}\right) = 35.87\ mm$$

Provide 20 mm ϕ Hysd steel bars at 30 mm spacing centre to centre near top surface.

Step 10. (E) Distribution reinforcement 8 mm ϕ Hysd steel bars shall be provide at 100 mm spacing c/c at top surface and at 50 mm also at the bottom surface.

Step 11. Design of shear key. Depth of shear key below the base shall be 1.426 m. Depth of shear key below the ground level shall be (1.426 + 1.2) = 2.626 m.

Consider 1 m length of the retaining wall.

Bending moment on shear key due to passive soil pressure

$$\int_{1.426}^{2.626} K_P \,.\, \gamma \,.\, H\,(H - 1.426)\, dH = 3 \times 18.85 \int_{1.426}^{2.626} \left(H^2 - 1.426\ H\right) dH$$

$$= 90.423\ \text{kN-m}$$

$$\text{Effective depth, } d = \left(\frac{90.423 \times 10^6}{0.9129 \times 1000}\right)^{1/2} = 314.72\ mm$$

Let the effective cover be 30 mm. The overall thickness shall be 344.72 mm. The overall thickness is kept as 400 mm. Then, the effective depth shall be 370 mm. Area of steel reinforcement

$$A_{st} = \left(\frac{90.423 \times 10^6}{230 \times 0.9038 \times 370}\right) = 1175.64\ mm^2$$

Let 12 m ϕ Hysd steel bars be provided, $A_\phi = 113.10\ mm^2$

$$\text{Spacing of bars, } s = \left(\frac{1000 \times 113.10}{1175.64}\right) = 96.25\ mm.$$

Provided bars at 90 mm spacing c/c. Shear shall be critical at a depth 370 mm below the base slab. Shear force shall be

$$V = \int_{1.57}^{2.626} K_P \,.\, \gamma \,.\, HdH$$

$$\text{or } V = 3.00 \times 18.85 \times \int_{1.57}^{2.626} HdH = 125.286 \ kN$$

$$\text{Nominal shear stress, } \tau_v = \left(\frac{125.286 \times 10^3}{1000 \times 370}\right) = 0.3386 \ N/mm^2$$

For $\left(\frac{100\ A_s}{bd}\right) = p_B = 0.4398$ percent and for M 20 grade of concert $\tau_c = 0.2806 N/mm^2$

Since, τ_v is more than τ_c, the depth is increased, $\left(370 \times \frac{0.3388}{0.2806}\right) = 446.49 \ mm.$

Let the effective cover be 30 mm. Let the overall thickness shall be 476.49 mm. Let the overall thickness be 480 mm. Distribution reinforcement

$$\left(\frac{0.12}{100} \times 1000 \times 480\right) = 576 \ mm^2$$

8 mm diameter Hysd steel bars shall be provided

$$\left(A_\phi = 50.26 \ mm^2\right)$$

$$\text{Spacing} = \left(\frac{1000 \times 50.26}{576}\right) = 87.266 \ mm$$

These bars are provided at 80 mm spacing centre to centre.

Step 12. Design of counterforts. The structural behaviour of counterforts is like T-beam. The vertical wall (stem acts as the flange and the counterforts acts as wab. Along the inclined edge of the counterfort, the bars shall be provided as tension reinforcement.

From Fig. 25.21,

$$\tan \phi = (3.12/8.8) = 0.3545, \ \therefore \phi = 19.52^\circ$$

perpendicular distance PQ

$$PQ = 8.8 \sin \phi = (8.8 \times 0.3342) = 2.941 \ m$$

Thickness of counterfort is 0.4 m clear spacing between the counterfort

$$S' = 2.893 \ m.$$

Maximum bending moment at the base of slab

$$M = \left(\frac{1}{6} \,.\, K_{AH} \,.\, \gamma \,.\, H^3 \,.\, s'\right)$$

$$=\left(\frac{1}{6}\times 0.333\times 18.85\times 8.8^3\times 2.893\right)=2062.53\ kN-m$$

Effective depth of beam required (even, in case, it is considered as rectangular beam)

$$d=\left(\frac{2062.53\times 10^6}{0.9129\times 370}\right)^{1/2}=2471\ mm$$

Let the effective cover be 30 mm. Then, the overall depth shall be 2501.08 mm

However, the overall depth is PQ = 2941 mm

Let the lever arm factor j for the T-beam be assumed as 0.90. Area of steel bars needed

$$A_{st}=\left(\frac{M}{\sigma_{st}\,.\,j.\,d.\cos\phi}\right)(\cos\phi=0.9425)$$

$$=\left(\frac{2062.53\times 10^6}{230\times 0.90\times 0.9425\times(2941-30)}\right)=3630.736\ mm^2$$

Provide 20 mm ϕ Hysd steel bars, A_ϕ = 314.159 mm²

$$\text{Number of bars}\left(\frac{3630.736}{314.159}\right)=11.557$$

Provide 12 bars of 20 mm ϕ Hysd steel along the inclined sum face of the counterfort

Step 12 (A) Horizontal ties. The vertical wall (stem) is firmly attached with the counterforts so that the vertical wall cannot separate out. Horizontal ties are provided for this purpose. Horizontal force which tends to separate the wall

$$K_A\,.\,\gamma\,.\,H\times s'=(0.333\times 18.85\times 8.80\times 2.893)=159.80\ kN$$

Area of 2-legged stirrups, (12 mm ϕ Hysd steel bars)

$$A_\phi=\left(2\times\frac{\pi}{4}\times 12^2\right)=266.19\ mm^2$$

$$\text{Area of ties required, }\left(\frac{159.80\times 10^3}{230}\right)=694.78\ mm^2$$

Vertical spacing of the horizontal ties shall be

$$s=\left(\frac{1000\times 226.19}{694.78}\right)=325.56\ mm$$

Provide the horizontal ties at 320 mm spacing. This spacing may be increased towards the top to 450 mm dc.

Step 12. (B) Vertical ties. Because of net load acting downward, the heel slab also tends to separate out from the counterforts. To check this, the vertical ties are provided. Net downward force per 1 m horizontal width from step 8.

$= 62.50 \text{ kN/m}^2$

Total downward load, $(62.50 \times 2.893) = 180.81\ kN$

Area of ties required, $\left(\dfrac{180.81 \times 10^3}{230}\right) = 786\ mm^2$.

Two legged 12 mm ϕ Hysd steel bars shall be provided for the vertical ties.

$A_\phi = 226.19 \text{ mm}^2$.

Spacing of the vertical ties, $s = \left(\dfrac{1000 \times 226.19}{786}\right) = 287.72\ mm$

The vertical ties are provided at a horizontal spacing of 280 mm c/c.

Step 12. (C) Check for shear. The counterforts may be checked for shear. Total shear force

$$V = \left(\frac{1}{2} K_{AH} \cdot \gamma \cdot H^2 \times 2.893\right) kN$$

$$V = \left(\frac{1}{2} \times 0.333 \times 18.85 \times 8.8^2 \times 2.893\right) = 703.136\ kN$$

$$\left(V' - \frac{M}{d}\tan\phi\right) = \left[703.136 - \frac{2062 \times 10^6}{2911 \times 1000} \times 0.3545\right] kN$$

$[703.136 - 251.11] = 452.03\ kN$

Nominal shear stress

$$\tau_v = \left(\frac{452.03 \times 10^3}{370 \times 2911}\right) = 0.4196\ N/mm^2,\ and\ \tau_c = 0.2806\ N/mm^2$$

$$V_c = \left(\frac{0.2806 \times 370 \times 2911}{1000}\right) = 302.22\ kN$$

Shear strength of horizontal ties

$$V_s = \left(\frac{230 \times 226.19 \times 2911}{320 \times 1000}\right) = 473.25\ kN$$

$$(V_c + V_s) = (302.22 + 473.25) = 775.42\ kN > 452.03\ kN$$

The counterfort is safe in shear.

Step 12 (D) Curtailment of inclined bars. The inclined bars carrying tension due to bending and provided near the inclined edge of the counterfort may be curtailed at higher elevations.

Half of the bars may be curtailed at a height from top

$$H = \left(\frac{H_1}{\sqrt{2}}\right) = \left(\frac{8.8}{\sqrt{2}}\right) = 6.22\ m$$

These remaining 6 bars shall be extended 12 times the diameter of the steel bars. Then, the actual point of cut-off from the top

$$\left(6.22 - \frac{12 \times 20}{1000}\right) = 5.98\ m$$

Shear force at this height

$$V' = \left(\frac{1}{2} K_{AH} \cdot \gamma \cdot H^2 \cdot 2.893\right) kN$$

$$\left(\frac{1}{2} \times 0.333 \times 18.85 \times 5.98^2 \times 2.893\right) = 324.696\ kN$$

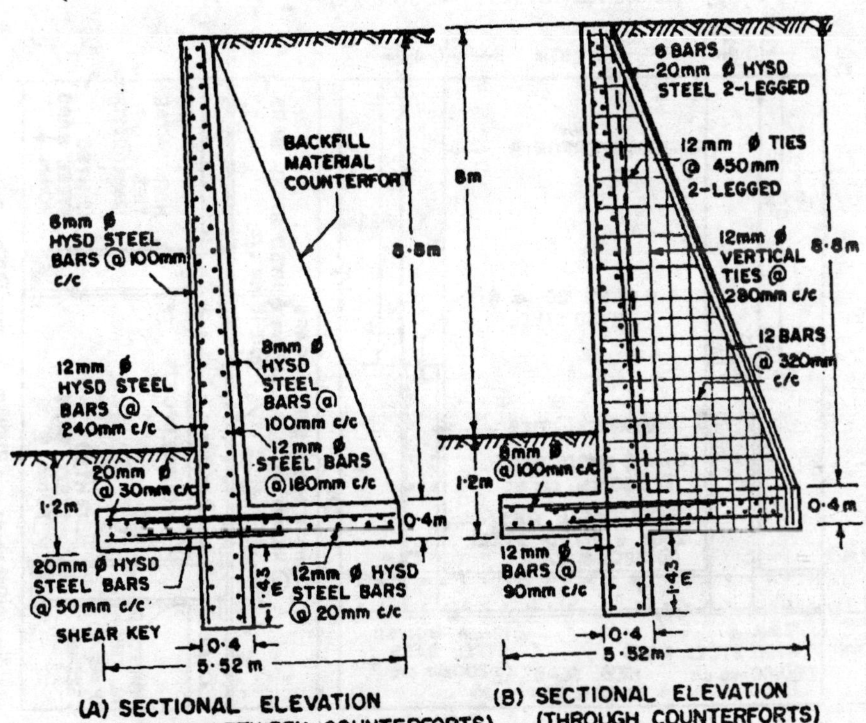

Fig. 25.22

Effective depth at this point (for T-beam)

$$(0.2 + 5.98 \tan 19.52°) = (0.2 + 5.98 \times 0.3545) = 2.32\ m$$

$$A_{st} = 6 \times \left(\frac{\pi}{4} \times 20^2\right) = 1884.96\ mm^2$$

$$\left(\frac{100\ A_{st}}{bd}\right) = \left(\frac{100 \times 1884.96}{370 \times 2320}\right) = 0.2196$$

From IS : 456-2001 for M 20 grade of concrete, $\tau_c = 0.22\ N/mm^2$

$$V_c = \tau_c \frac{bd}{1000} = \left(\frac{0.22 \times 370 \times 2320}{1000}\right) = 188.848\ kN$$

For 12 mm ϕ Hysd steel 2-legged

Horizontal ties at 320 mm spacing

$$V_s = \left(\frac{226.19 \times 230 \times 2320}{320 \times 1000}\right) = 377.17\ kN$$

$$(V_c + V_s) = (188.848 + 377.17) = 566.02\ kN$$

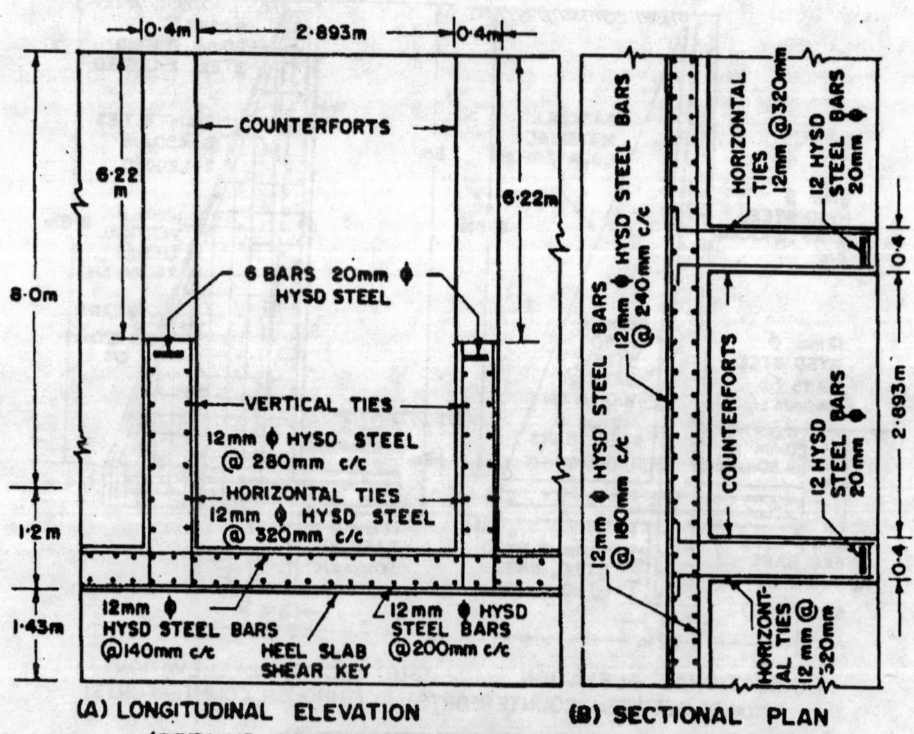

(DETAILS OF REINFORCEMENT) DEIGN OF COUNTERFORT RETAINING WALL

Fig. 25.23

Therefore, ($V_c + V_s$) is more than V'

$$\frac{2}{3}\left(V_c + V_s\right) = \frac{2}{3} \times 566.02 = 377.346\ kN$$

It is also more than V' = 324.696 kN.

Hence, the bars may be curtailed at this height.

The details of reinforcement for the counterfort retaining wall are shown in Figs. 25.22 and 25.23.

PROBLEMS

25.1 **A backfill material consisting of sand and gravel without fine particles of 3.20 m height is to be retained by a gravity retaining wall. A live load surcharge of 12 kN/m^2 shall be acting over the horizontal surface of the back. Design the gravity retaining wall.**

25.2 **Design a cantilever retaining wall to retain a backfill of sand and gravel without fine particles of 3.20 m height. A live load surcharge of 12 kN/m^2 shall be acting over the horizontal surface of the backfill M 20 grade of concrete and Hysd steel bars shall be used. The maximum bearing capacity of the soil is 160 kN/m^2.**

25.3 **Design a counterfort retaining wall to retain a backfill material consisting of sand and gravel without fine particles of 7.40 m height. A live load of 12 kN/m^2 shall be acting over the horizontal surface of the backfill. M 20 grade of concrete and Hysd steel bars shall be used. The safe bearing capacity of soil is 220 kN/m^2. Take foundation 1.00 m deep.**

26

Design of Domes

26.1 INTRODUCTION

A **dome** is a structure having thin curved surface formed by revolution of curved surfaced about a vertical axis. A vertical section through this axis in any direction is, as a rule, an arc of a curve. A **spherical dome** is obtained by revolving a segment of a circle about its vertical axis. A **conical dome** is formed by the revolution of a right angled triangle about its vertical axis. A **spheroid dome** is generated by revolving an elliptical curve through its axis of revolution. In general, the spherical domes are very commonly used.

Domes are needed to roof large spans, *for examples* auditoriums, assemble halls, temples, mosques, hangers, exhibition halls, circular tanks, bins and bunkers, . . . etc. Domes have been built since the earliest times. Almost all the schools of architecture (modern, ancient, western, oriental) have recognized the innate beauty of domes. In early days, the **masonry domes** were built. The various examples of the *maxonry domes* are pantheon dome in Rome, St. Sophia's domes in Constantinople, St. Paul's dome in London (in western world) and dome of Taj Mahal at Agra and the Gol Gumbaz at Bijapur in India). Dome of the civic centre in Charlotte, U.S.A. built in 1954, is built in *steel.* Dome of discovery constructed in London for the festival of Britain in 1951 (which was later dismantled after the festival was over) was built in *aluminium.* Domes built in concrete and suitably reinforced are advantageous, since the wet concrete readily takes up the curvature required in two directions.

From the practical point of view, the curved shuttering becomes expensive and there are difficulties for the accurate erection and placing of concrete on both faces where the inclination to the horizontal becomes more than 30 degrees. The erection cost becomes very high. However it becomes economical, in case, the shuttering is repetitively used.

As compared to the various other dimensions of dome, the thickness of dome is very small. It may vary from 75 mm to 150 mm. The domes may be designed with the normal dimensions in case the loading is symmetrical about the axis of revolution and the dome is supported in a plane at right angles to its axis of revolution and the supports being level all round the edge. In case, these specified conditions are satisfied, only direct compressive forces are needed to maintain the stability.

26.2 DEFINITIONS OF TERMS RELATED WITH DOMES

The definitions of following terms related with the domes have been given so that the domes may be studied with convenience. Consider a frustrum of a spherical dome as shown in Fig. 26.1 (A) and (B). Hypothetically, the dome may be considered to consist of number of rings (horizontal slices) of uniform heights and decreasing diameters and placed one above the other. Each ring supports the load of all the rings above it and transfers the load to the ring immediately below it.

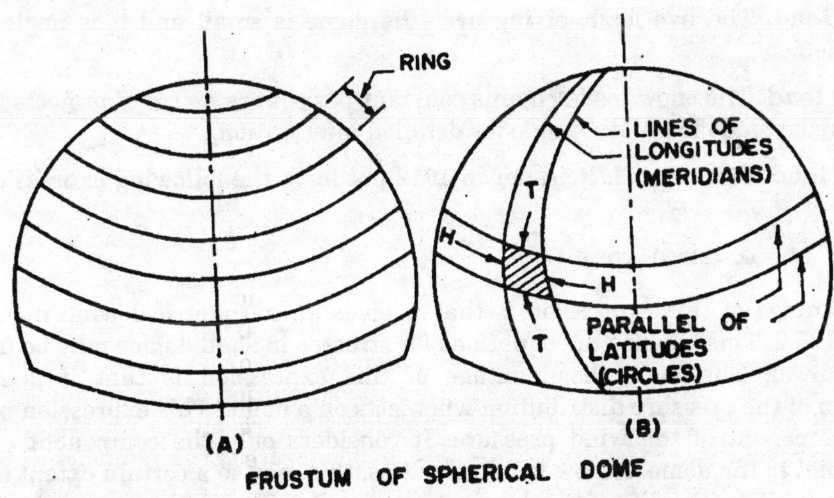

Fig. 26.1

1. Latitudes and longitudes. The **latitude** is the circle of a ring. In case, the circles are drawn through top and bottom points dimetrically opposite to each other, then, these are called as **circles of meridians**. A line corresponding to each circle of meridian is called as *longitude*. The latitude and longitude are used to represent two directions. Mathematically, these are used to represent two angular directions. Geographically these terms are used to represent the position of a place. The longitude is an arc of the equator between the meridian of a place and a standard meridian, where as the latitude is the angular distance from the equator.

The top most point of the spherical dome is called **crown.** The meridian circles converge at the crown.

2. Meridianal thrust. The loads coming from all the above rings over a ring acting down the lines of longitudes, Fig. 26.1 (B) and the reaction from the bottom provides direct compression over the ring. This direct compressive force acting along the meridians is called as *meridianal thrust,* represented by letters, T and T.

3. Hoop compression. It is seen that the dome shown in Fig. 26.1 (B) remains in equilibrium by virtue of top part of the wedge shaped sides of each sector pressing one another. The top of each sector tends to fall inwards.

4. Hoop tension. Towards the support, these lateral reactions becomes **circumferential tension.** The circumferential tensions are provided by the circumferential bands or circular ties near the bottom.

26.3 VARIOUS LOADS ACTING ON DOMES

The various loads acting on the domes are as under :

1. Self-weight 2. Live load 3. Snow load 4. Wind load

1. Self-weight. The self-weight of dome is a very variable factor. It depends upon the type of dome provided. However, a rough figure of 0.24 to 0.58 kilo-Newton per square metre of projected plan area may be assumed.

2. Live load. The live load acting over the dome is small and it is neglected in the design of dome.

3. Snow load. The snow load remains constant per square metre of projected plan area. Reference may be made to building code for detailed information.

4. Wind load. Professor H. Reissner in 1912 put forth the following expression for wind distribution

$$p = \phi_o \cdot \sin\phi \cdot \cos\phi \qquad \text{(i)}$$

The advantage of this expression is that it gives an antimetrical wind distribution as shown in Fig. 26.2. This enables the equations for stresses in shell domes may be framed very easily and solved. The main disadvantage of this expression is that it is not a true representation of the pressure distribution when acts on a dome. This expression neglects the tangential component of the wind pressure. It considers only the component of the wind pressure normal to the dome. At low wind velocities, this may to a certain extent be true, but at high wind velocities, the life of the dome may seriously affect the pressure distribution.

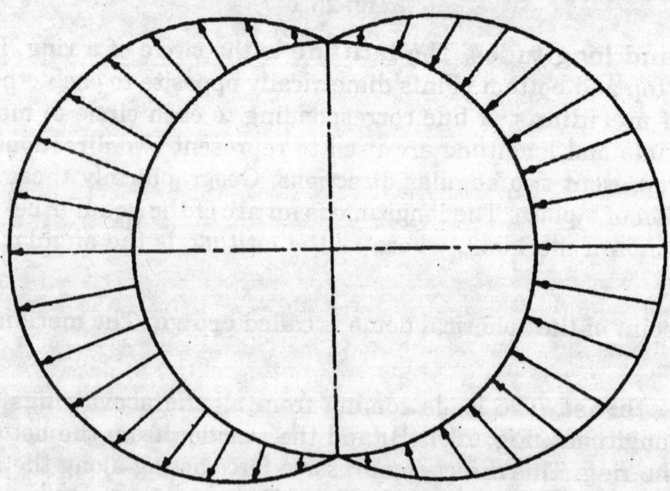

P = Po Sin Ø Cos Ø

ANTIMETRICAL DISTRIBUTION OF WIND PRESSURE

Fig. 26.2

Stresses due to wind. It is difficult to estimate the stresses due to wind, temperature and shrinkage. The analysis of dome subjected to these effects is complicate. However, the stresses due to these effects are small as the stresses due to dead load themselves are small. The horizontal pressure due to wind from one direction tends to cant the otherwise horizontal circumferential lines of force round the dome to a small angle with the horizontal and at the same time slightly increase the meridional thrust in the leeward side. The effects are accounted for by the addition of reinforcement in both the directions to the extent of about 0.2 percent of the cross-sectional area.

Professor F. Dischinger put the better representation of the wind pressure distribution diagram on the structure as below :

$$p = p_o \sin^2 \phi \,.\, \sin^2 \psi, \quad where\ \psi = (90 - \theta) \tag{ii}$$

Experimental tests (conducted by F. Dischinger) gave results closer to those obtained by the expression (ii) than to those given by the expression (i).

The disadvantage of Dischinger's expression is that is does not give an antimetrical loading on the dome. To overcome this difficulty, Professor F. Dischinger replaced $\sin^2 \psi$ *by* $\sin \psi$, *and* $\sin^3 \psi$

So that

$$p = p_o \sin^2 \phi \left|c \,.\, \sin \psi + (1 - c) \sin^3 \psi\right| \tag{iii}$$

where, c is a constant determined from the consideration that the horizontal load on the dome must in both formulate be the same. The value of c was suggested as 0.4. Therefore,

$$\sin^2 \psi = (0.4 \sin \psi + 0.6 \sin^2 \psi) \tag{iv}$$

$$\sin^3 \psi = (0.75 \sin \psi - 0.25 \sin 3\psi) \tag{v}$$

$$\text{Then } p = p_o \sin^2 \phi \left[0.85 \sin \psi - 0.15 \sin 3\psi\right] \tag{vi}$$

26.4 SPHERICAL DOME CARRYING UNIFORMLY DISTRIBUTED LOAD

The spherical dome of uniform thickness carrying uniformly distributed load may be analysed and hoop compression and meridional thrust at any fibre of the dome may be found.

The weight of covering (if any) for example, tiles, sheet metal or asphalt, and suspended ceiling (if provided) shall be considered in addition to the self-weight of the structural concrete. The weight of covering remains proportional to the surface of the dome. Whereas, the weight of the ceiling remains in proportion to that of the projected area. The sum of those may be adopted as 0.96 kN/m². It is added to the self-weight of the concrete. It shall be the uniformly distributed load per unit area. Let the uniformly distributed load acting on the spherical dome be w per unit area.

Fig. 26.3 shows a section through the vertical axis of revolution of a thin spherical dome of thickness, *t* and the radius, *r*. An elemental ring bounded by two horizontal planes CD and EF close together is considered. The positions of the ring are defined by angles q and $(\theta + \Delta\theta)$ with the vertical. The meridional thrust and the hoop compression at any fibre of the concrete may be determined as under :

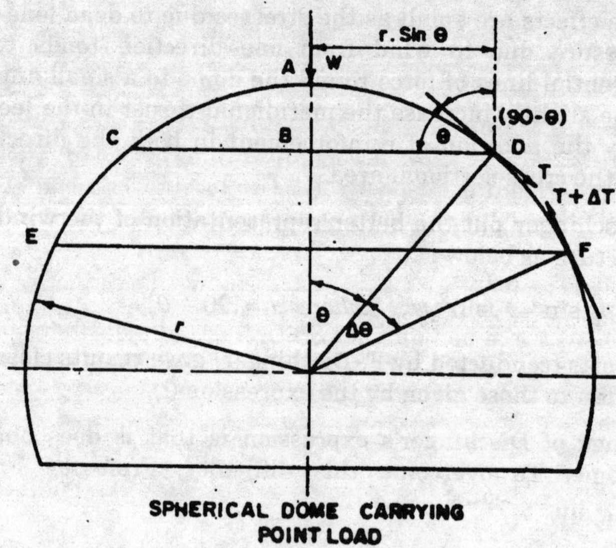

Fig. 26.3

1. Meridional thrust. The elemental ring DEFE is in equilibrium under meridional thrust, T per unit length of the circle of latitude CD down the lines of longitude tangential to the surface and a similar horizontal thrust (T + Δ T) per unit length acting up the lines of longitude also tangential to the surface round the circle of latitude, EF and the weight of ring acting vertically downward. Two meridional thrusts T and $(T + \Delta T)$ are not equal. These thrusts act at a small angle $\Delta\theta$ to one another between them. The meridional thrust (T + ΔT) acting upward is more than the meridional thrust, T.

$$(T + \Delta T) - (T) = \Delta T \quad \text{(i)}$$

The meridional thrust T is due to weight of the dome portion CAD. The surface area of portion of dome CAD.

$$= 2\,\pi\,\gamma\,.\,AB \quad \text{(ii)}$$

Let w be weight per unit area, then, the total weight

$$= w\,.\,(2\,\pi\,\gamma.\,AB) \quad \text{(iii)}$$

From Fig. 26.3, $AB = (AO - BO) = \gamma\,(1 - \cos\theta)$ (iv)

Therefore, the weight of portion of dome CAD $= w\,.\,2\,\pi r^2\,.\,(1 - \cos\theta)$ (v)

The sum of the vertical components of all the thrust T per unit length acting round the circumference of circle, CD is equal to the above wt.

$$T\,.2\pi\,.\,BD\,.\cos(90 - \theta) = w\,.\,2\,\pi r^2\,(1 - \cos\theta)$$

$$\text{or } T\,.\,2\,\pi\,(\gamma.\sin\theta).\sin\theta = w\,.\,2\,\pi r^2\,(1 - \cos\theta)$$

$$\therefore \quad T = w.r.\left(\frac{1-\cos\theta}{\sin^2\theta}\right) \tag{26.1}$$

2. Hoope compression. The hoop compression, H develops due to the thrust T acting downward and thrust (T + ΔT) acting upward. These two thrusts act at angles θ and $(\theta + \Delta\theta)$ with the horizontal, respectively. The breadth of elemental ring, DF measured on great circle are (viz., a line of longitude through the crown, A) is $r\Delta\theta$. The hoop force

$$= -H \,.\, r\Delta\theta \tag{vi}$$

where, H is the hoop force per unit length of surface measured on the great circle (already defined above). The component of the meridional thrust, T acting horizontally and radially (outward)

$$= T.\cos\theta \tag{vii}$$

This horizontal component develops hoop tension and tends to increase the diameter of this ring.

Hoop tension = T. cos θ . (radius of ring, *CD*)

$$= T\,.\cos\theta\,.\,r\sin\theta$$

The component of the meridoinal thrust (T + ΔT) acting horizontally and radially (inward)

$$= (T + \Delta T)\,.\cos(\theta + \Delta\theta) \tag{ix}$$

This horizontal component develops hoop compression and tends to decrease the diameter of this ring. Hoop compression

$$= (T + \Delta T)\,.\cos(\theta + q\Delta)\,.\,r\ \sin\ (\theta + \Delta\theta) \tag{x}$$

The difference between the two expressions (viii) and (ix) gives the hoop compression or tension (whichever is more) on the elemental ring. The difference is obtained as below :

$$(T + \Delta T)\,.\,[\cos(\theta + \Delta\theta)\,.\,r\sin(\theta + \Delta\theta)] - T\cos\theta\,.\,r\sin\theta$$

$$= (T + \Delta T)\,.\,r.\,[\cos\theta\,.\cos\Delta\theta - \sin\theta\,.\sin\Delta\theta) \times (\sin\theta\cos\Delta\theta + \cos\theta\,.\sin\Delta\theta)]$$

$$- T\cos\theta\sin\theta$$

$$= (T + \Delta T)\,.\,r\,[(\cos\theta\,.\sin\theta\,.\cos^2\theta\,\Delta\theta\,.\sin^2\theta\,.\sin\Delta\theta\,.\cos\Delta\theta + \cos^2\theta\,.$$

$$\cos\Delta\theta\sin\Delta\theta - \cos\theta\,.\sin\theta\,.\sin^2\Delta\theta) - T\,.\,r.\cos\theta\sin\theta \tag{xi}$$

When Δθ tends to zero, cos Δθ becomes *unity* and sin Δθ becomes zero. Therefore, above difference may be written

$$(T + \Delta T)\,.\,r\cos\theta\,.\sin\theta - T.\,r.\cos\theta\sin\theta = \Delta(T\,.\,r.\cos\theta\,.\sin\theta) \tag{xii}$$

Hoop force

$$-H.r.\,\Delta\theta = \Delta\,.\,(T.\,r.\cos\theta\,.\sin\theta) \tag{xiii}$$

From Eq. 26.1, $T = \left(\dfrac{w.r\,(1-\cos\theta)}{\sin^2\theta}\right)$ (xiv)

Therefore, $-H = \dfrac{1}{2}\dfrac{\Delta}{\Delta\theta}\left[w.r.\dfrac{(1-\cos\theta).\,r}{\sin^2\theta}\cos\theta\sin\theta\right]$ (xv)

The expression (xv) may be written in the form of differentiation as below :

$$-H = w.r.\frac{d}{d\theta}\left(\frac{\cos\theta}{\sin\theta}-\frac{\cos^2\theta}{\sin\theta}\right) \quad \text{(xvi)}$$

The rules for differentiation for a product and a quotient may be applied.

Therefore,

$$(-H) = w\,.\,r.\left[\frac{-\sin^2\theta-\cos^2\theta}{\sin^2\theta}-\left(\frac{-2\sin^2\theta\,.\cos\theta-\cos^3\theta}{\sin^2\theta}\right)\right]$$

$$= w.r.\left[\frac{-1+2\sin^2\theta\,.\cos\theta+\cos^3\theta}{\sin^2\theta}\right]$$

$$= w.r.\left[\frac{-1+2\cos\theta-2\cos^3\theta+\cos^3\theta}{\sin^2\theta}\right]$$

$$(-H) = w.r.\left[\frac{-1+2\cos\theta-\cos^3\theta}{\sin^2\theta}\right] = -w.r.\left[\frac{1-2\cos\theta+\cos^3\theta}{\sin^2\theta}\right]$$

$$\therefore H = w.r.\left[\frac{1-2\cos\theta+\cos^2\theta-\cos^2\theta+\cos^3\theta}{1-\cos^2\theta}\right]$$

$$\therefore H = w.r.\left[\frac{(1-\cos\theta)^2-\cos^2\theta\,(1-\cos\theta)}{(1+\cos\theta)\,(1-\cos\theta)}\right]$$

$$H = w.r.\left[\frac{1-\cos\theta-\cos^2\theta}{1+\cos\theta}\right] \quad (26.2)$$

At A, Fig. 26.3, (viz., crown of the dome) the angle θ shall be zero. From Eq. 26.2,

$$H = \left(-\frac{w.r}{2}\right) \quad \textit{(i.e., compressive)} \quad (26.3)$$

The intensity of hoop stress shall be

$$\sigma_H = \left(\frac{H}{t}\right) = \left(-\frac{w.r}{2t}\right) \quad (26.4)$$

The hoop stress is compressive and maximum at the crown.

The hoop compressive force shall be zero when the right side of Eq. 26.2 disappears. That is

$$w.r.\left[\frac{1-\cos\theta-\cos^2\theta}{1+\cos\theta}\right]=0$$

or $(1 - \cos\theta - \cos^2\theta) = 0$

then $\cos\theta = 0.618$, i.e., $\theta = 51°48'$

Where there is no hoop compression and no hoop tension, the plane through the circle is called as the **plane of rupture.** When the angle q exceeds 51 degrees and 48 minutes, the hoop force changes its sign. It becomes tensile. The variation of hoop stress is as shown in Fig. 26.4.

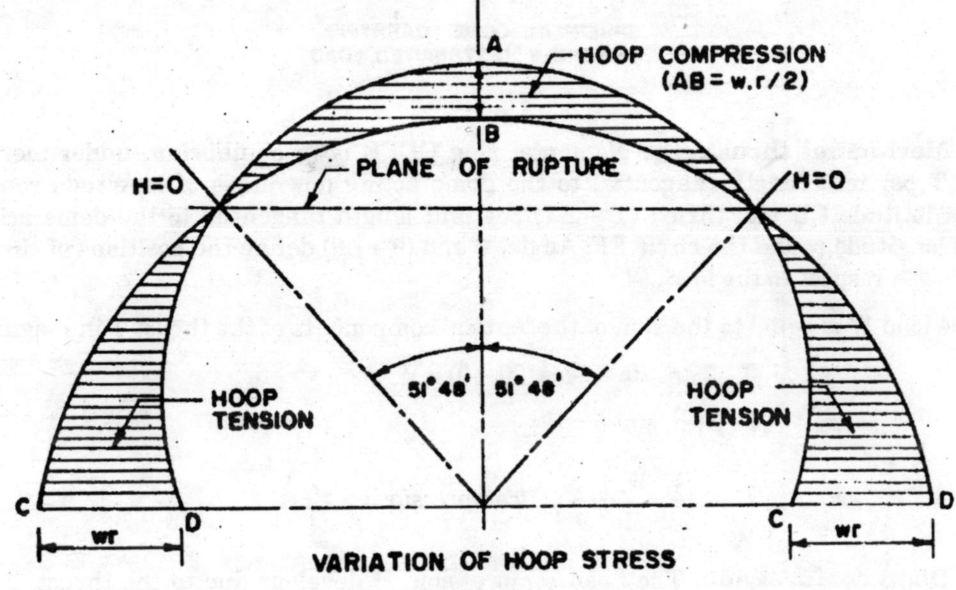

Fig. 26.4

26.5 SPHERICAL DOME CARRYING CONCENTRATED LOAD AT CROWN

A spherical dome of uniform thickness carrying a point load W at the crown is shown in Fig. 26.5, through the vertical axis of revolution. The meridional thrust and hoop force due to self-weight of the dome shall be calculated as described in Art. 26.4. The stresses in dome due to point load are determined separately. The resultant stresses are obtained by adding the stresses calculated for the two cases.

An elemental ring bounded by two horizontal planes CD and EF close together is considered. The positions of the ring are defined by angle θ and (θ + Δθ) with the vertical axis. The meridional thrust and the hoop force at any fibre of the concrete may be obtained as under :

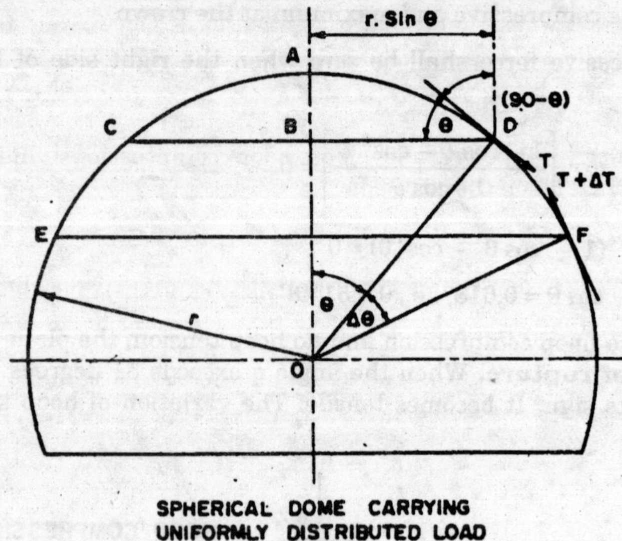

Fig. 26.5

1. Meridional thrust. The elemental ring CDFE is in equilibrium under meridional thrust, T per unit length tangential to the dome acting downlines of longitude round the circle of latitude CD and thrust (T + ΔT) per unit length tangential to the dome acting up lines of longitude round the circle EF. Angles θ and (θ + Δθ) define the positions of circles CD and EF with respect to the load, W.

The load W is equal to the sum of the vertical components of the thrust T in magnitude

$$\therefore \quad T \,.\, 2\pi r \,.\, \sin\theta \,.\, \cos(90-\theta) = W$$

or $$T \,.\, 2\pi r \,.\, \sin^2\theta = W$$

or $$T = \left(\frac{W}{2\pi r.\sin^2\theta}\right) \quad (compression) \tag{26.5}$$

2. Hoop compression. The hoop compression, H develops due to the thrust, T acting downward and thrust (T + ΔT) acting upward. These two thrusts act at angles θ and (θ + Δθ) with the horizontal, respectively. The breadth of elemental ring DF measured on great circle are (viz., a line of longitude through crown A) is rΔθ. The Hoop force

$$= (-H \,.\, r\,\Delta\theta) \tag{i}$$

where, H is the hoop force per unit length of surface measured on great circle. The component of meridional thrust, T acting horizontally and radially (outward)

$$= T \,.\, \cos\theta \tag{ii}$$

This horizontal component develops hoop tension and tends to increase the diameter of this ring. Hoop tension

$$T \,.\, \cos\theta \text{ (radius of ring CD)} = T \,.\, \cos\theta \sin\theta \tag{iii}$$

The component of the meridional thrust (T + ΔT) acting horizontally and radially (inward)

$$= (T + \Delta T) \,.\, \cos(\theta + \Delta\theta) \qquad \text{(iv)}$$

This horizontal component develops hoop compressions and tends to decrease the diameter of this ring. Hoop compression

$$= (T + \Delta T)\ \cos(\theta + \Delta\theta).\, r \sin(\theta + \Delta\theta) \qquad \text{(v)}$$

The difference between the two expressions (v) and (iii) gives hoop force. Therefore,

$$-H\,.\,r\,.\,\Delta\,\theta = \Delta\,(T\,.\,r\,.\cos\theta\,.\sin\theta) \qquad \text{(vi)}$$

$$\text{From Eq. 26.5, } T = \left(\frac{W}{2\pi r\,.\sin^2\theta}\right) \qquad \text{(vii)}$$

$$\therefore\ -H\,.r.\,\Delta\theta = \Delta.\left(\frac{W}{2\pi\gamma\sin^2\theta}.\,r\,.\,\cos\theta\sin\theta\right)$$

$$\text{or}\ \ -H = \frac{W}{2\pi\gamma}.\frac{\Delta}{\Delta\theta}\left(\frac{\cos\theta}{\sin\theta}\right) \qquad \text{(viii)}$$

The expression (viii) may be written as

$$(-H) = \frac{W}{2\pi r}\times\left[\frac{d}{d\theta}\left(\frac{\cos\theta}{\sin\theta}\right)\right] = \frac{W}{2\pi r}.\left(\frac{-\sin\theta\sin\theta - \cos\theta\cos\theta}{\sin^2\theta}\right)$$

$$H = \frac{W}{2\pi r}.\left(\frac{\sin^2\theta + \cos^2\theta}{\sin^2\theta}\right) = \left(\frac{W}{2\pi r}\right).\,\text{cosec}^2\,\theta\ (tension) \qquad (26.6)$$

At *crown* the angle θ shall be zero and cosec θ shall be infinity. Therefore, the value of H shall be *infinite.* Therefore, when the lantern is attached, the lantern load shall be acting as concentrated load at angle θ equal to zero. This lantern load is to be spread over an appreciable area and if necessary, the dome must be thickened, reinforced strongly or both.

The values of various coefficients of the hoop stresses are given for a thin spherical dome of uniform thickness for uniformly distributed load and for concentrated load at the crown for different values of angle q in Table 26.1.

The coefficients of w. r/ (W/r) for meridional thrust, T and the hoop stresses, H are given for uniformly distributed load and for point load at the crown for the segmental dome of uniform thickness, separately.

However, for any design, the effect of both the loadings must be considered simultaneously at any level. The level of zero hoop force or the joint of rupture depends on the relative magnitude of the unit weight w of the dome and the concentrated load, W and on the radius, r. It shall be no longer occur at angle of 51 degree and 48 minutes. The value of θ at the joint of rupture for the combined loading, the whole expression may be equated to zero. That is,

$$\left[w.r.\left(\frac{1-\cos\theta-\cos^2\theta}{1+\cos\theta}\right)+\frac{W}{2\pi r}.\operatorname{cosec}^2\theta\right]$$ *is equated to zero.*

26.6 DESIGN OF DOMES

The thickness of dome needed from the consideration of stresses developed due to load is very small in comparison with the other dimensions. The thickness of dome is kept at 75 mm minimum and it may be 150 mm. For water tanks the ratio of span to rise of the dome may be kept as 4 to 6.

TABLE 26.1. COEFFICIENTS OF w.r/(W/r) FOR MERIDIONAL THRUST T AND HOOP FORCES, H

θ (deg.)	Uniformly distributed load		Point load W at crown	
	Meridional thrust T coeff. of w. γ	Hoop force H coeffi. w.r.	of Meridional thrust R coeffi. of W/γ	Hoop force n coeffi. of W/γ
	$\left(\frac{1-\cos\theta}{\sin^2\theta}\right)$	$\left(\frac{1-\cos\theta-\cos^2\theta}{1+\cos\theta}\right)$	$\left(\frac{1}{2\pi.\sin^2\theta}\right)$	$\left(\frac{1}{2\pi}.\frac{1}{\sin^2\theta}\right)$
0	0.5	–0.5	∞	∞
5	0.500	–0.496	21.000	21.000
10	0.505	–0.480	5.300	5.300
20	0.556	–0.425	1.370	1.370
30	0.537	–0.330	0.640	0.640
40	0.566	–0.200	0.380	0.380
50	0.608	–0.034	0.270	0.270
51°48′	0.618	0.000	0.260	0.260
60	0.667	+0.167	0.210	0.210
70	0.747	+0.402	0.180	0.180
80	0.838	+0.680	0.160	0.160
90	1.000	+1.000	0.160	0.160
100	1.110	+1.380	0.160	0.160
110	1.520	+1.860	0.180	0.180
120	2.000	+2.500	0.210	0.210
130	2.790	+3.450	0.270	0.270
140	4.270	+5.050	0.380	0.380
150	7.480	+ 8.330	0.640	0.640
160	16.600	+17.500	1.370	1.370
170	66.000	+66.000	5.300	5.300
180	∞	∞	∞	∞

(**Reference.** Terrington, J.S., `Design of Domes' Published by Concrete Publications Limited 14 Dartsmouth Street, London, S.W.1)

Note : Positive sign is for tensile stress.

Ring beam. Mostly of the domes used to be segmental. At the free edge of the segmental dome, the meridional compression shall be inclined with the horizontal. The horizontal component of the meridional compression provides an outward thrust. In order to resist this outward thrust a *ring beam* is provided at the base of the dome. The reinforcement provided in the ring beam resists hoop tension. Only vertical component is transmitted to the supporting wall. In hemispherical dome, the hoop compression remains vertical.

Reinforcement. A minimum reinforcement of 0.2 percent of the cross-sectional area of the dome is provided in both the directions (viz., latitudes and longitudes). In order to avoid congestion of the steel bars, the steel bars are provided upto certain latitude and mesh reinforcement is provided at the crown.

Example 26.1. The inside diameter of a circular room is 9 m. Design a spherical dome to carry a uniformly distributed live load of 1.50 kN/m². The dome is to support a lantern of load 18 kN as point load from a circular opening of 1.8 m at the crown. M 20 grade of concrete and Hysd steel of grade Fe 415 shall be used.

Design : The modular ratio of M 20 grade of concrete

$$m = \left(\frac{280}{3 \times 7}\right) = 13.33$$

Step 1 : Design constants. For M 20 grade of concrete and Hysd-steel bars of grade Fe 415, for the balanced section, the design constant are as follows.

The ratio of limiting value of depth of neutral axis to the effective depth

$$k_B = 0.2886,\ j_B = 0.9038,\ R_B = 0.9129,\ p_B = 0.4398$$

Step 2 : Dimension of dome. The dimension of dome are as shown in Fig. 26.6. The inside diameter of dome is 9 m. The span to rise ratio is kept 4 to 6. Therefore, the rise of dome

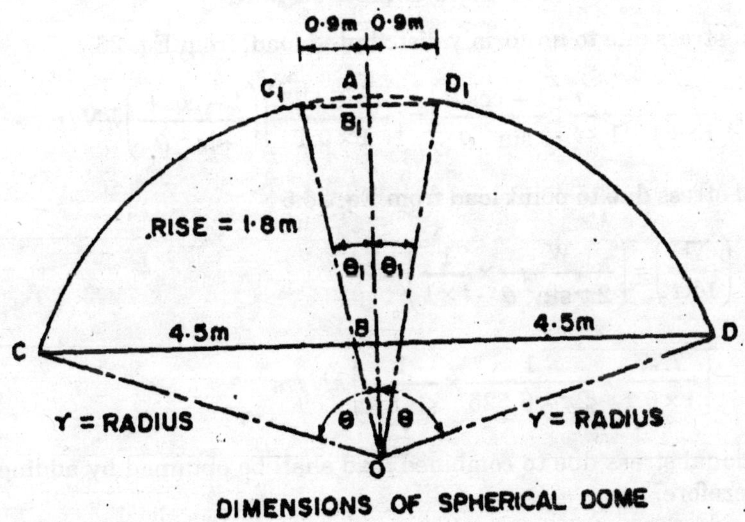

Fig. 26.6

(9/4) = 2.25 m, and (9/6) = 1.50 m (i)

Provide a rise of 1.8 m, Let r be the radius of dome. Then, Rise. (2r –rise) = 4.5^2 (ii)

$$1.8\times(2r-1.8)=20.25, \textit{therefore}, r=6.525\ m$$

$$\sin\theta=(4.5/6.525)=0.6897 \text{ and } \cos\theta=0.7241 \text{ and } \theta=43.603^\circ \quad \text{(iii)}$$

Similar to the expression (i) for the opening of 1.8 m, AB_1 may be calculated

$$AB_1\times(2r-AB_1)=(0.9\times0.9)=0.81$$

$$\text{or } AB_1\times(2\times6.525-AB_1)=0.81, \textit{therefore } AB_1=0.064\ m \quad \text{(iv)}$$

$$\sin\theta_1=(0.9/6/525)=0.1379 \text{ and } \cos\theta_1=0.9904, \text{ and } \theta_1=7.928^\circ$$

Step 3 : Loads. It is to note that in Arts. 26.4 and 26.5, therefore was no opening in the dome. In this case, an opening of 1.8 m, diameter has been provided. It may be assumed that there is no opening. The effective weight of lantern may be considered to consider the effect of opening.

Let the thickness of dome be 100 mm. The uniformly distributed load per square metre of surface area due to

$$\text{Self-wt.} = 0.1\times25=2.5\ kN/m^2, \quad \textit{Live load} = \ =1.5\ kN/m^2$$

Total = 4 kN/m²

Weight of the portion of dome C_1AD_1

$$(4\times2\pi r\times AB_1)=(4\times2\pi\times6.525\times0.0624)=10.23\ kN$$

Effective weight of lantern, = (18 – 10.23) = 7.77 kN

Step 4 : Meridional stress due to combined load

Meridional stress due to uniformly distributed load, from Eq. 26.1

$$\left(\frac{T}{1\times t}\right)=\frac{w\,.\,r}{1\times t}\left(\frac{1-\cos\theta}{\sin^2\theta}\right)=\left(\frac{4\times6.525}{1\times0.1}\right)\left(\frac{1-\cos\theta}{\sin^2\theta}\right)kN/m^2 \quad \text{(v)}$$

Meridional stress due to point load from Eq. 26.5

$$\left(\frac{T}{1\times t}\right)=\left(\frac{W}{2\pi r\sin^2\theta}\times\frac{1}{t\times1}\right)$$

$$=\left(\frac{7.77}{1\times0.1}\times\frac{1}{2\pi\times6.525}\times\frac{1}{\sin^2\theta}\right)kN/m^2 \quad \text{(vi)}$$

The meridional stress due to combined load shall be obtained by adding the expressions (v) and (vi). Therefore,

$$= \left[\frac{4 \times 6.525}{1 \times 0.1}\left(\frac{1 - \cos\theta}{\sin^2\theta}\right) + \frac{7.77}{1 \times 0.1 \times 2\pi \times 6.525}\frac{1}{\sin^2\theta}\right]$$

$$= \left[261\left(\frac{1 - \cos\theta}{\sin^2\theta}\right) + \frac{11.91}{2\pi}\left(\frac{1}{\sin^2\pi}\right)\right] \quad \text{(viii)}$$

The values of meridional stress for combined loads for the different values of angle θ have been tabulated in Table 26.2 given below :

Step 5 : Hoop stress due to combined load

Hoop stress due to uniformly distributed load, from Eq. 26.2,

$$\left(\frac{H}{t \times 1}\right) = \frac{w \, . \, r}{t \times 1}\left[\frac{1 - \cos\theta - \cos^2\theta}{1 + \cos\theta}\right]$$

$$\left(\frac{4 \times 6.525}{0.1 \times 1}\right)\left[\frac{1 - \cos\theta - \cos^2\theta}{1 + \cos\theta}\right] = 261\left(\frac{1 - \cos\theta - \cos^2\theta}{1 + \cos\theta}\right) \quad \text{(ix)}$$

Hoop stress due to point load, from Eq. 26.6

$$\left(\frac{H}{t \times 1}\right) = \left(\frac{w}{2\pi r} \cdot \frac{1}{t \times 1} \cdot \operatorname{cosec}^2\theta\right)$$

$$\left(\frac{H}{t \times 1}\right) = \left(\frac{7.77}{2\pi \times 6.525} \times \frac{1}{0.1 \times 1} \times \frac{1}{\sin^2\theta}\right) = \left(1.895\frac{1}{\sin^2\theta}\right) kN/m^2 \quad \text{(x)}$$

The value of hoop stress due to combined load shall be obtained by adding the expressions (ix) and (x). Therefore,

$$= \left[261\left(\frac{1 - \cos\theta - \cos^2\theta}{1 + \cos\theta}\right) + \frac{11.91}{2\pi}\left(\frac{1}{\sin^2\theta}\right)\right] kN/m^2 \quad \text{(xi)}$$

The values of hoop stress for combined loads for the different values of angle θ have been calculated using Table 26.1 and tabulated in Table 26.2, given below :

TABLE 26.2

Angle, q	Meridional stress, kN/m²			Hoop stress, kN/m²		
	Due to UDL	Due to POINT LOAD	Total	Due to UDL	Due to POINT LOAD	Total
5°	130.50	250.11	–	–129.46	250.11	–
7.928°	131.27	140.56	271.83	–127.01	140.56	+13.55
10°	131.81	63.12	194.93	–125.28	63.12	–62.16
20°	134.68	16.32	151.00	–110.93	16.32	–94.61

30°	140.16	7.62	147.78	–86.13	7.62	–78.51
40°	147.73	4.53	152.26	–52.29	4.53	–47.67
43.603°	151.68	4.06	155.74	–*39.98	*4.6	–35.92
50°	158.688	3.22	–	–8.874	3.22	–

*This value is calculated by extrapolation instead of intrapolation.

Step 6 : Hoop stress due to combined loading (without live load). The hoop stress is determined for the combined loading without live load as the hoop tensile stress near the lantern opening is likely to be more. Self-weight of the dome

$$(0.1 \times 1 \times 25) = 2.5\ kN/m^2$$

Weight of the portion of dome C_1AD_1

$$(2.5 \times 2\,\pi \times 6.525 \times AB_1) = 2.5 \times 2\pi \times 6.525 \times 0.0624 = 6.40\ kN$$

Effective weight of lantern

$$(18 - 6.40) = 11.60 \text{ kN}$$

Hoop stress due to combined loading without live load shall be (H/t x 1)

$$\left[\frac{2.5 \times 6.525}{0.1 \times 1}\left(\frac{1 - \cos\theta - \cos^2\theta}{1 + \cos\theta}\right) + \frac{11.60}{6\pi \times 6.625} \times \frac{1}{0.1 \times 1} \times \frac{1}{\sin^2\theta}\right] kN/m^2$$

$$= \left[163.125\left(\frac{1 - \cos\theta - \cos^2\theta}{1 + \cos\theta}\right) + \frac{17.77}{2\pi \sin^2\theta}\right] kN/m^2$$

The values of hoops stress for combined load without live load shall be calculated by decreasing values of uniformly distributed load in the ratio (163.125/261) = 0.625 and increasing the values of those due to point load in the ratio of (17.77/11.91) = 1.492 and these values have been tabulated in Table 26.3.

TABLE 26.3

	Hoop stress, kN/m²		
	Due to UDL	Due to point load	Total
7.928°	– 79.38	209.72	130.34
10°	– 78.30	94.18	15.88
20°	– 69.33	24.35	–44.98
30°	– 53.83	11.37	–42.46
40°	– 32.63	6.77	–25.86
43.603°	– 24.99	6.06	–18.93

From Table 26.3 maximum hoop tension occurs at 7.9280° (i.e., at the opening)

130.34 kN/m² = 0.130 N/mm²

Step 7 : Reinforcement

Maximum meridional stress (compression), ∴ 271.83 kN/m² = 0.272 N/mm²

Maximum hoop tension occurs at 7.928° is 0.130 N/mm²

Maximum hoop tension for 1 m length

$$\left(\frac{(0.1310 \times t \times 1000)}{1000}\right) = \left(\frac{0.130 \times 100 \times 1000}{1000}\right) = 13\ kN/m$$

Cross-sectional area of reinforcement (Hysd steel)

$$\left(\frac{13 \times 1000}{230}\right) = 56.52\ mm^2/m$$

$$\text{Distribution reinforcement, } \left(\frac{0.12}{100} \times 100 \times 1000\right) = 120\ mm^2/m$$

Total reinforcement, = 176.52 mm²/m

Provide 8 mm ϕ Hysd steel bars, $A_\phi = 50.26\ mm^2$

$$\text{Spacing of the steel bars, } \left(\frac{100 \times 50.26}{176.52}\right) = 284.758\ mm$$

8 mm ϕ Hysd steel bars shall be provided at 200 mm c/c, where, hoop tension exists. In case, there is no hoop tension 0.12 percent viz., 120 mm² reinforcement shall be provided at 400 mm c/c spacing.

The minimum reinforcement of 0.12 percent viz., 120 mm² shall also be provided in the direction of the meridian (i.e., at 400 mm c/c spacing).

Step 8 : Ring beam at bottom

The meridional thrust at the bottom

155.74 kN/m² = 0.156 N/mm²

Meridional thrust per 1 m length

$$\left(\frac{0.156 \times (t \times 1000)}{1000}\right) = \left(\frac{0.156 \times 100 \times 1000}{1000}\right) = 15.6\ kN/m$$

Horizontal component

15.6 cos (43.603) = (15.6 × 0.7421) = 11.297 kN/m

This horizontal component tends to rupture the dome

$$11.297 \times \left(\frac{2 \times 6.525}{2}\right) = 73.71\ kN/m$$

Cross-sectional area of steel, $\left(\dfrac{73.71 \times 1000}{415}\right) = 737.10\ mm^2$

Provide 10 mm ϕ Hysd steel bars.

Number of rings

(737.10/78.54) = 9.385

Provide 10 rings of 10 mm ϕ bars.

The equivalent cross-section are of ring beam

$$(A + (m-1)\,A_{st}) = (A + (13.33 - 1) \times 10 \times 78.54) = (A + 9683.98)\,mm^2$$

Tensile stress of 1.68 N/mm², may be allowed in M 20 grade of concrete. Then

$$\left(\frac{73.71 \times 10^3}{A + 9683.98}\right) = 1.68$$

$$\text{or}\ \ 73.71 \times 10^3 = \left[1.68 \times A + 16.27 \times 10^3\right]$$

$$A = \left(\frac{57.44}{1.68} \times 1000\right) = 34190\ mm^2$$

Size of ring beam = 200 mm x 200 mm. 6 mm ϕ ties are provided at 150 mm c/c.

Step 9 : Ring beam at top. Meridional thrust at top (viz., at θ = 7.928°)

= 271.83 kN/m² = 0.272 N/mm²

Horizontal component 0.276 cos (7.928)°

= 0.272 × 0.9904 = 0.269 N/mm²

Horizontal component per 1 m length

$$\left(\frac{0.269 \times (t \times 1000)}{1000}\right) = \left(\frac{0.269 \times 100 \times 1000}{1000}\right) = 26.9\ kN/m$$

Hoop compression

$$26.9 \times (Diameter/2) = \left(\frac{26.9 \times 1.8}{2}\right) = 24.246\ kN/m$$

A ring shall be provided which shall be resisting the compression. It will distribute the lantern load to the dome.

A ring beam of 150 mm x 150 mm may be provided.

$$\left(\frac{24.246 \times 1000}{150 \times 150}\right) = 1.078\ N/mm^2$$

It is safe. 8 mm f rings shall be extended within the ring beam. The details of reinforcement is shown in Fig. 26.7.

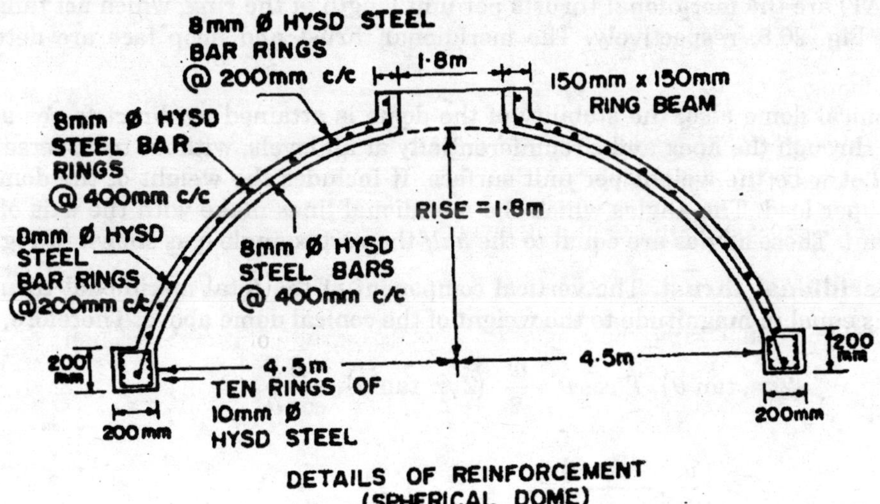

Fig. 26.7

26.7 CONICAL DOME

A conical dome is a surface of revolution generated by revolving a right angled triangle about its vertical axis. A conical dome may also be used to support a load at the crown. A section through the vertical axis of the dome is shown in Fig. 26.8.

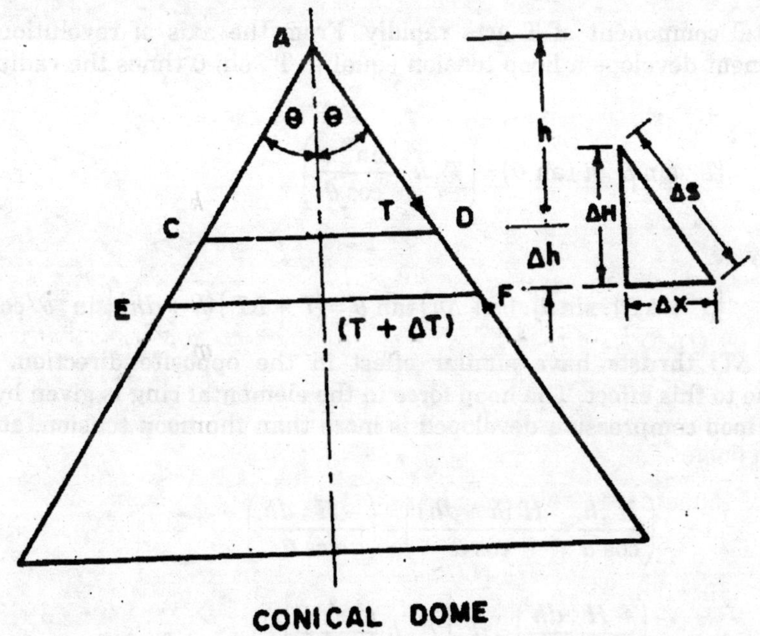

Fig. 26.8

An elemental ring CDFE bounded by two horizontal planes CD and EF close together is considered. The depths of horizontal plane CD and EF are h and (h + Δh) from the crown A. T and (T + ΔT) are the meridional thrusts per unit length of the ring, which act tangentially as shown in Fig. 26.8, respectively. The meridional thrust and hoop face are determined as follows.

In conical dome also, the stability of the dome is attained by direct forces acting down the dome through the apex and circumferentially at all levels, without transverse or bending stresses. Let w be the weight per unit surface. It includes the weight of the dome, covering and the super load. The angles which the meridional lines make with the axis of revolution are constant. These angles are equal to the *half* the vertex angle q as shown in Fig. 26.8.

1. Meridional thrust. The vertical component of the total meridional thrust T at the level CD is equal in magnitude to the weight of the conical dome above. Therefore,

$$(2\pi r \cdot \tan\theta) \cdot T \cos\theta = \frac{w}{2} \cdot (2\pi r \tan\theta) \cdot \frac{h}{\cos\theta}$$

$$\therefore \quad T = \left(\frac{w}{2} \cdot \frac{h}{\cos^2\theta}\right) \tag{26.7}$$

At the level of EF of the elemental ring, the meridional thrust increases to (T + ΔT).

2. Hoop force. The small increase in meridional thrust, ΔT causes hoop compression around the circles of latitude. The value of compression, H may be found as below :

$$(H \cdot 2\pi r \cdot \tan\theta \cdot ds) = \left(H \cdot 2\pi r \cdot \tan\theta \frac{dh}{\cos\theta}\right) \tag{i}$$

Horizontal component of T acts rapidly. From the axis of revolution is T-sin θ. This radial component develops a hoop tension equal to T . sin θ times the radius of the ring at a depth i.e.,

$$(T \cdot \sin\theta \cdot h \tan\theta) = \left(T \cdot h \cdot \frac{\sin^2\theta}{\cos\theta}\right) \tag{ii}$$

Similarly,

$$[(T + \Delta T) \cdot \sin\theta] \cdot (h + \Delta h) \tan\theta = (T + \Delta T)(h + dh)(\sin^2\theta / \cos\theta) \tag{iii}$$

(T + ΔT) thrusts have similar effect in the opposite direction. Compressions is developed due to this effect. The hoop force in the elemental ring is given by the difference of two effects. Hoop compression developed is more than the hoop tension, at all the points on the surface

$$\left(\frac{H \cdot h}{\cos\theta} - \frac{H(h + dh)}{\cos\theta}\right) = \left(\frac{-H \cdot dh}{\cos\theta}\right) \tag{iv}$$

$$\left(\frac{-H \cdot dh}{\cos\theta}\right) = d \cdot \left[T \cdot h \frac{\sin^2\theta}{\cos^2\theta}\right]$$

$$\text{or} \quad \left(\frac{-H\,.\,dh}{\cos\theta}\right) = d\,.\left[\frac{w}{2}\,.\,\frac{h}{\cos^2\theta}\,.\,\frac{h\sin^2\theta}{\cos\theta}\right]$$

$$(-H) = \left(\frac{w}{2}\,\frac{\cos\theta\,.\,\sin^2\theta}{\cos^2\theta}\,.\,d\left(h^2\right)\right) = \left(\frac{w}{2}\,.\,\frac{\sin^2\theta}{\cos^2\theta}\,.\,2h\right)$$

$$\text{or} \quad H = \left(-w\,.\,\tan^2\theta\,.\,h\right) \tag{26.8}$$

The negative sign indicates that the cone is in hoop compression at all the levels. The supports are to resist the inclined thrust. It gives a horizontal reaction

$$T\,.\sin\theta = \left(\frac{w\,.\,h\,.\sin\theta}{2\,.\cos^2\theta}\right) \tag{26.9}$$

This force may be taken by a band of reinforcement in ring tension. The hoop stress in dome shall be H/(t x 1).

Example 26.2. The inside diameter of a circular room is 16 m. The uniformly distributed load is 1.8 kN/m². Design the conical dome. M 20 grade of concrete and Hysd steel of grade Fe 415 shall be used.

Design : The modular ratio of M 20 grade of concrete

$$m = \left(\frac{280}{3\times 7}\right) = 13.33$$

Step 1: Design constants. For M 20 grade of concrete and Hysd steel bars of grade Fe 415, the design constants are as follows :

$k_B = 0.2886$, $j_B = 0.9038$, $R_B = 0.9129$, $p_B = 0.4398$

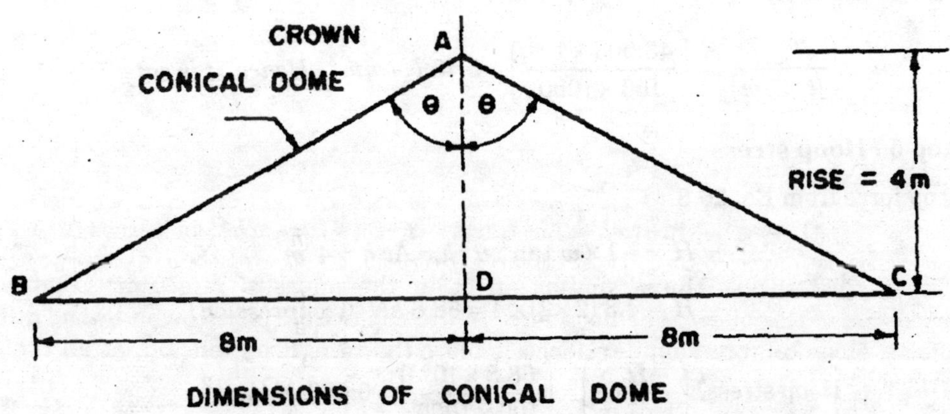

Fig. 26.9

Step 2 : Dimensions of dome. The dimensions of dome are as shown in Fig. 26.9. The inside diameter of dome is 16 m. The span to rise ratio is kept 4 to 6. Therefore, the rise of dome

(16/4) = 4 m (i)

Height of dome = Rise = 4 m

$$\tan\theta = \left(\frac{8}{4}\right) = 2, \ \textit{therefore}\ \theta = 63.435^{\circ}$$

$\sin\theta = 0.8944$, and $\cos\theta = 0.4472$

Step 3 : Loads. Uniformly distributed loads = 1.8 kN/m^2

Let the thickness of dome be 100 mm. Self-weight of the dome per square metre of surface area

$$\left(\frac{100}{1000}\times 1\times 1\times 25\right) = 2.5\ kN/m^2$$

Total uniformly distributed load

w = 4.3 kN/m^2

Step 4. Meridional stress. Meridional force Eq. 26.7

$$T = \left(\frac{w}{2}\cdot\frac{h}{\cos^2\theta}\right) \quad \text{(i)}$$

From the expression (i), it is seen that the height (depth) of cone from the crown. The meridional force is maximum at a = 4 m. Therefore,

$$T_{max} = \left(\frac{4.3}{2}\times\frac{4}{(0.4472)^2}\right) = 43.003\ kN$$

∴ Meridional stress

$$\left(\frac{T_{max}}{(t\times 1\,m)}\right) = \left(\frac{43.003\times 10^3}{100\times 1000}\right) = 0.43\ N/mm^2 \ \ \textit{Hence, it is safe.}$$

Step 5 : Hoop stress

Hoop force from Eq. 26.8

$$H = -1\times w\tan^2\theta\,.\,h. \ \ At\ h = 4\ m$$

$$H = 4.3\,(2\times 2)\times 4 = 68.8\ kN\ \ (compression)$$

$$\text{Hoop stress, } \left(\frac{H}{t\times 1\,m}\right) = \left(\frac{68.8\times 10^3}{100\times 1000}\right) = 0.688\ N/mm^2$$

Hence, it is safe.

Step 6: Reinforcement. Distribution reinforcement shall be provided in both the directions (viz., along the latitudes and longitudes).

$$A_d = \left(\frac{0.12}{100} \times 100 \times 1000\right) = 120\ mm^2$$

8 mm ϕ Hysd steel bars shall be provided. $A\phi_d$ = 50.26 mm². The spacing of steel bars

$$s_d = \left(\frac{1000 \times 50.26}{120}\right) = 418.88\ mm$$

8 mm ϕ Hysd steel bars shall be provided at 400 mm spacing c/c. To avoid congestion of the steel bars, a wire mesh is provided at the crown of the dome.

Step 7 : Ring beam. An outward force acts at the bottom of dome due to horizontal component of the meridional thrust, T.

Hoop component of meridional thrust acting radially outward

$$T \cdot \sin\theta = (43.003 \times 0.8944) = 38.462\ kN/m$$

Total tension = intensity of radial pressure times radius of dome

$$\therefore \quad \text{total tension} = (38.462 \times 8) = 307.70\ \text{kN}$$

$$\text{Area of steel, } A_{st} = \left(\frac{307.70 \times 10^3}{230}\right) = 1337.80\ mm^2$$

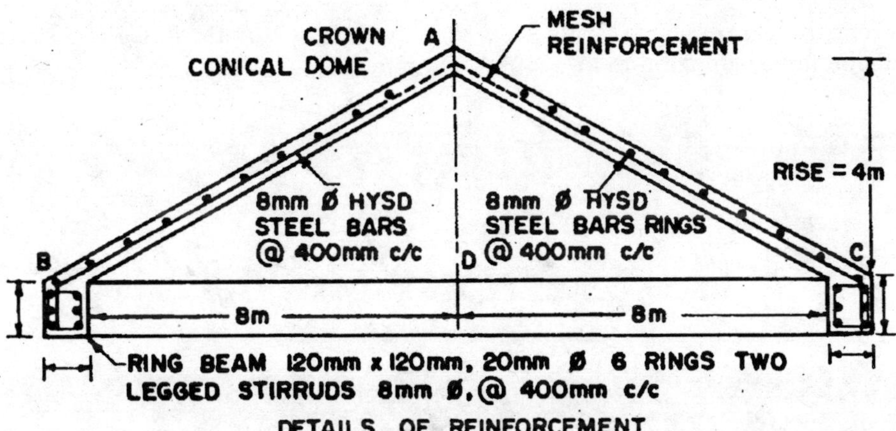

Fig. 26.10

Provided 20 mm ϕ Hysd steel bars, $A_\phi = 314.159\ mm^2$

$$\text{Number of steel bars, } \left(\frac{1337.80}{314.159}\right) = 4.258 = 6\ bars$$

Area of steel provided

$$(6 \times 314.159) = 1884.66\ mm^2$$

Two legged stirrups of 8 mm ϕ Hysd steel bars may be provided at 400 mm c/c. Equivalent area of ring beam

$$(A + 13.33 \times 1884.96) = (A + 25126.46)\, mm^3$$

Let the allowable stress in tension in M 20 grade of concrete be 1.8 N/mm².

$$\left(\frac{25126.46}{A + 25126.46}\right) = 1.8, 13959.14 = A + 25126.46, A = 11167.36\, mm^2$$

Size of ring beam may be 120 mm x 120 mm. The details of reinforcement is shown in Fig. 26.10.

PROBLEMS

26.1 The inside diameter of a circular room is 10 mm. Design a spherical dome to carry a uniformly distributed load (live load) is 1.8 kN/m². The dome is to support a lantern of load 24 kN as point load from a circular opening of 2 m at the crown. M 20 grade of concrete and Hysd steel of grade Fe 415 shall be used.

26.2 The inside diameter of a circular room's 18 m. The uniformly distributed load is 2 kN/m². Design the conical dome. M 20 grade of concrete and Hysd steel bars of grade Fe 415 shall be used.

27

Design of Reinforced Brick Slabs and Hollow Tiles

27.1 INTRODUCTION

Reinforced brick and hollow tiles are needed and used where the light floor loads are carried on long spans and smooth ceilings are required. These are used to a great extent in school houses, hospitals, and office buildings, in combination with reinforced concrete. In the solid slab, the concrete is replaced by bricks and cement concrete in a solid slab is replaced by bricks as shown in Fig. 27.1. The bricks are placed side by side and the joints between the bricks are filled with cement mortar. The steel bars provided as reinforcement resists the tensile stresses as usual. The steel bars are placed inside the cement mortar used to fill the gaps between the bricks.

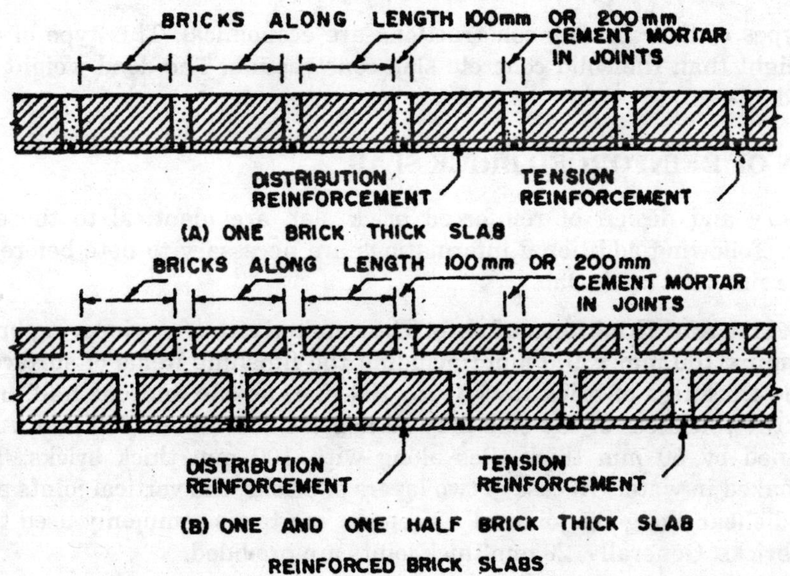

Fig. 27.1

In solid slab, a portion of concrete below the neutral axis being in tension zone is neglected. This portion of concrete does not contribute to the bending strength. This portion of concrete is replaced by hollow tiles blocks as shown in Fig. 27.2. The dead weight of slab is very much reduced. The hollow tile blocks are closely placed. The gaps between the blocks are filled with cement concrete which forms the ribs. These ribs are connected at the top by a thin concrete slab, which one is usually at *topping*. These types of slabs are referred as *ribbed slabs.*

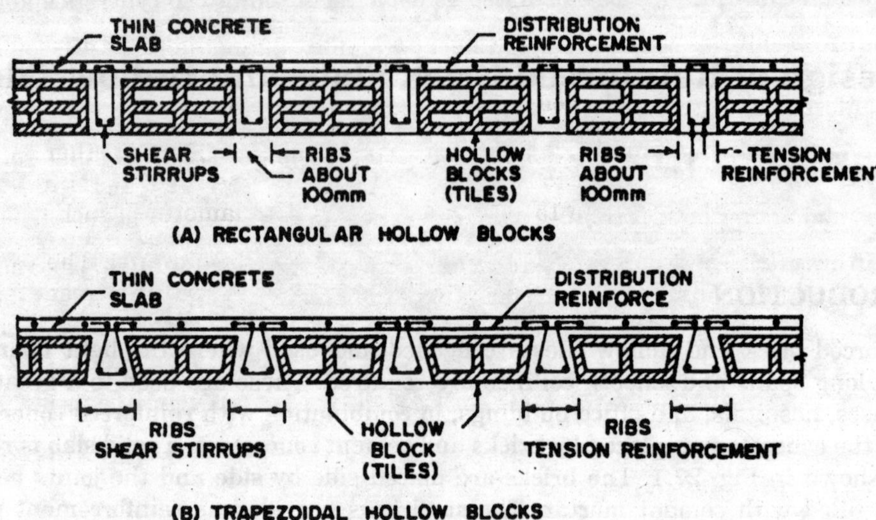

Fig. 27.2

These types of floor in slab constructions are economical. This type of construction is lighter in weight than the solid concrete slab construction. The dead weight of slab is very much reduced.

27.2 DESIGN OF REINFORCED BRICK SLAB

The theory and design of reinforced brick slab are identical to those of reinforced concrete slab. Following additional informations are necessary to note before to proceed for the design of reinforced brick slab.

1. Brick size and thickness of slab. The nominal size of engineering bricks as per IS : 1077 is 200 mm x 100 mm x 100 mm (including the finishing items of surface, e.g. plaster). Therefore, the thickness of brick thick (viz., 100 mm) or one and one-half brick thick (viz., 150 mm) or the thickness of the slab may be kept as 200 mm or 300 mm. A 150 mm thick slab is obtained by 50 mm thick tiles along with 100 mm thick bricks. The bricks are thoroughly soaked in water. In case of two layers of bricks, the vertical joints are scattered in order to avoid shear along the joints. 1 : 3 cement mortar is commonly used to fill the joints between the bricks. Generally, 20 mm thick joints are provided.

2. Permissible compressive strength of bricks and steel reinforcement bars. The permissible compressive bending stress in bricks may vary from 2 N/mm^2 to 3 N/mm^2. However, an average value of 2.5 N/mm^2 may be adopted in design. The modular ratio for reinforced brick slab may be adopted as 40.

As regards, the permissible stress in steel bars of mild steel, in tension may be adopted as 140 N/mm² for diameter of bars upto and including 20 mm. However, for the diameter of bars exceeding 20 mm, its value may be adopted as 130 N/mm². For, the Hysd steel bars, a value of 230 N/mm² is used.

The design constant may be calculated as usual as in reinforced concrete slabs.

3. Reinforcement. Small diameter steel bars upto 12 mm diameter are commonly used for the reinforced brick slabs. The steel bars are properly anchored. *Fifty* percent of the tension reinforcement provided is bent up at 0.1428 *times* the span.

4. Cover to reinforcement. For tension, compression or shear or other reinforcement in reinforced brick slabs, the cover to reinforced brick slabs, the cover to reinforcement (viz., clear cover) should not be less than 15 mm, nor less than the diameter of such bar.

5. Minimum reinforcement (viz., **distribution reinforcement).** The reinforcement in either direction in reinforced brick slabs should not be less than 0.15 percent for mild steel or 0.12 percent for Hysd steel of the total cross-sectional area.

6. Shear stress. The value of permissible shear stress in reinforced brick slabs varies from 0.07 N/mm² to 0.3 N/mm². However, an average of 0.1 N/mm² may be used.

7. Development length. An appropriate development length of the steel bars is provided, as it is provided in the reinforced concrete slab.

Example 27.1 A reinforced brick slab is to be provided over a room of 2.8 width. A 20 mm thick finishing coat shall be used over the slab. The floor of the slab shall be used for residential building. Hysd steel bars shall be provided. Design reinforced brick slab.

Design : The modular ratio for brick may be taken as 40.

Step 1 : Design constants. Permissible compressive strength for bricks and tensile stress for Hysd steel bars in bending are 2.5 N/mm², and 230 N/mm², respectively.

Factor for depth of neutral axis

$$k = \left(\frac{m.\ \sigma_{cbc}}{m\ \sigma_{cbc} + \sigma_{st}}\right) = \left(\frac{40 \times 25}{40 \times 2.5 + 230}\right) = 0.303$$

Factor for lever arm

$$j = \left(1 - \frac{k}{3}\right) = \left(1 - \frac{0.303}{3}\right) = 0.899$$

Factor for moment of resistance

$$R = \frac{1}{2}\,\sigma_{cbc}\ .\ k\ .\ j = \left(\frac{1}{2} \times 2.5 \times 0.303 \times 0.899\right) = 0.3405$$

Step 2 : Loads. The thickness of reinforced brick slab may be assumed as 200 mm.

Self-weight of reinforced brick slab, $(0.2 \times 1 \times 1 \times 22) = 4.4\ kN/m^2$

Weight of wearing coat, $\left(\frac{20}{1000} \times 1 \times 1 \times 24\right) = 0.48\ kN/m^2$

Total dead load + 4.88 kN/m²

The floor of reinforced brick slab is likely to be used for the residential building. Therefore, the intensity of live load = 2.5 kN/m²

Total load, (4.88 + 2.5) = 7.38 kN/m²

Step 3 : Effective span. The thickness of slab has been assumed as 200 mm. Let the effective cover be 20 mm. Then, the effective depth shall be 180 mm. The effective span of slab.

ES = (2.8 + 0.180) = 2.98 m.

Step 4 : Bending moment. The reinforced brick slab shall be acting as simply supported at both the ends.

$$\text{Maximum bending moment, } \left(\frac{w \cdot L^2}{8}\right) = \left(\frac{7.38 \times 2.98^2}{8}\right) = 8.192\ kN-m$$

Step 5 : Effective depth of slab

$$d = \left(\frac{M}{R \times 1000}\right)^{1/2} = \left(\frac{9.192 \times 10^6}{0.3405 \times 1000}\right)^{1/2} = 155.11\ mm$$

Let the effective cover be 20 mm. Then the overall thickness of slab shall be (15.11 + 20) = 175.11 mm. However, the overall thickness of 270 mm is provided keeping in view the size of bricks. Therefore, the effective depth of slab will be 180 mm.

Step 6 : Tension reinforcement. Let the cross-sectional area of the tension reinforcement be A_{st}.

$$A_{st} = \left(\frac{M}{\sigma_{st} \cdot j_B \cdot d}\right) = \left(\frac{8.192 \times 10^6}{230 \times 0.899 \times 180}\right) = 220.105\ mm^2.$$

Let the diameter of Hysd steel bars be 8 mm. Cross sectional area of one steel bar

$A_\phi = 50.265\ mm^2$

$$\text{Spacing of the steel bars } s = \left(\frac{1000 \times 50.265}{220.105}\right) = 228.37\ mm.$$

A gap of 20 mm will be left between the two adjacent joint. The modular dimension in each direction may taken as 10 mm less for 200 mm length of the brick, the modular dimension shall be 190 mm. With 20 mm gap on each side, the centre to centre distance between steel bars

= (190 + 10 + 10) = 210 mm

Fifty percent of the bars less shall be bent up a distance 0.1428 times span (viz., 0.1428 x 1000 = 400 mm) from the face of support.

Step 7 : Distribution reinforcement. 0.12 percent of the total cross-section of slab for Hysd steel shall be provided as the distribution reinforcement

$$A_d = \left(\frac{0.12}{100} \times 1000 \times 200\right) = 240\ mm^2$$

Let the diameter of Hysd steel bars be 6 mm. Cross-sectional area of one steel bar

$$A_{\phi d}\ 28.27\ mm^2.$$

Spacing of the distribution reinforcement, $s_d = \left(\frac{1000 \times 28.27}{240}\right) = 118\ mm$

A 20 mm gap will be left between two adjacent bricks. The modular dimension of brick shall be 90 mm. Then, the spacing of 6 mm bars

(90 + 10 + 10) = 110 mm.

Step 8 : Shear force. The maximum shear force for the uniformly distributed load

$$V = \left[\frac{wL}{2} - w\left(\frac{b_s}{2} + d\right)\right]$$

Wall thickness, b_s = 300 mm

$$\left[\frac{73.38 \times 2.8}{2} - 7.38\,(0.115 + 0.180)\right] = 8.155\ kN$$

Nominal shear stress

$$\tau_v = \left(\frac{V}{b\,.\,d}\right) = \left(\frac{8.156 \times 10^3}{1000 \times 180}\right) = 0.045\ N/mm^2.$$

The nominal shear stress is very small. Hence, it is safe.

Step 9 : Development length. At support, the development length

$$L_d < [(M_1/V) + L_o]$$

Moment of resistance of slab at support

$$M_1 = (\sigma_{st}\ .\ A_{st}\ .\ j_d)$$

$$M_1 = \frac{230}{1000} \times \left[\frac{1000 \times 50.265}{2 \times 220}\right] \times \left(\frac{0.899 \times 180}{1000}\right) = 4.252\ kN-m$$

$$V = \left(\frac{wL}{2}\right) = \left(\frac{7.38 \times 2.8}{2}\right) = 10.332\ kN$$

$$(M_1/V) = \frac{4.252}{10.332} = 0.412\ m.$$

$$L_o = \left[\frac{b_s}{2} - x' + 13\phi\right] = \frac{1}{1000}\left(\frac{300}{2} - 25 + 13 \times 8\right) = 0.229\,m$$

$$\left(\frac{M_1}{V} + L_o\right) = (0.412 + 0.229) = 0.641\ m.$$

For Hysd steel bars, $L_d = \left(\frac{8 \times 68.65}{1000}\right) = 0.548\ m$

$$L_d < \left(\frac{M_1}{V} + L_o\right) \text{ i.e., } 0.548\ m < 0.641\ m.$$

Hence, safe. The details of reinforcement are shown in Fig. 27.3.

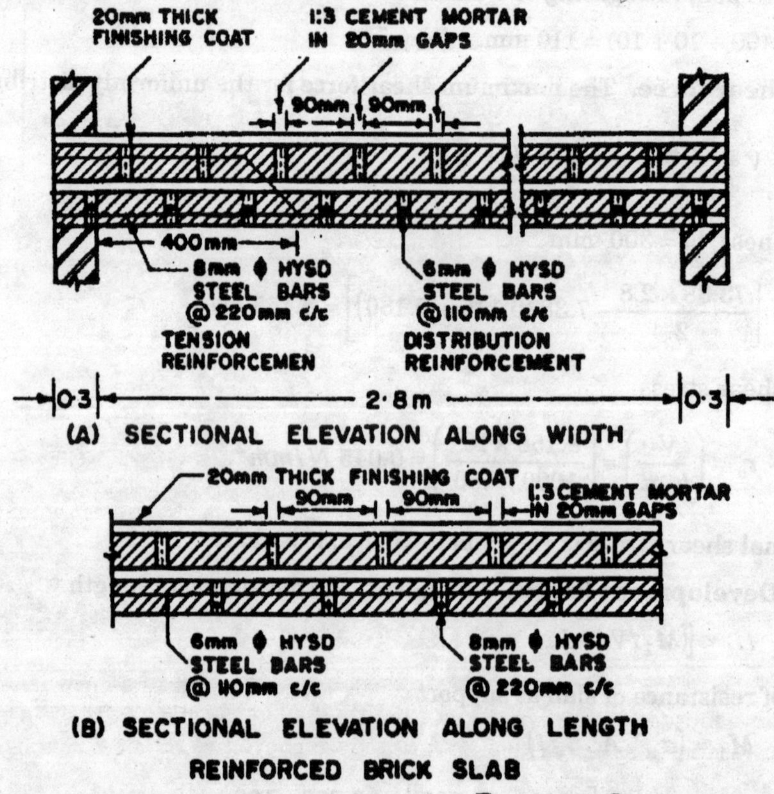

Fig. 27.3

27.3 HOLLOW CLAY TILES

In the solid slab spanning in one direction, the concrete below the neutral axis remains in tension. While calculating the strength of slab in bending, the portion of concrete in tension is neglected. In order to have the light weight roofs, the concrete in tension zone is not provided. Such a slab without concrete in tension zone appears as T-beam with series of small closely spaced ribs. In order to present a good appearance and to provide a plain ceiling, the spaces between ribs are filled with hollow blocks made of clay, concrete, fibre glass or P.V.C. as shown in Fig. 27.2. The slab is called as **hollow block slab or ribbed slab.**

In ribbed slab, the thickness of top slab used to be very small and the slabs have ribs in one direction as shown in Fig. 27.4. The concrete girders run in one direction. the construction consists of hollow tiles, reinforced concrete girders and concrete topping above the tiles. One way hollow tile slabs are suitable for moderate light and uniformly distributed loads.

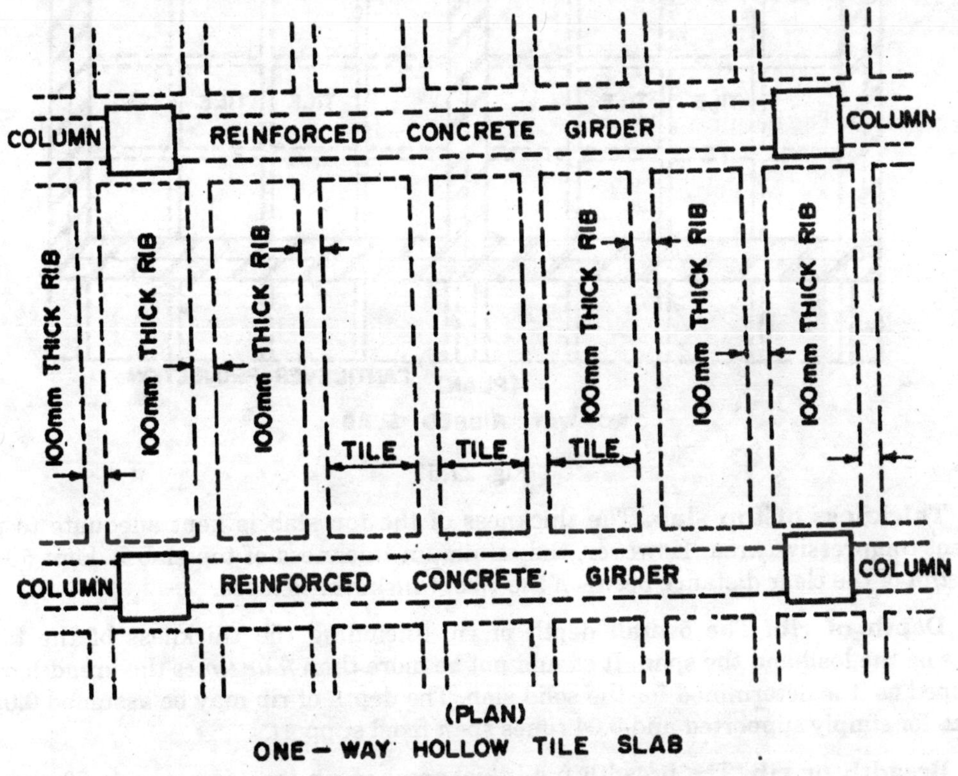

Fig. 27.4

Two-way hollow tiles having ribs in two directions are used as shown in Fig. 27.5 when the uniformly distributed loads are heavy or points loads are acting over the slab. In the two-way system, the supports are provided on four sides. There are series of ribs in two direction, and these ribs are closely spaced.

27.4 DESIGN OF RIBBED SLAB

The nominal size of tiles as per IS : 1077 is 200 mm x 100 mm x 50 mm. The modular dimensions are 10 mm less in each direction. The size of tiles or hollow blocks decides the spacing between the ribs. However, it should not exceed 500 mm. The ribbed slabs are used for spans upto 7 m. When the span is more than 5 m, ribs are also provided in the lateral direction also. The ribbed slabs are designed as follows :

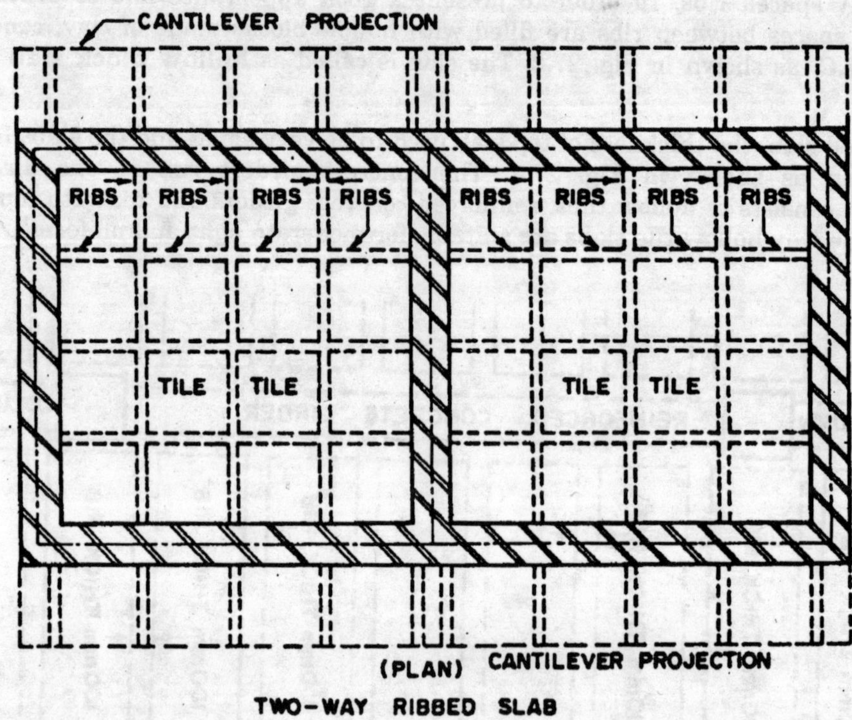

Fig. 27.5

1. Thickness of top slab. The thickness of the top slab is kept adequate to provide sufficient compressive area. However, the minimum thickness of top slab is kept 50 mm or *one-twelfth* of the clear distance between the ribs, whichever is more.

2. Depth of rib. The overall depth of rib (including the thickness of the top slab) depends on the load and the span. It should not be more than *four times* the breadth or rib. It is obtained as it is determined for the solid slab. The depth of rib may be assumed 0.05 *times* the span for simply supported and 0.04 times span fixed support.

3. Breadth or rib. The breadth (i.e., thickness) of rib is kept between 60 mm to 100 mm.

However, the breadth and effective depth of ribs also depend upon shear.

The ribs are provided parallel to the span in case of one-way slab and parallel to each edge in case of two-way slab.

4. Spacing between ribs. The spacing between two adjacent ribs should not be morethan 1.5 m.

5. Reinforcement. The design of top slab is not necessary as the span is small. The distribution reinforcement is provided (for temperature and shrinkage) in the slab. The tension reinforcement is provided at the bottom of the slab.

For tension, compression, shear, or other reinforcement in a slab, the clear cover to reinforcement should not be less than 15 mm nor less than the diameter of such bar.

One-way ribbed slab is designed as ordinary T-beam. However, the two-way ribbed slab may be designed as two-way solid slab.

6. Permissible stresses. The permissible stresses in the material shall be same as those for the different grades of concrete. The permissible stresses for steel bars shall be as those of mild steel or Hysd steel used. The modular ratio shall be calculated as usual.

Example 27.2. Design a one-way ribbed slab using hollow tile blocks for a residential building for room 4.2 m wide. M 15 grade of concrete and mild steel reinforcement shall be used for the slab. A finishing coat of 20 mm thickness of cement concrete should be provided.

Design : For M 15 grade of concrete, the permissible stress in concrete in bending in compression, σ_{cbc} is 140 N/mm². Modular ratio of concrete

$$m = \left(\frac{280}{3 \times 5}\right) = 18.67$$

Step 1 : Design constants. The permissible stress in mild steel for diameter of steel bars less than or equal to 20 mm, is 140 N/mm². Various design constants are as follows :

Depth for neutral axis

$$k = \left(\frac{m\ \sigma_{cbc}}{m\ \sigma_{cbc} + \sigma_{st}}\right) = \left(\frac{18.67 \times 5}{18.67 \times 5 + 140}\right) = 0.4$$

Lever arm factor, j = (1 – k/3) = (1 – 0.4/3) = 0.867

Factor for moment of resistance

$$R = \left(\frac{1}{2} \times \sigma_{cbc} \times k \times j\right) = \left(\frac{1}{2} \times 5 \times 0.4 \times 0.867\right) = 0.867.$$

Percentage of steel reinforcement, p = 0.71428 percent.

Step 2 : Effective span of the slab. The slab is shown in Fig. 27.6. Let the wall thickness be 300 mm. Then, the effective span

$$ES = \left(clear\ span + \frac{bs}{2} + \frac{bs}{}\right) = (4.2 + 0.150 + 0.150) = 4.5\ m$$

Let the effective depth of rib be 300 mm

ES = (4.2 + 0.3) = 4.5 m.

Effective span L = 4.5 m.

Step 3 : Loads. Overall depth of rib may be assumed as 240 mm and the thickness of slab may be assumed as 50 mm.

Self-weight of the slab, $\left(\frac{50}{1000} \times 1 \times 1 \times 25\right) = 1.25\ kN/m^2$

Weight of the wearing coat, $\left(\frac{20}{1000}\times1\times1\times24\right)=0.48\ kN/m^2$

Live load for the residential building, $(25\times1\times1)=2.5\ kN/m^2$

Weight of tiles (hole blocks) $(0.2\times1\times1)=0.2\ kN/m^2$.

For slab, total load = 4.43 kN/m².

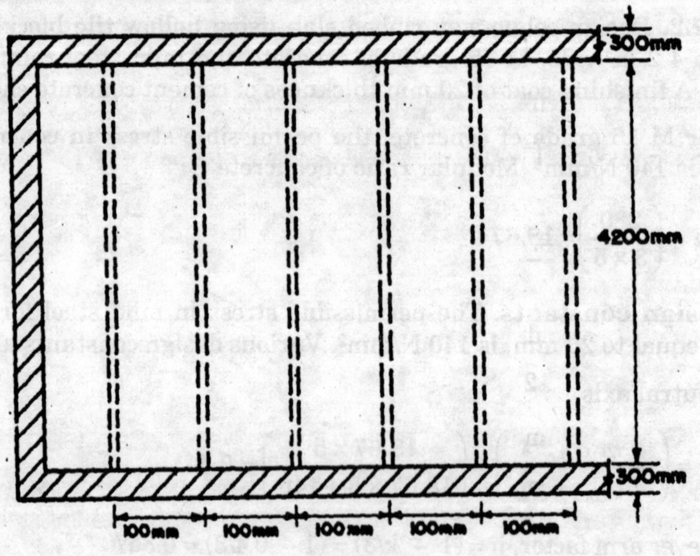

Fig. 27.6

Step 4 : Bending moment for slab. Consider 1 m width of the slab. Maximum bending moment (taking advantage of continuity of slab over number of ribs)

$$\left(\frac{wL^2}{10}\right)=\left(\frac{4.43\times1^2}{10}\right)=0.443\ kN-m$$

Step 5 : Effective depth of slab. Let the effective depth of slab be d.

$$d=\left(\frac{M}{R\times1000}\right)^{1/2}=\left(\frac{0.443\times10^6}{0.867\times1000}\right)^{1/2}=22.6\ mm.$$

Let the effective cover be 25 mm. Then, the overall depth shall be 47.60 mm. Let the overall depth of slab be 50 mm (as assumed), then, d = 25 mm.

Step 6 : Reinforcement in slab. Let the reinforcement in slab be A_{st}

$$A_{st}=\left(\frac{M}{\sigma_{st}.j.d}\right)=\left(\frac{0.443\times10^5}{140\times0.867\times25}\right)=145.99\ mm^2$$

Minimum reinforcement as per IS : 456-2001

$$A_{st.\min} = \left(\frac{0.15}{100} \times 50 \times 1000\right) = 75\ mm^2 \quad A_{st} > A_{st.\min}.\ (\textit{It is alright})$$

Let the diameter of mild steel bars be 5 mm. Cross-sectional area of one bar

$$A_\phi = 28.27\ mm^2$$

$$\text{Spacing the steel bars, } s = \left(\frac{1000 \times 28.27}{145.99}\right) = 193.65\ mm$$

$$\text{Maximum spacing, } (3d = 3 \times 25) = 75\ mm$$

Main reinforcement steel bars in tension are provided at 75 mm c/c. Spacing of distribution reinforcement

$$s_d = \left(\frac{1000 \times 28.75}{75}\right) = 376.93\ mm$$

Maximum spacing for the distribution bars

$$5d = (5 \times 25) = 125\ mm$$

The distribution reinforcement steel bars of 6 mm f are provided at 125 mm.

Design of rib. **Weight from slab = 4.43 kN/m². Ribs are also provided at 1 m spacing centre to centre. Therefore, the self-weight or rib (assuming thickness of slab as 150 mm)**

$$\left(\frac{150}{1000}\right) \times \left(\frac{250 - 50}{1000}\right) \times 1 \times 25 = 0.75\ kN$$

Total load = 5.18 kN/m², say = 5.02 kN/m

Step 7 : Bending moment in rib. **Maximum bending moment in the rib**

$$\left(\frac{5.20 \times 4.5^2}{8}\right) = 13.163\ kN - m$$

Step 8 : Effective width of flange. **The effective width of flange as per IS : 456-2000**

$$b_f = \left(\frac{l_o}{6} + b_w + 6\,D_f\right) = \left(\frac{4500}{6} + 150 + 6 \times 50\right) = 1200\ mm$$

The effective width of flange should not be more than

$$\left(b_w + \frac{1}{2}.\ \textit{clear span on left} + \frac{1}{2}\ \textit{clear span on right}\right)$$

$$= \left[150 + \frac{1}{2}(1000 - 150) + \frac{1}{2}(1000 - 150)\right] = 1000\ mm$$

Therefore, the effective width of T-beam shall be b_f = 1000 mm.

Step 9 : Effective depth of rib. (A) The effective depth of rib may be determined by equating the bending moment, M and the moment of resistance of T-beam section, ∴ M = MR

$$M = 0.45 \times b_f \times D_f \times \sigma_{cbc} \times \left(\frac{2kd - D_f}{k}\right)$$

$$\text{or } 13.163 \times 10^6 = 0.45 \times 1000 \times 50 \times 5 \times \left(\frac{2 \times 0.4\, d - 50}{0.4}\right)$$

$$\text{or } 1316.3 = (28.125 \times 0.8\, d - 28.125 \times 50)$$

$$\text{or } d = \left(\frac{1316.3 + 1406.25}{28.125 \times 0.8}\right) = 121\, mm$$

(B) The effective depth of rib may be calculated from the economical depth

$$d = \left(\frac{\gamma\, .\, M}{\sigma_{st}\, .\, j\, .\, b_w}\right)^{1/2} = \left(\frac{70 \times 13.163 \times 10^6}{140 \times 0.867 \times 150}\right)^{1/2} = 224.99\, mm.$$

Let the effective cover be 25 mm. Then, the overall depth of rib will be 249.99 mm. The overall depth of rib has been assumed as 250 mm. The effective depth of rib shall be 225 mm.

Step 10 : Reinforcement in rib.

$$A_{st} = \left(\frac{M}{\sigma_{st}\, .\, j\, .\, d}\right) = \left(\frac{13.163 \times 10^6}{140 \times 0.867 \times 225}\right) = 481.98\, mm^2.$$

Let the diameter of bars be 18 mm. Provide 2 (two) bars in one rib.

A_ϕ = 254.47 mm^2.

Step 11 : Stresses in concrete and steel (A) Neutral axis. Let the neutral axis lies below the slab, neglecting the portion of concrete in tension the depth of neutral axis may be calculated by equating the first moment of areas of concrete and steel about neutral axis

$$b_f\, .\, D_f\, .\left(n - \frac{D_f}{2}\right) = m\, A_{st}\, (d - n)$$

$$1000 \times 50\, \left(n - \frac{50}{2}\right) = 18.67 \times 508.94\, (225 - n)$$

$$(n - 25) = 0.19\, (25 - n)$$

$$n = \left(\frac{(25 + 0.19 \times 225)}{1.19}\right) = 56.94\, mm$$

(B) Distance of centroid of compressive force. Stress below the slab

$$\sigma_{cbc_1} = \left(\frac{n - D_f}{n}\right)\sigma_{cbc} = \left(\frac{56.94 - 50}{56.94}\right)\sigma_{cbc} = 0.122\ \sigma_{cbc}$$

Distance of line of action of resultant compression

$$\bar{y} = \frac{D_f}{3}\left(\frac{2\ _{cbc_1} + \sigma_{cbc}}{\sigma_{cbc_1} + \sigma_{cbc}}\right) = \frac{D_f}{3}\left(\frac{2 \times 0.122\ \sigma_{cbc} + \sigma_{cbc}}{0.122\ \sigma_{cbc} + \sigma_{cbc}}\right)$$

$$\text{or } \bar{y} = \frac{50}{3} \times \left(\frac{0.244 + 1}{0.122 + 1}\right) \cdot \left(\frac{\sigma_{cbc}}{\sigma_{cbc}}\right) = 18.48\ mm$$

Take the moment of resultant compressive force about the centre line of steel bars in tension.

Moment of resistance of T-beam, neglecting the portion of concrete in web in compression

MR = Resultant compressive force × Lever arm

$$MR = C_R \cdot (d - \bar{y}) = b_f \cdot D_f \cdot \left(\frac{\sigma_{cbc} + \sigma_{cbc_1}}{2}\right)(d - \bar{y})$$

$$13.163 \times 10^6 = 1000 \times 50 \times \left(\frac{\sigma_{cbc} + 0.112\ \sigma_{cbc}}{2}\right)(225 - 18.48)$$

$$\sigma_{cbc} = \left(\frac{13.163 \times 10^6 \times 2}{1000 \times 50 \times 1.112 \times 206.52}\right) = 2.293\ N/mm^2$$

The compressive stress in bending in concrete Scbc = 2.293 N/mm² is less than the permissible stress in compression in bending in concrete 5 N/mm².

Stress in steel bars provided as reinforcement in tension 140 N/mm².

$$\sigma_{st} = m \cdot \sigma_{cbc} \cdot \left(\frac{d - n}{n}\right)$$

$$\sigma_{st} = 18.67 \times 2.293 \times \left(\frac{225 - 56.94}{56.94}\right)\sigma_{st} = 126.36\ N/mm^2$$

It is less than the permissible stress in steel bars n tension 140 N/mm².

Hence, the design is safe for the bending stress.

Step 12 : Shear stress : Maximum shear force

$$V = \frac{wL}{2} - w\left(\frac{b_s}{2} + d\right)$$

$$\left[\left(\frac{4.43 \times 4.5}{2}\right) - 4.43\,(0.150 + 0.225)\right] = 8.31\ kN$$

$$\text{Nominal shear stress, } \tau_v = \left(\frac{8.31 \times 1000}{150 \times 225}\right) = 0.246\ N/mm^2$$

There are two mild steel bars 18 mm diameter which shall be provided as tension reinforcement

$$A_{st} = (2 \times 254.47) = 508.94\ mm^2,\ \left(\frac{100 \times 508.94}{150 \times 255}\right) = 1.51$$

and for M 15 grade of concrete permissible shear stress in concrete, τ_c = 0.42 N/mm²

τ_c = 0.42 N/mm² is more than τ_v = 0.296 N/mm²

Hence, the nominal shear stress is only need. The minimum shear reinforcement in the form of stirrups shall be provided such that

$$\left(\frac{A_{sv}}{b.s_v}\right) = \left(\frac{0.4}{f_y}\right)$$

$$\text{Provide two legged 6 mm } \phi \text{ stirrups, } A_{sv} = \left(2 \times \frac{\pi}{4} \times 6^2\right) = 56.55\ mm^2$$

$$s_v = \left(\frac{56.55 \times 250}{150 \times 0.4}\right) = 235.62\ mm$$

Maximum spacing of shear reinforcement measured along the axis of the member should not exceed 0.75 d or 450 mm for the vertical stirrup

0.75 d = (0.75 x 225) = 168.75 mm

Provide the nominal stirrups at 160 mm spacing centre to centre.

Step 13 : Distribution reinforcement.

Distribution reinforcement in slab

= 0.15 percent if total cross-sectional area

$$A_{st} = \left(\frac{0.15}{100} \times 150 \times 250\right) = 56.25\ mm^2$$

Provide 6 mm ϕ mild steel bars A_ϕ = 28.125 mm²

$$\text{Spacing of the distribution reinforcement, } s_d = \left(\frac{1000 \times 56.25}{28.125}\right) = 2000\ mm$$

The distribution reinforcement 6 mm ϕ shall be provided at 320 mm.

The detail of reinforcement is shown in Fig. 27.7.

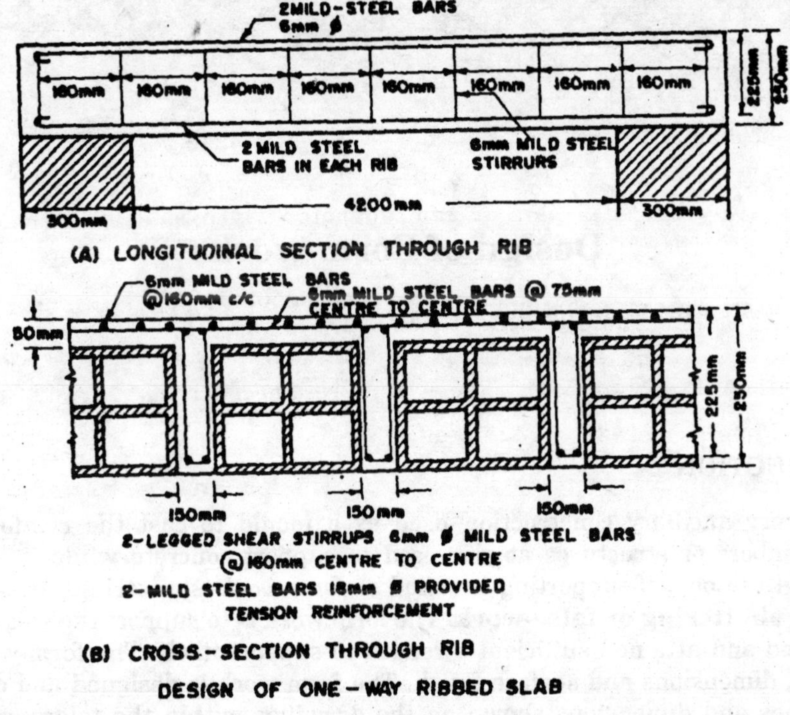

Fig. 27.7

PROBLEMS

27.1 A reinforced brick slab is to be provided over a room of 3.2 m. A 20 mm thick finishing coat shall be used over the slab. The floor of the slab shall be used for the residential building. Hysd steel bars be provided. Design the reinforced brick slab.

27.2 Design a one-way ribbed slab using hollow tile blocks for a residential building for room 4.8 m wide. M 15 grade of concrete and will steel reinforcement shall be used for slab. A finishing coat of 20 mm thickness of cement concrete shall be provided.

28

Design of Form-works

28.1 INTRODUCTION

A temporary ancillary construction used as a mould to cast the reinforced concrete structural members or structures at site used to support concrete while it is setting and gaining strength to be self-supporting is called as *form-work* (shuttering). It is also referred as **centering, shuttering or false-work**. The form work is to support the wet concrete until it has hardened and attained sufficient strength to support itself. The form-work is to give desired shape, dimensions and surface finish. The form work is designed and constructed to the shapes lines and dimensions shown on the drawings within the tolerances. The form-work is so constructed as to remain sufficiently rigid during the placing and the compacting of the concrete and shall be sufficiently tight to prevent loss of liquid from the concrete.

The tolerance on the dimensions of the structures is a practical necessity. The design provision accommodates some of these tolerances. The form-work shall be so constructed such that the internal dimensions are within the permissible tolerances specified in IS : 456-2000 and given in Table 28.1. The tolerances for bending dimensions for stirrups are shown in Fig. 28.1.

TABLE 28.1. TOLERANCE FOR DIFFERENT REINFORCED CONCRETE STRUCTURAL MEMBERS

S.No.	Structural members	Tolerance Deviation from specified dimension
1.	Cross-section of beams	– 6 mm
		+ 12 mm
2.	Cross-section of columns	– 6 mm
		+ 12 mm
3.	Footings (see note)	
	(i) Dimensions in plan	–52 mm
	(ii) Eccentricity	+ 50 mm
		0.02 times the width of the footing in the
	(iii) Thickness	direction of deviation but not more than 50 mm
		± 0.05 times the specified thickness

4.	For placing of reinforcement	
	(i) For effective depth 200 mm or less	± 10 mm
	(ii) For effective depth more tan 200 mm	± 15 mm

Note : (1) The tolerances apply to concrete dimensions only, not to positioning of vertical reinforcing steel or dowels; (2) The cover shall in no case be reduced by more than *one-third* of specified cover or 5 mm whichever is less.

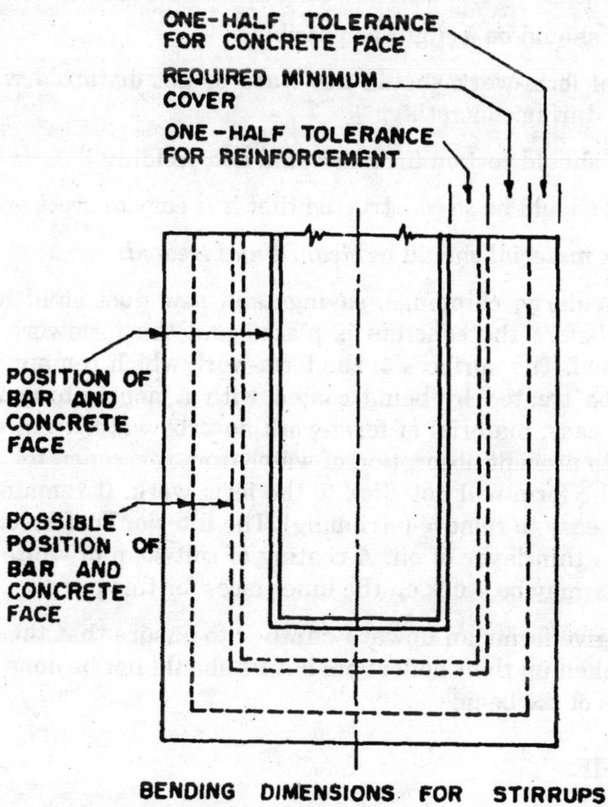

Fig. 28.1

28.2 REQUIREMENTS OF FORM-WORK

Following requirements should be satisfied for a good form-work.

1. The form-work should be *strong enough* to carry all types of dead and live loads (e.g., self-weight, weight of the reinforcement, weight of the wet concrete, loads due to workmen, construction equipment, other incidental loads and forces caused by dumping and consolidation of concrete) imposed upon it during and after casting of concrete.

2. The form-work should be *rigidly constructed* and efficiently propped and braced in horizontal as well as in vertical directions, so as to retain its shape without undue deflection.

3. The joints in the form-work should be tight against leakage of any liquid or cement grouting.
4. The form-work should be made of cheap material, which is easily available and should be suitable for re-use several times, if possible.
5. The form-work should have plane surfaces. It should be set accurately to the desired line and level.
6. The form-work should be as light as possible.
7. The material of form-work should not warp or get distorted when exposed to sun, rain or water during concreting.
8. The form work should rest on firm base and non yielding.
9. The form-work should be so constructed that it is easy to erect and strip (removed).
10. The form-work material should be *cleaned and treated.*

All rubbish, particularly, chippings, savings and saw dust shall be removed from the interior of the forms before the concrete is placed and the formwork in contact with the concrete shall be cleaned. The surfaces of the form-work which remain in contact with fresh (wet) concrete must be treated by being coated with a nonstaining mineral oil or other approved material. In case, material of form-work absorbs water from wet concrete, it shall be thoroughly wetted to prevent absorption of water from wet concrete. It is ensured that the concrete poured into the form will not stick to the form work. It remains easy to release the form-work. It remains easy to remove (stripping). The interior surface of the form-work shall be given a coating of a thin layer of oil. A coating of soft-soap of white-wash may be given. The vanvas or oil-paper may be stuck on the inner faces for this purpose.

It is desirable to give forms an upward **camber** to ensure that the beams do nɔt have a sag when they have taken up their deflection, but its should not be done unless allowed for in the design calculations of the beams.

28.3 STRIPPING TIME

The adequate time after which the form-work supporting the newly cast reinforced concrete is known as **stripping time,** IS: 456-2000 recommended that the form-work should not be removed until the concrete has reached a strength at least *twice* the stress which the concrete may be subjected at the time of removal of form-work. For this, it becomes necessary to estimate the strength of the concrete at the time of removal of form-work. The cubes are cast at the same time when the concrete is cast for the structural member/structure. The cubes shall be cast of concrete using the same cement and same aggregate with the same proportions and cured under conditions of temperature and moisture similar to those existing on the work. Where possible, the form-work shall be left longer as it would assist the curing.

In normal circumstances and where ordinary portland cement is used, form-work may generally be removed after the expiry of the following period (Table 28.2) as recommended in IS : 456-2001.

For rapid hardening cement, 3/7th of the period given for the ordinary portland cement shall be normally sufficient except that a minimum period of 24 hors is needed.

It is to note that the number of props left under, their size and disposition shall be such as to be able to safely carry full dead load of the slab, beam or arch as the case may be together with any live load likely to occur during further construction.

Where the shape of element is such that the form-work has re-entrant angles, the form-work shall be removed as soon as possible after the concrete has set, to avoid shrinkage cracking occurring due to the restraint imposed.

TABLE 28.2. PERIOD OF REMOVAL OF FORM-WORK FOR PORTLAND CEMENT

S.No.	Structural members	Period of removal of form-work
1.	Walls, columns and vertical faces of all structural members	24 to 48 hours as may be decided by the engineer in-charge
2.	Slabs (props left under)	3 days
3.	Beam soffits (props left under)	7 days
4.	Removal of props under slabs :	
	(i) Spanning upto 4.5 m	7 days
	(ii) Spanning over 6 m	14 days
5.	Removal of props under beams and arches	
	(i) Spanning upto 6 m	14 days
	(ii) Spanning over 6 m	21 days

Note : For other cements, the stripping time recommended for ordinary portland may be suitably modified.

All the form-work is removed carefully without shock or vibration as it would damage the reinforced concrete. Before, the soffit and the struts are removed, the concrete, surface shall be exposed, where necessary in order to ascertain that the concrete has sufficiently hardened. Proper precautions shall be taken to allow for the decrease in the rate of hardening that occurs with all the cements in cold weather.

Due regard must also be given to curing methods to be used before the form-work is removed.

28.4 ECONOMY OF FORM-WORK

The total cost of construction of reinforced concrete structures includes also the cost of form-work. The placing costs may vary depending upon good scheduling, proper crew planning and sequencing, efficient working spaces and adequate materials handling. All of these items are related to and dependent on forming system selected and on how the system is implemented. The form-work costs represent the major and most variable proportion of the cost of concrete in place, normally 33 to 55 percent.

The form-work system is designed and constructed such that the system is sufficiently economical. In case, the form-work may be used and reused number of times, the cost of form work will be reduced positively. The form-work and its components which may be reused number of times are called as *panel forms.* Other type of form-work is called *stationary forms.* The stationary forms are constructed for individual non-standard members. Such forms cannot be used for repetitive works. The reuse of form-work is connected with the stripping ability of the form-work. The economical reuse may be increased by designing and sequencing structural components so that the forms may be reused partially filled, bulk headed or slightly modified. The cheapest form-work available will usually be more expensive per square metre of the contact area when used only once as compared to very expensive forms having many reuses.

The cost of concrete formwork in place consists of two items. First is the make-up item, which is the cost of the form ready for erection at the job site. It is a one time cost. Therefore, the total form cost per square metre contact area is inversely proportional to that number of uses. The structure may be planned in such a way that there are minimum number of variations in the sizes of rooms, floor area etc. so as to allow the reuse of form-work moulds repetitively.

The second item is stripping time, cleaning, repairing, erection and expendable hardware. This item remains constant for ease of reuse upto a certain time, after which the cost of repairing and cleaning starts rising rapidly and the further reuse of the form system becomes uneconomical.

28.5 MATERIALS FOR FORM-WORKS

Timber, plywood, steel, reinforced concrete, plane, concrete, precast concrete, fibre glass, combined timber and steel are the various materials which are used to prepare the form-work. The nature of construction and the availability and cost of the material are the factors which decide the type of material to be used for the construction of the form-work. The suitability of different materials their advantages and their disadvantages are as under.

1. Timber. The form-work is very commonly made of timber in early stages, the form-work was made up of timber alone. The timber is economical material to be used for form work. The form-work of timber may be made of any shape and size easily as per the requirement. The form-work made up of timber may be easily erected. The timber used for form-work should be well seasoned light in weight, easily workable with nails without splitting and free from knots.

The possibility of warping, swelling and shrinkage of the timber form-works are the disadvantages. These defects may be reduced by covering the timber form-work elements by some water-impermeable coating. The form-work made up of timber is also used for repetitive uses.

The span of slab, or beam, floor to floor height and centre to centre spacing of the centring supports are the various factors which decide the size of timber sections for the different components of form-work for shuttering of 4.5 m span and 3.5 height are given in Table 28.3.

For other situations, the form-work should be properly designed in detail and the sizes of different components are determined. For the usual construction work, it is possible to have the repetitive use of shuttering. In such a case, the quantity of timber shuttering may be calculated on the assumption that a shuttering may be used 10 to 12 times.

TABLE 28.3. SIZES OF DIFFERENT COMPONENTS OF TIMBER FORM-WORKS

S.No.	Different components of form-work	Size (varies as per the, spacing of centring props from 1 m to 1.2 m)
1.	Sheeting beams and slabs, sides of the columns and bottom of beam	25 mm to 40 mm thick
2.	Joist	50 mm × 75 mm 50 mm × 150 mm
3.	Posts	75 mm × 100 mm 100 mm × 100 mm
4.	Ballies	At mid length ∢ 100 *mm* at thin-end ∢ 80 *mm*

Note : (i) For 1 m centre to centre spacing of the centring props, the minimum size s used.
(ii) For 1.2 m centre to centre spacing of the centring props, the maximum size is used.

The tongue and grooved joints are provided in the timber plants and the joints are ensured that they are adequately tight against leakage of cement grout.

The properties of some of timber species used in the form-work are listed in Table 28.4.

TABLE 28.4. PROPERTIES OF SOME OF TIMBER SPICES USED FOR FORM-WORK AS PER IS : 883-1970

S.No.	Properties of timber used for form-work	Spices of timber			
		Chir	Deoder	Fir	Kail
1.	Average unit weight at 12 percent of moisture content kN/m³	5.75	5.45	4.40	5.15
2.	Modulus of elasticity (all grades, all locations kN/m²	9.80	9.50	9.40	6.80
3.	**Permissible stresses** (N/mm²)				
(i)	*Bending and tension*				
	inside	8.4	10.2	7.8	6.6
	outside	7.2	8.8	6.6	5.6
	wet	6.2	7.0	5.6	5.0
(ii)	*Shear* (N/mm²)				
	horizontal	0.6	0.7	0.6	0.6
	along grain	0.9	1.0	0.8	0.8
(iii)	**Compression** (N/mm²)				
(A)	*Parellel to grain*				
	inside	6.4	7.8	6.0	5.2
	outside	5.6	7.0	5.2	4.6
	wet	4.6	5.6	4.1	3.8
(B)	*Perpendicular to grain*				
	inside	2.2	2.6	1.6	1.7
	outside	1.7	2.1	1.2	1.3
	wet	1.2	1.7	2.0	1.0

Note : 1. The values of horizontal shears are to be used for beams.
2. In all other cases, the shear along the grains are to be used.

2. Steel. The form-works which are to be used number of times should preferably be made of structural steel sections. The *panel forms* are fabricated from thin plates stifferened along the edges by small steel angles. The panel units are put together in position by or more clamps or bolts. The shuttering may be assembled and kept aligned by use of horizontal or vertical centring the panel forms may be made in large number in any desired modular shape or size (usually, rectangular). The panel forms are fabricated in 600 mm × 600 mm to 600 × 1200 mm in size.

The steel-forms are mostly employed in major construction projects. The panel-form may also be fabricated most suitably for circular, or curved shapes (e.g., circular tanks, columns, chimneys etc.). The erection and stripping of steel shutters are easy. These are repetitively used number of times. The contact surfaces of the structural components/ structures remain even and smooth. However, the cost involved for steel shutterings is high.

3. Plywood. Instead of timber, plywood is commonly used and becoming popular now-a-days. The panels of required size are made by attaching resin bonded plywood panels may be easily assembled by using bolts and nuts. The plywood shuttering gives quality surface finish. It is possible to have perfectly plain and smooth surface. The plywood shuttering may become comparatively economical because of reuse of shuttering number of times.

Fibre glass is mainly used for making the moulds for repetitive castings of pre-cast concrete products. For cast-in situation construction, (e.g., waffle slabs, or members having curved surfaces), the fibre glass, aluminium, or pre-cast concrete moulds may be used.

28.6 LOADS ON FORM-WORK

The form-work should be designed to support the following loads which are likely to affect them. The vertical loads and horizontal loads which are likely to act on the form-work are as follows :

Vertical loads. The vertical loads are as below :

1. The self-weight of the form-work themselves including their supports.

2. The dead load of wet concrete that is above them, using a unit weight of 25 kN/m³ to allow for the weight of some reinforcement.

3. The live load for the weight of a 0.8596 kN man standing in any probably position. This influences light forms for thin floors.

4. The weight of any equipment that may be placed upon the forms.

For Example, buggies full of concrete, specially the motorised ones that have been used so efficiently on some modern jobs, a concrete bucket that is likely to be set down temporarily.

5. The effect of impact due to vibrators.

It is difficult to find but may be minimized by proper supervision.

Horizontal loads. The horizontal loads/forces which are likely to act are as under :

6. The lateral (hydrostatic) pressure.

The intensity of the lateral pressure (hydro-static pressure) due to freshly poured concrete is uncertain. It is generally assumed to a limited extent to be hydrostatic in character. The rate of pouring of the fresh concrete, the temperature of the concrete, the nature of the coarse aggregate, the richness of the cement in concrete, the relative quantity of the fine aggregate, the slump, the use or absence of vibrators, and the use of retarding admixtures are the various factors which influence the unit pressure due to fresh wet concrete.

The fresh and wet concrete stiffens and it becomes a weak solid due to the setting of the cement (which one is about 45 minutes to one hour). Thereafter, the concrete will not act as fluid when concrete is deposited on it. Therefore, a height of concrete poured within 45 minutes to an hour may be taken while the pressure is calculated.

28.7 FORM-WORK FOR BEAM AND SLAB

In beam and slab construction, the beams are provided to support the slab and the slab remains continuous over the beam. In T-beam construction, the slab and beam (two components) jointly act together. In both these construction, form-work (viz., *shuttering*) is provided to support the slab as well as to support the beam, identically. The form-work for beam and slab is shown in Fig. 28.2.

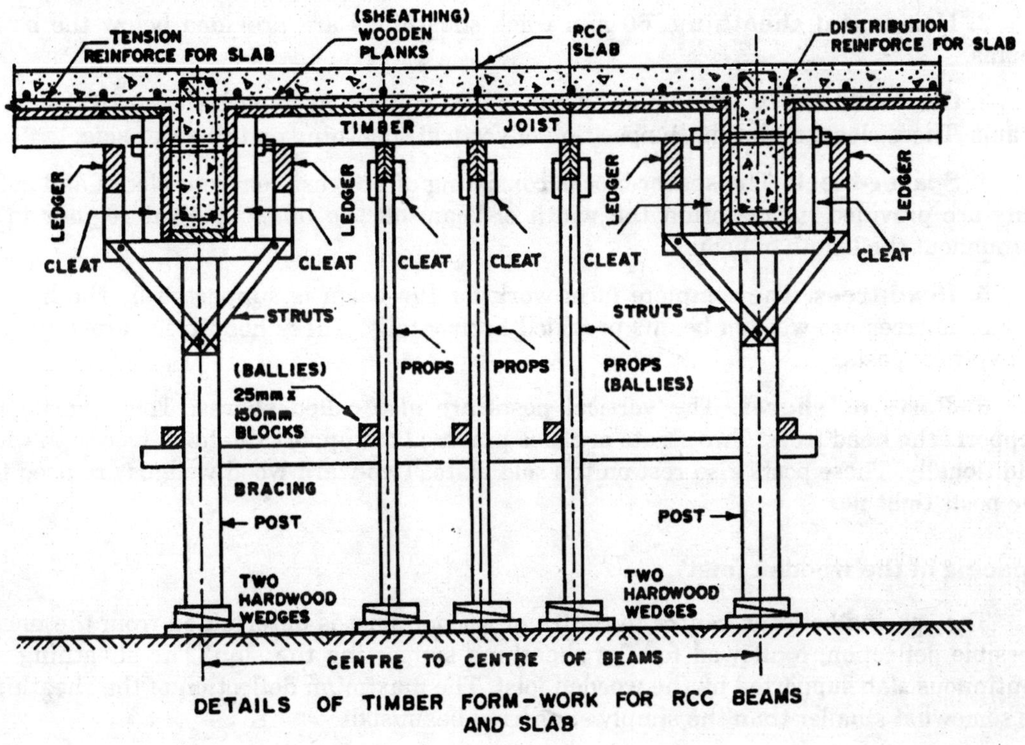

Fig. 28.2

The form-work consists of the following elements. These are described in the sequence in which the form-work is prepared.

1. Sheathing. The wooden planks 25 mm thick (also called as *sheathing*) are placed parallel to the beams to support the slab. The tongue and groove joints are provided to connect the different planks and the joints are made leakage proof, such that the cement grout does not come out.

2. Joists or battens. The *joists* are also called *battens.* The wooden joists of size 50 mm × 150 mm or 100 mm × 200 mm are provided below the planks at centre to centre spacing, s in mm (between the beams). The joists remain perpendicular to beams.

3. Props. The props made of ballies are fitted below the joists at regular intervals. At the bottom, the props are placed over sole plates and two hardwood wedges are used to keep the props tight. The props reduce the deflections of the various wooden joists.

4. Ledgers. The ledgers are horizontal wooden pieces. These are nailed to the ceats. The ledgers are provided at the top of props. The ledgers are also provided at the ends of the wooden joists near the beams.

The sheathing, joists or battens, props and the ledger form the shuttering for the slab. For the beams, the following elements are used to form the shuttering :

1. Vertical sheathing. 30 mm thick sheathings are used to support the vertical sides of the beams.

2. Horizontal sheathing. 60 mm thick sheathings are provided below the bottom of beams.

3. Cleats. 20 mm × 100 mm or 30 mm × 100 mm are fixed with the vertical sides of the beams. These cleats maintain the position of vertical sheathing on the outer side.

4. Spacer-block. The spacer-blocks consisting of pre-cast concrete blocks and bolts and nuts are provided (to maintain the width of beam on the inner side) at regular intervals throughout the length of beams.

5. Head-trees. The complete form work for the beam is supported by the head-trees. The head trees are wooden beams provided horizontally. These head-trees are supported on the vertical posts.

6. Posts or shores. The vertical posts are also called shores. These vertical posts supports the head trees. Two struts are also provided to support the head-tree with the posts additionally. These posts also rest on the sole plate. Two-ward-wood wedges are used to keep the posts tight fit.

Spacing of the wooden joist

The spacing between centre to centre of wooden joist is determined from the maximum possible deflection, δ allowed for the sheathing supporting the slab. The sheathing acts as continuous slab supported by the wooden joist. The maximum deflection of the sheathing will be somewhat smaller than the simply supported beam/slab.

The maximum allowable deflection

$$\delta = \frac{3}{384}\left(\frac{w}{E} \cdot \frac{s^4}{I}\right) \tag{28.1}$$

where, E is the Young's modulus of elasticity for the timber used for the sheathing, b is the width of sheathing and I is the moment of inertia of the sheathing cross-section.

$$\text{From Eq. 28.1, } s = \left[128 \, . \, EI \, . \left(\frac{\delta}{w}\right)\right]^{1/4} \qquad (28.2)$$

These calculations are also referred as the *design of decking.*

28.8 FORM-WORK FOR COLUMNS

The form-work for a square column is shown in Fig. 28.3. The form work consists of following elements :

1. Seathing. The wooden boarding 25 mm to 35 mm thick (i.e., timber plants) called as sheathing are provided vertically all along the perimeter of the column. These plants are nailed with the side yokes and end yokes.

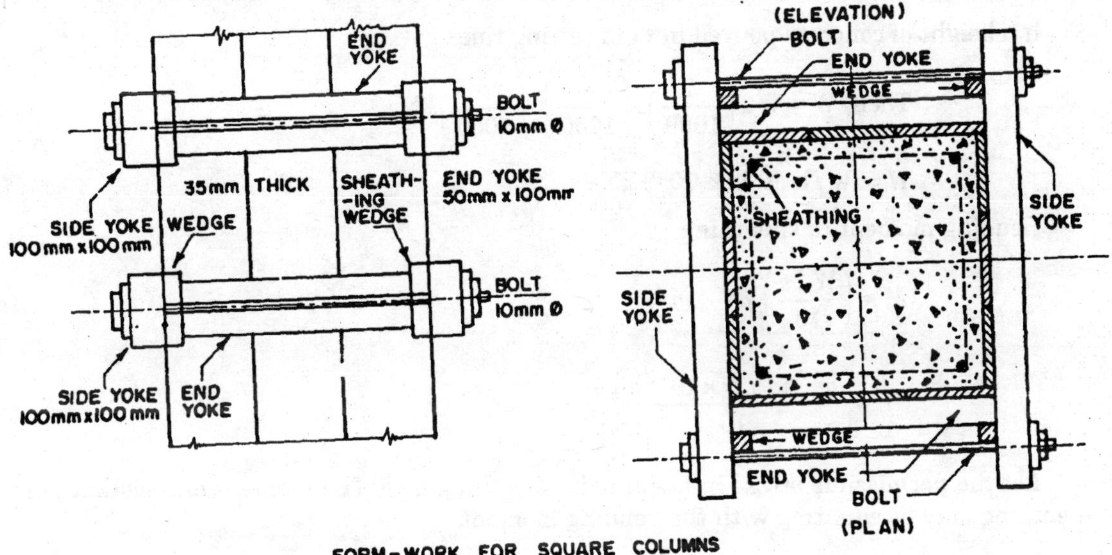

Fig. 28.3

2. Side and end yokes. A set of two side yokes and two end yokes 100 mm × 100 mm in size and 50 mm × 100 mm in size, respectively, is provided at one level of the column form-work. Such sets are suitably spaced along the height of the vertical sheathings. The sides yokes provided on opposite faces are inter-connected together by providing two long bolts.

3. Wedges. In order to tighten the various elements of the form-work, the four wedges are inserted (one at each corner) between the bolt and end yoke.

Spacing of yokes. The sets of yokes are provided at a spacing s in mm. The sheathing remains continuous over number of yokes. The wet concrete poured within the setting time, 45 minutes to an hour. The wet concrete will provide hydrostatic pressure over the sheathing. In case h (mm) is the height of the wet concrete poured within the setting time and w is the equivalent unit weight of concrete (kN/m³), the total uniformly distributed load acting across the width of column, b in mm shall be as under, as shown in Fig. 28.4.

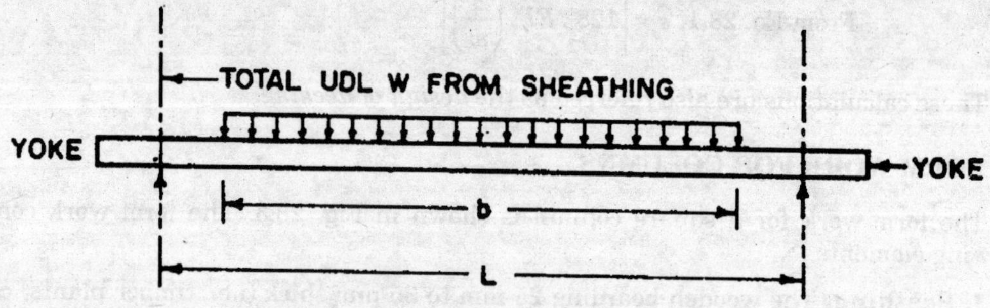

Fig. 28.4

where, w = equivalent hydrostatic pressure in kN/m³, (23 kN/m³)

h = height of concrete poured in m in setting time.

$$\text{Then, } W = W\,.\left(\frac{h}{1000}\right).\left(\frac{s}{1000}\right).\left(\frac{b}{1000}\right) kN \quad \text{(i)}$$

$$W = (w\,.\,h\,.\,s.\,b \times 10^{-9})\ kN \quad \text{(ii)}$$

Bending moment in sheathing

$$M = \left(\frac{W\,.\,s}{10}\right) kN - mm \quad \text{(iii)}$$

$$M = \left(\frac{w\,.\,h\,.\,s.\,b \times 10^{=9}.s}{10}\right) \quad \text{(iv)}$$

Let the permissible stress in bending in sheathing be f. The moment of resistance of the sheathing may be equating with the bending moment.

Then

$$f\,.\frac{1}{6}.\,b\,.\,d^2 = \left(\frac{w\,.\,h\,.\,s.b \times 10^{=9} \times s \times 10^{+3}}{10}\right) \quad \text{(v)}$$

where, b is the breadth of sheathing in mm and d is the depth (thickness) of the sheathing in mm. From the expression (v)

$$f\,.\frac{1}{6}.\,b\,.\,d^2 = \left(\frac{w\,.\,h\,.\,s^2.\,b \times 10^{-6}}{10}\right) \quad \text{(vi)}$$

$$\text{or } s^2 = \left(\frac{10}{6}\right)\left(10^6\right)\left(\frac{f}{wh}\right)d^2 \quad \text{(vii)}$$

$$s = 1291 \left(\frac{f}{wh} \right)^{1/2} .d. \tag{28.3}$$

The spacing, may also be determined from the strength of yoke in bending. Let f_1 be the permissible stress in the material of the yoke in bending and the section modulus of yoke be z mm³. Then,

$$\left(\frac{1000W}{2} \right) . \left(\frac{L}{2} - \frac{b}{4} \right) = f_1 Z \tag{viii}$$

or $$1000 . \frac{W}{8} (2L - b) = f_1 .Z$$

or $$\left(\frac{1000 w.h.s.b \times 10^{-9}}{8} \right) \times (2 L - b) = f_1 . Z$$

or $$\left(\frac{w . h . s . b. (2 L - b)}{8 \times 10^6} \right) = f_1 . Z$$

or $$s = \frac{8 \times 10^6 (f_1 . Z)}{w . h . b} \tag{28.4}$$

The spacing calculated from Eqs. 28.3 and 28.4, whichever is less, is adopted.

The form-work for the circular column is shown in Fig. 28.5. The form-work for octagonal column is shown in Fig. 28.6.

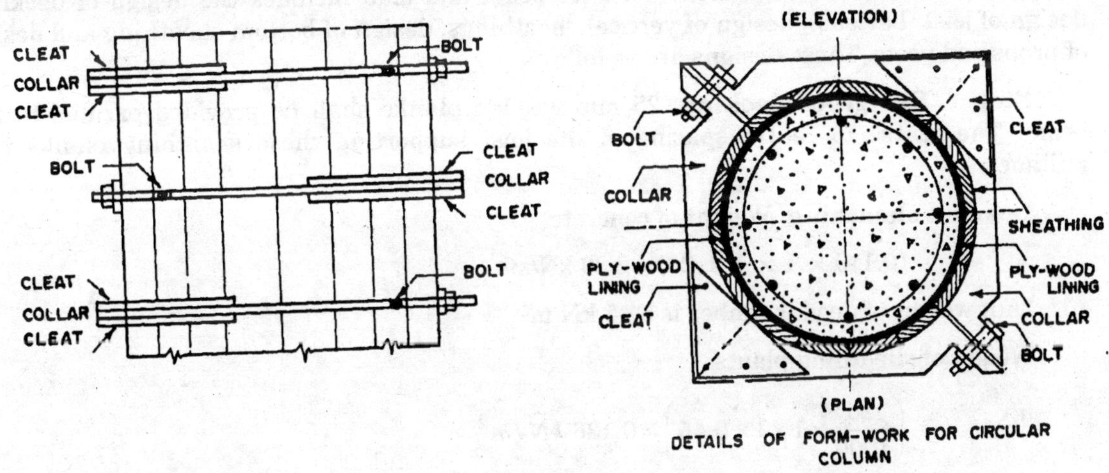

Fig. 28.5

Example 28.1. Design the form-work for beam and slab roof for the following :

Thickness of slab = 110 mm

Centre to centre spacing of beams = 3.6 m

Width of the beam = 320 mm

Depth of the beam below the slab = 440 mm

Height of ceiling of the roof = 3.40 m above floor.

Live load on sheathing is 4 kN/m^2 and unit weight of wet concrete is 26 kN/m^2. Use deoder timber for the form-work.

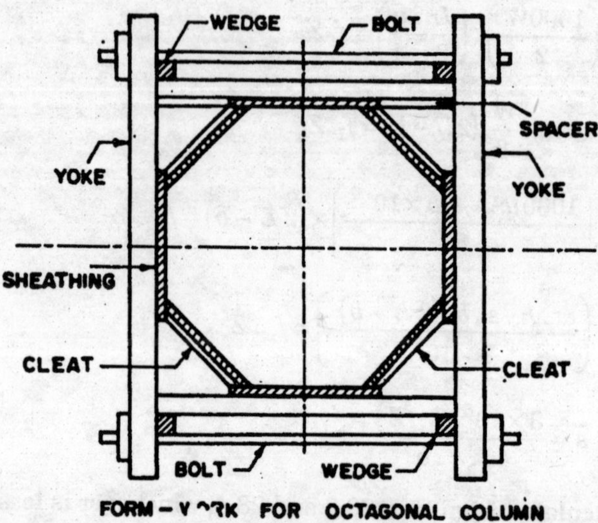

Fig. 28.6

Design : The design of form-work for beam and slab includes the design of decking, design of joist (battens), design of vertical sheathings, design of bottom sheathing and design of props and posts. These designs are as follows :

Step 1. Design of decking. 25 mm wooden planks shall be provided parallel to the beam. The centre to centre spacing of the joist supporting these sheathing planks is s millimetre.

Load on sheathing. Weight of concrete

$$(0.110 \times 1 \times 1 \times 26) = 2.86 \text{ kN/m}^2$$

Unit weight of deodar timber is 5.45 kN/m^2.

Weight of sheathing plants

$$\left(\frac{25}{1000} \times 1 \times 1 \times 5.45\right) = 0.136\ kN/m^2$$

Live load on sheathing = 4 kN/m^2

Total load on sheathing = 6.996 kN/m^2 = 7 kN/m^2 say = (7/1000) N/mm^2

Young's modulus of elasticity, E for timber is 9.5 kN/m^2 (i.e., 9500 N/mm^2).

Moment of inertia of sheathing (woodens plank) 1 mm width and 25 mm. Moment of inertia of sheathing

$$I = \left(\frac{1}{12} \times 1 \times 25^3\right) = 1302.08\ mm^4$$

$$EI = (9.5 \times 10^3 \times 1302.08) = 12.369 \times 10^6\ N/mm^2$$

3 mm deflection may be allowed for the sheathing. From Eq. 28.2.

(A) **Spacing of joist from deflection of planks**

$$s = \left[128\ .\ EI\ .\left(\frac{d}{w}\right)\right]^{1/4} = \left[\frac{128 \times 12.3698 \times 10^6 \times 3 \times 1000}{7}\right]^{1/4}$$

$$= [6785.7 \times 10^8]^{1/4} = 907.6\ \text{mm}$$

(B) **Spacing of joist from its bending moment.** Sheathing remains continuous over joist. The maximum bending moment for plank

$$M = \left(\frac{w.\ s^2}{10}\right) = \left(\frac{7}{1000} \times \frac{s^2}{10}\right) N - mm$$

Section modulus of section for 1 mm

$$Z = \left(\frac{1}{6} \times 1 \times 24^2\right) = 104.167\ mm^3$$

Permissible bending stress for deodar is 10.2 N/mm²

$$\therefore \left(\frac{7}{1000} \times \frac{s^2}{10}\right) = (104.167 \times 10.2)$$

$$\therefore \text{ Spacing } s = (12.32 \times 100) = 1232\ \text{mm.}$$

The spacing of the joists shall be, whichever is small. Therefore, the spacing of joist shall be 900 mm.

Step 2 . Design of joists (battens). The spacing of joists centre to centre is 900 mm. The span of joist

$$(3.6 - 0.320) = 3.28\ \text{m}$$

Total uniformly distributed load on the joist

$$\left(\frac{7}{1000} \times 900\right) = 6.3 N/mm$$

Maximum bending moment

$$M_1 = \left(\frac{w\ .\ l^2}{8}\right) = \left(\frac{6.3 \times 3280 \times 3280}{8}\right) = 8.47 \times 10^6\ N - mm$$

Equate moment of resistance and the bending moment. The width of joist is assumed 100 mm

$$10.2\times\frac{1}{6}\times 100\times d^2 = 8.47\times 10^6, therefore,\ d = 223.24\ mm$$

The size of wooden joist (deodar timber) may be kept as 100 mm × 250 mm.

Check for shear stress. The cross-section of wooden joist is rectangular. The maximum shear stress

$$\left(\frac{5}{8}w.l\times\frac{3}{2}\times\frac{1}{b.d}\right)=\left(\frac{5}{8}\times\frac{7\times 900}{1000}\times\frac{3280\times 1.5}{100\times 250}\right)=0.775\ N/mm^2$$

The maximum shear stress 0.775 N/mm² is less than the permissible shear stress, 1 N/mm².

Step 3. Design of vertical sheathings. The hydrostatic pressure of wet concrete poured within the setting time (45 minutes to an hour) shall be acting on the vertical sheathing. The equivalent fluid pressure may be assumed as 23 kN/m³. Overall depth of beam, (110 + 440) = 550 mm

Depth of vertical sheathing, (550 – 110 – 25) = 415 mm

Pressure at the top of vertical sheathing, $\left(23\times\frac{135}{1000}\right)=3.105\ kN/m^3$

Pressure at the bottom of vertical sheathing, $23\times\left(\frac{135+415}{1000}\right)=12.65\ kN/m^2$

Average pressure on the average sheathing, $\left(\frac{3.105+12.65}{2}\right)=7.88\ kN/m^2$

Total load per 1 m is 7.88 kN/m

The span of vertical sheathing shall be equal to that of wooden planks below the slab, (viz., 0.9 m = 900 mm).

$$\text{Bending moment, } M_2=\left(\frac{7.88\times 0.9^2}{10}\right)=0.6383\ kN-m$$

The thickness of vertical sheathing may be assumed as 35 mm.

Section modulus of the vertical sheathing

$$\left(\frac{1}{6}\times b\times d^2\right)=\left(\frac{1}{6}\times 415\times 35^2\right)mm^2$$

Bending stress in the vertical sheathing

$$\left(\frac{M}{Z}\right)=\left(\frac{0.6386\times10^6\times6}{415\times35\times3}\right)=7.53\ N/mm^2$$

Actual bending stress 7.53 N/mm^2 is less than the permissible bending stress for the deodar timber viz., 10.2 N/mm^2. Hence, it is safe.

Step 4. Design of horizontal sheathing below the beam :

Dead of wet concrete (including the thickness of slab)

$$\left(\frac{110+440}{100}\right)\times\left(\frac{320}{1000}\times1\times26\right)=4.576\ kN/m$$

Live load, (4 × 0.320 × 1) = 1.280 kN/m

Self-weight or horizontal sheathing may be assumed = 0.4 kN/m

Total load = 7.536 kN/m

Horizontal sheathing is supported on head free, which one is supported on the posts at a spacing equal to twice the spacing of ordinary wooden planks for the slab (viz., 2 × 900 = 1800 mm).

Bending moment for horizontal sheathing, $\left(\frac{7.536\times1.8^2}{8}\right)=3.052\ kN-m$

Width of bottom sheathing, (320 + 2 × 35) = 390 mm

Moment of resistance of the bottom horizontal sheathing may be equated to bending moment. Then,

$$10.2\times\frac{1}{6}\times390\times d_2^2=3.052\times10^6$$

$$d_2^2=\left(\frac{3.052\times10^6\times6}{10.2\times390}\right), therefore,\ d_2=67.85\ mm$$

The thickness of bottom horizontal sheathing below the beam may be kept as 70 mm.

Step 5. Design of posts. The vertical load on each post

= (7.536 × 1.8) = 13.565 kN

Load from the slab, 7 × (3.6 – 0.32) × 1.8 = 41.028 kN

Total load = 54.893 kN

Let the axial load on the post be 56 kN. Permissible direct stress in compression in deodar timber ballies along the grain for outside location is 7 N/mm^2.

Cross-sectional area required

$$\left(\frac{56\times1000}{7}\right)=8000\ mm^2$$

Size of ballies = 89.44 mm.

The size of ballies (posts) may be kept as 130 mm × 130 mm square size. The posts may be checked against buckling. Safe stress is given by [height of prop is (3.40 – 0.40) = 3 m]

$$f\left(1-\frac{h}{50d}\right)=7\left(1-\frac{3}{50\times 0.13}\right)=3.769\ N/mm^2$$

Allowable load

$$\left(\frac{3.769\times 130\times 130}{1000}\right)=63.7\ (>56\ kN)$$

Hence, the props/posts of size 130 mm × 130 are safe.

Example 28.2 A square reinforced concrete column is to be 380 mm × 380 mm in size. The height of column is 3.2 m. The complete concrete is to be poured for 3.2 m height in one stage. Design the form-work for the column. Deodar timber may be used for form-work.

Design : The design of form-work for the column consists of design of the vertical sheathing, and the design of yoke and designed of bolts. These are designed as under :

Step 1 : Design of vertical sheathing. Equivalent fluid weight for 1.5 m pour of concrete = 23 kN/m³ (i)

Equivalent fluid weight for 6.0 m pour of concrete = 12 kN/m³ (ii)

Equivalent fluid weight for 3.2 m pour of concrete

$$=\left[23-\left(\frac{23-12}{4.5}\right)(3.2-1.5)\right]=18.09\ kN/m^3$$

The centre to centre spacing of yoke is s mm. For deodar timber, the permissible bending stress is 10.2 N/mm². 25 mm thick wooden plans may be provided for the vertical sheathing. The spacing of sheathing (w = 18090 N/mm³)

$$s=1291\,d\,.\left(\frac{f}{w.h}\right)^{1/2}=1291\times 25\times\left(\frac{10.2}{18090\times 3.2}\right)^{1/2}$$

$$s=\left(\frac{1291\times 25\times 3.194}{240.6}\right)=428.46\ m. \qquad \text{(iii)}$$

The centre to centre spacing of yoke may be kept as 4.25 mm.

1. (A) **Check for shear.** Shear force

$$V=\left(\frac{1}{2}w\,.\,h\,.\,b\times\frac{425}{1000}\right)=\left(\frac{1}{6}\times 58.09\times 3.2\times 0.425\right)=4.674\ kN$$

The vertical sheathing is rectangular in section. The distribution of shear stress across a rectangular section is parabolic with maximum shear stress at the centre.

$$\text{Shear stress, } \left(\frac{3}{2}\times\frac{V}{b\,.\,d}\right)=\left(\frac{3}{2}\times\frac{4.674\times10^{3}}{380\times25}\right)=0.7381\ N/mm^{2}$$

Maximum permissible shear stress in deodar timber is 1 N/mm² is more the actual maximum shear stress. Hence, it is safe.

1. (B) **Check for deflection**. The permissible deflection for the vertical is 3 mm. Actual deflection is given by

$$y=\left(\frac{3}{384}\times\frac{w\,.\,h\,.\,b\,.\,s^{4}}{EI}\right)\ and\ E=9500\ N/mm^{2} \quad \text{(iv)}$$

$$I=\left(\frac{1}{12}\times380\times25^{3}\right)=49.48\times10^{4}\ mm^{4} \quad \text{(v)}$$

$$EI=(9.5\times10^{3}\times49.48\times10^{4})=4.7\times10^{9}\ \text{N-mm}^{2}$$

Load per 1 mm run of the sheathing

$$w\,.\,h\,.\,b=\left(18.09\times1000\times3.2\times0.380\times\frac{1}{1000}\right)=21.997\ N/mm$$

$$\therefore y=\left[\frac{3}{384}\times\frac{21.997\times(425)^{4}}{4.7\times10^{9}}\right]=1.193\ mm$$

It is less than the permissible deflection. Hence, it is safe.

Step 2 : Design of yoke. The distance between the centre to centre of bolts may be kept as 620 mm.

Load on the yoke

$$W=(w\,.\,h\,.\,b\,.\,s)=(18.090\times3.2\times0.380\times0.425)=9.349\ \text{kN}$$

This total load shall be acting as uniformly distributed load as shown in Fig. 28.4. Maximum bending moment

$$M=\frac{W}{8}(2L-b)=\frac{9.349}{8}(2\times0.62-0.380)=1.005\ kN-m$$

Permissible bending stress for deodar timber in bending is 10.2 N/mm². Section modulus required

$$\left(\frac{1.005}{10.2}\times10^{6}\right)=\left(0.09853\times10^{6}\right)-98.53\times10^{3}\ mm^{3}$$

Let the size of yoke be 80 mm × 100 mm. Section modulus

$$Z=\frac{1}{2}\times b\times d^{2}=\left(\frac{1}{2}\times80\times100^{2}\right)=133.33\times10^{3}\ mm^{3}$$

Actual section modulus is more than that required. Hence, it is safe.

Check for deflection

$$y = \left(\frac{5}{384} \times \frac{w \cdot l^3}{EI}\right) = \left(\frac{5}{384} \times \frac{9.349 \times 10^3 \times (620)^3}{9500 \times \frac{1}{12} \times 80\,(100)^3}\right) = 0.458\ mm$$

The permissible deflection is 3 mm. It is more than that of actual deflection. Hence, it is safe.

Step 3. Design of bolts. Actual tension in the bolt is equal to w = 9.349 kN

Tension in each bolt shall be 4.6745 kN.

The distance of wedge from the bolt is assumed as 45 mm. The bolt is also subjected to bending moment in addition to axial tension. Let 25 mm φ high strength bolt be used. The cross-sectional area of bolt

$$A = \left(\frac{\pi}{4} \times 25^2\right) = 490.87\ mm^2$$

$$\text{Direct tensile stress, } \left(\frac{4.6745 \times 1000}{490.87}\right) = 9.523\ N/mm^2$$

$$\text{Bending moment, } (4.6745 \times 45) = 210.35\ kN - mm$$

$$\text{Bending stress, } \left(\frac{210.35 \times 10^3}{\frac{\pi}{64} \times (25)^4} \times \frac{25}{2}\right) = 137.129\ N/mm^2$$

Total tension in bolt = (9.23 + 137.129) = 146.65 N/mm².

It is less than the permissible stress in tension 190 N/mm². Hence, it is safe.

PROBLEMS

28.1 Design the form-work for beam and slab roof for the following :
Thickness of slab = 130 mm
Centre to centre spacing of beams = 3.8 m
Width of beam = 380 mm
Depth of the beam below the slab = 480 mm
Height of ceiling of the roof above floor = 3.60 m
Live load on sheathing = 4 kN/m³
Use fir wood for the form-work.

28.2 A square reinforced concrete column is to be 360 mm × 360 mm in size. The height of column is 3.6 m. The total concrete is to be poured in one stage. Design the form-work for the column, air wood timber may be used for the form-work.

PART VIII

29 CONCRETE MIX AND LABORATORY TESTS

Design of Concrete Mix

29.1 INTRODUCTION

Portland cement, fine aggregate coarse aggregate and water fishly mixed together form wet concrete. The net concrete consolidates into a hard mass due to chemical reaction between the cement and water. The coarse aggregate works as a filler. The fine aggregate fills up the void in the coatrse aggregate. The cement in conjunction with the water acts as a binder. The cement paste and fines (and occasionally by admixtures) provide mobility.

In the design of concrete mix, a rational approach is adopted to determine the optimum proportions of the given ingredients (namely, cement, fine aggregate, coarse aggregate and water and admixtures, if necessary) to produce concrete possessing specified properties both in the fresh (viz., plastic) state and hardened state with the maximum overall economy. The condition of fresh concrete should be satisfactory. Therefore, the workability is considered as an important property for the fresh concrete. The ease of placement of concrete is controlled by the workability. The satisfactory compaction may be obtained only, in case, the concrete is sufficiently workable for the method of compaction to be used. Therefore, the workability necessary for the fresh concrete is specified. The condition of hardened concrete should also be satisfactory. The compressive strength and the durability are the important properties for the hardened state, so that the hardened concrete is not impaired. The durability is related primarily to the water-cement ratio. Both these properties are the pivotal points of proporting all the four constituents of the concrete. *The design of concrete mix is d one for a particular compressive strength of concrete with adequate workability so that the fresh concrete my be properly placed and compacted, and the required durability is obtained.* Professor Shack-lock (B.W.) in this publication `concrete constituents and mix proportions' in cement and concrete Association, 1914, and Professors Ghosh (R.K.), Chatterjee (M.R.) and Ram Lal in their publication `Flexural strength of concrete, its variations, relations, relationship with compressive strength and use in concrete mix design published in Road Research Bulletin No. 16, Indian Road Congress 1972 have suggested that, in special situations, the concrete may be designed for flexural strength or so that matter for any other specific property of concrete.

The grade designation of the concrete mix, type of cement, the maximum nominal size of aggregate, the water-cement ratio, workability and the minimum cement content are the *various factors in the choice of mix design.* The maximum nominal size of aggregates and the requirement of cement may jointly be put as aggregate-cement ratio. In addition to these above factors, *aggregate particle shape* and *texture* and *quantity of entrained air are more* additional points in the choice of mix design. Out of all these factors, the most critical factor in concrete mix design is water cement ratio. Two basic assumptions are made in the design of fresh (viz., plastic) concrete. The compressive strength of concrete is controlled by its water-cement ratio. And for a given aggregate characteristics, the workability of concrete is controlled by its water content. The purpose of design of concrete mix is to obtain the estimated minimum strength in the most economical manner. Total cost consists of cost of materials and cost of labour. The cost of labour includes cost of form-work, batching, mixing, transporting and curing and remains constant for every mix. The cost of cement is *many times* more than the cost of other constituents. The minimum cement content (consistent with strength and durability) is given all attention while designing a concrete mix. Considerable efforts have been done regularly since long and the effects of these factors on the properties of concrete have been studied. Still, the efforts are continuous to establish the precise relationship. Therefore, the design of concrete mix is an art rather than a science. The design of concrete mix is carried out by experience and trial and error, and available data for selecting mix proportions serve as a guide only. The mix proportions are adjusted and refinements are done in the observations made as trial mixes.

29.2 CONCRETE MIX PROPORTIONING

IS : 456-2000 recommends that the proportions of cement, aggregate and water may be determined to attain the required strength may be made *either by* designing the concrete mix (such concrete is called as `Design mix concrete') *or* by adopting the nominal concrete (such concrete is called `nominal mix concrete'). The designed mix concrete is preferred to nominal mix.

Various factors which influence the design of concrete mix have been narrated in Art. 29.1. Before designing a concrete mix, the information about these factors may be collected.

The **grade designation** (e.g., M 15, M 20, ... M 40) specifies the characteristic strength of 150 mm cube at 28 days expressed as N/mm^2. The concrete mix is to be designed for a target mean strength (somewhat more than the characteristic strength).

The **type of cement** (viz., ordinary Portland cement, Portland pozzalana. . . etc.) to be used is essentially asked as it is important mainly through its influence on the rate of development of compressive strength of concrete as well as durability under aggressive environments. The requirements of durability may be obtained by limitations in terms of minimum cement content, the type of cement and water-cement ratio.

The **maximum nominal size of aggregate** to be used in concrete is controlled by the type of structural member (e.g., for slabs, 10 mm, 12 mm aggregate, for beams and columns, 16 mm, 18 mm, 20 mm aggregates), the size of the section and the diameter and spacing of reinforcement steel bars IS: 456-2000 specifies that the nominal size of coarse aggregate should not be greater then one-fourth of the minimum thickness of the member, and it should be restricted to 5 mm less than the minimum clear distance between the main bars or 5 mm less than the minimum cover to the reinforcement and 5 mm less than the spacing between the cables, strands or sheathing in case of prestressed concrete. It is seen that the larger the maximum size of aggregate, the smaller is the cement requirement for a particular water-cement ratio as the workability of concrete increases with the increase in maximum size of aggregate. The type and size of aggregate (e.g., natural sand, crushed stone, gravel) along with source of supply as per IS : 383-1970 should also be asked.

The **water-cement ratio** is the main parameter which influences the strength of concrete and it has been described in Art. 2.15. It is seen that given sound aggregate and a fully compacted concrete (viz., other conditions being equal for workable mixes), the compressive strength of concrete varies as an inverse function of water-cement ratio. The exact relationship between the compressive strength and the free water-cement ratio has been stated by Abram's law described in Art. 2.15 by Eq. 2.3 and 2.4 and shown in Fig. 29.3.

The main effort in proportioning the different ingredients of a concrete mix is to use a minimum quantity of the paste, (that is, cement) which will lubricate the mass while the concrete is fresh and after hardening, it will bind the aggregate particles together and fill the space between them. The lubricating effect of paste provides the workability of the concrete mass, which is affected by the dilution of paste. In general, the aggregates used to be stronger than the paste compound, the strength of concrete is controlled by the strength of paste. It is therefore, necessary to consider the structure of the paste more closely. Instead of a solution of cement and water, a fresh paste in the form of suspension is needed. The dilute paste keeps greater spacing between the cement particles and makes the ultimate paste structure weak. The quantity of water needed also depends upon the quantity of paste. Therefore, it is essential that as small paste as possible should be paste. Therefore, it is essential that as small paste as possible should be used. The cost becomes more, in case the paste is in excess. The excess paste also increases the drying shrinkage of concrete, and susceptibility to percolation of water. The chances of attack by aggressive water and weathering action also increase.

Professors Cordon, W.A. and Gillespies, H.A. in their paper titled as `Variation in concrete aggregate and portland cement paste which influence the strength of concrete published in Journal of the American concrete Institute, August 1963 have shown that the water-cement ratio law is not final. The cement content for a specific strength changes depending upon the maximum size of aggressive. For the same water-cement ratio, different ranges of strength are obtained for the different size of aggregate.

The **workability of concrete** for satisfactory placing and compaction is related to the size and shape of the section to be concreted and the methods to be used for transportation, placing and compaction of concrete. The workability property of concrete has been described in Art. 1.12 along with the various properties of concrete. The workability of the concrete mass is provided by the lubricating effect of the water-cement paste and it is affected by the quantity of dilution of paste. The workability for a concrete mix is controlled by water-cement ratio as well as by aggregate cement ratio for a given type and grading of aggregate. The workability is also influenced by the maximum size of aggregate. The total surface area of aggregate becomes less as the size of aggregate increases.

The **durability of concrete** is also an important property of the concrete. It has been described in Art. 1.12. The durability of concrete depends essentially on the porosity. The voids in concrete are bound to exist in the paste, in case, the volume of the hardened paste is less than the combined volumes of cement and water that react chemically. Increase in the water used for preparing the concrete mix increases the voids. The concrete will become less durable and less impermeable with the increase in water-cement ratio. The value of water cement ratio has to be restricted to a maximum value from the consideration of the durability of the concrete.

Similarly, the maximum limit of **cement content** in the concrete may also have to be specified in appropriate circumstances. The shrinkage, cracking and creep of concrete increase with the cement paste content (i.e., with high cement content) when the thickness of concrete in structural members is thick and the member are restrained against movements, high cement content causes excessive cracking because differential thermal stresses due to hydration of cement in fresh concrete. For overall economy, the maximum cement content in the concrete mix is limited to 530 kg per m^3 for prestressed concrete structures (IS : 1343-1980) and for reinforced concrete liquid retaining structures [IS : 3370 (Part I) - 1965]. For high strength concrete, increasing cement content beyond a certain value of the order of 550 kg/m^3 or so, may not increase the compressive strength.

The concrete mix is to be proportioned to give an average strength which will be higher than the specified strength by an amount which will ensure that not more than the acceptable percentage of results will fall below the specified strength.

With the study of the various above properties of the materials of concrete, plastic (fresh) concrete and hardened concrete, a concrete mix for a particular strength may be designed efficiently.

The concrete mix proportioning is done by following the basic steps involved as mentioned below :

1. The mean target strength (defined in Art. 20.7) is arrived from the characteristic strength specified and the level of quality control.
2. The water-cement ratio is chosen for mean target strength and the requirements of durability are checked.
3. The water content for the required workability is arrived.
4. The cement content is calculated and the requirements of durability are also checked.
5. The relative proportion of the fine and coarse aggregates are choosen from the characteristics of coarse and fine aggregates.
6. The concrete mix proportions for the trial mix are arrived.
7. Suitable adjustments are made till the final mix composition is arrived at.

29.3 NOMINAL MIX CONCRETE

IS : 456-2000 recommends that the nominal mix concrete may be used for concrete of grades M 5, M 7.5, M 10, M 15 and M 20. The proportions of materials for nominal mix concrete has been given in Art. 2.14, Table 2.4. The proportions of nominal mix, that is, the quantities of cement and aggregates are given in terms of weight. The batching by weight has been preferred to the batching by volume (as it was specified in the earlier version of this code).

A rough guide for the nominal mix proportion by volume will be 1 : 3 : 6, 1 : 2 : 4 and 1 : 1 1/2 : 3 for M 10, M 15 and M 20 grades of concrete, respectively, provided that the nominal maximum size of aggregate is 20 mm. For other sizes of aggregates, adjustments in the ratios of the weight of coarse and fine aggregates are necessary as indicated in the note in Table 2.4. In all cases, the fine aggregates should conform to the grading of zone II or zone III of IS : 383-1970.

The cement content of the mix specified in Table 2.4 for any nominal mix shall be proportionately increased if the quantity of water in a mix has to be increased to overcome the difficulties of placement and compaction so that the water-cement ratio as specified is not exceeded.

It is to note that in case of vibrated concrete, the limit specified may be suitably reduced to avoid segregation.

It is to further note that the quantity of water used in the concrete mix for reinforced concrete work should be sufficient, but not more than sufficient to produce a dense concrete of adequate workability for its purpose, which will surround and properly grip all the reinforcement. The workability of the concrete should be controlled by maintaining a water content that is found to give a concrete which is just sufficient wet to be placed and compacted without difficulty with the means available.

If the nominal mix concrete in concrete made in accordance with the proportions given for a particular grade does not yield the specified strength, such concrete shall be classified as belonging to the appropriate lower grade. The nominal mix concrete proportioned for a given grade in accordance with Table 2.4 shall not, however, be placed in higher grade on the ground that the test strength are higher than the minimum specified.

The nominal mix concrete may be restricted to works of minor nature in which the strength of concrete is not critical.

29.4 COMMON TERMS

Following common terms are used in the statistical quality control of concrete :

1. Mean strength. The average strength determined by dividing the sum of strength of all the cubes by the number of cubes is known as the mean strength, σ_{mean}

$$\sigma_{mean} = \left(\sum \sigma / n\right) \tag{29.1}$$

where, $\sum \sigma$ is the sum of the strength of n number of cubes.

The average or mean strength gives the measure of central tendency.

2. Variance. The difference between any single observed data and the mean strength is called as *variance.*

$$\text{Varance} = \left(\sigma - \sigma_{mean}\right) \tag{29.2}$$

3. Coefficient of variation. The results of the variation may alternatively be expressed by *coefficient of variation.* The ratio of the standard deviation, s to the mean strength, σ_{mean} is called as coefficient of variation.

$$V_{coeff.} = \left(\frac{s}{\sigma_{mean}}\right) \times 100 \tag{29.3}$$

It is a non-dimensional measure of variation.

29.5 STATISTICAL QUALITY CONTROL OF CONCRETE

In order to derive the optimum use from the materials used, the importance of *quality control* of the concrete is being increasing realised. The quality control does not simply means the testing of concrete cubes at 28 days. The quality control means to see to all aspects of the choice of the materials, design and workmanship. It begins much earlier than the any concrete is available for testing at 28 days.

The variability both in materials and constructional methods is always seen in concrete upto certain extent. Due to this, the variation of strength from batch to batch and even within the batch is observed. The assessment of strength of the final product is very difficult. A large number of destructive tests cannot be done to evaluate the strength of final products. Therefore, the sample tests are always done. Too few tests may lead to erroneous situation and cannot be relied. Very rigid criteria to accept or not to accept a structure will be very costly.

In general, the specifications provide the condition for 28-day compressive strength requirement. It is useful to provide the criteria. However, the limitations of compressive strength data should be considered. Instead of actual strength of a structure, the potential strength is indicated by the test specimen. The decrease in strength of concrete due to poor workmanship in placing and curing is not reflected by test.

A reasonable control of concrete work may be provided by ensuring that the probability of test result falling below the design strength remains less than the specified tolerance level. The variability may be limited by quality control. A scientific approach is provided by the statistical quality control to understand the realistic variability of the materials. In the light of this, the design specifications may be prepared so that the unavoidable variations may also be included.

Many problems shall be introduced by the quality control as the strength of cube shall not be known until 28 days after placing the concrete. A minimum strength requirement is specified by most of the specifications. Statistical analysis shows that such specifications are difficult to be satisfied. By planning a proper sampling plan, it is possible to ensure a certain quality at a specified risk. The statistical method provides a scientific basis of acceptance which is realistic and restrictive as required by the requirements for the design for the concrete construction.

For large contracts, the quality control of concrete is of immense value. The efforts made will be more than the returns by the saving resulted in the overall concreting.

29.6 STANDARD DEVIATION

The difference between the individual test strength, σ from the average (mean) strength, σ_{mean} of n samples is called as *deviation.* It is denoted by Δ.

$$\Delta = (\sigma - \sigma_{mean}) \tag{29.4}$$

The *standard deviation* is defined as the root mean square deviation of all the results, the numerical value of the estimated standard deviation, *s* may be calculated as below :

$$s = \left(\frac{\Sigma \Delta^2}{(n-1)} \right)^{1/2} \tag{29.5}$$

or
$$s = \left(\frac{\Sigma\, (\sigma - \sigma_{mean})^2}{(n-1)} \right)^{1/2}, \tag{29.6}$$

where, n is the number of sample test results.

The standard deviation increases with increasing variability. The standard deviation of the results from their mean value is regarded as an index of the scatter and consequently of the degree of site control.

To constitute an acceptable record for calculation of the standard deviation, the number of test results should be atleast 30. The efforts may be made to obtain 30 test results, as early as possible, when a mix is used for the first time. The calculation of the standard deviation shall be brought up to date after every change of mix design and atleast once a month . The concrete of each grade shall be analysed separately to calculate its standard deviation. Where sufficient test results for a particular grade of concrete are not available, the value of standard deviation may be assumed as given in Table 29.1 as per IS : 456-2001.

However, when adequate past records for a similar grade exist and justify the value of standard deviation different from that given in Table 29.1, it shall be permissible to use that value.

When a large number of samples of a given grade of concrete manufactured even under stable conditions of control are tested, the results will show some variation. In case, the compressive strength of concrete is potted along x axis against the frequency (number of samples) occurrence, along y-axis, a bell shaped curve (viz., normal distribution curve or Gaussian distribution curve) as shown in Fig. 29.1 shall be plotted.

TABLE 29.1. ASSUMED STANDARD DEVIATION

Grade of concrete	Assumed standard deviation (N/mm^2)
M 10	2.3
M 15	3.5
M 20	4.6
M 25	5.3
M 30	6.0
M 35	6.3
M 40	6.6

The characteristic strength of concrete, f_{ck} (which is less than the mean strength, that is, target strength) is taken as reference for designing the grade of concrete in IS : 456-2001.

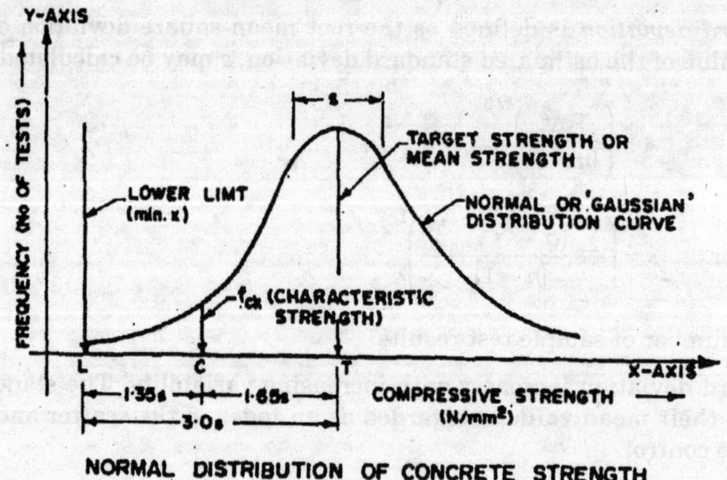

Fig. 29.1

A standard deviation shows the measure of dispersion. A small value of standard deviation shows a curve with dominant peak as shows in Fig. 29.2. A large value shows a flatter curve depending upon the level of control exercised in the manufacture of concrete.

The standard deviation calculations should be done a fresh and not brought up-to-date, whenever there is a change in mix design.

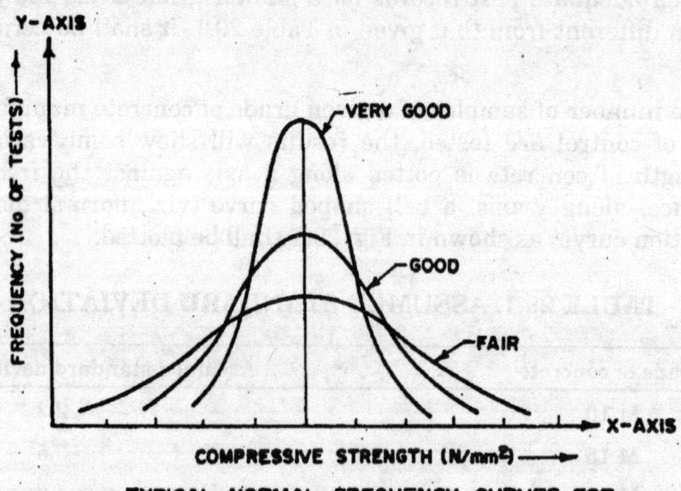

Fig. 29.2

29.7 TARGET STRENGTH

The strength requirement of a concrete mix has been specified in IS : 456-2000 in terms of the characteristic strength and the standard deviation. For specification and design, the characteristic strength of concrete has been used as a reference value. In case the inherent variability of concrete strength during production is considered, it is necessary to know the term target strength. When a concrete mix is designed, it should have target mean strength. The target mean strength is higher than the characteristic strength by a suitable margin. The target mean strength, f_t may be obtained as below :

Target strength = (characteristic strength + K . times standard deviation)

$$f_t = (f_{ck} + K . s) \tag{29.7}$$

where, s is the standard deviation, and K is a constant. Its value is equal to 1.65 as per the definition of the characteristic strength where not more than 5 per cent of the test results are expected to fall below the characteristic strength. Therefore,

$$f_t = (f_{ck} + 1.65\ s) \tag{29.8}$$

29.8 ACCEPTANCE CRITERIA

IS : 456-2000 specifies two sets of criteria for supplied concrete which may be considered to comply with the strength requirement and which do not comply with the strength requirement. These criteria are described here.

1. Supplied concrete deemed to comply with strength requirement. The supplied concrete may be considered to be satisfactory (i.e., acceptable) in case :

(a) every sample has a test strength *not less than* the characteristic strength of concrete, or

(b) the strength of one or more samples though less than the characteristics value, is in each case *not less* than the greater of the following

(i) the characteristic strength minus 1.35 times the standard deviation; that is,

$$\sigma < (f_{ck} - 1.35\ s) \tag{29.9}$$

(ii) 0.80 times the characteristic strength, that is,

$$\sigma \nless (0.80\ f_{ck}) \tag{29.10}$$

whichever is more in Eqs 29.9 and 29.10 and the average strength of all the samples is not less than the following :

$$\sigma \nless f_{ck} + \left[1.65 - 1.65 / (n)^{1/2}\right]. s, \qquad where \tag{29.11}$$

σ_{mean} = Average or mean strength of samples of concrete supplied

f_{ck} = characteristic strength

n = number of samples, (minimum = 30) and

s = standard deviation.

2. Supplied concrete deemed not to comply with strength requirement. The supplied concrete shall not be satisfactory (i.e., not acceptable) in case,

(a) the strength of concrete is less than the greater of the following

(i) the characteristic strength minus 1.35 times the standard deviation that is,

$\sigma < (f_{ck} - 1.35 \cdot s)$ (29.12)

(ii) 0.80 times the characteristic strength

$\sigma < (0.80\, f_{ck})$ 29.13)

whichever is more in Eqs. 29.12 and 29.13,

(b) the average strength of all the samples is less than the following :

$$\sigma_{mean} < \left(f_{ck} + \left[1.65 - 3/(n)\right]^{1/2}\right] . s\right) \quad (29.14)$$

The test strength of a sample shall be the average of the strength of three specimens. IS : 456-2000 recommends that the individual variation should not be more than ± 15 percent of the average, that is

$$(\sigma - \sigma_{mean}) \not> 0.15 \cdot \sigma_{mean} \quad (29.15)$$

Eq. 29.11 may be used to calculate average minimum field strength of concrete for a given concrete mix and a given value of the standard deviation. The average field strength may be increased by 15 percent to find the corresponding values of a laboratory strength. Table 29.2 has been prepared for the guidance as per IS : 456-2000 for the laboratory design strength.

TABLE 29.2. LABORATORY DESIGN STRENGTH

Description	Grades of concrete					
	15	20	25	30	35	40
1. Assumed standard deviation (IS : 456-2000 (N/mm^2)	3.5	4.6	5.3	6.0	6.3	6.6
2. Average field strength (N/mm^2)	19.72	26.20	32.148	38.09	43.497	48.902
3. Laboratory design strength N/mm^2	22.678	30.13	36.97	43.80	50.02	56.237

(Number of samples n = 30)

A random sampling procedure shall be adopted to ensure that each concrete batch shall have a reasonable chance of being tested that is, the sampling should be spread over the entire period of concreting and cover all mixing units. The minimum frequency of sampling of concrete of each grade shall be as recommend in IS: 456-2000 and given in Table 29.3, below :

TABLE 29.3. NUMBER OF SAMPLES

Quantity of concrete in the work, m^3	Number of samples
1–5	1
6–15	2
16–30	3
31–50	4
51 and above	4 plus one additional sample for 50 m^3 or part thereof

Note : Atleast one sample shall be taken from each shift.

IS: 456-2001 further specifies that the concrete of each grade shall be assessed separately. The concrete shall be assessed every day for compliance. The concrete is liable to be rejected if it is porous or honey-combed; its placing has been interrupted without providing a proper construction joint, the reinforcement has been displaced beyond the tolerance specified; or the construction tolerance has not been met. However, the hardened concrete may be accepted after carrying out suitable remedial measures.

The main statistical features of the acceptance criteria are as below :

(i) The minimum strength is mentioned by the characteristic strength below which certain test results are allowed to fall. Such `low' results, which are inevitable to occur, not regarded as failures and the concrete in the structure that they represent should not be considered automatically for rejection.

(ii) Notwithstanding such statistical concept for the minimum strength, the absolute value is also specified as in Eq. 29.10.

(iii) For the purpose of acceptance each sample is expected to be equal to or exceed the characteristic strength required.

(iv) Because of the random nature of strength, a sample may have strength lower than the characteristics strength although expectedly not lower than 3 times standard deviation below the mean, that is, 1.35 times the standard deviation below the characteristic strength but by the same logic, there should not be other samples whose strength should have exceeded the characteristic strength. The average of all such samples should then be not less than that given by Eq. 29.11.

It is to note that the standard deviation of average of n samples is $[1/(n)^{1/2}]$ times that of the individual samples, when n is sufficiently large, this approaches the mean or target strength, as it should be.

The acceptable is thus on the basis of the average of all samples tested till that time rather than on individual low results. Thus a larger set of test data is utilised in decision making.

(v) However, when the average of n such samples is less than that given by Eq. 29.14, the concrete is deemed not to have compiled with the acceptance criteria and the structural adequacy is to be investigated.

29.9 METHODS OF CONCRETE MIX DESIGN

Empirical relationships, charts and graphs developed from extensive experimental investigations are used in different countries as methods for the design of concrete mix. The introduction of reinforced cement concrete as an alternative for the steel constructions, in the beginning of *twentieth century,* necessitated the development and use of low and medium strength concretes. Professor Freyssinet introduced the prestressed concrete in 1930. Since than the importance of using high strength concrete to withstand the high compressive stresses developed due to prestressing was recognised. Early methods were base on obtaining the desired strength and developed the concrete with compressive strength ranging from 30 to 70 N/mm^2 by proper selection of the ingredients and improved methods of manufacture of the concrete. The increase in quantity of cement and with better compaction, the strength of concrete was further increased. The role of mixing water was not clearly under stood. The use of aggregates have less voids resulted in stronger concrete. Some of the earlier methods of concrete mix design are based on the principles of minimum voids and maximum density.

Recent developments in the field of concrete mix design have shown that the concrete of ultra-high strength with compressive strength of cube desired at 28 days ranging from 70 to 100 N/mm^2, without adopting any unusual material or processing and without incurring any significant technical difficulties may be produced. A concrete of compressive strength of about 110 N/mm^2 may be produced as a limit. Above this limit, it becomes essential to adopt special materials and sophisticated manufacturing technique like polymerization. The concepts of workability, durability and the strength and other factors influencing the concrete mix proportions are of comparatively recent origin.

With the present state of knowledge in the field of concrete mix design, it is possible to prepare and design concrete capable of resisting heat, sea water, frost and chemical attack arising out of industrial effluents. Consideration developments have been obtained in the design use of structural light weight concrete. The light weight concrete has the advantage of reduced density as well as increased thermal insulation. At present in nuclear age, high density concrete has been successively prepared for radiation shielding of nuclear reactors.

Following are the different methods of concrete mix design.

1. Minimum voids method
2. Maximum density method
3. Fineness modulus method
4. RRL method (the Road Research Laboratory, London method)
5. ACI method (The American Concrete Institute method)
6. British method
7. USBR method
8. The Indian Road Congress method
9. Indian standard method

These methods of concrete mix design have been described in the subsequent sections.

29.10 MINIMUM VOIDS METHOD

A concrete mix is designed to have minimum voids and thereby maximum density. The voids in coarse aggregate are filled by fine aggregate in enough quantity. The voids in fine aggregate are filled by cement paste in sufficient quantity. In order to give mobility to the mix and to allow the additional voids created by wedging acting of the fine aggregate particles on the coarse aggregate and similarly that of the cement particles on the fine aggregate, extra fine aggregate and extra cement paste are mixed.

The minimum voids method is based on experimental approach. The proportion of fine aggregate and coarse aggregate is first found and then the cement paste is mixed to get the concrete of the desired consistency.

The step by step procedure to design the concrete mix by the minimum voids method is as under :

1. The dry rodded unit weights of coarse aggregate and fine aggregate are found.

2. The proportions of graded coarse aggregate and fine aggregate are kept as 20 : 80, 30 : 70, 40 : 60, 50 : 50, 60 : 40, 70 : 30, 80 : 20 and these are mixed thoroughly. The dry robbed unit weights of the combined aggregate with each proportion of coarse aggregate and fine aggregate are found.

3. The percentage of voids in each of the proportion are found knowing the unit weights of coarse aggregate and fine aggregate.

4. The percentage of voids is plotted against the proportion of coarse aggregate and fine aggregate. The proportion which gives minimum percentage of voids is then found from plot and it is used in preparing a trial mix.

5. A mix of coarse aggregate and fine aggregate is prepared in the proportion corresponding to the minimum voids knowing the weights of coarse and fine aggregates.

6. A cement paste of suitable water-cement ratio is prepared separately.

7. This paste is added slowly in sufficient quantity to the mix of aggregates and simultaneously mixed so that a mix of desired consistency (workability) is obtained.

8. The properties of the concrete mix are checked by testing it in accordance with the relevant IS specifications. If necessary, another trial mix with suitable modifications is prepared.

The minimum voids method should be used carefully. This method gives widely varying results. The coarse aggregate particles separate out when the cement paste and the fine aggregates are added. The voids will increase as a result of this.

Example 29.1. In case, the coarse aggregate and the fine aggregate contain 44 percent and 30 percent voids, respectively, design the concrete mix.

Design. Let the quantity of coarse aggregate be 100 litres. The voids oc coarse aggregate are filled with 10 percent extra fine aggregate. The quantity of fine aggregate needed.

$$1.10\left(\frac{44}{100}\times 100\right) = 48.4 \textit{ litres}$$

The voids of fine aggregate are filled with 15 percent extra portland cement. The quantity of cement paste required,

$$1.15\left(48.4\times\frac{30}{100}\right) = 16.698 \textit{ litres}$$

Approximate quantity of cement may be taken as 20 percent extra.

$$(16.698\times 1.20) = 20.04 \textit{ litres}$$

Proportions of cement : fine aggregate : coarse aggregate by volume

$$(20.04 : 48.4 : 100) \text{ That is, } (1 : 2.415 : 4.99)$$

29.11 MAXIMUM DENSITY METHOD

Professor Fuller proposed maximum density method as an improvement over the minimum void method. It is a theoretical approach to find the grade of aggregate to determine the maximum density. In the maximum density method, it is assumed that the greater quantity of solid particles that may be packed in a given volume of concrete to give maximum density of the concrete, so that the strength of the concrete shall be higher. The graded aggregate is used to get maximum density for the concrete mix. Professor Fuller and Thompson gave ideal curve based on the empirical expression as below :

$$p_d = 100 \,.\, (d/D)^{1/2}, \text{ where} \qquad (29.16)$$

p_d = percentage of particle size by weight smaller than the size, *d*

D = maximum size of the coarse aggregate

d = maximum size of the fine aggregate

In case, it is needed to grade a mix with 30 mm maximum size of coarse aggregate, the percentage of particle sizes finer than 20 mm, 10 mm and 4.75 mm are as under : From Eq. 29.16

$$p_d = 100 \times (20/30)^{1/2} = 81.65 \text{ percent}$$

$$p_d = 100 \times (10/30)^{1/2} = 57.74 \text{ percent}$$

$$p_d = 100 \times (4.75/30)^{1/2} = 39.79 \text{ percent}$$

For the percentage of particle size finer than 4.75 mm, the quantity of cement and fine aggregate is 39.79 percent and that of coarse aggregate is 60.21 percent. In addition, the quantity of particles of various intermediate sizes should correspond with Fuller's ideal curve.

In this method, the effect of particle interference is ignored. In practice, it was seen that the aggregate graded to give maximum density results in a harsh and unworkable mix, since the workability is improved only when there is excess of paste above that needed to fill the voids in the fine aggregate and also in excess of mortar above that needed to fill the voids in the coarse aggregate.

Now, the concept of an ideal grading curve is not credited. The concrete may be prepared successfully from aggregates having gradings much different from the ideal curve.

The step by step procedure for the design of a concrete mix following Fuller's method is as under :

1. Depending on the maximum size of coarse aggregate, a suitable aggregate cement ratio may be assumed.
2. The percentage of particles smaller than 4.75 mm are determined. Eq. 29.16 is used for this purpose.
3. The quantity of fine aggregate is determined.
4. The quantity of coarse aggregate is also determined.
5. A suitable water-cement ratio is chosen to get a concrete mix of desired strength.
6. The proportions of the concrete mix are checked by testing it is in accordance with the relevant IS specification. If necessary, another trial is made with suitable modifications.

Example 29.2. The maximum size of coarse aggregate is 30 mm and 20 N/mm² is expected to be the characteristic strength at 28 days. Use ordinary portland cement. Design the concrete mix by maximum density method.

Design : Let the aggregate: cement ratio be 6 : 1. The quantity of portland cement shall be 14.286 kg in a total mix of 100 kg.

Percentage of particles smaller than 4.75 mm for 30 mm maximum size of coarse aggregate, from Eq. 29.16.

$$p_d = 100\,.(4.7/30)^{1/2} = 39.79 \text{ percent}$$

The quantity of particles smaller than 4.75 mm is 39.79 percent. It gives the combined percentage of cement and fine aggregate as 39.79 percent, that is 39.79 kg.

Quantity of cement = 14.286 kg

Quantity of fine aggregate (39.79 – 14.286) = 25.504 kg

Quantity of coarse aggregate (100 – 39.69) = 60.21 kg

Characteristic strength needed is 20 N/mm²

As per Table 29.2 and IS: 456-1978, the laboratory strength for M 20 is to be 30.13 N/mm². From strength: water-cement ratio curve, Fig. 2.3 is 0.58 corresponding to 30.13 N/mm². The proportion of concrete mix 14.286 : 25.504: 60.21, that is 1 : 1.785 : 4.215 with water-cement ratio as 0.58.

29.12. FINENESS MODULUS METHOD

Professor Abrams, D.A. evolved water-cement ratio law in 1918 and published a paper titled as `Design of concrete mixtures' in Structural Materials Research Laboratory', Lewis Institutive, Chicago, 1918, Bulletin No. 1 and since, then, it was referred as Abrams' law of water-cement ratio. It proved to be a significant advance for design of concrete mix.

Abram's law states that when the concrete is fully compacted (that is, there are no voids and no entrapped air in the concrete mix), its compressive strength (CS) is inversely proportion to water-cement ratio (w/c), that is

$$CS \propto [1/(w/c)] \tag{i}$$

In Art. 2.15, it is stated by Eqs. 2.3 and 2.4 as below

$$CS = K_1/(K_2)^{w/c} \tag{ii}$$

$$\text{or } \log_e (CS) = \log_e K_1 - (w/c) \log_e K_2 \tag{iii}$$

According to Abram's law, the compressive strength of concrete at 28 days strength is given by Eqs. 2.3 and 2.4 as above. In addition to water-cement ratio, the compressive strength also depends on relative proportions of fine and coarse aggregate such that the combined aggregate require minimum quantity of cement paste for full compaction. The cement paste needed will be minimum for given workability and the concrete shall be economical for given strength.

Professor Abram's assessed workability in relation of fineness modulus has been described in Art. 1.8. Fineness modulus was used as a single parameter to express the grading of aggregate. *The fineness modulus of the concrete is a measure of the mean size of particles in total quantity of the aggregate.* The fineness modulus is determined as the sum of the cumulative percentages retained on a set of ten IS : sieves (80 mm, 40 mm, 20 mm, 10 mm, 4.75 mm, 2.36 mm, 1.18 mm, 600 micron, 300 micron and 150 micron) and divided by 100. The smallest size of sieve is 150 micron; and the larger designations are such that a particular size is *twice* that of the preceding designation.

The surface area of coarse particles is less and therefore, less cement paste is needed to coat the surface. While the surface area of fine particles is more and hence more cement paste is required. The minimum surface area and minimum voids (that is, the maximum density) indicates that the aggregate is nicely graded. Professor Abram recommended that ideal grading of concrete as given in Table 29.4. It is not possible to get aggregate conforming to ideal grading.

Let p_{CA} be the fraction of coarse aggregate and its fineness modulus be FM_{CA}. Let p_{FA} be the fraction fine aggregate and its fineness modulus be FM_{FA} = Let FM_{MA} be the fineness modulus derived by mixing the coarse aggregate and fine aggregate. Then,

$$(p_{CA} + p_{FA}) = 1 \quad \text{(iv)}$$

$$FM_{MA} = (p_{CA} .\ FM_{CA} + p_{FA} .\ FM_{FA}) \quad \text{(v)}$$

Therefore, by solving the expression (iv) and (v)

$$p_{FA} = \left(\frac{FM_{CA} - FM_{MA}}{FM_{CA} - FM_{FA}} \right) \quad \text{(vi)}$$

Percentage of fine aggregate

$$p_{FA} = 100 . \left(\frac{FM_{CA} - FM_{MA}}{FM_{CA} - FM_{FA}} \right) \quad (29.17)$$

TABLE 29.4. IDEAL GRADING OF AGGREGATE (AS RECOMMENDED BY ABRAM)

IS : Sieve Designation	Percentage passing			
	Maximum size of aggregate			
	150 mm	75 mm	40 mm	20 mm
150 mm	100.00	–	–	–
75 mm	70.00	100.00	–	–
40 mm	50.00	70.00	100.00	–
20 mm	36.00	50.00	70.00	100.00
10 mm	25.00	36.00	50.00	70.00
4.75 mm	18.00	25.00	36.00	50.00
2.36 mm	12.50	18.00	25.00	36.00
1.18 mm	9.00	12.50	18.00	25.00

600 μ	6.25	9.00	12.50	18.00
300 μ	4.50	6.25	9.00	12.50
150 μ	3.00	4.50	6.25	9.00
cumulative percentage passing	334.25	331.25	326.75	320.50

(Note : This grading also follows Fuller's grading).

Practically, it was not possible to get aggregate conforming to the ideal grading. Professor Abram specified maximum and minimum limits for the fineness modulus as given in Table 29.5.

The step by step procedure to design concrete mix by fineness method is as below :

1. Knowing the minimum compressive strength and workability needed, the sizes of available coarse aggregate and fine aggregate, the grading of both aggregates, their weights, their bulking percentage, the water-cement ratio depending upon the strength or durability is found from Fig. 2.4.
2. The maximum size of coarse aggregate is decided depending on the minimum size or thickness of a structural member concrete section.
3. The grading of aggregate and fineness modulus of these aggregates are found.

Table 29.5. LIMITS OF FINENESS MODULUS (SPECIFIED BY ABRAM)

Type of Aggregate	Maximum size of Aggregate (mm)	Limits of fineness modulus	
		Maximum	Minimum
Fine	–	3.5	2.0
Coarse	20	6.9	6.0
	40	7.5	6.9
	75	8.0	7.5
	150	8.5	8.0
Combined	20	5.1	4.7
	25	5.5	5.0
	40	5.9	5.4
	75	6.3	5.8
	150	7.0	6.5

4. The weights of cement, and water needed for 100 cubic metre are found.
5. The proportions of cement, fine aggregate and coarse aggregate and their absolute weights are found.
6. The properties of trial mix of concrete is checked by testing in accordance with relevant IS specifications. If necessary, another trial mix is prepared with suitable modifications.

In this method, it is possible for aggregate to have substantially different gradings for the same fineness modulus. It needs number of trial mixes to obtain the desired workability.

The minimum voids method, maximum density method and fineness modulus method are referred as early mix design methods. These methods paved the way for better understanding of design of concrete mix. These methods indicated the importance of various parameters which influence the concrete mix.

Most of the methods used for the design of concrete mix in present time are based on extensive experimental investigations.

Example 29.3. The sieve analysis of 10 kg of coarse aggregate was carried out. The quantities of coarse retained over different sieves are shown in column 2 of table given below. Find the fineness modulus of coarse aggregate.

Solution. Quantity of coarse aggregate is 10 kg. The quantities of coarse aggregate retained over different sieves are as under :

IS : sieves	Mass of aggregate retained (kg)	Cumulative mass retained (kg)	Percentage cumulative mass retained (per cent)
150 mm	0	0	0
75 mm	0	0	0
40 mm	2.4	2.4	24
20 mm	3.2	5.6	56
10 mm	1.8	7.4	74
4.75 mm	2.3	9.7	97
2.36 mm	0.3	10.0	100
1.18 mm	–	10.0	100
600 m	–	10.0	100
300m	–	10.0	100
150 m	–	–	–
Total	10 kg	–	651

Fineness modulus of coarse aggregate

$$FM_{CA} = \left(\frac{Cumulative\ percentage\ retained}{100}\right) = \left(\frac{651}{100}\right) = 6.51$$

Example 29.4. The sieve analysis of 1 kg fine aggregate was carried out. The qualities of fine aggregate retained over different sieves are shown in column 2 of table below. Find the fineness modulus of fine aggregate.

Solution. Quantity of fine aggregate is 1 kg. The quantities of fine aggregate related over different sieves are as below :

IS : sieves	Mass of aggregate retained (kg)	Cumulative mass retained (kg)	Percentage cumulative mass retained (per cent)
150 mm	–	–	–

75 mm	–	–	–
40 mm	–	–	–
20 mm	–	–	–
10 mm	–	–	–
4.75 mm	–	–	–
2.36 mm	0.18	0.18	18
1.18 mm	0.22	0.40	40
600 m	0.24	0.64	64
300m	0.12	0.76	76
150 m	0.24	1.00	100
Total	1.00	–	298

Fineness modulus of fine aggregate

$$FM_{FA} = \left(\frac{Cumulative\ percentage\ retained}{100}\right) = \left(\frac{298}{100}\right) = 2.98$$

Example 29.5. The characteristic strength of concrete is to be 20 N/mm² at 28 days. The type of exposure is moderate. The fineness module of coarse aggregate and fine aggregate at 6.51 and 2.20, respectively. The voids in coarse aggregate and fine aggregate are 32 percent and 42 percent, respectively. Find the proportions of cement, fine aggregate and coarse aggregate for the concrete mix.

Design. The field strength of concrete mix is 20 N/mm² at 28 days. The laboratory design strength (from Table 29.2) of concrete mix shall be 30.13 N/mm².

For the strength requirement, water-cement ratio from Fig. 29.3, = 0.58

For the durability for moderate exposure, the water-cement ratio, = 0.55

The water-cement ratio may be kept as 0.57. The minimum cement content recommended for moderate exposure is 290 kg per m³. The quantity of water per m³ of concrete

$$(290 \times 0.57) = 165.3\ kg$$

Maximum size of coarse aggregate may be kept as 20 mm. The minimum limit of fineness modulus as recommended by Abram from Table 29.5 is 4.70 for mixed aggregate. The percentage of fine aggregate from Eq. 29.17

$$p_{FA} = 100 \cdot \left(\frac{FM_{CA} - FM_{MA}}{FM_{CA} - FM_{FA}}\right)$$

$$p_{FA} = 100 \cdot \left(\frac{6.51 - 4.70}{6.51 - 2.20}\right) = 41.76\ per\ cent$$

Then, percentage of coarse aggregate shall be 58.24 percent.

For combined mix of aggregate of 100 kg, the fine and coarse aggregates shall be 41.76 kg and 58.24 kg. respectively.

The densities of cement, fine and coarse aggregates are 1500 kg/m³, 1700 kg/m³ and 1600 kg/m³, respectively.

$$\text{Volume of fine aggregate, } \left(41.76 \times \frac{1}{1700}\right) = 24.565 \times 10^{-3}\ m^3$$

$$\text{Volume of coarse aggregate, } \left(58.24 \times \frac{1}{1600}\right) = 36.4 \times 10^{-3}\ m^3$$

Ratio of volumes of fine aggregates to coarse aggregates

$$\left(1 : \frac{36.4 \times 10^{-3}}{24.565 \times 10^{-3}}\right) = 1 : 1.482$$

$$\text{Apparent volume of cement, } \left(\frac{2900}{1500}\right) = 0.193\ m^3$$

$$\text{Volume of water, } \left(\frac{165.3}{1000}\right) = 0.1653\ m^3$$

Let the voids in cement be 60 percent.

Then, the solid volume of cement, $(0.193 \times 0.40) = 0.0772$

Solid volume of aggregate, $(1 - 0.0772 - 0.1653) = 0.7575$

$$\text{Solid volume of fine aggregate, } 0.7575 \times \left(\frac{1}{1 + 1.482}\right) = 0.3052\ m^3$$

Voids in fine aggregate is 42 percent

$$\text{Apparent volume of fine aggregate, } \left(0.3052 \times \frac{1}{0.58}\right) = 0.5262\ m^3$$

$$\text{Solid volume of coarse aggregate, } 0.7575 \times \left(\frac{1.482}{1 + 1.482}\right) = 0.4523\ m^3$$

Voids in coarse aggregate is 32 percent. Apparent volume of coarse aggregate

$$\left(0.4523 \times \frac{1}{0.68}\right) = 0.6652\ m^3$$

Proportion of cement, fine aggregate and coarse aggregate in concrete mix by volume

$$(0.193 : 0.5262 : 0.6652) = (1 : 2.726 : 3.447)$$

Therefore, the proportions of cement, fine aggregate, and coarse aggregate by mass

$$= (1 \times 1500) : (2.726 \times 1700) : 3.447 \times 1600$$

= 1 : 3.089 : 3.677, and water-cement ratio is 0.57.

29.13 ROAD RESEARCH LABORATORY METHOD

Professor Galnville, W.H. Collins, A.R. and Mathews, D.D. Published a paper titled as `The grading of aggregate and workability of concrete, 2nd revised Edition, Road Research Technical paper No. 4, HMSO, London, U.K., 1950. Based on the experimental investigations, exhaustive study and results in Road Research Laboratory (RRL) published in Road Note No. 4, the method of design of concrete mix proportions. This method is briefly referred as RRL-method. U.K., India and other countries have extensive application of this method. The designed concrete mix to be satisfactory in plastic (fresh) state and so also in hardened state. The design of concrete mix by RRL-method depends upon the parameters mentioned below :

1. Nominal maximum size of aggregate. Aggregate of 10 mm, 20 mm and 40 mm nominal sizes are used for normal reinforced concrete works. Aggregates of 75 mm and 80 mm sizes are used for leaner mix concrete.

2. Types of aggregate. Rounded, irregular or crushed rock type (any of these three type of aggregate) may be used for the concrete.

3. Grading of aggregate. RRL method recommends four types of the gradings (for each of the three sizes, viz., 10 mm, 20 mm and 40 mm nominal sizes) as shown in Fig. 29.3, Fig. 29.4 and Fig. 29.5 may be used in reinforced concrete works. Four types of grades for 10 mm, 20 mm or 40 mm nominal size of aggregate are recommended by RRL-method, in Tables 29.6, 29.7 and 29.8, respectively.

The curves shown in Figs. 29.3, 29.4 and 29.5 are not ideal curves. However, those figs. 29.3, 29.4 and 29.5 show the gradings used in RRL tests. For these nominal sizes of aggregate, four types of gradings of aggregate are given. Higher proportions of fines are represented by higher number.

TABLE 29.6. TYPES OF GRADINGS FOR 10-MM MAXIMUM SIZE OF AGGREGATE (FIG. 29.3)

IS-sieve designation	Percentage passing			
	Grading types			
	1	2	3	4
10 mm	100	100	100	100
4.75 mm	30	45	60	75
2.36 mm	20	33	46	60
1.18 mm	16	26	37	46
600 μ	12	19	28	34
300 μ	4	8	14	20
150 μ	0	1	3	6

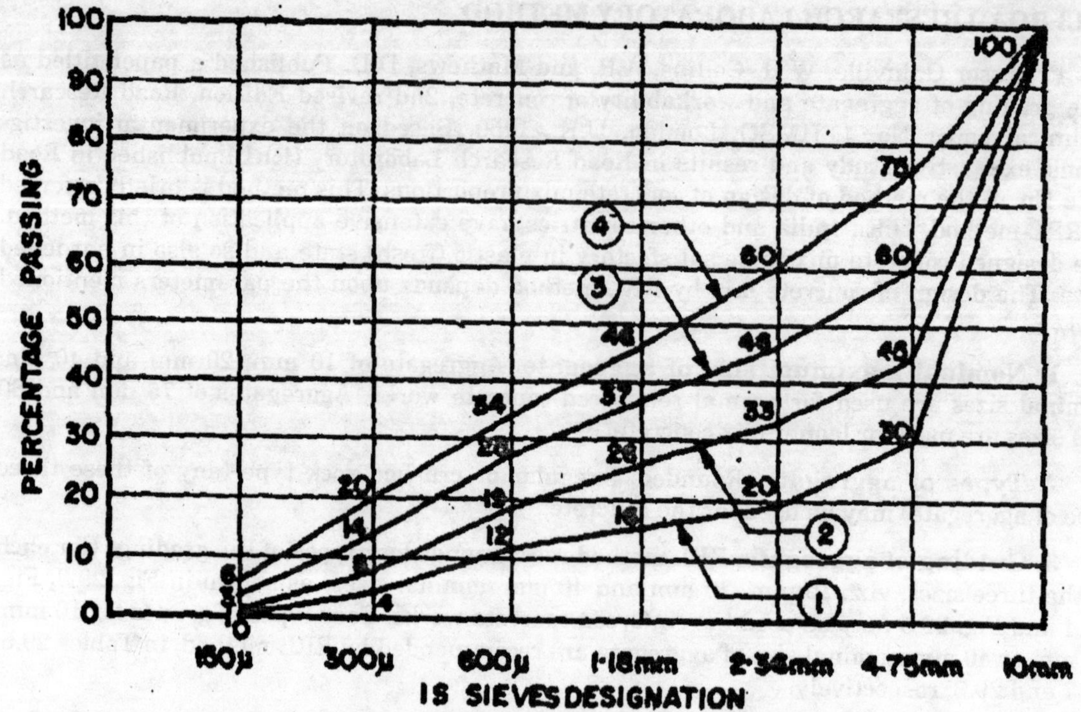

GRADING CURVES FOR
10–mm MAXIMUM – SIZE AGGREGATE

Fig. 29.3

TABLE 29.7. TYPE OF GRADINGS FOR 20-MM MAXIMUM SIZE OF AGGREGATE (FIG. 29.4)

IS-sieve designation	Percentage passing			
	Grading types			
	1	2	3	4
20 mm	100	100	100	100
10 mm	45	55	65	75
4.75 mm	30	35	42	48
2.36 mm	23	28	35	42
1.18 mm	16	21	28	34
600 µ	9	14	21	27
300 µ	2	3	5	12
150 µ	0	0	0	1.5

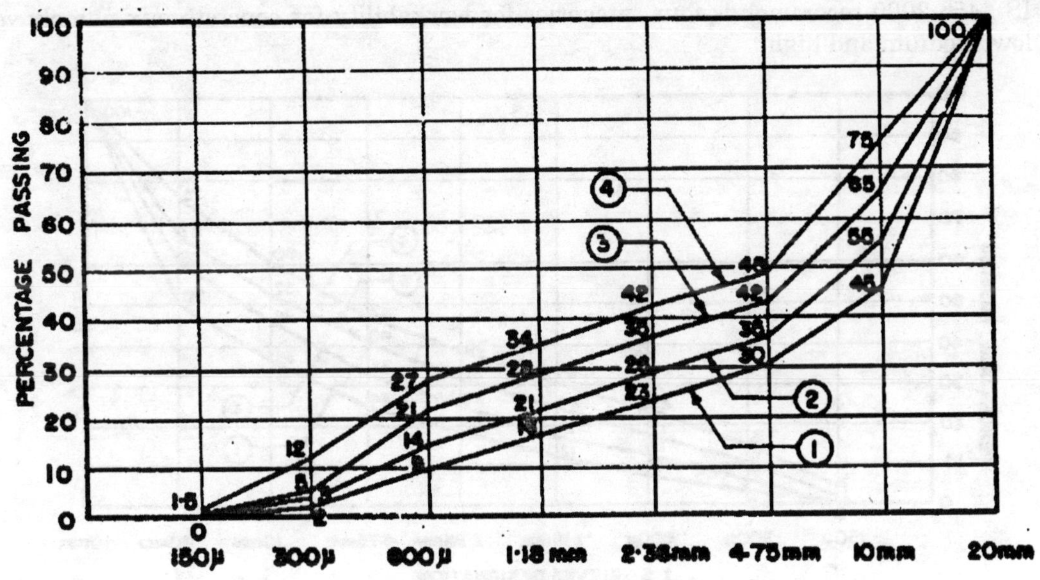

Fig. 29.4

Table 29.8. TYPE OF GRADINGS FOR 40-MM MAXIMUM SIZE OF AGGREGATE (FIG. 29.5)

IS-sieve designation	Percentage passing			
	Grading types			
	1	2	3	4
40 mm	100	100	100	100
20 mm	50	59	67	75
10 mm	36	44	52	60
4..75 mm	24	32	40	47
2.36 mm	18	25	31	38
1.18 mm	12	17	24	30
600 μ	7	12	17	23
300 μ	3	7	11	15
150 μ	0	0	2	5

4. Workability of concrete mix. The workability of concrete mix has been placed into five categories namely extremely low, very low, medium and high.

IS : 456-2000 recommends four categories for workability for concrete mix namely very low, low, medium and high.

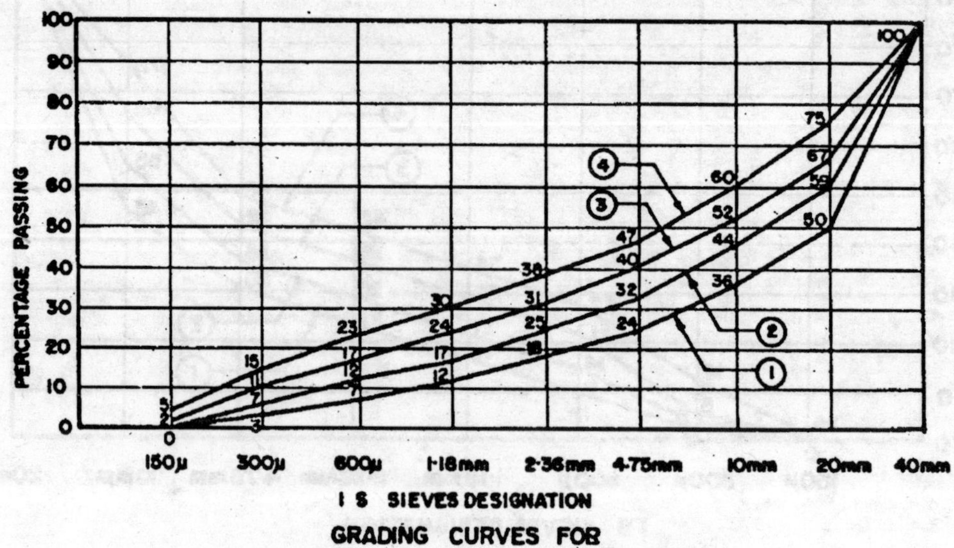

GRADING CURVES FOR
40-mm MAXIMUM- SIZE AGGREGATE

Fig. 29.5

The category of workability is chosen to suit the placing condition, type of construction, and the method of compaction. The values of slump and compaction factor of the concrete mix measured at the work site roughly indicate the degree of workability.

At the Road Research Laboratory (RRL), UK, extensive investigations were carried out for aggregate-cement ratio, water-cement ratio, degree of workability, grading f aggregate, shape and size of aggregates.

5. Average compressive strength. The characteristic strength of concrete and standard deviation were also used in RRL-test.

In addition to the above parameters considered in the design of concrete mix, the type of cement and durability requirements are also considered.

Varying above parameters one by one, the experiments were conducted and Table 29.9 (A) to 20.11 (C) were prepared to give aggregate-cement ratios for different water-cement ratio, different degrees of workability, and different types of aggregate. The fine and coarse aggregate are kept of the same type for these tables. The curves 1, 2, 3 and 4 in Figs. 29.3, 29.4 and 29.5 represent the grade numbers 1, 2, 3 and 4 in these tables. Thus, a trial mix is prepared. In case, the mix is found unsatisfactory, the aggregate-cement ratio is suitably adjusted and a new trial mix is prepared.

The free water-cement ratio needed to get the average compressive strength of concrete is found from Fig. 29.3 (chapter 2). In Fig. 29.3, the values of the compressive strength given are for fully-compacted devise-aggregate concrete of medium strength prepared with portland cement of average quality and cured at the same temperature.

The free water in the fine and coarse aggregate and the water added to the mixer are included in the effective water content. However, the water absorbed by the aggregate is not included. The free water in the aggregate is deducted. when the quantity of water needed in the mix is estimated and the quantity of water to be added at the mixer is determined. It is a matter of site control to find the free water in the aggregate.

The concrete mix is designed by RRL method by the following step by step procedure described as under :

Step 1. The maximum size of aggregate is decided depending on the minimum thickness of the concrete section.

Step 2. The degree of workability is also decided.

Step 3. The average compressive strength is determined from the characteristic strength of concrete by assuming a suitable standard deviation, s.

Step 4. The water-cement ratio is found from the consideration of strength and/or durability of concrete.

Step 5. The gradings of different size of aggregate are determined.

Step 6. The proportion of different size of is determined so that their combined grading matches reasonably with one of the type of gradings shown in the grading curves in Figs. 29.3, 29.4 and 29.5.

Step 7. The aggregate cement ratio is then determined from the respective Tables 29.9 (A) to 29.11 (C).

Step 8. A trial mix is then prepared and checked. In case, the trial mix is not satisfactory, a different aggregate cement ratio is used, and an another trial mix is made.

It is to note that the appropriate corrections for specified gravity and type of aggregate is made before the aggregate cement ratio is found.

TABLE 29.9 (A). AGGREGATE-CEMENT RATIO (40 mm rounded gravel aggregate)

Degree of workability	Aggregate cement-ratio by weight					
	Water-cement ratio by weight					
	0.35	0.40	0.45	0.50	0.55	0.60
Very low						
Grade No. 1	5.0	7.0	8.9	–	–	–
2	4.5	6.5	8.6	–	–	–
3	3.9	5.7	7.7	–	–	–
4	3.4	4.9	6.5	8.0	–	–
Low						
Grade No. 1	4.3	5.9	7.6	–	–	–
2	3.9	5.6	7.4	–	–	–

3	3.5	5.0	6.7	8.2	–	–
4	3.1	4.4	5.8	7.2	8.4	–
Medium						
Grade No. 1	3.4	4.7	6.0	7.5	–	–
2	3.1	4.6	6.1	7.6	8.9	–
3	2.9	4.3	5.7	7.1	8.1	–
4	2.7	3.8	5.0	6.3	7.3	–
High						
Grade No. 1	–	4.1	5.2	6.3	S	–
2	–	4.0	5.3	6.5	7.7	–
3	–	3.9	5.0	6.2	7.4	–
4	–	3.5	4.6	5.7	6.7	7.6

Note : 1. Letter S indicates that the mix would segregate.

2. These values have been obtained by extrapolation of other data and not based directly on the results of trial mixes.

TABLE 29.9 (B). AGGREGATE-CEMENT RATIO (40 mm irregular gravel aggregate)

Degree of workability	Aggregate-cement ratio by weight								
	Water-cement ratio by weight								
	0.35	0.40	0.45	0.50	0.55	0.60	0.65	0.70	0.75
Very low									
Grade No. 1	4.0	5.3	6.6	7.8	–	–	–	–	–
2	3.9	5.2	6.5	7.7	–	–	–	–	–
3	3.6	4.8	6.0	7.1	8.1	–	–	–	–
4	3.2	4.3	5.3	6.3	7.3	–	–	–	–
Low									
Grade No. 1	3.4	4.5	5.6	6.6	7.6	–	–	–	–
2	3.3	4.5	5.6	6.6	7.6	–	–	–	–
3	3.2	4.2	5.3	7.2	7.2	–	–	–	–
4	2.9	3.8	4.8	5.7	6.6	7.4	8.1	–	–
Medium									
Grade No. 1	–	3.8	4.6	5.5	6.2	7.0	7.8	–	–
2	–	3.8	4.7	5.7	6.5	7.3	8.1	–	–
3	–	3.7	4.6	5.5	6.3	7.1	7.8	–	–
4	–	3.4	4.3	5.1	5.9	6.6	7.3	7.9	–
High									
Grade No. 1	–	3.4	4.1	4.8	S	–	–	–	–
2	–	3.5	4.4	5.2	5.9	S	–	–	–
3	–	3.3	4.3	5.1	5.9	6.7	7.3	–	–
4	–	3.1	4.0	4.8	5.5	6.3	6.9	7.4	8.0

Note : Letter S indicates that the mix would segregate.

TABLE 29.9 (C). AGGREGATE CEMENT RATI (40 mm crushed rock aggregate)

Degree of workability	Aggregate-cement ratio by weight								
	Water-cement ratio by weight								
	0.35	0.40	0.45	0.50	0.55	0.60	0.65	0.70	0.75
Very low									
Grade No. 1	3.4	4.9	6.0	7.2	8.1	–	–	–	–
2	3.4	4.6	5.7	6.8	7.7	8.6	–	–	–
3	3.2	4.2	5.2	6.2	7.1	8.0	8.8	–	–
4	2.9	3.8	4.7	5.6	6.4	7.2	7.9	8.6	–
Low									
Grade No. 1	–	4.0	4.9	5.8	6.6	7.4	8.1	–	–
2	–	3.8	4.7	5.6	6.4	7.2	7.9	8.5	–
3	–	3.6	4.4	5.3	6.1	6.9	7.6	8.3	–
4	–	3.3	4.2	5.0	5.8	6.6	7.9	7.9	8.5
Medium									
Grade No. 1	–	3.3	4.1	4.8	5.5	6.1	S	–	–
2	–	3.3	4.1	4.8	5.5	6.2	6.9	7.5	–
3	–	3.2	3.9	4.7	5.4	6.1	6.8	7.5	–
4	–	3.0	3.8	4.6	5.3	6.0	6.6	7.3	7.8
High									
Grade No. 1	–	3.1	3.7	4.4	S	–	–	–	–
2	–	3.1	3.8	4.5	5.2	S	–	–	–
3	–	2.9	3.7	4.5	5.2	5.9	6.5	7.1	–
4	–	2.7	3.4	4.2	4.8	5.6	6.2	6.8	7.4

Note : 1. Letter S indicates that the mix would segregate.
2. These values have been obtained by extra-polation of other data and not based directly on the results of trial mixes.

TABLE 29.10 (A). AGGREGATE CEMENT-RATIO (20 mm rounded aggregate)

Degree of workability	Aggregate-cement ratio by weight								
	Water-cement ratio by weight								
	0.35	0.40	0.45	0.50	0.55	0.60	0.65	0.70	0.75
Very low									
Grade No. 1	4.5	6.6	8.1	–	–	–	–	–	–
2	4.2	6.1	7.6	–	–	–	–	–	–
3	3.7	5.4	6.7	8.0	–	–	–	–	–

4	3.2	4.5	5.8	7.0	8.1	–	–	–	–
Low									
Grade No. 1	3.8	5.3	6.9	8.2	–	–	–	–	–
2	3.6	5.1	6.6	8.0	–	–	–	–	–
3	3.3	4.6	5.9	7.0	8.2	–	–	–	–
4	3.0	4.1	5.1	6.0	6.9	7.7	8.4	–	–
Medium									
Grade No. 1	3.1	4.2	5.3	6.3	7.3	–	–	–	–
2	3.0	4.2	5.3	6.3	7.3	–	–	–	–
3	2.8	3.9	5.0	6.0	7.0	8.0	–	–	–
4	2.6	3.6	4.6	5.5	6.3	7.1	7.8	–	–
High									
Grade No. 1	–	3.7	4.6	5.5	6.3	S	–	–	–
2	–	3.8	4.8	5.7	6.5	7.2	7.7	–	–
3	–	3.6	4.5	5.4	6.1	6.8	7.4	7.9	–
4	–	3.3	4.1	4.8	5.5	6.1	6.6	7.1	7.6

Note : Letter S indicates that the mix would segregate.

TABLE 29.10 (B). AGGREGATE CEMENT RATIO (20 mm irregular gravel aggregate)

Degree of workability	Aggregate-cement ratio by weight									
	Water-cement ratio by weight									
	0.35	0.40	0.45	0.50	0.55	0.60	0.65	0.70	0.75	0.80
Very low										
Grade No. 1	3.6	4.9	6.0	7.2	8.3	9.4	–	–	–	–
2	3.6	4.8	5.8	6.8	7.8	8.7	–	–	–	–
3	3.5	4.6	5.5	6.4	7.3	8.1	–	–	–	–
4	3.0	4.1	5.0	5.9	6.7	7.4	8.0	–	–	–
Low										
Grade No. 1	3.0	3.9	4.8	5.5	6.2	6.9	7.5	8.0	–	–
2	3.0	3.9	4.8	5.5	6.2	6.9	7.5	8.0	–	–
3	3.0	3.9	4.7	5.4	6.1	6.7	7.3	7.8	–	–
4	2.7	3.5	4.3	5.0	5.7	6.3	6.8	7.3	7.9	
Medium										
Grade No. 1	–	3.3	4.0	4.6	S	–	–	–	–	–
2	–	3.4	4.1	4.8	5.4	6.0	S	–	–	–

3	–	3.4	4.1	4.8	5.3	5.9	6.4	6.8	7.2	7.5
4	–	3.2	3.9	4.5	5.1	5.6	6.1	6.6	7.0	7.4
High										
Grade No. 1	–	3.1	S	–	–	–	–	–	–	–
2	–	3.2	3.8	4.4	4.9	S	–	–	–	–
3	–	3.2	3.8	4.4	4.9	5.4	5.8	6.2	6.6	S
4	–	2.9	3.5	4.1	4.7	5.2	5.7	6.1	6.5	7.0

Note : Letter S indicates that the mix would segregate.

TABLE 29.10 (C). AGGREGATE CEMENT RATIO (20 mm crushed rock aggregate)

Degree of workability	Aggregate-cement ratio by weight								
	Water-cement ratio by weight								
	0.40	0.45	0.50	0.55	0.60	0.65	0.70	0.75	0.80
Very low									
Grade No. 1	4.5	5.5	6.5	7.2	7.8	8.3	8.7	–	–
2	4.1	5.0	5.9	6.6	7.2	7.7	8.2	–	–
3	3.8	4.6	5.4	6.0	6.6	7.2	7.7	8.2	–
4	3.5	4.3	5.0	5.7	6.2	6.9	7.5	8.0	–
Low									
Grade No. 1	3.5	4.3	5.0	5.7	6.3	6.0	7.4	7.9	–
2	3.5	4.2	4.9	5.5	6.0	6.5	7.0	7.5	–
3	3.2	3.9	4.5	5.0	5.6	6.1	6.6	7.0	7.4
4	3.0	3.7	4.3	4.8	5.3	5.8	6.3	6.7	7.2
Medium									
Grade No. 1	–	3.7	4.2	4.7	S	–	–	–	–
2	–	3.7	4.2	4.7	5.2	5.7	6.2	S	–
3	–	3.4	3.9	4.5	4.9	5.4	5.8	6.2	6.6
4	–	3.3	3.8	4.3	4.8	5.2	5.7	6.1	6.5
High									
Grade No. 1	–	3.5	S	–	–	–	–	–	–
2	–	3.5	3.9	S	–	–	–	–	–
3	–	3.2	3.8	4.3	–4.7	5.2	5.5	5.8	6.1
4	–	3.1	3.5	4.0	4.5	4.9	5.3	5.7	6.0

Note : Letter S indicates that the mix would segregate.

TABLE 29.11 (A) AGGREGATE CEMENT RATIO (10-mm rounded gravel aggregate)

Degree of workability	Aggregate-cement ratio by weight								
	Water-cement ratio by weight								
	0.40	0.45	0.50	0.55	0.60	0.65	0.70	0.75	0.80
Very low									
Grade No. 1	5.6	7.2	–	–	–	–	–	–	–

2	5.0	6.4	7.8	–	–	–	–	–	–
3	4.0	5.3	6.4	7.5	–	–	–	–	–
4	3.2	4.1	4.9	5.7	6.5	7.2	–	–	–
Low									
Grade No. 1	4.5	5.5	6.5	7.4	–	–	–	–	–
2	3.9	4.9	5.8	6.7	6.7	–	–	–	–
3	3.3	4.1	4.9	5.7	6.4	7.1	7.7	–	–
4	2.6	3.2	3.8	4.4	5.0	5.6	6.2	6.7	7.2
Medium									
Grade No. 1	3.9	4.7	5.4	6.1	6.7	7.3	7.9	–	–
2	3.5	4.3	5.0	5.7	6.3	6.9	7.5	–	–
3	3.0	3.7	4.3	4.9	5.5	6.1	6.7	7.3	7.7
4	2.4	3.0	3.5	4.0	4.5	5.3	5.5	5.9	6.3
High									
Grade No. 1	3.5	4.2	4.8	5.3	5.8	S	–	–	–
2	3.2	3.9	4.5	5.1	5.6	6.1	6.6	7.1	7.6
3	2.8	3.4	4.0	4.5	5.0	5.5	6.0	6.5	6.9
4	2.8	2.9	3.4	3.9	4.3	4.7	5.1	5.5	5.9

Note : Letter S indicates that the mix would segregate.

TABLE 29.11 (B) AGGREGATE CEMENT RATIO (10 mm irregular gravel aggregate)

Degree of workability	Aggregate-cement ratio by weight								
	Water-cement ratio by weight								
	0.40	0.45	0.50	0.55	0.60	0.65	0.70	0.75	0.80
Very low									
Grade No. 1						–	–	–	–
2	4.1	5.1	6.1	7.0	7.9	–	–	–	–
	3.8	4.8	5.8	6.7	7.6				
3	3.3	4.3	5.2	6.1	7.0	7.8	–	–	–
4	2.8	3.6	4.4	5.2	6.0	6.8	–	–	–
Low									
Grade No. 1	3.3	4.1	4.8	5.5	S	–	–	–	–
2	3.1	3.9	4.6	5.3	6.0	6.0	7.2	7.8	–
3	2.8	3.5	4.2	4.2	5.6	5.6	6.8	7.4	8.0
4	2.3	3.0	3.7	4.3	4.9	4.9	5.5	6.7	7.3
Medium									
Grade No. 1	–	3.5	4.2	S	–	–	–	–	–
2	–	3.4	4.1	4.7	5.3	5.9	6.4	6.9	7.4
3	–	3.2	3.8	4.4	5.0	5.6	6.1	6.6	7.1
4	–	2.8	3.4	4.0	4.5	5.0	5.5	6.0	6.4

water-cement ratio, this method considers the compressive strength of cement received from factories. The proportioning of the concrete mix is done by following the step-by-step procedure as under :

TABLE 29.21. AVERAGE DESIGN STRENGTH OF CONCRETE FOR DIFFERENT DEGREES OF QUALITY CONTROL AND TOLERANCE LEVELS (AS PER IRC 15-1970)

Degree of quality control	Minimum specified concrete strength at 28 days (N/mm²)								
	20 N/mm²			27.5 N/mm²			35 N/mm²		
	A	B	C	A	B	C	A	B	C
(1)	(2)	(3)	(4)	(5)	(6)	(7)	(8)	(9)	(10)
Very good	1 in 15	10	23.5	1 in 15	7	31.0	1 in 20	7	40.0
Good	1 in 10	15	25.0	1 in 15	10	32.5	1 in 15	14	41.5
Fair	1 in 15	20	27.0	1 in 10	15	34.0	–	–	–

where, A = Tolerance level, B = Coefficient of variation, C = Average design strength (N/mm²)

Note. Average design strengths given in this above Table are for tolerance factors corresponding to infinite samples as given in Table 29.22.

Step 1. In order to design the concrete mix by Indian Road Congress Method, following needed information is collected.

(i) The compressive strength of cement at 7 days, the minimum compressive or flexural field strength of concrete at 28 days and the specific gravity of cement are noted.

(ii) The type of aggregates, its pattern of grading and maximum size and the specific gravity of fine and coarse aggregate, the percentage of water absorption or free moisture in the aggregate are mentioned. The sieve analysis of fine aggregate is carried out and its fineness modulus is also noted.

(iii) Looking to the type of construction work, and the mode of compaction, the degree of workability is mentioned by specifying value of slump.

(iv) The degree of control likely to be exercised, is mentioned in terms of variation. The accepted tolerance is also indicated.

Step 2. Depending on the magnitude of concreting work and the specified tolerance level, the tolerance factor is noted corresponding to the number of samples from Table 29.14.

Step 3. The average design strength is determined for the noted values of coefficient of variation and tolerance level using Table 29.21.

Step 4. The water cement ratio is determined for the seven days compressive strength of cement likely to be used and 28 days average compressive strength of concrete from Fig. 29.6.

Step 5. A combined grading curve, which one is very close to the suggested coarse aggregate grading is found for the known sieve analysis of the coarse aggregate given in Table 29.23, either by analytical method or by graphical method.

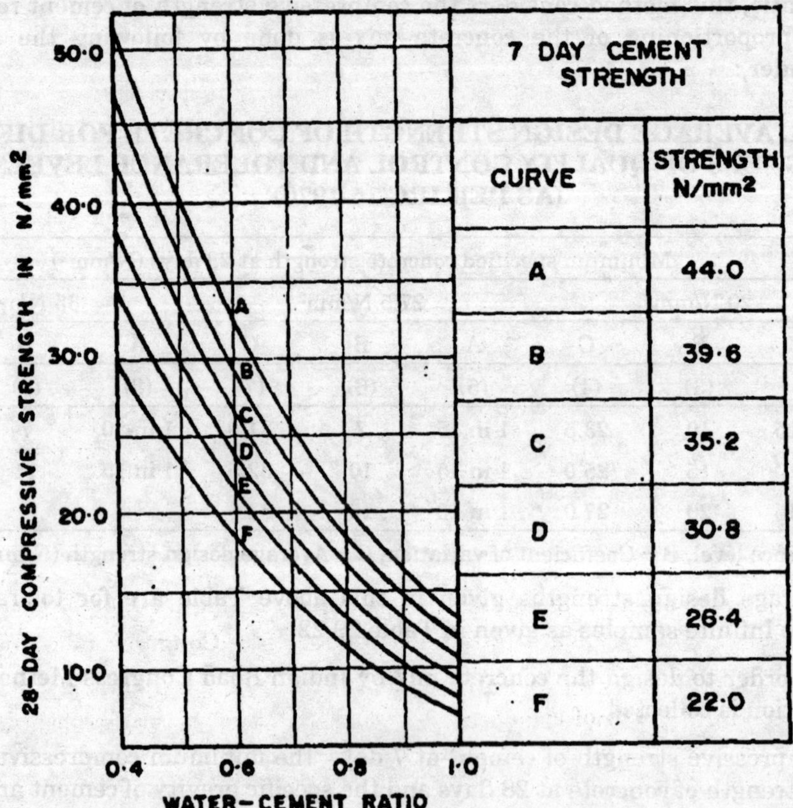

Fig. 29.6

Step 6. The water content in kg per cubic metre of concrete is determined for the maximum size of the aggregate. The sand content in percent of total aggregate by absolute volume is also determined using Table 29.24.

TABLE 29.22. VALUES OF TOLERANCE FACTOR (AS PER IRC 15-1970)

Tolerance level	Number of samples			
	10	20	30	Infinite
1 in 10	1.37	1.32	1.31	1.28
1 in 15	1.65	1.58	1.54	1.50
1 in 20	1.81	1.72	1.70	1.64
1 in 40	2.23	2.09	2.04	1.96
1 in 100	2.76	2.53	2.46	2.33

Note. Under conditions of a major concrete job, where large number of samples shall be tested, it would be appropriate to adopt a tolerance factor corresponding to infinite number of samples.

TABLE 29.23 (A). SUGGESTED COARSE AGGREGATE GRADING (MAXIMUM SIZE OF AGGREGATE 20 MM) (AS PER IRC 15-1970)

Zone	Percentage passing			
	Sieve sizes			
	40 mm	20 mm	10 mm	4.75 mm
A	100	100	21–32	0
B	100	100	31–40	0
C	100	100	40–52	0

Note. Allowance for over size in the nominal maximum size of aggregate shall be limited to 5 percent.

TABLE 29.23 (B) SUGGESTED COARSE AGGREGATE GRADING (MAXIMUM SIZE OF AGGREGATE 20 MM) (AS PER IRC 15-1970)

Zone	Percentage passing			
	Sieve sizes			
	40 mm	20 mm	10 mm	4.75 mm
A	100	30–40	16–18	0
B	100	40–45	18–20	0
C	100	45–53	20–25	0

Note : Allowance for over size in the nominal maximum size aggregate shall be limited to 5 percent.

TABLE 29.24. APPROXIMATE SAND AND WATER CONTENTS PER CUBIC METRE OF CONCRETE (As per IRC 15-1970)

Maximum size of aggregate (mm)	water content of saturated surface dry basis per cubic metre of concrete *(kg)	Sand in percent of total aggregate by absolute volume (%)
80 mm	172.0	28.0
60 mm	175.0	33.5
25 mm	177.5	38.0
20 mm	178.0	40.0

Note. By saturated surface dry conditions, it is understood that the aggregates are fully saturated but there is no free moisture at the surface, if the aggregates are not in this condition, water to be added in the mixer shall be required to be increased or decreased to make up difference depending on whether the aggregates are dry or wet.

TABLE 29.25. ADJUSTMENT OF VALUES IN WATER AND SAND CONTENT FOR OTHER CONDITIONS (As per IRC 15-1970)

Changes in conditions stipulated for Table 29.24	Adjustment required in	
	Water content %	sand content %
Each 0.05 increase or decrease in water-cement ratio	0	± 1
Each 0.1 increase or decrease in F.M. of sand	0	0 ± 0.5

Each 25 mm increase or decrease in slump	± 4	0
Rounded aggregate	litres (adhoc)	

Step 7. The values determined at Step 6 from Table 29.24 are for angular aggregate, water-cement ratio of 0.5, slump of 25 mm and fineness modulus of 2.60. For other different values, the adjustments are needed as mentioned in Table 29.25.

Step 8. The cement content is calculated from the water cement ratio and water content.

Step 9. The volume of entrained air corresponding to the maximum size of the aggregate are obtained Table 29.14 (B).

Step 10. The total aggregate content per unit volume of concrete is found from Eq. 29.19.

$$V = \left[W_W + \frac{W_C}{S_C} + \frac{P \cdot W_A}{S_S} + \frac{(1-P) \cdot W_A}{S_A} \right] \times 10^{-3}, \textit{where} \quad (29.19)$$

V = absolute volume of the wet mix

(Gross volume 1 m^3 – volume of entrained air)

W_W = weight of water in kg/m^3 of concrete

W_C = weight of cement in kg/m^3 of concrete

W_A = weight of total aggregate in kg/m^3 of concrete

P = proportion of sand in total aggregate S_C . S_S and S_A are the specific gravities of cement, sand and aggregate respectively are known.

Step 11. The proportion of concrete mix becomes known, once water-cement ratio, cement, sand and aggregate, respectively, are known

Step 12. The proportion of mix is adjusted, if the aggregate is moisted.

Example 29.8 Design the concrete mix by Indian Road Congress method for the following :

(i) Minimum compressive field strength at 28 days is 35 N/mm^2.

(ii) Maximum size of angular shape coarse aggregate is 40 mm.

(iii) Degree of workability is 25 mm slump.

(iv) Degree of quality control is good (coefficient of variation is 10 percent).

(v) Accepted tolerance level is 1 in 15.

(vi) A major concrete construction work shall be done. It needs testing of large number of samples : that is, tolerance factor is 1.50.

(vii) Specific gravity of cement = 3.15, fine aggregate = 2.66, and coarse aggregate = 2.64

(viii) Percentage of water absorption for

(a) coarse aggregate = 0.3 percent, (b) fine aggregate = 0.5 percent.

(ix) Free (surface) moisture

(a) coarse aggregate = nil, (b) fine aggregate = 2 percent

(x) 7 days compressive strength of cement is 28 N/mm³.

Sieve size (mm)	Fraction of coarse aggregate Percent passing Grade I	Grade II
40 mm	100	100
20 mm	16	100
10 mm	10	32
4.75 mm	03	05

Sieve sizes	Fine aggregate	Cumulative percent retained
4.75 mm	100	00
2.36 mm	97	03
1.18 mm	88	12
600 micron	52	48
300 micron	10	90
150 micron	04	96

$$\text{Fineness modulus} \left(\frac{249}{100}\right) = 2.49$$

Combination of fraction of coarse aggregate (Analytical Method)

Grade of aggregate	Percentage passing IS : Sieve sizes 40 mm	20 mm	10 mm	4.75 mm
Available I	100	16	8	02
Available II	100	100	32	05
Combined				
Ref. : Table 29.23	100	42	20	0
First trial				
70 percent I	70	11.2	7	2.1
30 percent II	30	30.0	9.6	1.5
Combined	100	41.2	16.6	3.6
Second trial				
68 percent I	68	10.88	6.8	2.04
32 percent II	32	32	10.24	1.60
Combined	100	42.88	17.04	3.64

Illustration. *For grade I*

$$(100 \times 70\% = 70), (16 \times 70\% = 11.2), (10 \times 70\% = 7), (3 \times 70\% = 2.1)$$

For grade II

$$100 \times 30\% = 30, 100 \times 30 = 30, 32 \times 30\% = 9.6, 5 \times 30\% = 1.5.$$

Percentage passing for trial grade (second trial combined) is close to the suggested grade of coarse aggregate as per Table 29.23.

Therefore, the aggregate may be combined in the proportion 68 : 32. Further procedure for the design of concrete mix is as follows :

Step 1. Tolerance factor is 1.50 and the coefficient of variation is 90 percent.

Average strength

$$\left(\frac{S_{min}}{1-\frac{t\,.\,v}{100}}\right) = \left(\frac{35}{1-\frac{1.5 \times 10}{100}}\right) = 41.18\ N/mm^3\ (Say\ 42\ N/mm^2)$$

Step 2. Water-cement ratio is determined. From Fig. 29.6. Water cement ratio corresponding to 7 days compressive strength (28 N/mm²) of cement and design strength of concrete 42 N/mm² is 0.4.

Step 3. From Table 29.24, approximate contents of sand and water per m³ of concrete corresponding to

Water-cement ratio = 0.50, Slump = 25 mm

Fineness modulus = 2.60, Type of aggregate = angular

Water content per m³ of concrete is = 175 kg and Sand content per m³ of concrete is = 33.5%.

Step 4. Adjustments of values in contents of water and sand are needed as the actual values of water-cement ratio, fineness modulus of sand and slump are different.

Difference in conditions	Adjustments needed in	
	Water content	Sand content
(i) Decrease in w-c ratio (0.50–0.40) = 0.10	0	– 2.0%
(ii) Each 0.1 increase or decrease in FM of sand (2.60 – 2.49) = 0.11	0	– 0.555
(iii) Slump (25 mm – 25 mm) = 0	0	0%
Total	0	–2%

Water-cement per m³ of concrete 175 kg/m³ remains unchanged. Sand content per m³ of concrete

$$P = (33.5 - 2.0) = 31.5 \text{ or } 0.315 \text{ percent.}$$

Step 5. Cement content is calculated as below. Water-cement ratio

$(W_w/W_c) = 0.40$

Content of water = 175 kg/m³ of concrete.

Content of cement

$(175/W_C) = 0.40$, therefore, $W_C = (175/0.40) = 437.5$ kg

Step 6. Total aggregate content per unit volume of concrete is determined as under. The specified maximum size of aggregate is 40 mm. For air-entrained concrete. Table 29.14 (A), approximate entrained air content is 1 percent

$W_W = 175$ kg

Cement content $W_C = 437$ kg

Weight of total aggregate $W_A = W_A$

Proportion of sand in total aggregate, $P = 0.315$

Specific gravity of

Cement $S_C = 3.15$, Sand $S_S = 2.65$, Aggregate $S_A = 2.64$.

Absolute volume of the wet mix

$V = (1 - 0.01) = 0.99 \text{ m}^3$.

From Eq. 29.19,

$$V = \left[W_W + \frac{W_C}{S_C} + \frac{P \,.\, W_A}{S_s} + \frac{(1-P).\,W_A}{S_A} \right] \times 10^{-3}$$

or $$V = \left[175 + \frac{437}{3.15} + \frac{0.315\, W_A}{2.66} + \frac{(0.865).\, W_A}{2.64} \right] \times 10^{-3}$$

or $V = [175 + 138.73 + 0.118\, W_A + 0.26\, W_A] \times 10^{-3}$

$0.378\, W_A = (0.99 \times 1000 - 313.73)$

$W_A = (676.27/0.378) = 1789.10$ kg

Therefore, the proportion of concrete mix

Water	Cement	Aggregates
175 kg	437 kg	1789.10 kg
1	2.497	10.223
0.40	0.9988 = 1.0	4.10 (by weight)

Step 7. For one bag of cement, that is 50 kg, actual qualities of

(i) Cement = 50 kg, (ii) Aggregate = 205 kg.

(iii) Fine aggregate $\left(\frac{32}{100} \times 205\right) = 65.6\ kg$

(iv) Coarse aggregate $\left(\frac{68}{100} \times 205\right) = 139.4\ kg$

Type I fraction is 68% of 139.4 kg is 94.792 kg

Type II fraction is 32% of 139.4 kg is 44.608 kg

(v) Water :

(a) For water-cement ratio of 0.40 = 20 litres

(b) Extra water to be added to account for absorption in case of coarse aggregate, at 0.3 percent by weight is = 0.418 litres

(c) water to be reduced for free moisture in fine aggregate 2 percent by weight

$$\left(65.6 \times \frac{2}{100}\right) = 1.312 \textit{ litres}$$

(d) Net quantity of water

$$(20 + 0.418 - 1.312) = 19.11 \text{ litres}$$

(vi) Net weight of fine aggregate needed keeping allowance for free moisture

$$= (65.6 + 1.312) = 66.912 \text{ kg}$$

(vii) Net weight of coarse aggregate needed

$$\text{Type I fraction, } \left(94.792 + 94.792 \times \frac{3}{100}\right) = 97.64\,kg$$

$$\text{Type II fraction } \left(44.608 + 44.608 \times \frac{5}{100}\right) = 46.84\,kg$$

Actual quantities of different ingredients needed for the mix are as below :

Water = 19.11 litres and Cement = 50 kg

Fine Aggregate = 66.912 kg

Coarse aggregate, Type I = 97.64 kg and Type II = 46.84 kg

29.18 INDIAN STANDARD MIX DESIGN METHOD

IS : 10262-1962 recommended guidelines for concrete mix design includes the design of normal concrete mixes (non-air-entrained). This method is applicable to both medium and high strength concrete. For a design of a concrete mix, following data needed are specified.

(i) The characteristic compressive strength of a concrete mix at 28 days.

(ii) Expected degree of workability (for guidance, see Table 1.9, Chapter 1).

(iii) To ensure adequate durability for the type of exposure, the limitations on the minimum cement content (see Table 29.15 and Table 29.16).

(iv) Standard deviation, s for the compressive strength of concrete.

The standard deviation, s is determined from the results of tests. In case the results of sufficient member of test under site conditions and for the grade of concrete are not available, the values of standard deviation for different degree of control from Table 29.26 may be used. Depending upon the infrastructure and practices adopted at the construction site, the degree of quality control as desired may be seen from Table 29.27 for guidance.

The proportioning of the concrete mix is done by following the step by step procedure as under :

TABLE 29.26. SUGGESTED VALUES OF STANDARD DEVIATION (AS PER SP 23-1982)

Grade of aggregate	Standard deviation for different degree of control, N/mm²		
	Very good	Good	Fair
M 10	2.0	2.3	3.3
M 15	2.5	3.5	4.5
M 20	3.6	4.6	5.6
M 25	4.3	5.3	6.3
M 30	5.0	6.0	7.0
M 35	5.3	6.3	7.3
M 40	5.6	6.6	7.6
M 45	6.0	7.0	8.0
M 50	6.4	7.4	8.4
M 55	6.7	7.7	8.7
M 60	6.8	7.8	8.8

TABLE 29.26. DEGREE OF QUALITY CONTROL UNDER DIFFERENT SITE CONDITIONS (AS PER SP 23 - 1982)

Grade of control	Conditions of productions
Very good	Fresh cement from single source and regular tests, weigh batching of all materials, aggregates supplied in single sizes, control of aggregate grading and moisture content, control of water added, frequent supervision, regular workability and strength tests, field laboratory facilities.
Good	Carefully stored cement and periodic tests, weight batching of all materials controlled water, graded aggregate supplied, occasional grading and moisture test, periodic check of workability and strength, intermittent supervision, experienced workers.
Fair	Proper storage of cement, volume batching of all aggregates allowing for building of sand, weight batching of cement, water-content controlled by inspection of mix, occasional supervision and tests.

Step 1. The target mean strength for mix design is calculated as below

$$f_t = (f_{ck} + s \ . \ K) \qquad \text{(i)}$$

where, f_t = target mean compressive strength at 28 days

f_{ck} = characteristic compressive strength at 28 days

s = standard deviation

K = a statistical value depending upon the accepted proportion of low results (risk factor) and number of test. In such case K = 1.65, then, the values of K are given in Table 29.28.

$$f_t = (f_{ck} + 1.65\ s) \qquad \text{(ii)}$$

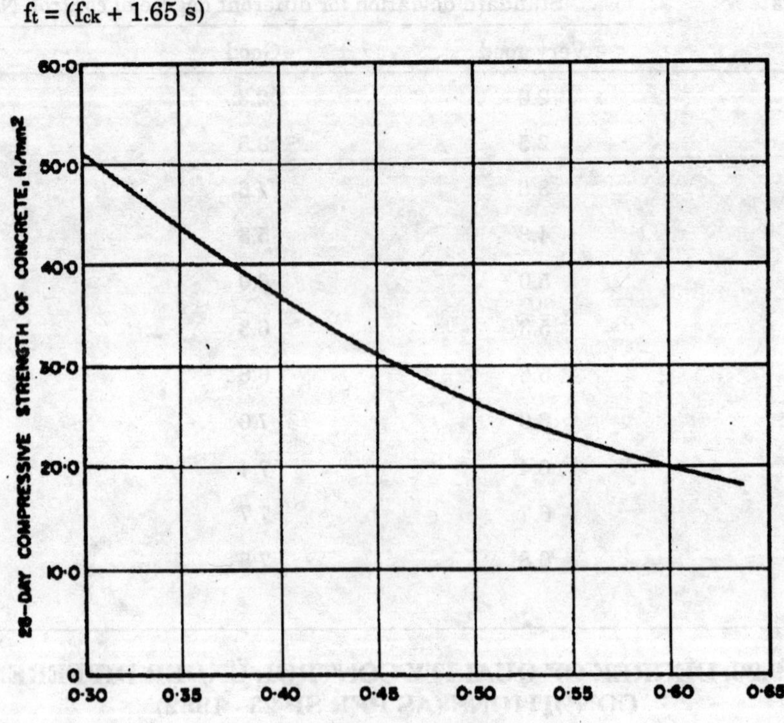

GENERALIZED RELATIONSHIP BETWEEN FREE WATER-CEMENT RATIO AND COMPRESSIVE STRENGTH OF CONCRETE

Fig. 29.7

[Note. Fig. 29.7 is prepared for a large number of results under Indian conditions. For a given situation, it requires slight modification depending upon the characteristics of cement available (Reference : Mallick, A.K., Iyer, SR, and Babu, K.H. variability of cements and adjustments in concrete mixes by accelerated tests, VII ERMCO Congress London, 1983). as such it is used more as a guide and actual water-cement ratio is determined by means of trial mixes as described in IS : 10262-1982].

Step 2. Water cement ratio is determined for the target mean strength from Fig. 29.7 and it is checked against the limiting water-cement ratio for the requirement of durability (see Table 29.15 and Table 29.6) and the lower of two values shall be used. Fig. 29.8 may be used to find a more precise estimate of the preliminary water cement ratio corresponding to the target mean strength for the curves corresponding to the 28 days compressive strength of cement.

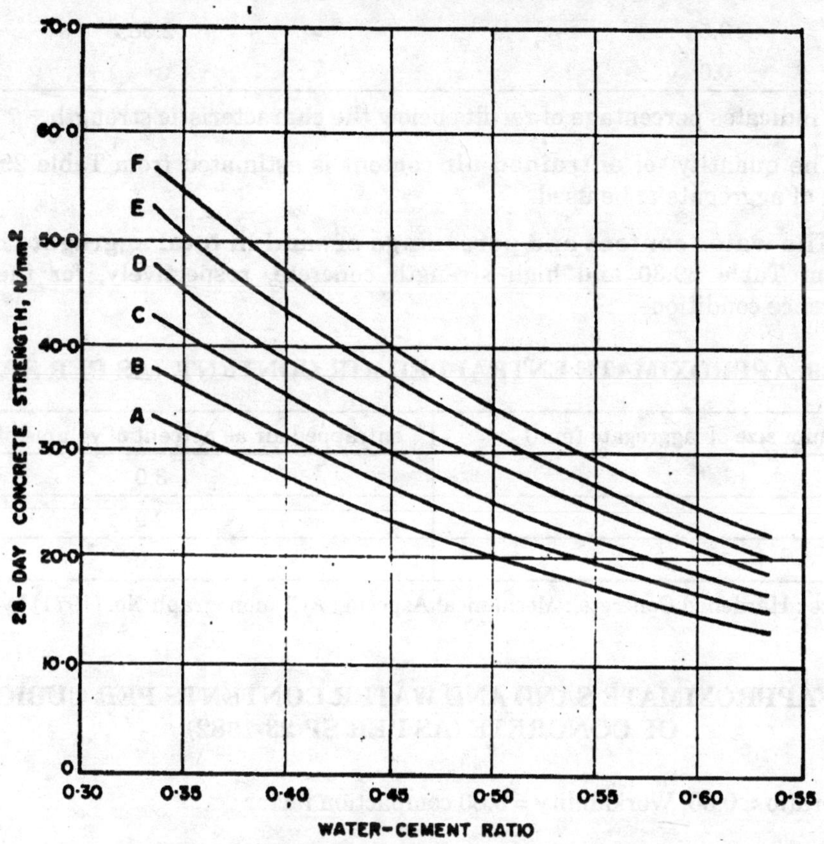

Fig. 29.8

TABLE 29.28. VALUES OF K, (AS PER SP 23-1982)

PR	K
50	0.00
16	1.00
10	1.28
5	1.65
2.5	1.96
1.0	2.33

0.5	2.58
0.0	∝

Note. PR indicates percentage of results below the characteristic strength.

Step 3. The quantity of **entrained air** content is estimated from Table 29.29 for the maximum size of aggregate to be used.

Step 4. The water content and percentage of sand in total aggregate by volume are noted from Table 29.30 and high strength concrete, respectively, for the following standard reference conditions.

TABLE 29.29. APPROXIMATE ENTRAPPED AIR CONTENT, (AS PER SP 23-1982)

Nominal maximum size of aggregate (mm)	entrapped air as percent of volume of concrete
10	3.0
20	2.0
40	1.0

Note. (Reference : Hardened Concrete : Mechanical Aspects : ACI monograph No. (1971) by A.M. Neville).

Table 29.30. APPROXIMATE SAND AND WATER CONTENTS PER CUBIC METRE OF CONCRETE (AS PER SP 23-1982)

Water cement ratio = 0.60, Workability = 0.80 compaction factor

(Applicable for concrete upto grade M 15)

Maximum size of aggregate (mm)	Water-content *per m^3 of concrete (kg)	Sand as percent of total aggregate by absolute volume
10	208	62
20	186	45
30	165	40

(Note. Water content corresponding to saturated surface dry aggregate)

Table 29.31. APPROXIMATE SAND AND WATER CONTENTS PER M³ CONCRETE (AS PER SP 23 - 1982)

Water cement ratio = 0.35, Workability = 0.80 compaction factor

(Applicable for concrete above M 15 grade)

Maximum size of aggregate (mm)	Water-content *per m^3 of concrete (kg)	Sand as percent of total aggregate by absolute volume
10	200	26
20	190	23

TABLE 29.32. ADJUSTMENT OF VALUES IN WATER-CONTENT AND SAND PERCENTAGE FOR OTHER CONDITIONS (As per SP 23-1982)

S.No.	Changes in conditions	Adjustments required	
	Stipulated for Tables 29.29 and 29.30	Water content	Percent sand in total aggregate
1	For sand conforming to grading zones I, III or IV of Table 4 (IS 383-1970)	0	+ 1.5 percent for zone I – 1.5 percent for zone III – 3.0 percent for zone IV
2	Increase or decrease in the value of compaction factor by 0.1	± 3 per cent	0
3	Each 0.05 increase or decrease in water content	0	± 1 percent
4	For rounded aggregate	– 15 kg/m³	– 7 percent

(i) Crushed (angular) coarse aggregate conforming to IS : 383-1970.

(ii) Fine aggregate consisting of natural sand conforming to grading zone II of Table 4, IS : 383-1970, in saturated surface dry condition.

(iii) Water cement ratio of 0.60 and 0.35 for medium and high strength concrete, respectively, and

(iv) Workability corresponding to compaction factor of 0.80

Adjustments in water content and percentage of sand in total aggregate are made as per Table 29.32 for any departure from above conditions.

Step 5. The cement content per unit volume of concrete may be determined from free water-cement ratio and the final water content after adjustment.

The cement content so determined is checked against the minimum cement content from the requirements of durability (see Table 29.15 and Table 29.16) and the greater of two values shall be adopted.

Step 6. The coarse and fine aggregate content per unit volume of concrete and determined from the following :

$$V = \left(W_W + \frac{W_C}{S_C} + \frac{1}{P} \cdot \frac{W_{FA}}{S_{FA}} \right) \times 10^{-3} \quad (29.20)$$

and $$V = \left(W_W + \frac{W_C}{S_C} + \frac{1}{(1-p)} \cdot \frac{W_{CA}}{S_{CA}} \right) \times 10^{-3}, \text{ where} \quad (29.21)$$

V = absolute volume of fresh concrete (which one is equal to gross volume, m³ minus the volume of entrained air)

W_W = mass of water, kg per m³ of concrete

W_C = mass of cement, kg per m³ of concrete

p = ratio of fine aggregate to total aggregate by absolute volume

W_{FA}, W_{CA} = total masses of fine aggregate and coarse aggregate, kg per m³ of concrete, respectively.

S_C = specific gravity of cement

S_{FA} = specific gravity of fine aggregate, and

S_{CA} = specific gravity of coarse aggregate

Checks for workability, freedom from segregation, bleeding and its finishing property are applied for the trial mix so arrived.

In case the measured workability is different from the assumed in the calculation, an adjustment in the water content is done from Table 29.32 and the complete mix design is to be redone and the water-cement ratio remains unchanged. Minor modifications in the aggregate may be done to improve the finishing quality or freedom from segregation and bleeding, the water-cement ratio is now changed by ± 10 percent of previously assumed value and the proportions of mix are again determined. Three trial mixes are simultaneously tested. Sufficient information is received from these tests. These tests shall also provide relationship between the compressive strength and the water-cement. The mix proportions for field trials may then be finalised.

Example 29.9 Design the concrete mix by Indian standard method for the following :

(A) Design stipulations

(a) Characteristic compressive strength needed in the field at 28 days. 20 N/mm²

(b) Maximum size of aggregate 20 mm (angular)

(c) Degree of workability compaction factor 0.90

(d) Degree of quality control good

(e) Type of exposure mild

(B) Test data for materials

(a) Cement used-Ordinary portland cement conforming to IS : 269-1976

(b) Specific gravity of cement 3.15

(c) Specific gravity of

(i) coarse aggregate 2.64 (ii) fine aggregate 2.66

(d) Percentage of water absorption

(i) coarse aggregate 0.3 percent and (ii) fine aggregate 0.5 percent

(e) Free (surface) moisture

(i) coarse aggregate = nil and (ii) fine aggregate = 2 percent

(f) 7 days compressive strength of cement satisfies the requirement of IS : 269-1976

(g) Sieve analysis - coarse and fine aggregate as under.

1. Coarse aggregate

Grade of aggregate	Percent passing			
	IS : Sieve size			
	20 mm	10 mm	4.75 mm	2.36 mm
Available I	100	0	–	–
Available II	100	71.20	9.40	0
Percentage of different fractions				
I 60 percent	60	0	–	–
II 40 percent	40	28.5	3.7	–
Combined 100 percent	100	28.5	3.7	–

(**Remarks.** Conforming to Table 2 of IS : 383-1970)

2. Fine-aggregate

Grade of aggregate	Percent passing					
	IS : Sieve size					
	4.75 mm	2.36 mm	1.18 mm	600 m	300 m	150 m
Available	100	100	93	60	12	2

(**Remarks.** conforming to grading zone III of Table, 4, IS : 383-1970)

Design. The concrete mix is designed by Indian standard method as under :

Step 1. Target mean strength of concrete. Characteristic compressive strength required in the field at 28-days

$$f_{ck} = 20 \text{ N/mm}^2 \qquad \text{(i)}$$

Using Table 29.26, the value of standard deviation for a tolerance factor of 1.65

$$S = 4.6 \text{ N/mm}^2 \qquad \text{(ii)}$$

Target mean strength, $f_t = (f_{ck} + K.s.)$

$$f_t = (20 + 1.65 \times 4.6) = 27.6 \text{ N/mm}^2$$

Step 2. Water-cement ratio. From Fig. 29.7, the water cement ratio needed for the target mean strength of 27.6 N/mm²

$$w - c \text{ ratio} = 0.50 \qquad \text{(iv)}$$

From Table 29.15, the maximum water-cement ratio for mild exposers

$$\text{Max . } w - c \text{ ratio} = 0.65 \qquad \text{(v)}$$

Step 3. Water and sand contents. From Table 29.30, for 20 mm nominal maximum size aggregate and sand conforming to grading zone II, water content per m³ of concrete

$$W_c = 186 \text{ kg} \qquad \text{(vi)}$$

Sand content as percentage of total aggregate by absolute volume p = 35 percent (vii)

Following adjustment is needed for the change in value in water cement ratio, compactor factor, for sand belonging to zone III, (see Table 29.32).

Change in conditions	Adjustment required in	
	water content	Percent sand in total aggregate
1. For decrease in water-cement ratio (0.60 – 0.50) = 0.1	0	–0.20
2. For increase in compaction factor (0.9 – 0.8) = 0.1	+ 3	0
3. For sand conforming to zone III of Table 4 of IS : 383-1970	0	–1.5
Total	+ 3	– 3.5

Therefore, sand content needed as percent of total aggregate by absolute volume

(35 – 3.5) = 31.5 percent

Water content needed, 186 + (186 × 3/100) = 191.6 litres/m³

Step 4. Cement content

Water-cement ratio = 0.50, water = 191.6 litres/m³,

$$\text{cement} = \left(\frac{191.6}{0.5}\right) = 383\ kg/m^3$$

The cement is sufficient for *mild exposure* (see Table 29.15).

Step 5. Contents of coarse and fine aggregates. Maximum size of coarse aggregate is 20 mm. From Table 29.29, the quantity of entrained air in the net concrete = 2 percent.

From Eq. 29.20,

$$V = \left(W_W + \frac{W_C}{S_C} + \frac{1}{p}\cdot\frac{W_{FA}}{S_{FA}}\right)\times 10^{-3}$$

$$(1-0.02) = \left(191.6 + \frac{383}{3.15} + \frac{1}{0.315}\cdot\frac{W_{FA}}{2.66}\right)\times 10^{-3}$$

$$0.98 = [191.6 + 121.59 + 1.19\ W_{FA}]\times 10^{-3}$$

$$W_{FA} = \left(\frac{980 - 313.19}{1.19}\right) = \left(\frac{668.81}{1.19}\right) = 560.34\ kg/m^3$$

From Eq. 29.21, $V = \left(W_W + \frac{W_C}{S_C} + \frac{1}{(1-p)}\cdot\frac{W_{CA}}{S_{CA}}\right)\times 10^{-3}$

$$(1-0.02) = \left(191.6 + \frac{383}{3.15} + \frac{1}{0.685}\cdot\frac{W_{CA}}{2.64}\right)\times 10^{-3}$$

$$0.98 = [191.6 + 121.59 + 0.553\, W_{CA}] \times 10^{-3}$$

$$W_{CA} = \left(\frac{982 - 313.19}{0.553}\right) = 1209.42\ kg/m^3$$

Step 6. Mix proportions

Water	Cement		Sand	Coarse aggregate
191.6 litres :	383 kg :		560 kg :	1209.42 kg
0.50 :	1 :		1.46 :	3.158
For 50 kg of cement				
25 litres :	50 kg :	73 kg :		157.9 kg

Fraction I = 94.74 kg and Fraction II = 63.16 kg

Step 7. Water content (corrected)

(i) Water content for water-cement ratio 0.50 - 25 litres

(ii) Extra water to be added for absorption in case of coarse aggregate at 0.3 percent by mass

$$\left(\frac{0.3 \times 157.9}{100}\right) = 0.474\ litres$$

(iii) Water to be reduced for free moisture present in sand

$$\left(\frac{2}{100} \times 73\right) = 1.46\ 24.01\ litres$$

(iv) Actual quantity of water to be added

(25 + 0.474 – 1.46) = 24.01 litres.

Step 8. Quantity of coarse aggregate. Actual quantity of coarse aggregate needed

(i) Fraction I = (94.74 – 0.474 × 0.6) = 94.46 kg

(ii) Fraction II = (63.16 – 0.474 × 0.4) = 62.97 kg

Step 9. Actual quantities of various ingredients of the concrete mix are as below :

water = 24.01 litres, cement = 50 kg, sand = 74.46 kg,

coarse aggregate and Fraction I = 94.46 kg, Fraction II = 62.97 kg

The outline of design of concrete mix may be schematically represented as shown in Fig. 29.9.

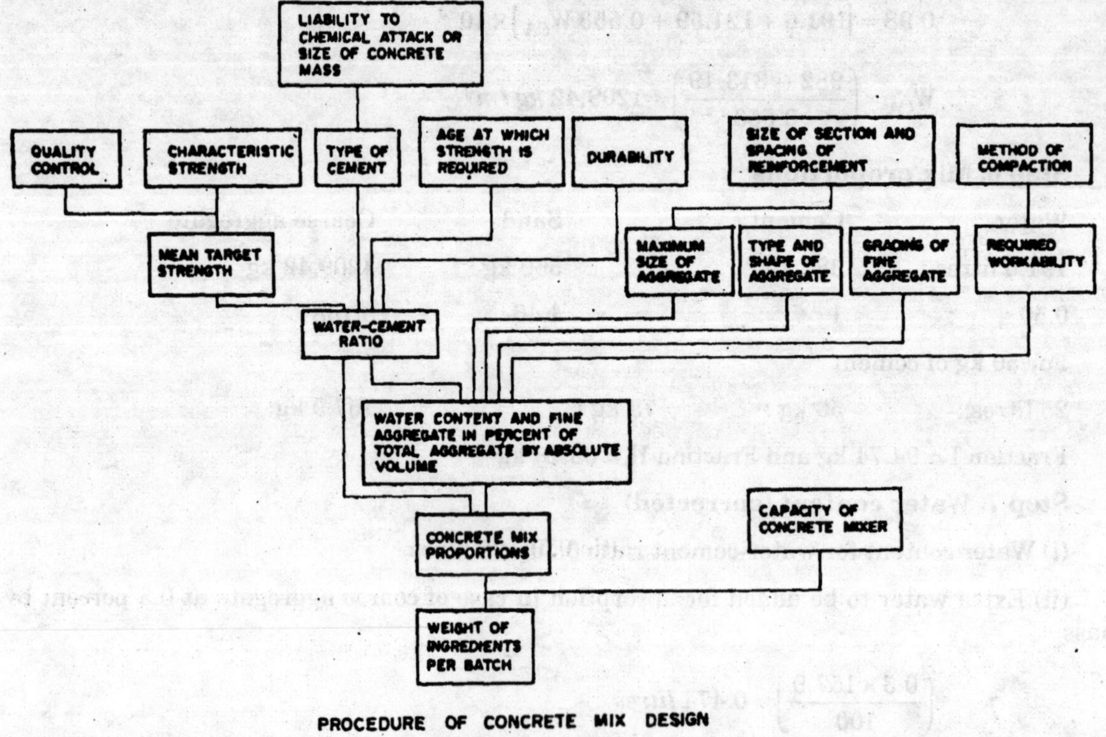

PROCEDURE OF CONCRETE MIX DESIGN

Fig. 29.9

PROBLEMS

29.1 Design a concrete mix to have a characteristic compressive field strength at 28 days of 25 N/mm². Use appropriate data for the design of concrete mix. Adopt the minimum voids method.

29.2 Design the concrete mix in Problem 29.1. Adopt the maximum density method.

29.3 Design the concrete mix in Problem 29.1 Adopt the fineness modulus method.

29.4 Design the concrete mix in Problem 29.1. Adopt RRL-method.

29.5. Design the concrete mix in Problem 29.1. Adopt ACI-method.

29.6 Design the concrete mix in Problem 29.1. Adopt British method.

29.7. Design the concrete mix in Problem 29.1. Adopt USBR method.

29.8 Design the concrete mix in Problem 29.1. Adopt IRC method.

29.9 Design the concrete mix in Problem 29.1. Adopt Indian Standard method.

30

Laboratory Tests for Cement and Concrete

30.1 INTRODUCTION

The cements are manufactured in various factories. The specifications have been specified in the various standards as regards the physical requirements of cement and the chemical composition. The large variations in the quality and the properties of cement are possible. It is seen that the different results are obtained for the same materials, when the materials are tested. *To ensure the suitability and the standard quality, the cements are tested in the laboratories.* The object of these tests is to fix values of certain readily found physical and chemical properties for cements. It may be verified that the cement to be used in practice is satisfactory and the inferior materials may be detected. The fineness, chemical composition, compressive strength, tensile strength, setting time, soundness, heat of hydration. etc., are the various tests which are carried out in accordance with respective Indian Standards. The methods of testing are implicity followed so that the reliable results are obtained. Some of the tests of cement have been described in Article 1.4 (Chapter 1).

The representative samples of aggregate brought from any pit, quarry or other source of supply for use in concrete are obtained. *To ensure the suitability of aggregates for concrete work, these samples are tested.* The sieve analysis, determination of clay, silt and fine dust, the determination of organic impurities, the specific gravity and absorption of aggregate coarser that the specified one, sand finer than the specified one, the aggregate crushing value, bulk density (or unit weight), determination of voids, determination of suitability of aggregates by reference to the strength of test pieces of concrete made with them are the various tests which are carried out in accordance of Indian Standards.

It is necessary to test the concrete in both fresh and hardened states. To find the capacity of concrete for satisfactory placing, the fresh concrete is tested for workability. The information about the consistency of the concrete mix such that the concrete may be transported, placed, compacted and finished easily and without segregation is obtained from the workability test. The analysis of fresh concrete is needed to judge the stability that is to indentify segregation of the concrete mix, uniformity of the concrete mix and to determine the proportions of ingredients of concrete actually used. To find the time upto which the concrete may be vibrated and to decide the rate of movement of formwork (for example, in slip from construction), the tests for setting time of concrete is needed.

To check the quality and compliance with the specifications, the hardened concrete specimens are tested. The concrete cubes of 150 mm size are tested to find the compressive strength at 28 days. The concrete is also tested for the flexural strength.

Test specimens are tested the laboratory and crushed (i.e., till the failure of the specimens). These tested are conducted for the last many years. These tests are essential and these are conventional and traditional. However, these tests have not been referred as **destructive test,** since the tests are conducted for the constructive purpose. In the recent past, **non-destructive tests** have been developed. These tests are also conducted for constructive test purpose. However, the test specimens are neither crushed nor loaded till their failure occurs. Incase, there specimens are commercial products, then, these may be still sold in the market. These test specimens are not damaged in any manner.

30.2 FINENESS OF CEMENT

The test for fineness of cement has been briefly described in Art. 1.4 (Chapter 1). Indian Standard Specifications prescribe dry sieve analysis method and Blaine air permeability method for determining the fineness of cement.

1. Dry sieve analysis method. In this method, the fineness of cement is represented by the weight of the residual left on a standard 90-micron IS : 460-1972. A 100 gm of cement is placed on 90-micron IS : Sieve. Holding the sieve in both hands, the sieving is done with a gently wrist motion. The continuous rotation of the sieve shall be done throughout the sieving. After 15 minutes of sieving the residue left over is weighed. This weight should not exceed the limit specified in the respective standard specification. The method of finding the fineness modulus of cement in not satisfactory.

2. Blaine air permeability method. Blaine's air permeability apparatus is shown in Fig. 30.1. The permeability cell as shown in Fig. 30.1 (C), plunger and the monometer tube assembly are the three essential parts of this apparatus which are described below :

A definite quantity of air is drawn through a prepared bed of cement of definite porosity. The size of particles governs the number and size of the pores in a prepared bed of cement. The rate of air flow through the bed's determined.

(i) Permeability cell. A rigid cylinder (12.7 ± 1) mm in inside diameter constructed of glass or non-corroding metal forms the cell. The top of the cell shall be at right angles to the principal axis of the cell. The bottom of cell provides in air tight connection at the top of a monometer. The cell is connected with the monometer tube assembly. A ledge of (0.5 to 1) mm in width shall be an integral part of the cell or firmly fixed in the cell (50 ± 15) mm from the top of the cell for support of the perforated metal disc. This noncorrodic metallic disc is 0.9 ± 0.1) mm in thickness. It has 30 to 40 perforated holes of 1 m diameter equally distributed over its area. The disc fits the inside of the cell snugly.

(ii) Plunger. The permeability cell carries a plunger as shown in Fig. 30.1 (B) fitted into it with a clearance of not more than 0.1 mm. The plunger carries a sharp square edge at its bottom remaining at right angles to the principal axis. The plunger has also an air vent either at the centre or on one side of the plunger. A collar is provided at the top of the plunger such that when the plunger is placed in the cell and the collar is brought in contact with the top of the cell, the distance between the bottom of the plunger and the top of the perforated disc shall be (15 ± 1.0) mm. A filter paper in the form of circular disc with smooth edge and having the same diameter as the inside of the cell shall be used as medium retentive.

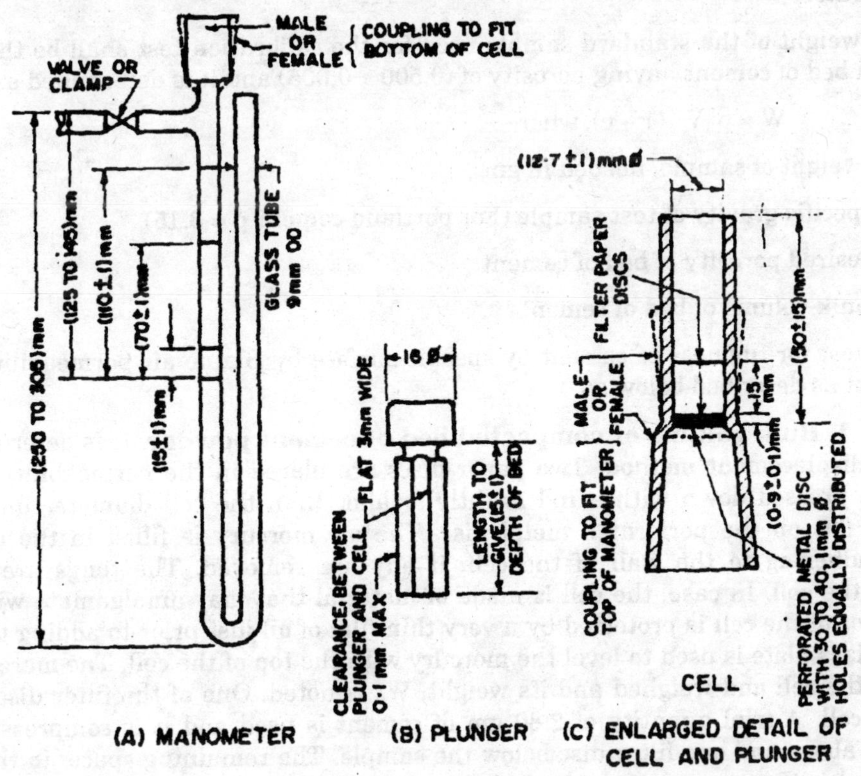

Fig. 30.1

(iii) Monometer. A U-tube manometer as shown in Fig. 30.1 (A) consists of nominal 9 mm outside-diameter (standard wall glass tubing). An air-tight connection with the permeability cell is kept with the top of one arm of the monometer. The monometer arm connected to the permeability cell shall have a line etched around the tube at (125 to 145) mm below the top of the inside outlet and also others at distance of (15 ± 1) mm, (70± 1) mm and (110 ± 1) mm above that line. For evacuation, a side outlet is provided at (250 to 305) mm above the bottom of the manometer. A positive air-tight valve o clamp is provided on the side outlet not more than 50 mm from the monometer arm. The monometer is firmly attached and its arms are kept vertical.

Sample. Air permeability apparatus is first calibrated by using *National Bureau of Standards Sample No. 114* or any other cement sample having the specific surface equal to the appropriate certificate value of NBS Standard Sample No. 114. The sample shall be at room temperature when tested. It is to note that in the absence of NBS Standard Sample No. 114, samples of Indian Standard cements having the specific surface equivalent to NBS Standard No. 114 may be used. Such samples may be brought from the ACC Central Research Station, Bombay-Agra Road, P.O. Wagle Industrial Estate, Thana, (Maharashtra).

The contents of a vial of the standard cement sample shall be enclosed in a 125 gm jar and shaken vigorously for 2 minutes to fluff the cement and break up lumps or agglometrates.

The weight of the standard sample used for the calibration test shall be that needed to produce a bed of cement having porosity of (0.500 ± 0.005) and it is determined as below :

$$W = p\ V . (1 - e), \text{ where} \tag{30.1}$$

W = weight of sample, needed in gm

p = specific gravity of test sample (For portland cement p = 3.15)

e = desired porosity of bed of cement

V = bulk volume of bed of cement

The test for fineness of cement by specific surface by Blaine air permeability method is carried out as described below :

Step 1. Bulk volume of compacted bed of cement powder. It is determined by the mercury displacement method. Two paper discs are placed in the permeability cell and its edges are pressed down with a rod slightly smaller than the cell diameter until the filter discs are flat on the perforated metal disc. The air mercury is filled in the cell. The air bubbles adhering to the wall of the cell, if any are removed. The tongs are used while handling the cell. In case, the cell is made of material that may amalgamate with mercury, the interior of the cell is protected by a very thin film of oil just prior to adding the mercury. A small glass plate is used to level the mercury with the top of the cell. The mercury is taken out from the cell and weighed and its weight, W_A is noted. One of the filter disc is removed from the cell. A trial quantity of 2.80 gm of cement is used and it is compressed with one filter disc above and one filter disc below the sample. The remaining space in the top of the cell is filled with mercury, entrapped air is removed and mercury is leveled with the top as before. The mercury is again taken out and weighed and its weight, W_B is also recorded. The volume occupied by the cement is found from Eq. 30.2, below :

$$V = (W_A - W_B)/D \tag{30.2}$$

where, D is the density of mercury at temperature of test in gm/cm^3. At least, the bulk volume of cement sample shall be determined twice.

It is to note that the bed of cement is prepared by seating a perforated disc on the ledge in the permeability cell. A filter paper is placed on the metal disc and the edges are pressed down with a rod. A weight quantity of cement is placed in the cell. In order to level the bed of cement, the side of the cell is tapped lightly. A filter disc shall be placed on the top of the cement and the cement is compressed with the plunger until the plunger collar is in contact with the top of the cell. The plunger is removed slowly.

Step 2. Permeability test. The monometer tube is attached with the permeability cell. The connection is made air-tight. It is seen that the prepared bed of cement is not distributed.

One arm of the monometer U-tube is slowly evacuated of air until the liquid reaches the top mark and valve is then closed tightly. When the bottom of the meniscus of the monometer liquid reaches the second (next to the top) mark, the stopwatch is started. The stopwatch shall be stopped as the bottom of meniscus of the liquid reaches the third (next to the bottom) mark. The measured interval of time is noted in second. The temperature of test is also noted in degrees centigrade.

Step 3. Procedure of test. The cement sample shall be at room temperature when it is tested. The weight of the cement sample used for the test shall be the same as that used in the calibration test on the standard sample. Instead of porosity of (0.500 ± 0.005) for bed of the cement in standard calibration test, the weight of sample shall be that needed to produce a test bed having a porosity of (0.35 ± 0.005) while determining the fineness of high-early strength cement. The bed of test sample is prepared as it is prepared for bed of cement for standard test. The permeability test is also carried out as it is done for standard sample. For one prepared bed of actual cement, only one time of flow determined is made instead of three separately prepared beds of standard sample.

Step 4. Calculations. The values of specific surface as determined using Eq. 30.3 as below :

$$S = S_g \cdot (T/T_g)^{1/2}, \text{ where} \qquad (30.3)$$

S = specific surface in square cm per gram of the test sample

S_g = specific surface in square cm per gram of the standard sample used in calibration of the apparatus.

T and T_g are measured time-interval, in seconds, of monometer drop for test sample and standard sample used in calibration of the apparatus, respectively.

Eq. 30.3 is used in case the temperature of test of the test sample is within ± 3°C of the temperature of calibration test of the temperature of test is outside this range, Eq. 30.4 as below shall be used :

$$S = S_g \cdot \left[\left(\frac{T}{T_g}\right)\cdot\left(\frac{n_g}{n}\right)\right]^{1/2} \qquad (30.4)$$

where n_g and n are the viscosities of air in poises at the temperature of test of the standard sample used in calibration of the apparatus and test of the test sample, respectively. Eq. 30.5 is used if the temperature of test of the test sample is within ± 3°C of the temperature of calibration test of the standard fineness sample

$$S = S_g \cdot \left(\frac{1-e_g}{1-e}\right)\left[\left(\frac{e^3}{e_g^3}\right)\left(\frac{T}{T_g}\right)\right]^{1/2} \qquad (30.5)$$

where e_g and e are the porosities of prepared bed of standard sample used in calibration of apparatus and bed of test sample, respectively. Eq. 30.6 as below is used if the temperature of test is outside this range.

$$S = S_g \cdot \left(\frac{1-e_g}{1-e}\right)\left[\left(\frac{e^3}{e_g^3}\right)\cdot\left(\frac{T}{T_g}\right)\left(\frac{n_g}{n}\right)\right]^{1/2} \qquad (30.6)$$

It is to note that Eqs. 30.3 and 30.4 are used in the calculations of fineness of portland cements compacted to the sample porosity as that of the standard sample.

Whereas, Eqs. 30.5 and 30.6 are used in calculation of fineness of portland cements compacted to a porosity other that of the standard fineness sample used in the calibration test.

TABLE 30.1. VALUES OF DENSITY OF MERCURY AND VISCOSITY OF AIR (N) AND (N)^1/2 (AS PER IS : 456-2001)

Room temperature 0°C	Density of mercury* gm/cm³	Viscosity of air (poises) n	$(n)^{1/3}$
16	13.56	178.8×10^{-6}	13.37×10^{-3}
18	13.55	179.8×10^{-6}	13.41×10^{-3}
20	13.55	180.8×10^{-6}	13.44×10^{-3}
22	13.54	181.8×10^{-6}	13.48×10^{-3}
24	13.54	182.8×10^{-6}	13.52×10^{-3}
26	13.53	183.7×10^{-6}	13.55×10^{-3}
28	13.53	184.7×10^{-6}	13.59×10^{-3}
30	13.52	185.7×10^{-6}	13.62×10^{-3}
32	13.52	186.7×10^{-6}	13.66×10^{-3}
34	13.51	187.6×10^{-6}	13.69×10^{-3}

(*The density of mercury is determined separately).

Values of $(e^3)^{1/2}$ and $(T)^{1/2}$ may be taken from IS : 4031–1968.

In case, the fineness of materials other than portland cement are calculated, Eqs. 30.7 and 30.8 below are used. Out of these equations, Eq. 30.7 as below is used when the temperature of test of the sample is within ±3°C of the temperature of calibration test of the standard fineness sample.

$$S = S_g \cdot \left(\frac{p_g}{p}\right) \cdot \left(\frac{1-e_g}{1-e}\right) \cdot \left[\left(\frac{e^3}{e_g^3}\right) \cdot \left(\frac{T}{T_g}\right)\right]^{1/2} \tag{30.7}$$

Eq. 30.8 as below is used if the temperature of the test is outside of this range.

$$S = S_g \cdot \left(\frac{p_g}{p}\right) \cdot \left(\frac{1-e_g}{1-e}\right) \cdot \left[\left(\frac{e^3}{e_g^3}\right) \cdot \left(\frac{T}{T_g}\right)\left(\frac{n_g}{n}\right)\right]^{1/2} \tag{30.8}$$

where, p_g and p are the specific gravities of the standard sample used in the calibration of apparatus (assumed to be 3.15) and the test simple, assumed to be 3.15 for the portland cement respectively.

Values of density of mercury viscosity of air, n and $(n)^{1/2}$ at different room temperatures may be adopted from Table 30.1.

30.3 SOUNDNESS OF CEMENT

The test for soundness of cement has been briefly described in Art. 1.4 (chapter 1). IS : 4031-1968 specifies **Lechatelier** method. The apparatus used is shown in Fig. 30.2. It consists of a small split cylinder of spring brass, or other suitable metal of 0.5 mm thickness, forming a mould 30 mm internal diameter and 30 mm high. On either side of the split are attached two indicators with pointed ends AA, the distance from these ends to the centre of cylinder being 165 mm.

The procedure for determining the soundness of cement by Lechatelier apparatus is as follows :

The mould is placed on a glass sheet and it is filled with cement paste (prepared from about 100 gm of cement) formed by gauging cement with 0.78 *times* the water needed to give a paste of standard consistency. Another piece of glass plate is used to cover the mould. A small weight is placed on this covering glass sheet and immediately the whole assembly is submerged in the water at a temperature (27 ± 2)°C. The assembly is kept in water for 24 hours. After 24 hours, the assembly is brought out and the distance separating the indicator points is measured. Again, the mould is submerged in water at temperature (27 ± 2)°C.

The water is boiled with submerged mould for 25 to 30 minutes. The water is kept boiling for three hours. The mould is brought out from the water and it is allowed to cool. The distance between the indicator points is then, measured. The difference between two measurements represents the expansion of the cement.

This expansion for unaerated low heat portland cement and ordinary rapid hardening cement should not exceed 10 mm.

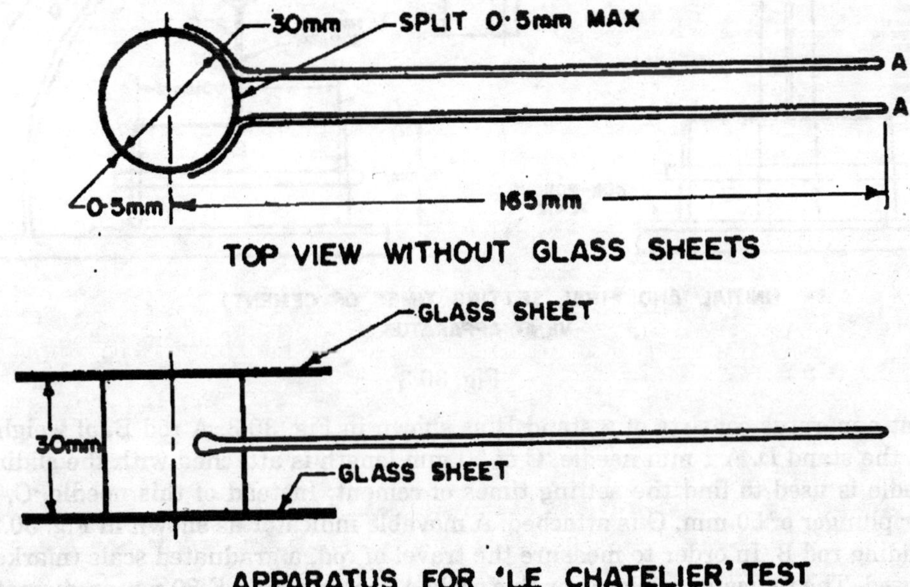

Fig. 30.2

30.4 CONSISTENCY OF STANDARD CEMENT PASTE

The object of this test is to find the quantity of water needed to produce a cement paste of *standard consistency*. The Vicat apparatus as shown in Fig. 30.3 is used.

The standard consistency of a cement paste is defined as that consistency which allows the Vicat plunger shown in Fig. 30.3 to penetrate to a point 5 to 7 mm from the bottom of the Vicat mould.

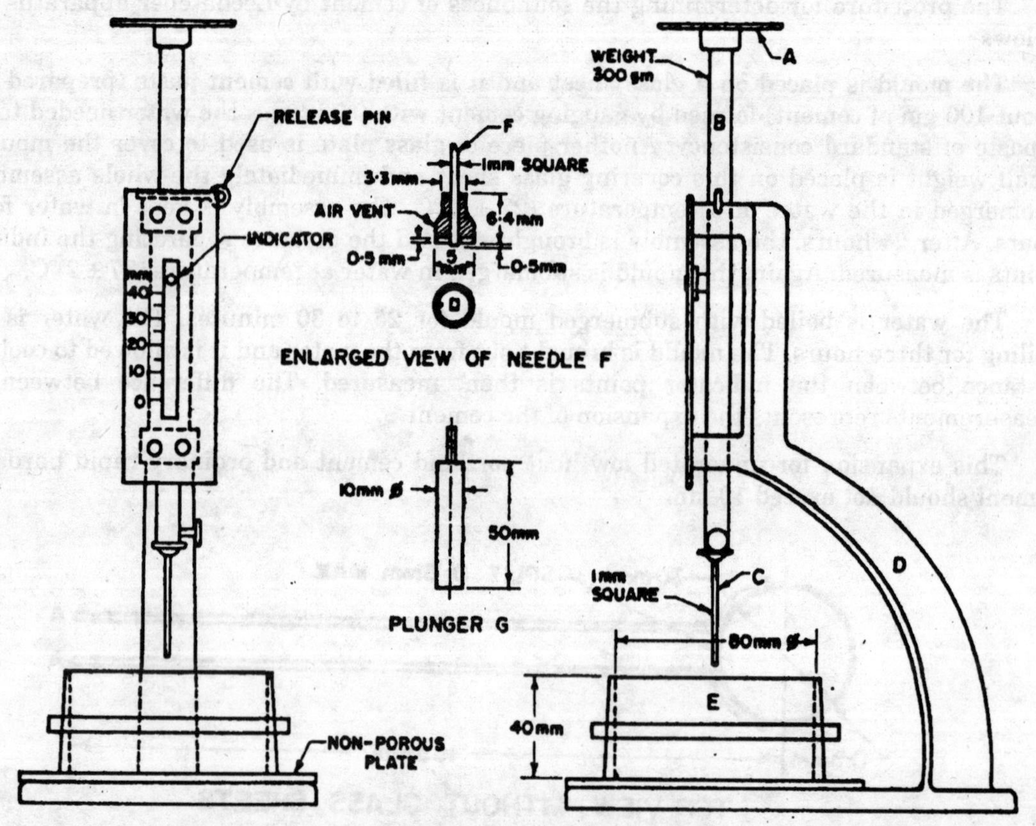

Fig. 30.3

Vicat apparatus consists of a stand D as shown in Fig. 30.3. A rod B, of weight 300 gm slides in the stand D. A 1 mm needle, C of 50 mm length is attached with the sliding rod, B. This needle is used to find the setting times of cement. Instead of this needle, C, a 10 mm diameter plunger of 50 mm, G is attached. A movable indicator as shown in Fig. 30.3 is fitted to the sliding rod B. In order to measure the travel of rod, a graduated scale (marked 0 to 40 mm) is used. The indicator moves over this scale A Vicate mould E 80 mm in diameter and of 40 mm height is used for cement paste sample. This mould is placed over a non-porous plate.

The test procedure to find the consistency of standard cement paste is as follows :

Step 1. A paste of weighed quantity of cement is prepared with a weighed quantity of water. The time for gauging is kept more than 3 minutes and less than 5 minutes. The gauging is done with spatula. The gauging is completed before any sign of setting occurs. The gauge time is measured from time of adding water to the dry cement upto beginning of filling of cement paste in the mould.

Step 2. The Vicat mould E is filled with cement paste. This mould E rests on a non-porous plate. The surface of paste is smoothed off after completely filling the mould. The cement paste is levelled with the top of the mould. While filling the mould, the operator's hand and the blade of the gauge trowel alone are used. The trowel weights about (210 + 10) gm. The clean appliances are used while gauging. While performing the above operations, the temperature of cement, water and that of test room shall be (27 ± 2)°C.

Step 3. Immediately after filling the mould, the mould together with non-porous plate are placed under the rod bearing the plunger. The plunger is gently lowered to touch the surface of the test block. The rod is quickly released and allowed to sink into the paste. The indicator-reading over the graduated scale is recorded.

Trial pastes with varying percentages of water prepared and tested as described above are tested until the quantity of water needed for making up the standard consistency is determined.

The quantity of water is expressed as a percentage by weight of the dry cement.

30.5 INITIAL AND FINAL SETTING TIMES OF CEMENT

The initial setting time is the time elapsed after initial contact of cement and water needed for the mortar to reach a penetration resistance of 3.43 N/mm².

The final setting time is the time elapsed after initial contact of cement and water needed for the mortar to reach a penetration resistance of 26.97 N/mm². The initial and final setting times of different cements have been discussed in Art. 1.4 (Chapter 1) and are given in Table 1.2.

Vicat apparatus as shown in Fig. 30.3 and described for consistency test in Art. 30.4 is used to find the initial and final setting times. The procedure to find the initial and final setting times of a cement is as follows :

Step 1. Preparation of test block. A neat cement paste is prepared by gauging the cement with 0.85 times the water needed to give a paste of standard consistency. As soon as, the water is added to the cement, the stopwatch is started. The gauged cement paste is filled in the Vicat mould resting on a non-porous plate. The Vicat mould is filled fully, the surface of the cement paste is smoothed off and the surface is levelled with top of the mould. Thus a test block (cement block) is prepared. The appliances are kept clean while gauging the paste. The temperature of water, cement and that of the test room at the time of gauging is kept within (27 ± 2)°C.

Step 2. Initial setting time. The mould along with cement block resting on the non-porous plate is placed under the sliding rod. A needle, C of one mm square size is fitted with the sliding rod. The sliding rod is lowered down and the needle in contact with the surface of cement block gently and quickly released. The needle is permitted to penetrate into the cement block. The needle completely penetrates the cement block in the beginning. This procedure is repeated. At some stage, the needle fails to pierce the block for (5 ± 0.5) mm measured from the bottom of the mould. The initial setting time [measured from the time, of adding water upto the time, the needle fails to pierce the cement block by (5 ± 0.5) mm] is obtained.

Step 3. Final setting time. The needle, C of the Vicat apparature, as shown in Fig. 30.3 is replaced by the needle having annular attachment, F. The size of needle is one mm square, and it projects 0.5 mm out of the attachment as shown in Fig. 30.3.

The cement is considered as finally set when upon applying the needle gently to the surface of cement block the needle makes an impression thereon, while the attachment fails to do so.

The period elapsed between the time when water is added to the cement and the time at which the needle makes an impression on the surface of cement block while the attachment fails to do so shall be final setting time.

In case scum is formed on the surface of the cement block, the lower surface of the cement may be used to find the final setting time.

30.6 COMPRESSIVE STRENGTH OF CEMENT

The test for compressive strength of cement has been briefly described in Art. 1.4 (Chapter 1). The physical requirements of various types of cements are also given in Table 1.2.

The cement mortar cubes are cast and vibrated by means of a standard vibration machine for compaction. The standard Leighton Buzzard sand is used in preparing the cement mortar conform to IS : 650-1966.

The procedure to find the compressive strength of cement is as follows :

Step 1. Mix proportions and mixing. 200 gm of cement and 600 gm of standard sand and water are mixed. The quantity of water shall be [P/4 + 3.0] percent of combined weight of cement and sand, where P is the percentage of water needed to prepare a paste of standard consistency.

In case, the standard sand is not available, the single sized **Enore Sand** as an alternative to the use of regraded Enore Sand conforming to IS : 650-1966 may be used. The single graded Enore Sand shall pass through 850-micron IS : sieve and not more than 10 percent by weight shall pass through 600-micron IS : sieve. In such a case, 185 gm of cement 555 gm of standard sand and water are mixed. The quantity of water shall be [(P/4) + 3.5] percent of combined weight of cement and sand.

The mixture of cement and sand (in proportion 1 : 3) by weight is placed over a non-porous plate. It is mixed dry with a trowel for one minute and then with water until the uniform colour is seen. The time of mixture shall in any event be not less than 3 minutes. In case, the time taken to get a uniform colour exceeds 4 minutes, the mixture is rejected and a fresh quantity of cement, sand and water are taken and the operation is repeated.

Step 2. Moulding specimens. The metallic moulds for the cube specimens of 50 cm^2 face area shall be used. The metal should be such that it is not attacked by cement mortar, and there shall be sufficient material in the sides of the mould to prevent spreading and warping. The split moulds are used, which facilitate the removal of the mould specimen without damage. The assembled parts of the mould are positively held together. The joints between the halves of the mould are covered with a thin film of petroleum jelly and a similar jelly is also applied between the contact surfaces of the bottom of the mould and its base plate in order to ensure that no water escapes during vibration. The interior faces of mould are treated with a thin coating of mould oil.

The assembled mould is placed on the vibration table and it is firmly held is position by means of a suitable clamp. The mortar is filled in the cube mould immediately after the mortar is prepared. The mortar is prodded with the pocking rod made of nonabsorptive. nonabrasive and non-brittle material (such as rubber compounding). The mortar is prodded 20 times in 8 seconds and it is ensured that the entrained air and honey combing is eliminated. The remaining quantity of mortar is placed in the hopper of the cube mould. It is again prodded as specified for the first layer. The mortar is compacted by vibration machine for two minutes as the specified speed of (12,000 ± 400) vibrations per minute. The mould together with the base place is removed at the end of vibrations. The top surface of cement cube in the mould is finished, the surface is smoothed with the blade of a trowel and levelled. Such cubes are prepared and compacted in different mould. For each mould, the cement, sand and water are mixed separately.

Step 3. Curing specimens. The filled moulds are placed at a temperature of (27 ± 2)°C in an atmosphere of at least 90 percent relative humidity for 24 hours after the completion of vibration. The moulds are removed at the end of that period from the moulds. The cubes are immediately submerged in clean fresh water. The cubes are kept in water and taken out just prior to breaking.

Step 4. Testing of cubes for compressive strength. On the completion of specified period, there cubes are tested for the compressive strength. Three cubes are tested after curing for 7 days and three cubes are tested after curing for 28 days. The packing shall not be used between the cube and the steel plattens of the testing machine. The cubes are tested on their sides. One of the plattens shall be carried on a base and shall be self-adjusted and the load shall be steadily and uniformly applied, beginning from zero at a rate of 35 N/mm^2/minute.

The compressive strength is determined from the crushing load and the average area over which the load is applied. The result is expressed in N./mm^2. The compressive strength shall be average of the strength of three cubes for each period of curing.

30.7 TENSILE STRENGTH OF CEMENT

The test for tensile strength of cement has been briefly described in Art. 1.4 (Chapter 1). The physical requirements of various types of cements are also given in Table 1.2. The tensile strength of cement is found by calculating the tensile strength of cement-sand mortar by following the procedure described as under :

Step 1. Mix proportions and mixing. 250 gm of cement and 750 gm of **Standard Leighton Buzzard Sand** (viz., 1 : 3 parts) by mixing the water. The quantity of water to be used for making the mix is taken as [0.5 P + 2.5] percent of total weight of cement and sand, where, P is the percent of water needed for standard consistency. The mortar is produced in the manner identical to that for the test for the compressive strength.

Step 2. Preparation of specimen. For conducting the test, the briquettes of standard size, as shown in Fig. 30.4, are prepared in standard briquette moulds resting on non-porous plate. Split type moulds are used, which may open on both the faces, so that the prepared specimen may be taken out without damage. The minimum width and thickness of briquette specimen are 25.4 mm each so that the cross-sectional area at this minimum section is 645 mm^2.

The grease is applied on the two halves of the mould and oil is coated on the inner faces. The moulds are filled with the cement mortar prepared. The mortar is heaped over the mould and with the help of standard spatula, the mortar is beaten down so that surface water appears. The surface of the mould is smoothed and levelled with the top. The other face of the briquette is also prepared similarly. At least 12 briquette specimens are prepared. The briquettes are placed in room for 24 hours at (27 ± 2)°C temperature and 90 percent humidity. The briquettes so prepared are brought out from the moulds and immersed in water tank for curing.

Step 3. Testing of briquettes. Six briquettes are tested after 3 days curing and six briquettes are tested after 7 days curing in a special testing machine. The tensile load is applied by a compound lever at the rate of 3.5 N/mm^2 of the section per minute. Special shaped jaws hold the briquette specimen. The lead shots drop from a hopper in a pan and load is applied. The briquettes finally fail in ultimate tension. the lead shots automatically discontinue as soon as the briquette fails. The lead shots of pan are weighed in the balanced.

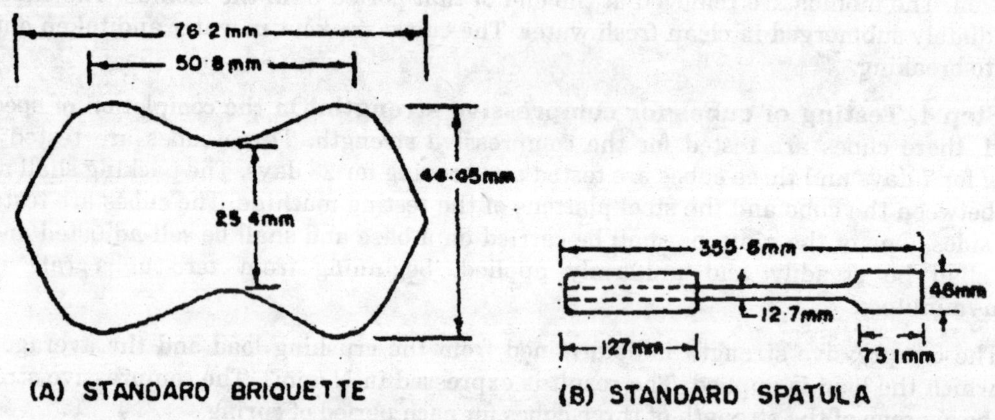

Fig. 30.4

The tensile strength at the failure of briquette

$$= \left(\frac{\textit{Ultimate tensile load}}{\textit{Min. sectional area of briquette}} \right)$$

The tensile strength at failure after 3 days for ordinary portland cement shall not be less than 2 N/mm^2 and after 7 days it should not be less than 2.5 N/mm^2. For rapid hardening cement, the stress at failure after 1 day, should not be less than 2 N/mm^2 and that after 3 days, it should not be less than 3 N/mm^2.

30.8 SLUMP TEST

The slump test has been briefly described in Art. 1.12 (Chapter 1). The slump test is conducted to measure the workability of fresh concrete mix. At actual site conditions, the slump test is more common for medium strength concretes. IS : 1199-1959 (Methods of sampling and analysis of concrete) specifies the method of conducting slump test. The procedure to find the slump is as follows :

Step 1. Apparatus. A truncated cone having 100 mm diameter at top and 200 mm diameter at bottom and 300 mm in height, as shown in Fig. 30.5, is used to conduct the slump where the maximum size of aggregate is not more than 38 mm in size. Suitable foot pieces are attached with the cone. The cone may be lifted vertically as needed during the test with the help of two handles fitted with it. A suitable guide attachment is also provided with the mould. A tamping rod 16 mm diameter and 600 mm long made of steel is used to tamp the concrete when it is filled in the slum cone.

Step 2. Test specimen. The slump cone mould is cleaned and placed on a smooth horizontal, rigid, non-absorbant base. The freshly mixed concrete is poured into the slump mould in three layers and each layer is tamped 25 strokes uniformly with tamping rod throughout its depth. The strokes of tamping rod pierce in the underlying layer when the second and subsequently layers are poured. After rodding the top layers, the concrete surface is smoothed and levelled with a trowel. The slump cone is, then, gently and vertically raised. The concrete settles down under its own weight. The vertical distance between its original level to the new level after subsidence (which one is called **slump**) is measured in mm as shown in Fig. 30.5.

Following three types of slumps as shown in Fig. 30.5 are seen while conducting the slump test.

(i) True slump. The concrete mixes subside uniformly and cohesively. Rich mixes and higher proportion of fine aggregate show this type of slump.

(ii) Shear slump. Half of the cone shears off along the inclined plane and half of the other slump remains true. The linear mixes, e.g. 1 : 6 of 1 : 8 show this type of slump. In case, a shear slump occurs in first instance, the test is repeated. In such cases, the slump is measured from the level of the cone to the centre of the shear plane as shown in Fig. 30.5.

(iii) Collapse slump. The concrete just collapses and spreads over a large area. Very wet mixes show this type of slump.

The mixes have a tendency to segregate (during slump test) by gently tapping the side of the subsided cone of the mix with the tamping rod. The cone subsides uniformly with the gentle tappings, without the coarse particles trying to separate from the mix. The mix is said to be cohesive. Lightly trowelling the mix and observing the smoothness of the surface, the harshness of the mix may be checked.

In case, the slump test indicates the marked change in the slump, two features are reflected from this.

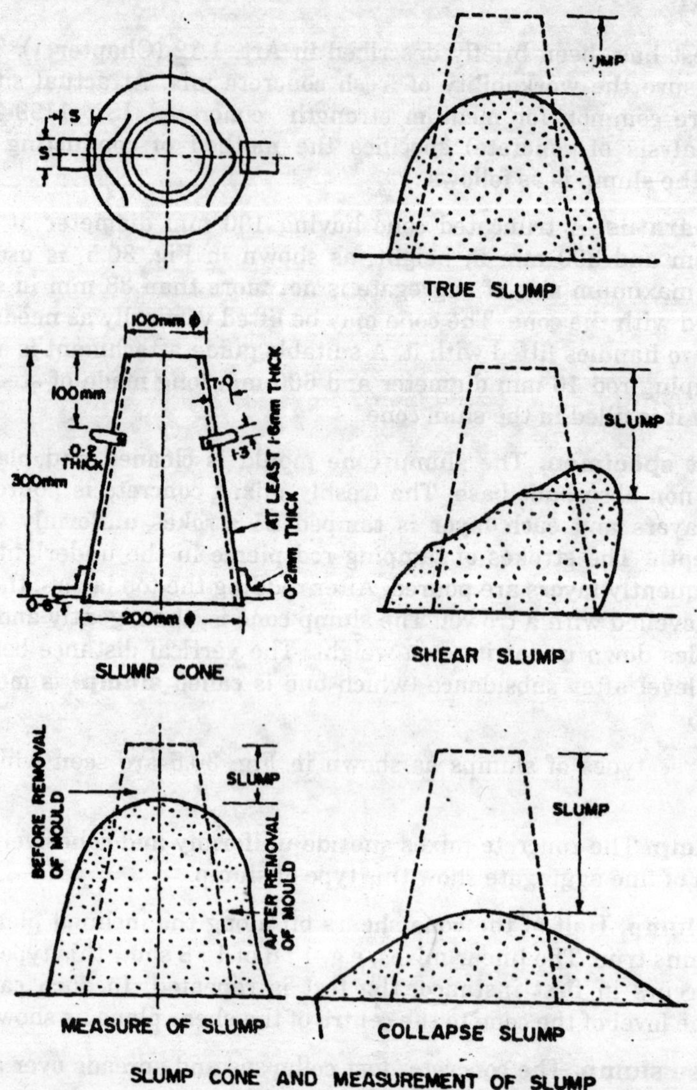

Fig. 30.5

(i) There is appreciable increase or decrease of the natural moisture content in the fine aggregate. The water-content is then, adjusted.

(ii) In case, there is no possibility of change in moisture content in the aggregates, it is likely that there is appreciable change in the granding of aggregate. In case, the slump found is less than the one expected, then, there is increase in the fine particles. The aggregate grading coarser if it is seen that the slump is more.

The proportion of the fine aggregate to the coarse aggregate is adjusted to obtained the expected slump.

The slump test provides only a rough check for the quality of the concrete.

Limitation of the slump test

There is no relation to useful internal work necessary for complete compaction. The slump test becomes ineffective if zero slump for the concrete mix for very low degree of workability are seen.

30.9 COMPACTING FACTOR TEST

The compacting factor test has been briefly described in Art. 1.12 (Chapter 1). The compacting factor test is more commonly used for high-strength concrete. This method is also used for low workability concrete, and very dry mixes. This test was developed at Road Research Laboratory, London. In this method, *the external work applied to a given quantity of concrete and to reduce to a minimum the work used in overcoming the friction between the concrete and the contacting surfaces.* The work applied includes the work done against the surface friction but this is reduced to a minimum, although probably the actual friction varies with the workability of the mix. IS : 1199-1959 (Methods of sampling and analysis of concrete) specifies the method of conducting compacting factor test. The procedure for conducting the compacting factor test is as follows :

Step 1. Apparatus. A compacting factor apparatus is shown in Fig. 30.6. Two conical hoffers A and B are mounted above a cylindrical mould with a rigid frame. These three components may also be easily detached from the frame. These are made of brass or bronze or thick sheet of steel. Both the hoppers have trap doors at their bottom. The trap doors are closed tight fit. The doors move about the hinge. The trap doors may be quickly released from catches. Two ordinary bricklayers trowels, one hand scoop and a rod of steep are used as appliances with the apparatus. A balance is used to weight the materials.

Step 2. Test. The upper hoper A is filled with freshly mixed concrete using the hand scoop. The surface of concrete is smoothed and levelled with the top of the hooper. The trap door of the upper hopper, A is opened. The concrete drops into the lower hopper, B, A rod is used to push the sickened concrete (if any). As soon as the concrete is seen at rest, the trap door of lower hopper, B is immediately opened. The concrete drops into the cylinder, C. The excess of concrete above the top of the cylinder is removed by trowels in both the hands. The blades are kept horizontal. The trowels are moved simultaneously from both the sides over the top of the cylinder. The blades of trowels remain pressed with the top of cylinder. The cylinder is also cleaned from the outside. Te concrete in the cylinder is weighed. This weight, W_{PC} is called as *weight of partially compacted concrete.* The fresh concrete of some sample is refilled in layers of approximately 50 mm depth. Each layer of concrete is heavily rammed. Alternately each layer may be vibrated to get complete vibration. The excess of completely compacted concrete is struck off with the level of top of the cylinder. Let the weight of completely compacted concrete be W_{CC}. *The ratio of weight of partially compacted concrete and that of completely compacted concrete is called as* **compacting factor.** Therefore,

Compacting factor = (W_{PC}/W_{CC})

It is expressed nearest to the second decimal place.

The weight of completely compacted concrete may also be obtained from the specific gravity of the materials. The dry mixes have low value of compacting factor and therefore, a low degree of workability. The wet mixes have a high value of compaction factor and high degree of workability. The suggested values of workability of concrete for some placing conditions measured in terms of slump compacting factor and Vee-bee time have been given in Table 1.9.

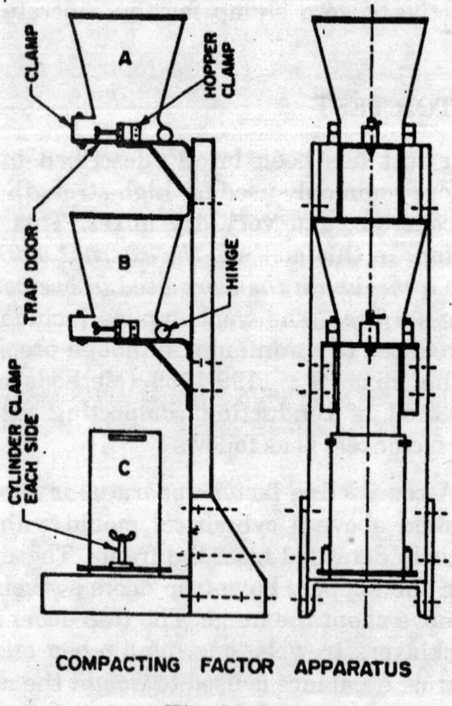

Fig. 30.6

30.10 COMPRESSIVE STRENGTH OF CONCRETE

In order to control and confirm the quality of cement concrete work, *it is important to test the hardened concrete.* It confirms that the fresh concrete used at site will develop the needed strength. The tests for compressive strength of concrete are conducted on the samples obtained from the frequency of sampling. These samples are tested at 28 days (after hardening of the concrete). Therefore, the compressive strength of sample shall be known after 28 days. It is, of course, the disadvantage of the conventional test. In case, the different ingredients are mixed and effective control is observed normally the concrete will develop the needed strength. Although, the results of testing of samples (viz., the compressive strength) shall be known, the proportions of materials may be adjusted for further production of concrete, if necessary. The shape and size of samples for conducting the test for compressive strength are as follows :

Shape and size of samples

The test for compressive strength of hardened concrete are conducted as per IS : 516-1959 on samples of cubical shape or cylindrical shape. The tests carried out on the samples prepared in the laboratory are termed as *preliminary tests.* The test carries out on the samples prepared in the field from the fresh concrete being mixed in the mixer are termed as *work test.*

Samples of cubical shape. In case, the maximum nominal size of aggregate is 30 mm then the cube samples of size 150 mm x 150 mm x 150 mm are prepared. However, in case the maximum nominal size of aggregate is 20 mm, then, the cube samples of size 100 mm x 100 mm x 100 mm are prepared.

Samples of cylindrical shape. The length of cylindrical shape samples is kept *twice* the diameter. The cylindrical shape samples of 150 mm diameter and 300 mm length are prepared. However, samples of smaller size may be prepared. But the ratio of diameter of the sample to the maximum nominal size of the aggregate 3 : 1 is to be maintained.

The moulds are used to prepare the samples.

Cube moulds. The split type metallic moulds made of steel or cast iron with metallic base are used to prepare cubical samples. The moulds are made stout enough, so that the distortion does not occur. The moulds are made in such a manner as to facilitate the removal of the mould samples without damage.

These moulds are so machined that when these are assembled ready for use, the dimensions and the internal faces are needed to be accurate. IS : 516-1959 specifies that the distance between the opposite faces and height of mould should be of the specified size ± 0.2 mm. The adjacent internal faces and top and bottom should have $(90 \pm 0.5)^{\circ}$ angles. The interior plane surfaces may have permissible variation of 0.03 mm. The dimensions of the base plate should be such that it supports the mould during filling without leakage. The base plate is attached with the mould by screws or springs. The assembled parts of the mould are positively and rigidly held together. The joints of the mould are thinly coated with mould oil. The contact surfaces of the assemble mould and base plate are also coated with mould oil. The internal faces of the mould are also coated with this oil so that adhesion of concrete does not occur.

Cylindrical mould. The metallic wall of the cylindrical mould shall be not less than 3 mm thickness. Each mould is made such that it may be opened longitudinally to facilitate the removal of sample. The ends of mould are made to be in plane surface perpendicular to the axis of the mould within ± 0.05 mm. The mean internal diameter of the assembled mould should be (150 ± 0.2) mm. The internal diameter of the mould should not be less than 149.5 mm or more than 150.5 mm. The height of mould should be (300 ± 0.1) mm. There must be a metallic base plate and a capping plate of glass with each mould. These plates should be 6.5 mm in thickness. These plates should have plane surface within ± 0.02 mm. The base plate should be such that it supports the mould during filling without leakage. The base plate is rigidly attached with the mould. The base plate and mould are coated with thin film of mould oil before use so that adhesion of concrete does not occur.

Alongwith the mould a tamping rod of steel of 16 mm diameter and 600 mm length with a bullet point at the lower end is used for compaction.

Preparation of sample. After mixing all the ingredients in their correct proportions, the fresh concrete is filled in the mould and *cube samples are prepared* as soon as practicable in such a way as to produce full compaction of the moulds in layers of approximately 50 mm depth. Each layer is either vibrated or compacted by hand. A standard tamping rod is used for compaction by hand. The strokes are applied evenly distributed. The number of strokes applied should not be less than 35 per layer for 150 mm cubes or 25 per layer of 100 mm cubes. The strokes are applied such that the rod penetrates the underlying layer. The sides of mould are tapped gently to close the voids, (if any). After compacting the top surface of the concrete, concrete surface is finished smooth and levelled with the top of mould using a trowel. The top surface is covered with a glass or metal plate to prevent evaporation.

For cylindrical samples, the number of strokes to be applied are not less than 30 per layer. After setting of concrete in the moulds for two to four hours a thin layer of stiff and neat portland paste is applied to both the ends of the samples of capping.

After completion of moulding, the mould is placed in place free from vibration and having 90 percent relative humidity and at room temperature of (27 ± 2)°C for 24 hours. The moulds are then removed and the cube/cylinder samples are immediately placed in clean and fresh water. These samples are kept submerged in water till these are tested in the compression testing machine.

The procedure for testing the sample is as follows :

Testing of samples. These samples (cubes/cylinders) prepared for compressive strength are tested in a compression testing machine. The bearing surface (platens) of the testing machined are wiped clean and any loose sand or other material is removed from the surfaces of the samples, which shall remain in contact with the platens of the machine. These samples are mostly tested for 7 days and 28 days compressive strength. These samples are immediately tested after taking out from the water and while these are wet. The axis of each sample is carefully aligned with the line of thrust of the spherically seated platen. The packing should not be used between the faces of the samples and the steel platens of the testing machine. The spherically seated block is brought to bear on the sample. In order to have uniform seating, the movable portion is rotated gently by hand. The thrust is gradually applied as approximate rate of 14 N/mm^2 per minute without shock. The thrust is continuously increased until the resistance of specimen to the increased load breaks down and no increased load may be sustained. The maximum load applied for the failure of the sample is noted. The appearance of the concrete and any unusual features in the type of failure are also recorded.

The cube samples are tested on the sides, that is, at right angles to the direction of placement (not to the top and bottom). In case of cube samples, it is seen that the tangential stresses are set up and there it leads to higher compressive strength. The cubes also have higher compressive strength as its height and side ratio is *unity* only. The advantage of cube samples test is that its shape resembles the shape of the structural members. The cube samples do not need capping.

The cylindrical samples are tested in the same direction in which these are moulded. The samples are crushed in the direction of placement. The central zone of cylindrical samples remain free from tangential force. These samples are subjected to uniform loading. These samples are less affected by the end restraints. The restraining effect of the platens of the testing machine extends over the entire height of the cube. A part of a test cylinder is left unaffected because of the greater height. The casting and loading of the structural members are identical to that of the cylindrical sample. These samples simulates the actual condition of the structural member in the field in respect of direction of load. The use of cylindrical samples is becoming more popular, particularly in the research laboratory.

Because of the restraint conditions, the strength of the cube prepared from the same concrete shall be different from the strength of the cylinder. Normally the strength of cylinder is taken as 0.8 times the strength of the cubes.

However, the experiments have shown that there is no unique relation between the strength of these two samples.

The maximum load applied to the sample upto failure during the test divided by the cross-sectional area (determined from the mean dimensions of the section) gives the compressive strength of the concrete nearest N/mm². Three samples are tested as the representative of the batch (provided the individual variation is not more than 15 percent of the average). Average compressive strength is then found. Otherwise, the tests are repeated by taking other samples.

30.11 FLEXURAL STRENGTH OF CONCRETE

The flexural strength of concrete is the resistance of concrete to tension. In the fields of design of concrete roads, design of air field pavements and the design of concrete pipes, the tensile stresses are important. The tensile stresses develop in concrete due to drying shrinkage, rusting of steel reinforcement, temperature gradients and many other reasons. The flexural tensile strength of concrete is about *one-eighth* to *one-tenth* of the flexural compressive strength. The shape and texture of the aggregate have more influence on the flexural tensile strength than on the compressive strength. The angular aggregates give higher strength values than the rounded aggregates.

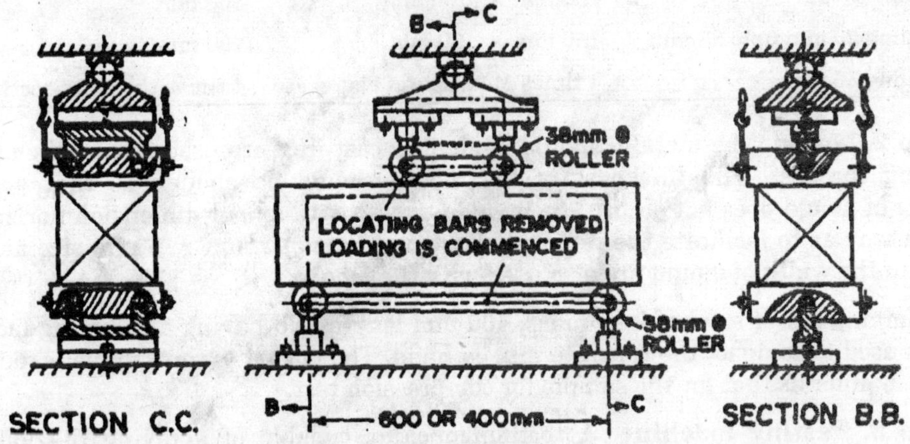

ARRANGEMENT FOR LOADING OF FLEXURAL TEST SPECIMEN AS PER I.S.

Fig. 30.7

The concrete slab roads are to resist the tensile stresses due to wheel loads and volume change in the concrete. Where there is insufficient subgrade support, the wheel loads develop high tensile stresses due to bending. The changes in temperature and moisture cause volumetric changes and as a result of this, the tensile stresses are produced due to warping and due to the movement of the slab along the subgrade and these stresses may be high. The restraint and temperature warping may cause longitudinal tensile stresses in the bottom of the pavement to the extent of 2.5 N/mm² at certain periods of the year and in transverse direction of the extent of 0.9 N/mm². All these stresses are additive to those developed due to wheel loads.

It is difficult to perform a test for direct tension. Therefore, the flexural strength test is conducted. The flexural strength property, that is, the extreme fibre stress calculated at failure (viz., *modulus of rupture*) is measured. The modulus of rupture depends on the cross-sectional size of the beam and the manner of the loading. The central point loading and the third point loading are the two different systems of loadings. In the central point loading of a simply supported beam, the load is applied at the centre and the maximum bend-moment also occurs at the centre. In case of third-point loading two loads are applied symmetrically at *one-third* span from the ends. The failure of beam may occur or a crack may develop at any section, which one is not strong to resist the stress with the middle third span in which portion, the bending moment is maximum. From the test, it may be seen that two point loading will give smaller value of the modulus or rupture than the central point loading. IS : 516-1959 specifies two point loading.

The procedure for conducting the flexural strength test is as follows.

Step 1. Test samples. The rectangular beams with square cross-sections are used as test samples for conducting the flexural strength. The standard size of sample :

Max. size of aggregate	Cross-sectional size	Length of beam
(i) Less than 20 mm	100 mm × 100 mm	500 mm
(ii) Exceeding 20 mm upto 38 mm	150 mm × 150 mm	700 mm
(ii) Exceeding 38 mm	4 times Max. size of aggregate	4 times the cross-sectional size

Step 2. Mould. The metallic mould of steel or cast-iron are used to cast the sample for conducting the test. The thickness of the mould should be enough so that warping or spreading of mould does not occur. The mould is made with longer dimension horizontal and in such a way as to facilitate the removal of the test for to 40 mm × 6 mm size and 50 mm wider than the width of mould are also used with the mould.

A tamping bar of steel of 2 kg mass, 400 mm length and having a ramming face 25 mm square is used to compact the concrete mix by hand. The curing of the sample is done in the identical manner as that for the sample for compression test.

Step 3. Testing machine. A testing machine capable of applying the load at the specified rate and of sufficiently capacity is used to test the sample of the beam cast. Where a high degree of accuracy is needed, the applied load should be accurate upto ± 0.5 percent. Two rollers 38 mm diameter are provided with the bed of testing machine. The beam sample is supported on the rollers as shown in Fig. 30.7. The centre to centre distance between the rollers is kept 600 mm for 150 mm sample and 400 mm for 100 mm sample. Two loads are applied through two similar rollers placed at 200 mm for 700 long beam sample and 133 mm for 500 mm long beam sample. The axes of sample and loading device are aligned.

Step 4. Testing of sample. The pacing are not used between the sample and the rollers. The rollers are cleaned before placing the sample. The loads are applied without shock and increasing continuously at a rate such that the extreme fibre stress increases at approximately 0.7 N/mm^2 per minute, that is at the rate of 4 kN per minute for 150 mm sample and at the rate of 1.8 kN per minute for 100 mm samples. The loads are increased until the failure of sample tąkes places. The maximum loads applied corresponding to the failure are noted. The appearance of the fractured faces of concrete and any unusual features in the type of failure are also recorded.

Step 5. Calculation of flexural strength. The flexural strength of the sample is expressed as the *modulus of rupture.* f_b. In case a equals the distance between the line of fracture and the nearer support, measured on the tensile side of the sample in mm, it is calculated nearest of 0.05 N/mm^2 as under

$$f_b = \left(\frac{P \cdot l}{b \cdot d^2}\right) \tag{30.8}$$

In case, a is less than 200 mm but more than 170 mm for 150 mm sample or less than 133 mm but more than 110 mm for 100 mm sample, the modulus or rupture is determine from

$$f_b = \left(\frac{3 \cdot P \cdot a}{b \cdot d^2}\right), \qquad \textit{where,} \tag{30.9}$$

b = measured width of the sample.

d = measured depth of the sample at the point of fracture

l = length in mm of the span on which the sample is supported.

P = maximum load applied to the sample in Newton.

In case a is less than 170 mm for a 150 mm sample or less than 110 mm for a 100 mm sample, the results of the test are discarded.

31

NON-DESTRUCTIVE TESTS EVALUATION (NDT/NDE) [NON-MECHANICAL TEST METHODS]

31.1 INTRODUCTION

Concrete (made either under controlled factory conditions or under site conditions) is a mixture of cement, fine aggregate, coarse aggregate, water and admixtures. These constituent materials have inherent variability in their characteristics. Depending upon the practices and level of workmanship employed in batching, mixing, transportation, placing, compaction and curing of concrete, further variability is induced in the final product (i.e., the concrete used in the structure). The properties of concrete are also dependent on the complete hydro-thermal (moisture - heat) history during the service life. Keeping in view of the variability of the various constituents, workmanship and atmosphere, it is essentially necessary to keep knowledge of the quality, strength and other performance characteristics of concrete. At the same time, it is almost impossible to have a one-to-one correspondence with the inputs. **Some kind of testing and evaluation** of the concrete becomes necessary to ensure that the concrete is safe and in acceptable state.

Compressive strength of concrete is a pre-dominant property among its various properties. The compressive strength of the concrete gives the nearest estimation to load carrying capacity of the structure. At the same time, many other properties of concrete are considered to the directly or indirectly related to its compressive strength.

At the same time, the comparison of the strength of concrete at the site gives a valuable tool for quality control by comparing it with the targeted compressive strength that should be achieved given the quality and quantity of ingredients, as estimated by the trial mixes.

In view of the unavoidable variability in the properties of concrete, some measures of strength is looked for, especially when the control cube specimens give expected results. Such variability is seen between different construction, different components of one structure and different portions of the same structure. It is not practicable to conduct full-scale testing of all the elements. Once a structure has been constructed, load-testing for its strength or load-carrying capacity, without damaging the structure is not always possible.

Non-destructive evaluation (NDE) methods provide a thought about inherent performance characteristics of the structures offer an attractive proposition. It is common in concrete construction throughout the world including our country.

A condition evaluable to find the pressure conditions and nature and causes of observed distress must be conducted earlier than attending repairs to the concrete. Otherwise, the absence of this information may lead to a high risk of failure of the repair.

The extent of the condition evaluation depends largely on the objective of the repairs. The factors which influence the repair objectives include safety and structural integrity the desired service-life extension change in intended use or loading requirements, serviceability, aesthetics and cost. ASCE and ACI (internationally recognized organisations, established in 1990, 1993) have published guidelines for the condition evaluation of structures.

In 1987, the Strategic Highway Research Program (SHRP) was established by Congress to improve the performation and durability of the nations highway system [Transport Research Board, 1986]. As part of this effort, a number of non-destructive tests/evation (NDT/NDE) tools were developed which are currently being implemented by the states and local government agencies, industry and private sectors. Some of methods developed by SHRP include alkali-silica reactivity (ASR) test methods and a rapid method for finding chloride content and rate of corrosion of steel reinforcement in hardened concrete. These alongwith other existing and emerging NDE techniques have been described in the subsequent sections.

31.2. NON-DESTRUCTIVE TESTS/EVALUATION (NDT/NDV) [NON MECHANICAL TESTS METHODS]

Mechanical methods of testing are in common use. However, there are, in addition , various testing and inspection methods of **non-mechanical** nature which are used use of x-rays and gamma rays, electron differaction and electron microscope, ultra-sonic insepction techniques, magnetic crack detection, flourescent inspection, and certain other inspection methods depending upon the magnetic or electrical properties. These methods are of recent origin, and play an increasing important part in the field of testing and studying the structures.

Non destructive tests are defined as indirect methods by which the property or characteristics of primary interest is concluded by measuring other properties or characteristic of concrete. These methods are commonly referred as non-destructive test/non-destructive evaluation (NDT/NDE) methods. These methods do not damage the function of the structure and allow retesting at the same location to evaluate changes in the properties with time. NDT methods for concrete progressed at slower pace as compared with steel. Concrete is highly heterogenous on a macroscopic scale. It is electrically non-conductive but commonly contains significant quantity of steel reinforcement. Concrete s commonly used in thick members. In addition, there remained little interest in the traditional NDT community (physicists, electrical engineers and mechanical engineers) to develop test methods for concrete.

Definition : of non-destructive tests

There is no standard definition for non-destructive tests as used for the concrete. For some persons, these are tests, which do not change the concrete. Whereas, for others, these are simply tests which do not damage the function of a structure, in which case, the drilling of cores is treated to be a NDT. For still others, these are tests, which do less damage to the structure than does the drilling of cores. This chapter deals with methods which either do not change the concrete or which result only in superficial local damage.

Non-destructive tests are defined as indirect methods by which the property or characteristics of primary interest is inferred by measuring other properties or characteristics.

These tests may also be defined as tests or methods that yield information regarding the quality of a structure or portion, that does not impair the serviceability of the element or structure.

Fundamentals of science and the inherent heterogeneity of concrete are the broad principles of application. More than the heterogenety, the larger problem is the fact that through NDE techniques, most of time, some other properties of concrete are measured. To know the strength of concrete, certain techniques of NDE related to the modulus of elasticity, density, moisture content and surface hardness of concrete and existence and absence of cracks, flaws etc are studied. The compressive strength of concrete is estimated from such measurements though calibration curves, which may never be very exact.

Besides limitations, the **Non-destructive evaluation** (**NDE**) techniques are increasingly chosen for the following purposes :

1. To estimate the compressive strength of concrete
2. To establish the uniformity and homogeneity of concrete
3. To establish the quality of concrete in to standard requirements
4. To access the quality of one component of concrete in relation with another.
5. To detect the existence of cracks, voids and other imperfections.
6. To monitor changes in the structure of the concrete which occur with time

31.3 COMPREHENSIVE (BROAD) CLASSIFICATION OF NDT TECHNIQUES

Various methods of NDT techniques may classified in two following different groups.

1. These techniques of which, the main purpose is to estimate **compressive strength**
2. Those techniques of which the main purpose is to evaluate conditions other than the **compressive strength** (i.e., to evaluate the integrity

The most reliable tests to estimate the compressive strength are those which result in superficial local damage. Nicholas J. Carino (author of chapter, 19 `Non-destructive Test Methods (Research structural engineer in he structure Division at the National Institute of Standards and Technology and Gaithersburg MD (expert in non-destructive test methods for concrete [Concrete construction Engineering Handbook, Editor-in-chief, Edward G. Nawy, CRC-Press, LLC Publication] prefer. The term **In-place tests** for this category.

The purpose of this chapter is to provide an introduction to commonly used NDT-methods for concrete. The principles underlying the various methods have been emphasized so that the students may understand their advantages and inherent limitations.

ACI 228.1 R (1995), Malhotra and Carino 1991, and Bungey (1989) are the additional references for more information on the application of these methods may be referred.

Various test methods (**In-place tests** to estimate the **compressive strength** are as under :

1. Rebound hammer	2. Ultra-sonic pulse velocity
3. Probe penetration	4. Pull out
5. Break off	6. Maturity method.

Various **non-destructive tests for the integrity,** are as below :

1. Visual Inspection	2. Stress wave propagation methods
3. Ground penetrating radar	4. Electrical/magnetic methods
5. Nuclear methods	6. Infrared thermography.

In case the principles of mechanism are looked into them, the classification may be of the following types.

1. Absorption and recovery of mechanical energy: (Test based on the rebound of a tool): Standard impact, velocity and attenuation of elastic waves set-up by a single or repeated broad spectrum impact or bursts of sonic or ultrasonic excitation, frequency and damping characteristics under free or forced vibration and other similar tests.

2. Local distress/fracture: In dentation, pullout of anchored inserts, driving in penetration devices and other similar tests, which result in local failure under controlled conditions are grouped in this category.

3. Electrical resistivity/Potential: The electrical resistivity, electrical self-potentials and their spatial and temporal variations may reflect a variety of phenomena including setting and hardening process of cement, possibility, extent and progress of corrosion in embedded reinforcement or metallic inserts and other physico-chemical interactions occurring in concrete.

4. **Magnetic/electro-magnetic effects**: Methods based on magnetic attraction, electro-magnetic induction, eddy current influences, etc. have been developed for the existence, size, location and depth of reinforcement and other metallic components embedded in concrete.

5. Acoustic emission: Many materials including concrete emit characteristic noise under stress, which may be detected, amplified, monitored and recorded. Any abnormal change in intensity or other parameters of the emission pattern may indicate significance changes in the state of stress.

6. Radiography: Transmission, absorption and back-scatter techniques using X-rays, gamma rays and neutrons sources may allow detection and mapping of density changes, reinforcement, disposition, inclusion, flaws, cracks, and moisture status in concrete.

Another classification may be on the basis of **fundamental properties** measured. Accordingly, the non-destructive techniques have been divided into following two groups.

Group 1 : It consists of those methods which measure some property of concrete from which an estimate of its strength, its durability, and its elastic parameters are found.

Some such properties of concrete are as below :

1. Its hardness
2. Its resistance to penetration, and
3. Its ability to allow **ultra-sonic** pulses to propagate through it.

Group 2 : It consists of those methods which may find the following :

1. Position and size of reinforcement	2. Areas of poor consolidation,
3. Voids	4. Cracks
5. Honey-cobing	6. Moisture content of in-place concrete

In the above two groups, a third group may be added.

Group 3 : It measures characteristics having effects on the durability of concrete. That is,

1. The permeability of concrete, which allows diffusion of gases or liquids containing harmful chemical ions,
2. Detection of existence of chemicals
3. Cement content, and
4. Depth of carbonation, etc.

31.4 HISTORICAL BACKGROUND (METHODS TO ESTIMATE IN-PLACE STRENGTH)

Adaptations of the Brinell hardness test for metals was first method to evaluate the in-place strength of concrete. In this method, a high-strength steel-ball is pushed into the test-piece under a given force and measures the area of the indentation (i.e., the depth of penetration caused by a specific indent or under a specific load). Therefore, this test is also referred as **indentation test.**

This type of test was modified for concrete structures. In 1934, Professor K. Gaede (Hanover, Germany) reported on the use of a spring-driven impact or to supply the force to drive a steel ball into the concrete [Malhotra, 1976].

A non-linear, empirical relationship was found between the cube strength and indentation diameter.

In 1936, J.P. Williams (England) report on a spring-loaded, pistol-shaped device, in which a 4 mm ball was attached to a plunger [Malhotra, 1976]. The spring was compressed by turning a screw, a trigger released the compressed spring, and the plunger was propelled toward the concrete. The diameter of the indentation produced by the ball was measured with a magnifying scale.

D.G. Skratajeb, of the central Institute of Industrial Building Research, Moscow published a landmark paper in 1938. 14 different techniques were summarised. Out of these, 10 were developed in the sovieth union, for measuring the in-place strength of concrete. It is

suggested that this paper may be studied by every student of non-destructive testing for its historical content. Skramtajev classified the test into two categories.

1. Those which needed installation of test hardware prior to placement of concrete, and

2. those which did not need preinstallation of hardware.

Following methods were included which were described by Skramtajev

1. Molds placed in the structure to form in-place test-specimens;

2. Pull out tests of embedded bars;

3. An in-place punching shear test;

4. An in-place fracture test using a pincer device

5. Penetration with a chisel driven by a hammer blows;

6. Guns that fires indentors into the concrete; and

7. Penetration with a ball powdered by a spring driven apparatus.

The variations of methods were suggested over one-half century ago.

The need for in-place testing was also commented by Skramtajev noted the following in 1938.

1. The curing conditions of standard test specimen are not representative of the concrete in the structure.

2. The number of standard test specimens is insufficient to assure the sufficiently of all members in a structure.

3. Standard test specimens which are tested at an age of one month do not provide information on the later-age strength of the concrete in the structure.

4. Surface tests do not provide an indication of the actual concrete strength owing, to the influences of carbonation, laitance, and moisture conditions.

5. Methods needing preplacement of hardware tend to provide more precise estimates of strength than these that do not need preplacement of hardware, but these lack flexibility for use at any desired location in an existing structure.

Same arguments are quoted even 50 later in relation to in-place testing. [ACI 228.1 R, 1995].

Above mentioned in-place tests and non-destructive tests have been described in the subsequent articles.

31.5 VISUAL INSPECTION

A condition evaluation of a concrete structure generally mean a visual examination of the structure with select coring of various concrete elements for compressive strength testing, perhaps a petrographic analysis, and sounding by chaining. Visual inspection is an integral part of a condition evaluation and, infact, is the most widely used of NDE.

The forms of distress and deterioration that is recorded in the field assessment and some of the associated causes cracking [possible causes : (Plaster shrinkage, drying shrinkage restraine, subgrade support deficiencies vapour barriers, expansion, corrosion of reinforcing steel/prestressing steel or other embedded metal components, thermal loading, vehicular impact, over loading, aggregate reaction)] scaling [possible causes : (inadequate air content, finishing problems, freezing-thawing cycling, chemical deicers)], spalling [possible causes : (aggregate reaction, corrosion, freezing-thawing cycling, construction (poor preparation of construction joints, early age loading)], disintegration [possible causes : (forzen concrete, freezing thawing cycling, low strength, chemical attack, sulphate attack)], honey combing and surface voids [possible causes : (poor placement, poor consolidation, congested reinforcement], discoloration and staining [possible causes : (different cement production, different water-cement ratios, corrosion, aggregate, use of calcium chlorides, curing, finishing non-uniform absorption of forms)] efflorescence [Possible cause : (Calcium carbonate and other mineral deposits)].

In some cases, the assessment may be conducted with a simple visual inspection.

However, in more cases, than not, some form of non-destructive testing (NDT) is needed to find the cause(s) of observed deterioration, assess its implication and the extent to which it influences structural integrity, and develop appropriate repair and maintenance strategies to meet the budget.

31.6 SONIC TESTS

Sound waves or waves of similar nature are involved in these non-destructive tests of concrete. These tests have important position. It is mainly because of two reasons. The whole mass of concrete is involved in the measurement.

Definition : Term **sonic** indicates that the waves remain within the normal audible frequency (Generally 30 to 17000 Hz).

The waves are used essentially in two modes of testing

1. To cause vibration

2. To pass a pulse of waves through the concrete and measure its velocity

In either case, fundamental relationships between resonance frequency velocity of pulse through the concrete and its material properties (e.g., density, modulus of elasticity and poisson's ratio) are derived from basic principles of physical sciences.

From the description, it will be apparent that these tests are not directly related to the compressive strength of concrete and estimation of concrete strength from **sonic test** could, at best, be approximate.

Such tests are, however, very useful in detection of overall quality and uniformity of concrete comparison of concrete in different portions of a structure, monitoring differences of concrete with time or in aggressive environments, cracks, voids and discontinuties in concrete structures etc.

In the sonic test and the ultrasonic tests, the principles of the test and equipment are essentially similar, the resonance frequency of transducers used in ultrasonic testing are considerably higher than those used in sonic testing.

31.7 SOME TERMS (ELECTRIC-GRATING AND ELECTRON DIFFRACTIONS) USED IN ULTRA SONIC METHOD

The wave lengths of X-rays are shorter than light waves. Therefore, the X-rays are invisible. X-rays, however, possess some of the properties of visible light. The light waves (or rays of light) may be reflected, refracted or diffracted. All are familiar with **reflection** and **refraction.** Breaking up of light ways into its seven separate colours of the spectrum by means of a prism is known as **disperson** of light waves. Another method is to rule fine parallel lines on glass to form a grating.

Grating : may be defined as frame-work of parallel beams. It is either transparent or reflecting. The distance between each pair of lines is of the same order of magnitude as that of the wave length (wave lengths of X-rays) to measured.

In 1912 Von-laue conceived the idea of using crystals, in which the molecules or atoms were regularly spaced. Von-Laue tried the experiment. A beam of X-rays were passed through a thin crystal plate. The rays were allowed to diffract (break up of beam of light into series of dark and light bands) and to strike a photographic plate. Fig. 31.1 shows X-rays beam with fine parallel lines on glass forming a **grating.** These parallel lines reflect as a single beam.

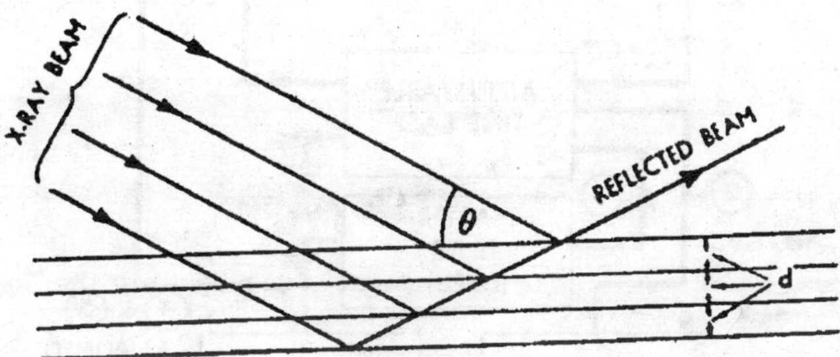

Fig. 31.1. Diffraction (Bragg Law)

There is this difference, however, the reflected rays from the different layers interfere with each other and a fuzzy, (indistinct) photograph is this obtained unless the relationship.

$$2d \sin \phi = n - \lambda, \tag{31.1}$$

is fulfilled, d is the distance between two successive layers of atomic planes, ϕ is the angle (the glanding angle) which the x-ray beam makes with the surface of the alloy under examination λ is the wave length of the x-ray and n is integer (as a whole number). This relationship is known as the Bragg law.

In 1925, it was suggested that the moving particles such as electrons should possess some of the qualities of wave-motion. In 1927, it was shown that the beams of electron could be diffracted by a crystal. This property was put to practical purpose. A beam of electrons is directed at an extremely thin specimen behind which a photographic plate is placed (transmission method) or the beam is directed so that it makes a small angle with the surface under examination and the photographic plate is placed to receive the electrons that bounce-off (reflection method).

The diffraction of X-rays is a consequence of scattering the X-ray beam by electrons belonging to atoms in the path of the beam.

Electron diffraction, however, takes place because the lattice structure in a crystalline substance has a periodic potential variation associated with it, thus forming an electric grating capacity of diffracting electron beams.

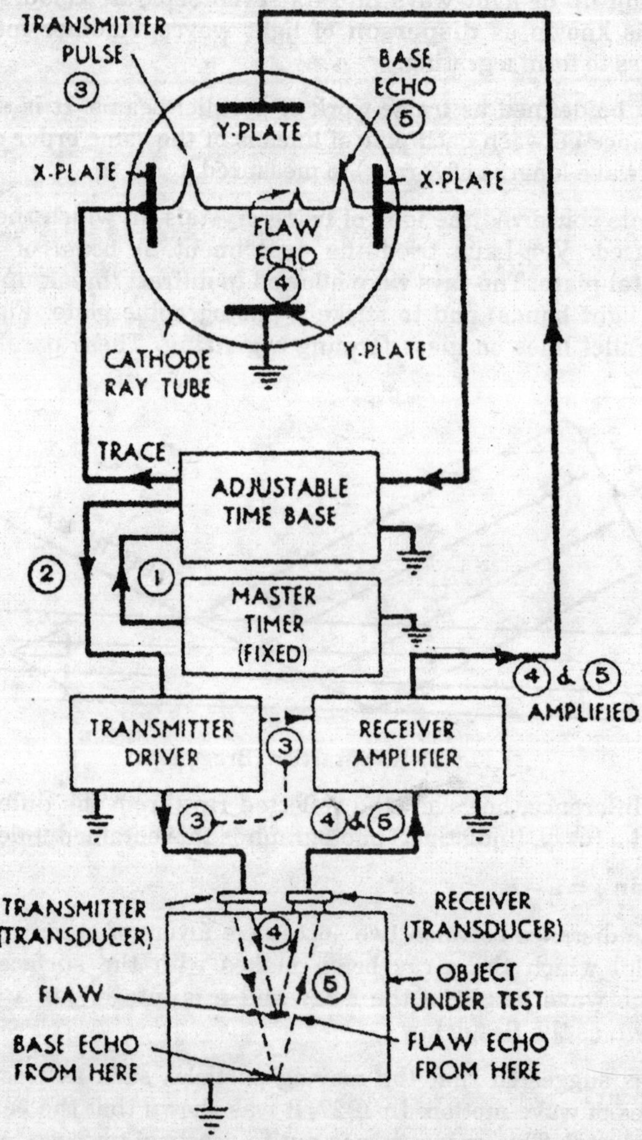

Fig. 31.2. Principal components of an ultra-sonic flaw detecter

There is difference, which should be noticed. The WAVE, length attributed to electron beams is shorter than that of the X-rays used for diffraction work. The wave lengths of such X-rays (produced at about 25000 volts to 75,000 volts) range from 0.50 A to 0.17 A (where, 1

Arigstrom unit = 10^{-7} mm = 1A) where as, corresponding wave lengths of the electron beams are of the order of 0.05 A. One consequence of this short wave is that θ may be used instead of sin θ in the Bragg formula quoted above.

The work must be carried out in vacuum because the range of an electron beam in air is short. The high stopping power of matter for the electrons also means that the specimens for examination by transmission must be very thin indeed, of the order of one-millionth of an inch.

Electron diffraction renders possible the examination of thin surface layers by means of a reflection of grazing incidence technique; that is, a beam of electrons directed so as just to graze, the surface. Such examination is not possible with the X-ray diffraction technique, because, as has been mentioned earlier, X-rays penetrate the matter beyond these thin surface layers.

31.8 ULTRASONIC (SUPERSONIC) PULSE VELOCITY (UPV) TEST

The range of sound frequencies known as **ultrasonic (or supersonic**) waves has been puts to practical use in recent year in widely varying applications. One such application is the detection of concealed flaws in objects by directing a concentrated beam of the vibrations against the test-piece and observing any reflections returned by the flaws.

The ultrasonic pulse velocity is used for the last five decades. This technique is truly non-destructive. It results in no damage. Soniscope was used by Leslie and Cheesman during World-war second in Canada.

The ultrasonic pulse velocity method is a stress wave propagation method, that involved measuring the travel time, over a known path length, of a pulse of ultrasonic compressional waves (these waves are associated with normal stress). The pulses are introduced into the concrete by a piezo-electric acostic transducer, and a similar transducer acts as a receiver to monitor the longitudinal vibration caused by the arrival of the pulse. Three different kinds of waves (longitudinal, shear (transverse), and surface pay length-waves) are generated. In general, the transducers are designed to generate longitudinal pulse. The velocity of longitudinal waves are more and their exact time of anvil may be picked up easily. The speed of compressional waves in a solid is related to the elastic constants (modulus of elasticity Poisson's ratio) and the density. By conducting tests at various points on a structure, lower quality concrete may be identified by its lower pulse velocity.

Definition : The term **ultrasonic** (above sound) arises from the fact that these waves or vibrations are much too high pitched to be audible to the human ear; or in other words, their frequencies are higher and outside the audio range to which our ears respond.

These are, nonetheless, a form of mechanical energy (vibrations of the air) and not to be confused with waves or rays of electro magnetic origin (radio, light, radiant heat, X-ray etc).

As the pitch or frequency of a sound is raised, so its wave-length diminishes. Thus, the wave-length (in air) of A above middle C is approximately 0.762 m, while the A at the top of the piano key-board (three octaves higher), has a wavelength of only about 95.25 mm. Now in case, one is to use a beam of sound waves to locate hidden flaws, it must be a narrow one, and to obtain narrow sound beams, the sound generator must have a diameter several times greater than the wave-length of the sound it is to radiate.

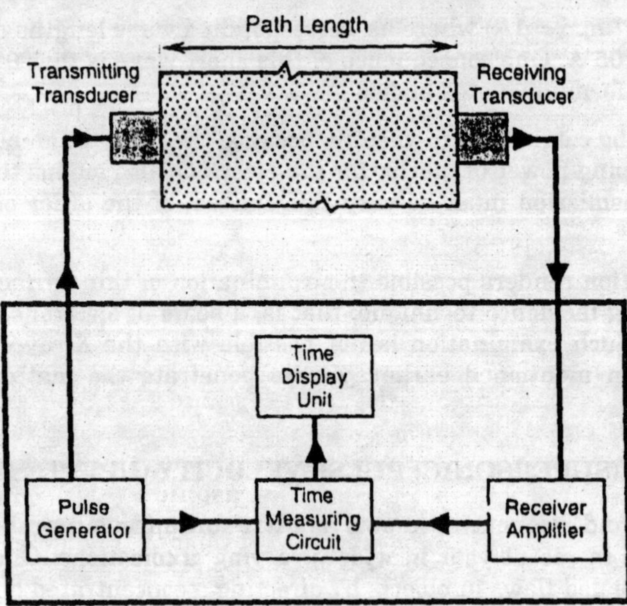

Fig. 31.3. Schematic of ultrasonic pulse velocity method.

For ordinary sound frequencies, such a device would be too large to be practicable. Further, the apparatus would be insensitive to small flaws.

Therefore, ultra-sonic waves (waves of very high-frequencies) are used. There are transmitted into the object under test by means of a transducer (a device for converting electrical energy into mechanical motion and vice-versa.

The ultrasonic waves reflected b flaws and other features of the object are also picked up by a transducer and after amplification are caused to influence some detection device, such as a cathode-ray tube, so that visible signals related to the position and magnitude of the flaws, .. .etc may be revealed.

In Fig. 31.2, the essential features of an ultrasonic flaw-detection apparatus are shown. The cathode ray tube is a familiar domestic object and now a days needs no introduction. It is only necessary that the interval beam shall rule on the screen a horizontal straight line as a base upon which echoes may be shown as blips or peaks. Any trace or line built-up on the screen is in reality the successive positions of a single light spot moving at high speed. The eye does not immediately lose the impression of any thing seen, there being some persistence on the ratna for about 1/20th second. In case, the trace, therefore, is repeated faster than about 20 times per second it will appear to the eye as a continuous line.

The master timer shown in the centre of the diagram sends out an electrical pulsed (marked 1) with a frequency ranging from every fiftieth to about every two-hundredth of a second, this timer is marked `fixed' because it is pre-set and needs no adjustment. In some sets, the mains frequencies is used for master-timing, so that the pulses are only every fiftieth of a second.

The pulse I travels to the component marked `adjustable time base' (adjustable to accommodate different object thickness) and ``triggers'' the time base, so that, the light spot is set in motion across the screen (the path of the spot, or trace, is shown diagramatically between the two X-plates). The time base itself, by the pulse marked 2, triggers the transmitter driver, which sends a very strong pulse 3 to the transmitting transducer, which is usually a quartz crystal or similar device. Pulse 3 is of quite a different character from 2 and 1. Instead of being a single short pulse of voltage, it is a short ``burst" of electrical vibrations which are converted into mechanical ultrasonic energy or waves by the transducer.

In case, the transducer (also known a probe) is coupled to the test object by thin film of oil or other liquid, these waves may be transmitted into the object in a reasonably narrow beam and as shown in the diagram, may be reflected.

Two reflections are shown 4 being from a flaw and 5 being the ``base echo". As 4 transverse a shorter path, it will arrive at the receiver before 5. At the receiver (also coupled to the object by means of a liquid) both echoes are turned back into electrical impulses which are fed into the receiver amplifier, where they are greatly magnified before being passed on the y-plate of the cathode-ray tube.

The events on the screh of the tube now by considered in slow motion. Every two-hundredth of a second or so, a spot of light starts its journey across from left to right and appears to the human eye as a straight line. When it begins to travel, the transmitter is set going and a very small fraction of its ultrasonic pulse 3 leaks, as shown by the dotted line, to the receiver and thence to the tube, so that the light spot is pulled upwards from its path to show the left hand blip 3. The light spot returns to its normal path until similarly deflected by the arrival of 4 and 5. It will be aparent that the nearer the flaw is the top surface, the nearer the blip 4 will be to the left-hand end of the trace.

Interpretation of the traces on the screen needs skill and experience, and reference samples or standards should always be available for checking observations. A large flaw may completely supress the base echo, for example, and severe porosity may prevent any transmission at all.

Frequencies, used in ultrasonic testing range from about half an megacycle (500,000 = 0.5 x 10^6 vibrations per second) to 20 megacycles per second. At the commonly used frequency of 2 M-c/s, the wave in steel (velocity 5050 metre per second) is approximately 2.54 mm, so the deflects of this order of size are easily found.

Whitehurst (1951) in his summary of industry's experience in the United States published following tentative classification for using pulse velocity as an indicator of quality.

Pulse velocity m/sec	condition
1. Above 4570	excellent
2. 3660 to 4570	generally good
3. 3050 to 3660	questionable
4. 2130 to 3050	generally poor
5. Below 2130	very poor

This table was quoted in many subsequent publications. Whitehurst reported that these values were established on the basis of tests of normal concrete having a density of about 2400 kg/m³.

There are following three ways (Direct, semi-direct and indirect transmissions) of measuring pulse velocity through concrete.

1. **Direct transmission** : In this, the transducers are held on opposite faces of the concrete In this, the path is well defined.

2. **Semi direct transmission** : When a test path extends across a corner of a large mass of concrete, then, the concrete surface needs this method.

3. **Surface transmission** : When only one face of the concrete surface is assessable (e.g., concrete roads, water tanks). This method is used to find dynamic modulus of elasticity). It is least satisfactory. The maximum energy of the pulse is directed into the concrete pulse velocity measurement indicates the quality of concrete only near the surface and not about deeper layer.

For dealing with very large components, such as aeroplane wing spars, the component is completely submerged in water tank. While the transducer automatically scans the object and any defects are recorded on a chart.

The transducers used in hand scanning may be adopted to exite different types of waves longitudinal, transverse, shear or surface waves, according to the nature and orientation of the flaws that are expected. With some apparatus, two separate probes are used; in other designs two transducers are contained in a single probe or one transducer is designed to perform the functions of both transmission reception.

Ultrasonic pulse velocity method is a relatively simple test to perform on set provided it is possible to gain access to both the sides of member. White tests may be performed with the transducers placed on the same surface, the results are not easy to interpret. Care must be exercised to assure that good and consistent coupling with the concrete surface is achieved. Other important factors besides concrete strength, that may be influenced the measured ultrasonic pulse velocity (UPV) and that should be considered in the report of ACI Committee [ACI 228.1 R, 1995]. These include

1. Moisture content - an increase in moisture content increases the pulse velocity
2. Presence of reinforcement oriented parallel to the pulse propagation direction - the pulse may propagate through the bars and result in an apparent pulse velocity that is higher tan that propogating through concrete; and
3. Presence of cracks and voids these may increase the length of travel path and result in long travel time

Because of these factors, the ultrasonic pulse velocity should be used for estimating concrete strength only by experienced individuals

Limitations and usefulness

1. This techniques provides an excellent means for establishing uniformity of concrete and helps in quality control operations.
2. This test may be carried out both in the laboratory and at site. Surface and internal cracks, both, are delineated.

3. Since large number of factors influence the relation between strength of concrete and its pulse velocity. Therefore, its use to predict flexural and compressive strength is not recommended.

31.9 SCHMIDT REBOUND HAMMER

A swiss engineer Erns Schmidt developed a device for testing the concrete based upon the rebound principle in 1948 [Malhotra, 1976, 1991]. the motivation for this new device came from tests developed to measure the hardness of metals (as was the case with earlier indentation test). This new device was an outcome of the Sclero-scope [In Greek, the word Sklero measn hard]. The **principle** of this test is based on measuring on rebound height of a diamond tipped hammer or mass, that is dropped from a fixed height above the test surface. Depending upon the impact energy, Schmidt's test hammers are categorised into four types (i.e., N, N. M and P - types). Type N test hammer is suitable for normal grade of concrete. Type L hammer hs smaller (low) impact energy. It is suitable for light weight concrete or small and impact sensitive part of concrete. Type M hammer has comparatively larger impact energy. Therefore, it is suitable for testing mass concrete Type P is used for cement mortar and cement plaster.

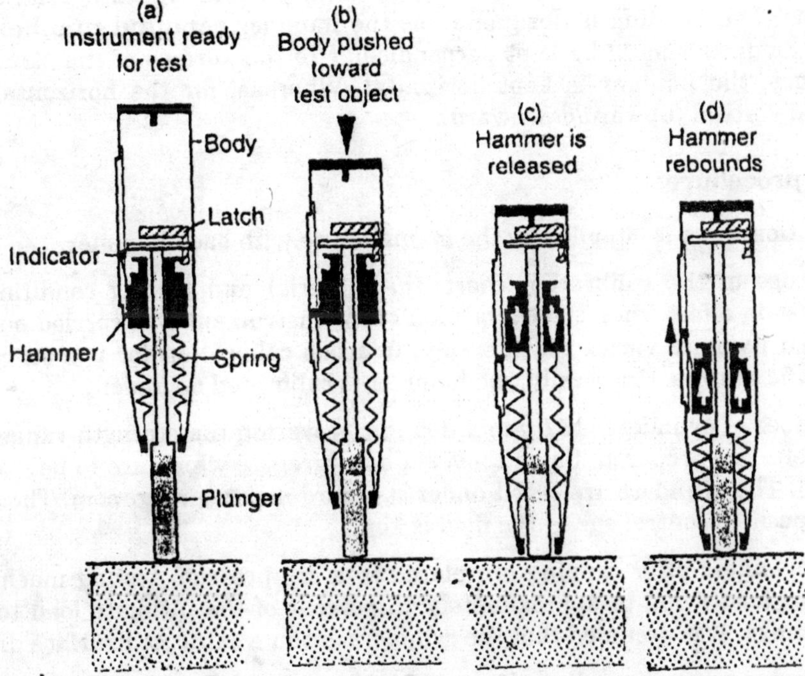

Fig. 31.4. Schematic cross of SCHMIDT rebound hammer showing principle of operation.

Kolek (1958) noted that, when concrete is struct by a hammer, the degree of rebound is an indicator of the hardness of concrete Schmidt standardized the hammer blow. A schematic cross-section of a rebound hammer is shown in Fig. 31.3.

The essential parts of the schmidt rebound hammer are (i) the outer body, (ii) the hammer (iii) the plunger, (iv) the spring and (v) the slide indicator.

Method of testing

The plunger is extended from the body of the instrument, which causes a latch mechanism to grab hold of the hammer (i.e., the plunger is locked) to perform the test, Fig. 31.3(A). The body of instrument is then pushed toward the concrete surface, which stretches are spring attached to the hammer and the body (Fig. 31.3(B)). When the body is pushed to the limit, the latch is released and the hammer is propelled toward the concrete by a combination of gravity and spring force, Fig. 31.3(C). The hammer strikes the sholder of the plunger and it rebounds, Fig. 31.3(D). The rebound distance (height) is measured on a scale by a slide indicator. The rebound distance is expressed as a rebound number, which is the percentage of the initial extension of the spring [Kolek, 1958]. Currently, different models of the instrument are available, which differ in mass of the hammer and the stiffness of the spring. Thus, different energies may be used for different materials.

The results of rebound give a measure of relative hardness and within limits, empirical correlations have been established between strength properties and the rebound number. Energy is lost in impact due to localised crushing of the concrete and the internal friction within the body of the concrete, which correlates strength property.

While the hammer is still in its testing position, the slide index is read to the nearest whole number. This reading is designated as the hammer **rebound number/or rebound index.** The hammer should be kept perpendicular to the surface of the structure. For the vertical surface, the laminar is kept horizontal. Whereas, for the horizontal surface, the hammer is old vertical (upward/downward).

Calibration procedure

A calibration chart is supplied by the manufacture with each hammer.

For the use of the calibration chart, the material and testing conditions should be similar to those in effect when the calibration of the instrument was carried out. The performation of each hammer varies considerably. It needs calibration for use on concrete made with the specific sources. Practically, the hammer is calibrated as under :

1. A number of cylinders (152 mm x 305 mm) covering the strength range to be met at the site are prepared. Same convert and aggregate which are to be used on site are used. The cylinders are cured under standard moist curing room. The curing period as specified control age in the field is kept.
2. The cylinders after capping are placed in a compression testing machine under an initial load equal to approximately 15 percent of the ultimate load to restrain the specimen. It is assumed that the cylinders are in a saturated surface dry condition.
3. 15 hammer rebound readings (5 on each of 3 vertical lines 120 degree apart) are made against the sde surface in the middle two thirds of each cylinder. Hitting the same spot twice is avoided. 5 readings on each of four (4) moulded faces without hitting the same spot twice is taken for cubes.
4. Average of the reading are found. It is called as **rebound number** for the cylinder (or cube) under test.
5. For all the cylinders (or cubes), this procedure is repeated.

6. The test cylinders (or cubes) are tested to failure in compression. The rebound numbers are the compressive strength in N/mm^2 are plotted on a graph.

7. A curve or a line is fitted by the method of least squares.

Limitation : There are serious limitations, although the rebound hammer provides a quick inexpensive means of checking uniformity of concrete. The results obtained by the Schmidt rebound hammer are influenced by the following :

1. The surface texture influences the rebound number, with smooth hard troweled surface giving higher values than rough-textured surface.

2. The presence of a surface layer of carbonation increases the rebound number

3. The surface and the internal moisture influences the rebound number. A dry surface results in a higher rebound number

4. The rebound number is influenced by the orient of the instrument in relation to the direction of gravity (Approximate correction factors are available.

5. Size, shape, rigidity of the specimen and its age influence the rebound number

6. Types of cement, coarse aggregate and the moulds influence the rebound number.

The rebound number method is recognized as a useful tool for performing quick surface to evaluate the uniformity of concrete. It is not generally recommended where accurate strength estimates are required, since, the method is not very accurate. The probable accuracy of strength of the concrete in a structure is ± 25 per cent.

31.10 COMBINED USE OF REBOUND HAMMER AND ULTRA-SONIC PULSE VELOCITY METHOD

Each method has its limitations for predicting the strength of strength of concrete. In view of this, both ultrasonic plus velocity (UPV) and rebound hammer methods are sometimes used in combination. The errors arising of the effects of material, mix and environmental parameters on the respective measurements may be lessened.

Relationships between UPV, rebound number and compressive strength are found by the multiple regression of measured values on laboratory test specimen. Combined use of two methods, give better accuracy to estimate the strength of concrete. However, there are limitations in this combined approach also.

Limitations

1. Co-relations between UPV, rebound number and compressive strength are valid only for the materials and mix proportions used in the trials.

2. The inherent difference between the laboratory test specimens and in-situ concrete (e.g., surface texture, moisture conditions, presence of reinforcement) also influence the accuracy of the test results.

3. The co-relation is valid only within the range of values of pulse velocity, rebound number and the compressive strength used and any extrapolation beyond this is

questionable. (e.g., for structures where the pulse velocity may be lower due to poorer workmanship or existence of cracks, voids etc).

Method : Combined use of UPV and rebound hammer methods are used for the evaluation of the quality and compressive strength of in-situ concrete. It is taken to be indicative of the whole mass only when the overall quality of concrete is examined by the UPV is good. When the quality evaluated is medium the estimation of compressive strength by rebound indices is extended to the whole mass only on the basis of other co-lateral measurement (e.g., strength of control cube specimens, cement content of hardened concrete by chemical analysis or concrete core testing). When the quality of concrete is poor, no evaluation of the strength of concrete is done from the rebound indices.

31.11 ELECTRICAL AND MAGNETIC METHODS FOR REINFORCEMENT

To assess the strength of reinforced concrete members, the information about the quantity, location, and condition of the embedded steel reinforcement and concrete cover. Following electrical and magnetic methods (devices) are used for this purpose.

Cover-metres. These are portable battery operated magnetic devices to locate reinforcing bars and to estimate the diameter and depth of concrete cover. Such cover-meters are called **Pacho-meters** or **Profo-meters**. These are portable and battery-operated devices.

These devices work on the principle of inter-actions between the steel bars (magnetic object) and the low-frequency electro-magnetic fields, change in the magnetic flux in the field of an electromagnet due to existence of steel reinforcement. The physical principle that is used is that of **electro magnetic induction** whereby an alternatic magnetic field induces an electrical potential in an electrical circuit intersected by the field. Commercial cover-meters may be divided into two-classes. Those based on

1. the principle of **magnetic reluctance** and
2. the principle of **eddy-currents**

Magnetic reluctance meters

A probe unit consists of a highly permeable U-shaped magnetic core on which two coils are mounted as shown in Fig. 31.4.

When an alternating current flows through an electrical coil, a magnetic field is created and there is a flow of magnetic flux between poles analogues to the flow of current in an electrical circuit [Fitzerald *et al.*, 1967]. The resistance to flow of magnetic flux is called **reluctance** which is analogus to the resistance to flow of current in an electric circuit.

The convertor is based on changes in the reluctance of a magnetic circuit caused by the presence or absence of a bar within the surrounding of the search head.

The seach-head consists of a ferro-magnetic U-shaped core, (yoke) an excitation coil, and a sensing coil. When an alternating current (less than 100 Hz) is applied to the excitation coil, an alternating magnetic field is created and magnetic flux flows between the poles of the yoke. In the absence of a bar, Fig. 31.4(A), the magnetic circuit, consisted of the yolk end the concrete between the ends of the yoke, has a high reluctance and the alternating magnetic flux flowing between the poles will be small. The alternating flux induces a small, secondary current in the sensing coil.

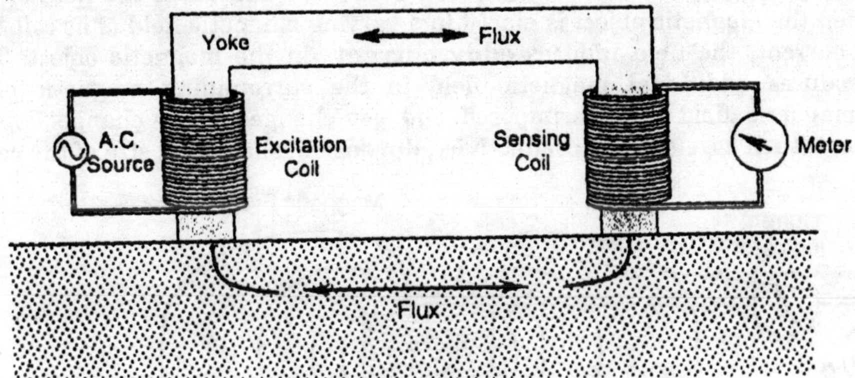

(a) Small current induced in sensing coil when no bar is present

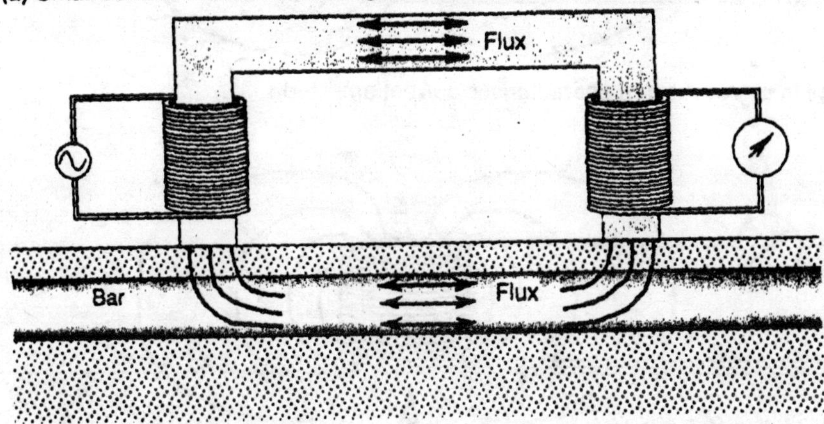

(b) Presence of bar increases flux and increases current in sensing coil

Fig. 31.5. Covermeter based on the principle of magnetic reluctance (adapted from Carino, 1992b).

In case, a ferro magnetic bar exists, Fig. 31.4(B), the reluctance decreases, the magnetic flux amplitude increases, and the current in the sensing-coil increases. Thus, the existance of steel bar is indicated by a change in the output from the sensing coil.

For a given reinforcing steel bar, the reluctance of the magnetic circuit depends strongly on the distance between the bar and the poles of the yoke.

An increase in concrete cover increases the reluctance and reduces the current in the sensing coil.

In case, the meter out put were plotted on a function of the cover a calibration of relationship may be established. That may be used to measure the cover. Since the size of the bar influences the reluctance of the magnetic circuit, there would be a separate relationship for each bar size.

Magnetic eddy-current meters Reinforcement bar scanner

The principle of this device is existence of steel bars influences the field of an electromagnet. When the magnetic object is placed in a varying magnetic field of as coil carrying an alternating current, the field induces **eddy current** in the magnetic object. These eddy currents produces additional magnetic field in the surrounding magnetic object. This additional magnetic field is superimposed and get changed. This change depends upon various factors (such as, electrical conductivity, dimension, shape and size of the coil.

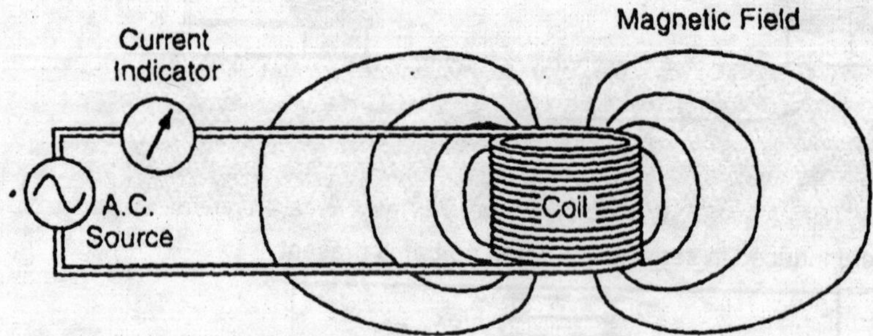

(a) Coil in air results in a characteristic current amplitude

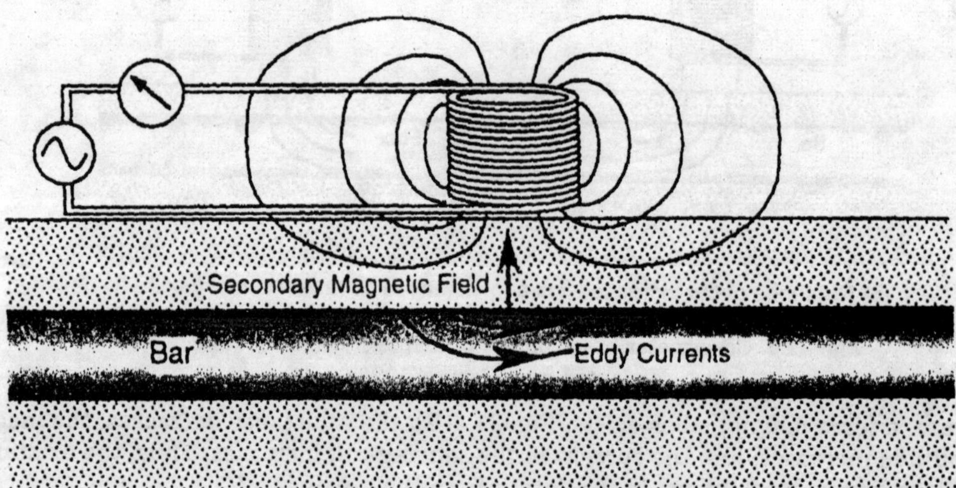

(b) Interaction with reinforcing bar causes changes in coil impedance and current amplitude

Fig. 31.6 Covermeter based on eddy current principle (adapted from Carino, 1992 b).

Any current flow gives rise to a magnetic field, reddy current produce a secondary magnetic field that interacts with the field of the coil. The second class of covermeters is based on monitoring the influences of the eddy currents induced in a reinforcing bar. Schematic diagram of a continuous eddy current covermeter is shown in Fig. 31.5.

In the absence of a reinforcing bar, the magnitude of the alternating current (usually at about 1000 Hz) in the coil depends on the coil impedance.

In case, the coil is brought near a reinforcing bar, alternating eddy currents are established within the surface of the bar. The eddy currents give rise to an alternating secondary magnetic field that induces a secondary current in the coil. In accordance with Lenz's law. [Serway, 198] the secondary currents opposes the primary current. As a the net current flowing through the coil, is reduced, and the apparent impedance of the coil increased [Hagemair, 1990]. Thus, the existence of the bar s inferred by monitoring the change in current flowing through the coil.

31.12 HALF CELL POTENTIAL METHOD

In order to assess the corrosion activity of steel reinforcement bars electrical methods are used. An understanding of the underlying principles of these electrical method is necessary to obtain meaningful results. Corrosion is an electro-chemical process involving the flow of changes (electrons and ions) Galvonic corrosion cells are formed in the concrete when corrosion of reinforcement takes place. At active sites on the steel bars, called **anodes,** iron atoms lose electrons and these electrons move into surrounding concrete as ferrous ions. This process is called **a half-cell oxidation reaction** or the **anodic-reaction.** This reaction may be represented as under :

$$Fe = (Fe^{2+} + 2e^{-})$$

The electrons remain in the steel bar and flow to sites, called **cathodes.** At cathodes, the electrons combine with water and oxygen which are present in the concrete. The reaction at the cathode is called **a half-cell reduction reaction.** It is represented as under

$$\left(2H_2O + O_2 + 4\,\bar{e} \rightarrow 4\,O\overline{H}\right)$$

The ferrous ions moves through the concrete to these cathodic sites to maintain electrical neutrality. At cathode sites, these ions combine with water and oxygen to form hydrated iron-oxide that is, rust. When the bar is corroding there is a flow of electrons through the bar and a flow of ions through the concrete. When the bar is not corroding, there is no flow of electrons and ions.

As the ferrous ions move into the surrounding concrete, the electrons which are left behind in the bar give the bar a negative charge. This negative charge is detected by **the half-cell potential method** and the corrosion activity is indicated.

The corrosion of steel reinforcement bar may be detected by measuring the **half-cell potential** (also called **corrosion potential**) of the reinforcement with reference to a reference electrode which is inserted in the galvanic corrosion. A schematic diagram of the apparatus is shown in Fig. 31.6. The apparatus includes a copper-copper sulphate half-cell connecting wires and a high impedance voltmeter. Instead of copper-copper sulphate, silver-silver chloride are commonly used. It is composed of a copper bar immersed n a saturated copper sulphate solution. The positive terminal of the voltmeter is attached to the reinforcement and negative terminal is attached to the copper-copper sulphate half-cell. A high impedance voltmeter is used so that with little current flows through the circuit. Half-cell makes electrical contact with the moist concrete by measn of a porous plug and a spong (i.e., moistened with a wetting solution, such as liquid detergent). The moist concrete functions as an electrolyte.

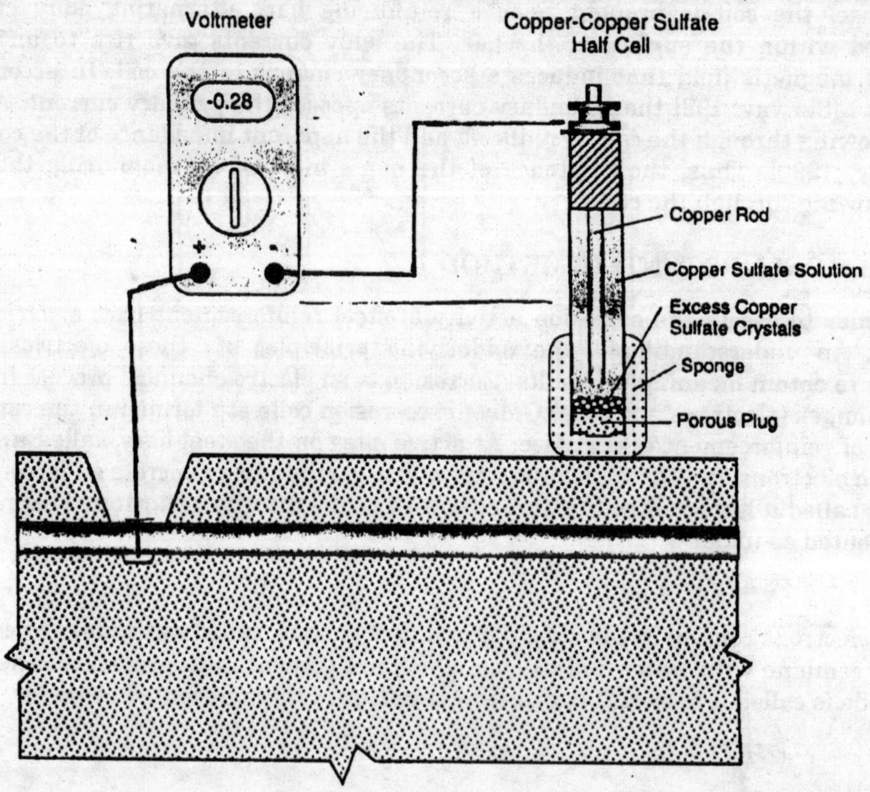

Fig. 31.7 Apparatus for half-cell potential method described in ASTM C876.

In case, the bar is corroding electron would tend to flow from the bar to the half-cell. At the half, the electrodes are consumed in a reduction reaction that transforms copper ions in the copper sulphate solution into copper atoms deposited on the rod. Because of the manner in which the terminals of the voltmeter are connected, the voltmeter would show a negative value. Higher negative voltage reading shows the higher possibility of the bar is corroding. It is an open-circuit potential because it is measured under the condition of no current in the measuring circuit [ASTM G 15]

Half-cell potential measurement is suitable in the field. It is applicable to concrete members in respective of their size or depth of concrete cover. In practice, the measurements are taken an predetermined grids over the surface of the concrete. The equipment is portable.

Plots of equipotential contours are constructed to locate areas of possible corrosion activity in a concrete member.

The detailed procedure of half-cell potential measurement is given in ASTM C 870. (Test method for Half-cell potential of Reinforcing steel Bars in concrete). Access to the reinforcement is to be provided. This method cannot be applied to concrete with expos-coated reinforcements. It does measure the **rate of corrosion.** It only shows the possibility is that occurring corrosion activity at the time measurement in done.

31.13. MATURITY METHOD (TEMPERATURE TIME FACTOR)

The compressive strength of well-cured concrete increases with time. The temperature of curing influences this increase. The product of time, and temperature is considered as **maturity,** with datum temperature +10C. Below this temperature, cement does not hydrate.

The **maturity method** is a technique which may be used for estimating the compressive strength developed during its curing period by measuring temperature of the concrete. It is monitored by thermo-couples of by the instrument `maturity meter'. Established co-relation graph between maternity and in-place compressive strength of concrete had been prepared.

History : Carino (1991) provided the a comprehensive review of the history of the method and some of its applications. Its origin is traced to a series of papers from England. These papers (McIntosh, 1949; Nurse, 1949; Saul, 1951) deal with accelerated curing methods. Nurse-saul maturing function may be used to find maturity index

$$MI = \sum_{o}^{t} (T - T_o) . \Delta t, \quad \text{where,} \tag{31.1}$$

MI = maturity index (degree calcium-hours (or degree Calcium-days)

T = average temperature of concrete (Degree calcium during time interval ΔT

To = datum temperature. 10°C

The index determined by using above expression is called as maturity index. However, the current terminology is the **temperature time factor.** (ASTMC 1074). Following principle was presented by Saul (1951) and it was called as maturity rule.

Concrete of the same mix at the same maturity (calculated in temperature-time) has approximately the same strength, whatever combination of temperature and time to go to make up that maturity. Eq. 31.1 is based on the assumption that rate of strength gain is a linear function of temperature.

It was soon realised that above approximation may not be valid. As a result, a series of alternatives to the Nurse-saul function were proposed by other researchers [Malhotra, 1971]. However, the proposed alternatives did not receive wide acceptance. Eq. 31.1 was used world-wide until improved function was proposed in 1970s.

A new function was proposed to calculate a maturity index in 1977, from the recorded temperature history of the concrete [Freiesleben Hansen and Pedersen, 1977]. The basis of this function was Arrhenium equation [Brown and Lemay, 1988]. It described the effect of temperature on the rate of chemical reaction.

The equivalent age of concrete was determined from new function

$$t_e = \sum_{o}^{t} e^y \tag{31.2A}$$

$$y = \left(-\frac{E}{R} \left(\frac{1}{T} - \frac{1}{T_r} \right) \right) . \Delta t \tag{31.2B}$$

where,

te = the equivalent age at the reference temperature

E = apparent activation energy, J/mol.

R = universal gas constant, 8.314 J/mol-K

T = average absolute temperature of concrete during time internal Δt, degrees kelvin

T_r = absolute reference temperature, degrees, kelvin

Actual age of the concrete is converted to its equivalent age by using Eq. 31.2 in terms of gain in strength at the reference temperature Eq. 31.2 considers non-linear relationship between the rate of strength development ad curing temperature. Eq.31.2 is superior to Eq. 31.1.

Prior placing of the maturity meters in the form work at the time of construction is used to monitor the strength of early age of concrete. The form-work may then, be removed confidently. These are the advantages of maturity concept.

31.14 PERMEABILITY TEST FOR CONCRETE

Significance of permeability in addition to compressive strength in evaluating the quality of concrete was realized looking to the increasing instances of corrosion in reinforced concrete. While considering the durability of concrete, it has been seen that the permeability is the most critical factor. It is necessary to monitor the permeability in the existing structure and so also during the process of new construction. Following cases of permeability of concrete have considered.

1. Permeability to air

In case the concrete is permeable to atmospheric air, then concrete will release its free water from the layers next to the surface. During evaporation, the pore water in the concrete is replaced by air. It allows the carbon di oxide (CO_2) from air to penetrate the concrete. The reaction between CO_2 and the alkali compounds of the concrete takes place. This process between CO_2 and the hydration products of the hardened paste is called **carboration**.

Holes are drilled in concrete as shown in Fig. 31.7 (which also shows the test apparatus). After evacuating the holes upon a certain level of mercury head, the time taken by the mercury head to rise by 60 mm (in general) is measured by the stop watch. This gives an idea of the rate at which air may flow through concrete into the hole. The permeability rate is given by the rise of mercury head divided by the time taken for the rise. The recovery speed is influenced by several factors (e.g., the length of nose, depth of the test-hole, presence of holes in the neighbourhood and the plugging). Relationships have been developed to correlate various parameters with the recovering speed of the manometer head. Presence of moisture, curing period and location of testing do influence the results.

Above described method was originally proposed by John Figg and lateron, it has been modified by other investigators.

2. Permeability to gases

Concrete storage vessels should be gas tight. To quantify this property, various methods of gas permeability technique have been suggested. However, the methods are restricted to laboratories only. Low-air pressure is applied to the surface of concrete. For the specific depth, the pressure will be increased with time and it is monitored. It will given an idea of the permeability to air.

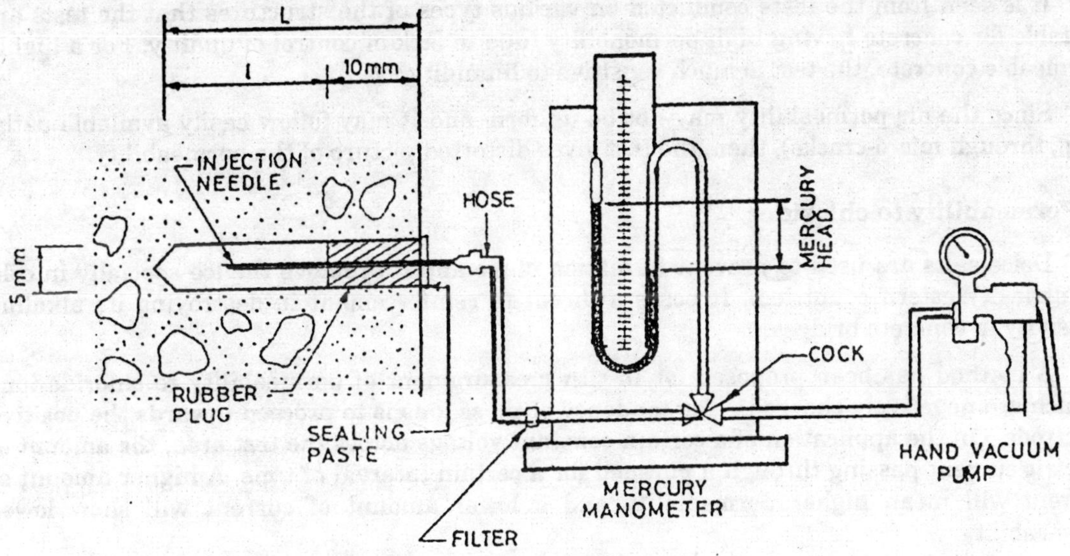

Fig. 31.8. Rapid Test Apparatus for Air Permeability of Concrete

The apparatus and arrangements of the test are shown in Fig. 31.8. It is seen that while increase in pressure over time in quite fast in case of permeable concrete, the pressure increase is almost negligible for practically impermeable concrete.

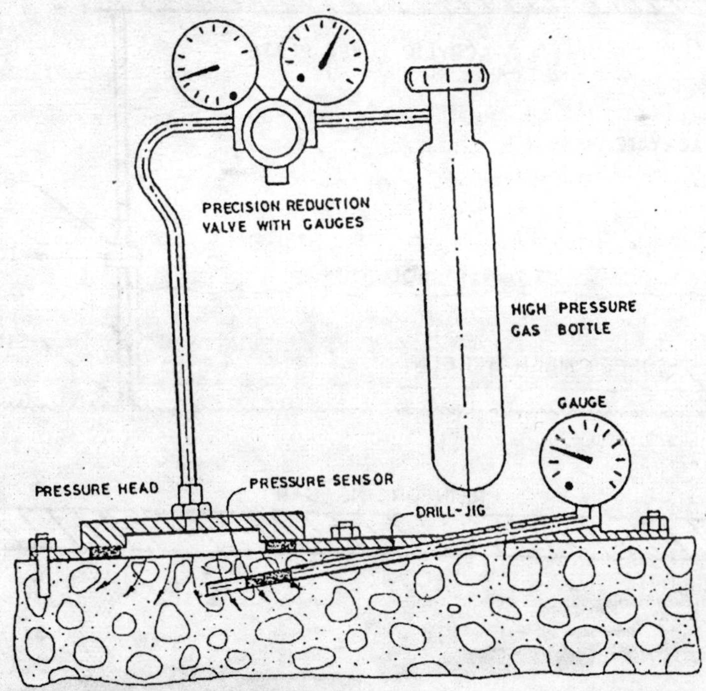

Fig. 31.9. Apparatus for Determination of Air Permeability

It is seen from the tests conducted on various types of the structures that the tests are suitable for concrete having high permeability (due to lack of control of quality. For a highly permeable concrete, the test is much sensitive to humidity.

Since the air permeability may not be uniform and it may follow easily available paths (e.g., through micro-cracks), then, the test gives distorted picture of the permeability.

3. Permeability to chlorides

Deice salts are used to prevent formation of ice and to remove the ice specially in cold countries (Western countries). It poses a threat to reinforcement in destroying its alkaline passivity in concrete bridges.

A method has been proposed for in-situ measurement of permeability to chloride ions (which are negatively charged). The tendency of these ions is to proceed towards the positive electrode. On the application of a certain constant voltage across the test area, the amount of electric current passing through a pursued for a certain interval of time. A higher amount of current will mean higher permeability and a lower amount of current will show lower permeability.

Reduction in the strength of sodium chloride (NaCl) solution measures the penetration of chloride into concrete. The test arrangement is shown in Fig. 31.9.

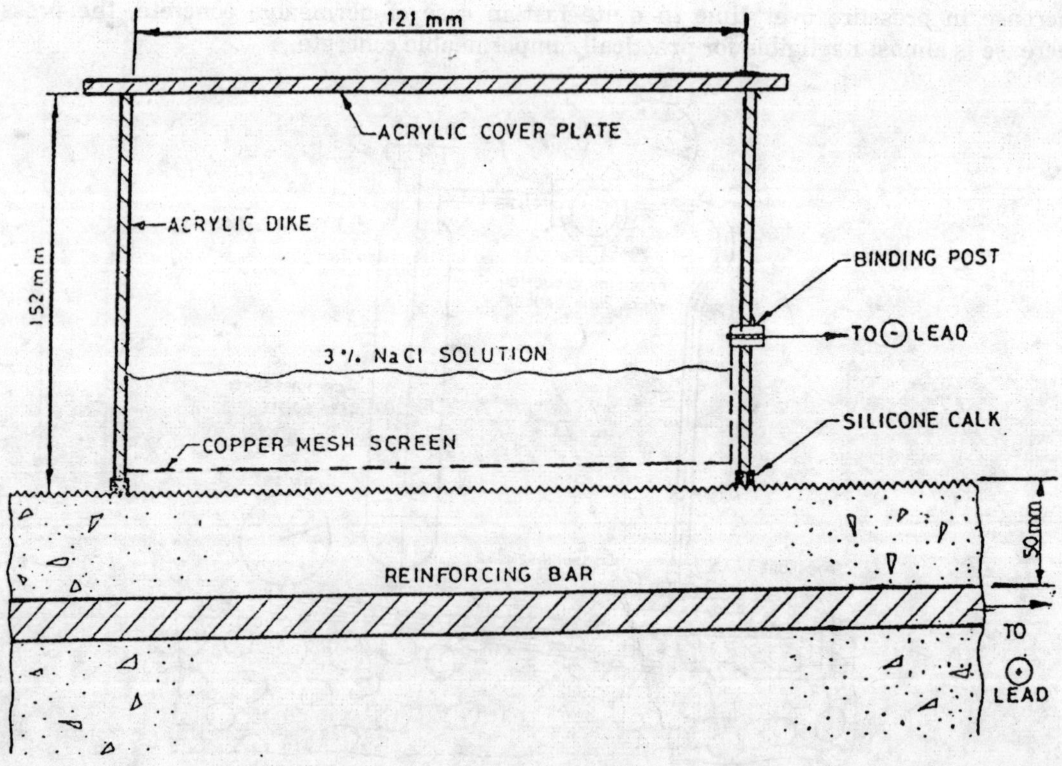

Fig. 31.10. Applied voltage apparatus used on slab specimens

31.15 RESONANCE FREQUENCY METHOD

Fundamental property, namely **Resonance Frequency of vibration** of a material is related by the basic relationship.

$$n = \frac{K \cdot m^2}{2\pi L^2}\sqrt{\frac{E}{\mu}}, \quad \text{where} \tag{31.3}$$

n = resonance frequency of vibration.

K = radiation of gyration of the section.

L = length of the specimen

E = modulus of elasticity

μ = Poisson ratio

m = a constant

Eq. 31.3 is applicable to many materials. Three modes of vibrations are used for its use in concrete.

1. Longitudinal, 2. transverse and 3. torsional.

Fig. 31.10 shows the arrangement of driving circuit. Oscillator and receiving (pick-up) circuit transducers are placed on the ends of the specimen. Specimens 25 mm x 25 mm x 275 mm, 50 mm x 275 mm, 150 mm x 150 mm x 750 mm and 100 mm x 100 mm x 500 mm size used.

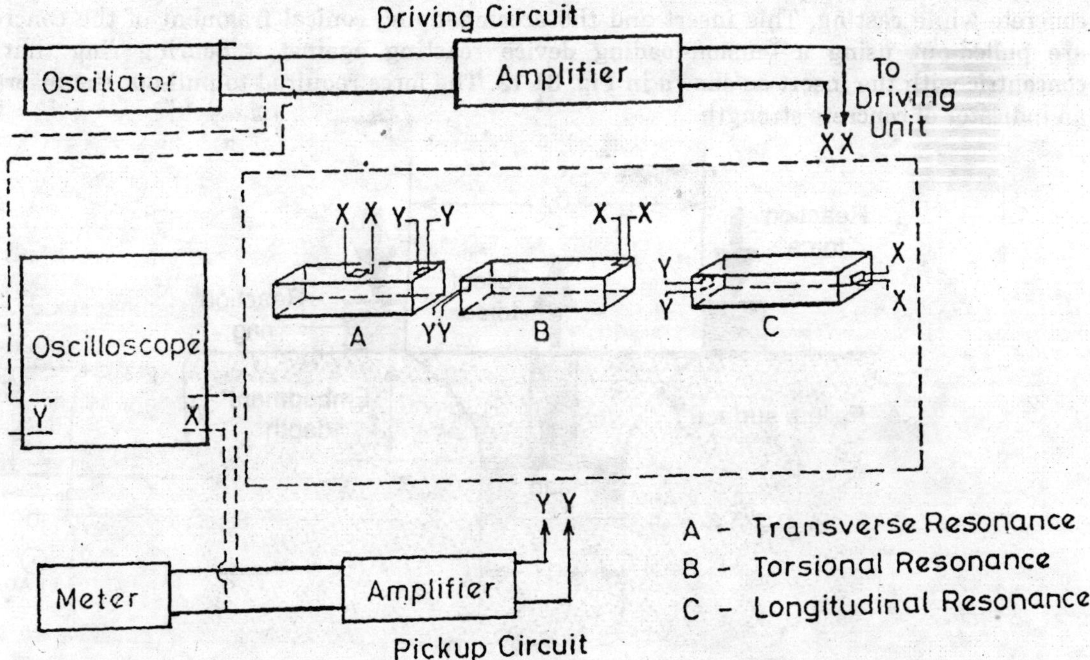

Fig. 31.11. Block diagram of apparatus for resonant frequency determination

Dynamic modulus of elasticity is found from the following by conducting the test.

$$E = (C \cdot W \cdot n^2) \quad \text{where} \tag{31.4}$$

W = weight of the specimen, and

C = a constant which depends upon the mode of vibration and the linear dimensions of the specimen.

So calculated dynamic modulus elasticity is free from the influences of creep. It is greater than the static modulus. Ratio of dynamic modulus and static modulus depends upon and mix parameters. Its value is of the order 1.3 to 1.7.

Damping constant Q may be determined from the following:

$$Q = f_o/(f_2 - f_1), \quad \text{where,} \tag{31.5}$$

f_o = resonance frequency of vibration.

f, f_2 = frequencies on either side of resonance at which the amplitude is equal to 0.707 **times** that at resonance.

It is a common method of evaluating durability of concrete (e.g., in freezing and thawing tests). The decreasing resonance frequency, by 50 (fifty percent of the original value, the durability factor (given by number of cycles is calculated).

31.16. PULL-OUT TEST

In this method, a specially shaped steel rod with enlarged head on insert is embedded in concrete while casting. This insert and the accompanying conical fragment of the concrete are pulled-out using a tension-loading device reacting against a bearing ring that is concentric with the insert as shown in Fig. 31.12. The force required to pull-out the insert is an indicator of concrete strength.

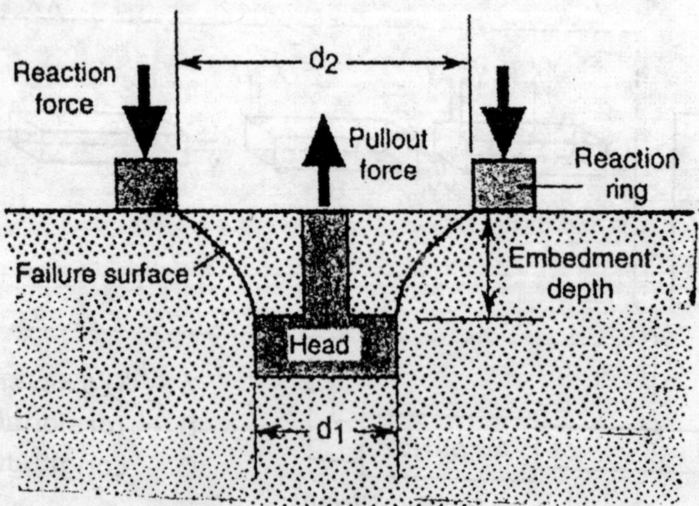

Fig. 31.12A. Schematic of cast-in-place pullout test.

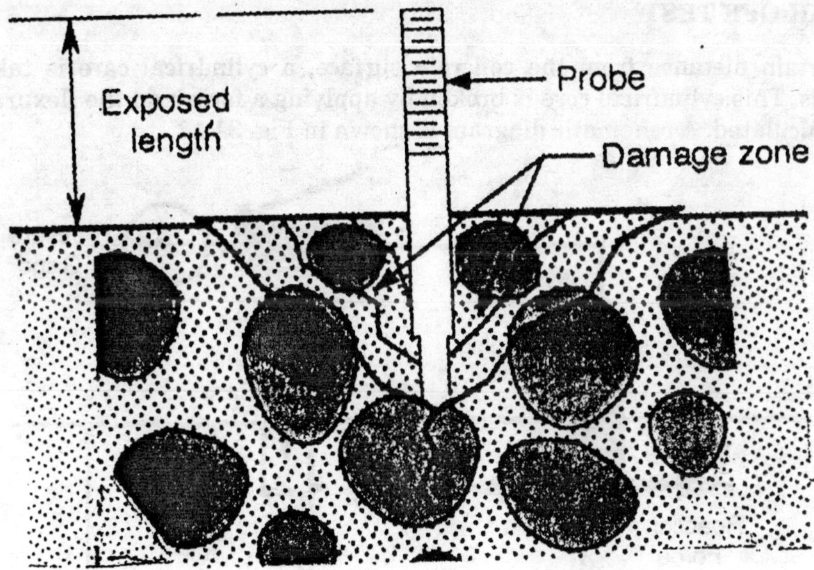

Fig. 31.12B. Schematic of conical failure zone during problem penetration test.

The pull-out strength is of the same order of magnitude as the direct shear strength of the concrete. It is 10 percent to 30 percent of the compressive strength.

Details of test method is given in ASTM standard C900. In Denmark, the pull-out test is called as `Lok-test'. A disc embedded in concrete is pulled-out through a cylindrical counter-pressure member.

For a given strength the ratio of pull-out strength to compressive strength is independent of age. However, it decreases with the increase in the level of strength.

It is not required always. It is sufficient to apply pre-calculated force to the embedded rod. In case the rod is not pulled out, it is assumed that the given strength exit.

Cast-in-place pull-out test is most reliable technique for estimating the in-place strength of concrete during construction.

Advantages: In this test, in-situ strength of concrete is measured. This test may be easily carried out in the field. The results of this test are re-producable. This test is superior to re-bound hammer test and to the penetration resistance test, because a large volume and a deeper depth of concrete are involved in this test.

Dis-advantage: It damages to the concrete. The damage is required to be repaired. Pull-out rods are be inserted at the time of casting of the concrete. It has to be pre-planned unlike other in-situ tests, this test cannot be performed at random after the concrete has hardened.

There are other variants in this test. Holes are drilled in the hardened concrete. Split-sleeves assemblies or wedge anchors are installed. At the time of pulling out the rods, the internal cracking of the concrete takes place. Alternatively, a bolt to may be set and later-on, the bolt is pulled out.

31.17. BREAK-OFF TEST

At a certain distance from the concrete surface, a cylindrical care is taken from the concrete mass. This cylindrical core is broken by applying a force and the flexural strength of concrete is calculated. A schematic diagram is shown in Fig. 31.13.

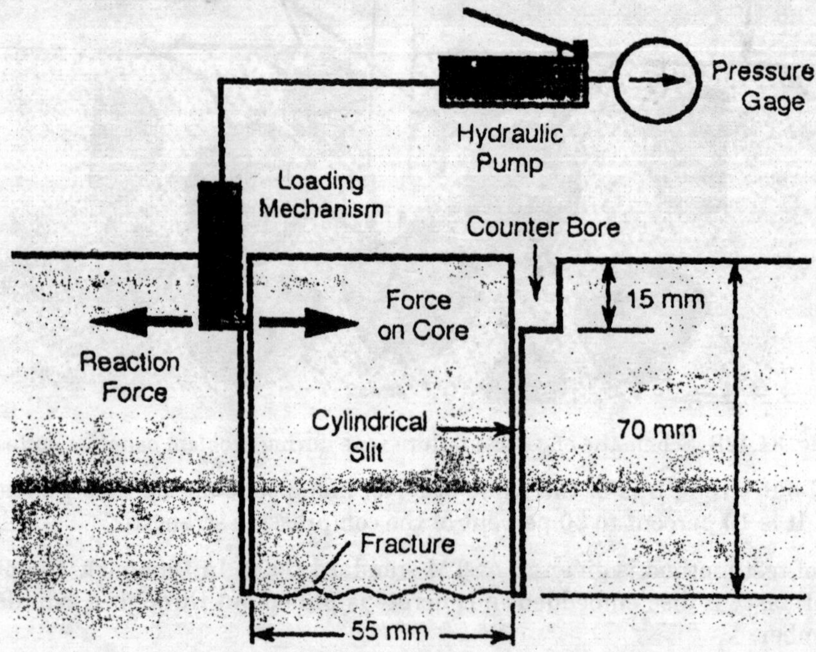

Fig. 31.13. Schematic of break-off test.

In this method, tubular disposable forms plastic sleevers are placed in the fresh concrete. These inserts are removed at the time of testing. The concrete core is broken-off at the bottom. The force is applied at the top perpendicular to the axis of the core. A special, hand-operatured, hydraulic loading jack is placed into the counter bore. The hydraulic fluid pressure is monitored with a pressure gauge, and the maximum pressure gauge readings in units of bar [1 bar = 0.1 MPo (0.1 N/mm^2)] is referred to as the **break-off number** of the concrete. Force is applied slowly and the static strength property of the concrete is measured. The core is loaded as a cantilever the concrete at the base of the core is subjected to a combination of bending and shearing stresses. The pressure gauge readings are related directly to the compressive strength.

A typical co-relation, prepared between break-off strength and the modulus of rupture of concrete. The co-relation between break-off strength and the compressive strength have been seen to be non-linear as per the usual practical of relating the nodulus of rupture of the concrete to the square-root of compressive strength. It is also seen that the co-relation between break-off strength and modulus of rupture may be more uncertain than that between break-off strength and compressive strength.

Failure occurs by fracture at the base 55 mm diameter-core. The crack initiates at the most highly stressed point. It then propagates throught he mortar and in most cases, around the coarse aggregate particles located at the base of core. The particular arrangement of aggregate particles within the failure region would be expected to influence the ultimate load. The distribution of aggregate particles will be different at different test location.

The break-off test is seen promising for certain applications (e.g., air-port and high-way parements). This test needs pre-planning and within-test, the variations are higher than the other standard test for strength. This test is very useful for fresh concrete. There are some difficulties in placing the tube into the stiff concrete mix. This method is rapid and simple. The test results are not influenced by the surface conditions, or by temperature and shrinking effects.

31.18. PROBE - PENETRATION METHOD

There are two techniques for the evaluation of hardness of concrete by probing.

1. A hammer known as `simbi' is used for perforating concrete and the depth of core hole was co-related to the compressive strength of concrete.
2. A gun is used to drive a hardened steel rod (pins) into the concrete and measures the exposed length of the probe. The depth of penetration is co-related with the compressive strength of concrete.

In principle, as the strength of the concrete increases, the exposed probe length also increases.

By means of suitable correlation, the exposed length may be used to estimate the compressive strength.

Windsor probe test method was adopted as a tentative ASTM standard (C 803) in 1975. This standard was modified in 1990 to include a pin-penetration device, in which a small pin is forced into the concrete using a spring loaded driver [ACI 228. 1 R, 1995, Malhotra and Caretta, 1991; Nasser and Al-Manseer, 1987].

In one of the probe test, the probe consists of a powdered activated gun or a driver, hardened alloy probe, loaded cartridges and a depth gauge for measuring the penetration. The diameter and the length of the probe are 6.3 mm and 79.5 mm, respectively. There is a frusto-conical point on the front. The rear of the probe is threaded and screwed into a probe driving head. The head fits smugly into the bore of the driver. This driver is used to fire a probe into the concrete using a single probe templet. The exposed length of the probe is co-related with the compressive strength.

Underlying principle is that for standard test condition, the penetration is inversely proportional to the compressive strength. Deeper the exposed length of the probe, greater is the compressive strength.

Windsor probe is primarily a hardness tester and provides an excellent means of calculating the relative strength in different structures. Probe is not expected to reveal the absolute values of the strength of concrete.

The precision of Windser probe measurement has been seen to vary with the maximum size of aggregate in concrete. It is influenced by the hardness of the aggregate. A separate callibration curve should be for the type of concrete under investigation.

Test is almost non-destructive as the damage to the concrete made by 6 mm probes is only local.

31.19. NUCLEAR (RADIO ACTIVE) METHODS

Nuclear (or radioactive) methods for non-destructive evaluation of concrete involve the use of high-energy electro-magnetic radiation to get information about the internal structure of the test object. A source of penetrating electromagnetic radiation and a sensor to measure the intensity of the radiation after it has moved through the object are involved. Two techniques involved are as under:

1. **Radiometry:** In this, the sensor is an electronic device that converts the incident radiation into electrical pulses.

2. **Radiography:** In case the sensor is in the form of special photographic film.

Malhotra (1976) and Mitchell (1991) presented review of the early developments in the use of nuclear methods.

Two non-destructive techniques used (which depend upon form of penetrating radiation) are the sources used to generate them and their penetrating ability.

X-Rays produce radiograph that reveal the internal structure of concrete (initial work was focused in late 1940). X-rays are produced by high-voltage electronic devices. The penetrating ability of X-rays depends on the voltage of the generating instrument. X-rays have been described as the `all-seeing eye' of science by means of which the inner structures of many substances may be observed X-rays are used for two distinct purposes (i) to determine the atomic structure, (ii) to detect internal flaws such as voids, segregates (concentration of impurities, internal cracks and so on).

These remarkable rays were discovered in 1985 by Rontger. These are a form of radiation similar in character to radiant heat, light, radio and electromagnetic waves. These waves of exceedingly short wave-lengths (the distance between crest to crest). Variation in wave-length is the fundamental difference between x-rays, light waves, radio waves and the other member, of the electro-magnetic family. X-ray wave-lengths are, in fact of the order of 10^{-7} millimetre.

The wave-length of X-rays are thus, shorter than those of light waves X-rays are invisible. However, these rays passes some properties of the visible light (these rays may be reflected refracted or diffracted). This property of diffraction is particularly useful to us.

Diffraction means dispersion or breaking-up of waves. When a light passes through a prism, it disperses into separate colours of spectrum. **Grating** that is, either transparent or reflecting is one another process, in which the light is to rule fine parallel lines on glass. The distance between each pair of parallel lines s of the same order of magnitude as that of the wave-length measured. In working with the ordinary visible light-rays, this needs several thousands of lines to each centimetre, so that, in case of x-rays of much smaller wave lengths the use of a diffraction grating appeared to be out of the question.

This method is not very popular for field use because of its limitations of high cost and immobility of 15 equipment requiring high voltage.

In the 1950s, attention turned to the use of **gamma rays**. Effects of using gamma-rays were made in Great Britain. These methods were used to locate reinforcement bars, measure density and locate voids in grouted post-tensioning ducts [Forester 1970]. Eastern Europe and soviet Union also conducted early studies the eventually led to the development of portable density meters for concrete. Gamma rays are produced by the byproducts of the disintegration of radioactive isotopes. The penetrating ability of gamma rays depends on the radio-active isotope and its age.

Fortunately, the limitations as regards thickness do not apply to the same extent when radiation from radioactive isotopes is used instead of X-rays. Thus, **iridium 192, tantalum 182, caeslum 167** and **cobalt** 60 have all been used as a source for this purpose. All emit the powerful, penetrating gamma-rays, with the aid of which defects may be recorded on photographic film just as with the X-rays.

31.20. RADIOMETRIC METHODS

In non-destructive testing of concrete, there two basic radiometric methods

1. In one method *x-rays* are used while,
2. In other method *gamma-rays* are used

Radioactive methods are being used more and more for concrete structures in recent times

In the **transmission, method** the amplitude of the radiation passing through a member is measured. As the radiation passes through a member, the attenuation is dependent on the density of the material and the path-length from the source to the sensor. Direct transmission techniques may be used to detect the reinforcement. The main use of the technique is to measure the inplace density, both in fresh and hardened concrete. The structures of heavy weight and roller-compacted concrete are the cases where the method is of particular value. For such cases (applications), the radioactive source is contained in a tube that is pushed into the fresh concrete and the detector is set on the surface of concrete. This density meter developed at the Technical university of Brno (Czechoslovakia) is an example of such a device.

The source may be lowered upto a depth of 200 mm into a hollow steel needle that is pushed into the fresh concrete. A spherical lead shield supresses the radiation when the source is in this retracted position. Detectors are located beneath the treads used to push the needle into concrete.

It is claimed that the instrument has a resolution of 10 kg/m³ [Honig, 1984].

ASTM C 1040 provides procedures for using nuclear methods to measure the in-place density of fresh or hardened concrete. The key element of the procedure is development of the calibration curve for the instrument. This is accomplished by making test specimens of different densities and calculating the gauge out-put for each specimen. The gauge out-put is plotted as a function of the density, and a best fit curve is found.

Radiometry methods uses a concentrated source and a detector to measure the received emissions at a point on the member Two types of arrangements (i.e., direct transmission and backscatter are adopted for locating source and detector in this method.

31.21 BACK-SCATTER METHOD

This process is known as `Compton scattering [Mitchell, 1991]. Back-scatter techniques are particularly suitable for applications where large number of in-situ measurements are needed. Since, the back-scatter measurements are influenced by the top 40 mm to 100 mm, the method is suitable for measurement of the surface zone of a concrete element. A good example of the use of this method is the monitoring of the density of bridge deck overlays. Non-contacting equipment has been developed that is, used for continuous monitoring of concrete pavement density slip-form operations [Mitchell *et al.*, 1979].

A radioactive source is used to supply **gamma-rays**, and a detector close to the source is used to measure the backscattered rays. The scattered rays are lower energy than the transmitted ones and are produced when a photon collides with an electron in an aton. Part of the photon energy is given to the electron, and a new photon come out moving in a new direction with lower energy.

In ASTM C 1040, procedures for using backscatter methods to measure concrete density are given.

It is necessary to establish a calibration curve prior to using a nuclear backscatter gauge to measure in place density. The inherent precision of backscatter density gauge is less than that of direct transmission devices.

ASTM C 1040 needs that a suitable backscatter gauge for density measurement result in a standard deviation of less than 16 kg/m^3, while the standard deviation should be less than 8 kg/m^3 for a direct transmission gauge. According to ASTM C 1040, backscatter gauges are typically affected by the top 75 mm to 125 mm of material. The top 25 mm calculates 50 percent to 70 percent of the count rate, and the top 50 mm calculates 80 percent to 95 percent of the count rate.

This method is also referred as **Neutron scattering method** for finding the moisture content of Hydrogen present in water in concrete slow down the energy of fast neutron. The number of the slowed down neutrons resulting from the interaction of **fast neutrons** with the matrix containing hydrogen are countered, which gives measure of the hydrogen and hence the moisture content of the concrete. For generating fast neutrons, radium with beryllium is generally used in commercially available equipment.

31.22 RADIOGRAPHIC METHODS

The British standard Institute has adopted a standard for radiographic testing of concrete BS : 1881 : Part 205 (Recommendations for radiography of concrete. [Mitchell, 1991]. There are two types of radiographic techniques. The selection of a source depends on the density and thickness of the test object and the exposure time that may be tolerated. There are limits on the thickness member that may be tested by the radiographic methods.

1. X-ray radiography, and
2. Gamma radiography

In X-ray radiography, the radiation is produced by an X-ray source tube [Mitchell, 1991]. The penetrating ability of the X-rays is dependent on the operating voltage of the X-ray tube.

A system, known as scorpion IT, was developed in France that uses a linear accelerator the produce **very high energy** X-rays than may penetrate upto 1 m of concrete. This system was developed for the inspection of prestressed members to establish the condition and location of prestressing strands and to find the quality of grouting in tendon ducts [Mitchell, 1991].

In gamma radiography, a radio active istope is used as the radiation source. The most intense source is cobalt-60, which may be used to penetrate upto 500 mm of concrete. For members of thickness of 150 mm or less, iridium-192 or cesium 137 may be used [Mitchell, 1991]. The film type will depend on the thickness and density of the member being tested. This radiography is used to find the position and condition of the reinforcement in concrete, voids in the concrete and grouting of post-tension cable in structures BS : 4408 (Part 3) non-destructive methods of test for concrete specifies details on suitability of sources of radiation test procedures and safety precautions.

Most field applications have used radioactive sources because of their greater penetrating ability (higher energy radian) compared with X-rays. The radiographic methods consist of taking photographs of the interior of the concrete member.

31.23. COMMENTS ON NUCLEAR METHODS

The nuclear methods have the **ability to see** into the concrete. However, the nuclear methods are combersome and need trained and licensed personal [Mitchell, 1991]. The testing across the full thickness of a concrete element is particularly hazardous and needs extensive precautions skilled personnel, and highly specialized equipment.

Radiographic procedures are costly and need evacuation of the structure by the persons not involved in the actual testing.

The **use of X-ray** equipment poses an additional danger owing to **high voltage** that are used.

The **use of gamma-rays radiography** are limited according to the thickness of member. Thick members (more than 500 mm thickness) need unacceptable long exposure times.

Radiography is not very useful for locating crack places perpendicular to the radiation beam. Because of these major draw bracks, the radiographic methods are not used for flaw detection.

IIowever, there may bc situations, where the ability to see the internal structure of the member surpasses these draw backs.

31.24 COMMENTS FOR USE AND APPLICATIONS OF NONDESTRUCTIVE TESTS/EVALUATIONS

Available non-destructive techniques for evaluating the properties and the condition of concrete have been divided in following two categories.

Underlying principles and some of their inherent limitations have been described. ACI relevant ASTM standards may be referred for additional information on using these methods.

1. For estimating the in-place compressive strength

Main principle of the methods for evaluating in-place strength is the compressive strength relationship which correlates the concrete strength to the results of the in-place tests. The strength relationship is developed experimentally before using the method to assess in-place strength.

For new construction, the test-specimens are made of concrete similar to those, which will be used in the structure. The companion in-place tests and standard strength test are carried out on specimen of the equal maturity at strength level.

For the existing construction, it is necessary to carry out in-place tests and get cores at different locations so that a wide range of concrete strengths may be used to develop the strength relationship. After establishing the strength relationships, in-place tests are carried out on the structure, and statistical methods are used to convert the average of in-place test results to a reliable estimate of in-place strength. Generally, in-place test methods that result in local failure of the concrete the more reliable than those within are totally non-destructive.

2. Flow flaw detection and condition assessment

Most of these methods are based on monitoring the response of the structure, when it is subjected to some type of disturbance.

Two broad group of non-destructive methods are those based on (i) stress-wave propagation and (ii) electromagnetic wave propagation.

Other than the visual inspection methods generally need sophisticated instrumentation. There are inherent advantages and disadvantages of all these non-destructive test methods. It is better to use more than one method to make the evaluation.

(i) Methods based on stress-wave propagation are suitable to identify the existence of internal concrete air interfaces, such as due to cracking or voids.

(ii) Electrical methods are well suitable for getting information about embedded reinforcement, such as location, approximate size, and whether the active corrosion exists.

(iii) Radar is approximate for finding deep metallic embedments. It is also sensitive to the presence of moisture. Large portion of a structure may be scanned in a short time.

Non-destructive test methods may be used by properly trained operators. In that case, these methods offered technical and economic advantages in comparison with other destructive sampling technique.

There are following two types of neutron activation analysis methods for finding cement content of concrete.

(i) In thermal neutron activation method, the sample is subjected to impact of electrons with low-kinetic energy neutron from an isotopic neutron source or from a nuclear reactor. The elements as a result of such an impact become radioactive and the radiation energy released upon their decay is monitored to measure the quantity of desired element in the sample after comparison with standard specimen subjected to similar impact.

Since, the calcium and silicon contents of portland cements are relative high compared to other constituents. These elements are detected easily by this technique in which Ca49 and A128 isotopes are formed from Ca and Si constituents of the concrete.

This method is however not suitable for calcaious aggregate.

(ii) Fast neutron activation method is the second type of neutron activation analysis. In this technique the accelerator produces neutrons of high energy by deutron (heavy hydrogen nuclel, ^{2}H) into a target containing tritium (heavy hydrogen with a mass 3 or ^{3}H). These fast neutrons are capable of producing isotopes from elements. The silicon is found to be very sensitive for this technique and forms the basis for the analysis. The calcium does not respond.

REFERENCES

1. `Concrete Construction Engineering Hand-book' by Edward G. Nawy (Editor-in-Chief), 1997 Edition Non-destructive Test Methods by Nicholas J. Carino [PP (19-1) to (19-59)].
2. `The Structure of Steel' by Edwin Gregory, Eric N. Simons, Third Edition, 1957, Odnams Press Limited, Non-mechanical test methods (pp 100-113).
3. `Non-destructive testing of RCC structures' by Sanjay Sharma Department of Civil Engineering National Institute of Technical Teachers & Research, Sector 26, Chandigarh, 1600190, March 2005 (Cyclostyled - notes).

32

FUTURISTIC CONSTRUCTION MATERIALS AND TECHNIQUES (FCMT)

32.1 INTRODUCTION

In ancient time (i.e., prehistoric age), it was stone age. Development of civilization began and the buildings were built for shelter, security and residence with stone and mud was used as mortar. Timber (wooden scantlings) was also used for the construction for the residence and some protection. Kacha bricks and in due course of time, pucca bricks were used to build buildings. As early as 3000 BC, specially in the Middle East clay bricks using moulds, used to be manufactured, from clay, clayey soils or soft slate or sale containing about 30 percent of sand. Clays were first ground or crushed in mills and mixed with water, as needed to obtain the desired consistency for forming. These bricks were generally sun dried. King Nebuchadnezzar's palace in Babylon was built in the River Euphrates in Iraq, about 600 BC. It is an example of early brick construction.

The Greek civilisation used the same techniques as the Egyptians, using **stone walls** Piers and lintels were built in sophisticated palaces and temples. The Roman, and their successors as the dominant civilisation in the Middle East and Western Europe were even greater builders. There are examples of the buildings and structures for almost every purpose. **Stone and fire clay bricks** were used for the construction of buildings for long time and these are still in use even now. Roman building continued. The Byzantine or Eastern Empire, specially with the construction of Churches, some on a very large scale, such as Hagia Sophia in Constantinople (now, Istanbul) in the 6th century AD. Though in the Middle Ages, the most notably masonry buildings in Europe were chur ches, cathedrals and Castles, latter ranged from vast complexes with formidable defensive walls to relatively small towers in the disputed areas. An example of a border tower is Borthwick castle near Edinburgh built in 1430 with walls over 4 m thick.

Cathedrals built between the 11th and 14 centuries reached an extremely high technical and aesthic level of buildings using stone masonry, for example, Durham Cathedral. The Romans were known to be the best bridge builders between 200 BC and 260 BC and some of the **masonry arch** bridges built by them still exist. The arch bridge known as Pont-Du-Gard was built in Frame in 14 AD and due to good preservation and maintenance, the bridge is still in a fairy good condition.

After the **stone age,** the civilisation reached **iron age,** which began somewhat 1000 BC in Western Asia and Egypt. Iron occurs in nature (as raw material) in combined state. Weapons and arms were primarily made of iron. These were used for hunting and for fighting wars.

In our country, there are many examples of use of iron which may be cited from Ramayan and Mahabharat.

Some of the structures were converted into residences, (e.g., Dalhousie Castle), also near Edinburgh, which was built in the 12th century and converted into a country house in the 19th century and it is still in use as a hotel. Most often, the land-owners built new mansions or palaces on or close to the site of the original castle, leading to a new style of **masonry building,** sometimes on a very grand scale. Many `stately homes' of this kind were built between the 16th and 19th centuries and are valued as part of National heritage.

Valley civilisation, there is archeae logical evidence of use of **iron.** In the fifth century, **an iron Isthamb** was made in **Mehrauli village, Delhi** within a small distance, from **QutubMinar.** An iron pillar in **Kodachadri Village** in Karnataka is seen. It does not rust for nearby 1500 years. In the eighteen century, famous bridge engineers began the use of iron. Initially, a number of **arch bridges,** were built using **casterion** in the 18th century. First iron bridge of 30.5 m span was built in 1779 over the river seven in Coalbrookdale at stropshire, England. Cast-iron was gradually replaced by wrought iron dominated the field until close to the end of the 19th century which in time was also replaced by steel. First major structure was the Eads Bridge built using carbon steel (st. Louis Bridge over the river Missouri). It was completed in 1874. (This was a steel `bridge of three spans of 153 m. First major **steel structure** in UK was the fourth bridge completed in 1889. Which in time (midnineteenth century) was replaced by steel when Bessemier Process of steel (BPS) making was introduced. With the introduction of steel as the bridge material in place of iron and with the improvement of the quality of steel, these was rapid development in the construction of large span steel bridges, particularly the suspension as well as the large steel cantilevers.

Steel was invented and patented in 1885 in England by Sir Henery Bessemier. At about the same time, the technique of making steel was developed by William Kelly of USA. In the heating process, the undesirable elements including excessive quantities of carbon were removed. Subsequently, the desirable quantities of carbon and manganese were added. It became the **beginning of steel age.** Use of steel became popular in nineteenth century Siemens Martin Process of open-hearth (OHP) technique (oldest) was developed along with Bessemier Process. Open-Hearth Process (OHP) was replaced by the basic oxygen system (briefly, referred as BOS) and the electric arc method. More recently, a new process has been devised. It is continuous casting process (briefly) referred as CONCAST).

In our country, Howrah Bridge (Rabindra Sethu) in Kolkata is a marvellous structure built in steel in 1943. It is a balanced cantilever bridge over river Hoogly having 457 m span. It is fourth largest cantilever bridge in the world. Recently, a cable-stayed bridge (called as second Hoogly Bridge i.e., Vidhya Sagar Sethu) has been constructed over this river Hoogly. It is completed in 1992. A huge steel construction may be seen in the Jogighopa bridge over river Brahmputra. In our country, there are many bridges built in steel for railways.

In 1824, Joseph Aspdin, a British stone mason obtained a potent for **cement** produced in his kitchen. Aspdin heated a mixture of finely ground limestone and clay and ground the mixture into a power, which created a hydraulic cement, one that hardens with the addition of water. Aspin named this product as **Portland cement** because it resembled a stone quarried on the Isle of Portland of the British Coast. The Greeks and Romans were aware of the fact that certain **volcanic ash and tuff** wen mixed with lime and sand yielded mortar possessing superior strength and better durability in fresh or salt water. This volcanic tuff was found near **pozzuoli** village near Mount Vesuvius in Italy, thus acquired the name **pozzolana.** Later on **Pozzolana** name was applied to any other material matural or artificial which at ordinary temperature, react with lime in presence of water, resulting in cementing materials. Diatomaceous earth or material processed by calcization of the soil (e.g., burnt day) or artificial material such as fly-ash, burnt clay and pumicite are used as **pozzolana**. The term pozzolana means silicious material, having nearly the same composition as that of volcanic tuff or ash at **Pozzuoli.**

In 18th and 19th centuries, the Industrial Revotion took place in Europe, the population expanded rapidly and town and cities grew proportionately. This resulted in building, this was almost entired in masonry and timber until near the end of 19th century. A great deal of construction was required for new industrial development and transport infrastructure through this period, again built largely in masonry. Earliest high rise buildings in fact built in masonry. One of the early `sky-scrapers' in Chicago, was the Monadnock building, now preserved for its historic interest. This building is 16 storeys in height and of brick work.

Reinforced concrete bridges gained popularity in the 20th century because of their versatility in construction and economy in cost and maintenance. These reinforced concrete bridges may be cast in any convenient shapes and forms to meet architectural requirements. One of the longest span reinforced concrete bridges is the Sando Bridge built in Sweden in 1943 with a span of 264 m. Only about a decade ago, the new Sydney Harbour Bridge (which is an R.C.C. arch bridge having a span of 305 m had been constructed.

Manufacture of **high strength concrete** and **prestressing** the same by the use of **high tensile steel wires** further improved the construction of concrete bridges. One of the **early prestressed concrete** bridges is the Marine Bridge built in France. First **prestressed concrete** bridge built in our country (Tamilnadu state) is the Polar Bridge many long span prestressed concrete bridges in our country have been constructed. Some of these bridges are Barak Bridge at Silchar, Assam with a central span of 122 m, Bassein Creek Bridge near Bombay with two central spans of 115 m each, Ganga bridge at Patna with a number of central spans of 121 m each Longest span prestressed concrete bridge in our country is the Lubha Bridge in Assam with a central span of 130 m.

All the conventional materials bricks, stone, timber, iron and steel, and concrete described above are likely to be exhausted in some small or long period of time. A small period of time span may be referred by the word **future**. Where as the word **futuristic** is used to a very long period of time span. **Future** is predictable based on statistics, while **futuristic** is a fantasy dream and revolutionary. It is therefore essential to make efforts to find new materials and techniques for their use in construction.

32.2 DESIRED REQUIREMENTS FOR FUTURISTIC CONSTRUCTION MATERIALS (FCMT)

Industrial revolution took place in Europe in the 18th and 19th centuries. The population expanded rapidly in towns and cities at a very fast pace. There are many reasons for this. Lack of infrastructure facilities and employment opportunities in villages are main reasons among them. Urban population continues to increase in geometric progression.

However, majority of population in our country lives in areas. Special attention is needed for their housing. New materials and technology is required to be developed, which one should be friendly to the people. New materials found may have **desired requirements.**

(1) Materials should have minimum cost and require minimum maintenance.

(2) Building materials should be sustainable. Ecological balance should not be disturbed due to such materials.

(3) The materials should be light in weight and seismic resistant

(4) Self-help type technologies may be developed

Efforts may be made to discontinue migration of people from rural areas to urban areas. Some of such materials were found some years back, while the use of some of some materials started recently.

These materials are as below

(1) Self-compacting concrete (SCC)

(2) High Performance concrete (HPC)

(3) Concrete Nano-technology.

(4) Polymer-modified cement, mortar (PCM) and concrete (PCC)

(5) Polymer-mortar (PM) and Polymer-concrete (PC)

(6) Fibre reinforced polymer (FRP)

32.3 SELF-COMPACTING CONCRETE (SCC)

(*References has been cited at the end of this articles)

Definition

It is the concrete that is able to flow under its own Weight and completely fill the form-work even in the present of congestion of reinforcement without the need of any vibration, whilst maintaining homogeneity.

Since, the early 1990's, Japan has used **self-compacting concrete** (SCC) in bridges, buildings and tunnel constructions. A number of SCC bridges have been constructed in Europe. The application of SCC in high-way bridges construction is very limited at this time in the United states.

The precast concrete industry in the U.S. however, is beginning to apply technology to architectural concrete.

Placement of seal concrete underwater is done by the use of a tremie without vibration, mass concrete has been placed vibration, and shaft concrete may be successfully placed without vibration. These seal mass and shaft concrete are the examples of SCC. This concrete has generally low strength (less than 34.5 N/mm^2). Its application is focussing on high performance, better and more reliable quality, dense and uniform surface texture and improved.

Presently used concrete lacks in uniformly and complete compaction even after using the vibration. This fact was recognized Researchers at the University of Tokya, Japan began to develop **self compacting concrete** (SCC) in late 1980's. The concept of self-compacting concrete was proposed by Okamura, H. in 1986. Its prototype of was first completed in 1988. In Japan, self-compacting concrete was developed in the early 1990's. This concrete does not require vibration to obtain full compaction. Many of it applications of its construction have been reported there. Quantity of SCC used for it ready mixed concrete (cast-in-place) and prefabricated products (precast member) was about 4,00,000 m³ upto 2000.

Recent remarkable changes in the construction environment, are demanding improved technology for the produ of **high performance concrete** with for greater workability, high strength and durability. Self-compacting concrete has excellent deformability and resistance to segregation. This concrete may be filled in heavily reinforced form-work without vibration. The definition of this concrete may be stated as under. (see page 1) durability. It is difficult to attain consistent quality. The construction using this concrete is faster. This concrete has high potential for wider structural applications, specially in highway bridge construction. Self-compating concrete contains (i) cement (ii) Aggregate (iii) Mixing water (iv) Admixtures (super-plasticizers (SP), Viscosity, modifying agant (VMA)] (v) Additions including minerals fillers (semi-inert, pozzolanic or latent hydraulic) (vi) Fibres and (vii) Pigments.

Required workability for casting depends on several factors, such as type of construction, the selected placement and consolidation methods, the shape of form-work and the nature of congestion of the reinforcement. The congestion in the reinforcement in mat foundation and moment-resisting concrete structures is increasing. The interest in highly flowable concrete is high growing.

Several chemical and mineral admixtures are often incorporated in SCC. [In particular a super-plasticizers (SP) and a viscosity modifying admixtures (VMA)]. Admixture SP is used to insure high-fluidity and to reduce the water powder ratio (W/P ratio). Admixture VMA is incorporated to increase the yield value and viscosity of the fluid mixture. Therefore, the bleeding, segregation, and settlement are reduced. High-range water reducing admixtures (HRWR) are also added in concrete mixes.

In major, civil engineering works, in Japan used SCC (e.g., Massive LPG storage tank and huge anchorages for 2 kg-span Akashi-Kaikyo suspension bridge. These anchorages contain complex mixture of dense reinforcement and cabling that had to be encased in 512 x 1000 m³ and 250 x 1000³ of SCC each, Japan, Europe and the United States.

Self-compacting concrete is a considerable step forward with respect to durability. Mixing and casting are easier. There are many **advantages** (which are offered by self compacting concrete) for the precast, pre-stressed concrete industry and for cast-in-situation construction.

1. Improved durability
2. Higher strength
3. Better surface finish and product quality
4. Faster construction

5. Easier placing

6. Greater freedom in design

7. Thinner concrete sections

8. Saving in lavour, vibrators and energy solution of site problems increased speed of work.

9. Low noise-level in the plants and construction site

10. Safer working environment.

Professional troubles of arrangement of labours (workers) are reduced. Difficulties experienced in conventional concrete increased the interest in self-compacting concrete.

1. **Poor vibration**: Which leads to segretion of the aggregates, greater settlement, fluctuations in behaviour while setting and the presence of voids, high permeability and poor distribution of pores in case the vibration is insufficient.

In addition to these, poor surface appearance with regard to colour and sand striking poor impermeability due to bleeding around the aggregates, poor quality of the concrete skin, in other words of its surface and lack of homogenecity with the concrete which leads to vibrations in strength. Honey combs and cold joints also create difficulties.

2. **Retempering**: Loads to increase in water cement ratio and therefore decrease in strength, less cohesively, low homogenecity and poor durability.

32.4. PROPERTIES OF SELF-COMPACTING CONCRETE

Properties of self-compacting concrete in **its fresh state** are its fluidity, deformability and its resistance against segregation, which are important **requirements.**

1. **Fluidity of the concrete**: It is defined as its ability to flow (under its own weight without vibration) into the moulds or forms, and through heavily congested reinforcement (even, the presence of obstackles, which many interfere with its path.

2. **Deformability of the concrete:** It is defined as the capacity to alter shape and to pass through different sections without loosing cohesively.

3. **Segregation of the concrete** : It is the tendency of the coarse aggregate to separate from the mortar. This segregation does not occur in this concrete. A concrete mix may be **classified** as self-compacting concrete incase above stated requirements are fulfilled. The **mechanisms** of self-compaction in divides (i) high quantity fines (550-550 kg/ m^3) (ii) Super plasticizers with high water reduction and (iii) Visco modifying agents (VMA).

Properties of self-compacting concrete in **its hardened state** are similar to those obtainable in the conventional HPC vibrated concrete; which include the following structural properties

1. **Compressive strength of SCC** are comparable to those of conventional vibrated concrete made with similar mix proportions and water/cement ratio

(i) At 28 days - 40 to 80 N/mm^2

(ii) At 91 days - 55 to 100 N/mm^2

There is no difficulty in producing SCC with compressive strengths upto 60 N/mm^2.

2. **Tensile strength of SCC** (at 28 days 2.4 to 4.8 N/mm^2) are based on the indirect splitting test on cylinders.

3. **Bond strength** between concrete and reinforcement has been found by conducting pull out tests for different diameters of bars.

In general, the SCC bond strengths expressed in terms of the compressive strengths are higher than those of conventional concrete.

4. **Modulus of elasticity.** SCC and conventional concrete bear a similar relationship between. Modulus of elasticity and characteristic compressive strength expressed in the form $E/(f_c)^{0.5}$.

where, E = Modulus of elasticity

f_c = Characteristic compressive strength.

It is similar to that recommended by ACI for conventional normal weight concrete.

Note : Water binder ratio used is 25 to 40 percent where as the air content is 4.5 to 6.0 percent

5. **Durability of SCC :** Durability of SCC and so also conventional concrete depends upon two important properties (namely, permeability and porosity). The intrinsic causes of deterioration are linked to the fact that the concrete often has weak points. Concrete is likely to deteriorate due to following.

1. **Physical causes** include freezing and thawing, shrinkage and high temperature.

2. **Chemical causes** include attacks of sulphates, organic acids and substances, chlorides (on the reinforcing bars) and others (ammonium salts, magnesium etc) carbon dioxide and inorganic, acids. For pH less tan 5, the concrete must be protected with coatings.

3. **Mechanical causes** include abrasion, erosion and cavitation.

4. **Intrinsic causes in the material** are as follows

(i) The composition of concrete (water/cement ration and aggregate/cement ration)

(ii) Aggregate, water, admixtures

(iii) Workability of the concrete at the time of casting

(iv) Curing of the concrete after casting.

Durability of concrete is defined as the capacity of a work to withstand aggregate agents of nature, keeping the mechanical and functional properties unchanged.

All efforts to make durable and reliable structures, which require insignificant maintenance work, then, it may be said, it has been worthwhile.

32.5 HIGH PERFORMANCE CONCRETE (HPC)

HPC may be **defined** as a concrete made with appropriate materials (such as super-plastizers, retarders, fly ash, blast furnace slag and silica fume) combined according to a selected mix design and properly mixed, transported, placed, consolidated, and cured to give excellent performance in some properties of concrete (such as high compressive strength, high density, low permeability, and good resistance to certain forms of attack.

In many applications, high performance concrete is used only because of it high durability quality rather than the need for the strength. HPC is often called **durable** concrete because its strength and impermeability to chloride penetration makes it, to last longer than conventional PCC.

It may be also be defined as **engineered concrete** made up of the classic elements of water, Portland cement and fine and coarse aggregates, but with added components. In HPC, materials and admixtures are carefully selected and proportioned (optimized) to form high early strength, high ultimate strength and high durability beyond conventional concrete.

It is mainly used for long span bridges and high-rise buildings. It is the latest development in concrete technology. It has replaced high strength concrete developed (in the early 1980s), over the last two decades.

HPC has an excellent frost and salt scaling resistance, practically useful for protection of reinforcing steel bars against corrosion. With low water cement ratios, super plasticizers, mix water conditioners, concrete densifiers sealers and appropriate aggregate, cement and mineral admixtures, it is possible to produce concrete with high compressive strength, very low porosity, and extremely low permeability.

In the conventional concrete, following are enhanced (i) ease of placement and completion without segregation, (ii) impermeability and high density, (iii) durability and toughness (iv) low heat of hydration, (v) flow ability and self-levelling capability, (vi) early age strength (vii) volume stability and (viii) long life in severe environment.

Different industrial waste materials are used to increase the desired properties of concrete. The properties of HPC have been listed in Table 32.15.

Table 32.15 PROPERTIES OF HPC

S.No.		
1.	High ultimate strength	Compressive strength at 28 or 56 days at 56 days
2.	High early strength	(18 to 24) Hours Compressive strength Flexural strength
3.	High modulus of elasticity	More than
4.	High durability	Protection against
	(i) Corrosion of	embedded steel
	(ii) other severe conditions	servere service
5.	High workability, palpability and finish ability	
	(i) Mid range concrete	150 mm to 200 mm slump
	(ii) Flowing conc.	More than 200 mm is slump (without segregation)
	(iii) Reduced pumping pressures	

(iv) Easier finishing

(v) Self consolition

6. Placed in cold weather

(i) Normal setting time

(ii) Accelerated strength gain

(iii) Low temperature freeze protection.

7. Placed in hot weather

(i) Normal slump retaction

(ii) Control of hydration

8. Controlled hydration

(i) Extended setting time as needed by project conditions

TABLE 32.2 ADMIXTURES FOR HPC

S.No.	Admixtures category	Applications/Uses Standards	Consideration for most effective use
1.	Conventional admixtures	water reducers (i) Retarders (ii) Accelerators (iii) Combinations	Allow use as needed by Producer or contractor
2.	Super-plasticizers	(i) More water reduction 12 to 304% (ii) Flowing concrete (iii) Increased pump ability (iv) Self-consolidation	Allow slumps in excess of 200 mm Allow free fall of 4.572 m, 7.620 m Allow redosing if necessary Allow batch plant addition
3.	Silica fume	(i) High strength (ii) High durability	Must include use of super-plasticizer specify amount Specify Amount suitable to the application
4.	Corrosion	Extended time to inhibitors Corrosion by deicing salts or marine exposure	Allow use as needed caused by conditions
5.	Non-chloride temperatures	(i) Improved setting time (ii) Improved strength gain characteristics	Reduce protection accelerators at low Allow early form removals
6.	Cold weature admixtures as low as 20ºF	(i) Allow concrete to set at ambient temperature (ii) Allow placement (iii) Reduce winter protection	Reduce winter heating concrete without protection

7.	Hydration control admixtures	(i) Extended setting time (ii) Long-haul Application	Allow use of treated conc. Allow use of treated conc. Allow increased haul-time

About 15 years before, the **concept of high performance concrete,** as a technology, emerged out of laboratory. However, actual research work began in the last decade only on the possible use of different materials. Fly-ash, slag from smelters and different types of other waste coming from the industries greatly helped in increasing the strength and other properties desired for particular applications. In fact, many materials are to be tested in order to increase the use of other materials in the field. Some of the admixtures, which have widely accepted for use in HPC along with their field of application are given in Table 32.

Following **technical problems** associated with the use of HPC **before and during construction are concerned.**

(i) Finish ability and slump

HPC is typically placed at relatively **high slumps** from 200 mm to 250 mm, because of the super-plasticizers needed for workability.

During finishing,

HPC with silica fume may be sticky and may lead to tears and pulls. Screeding operations must begin as soon as possible after placement.

(2) Lack of bleeding water

HPC forms lumps during its application. It is difficult to handle lumps. The **workability** and **bleeding of water** are reduced Admixture super plasticizers may be used in proper ratio to solve this problem to a certain extent

(3) Trial mixes In HPC jobs, the trial mixes in pre construction mocks-up or trial slabs are usually needed.

Even the contractors with prior HPC experience may benefit from tackling potential problems with the mix before beginning of the construction.

(4) Pre-construction meeting. HPC jobs are not routine. A pre-biomeeting is a must on HPC projects, participants are required to be aware of new and non standard caluses in the specifications. Innovative construction proposals from an experienced concrete contractor during pre construction meetings with project engineers may prevent expensive delays later.

(5) Testing : In order to check the required properties, testing of concrete for 28 days compressive strength has to be conducted. Match curing is often specified for HPC construction. Match curing uses a special test cylinder mould (generally 100 mm x 200 mm) with a thermal jacket, thermo-couple and controller.

High performance concrete has been **used** for bridges, tall buildings and pavements.

Use of high performance concrete in **bridges** enables the standard prestress concrete girders to span longer distances or to carry heavy loads.

HPC greatly helps in reducing the time needed for the construction of **tall buildings.**

There is potential economical benefit that may be derived from the early strength gain of HPC for **pavement** construction.

Some areas, which require research in **future** of HPC are as under :

(i) Self-curing (ii) Methods of triggering set (iii) durability without air entrainment, (iv) composites (for example, synthetic fibres), and (v) Reducing thermal curve by making concrete more conductive.

It is the need for different construction works, which has introduced the concept of HPC.

32.6 CONCRETE NANO-TECHNOLOGY

Nano refers to one thousand' millionth part (10^{-9}, a very smaller part, numerically. It is worth while to point out that the word **nano-technology** has become very popular. It is used to describe many types of research, where the characteristic dimensions are less than about 1000 nano-metres.

For example, continued improvements in lithography have in line widths which are less than one micron (one millionth of a metre (1/10,00,000), which one is denoted by μ). This work is often called **nano-technology**. Terms **molecular nano-technology** or **molecular manufacturing** are often used in the literature in the broad sense instead of nano-technology. This technology allows

(1) to get essentially every atom in the right place

(2) to make almost any structure consistent with the laws of physics which may specify in molecular detail

(3) to have manufacturing costs not greatly exceeding the cost of the required raw materials and energy.

In addition to above, two more concepts are commonly associated with the **nano-technology** [namely (i) position assembly, and (ii) massive parallelism]. In case, first three objects are achieved simultaneously by using any are method, it is sufficiently satisfactory. However, it seems difficult without using some form of **positional assembly** (that is, to get the right places) and some form of **massive parallelism** (to keep the costs down).

Portland cement concrete technology is brought into the age of **nano-technology** by Federal High-way Administration (FHWA). The research is conducted at the level of the atom and molecule. The scale of such a research is generally 10 times the diameter of a water molecule (i.e., one billionth of a meter, a **nano-metre.**

Study is under the way on a variety of nono-technology applications within high-ways and bridge industries under FHWA,s advanced infrastructures research programme

(1) Alkali-silica reactivity (ASR) and delayed ettringite formations, the bane of concrete highways and bridges are considered as ills of concrete. These are explored at the molecular level using neutron scattering technology and other processes.

(2) Using neutron scattering technology, the fundamental research into the inter-actions between fly-ash and nano-structure of portland cement gel is under way.

(3) **Nano-technology** is providing a close-up look at the hydration of cement grains, and the nano-structure of cement reactivity as hydrated surfaces develop on individual cement grains.

(4) These feasibility of Cyberst or smart aggregate, as wireless sensors embedded in concrete or soil is under examination.

Both asphalt and concrete are nano-materials Dr. Richard Livinston senior physical scientist at FHWA's advanced infra-structure research programme, said ``until now it is not clearly under stood, what is going on at that level but what happens there, affects the performance of those materials.

Nano-technology may be defined as under

It encompasses research development and manufacture that utilizes and manupulates the unique properties of matter existing at the **nano-scale.**

At this scale, approximately 1 to 100 nanometers, clusters of atoms and molecules exhibit properties quite different from those found at larger scales. In this way, nanoscale, science and engineering involve more than mere hyper miniaturization. Instead, the discipline represent an opportunity to gain unprecedented insight into the unique phenomens existing at the nano-scale and to use that knowledge to engineer materials/devices with novel characteristics. With nano-technology, super-small devices may be designed and manufactured to infinitesimal degree of tolerance. Thereby, **nano-technology is defined** as fabrication of devices with atomic or molecular scale precision. A nano-tech education and networking consultant firm, at such a small scale, physical forces different from those of the ordinary human dimensions are at play.

Concrete is a porous material characterized by air voids ranging in size to **nanometer-scale** pores produced by the cement water chemical reaction. The combined effects of concrete porosity and composition are examined. These **nano-scale** pores control the properties of the calcium-silicate hydration product. This is the main glue which holds concrete together. It was seen that the concrete is some-ways, a nano-scale material

Under the assault of deicing chemicals that penetrate concrete's porous structure and oxidize the reinforcing steel within, causing cracking and deterioration to the structure, concrete's microstructures needs more attention. It was realized that the concrete is a complex composite. The addition to concrete mixes of nanoscale silica fume, an industrial byproduct of glass manufacture, is increasingly recognized as a means to improve durability of concrete structures exposed to deicing salts.

Silica fume additive operates at a nanoscale and makes more durable concrete. In case, it is put too much in the concrete will become brittle. It needs to better understand how it al works together, especially now that all kinds of new materials (e.g., ground granulated blast furnance slag or super plasticizing chemicals are added.

Nuclear resonance reaction analysis (NRRA) is used to study cement hydration at a **nano scale** level. The action on the surface of the cement particles as it hydrates is studied. Reactivity of cement with water on a nano-scale is measured. How to control the timing of concrete setting is studied. NRRA employs a beam of nitrogen atoms to probe a reacting cement grain to locate hydrogen atoms, a necessary component of water (or reaction by products) Results of the probe are plotted to create a hydrogen depth profile (a graph showing the rate of water penetration as well as the arrangement of various surface layers formed during the reaction).

A 20-nanometer thick surface layer acts as a semi-permeable barrier that allows water to enter the cement grain and calcium ions to leach out. However, the larger silicate ions in the cement are trapped behind this layer. As the reaction continues, a silicate gel forms there, causing swelling within the cement grain and eventual breakdown of the countermost layer. The surface disintegration then releases accumulated silicate into the surrounding solution, where it reacts with calcium ions to form a calcium silicated hydrate gel, which birds cement grains together and sets the concrete.

The evolution of the hydrogen profile shows the timing of the surface layer's break-down. The concrete setting process as a function fo time, temperature, cement chemistry and other factors is studied. In addition to NRRA studies, concrete material research conducted by FHWA's advanced infrastructure research programme includes following on going projects:

(i) Colloidal chemistry of Alkalisilicate Reaction (ASR) gel (ii) Fly-ash reactivity characterization (iii) Aggregate ASR potential test (iv) Delayed ettringate formation (v) cement hydration kinetics. These projects are related with Numoscala attack on ASR.

Yet another FHWA - funded project involves research on Cyberliths and their radio communication properties. This study target the feasibility of using wireless sensors in highway construction materials. Electromagnetic finite element models are used to determine the range and spatial (area) resolution of wireless sensors embedded in concrete or soil under various conditions.

These smart aggregate sensors may be embedded throughout the a structure during construction, added to the mix right before placement. Comprising the system are smart aggregate and a data reader that may be mounted on a car or truck. The reader powers the aggregate as it passes over them and collects the sensor onto a PC.

Each smart Aggregate contains wireless power receiver and data transmission coils, incorporates ceramic hybrid integrated circuit technology to withstand mechanical stresses and concrete's high pH environment. The aggregates are built to have a life-time of 50 plus years. This wireless power transmission and data collection technique eliminates the necessity and potential problems of batteries, cables and connectors. Recently manufactured prototype smart Aggregates are now undergoing reliability measurements.

Strictly speaking, while neither cyberliths nor smart Aggregates constitute nontechnology due to their size of several millimeters and use of conventional electronics, these do not demonstrate the value of embedded sensors and augers what may be accomplished in the future, as the technology is further reduced in size.

32.7 POLYMER-MODIFIED CEMENT, MORTAR AND CONCRETE

(References have been given at the end of this article).

Civil engineers remained in search for alternative to steels and alloys to combat the high cost of repair and maintenance of structures damaged by corrosion and heavy use.

History of using polymers as construction material is old. It dates back to 3000 BC. Polymers were used with time-based mortar and concrete in the construction of important buildings to meet the requirements of strength, impermeability and durability in the ancient times. The polymers were used as plaster, binders, stabilizing material and water-proofing material.

Some of the natural materials were used as polymers in the construction. Asphalt and bitumen are the oldest natural polymers used in our country, in Egypt and Babylonia. Rice-paste, rice-huts, clay, coconut fibres, jute blood, fruit-juice and natural rubber. Some natural materials and organic matter used as polymers have been listed in Table 32.1 alongwith their dates of use. Polymers were used in construction of the some historical structures in our world. Name of the structures, dates of use and natural material used have been listed in Table 32.2.

Table 32.1. NATURAL MATERIALS, NM AND DATES OF USE

A.N.	Natural Materials	Date of use		Place
1	Asphalt	3000	B.C.	Indus Valley
2	Bitumen	2500	B.C.	Babylon
3	Albumen	150	B.C.	Egypt
4	Blood	150	B.C.	Egypt
5	Curdled Milk	46	B.C.	Vitrifies
6	Barley	23	B.C.	Piney
7	Fruit juice	1200	A.D.	—
8	Cheese	1703	A.D.	Never Moon
9	Cotton	1850	A.D.	Burn ell
10	Butter	1873	A.D.	Vicat Smith

TABLE 32.2. NATURAL MATERIAL AS POLYMER STRUCTURE AND DATE

A.N.	Natural material	Date of use	Name of structure
1	Bitumen	3000 BC	Great Bath of Mohan-Dorado
2	Asphalt and Bitumen	2000 BC	Walls and roads in Babylon
3	Rice-husk with lime	400 AD	Amanita caves India
4	Rubber latex	1300 AD	Walls at Chan Peru
5	Rice-paste with lime	1500 AD	Great wall of China

Since 1940s, composite materials, formed by the combination of two or more materials in a microscopic scale, have geared increasing popularity in the engineering field.

Cement is partially replaced by polymer latexes, liquid resins and monomers than the composites so formed is termed as **Polymer-modified Cement mortar** (PCM) and **Polymer-modified cement concrete (PCC)**. The matrix of composite gel so farmed is homogeneous consisting of cement and polymer matrix interpenetrating with each other.

Polymer composites are multi-phase materials produced by combining polymer resins such as polyster, vinylester and epoxy, with fillers and reinforcing fibres to produce a bulk materials with properties better than those of the individual base materials. Fillers are often used to provide bulk to the material, reduce cost, lower bulk-density or to produce aesthetic features. Fibres are used to reinforce the polymer and improve mechanical properties such as stiffness and strength. High strength fibres of glass aramid and carbon are used as the primary means of carrying-load, white the polymer resin protects the fibres and binds them into a cohesive structural unit. These and commonly called **fibre composite materials.** Properties outfresh and (2) hardened PCM and PCC are as under:

1. **Fresh PCM and PCC** : Posses improved consistency due to ball bearing actions of polymer particles and entrainer air. As such its **workability** is better than conventional mortar and concrete. Flow of polymer-modified mortar and concrete increases with increasing water cement ratio. Entrapped air content is normally higher than the conventional concrete. The durability against freezing and thawing improves but the strength is reduced due to excessive air.

Fresh PCM and PCC have excellent resistance to bleeding and segregation due to hydrophilic colloidal properties of polymers setting time is delayed to some extent depending upon polymer type and polymer cement ratio.

2. **Hardened PCM and PCC:** Following properties are affected:

(i) **Strength**: Polymer cement ratio, water cement ratio, binder void ratio, air content and condition of the curing are the various factors which influence the strength. Tensile, flexural and shear strengths are increased. However, the increase in compressive strength is less as seen in Table 32.3. High tensile strength by the polymers and overall improvement in cement aggregate bond due to polymers are the contributory factors which result in increase in strength.

Table 32.3. COMPARISON OF STRENGTH IN POLYMER MODIFIED CONCRETE REL. (RELATIVE)

	Types of concrete	Polymer cement Ratio	Water/ Cement Ratio	Rel. Comp.	Rel. Flexure	Rel. Tensile	Rel. shear
1	Unmodified	0	0.6	100	100	100	100
2	Styrene Bitadience Rubber	5	0.53	123	118	126	131
		10	0.48	134	129	154	144
		15	0.44	150	153	212	146
		20	0.40	146	178	236	149
3	Poly-acrylic Easters modified	5	0.43	159	127	150	111
		10	0.33	179	146	158	116

		15	0.31	157	143	192	126
		20	0.30	140	192	184	139
4	Poly-vinyla-acetale modified	5	0.51	98	95	112	102
		10	0.44	82	105	120	106
		15	0.42	55	80	90	88
		20	0.36	37	62	91	60

(ii) **Modulus of elasticity:** It is slightly higher. It is listed in Table 32.4.

Table 32.4. MODULUS OF ELASTICITY IN POLYMER MODIFIED CONCRETE (E - x 10^5 N/mm²)

S.No.	Types of concrete	Polymer cement ratio	Modulus of Elasticity in compr. values of x.
1	Unmodified	0	2.11
2	Styrene Buta diene rubber modified	5	2.27
		10	2.36
		15	2.30
		20	2.24
3	Poly-acrylic Easterns modified	5	2.28
		10	2.43
		15	2.42
		20	2.02

(iii) **Poisson's ratio:** Its value about 0.16 to 0.17 is almost same.

(iv) **Drying shrinkage:** It is less than conventional concrete. The decreases in its value are due to increase in polyment cement ratio.

(v) **Creep:** It less than conventional concrete. The decreases in its value are due to increase in polymer cement ratio.

(vi) **Durability:** The pores in the microstructure are filled with polymers. Total pre volume and pore radius are reduced. Therefore, the durability is increased. Chloride ion penetration diffusion as listed in Table 32.5 indicated the reduction in values as compared to unmodified concrete carborateion depth is greatly reduced.

Table 32.5. CHLORIDE ION DIFFUSION COEFFICIENT IN POLYMER MODIFIED CONCRETE

S.No	Type of concrete	Polymer-cement Ratio	Chloride coeff. 100mm²/sec	Diffusion
1	Unmodified	0	2.2 x 10	–x
2	Styrene	10	1.9 x 10	–x
	Butadine Rubber-modified	20	9.3 x 10'	y
3	Poly-acrylic	10	6.2 x 10	–y
	Easters modified	20	5.8 x 10	y

Principle of polymer modification is as follows

The polymer particles are dispersed in the cement paste when the polymer latexes are mixed in the preparation of PCM and PCC. The hydration of cement produces CSH gel and the polymer particles are gradually confirmed in the capillary pores. As the hydration of cement progresses, the capillary water and pores further reduce. The polymer particles begin to flocculate and form a continuous closely packed layer around the gel and aggregate particles. With complete withdrawal of water, a complete membrane is formed around gel and aggregates.

PCM and PCC Mixing of ingredients

The method of mixing in gradients is same as conventional mortar and concrete except polymeric modifier. Steps to prepare mix-design are almost same as these for the conventional concrete. Cement to sand ratio for PCM is 1:2 to 1:3, polymer to cement ratio 5 to 20 percent, water to cement ratio 0.3 to 0.6.

Polymer to cement ratio for PCC is 5 to 15 percent and water to cement ratio is 0.3 to 0.5.

The final misdoing needs a number of trials in specific construction. The air content is normally high. Therefore, air entraining agents should not be used the process technology of PCM and PCC is similar to conventional mortar and concrete.

Placing and curing are as under:

Before placing the modified concrete, the surface should be cleaved and free from dust and grease. In general, the modified mortar may be recommended for thickness of 30 mm or less, and the modified concrete for thickness exceeding 30 mm. The mortar and concrete should be placed in the temperature range of 5 - 30°C.

PCM and PCC, require different curing. However, the curing is done in a different way. These need moist curing for 2-3 days only and then the dry ambient temperature curing is sufficient.

High tensile and flexural strength, high water proof ness and good chemical resistance are the properties which are making PMC a better preposition. The corrosion of steel is greatly checked as the chloride induced penetration, oxygen diffusion and moisture penetration is considerably low.

32.8 POLYMER MORTAR (PM) AND POLYMER CONCRETE (PC)

Polymeric binders and liquid resins are used to be replaced completely the cement hydrate and **composites** are made for PM and PC. There is no cement hydrate phase. Thermo-setting resins, vinyl monomers and tar-modified resins are the examples of polymeric binders.

Polymeric binders strongly bound the aggregates to each other in polymer mortar and concrete. This provides the uniform matrix phase around aggregate. The bond of adhesion in PC and PM is different than cement hydrate. It is principle of polymer replacement.

PM and PC Mixing of Ingredients

Epoxy resin (EP), unsaturated polyster-resins (UP), vinyl ester resin (VE) methalcrylate monomer (MMA) are most commonly used for polymeric binders. The polymeric binders are toxic and flammable. Therefore, proper safety measures should be taken. The most essential properties of aggregates used in PC and PM are soundness, low water absorption and high strength.

River sand, gravels, crushed sand and light-weight aggregates are used in making PC and PM.

Calcium carbonate, silica flour and fly-ash are used as fillers to fill the very fine pores.

Better bonding between aggregates and binders may be obtained, in case coupling agents are used alongwith binders. Colour-pigments may be used for aesthetics.

In the structural application mild steel bars, fibre reinforced plastic, glass fibres or steel fibres may be used with PM.

The polymeric binder contents are normally 9 to 25 percent.

The common procedure for mix-proportioning requires aggregates to be selected in such a way that the void volume is minimum. The typical mix proportion for PM and PC is as shown in Table 32.6.

The mixing should be carried out in a mechanical mixer to obtain uniform mixing of polymers and aggregates. Aggregates and fillers should be dry mixed in the mixer. The polymeric binder should be added and mixed for 3 to 5 minutes. After mixing PM and PC should be removed immediately keeping in mind pot life of polymer same as he process technology of polymer mortar and concrete is that for the conventional concrete.

Table 32.6. MIX-PROPORTIONING IN POLYMER MORTAR (PM) AND CONCRETE (PC)

S.No	Type	Polymer	Fillers	Fine aggregate
1	Polymer mortar	1	0 to 1.5	3 to 7
2	Polymer concrete	1	1 to 1.5	8 to 8.5

Before **placing** the polymer binder concrete, the surfaced should be cleaned and free from dust and grease. For **curing**, the most important precaution is to centralise the moisture content of concrete constituents.

In general, the binder mortar may be recommended for thickness of 5-10 mm, and the polymer binder concrete for thickness 50-100 mm with maximum upto 300 mm.

The **placement** should be done in the temperature range 0°-50°C.

The polymer mortar and concrete are subjected to ambient temperature cure, heat cure or combined cure.

Properties of PM and PC

Polymer mortar and concrete are **fresh** they have good resistance to bleeding and segregation.

The setting shrinkage of PM or PC is large as compared to cement mortar or concrete.

Due to large setting, there is problem in moulding or placing.

The workability is normally less in polymer mortar and concrete because of the short pot-life of the polymer. Therefore, controlling pot-life by adding suitable initiators, promoters or hardener is important.

Hardened PM and PC following properties:

The **strength** of polymer mortar (PM) and polymer concrete (PC) depend on polymer aggregate ratio and strength of aggregates. It has been seen that strength of PM and PC is high as compared to cement mortar and cement concrete as listed in Table 32.7.

Table 32.7. COMPARISON OF STRENGTH IN POLYMER CONCRETE (PC)

S.No	Type of concrete	Compressive strength (N/mm²)	Tensile strength (N/mm²)	Flexural Strength (N/mm²)
1	Unpolymeric concrete (ordinary)	100-600	10-50	20-70
2	Epoxy-concrete	800-1200	100-110	170-310
3	Polyster concrete	800-1600	90-140	140-350
4	Polyurethane concrete	650-720	80-90	200-230

Modulus of elasticity of PM and PC is same or slightly less as compared to cement mortar and concrete as listed in Table 32.8.

Table 32.8. COMPARISON OF WATER-ABSORPTION IN POLYMER CONCRETE

S.No	Type of concrete	Water absorption (wt %)
1	Unpolymeric (ordinary) concrete	4.0-6.0
2	Expoxy concrete	0.05-0.2
3	Polyester concrete	0.05-0.2
4	Polyurethane concrete	0.3-1.0

The **creep** of polymer mortar and polymer concrete depends on type and quantity of polymeric binders and temperature. Creep of PM and PC is almost same as compared to cement mortar and concrete.

As regards **durability,** the most polymer mortar and concrete have such a micro-structure that makes it impermeable. Therefore, the penetration of moisture and air is greately reduced as listed in Table 32.9.

Table 32.9. COMPARISON OF FREEZE-THAW DURABILITY IN POLYMER-CONCRETE

S.No	Type of concrete	Number of cycle Off/T	Dynamic Modulus of Elasticity $(x \times 10^4 \text{ N/mm}^2)$	Flexural strength N/mm²
1	Epoxy concrete	0	28.3	173
		100	28.1	170
		300	25.6	170
2	Polyster concrete	0	33.6	227

		100	33.6	224
		300	33.1	—
3	Polyurethane concrete	0	18.3	193
		100	17.4	184
		300	16.9	179

The chemical resistance, abression resistance and impact resistance is improved in PM and PC as compared as compared to unpolymeric mortar or concrete.

It may be seen from Table 32.10 that the freeze-thaw durability is high as compared to impolymeric mortar and concrete.

Table 32.10. POLYMER IMPREGNATED CONCRETE STRENGTH

S.No	Type of concrete	Compressive strength	Tensile strength	Flexural Strength
1	Impregnated concrete	37.0	2.9	5.2
2	PIC-Radiation	103.4	8.5	16.8
3	PIC-Thermal	70.2	5.9	8.2
4	Polymethyl Methyl crylate	142.4	11.4	18.5
			Impregnated concrete Radiation	
5	Polymethyl Methyl Crylate	127.7	10.6	16.1
			Impregnated concrete Thermal	

Currently polymer mortar and concrete are used for finishing work and repair work. It has also been used for construction of high strength pre-cast panels, grouting, anti corrosive coatings and flooring. These are put to use in Europe, USA and Japan.

32.9. POLYMER IMPREGNATED MORTAR (PIM) AND POLYMER IMPREGNATED CONCRETE (PIC)

The polymer impregnation is **defined** as fully or partially filling the poes of conventional cement mortar and concrete by means of polymers.

History of PIM and PIC is very recent. The **process** is new and costly.

There is no accurate method available to measure the depth of impregnation and therefore, the quality control is difficult.

The process of impregnating mortar and concrete consists of interpenetrating micro-structure of cement paste, polymer phase and aggregate phase. The improved microstructure increases the strength water tightness, chemical resistance and durability. Polymer impregnation depth and polymer loading control the main properties of PIM and PIC.

Properties of PIM and PIC

Low viscosity polymers are impregnated in the concrete in case the initial mortar and concrete have same properties. Water filled pores are filled by polymers during impregnation.

Hardened PIM and PIC have the following properties. Increase in compressive, tensile and **strength** is seen as in Table 32.1.

Increase in values of **modulus elasticity** of PIM and PIC are seen as in Table 32.11.

Table 32.11. POLYMER IMPREGNATED MODULUS OF ELASTICITY

S.No	Type of concrete	Modulus of Elasticity ($x \times 10^5$ N/mm^2)
1	Un-impregnated	x = 25
2	PIC-Radiation	x = 54
3	PIC-thermal	x = 52
4	Polymethyl methyl crylate impregnated concrete-Radiation	x = 44
5	Polymethyl Methyl crylate Impregnated concrete - Thermal	x = 43

Durability increases as in the microstructure of PIM and PIC, the pores are filled with polymers parameters indicating durability (such as water absorption, water permeability, divisibility and acid resistance) show increased performance as seen in Table 32.13.

Table 32.12. POLYMER IMPREGNATED CONCRETE WATER ABSORPTION AND DIFFUSIBILITY

S.No	Type of concrete	Compressive strength (N/mm^2)	Diffusibility m2h
1	Unpregnated concrete	6.40	0.0036
2	PIC-Radiation	0.51	0.0037
3	PIC-Thermal	0.70	0.0038
4	Polymethyl Methyl Crylate Impregnated Concrete Radiation	1.08	0.0038
5	Polymethyl Methyl crylate Impregnated concrete thermal	0.34	0.0036

Table 32.13. POLYMER IMPREGNATED CONCRETE FREEZE THAW DURABILITY

S.No	Type of concrete	Number of cycles	Weight changes %
1	Impregnated Concrete	490	25.0
2	PIC-Radiation	620	6.5
3	PIC-Thermal	620	0.5
4	Polymethyl methyl crylate Impregnated concrete ration	750	4.0
5	Polymethyl Methyl crylate Impregnated concrete thermal	750	0.5

Freeze-thaw durability in PIM an dPIC is high as compared to unimpregnated mortar or concrete.

Mix proportioning methodology adopted in PIM and PIC and the **process technology** are completely different than the previous methods Monomeric impregnants, base mortar and base concrete are the various materials used for the impregnation vinyl monomers (such as methyl methacrylate and styrene) are the common materials used for impregnation. The polymer loading in PIM and PIC is about 5 to 15 percent by weight in full impregnation. Cement mortar and concrete, asbestos sheet and gypsum products are the materials used for base mortar and base concrete.

While **Placing and curing**, the dryness of existing mortar and concrete influences the properties of PIM and PIC to a large extent. Therefore, the surface should be made dry and clean. The polymer impregnation depth is about 20 mm to 30 mm and process may be carried out in the low temperature range also.

PIM and PIC are used in flooring, pavements, water-proofing runways and deck coverings in Japan and Germany. PIM and PIC are presently used for abrasion resistant bricks, pre-cast panels and acid-proof panels. There are in development stage for field use.

32.10. POLYMER COMPOSITES IN CIVIL ENGINEERING CONSTRUCTION

Civil engineers remained in search (for years) for alternative to steels and alloys to combat the high costs of repair and maintenance of the structures damaged by corrosion and heavy use. Since, 1940s, **composite materials** formed by the combination of two or more distinct materials in a microscopic scale.

Definition: Polymer composites are multiphase materials produced by combining polymer resins (such as polyster, vinylester and epoxy, with fillers and reinforcing fibres to produce a bulk material with properties better than those of the individual base materials.

Fillers are after used to provide bulk to the material, reduced cost, lower bulk density or to produce aesthetic features. Fibres are used to reinforce the polymer and improve mechanical properties such as stiffness and strength. High strength fibres of glass aramid and carbon are used as the primary means of carrying load, while the polymer resins protects the fibres and birds them into a cohesive structural unit. These are commonly called **fibre composite materials.**

In the recent times, fibre composite materials have been increasingly considered for structural load bearing applications by the construction industries and have established themselves as a viable and competitive option for rehabilitation and retrofit of existing civil structures, as a replacement for steel in reinforced concrete and to a lesser extent.

Fibre composites have potential benefits, which include high specific strength and specific stiffness, tailorable durability, good fatigue performance and the potential to reduce long-term costs.

Use of some fibre composite materials may potentially produce light structures and lead to cost saving.

32.11. FIBRE REINFORCED POLYMER (FRP) AND ALTERNATIVE CONSTRUCTION MATERIAL

(References have been given at the end of this article).

In the aerospace and automobile industries, the use **`Fibre Reinforced Polymer (FRP)'** is well established. In civil engineering, use of FRP is gaining momentum world-wide.

FRP composites present a new generation of energy effi-materials. In the view of its high strength and module, fatique and corrosion resistance and di-electric properties, it is almost dominating over the metallic materials. This material in use since 1990s. However, recently, it has attracted the attention of engineers working in the construction of civil structures.

FRP structural shapes and structural reinforcement, offer many potential and possible alternative to conventional reinforcement.

Fibre reinforced composites are composed of fibres embedded in matrix material. Its fibres are small in diameter and posses high strength of composites. Matrix material provides support for the fibres and carrying the loads and stability to composite material synthetic organic fibres [Aramid Fibres (Kevier™)] and inorganic fibres [Glass, boron and carbon ceramics] are available in the world-market.

Glass is an important engineering fibre. It possesses a high specific strength, at the same time, it cost is low. The specific stiffness and specific strength of a fibre may be 13 and 16 times that of structural metals. Through the use of advanced composites, there is the potential for large weight saving.

Polymers, metals and ceramics are used as matrix materials in FRP composites. The most commonly used matrix materials are as under:

1. **Polymeric matrix**: for example, Phenolics, polystes, Epoxies, Polyprophtene, Nylon, Poly-ether-sulphone.

2. **Metal matrix** for example, Almunium, Titanium, Copper.

3. **Ciramic matrix,** for example, Carbon, silicon carbide and silicon nitride.

These are most commonly used matrix materials.

The actual advantages of using FRP may be estimated from the following points of view:

1. **Technical point of view**: Composite materials have significant advantages

(i) These offer significant weight saving over existing metals about 25-45 percent (as shown in Fig. 32.1.)

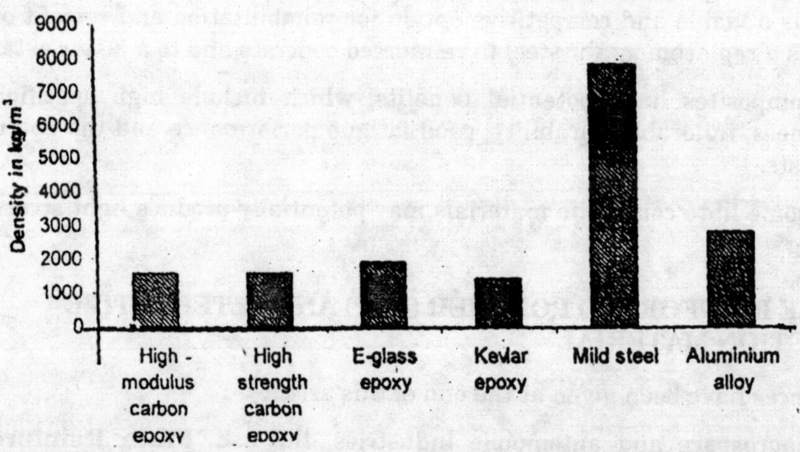

Fig. 32.1. Density of composite materials and conventional metals.

(ii) Undimensional fibre composite have specific tensile strength (ratio of material strength to density) about 6 to 14 times more than that of steel and aluminium (as shown in Fig. 32.2).

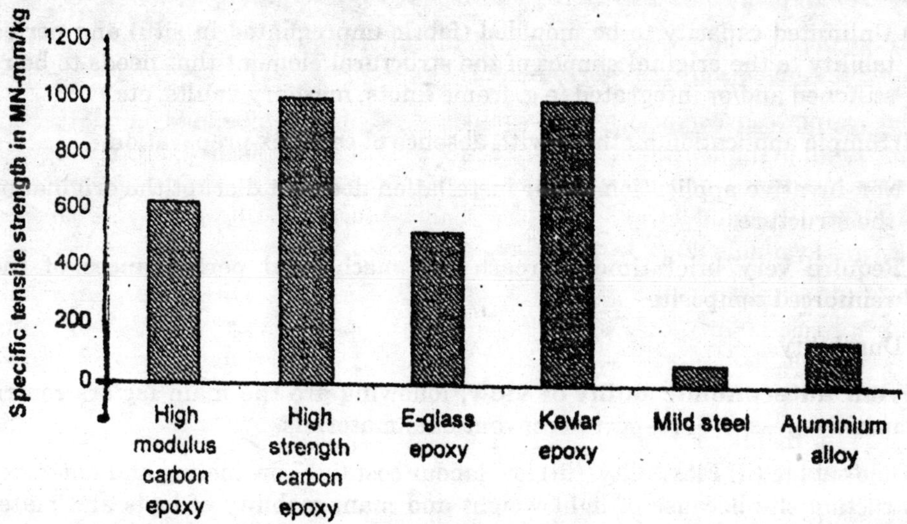

Fig. 32.2. Specific tensile strength of composite materials and conventional metals.

(iii) Undimensional fibre composites have specific modulus (ratio of materials stiffness to density) about 3 to 5 times more than that of steel and aluminium (as shown in Fig. 32.3).

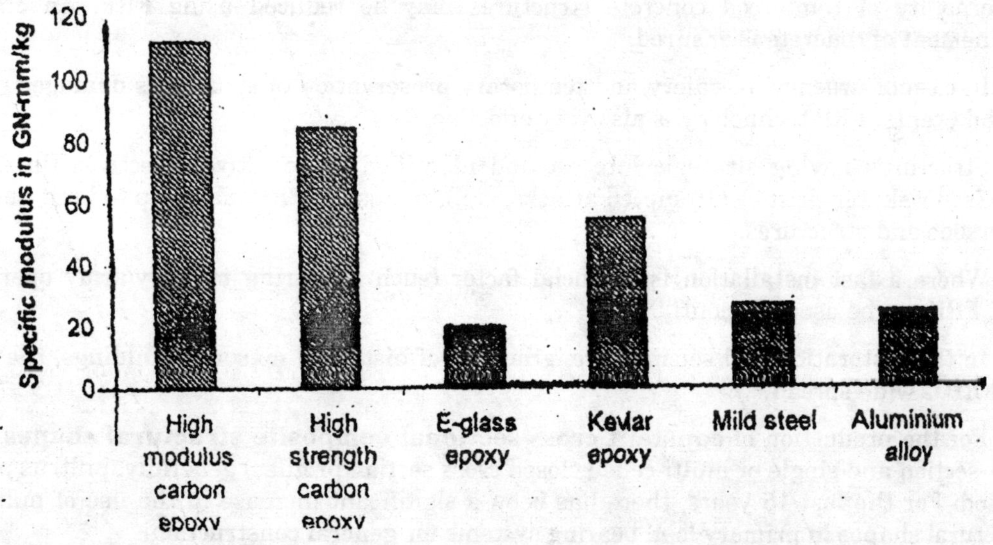

Fig. 32.3. Specific modulus of composite materials and conventional metals.

(iv) Fatigue endurance limit of composites may reach 60 percent of their ultimate tensile strength. This value is considerably lower for steel and aluminium.

(v) Corrosion resistance of fibre composites leads to reduced life cycle cost.

(vi) Fibre reinforced composites may be designed with excellent structural damping features. These are less noisy and provide lower vibration transmission than metals.

(vii) Unlimited capacity to be moulded (fabric impregnated in situ) and perfectly adoptability to the original shapes of the structural element that needs to be reinforced, stiffened and/or integrated (e.g. frame knots, masonry vaults, etc.)

(viii) Simple application methods with absence of complex preparation.

(ix) Non-invasive application. Their installation does not disrupt the original material of the structure.

(x) Require very brief time to reach full mechanical performances of the applied reinforced composite.

(xi) Durability.

2. **From an economic point of view,** following are the main factors contributing to their competitiveness with respect to conventional materials.

(i) Time saving (ii) Flexibility, (iii) low labour cost (iv) Low tooling and machinery cost on the construction site because of light weight and manageability of tools and material used. Advantages of strengthening reinforced concreted with FRP compared to conventional methods may be seen in 65 percent drop of total cost of application because of reduced labour needed and ease of installation in situ. (v) Possibility of restoring a structure without interrupting its utilization by users.

Use of FRP is more wide spread as an optimal innovative system. The seismic vulnerability of reinforced concrete structures may be reduced using FRP. An effective confinement of concrete is ensured.

In case of urgency, or safety and temporary preservation of structures damaged during special events, FRP technology is also very effective.

Structures having strategic interest and identified as sensitive objectives (in case of exploison risk, terrorist or attempted attacks, etc.), the use of FRP may help to limit damages of persons and structures.

Where a fast installation is a crucial factor (such as during military/army operations etc.), FRP can be used efficiently.

In the restoration and seismic strengthening of historical masonry buildings, the use of the FRP is wide spread.

For the production of **constant cross-sectional composite structural shapes,** (both open-section and single or multi-celled closed cross section profiles) generally, pultrus process is used. For the last 15 years, there has been a significant increase in the use of pultruded structural shapes in primary load bearing systems for general construction.

Prominent bridge and building structures have been designed and constructed using pultruded profiles, which include the following:

(1) Major pedestrian bridge 114 m long cable-stayed foot-bridge in Aberfeldy, Scotland. It was constructed in 1992 and designed by `Mausell Structural Plastics" (Beckenham, U.K.)

(2) 40 m long cable-stayed fibre-lined bridge in Kolding, Denmark. It was constructed in 1997 and designed by the Danish irm `RAMBOLL'.

(3) A six storey 19.4 m high load-bearing stair tower was recently constructed in Fort story, virginia of 250 mm x 250 mm pultruded I-section shapes manufactured by strongwell in Basel, Switzerland.

(4) A five-storey 15 m high residential/official building with a primary-load bearing structural frame of FRP shapes for the 1999 Swissbau fair.

In the past few years, FRP **bridge decks** have received maximum attention. It is due to their inherent advantages in strength and stiffness per unit weight as compared to traditional steel reinforced concrete decks. Weight of replacement decks in rehabilitation projects is reduced. It provides an opportunity for rapid replacement and reduction in dead load. Therefore, the live load rating of the structure is increased.

Using assemblies of adhesively banded pultruded shapes, the presently available commercial decks are constructed. Such shapes may be economically produced in continuous lengths by numerous manufacturing's using well-established processing methods design flexibility in this type of deck is obtained by changing the constituents of shapes (such as fibres and its orientations) and by changing the cross-sections of the shapes.

Externally bonded FRP composites are increasingly utilized for strengthening and retrofit techniques. These offers unique properties in terms of strength, lightness, chemical resistance, and ease of application. This meets the need for upgrading necessary due to the aging of infrastructure.

Various types of structural elements have been strengthed, including beams, columns, slabs, shear walls, joints, chimneys, domes and trusses. Flexural strengthening of RC elements may be provided by epoxy bonding the materials to proportions of the elements in tension with fibres parallel to the principal stress directions. Shear strengthening of RC elements may be provided by epoxy bonding FRP materials with fibres as parallel as practically possible to the principal tensile stresses.

Use of FRP reinforcement for reinforced and prestressed concrete [RC and PC] is commercially much less established than that of externally bonded reinforcement. Low-cost E-glass FRPs are generally chosen for non-prestressed applications, whereas high-strength carbon and aramid fibre are preferred for prestressed applications. Thermosetting resins such as vinyle stear and epoxy are the pre dominant polymers chosen for **internal reinforcement.**

REFERENCES

I. Self compacting concrete

1. Shikoku Island Concrete Research Association : Report by Self-Compacting Concrete Research Committee, ``Self-Compacting Concrete in Shikoku Island" 2000 to 2002, 2002.

2. Japan Society of Civil Engineers: Concrete Library 93, High-fluidity Concrete Construction Guideline, 1999.

3. Bernabeu and Laborde, ``SCC Production System for Civil Engineering", Final Report of Task 8.3, Brite EuRam Contract No. BRPR - CT96-0366.

4. Precase, Prestressed Concrete Institute, ``Interim Guidelines for the Use of Self-Consolidating Concrete in Precast, Prestressed Concrete Institute Member Plante", TR-6-03.

5. Okamura, H., Maekawa, K., Ozaqa, K., ``High-Performance Concrete", Gihodo Publishing, 1993.

6. Nakajima, Y., Nakazono, A., Mori, S., ``High Strength Self-Compacting Colored Concrete for Ritto Bridge Substructure (New Meishin Expressway)", Proceedings of the first fib Congress 2002, pp. 137-146. October, 2002.

7. Taniguchi, H., Tanaguchi, K., Uechi, H., Akizuki, S. ``Fabrication of Prestressed Concrete Composite Girders by Self-Compacting Concrete using Fly Ash", Technical Report of Sumitomo Construction Co., Vol. 120, 2002.

8. Billberg, P. Osterberg, T. Self-Compacting Concrete, Technique of use, CBI report 2 : 2002, Stockholm 2002 (In Swedish).

9. UGC SPONSORED FCMT, 2005. NATIONAL SEMINAR, February 21-22, 2005 Futuristic construction Materials and Techniques' organised by Department of structural Engineering Jai Narain Vyas University, Jodhpur 342001 (Raj.) India

II. (High Performance Concrete)

1. Aticin, P.C. and Neveille, A. (1993). ``High Performance Concrete Demystified". Concrete International, Jan, Vol. 15, No. 1, pp. 21-26.

2. Froster, S.W. (Oct. 1994). ``High Performance Concrete-Stretching the Paradigm". Concrete internation, Vol. 16, No. 10, pp. 33-34.

3. Focus Newsletter (May 1996). ``Workshop Showcases High Performance Concrete Bridges".

4. Goodspeed, C.H., Vanikar, S. and Cook, R.A. (Feb. 1996). ``High Performance Concrete Defined for Highway Structures", Concrete International, Vol. 18, No. 2, pp 62-67.

5. Papadakis, V. (1999). ``Experimental Investigation and Theoretical Modelling of Silica Fume Activity in Concrete", Cement and Concrete Research, Vol. 29, pp. 79-86.

6. Richard, Peirre and Cheyrezy, Marcel (1995). ``Composition of Reactive Powder Concretes", Cement and Concrete Research, Vol. 25, No. 7, pp. 1501-1511.

7. Zia, P., Leming, M.L. and Ahmad, S.H. (1999). ``High Performance Concrete-A State of the Art Report", SHRP-C/91-103, Strategic Highway Research Programme, National Research Council, Washington, D.C.

III. (Concrete Nano-technology)

1. Arun Parihar, Ajai Sharma : ``concrete nano-technology) A national Seminar `Future construction Materials and Techniques (February 21-22, 2005). Department of Structural Engineering, Faculty of Engineering MBM. Engineering College, Jai Narain Vyas, University of Jodhpur (342001), Rajasthan.

IV. Polymer modified Mortar PCM and Polymer Modified Cement PCC

* Neviile and Brooke, 1999, Concrete Technology, Addison and Wesley publication.
* Mehta and Monteiro, 1999, Concrete, ICI Indian Edition.
* Gambhir, 1998, Concrete Technology, Tata McGraw Hill Pub. Co. Ltd., New Delhi.
* Shetty, 2003, Concrete Technology, S. Chand and Comparny, New Delhi.
* Vance Dodson, Concrete Admixtures, Vein Nostrad Reinhold, New York.
* Chandra and Ohama, 1994, Polymers in Concrete, CRC Press.
* ACI Manul of Concrete Practice, Part 5, ACI publication.

REFERENCES :

1. Bank, L.C. and Gentry, T.R. 2000. ``Composite Materials for Roadside Safety Structute", Proceedings of ACUN-II International Composites Conference, University of New South Wales, Australia.
2. Barbero, E., Gango Rao, H.V.S., 1991. ``Structural Application of Composites in Infrastructure", SAMPE Journal, Vol. 27, No. 6 SAMPE, USA.
3. Carse, A. 1997. ``The Asset Management of a Long Bridge Structure Affected by Alkali-Silica Reaction". Proceedings of Concrete 97 Conference, Concrete Institute of Australia, Adelaide.
4. FHWA, ``Fibre Reinforced Polymer Composites Bridge Technology", US Department of Transport Federal Hyway Administration, Available March 2002.
5. Gowripalan, N., 1999. ``Fibre reinforced Polymer (FRP) Applications for Prestressed concrete Bridges", Proceedings of ACUN1 Conference Composites : Innovations and Structural Applications, UNSW, Australia.
6. Gowripalan, N. 2000. ``Design Consideration for Prestressed Concrete Beams with Fibre Reinforced Polymer (FRP) Tendons", Fibre Composites : Who Needs Them? A Workshop for Civil and Structural Engineers, University of Southern Queensland, Australia.
7. Hollaway, L., 2002. ``The Development and the Future of Advanced Polymer Composites in the Civil Infrastructure", Proceedings of the international Conference on Advanced Composites in Construction, University of Southampton, Available November, 2000.
8. Ko, K., *et al.*, 1997. ``Fibre Architecture Based Design of Ductile Composite Rebar for Concrete Structures", Proceedings of ICCM-11 : International Conference on Composite Materials, Australian Composites Structures Society, Australia.
9. UGC SPONSORED FCMT 2005 NATIONAL SEMINAR February 21-22, 2005. `Futuristic construction materials and Techniques' Organised by Department of Structural Enginnering Jai Narain Vyas University, Jodhpur 34200 (Raj.) India.

VII (Composite Materials)

1. Bank, L.C. and Gentry, T.R. 2000. ``Composite Materials for Roadside Safety Structute", Proceedings of ACUN-II International Composites Conference, University of New South Wales, Australia.
2. Barbero, E., Gango Rao, H.V.S., 1991. ``Structural Application of Composites in Infrastructure", SAMPE Journal, Vol. 27, No. 6 SAMPE, USA.

3. Carse, A. 1997. ``The Asset Management of a Long Bridge Structure Affected by Alkali-Silica Reaction". Proceedings of Concrete 97 Conference, Concrete Institute of Australia, Adelaide.

4. FHWA, ``Fibre Reinforced Polymer Composites Bridge Technology", US Department of Transport Federal Hyway Administration, Available March 2002.

5. Gowripalan, N., 1999. ``Fibre reinforced Polymer (FRP) Applications for Prestressed concrete Bridges", Proceedings of ACUN1 Conference Composites : Innovations and Structural Applications, UNSW, Australia.

6. Gowripalan, N. 2000. ``Design Consideration for Prestressed Concrete Beams with Fibre Reinforced Polymer (FRP) Tendons", Fibre Composites : Who Needs Them? A Workshop for Civil and Structural Engineers, University of Southern Queensland, Australia.

7. Hollaway, L., 2002. ``The Development and the Future of Advanced Polymer Composites in the Civil Infrastructure", Proceedings of the international Conference on Advanced Composites in Construction, University of Southampton, Available November, 2000.

8. Ko, K., *et al.*, 1997. ``Fibre Architecture Based Design of Ductile Composite Rebar for Concrete Structures", Proceedings of ICCM-11 : International Conference on Composite Materials, Australian Composites Structures Society, Australia.

9. UGC SPONSORED FCMT 2005 NATIONAL SEMINAR February 21-22, 2005. `Futuristic construction materials and Techniques' Organised by Department of Structural Enginnering Jai Narain Vyas University, Jodhpur 34200 (Raj.) India.

VIII. (Fibre Reinforced Plastics)

1. Alampalli, S., *et al.*, (1999). ``Fiber-reinforced Plastics for Bridge Construction and Rehabilitation in New york." Materials and Construction : Exploring the connection, Proc., 5th Materials Enginnering Congress, L.C. Bank, ed., ASCE, Reston, Va. 344-350.

2. Bakis, C.E., *et al.*, (2002). ``Fibre-Reinforced Polymer Compsites for Construction State of the Art Review." J. Compos. Constr. 6 (2), 73-87.

3. Balasamo, A, *et al.*, (2001). ``FRP in Construction : Applications, Advantages, Barriers and Perspectives." Proc., of the Int. Workshop, Compsites in Construction a reality, ASCE, Capri, Italy, 58-64.

4. Chandak, R., Upadhyay, A. and Bhargava, P. (2004). ``Shear Lag Effect in Laminated Composite Box beams". Int. Conference on Recent Advances in Composite Materials. IT-BHU, Varanasi.

5. Davalos, J. F. *et al.*, (2001). ``Modeling and Characterization o Fiber-reinforced Plastic Honeycomb sandwich Panels for Highway Bridge Applications". J. Compos. Constr. 52 (3-4), 441-452.

6. GangaRao, H.V.S., *et al.*, (1999). ``Development of Glass Fiber Reinforced Polymer Composite Bridge Deck." SAMPE J., 35 (4), 12-24.

7. Gupta, L. (1998). Advanced Composite Materials, Himalayan Books, New Delhi.

8. Kaw, A.K. (1997). ``Mechanics of Composite Materials." CRC Press, New Delhi.

9. Lesko, J.J. *et al.*, (2001). ``Characterization and Durability of FRP Structural Shapes and materials." Proc. of the Int. Workshop, Composites in Construction a reality, ASCE, Capri, Italy, 110-120.

10. Nanni, A., *et al.* (1993). Fiber-reinforced plastic for Concrete Structures : properties and Applications, ElsevierScience, Amsterdam.

11. Neale, K.W. *et al.*, (2001). ``FRP's for Strengthening and Rehabilitation: Durability Issues." Proc. of the Int. Workshop, Compsites in Construction a reality, ASCE Capri. Italy, 102-109.

12. Pecce, M. (2001). ``Structural behaviour of FRP profiles." Proc. of the Int. Workshop. Composites in Construction a reality, ASCE, Capri, Italy, 241-249.

13. Taerwe, L. and Matthys, S. (2001). ``Survey of European Guidelines for the Use of FRP in Structural Applications." Proc. of the Int. Workshop, Composites in Construction a reality, ASCE, Capri, Italy, 19-27.

14. Tan, K.H. (2001). ``Unlifted Code on structural use of FRP : Suggestions for Strengthening Design." Proc. of the Int. Workshop, Composites in Construction a reality. ASCE, Capri, Italy, 28-35.

15. Upadhyay, A., Kalyanaraman, V. (2003). ``Simplified analysis of FRP box-girders". J. Composite Struct., Elsevier, 59, 217-25.

16. Vinson, J.R., and Sierakowski, R.L. (1986). ``The Behaviour of Structures Composed of Composite Materials." Martinus Nijhoff.

APPENDIX - A

EARTHQUAKES
(KEEP AWARENESS ABOUT EARTHQUAKES)

A.1 INTRODUCTION

There is a need to create, to keep awareness and to educate the persons of our country about `Earthquake'. Wide spread rumours and misconceptions about tremors remained a daily event. Grief sticken people of Gujarat still prefer to sleep out in the open, and have developed a deep aversion (great dislike and hatredness) for multi-storeys. Among the many rumours being spread, three are particularly dangerous (viz., earthquakes may be predicted, aftershocks are less severe and all concrete buildings kill).

S.K. Srivastava, additional director general of Indian Meterological Department, said that any one may predict that an earthquake is going to happen in two hours, simply because there are 10,000 earthquakes of magnitude 03 (three) on Ritcher scale happening every-year. There are 3000 earthquakes of magnitude 05 (five) and above every-year, that is, three earthquakes per day on an average.

A.J. Lal, deputy advisor, ministry of statics and programme implementation, echoes his view point. About 1,00,000 (one lakh) earthquakes occur every-year, majority of 04 (four) on the Ritcher-scale happen every year.

Two-thirds of the country lies in seismic zones of moderate to severe intensity. Himalayan, Naga Lushai region, Indo-Gangetic plains, Western India, Kutch and Kathiawar regions are geological unstable regions.

According to A.R. Chandrashekharan, formerly professor of department of earthquake engineering at university of Roorkee, the idea of earthquake prediction has undergone major changes over time.

One should be able to say, when, where, and of what size would an earthquake is likely to occur. IS : 1893-2000, our country is divided into five seismic zones. (Kutch, North East India and Some Himalayan regions as belonging to high-intensity seismic zones). Even this draft revision cannot predict the activities of tectonic-plates.

In our country, there is no region which is earthquake free. But in the world, no earthquake goes unnoticed. Such is the sensitivity of the observatory net-work this helps in predicting after-shocks. Many take after shocks lightly. But these could be severe and as devastating as the first earthquake.

One can certainly predict the after shocks. In Later, many people died of. But in Gujarat, on January 27 itself, proper warning about after-shocks of intensity ranging from 5 to 7 was given. A red diet was declared and 14 lakh people came-out of their houses, thus avoiding major trajedy. Srivastava has said, `Science saved people's lives by preciting after-shocks.

Scientists studying the Garhwal earthquake were worried about the western coast of India that includes the highly active seismic zone of Ahmedabad, Ajmer, Delhi (Aravali-range) and had predicted that there would be stronger earthquake in the region. Their belief was based on the ongoing paleo-seismological investigations along Himalayan region. Knowing the vulnerability of this region to large earthquakes, the necessity for adoption of an altogether new strategy was emphasised.

Studying Chamoli earthquake and Uttarkashi in 1991 in detail, most geologists were alarmed over the high seismicity along the main boundary-thrust and Main central thrust due to rapid rate of convergence between the Indian and Eurasian plates, National and International Workshops from time to time at the Wadia Institute of Himalayan Geology here have expressed concern over the neo-tectonic activities in the region. any geologist attending a work shop at the Institute inferred that a great detachment underlying the sedimentary and metamorphic cover of the Himalayas, which raptures from time to time leads to great earthquakes in the region. During the past 100 year apart from Uttarkashi and Chamoli earthquakes, which hit the region within the past one decade, four major earthquakes measuring above seven on the Ritche-scale, shock the sub-Himalayan region causing a great loss of life and property. These big tremors hit the region in 1897, 1905, 1934 and 1950.

V.C. Thakur, former director of WIHG, said that there were inherent similarities in source processes of these earthquakes which point to their occurrence on the sub-horizontal plate boundary faults. The diverse data, according to other Himalayan geologist, K.N. Khatri, CSIR emeritus scientists, provide ample evidence that a uniform type of deformation exists in the Himalayan region atleast for the occurrence of great earthquakes. Khatri said `thrust type earthquake hypo centre define the plate boundary fault and sense of slip on it. Hence, the future earthquakes are expected to occur in a similar manner. The physical processes involved in the evolution of the Himalayas are well understood under the theory of plate-tectonics.

According to another senior scientist at the WIHG, Devendra Pal, a major fundamental fracture which divides the Himalayas in two parts (eastern and western), is very active scismically. It has been named as `Delhi-Haridwar-Harsil' ridge' (DHH). While the western part of this ridge comprises Himachal, Jammu and Kashmir, the eastern part comprises eastern Garhwal, Nepal and Kumaoun. This underground ridge is 65 kM to 70 kM in width between Ganga and Yamuna across Dehradun and is expected at a depth of 2.5 km.

The structures of east and west are totally different. Pal gives a detailed scientific explanation of the difference between the two structures. It is said that DHH forms a western limit of newly created India-China plate.

The ridge originates from Cariberg ridge in the Indian Ocean, runs as submarine ridge via chagos archipelago. Male, Lakshwa deep and Gulf of Cambay along the western coast of India, Ahmedabad, Ajmer, Delhi (Aravali-range) and continues in north-north eastern direction as an undergound ridge between Delhi and Haridwar and further between Himalayas in Harshi in Uttarkashi.

Robert S yeast, a scientist from Oregon State University, said at an international Workshop at WIHG some months ago that there was a consensus among the scientists that future great earthquakes will strike on the Himalayan Front. Some of these may be magnitude 8 in the Ritcher-Scale. He said because of great increases in population in the region, loss of life and property in the impending earthquakes are expected to be catastrophic. It is however, well understood among the scientists that seismic hazard assessment is quite of risky job because our understanding of seismic processes resulting in earthquakes is very limited.

Having found the region highly earthquake prone, some scientists have begun to talk about an adhoc disaster management plan and need for evolving a definite methodology for the prediction of major earthquakes in the area. Such a plan, according to them, would also have to address the problem of seismic safety on top priority basis. The earthquake strikes without warning. This fact calls for the state of preparedness at every level. Past failures in managing the past-earthquake disaster in uttarkashi and Chamoli further demands a serious planning and strategy for disaster management in the Himalayan region.

A.2 POSSIBILITY OF MANY MORE EARTHQUAKES

While, Gujarat grapple with (struggle and try to conquer) the horific devastation, for the rest of the country, seismic zones in particular, the hard times are not yet over.

Knowing the fact that our country (India) is moving north towards the Himalayas at the rate of 76 mm per year, the enormous stress produced in the Indian tectonic plate due to collosion in the Indian tectonic plate due to collosion with the Himalayas, could lead to many more earth-quakes. However, what is more disheartening is the fact that while other countries with high earthquake-prone. Zones are making use of the available technology to make estimated fore casts about areas and time frame of large earthquakes, our government appears to remain legged behind.

Countries like Japan, China, Turkey, U.S. (California is a good example) and those of European Union are using Global Positioning System (GPS) satellites these orbit the earth and monitor its surface to collect data. These satellites provide data on the rates at which plates are moving and the crust is deformed along plate boundaries and faults. This data helps to estimate size, frequency and probability of occurrence of future earthquakes.

Veteran seismologist at the Massachusetts Institutes of Technology, Cambridge, US Professor M. Nafi Toksoz, also the founder of Earth Resources Laboratories here, says the technology is available anyone may purchase it. ``It's matter of commitment of resources on the part of any government. Indian londmass is moving faster than the plate boundary on its western part and no one may say whether or not another earthquake will hit other high seismic zones in near future.

With the GPS Satellite orbiting the earth, special sensor, called GPS receivers, are placed on the ground close to known fault lines. These sensors may measure ground movements as small as one mm. Acknowledging that there is no credible way by which the location, time and magnitude of a large earthquake may be predicted with some measure of certainty. Professor Toksoz says `Scientists simply do not understand the physics of how a big earthquake occur. He however there are three ways in which the problem may be countered.

``The public should be educated that they live in a seismic zone and that if an earthquake occurs the ground will move to a certain degree. New buildings should be earthquake resistant and old ones, especially those of critical facilities, should be strengthened. Finally local authorities should train people for response and have logistical arrangements. If the public is aware, it will pressure the government to do something.

Emphasizing stringnet (severe/strict) construction rules and enforcement, scientists here cite two examples earthquake in North-ridge, California, in 1994 (Moment magnitude 6.8) and that in Kube, Japan, in 1995 (of moment magnitude (7.2). In many ways, two earthquakes were similar, the devastation differed widely. In the former, 57 persons were killed, 9300 injured, 13000 buildings destroyed and total damage was estimated of $13 billion. In Kube, about $200 persons died, 26000 were injured, 50,000 buildings were destroyed and the economic loss was a staggering $200 billion. Investigators attributed the difference in loss to varied building construction. Recent buildings bylaws in California require structures to be set back by a certain minimum distance from active faults and those houses on sediments that may liquefy must use specific building techniques. Stressing enforcement professor Toksoz says through Turkey had a large earthquake as recent as the 1970 and had building codes n place but poor enforcement led to heavy casualities in 1999.

No such reason holds good for our country except that lackadaiscal (to become dull) government officials quoted as say `they were not expecting this,' which make them the laughing stock of global community.

Earthquakes are classified as **tectonic earthquakes**. It is called so to designate its intimate relationship to the structural changes in the crust and upper mantle. Two other groups of earthquakes may be distinguished which owe their origin to volcanism and karst phenomena. However, these occur seldom and are rather weak.

Earthquakes are frequently preceded by weak **fore-shocks.** Fore-shocks may presumable be accounted for by the overstress and failures at some weak locations, which however, do not yet cause a major rupture.

The unreleased energy continues to develop stresses in the new bonds, which after a while exceed the breaking strength of some bonds, resulting in a new rupture and an **after shock** a new shock.

A.3. PREDICTION OF EARTHQUAKES

``Earthquakes cannot be predicted by any one, anywhere, in any-country. This is a scientific truth", despite what some my claim.

Gujarat earthquake was in terms of magnitude, the worst the country has experienced in more than a decade. Though the country has observed and surveyed, bigger ones in the past one in Assam, on Independent day in 1950, measured 8.5 on the Richter scale. A way back in June 1819, Gujarat's Kuchch area, seismically active, saw a big one measuring 8 on the scale. Maharashtra, which falls in the least earthquake prone, `Zone I' area of the country, saw a devastating earthquake in Latur and Osmanabad districts in September,

1993. That earthquake measured 6.4 on the Richter scale and took the lives of more than 7,500 people.

Our country's most active regions seismically, where earthquakes of magnitude 8 or more could occur are the Himalayan belt, the North-east, and Kuchch. Delhi incidently, falls in the next most active area category.

Scientists can tell us this much. But there is one thing that has not changed in all these centuaries, it is acknowledged inability to predict an earthquake, say R.R. Kelkar, Director - general of the India, Meterological Department (IMD).

Bhavnagar, for instance, had been felling the tremors for some time. These could be observed but could not, says Kelkar, be taken as hearlding a big earthquake. So what can residents of earthquake prone areas do? Re-engineering homes to survive earthquakes big and small.

How much damage an earthquake does may depend on more than one factor. For instance, scientists say, the depth of earthquake's epicentre may make a difference. Earthquakes may be upto 700 km deep, if this is so, the surface damage may be less. But, Gujarat earthquake relatively shallow, just 15 km deep. The earthquakes in Garhwal in March, 1999 which killed 103 persons was 25 km deep.

Nor do scientists believe, there is anything very unusual about the recurrence of such unheavals. On an average, they say, the world sees 18 major earthquakes measuring 7 to 7.9 on the Richert-scale every year. And about a thousand moderate earth quakes a year measuring 5 to 5.9 on the Richert scale. Very often, scientists say, these occur in un-inhabited areas and do not make such an impact. There has already been one big earthquake this year, measuring 7.6 on the Richert scale, it left 700 persons died in EI salvodor.

But while scientists take most earthquakes as usual, there are some which even they find unusual. For example, three earthquakes exceeding a magnitude of 8 occurred on December, 16, 1811, and February 7 and 23, 1812 in New Madrid (near Missouri in the US). More recently in January 22, 1988, three earthquakes of a magnitude of more than 6 occrred in a 12-hour span in Tennant Creek, Australia. These were out of the ordinary, most are not Seismologis in many countries of the world are on lookout for reliable techniques that would enable them to predict the location, time of occurrence and severity of earth-quakes. In case, the earthquakes may be forecast many eventualities may be avoided. In case the duration and return periods of the earthquakes become known, the designers would be able to decide actual need to render buildings and other structures earthquake-proof (earthquake-resistant). The cost of antiseismic measures may be cut-down and unnecessary strengthening of some solid structures may be avoided. In fact, the mss evacuation of people and valuable to safe places is obviously quite a big problem in itself, especially in cities and towns. It can hardly be regarded as a way out in case the nearby town is nearly demolished, which is inevitable in case the buildings have not been designed to carry seismic loads.

However, efforts may be made to solve the problem. S.V. Polyakov of USSR [Reference Design of Earthquake Resistant Structures, Mir Publishers Moscow, in second edition, 1985] in his book suggested the lines on which attempt may be made for the perdition of earthquakes.

Strain accumulation that takes place in rock before sudden rupture may be observed. This process is accompanied by variations in magnetic fields, bench marks and other factors, which once measured, may indicate that the rock in that locality is about the exceed a limit.

The relative velocities at which P and S-waves arrive at seismograph stations first decrease and then increase. The velocity changes are brought about by stress build-up in, and the ensuing micro-cracking of the rock. The interval between the instant when the difference in velocity between P- and S-waves. begings to rise and the instant at which the earthquake occurs makes up about 10 percent of the time during which the velocity changes are observed.

E. Savarensky and I. Nersesove of the USSR, in 1976, have submitted their report at the UNESCO conference. It was also submitted that the efforts may be made to solve the earthquake prediction problem.

It may consist in recording the changes in void pressure in wells and measuring crust deformations by strain-members and clinometers, using geomorphological and geodetic techniques observations were made in Japan in 1964, when in few hours before the Niigata earthquake vertical strain sensitive devices recorded abnormal tensile strains in the range of 10^{-5} along a distance of 10 km.

Otherway is to study the progress of minor earthquakes, which may frequently undergo noticeable changes sometime before a major earthquake. The relative velocities of man-made, explosion induced P and S-waves may like wise change during that period if the waves happen to cross an earthquake focus on their way to the station. Changes in the vibration spectra may take place too.

Quite a reliable approach may be to study changes in geoelectric and geo-magnetic fields in seismic regions. In Japan, geo-magnetic survey had been carried out in 1912. Well defined changes in geo-magnetic field were observed during the Nankaido earthquake of December 21, 1946.

A promising approach in the earthquake prediction may be to study geo-chemical precursors of an earthquake From Tashkent, specialised worked out on this approach. It relies upon the fact that the isotopic analysis of the inert gases and other chemical elements in wells undergoes a noticeable changes before major earthquake. In 1966 during the Tashkent earthquake observations were made on these lines. Further work is being done in ascertaining its applicability.

According to the results of the statistical analysis carried out by S. Fedotov, seismic events taking place in the same area keep within a definite temporal `seismic cycle'. Using this approach, Sovient seismo-logists are trying to work-out an earthquake forecast for the next 30 years for the Kuril-Kamchatka region.

Above approaches need more efforts.

A4. DISASTER MANAGEMENT CENTRE, YASHDA, (MAHARASHTRA)

The number of casualities resulting from the earthquake in Gujarat (January, 26, 2007) would have been far less had the Gujarat government followed the Maharashtra government's example of setting-up a full fledged disaster management centre.

From the devastations due to earth-quake in Latur in 1993 the Maharashtra government established a fully functional disaster management centre and control-room at Yashda (Yeshwant Rao Chavan Development Academy), Pune.

This centre has developed a comprehensive disaster management plan for Maharashtra and has created a state-wide wire-less and VSAT link between the state head quarters and Mumbai and district collectors and tehsildars across Maharashtra.

This centre has also created district wise data bases listing of the location of heavy earth-moving equipment, blood-banks hospitals, doctors, specialists and others, says Dr. Vinod Menon, Professor of disaster management at Yashda's centre for Disaster Management Dr. Menon said `While one cannot prevent earthquakes, one can definitely minimise the impact of the disaster if one knows where the resources are located and how rapidly these can be deployed.

It was pointed out that the centre located in Pune was linked via Internet, wire-less and Intenor to district head quarters across the state.

Using Geographical Information System (GIS) technology, the centre had also created a `disaster management information system for Maharahs' which comprised digitised mapping of all tehsils. The digitised computer-based maps enable the pin-pointing of location of villages, availability of disaster management facilities and alternative routes to reach the villages via the state and national high-ways.

Emphasising that earthquake-resistant construction techniques were extremely important to minimise the number of casualties during earthquakes, Menon said `Civic authorities in urban cėntres located in earthquakes prone areas have to seriously enforce building regulations and adopt earthquake resistant techniques, such as Lintel beams, retrofitting and seismic strengthening.

Menon pointed out that the loss of life in the Taiwan earthquake of September, 1999 was far lesser than that in the Turkey earthquake of August 1999 due to the strict enforcement of the building code.

A5. NECESSITY OF STUDIES

A three-year study of the Kutch region by the Centre for Earth-Sciences in Thiruvananthapuram had indicated in November, 2000 that the seismogenic fault of Kutch, a boundary fault of the rift system appears to be capable of generating larger earthquake of 7 or more on the Richter-scale.

The seismologist, C.P. Rajendran, pointed out that the study commissioned by the department of science had closely scrutinised the rupture zone of 1819 earthquake scrap with 4.3 m height was activated by the reverse stress activated by Himalayas. A reverse stress is when a part of surface moves up relative to the foot-wall that makes two uneven blocks.

The earthquake of 1819 had devasted Bhuj and caused heavy devastations in the whole terrain.

The prevailing theory of Indian Sub-continent tells us that the Indian plate severed from African Gondwana region floated around for pretty long time before collding with the Asian mainland nearly 40 million years ago. The hit produced Himalayas and the stress or compression of the process continues.

It is also a scientific conjecture that the southern part of the country is pressing further to the north, which will generate tremors of varied intensity. According to Kusala Rajindran, senior scientists of CESS, the CESS has documented the Kutch part well. It has shown that there existed more tan three slightly curvy rifts running horizontally to a 100 kM stretch. These become dormant or active, according to the stress generated.

Another major rift, called the Gult of Kutch, exists to the South. Apart from this, there are two semi-vertical rifts stretching from Rajasthan to Broach.

The scientists had tired to answer why rift-related tremors in Kutch, which formed the stable continental Regions was in surprisingly quick frequency. The earthquake of 1819 was followed by the 1956 Anjar tremor. The gap between 1956 and 2001 is pretty quick." The relative proximity of the collision boundary may be a possible cause." It was suggested a detailed geological geophysical and geodetic studies to determine the causes and precautionary measures may be made.

The tremors in Kerala and Tamil Nadu in the month of December, 2000, however, have no possible connection with the Gujarat earthquake of Sunday, January 26, 2001.

C.P. Rajendran has said while human knowledge is not matured enough to forecast an earthquake like weather. The scientists have fairly good information of the tremors in SCRs. It makes the authorities duty bound to take two precautionary measures (modification in existing building and sufficient medical facilities for a calamity, if it happens).

A6. DEVISE ELABORATE CRISIS MANAGEMENT SYSTEM

It is not easy to live on the back of a monster.

The residents of sun-shine state, California have learnt to live. In California, the San Andreas fault lies on the top (An intricate network more than 1312 km long). Literally thousands of earthquakes (big and small) occurs in the state each year.

In order to cope, it has devised an elaborate crisis management system. For one thing, the government is actively involved in pro-active planning. The Division of Mines and Geology develops state-wide earthquake risk maps using information abut California's regional geology and earthquake history.

To record how the ground and the structures shake during earthquakes, the California strong motion Instrumentation Programme installs and maintains instruments in and outside buildings, on dams and bridges and in natural sites.

Seismologists and engineers analyse the data, compare it with information about how the buildings and ground behaved when shaken and use the information to improve building codes and design safer buildings. In San Francisco, containers of layered steel and rubber are situated below the foundation of the buildings to act as shock absorbers. In several cities, the programmes are underway to strengthen a tear down old buildings most likely to collapse during earthquakes.

Under the seismic hazards maping act, the state geologist is required to establish regulatory zones and issue appropriate seismic hazard maps. The maps are distributed to cities, countries, and state agencies for planning and controlling construction and development. Before a development permit can be issued or a sub-division approved, a site-specific investigation is required to determine whether a significant hazard exists at the site and, if so, recommended measures to reduce the risk to an acceptable level. Stringent disclosure norms have also been put in place, if a property is located in a seismic hazard zone, the seller or the seller's agent must disclose this fact to a potential buyer.

Local emergency management offices also regularly keep in touch with the community and disseminate information about how to handle earthquakes. Thus, citizens are regularly advised to fasten shelves securely to walls, breakable items in low, closed cabinets with latches, hang heavy items such as pictures and mirrors away from beds, couches, and anywhere people sit, ...etc.

Families are also urged to keep handy dissaster supplies like flash-lights and extra batteries, portable battery, battery operated radio, a first aid kit and manual. Apart from that, they are encouraged to develop an emergency communication plan in case, they get separated.

The preparation is not restricted to minimising before the event. The entire state machinery swings into gear the moment an earthquake occurs. Scientists and emergency managers are constantly searching ways to improve relief efforts. One of these techniques is the use of seismograph networks.

In 1994, the Northridge earthquake in southern California left more than 40,000 people homeless and 16,000 residents unsafe and occupation in Los Angles. Scientists and engineers instantly began to prepare a shaking intensity map for all parts of the greater Los Angles area.

Although damage typically decreases with distance from the earthquake source local geologic conditions can alter this pattern by amplifying shaking in specific areas. By showing the estimated severity of shaking and the level of damage associated with such shaking, a shaking intensity map becomes a valuable tool for emergency managers.

The North ridge earthquake was the first instance of such a map being used during a disaster. It was finished long before a complete picture of the damage could be assembled. Using the map, emergency managers located the hardest-hit areas and rushed relief teams there.

The map also helped emergency managers to reduce the time for delivery of relief money to individuals from weak to hours. It has now become a standard feature of post-earthquake relief efforts in California and provides a classic example of how technology and common sense can be used in conjunction to cope with crises.

A7. EARTHQUAKE ALARMING INSTRUMENTS

After having devastation due to earthquake in Kutch, Saurashtra and in Ahmedabad (Gujarat) on the Republic Day (January, 26, 2001) a local firm has entered into a tie-up with an American firm for marketing of an earthquake alarming instrument (an earthquake indicator) developed by the California based company.

This local company is first in the country to place an order of more than 2000 such instruments the first consignment of which might have been received within a week after the said earthquake. Several people placed their orders to purchase, the alarms in about Rs 2500/- in the local market.

This alarm provides early warming by detecting primary (P) - waves of an earthquake before the destructive shear (s) - waves, which come later, strike. It has been designed to provide instant warning of seismic activity.

It is a small device not much bigger than a thermo-stat that can be mounted on a wall with adhesive velcro-strips and operates on a nine-volt battery. The internal components are made of a patented reverse pendulum system that is sensitive to p-waves that precede an earthquake. This pendulum moves, tripping a switch circuit which sounds the loud buzzar.

This alarm gives about 45 to 50 seconds time to help children or others find a safe place such as under a table and the buzzer is loud enough to wake you up. This earthquake alarm was demonstrated before scribes by directors of the local company (Kalpak Shah and Jagdeep Shah).

According to them, a survey conducted by the United states Geological department has shown that S-waves travel 1.75 times slower than P-waves persons should use this earthquake alarm because it can wake you up at night and at the time of the start of an earthquake. Besides, it confirms that what you might be feeling is in fact the tremors of earthquake.

Jagdeep Shah said `An earthquake produces many waves caused by the plate or rock movements beneath the surface. Two main waves are `p' compression waves which are moving fastest. The earthquakes alarm is sensitive enough to detect. This wave movement and sounds the alarm before the S - or shear waves hit. And the time gap between the two gives few seconds to react.

A.8. GLOSSARY OF TERMS RELATED TO EARTHQUAKES

Following terms are given in order to have better understanding and after learning these terms, to appreciate nicely.

(1) Earthquake : The release of sudden energy caused by sudden fracture and movement of rocks inside the earth. Part of energy released produces seismic waves that travel outward in all directions from the point of initial rupture these waves shake the ground as they pass by. An earthquake is felt if the shaking is strong enough to cause ground accelerations exceeding approximately 10 mm per square second huge amount of water is activated and the large tidal waves [(Tsunami), meaning harbour waves] are produced.

The origion of the earthquakes is confined to definite geographical regions, known as **earthquake belts**, (which are relatively narrow with respect to their lengths. It has been found that the location of the belts coincides with the present day location of major foldings. The earthquakes belts may be distinguished into three groups namely the **Circum-pacific** (also called the great earthquake belt) which surrounds the stable mass of the Pacific-ocean, the **Alpide belt** (sometimes referred to as the trans Asiatic earthquake zone), and the **secondary belts** which include the Arctic ocean belt, the West-Indian Oceanic belt, and the **East-African belt**.

The mechanism of an earth quake, is explained by most of the smologists by **elastic release theory**. This connects the generation of an earthquake with a sudden release of elastic strain energy. The earthquake shock is usually due to sudden failure of the crust caused by overstress. The stress at any of these locations may have been building through centuries because of a slow distorsion of the crust which gradually stored up an enormous elastic strain energy as an overload spring. Such a system slowly reaches its ultimate capacity for sustrained load, after which a small addition causes it to fail suddenly and violently by rupture or shearing either along an old fault where the rock is weakened by previous deformations or along a newly formed fault. The shock of this suddenly released energy radiations in all directions in the form of seismic waves.

During an earthquake, the very slow tectonic movements transform into seismic movements howing a high velocity owing to sudden release of the stored energy. This energy is released within a mere 10 seconds to 15 seconds (although period of 40 seconds to 60 seconds have also been observed on some rare occasions. The largest of all the recorded amounts of energy released during a major earthquake has been as high as 10^{18} Joules.

The damages caused by earthquakes have been classified in five categories.

1. **First degree (Small) damage** : Small cracks in walls, flaking of plaster and strucco.
2. **Second degree (Moderate) damage** : Small cracks in walls and the joints between panels, flaking of large pieces of plaster and stucco (fine plaster or cement), fall of roof tiles, cracks in chimneys, fall of some chimneys.
3. **Third degree (severe) damage** : Large deep and through cracks in walls and joints between ponds, fall of chimneys.
4. **Fourth degree (destruction) damage** : Fall of interior walls, filler-wall panels thrown out of frames, breaks in walls, partial fall of buildings, destruction of braces between individual parts of buildings.
5. **Fifth degree (collapse) damage :** Total destruction of buildings.

During the period of last 100 years, had been following major earthquakes which have been listed alphabetically) along with. Total depth (fd) and their intensity on Richter-scale (RS) and their year of occurrence

1. Assam earthquake in 1897 (RS 8.7)
2. Assam-Tibet earthquake in 1950 (RS 8.5 and fd = 15 km)
3. Bihar-Nepal earthquake in 1934 (RS 8.4)
4. Bihar-Nepal earthquake in 1988 (RS 6.6 and fd = 71 km)
5. Bhuj (Gujarat) earthquake in 1819
6. Bhuj (Gujarat) earthquake in 2001 (RS 6.9)
7. Earth tremor at Idukki Kottayam (Kerala) in 2001 (RS 4.8)
8. Chamoli (Uttar Pradesh) earthquake in 1999 (RS 6.8 and fd = 21 km)
9. Jabalpur earthquake in 1997 (RS 6.0 and fd= 35 kM)
10. Koyna earthquake in 1967 (RS 6.5 and fd = 8 km)
11. Latur earthquake n 1993 (RS 6.4 and fd = 12 km)

12. Quetta earthquake in 1935 (RS 7.6)

13. Uttarkashi earthquake in 1991 (RS 6.6 and fd = 12 km)

Epicentre : The location on the surface of the earth directly above the focus or place where the earthquake originates.

Fault : A break in the Earth along which movement occurs. Sudden movements along a fault line produce an earthquake. Slow movements produce seismic creep.

Seismology experts have established that the epicentre of 26 January, (Republic day of our country), earthquake was Amritsar, about 8 kM to 10 kM north of Bhachau, and not Lodai near Bhuj as is commonly believed.

A team of geologist, led by seismology expert Professor R-V. Karanth of the M.S. University, Vadodara, after conding extensive study of the ground deformities, ruptures and fissueres along the Kutch main-land fault have arrived at a conclusion that the epicentre seemed to be Amarsar.

The ground deformity and ruptures are more on the east of Bhuj along the Kutch main-land fault than on the western side. Their study indicated the deformation is extensive near Amarsar, which suggests that Amarsar might be epicentre and Ladai.

It was noted that the epi-centre was not one particular point. It extends all along the Kutch main-land fault and more expressive between Lodai and Rapar taluka and land expression was more evident near Amarsar.

Karanth and his team had moved the course of Kutch main-land fault and physically checked all strategic areas lying on it. He reached Kutch a couple of hours after the earthquake struck on January 26. It was his second visit and the team travelled along the fault beforing reaching the conclusion.

The extensive damage to Bhachau taluka and other villages on the east side of Bhuj might be because their proximity to Amarsar, the probably epicentre. The damage to life and property in villages and other areas on the west of Bhuj towards Nakhtaran and Lakhpat is comparatively less. The ground deformation and number of ruptures are also less in the said regions Karanth observed that it was imporative to have an Institute of Earthquake studies in Kutch. To study tectonics, seismology and technology.

This zone remained geologically active always.

Earthquakes are classified on the basis of magnitude into four categories. One of magnitude upto 4.9 on the Richart-scale would be considered `slight, one between 5 and 6.9 `moderate', one between 7 and 7.9 great and one of magnitude 8 or more very great.

The depth of occurrence of the **earthquake focus** provides a suitable basis for classification.

1. **Shallow-focus on normal earthquakes** occur at depths down to 70 kM. Most f the earthquakes have shallow foci.

2. **Intermediate focus earthquakes** have their foci at 70 km to 300 km below the Earth's surface.

3. **Deep focus earthquakes** occur at depths over 300 km. These earthquakes releasing a vast amount of energy happen seldom and then predominantly at or near ocean trenches.

Measurement of severity of skaking at a particular site. It is usually estimated from the descriptions of damage to buildings and terrain and structures. The intensity is often greater near the epicentre of earthquake. The magnitude of the earthquake is usually expressed in terms of Richter-scale. It does not have a mathematical basis, but it is based on observed effects.

Hypo-centre : Earthquakes occur locally and this is another aspect in which seismic movement differ from tectonic process. Apparently, the generation of an earthquake is due to rock failure within a confined area of some depth below the Earth's surface. Since the surrounding is weakened by the failure, the disturbance propogates through a certain area inside the Earth, which is called focal or the hypo-central region of the earthquake.

The paint from which the earthquake originate is termed as the **focus** or the hypo-centre.

4. **Richter scale** : The magnitude of earthquake is usually expressed interims of the Richter scale. It was developed in 1935 by Charles Francis Richter of the California-Institute of Technology as a mathematical device to compare the size of earthquakes. The magnitude of earthquakes is determined from the logarithm of the amplitude of waves recorded by the seismographs with the instruments which automatically detect and record the intensity, direction and duration of movement of the ground. It is not a physical device. The scale measures the earthquakes interims of energy released starting from one and open ended does not have an upper limit. Because of the logarithmic basis of the scale, each whole number increases in magnitude represents a ten-fold jump (times greater than previous one) increase (31 times more energy amplitude.

Seismologists say `magnitude' and `intensity' are the two ways in which the strength of earthquake is generally expressed.

5. To handle an earthquake

5 (i) **Before an earthquake :** A family plan may be developed and the following may be located :

(a) Safest places in the house which exist
(b) Alternative places
(c) Utility valves may be shut-off
(d) Flash lights and batteries
(e) First aid kit
(f) Fire-extinguishers
(g) Food and water supplies
(h) Batteries and transistor radios

5 (ii) During an earthquakes :

(a) Rush to the ground-floor whatever the height of the building.

(b) In case, one is unable to leave the room, take shelter under a table since the table absorbs the effect of earthquake

(c) Stay-away from the buildings

(d) stay-away from trees

(e) Take shelter in the structures built in with reinforced cement concrete

(f) If the building has already developed cracks do not kenture back till the tremors have subsided.

5 (iii) **After an earthquake :** One of the many earthquakes that often occur during the days and months that follows some large earthquake. After shocks occur in the same general region as the main shock and are believed to be the result of minor readjustment of stress at place in the fault-zone. Generally, major earthquakes are followed by a large number of aftershocks, decreasing in frequency with time.

(a) Injuries, if received, may be checked. Give first aid, cover seriously injured with blankets to prevent shock.

(b) To do not attempt to drive anywhere Roads may be damaged.

(c) Battery-operated or car-radio may be turned and listen for information on what to do.

(d) Plan for the long after shocks.

(e) stay out of damaged and already weakened homes

(f) Do not touch down power-lines or objects in contact with power lines.

(g) Do not loose hope.

6. Frequency of earthquakes : Earthquakes occur every day around the world. Each day, there are about 1000 (one thousand) small earthquakes measuring 1 to 2 on the Richter scale. Approximately, there is one every 87 seconds. Annually on average, there are 800 earthquakes capable of causing damage with a magnitude of 5 to 5.9 and 18 major ones with a magnitude of 7 or more.

7. Intensity Scale (Modified Mercalli)

Efforts have been done to devise a scale measuring the violence of the ground motion during an earthquake, ever since earthquakes have been studied.

In various countries, about 50 intensity scales have been proposed for the purpose. Most of them, give appropriate grades of intensity expressing the force of ground motion in a particular area interms of the observed effects on the structures and people.

Amongst the oldest intensity scales are the Rossi-Forel scale (1883) and the Mercalli-Cancani-Sieberg Scale (1917). Latter one is used even at the present time in number of European Countries.

Modified Mercalli Scale (MM-Scale)

In the USA, the most commonly quoted intensity scale is the modified Mercalli Scale (MM-Scale).

This MM-scale has 12 grades of intensity. It was adopted with modifications to suit conditions in America, by H.O. Wood and F. Neumann in 1931 from a proposal originally devised by G. Mercalli for use in Italy.

A team of experts A.S. Arya, National Seismic advisor, D.K. Paul, Head of Department of Earthquake Engineering at Indian Institute of Technology (I.I.T., Roorkee) and Pankaj Aggarwal organised a special training programmes for PWD engineers in Jammu and Kashmir from October 22 to 26, 2005 at Baramulla, Uri and Kupwara.

It is said earthquake resistant designs could withstand tremmors to an intensity 6 mm (modified mercalli). Upto 6 mm intensity, only cracks cracks would develop in the earthquake resistant houses. In 6-9 mm, there would be some more damage when during 9 mm and above would the house collapses.

A.9. ORIGIN AND CAUSES

Seismology is defined as the study of earthquakes and the internal structure of the Earth. The term seismology is derived from the Greek word **seismos** (which measn earthquake). An Irish scientist and engineers, Robert Malet first used this term in about 1880. In Russia, a major contribution to the science was made by A. Orlov (1840-1889), I. Mushketove (1850-1902), B. Golitsyn (1862-19160 and other eminent scientists. **Structural seismology** is the part of seismology concerned with the study of seismic phenomena as applied to earthquake resistant structures. Among the other things, it deals with the location of seismically active areas and prediction of the ground motion character.

I. Mushketov, one of the founders of seismology in Russia wrote in 1888 ``Earthquake have always mankind's scourge inspiring panic terror, which is why man has always been prone to regard them as a heavenly punishment for his sins, or a kind of devil's witch craft, or a whim of an underground monster".

M. Lomonsove, (a Russian scientist in 1711-1765) was one of the first to associate earthquakes with processes within the Earth.

Lomonosov established the relationship between earthquake and orogeny (a period of mountain-building, during, which rocks are severely folded, metamorphosed; and uplifted), a process which is an outcome of never ending activities deep inside the Earth.

Present day drilling techniques do not allow the research workers to take samples from deeper than 10 km of the Earth's body. (on December 27, 1983 a bore hill drilled in the Kola Peninsula reached a depth of 12 km). The distance from the surface to the centre of the globe (Earth) is about 6370 km. This implies that almost all of the Earth's body is left unexplored.

One of the most reliable geophysical methods is **seismic surveying**. It consists essentially in measuring the velocities at which elastic waves propagates in various Earth's layers, which depends on the type of wave and the density and hardness of the media in which waves propogate. Elastic waves inside the Earth may be excided and the propagation time may be measured (forward and backward reflected) and the thickness of layers through which the waves have passed may be evaluated. In addition to seismic surveying, geophysics has at its uioposal some other methods, including gravimetry and magnometry, which are likewise useful in exploring the Earth's interior are frequently employed in conjuction with seismic surveying to supplement and check are another.

Earth is a multi-layered body. It consists of a series of shells differing from one another in mineralogy, density, elastic properties, temperature, and pressure. Fig. 2.1 (A) schematically shows a section through part of the globe (Earth) and the characteristics of the constituent shells. Outer-shell (**crust**) itself consists of several layers. Upper layer (Stratified sedimentary rock) is several kilometers thick on the average. Below which a granite layer lies, which sometimes comes up to the surface. The granite layer is about 10 km thick in valleys and upto 40 km thick beneath mountain ridges. Further downward, there is a layer of **basalt** (30 km under-valleys and into 20 km under ridges). All in all, the crust is 35 km thick (on the average) under continental shields, being 60 to 70 km thick under high mountains. In contrast to its continented counter part, the oceanic crust is must thinner (6 km to 8 km), the basalts lying right under the sedimentary rock. This outer crust is separated from the underlying stratified material referred to as the **mantle**), by a surface called the **Mohorovicic discontinuity (or moho for short**) after the Jugoslav seismologist who discovered it. At this, the velocity of propagation suddenly changes (which indicates difference in density and elastic properties between the crust and upper matatal materials.

The Earth's rocks, especially those forming the crust, contain radio-active elements, which decay and emit heat. At a depth of 50 km to 100 km below the surface, there is a well pronounced layer of **asteno-sphere** (or teto-sphere) about 200 km thick, where the velocity of transverse waves suddenly falls. It shows that the materials is less hard than elsewhere above and that some of it is probably molten. The molten materials is concentrated, within isolated portions in what may be called **lenses,** rather than forming continuous shell. Those portions are directly responsible for **volcanism** and that these are located near the crust areas where intensive folding takes place. There, the concentration of radio-active elements, and as a consequence the amount of heat emitted is considerable. Below the lower matle lies the **core** whose outer portions is believed to be in molten state and whose central.

The surface of Earth remains in the state of continuous motion with some portions of the crust sinking and some going up. The crust executes horizontals motion. These processes are referred to as **diastrophism** or **tectonic movement** (from the Greek work tektonikos' meaning a builder.

In 1852, L. Elie de Beaumont suggested that the Earth cools and, in consequence, contracts. But this theory was neglected.

V. Obruchev and M. Usov of the USSR and W. Bocher of the USA put forward the **pulsation hypothesis,** according to which the Earth continually experiences alternate contractions and expansions. Due to contractions, the crystal movement occurs mainly in the areas of folding. The expansions give rise to crustal ruptures.

Few years earlier to this, in 1922, it was suggested that **convection currents** take place inside the Earth. By this hypothesis the tectonic movement was accounted for by the transportation of masses under the gravitational differentiation of materials of different density, caused by the difference in temperature. The convection in the mantle converts heat into the mechanical work of continental migration and orogeny.

It was observed long ago that the eastern margins of North and South America and the western margin of Europe and Africa are remarkably similar in shape and fit together like two jigsaw pieces. About 200 million years ago, these formed one vast mega-continental cluster called Pangea surrounded by a single ocean called the Tethys sea and later (about 135 million years ago) Pangea broke apart to form a cluster of north continents, named Laurasia, and a cluster of southern continents, named Gondwanalan. These two continents moved away from each other and the gap between them was filled with the ocean. Interesting this slow continental migration continues even to day In 1915, A. Wegener of Austria substantiated the **theory of continental drift or mobilism.** Some of this concepts were found to be erroneous and the theory fell into oblivion (forgotton) for many years. The reasons of continental drift (mobilism) were not given. According to wegener, the continuental migration takes place in subcrustal layers and oceanic crust and owes its existence to the axial rotation of the Earth, which gives rise to forces moving the colossal contents over the ocean floor. This explanation remained far from true. These forces brought about by the rotation of the Earth would be too weak for the job Even if those forces might be adequately large, the continents would drift in the same direction, rather than in the opposite direction as is the case in reality.

The hypothesis of convection current (put forward by A. Holmes) produced by the radiation heating of the subcrustal layers opened up a new promising prospect in substantiating the theory of continental drift, which has been given a new lease of life by the findings of geophysicist in the recent decades.

It is seen that the mountains ridges (upto 4 km wide and 6 km high). Pass through all the central regions of the oceans. In contrast to their continent counter-parts, these are made up of high basalt benches. There are rift valleys between them, which are, in effect large fault along the ridges. **A. fault** may be defined as a zone where the crust (and sometimes of the upper mantle) is fractured to a great extent and may, therefore, undergo considerable deformation. The ridges are cut by quite a that are called transform fault and deep trenches sometimes several hundred kilometers long. The surface temperature here is higher than elsewhere on the Earth.

Scientist formulated **plate tectonics** on the basis of new data available. By this theory, the crust and tie underlying upper mantle (which forms between them litho-sphere (rocky crust of the earth) are divided together with the overlying continents and ocean into several colossal plates) jointed. Under the action of convention currents, the plates continually move relative to one another, floating as it were on the heated layer of astheno-sphere (a hypohetical sphere lying between the crust and the solid nucleus of the earth). The distance between London and New York increases at a velocity of about 10 mm per year. **In fact two plates are in a state of collision which creates mountain belts as one plate overrides the other, giving rise to earthquakes** from time to time. The plates may also move laterally, thereby creating fractures and faults, which may be traced on the continents, too. The process of movement of plates is accompanied by considerable friction which heats up the rock and makes for volcanism. It hs been found that it is along ocean trenches that the deep foci of major earthquakes are located. The movement of plates away from each other, the movement of plates towards each other and the movement of plates in the lateral direction are three types of plate movements.

All the three types of plate motion are responsible for the creation of earth-quake foci which lie very deeply along ocean trenches. It is therefore legimate to suppose that the seismic activity s most pronounced along the plate boundaries. By plate tectonics theory, the inner plate areas may suffer surface deformation, too. However, these deformations are insignificant as compared with those along plate boundaries.

It is not necessary that a fault may be traced on the surface. Some may remain inactive for quite a long time. There may be some faults unknown to the scientists yet, which may lie deep inside the crust and the upper mantle and may never come up to the surface.

A.10. EARTHQUAKES UNDER SEA

Earthquakes which occur under the sea are known as **sea earthquakes** (In Japan, these are called Tsunami - (killer waves). In case the magnitude of earthquakes (MOE) is more than seven, and with shallow focal length these produce large tidal. A huge energy is transferred to the water. As there is no shear-resistance in the water, it travels in small waves (450 mm to 600 mm in height) in the deep oceans with speeds as high as 900 kmph towards the coast. The waves have large kinetic energy. These waves get shear resistance when they meet the coast. The shear energy is converted to the potential energy. These waves rise up to considerable height. In April, 1946, near Alaska Aleutin islands, sea earthquake of magnitude MOE = 7.8 produced Tsunami waves of 30 m height. At the coast, commonly these sea waves have heights of about 13 m to 16 m. In the Pacific sea, these are more frequent. In these regions on early warming system has been established.

On December 26, 2004, the sea earthquake occurred in the Indian Ocean near Indonesia on a known fault more than 1000 km long along a line from this place to Burma. This sea-earthquake was of MOE9. The ocean flow had a rise more than 10 m height along the fault. Tsunami (tital waves) was higher than a two storey building at places. This created homoc in the various countries (viz., Indonesia, Thailand, Sri Lanka and India (mainly Andaman and Nicobar Islands, Tamil Nadu and Kerala states) of the Indian Ocean and some damage to few countries of Africa.

An TSUNAMI (Killer-wave) expert (Condian - Indian- Todepally Murthy ala Tad Murthy, adviser to Government of Gujarat on the development of a warning system) submits on the basis of the past record of Tsunami striking the Indian region is likely to take place in the Arabian-sean any time in the next 60 years.

He has pointed out that the first major Tsunami to strike the Indian region happened on August 27, 1883 in the Bay of Bengal. The Second major one struck the Arabian sea on November, 26 1945. And the third (latest) was December 26, 2004 Killer-way in the Bay-Bengal, plus eastern coast including madras.

It shows that there had been an internal of nearly 60 years between the two successive disasters during which , the strain built-up in the earth crust. As such , it is expected that a Tsunami to hit the west coast in next 60 years.

Dr. Murthy rated India's cyclone evacuation plans superior to the US capabilities in that respect." India has had a good cyclone warning system in place for 50 years now". The US has a vertical mode of evacuation in which all the cyclone affected people are driven to a cyclone shelter a long way in the interior. India follows a horizontal evacuation plan where people are moved to cyclone shelter within walking distance of their homes. This saves a lot to time and effort. Persons are brought to safety faster.

As a member of Canadian delegation to the Indo-Canadian Science and Technology, Dr. Murthy Visits summit meetings held in Bangalore, Delhi and Chennai.

Dr. Murthy (AP-born scientist) rated India's disaster management system among the top ten in the world as was evident from the speed with which the country is setting up tsunami warning systems since last December's tidal wave disaster. India's Meterological Department has been rated as top five in the world.

A11. ANCINET AND ASTRONOMICAL VIEWS ABOUT EARTHQUAKES

Even as seismologists, geologist and scientists research about earthquakes, their causes and their effects on human and animal life, an entire research on earthquake is published in the **Brahat Samihita** (a manuscript written by an eminent astrologer Varamihir datng back to 500 AD.

The manuscript contains a special chapter on earthquake and describes four types of earthquakes, ways to judge its advent and the reasons for their occurrence.

Lecturer in Vedic Astrology at the M.S. University Sanskrit Maha Vidhyalaya, Rajendra Vyas tells about **Bhukamp Lakshan** (meaning signs indicating earthquakes before it strikes and after it happens). In above mentioned manuscript in Chapter 32, four types of earthquakes are

1. Vayu andal, 2. Agnimandal

3. Indramandal, and 4. Varunmandal

1. The Indramandal type of earthquake (that occurred on January 26 in Kutch-Bhuj) is triggered due to certain planetary conditions and occurs when the earths enters one of the Nakshatras that fall between Abhijeet and Anuradha. On January 26, it was Ghanishta Nakshatra between the two coupled with particular condition that caused the earthquake (says, Vyas).

An interesting fact that varamihir has written is a shloka describing the **Indramandal Bhukamp.** It says, this earthquake occurs in `Arbu-Surastra, Malav, Karmishti, Vrushti Karam Manav avanchit" (Abu, Saurastra and Malav-ancient name of Kutch). It brings with its beneficial rains and leads to springing up water benefiting the society. This time therefore, astrology predicts that now rains would bless Kutch and this area would get enough water' (adds Vyas).

Varamihir writes that the certan physiological and psychological effects are seen around 14 days before the Indramandal earthquake. It speaks of animal behaving strangely and about the deterioration of health of human beings. Infections or irritation of the mouth, throat and loose motions are common.

According to yas, these symptoms were seen among people during **Uttarayan** when many people had complained to doctors about such sumptoms. `I undertook a study to find out if what Varahmihir wrote about earthquakes matches with scientific research and facts seen now' (says Vyas).

`This research proves that the ancient astrologers were scientific in their analysis and the persons could not take ancient manuscript seriously. These rich sources of scientific information should be taken up seriously so that the earthquake possibility may be understood before it happens.

`Varahmihir writes that earthquakes occur due to the peculiar tilt of **prithvi** (earth) towards the northern side. To-day modern astronomers have proved that the earth is tilted on its axis. Varamihir further writes that due to this tilt, the northern-hemisphere of the earth is more likely to be earthquake prone. Historically, research has shown that 80 percent of the total earthquakes take place in the northern half of the earth.

Varahmihir has written that human beings are animals and just like animals, human beings too show signs of an impending earthquake. ``But due to **aamkar** (pride) about their intellect, human beings do not recognize signs that animals who are more developed feel". Today, many scientists have found that an earthquake ravaged areas, change in animal and human health are due to certain positively and negative charged ions that are released into the air.

Describing other earthquakes, Varamihir has said that the **Vayumandal earthquake** is preceded by peculiar smoke emissions in the sky, strong winds that uproot trees, destroy jungls. It is also characterised by blunted rays of sun.

The **Agnimandal earthquake** may be predicted if indicators like `ulka varsha' (i.e., falling stars) are visible in the sky. If the temperature soars and clouds are reduced, water tables deplete and intensity of the sun increases.

APPENDIX -B

SERVICE LIFE OF CONCRETE STRUCTURES

B.1 INTRODUCTION

Professor A.W. BEEBY presented a report at Building Research Establishment (BRE), seminar on `**Design for Durability**', December, 1990. It was prepared by the British Cement Association for BRE as part of a major collaborative project. Second CEB/RILEM International Workshop on the Durability of Concrete Structures provided technical functional and economic definitions of service life as under :

Technical service life is defined as the time in service until a defined unacceptable state is reached. **Functional service Life** is defined as the time in service until the structure becomes obsolete due to change in functional requirements. And, **economic service life** is the time in service until replacement of the structure is economically more advantageous than keeping the structure in service.

The performance of the structure may be considered to reduce with time. It remained thinking behind the technical service life, while the recognition that, in fact, the structures more commonly reach the end of life (i.e., these are dismantled and the site is redeveloped), remained the consideration behind the definitions of functional service life and economic service life. As the structures become out of date, than because of an inability to perform in the manner desired at the design stage. As a discussion point, the definition serves some purpose, functional service life may be probably be considered to fall within the definition of economic service life. In reality, except for certain special circumstances, the technical service life would also fall within this definition. Most technical, shortcomings may be corrected by repair and where repairs are done, or the elements in question or the structure replaced will depend upon the relative economics of two options. In fact, the only circumstance that comes easily to mind where service life is not defined by economic condition is, if the structure collapses without warming.

In all other circumstances, where time exists to make a decision about the fate of a structure, a balance will be struct between the costs of repair renovation, modification, the value of the upgraded structure and the costs of replacement. The value may not be exclusively monetary as in case of ancient monuments or cathedrals, where the value is a highly subjective quantity.

Practically, the **end of life** may be considered a stage in time, when it is concluded that the structure may no longer be economically capable to survive. In principle, the cost of whole life of the structure may be considered. BS : 5400 (Code of Practice for the design of concrete bridges), Part I defines **the design life** as 120 years. This code emphasises that this does not necessarily mean that a bridge designed in accordance with it will no longer be fit for its after 120 years nor that, it will continue to be serviceable for that period of time, without adequate and regular inspection and maintenance. After defining a suitable design life from economic consideration, it would then be necessary to ensure that the technical design life exceeds this figure. In Practice, detailed studies to define a design life would only rarely be appropriate whatever will be provided, shall be an arbitrary figure such as mentioned above. A different definition is needed, which should be based on Technical service life. A possible definition for this could be as under :

Design Service Life (DSL) : Any definition must contain the following elements, which are essential to the definition of a logical design approach. These are (i) a specific environment, (ii) a definite criterion of failure (iii) the statement must be in specified terms and (iv) an assumed level of maintenance. Then, the DSL may be defined as anticipated time in service in the anticipated environment until a defined acceptability that the structure would have reached an unacceptable state is exceeded, for a defined level of maintenance.

B.2. END OF SERVICE LIFE

It is necessary to clarify the end point of the technical design life so that it becomes a reality to specify the design life. The end of service life has been mentioned as a point when the defined unacceptable state is reached.

The Institution of structural Engineers considered the acceptable level of safety in relation to the appraisal of existing structures. In its report allowed the reduction in safety factors to be used in assessment situation.

In Limit state design, two partial factors are used (r_{FL}, partial factor of safety for loads, and r_{ms}, partial factor of safety for strength of materials). The factor of safety on the dead loads may be reduced in case the member sizes are actually measured similarly, the factor of safety for the strength of concrete may be reduced in case the actual strength of the concrete has been assessed from cores, UK and European design documents does not allow any reduction in the definition of r_{ms}. In the definition of r_{ms}, the allowance for the deterioration in resistance with time may be included.

In principle, r_{ms} may be divided in two parts such that

$$r_{ms} = (r_{msu} \, . \, r_{ms.d}), \quad \text{where}$$

$r_{ms.u} = r_{ms}$, partial factor of safety for the material strength as presently defined.

$r_{ms.d}$ = Allowance for possible reduction in the resistance due to the influences of the environment over the design life.

The current design philosophy does not allow any possibility of degradation in strength (i.e., which influences the ultimate Limit State) with time to be considered in design. It is possible that there may be degradation in serviceability limit state (i.e., cracking, or possibly spalling of concrete, which one is very unsightly. In many circumstances, the spalled lumps of concrete may fall down and potentially cause damage. In columns and other members, the concrete remains in compression. Concrete lumps get removed by spalling. The concrete was design to carry stress and the safety is reduced. The spalling of concrete lies beyond the limit of degradation.

From the serviceability point of view, it is therefore reasonable to define the limit to allowable degradation as development of cracks in excess of 0.3 mm. This may be seen satisfactory provided the appearance of such crack is not accompanied by any significant reduction in strength.

In other manner, the end of service life may be defined by considering the corrosion of the steel reinforcement bars. It may be stated that the reinforcement steel bars are so properly treated that there occurs no corrosion during the design life. Therefore, the end of the service life corresponds to the moment when either carbonation or sufficient chlorides to depassivate the steel, first reaches the reinforcement.

Two alternatives for defining the end of service life may be compared. In comparison, as compared to use limiting crack width, it is more conservative definition which considers corrosion. There is advantage of this definition. At present, far more is known about the rates of carbonation than, rates of corrosion. Present level of knowledge may allow to formulate reasonable design rules for meeting this criterion. For formulating rational rules for the cracking approach substantial more knowledge is needed. Furthermore, a reasonable prediction of the time for the carbonation to reach the reinforcement may be made from the measurements made early in the life of the structure. This approach is probably not excessively conservative for field environments where the corrosion may be rapid once it begings. There is a disadvantage in this approach. The design shall be unnecessarily conservative.

In defining the end service life of a structure, the quantity of concrete round the reinforcement would be carbonated and the probability of this occurring are to be mentioned.

The carbonation may extend down the faces of cracks to the reinforcement where the bars intersect the cracks in a short time.

Some corrosion had been seen on bars with 25 mm cover at the root of cracks as 0.1 mm after four years of exposure. However, tests over 10 years showed this corrosion at the root of small cracks did not develop. Experience generally suggests that the very limited areas of carbonation generated by cracks, where the bars intersect them do not generally sustained corrosion. The end point of service life must therefore. Correspond to considerably larger areas of carbonated concrete existing in contact with the reinforcement than that expected at the root of cracks.

Siemes, Vroumenvelder and Vanden Beukel carried out probabilistic studies of durability problems which led to distributions of service life. In the particular case, their studies gave the following approximate results for characteristics and mean lives for two situations.

	Characteris life	Mean Life
15mm cover	10 years	34 years
30 mm cover	40 years	123 years

It was seen that the scatter was very large and the mean life was roughly three times the characteristics life.

As such, the statement mentioned in BS : 5400, about 120 years of design life does not seem to be reasonable. The unacceptable performance of the structure must be mentioned. The probability of this occurring should be aimed at in the design.

Reliability and issues of service life

By increasing the cover for the reinforcement or by having concrete mix properly designed, the rate of degradation may be controlled. The durability failures occurring in practice may be avoided. Though, this is a realistic model in principle, it is only rarely relevant to real failure situations.

The problem of durability may be classified under two titles. `Life problems and Reliability problems.

Life problems are those where the design concept is useful. Whereas, the reliability problems are those where durability failure occur due to system failure. In case, the reinforcement steel bars are nicely coated by an appropriate coating, the coating remains undamaged then the life of a structure may be 150 years.

In order to solve, the problems of durability the following possible strategies may be adopted (i) Avoid degradation by use of non-reactive materials.

(ii) Concrete itself may be used as protection (e.g., provision of adequate cover may be specified in the code of practice).

(iii) Block ingress of deleterious substances by a barrier

(iv) Inhibit degradation reactions

(v) Arrange for planned replacement

(vi) combined solutions.

Choice of structural form and workmanship are the two aspects, which influence the durability of concrete.

Structural form

Leakage of water contaminated with de-icing salt through the joints onto the bearing shelf in bridges is a major cause of corrosion problems. This is completely a problem of choice of an appropriate structural arrangement. In case, the problem had been foreseen it could have been totally avoided, possibly by arranging for proper, maintenable drainage below the joints or possibly by water-proofing the bearing seats or by using a structural form which would have avoided the necessity of a joint at this point.

Many recent codes (inclusive of IS : 456-2000) and manual now attach great emphasis on the positive consideration about durability. The CEB Manual on design for durability and the draft CEB Model Code'90 and given importance to durability. The majority of which is concerned with general advice of cover and quality of concrete.

Workmanship : It has been seen in wide studies that the insufficient cover had been the major cause of problems in the structures. It had been noticed that the insignificantly less cover had been provided than specified.

The detailing and site control raised are similar to the example of coating described above. However, the issues are complex. The desired design life could be achieved by specifying design rules which mentions covers and quality of concrete, and which are followed correctly. Factors normally expected (e.g., variations in materials, environment and workmanship) have been taken into consideration.

The quality of the design input and the quality of the workmanship are of fundamental importance in ensuring durable structures. Durability is a major issue. Code committees (so also, IS 456-2000) have given emphasis on providing cover for the reinforcement steel bars. Formerly, a cover which was 12 mm recommended, is now specified as 40 mm with a concrete strength of roughly 25 N/mm^2. BS 8110 [13] mentions a cover of 40 mm for a grade of concrete of 40 N/mm^2 for the same environment.

The great majority of durability problems arise from lapses in the quality of design or the construction and are reliability related failure Better quality of concrete, better supervision, better trained and educated staff may result in more durable concrete structures.

Conclusion : It is expected that the construction will last without undue deterioration for its projected useful life. The necessity of introducing a life time or design life explicity into the design process is to be considered stonework or brickwork or mass concrete are considered to last effectively indefinitely in case properly specified and constructed. Life of Steel structures (in case properly, protected and protection is maintained) is not threatened.

Reinforced or Prestressed concrete seems unique in the requirement for a long life (say, 50 years. Specific cause of deterioration, corrosion of reinforcement or prestressing tendons and other forms of deterioration are treated by avoiding the reasons.

Mix ingredients should be selected so that the deleterious alkali-silica reactions do not occur Frost attack and Sulphate attack should be properly treated and dealt. In case, these shorts of deterioration occur, then, it is a reliability system failure.

The reinforcing steel bars and the prestressing tendons should be inspected and replaceable to achieve design life.

In the relatively, near future, it will be possible to select appropriate covers and mix designs to reach foreseen design Life.

The improvement of design for durability in two stages may lead to have long life.

Improved education and training of all involved in the construction process may provide good workmanship and there are no errors in the design. Improved methods of design (which are directly related to design life) may be adopted.

REFERENCES

1. A.W. BEEBY : Design for durability report presented at BRE seminar design for durability, December, 1990.

2. A report produced by the British Cement Association for the Building Research Establishment.

INDEX

Abram's law 93
Active earth pressure 749
Admixture 18
  Accelerating 18
  Air-entraining 19
  Plasticizing 19
  Retarding 18
  Water-reducing 19
Aggregates 09
All-in aggregate 10
  Coarse 10
  Classification 14
  Fine 10
  Fineness modulus 15
  Grading 11
  Heavy-weight 14
  Impurities 13
  Light-weight 14
  Normal-weight 14
  Oter-types 14
  Size 13
  Storage 24
Aggregate-cement ratio 94
Anchorage
  Plain bars 180
    Compression 180
    Tension 180
  Reinforcement bars 180
  Bends 180
  Hooks 180
  Value 181
Arrangement of live load 279, 298
  BM 279
  SF 298
Balanced section 115
Bars
  Anchorage 180
    Plain bars 180
    Deformed bars
    Splice 187
    Twisted bars 23
Beams
  Singly-reinforced 107
  Continuous 272
  Doubly-reinforced 193
  Flanged 224
  Inclined bars (stirrups) 160
  L-beams 224
  Tee-beams 224
  Vertical stirrups 156
  With shear reinforcement 156
  Without shear reinforcement 152
Behaviour
  Beams (reinforcement concrete) 108
  Diagonally craked beam 153
  Flat slab 376
  Retaining wall 763
  Slabs
    Spanning in one direction 302
    Spanning in two direction 333
Bent-up bars 159
Blaine air permeability method 933
Bond 176
  Permissible stress 180
  Stress 177
Brick (reinforcement) 833

Design 833, 834
Development length 835
Effective span 833
Permissible stress
Compressive 834
Shear 834
Reinforcement 835
Cover 835
Minimum 835
Shear stress 835
Cantilever
Beams 107
Retaining wall 767, 769
Cement 04
Fineness 15, 1026
33-grade 06
43 grade 06
53 grade 06
Ordinary portland 04
Chemical composition 05
Mineral compounds 05
Specifications 13
Storage 24
Test 06, 931
Compressive strength 08
Consistency 938
Standard 938
Soundness 09, 937
Tensile strength 09, 1043
Circular slabs 427
Edges 428, 430
Simply supported 428
Fixed 430
Coarse aggregate 10
Columns 453, 454
Composite 483
Design 481
Distribution of stress 511, 513, 516
Eccentricity 469
Compressive force 508
Intermediate 529
Large 497
Less than 0.25 D, 511
Minimum 469
More than 1.5 D 513
Tensile force 543
Intermediate 563
Less than 0.25 D, 557
More than 1.5 D, 544
Equivalent
Moment of inertia 469
Sectional area 461
Footing 475
Lateral reinforcement 468
Spiral 487
Ties 474, 497, 487
Length
Actual length 457
Effective length 456, 458
Unsupported length 465
Loaded
Axially 453, 454
Eccentrically 498
Circular 543
Rectangular 508
Long 457
Permissible stresses
Concrete 479
Steel 479
Reduction factor 483
Reinforcement 483
Clear cover 470
Longitudinal 483
Main 483
Transverse 485
Colum (continued)
Radius of gyration 462
Section 504
Area 504
Cracked 504
Gross 504
Transformed 504
Uncracked 502
Short 463
Slender 463
Slenderness ratio 463
Limit 464
Strength 471

Composite 483
Long 473
Short 471
Slender 473
Combined footing 640
Compaction factor 31
Compaction factor test 25
Concrete 03
Aggressive soils 31
Mix proportion 528
Sea water 32
Fire resistance 42
Building Frame 276
Fresh 32
Foundation 38
Grade 87
Hardened 93
Properties 32
Durability 27
Workability 25
Compacting factor 25
Slump test 25
Vee-bee test 26
Proportioning 937
Strength 63, 05
Compressive 09
Tensile strength 10
Test 25, 931
Compacting factor 945
Compressive strength 946
Cubical sample 947
Cylindrical sample 947
Flexural strength 949
Slump 943
Collapse 945
Shear 945
True 945
Concrete mix 931
Acceptance criteria 875
Common terms 871
Coefficient of variation 871
Mean strength 871
Target strength 874
Variance 871
Laboratory design strength 876
Nominal mix 870
Number of samples 876
Proportioning 868
Standard deviation 872
Standard deviation (assumed) 873
Methods of mix design 867, 877
ACI 901
British 909
Fineness modulus 881
IRC 912
ISI 920
Limits of fineness modulus 883
Maximum density 879
Minimum voids 878
RRL 887
USBR 910
Conditions of curtailment of bars 184
Consistency test 938
Construction joints,
Continuous beams 272
Bending of bars 269
Coefficients
Moment 275
Shear 297
Design 286
Effective span 273
Haunches 285
Control of deflection 191
Core diameter 472
Cracked beams 110
Diagonally cracked 153
Creep 76
Critical neutral axis
Critical section for shear force 165
Critical section for moment 274
Bending moment (column footing) 566
Development length, 566
Shear 567
Beam shear 567
Flat slab shear 568
One-way shear 567
Two-way 408, 568
Shear force 146, 567

Curing 37
Curtailment of reinforcement 182
  Conditions 184
  Special requirement 185
  Tension 182
Dead loads 57
Deflection 191
  Control 191
Deformed bars 23
Depth of neutral axis 111
Design
  Beams
    Continuous 249
    Doubly reinforced 204
    Effective span 128
    Flanged 224, 245
    L-beams 260
    Singly reinforced 128
    Tee-beams 245
  Codes 47
  Columns 557
  Cncrete mix 867
    Domes 820
    Footings 557
    Formwork 848
    Hollow tiles 833
    Reinforced brick 833
    Retaining walls 769, 785
    Singly reinforced 107
    Slabs 302, 304
    Shear reinforcement 308
    Tiles 839
Development length 179
  Beams
    Compression 220
    Tee beam 271
    Slab tension 352
  Diagonal tension 150
  Direct and bending stress 498, 499
  Distribution 307
    Reinforcement 308
    Spacing 308
    Stress 310
Domes 810
  Conical 827
  Definition of terms 869
  Design 820
  Hoop compression 811
  Hoop tension 811
  Latitudes 811
  Longitude 811
  Meriodional thurst 811, 828
    Concentrated load 817
    Uniformly dist. load 813
  Reinforcement . 821
  Spherical 813
  Various loads 812
    Live load 812
    Self-weight 812
    Snow load 812
    Wind load 812
Doubly reinforced beams 193
  Bending strength 196
  Design 204
  Location of NA 195
  Nominal shear stress 146
  Shear stresses 144
Durability 38
Earth pressure
  Active 749
  Passive 752
Eccentrically loaded columns 497
Economical depth of T-beam 244
Effects
  Patterns loadings 383
Effective depth
  Singly reinforced beam 109
  Doubly reinforced beam 194
  Minimum 220
  T-bemm 243
Effective span 128
  Beam 128
  Cantilever 128
  Continuous 272
  Simply supported 107
  Slab
    Cantilever 305
    Continuous 305

One-way 302
Simply supported 304
Equivalent section 394
Area 108
Column 394
Erection loads 82
Exposure conditions 33
Environment 33
Factors influencing
Strength of concrete 92
Final setting time 05, 939
Fine aggregate 10
Fineness modulus 15
Flanged beams 224
Bending strength 224, 230
Design 224, 245
Effective width 225
Lever arm 228
Location of neutral axis 226
Stresses 239
Concrete 239
Steel, 239
Flat slabs 369
Behaviour 376
Bending 376
Shear 407
Column head 382
Design 377
Direct method 377
Equivalent Frame method 384
Development 371
Drop Panel 371, 372
Elements 372
Effects of pattern of loading 383
Equivalent column 394
Loading pattern 398
Moment in columns 402
Distribution 379, 381
Negative BM 379
Positive BM 379
Opening 379
Reinforcement 403
Area 404
Spacing 405
Relative stiffness of slabs 385
Shear 407, 408
Beam type 407
One-way 407
Two-way 408
Statics 373
Thickness 371
Torsional member 395
Total design moments 378
Strips of flat slabs 373
Column strip 373
Middle strip 373
Stiffness factors for slabs 387
Footings 557
Axially loaded 557
Balanced footing pressure 621
Basic requirement 571
Bearing pressure 559, 560
Combined 640
Approximate method 653
Cantilever 642
Continuous strip 644
Grid 646, 647
Mat 647, 648
Piles 480
Rectangular 648
Strap footing 642
Trapezoidal 651, 641
Depth of footing 562
Design 557
Eccentrically loaded 640, 622
Independent 557
Circular 611, 612
Rectangular 577, 578
Sloped 582, 583
Square 575, 576
Stepped 580, 581
Strip 644
Wall 572
Isolated 557, 571
Kern distance 625, 626
Symmetrical 628
Thickness at edge 562
Unsymmetrical 629

Form-work 848
Beam 855
Columns 857
Economy 851
Loads on form work 854
Horizontal 854
Vertical 854
Material 852
Plywood 854
Steel 854
Timber 852
Requirements 849
Slab 855
Stripping time 850
Tolerance 848
Foundation 557
Deep 558
Shallow 558
Futuristic constr. materials (FCM) 990
Fibre Reinforced polymer (FRP) 1009
Ciramic matrix 1010
Externally bended 1013
Metal matrix 1010
Polymeric matrix 1010
High performance conc (HPC) 992
Properties 993
High durability 993
High early str. 995
High modulus of elas. 995
High ultimate strength 995
High workability 995
HPC (admixtures) 996
Technical problems 997
Self-compacting conc (scc) 991
Properties 993
Bond strength 994
Compr. strength 993
Deformability 993
Durability 994
Fluidity 993
Modulus of elas. 994
Segregation 993
Polymer-modified cement (PC) 1000, 1004
Polymer-modified concrete (PCC) 1000, 1002
Polymeric impregnated conc (PIC) 1007
Polymeric-modified mortar (PIC) 1000 (PM)
History of use 1001
PC, PM 1004
Mixing of ingradients 1005
Properties 1005
PIC, PIM (properties) 1008
Polymer composites in CE 1009
Concrete Nano-technology (FCT) 998
Definition 999
Nano-meter 999
Nano-scale 999
Nuclear analysis 999
Grade of aggregate 11
Grade of concrete 87
Grashoff-Rankine method 334
Gross cross-section 44
Hardening
Cement 05
Concrete 32
Heat of hydration 04
Heel slab 764
Helical reinforcement 468, 471, 485
Hiley's formula 697
Hollow tiles 839
Hook 180
Hoop stresses in domes 811
Hydration 04
Hydrolysis 92
Impact 82
Inclined bars in shear 160
Initial modulus of elasticity 95
Initial setting time 05, 939
Isolated footing 558
Isolated L-beams 226
Isolated tee-beams 226
IS:method (Design of slabs) 349
Lap Splice 187
Compress. reinforcement 221
Tension reinforcement 188
Lateral girder effects 162
L-beam 260
Torsion 261, 264
Lever arm 113

Factor 114
T-beam 228
Live loads 61
Arrangement for BM 297
Arrangement for SF 298
Percentage reduction 68
Links 485
Columns 485
Piles 695
Loads 56
Dead loads 57
Unit weight of materials 57
Erection loads 82
Live loads 49, 870
Domes 810
Floors 63
Loads 56
Seismic loads 71
Snow 71
Wind load 68
Loading pattern 398
Locations of NA 457
Beams
Doubly reinforced 195
Flanged 203
Singly reinforced 111
Long columns 454
Marcus method 339
Maximum reinforcement
Beams (tension) 132
Columns (compression) 483
Methods of design 102
Limit state 103
Working stress 103
Ultimate 103, 104
Minimum reinforcement
Beam for BM (tension) 132
Beam for shear 143
Column 483
Modular ratio
Concrete in compression 215
Concrete in tension 111, 112
Moment of resistance 115
Balanced section 116
Doubly reinforced beam 196
Over reinforced 117
Singly reinforced beam 114
Tee-beam 207
Under reinforced 116
Non-destructive evaluation (NDE) 952
Non-destructive tests (NDT) 952
Non-mechanical tests 953
Classification of NDT Tech 954
Acoustic emission 955
Electro-magnetic 955
Magnetic effects 955
Mechanical energy 955
Absorption 955
Recovery 955
Sonic test 958
Super sonic PVT 961
ultra-sonic 959
Comments for use & application 985
Condition assessment 986
Electrical 986
Radar 986
Stress-propogation 986
Estimating in place Compr. Stren. 986
Existing construction 986
New construction 986
Flow detection 986
Historical back-ground 956
Methods of testing 966
Combined use 965
(ultra-sonic + Rebound. H)
Electrical methods 968
Magnetic methods
Combined (Elect. + Magnetic) 968
Schmidt Rebound Hammer 965
Methods
Back-scatter 984
Half-cell (potential) 971
Maturity method 973
Nuclear methods 982
Probe penetration 981
Radio-active method 982
Radio-graphic method 984
Gamma-rays 985

X-ray 984
Radio-metric methods 983
Reinforcement 968
Combined (Elec + Mag) 968
Electrical 968
Magnetic 968
Meters 968
Cover meters 968
Pacho-meters 968
Profo-meters 968
Magnetic eddy-current 968
Magnetic reluctance 968
Tests 974, 980
Break-off 980
Permeability (conc) 974
Air 974
Chlorides 976
Gases 974
Visual inspection 957
Neutral axis of beam
Doubly reinforced 195
Flanged 226
Singly reinforced 111
Nominal mix
Proportions 28
Nominal shear stress 146
Beams
Doubly reinforced 215
Singly reinforced 146
Tee-beams 256
Number of Piles 695
One-way slab 302, 304
Cover reinforcement 306
Distribution reinforcement 307
Effective span 304
Main reinforcement 306
Spacing of reinforcement 308
Opening in domes 820
Open newel stair 718
Open well star 718
Ordinary portland cement 4
Over-reinforced beams 117
Other forces and effects 81
Elastic axial shortenings 81
Flatigue 82
Foundation movements 81
Impact 82
Solid and fluid pressure 81
Vibration 82
Permissible stresses
Bending 119
Bond 215
Direct compression 479
Shear (concrete) 119, 167
Steel 94
Compression 173
Tension 94
Pigead's coefficient method 344
Piles 694
Bearing piles 694
Caps 702, 703
Analysis 704
Cap design 703
Cast-in place 695
Design 707
Erection 702
Footing 654
Formula 697
Friction 688
Load capacity 696
Number 654
Precast 695
Reaction 655
Reinforced concrete 695
Stresses 700
Driving 702
Erection 702
Handling 700
Properties of concrete 24
Compaction 37
Curing 37
Permeability 36
Proportioning
Aggregate 1005
Concrete mix 867
Raft foundation 647
Rankine-Grashoff method 334
Reinforcement 21

Bending (tension) 131, 132
Circular slabs 443
Compression reinforcement
Beam (Doubly Reinf) 193
Column
Maximum 483
Minimum 483
Cover for reinforcement
Beam 130
Columns 470
Slabs 306
Design 272
Extrension 272
General requirements 32
Shear 152
Side face 130
Spacing of Main reinforcement
Beam 132
Slabs 306
Column (longitudinal) 483
Column (Ties, Spiral) 497, 485
Spacing of shear reinforcement
Maximum 158
Minimum 158
Reinforcement (continued)
Type 22
Cold-twisted 23
Deformed bars 23
Hard-drawn 24
Hot-rolled 22
Mild steel bars 22
Plain steel bars 22
Rolled steel bars 23
Relative stiffness 381
Columns 390
Slabs 381
Requirements of in flat slab 390
Durability 28
Sliding 44
Overturning 44
Stability 43
Stiffness 44
Strength 43
Restrained slabs 337
Main reinforcement 352
Torsional reinforcement 354
Retaining wall 748
Behaviour 763, 786
Coefficient 766
Active earth pressure 749
Earth pressure 754, 755, 756
Design 769, 787
Cantilever 769
Counterfort 785
Active earth press 749
Common conditions 805
Passive earth press 752
Heel slab 769
Toe slab 769
Effective of surcharge 753
Forces 757
Proportioning (preliminary) 765
Cantilever 765
Stability 758
Overturning 758
Sliding 758, 760, 767
Subsidence 758, 761
Strength 758
Types 748
Buttress 749
Cantilever 748
Counterfort 749
Gravity 748
Ribbed slab 839
Breadth 839
Deoth 841
Design 839
Ribs 840
One-way 842
Two-way 843
Spacing of ribs 840
Thickness of slab 842
Service life of concrete 45
Seismic oad 71
Setting time 05, 939
Final 105, 939
Initial 05, 939
Secant modulus of elasticity 95

Shear reinforcement
Design 168
Function 161
Truss analogy 162
Shear stress
Beam of varying depth 147
Continuous beams 297, 298
Distribution 144
Doubly reinforced beam 230, 215
Singly reinforced 144
Shrinkage 72
Singly reinforced beam 107, 143
Balanced section 115
Behaviour 108
Bending strength 110
Bent-up bars 159
Cover 130
Design 128
Effective span 128
Equivalent beam section
Cracked 110
Uncracked 109
Fundamental assumptions 111
Inclined bars 160
Maximum reinforcement 132
Minimum reinforcement 131
Moment of resistance 126
Neutral axis 110
Over-reinforced 115, 116
Percentage of steel reinforcement 113
Permissible stresses 119
Reinforcement 21
Maximum reinforcement 132
Minimum reinforcement 131
Side face 130, 132
Shear reinforcement 156
Shear stresses 144
Slenderness limits 136
Spacing of reinforcement 132
Minimum distance 132
under reinforced 115, 116
Varying depth 138
Vertical stirrups 156
Spacing 157
Maximum 158
Minimum 158
Slabs 302, 333
Cantilever 312
Carrying concenrated loads 310
Circular 427
Continuous 320
Design 302
Estimation of thickness 314
Flat slabs 369
One way 302
Restrained slabs 357
Spanning in
One direction 302
Two direction 333
Two-way 333
Main reinforcement 287
Slenderness-limits (Beams) 192
Slump test 25, 943
Snow loads 71
Soundness of cement 9, 937
Stairs 717
Types 718
Circular 719
Dog-lagged 719
Distribution of loads 723
Load on
Geometric 719
Open newel 719
Open well 719
Spiral 719
Straight 718
Effective span 720
Spanning horizontally 720
Spanning longitudinally 721
Steel beam theroy 208
Steel reinforcement
Centre (Tension steel)
Hysd steel 23
Mild steel 22
Support 281, 283
Compression 281
Tension 283
Stiffness factors in slabs 385, 387

Stirrups 157
Bent-up bars 159
Four legged 156
Inclined 160
Single legged 157
Six legged 157
Two legged 157
Vertical 156
Strap (cantilever) footings 673
Storage of materials 24
Aggregate 24
Cement 24
Reinforcement bars 24
Strength
Beams 102
Cement (Tensile) 09
Columns (axially loaded) 453
Concrete compressive 95
Factors influencing 92
Increase in strength 89
Stresses 113
Beams
Bending (concrete) 146
Shear (concrete) 152
Bond 177
Concrete 197
Permissible 119, 167, 180, 194
Shear
Actual 151
Distribution 146, 144
Nominal shear 146
Steel 197
Stress-strain relationship
Concrete 95, 96
Steel 99
Deformed (Hysd) 101
Mild 99
Structural safety 83
Structural failures 84
Tangent modulus of elasticity 95
Tee-beams 224
Bending strength 230
Design 224, 245
Effective width of flange 225
Elements 224
Flange 224
Isolated 206
Lever arm 228
Location of NA 228
Proportioning
Depth 243
Economical 245
Effective 243, 244
Minimum 243
Overall 246
Rib 246
Web 246
Width 246
Web 243
Stresses in conc. and steel 239
Temperature 308
Effects 72
Reinforcement 308
Temperature and shrinkage 72
Reinforcement 308
Tensile strength of cement 941
Toe slab 764
Torsion
L-beams 260, 262, 264
Reinforcement 264
Slabs 354, 355
Torsional member 395
Torsion and shear 245
Truss analogy (shear) 162
Two-way slabs 333
Behaviour 333
Design 356
Design method 334
Grashoff-Rankine 334
IS : 456-1978, 349
Marcus, 339
Pigeud's 344
Westergaard's 345
Two-way shear 407
Ultimate method 104
Vertical stirrups 156
Design 168, 169
Spacing 157

Maximum 158
Minimum 158
Wall 768, 769
Vicat's apparatus 938
Water 16
Solids (present) 17
Water-cement ratio 93
Westergaard's method 345
Wind loads 44
Workability 25
Working stress method 103